HANDBOOK OF QUANTUM LOGIC AND QUANTUM STRUCTURES

QUANTUM LOGIC

HANDBOOK OF QUANTUM LOGIC AND QUANTUM STRUCTURES

QUANTUM LOGIC

Edited by

KURT ENGESSER
Universität Konstanz, Konstanz, Germany

DOV M. GABBAY
King's College London, Strand, London, UK

DANIEL LEHMANN
The Hebrew University of Jerusalem, Jerusalem, Israel

ELSEVIER

Amsterdam • Boston • Heidelberg • London • New York • Oxford
Paris • San Diego • San Francisco • Singapore • Sydney • Tokyo
North-Holland is an imprint of Elsevier

N·H

North-Holland is an imprint of Elsevier

Radarweg 29, PO Box 211, 1000 AE Amsterdam, The Netherlands
The Boulevard, Langford Lane, Kidlington, Oxford OX5 1GB, UK

First edition 2009

British Library Cataloguing-in-Publication Data
A catalogue record for this book is available from the British Library

Library of Congress Cataloging-in-Publication Data
A catalog record for this book is available from the Library of Congress

ISBN: 978-0-444-52869-8

For information on all North-Holland publications
visit our website at elsevierdirect.com

Printed and bound in Hungary

09 10 11 12 10 9 8 7 6 5 4 3 2 1

CONTENTS

FOREWORD

More than a century ago Hilbert posed his unsolved (and now famous), 23 problems of mathematics. When browsing through the Internet recently I found that Hilbert termed his Sixth Problem non-mathematical. How could Hilbert call a problem of mathematics non-mathematical? And what does this problem say?

In 1900, Hilbert, inspired by Euclid's axiomatic system of geometry, formulated his Sixth Problem as follows: *To find a few physical axioms that, similar to the axioms of geometry, can describe a theory for a class of physical events that is as large as possible.*

The twenties and thirties of the last century were truly exciting times. On the one hand there emerged the new physics which we call quantum physics today. On the other hand, in 1933, N. A. Kolmogorov presented a new axiomatic system which provided a solid basis for modern probability theory. These milestones marked the entrance into a new epoch in that quantum mechanics and modern probability theory opened new gates, not just for science, but for human thinking in general.

Heisenberg's Uncertainty Principle showed, however, that the micro world is governed by a new kind of probability laws which differ from the Kolmogorovian ones. This was a great challenge to mathematicians as well as to physicists and logicians. One of the responses to this situation was the, now famous, 1936 paper by Garret Birkhoff and John von Neumann entitled "The logic of quantum mechanics", in which they suggested a new logical model which was based on the Hilbert space formalism of quantum mechanics and which we, today, call a quantum logic. G. Mackey asked the question whether every state on the lattice of projections of a Hilbert space could be described by a density operator; and his young student A. Gleason gave a positive answer to this question. Although this was not part of Gleason's special field of interest, his theorem, now known as Gleason's theorem, had a profound impact and is rightfully considered one of the most important results about quantum logics and structures. Gleason's proof was non-trivial. When John Bell became familiar with it, he said he would leave this field of research unless there would be a simpler proof of Gleason's theorem. Fortunately, Bell did find a relatively simple proof of the partial result that there exists no two-valued measure on a three-dimensional Hilbert space. An elementary proof of Gleason's theorem was presented by R. Cooke, M. Keane and W. Moran in 1985.

In the eighties and nineties it was the American school that greatly enriched the theory of quantum structures. For me personally Varadarajan's paper and subsequently his book were the primary sources of inspiration for my work together with

Gleason's theorem. The theory of quantum logics and quantum structures inspired many mathematicians, physicists, logicians, experts on information theory as well as philosophers of science. I am proud that in my small country, Czechoslovakia and now Slovakia, research on quantum structures is a thriving field of scientific activity.

The achievements characteristic of the eighties and nineties are the fuzzy approaches which provided a new way of looking at quantum structures. A whole hierarchy of quantum structures emerged, and many surprising connections with other branches of mathemtics and other sciences were discovered. Today we can relate phenomena first observed in quantum mechanics to other branches of science such as complex computer systems and investigations on the functioning of the human brain, etc.

In the early nineties, a new organisation called *International Quantum Structures Association (IQSA)* was founded. *IQSA* gathers experts on quantum logic and quantum structures from all over the world under its umbrella. It organises regular biannual meetings: Castiglioncello 1992, Prague 1994, Berlin 1996, Liptovsky Mikulas 1998, Cesenatico 2001, Vienna 2002, Denver 2004, Malta 2006.

In spring 2005, Dov Gabbay, Kurt Engesser, Daniel Lehmann and Jane Spurr had an excellent idea — to ask experts on quantum logic and quantum structures to write long chapters for the *Handbook of Quantum Logic and Quantum Structures.* It was a gigantic task to collect and coordinate these contributions by leading experts from all over the world. We are grateful to all four for preparing this monumental opus and to Elsevier for publishing it.

When browsing through this Handbook, in my mind I am wandering back to Hilbert's Sixth Problem. I am happy that this problem is in fact not a genuinely mathematical one which, once it is solved, brings things to a close. Rather it has led to a new development of scientific thought which deeply enriched mathematics, the understanding of the foundations of quantum mechanics, logic and the philosophy of science. The present Handbook is a testimony to this fact. Those who bear witness to it are Dov, Kurt, Daniel, Jane and the numerous authors. Thanks to everybody who helped bring it into existence.

Anatolij Dvurečenskij, President of IQSA
July 2006

EDITORIAL PREFACE

There is a wide spread slogan saying that Quantum Mechanics is the most successful physical theory ever. And, in fact, there is hardly a physicist who does not agree with this. However, there is a reverse of the medal. Not only is Quantum mechanics unprecedently successful but it also raises fundamental problems which are equally unprecedented not only in the history of physics but in the history of science in general.

The most fundamental problems that Quantum Mechanics raises are conceptual in nature. What is the proper interpretation of Quantum Mechanics? This is a question touching on most fundamental issues, and it is, at this stage, safe to say that there is no answer to this question yet on which physicists and philosophers of science could agree. It is, moreover, no exaggeration to say that the problem of the conceptual understanding of Quantum Mechanics constitutes one of the great intellectual puzzles of our time.

The topic of the present Handbook is, though related to this gigantic issue, more modest in nature. It can, briefly, be described as follows. Quantum Mechanics owes is tremendous success to a mathematical formalism. It is the mathematical and logical investigation of the various aspects of this formalism that constitutes the topic of the present Handbook.

This formalism the core of which is the mathematical structure of a Hilbert space received its final elegant shape in John von Neumann's classic 1932 book "Mathematical Foundations of Quantum Mechanics". In 1936 John von Neumann published, jointly with the Harvard matthematician Garret Birkhoff, a paper entitled "The logic of quantum mechanics". In the Introduction the authors say: "The object of the present paper is to discover what logical structure we may hope to find in physical theories which, like quantum mechanics, do not conform to classical logic". The idea of the paper, which was as ingenious as it was revolutionary, was that the Hilbert space formalism of Quantum Mechanics displayed a logical structure that could prove useful to the understanding of Quantum Mechanics. Birkoff and von Neumann were the first to put forward the idea that there is a link between logic and (the formalism of) Quantum Mechanics, and their now famous paper marked the birth of a field of research which has become known as Quantum Logic. The Birkhoff-von Neumann paper triggered, after some time of dormancy admittedly, a rapid development of quantum logical research. Various schools of thought emerged. Let us, in this Introduction, highlight just a few of the milestones in this development.

In his famous essay "Is logic empirical?" Putnam put forward the view that the role played by logic in Quantum Mechanics is similar to that played by geometry in the theory of relativity. On this view logic is as empirical as geometry. Putnam's revolutionary thesis triggered a discussion which was highly fruitful not only for Quantum Mechanics but for our views on the nature of logic in general. The reader will find a discussion of Putnam's thesis in this Handbook.

Another school of thought was initiated by Piron's "Axiomatique quantique". This school, which has become known as the Geneva school, aimed at reconstructing the formalism of Quantum Mechanics from first principles. It was Piron's student Diederik Aerts who continued this in Brussels. The achievements of the Geneva-Brussels school are reflected in various chapters witten by Aerts and former students of his.

In Italy it was Enrico Beltrametti and Maria Luisa Dalla Chiara, just to mention two names, who founded another highly influential school which is well represented in this Handbook.

Highly sophisticated efforts resulted in linking the logic of Quantum Mechanics to mainstream logic. Just to give a flavour of this, let us mention that Nishimura studied Gentzen type systems in the context of Quantum Logic. Abramsky and Coecke in Oxford and Sernadas in Lisbon as well as others established the connection with Categorial Logic and Linear Logic, and the connection with Non-Monotonic Logic was made by the editors.

Prior to this, another direction of research had focused on the lattice structures relevant to Quantum Logic. Essentially, this field of research was brought to fruition in the USA by the pioneering work of Foulis and Greechie on orthomodular lattices.

Moreover, we have to mention the Czech-Slovak school which was highly influential in establishing the vast field of research dealing with the various abstract Quantum Structures which constitute the topic of a whole volume of this Handbook. Let us in this context just mention the names of Anatolij Dvurecenskij and Sylvia Pulmannova in Bratislava and Pavel Ptak in Prague.

The editors are happy and grateful to have succeeded in bringing together the most eminent scholars in the field of Quantum logic and Quantum Structures for the sake of the present Handbook. We cordially thank all the authors for their contributions and their cooperation during the preparation of this work. Most of these authors are members of the Internatonal Quantum Structures Association (IQSA). We would like to express our deep gratitude to IQSA and in particular to its President, Professor Anatolij Dvurecenskij, for cooperating so closely with us and supporting us so generously during the preparation of this Handbook.

The present Handbook is an impressive document of the intellectual achievements which have been made in the study of the logical and mathematical structures arising from Quantum Mechanics. We hope that it will turn out to be a milestone on the path that will ultimately lead to the solution of one of the great intellectual puzzles of our time, namely the understanding of Quantum Mechanics.

The Editors: Kurt Engesser, Dov Gabbay and Daniel Lehmann
Germany, London, and Israel
May 2008

Edited by K. Engesser, D. M. Gabbay and D. Lehmann

THE BIRKHOFF–VON NEUMANN CONCEPT OF QUANTUM LOGIC

Miklós Rédei

1 INTRODUCTION

Quantum logic was born with the following conclusion of Garrett Birkhoff and John von Neumann in their joint paper (henceforth "Birkhoff-von Neumann paper") published in 1936:

> Hence we conclude that the *propositional calculus of quantum mechanics has the same structure as an abstract projective geometry.* [Birkhoff and von Neumann, 1936] (Emphasis in the original)

This was a striking conclusion in 1936 for two reasons, one ground breaking and one conservative: ground breaking because it opened the way for the development of algebraic logic in the direction of non-classical algebraic structures that have much weaker properties than Boolean algebras. Conservative because an abstract projective geometry is an orthocomplemeneted, (non-distributive), *modular* lattice; however, the non-Boolean algebra that seemed in 1936 to be the most natural candidate for quantum logic was the *non-modular*, orthomodular lattice of all projections on an infinite dimensional complex Hilbert space. Indeed, subsequently quantum logic was (and typically still is) taken to be an orthocomplemented, *non-modular*, orthomodular lattice. Hence, the concept of quantum logic proposed by Birkhoff and von Neumann in their seminal paper differs markedly from the notion that became later the standard view – it is more conservative than one would expect on the basis of later developments.

There are not many historical investigations in the enormous quantum logic literature [Pavicic, 1992] that aim at scrutinizing the Birkhoff-von Neumann notion of quantum logic, and especially the discrepancy between the standard notion and the Birkhoff-von Neumann concept: [Bub, 1981b], [Bub, 1981a], [Rédei, 1996], Chapter 7. in [Rédei, 1998], [Rédei, 2001], [Dalla-Chiara *et al.*, 2007] (see also [Popper, 1968] and [Scheibe, 1974]). The recent discovery and publication in [Rédei, 2005] of von Neumann's letters to Birkhoff during the preparatory phase (in 1935) of their 1936 paper have made it possible to reconstruct in great detail [Rédei, 2007] the conceptual considerations that culminated in the 1936 paper's main conclusion cited above. As a result of these historical studies we now understand quite well why Birkhoff and von Neumann postulated the "quantum propositional calculus" to be a modular lattice and rejected explicitly the idea that quantum propositional calculus can be identified with the non-modular, orthomodular lattice of all closed linear subspaces of an infinite dimensional complex Hilbert space (Hilbert lattice).

The aim of this review is to recall the Birkhoff-von Neumann concept of quantum logic together with the pertinent mathematical notions necessary to understand the development of their ideas. Special emphasis will be put on the analysis of the difference between their views and the subsequent standard notion of quantum logic. The structure of the review is the following. For reference, and in order to place the Birkhoff-von Neumann concept in appropriate context, section 2 recalls briefly the standard notion of quantum logic in terms of algebraic semantics. Based on excerpts from the recently discovered and published letters by von Neumann to Birkhoff, section 3 reconstructs the main steps of the thought process that led Birkhoff and von Neumann to abandon Hilbert lattice as quantum logic and to propose in their published paper an abstract projective geometry as the quantum propositional system. Section 4 argues that von Neumann was not satisfied with their idea after 1936: He would have liked to see quantum logic worked out in much greater detail – he himself tried to achieve this but did not succeed. Section 4 also attempts to discern the conceptual obstacles standing in the way of elaborating quantum logic along the lines von Neumann envisaged. The concluding section 5 summarizes the main points and makes some further comments on the significance of the Birkhoff-von Neumann concept.

2 QUANTUM LOGIC: LOGICIZATION OF NON-BOOLEAN ALGEBRAIC STRUCTURES. THE STANDARD VIEW.

It is well known that both the syntactic and the semantic aspects of classical propositional logic can be described completely in terms of Boolean algebras: The Tarski-Lindenbaum algebra $\mathcal{A}$ of classical propositions is a Boolean algebra and a deductive system formulated in a classical propositional logic can be identified with a filter in $\mathcal{A}$. The notions of syntactic consistency and completeness correspond to the filter being proper and being prime (equivalently: maximal), respectively. The notion of interpretation turns out to be a Boolean algebra homomorphism from $\mathcal{A}$ into the two element Boolean algebra, and all the semantic notions are defined in terms of these homomorphisms. All this is expressed metaphorically by Halmos' famous characterization of the (classical) logician as the dual of a (Boolean) algebraist [Halmos, 1962, p. 22], a characterization which has been recently "dualized" by Dunn and Hardegree: "By duality we obtain that the algebraist is the dual of the logician." [Dunn and Hardegree, 2001, p. 6].

The problem of quantum logic can be formulated as the question of whether the duality alluded to above also obtains if Boolean algebras are replaced by other, typically weaker algebraic structures arising from the mathematical formalism of quantum mechanics. It turns out that formal logicization is possible for a large class of non-Boolean structures. Following Hardegree [Hardegree, 1981b], [Hardegree, 1981a] the standard (sometimes called "orthodox") concept is described below, and it is this concept with which the Birkhoff von Neumann concept will be contrasted.

Standard quantum logic comes in two forms: abstract (also called orthomodular) quantum logic and concrete (also called Hilbert) quantum logic. The semantics is similar in both cases, but the latter determines a stronger logic.

Let $\mathcal{K} = \{P, \&, \sqcup, \sim\}$ be a zeroth order formal language with the set P of sentence variables $p, q \ldots$, two place connectives & (*and*), $\sqcup$ (*or*), negation sign $\sim$, parentheses (,), and let $\mathcal{F}$ be the set of well formed formulas in $\mathcal{K}$ defined in the standard way by induction from P: $\mathcal{F}$ is the smallest set for which the following two conditions hold:

$$P \subset \mathcal{F} \tag{1}$$

$$\text{if} \quad \phi, \psi \in \mathcal{F} \quad \text{then} \qquad (\phi \& \psi), (\phi \sqcup \psi), (\sim \phi) \in \mathcal{F} \tag{2}$$

Let $(\mathcal{L}, \vee, \wedge, \perp)$ be an orthomodular lattice. Orthomodularity of $\mathcal{L}$ means that the following condition holds:

(3) orthomodularity: If $A \leq B$ and $A^{\perp} \leq C$, then $A \vee (B \wedge C) = (A \vee B) \wedge (A \vee C)$

Orthomodularity is a weakening of the modularity law:

(4) modularity: If $A \leq B$, then $A \vee (B \wedge C) = (A \vee B) \wedge (A \vee C)$

which itself is a weakening of the distributivity law:

(5) distributivity: $A \vee (B \wedge C) = (A \vee B) \wedge (A \vee C)$ for all A,B,C

The set $\mathcal{G}_Q \subseteq \mathcal{L}$ in an orthomodular lattice is called a *generalized filter* if

$$I \in \mathcal{G}_Q \tag{6}$$

$$\text{if } A \in \mathcal{G}_Q \text{ and } A^{\perp} \vee (A \wedge B) \in \mathcal{G}_Q \text{ then } \quad B \in \mathcal{G}_Q \tag{7}$$

Given a pair $(\mathcal{L}, \mathcal{G}_Q)$ the map $i \colon \mathcal{F} \to \mathcal{L}$ is called an $(\mathcal{L}, \mathcal{G}_Q)$-interpretation if

$$i(\phi \& \psi) = i(\phi) \wedge i(\psi) \tag{8}$$

$$i(\phi \sqcup \psi) = i(\phi) \vee i(\psi) \tag{9}$$

$$i(\sim \phi) = i(\phi)^{\perp} \tag{10}$$

Each interpretation i determines a $(\mathcal{L}, \mathcal{G}_Q)$-valuation v_i by

$$(11) \quad v_i(\phi) = \begin{cases} 1 \ (true) & \text{if} \quad i(\phi) \in \mathcal{G}_Q \\ 0 \ (false) & \text{if} \quad i(\sim \phi) \in \mathcal{G}_Q \\ undetermined & \text{otherwise} \end{cases}$$

If $V(\mathcal{L})$ denotes the set of all $(\mathcal{L}, \mathcal{G}_Q)$-valuations and V is the class of valuations determined by the class of orthomodular lattices, then $\phi \in \mathcal{F}$ is called valid if $v(\phi) = 1$ for every $v \in V$, and a class of formulas Γ is defined to entail ϕ if $v(\psi) = 1$ for all $\psi \in \Gamma$ implies $v(\phi) = 1$. One can define the quantum analog $\rightarrow_Q$ of the classical conditional by

(12) $\phi \rightarrow_Q \psi = \sim \phi \sqcup (\phi \& \psi)$

and one can formulate a deduction system in $\mathcal{K}$ using $\rightarrow_Q$ such that one can prove soundness and completeness theorems for the resulting quantum logical system (see [Hardegree, 1981b],[Hardegree, 1981a]).

A specific class of orthomodular lattices is the category of Hilbert lattices: A Hilbert lattice $\mathcal{P}(\mathcal{H})$ with lattice operations $\wedge, \vee, \perp$ is the set $\mathcal{P}(\mathcal{H})$ of all projections (equivalently: closed linear subspaces) of a complex, possibly infinite dimensional Hilbert space $\mathcal{H}$, where the lattice operations $\wedge, \vee$ and $\perp$ are the set theoretical intersection, closure of the sum and orthogonal complement, respectively. Note that the Hilbert lattice $\mathcal{P}(\mathcal{H})$ is not only non-distributive but it also is non-modular if the dimension of the Hilbert space is infinite [Rédei, 1998].

It is important that while all the definitions and stipulations made above in connection with orthomodular lattices are meaningful for Hilbert lattices, no completeness results are presently known for the semantics determined by Hilbert lattices: The deduction system that works in the case of (abstract) orthomodular lattices is not strong enough to yield all statements that are valid in Hilbert lattices: the "ortho-arguesian law", which is valid in Hilbert lattices, does not hold in every orthomodular lattice (see [Kalmbach, 1981], [Dalla-Chiara and Giuntini, 2002]).

3 THE BIRKHOFF-VON NEUMANN CONCEPT OF QUANTUM LOGIC

The Birkhoff-von Neumann paper can be viewed as the first paper in which the suggestion to logicize a non-Boolean lattice appears. There are however several types of non-Boolean lattice. Which one is supposed to be logicized?

3.1 Which non-Boolean lattice to logicize?

At the time of the birth of quantum logic the notion of an abstract orthomodular lattice did not yet exist; however, the canonical example of non-distributive, orthomodular lattices, the Hilbert lattice $\mathcal{P}(\mathcal{H})$, was known already, and, since this structure emerges naturally from the Hilbert space formalism of quantum mechanics, $\mathcal{P}(\mathcal{H})$ was the most natural candidate in 1935 for the propositional system of quantum logic. Indeed, Birkhoff and von Neumann did consider $\mathcal{P}(\mathcal{H})$ as a possible propositional system of quantum logic; yet, this lattice was not their choice: The first indication that $\mathcal{P}(\mathcal{H})$ may not be a suitable candidate for a quantum propositional system is in von Neumann's letter of (January 19, 1935). Von Neumann writes:

Using the operator-description,
$a \vee b$, $a \wedge b$ can be formed, if the
physically significant operators form a *ring*. (← I believe this).
This, I think should be assumed anyhow,
even if one does not require that
all operators are phys.[ically] significant. (← but I am rather doubting lately this.)

But we need probably not insist on this point too much.
(von Neumann to Birkhoff, January 19, 1935 ?[1]), [Rédei, 2005, p. 51]

3.2 Von Neumann algebras

A "ring of operators" von Neumann is referring to in the above quotation is known today as a von Neumann algebra: a set $\mathcal{N}$ of bounded operators on a Hilbert space $\mathcal{H}$ is a von Neumann algebra if it contains the unit, is closed with respect to the adjoint operation and is closed in the strong operator topology. The latter requirement means that if Q_n is a sequence of operators from $\mathcal{M}$ such that for all $\xi \in \mathcal{H}$ we have $Q_n\xi \to Q\xi$ for some bounded operator Q on $\mathcal{H}$, then $Q \in \mathcal{M}$ (see [Takesaki, 1979] for the theory of von Neumann algebras).

If S is any set of bounded operators on $\mathcal{H}$, then its (first) commutant S' is the set of bounded operators that commute with every element in S i.e.:

$$S' \equiv \{Q : QX = XQ, \text{ for all } X \in S\}$$

The operation of taking the commutant can be iterated: $S'' \equiv (S')'$, and it is clear that S is contained in the second commutant, so the second commutant S'' is an extension of S. How much of an extension? The answer to this question, von Neumann's double commutant theorem, is the most fundamental result in the theory of von Neumann algebras:

PROPOSITION 1 Double commutant theorem. *S is strongly dense in S''.*

Proposition 1 implies that a $*$-algebra of bounded operators on a Hilbert space that contains the unit is a von Neumann algebra if and only if it coincides with its second commutant. A von Neumann algebra $\mathcal{N}$ is called a *factor* if the only elements in $\mathcal{N}$ that commute with every other element in $\mathcal{N}$ are the constant multiples of the identity, i.e. if

(13) $\mathcal{N} \cap \mathcal{N}' = \{\lambda I : \lambda \text{ a complex number}\}$

An immediate corollary of the double commutant theorem is the characteristic property of von Neumann algebras that they contain many projections; in fact, they contain enough projections for the set of projections $\mathcal{P}(\mathcal{M})$ to determine the von Neumann algebra completely in the sense

[1]Von Neumann's letters are not always properly dated: the year is occasionally missing. If this is the case, we put a question mark after the year – the context makes it clear that the year of writing is 1935 in cases of all the letters quoted here.

(14) $\mathcal{M} = (\mathcal{P}(\mathcal{M}))''$

Moreover, we have

PROPOSITION 2. *The set $\mathcal{P}(\mathcal{M})$ of projections of a von Neumann algebra $\mathcal{M}$ is a complete, orthomodular lattice (called von Neumann lattice) with respect to the lattice operations inherited from $\mathcal{P}(\mathcal{H})$.*

In particular the Hilbert lattice $\mathcal{P}(\mathcal{H})$ of *all* projections on a Hilbert space is a complete orthomodular lattice, since the set $\mathcal{B}(\mathcal{H})$ of all bounded operators on $\mathcal{H}$ is a von Neumann algebra. Specifically the spectral projections of the set of *all* (not necessarily bounded) selfadjoint operators coincides with the set of *all* projections.

It is very important that while all von Neumann lattices are orthomodular, some have the stronger property of modularity. There is a subtle connection between the type of a von Neumann algebra (in the sense of the Murray-von Neumann classification theory) and the modularity of its projection lattice. We shall return to this issue later.

3.3 Non-modularity of Hilbert lattice

While in January 1935 von Neumann did not intend to insist on restricting the set of physical quantities to a proper subset of all possible operators, by November 1935 he changed his mind:

> I am somewhat scared to consider all physical quantities = bounded self-adjoint operators as a lattice.
> (von Neumann to Birkhoff, November 6, 1935 ?), [Rédei, 2005, p. 53]

The reason why he changed his mind was the realization that the Hilbert lattice $\mathcal{P}(\mathcal{H})$ is not modular if the Hilbert space is infinite dimensional (note that "B-lattice" means modular lattice in the next quotation):

> In any linear space H the linear subspaces $K, L, M, \ldots$ form a B-lattice $\mathcal{L}$ with the
>
> "meet" $K \cap L$: intersection of K and L in the sense of set theory
>
> "join" $K \cup L$: linear sum of K and L, i.e. set of all $f+g$, $f \in K$, $g \in L$
>
> (Proof obvious.) But in a *metric-linear space* $\overline{H}$ the lattice $\overline{\mathcal{L}}$ of all *closed-linear subspaces* $\overline{KLM}, \ldots$, for which the "join" is defined as
>
> "join" $\overline{K} \vee \overline{K}$: closure of the linear sum of K and L, i.e. the set of all condensation points of the $f+g$, $f \in K$, $g \in L$

> while the "meet" is as above, is *not necessarily a B-lattice. This is in particular the case in Hilbert space.* ($\overline{K} \cup \overline{L}$ and $\overline{K} \vee \overline{L}$ are identical if $\overline{K}, \overline{L}$ are both closed and *orthogonal* to each other, but not for any two closed $\overline{K}, \overline{L}$!)
>
> In fact, it is possible to find three closed-linear subspaces $\overline{K}$, $\overline{L}$, $\overline{M}$ of Hilbert space $\overline{H}$, for which
>
> (15) $\overline{K} \subsetneqq \overline{M}, \quad (\overline{K} \vee \overline{L}) \cap \overline{M} \supsetneqq \overline{K} \vee (\overline{L} \cap \overline{M})$
>
> (von Neumann to Birkhoff, November 6, 1935 ?), [Rédei, 2005, p. 54], emphasis in original.

This letter contains a detailed proof that subspaces $\overline{K}$, $\overline{L}$, $\overline{M}$ exist that satisfy (15) (and thereby violate modularity (4)). (The Birkhoff-von Neumann paper just states this fact without detailed argument.) Von Neumann's proof makes use of the theory of *unbounded* selfadjoint operators, utilizing the fact that one can find two unbounded selfadjoint operators X and Y on an infinite dimensional Hilbert space such that the intersection of their domains is empty. Von Neumann emphasizes this feature of his proof:

> Examples could be constructed which make no use of operator theory, but I think that this example shows more clearly "what it's all about": It is the existence of "pathological" operators – like X, Y above – in Hilbert space, which destroys the B-lattice character.
> (von Neumann to Birkhoff, November 6, 1935 ?), [Rédei, 2005, p. 55]

Von Neumann regarded this pathological behavior of the set of all unbounded operators on a Hilbert space a very serious problem because it prohibits adding and composing these operators in general, which entails that these operators do not form an algebra. In von Neumann's eyes this was a great obstacle to doing computations with those operators, and since the selfadjoint operators are representatives of quantum physical quantities, it appeared unnatural to him that they behave so irregularly that forming an algebra from them was not possible. He pointed out this pathology several times in his published papers (see e.g. paragraph 6. of Introduction in [Murray and von Neumann, 1936]), and the pathological character of the set of all selfadjoint operators was one of the main reasons why he hoped as late as in his famous talk on "Unsolved Problems in Mathematics" in 1954 (see [von Neumann, 2001] and [Rédei, 1999]) that a restricted set of operators, and therefore a specific von Neumann algebra, the type II_1 factor (see below) would be a more suitable mathematical framework for quantum mechanics than Hilbert space theory.

In this situation von Neumann saw two options:

> **(I)** *Either* we define the "join" by $\cup$ (as a honest linear sum), then the lattice is B, but we must admitt all (not-necessarily-closed-) linear subspaces,
>
> **(II)** *or* we define the "join" by $\vee$ (closure of the linear sum), then the B-character is lost.
>
> (von Neumann to Birkhoff, November 6, 1935 ?), [Rédei, 2005, p. 55]

Since $\mathcal{P}(\mathcal{H})$ is not modular and given that von Neumann wished to preserve modularity as a property of the quantum propositional system, one would expect von Neumann to choose option **(I)**. But this is not the case. Von Neumann thinks through the consequences of choosing option **(I)** first:

> Let us first consider the alternative **(I)**. The orthogonal complement K' still has the property $K' \cup L' = (K \cap L)'$, but $K' \cap L' = (K \cup L)'$ and $K'' = K$ are lost. We have $K \cap K' = 0$, while $K \cup K'$ is everywhere dense, but not necessarily I. There is a funny relationship between K and its "closure" K''. (For instance: All probabilities in the state K are equal to those in the state K'', but "meets" ($K \cap L$ and $K'' \cap L''$, even for closed L's) may differ.)
>
> The situation is strongly reminiscent of the "excluded middle" troubles, although I did not yet compare all details with those of the class-calculus in "intuitionistic" logics.
>
> After all it is so in normal logics, too, that these troubles arise as soon as you pass to infinite systems, although I must admitt, that the difficulties there are more "optional" then[2] here.
>
> It has to be said, finally, that even in this case **(I)** complements exist, i.e., that for every K there exists K^*'s for which $K \cup K^* = I$, $K \cap K^* = 0$, but one needs the Hammel-basis-construction to get them.
> (von Neumann to Birkhoff, November 6, 1935 ?), [Rédei, 2005, p. 55]

So, while von Neumann evaluates alternative **(I)** as representing an option which cannot be excluded on the grounds of being either algebraically or logically extremely weird (although it is clear from the above that he did not like the asymmetric failure of De Morgan's law), he prefers option **(II)** in spite of its being seemingly counterintuitive. Here is why:

> Alternative **(II)** seems to exclude Hilbert space, if one sticks to B-lattices.[3] Still one may observe this:
>
> Consider a ring $\mathcal{R}$ of operators in Hilbert space. The idempotents of $\mathcal{R}$ form a lattice $\mathcal{L}_{\mathcal{R}}$. One sees easily, that $\mathcal{L}_{\mathcal{R}}$ is irreducible (=no direct sum), if and only if the center of $\mathcal{R}$ consists of the αI (α=complex

[2]Spelling error, should be "than".

[3]Recall that B-lattice means modular lattice.

number) only, i.e. if $\mathcal{R}$ is a ring of the sort which Murray and I considered. (We called them "factors".) $\mathcal{L}_{\mathcal{R}}$ contains $0, I$ and a complement which dualises $\cup$ and $\cap$. (Now $\cup$ corresponds to what I called $\vee$, case **(II)**.) One may ask: When is $\mathcal{L}_{\mathcal{R}}$ a B-lattice? The answer is (this is not difficult to prove): If and only if the ring $\mathcal{R}$ is finite in the classification Murray and I gave. I.e.: $\mathcal{R}$ must be isomorphic:

1. either to the full matrix-ring of a finite-dimensional Euclidean space (say n-dimensional, $n = 1, 2, \ldots$),
2. to one of those of our rings, in which each idempotent has a "dimensionality", the range of which consists of all real numbers $\geq 0, \leq 1$, and which is uniquely determined by its formal properties.

We called 1. "Case I_n" and 2. "Case II_1".

Thus for operator-lattices the B-lattice axiom

$$a \leq b \quad \rightarrow \quad (a \cup b) \cap c = a \cup (b \cap c)$$

leads directly to the cases $\mathrm{I}_1, \mathrm{I}_2, \ldots$ *and* II_1!
(von Neumann to Birkhoff, November 6, 1935 ?), [Rédei, 2005, p. 56]

3.4 Types of von Neumann algebras

Von Neumann refers here to the Murray-von Neumann classification theory of factors, which was worked out by him in collaboration with F.J Murray precisely at the time (1934-1935) when he was working with G. Birkhoff on quantum logic [Murray and von Neumann, 1936]. Von Neumann (partly in collaboration with F.J. Murray) published four major papers on the theory of von Neumann algebras [Murray and von Neumann, 1936], [Murray and von Neumann, 1937], [von Neumann, 1940] and [Murray and von Neumann, 1943]. The first paper's main result was a classification theory of von Neumann algebras that are irreducible in the sense of not containing non-trivial operators commuting with every other operator in the algebra (i.e. "factors"). The set $\mathcal{B}(\mathcal{H})$ of *all* bounded operators is clearly a factor and it turned out that there are five classes of factors, the different types are denoted by von Neumann as I_n, I_∞, II_1, II_∞ and III_∞. The classification of factors was given in terms of a (relative) dimension function d defined on the lattice of projections $\mathcal{P}(\mathcal{M})$ of a von Neumann algebra $\mathcal{M}$. The map d from $\mathcal{P}(\mathcal{M})$ into the set $\mathbb{R}^+ \cup \{\infty\}$ is a dimension function if

(i) $d(A) > 0$ if and only if $A \neq 0$,

(ii) $d(A) = d(B)$ if there exists an isometry $U \in \mathcal{M}$ between ranges of the projections A and B,

(iii) $d(A) + d(B) = d(A \vee B) + d(A \wedge B)$.

The (relative) dimension function d on an arbitrary factor is a generalization of the ordinary linear dimension of the closed linear subspace a projection projects to, and the ordinary dimension takes on the positive integer values $0, 1, 2, \ldots, n$ and $0, 1, 2, \ldots$, respectively, in the two well-known cases of the set of all bounded operators on a finite, n-dimensional (respectively infinite) dimensional Hilbert space. In the cases II_1, II_∞ and III_∞ the ranges of the dimension function are, respectively, the following: the unit interval $[0,1]$, the set of non-negative real numbers $\mathbb{R}^+$ and the two element set $\{0, \infty\}$. (See [Takesaki, 1979] for a systematic treatment, or [Rédei, 1998], [Petz and Rédei, 1995] for a brief review of the Murray-von Neumann classification theory). The result of the classification theory can thus be summarized in the form of the following table:

range of d	type of factor $\mathcal{N}$	the lattice $\mathcal{P}(\mathcal{M})$
$\{0, 1, 2, \ldots, n\}$	I_n	modular
$\{0, 1, 2, \ldots, \infty\}$	I_∞	orthomodular, non-modular
$[0, 1]$	II_1	modular
$\mathbb{R}^+$	II_∞	orthomodular, non-modular
$\{0, \infty\}$	III_∞	orthomodular, non-modular

3.5 The type II_1 factor

Thus the significance of the existence of type II_1 factors is that their projection lattices are modular. (Accordingly, the set of all (not necessarily bounded) selfadjoint operators that they determine are free of the pathologies which von Neumann considered undesirable.)

Von Neumann's conclusion:

> This makes me strongly inclined, therefore, to take the ring of all bounded operators of Hilbert space ("Case I_∞" in our notation) less seriously, and Case II_1 more seriously, when thinking of an ultimate basis of quantum mechanics.
> (von Neumann to Birkhoff, November 6, 1935 ?), [Rédei, 2005, p. 56]

As can be inferred from von Neumann's letter to Birkhoff (November 6, 1935 ?), [Rédei, 2005, p. 59-64], Birkhoff suggested another idea to save the modularity of the lattice formed by some subspaces of a Hilbert space: by restricting the linear subspaces to the finite dimensional ones. Von Neumann did not consider this idea in detail, but thought that it was not an attractive one:

> Many thanks for your letter. Your idea of requiring $a \leq c \rightarrow a \cup (b \cap c) = (a \cup b) \cap c$ in Hilbert space for the finite a, b, c only is very interesting, but will it permit to differentiate between Hilbert-space and other Banach-spaces?
> (von Neumann to Birkhoff, November 13, 1935 ?), [Rédei, 2005, p. 59]

Rather than answering this rhetorical question, von Neumann makes his famous confession (quoted in part by Birkhoff in [Birkhoff, 1961]), reaffirming that the operator algebraic results related to classification theory of von Neumann algebras reduce the privileged status of Hilbert space quantum mechanics:

> I would like to make a confession which may seem immoral: I do not believe absolutely in Hilbert space any more. After all Hilbert-space (as far as quantum-mechanical things are concerned) was obtained by generalizing Euclidean space, footing on the principle of "conserving the validity of all formal rules". This is very clear, if you consider the axiomatic-geometric definition of Hilbert-space, where one simply takes Weyl's axioms for a unitary-Euclidean-space, drops the condition on the existence of a finite linear basis, and replaces it by a minimum of topological assumptions (completeness + separability). Thus Hilbert-space is the straightforward generalization of Euclidean space, if one considers the *vectors* as the essential notions.
>
> Now we[4] begin to believe, that it is not the *vectors* which matter but the *lattice of all linear (closed) subspaces*. Because:
>
> 1. The vectors ought to represent the physical *states*, but they do it redundantly, up to a complex factor, only.
> 2. And besides the *states* are merely a derived notion, the primitive (phenomenologically given) notion being the *qualities*, which correspond to the *linear closed subspaces*.
>
> But if we wish to generalize the lattice of all linear closed subspaces from a Euclidean space to infinitely many dimensions, then one does not obtain Hilbert space, but that configuration, which Murray and I called "case II_1." (The lattice of all linear closed subspaces of Hilbert-space is our "case I_∞".) And this is chiefly due to the presence of the rule
>
> $$a \leq c \rightarrow a \cup (b \cap c) - (a \cup b) \cap c$$
>
> This "formal rule" would be lost, by passing to Hilbert space!
> (von Neumann to Birkhoff, November 13, 1935 ?), [Rédei, 2005, p. 59]

3.6 *From the type II_1 factor to abstract continuous geometry*

So it would seem that the modular lattice of the type II_1 factor von Neumann algebra emerges as the strongest candidate for logicization, and so one would expect this lattice to be declared in the Birkhoff-von Neumann paper to be the propositional system of quantum logic. But this is not the case; in fact, the published paper does *not* at all refer to the results of the Murray-von Neumann classification theory of von Neumann algebras to support the modularity postulate.

[4]With F.J. Murray, von Neumann's coauthor.

Why? Von Neumann's letters to Birkhoff also contain clues for the the answer to this question, and the answer is that von Neumann's mind moved extremely quickly from the level of abstractness of von Neumann algebras to the level of abstractness represented by continuous geometries — and this move was taking place precisely during the preparation of the quantum logic paper: in his letter to Birkhoff (November 6, 1935 ?) von Neumann writes:

> Mathematically – and physically, too – it seems to be desirable, to try to make a general theory of dimension in *complemented, irreducible B-lattices*, without requiring "finite chain conditions". I am convinced, that by adding a moderate amount of continuity-conditions, the existence of a numerical dimensionality could be proved, which
>
> 1. is uniquely determined by its formal properties,
> 2. and after a suitable normalisation has either the range $d = 1, 2, \ldots, n$ ($n = 1, 2, \ldots,$ finite!) or $d \geq 0, \leq 1$.
>
> I have already obtained some results in this direction, which connect the notion of dimension in a very funny way with the perspectivities and projectivities in projective geometry.
>
> It will perhaps amuse you if I give some details of this. Here they are: (von Neumann to Birkhoff, November 6, 1935 ?), [Rédei, 2005, p. 56]

And there follows in the letter a three page long exposition of the theory of continuous geometries, which is not reproduced here. In his letter written a week later (November 13, 1935 ?), [Rédei, 2005, p. 59-64], von Neumann gives an even more detailed description of his results on continuous geometry, which confirm the two conjectures 1. and 2. above completely: on every projective geometry there exists a dimension function d having the properties

$$0 \leq \quad d(A) \quad \leq 1 \tag{16}$$

$$d(A) + d(B) \quad = \quad d(A \vee B) + d(A \wedge B) \tag{17}$$

and having discrete or continuous range.[5] These results do not appear in the Birkhoff-von Neumann paper on quantum logic, von Neumann published them separately [von Neumann, 1936] (cf. footnote 33 in the Birkhoff-von Neumann paper [Birkhoff and von Neumann, 1936]).

Thus by the time it came to the final version of the quantum logic paper, von Neumann knew that the projection lattice of a type II_1 von Neumann algebra is just a special case of more general continuous geometries that admit well-behaving probability measures, and this explains why the major postulate in the Birkhoff-von Neumann paper is formulated in the section entitled "Relation to abstract projective geometries" and reads:

[5]The discrepancy between Eq. (16) and the ranges mentioned under 2. above are due to different normalizations.

Hence we conclude that the *propositional calculus of quantum mechanics has the same structure as an abstract projective geometry.* [Birkhoff and von Neumann, 1936] (Emphasis in the original)

3.7 Probability and quantum logic

The finite dimension function on a projective geometry, in particular the dimension function with the range [0, 1] on the continuous projective geometry, was for von Neumann crucially important in his search for a proper quantum logic: he interpreted the dimension function as an *a priori* probability measure on the modular lattice of the quantum propositional system. Thus by requiring quantum logic to be an abstract continuous geometry with a dimension function, Birkhoff and von Neumann created an analogy with classical logic and probability theory, where a Boolean algebra is both a propositional system and a random event structure on which probability measures are defined. While there is no detailed discussion of this aspect of the dimension function in the von Neumann-Birkhoff correspondence, the Birkhoff-von Neumann paper points out that properties (16)-(17) of the dimension function describe the formal properties of probability. Von Neumann regarded it as another pathology of the total Hilbert lattice $\mathcal{P}(\mathcal{H})$ that there exists no probability measure on it that satisfies conditions (16)-(17. This is because one has the following theorem:

PROPOSITION 3. *Let $\mathcal{L}$ be a bounded lattice. If there exists a finite dimension function d on $\mathcal{L}$ (i.e. a map d from $\mathcal{L}$ into the set of real numbers that has the properties (16)-(17)), then the lattice is modular.*

It is very easy to see that subadditivity (property (17)) is a necessary condition for a measure to be interpreted as probability understood as relative frequency in the sense of von Mises [von Mises, 1919], [von Mises, 1981]:

Assume that the probability $p(X)$ $(X = A, B, A \wedge B, A \vee B)$ is to be interpreted as relative frequency in the following sense:

1. There exists a fixed ensemble $\mathcal{E}$ consisting of N events such that
2. for each event X one can decide unambiguously and
3. without changing the ensemble whether X is the case or not;
4. $p(X) = \frac{\#(X)}{N}$ where $\#(X)$ is the number of events in $\mathcal{E}$ for which X is the case.[6]

Under the assumptions 1.-4. it trivially follows that (17) holds since one can

[6]Strictly speaking one should write $p(X) = \lim_{N\to\infty} \frac{\#(X)}{N}$; however, the limit is not important from the point of view of the present considerations, so we omit it.

write

$$
\begin{aligned}
\frac{\#(A \cup B)}{N} + \frac{\#(A \cap B)}{N} &= \\
\frac{\#((A \setminus A \cap B) \cup (B \setminus A \cap B) \cup A \cap B))}{N} + \frac{\#(A \cap B)}{N} &= \\
\frac{\#(A \setminus A \cap B) + \#(B \setminus A \cap B) + \#(A \cap B) + \#(A \cap B)}{N} &= \\
\frac{\#(A) + \#(B)}{N}
\end{aligned}
$$

which is the subadditivity. Thus, if a map d on a lattice does *not* have subadditivity (17) then the probabilities $d(X)$ *cannot* be interpreted as probabilities in the sense of relative frequency formulated above via 1.-4.; consequently, the lattice cannot be viewed as representing a collection of random events in the sense of a relative frequency interpreted probability theory specified by 1.-4. (with the understanding that $A \wedge B$ denotes the joint occurrence of events A and B). Since von Neumann embraced the frequency interpretation of probability in the years 1927-1935, this makes understandable why he considered the subadditivity (17) a key feature of probability and, consequently, modularity an important condition to require.

Thus it would seem that within the mathematical framework of continuous geometry, especially within the theory of the type II_1 von Neumann algebras, the Birkhoff and von Neumann concept of quantum logic could restore the harmonious classical picture: random events can be identified with the propositions stating that the event happens, and probabilities can be viewed as relative frequencies of the occurrences of the events. But this restored harmony is illusory for the following reason: von Neumann and Murray showed that a dimension function d on the projection lattice $\mathcal{P}(\mathcal{N})$ of a type II_1 algebra $\mathcal{N}$ can be extended to a trace τ on $\mathcal{N}$. The defining property of a trace τ is

(18) $\tau(XY) = \tau(YX) \qquad \text{for all} \quad X, Y \in \mathcal{N}.$

That is to say, the trace is exactly the functional which is *insensitive* (in the sense of (18)) for the non-commutativity of the algebra. On the other hand, it can be shown that a normal state ϕ on a von Neumann lattice satisfies the additivity (17) if (and only if) it is a trace [Petz and Zemanek, 1988]. Thus, the only quantum probability measures that mesh with the relative frequency interpretation via 1.-4. are the ones given by the trace.

Behind the mathematical fact that only traces satisfy subadditivity lies the conceptual difficulty that assumptions 2. and 3. of the frequency interpretation of probability cannot be upheld in interpreting the elements of a von Neumann lattice as random quantum events and the lattice operation $A \wedge B$ as the joint occurrence of A and B: assumption 3. fails if "deciding" means "measuring", since measuring disturbs the measured system, hence also the ensemble; therefore, there is no single, fixed, well-defined ensemble in which to compute as relative frequencies the probabilities of *all* projections representing quantum attributes.

Von Neumann was fully aware of this difficulty: One of his arguments against hidden variables is essentially the argument that if hidden parameters did exist, then it should be possible to resolve any ensemble into subensembles that are dispersion-free, but this is not possible if "resolving" means selecting subensembles by measurement, since if one selection ensures dispersion-freeness with respect to observable Q_1, the subsequent selection by measurement of this subensemble into a further subensemble in which another observable Q_2 has sharp value destroys the result of the first step if Q_1 and Q_2 are incompatible (see [von Neumann, 1932, p. 304]). Yet, in his 1932 book von Neumann thought to be able to maintain an ensemble interpretation of quantum probability by getting around the problem that quantum measurements disturb the ensemble:

> Even if two or more quantities R, S in a single system are not simultaneously measurable, their probability distributions in a given ensemble $[S_1, \dots S_N]$ can be obtained with arbitrary accuracy if N is sufficiently large.
>
> Indeed, with an ensemble of N elements it suffices to carry out the statistical inspections, relative to the distribution of values of the quantity R, not on all N elements $[S_1, \dots S_N]$, but on any subset of M ($\leq N$) elements, say $[S_1, \dots S_M]$ – provided that M, N are both large, and that M is very small compared to N. Then only the M/N-th part of the ensemble is affected by the changes which result from the measurement. The effect is an arbitrary small one if M/N is chosen small enough – which is possible for sufficiently large N, even in the case of large M... In order to measure two (or several) quantities R, S simultaneuosly, we need two sub-ensembles, say $[S_1, \dots S_M]$, $[S_{M+1}, \dots S_{2M}]$ ($2M \leq N$), of such a type that the first is employed obtaining the statistics of R, and the second in obtaining those of S. The two measurements therefore do not disturb each other, although they are performed in the same ensemble $[S_1, \dots S_N]$ and they can change this ensemble only by an arbitrarily small amount, if $2M/N$ is sufficiently small – which is possible for sufficiently large N even in the case of large M ...
>
> [von Neumann, 1932, p. 300]

Implicit in this reasoning is the assumption that the subensembles are representative of the large ensemble in the sense that the relative frequency of every attribute is the same both in the original and in the subensemble. This non-trivial assumption, known in von Mises' theory as the requirement of "randomness" concerning the ensembles that can serve as ensembles to compute probabilities as frequencies, is crucial in von Mises' theory, and von Mises takes pains in giving it a precise formulation (see "Forderung II" in [von Mises, 1919], [von Mises, 1981, p. 61]). Von Neumann does not elaborate on the details and significance for his interpretation of quantum probability of the randomness requirement; apparently he did not see any problem with taking advantage of this non-trivial (and controversial) feature of von Mises interpretation.

However, even granting that an ensemble interpretation remains meaningful if one relaxes 3. in the specification of the frequency interpretation of probability in the way von Neumann does, the problem remains for von Neumann that 2. does not make sense at all in quantum mechanics if one takes the position that (i) $A \wedge B$ represents the joint occurrence of A and B, and (ii) the joint occurrence cannot be checked by measurement at all on whatever ensemble if A and B are not simultaneously measurable.

In sum: There are no "properly non-commutative" probability spaces – as long as one insists on the frequency interpretation of probability; hence, if one wants to maintain the idea of non-commutative probability spaces, with a non-distributive lattice taking the place of Boolean algebra, the frequency view has to go.

4 VON NEUMANN'S POST 1936 STRUGGLE WITH QUANTUM LOGIC

It did: von Neumann abandoned the frequency interpretation after 1936. In an unfinished manuscript written about 1937 and entitled "Quantum logic (strict- and probability logics)" he writes:

> This view, the so-called 'frequency theory of probability' has been very brilliantly upheld and expounded by R. von Mises. This view, however, is not acceptable to us, at least not in the present 'logical' context. [von Neumann, 1961]

Instead, von Neumann embraces in this unfinished note a "logical theory of probability", which he associates with J. M. Keynes, but which he does not spell out in detail in the unfinished note. The only formulation of this idea can be found in his talk delivered years later at the International Congress of mathematicians in Amsterdam 1954:

> Essentially if a state of a system is given by one vector, the transition probability in another state is the inner product of the two which is the square of the cosine of the angle between them. In other words, probability corresponds precisely to introducing the angles geometrically. Furthermore, there is only one way to introduce it. The more so because in the quantum mechanical machinery the negation of a statement, so the negation of a statement which is represented by a linear set of vectors, corresponds to the orthogonal complement of this linear space.
>
> And therefore, as soon as you have introduced into the projective geometry the ordinary machinery of logics, you must have introduced the concept of orthogonality. ... In order to have probability all you need is a concept of all angles, I mean angles other than 90°. Now it is perfectly quite true that in geometry, as soon as you can define the right angle, you can define all angles. Another way to put it is

> that if you take the case of an orthogonal space, those mappings of this space on itself, which leave orthogonality intact, leave all angles intact, in other words, in those systems which can be used as models of the logical background for quantum theory, it is true that as soon as all the ordinary concepts of logic are fixed under some isomorphic transformation, all of probability theory is already fixed.
>
> What I now say is not more profound than saying that the concept of a priori probability in quantum mechanics is uniquely given from the start. ... This means, however, that one has a formal mechanism, in which logics and probability theory arise simultaneously and are derived simultaneously.
> [von Neumann, 2001, p. 244-245]

Von Neumann was intrigued by the determination of probability by logic in the quantum context but he did not consider this "logical theory" (interpretation) of probability as fully understood: he mentions the need for an axiomatic investigation of this issue in his address to the International Congress of Mathematicians (Amsterdam, 1954) [von Neumann, 2001] as one of the open problems in mathematics: he thought that it would shed

> ... a great deal of new light on logics and probably alter the whole formal structure of logics considerably, if one succeeds in deriving this system from first principles, in other words from a suitable set of axioms.
> [von Neumann, 2001, p. 245]

It seems that he worked on this problem and tried to work out a systematic theory of quantum logic after 1936, but he did not succeed. The recently published series of letters he wrote to F.B. Silsbee in [Rédei, 2005], prove this in a remarkable way. The correspondence between von Neumann and Silsbee, then the president of the Washington Philosophical Society, starts with Silsbee's letter (October 31, 1944, unpublished, Library of Congress, Washington D.C.) inviting von Neumann to deliver the Fourteenth Joseph Henry Lecture scheduled for March 17, 1945. In his reply to Silsbee (November 3, 1944) [Rédei, 2005, p. 216] von Neumann accepted the invitation, promising at the same time to write up the lecture in a paper. In his second letter to Silsbee (February 14, 1945) [Rédei, 2005, p. 217] von Neumann specifies the problem of relation of logic and probability in quantum mechanics as the topic of the lecture. Von Neumann did deliver his talk as planned; however, as his letter to Silsbee (June 11, 1945) [Rédei, 2005, p. 218] shows, he did not meet the first deadline of submitting the manuscript of the planned paper entitled "Logic of quantum mechanics". He promises at the same time to deliver the paper by July 7, the latest. But he did not meet this second deadline either; in fact, the promised paper was never written, and von Neumann's subsequent letters to Silsbee (July 2, 1945, October 22, 1945, April 20, 1946 and December 23, 1946, all in [Rédei, 2005]) show von Neumann agonizing over this project. Most revealing is his letter of July 2, 1945 [Rédei, 2005, p. 218-220]. Von Neumann confesses that even if he had not been disrupted by war-work, he might not have been able to write the promised paper on quantum logic:

> It is with great regret that I am writing these lines to you, but I simply cannot help myself. In spite of very serious attempts to write the article on the "Logics of quantum mechanics" I find it completely impossible to do it at this time. As you may know, I wrote a paper on this subject with Garrett Birkhoff in 1936 ("Annals of Mathematics", vol. 37, pp. 823-843), and I have thought a good deal on the subject since. My work on continuous geometries, on which I gave the Amer. Math. Soc. Colloqium lectures in 1937, comes to a considerable extent from this source. Also a good deal concerning the relationship between strict and probability logics (upon which I touched briefly in the Henry Joseph Lecture) and the extension of this "Propositional calculus" work to "logics with quantifiers" (which I never so far discussed in public). All these things should be presented as a connected whole (I mean the propositional and the "quantifier" strict logics, the probability logics, plus a short indication of the ideas of "continuous" projective geometry), and I have been mainly interrupted in this (as well as in writing a book on continuous geometries, which I still owe the Amer. Math. Soc. Colloqium Series) by the war. To do it properly would require a good deal of work, since the subjects that have to be correlated are a very heterogenous collection — although I think that I can show how they belong together.
>
> When I offered to give the Henry Joseph Lecture on this subject, I thought (and I hope that I was not too far wrong in this) that I could give a reasonable general survey of at least part of the subject in a talk, which might have some interest to the audience. I did not realize the importance nor the difficulties of reducing this to writing.
>
> I have now learned — after a considerable number of serious but very unsuccessful efforts — that they are exceedingly great. I must, of course, accept a good part of the responsibility for my method of writing — I write rather freely and fast if a subject is "mature" in my mind, but develop the worst traits of pedantism and inefficiency if I attempt to give a preliminary account of a subject which I do not have yet in what I can believe to be in its final form.
>
> I have tried to live up to my promise and to force myself to write this article, and spent much more time on it than on many comparable ones which I wrote with no difficulty at all — and it just didn't work. Perhaps if I were not continually interrupted by journeys and other obligations arising from still surviving war work, I might have been able to do it — although I am not even sure of this.
>
> (von Neumann to Silsbee, July 2, 1945), [Rédei, 2005][p. 218-219]

5 SUMMARY AND CONCLUDING REMARKS

As we have seen, the modularity postulate of the 1936 Birkhoff-von Neumann concept of quantum logic was motivated by the need to establish conceptual coherence between different things "quantum": (quantum) logic, (quantum) probability and quantum mechanics. Thus to understand the Birkhoff-Von Neumann concept of quantum logic one has to keep in mind that while searching for quantum logic, von Neumann did not just want to create a non-classical logic: he was motivated by the desire of creating the non-classical (non-commutative) analogy of the classical situation where a Boolean algebra plays two roles at the same time: it represents the propositional algebra of a classical propositional calculus and it also represents the set of random events on which probability measures are defined. Moreover, probabilities given by a classical probability measure can in principle be interpreted as relative frequencies in the sense of von Mises (the well-known difficulties of the frequency interpretation notwithstanding).

Von Neumann had realized that the failure of modularity in the Hilbert lattice prohibits the interpretation of the Hilbert lattice as an event structure for a relative frequency interpreted non-commutative probability theory and, consequently, he abandoned the Hilbert lattice as quantum logic — thereby also abandoning Hilbert space quantum mechanics as well. He could do this in 1936 only because by then he was in the position to suggest another mathematical framework for quantum theory, a framework that he hoped would be conceptually more suitable than Hilbert space quantum theory: at the time of cooperating with Birkhoff on the 1936 quantum logic paper, he also was working on the theory of "rings of operators" ("von Neumann algebras"), and in the year of the publication of the Birkhoff-von Neumann paper on quantum logic von Neumann also published a joint paper with J. Murray that established the classification theory of von Neumann algebras. One result of this classification theory was the discovery of a specific type of a von Neumann algebra, the so-called type "II_1 factor". The projections of this algebra form a modular lattice that von Neumann hoped would be a proper quantum logic. The lattice of a type II_1 factor is a special case of a continuous geometry.

Thus the 1936 Birkhoff-von Neumann concept of quantum logic is related to deep mathematical discoveries in the mid thirties and to conceptual difficulties in connection with the frequency interpretation of quantum probability. So the issue is a convoluted one. The complexity of the problem is also reflected by the fact that, as we have seen, von Neumann himself was never quite satisfied with how he had worked out quantum logic: It turned out that even the modular lattices of type II_1 von Neumann algebras are too week algebraically to regard them as event structure for a truly non-commutative probability theory — if the probabilities are to be interpreted as relative frequencies. This was likely the main reason why von Neumann abandoned the frequency interpretation of quantum probability after 1936 in favor of a "logical interpretation" of probability, which von Neumann did not regard as very well developed and understood, however.

While the conceptual coherence between quantum logic, quantum probability and quantum mechanics cannot be achieved in the way Birkhoff and von Neumann seem to have envisaged, and the interpretation of quantum probability remains a much debated issue even today, their challenging the Hilbert space formalism was a very significant move — albeit one not widely noticed at the time. Through the type II_1 factor and its projection lattice the Birkhoff-von Neumann concept of quantum logic got related to the classification theory of von Neumann algebras, which opened up the possibility that types *other* than II_1 and the I_∞ (= Hilbert space quantum mechanics) might also be utilized in applications to quantum systems. The subsequent development of quantum theory proved that von Neumann algebras of *all* the types discovered by Murray and von Neumann in 1935 are needed in modelling quantum systems (see [Rédei and Summers, 2007] for a review of the role of type in quantum theory). The projection lattices of such von Neumann algebras all represent non-classical (quantum) logics, and the specific type-related features of these projection lattices are crucial in understanding the behavior of the quantum world. Detailing these latter issues is beyond the scope of the present review however. Nor were they discussed in the Birkhoff-von Neumann paper: Birkhoff and von Neumann were fully aware that their work was just the beginning of a rich field to be developed further:

> Your general remarks, I think, are very true: I, too, think, that our paper will not be very exhaustive or conclusive, but that we should not attempt to make it such: The subject is obviously only at the beginning of a development, and we want to suggest the direction of this development much more, than to reach "final" results. I, for one, do not even believe, that the right formal frame for quantum mechanics is already found.
> (von Neumann to Birkhoff, November 21, 1935 ?), [Rédei, 2005, p. 65]

ACKNOWLEDGEMENTS

This paper was written while I was staying as Donders Professor in the Department of History and Foundations of Mathematics and Science, Utrecht University, The Netherlands. I wish to thank Utrecht University and especially, Professor D. Dieks for their hospitality. Work also supported by OTKA (contract number: T 43642)

BIBLIOGRAPHY

[Birkhoff and von Neumann, 1936] G.D. Birkhoff and J. von Neumann. The logic of quantum mechanics. *Annals of Mathematics*, 37:823–843, 1936. Reprinted in [Taub, 1961a], No. 7.

[Birkhoff, 1961] G. Birkhoff. Lattices in applied mathematics. In R.P. Dilworth, editor, *Lattice Theory: Proceedings of the Second Symposium in Pure Mathematics of the American Mathematical Society, Monterey, April, 1959*, pages 155–184. American Mathematical Society, Providence, 1961.

[Bub, 1981a] J. Bub. Hidden variables and quantum mechanics — a sceptical review. *Erkenntnis*, 16:275–293, 1981.

[Bub, 1981b] J. Bub. What does quantum logic explain? In E. Beltrametti and B.C. van Fraassen, editors, *Current Issues in Quantum Logic*, pages 89–100. Plenum Press, New York, 1981.

[Dalla-Chiara and Giuntini, 2002] M.L. Dalla-Chiara and R. Giuntini. Quantum logic. In D.M. Gabbay and F. Guenthner, editors, *Handbook of Philosophical Logic*, volume 6, pages 129–228. Kluwer Academic Publishers, Dordrecht, 2002.

[Dalla-Chiara *et al.*, 2007] M.L. Dalla-Chiara, R. Giuntini, and M. Rédei. The history of quantum logic. In D.M. Gabbay, editor, *Handbook of History of Logic*. Kluwer Academic Publishers, Dordrecht. Vol. 38 (2007), 390–417.

[Dunn and Hardegree, 2001] J. Michael Dunn and Garry M. Hardegree. *Algebraic Methods in Philosophical Logic*. Claredon Press, Oxford, 2001.

[Halmos, 1962] P.R. Halmos. *Algebraic Logic*. Chelsa Publishing Company, New York, 1962.

[Hardegree, 1981a] G.M. Hardegree. An axiom system for orthomodular quantum logic. *Studia Logica*, 40:1–12, 1981.

[Hardegree, 1981b] G.M. Hardegree. Some problems and methods in formal quantum logic. In E. Beltrametti and B.C. van Fraassen, editors, *Current Issues in Quantum Logic*, pages 209–225. Plenum Press, New York, 1981.

[Kalmbach, 1981] G. Kalmbach. Omologic as Hilbert type calculus. In E. Beltrametti and B.C. van Fraassen, editors, *Current Issues in Quantum Logic*, pages 333–340. Plenum Press, New York, 1981.

[Murray and von Neumann, 1936] F.J. Murray and J. von Neumann. On rings of operators. *Annals of Mathematics*, 37:116–229, 1936. Reprinted in [Taub, 1961b] No. 2.

[Murray and von Neumann, 1937] F.J. Murray and J. von Neumann. On rings of operators, II. *American Mathematical Society Transactions*, 41:208–248, 1937. Reprinted in [Taub, 1961b] No. 3.

[Murray and von Neumann, 1943] F.J. Murray and J. von Neumann. On rings of operators, IV. *Annals of Mathematics*, 44:716–808, 1943. Reprinted in [Taub, 1961b] No. 5.

[Pavicic, 1992] M. Pavicic. Bibliography on quantum logic. *International Journal of Theoretical Physics*, 31, 1992.

[Petz and Rédei, 1995] D. Petz and M. Rédei. John von Neumann and the theory of operator algebras. In F. Brody and T. Vámos, editors, *The Neumann Compendium*, volume I of *World Scientific Series of 20th Century Mathematics*, pages 163–181. World Scientific, Singapore, 1995.

[Petz and Zemanek, 1988] D. Petz and J. Zemanek. Characterizations of the trace. *Linear Algebra and its Applications*, 111:43–52, 1988.

[Popper, 1968] K.R. Popper. Bikhoff and von Neumann's interpretation of quantum mechanics. *Nature*, 219:682–685, 1968.

[Rédei and Summers, 2007] M. Rédei and S.J. Summers. Quantum probability theory. *Studies in the History and Philosophy of Modern Physics*, 2007. forthcoming.

[Rédei, 1996] M. Rédei. Why John von Neumann did not like the Hilbert space formalism of quantum mechanics (and what he liked instead). *Studies in the History and Philosophy of Modern Physics*, 27:1309–1321, 1996.

[Rédei, 1998] M. Rédei. *Quantum Logic in Algebraic Approach*, volume 91 of *Fundamental Theories of Physics*. Kluwer Academic Publisher, 1998.

[Rédei, 1999] M. Rédei. 'Unsolved problems in mathematics' J. von Neumann's address to the International Congress of Mathematicians Amsterdam, September 2-9, 1954. *The Mathematical Intelligencer*, 21:7–12, 1999.

[Rédei, 2001] M. Rédei. Von Neumann's concept of quantum logic and quantum probability. In M. Rédei and M. Stöltzner, editors, *John von Neumann and the Foundations of Quantum Physics*, Institute Vienna Circle Yearbook, pages 153–172. Kluwer Academic Publishers, Dordrecht, 2001.

[Rédei, 2005] M. Rédei, editor. *John von Neumann: Selected Letters*, volume 27 of *History of Mathematics*, Rhode Island, 2005. American Mathematical Society and London Mathematical Society.

[Rédei, 2007] M. Rédei. The birth of quantum logic. *History and Philosophy of Logic*, 28:107–122, May 2007.

[Scheibe, 1974] E. Scheibe. Popper and quantum logic. *The British Journal for the Philosophy of Science*, 25:319–342, 1974.

[Takesaki, 1979] M. Takesaki. *Theory of Operator Algebras*, volume I. Springer Verlag, New York, 1979.

[Taub, 1961a] A.H. Taub, editor. *John von Neumann: Collected Works*, volume IV. Continuous Geometry and Other Topics, New York and Oxford, 1961. Pergamon Press.

[Taub, 1961b] A.H. Taub, editor. *John von Neumann: Collected Works*, volume III. Rings of Operators, New York and Oxford, 1961. Pergamon Press.

[von Mises, 1919] R. von Mises. Grundlagen der Wahrscheinlichkeitsrechnung. *Mathematische Zeitschrift*, 5:52–99, 1919.

[von Mises, 1981] R. von Mises. *Probability, Statistics and Truth.* Dover Publications, New York, 2nd edition, 1981. Originally published as 'Wahrscheinlichkeit, Statistik und Wahrheit' (Springer, 1928).

[von Neumann, 1932] J. von Neumann. *Mathematische Grundlagen der Quantenmechanik.* Springer Verlag, Berlin, 1932. English translation: Mathematical Foundations of Quantum Mechanics (Princeton University Press, Princeton) 1955.

[von Neumann, 1936] J. von Neumann. Continuous geometry. *Proceedings of the National Academy of Sciences*, 22:92–100, 1936. Reprinted in [Taub, 1961a] pp. 126-134.

[von Neumann, 1940] J. von Neumann. On rings of operators, iii. *Annals of Mathematics*, 41:94–161, 1940. Reprinted in [Taub, 1961b] No. 4.

[von Neumann, 1961] J. von Neumann. Quantum logic (strict- and probability logics). In A.H. Taub, editor, *John von Neumann: Collected Works*, volume IV. Continuous Geometry and Other Topics, pages 195–197. Pergamon Press, New York and Oxford, 1961. Unfinished manuscript, reviewed by A.H. Taub.

[von Neumann, 2001] J. von Neumann. Unsolved problems in mathematics. In M. Rédei and M. Stöltzner, editors, *John von Neumann and the Foundations of Quantum Physics*, Institute Vienna Circle Yearbook, pages 231–245. Kluwer Academic Publishers, Dordrecht, 2001.

Edited by K. Engesser, D. M. Gabbay and D. Lehmann

IS QUANTUM LOGIC A LOGIC?

Mladen Pavičić and Norman D. Megill

1 INTRODUCTION

Thirty-seven years ago, Richard Greechie and Stanley Gudder wrote a paper entitled *Is a Quantum Logic a Logic?* [Greechie and Gudder, 1971] in which they strengthen a previous negative result of Josef Jauch and Constantin Piron. [Jauch and Piron, 1970]

"Jauch and Piron have considered a possibility that a quantum propositional system is an infinite valued logic... and shown that standard propositional systems (that is, ones that are isomorphic to the lattice of all closed subspaces of a Hilbert space) are not conditional and thus cannot be logic in the usual sense." [Greechie and Gudder, 1971] A *conditional* lattice is defined as follows. We define a valuation $v[a]$ as a mapping from an element a of the lattice to the interval $[0, 1]$. We say that two elements a, b are conditional if there exists a unique c such that $v[c] = min\{1, 1 - v[a] + v[b]\}$. We call c the *conditional* of a and b and write $c = a \to b$. We say that the lattice is conditional if every pair a, b is conditional. Greechie and Gudder then proved that a lattice is conditional if and only if it contains only two elements 0 and 1.[1] This implies that [0,1] reduces to $\{0, 1\}$ and that the lattice reduces to a two-valued Boolean algebra. In effect, this result shows that one cannot apply the same kind of valuation to both quantum and classical logics.

It became obvious that if we wanted to arrive at a proper quantum logic, we should take an axiomatically defined set of propositions closed under substitutions and some rules of inference, and apply a model-theoretic approach to obtain valuations of every axiom and theorem of the logic. So, a valuation should not be a mapping to [0,1] or $\{0, 1\}$ but to the elements of a model. For classical logic, a model for logic was a complemented distributive lattice, i.e., a Boolean algebra. For quantum logics the most natural candidate for a model was the orthomodular lattice, while the logics themselves were still to be formulated. Here we come to the question of *what logic is.* We take that logic is about propositions and inferences between them, so as to form an axiomatic deductive system. The system always has some algebras as models, and we always define valuations that map its propositions to elements of the algebra — we say, the system always has its semantics — but our definition stops short of taking semantics to be a part of the

[1] We define 0 and 1 in a lattice in Section 2.

system itself. Our title refers to such a definition of logic, and we call quantum logic so defined *deductive quantum logic.*[2] Classical logic is deductive in the same sense.

In the early seventies, a number of results and a number of predecessors to deductive quantum logics were formulated. Jauch, Piron, Greechie, and Gudder above assumed the conditional — from now on we will call it *implication* — to be defined as $a \to_0 b = a' \cup b$ (see Section 2 for notation). However, it was already then known that in an orthomodular lattice[3] an implication so defined would not satisfy the condition $a \to b = 1 \ \Leftrightarrow \ a \le b$, which holds in every Boolean algebra and which was considered plausible to hold in an orthomodular lattice too. In 1970, the following implication was found to satisfy this condition: $a \to_1 b = a' \cup (a \cap b)$ (the so-called *Sasaki hook*[4]) by Peter Mittelstaedt [Mittelstaedt, 1970] and Peter Finch [Finch, 1970]. The Sasaki hook becomes equal to $a' \cup b$ when an orthomodular lattice satisfies the distributive law, i.e. when it is a Boolean algebra. The Sasaki implication first served several authors simply to reformulate the orthomodular lattice in a logic-like way and call it "quantum logic." [Finch, 1970; Clark, 1973; Piziak, 1974] In 1974 Gudrun Kalmbach proved that in addition to the Sasaki hook, there are exactly four other "quantum implications" that satisfy the above plausible condition and that all reduce to $a' \cup b$ in a Boolean algebra.

In the very same year, four genuine (i.e. propositional) deductive quantum logics — using three different implications and none at all, respectively — were formulated by Gudrun Kalmbach [Kalmbach, 1974] (a standard propositional logic based on the *Kalmbach implication*,[5]) Hermann Dishkant [Dishkant, 1974] (a first-order predicate logic based on the *Dishkant implication*[6]), Peter Mittelstaedt [Mittelstaedt, 1974] (a dialog logic based on the Sasaki hook), and Robert Goldblatt [Goldblatt, 1974] (a binary logic with no implication — the binary inference '$\vdash$' represented the lattice '$\le$'). Several other quantum logics were later formulated by Maria Luisa Dalla Chiara [Dalla Chiara, 1977] (first-order quantum logic), Jay Zeman [Zeman, 1978] (*normal logic*), Hirokazu Nishimura [Nishimura, 1980] (Gentzen sequent logic), George Georgacarakos [Georgacarakos, 1980] (*orthomodular logics* based on *relevance*,[7] Sasaki, and Dishkant implications), Michael Dunn [Dunn, 1981] (predicate binary logic), Ernst-Walter Stachow [Stachow, 1976] (*tableaux calculus*, a Gentzen-like *calculus of sequents*, and a Brouwer-like logic), Gary Hardegree [Hardegree, 1981] (*orthomodular calculus*), John Bell [Bell, 1986] (quantum

---

[2]Note that many authors understand quantum logic as simply a lattice [Jauch, 1968] or a poset [Varadarajan, 1968; Pták and Pulmannová, 1991]. Quantum logics so defined do not have the aforementioned valuation and are not deductive quantum logics. Such a definition stems from an operationalist approach, which started with the idea that quantum logic might be empirical. It was argued that propositions might be measured and that properties such as orthomodularity for quantum systems or distributivity for classical ones can be experimentally verified. [Jauch, 1968]

[3]The lattice of all closed subspaces of a Hilbert space is an orthomodular lattice. See Section 2.

[4]The Sasaki hook is an orthocomplement to the *Sasaki projection* [Sasaki, 1964].

[5]Kalmbach implication is defined as $a \to_3 b = (a' \cap b) \cup (a' \cap b') \cup (a \cap (a' \cup b))$.

[6]Dishkant implication is defined as $a \to_2 b = b' \to_1 a'$.

[7]Relevance implication is defined as $a \to_5 b = (a \cap b) \cup (a' \cap b) \cup (a' \cap b')$.

"attribute" logic), Mladen Pavičić [Pavičić, 1987] (binary quantum logics with *merged implications*[8]), Mladen Pavičić [Pavičić, 1989] (unary quantum logic with *merged implications*),[9] Mladen Pavičić and Norman Megill [Pavičić and Megill, 1999] (unary quantum logics with *merged equivalences*[10]), etc. Logics with the $v(a) = 1$ lattice valuation corresponding to $\vdash a$ we call *unary* logics and logics with the $v(a) \leq v(b)$ lattice valuation corresponding to $a \vdash b$ we call *binary* logics.

Still, the parallels with classical logic were a major concern of the researchers at the time. "I would argue that a 'logic' without an implication ... is radically incomplete, and indeed, hardly qualifies as a theory of deduction" (Jay Zeman, 1978). [Zeman, 1978] So, an extensive search was undertaken in the seventies and eighties to single out a "proper quantum implication" from the five possible ones on purely logical grounds,[11] but none of the attempts proved successful.

In 1987 Mladen Pavičić [Pavičić, 1987; Pavičić, 1989] proved that there is no "proper quantum implication" since any one of the conditions $a \rightarrow_i b = 1 \Leftrightarrow a \leq b$, $i = 1, \ldots, 5$[12] *is* the very orthomodularity which, when satisfied by an orthocomplemented lattice (the so-called *ortholattice*), makes it orthomodular. In terms of a logic, the corresponding logical rules of inference turn any *orthologic* or *minimal quantum logic* into a quantum logic. He also proved that when we add the condition $a \rightarrow_0 b = 1 \Leftrightarrow a \leq b$ is saytisfied by an ortholattice, the lattice becomes a Boolean algebra.[13] A corresponding logical rule of inference turns any orthologic into a classical logic.

This finding was soon complemented by a proof given by Jacek Malinowski in 1990 that "no logic determined by any class of orthomodular lattices admits the deduction theorem," [Malinowski, 1990] where the *deduction theorem* says that if we can derive b from $S \bigcup \{a\}$ then we can derive $a \rightarrow b$ from S.[14] He also proved that no extension of quantum logic, i.e., no logic between the quantum and the classical one, satisfies the deduction theorem. [Mortensen, 1991] The conclusion was: "Since orthomodular logic is algebraically well behaved, this perhaps shows that implication is not such a desirable operation to have." [Mortensen, 1991]

The conjecture was confirmed by Mladen Pavičić in 1993 [Pavičić, 1993]. The above orthomodularity condition does not require implications. One can also have

---

[8]Under *merged implications* all six implications are meant; $a \rightarrow_i b$, $i = 0, 1, 2, 3, 5$ are defined above; $a \rightarrow_4 b = b' \rightarrow_3 a'$ is called *non-tollens implication*. In these logics of Pavičić, axioms of identical form hold for each of the implications yielding five quantum logics and one classical (for $i = 0$).

[9]Again, axioms of identical form hold for all implications.

[10]Merged equivalences, $a \equiv_i b$, $i = 0, \ldots, 5$, are explicit expressions (by means of $\cup, \cap, '$) of $(a \rightarrow_i b) \cap (b \rightarrow_j a)$, $i = 0, \ldots, 5$, $j = 0, \ldots, 5$, in any orthomodular lattice as given by Table 1 of Ref. [Pavičić and Megill, 1999]. In these logics, axioms of identical form hold for all equivalences.

[11]An excellent contemporary review of the state of the art was written in 1979 by Gary Hardegree [Hardegree, 1979].

[12]$a \rightarrow_i b$, $i = 1, \ldots, 5$ are defined above. See footnotes Nos. 8 and 9.

[13]In any Boolean algebra all six implications merge.

[14]It should be stressed here that the deduction theorem is not essential for classical logic either. It was first proved by Jaques Herbrand in 1930. [Herbrand, 1931] All classical logic systems before 1930, e.g., the ones by Whitehead and Russell, Hilbert, Ackermann, Post, Skolem, Łukasiewicz, Tarski, etc., were formulated without it.

it with an essentially weaker equivalence operation: $a \equiv b = 1 \Leftrightarrow a = b$, where $a \equiv b = (a \cap b) \cup (a' \cap b')$; we say a and b are *equivalent*. [Pavičić, 1993; Pavičić and Megill, 1999] As above, when this condition is satisfied by an ortholattice it makes it orthomodular.[15] Moreover in any orthomodular lattice $a \equiv b = (a \to_i b) \cap (b \to_i a)$, $i = 1, \ldots, 5$. The analogous classical condition $a \equiv_0 b = 1 \Leftrightarrow a = b$, where $a \equiv_0 b = (a' \cup b) \cap (a \cup b')$, amounts to distributivity: when satisfied by an ortholattice, it makes it a Boolean algebra. [Pavičić, 1998; Pavičić and Megill, 1999]

On the other hand, it turned out that everything in orthomodular lattices is sixfold defined: binary operations, unary operation, variables and even unities and zeros. They all collapse to standard Boolean operations, variables and 0,1 when we add distributivity. For example, as proved by Norman Megill and Mladen Pavičić [Megill and Pavičić, 2001] $0_{1(a,b)} = a \cap (a' \cup b) \cap (a \cup b')$,...,$0_{5(a,b)} = (a \cup b) \cap (a \cup b') \cap (a' \cup b) \cap (a' \cup b')$; $a \equiv_3 b = (a' \cup b) \cap (a \cup (a' \cap b'))$; etc. [Megill and Pavičić, 2002] Moreover, we can express any of such expressions by means of every appropriate other in a huge although definite number of equivalence classes. [Megill and Pavičić, 2002] For example, a shortest expression for $\cup$ expressed by means of quantum implications is $a \cup b = (a \to_i b) \to_i (((a \to_i b) \to_i (b \to_i a)) \to_i a)$, $i = 1, \ldots, 5$. [Megill and Pavičić, 2001; Megill and Pavičić, 2002; Pavičić and Megill, 1998a; Megill and Pavičić, 2003]

For such a "weird" model, the question emerged as to whether it is possible to formulate a proper deductive quantum logic as a general theory of inference and how independent of its model this logic can be. In other words, can such a logic be more general than its orthomodular model?

The answer turned out to be affirmative. In 1998 Mladen Pavičić and Norman Megill showed that the deductive quantum logic is not only more general but also very different from their models. [Pavičić and Megill, 1998b; Pavičić and Megill, 1999] They proved that

- Deductive quantum logic is not orthomodular.
- Deductive quantum logic has models that are ortholattices that are not orthomodular.
- Deductive quantum logic is sound and complete under these models.

This shows that quantum logic is not much different from the classical one since they also proved that [Pavičić and Megill, 1999]

- Classical logic is not distributive.[16]

[15] The same holds for $a \equiv_i b$, $i = 1, \ldots, 5$ from footnote No. 10, as well. [Pavičić and Megill, 1999]

[16] Don't be alarmed. This is *not* in contradiction with anything in the literature. The classical logic still stands intact, and the fact that it is not distributive is just a feature of classical logic that — due to Boole's heritage — simply has not occurred to anyone as possible and which therefore has not been discovered before. See the proof of Theorem 30, Theorem 45, Lemma 50, and the discussion in Section 10.

- Classical logic has models that are ortholattices that are not orthomodular and therefore also not distributive.
- Classical logic is sound and complete under these models.

These remarkably similar results reveal that quantum logic is a logic in the very same way in which classical logic is a logic. In the present chapter, we present the results in some detail.

The chapter is organized as follows. In Section 2, we define the ortholattice, orthomodular lattice, complemented distributive lattice (Boolean algebra), weakly orthomodular lattice WOML (which is not necessarily orthomodular), weakly distributive lattice WDOL (which is not necessarily either distributive or orthomodular), and some results that connect the lattices. In Section 3, we define quantum and classical logics. In Sections 4 and 5, we prove the soundness of quantum logic for WOML and of classical logic for WDOL, respectively. In Sections 6 and 7, we prove the completeness of the logics for WOML and WDOL, respectively. In Sections 8 and 9, we prove the completeness of the logics for OML and Boolean algebra, respectively, and show that the latter proofs of completeness introduce hidden axioms of orthomodularity and distributivity in the respective Lindenbaum algebras of the logics. In Section 10, we discuss the obtained results.

2 LATTICES

In this section, we introduce two models for deductive quantum logic, orthomodular lattice and WOML, and two models for classical logic, Boolean algebra and WDOL. They are gradually defined as follows.

There are two equivalent ways to define a lattice: as a partially ordered set (poset)[17] [Maeda and Maeda, 1970] or as an algebra [Birkhoff, 1948, II.3. *Lattices as Abstract Algebras*]. We shall adopt the latter approach.

DEFINITION 1. An *ortholattice*, OL, is an algebra $\langle \mathcal{OL}_0, ', \cup, \cap \rangle$ such that the following conditions are satisfied for any $a, b, c \in \mathcal{OL}_0$ [Megill and Pavičić, 2002]:

$$a \cup b = b \cup a \tag{1}$$
$$(a \cup b) \cup c = a \cup (b \cup c) \tag{2}$$
$$a'' = a \tag{3}$$
$$a \cup (b \cup b') = b \cup b' \tag{4}$$
$$a \cup (a \cap b) = a \tag{5}$$
$$a \cap b = (a' \cup b')' \tag{6}$$

In addition, since $a \cup a' = b \cup b'$ for any $a, b \in \mathcal{OL}_0$, we define:

$$1 \overset{\text{def}}{=} a \cup a', \qquad 0 \overset{\text{def}}{=} a \cap a' \tag{7}$$

[17]Any two elements a and b of the poset have a least upper bound $a \cup b$ — called *join* — and a greatest lower bound $a \cap b$ — called *meet*.

and

$$a \le b \quad \overset{\text{def}}{\Longleftrightarrow} \quad a \cap b = a \quad \Longleftrightarrow \quad a \cup b = b \tag{8}$$

Connectives $\to_1$ (*quantum implication, Sasaki hook*), $\to_0$ (*classical implication*), $\equiv$ (*quantum equivalence*), and $\equiv_0$ (*classical equivalence*) are defined as follows:

DEFINITION 2. $a \to_1 b \overset{\text{def}}{=} a' \cup (a \cap b)$, $\qquad a \to_0 b \overset{\text{def}}{=} a' \cup b$.

DEFINITION 3.[18] $a \equiv b \overset{\text{def}}{=} (a \cap b) \cup (a' \cap b')$.

DEFINITION 4. $a \equiv_0 b \overset{\text{def}}{=} (a \to_0 b) \cap (b \to_0 a)$.

Connectives bind from weakest to strongest in the order $\to_1$ ($\to_0$), $\equiv$ ($\equiv_0$), $\cup$, $\cap$, and $'$.

DEFINITION 5. (Pavičić and Megill [Pavičić and Megill, 1999]) An ortholattice that satisfies the following condition:

$$a \equiv b = 1 \qquad \Rightarrow \qquad (a \cup c) \equiv (b \cup c) = 1 \tag{9}$$

is called a *weakly orthomodular ortholattice*, WOML.

DEFINITION 6. (Pavičić [Pavičić, 1993]) An ortholattice that satifies the following condition:

$$a \equiv b = 1 \qquad \Rightarrow \qquad a = b, \tag{10}$$

is called an *orthomodular lattice*, OML.

Equivalently:

DEFINITION 7. (Foulis [Foulis, 1962], Kalmbach [Kalmbach, 1974]) An ortholattice that satisfies either of the following conditions:

$$a \cup (a' \cap (a \cup b)) = a \cup b \tag{11}$$

$$a\,\mathcal{C}\,b \quad \& \quad a\,\mathcal{C}\,c \quad \Rightarrow \quad a \cap (b \cup c) = (a \cap b) \cup (a \cap c) \tag{12}$$

where $a\,\mathcal{C}\,b \overset{\text{def}}{\Longleftrightarrow} a = (a \cap b) \cup (a \cap b')$ (*a commutes* with b), is called an *orthomodular lattice*, OML.

DEFINITION 8. (Pavičić and Megill [Pavičić and Megill, 1999]) An ortholattice that satisfies the following condition:[19]

$$(a \equiv b) \cup (a \equiv b') = (a \cap b) \cup (a \cap b') \cup (a' \cap b) \cup (a' \cap b') = 1 \tag{13}$$

is called a *weakly distributive ortholattice*, WDOL.

---

[18] In every orthomodular lattice $a \equiv b = (a \to_1 b) \cap (b \to_1 a)$, but not in every ortholattice.

[19] This condition is known as *commensurability*. [Mittelstaedt, 1970, Definition (2.13), p. 32] Commensurability is a weaker form of the commutativity from Definition 7. Actually, a metaimplication from commensurability to commutativity is yet another way to express orthomodularity. They coincide in any OML.

DEFINITION 9. (Pavičić [Pavičić, 1998]) An ortholattice that satisfies the following condition:

$$a \equiv_0 b = 1 \qquad \Rightarrow \qquad a = b \tag{14}$$

is called a *Boolean algebra.*

Equivalently:

DEFINITION 10. (Schröder [Schröder, 1890]) An ortholattice to which the following condition is added:

$$a \cap (b \cup c) = (a \cap b) \cup (a \cap c) \tag{15}$$

is called a *Boolean algebra.*

The opposite directions in Eqs. (10) and (14) hold in any OL.

Any finite lattice can be represented by a Hasse diagram that consists of points (*vertices*) and lines (*edges*). Each point represents an element of the lattice, and positioning element a above element b and connecting them with a line means $a \leq b$. For example, in Figure 1 we have $0 \leq x \leq y \leq 1$. We also see that in this lattice, e.g., x does not have a relation with either x' or y'.

Definition 11 and Theorems 12 and 14 will turn out to be crucial for the completeness proofs of both quantum and classical logics in Sections 6 and 7.

DEFINITION 11. We define O6 as the lattice shown in Figure 1, with the meaning $0 < x < y < 1$ and $0 < y' < x' < 1$,

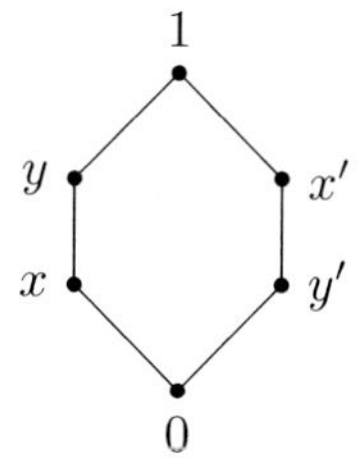

Figure 1. Ortholattice O6, also called *benzene ring* and *hexagon.*

THEOREM 12. *An ortholattice is orthomodular if only if it does not include a subalgebra isomorphic to the lattice* O6.

Proof. Samuel Holland [Holland, 1970]. See also Gudrun Kalmbach [Kalmbach, 1983, p. 22]. ■

COROLLARY 13. O6 *violates the distributive law.*

Proof. Distributivity implies orthomodularity. We can also easily verify on the diagram: $y \cap (x \cup x') = y \cap 1 = y$, but $(y \cap x) \cup (y \cap x') = x \cup 0 = x$. ■

THEOREM 14. *All conditions of* WOML *and* WDOL *hold in* O6.

Proof. As given by Mladen Pavičić and Norman Megill. [Pavičić and Megill, 1998b; Pavičić and Megill, 1999] It boils down to the fact that O6 violates none of the conditions given by Eqs. (1-6), (9), and (13) ■

THEOREM 15. *There exist* WDOL *lattices that are not orthomodular and therefore not distributive,* WOML *lattices that are not orthomodular, ortholattices that are neither* WOML *nor* WDOL, *and there are* WOML *lattices that are not* WDOL.

Proof. As given by Mladen Pavičić and Norman Megill. [Pavičić and Megill, 1998b; Pavičić and Megill, 1999]. ■

On the one hand, the equations that hold in OML and Boolean algebra properly include those that hold in WOML and WDOL, since WOML and WDOL are strictly more general classes of algebras. But on the other hand, there is also a sense in which the equations of WOML and WDOL can be considered to properly include those of OML and Boolean algebra, via mappings that the next theorems describe.

THEOREM 16. *The equational theory of* OML*s can be simulated by a proper subset of the equational theory of* WOML*s.*

Proof. The equational theory of OML consists of equality conditions, Eqs. (1)–(6) together with the orthomodularity condition Eq. (11) (or Eq. (10) or Eq. (12)). We construct a mapping from these conditions to WOML conditions as follows. We map each of the OML conditions, which is an equation in the form $t = s$ (where t and s are terms), to the equation $t \equiv s = 1$, which holds in WOML. Any equational proof in OML can then be simulated in WOML by replacing each axiom reference in the OML proof with its corresponding WOML mapping [Pavičić and Megill, 2008]. Such a mapped proof will use only a proper subset of the equations that hold in WOML: any equation whose right-hand side does not equal 1, such as $a = a$, will never be used. ■

COROLLARY 17. *No set of equations of the form $t \equiv s = 1$, where t and s are terms in* OML *and where $t = s$ holds in* OML, *determines* OML *when added to the conditions for ortholattices.*

Proof. Theorem 16 shows that all equations of this form hold in a WOML and none of WOML conditions given by Eqs. (1-6,9) is violated by O6. Hence, Theorem 12 completes the proof. ■

THEOREM 18. *The equational theory of Boolean algebras can be simulated by a proper subset of the equational theory of* WDOL*s.*

Proof. The equational theory of Boolean algebras consists of equality conditions Eqs. (1)–(6) together with the distributivity condition Eq. (15). We construct a mapping from these conditions into WDOL as follows. We map each of the Boolean algebra conditions, which is an equation in the form $t = s$ (where t and s are terms), to the equation $t \equiv_0 s = 1$, which holds in WDOL. Any equational proof in a Boolean algebra can then be simulated in WDOL by replacing each condition reference in the Boolean algebra proof with its corresponding WDOL mapping. [Pavičić and Megill, 2008] Such a mapped proof will use only a proper subset of the equations that hold in WDOL: any equation whose right-hand side does not equal 1, such as $a = a$, will never be used. ■

COROLLARY 19. *No set of equations of the form $t \equiv_0 s = 1$, where t and s are terms in any Boolean algebra and where $t = s$ holds in the algebra, determines a Boolean algebra when added to an ortholattice.*

Proof. Theorem 18 shows that all equations of this form hold in a WDOL and none of WDOL conditions given by Eqs. (1-6,8) is violated by O6. Hence, Corollary 13 completes the proof. ■

3 LOGICS

Logic, $\mathcal{L}$, is a language consisting of propositions and a set of conditions and rules imposed on them called axioms and rules of inference.

The propositions we use are well-formed formulas (wffs), defined as follows. We denote elementary, or primitive, propositions by $p_0, p_1, p_2, \ldots$, and have the following primitive connectives: $\neg$ (negation) and $\vee$ (disjunction). The set of wffs is defined recursively as follows:

p_j is a wff for $j = 0, 1, 2, \ldots$

$\neg A$ is a wff if A is a wff.

$A \vee B$ is a wff if A and B are wffs.

We introduce conjunction with the following definition:

DEFINITION 20. $A \wedge B \overset{\text{def}}{=} \neg(\neg A \vee \neg B)$.

The statement calculus of our metalanguage consists of axioms and rules from the object language as elementary metapropositions and of compound metapropositions built up by means of the following metaconnectives: $\sim$ (*not*), & (*and*), $\veebar$ (*or*), $\Rightarrow$ (*if…, then*), and $\Leftrightarrow$ (*iff*), with the usual *classical* meaning. Our metalanguage statement calculus is actually the very same classical logic we deal with in this chapter, only with the {0,1} valuation. We extend the statement calculus of the metalanguage with first-order predicate calculus — with quantifiers $\forall$ (*for all*) and $\exists$ (*exists*) — and informal set theory in the usual way.

The operations of implication are the following ones (classical, Sasaki, and Kalmbach) [Pavičić, 1987]:

DEFINITION 21. $A \to_0 B \stackrel{\text{def}}{=} \neg A \vee B.$

DEFINITION 22. $A \to_1 B \stackrel{\text{def}}{=} \neg A \vee (A \wedge B).$

DEFINITION 23. $A \to_3 B \stackrel{\text{def}}{=} (\neg A \wedge B) \vee (\neg A \wedge \neg B) \vee (A \wedge (\neg A \vee B)).$

We also define the *equivalence* operations as follows:

DEFINITION 24. $A \equiv B \stackrel{\text{def}}{=} (A \wedge B) \vee (\neg A \wedge \neg B).$

DEFINITION 25. $A \equiv_0 B \stackrel{\text{def}}{=} (A \to_0 B) \wedge (B \to_0 A).$

Connectives bind from weakest to strongest in the order $\to$, $\equiv$, $\vee$, $\wedge$, $\neg$.

Let $\mathcal{F}^\circ$ be the set of all propositions, i.e., of all wffs. Of the above connectives, $\vee$ and $\neg$ are primitive ones. Wffs containing $\vee$ and $\neg$ within logic $\mathcal{L}$ are used to build an algebra $\mathcal{F} = \langle \mathcal{F}^\circ, \neg, \vee \rangle$. In $\mathcal{L}$, a set of axioms and rules of inference are imposed on $\mathcal{F}$. From a set of axioms by means of rules of inference, we get other expressions which we call theorems. Axioms themselves are also theorems. A special symbol $\vdash$ is used to denote the set of theorems. Hence $A \in \vdash$ iff A is a theorem. The statement $A \in \vdash$ is usually written as $\vdash A$. We read this: "A is provable" since if A is a theorem, then there is a proof for it. We present the axiom systems of our propositional logics in schemata form (so that we dispense with the rule of substitution).

3.1 *Quantum Logic*

All unary quantum logics we mentioned in the Introduction are equivalent. Here we present Kalmbach's quantum logic because it is the system which has been investigated in the greatest detail in her book [Kalmbach, 1983] and elsewhere [Kalmbach, 1974; Pavičić and Megill, 1998b]. Quantum logic, $\mathcal{QL}$, is defined as a language consisting of propositions and connectives (operations) as introduced above, and the following axioms and a rule of inference. We will use $\vdash_{\mathcal{QL}}$ to denote provability from the axioms and rule of $\mathcal{QL}$ and omit the subscript when it is clear from context (such as in the list of axioms that follow).

Axioms

A1 $\quad \vdash A \equiv A \qquad (16)$

A2 $\quad \vdash A \equiv B \to_0 (B \equiv C \to_0 A \equiv C) \qquad (17)$

A3 $\quad \vdash A \equiv B \to_0 \neg A \equiv \neg B \qquad (18)$

A4 $\quad \vdash A \equiv B \to_0 A \wedge C \equiv B \wedge C \qquad (19)$

A5 $\quad \vdash A \wedge B \equiv B \wedge A \qquad (20)$

A6 $\quad \vdash A \wedge (B \wedge C) \equiv (A \wedge B) \wedge C \qquad (21)$

A7 $\quad \vdash A \wedge (A \vee B) \equiv A \qquad (22)$

A8	$\vdash \neg A \wedge A \equiv (\neg A \wedge A) \wedge B$	(23)
A9	$\vdash A \equiv \neg\neg A$	(24)
A10	$\vdash \neg(A \vee B) \equiv \neg A \wedge \neg B$	(25)
A11	$\vdash A \vee (\neg A \wedge (A \vee B)) \equiv A \vee B$	(26)
A12	$\vdash (A \equiv B) \equiv (B \equiv A)$	(27)
A13	$\vdash A \equiv B \rightarrow_0 (A \rightarrow_0 B)$	(28)
A14	$\vdash (A \rightarrow_0 B) \rightarrow_3 (A \rightarrow_3 (A \rightarrow_3 B))$	(29)
A15	$\vdash (A \rightarrow_3 B) \rightarrow_0 (A \rightarrow_0 B)$	(30)

Rule of Inference (*Modus Ponens*)

$$\text{R1} \qquad \vdash A \quad \& \quad \vdash A \rightarrow_3 B \quad \Rightarrow \quad \vdash B \tag{31}$$

In Kalmbach's presentation, the connectives $\vee$, $\wedge$, and $\neg$ are primitive. In the base set of any model (such as an OML or WOML model) that belongs to OL, $\cap$ can be defined in terms of $\cup$ and $'$, as justified by DeMorgan's laws, and thus the corresponding $\wedge$ can be defined in terms of $\vee$ and $\neg$ (Definition 20). We shall do this for simplicity. Regardless of whether we consider $\wedge$ primitive or defined, we can drop axioms A1, A11, and A15 because it has been proved that they are redundant, i.e., can be derived from the other axioms. [Pavičić and Megill, 1998b] Note that A11 is what we would expect to be *the* orthomodularity[20] — see Eq. (37) and the discussion following the equation.

DEFINITION 26. For $\Gamma \subseteq \mathcal{F}^\circ$ we say A is derivable from Γ and write $\Gamma \vdash_{\mathcal{QL}} A$ or just $\Gamma \vdash A$ if there is a sequence of formulas ending with A, each of which is either one of the axioms of $\mathcal{QL}$ or is a member of Γ or is obtained from its precursors with the help of a rule of inference of the logic.

3.2 Classical Logic

We make use of the PM classical logical system $\mathcal{CL}$ (Whitehead and Russell's *Principia Mathematica* axiomatization in Hilbert and Ackermann's presentation [Hilbert and Ackermann, 1950] but in schemata form so that we dispense with their rule of substitution). In this system, the connectives $\vee$ and $\neg$ are primitive, and the $\rightarrow_0$ connective shown in the axioms is implicitly understood to be expanded according to its definition. We will use $\vdash_{\mathcal{CL}}$ to denote provability from the axioms and rule of $\mathcal{CL}$, omitting the subscript when it is clear from context.

Axioms

A1	$\vdash A \vee A \rightarrow_0 A$	(32)
A2	$\vdash A \rightarrow_0 A \vee B$	(33)
A3	$\vdash A \vee B \rightarrow_0 B \vee A$	(34)
A4	$\vdash (A \rightarrow_0 B) \rightarrow_0 (C \vee A \rightarrow_0 C \vee B)$	(35)

[20]Cf. Definition (7), Eq. (11)

Rule of Inference (*Modus Ponens*)

$$\text{R1} \qquad \vdash A \quad \& \quad A \to_0 B \quad \Rightarrow \quad \vdash B \tag{36}$$

We assume that the only legitimate way of inferring theorems in $\mathcal{CL}$ is by means of these axioms and the Modus Ponens rule. We make no assumption about valuations of the primitive propositions from which wffs are built, but instead are interested in wffs that are valid in the underlying models. Soundness and completeness will show that those theorems that can be inferred from the axioms and the rule of inference are exactly those that are valid.

We define derivability in $\mathcal{CL}$, $\Gamma \vdash_{\mathcal{CL}} A$ or just $\Gamma \vdash A$, in the same way as we do for system $\mathcal{QL}$.

4 THE SOUNDNESS OF $\mathcal{QL}$: ORTHOMODULARITY LOST

In this section we show that the syntax of $\mathcal{QL}$ does not correspond to the syntax of an orthomodular lattice. We do this by proving the soundness of $\mathcal{QL}$ for WOML. To prove soundness means to prove that all axioms as well as the rules of inference (and therefore all theorems) of $\mathcal{QL}$ hold in its models. Since by Theorem 16 WOML properly includes OML, proving the soundness of $\mathcal{QL}$ for OML would not tell us anything new, and we can dispense with it.

DEFINITION 27. We call $\mathcal{M} = \langle \mathcal{L}, h \rangle$ a model if $\mathcal{L}$ is an algebra and $h : \mathcal{F}^\circ \longrightarrow \mathcal{L}$, called a valuation, is a morphism of formulas $\mathcal{F}^\circ$ into $\mathcal{L}$, preserving the operations $\neg, \vee$ while turning them into $', \cup$.

Whenever the base set $\mathcal{L}$ of a model belongs to WOML (or another class of algebras), we say (informally) that the model belongs to WOML (or the other class). In particular, if we say "for all models in WOML" or "for all WOML models," we mean for all base sets in WOML and for all valuations on each base set. The term "model" may refer either to a specific pair $\langle \mathcal{L}, h \rangle$ or to all possible such pairs with the base set $\mathcal{L}$, depending on context.

DEFINITION 28. We call a formula $A \in \mathcal{F}^\circ$ valid in the model $\mathcal{M}$, and write $\models_{\mathcal{M}} A$, if $h(A) = 1$ for all valuations h on the model, i.e. for all h associated with the base set $\mathcal{L}$ of the model. We call a formula $A \in \mathcal{F}^\circ$ a consequence of $\Gamma \subseteq \mathcal{F}^\circ$ in the model $\mathcal{M}$ and write $\Gamma \models_{\mathcal{M}} A$ if $h(X) = 1$ for all X in Γ implies $h(A) = 1$, for all valuations h.

For brevity, whenever we do not make it explicit, the notations $\models_{\mathcal{M}} A$ and $\Gamma \models_{\mathcal{M}} A$ will always be implicitly quantified over all models of the appropriate type, in this section for all WOML models $\mathcal{M}$. Similarly, when we say "valid" without qualification, we will mean valid in all models of that type.

We now prove the soundness of quantum logic by means of WOML, i.e., that if A is a theorem in $\mathcal{QL}$, then A is valid in any WOML model.

THEOREM 29. [Soundness] $\quad \Gamma \vdash A \quad \Rightarrow \quad \Gamma \models_{\mathcal{M}} A$

Proof. We must show that any axiom A1–A15, given by Eqs. (16–30), is valid in any WOML model $\mathcal{M}$, and that any set of formulas that are consequences of Γ in the model are closed under the rule of inference R1, Eq. (31).

Let us put $a = h(A)$, $b = h(B)$, ...

By Theorem 16, we can prove that WOML is equal to OL restricted to all orthomodular lattice conditions of the form $t \equiv s = 1$, where t and s are terms (polynomials) built from the ortholattice operations and $t = s$ is an equation that holds in all OMLs. ■

Hence, mappings of $\mathcal{QL}$ axioms and its rule of inference can be easily proved to hold in WOML. Moreover, mappings of A1,A3,A5–A13,A15 and R1 hold in any ortholattice. In particular, the

$$\text{A11 mapping}: \qquad (a \cup (a' \cap (a \cup b))) \equiv (a \cup b) = 1 \tag{37}$$

holds in every ortholattice and A11 itself is redundant, i.e., can be be inferred from other axioms. Notice that by Corollary 17, $a \equiv b = 1$ does not imply $a = b$. In particular, Eq. (37) does not imply $(a \cup (a' \cap (a \cup b))) = (a \cup b)$

5 THE SOUNDNESS OF $\mathcal{CL}$: DISTRIBUTIVITY LOST

In this section we show that the syntax of $\mathcal{CL}$ does not correspond to the syntax of a Boolean algebra. In a way analogous to the $\mathcal{QL}$ soundness proof, we prove the soundness of $\mathcal{CL}$ only by means of WDOL.

Recall Definitions 27 and 28 for "model," "valid," and "consequence."

We now prove the soundness of classical logic by means of WDOL, i.e., that if A is a theorem in $\mathcal{CL}$, then A is valid in any WDOL model.

THEOREM 30. [Soundness] $\qquad \Gamma \vdash A \quad \Rightarrow \quad \Gamma \vDash_{\mathcal{M}} A$

Proof. We must show that any axiom A1–A4, given by Eqs. (32–35), is valid in any WDOL model $\mathcal{M}$, and that any set of formulas that are consequences of Γ in the model are closed under the rule of inference R1, Eq. (36).

Let us put $a = h(A)$, $b = h(B)$, ...

By Theorem 18, we can prove that WDOL is equal to OL restricted to all Boolean algebra conditions of the form $t \equiv_0 s = 1$, where t and s are terms and $t = s$ is an equation that holds in all Boolean algebras. Notice that according to Corollary 19, $t \equiv_0 s = 1$ is not generally equivalent to $t = s$ in WDOL. For example, the mappings of A1–A3 and R1 hold in every ortholattice, and the ortholattice mapping of A4 does not make the ortholattice even orthomodular let alone distributive. In other words,

$$(a \cap (b \cup c)) \equiv_0 ((a \cap b) \cup (a \cap c)) = 1 \tag{38}$$

does not imply $(a \cap (b \cup c)) = ((a \cap b) \cup (a \cap c))$, and therefore we cannot speak of distributivity within $\mathcal{CL}$. ■

6 THE COMPLETENESS OF $\mathcal{QL}$ FOR WOML MODELS: NON-ORTHOMODULARITY CONFIRMED

Our main task in proving the soundness of $\mathcal{QL}$ in the previous section was to show that all axioms as well as the rules of inference (and therefore all theorems) from $\mathcal{QL}$ hold in WOML. The task of proving the completeness of $\mathcal{QL}$ is the opposite one: we have to impose the structure of WOML on the set $\mathcal{F}^\circ$ of formulas of $\mathcal{QL}$.

We start with a relation of congruence, i.e., a relation of equivalence compatible with the operations in $\mathcal{QL}$. We make use of an equivalence relation to establish a correspondence between formulas of $\mathcal{QL}$ and formulas of WOML. The resulting equivalence classes stand for elements of a WOML and enable the completeness proof of $\mathcal{QL}$ by means of this WOML.

Our definition of congruence involves a special set of valuations on lattice O6 (shown in Figure 1 in Section 2) called $\mathcal{O}6$ and defined as follows. Its definition is the same for both the quantum logic completeness proof in this section and the classical logic completeness proof in Section 7.

DEFINITION 31. Letting O6 represent the lattice from Definition 11, we define $\mathcal{O}6$ as the set of all mappings $o : \mathcal{F}^\circ \longrightarrow \mathrm{O6}$ such that for $A, B \in \mathcal{F}^\circ$, $o(\neg A) = o(A)'$, and $o(A \vee B) = o(A) \cup o(B)$.

The purpose of $\mathcal{O}6$ is to let us refine the equivalence classes used for the completeness proof, so that the Lindenbaum algebra will be a proper WOML, i.e. one that is not orthomodular. This is accomplished by conjoining the term $(\forall o \in \mathcal{O}6)[(\forall X \in \Gamma)(o(X) = 1) \Rightarrow o(A) = o(B)]$ to the equivalence relation definition, meaning that for equivalence we require also that (whenever the valuations o of the wffs in Γ are all 1) the valuations of wffs A and B map to the same point in the lattice O6. For example, the two wffs $A \vee B$ and $A \vee (\neg A \wedge (A \vee B))$ will become members of two separate equivalence classes by Theorem 37 below. Without the conjoined term, these two wffs would belong to the same equivalence class. The point of doing this is to provide a completeness proof that is not dependent in any way on the orthomodular law, to show that completeness does not require that the underlying models be OMLs.

THEOREM 32. *The relation of* equivalence $\approx_{\Gamma,\mathcal{QL}}$ *or just* $\approx$, *defined as*

$$A \approx B \stackrel{\text{def}}{=} \Gamma \vdash A \equiv B \ \& \ (\forall o \in \mathcal{O}6)[(\forall X \in \Gamma)(o(X) = 1) \Rightarrow o(A) = o(B)], \tag{39}$$

is a relation of congruence in the algebra $\mathcal{F}$, *where* $\Gamma \subseteq \mathcal{F}^\circ$

Proof. Let us first prove that $\approx$ is an equivalence relation. $A \approx A$ follows from A1 [Eq. (16)] of system $\mathcal{QL}$ and the identity law of equality. If $\Gamma \vdash A \equiv B$, we can detach the left-hand side of A12 to conclude $\Gamma \vdash B \equiv A$, through the use of A13 and repeated uses of A14 and R1. From this and commutativity of equality, we conclude $A \approx B \Rightarrow B \approx A$. (For brevity we will not usually mention further uses of A12, A13, A14, and R1 in what follows.) The proof of transitivity runs as follows.

$$\begin{aligned} A \approx B \quad & \& \quad B \approx C \\ & \Rightarrow \Gamma \vdash A \equiv B \quad \& \quad \Gamma \vdash B \equiv C \\ & \& \ (\forall o \in \mathcal{O}6)[(\forall X \in \Gamma)(o(X) = 1) \ \Rightarrow \ o(A) = o(B)] \\ & \& \ (\forall o \in \mathcal{O}6)[(\forall X \in \Gamma)(o(X) = 1) \ \Rightarrow \ o(B) = o(C)] \\ & \Rightarrow \Gamma \vdash A \equiv C \\ & \& \ (\forall o \in \mathcal{O}6)[(\forall X \in \Gamma)(o(X) = 1) \ \Rightarrow \ o(A) = o(B) \ \& \ o(B) = o(C)]. \end{aligned} \tag{40}$$

In the last line above, $\Gamma \vdash A \equiv C$ follows (see Sec. 3.1) using A2, A14 twice, and R1 six times, and the last metaconjunction reduces to $o(A) = o(C)$ by transitivity of equality. Hence the conclusion $A \approx C$ by definition.

In order to be a relation of congruence, the relation of equivalence must be compatible with the operations $\neg$ and $\vee$. These proofs run as follows.

$$\begin{aligned} A \approx B & \\ & \Rightarrow \Gamma \vdash A \equiv B \\ & \& \ (\forall o \in \mathcal{O}6)[(\forall X \in \Gamma)(o(X) = 1) \ \Rightarrow \ o(A) = o(B)] \\ & \Rightarrow \Gamma \vdash \neg A \equiv \neg B \\ & \& \ (\forall o \in \mathcal{O}6)[(\forall X \in \Gamma)(o(X) = 1) \ \Rightarrow \ o(A)' = o(B)'] \\ & \Rightarrow \Gamma \vdash \neg A \equiv \neg B \\ & \& \ (\forall o \in \mathcal{O}6)[(\forall X \in \Gamma)(o(X) = 1) \ \Rightarrow \ o(\neg A) = o(\neg B)] \\ & \Rightarrow \neg A \approx \neg B \end{aligned} \tag{41}$$

$$\begin{aligned} A \approx B & \\ & \Rightarrow \Gamma \vdash A \equiv B \\ & \& \ (\forall o \in \mathcal{O}6)[(\forall X \in \Gamma)(o(X) = 1) \ \Rightarrow \ o(A) = o(B)] \\ & \Rightarrow \Gamma \vdash (A \vee C) \equiv (B \vee C) \\ & \& \ (\forall o \in \mathcal{O}6)[(\forall X \in \Gamma)(o(X) = 1) \ \Rightarrow \ o(A) \cup o(C) = o(B) \cup o(C)] \\ & \Rightarrow (A \vee C) \approx (B \vee C) \end{aligned} \tag{42}$$

In the second step of Eq. 41, we used A3. In the second step of Eq. 42, we used A4 and A10. For the quantified part of these expressions, we applied the definition of $\mathcal{O}6$. ■

DEFINITION 33. The equivalence class for wff A under the relation of equivalence $\approx$ is defined as $|A| = \{B \in \mathcal{F}^\circ : A \approx B\}$, and we denote $\mathcal{F}^\circ/\approx\ = \{|A| : A \in \mathcal{F}^\circ\}$. The equivalence classes define the natural morphism $f : \mathcal{F}^\circ \longrightarrow \mathcal{F}^\circ/\approx$, which gives $f(A) \ =^{\text{def}} \ |A|$. We write $a = f(A)$, $b = f(B)$, etc.

LEMMA 34. *The relation* $a = b$ *on* $\mathcal{F}^\circ/\approx$ *is given by:*

$$|A| = |B| \qquad \Leftrightarrow \qquad A \approx B \tag{43}$$

LEMMA 35. *The Lindenbaum algebra* $\mathcal{A} = \langle \mathcal{F}^\circ/\approx, \neg/\approx, \vee/\approx\rangle$ *is a* WOML, *i.e., Eqs. (1)–(6) and Eq. (9) hold for* $\neg/\approx$ *and* $\vee/\approx$ *as* $'$ *and* $\cup$ *respectively* [where — for simplicity — we use the same symbols ($'$ and $\cup$) as for O6, since there are no ambiguous expressions in which the origin of the operations would not be clear from the context].

Proof. For the $\Gamma \vdash A \equiv B$ part of the $A \approx B$ definition, the proofs of the ortholattice conditions, Eqs. (1)–(6), follow from A5, A6, A9, the dual of A8, the dual of A7, and DeMorgan's laws respectively. (The duals follow from DeMorgan's laws, derived from A10, A9, and A3.) A11 gives us an analog of the OML law for the $\Gamma \vdash A \equiv B$ part, and the WOML law Eq. (9) follows from the OML law in an ortholattice. For the quantified part of the $A \approx B$ definition, lattice O6 is a WOML by Theorem 14. ■

LEMMA 36. *In the Lindenbaum algebra* $\mathcal{A}$, *if* $f(X) = 1$ *for all* X *in* Γ *implies* $f(A) = 1$, *then* $\Gamma \vdash A$.

Proof. Let us assume that $f(X) = 1$ for all X in Γ implies $f(A) = 1$ i.e. $|A| = 1 = |A| \cup |A|' = |A \vee \neg A|$, where the first equality is from Definition 33, the second equality follows from Eq. (7) (the definition of 1 in an ortholattice), and the third from the fact that $\approx$ is a congruence. Thus $A \approx (A\vee\neg A)$, which by definition means $\Gamma \vdash A \equiv (A\vee\neg A)$ & $(\forall o \in \mathcal{O}6)[(\forall X \in \Gamma)(o(X) = 1) \Rightarrow o(A) = o((A\vee\neg A))]$. This implies, in particular, $\Gamma \vdash A \equiv (A\vee\neg A)$. In any ortholattice, $a \equiv (a\cup a') = a$ holds. By analogy, we can prove $\Gamma \vdash (A \equiv (A \vee \neg A)) \equiv A$ from $\mathcal{QL}$ axioms A1–A15. Detaching the left-hand side (using A12, A13, A14, and R1), we conclude $\Gamma \vdash A$. ■

THEOREM 37. *The orthomodular law does not hold in* $\mathcal{A}$.

Proof. This is Theorem 3.27 from [Pavičić and Megill, 1999], and the proof provided there runs as follows. We assume $\mathcal{F}^\circ$ contains at least two elementary (primitive) propositions $p_0, p_1, \ldots$. We pick a valuation o that maps two of them, A and B, to distinct nodes $o(A)$ and $o(B)$ of O6 that are neither 0 nor 1 such that $o(A) \le o(B)$ [i.e. $o(A)$ and $o(B)$ are on the same side of hexagon O6 in Figure 1 in Section 2]. From the structure of O6, we obtain $o(A) \cup o(B) = o(B)$ and $o(A)\cup(o(A)'\cap(o(A)\cup o(B))) = o(A)\cup(o(A)'\cap o(B)) = o(A)\cup 0 = o(A)$. Therefore $o(A)\cup o(B) \ne o(A)\cup(o(A)'\cap(o(A)\cup o(B))$, i.e., $o(A\vee B) \ne o(A\vee(\neg A\wedge(A\vee B)))$. This falsifies $(A \vee B) \approx (A \vee (\neg A \wedge (A \vee B))$. Therefore $a \cup b \ne a \cup (a' \cap (a \cup b))$, providing a counterexample to the orthomodular law for $\mathcal{F}^\circ/\approx$. ■

LEMMA 38. $\mathcal{M} = \langle \mathcal{F}/\approx, f\rangle$ *is a* WOML *model.*

Proof. Follows from Lemma 35. ■

Now we are able to prove the completeness of $\mathcal{QL}$, i.e., that if a formula A is a consequence of a set of wffs Γ in all WOML models, then $\Gamma \vdash A$. In particular, when $\Gamma = \varnothing$, all valid formulas are provable in $\mathcal{QL}$. (Recall from the note below Definition 28 that the left-hand side of the metaimplication below is implicitly quantified over all WOML models $\mathcal{M}$.)

THEOREM 39. [Completeness] $\quad \Gamma \vDash_{\mathcal{M}} A \quad \Rightarrow \quad \Gamma \vdash A.$

Proof. $\Gamma \vDash_{\mathcal{M}} A$ means that in all WOML models $\mathcal{M}$, if $f(X) = 1$ for all X in Γ, then $f(A) = 1$ holds. In particular, it holds for $\mathcal{M} = \langle \mathcal{F}/\approx, f\rangle$, which is a WOML model by Lemma 38. Therefore, in the Lindenbaum algebra $\mathcal{A}$, if $f(X) = 1$ for all X in Γ, then $f(A) = 1$ holds. By Lemma 36, it follows that $\Gamma \vdash A$. ■

7 THE COMPLETENESS OF $\mathcal{CL}$ FOR WDOL MODELS: NON-DISTRIBUTIVITY CONFIRMED

In this section we will prove the completeness of $\mathcal{CL}$, i.e., we will impose the structure of WDOL on the set $\mathcal{F}^\circ$ of formulas of $\mathcal{CL}$.

We start with a relation of congruence, i.e., a relation of equivalence compatible with the operations in $\mathcal{CL}$. We have to make use of an equivalence relation to establish a correspondence between formulas from $\mathcal{CL}$ and formulas from WDOL. The resulting equivalence classes stand for elements of a WDOL and enable the completeness proof of $\mathcal{CL}$.

THEOREM 40. *The relation of* equivalence $\approx_{\Gamma,\mathcal{CL}}$ *or just* $\approx$*, defined as*

$$A \approx B \tag{44}$$
$$\stackrel{\text{def}}{=} \Gamma \vdash A \equiv_0 B \;\&\; (\forall o \in \mathcal{O}6)[(\forall X \in \Gamma)(o(X) = 1) \Rightarrow o(A) = o(B)],$$

is a relation of congruence in the algebra $\mathcal{F}$.

Proof. The axioms and rules of $\mathcal{QL}$, A1–A15 and R1, i.e., Eqs. (16)–(31), are theorems of $\mathcal{CL}$, A1–A4 and R1, i.e. Eqs. (32)–(36). To verify this we refer the reader to *Principia Mathematica* by Alfred Whitehead and Bertrand Russell [Whitehead and Russell, 1910], where the $\mathcal{QL}$ axioms either will be found as theorems or can easily be derived from them. For example, axiom A1 of $\mathcal{QL}$ is given as Theorem *4.2 [Whitehead and Russell, 1910, p. 116] after using Theorem *5.23 to convert from $\equiv_0$ to $\equiv$. This will let us take advantage of parts of the completeness proof for $\mathcal{QL}$, implicitly using Theorem *5.23 [Whitehead and Russell, 1910, p. 124] in either direction as required.

With this in mind, the proof that $\approx$ is an equivalence and congruence relation becomes exactly the proof of Theorem 32. ■

DEFINITION 41. The equivalence class for wff A under the relation of equivalence $\approx$ is defined as $|A| = \{B \in \mathcal{F}^\circ : A \approx B\}$, and we denote $\mathcal{F}^\circ/\approx = \{|A| \in \mathcal{F}^\circ\}$. The equivalence classes define the natural morphism $f : \mathcal{F}^\circ \longrightarrow \mathcal{F}^\circ/\approx$, which gives $f(A) \ =^{\mathrm{def}}\ |A|$. We write $a = f(A)$, $b = f(B)$, etc.

LEMMA 42. *The relation $a = b$ on $\mathcal{F}^\circ/\approx$ is given as:*

$$|A| = |B| \qquad \Leftrightarrow \qquad A \approx B \tag{45}$$

LEMMA 43. *The Lindenbaum algebra $\mathcal{A} = \langle \mathcal{F}^\circ/\approx, \neg/\approx, \vee/\approx, \wedge/\approx \rangle$ is a* WDOL, *i.e., Eqs. (1)–(6) and Eq. (13), hold for $\neg/\approx$ and $\vee/\approx$ as $'$ and $\cup$ respectively.*

Proof. For the $\Gamma \vdash A \equiv_0 B$ part of the $A \approx B$ definition, the proofs of the ortholattice axioms are identical to those in the proof of Lemma 35 (after using using Theorem *5.23 on p. 124 of Ref. [Whitehead and Russell, 1910] to convert between $\equiv_0$ and $\equiv$). The WDOL law Eq. (13) for the $\Gamma \vdash A \equiv_0 B$ part can be derived using Theorems *5.24, *4.21, *5.17, *3.2, *2.11, and *5.1 [Whitehead and Russell, 1910, pp. 101–124]. For the quantified part of the $A \approx B$ definition, lattice O6 is a WDOL by Theorem 14. ■

LEMMA 44. *In the Lindenbaum algebra $\mathcal{A}$, if $f(X) = 1$ for all X in Γ implies $f(A) = 1$, then $\Gamma \vdash A$.*

Proof. Identical to the proof of Lemma 36. ■

THEOREM 45. *Distributivity does not hold in $\mathcal{A}$.*

Proof. $(a \cap (b \cup c)) = ((a \cap b) \cup (a \cap c))$ fails in O6. Cf. the proof of Theorem 37. ■

LEMMA 46. *$\mathcal{M} = \langle \mathcal{F}/\approx, f \rangle$ is a* WDOL *model.*

Proof. Follows Lemma 43. ■

Now we are able to prove the completeness of $\mathcal{CL}$, i.e., that if a formula A is a consequence of a set of wffs Γ in all WDOL models, then $\Gamma \vdash A$. In particular, when $\Gamma = \varnothing$, all valid formulas are provable in $\mathcal{QL}$.

THEOREM 47. [Completeness] $\qquad \Gamma \vDash_{\mathcal{M}} A \quad \Rightarrow \quad \Gamma \vdash A$

Proof. Analogous to the proof of Theorem 39. ■

8 THE COMPLETENESS OF $\mathcal{QL}$ FOR OML MODELS: ORTHOMODULARITY REGAINED

Completeness proofs for $\mathcal{QL}$ carried out in the literature so far — with the exception of Pavičić and Megill [Pavičić and Megill, 1999] — do not invoke Definition 11 and Theorem 14, and instead of Theorem 32 one invokes the following one:

THEOREM 48. *Relation $\approx$ defined as*

$$A \approx B \quad \stackrel{\text{def}}{=} \quad \Gamma \vdash A \equiv B \tag{46}$$

is a relation of congruence in the algebra $\mathcal{F}$.

Instead of Definition 33 one has:

DEFINITION 49. The equivalence class under the relation of equivalence is defined as $|A| = \{B \in \mathcal{F}^\circ : A \approx B\}$, and we denote $\mathcal{F}^\circ/\approx\, = \{|A| \in \mathcal{F}^\circ\}$ The equivalence classes define the natural morphism $f : \mathcal{F}^\circ \longrightarrow \mathcal{F}^\circ/\approx$, which gives $f(A) \;=^{\text{def}}\; |A|$. We write $a = f(A)$, $b = f(A)$, etc.

And instead of Lemma 34 one is able to obtain:

LEMMA 50. *The relation $a = b$ on $\mathcal{F}^\circ/\approx$ is given as:*

$$a = b \quad \Leftrightarrow \quad |A| = |B| \quad \Leftrightarrow \quad A \approx B \quad \Leftrightarrow \quad \Gamma \vdash A \equiv B \tag{47}$$

Hence, from the following easily provable theorem in $\mathcal{QL}$:

$$\vdash (A \equiv B) \equiv (C \vee \neg C) \quad \Rightarrow \quad \vdash A \equiv B \tag{48}$$

one is also able to get:

$$a \equiv b = 1 \quad \Rightarrow \quad a = b \tag{49}$$

in the Lindenbaum algebra $\mathcal{A}$, which is the orthomodularity as given by Definition 6. [Pavičić, 1998]

The point here is that Eq. (49) has nothing to do with any axiom or rule of inference from $\mathcal{QL}$ — it is nothing but a consequence of the definition of the relation of equivalence from Theorem 48. Hence, the very definition of the standard relation of equivalence introduces a hidden axiom — the orthomodularity — into the Lindenbaum algebra $\mathcal{A}$, thus turning it into an orthomodular lattice. Without this hidden axiom, the Lindenbaum algebra stays WOML as required by the $\mathcal{QL}$ syntax. With it the Lindenbaum algebra turns into OML as follows.

LEMMA 51. *In the Lindenbaum algebra $\mathcal{A}$, if $f(X) = 1$ for all X in Γ implies $f(A) = 1$, then $\Gamma \vdash A$.*

Proof. In complete analogy to the proof of Theorem 36. ■

THEOREM 52. *The orthomodular law holds in $\mathcal{A}$.*

Proof. $a \cup (a' \cap (a \cup b)) = a \cup b$ follows from A11, Eq. (26) and Eq. (49). ■

LEMMA 53. $\mathcal{M} = \langle \mathcal{F}/\approx, f\rangle$ *is an* OML *model.*

Proof. Follows from Lemma 51. ■

Now we are able to prove the completeness of $\mathcal{QL}$, i.e., that if a formula A is a consequence of a set of wffs Γ in all OML models, then $\Gamma \vdash A$.

THEOREM 54. [Completeness] $\qquad \Gamma \vDash_{\mathcal{M}} A \quad \Rightarrow \quad \Gamma \vdash A$

Proof. Analogous to the proof of Theorem 39. ■

9 THE COMPLETENESS OF $\mathcal{CL}$ FOR BOOLEAN ALGEBRA MODELS: DISTRIBUTIVITY REGAINED

The completeness proof carried out in almost all logic books and textbooks do not invoke Definition 11, Theorem 14, and Theorem 40. An exception is the *Classical and Nonclassical Logics* by Eric Schechter [Schechter, 2005, p. 272] who adopted them from Pavičić and Megill [Pavičić and Megill, 1999] and presented in a reduced approach which he called the *hexagon interpretation*. Other books, though, are based on:

THEOREM 55. *Relation $\approx$ defined as*

$$A \approx B \quad \stackrel{\text{def}}{=} \quad \Gamma \vdash A \equiv_0 B \tag{50}$$

is a relation of congruence in the algebra $\mathcal{F}$.

Instead of Definition 41 one has:

DEFINITION 56. The equivalence class under the relation of equivalence is defined as $|A| = \{B \in \mathcal{F}^\circ : A \approx B\}$, and we denote $\mathcal{F}^\circ/\approx \;=\; \{|A| \in \mathcal{F}^\circ\}$ The equivalence classes define the natural morphism $f : \mathcal{F}^\circ \longrightarrow \mathcal{F}^\circ/\approx$, which gives $f(A) \;=^{\text{def}}\; |A|$. We write $a = f(A)$, $b = f(A)$, etc.

And instead of Lemma 42 one is able to obtain:

LEMMA 57. *The relation $a = b$ on $\mathcal{F}^\circ/\approx$ is given as:*

$$a = b \quad \Leftrightarrow \quad |A| = |B| \quad \Leftrightarrow \quad A \approx B \quad \Leftrightarrow \quad \Gamma \vdash A \equiv_0 B \tag{51}$$

Hence, from the following easily provable theorem in $\mathcal{CL}$:

$$\vdash (A \equiv_0 B) \equiv_0 (C \vee \neg C) \quad \Rightarrow \quad \vdash A \equiv_0 B \tag{52}$$

one is also able to get:

$$a \equiv_0 b = 1 \quad \Rightarrow \quad a = b \tag{53}$$

in the Lindenbaum algebra $\mathcal{A}$, which is the distributivity as given by Definition 9. [Pavičić, 1998] The point here is that Eq. (53) has nothing to do with any axiom or rule of inference from $\mathcal{CL}$ — it is nothing but a consequence of the definition of the relation of equivalence from Theorem 55. Hence, the very definition of the standard relation of equivalence introduces the distributivity as a hidden axiom into the Lindenbaum algebra $\mathcal{A}$ and turns it into a Boolean algebra.

THEOREM 58. [Completeness] $\quad \Gamma \vDash_{\mathcal{M}} A \quad \Rightarrow \quad \Gamma \vdash A$

Proof. Analogous to the proof of Theorem 47. ∎

10 DISCUSSION

In the above sections, we reviewed the historical results that we considered relevant to decide whether quantum logic can be considered a logic or not. In the Introduction, we showed that many authors in the past thirty years tried to decide on this question by starting with particular models and their syntax — the orthomodular lattice for quantum logic and Boolean algebra for classical. They compared the models and often came to a conclusion that since they are so different, quantum logic should not be considered a logic. This was, however, in obvious conflict with the growing number of well-formulated quantum logic systems over the same period. We mentioned some of them in the Introduction.

Orthomodular lattices and Boolean algebras *are* very different. As reviewed in the Introduction, in any orthomodular lattice all operations, variables, and constants are sixfold defined (five *quantum* and one *classical*), and in a Boolean algebra they all merge to classical operations, variables, and constants (0,1). Both an orthomodular lattice and a Boolean algebra can be formulated as equational systems — as reviewed in Section 2. Such equational systems can mimic both quantum and classical logics and show that one can formulate the Deduction Theorem in a special orthomodular lattice — a distributive one, i.e., a Boolean algebra — but cannot in a general one. As a consequence, the operation of implication — which the Deduction Theorem[21] is based on — plays a special unique role in classical logic and does not in quantum logic. Also, the Boolean algebra used as a model for classical logic is almost always two-valued, i.e., it consists of only two elements 0 and 1, and an orthomodular lattice, according to the Kochen-Specker theorem, cannot be given a $\{0,1\}$ valuation.[22]

So, recently research was carried out on whether a logic could have more than one model of the same type, e.g., an ortholattice, with the idea of freeing logics from any semantics and valuation. The result was affirmative, and a consequence was that quantum logic can be considered a logic in the same sense in which classical logic can be considered a logic. The details are given in Sections 3–9, where we chose Kalmbach's system to represent quantum logic in Section 3.1 and Hilbert and Ackermann's presentation of *Principa Mathematica* to represent classical logic in Section 3.2 (although we could have chosen any other system mentioned in the Introduction or from the literature).[23]

In Sections 4 and 6, we then proved the soundness and completeness, respectively, of quantum logic $\mathcal{QL}$ for a non-orthomodular model WOML and in Sections

[21]See footnote No. 14.

[22]In 2004 Mladen Pavičić, Jean-Pierre Merlet, Brendan McKay, and Norman Megill gave exhaustive algorithms for generation of Kochen-Specker vector systems with arbitrary number of vectors in Hilbert spaces of arbitrary dimension. [Pavičić *et al.*, 2004; Pavičić *et al*, 2005; Pavičić, 2005] The algorithms use MMP (McKay-Megill-Pavičić) diagrams for which in 3-dim Hilbert space a direct correspondence to Greechie and Hasse diagrams can be established. Thus, we also have a constructive proof within the lattice itself.

[23]Quantum logics given by Mladen Pavičić [Pavičić, 1989] and by Mladen Pavičić and Norman Megill [Pavičić and Megill, 1999] are particulary instructive since they contain only axioms designed so as to directly map into WOML conditions.

5 and 7 the soundness and completeness, respectively, of classical logic $\mathcal{CL}$ for a non-distributive model WDOL. Hence, with respect to these models, quantum logic $\mathcal{QL}$ cannot be called orthomodular and classical logic $\mathcal{CL}$ cannot be called distributive or Boolean. Also, neither $\mathcal{QL}$ nor $\mathcal{CL}$ can have a numerical valuation in general, since the truth table method is inapplicable within their OML, WOML, and WDOL models.

One might be tempted to "explain" these results in the following way. "It is true that WOML and WDOL obviously contain lattices that violate the orthomodularity law, for example the O6 hexagon (shown in Figure 1 in Section 2) itself, but most probably they also *must* contain lattices that pass the law and that would, with reference to Theorem 16, explain why we were able to prove the completeness of quantum and classical logic for WOML and WDOL." This is, however, not the case. We can prove the soundness and completeness of quantum and classical logics using a class of WOML lattices none of which pass the orthomodularity law [Pavičić and Megill, 2008]. Moreover, Eric Schechter has simplified the results of Pavičić and Megill [Pavičić and Megill, 1999] to the point of proving the soundness and completeness of classical logic for nothing but O6 itself. [Schechter, 2005, p. 272]

One of the conclusions Eric Schechter has drawn from the unexpected non-distributivity of the WDOL models, especially when reduced to the O6 lattice alone, is that all the axioms that one can prove by means of $\{0, 1\}$ truth tables, one can also prove by any Boolean algebra, and by O6. So, logics are, first of all, axiomatic deductive systems. Semantics are a next layer that concern models and valuations. Quantum and classical logics can be considered to be two such deductive systems. There are no grounds for considering any of the two logics more "proper" than the other. As we have shown above, semantics of the logics that consider their models show bigger differences between the two aforementioned classical models than between two corresponding quantum and classical models.

Whether we will ever use O6 semantics of classical logic or WOML semantics of quantum logic remains an open question, but these semantics certainly enrich our understanding of the role of logics in applications to mathematics and physics. We cannot make use of bare axiomatics of logic without specifying semantics (models and valuations) for the purpose. By making such a choice we commit ourselves to a particular model and disregard the original logical axioms and their syntax. Thus we do not use quantum logic itself in quantum mechanics and in quantum computers but instead an orthomodular lattice, and we do not use classical logic in our computers today but instead a two-valued Boolean algebra (we even hardly ever use more complicated Boolean algebras). We certainly cannot use O6 semantics to build a computer or an arithmetic; however, one day we might come forward with significant applications of these alternative semantics, and then it might prove important to have a common formal denominator for all the models — logics they are semantics of. We can also implement an alternative scenario — searching for different ortholattice semantics of the same logics [Pavičić and Megill, 2008].

Whatever strategy we choose to apply, we should always bear in mind that the syntaxes of the logics correspond to WOML, WDOL, and O6 semantics (models) while OML and Boolean algebra semantics (models) are imposed on the logics with the help of "hidden" axioms, Eqs. (49) and (53), that emerge from the standard way of defining the relation of equivalence in the completeness proofs, Theorems 48 and 55, of the logics for the latter models.

ACKNOWLEDGEMENT

We acknowledge the support of the *Ministry of Science, Education and Sport of Croatia* through teh project No. 082-0982562-3160.

BIBLIOGRAPHY

[Bell, 1986] J. L. Bell, A New Approach to Quantum Logic, *Brit. J. Phil. Sci.* **37**, 83–99 (1986).

[Birkhoff, 1948] G. Birkhoff, *Lattice Theory*, volume XXV of *American Mathematical Society Colloqium Publications*, American Mathematical Society, New York, 2nd (revised) edition, 1948.

[Clark, 1973] I. D. Clark, An Axiomatization of Quantum Logic, *J. Symb. Logic* **38**, 389–392 (1973).

[Dalla Chiara, 1977] M. L. Dalla Chiara, Quantum Logic and Physical Modalities, *J. Phil. Logic* **6**, 391–404 (1977).

[Dishkant, 1974] H. Dishkant, The First Order Predicate Calculus Based on the Logic of Quantum Mechanics, *Rep. Math. Logic* **3**, 9–18 (1974).

[Dunn, 1981] J. M. Dunn, Quantum mathematics, in *PSA 1980: Proceedings of the 1980 Biennial Meeting of the Philosophy of Science Association*, edited by P. D. Asquith and R. N. Giere, volume 2, pages 512–531, Philosophy of Science Association, East Lansing, Michigan, 1981.

[Finch, 1970] P. D. Finch, Quantum Logic as an Implication Algebra, *Bull. Austral. Math. Soc.* **2**, 101–106 (1970).

[Foulis, 1962] D. J. Foulis, A Note on Orthomodular Lattice, Portugal. Math. **21**, 65–72 (1962).

[Georgacarakos, 1980] G. N. Georgacarakos, Equationally Definable Implication Algebras for Orthomodular Lattices, *Studia Logica* **39**, 5–18 (1980).

[Goldblatt, 1974] R. I. Goldblatt, Semantic Analysis of Orthologic, *J. Phil. Logic* **3**, 19–35 (1974).

[Greechie and Gudder, 1971] R. J. Greechie and S. P. Gudder, Is a Quantum Logic a Logic?, *Helv. Phys. Acta* **44**, 238–240 (1971).

[Hardegree, 1979] G. M. Hardegree, The Conditional in Abstract and Concrete Quantum Logic, in *The Logico-Algebraic Approach to Quantum Mechanics*, edited by C. A. Hooker, volume II. Contemporary Consolidation, pages 49–108, D. Reidel, Dordrecht, 1979.

[Hardegree, 1981] G. M. Hardegree, An Axiomatic System for Orthomodular Quantum Logic, *Studia Logica* **40**, 1–12 (1981).

[Herbrand, 1931] J. Herbrand, Recherches sur la théorie de la démonstration, *Travaux de la Société des Sciences et des Lettres de Varsovie, Classe III sci. math. et phys.* **24**, 12–56 (1931), Translated into English in Jean van Heijenoort (ed.), *From Frege to Gödel: A Source Book in Mathematical Logic, 1879-1931*, Harvard University Press, Cambridge, Mass. 1967, under the title: "Investigations in Proof Theory: The Properties of True Propositions," pp. 525–581.

[Hilbert and Ackermann, 1950] D. Hilbert and W. Ackermann, *Principles of Mathematical Logic*, Chelsea, New York, 1950.

[Holland, 1970] S. S. Holland, JR., The Current Interest in Orthomodular Lattices, in *Trends in Lattice Theory*, edited by J. C. Abbot, pages 41–126, Van Nostrand Reinhold, New York, 1970.

[Jauch, 1968] J. M. Jauch, *Foundations of Quantum Mechanics*, Addison-Wesley, Reading, Massachusetts, 1968.

[Jauch and Piron, 1970] J. M. Jauch and C. Piron, What is 'Quantum Logic'?, in *Quanta*, edited by P. G. O. Freund, C. J. Goebel, and Y. Nambu, pages 166–181, The University of Chicago Press, Chicago and London, 1970.

[Kalmbach, 1974] G. Kalmbach, Orthomodular Logic, *Z. math. Logik Grundl. Math.* **20**, 395–406 (1974).

[Kalmbach, 1983] G. Kalmbach, *Orthomodular Lattices*, Academic Press, London, 1983.

[Maeda and Maeda, 1970] F. Maeda and S. Maeda, *Theory of Symmetric Lattices*, Springer-Verlag, New York, 1970.

[Malinowski, 1990] J. Malinowski, The Deduction Theorem for Quantum Logic — Some Negative Results, *J. Symb. Logic* **55**, 615–625 (1990).

[Megill and Pavičić, 2001] N. D. Megill and M. Pavičić, Orthomodular Lattices and a Quantum Algebra, *Int. J. Theor. Phys.* **40**, 1387–1410 (2001).

[Megill and Pavičić, 2002] N. D. Megill and M. Pavičić, Deduction, Ordering, and Operations in Quantum Logic, *Found. Phys.* **32**, 357–378 (2002).

[Megill and Pavičić, 2003] N. D. Megill and M. Pavičić, Quantum Implication Algebras, *Int. J. Theor. Phys.* **42**, 2825–2840 (2003).

[Mittelstaedt, 1970] P. Mittelstaedt, Quantenlogische Interpretation orthokomplementärer quasimodularer Verbände, *Z. Naturforsch.* **25**, 1773–1778 (1970).

[Mittelstaedt, 1974] P. Mittelstaedt, Quantum Logic, in *PSA 1974, Proceedings of the 1974 Biennial Meeting of the Philosophy of Science Association*, edited by R. S. Cohen, C. A. Hooker, A. C. Michalos, and J. W. van Evra, volume 101 of *Synthese Library*, pages 501–514, D. Reidel, Dordrecht–Holland, 1974.

[Mortensen, 1991] C. Mortensen, "Malinowski, Jacek, The Deduction Theorem for Quantum Logic — Some Negative Results," *Mathematical Review* **1056375**, 91g:03124 (1991).

[Nishimura, 1980] H. Nishimura, Sequential method in quantum logic, *J. Symb. Logic* **45**, 339–352 (1980).

[Pavičić, 1987] M. Pavičić, Minimal Quantum Logic with Merged Implications, *Int. J. Theor. Phys.* **26**, 845–852 (1987).

[Pavičić, 1989] M. Pavičić, Unified Quantum Logic, *Found. Phys.* **19**, 999–1016 (1989).

[Pavičić, 1993] M. Pavičić, Nonordered Quantum Logic and Its YES–NO Representation, *Int. J. Theor. Phys.* **32**, 1481–1505 (1993).

[Pavičić, 1998] M. Pavičić, Identity Rule for Classical and Quantum Theories, *Int. J. Theor. Phys.* **37**, 2099–2103 (1998).

[Pavičić, 2005] M. Pavičić, *Quantum Computation and Quantum Communication: Theory and Experiments*, Springer, New York, 2005.

[Pavičić and Megill, 1998a] M. Pavičić and N. D. Megill, Quantum and Classical Implication Algebras with Primitive Implications, *Int. J. Theor. Phys.* **37**, 2091–2098 (1998).

[Pavičić and Megill, 1998b] M. Pavičić and N. D. Megill, Binary Orthologic with Modus Ponens Is either Orthomodular or Distributive, *Helv. Phys. Acta* **71**, 610–628 (1998).

[Pavičić and Megill, 1999] M. Pavičić and N. D. Megill, Non-Orthomodular Models for Both Standard Quantum Logic and Standard Classical Logic: Repercussions for Quantum Computers, *Helv. Phys. Acta* **72**, 189–210 (1999).

[Pavičić and Megill, 2008] M. Pavičić and N. D. Megill, Standard Logics are Valuation-Monotonic, to appear in the *Journal of Logic and Computation*, (2008).

[Pavičić *et al.*, 2004] M. Pavičić, J.-P. Merlet, and N. D. Megill, Exhaustive Enumeration of Kochen–Specker Vector Systems, The French National Institute for Research in Computer Science and Control Research Reports **RR-5388** (2004).

[Pavičić *et al*, 2005] M. Pavičić, J.-P. Merlet, B. D. McKay, and N. D. Megill, Kochen–Specker Vectors, *J. Phys. A* **38**, 497–503 (2005), Corrigendum, *J. Phys. A* **38**, 3709 (2005).

[Piziak, 1974] R. Piziak, Orthomodular Lattices as Implication Algebras, *J. Phil. Logic* **3**, 413–438 (1974).

[Pták and Pulmannová, 1991] P. Pták and S. Pulmannová, *Orthomodular Structures as Quantum Logics*, Kluwer, Dordrecht, 1991.

[Sasaki, 1964] U. Sasaki, Orthocomplemented Lattices Satisfying the Exchange Axiom, *J. Sci. Hiroshima Univ. A* **17**, 293–302 (1964).

[Schechter, 2005] E. Schechter, *Classical and Nonclassical Logics: An Introduction to the Mathematics of Propositions*, Princeton University Press, Princeton, 2005.

[Schröder, 1890] E. Schröder, *Vorlesungen über die Algebra der Logik (exacte Logik)*, volume 1, Leipzig, 1890.
[Stachow, 1976] E. Stachow, Quantum Logical Calculi and Lattice Structures, *J. Phil. Logic* **6**, 347–386 (1976).
[Varadarajan, 1968] V. S. Varadarajan, *Geometry of Quantum Theory, Vols. 1 & 2*, John Wiley & Sons, New-York, 1968, 1970.
[Whitehead and Russell, 1910] A. N. Whitehead and B. Russell, *Principia Mathematica*, Cambridge University Press, Cambridge, 1910.
[Zeman, 1978] J. J. Zeman, Generalized Normal Logic, *J. Phil. Logic* **7**, 225–243 (1978).

Edited by K. Engesser, D. M. Gabbay and D. Lehmann

IS LOGIC EMPIRICAL?

Guido Bacciagaluppi

1 INTRODUCTION

In 1968 Hilary Putnam published a well-known paper on the question 'Is logic empirical?' (Putnam 1968), which gave rise to much controversy in the 1970s and 1980s. The main claims of Putnam's paper (repeated in Putnam 1974) can be paraphrased as follows:

(a) Quantum mechanics prompts us to revise our classical logical notions in favour of 'quantum logical' ones. This is explained by analogy to geometry, in the sense that also general relativity prompts us to revise our Euclidean (or rather Minkowskian) geometrical notions in favour of Riemannian (or rather pseudo-Riemannian) geometrical notions.

(b) This revision of logic is not merely local, i. e. not merely an instance of a logical system especially suited to a particular subject matter, but it is truly global. Quantum logic is the 'true' logic (just as the 'true' geometry of space-time is non-Euclidean). Indeed, we have so far failed to recognise that our usual logical connectives *are* the connectives of quantum logic.

(c) Recognising that logic is thus quantum solves the standard paradoxes of quantum mechanics, such as the measurement problem or Schrödinger's cat.

Of these truly ambitious and indeed exciting claims, the third claim (c) in particular was discussed extensively, and an almost universal consensus was reached (now shared by Putnam, 1994) that a move to quantum logic, even were it otherwise justified, would not resolve the puzzles of quantum mechanics. There have been notable reactions also to Putnam's first two claims. Yet, with few exceptions (one needs only to recall the masterly paper by Michael Dummett, 1976), the topic seems to be riddled with misunderstandings. Indeed, very few philosophers appear to still consider seriously the possibility that quantum mechanics might have something to say about the 'true' logic (I know of only one recent attempt to resurrect this idea, namely by Michael Dickson, 2001, on whose views more below). This chapter aims at clearing such misunderstandings, and at providing a much-needed overall assessment of Putnam's claims, by updating the debate in the light of the current state of the art in the foundations of quantum mechanics.

As regards Putnam's claim (a), I take it that it is indeed justified, at least provided one takes 'quantum logic' as a *local* logic, suitable to describing a class of propositions in the context of quantum mechanical experiments (or the corresponding class of propositions about properties of quantum mechanical systems).

This claim is analogous to the claim that intuitionistic logic is indeed suitable to describing a class of propositions dealing with mathematical constructions. This is distinct from the claim that intuitionistic logic is in fact the logic that underlies all rigorous human thought (and is thus the 'true' logic). Claim (a) understood in this sense, I should think, is relatively uncontroversial, and shall be taken as such for purposes of further discussion. The explanation that quantum logic, suitably defined, has all the main formal properties required of a 'good' logic will also fall into this part of the discussion.

Claim (b) is the most controversial one, and its assessment will therefore need the most care. There are two points at issue (both well emphasised already by Dummett). The first point is that motivating a revision of logic does not only require motivating the introduction of some non-classical connectives. An advocate of a revision of logic must show why these connectives do not merely sit alongside the classical connectives, but actually *replace* them. The second point is a Quinean one: such a revision of logic means that, as part of the various revisions to our network of beliefs prompted by the empirical consideration of quantum phenomena, it is possible to choose to make some revisions in our conception of logic. But it is clear that empirical considerations alone cannot *force* us to revise our logic: a distinctly philosophical component will be needed in order to justify whether a revision of logic, as opposed to a revision somewhere else in our network of beliefs, might be desirable. (In the case of geometry, this is the same situation we have known ever since Poincaré. And indeed, we shall note in section 7 that in the interpretation of quantum mechanics one can find a rather close analogy to issues about conventionalism in physical geometry.)

An aspect of claim (b) that is of special importance is the subsidiary claim that the quantum logical connectives are not new connectives that can be defined in terms of the classical ones (and of some additional physical concepts), but that the classical connectives *are* in fact the quantum logical ones in disguise. We shall therefore have to discuss in depth whether there is a sense in which the classical connectives might be reducible to the quantum logical ones, either in some strict formal sense, or in some physical limit. In this context, as we shall see, questions of interpretation in quantum mechanics play an important role. Indeed, most discussions of quantum logic as the 'true' logic have taken place, at least implicitly, in the context of the so-called standard interpretation of quantum mechanics. This, however, is the interpretation that is riddled with the usual paradoxes. As we shall see, which alternative approach to the foundations of quantum mechanics one accepts might influence the assessment of whether a global revision of logic is acceptable. Conversely, one might add, one's views on whether a global revision of logic is acceptable might influence the assessment of which approach to the foundations of quantum mechanics is most appealing.

We shall proceed as follows. In section 2, we sketch a few basic elements of quantum mechanics that will be needed later. In section 3, we introduce quantum logic (in its lattice-theoretic form) as a local logic of certain experimental propositions; we further discuss the formal properties of such a logic, and mention a few alternative forms of quantum logic. Section 4 introduces the so-called standard interpretation of quantum mechanics, and section 5 assesses Putnam's claims in the context of this interpretation. The claims are found hard to defend, but the standard interpretation itself is not a believable interpretation because it gives rise to the usual paradoxes, at least if one applies it to standard quantum mechanics. Putnam's claims are thus reassessed first, in section 6, in the context of more general quantum-like theories (based on von Neumann algebras), where the classical connectives seem indeed to be reducible to the quantum ones. Then, in section 7, we shall reassess Putnam's claims in the context of the main current approaches to the foundations of quantum mechanics that explicitly address the paradoxes of the standard theory. Our conclusions will be that, while in the case of the approaches known as de Broglie-Bohm theory and as spontaneous collapse theories quantum logic at most can be introduced alongside classical logic, and thus in no way can be construed as replacing it, in the case of the Everett (or many-worlds) approach a case can indeed be made that the classical connectives emerge from the quantum ones.

Before proceeding, I should emphasise that although the title of this chapter may suggest a general treatment of the question of whether logic is empirical, it will deal only with the question of whether considerations related to quantum mechanics may provide an argument for the general claim. (Putnam's original paper, 1968, does the same.[1]) Of course, if quantum logic provides us with an intelligible global alternative to classical logic, the case for logic being empirical will be strengthened. However, I believe that a comprehensive assessment of the question of whether empirical considerations might prompt us to revise our logic will depend less on the details of the physics and more on the largely conceptual question of whether the notion of logical consequence is *a priori* or is an abstraction from what appear to be valid inferences in our practical use of language.[2] Indeed, unless one tends towards the latter position, i. e. unless one thinks that classical logic is already an abstraction from the classical empirical world around us (thus already conceding that 'logic is empirical'), one will be disinclined to take discoveries in microphysics to be relevant at all to the revisability of logic. We shall not attempt to address this more general question.

2 QUANTUM MECHANICS IN A NUTSHELL

In the interest of a self-contained presentation, I summarise a few essentials about quantum mechanics that will be needed below. (This section will be rather abstract but elementary.)

In classical mechanics, the state of a system can be represented by a point

[1]As a matter of fact, Putnam's paper was later reprinted with the modified title 'The logic of quantum mechanics'.

[2]For a recent discussion of the apriorism issue in logic, see e. g. Bueno and Colyvan (2004). Note also that one could very well conceive adopting an apriorist position with regard to quantum logic rather than classical logic.

in (or a subset of, or a probability distribution over) a set called phase space, encoding the positions and momenta of all the particles forming the system. In quantum mechanics, instead, the state of a system is represented by an element in a complex *Hilbert space* (which is a vector space, equipped with a scalar product, that is complete in the norm induced by the scalar product). In particular, this means that for any two states (e.g., for a spin-1/2 system, the states of spin-up and spin-down in direction x), an arbitrary linear combination (or 'superposition') is also a possible state:

$$(1) \quad |\varphi\rangle = \alpha|+_x\rangle + \beta|-_x\rangle \ .$$

Note that the same vector can always be expressed as an appropriate linear combination of vectors in any other basis:

$$(2) \quad |\varphi\rangle = \gamma|+_y\rangle + \delta|-_y\rangle \ .$$

In quantum mechanics, overall scalar factors do not count, i. e. the vectors $|\varphi\rangle$ and $\varepsilon|\varphi\rangle$ for arbitrary complex ε represent the same state, and by convention all states are usually normalised, i. e. have length 1. Therefore, if the basis vectors are normalised and orthogonal, as in the example above, one has $|\alpha|^2 + |\beta|^2 = |\gamma|^2 + |\delta|^2 = 1$.

A second crucial distinction between classical and quantum mechanics is that, when describing composite systems in quantum mechanics, instead of taking the Cartesian product of the given phase spaces as in classical mechanics, one has to take the *tensor product* of the given Hilbert spaces. For instance, for two spin-1/2 subsystems with Hilbert spaces generated (spanned) by

$$(3) \quad \{|+^1_x\rangle, |-^1_x\rangle\}, \quad \{|+^2_x\rangle, |-^2_x\rangle\} \ ,$$

one takes the Hilbert space generated by a basis given by the products of the basis vectors:

$$(4) \quad \{|+^1_x\rangle \otimes |+^2_x\rangle, \ |+^1_x\rangle \otimes |-^2_x\rangle, \ |-^1_x\rangle \otimes |+^2_x\rangle, \ |-^1_x\rangle \otimes |-^2_x\rangle\}$$

(this construction is independent of the bases chosen for the subsystems).

The fundamental consequence of taking tensor products to describe composite systems is that some states of the composite are not product states, e. g. the so-called singlet state of two spin-1/2 systems:

$$(5) \quad \frac{1}{\sqrt{2}}\Big(|+^1_x\rangle \otimes |-^2_x\rangle - |-^1_x\rangle \otimes |+^2_x\rangle\Big) \ .$$

Such non-factorisable states are called *entangled* (the property of being entangled is also independent of the bases chosen for representation in the component systems[3]). If the state of a composite system is entangled, then the subsystems are evidently not described separately by vectors in their respective Hilbert spaces.

[3]Incidentally, it is not independent of the choice of the *subsystems* into which the system is decomposed.

This is a characteristic trait of quantum mechanics (Schrödinger, 1935, p. 555, called it 'not .... *one* but rather *the* characteristic trait of quantum mechanics'), and it is related to the Einstein-Podolsky-Rosen paradox, the Bell inequalities, quantum non-locality *et cetera.*

How can this be? The key, and the third crucial ingredient in our brief summary of quantum mechanics besides Hilbert spaces and tensor products, is the *phenomenology of measurement.* In classical mechanics one can idealise measurements as testing whether a system lies in a certain subset of its phase space. This can be done in principle without disturbing the system, and the result of the test is in principle fully determined by the state of the system. In quantum mechanics we are empirically confronted with the following situation. (i) Measurements can be idealised as testing whether the system lies in a certain (norm-closed) subspace of its Hilbert space — a subset which, in particular, is closed under linear combinations. (ii) A measurement in general disturbs a system: unless the state of the system is either contained in or orthogonal to the tested subspace, the state is projected ('collapsed') onto either the tested subspace or its orthogonal complement. (c) This collapse process is indeterministic, and the relevant probabilities are given by the squared norms of the projections of the state on the given subspace and its orthogonal complement, respectively.

For example, take the initial state (1) and test for spin-up in direction x (test for the subspace P_{+_x}): the final state will be $|+_x\rangle$ with probability $|\alpha|^2$, or $|-_x\rangle$ with probability $|\beta|^2$. Now take the singlet state (5) as the intial state and test for $P^1_{+_x} \otimes P^2_{+_x}$: the test will come out negative with probability 1, and the state will be undisturbed, since it lies in a subspace orthogonal to the tested one. (The same will be the case if one tests for $P^1_{-_x} \otimes P^2_{-_x}$.) Test instead for $P^1_{+_x} \otimes P^2_{-_x}$ (or for $P^1_{-_x} \otimes P^2_{+_x}$): the result (in both cases) will now be $|+^1_x\rangle \otimes |-^2_x\rangle$ or $|-^1_x\rangle \otimes |+^2_x\rangle$, each with probability 1/2. Weaker correlations will be observed if spin is measured along two different directions on the two subsystems. Entanglement thus introduces what appear to be irreducible correlations between results of measurements (even carried out at a distance), and this for a generic pair of tests.

The last two elements of quantum mechanics that we shall also refer to are the Schrödinger equation and the notion of (self-ajoint) operator as an observable quantity.

The Schrödinger equation describes the time evolution of quantum state vectors. It is a linear and unitary equation, i. e. it maps linear combinations into linear combinations, and it preserves the norm (length) of vectors. In its most familiar form, it is a differential equation for the quantum states represented as complex (square-integrable) functions on configuration space (the space of positions of all particles), the so-called Schrödinger waves or wave functions.

Operators, specifically self-adjoint operators (which by the spectral theorem can be decomposed uniquely — in the simplest case — into a real linear combination of projectors onto a family of mutually orthogonal subspaces) play two roles in quantum mechanics. On the one hand they mathematically generate Schrödinger-type evolutions, on the other hand they can be conveniently used to classify simultaneous experimental tests of families of mutually orthogonal subspaces. A system will test positively to only one of these tests, and to this test will be associated the measured value of the corresponding observable. Instead of being understood as specifying probabilities for results of tests, quantum states can be thus equivalently understood as specifying expectation values for observables.

3 QUANTUM LOGIC IN A NUTSHELL

3.1 Quantum logic as a logic of experimental propositions

The easiest way to introduce the concepts of quantum logic is in terms of a logic of 'experimental propositions'. That is, one can define explicitly some non-classical connectives for a certain special class of propositions, relating to idealised quantum mechanical tests. These connectives will be arguably well suited for the limited subject matter at hand. If as a result one obtains a logical system satisfying certain formal requirements, we shall say that one has introduced a local non-classical logic. This is meant to be uncontroversial. Indeed, it should be relatively uncontroversial that (provided the formal requirements are indeed met) such a procedure is legitimate, although there may still be scope for disagreement as to how useful the introduction of such a logic is. In order to go on to assess Putnam's further claims it is essential, at least for the sake of argument, that one accept that in this sense different logics may be better adapted to different subject matters.[4]

The prime example for such a procedure is Kolmogoroff's (1931) interpretation of intuitionistic logic as a calculus of mathematical tasks (*Aufgabenrechnung*). In this framework, each mathematical proposition p stands for solving the corresponding mathematical task. The classical negation of p (not solving the task) is not itself a mathematical task, so the chosen set of propositions is not closed under classical negation. Instead, showing that a task is impossible to solve is again a mathematical task. This justifies introducing a *strong* negation, for which the law of excluded middle $p \vee \neg p$ breaks down. On this basis, one can set up a logical system, which is just the system of intuitionistic logic. More radical claims are not engaged with at this stage. (Indeed, one can argue that this is the correct and only way of interpreting intuitionistic logic, thus safeguarding the primacy of classical logic.)[5]

In the quantum context, let us define experimental propositions as (suitable equivalence classes of) statements of the form: 'The system passes a certain test with probability 1'. From the discussion in the previous section, we recognise that these propositions are in bijective correspondence to closed subspaces of the

[4]What we sketch here is quantum logic as descriptive of the empirical behaviour of certain experiments (albeit idealised ones). One can of course also introduce quantum logic abstractly and axiomatically based on the notion of a 'yes-no' test. This is the approach of the so-called 'Geneva school' of quantum logic (see e. g. Jauch and Piron 1969).

[5]There are further analogies between intuitionistic logic and quantum logic that could be brought to bear on the issue of the revision of logic. Both logics, for instance, allow for classical modal translations (see, respectively, Gödel 1933, and Dalla Chiara 1981).

Hilbert space of the system. The classical negation of such a proposition is not an experimental proposition in this sense. Instead, the proposition stating that the system passes with probability 1 the test corresponding to the orthogonal complement of the given subspace is an experimental proposition. This, again, can be taken to define a strong negation. In the quantum case, however, we go further than in the case of Kolmogoroff's task logic. Indeed, even the classical disjunction of two experimental propositions p and q, corresponding to the set-theoretic union of the two subspaces, is not itself a subspace in general, thus it is not an experimental proposition. Instead, the proposition corresponding to the (closed) span of the two subspaces P and Q (the smallest closed subspace containing both the subspaces P and Q) is an experimental proposition, and we can introduce a corresponding 'quantum logical' disjunction. This proposition corresponds to the most stringent test that will be passed with probability 1 if the tests corresponding to P and Q will. The classical conjunction of p and q, corresponding to the intersection of the two subspaces P and Q, is itself an experimental proposition, so in this sense there is no need to introduce a separate quantum logical conjunction. The closed subspaces of a Hilbert space are ordered by inclusion and form a lattice (i. e. suprema and infima are pairwise always well defined), which is further orthocomplemented under the orthogonal complement defined via the scalar product. The quantum logical connectives correspond to the supremum, infimum and orthocomplement in this lattice.

As a consequence of the introduction of the quantum logical connectives, it is not the law of excluded middle that fails, but (one half of) the *distributive law*: the proposition $p \wedge (q \vee r)$ is generally weaker than the proposition $(p \wedge q) \vee (p \wedge r)$. This can be trivially seen by taking the subspaces Q and R to be two rays spanning a plane, and P to be a ray lying in the same plane but non-collinear with either Q or R. In that case, $p \wedge (q \vee r)$ corresponds to the same subspace as p, but both $p \wedge q$ and $p \wedge r$ correspond to the zero subspace, and so does their quantum logical disjunction. Propositions that thus engender violations of distributivity are called *incompatible*; more precisely, two propositions p and q are called compatible iff the sublattice generated by (the subspaces corresponding to) p, $\neg p$, q and $\neg q$ is distributive.[6]

3.2 Formal properties of the logic

So far, what we have described is a *semi-interpreted language* (Van Fraassen, 1970). We have taken a propositional language, and we have fixed a class of structures that are intended as models of the language, namely the class of lattices of subspaces of Hilbert spaces (henceforth: Hilbert lattices). A model in this sense will be a mapping of the propositions onto the subspaces of some Hilbert space, such that (syntactic) conjunctions shall be mapped to intersections, disjunctions to (closed)

[6]There is more than one definition of compatibility in the literature, but this is immaterial for the purposes of this paper. Furthermore, they all coincide in the important case of orthomodular lattices.

spans and negations to orthogonal complements of the corresponding subspaces.

In order to say that we are introducing a logic in the formal sense (even a local one), we must have at least also a notion of logical consequence and of logical validity, and presumably other formal properties as well, such as soundness and completeness results for some appropriate logical calculus.

With this in mind, let us return to the classical case. Also in the classical case, we could define a semi-interpreted language by defining a model of the language in terms of subsets of some set, and mapping the logical connectives to the corresponding set unions, set intersections, and complements within the set (these are the lattice operations for the subset ordering relation). Every such lattice of subsets is a distributive lattice (also called a Boolean lattice or Boolean algebra), and conversely every distributive lattice is representable as the lattice of subsets of some set.

One can turn this semi-interpreted language into a logic by defining truth valuations as (orthocomplemented-lattice) homomorphisms from an arbitrary Boolean algebra onto the two-element algebra $\{0, 1\}$, and defining the notion of logical validity by taking the class of all Boolean algebras as reference class. That is, a sentence in the language will be a logical truth, iff it is true under every truth valuation of every model. The logic characterised by this notion of logical validity can be axiomatised, is sound and complete, and is of course the usual classical logic.

In order to extend this treatment to quantum logic, we need to extend the notion of a truth valuation to non-distributive lattices. Homomorphisms of the entire lattice onto $\{0, 1\}$ will not do, because in general there are no such total homomorphisms (Jauch and Piron 1963). More precisely, Jauch and Piron show that any so-called orthomodular lattice (in particular any Hilbert lattice) admits total homomorphisms onto $\{0, 1\}$ iff it is distributive.[7] Note that this means that any form of quantum logic *must give up bivalence.* Thus, to insist that every proposition is indeed always true or false (as a matter of logic!) would be question-begging, and, at least for the sake of argument, the failure of bivalence must not be taken as a reason for rejecting the whole framework out of hand.

Instead, one can define workable truth valuations as *partial homomorphisms* onto $\{0, 1\}$, i. e. homomorphisms q defined on some proper (orthocomplemented) sublattice $\mathcal{Q}$ of a given lattice $\mathcal{L}$, provided one requires also that such a partial homomorphism be *filtered*, i. e. for all $a \in \mathcal{Q}$ and $b \in \mathcal{L}$,

$$(6) \quad a < b,\ q(a) = 1 \quad \Rightarrow \quad b \in \mathcal{Q} \text{ and } q(b) = 1\ ,$$

and *maximal*, i. e. have no proper extensions. The intuition behind these properties is that as many propositions as possible should be true or false under a truth valuation (maximality) and, in particular, a proposition that is weaker than a true proposition should also be true (filtering). Both properties are of course trivial for total homomorphisms on Boolean lattices. Note also that any partial homomorphism has both a canonical filtered extension and — by an application

[7]A quick proof for the special case of Hilbert lattices is given in Bell (1987, pp. 5–6).

of Zorn's lemma — a maximal extension. A maximal partial homomorphism is always filtered.

A useful characterisation of truth valuations is the following. For any partial homomorphism q, let $\mathcal{S}$ denote the subset of all $s \in \mathcal{Q}$ such that $q(s) = 1$ (the set of all true propositions). The set $\mathcal{S}$ is a non-empty proper subset of $\mathcal{Q}$, closed under conjunctions. Together with property (6), this means that it is a so-called *filter*; and maximality of q means that $\mathcal{S}$ is a maximal filter, so-called *ultrafilter*. Truth valuations q are thus in bijective correspondence with ultrafilters $\mathcal{S}$ on the lattice. Note that $\mathcal{S}^{\perp}$ is the set of all false propositions, and $\mathcal{Q} = \mathcal{S} \cup \mathcal{S}^{\perp}$.

Given the above definition of truth valuation, one can now proceed with quantum logic as with classical logic and define a notion of logical validity and logical consequence by fixing a suitable reference class of non-distributive lattices. Quantum mechanics (if assumed to be strictly true) tells us that the world is one specific (only partially known) Hilbert lattice, but the corresponding notion of logic will need to be general enough to cover all possible Hilbert lattices.[8] Admittedly, the choice of reference class is not as obvious as in the case of Boolean algebras, and there is some trade-off involved in the choice. One could choose the class of all Hilbert lattices, but it is unclear to date whether the resulting logic is axiomatisable. On the other hand, one can choose more general classes of lattices as reference class, for instance the class of all orthocomplemented lattices or the more restrictive class of all orthomodular lattices. These yield axiomatisable logics that are both sound and complete (see e. g. Dalla Chiara and Giuntini 2002, section 6). Note that the logic of all Hilbert lattices, the logic of all orthomodular lattices and the logic of all orthocomplemented lattices are indeed all distinct, i. e. they have different sets of logical truths.

Choosing the logic of all Hilbert lattices would more properly characterise the 'logic of quantum mechanics'. On the other hand, even if one takes a reference class more general than that of all Hilbert lattices, one can still argue that quantum phenomena have prompted the adoption (at least locally) of a non-classical logic. (Also, as mentioned in section 6, quantum theories of systems with infinitely many degrees of freedom seem to require a larger reference class.) The choice of orthomodular lattices seems particularly attractive, since in an orthomodular lattice there is a unique conditional reducing to the standard conditional for compatible propositions; the resulting connective has some unusual features, but these can be explained in analogy to counterfactual connectives, as is reasonably intuitive in a logic that gives up bivalence (Hardegree 1975).

In any case, the resulting logic is strictly weaker than classical logic, since the reference class that defines logical validity is extended beyond the class of Boolean algebras. Irrespectively of the details of the choice, we shall take it that such a notion of quantum logic provides us with a basis for discussing Putnam's claims, the interest of which after all lies primarily in the idea that empirical considerations

[8]Similarly, general relativity (if assumed to be strictly true) tells us that the world is one specific (only partially known) Lorentzian manifold, but the corresponding notion of geometry will cover all possible Lorentzian or pseudo-Riemannian manifolds.

might force us to give up classical logic, and not (or only in the second place) in the details of which logic should replace it.

3.3 Alternative frameworks

As an aside, let us remark that we have presented above merely one possible framework for introducing a quantum logic, and that others have been proposed. We should mention two in particular.

First, one could choose a different idealisation for quantum mechanical experiments, in order to include more realistic measurements (described technically by positive-operator-valued measures rather than projection-valued measures). This leads one to consider, instead of the lattice of projections (equivalent to the lattice of subspaces), the poset (partially ordered set) of positive operators. This in turn prompts the introduction of fuzzy quantum logics and other quantum logics that generalise the lattice-theoretic approach (see e. g. Dalla Chiara and Giuntini 2002, sections 11–16). More general poset-theoretical structures arise also as the logics associated with theories of quantum probability, as in the test space approach of Foulis and Randall (1981).

Second, one can focus on a different general aspect of quantum mechanical experiments, namely their incompatibility; and instead of introducing apparently new logical connectives, one can restrict the use of the usual connectives to pairs of compatible propositions. This is the partial Boolean algebra approach to quantum logic (Kochen and Specker 1965a, b, 1967), which also gives rise to logical systems with nice formal properties. The partial Boolean algebra approach and the poset-theoretical approach overlap, unsurprisingly, in that so-called transitive partial Boolean algebras are canonically equivalent to so-called coherent orthomodular posets (Finch 1969, Gudder 1972), so that the corresponding logics are the same.

Note that the partial Boolean algebra approach may present advantages to the advocate of a global revision of logic, because the implied revision of logic appears to be more modest (although in a sense equivalent), and because it is easier to argue that the meaning of the logical connectives has remained the same. One does not construct new connectives that must somehow turn out to be the usual ones in disguise. One merely needs to argue that our usual connectives can be applied only to propositions that are compatible, and that it is an empirical matter, settled in the negative by quantum mechanics, whether all propositions are indeed so. We shall not attempt to develop here this line of argument, merely note that Putnam himself switched to using at least the formalism of partial Boolean algebras in some later publications (notably Friedman and Putnam 1978). We shall keep to talking of quantum logic in the lattice-theoretic approach, because most of the discussion about Putnam's suggested revision of logic has been in the context of this approach and of the corresponding failure of distributivity.

4 STANDARD INTERPRETATION AND MEASUREMENT PROBLEM OF QUANTUM MECHANICS

It is certainly an empirical fact that, if one defines experimental propositions as in the previous section, the resulting lattice fails to be distributive, and a fact that is characteristic of quantum mechanics. If all physics were classical, then the lattice of experimental propositions defined in this way would be distributive. It may also be reasonable to want to define a local non-distributive logic for dealing with such experimental propositions. However, it is not clear at this stage why this logic should be even a candidate for a revised global logic. If one takes a 'naive' *instrumentalist* position, then quantum mechanics just provides us with the means of calculating the probabilities for the results of our experiments. The resulting procedure is certainly different from that in any classical framework, but there seems to be little need to revise anything but our algorithmic procedures for predicting experimental results. If one adopts a subtler Bohrian position, then the language of classical physics becomes a prerequisite for the description of quantum experiments, so that the very formulation of quantum mechanics would seem to require classical logic. Clearly, more than empirical considerations are needed in order to mount a case for the revision of logic at the global level. In particular, a strong *opposition* to the instrumentalist or Bohrian position is necessary in order to reject the overall package that includes classical logic and an instrumentalist or Bohrian reading of quantum mechanics.

In this section we shall sketch the 'naive' *realist* interpretation of quantum mechanics. This interpretation, variously referred to as 'standard' or 'orthodox' or 'von Neumann-Dirac', is problematic, because it gives rise to the usual paradoxes, but it is usually taken as the starting point for further discussion and elaboration of other subtler approaches to quantum mechanics. It is thus, so to speak, the default realist position in the foundations of quantum mechanics. And in fact, it is the interpretation of choice (at least implicitly) also for discussions of Putnam's claims on the revision of logic. (Other realist approaches, and their implications for Putnam's claims, will be discussed in section 7.)

The standard interpretation consists in the following assignment of (intrinsic) *properties* to quantum systems. A quantum system has a certain property iff it passes with probability 1 a corresponding experimental test (in the sense of the previous section). Properties assigned in this way are thus in bijective correspondence to the closed subspaces of Hilbert space. What can it mean to assign such properties to a physical system?

The case of one-dimensional subspaces is relatively straightforward: a one-dimensional subspace (ray) is the set of all scalar multiples of a given vector, and these all describe the same quantum state. So, saying that a quantum system has a certain one-dimensional property corresponds to saying that its state is a certain vector in the Hilbert space.

The case of multi-dimensional properties is more difficult, but it is also quite crucial. In this case, one should think of entangled systems, where there is a vector describing the composite system but no vector describing each subsystem separately. The composite system will thus be assigned a one-dimensional property, but not the subsystems. Nevertheless, if two systems are entangled there are always *multi*-dimensional tests (in general non-trivial) on the individual subsystems, for which the subsystems will test positively with probability 1. Therefore, according to the standard interpretation, the subsystems are assigned the corresponding multi-dimensional properties. Unless one accepts some form of holism, in which only the composite system is assigned intrinsic properties, one is forced to generalise the notion of properties for the individual subsystems to include also multi-dimensional ones.

The motivation behind the standard interpretation can be phrased in the language of dispositions. Quantum mechanical systems exhibit a range of dispositional properties in the context of experimental tests, some of which are *sure-fire* dispositions. The standard interpretation suggests that these sure-fire dispositions (whether one-dimensional or multi-dimensional) support an inference to real, objective properties of the quantum system, which are revealed by idealised tests.

But now, enter the paradoxes, specifically the measurement problem and Schrödinger's cat (which we shall take here as two examples of the same problem). If we take quantum mechanics to be correct and universally valid, then one can easily construct examples in which the dynamics of the theory, the Schrödinger equation, will lead to entanglement between microscopic and macroscopic systems, e. g. between a quantum system being measured and the corresponding measuring apparatus, or between a microscopic system and a cat, in such a way that macroscopically distinguishable states of the apparatus (different readings) or the cat (alive or dead) are correlated with different states of the microscopic system. In such cases, on the standard interpretation, only the multi-dimensional subspaces spanned by the macroscopically distinguishable states correspond to properties of the macroscopic system, and these do not correspond to the macroscopic states we witness (the different readings, the cat alive or dead).[9]

One could say, paradoxically, that the cat is neither alive nor dead, but this formulation trades on an ambiguity: this statement would be paradoxical if 'dead' were understood as 'not alive' in the classical sense, but if it is understood as 'not alive' in the sense of the strong negation of quantum logic (assuming that the live and dead states span the Hilbert space of the cat), then the statement makes perfect sense, since in this case 'dead' is strictly stronger than the classical 'not alive'.

[9]Furthermore, this problem cannot be lifted by modelling the states of the apparatus as statistical distributions over microscopic states. If the dynamics is the unitary Schrödinger dynamics, one cannot reproduce the correct measurement statistics for all initial states, unless the state of the apparatus depends on the state of the system to be measured. This result was known already to von Neumann — indeed it prefaces his discussion of measurement in quantum mechanics (von Neumann 1932, section VI.3). For a modern, more general discussion, see e. g. Brown (1986) and references therein.

5 QUANTUM LOGIC AND THE STANDARD INTERPRETATION OF QUANTUM MECHANICS

5.1 Quantum logic as a logic of properties

The obvious interest of the standard interpretation, from the point of view of quantum logic, is that it allows one to apply the quantum logical structures introduced for experimental propositions also to propositions about intrinsic properties of a quantum system. Thus one speaks of the property lattice of the system, or of the lattice of 'testable' propositions about the system.

This move from experimental propositions to properties of a system is explicitly made for instance by Jauch and Piron (1969), who further propose that a quantum state should be understood as the set (in fact the ultrafilter) of true properties about the system. They thus propose, in effect, that a quantum state should be understood as a truth valuation on the lattice of properties of a quantum system. And, indeed, quantum states in the sense of rays in the Hilbert space are in bijective correspondence to the ultrafilters of true propositions they generate (by assigning them probability 1 upon measurement). Thus, truth valuations on the Hilbert lattice of quantum propositions encode all the information about quantum mechanical expectation values.

Abstractly, the introduction of quantum logic for testable propositions is possible simply because there is a bijective correspondence between the experimental propositions and the testable propositions (both being in bijective correspondence with the closed subspaces of the Hilbert space). A closed subspace of the Hilbert space will now represent a proposition about an intrinsic property of the relevant quantum system, and the closed span, intersection and orthogonal complement of such subspaces will correspond to the quantum logical disjunction, conjunction and negation of the respective propositions.

Concretely, the standard interpretation introduces properties corresponding to one-dimensional subspaces P, Q etc., and properties corresponding to multi-dimensional subspaces such as the span of P and Q. The novelty of these properties lies in the fact that under all possible truth valuations, whenever P obtains or whenever Q obtains, also the property corresponding to their span obtains. The interpretation of this property as a quantum logical disjunction $p \vee q$ allows one to interpret such relations between propositions as relations of *logical consequence*.

By considering quantum logic at the level of intrinsic properties of physical objects, we make a further step in the direction of Putnam's proposals. Indeed, at least as regards the more modest claim (a), the fact that this logic in general is a non-distributive lattice is clearly an empirical fact. The fact that it is best understood as a semi-interpreted language, and the fact that this language has a number of properties that justify calling it a logic in the formal sense, have been discussed above. In this sense thus, it should be relatively uncontroversial that quantum phenomena give us empirical grounds for introducing a logic adapted to the world of physics that is non-distributive and hence non-classical. I take it that what we have called above Putnam's claim (a) is thus both intelligible and justifiable.

The claim that is controversial is claim (b), that this gives us further reasons to revise logic *tout court*, i. e. that this logic of testable quantum mechanical propositions, or logic of quantum mechanics, is in fact the 'true' logic and that we have failed to recognise so far that our usual, apparently classical connectives are in fact the connectives of quantum logic. We turn at last to this controversial point, for the time being in the context of the standard interpretation of quantum mechanics.

5.2 *The revision of logic*

If we have successfully introduced the quantum logical connectives in the context of propositions about material properties of physical objects, is this not *ipso facto* saying that 'the logic' of the world is quantum? Surely quantum mechanics is a theory that applies to all material objects, so that the resulting quantum logic is not a local but a global logic?

As mentioned in the introduction, the fact that one may justify the introduction of non-classical connectives does not yet mean that logic has been revised. The crucial point is whether these connectives have been introduced *alongside* the classical connectives, or whether they *replace* them (in an appropriate sense). As we shall see now, the standard interpretation is neutral with regard to this question. Indeed, one has a choice between two opposing views.

On the one hand, it is perfectly possible to interpret the properties assigned to systems in the standard interpretation as elementary properties in the sense of classical logic. Indeed, a quantum logical disjunction $p \vee q$ classically must be an elementary proposition: it is not a classical disjunction of terms that include p and q, although one might be tempted to think that it is the disjunction of all one-dimensional subspaces contained in the span of P and Q. As a matter of fact, this is not true: if it were, in the case of entangled systems there would be a quantum state that describes an individual subsystem, contrary to what quantum mechanics says. On top of the elementary propositions, however, one can perfectly well consider complex ones, constructing them by applying the classical connectives to this new quantum set of elementary propositions, e. g. the disjunction of all one-dimensional subspaces contained in the span of P and Q can be considered alongside with the quantum disjunction $p \vee q$ itself. The quantum aspect is physical and lies in the determination of the elementary properties, while the logic remains classical.

This position has admittedly some disadvantages. Complex propositions in general are not directly testable, i. e. verifiable with probability 1. This is simply because neither the set union of P and Q nor the set complement of P are subspaces of the Hilbert space. This would thus be an empirical limitation characteristic of quantum mechanics. More importantly, perhaps, the above relation between p or q and the quantum logical $p \vee q$ cannot be analysed as logical consequence. That is, there are triples of propositions p, q, r such that whenever p or q hold, also r holds; but since r is elementary, this relation of consequence cannot be analysed as logical consequence.

This, however, may seem a small price to pay in order to refrain from revising our logic. And in fact, we shall see in section 7 that this is arguably the position most naturally associated with the approaches to quantum mechanics known as spontaneous collapse theories.

The opposite position consists in maintaining that the properties assigned in the standard interpretation are *all* the possible properties of a physical system: there is *no* property corresponding to the classical disjunction of p and q, or to the classical negation of p. An equivalent way of saying this is that if p and q are the propositions that some physical quantities take certain values, say '$A = 4$' and '$B = 9$', then there is no meaningful physical quantity that can encode the classical disjunction '$A = 4$ or $B = 9$'. Note that there is no quantum mechanical observable that encodes it. Indeed, in disanalogy to classical physics, the operator $(A - 4)(B - 9)$ in general does not represent a quantum mechanical quantity, because in general the operators A and B do not commute (so that $(A - 4)(B - 9)$ is not self-adjoint). But if there is no meaningful physical quantity whatsoever that represents a classical disjunction, insisting that the properties of the standard interpretation are elementary would mean that the vast majority of complex propositions constructed from elementary propositions about quantum systems are meaningless.

If one drops altogether the possibility of using the classical connectives to form complex propositions, one can instead interpret some of the testable propositions as complex propositions in the sense of quantum logic. In so doing one removes the mismatch between logical propositions and physical propositions (indeed, all propositions are testable propositions), and one ensures that the consequence relation described above between testable propositions and their quantum disjunctions is indeed a relation of logical consequence.

This is presumably the best case that can be made for a revision of logic in the context of the standard interpretation of quantum mechanics. It is not made explicitly by Putnam, although some of it must be implicit in his discussion; it is present more or less in Dickson (2001), who explicitly denies the 'empirical significance' of classical disjunctions and negations. Still, it appears that if one follows this line of argument, the quantum logical connectives have *supplanted* the meaningless classical ones. What about the claim that the quantum logical connectives are the *same* as the classical connectives? Indeed, since every physical system is a quantum system, we seem to have arrived at the conclusion that an 'everyday' disjunctive proposition about any physical object whatsoever is meaningless. But Putnam's claim that the classical and quantum connectives are the same is surely meant in the sense that we should be able to gain a better understanding of our usual everyday classical connectives by realising that they are indeed quantum logical.

What is missing from the above is an explanation of why classical logic appears to have been so effective until now. One needs to explain how, if the true logic is non-distributive, it is still possible for the connectives to behave *truth-functionally* in special cases. This would give rise to the possibility of abstracting classical logic (empirically!) from our everyday use, and of applying it in the appropriate circumstances (as Putnam undoubtedly did in the act of writing his famous paper).

Putnam's claim (c), that adopting quantum logic will solve the paradoxes of quantum mechanics, can be understood as an attempt to fill this gap: the quantum logical point of view does indeed explain, according to Putnam, why the world appears classical to us. Indeed, for Putnam the main advantage of a revision of logic is precisely that it will solve the paradoxes of quantum mechanics. We shall briefly discuss now how Putnam argues for this point and why his arguments are justly regarded as flawed. One other author at least, namely Dickson, attempts to argue that, although the quantum connectives are the true connectives, they behave classically when applied to the everyday, macroscopic realm. As we shall also see, his attempt appears to fail on ultimately similar grounds.

If this is so, then we are left with the following situation. There is a coherent, perhaps even a reasonably convincing case to be made that a non-classical logic is well adapted to a world in which quantum mechanics under the standard interpretation is true. But this world is hugely different from our own. This is precisely what the measurement problem and Schrödinger's cat highlight. Indeed, in such a world it would not seem possible for any intelligent beings to develop at all, let alone beings capable of formulating any kind of logic (let alone quantum mechanics). If the argument does not apply to our world (or at least to a possible world similar to ours), then it loses most of its interest.

5.3 Putnam and the paradoxes

The seemingly logical paradox of Schrödinger's cat, that the cat is neither alive nor dead, trades as we have mentioned on the ambiguity between classical and quantum logical terms. Putnam's way of resolving the paradox is to choose a strictly quantum logical reading: 'dead' is interpreted as 'not alive' in the quantum logical sense of orthocomplementation in the lattice, and the cat is then indeed alive or dead, but in the sense of the quantum logical disjunction. Putnam, however, seems to want to go further, namely he claims that, since the cat is alive or dead (quantum logically), there is a *matter of fact* about the biological state of the cat.

To make the point clearer, let us take an example adapted from Putnam himself (1968, pp. 184–185). Consider an n-dimensional Hilbert space and take an orthonormal basis $|x_1\rangle, \ldots, |x_n\rangle$ in the Hilbert space, which one can associate with a family of tests, or equivalently with some observable X. Denoting the propositions corresponding to the one-dimensional projectors onto the basis vectors as $x_1, \ldots, x_n$, the following is a true proposition under all truth valuations:

(7) $x_1 \vee \ldots \vee x_n$.

Its truth, however, is understood by Putnam as meaning that the observable X has indeed a value corresponding to one of the x_i. As the reasoning is independent of the particular choice of basis, Putnam concludes that the system possesses values for all such observables. He then interprets measurements as simply revealing those preexisting values, thus proposing that the measurement problem of quantum mechanics is solved by a move to quantum logic.

This is rather bewildering, since, as we have seen in section 3, quantum logic comes equipped with a well-defined semantics, which underlies the quantum logical notion of consequence. And we have seen that truth valuations in this semantics are such that the proposition $x_1 \vee \ldots \vee x_n$ can be true without any of the x_i being true. *Any* quantum state that is a non-trivial linear combination of the basis vectors will define such a truth valuation; and in the case of entangled systems, we have seen that a quantum logical proposition can be true *without* any of the one-dimensional projections spanning it being true. To be fair, at the time of Putnam's 1968 paper, the semantics of quantum logic was not fully developed as yet, but the reasoning implied in the paper seems to be technically in error, since he appears to be using a different semantics from that required in quantum logic.

A more charitable reading (perhaps more in line with his later papers, e. g. Putnam 1981), takes Putnam as distinguishing between a quantum level, obeying quantum logic, and a 'hidden' level obeying classical logic. It has in particular been suggested that Putnam's proposals can be analysed in terms of a so-called non-contextual hidden variables theory (Friedman and Glymour 1972), which however confronts them with the standard problems facing such approaches, notably the no-go theorem by Kochen and Specker (1967). Perhaps more plausibly, it has also been suggested to analyse Putnam's proposals in terms of a so-called contextual hidden variables theory (Bacciagaluppi 1993), which however confronts them with the proofs of non-locality for this kind of approaches, specifically those by Heywood and Redhead (1983) and by Stairs (1983). In either case, however, Putnam would seem to be backing away from the proposal that quantum logic is the global logic. (For Putnam's most recent views on the subject, see Putnam 1994.) In section 7, we shall return to the issue of quantum logic in hidden variables approaches, namely in the context of the most successful of these, pilot-wave or de Broglie-Bohm theory.

Dickson's (2001) attempt to explain how classical logic is effective despite quantum logic supposedly being the global logic, proceeds along slightly different lines. Dickson points out that in the macroscopic realm, when talking about measurement results or cats, we apply logic always to a *distributive sublattice* of all (quantum logical) propositions. As it stands, however, this argument is inconclusive. The sublattice generated by the propositions $x_1, \ldots, x_n$ in Putnam's example is distributive, but this fact does not guarantee that the logical connectives will behave truth-functionally, and that is what is at stake. Again to be fair, Dickson suggests that the proper framework for discussing Putnam's claims is that of the more general quantum-like theories based on the formalism of von Neumann algebras. And we shall see in the next section that in that framework the connectives can indeed be shown to behave truth-functionally in certain cases.

As far as the claim concerns the usual formalism of quantum mechanics, however, it may be that Dickson falls prey to a common fallacy. Admittedly, it is a fact that we cannot construct in practice an experiment that would test for a macroscopically entangled state (in particular because of the phenomenon called decoherence), and that at the macroscopic level the only tests we have available are all compatible (so that the corresponding experimental propositions form a distributive lattice). And this fact has often been trumpeted as showing that the measurement problem does not arise. But this practical impossibility is irrelevant to the point that macroscopically entangled states (under the standard interpretation) are incompatible with macroscopic objects having the properties they appear to have, nor does it show that such states do not arise in practice. It thus seems that Dickson's argument fails to improve on Putnam's attempt.

6 QUANTUM LOGIC AND CLASSICAL PROPOSITIONS

Before proceeding further and enquiring into the status of quantum logic in realist approaches to quantum mechanics other than the standard interpretation, let us dwell in more detail on the question of what it could mean for the quantum logical connectives to be the same as the classical connectives.

There is an interesting way of making the case that the meaning of the connectives is indeed the same in classical and quantum logic, namely to argue that it is always given in terms of the supremum, infimum and orthocomplement of the lattice: the conjunction of two propositions is the weakest proposition that implies both propositions, their disjunction is the strongest proposition that is implied by either, and the negation of a proposition is its orthocomplement in the lattice. Until empirical evidence for quantum mechanics was obtained, we used to believe that all lattices of propositions we could ever consider would be distributive. We used to believe that the universe of sets was the correct framework for abstract semantics, because we believed it was rich enough to describe the physical world. But, so the argument goes, it has turned out that it is only the 'universe of Hilbert spaces' that is rich enough for that purpose. (This line of thought presupposes of course that one has already accepted that the logic should be read off the structure of the lattice of empirical propositions.)

The trouble with this suggestion is that, although at this more general level the quantum and classical connectives can thus be said to be the same, still, if the actual lattice of properties is a Hilbert lattice (of dimension greater than 1), the connectives will just not behave truth-functionally, so that the quantum connectives do not seem to reduce to the classical ones in everyday macroscopic situations. This is precisely the problem facing Dickson: can one have the (unique) logical connectives behave truth-functionally when applied to some propositions in the lattice but not to others? We shall see in the present section that, if, as is quite standardly done, one defines logical consequence through a reference class of lattices that is larger than the class of all Hilbert lattices (which, as noted above, is not known to lead to an axiomatisable logic), in particular if one considers quantum logic to be the logic of all orthocomplemented lattices or of all orthomodular lattices, then there is a rigorous sense in which the connectives interpreted in these non-distributive lattices (i. e. the standard quantum logical connectives) can behave truth-functionally in certain cases. Thus, at least in this more abstract setting, there are situations in which one could arguably 'mistake' the logic to be classical.

Recall that two propositions p and q are compatible iff the lattice generated by p, $\neg p$, q and $\neg q$ is distributive. For a subset $\mathcal{A}$ of an orthocomplemented lattice $\mathcal{L}$, denote by $\mathcal{A}^c$ the set of propositions compatible with all propositions in $\mathcal{A}$. If one considers lattices $\mathcal{L}$ more general than Hilbert lattices, the set $\mathcal{L}^c$ (the so-called centre of the lattice) may be non-trivial, i. e. there may exist propositions (other than the trivially true and false propositions) that are compatible with all propositions in the lattice. Such propositions are called *classical* propositions. Now, it is a theorem that under any truth valuation on $\mathcal{L}$, a classical proposition is *always* true or false.

Indeed, let q be a truth valuation from $\mathcal{L}$ onto $\{0, 1\}$, defined on an orthosublattice $\mathcal{Q} = \mathcal{S} \cup \mathcal{S}^\perp$ of $\mathcal{L}$, where $\mathcal{S}$ is the ultrafilter of propositions made true by q. For any ultrafilter $\mathcal{S}$ in $\mathcal{L}$,

(8) $(\mathcal{S} \cup \mathcal{S}^\perp)^c \subset (\mathcal{S} \cup \mathcal{S}^\perp)$

(Raggio 1981, Appendix 5, Proposition 3). A classical proposition, being compatible with any $a \in \mathcal{L}$, is obviously contained in $(\mathcal{S} \cup \mathcal{S}^\perp)^c$ for any set $\mathcal{S}$. Therefore, for any truth valuation q, $c \in \mathcal{Q}$, i. e. $q(c) = 1$ or $q(c) = 0$. QED.

It now follows, just as in the classical case, that if a lattice contains classical propositions, the lattice-theoretical connectives applied to the classical propositions will behave truth-functionally, in particular for any two classical propositions a and b, and any truth valuation q that makes $a \vee b$ true, q will make a true or b true.

Indeed, let q be any truth valuation with $q(a \vee b) = 1$. Since a and b are classical, by the above they are both either true or false under q. But if $q(a) = 0$ and $q(b) = 0$, then, since q is homomorphic, $q(a \vee b) = 0$, contrary to assumption. Therefore, if a and b are classical,

(9) $q(a \vee b) = 1 \quad \Rightarrow \quad q(a) = 1 \text{ or } q(b) = 1$,

for any truth valuation q. QED.

Note that the fact that a certain proposition a is classical depends on the lattice $\mathcal{L}$ chosen as a model of the logic. Specifically, it depends on the relation of a with all the other propositions in the chosen model. It thus depends on the *meaning* of a. We see that the quantum logical connectives can indeed behave truth-functionally in certain models, but depending on the meaning of the propositions involved. Classical logic appears to be valid in special cases, but the additional inferences one can make in these cases are not logical inferences: they are not based on the propositional *form* of the statements involved, they are based instead on the fact that the statements have a *classical content*.

If the lattice of properties in our world is the lattice of projections of some Hilbert space, our world does not contain classical propositions (*pace* Putnam and Dickson). On the other hand, at least some lattices that are more general than Hilbert lattices appear to be physically motivated. Indeed, generalisations of quantum mechanics that allow in general for classical propositions exist, and are required to treat systems with infinitely many degrees of freedom, such as in quantum field theory or in quantum statistical mechanics (when taking thermodynamic limits).

Mathematically, these theories are based on more abstract algebras of observables than the algebra of (self-adjoint) operators on a Hilbert space. For the purposes of quantum logic, the most interesting class of such algebras is that of so-called von Neumann (or W^*-) algebras, which can be represented as certain subalgebras of operators on Hilbert space. Von Neumann algebras can be generated by their projections, so that one can again reduce all statements about observables to statements about projections (i. e. to yes-no tests).[10] The lattices of projections of von Neumann algebras are always orthomodular lattices. (Indeed, historically, the study of orthomodular lattices developed out of the study of von Neumann algebras.) Therefore, unless one insists on characterising quantum logic by the class of all Hilbert lattices, lattices of projections of von Neumann algebras are already included in the models of the most usual varieties of quantum logic, and they are thus a *bona fide* source of examples for the behaviour of the usual quantum logical connectives. Incidentally, we note that J. von Neumann is also associated both with the standard interpretation of quantum mechanics (rightly or wrongly), through his book *Mathematische Grundlagen der Quantenmechanik* (von Neumann 1932), and with the first proposal that quantum mechanics should be interpreted in terms of a non-distributive logic, clearly stated in his paper with Birkhoff four years later (Birkhoff and von Neumann 1936).

When we say that general lattices of projections of von Neumann algebras include classical propositions, the intuition behind it is that there is a breakdown in the linear structure of the state space of a physical system. Indeed, defining a classical observable as an observable C such that propositions of the form 'C has value α' are classical, superpositions of states in which a classical observable has different values simply do not exist (one says that such states are separated by a superselection rule).

The framework of von Neumann algebras is general enough to include both quantum and classical physics, and intermediate theories besides. For instance, one can build algebras that are tensor products of a standard quantum system and a purely classical system, and for which there are no states entangling the quantum system and the classical system (Raggio 1988, see also Baez 1987). Although it is generally believed that such theories would be rather *ad hoc*, they do allow one to describe a world in which the measurement problem of quantum mechanics does not arise, a world in which all measuring apparatuses (as well as cats) are made

[10]Indeed, Raggio (1981) has proved that if $\mathcal{L}$ is the projection lattice of a W^*-algebra $\mathcal{M}$, there is a bijective correspondence between truth valuations on $\mathcal{L}$ and pure normal states on $\mathcal{M}$, in the sense of normalised positive linear functionals. That is, truth valuations indeed encode all the information about expectation values of observables in the algebra.

out of classical observables.

It is instructive to see explicitly how the truth-functionality of the quantum logical connectives would apply to a measurement scenario if the 'pointer' observable of a measuring apparatus were assumed to be a classical observable C. Suppose the apparatus measures a non-classical observable B taking, say, the two values ± 1. Now assume that at the end of an (ideal) measurement the following proposition is true (in obvious notation):

(10) $(B = 1 \wedge C = 1) \vee (B = -1 \wedge C = -1)$,

where $\wedge$ and $\vee$ denote the infimum and supremum in the lattice. We can now show from the fact that C is classical that the disjunction in (10) is truth-functional, i. e. under any truth valuation q that makes (10) true, $(B = 1 \wedge C = 1)$ or $(B = -1 \wedge C = -1)$ are also true.

Let q be such a truth valuation, i. e.

(11) $q\Big((B = 1 \wedge C = 1) \vee (B = -1 \wedge C = -1)\Big) = 1$.

We need to show that

(12) $q(B = 1 \wedge C = 1) = 1 \quad \text{or} \quad q(B = -1 \wedge C = -1) = 1$.

Because q is filtered, we have

(13) $q(C = 1 \vee C = -1) = 1$.

Since C is classical,

(14) $q(C = 1) = 1 \quad \text{or} \quad q(C = -1) = 1$,

by (9). Suppose for instance that $q(C = 1) = 1$. Since q is a homomorphism, we have that

(15) $q\Big[\Big((B = 1 \wedge C = 1) \vee (B = -1 \wedge C = -1)\Big) \wedge C = 1\Big] = 1$.

But now, the propositions $B = \pm 1$, $C = \pm 1$ are all mutually compatible, so that we can distribute over $\vee$ in (15), yielding

(16) $q(B = 1 \wedge C = 1) = 1$.

Analogously, if $q(C = -1) = 1$ we obtain $q(B = -1 \wedge C = -1) = 1$. QED.

Note in particular that the truth-functionality has spread to propositions that include non-classical terms. (This appears to be related to the results by Bub and Clifton (1996) on maximal truth-value assignments in a Hilbert lattice compatible with a certain 'preferred' observable being assigned definite values.)

If such examples do not describe the actual physics, however, what have we gained in showing that the connectives can sometimes behave truth-functionally? We should perhaps distinguish two questions: (i) Can we envisage worlds, perhaps merely inspired by quantum mechanics and sufficiently close to our own, in which we would consider revising our logic? (ii) Is our world such a world?

In a world as the above, one could indeed maintain that the only meaningful propositions are the propositions in the lattice, since the lattice is general enough to include propositions for which classical logic holds, and a generalised quantum mechanics together with the standard interpretation could arguably meet the objections detailed in the previous section against a revision of logic. We can thus make a case that logic is empirical because there is a *possible* world in which we might be prompted by empirical considerations to revise our logic (question (i)). This is different from establishing that in *our* world we may have good reasons for a revision of logic (question (ii)). Note that while Putnam's ultimate aim was to show that logic is indeed empirical, his actual claim was that we have reasons to revise our logic in this world.

In order to proceed further with question (ii), and thus address the revision of logic in Putnam's own terms, we shall have to return to standard quantum mechanics. However, we shall have to consider approaches to the foundations of quantum mechanics other than the standard interpretation, in particular approaches that have some credible claim to providing solutions to the standard puzzles.

7 QUANTUM LOGIC IN OTHER APPROACHES TO QUANTUM MECHANICS

We now leave the standard interpretation of quantum mechanics. While reverting to the standard formalism of quantum mechanics in this section, we shall discuss the status of the claims about quantum logic in the context of other approaches to the foundations of quantum mechanics, approaches that do propose solutions to the puzzles presented to us by quantum mechanics and in particular propose to explain why classical logic is effective (whether or not it be the true logic) in a world in which quantum mechanics is indeed true. The approaches we shall discuss in turn are (i) the pilot-wave theory of de Broglie and Bohm, (ii) spontaneous collapse theories, and (iii) the Everett or many-worlds interpretation. (The presentation of these approaches will necessarily be rather condensed.)

7.1 de Broglie-Bohm theory

The pilot-wave theory of de Broglie and Bohm is a very well-known and well-understood approach to the foundations of quantum mechanics. The theory, as presented by Louis de Broglie at the fifth Solvay conference in October 1927 (de Broglie 1928), is a new dynamics for n-particle systems, described in configuration space (which encodes only the positions of the particles) rather than in phase space. The motion of the particles is determined by a field of velocities defined by the phase S of the complex wave function. At least as regards particle detections, the theory can clearly predict both interference and diffraction phenomena: around the zeros of the wave function, the phase S will behave very

makes sense to identify these components as quasi-separate 'worlds', and to define an internal perspective as centred on each such world.

When a measurement occurs, each observer develops into generally many successors, indexed by their different measurement results. So, which measurement result obtains is a matter of perspective: from the perspective of the live cat, the atom has not decayed and thereby triggered the smashing of the phial of cyanide; from the perspective of the dead cat, it has.

Further recent work pioneered by Deutsch (1999) and perfected in particular by Wallace (2007) has sought to justify the use of the usual quantum probabilities on the basis of rational decision theory as adapted to such a 'splitting' agent. If one accepts Lewis's Principal Principle as the definition of objective chances, the Deutsch-Wallace results imply that the quantum probabilities are indeed objective in each world.

What about logic? Note first of all that, from the perspective of each world, the standard interpretation of quantum mechanics can be applied, taking the relevant component of the universal wave function to be the quantum state for that world. Note also that, although the description of a world given by the relevant component of the wave function is perspectival, it is no less objective than the description of the universe as given by the total wave function.

Thus, again from the perspective of each world (which is the only perspective that makes sense empirically), quantum logic is well adapted to describe the intrinsic properties of physical systems. The question, as we know by now, is whether classical logic is required separately to make sense of the effectiveness of classical logic on the macroscopic scale, or whether there is a sense in which quantum logic can explain how classical logic can be effective in everyday cases, and therefore how we may have arrived to our classical conception of logic by abstraction from the everyday world.

In the case of the Everett interpretation it now seems that this challenge is met. Indeed, while in general a quantum disjunction does not behave truth-functionally (because the different components of the wave function do not decohere, thus all belong to the same world), there are cases in which it does (because the different components do decohere and thus belong to different worlds). In such cases, from the perspective of each world, the disjuncts behave like classical alternatives, one of which is actual, the others counterfactual. Although in every world the properties of all physical systems are in bijective correspondence with subspaces of the Hilbert space, *de facto* unobserved macroscopic superpositions are not the kind of properties that appear in a typical world, unlike the case of the standard interpretation. And this is because interference between different components becomes negligible, and an effective superselection rule arises between the different non-interfering components, thus mimicking the case of von Neumann algebras discussed in section 6. The relation between the quantum and the classical connectives is not a formal relation as we had in the case of von Neumann algebras, but the connectives behave classically in a suitable physical limit.

Thus, while the structure of the intrinsic properties of physical systems supports a non-distributive logic at the fundamental level (even in the individual worlds), one can claim that, unlike the case of pilot-wave theory or spontaneous collapse, the perspectival element characteristic of the Everett interpretation introduces a genuine emergence of the classical connectives from the quantum connectives. In this sense, it is only the Everett interpretation, among the major approaches to quantum mechanics, that is compatible with a revision of logic. One is not forced to accept the overall package, but, while perhaps not entirely as Putnam had articulated it, there is an intelligible sense in which (standard) quantum mechanics may suggest that logic be revised.

8 CONCLUSION

We hope to have clarified in what sense empirical considerations of quantum phenomena may have a bearing on the issue of the 'true' logic. Some of Putnam's (1968) claims in this regard can be justified, but with qualifications.

What can be said about the status of quantum logic in our world, assuming current approaches to the foundations of quantum mechanics, depends on the details of the chosen approach. In particular, one might justify a revision of logic at most if one chooses an Everett interpretation. Indeed, it is a general lesson in the philosophy of physics, confirmed in the present case, that bold philosophical claims made on the basis of quantum mechanics turn out to be highly dependent on the interpretational approach one adopts towards the theory.

The scenario in which consideration of quantum or quantum-like phenomena might make a revision of logic most appealing is possibly that of von Neumann algebras — thus perhaps vindicating Dickson's (2001) intuition —, where there is a rigorous sense in which the quantum and classical connectives can be said to be the same and to behave truth-functionally or not according to the meaning of the propositions involved. This possibility is presumably not realised in our world, but whether it is or not is itself an empirical issue, thus lending at least some support to the idea that logic is indeed empirical.

ACKNOWLEDGEMENTS

The ideas contained in this chapter were developed over many years, and it is impossible for me to remember all those who provided significant input, feedback or encouragement. Hans Primas and Ernst Specker certainly exerted an early and durable influence on my ideas. While at Berkeley, I had the opportunity to discuss aspects of this project with a number of colleagues and students, in particular John MacFarlane, Russell O'Connor, Chris Pincock and Zoe Sachs-Arellano. More recently, Huw Price and Ofer Gal gave me opportunities to present and discuss this material, and I am indebted for comments and discussions to Mark Colyvan, Stephen Gaukroger, Jason Grossman and, most particularly, to Sungho Choi. Finally, I wish to thank Kurt Engesser both for his encouragement and his patience as editor, without either of which this chapter would not have been written.

BIBLIOGRAPHY

[Bacciagaluppi, 1993] G. Bacciagaluppi. Critique of Putnam's quantum logic, *International Journal of Theoretical Physics* **32**, 1835–1846, 1993.

[Baez, 1987] J. Baez. Bell's inequality for C^* algebras, *Letters in Mathematical Physics* **13**, 135–136, 1987.

[Bell, 1987] J. S. Bell. *Speakable and Unspeakable in Quantum Mechanics* (Cambridge: Cambridge University Press), 1987.

[Beltrametti and van Fraassen, 1981] E. Beltrametti and B. C. van Fraassen, eds. *Current Issues in Quantum Logic* (New York: Plenum Press), 1981.

[Birkhoff and von Neumann, 1936] G. Birkhoff and J. von Neumann. The logic of quantum mechanics, *Annals of Mathematics* **37**, 823–843, 1936. Reprinted in Hooker (1975), pp. 1–26.

[Bohm, 1952] D. Bohm. A suggested interpretation of the quantum theory in terms of "hidden" variables, I and II, *Physical Review* **85**, 166–179 and 180–193, 1952.

[de Broglie, 1928] L. de Broglie. La nouvelle dynamique des quanta, in *Électrons et Photons: Rapports et Discussions du Cinquième Conseil de Physique Solvay* (Paris: Gauthier-Villars), pp. 105–132, 1928. Translated in G. Bacciagaluppi and A. Valentini, *Quantum Theory at the Crossroads: Reconsidering the 1927 Solvay Conference* (Cambridge: Cambridge University Press, forthcoming), pp. 375–399.

[Brown, 1986] H. Brown. The insolubility proof of the quantum measurement problem, *Foundations of Physics* **16**, 857–870, 1986.

[Bub and Clifton, 1996] J. Bub and R. Clifton. A uniqueness theorem for "no collapse" interpretations of quantum mechanics, *Studies in History and Philosophy of Modern Physics* **27B**, 181–219, 1996.

[Bueno and Colyvan, 2004] O. Bueno and M. Colyvan. Logical non-apriorism and the "law" of non-contradiction, in G. Priest, J. C. Beall and B. Armour-Garb (eds.), *The Law of Non-Contradiction: New Philosophical Essays* (Oxford: Oxford University Press), pp. 156–175, 2004.

[Dalla Chiara, 1981] M. L. Dalla Chiara. Some metalogical pathologies of quantum logic, in Beltrametti and Van Fraassen (1981), pp. 147–159.

[Dalla Chiara and Giuntini, 2002] M. L. Dalla Chiara and R. Giuntini. Quantum logics, in D. Gabbay and F. Guenthner (eds.), *Handbook of Philosophical Logic*, vol. 6 (Dordrecht: Kluwer), pp. 129–228, 2002.

[Deutsch, 1999] D. Deutsch. Quantum theory of probability and decisions, *Proceedings of the Royal Society of London* **A 455**, 3129–3137, 1999.

[Dickson, 2001] M. Dickson. Quantum logic is alive $\wedge$ (it is true $\vee$ it is false), *Philosophy of Science (Proceedings)* **68**, S274–S287, 2001.

[Dummett, 1976] M. Dummett. Is logic empirical?, in H. D. Lewis (ed.), *Contemporary British Philosophy*, 4th series (London: Allen and Unwin), pp. 45–68, 1976. Reprinted in M. Dummett, *Truth and other Enigmas* (London: Duckworth, 1978), pp. 269–289.

[Finch, 1969] P. D. Finch. On the structure of quantum logic, *Journal of Symbolic Logic* **34**, 275–282, 1969. Reprinted in Hooker (1975), pp. 415–425.

[Foulis and Randall, 1981] D. J. Foulis and C. H. Randall. What are quantum logics and what ought they to be?, in Beltrametti and Van Fraassen (1981), pp. 35–52.

[Friedman and Glymour, 1972] M. Friedman and C. Glymour. If quanta had logic, *Journal of Philosophical Logic* **1**, 16–28, 1972.

[van Fraassen, 1970] B. C. van Fraassen. On the extension of Beth's semantics of physical theories, *Philosophy of Science* **37**, 325–339, 1970.

[Friedman and Putnam, 1978] M. Friedman and H. Putnam. Quantum logic, conditional probability, and interference, *Dialectica* **32**, 305–315, 1978.

[Ghirardi *et al.*, 1995] G.C. Ghirardi, R. Grassi, and F. Benatti. Describing the macroscopic world: closing the circle within the dynamical reduction program, *Foundations of Physics* **25**, 5–38, 1995.

[Ghirardi *et al.*, 1986] G.C. Ghirardi, A. Rimini, and T. Weber. Unified dynamics for microscopic and macroscopic systems, *Physical Review* **D 34**, 470–491, 1986.

[Gödel, 1933] K. Gödel. Eine Interpretation des intuitionistischen Aussagenkalküls, *Ergebnisse eines mathematischen Kolloquiums* **4**, 39–40, 1933. Reprinted and translated in K. Gödel, *Collected Works*, vol. 1, ed. by S. Feferman, S. Kleene, G. Moore, R. Solovay and J. van Heijenoort (Oxford: Oxford University Press, 1986), pp. 300–303.

[Greaves, 2004] H. Greaves. Understanding Deutsch's probability in a deterministic multiverse, *Studies in History and Philosophy of Modern Physics* **35B**, 423–456, 2004.

[Gudder, 1972] S. Gudder. Partial algebraic structures associated with orthomodular posets, *Pacific Journal of Mathematics* **41**, 712–730, 1972.

[Hardegree, 1975] G. M. Hardegree. Stalnaker conditionals and quantum logic, *Journal of Philosophical Logic* **4**, 399–421, 1975.

[Heywood and Redhead, 1983] P. Heywood and M. L. G. Redhead. Nonlocality and the Kochen-Specker paradox, *Foundations of Physics* **13**, 481–499, 1983.

[Hooker, 1975] C. A. Hooker. *The Logico-Algebraic Approach to Quantum Mechanics*, vol. 1 (Dordrecht: Reidel), 1975.

[Jauch and Piron, 1963] J. M. Jauch and C. Piron. Can hidden variables be excluded in quantum mechanics?, *Helvetica Physica Acta* **36**, 827–837, 1963.

[Jauch and Piron, 1969] J. M. Jauch and C. Piron. On the structure of quantal proposition systems, *Helvetica Physica Acta* **43**, 842–848, 1969. Reprinted in Hooker (1975), pp. 427–436.

[Kochen and Specker, 1965a] S. Kochen and E. P. Specker. Logical structures arising in quantum theory, in L. Addison, L. Henkin and A. Tarski (eds.), *The Theory of Models* (Amsterdam: North-Holland), pp. 177–189, 1965. Reprinted in Hooker (1975), pp. 263–276.

[Kochen and Specker, 1965b] S. Kochen and E. P. Specker. The calculus of partial propositional functions, in Y. Bar-Hillel (ed.), *Logic, Methodology, and Philosophy of Science* (Amsterdam: North-Holland), pp. 45–57, 1965. Reprinted in Hooker (1975), pp. 277–292.

[Kochen and Specker, 1967] S. Kochen and E. P. Specker. The problem of hidden variables in quantum mechanics, *Journal of Mathematics and Mechanics* **17**, 59–88, 1967. Reprinted in Hooker (1975), pp. 293–328.

[Kolmogoroff, 1931] A. Kolmogoroff. Zur Deutung der intuitionistischen Logik, *Mathematische Zeitschrift* **35**, 58–65, 1931.

[von Neumann, 1932] J. von Neumann. *Mathematische Grundlagen der Quantenmechanik* (Berlin: Springer), 1932.

[Pearle, 1989] P. Pearle. Combining stochastic dynamical state-vector reduction with spontaneous localization, *Physical Review* **A 39**, 2277–2289, 1989.

[Putnam, 1968] H. Putnam. Is logic empirical?, in R. Cohen and M. Wartofsky (eds.), *Boston Studies in the Philosophy of Science*, vol. 5 (Dordrecht: Reidel), pp. 216–241, 1968. Reprinted as 'The logic of quantum mechanics' in H. Putnam, *Mathematics, Matter, and Method. Philosophical Papers*, vol. 1 (Cambridge: Cambridge University Press, 1975), pp. 174–197.

[Putnam, 1974] H. Putnam. How to think quantum-logically, *Synthese*, **29**, 55–61, 1974. Reprinted in P. Suppes (ed.), *Logic and Probability in Quantum Mechanics* (Dordrecht: Reidel, 1976) pp. 47–53.

[Putnam, 1981] H. Putnam. Quantum mechanics and the observer, *Erkenntnis* **16**, 193–219, 1981.

[Putnam, 1994] H. Putnam. Michael Redhead on quantum logic, in P. Clark and R. Hale (eds.), *Reading Putnam* (Oxford: Basil Blackwell), pp. 265–280, 1994.

[Raggio, 1981] G. A. Raggio. *States and Composite Systems in W^*-algebraic Quantum Mechanics*, Diss. ETH No. 6824, 1981.

[Raggio, 1988] G. A. Raggio. A remark on Bell's inequality and decomposable normal states, *Letters in Mathematical Physics*, **15**, 27–29, 1988.

[Saunders, 1993] S. Saunders. Decoherence, relative states, and evolutionary adaptation, *Foundations of Physics* **23**, 1553–1585, 1993.

[Schrödinger, 1935] E. Schrödinger. Discussion of probability relations between separated systems, *Proceedings of the Cambridge Philosophical Society* **31**, 555–563, 1935.

[Stairs, 1983] A. Stairs. Quantum logic, realism, and value-definiteness, *Philosophy of Science* **50**, 578–602, 1983.

[Wallace, 2003] D. Wallace. Everett and structure, *Studies in History and Philosophy of Modern Physics* **34B**, 87–105, 2003.

[Wallace, 2007] D. Wallace. Quantum probability from subjective likelihood: improving on Deutsch's proof of the probability rule, *Studies in History and Philosophy of Modern Physics*, **38**, 311–332, 2007.

Edited by K. Engesser, D. M. Gabbay and D. Lehmann

QUANTUM AXIOMATICS

Diederik Aerts

1 INTRODUCTION

Quantum axiomatics has its roots in the work of John von Neumann, in collaboration with Garett Birkhoff, that is almost as old as the standard formulation of quantum mechanics itself [Birkhoff and von Neumann, 1936]. Indeed already during the beginning years of quantum mechanics, the formalism that is now referred to as standard quantum mechanics [von Neumann, 1932], was thought to be too specific by the founding fathers themselves. One of the questions that obviously was at the origin of this early dissatisfaction is: 'Why would a complex Hilbert space deliver the unique mathematical structure for a complete description of the microworld? Would that not be amazing? What is so special about a complex Hilbert space that its mathematical structure would play such a fundamental role?'

Let us turn for a moment to the other great theory of physics, namely general relativity, to raise more suspicion towards the fundamental role of the complex Hilbert space for quantum mechanics. General relativity is founded on the mathematical structure of Riemann geometry. In this case however it is much more plausible that indeed the right fundamental mathematical structure has been taken. Riemann developed his theory as a synthesis of the work of Gauss, Lobatsjevski and Bolyai on nonEuclidean geometry, and his aim was to work out a theory for the description of the geometrical structure of the world in all its generality. Hence Einstein took recourse to the work of Riemann to express his ideas and intuitions on space time and its geometry and this lead to general relativity. General relativity could be called in this respect 'the geometrization of a part of the world including gravitation'.

There is, of course, a definite reason why von Neumann used the mathematical structure of a complex Hilbert space for the formalization of quantum mechanics, but this reason is much less profound than it is for Riemann geometry and general relativity. The reason is that Heisenberg's matrix mechanics and Schrödinger's wave mechanics turned out to be equivalent, the first being a formalization of the new mechanics making use of l_2, the set of all square summable complex sequences, and the second making use of $L_2(\mathbb{R}^3)$, the set of all square integrable complex functions of three real variables. The two spaces l_2 and $L_2(\mathbb{R}^3)$ are canonical examples of a complex Hilbert space. This means that Heisenberg and Schrödinger were working already in a complex Hilbert space, when they formulated matrix mechanics and wave mechanics, without being aware of it. This made it a straightforward choice for von Neumann to propose a formulation of quantum mechanics in an abstract complex Hilbert space, reducing matrix mechanics and wave mechanics to two specific cases.

One problem with the Hilbert space representation was known from the start. A (pure) state of a quantum entity is represented by a unit vector or ray of the complex Hilbert space, and not by a vector. Indeed vectors contained in the same ray represent the same state or one has to normalize the vector that represents the state after it has been changed in one way or another. It is well known that if rays of a vector space are called points and two dimensional subspaces of this vector space are called lines, the set of points and lines corresponding in this way to a vector space, form a projective geometry. What we just remarked about the unit vector or ray representing the state of the quantum entity means that in some way the projective geometry corresponding to the complex Hilbert space represents more intrinsically the physics of the quantum world as does the Hilbert space itself. This state of affairs is revealed explicitly in the dynamics of quantum entities, that is built by using group representations, and one has to consider projective representations, which are representations in the corresponding projective geometry, and not vector representations [Wigner, 1959].

The title of the article by John von Neumann and Garett Birkhoff [Birkhoff and von Neumann, 1936] that we mentioned as the founding article for quantum axiomatics is 'The logic of quantum mechanics'. Let us explain shortly what Birkhoff and von Neumann do in this article. First of all they remark that an operational proposition of a quantum entity is represented in the standard quantum formalism by an orthogonal projection operator or by the corresponding closed subspace of the Hilbert space $\mathcal{H}$. Let us denote the set of all closed subspaces of $\mathcal{H}$ by $\mathcal{L}(\mathcal{H})$. Next Birkhoff and von Neumann show that the structure of $\mathcal{L}(\mathcal{H})$ is not that of a Boolean algebra, the archetypical structure of the set of propositions in classical logic. More specifically it is the distributive law between conjunction and disjunction

$$(a \vee b) \wedge c = (a \wedge c) \vee (b \wedge c) \tag{1}$$

that is not necessarily valid for the case of quantum propositions $a, b, c \in \mathcal{L}(\mathcal{H})$. A whole line of research, called quantum logic, was born as a consequence of the Birkhoff and von Neumann article. The underlying philosophical idea is that, in the same manner as general relativity has introduced nonEuclidean geometry into the reality of the physical world, quantum mechanics introduces nonBoolean logic. The quantum paradoxes would be due to the fact that we reason with Boolean logic about situations with quantum entities, while these situations should be reasoned about with nonBoolean logic.

Although fascinating as an approach [Mittelstaedt, 1963], it is not this idea that is at the origin of quantum axiomatics. Another aspect of what Birkhoff and von Neumann did in their article is that they shifted the attention on the mathematical structure of the set of operational propositions $\mathcal{L}(\mathcal{H})$ instead of the Hilbert space $\mathcal{H}$ itself. In this sense it is important to pay attention to the fact that $\mathcal{L}(\mathcal{H})$ is the set of all operational propositions, *i.e.* the set of yes/no experiments on a quantum entity. They opened a way to connect abstract mathematical concepts of the quantum formalism, namely the orthogonal projection operators or closed subspaces of the Hilbert space, directly with physical operations in the laboratory, namely the yes/no experiments.

George Mackey followed in on this idea when he wrote his book on the mathematical foundations of quantum mechanics [Mackey, 1963]. He starts the other way around and considers as a basis the set $\mathcal{L}$ of all operational propositions, meaning propositions being testable by yes/no experiments on a physical entity. Then he introduces as an axiom that this set $\mathcal{L}$ has to have a structure isomorphic to the set of all closed subspaces $\mathcal{L}(\mathcal{H})$ of a complex Hilbert space in the case of a quantum entity. He states that it would be interesting to invent a set of axioms on $\mathcal{L}$ that gradually would make $\mathcal{L}$ more and more alike to $\mathcal{L}(\mathcal{H})$ to finally arrive at an isomorphism when all the axioms are satisfied. While Mackey wrote his book results as such were underway. A year later Constantin Piron proved a fundamental representation theorem. Starting from the set $\mathcal{L}$ of all operational propositions of a physical entity and introducing five axioms on $\mathcal{L}$ he proved that $\mathcal{L}$ is isomorphic to the set of closed subspaces $\mathcal{L}(V)$ of a generalized Hilbert space V whenever these five axioms are satisfied [Piron, 1964]. Let us elaborate on some of the aspects of this representation theorem to be able to explain further what quantum axiomatics is about.

We mentioned already that Birkhoff and von Neumann had noticed that the set of closed subspaces $\mathcal{L}(\mathcal{H})$ of a complex Hilbert space $\mathcal{H}$ is not a Boolean algebra, because distributivity between conjunction and disjunction, like expressed in (1), is not satisfied. The set of closed subspaces of a complex Hilbert space forms however a lattice, which is a more general mathematical structure than a Boolean algebra, moreover, a lattice where the distributivity rule (1) is satisfied is a Boolean algebra, which indicates that the lattice structure is the one to consider for the quantum mechanical situation. As we will see more in detail later, and to make again a reference to general relativity, the lattice structure is indeed to a Boolean algebra what general Riemann geometry is to Euclidean geometry. And moreover, meanwhile we have understood why the structure of operational propositions of the world is not a Boolean algebra but a lattice. This is strictly due to the fact that measurements can have an uncontrollable influence on the state of the physical entity under consideration. We explain this insight in detail in [Aerts and Aerts, 2004], and mention it here to make clear that the intuition of Birkhoff and von Neumann, and later Mackey, Piron and others, although only mathematical intuition at that time, was correct.

When Piron proved his representation theorem in 1964, he concentrated on the lattice structure for the formulation of the five axioms. Meanwhile much more research has been done, both physically motivated in an attempt to make the approach more operational, as well as mathematically, trying to get axiomatically closer to the complex Hilbert space. In the presentation of quantum axiomatics we give in this article, we integrate the most recent results, and hence deviate for this reason from the original formulation, for example when we explain the representation theorem of Piron.

Axiomatic quantum mechanics is more than just an axiomatization of quantum mechanics. Because of the operational nature of the axiomatization, it holds the potential for 'more general theories than standard quantum mechanics' which however are 'quantum like theories'. In this sense, we believe that it is one of the candidates to generate the framework for the new theory to be developed generalizing quantum mechanics and relativity theory [Aerts and Aerts, 2004]. Let us explain why we believe that quantum axiomatics has the potential to deliver such a generalization of relativity theory and quantum mechanics. General relativity is a theory that brings part of the world that in earlier Newtonian mechanics was classified within dynamics to the geometrical realm of reality, and more specifically confronting us with the pre-scientific and naive realistic vision on space, time, matter and gravitation. It teaches us in a deep and new way, compared to Newtonian physics, 'what are the things that exists and how they exist and are related and how they influence each other'. But there is one deep lack in relativity theory: it does not take into account the influence of the observer, the effect that the measuring apparatus has on the thing observed. It does not confront the subject-object problem and its influence on how reality is. It cannot do this because its mathematical apparatus is based on the Riemann geometry of time-space, hence prejudicing that time-space is there, filled up with fields and matter, that are also there, independent of the observer. There is no fundamental role for the creation of 'new' within relativity theory, everything just 'is' and we are only there to 'detect' how this everything 'is'. That is also the reason why general relativity can easily be interpreted as delivering a model for the whole universe, whatever this would mean. We know that quantum mechanics takes into account in an essential way the effect of the observer through the measuring apparatus on the state of the physical entity under study. In a theory generalizing quantum mechanics and relativity, such that both appear as special cases, this effect should certainly also appear in a fundamental way. We believe that general relativity has explored to great depth the question 'how can things **be** in the world'. Quantum axiomatics explores in great depth the question 'how can be **acted** in the world'. And it does explore this question of 'action in the world' in a very similar manner as general relativity theory does with its question of 'being of the world'. This means that operational quantum axiomatics can be seen as the development of a general theory of 'actions in the world' in the same manner that Riemann geometry can be seen as a general theory of 'geometrical forms existing in the world'. Of course Riemann is not equivalent to general relativity, a lot of detailed physics had to be known to apply Riemann resulting in general relativity. This is the same with operational quantum axiomatics, it has the potential to deliver the framework for the theory generalizing quantum mechanics and relativity theory, but a lot of detailed physics will have to be used to find out the exact way of doing this.

We want to remark that in principle a theory that describes the possible actions in the world, and a theory that delivers a model for the whole universe, should not be incompatible. It should even be so that the theory that delivers a model of the whole universe should incorporate the theory of actions in the world, which would mean for the situation that exists now, general relativity should contain quantum mechanics, if it really delivers a model for the whole universe. That is why we believe that Einstein's attitude, trying to incorporate the other forces and interactions within general relativity, contrary to common believe, was the right one, globally speaking. What Einstein did not know at that time was 'the reality of nonlocality in the micro-world'. Nonlocality means nonspatiality, which means that the reality of the micro-world, and hence the reality of the universe as a whole, is not time-space like. Time-space is not the global theatre of reality, but rather a cristallization and structuration of the macro-world. Time-space has come into existence together with the macroscopic material entities, and hence it is 'their' time and space, but it is not the theatre of the microscopic quantum entities. This fact is the fundamental reason why general relativity, built on the mathematical geometrical Riemannian structure of time-space, cannot be the canvas for the new theory to be developed. A way to express this technically would be to say that the set of events cannot be identified with the set of time-space points as is done in relativity theory. Recourse will have to be taken to a theory that describes reality as a kind of pre-geometry, and where the geometrical structure arises as a consequence of interactions that collapse into the time-space context. We believe that operational quantum axiomatics, as presented in this article, can deliver the framework as well as the methodology to construct and elaborate such a theory. In the next section we introduce the basic notions of operational quantum axiomatics.

Mackey and Piron introduced the set of yes/no experiments but then immediately shifted to an attempt to axiomatize mathematically the lattice of (operational) propositions of a quantum entity, Mackey postulating right away an isomorphism with $\mathcal{L}(\mathcal{H})$ and Piron giving five axioms to come as close as possible to $\mathcal{L}(\mathcal{H})$. Also Piron's axioms are however mostly motivated by mimicking mathematically the structure of $\mathcal{L}(\mathcal{H})$. In later work Piron made a stronger attempt to found operationally part of the axioms [Piron, 1976], and this attempt was worked out further in [Aerts, 1981; Aerts, 1982; Aerts, 1983], to arrive at a full operational foundation only recently [Aerts, 1999a; Aerts, 1999b; Aerts *et al.*, 1999; Aerts, 2002].

Also mathematically the circle was closed only recently. At the time when Piron gave his five axioms that lead to the representation within a generalized Hilbert space, there only existed three examples of generalized Hilbert spaces that fitted all the axioms, namely real, complex and quaternionic Hilbert space, also referred to as the three standard Hilbert spaces.[1] Years later Hans Keller constructed

[1]There do exist a lot of finite dimensional generalized Hilbert spaces that are different from the three standard examples. But since a physical entity has to have at least a position observable, it follows that the generalized Hilbert space must be infinite dimensional. At the time of Piron's representation theorem, the only infinite dimensional cases that were known are the three standard Hilbert spaces, over the real, complex or quaternionic numbers.

the first counterexample, more specifically an example of an infinite dimensional generalized Hilbert space that is not isomorphic to one of the three standard Hilbert spaces [Keller, 1980]. The study of generalized Hilbert spaces, nowadays also called orthomodular spaces, developed into a research subject of its own, and recently Maria Pia Solèr proved a groundbreaking theorem in this field. She proved that an infinite dimensional generalized Hilbert space that contains an orthonormal base is isomorphic with one of the three standard Hilbert spaces [Solèr, 1995]. It has meanwhile also been possible to formulate an operational axiom, called 'plane transitivity' on the set of operational propositions that implies Solèr's condition [Aerts and van Steirteghem, 2000], which completes the axiomatics for standard quantum mechanics by means of six axioms, the original five axioms of Piron and plane transitivity as sixth axiom.

2 STATE PROPERTY SPACES

In this section we introduce the basic notions and basic axioms for quantum axiomatics. We introduce notions and axioms that are as simple as possible, but each time show how the more traditional axioms of quantum axiomatics are related and/or derived from our set of axioms.

2.1 States and properties

With each entity S corresponds a well defined set of states Σ of the entity. These are the modes of being of the entity. This means that at each moment the entity S 'is' in a specific state $p \in \Sigma$. Historically quantum axiomatics has been elaborated mainly by considering the set of properties[2]. With each entity S corresponds a well defined set of properties $\mathcal{L}$. A property $a \in \mathcal{L}$ is 'actual' or is 'potential' for the entity S. To be able to present the axiomatisation of the set of states and the set of properties of an entity S in a mathematical way, we introduce some additional notions.

Suppose that the entity S is in a specific state $p \in \Sigma$. Then some of the properties of S are actual and some are not, hence they are potential. This means that with each state $p \in \Sigma$ corresponds a set of actual properties, subset of $\mathcal{L}$. This defines a function $\xi : \Sigma \rightarrow \mathcal{P}(\mathcal{L})$, which makes each state $p \in \Sigma$ correspond to the set $\xi(p)$ of properties that are actual in this state. With the notation $\mathcal{P}(\mathcal{L})$ we mean the 'powerset' of $\mathcal{L}$, i.e. the set of all subsets of $\mathcal{L}$. From now on we can replace the statement 'property $a \in \mathcal{L}$ is actual for the entity S in state $p \in \Sigma$' by '$a \in \xi(p)$'.

Suppose that for the entity S a specific property $a \in \mathcal{L}$ is actual. Then this entity is in a certain state $p \in \Sigma$ that makes a actual. With each property $a \in \mathcal{L}$

[2]In the original paper of Birkhoff and Von Neumann [Birkhoff and von Neumann, 1936], the basic notion is the one of 'operational proposition'. An operational proposition is not the same as a property [Randall and Foulis, 1983; Foulis *et al.*, 1983], but it points at the same structural part of quantum axiomatics.

we can associate the set of states that make this property actual, i.e. a subset of Σ. This defines a function $\kappa : \mathcal{L} \rightarrow \mathcal{P}(\Sigma)$, which makes each property $a \in \mathcal{L}$ correspond to the set of states $\kappa(a)$ that make this property actual. We can replace the statement 'property $a \in \mathcal{L}$ is actual if the entity S is in state $p \in \Sigma$' by the expression '$p \in \kappa(a)$'.

Summarising the foregoing we have:

$$\begin{aligned}&\text{property } a \in \mathcal{L} \text{ is actual for the entity } S \text{ in state } p \in \Sigma\\ &\Leftrightarrow a \in \xi(p)\\ &\Leftrightarrow p \in \kappa(a)\end{aligned} \tag{2}$$

This expresses a fundamental 'duality' between states and properties. We introduce a specific mathematical structure to represent an entity S, its states and its properties, taking into account this duality. First we remark that if Σ and $\mathcal{L}$ are given, and one of the two functions ξ or κ is given, then the other function can be derived. Let us show this explicitly. Hence suppose that Σ, $\mathcal{L}$ and ξ are given, and define $\kappa : \mathcal{L} \rightarrow \mathcal{P}(\Sigma)$ such that $\kappa(a) = \{p \mid p \in \Sigma, a \in \xi(p)\}$. Similarly, if Σ, $\mathcal{L}$ and $\kappa : \mathcal{L} \rightarrow \mathcal{P}(\Sigma)$ are given, we can derive ξ in an analogous way. This means that to define the mathematical structure which carries our notions and relations it is enough to introduce Σ, $\mathcal{L}$ and one of the two functions ξ or κ.

DEFINITION 1 State property space. Consider two sets Σ and $\mathcal{L}$ and a function

$$\xi : \Sigma \leftarrow \mathcal{P}(\mathcal{L}) \quad p \mapsto \xi(p) \tag{3}$$

then we say that $(\Sigma, \mathcal{L}, \xi)$ is a state property space. The elements of Σ are interpreted as states and the elements of $\mathcal{L}$ as properties of the entity S. For $p \in \Sigma$ we have that $\xi(p)$ is the set of properties of S which are actual if S is in state p. For a state property space $(\Sigma, \mathcal{L}, \xi)$ we define:

$$\kappa : \mathcal{L} \rightarrow \mathcal{P}(\Sigma) \quad a \mapsto \kappa(a) = \{p \mid p \in \Sigma, a \in \xi(p)\} \tag{4}$$

and hence for $a \in \mathcal{L}$ we have that $\kappa(a)$ is the set of states of the entity S which make the property a actual. The function κ is called the Cartan map of the state property space $(\Sigma, \mathcal{L}, \xi)$.

PROPOSITION 2. *Consider a state property space* $(\Sigma, \mathcal{L}, \xi)$, *and* κ *defined as in* (4). *We have:*

$$a \in \xi(p) \Leftrightarrow p \in \kappa(a) \tag{5}$$

There are two natural 'implication relations' on a state property space. If the situation is such that if '$a \in \mathcal{L}$ is actual for S in state $p \in \Sigma$' implies that '$b \in \mathcal{L}$ is actual for S in state $p \in \Sigma$' we say that the property a implies the property b. If the situation is such that '$a \in \mathcal{L}$ is actual for S in state $q \in \Sigma$' implies that '$a \in \mathcal{L}$ is actual for S in state $p \in \Sigma$' we say that the state p implies the state q.

DEFINITION 3 Property implication and state implication. Consider a state property space $(\Sigma, \mathcal{L}, \xi)$. For $a, b \in \mathcal{L}$ we introduce:

$$a \leq b \Leftrightarrow \kappa(a) \subseteq \kappa(b) \tag{6}$$

and we say that a 'implies' b. For $p, q \in \Sigma$ we introduce:

$$p \leq q \Leftrightarrow \xi(q) \subseteq \xi(p) \tag{7}$$

and we say that p 'implies' q[3].

DEFINITION 4 Equivalent properties and equivalent states. Consider a state property space $(\Sigma, \mathcal{L}, \xi)$. We call properties $a, b \in \mathcal{L}$ equivalent, and denote $a \approx b$ iff $\kappa(a) = \kappa(b)$. We call states $p, q \in \Sigma$ equivalent and denote $p \approx q$ iff $\xi(p) = \xi(q)$.

Let us give two important examples of state property spaces. First, consider a set Ω and let $\mathcal{P}(\Omega)$ be the set of all subsets of Ω, and consider the function $\xi_\Omega : \Omega \to \mathcal{P}(\mathcal{P}(\Omega))$, such that for $p \in \Omega$

$$\xi_\Omega(p) = \{A \,|A \in \mathcal{P}(\Omega), p \in A\} \tag{8}$$

The triple $(\Omega, \mathcal{P}(\Omega), \xi_\Omega)$ is a state property space. For $A \in \mathcal{P}(\Omega)$ we have $\kappa_\Omega(A) = \{p \,|p \in \Omega, A \in \xi(p)\} = \{p \,|p \in \Omega, p \in A\} = A$. This shows that $\kappa_\Omega : \mathcal{P}(\Omega) \to \mathcal{P}(\Omega)$ is the identity.

Second, consider a complex Hilbert space $\mathcal{H}$, and let $\Sigma(\mathcal{H})$ be the set of unit vectors of $\mathcal{H}$ and $\mathcal{L}(\mathcal{H})$ the set of orthogonal projection operators of $\mathcal{H}$. Consider the function $\xi_\mathcal{H} : \Sigma(\mathcal{H}) \to \mathcal{P}(\mathcal{L}(\mathcal{H}))$, such that for $x \in \Sigma(\mathcal{H})$

$$\xi_\mathcal{H}(x) = \{A \mid A \in \mathcal{L}(\mathcal{H}), Ax = x\} \tag{9}$$

The triple $(\Sigma(\mathcal{H}), \mathcal{L}(\mathcal{H}), \xi_\mathcal{H})$ is a state property space. For $A \in \mathcal{L}(\mathcal{H})$ we have $\kappa_\mathcal{H}(A) = \{x \,|x \in \Sigma(\mathcal{H}), Ax = x\}$.

The two examples that we propose here are the archetypical physics examples. The first example is the state property space of a classical physical system, where Ω corresponds with its state space. The second example is the state property space of a quantum physical system, where $\mathcal{H}$ is the complex Hilbert space connected to the quantum system.

DEFINITION 5 Pre-order relation. Suppose that we have a set Z. We say that $\leq$ is a pre-order relation on Z iff for $x, y, z \in Z$ we have:

$$\begin{aligned} &x \leq x \\ &x \leq y \text{ and } y \leq z \Rightarrow x \leq z \end{aligned} \tag{10}$$

For two elements $x, y \in Z$ such that $x \leq y$ and $y \leq x$ we denote $x \approx y$ and we say that x is equivalent to y.

[3]The state implication and property implication are not defined in an analogous way. Indeed, then we should for example have written $p \leq q \Leftrightarrow \xi(p) \subseteq \xi(q)$. That we have chosen to define the state implication the other way around is because historically this is how intuitively is thought about states implying one another.

It is easy to verify that the implication relations that we have introduced are pre-order relations.

PROPOSITION 6. *Consider a state property space* $(\Sigma, \mathcal{L}, \xi)$, *then* $\Sigma, \leq$ *and* $\mathcal{L}, \leq$ *are pre-ordered sets.*

We can show the following for a state property space:

PROPOSITION 7. *Consider a state property space* $(\Sigma, \mathcal{L}, \xi)$. *(1) Suppose that* $a, b \in \mathcal{L}$ *and* $p \in \Sigma$. *If* $a \in \xi(p)$ *and* $a \leq b$, *then* $b \in \xi(p)$. *(2) Suppose that* $p, q \in \Sigma$ *and* $a \in \mathcal{L}$. *If* $q \in \kappa(a)$ *and* $p \leq q$ *then* $p \in \kappa(a)$.

Proof. (1) We have $p \in \kappa(a)$ and $\kappa(a) \subseteq \kappa(b)$. This proves that $p \in \kappa(b)$ and hence $b \in \xi(p)$. (2) We have $a \in \xi(q)$ and $\xi(q) \subseteq \xi(p)$ and hence $a \in \xi(p)$. This shows that $p \in \kappa(a)$. ■

Suppose we consider a set of properties $(a_i)_i \subseteq \mathcal{L}$. It is very well possible that there exist states of the entity S in which all the properties a_i are actual. This is in fact always the case if $\cap_i \kappa(a_i) \neq \emptyset$. Indeed, if we consider $p \in \cap_i \kappa(a_i)$ and S in state p, then all the properties a_i are actual. If it is such that the situation where all properties a_i of a set $(a_i)_i$ and no other are actual is again a property of the entity S, we will denote this new property by $\wedge_i a_i$, and call it a 'meet property' of $(a_i)_i$. Clearly we have $\wedge_i a_i$ is actual for S in state $p \in \Sigma$ iff a_i is actual for all i for S in state p. This means that we have $\wedge_i a_i \in \xi(p)$ iff $a_i \in \xi(p)$ $\forall i$.

DEFINITION 8 Meet property. Consider a state property space $(\Sigma, \mathcal{L}, \xi)$ and a set $(a_i)_i \subseteq \mathcal{L}$ of properties. If there exists a property, which we denote by $\wedge_i a_i$, such that

$$\kappa(\wedge_i a_i) = \cap_i \kappa(a_i) \tag{11}$$

we call $\wedge_i a_i$ the 'meet property' of the set of properties $(a_i)_i$.

If we have the structure of a pre-ordered set, we can wonder about the existence of meets and joins with respect to this pre-order, or conjunctions and disjunctions with respect to the implication related to this pre-order. In relation with the meet property we can prove the following

PROPOSITION 9. *Consider a state property space* $(\Sigma, \mathcal{L}, \xi)$ *and a set* $(a_i)_i \subseteq \mathcal{L}$ *of properties. The property* $\wedge_i a_i$, *if it exists, is an infimum*[4] *for the pre-order relation* $\leq$ *on* $\mathcal{L}$.

Proof. We have $\kappa(\wedge_i a_i) = \cap_i \kappa(a_i) \subseteq \kappa(a_j)$ $\forall$ j, and hence $\wedge_i a_i \leq a_j$ $\forall$ j. Suppose that $x \in \mathcal{L}$ is such that $x \leq a_j \forall j$, then we have $\kappa(x) \subseteq \kappa(a_j) \forall$ j, and hence $\kappa(x) \subseteq \cap_i \kappa(a_i) = \kappa(\wedge_i a_i)$. As a consequence we have $x \leq \wedge_i a_i$. This proves that $\wedge_i a_i$ is an infimum. ■

[4]An infimum of a subset $(x_i)_i$ of a pre-ordered set Z is an element of Z that is smaller than all the x_i and greater than any element that is smaller than all x_i.

2.2 Tests

For the operational foundations of the state property space, we need to make explicit how we test whether for a physical entity a specific property is actual.

A test is an experiment we can perform on the physical entity under investigation with the aim of knowing whether a specific property of this physical entity is actual or not. We identify for each test two outcomes, one which we call 'yes' corresponding to the occurrence of the expected outcome, and another one which we call 'no' corresponding to the non occurrence of the expected outcome. However, if for a test the outcome 'yes' occurs, this does not mean that the property which is tested is actual. It is only when we can predict with certainty, i.e. with probability equal to 1, that the test would have an outcome 'yes', if we would perform it, that the property a is actual.

Let us consider the example of an entity which is a piece of wood. We have in mind the property of 'burning well'. A possible test for this property consists of taking the piece of wood and setting it on fire. In general, when we perform the test on a piece of dry wood, the piece of wood will be destroyed by the test. So the property of 'burning well' is a property that the piece of wood eventually has before we make the test. Of course it is after having done a number of tests with pieces of wood and having got always the outcome yes, we decide that the one new piece of wood, prepared under equivalent conditions, whereon we never performed the test, has actually the property of burning well. We will say that the test is 'true' if this is the case.

DEFINITION 10 Testing a property. Consider a physical entity with corresponding state property space $(\Sigma, \mathcal{L}, \xi)$. α is a test of the property $a \in \mathcal{L}$ if we have

$$a \in \xi(p) \ \Leftrightarrow \ \text{`yes}' \text{ can be predicted with certainty for } \alpha \text{ when } S \text{ is in state } p \quad (12)$$

Similarly with the pre-order relations on the sets of properties we have pre-order relations on the sets of tests.

DEFINITION 11 Test implication. We say that a test α is stronger than a test β and denote $\alpha \leq \beta$ iff whenever the physical entity is in a state such that α is true then also β is true.

PROPOSITION 12. *Consider a state property space* $(\Sigma, \mathcal{L}, \xi)$. *If the test* α *tests property* a, *and the test* β *tests property* b, *we have*

$$\alpha \leq \beta \Leftrightarrow a \leq b \tag{13}$$

Proof. Suppose that $\alpha \leq \beta$, and consider $p \in \Sigma$ such that $a \in \xi(p)$. This means that the test α gives with certainty outcome 'yes' if the entity is in state p. Hence also β gives with certainty 'yes' if the entity is in state p. This means that $b \in \xi(p)$. Hence we have proven that $a \leq b$. Suppose now that $a \leq b$, and suppose that the entity is in a state p such that α gives with certainty outcome 'yes'. This means that $a \in \xi(p)$. Then we have $b \in \xi(p)$. Since β tests b we have that β gives with certainty the outcome 'yes'. Hence we have proven that $\alpha \leq \beta$. ■

DEFINITION 13 Equivalent tests. We say that two tests α and β are equivalent, and denote $\alpha \approx \beta$ iff $\alpha \leq \beta$ and $\beta \leq \alpha$.

PROPOSITION 14. *Equivalent tests test equivalent properties, and tests that test equivalent properties are equivalent tests.*

Proof. Consider two equivalent tests $\alpha \approx \beta$ testing respectively properties a and b. Since we have $\alpha \leq \beta$ and $\beta \leq \alpha$ this implies that $a \leq b$ and $b \leq a$, and hence $a \approx b$. Consider two equivalent properties $a \approx b$ being tested respectively by tests α and β. Since we have $a \leq b$ and $b \leq a$ this implies that $\alpha \leq \beta$ and $\beta \leq \alpha$, and hence $\alpha \approx \beta$. ■

In general the outcomes of a test of one property are profoundly influenced by the testing of another property. In most cases it makes even no sense to perform two tests at once or one after the other on the entity. But still it is so that every entity can have several properties which are actual at once. There is indeed a way to construct a test that makes it possible to test the actuality of several properties at once, even if the tests corresponding to the different properties disturb each other profoundly. Let us illustrate this by means of the example of the piece of wood.

Consider the following two properties of the piece of wood: Property c 'the piece of wood burns well' and property d 'the piece of wood floats on water'. Suppose that γ is a test of property c which consists of setting the wood on fire and giving the outcome 'yes' if it burns well. The test δ consists of putting the wood on water and giving the outcome 'yes' if it floats, hence it is a test of property d. If we perform first the test δ, and put the piece of wood on water, we have changed the state of the wood in a state of 'wet wood' and as a result the wood will not burn well. On the other hand if we perform the test γ and burn the wood, it will no longer float on water. However we all know plenty of pieces of wood for which both properties c and d are actual at once. This means that the way in which we decide both properties to be actual for a specific piece of wood is not related to performing both tests one after the other. If we analyse carefully this situation we see that we agree for a piece of wood both properties c and d to be actual if which ever of the tests γ or δ is performed, the outcome 'yes' can be predicted with certainty for this test. Hence, to state this in a slightly more formal way: 'If we choose, or if some process external to us produces a choice, between one of the two tests γ or δ, and it is certain to obtain the outcome 'yes' no matter what is this choice, then we agree that both properties c and d are actual for the piece of wood'.

This leads us to the following. Given two tests γ and δ we define a new test which we denote $\gamma \cdot \delta$ and call the product test of γ and δ. The performance of $\gamma \cdot \delta$ consists of a choice being made between γ and δ, and then the performance of this chosen test, and the attribution of the outcome obtained in this way. As a consequence, we have $\gamma \cdot \delta$ is true iff γ is true 'and' δ is true, which shows that $\gamma \cdot \delta$ tests both properties c and d, or, it tests the conjunction of properties c and *d*. Remark that for the performance of the test $\gamma \cdot \delta$ only one test γ or δ has to be performed, and hence the definition of the product test is valid independent of the way in which tests disturb each other. The definition of product test is valid for any number of tests, which means that we have found a way to test any number of properties at once, or, to test the conjunction of any number of properties. Let us formally introduce the product test for an arbitrary number of properties.

Consider a family $(a_i)_i$ of properties a_i and tests α_i, such that α_i tests property a_i. A test which tests the actuality of all the properties a_i, and which we denote $\Pi_i \alpha_i$ and call the product of the α_i is the following:

DEFINITION 15 Product test. The performance of $\Pi_i \alpha_i$ consists of a choice between one of the tests α_i followed by the performance of this chosen test.

PROPOSITION 16. *For a set of tests $(\alpha_i)_i$ we have*

$$\Pi_i \alpha_i \leq \alpha_j \ \forall \ j \tag{14}$$

Let us prove that the product test tests the meet of a set of properties.

PROPOSITION 17. *Consider an entity with corresponding state property space $(\Sigma, \mathcal{L}, \xi)$ and a set of properties $(a_i)_i \subseteq \mathcal{L}$. Suppose that we have tests $(\alpha_i)_i$ available for the properties $(a_i)_i$, then the product test $\Pi_i \alpha_i$ tests a meet property $\wedge_i a_i$.*

Proof. Following the definition of 'meet property' given in Definition 8, to prove that $\Pi_i \alpha_i$ tests the meet property $\wedge_i a_i$ of the set of properties $(a_i)_i$, where α_i tests a_i, we need to show that 'yes can be predicted with certainty for $\Pi_i \alpha$ the entity being in state p' is equivalent to '$a_i \in \xi(p) \ \forall \ i$'. This follows from the definition of the product test. Indeed 'yes can be predicted with certainty for $\Pi_i \alpha_i$ the entity being in state p' is equivalent to 'yes can be predicted with certainty for $\alpha_i \ \forall \ i$ the entity being in state p'. ■

2.3 Orthogonality

Let us investigate the operational foundation of orthogonality.

DEFINITION 18 Inverse test. For a test α we consider the test that consists of performing the same experiment and changing the role of 'yes' and 'no'. We denote this new test by $\widetilde{\alpha}$, and call it the inverse test of α.

PROPOSITION 19. *Consider a test α and a set of tests $(\alpha_i)_i$, then we have*

$$\widetilde{\widetilde{\alpha}} = \alpha \tag{15}$$

$$\widetilde{\Pi_i \alpha_i} = \Pi_i \widetilde{\alpha_i} \tag{16}$$

Proof. Obviously if we exchange 'yes' and 'no' for an experiment corresponding to the test α, and then exchange 'yes' and 'no' again, we get the same test. The test $\widetilde{\Pi_i \alpha_i}$ consists of exchanging 'yes' and 'no' for the experiment corresponding to the test $\Pi_i \alpha_i$. This comes to exchanging 'yes' and 'no' after the choice of one of the α_j is made. The test $\Pi_i \widetilde{\alpha_i}$ on the contrary consists of exchanging the 'yes' and 'no' of each of the tests α_i, hence before the choice of one of the α_j is made. These are the same tests. ■

There is a fundamental problem with the inverse test which is the following. Suppose that $\alpha(a)$ tests the property a, and $\beta(a)$ also tests the property a, then $\widetilde{\alpha}$ and $\widetilde{\beta}$ in general test completely different properties. Let us show this by means of a concrete example. We introduce the test τ, which is the unit test, in the sense that it is a test which gives 'yes' as outcome with certainty for each state of the entity. The test τ tests the maximal property I. Obviously $\widetilde{\tau}$ is a test which gives with certainty outcome 'no' for each state of the entity, which means that it tests a property which is never actual. It can be shown that this property which is never actual can only be represented by the minimal property 0, hence $\widetilde{\tau}$ tests 0. Consider now an arbitrary property $a \in \mathcal{L}$, and a test α which tests property a. Let us suppose that $\widetilde{\alpha}$ tests a property b, and that both properties a and b can be actual. Consider the product test $\alpha \cdot \tau$. This is a test which also tests the property a, because indeed $\alpha \cdot \tau$ gives with certainty the outcome 'yes' iff α gives with certainty the outcome 'yes'. However $\widetilde{\alpha \cdot \tau} = \widetilde{\alpha} \cdot \widetilde{\tau}$ tests the property 0, and not the property b. Indeed $\widetilde{\alpha} \cdot \widetilde{\tau}$ gives with certainty the outcome 'yes' iff $\widetilde{\alpha}$ gives with certainty 'yes' and $\widetilde{\tau}$ gives with certainty 'yes'. This is never the case, which proves that it tests the property 0.

DEFINITION 20 Orthogonal states. If p and q are two states of S we will say that p is orthogonal to q, iff there exists a test γ such that γ is true if S is in the state p and $\widetilde{\gamma}$ is true if S is in the state q. We will denote then $p \perp q$

PROPOSITION 21. *For $p, q, r, s \in \Sigma$ we have*

$$p \perp q \ \Rightarrow \ q \perp p \tag{17}$$

$$p \perp q, \ r \leq p, \ s \leq q \ \Rightarrow \ r \perp s \tag{18}$$

$$p \perp q \ \Rightarrow \ p \wedge q = 0 \tag{19}$$

Proof. The orthogonality relation is obviously symmetric. If $r \leq p$ and $s \leq q$, and $p \perp q$, and γ is a test such that γ is true if the entity is in state p and $\widetilde{\gamma}$ is true if the entity is in state q, then we have that γ is true if the entity is in state r and $\widetilde{\gamma}$ is true if the entity is in state s. This proves that $r \perp s$. ■

DEFINITION 22 Orthogonal properties and states. We say that a state $p \in \Sigma$ is orthogonal to a property $a \in \mathcal{L}$ iff for every $q \in \Sigma$ such that $a \in \xi(q)$ we have $p \perp q$. We denote $p \perp a$. We say that two properties $a, b \in \mathcal{L}$ are orthogonal iff for every $p, q \in \Sigma$ such that $a \in \xi(p)$ and $b \in \xi(q)$ we have $p \perp q$. We denote $a \perp b$.

PROPOSITION 23. *For $a, b, c, d \in \mathcal{L}$ and $p, r \in \Sigma$ we have*

$$p \perp a,\ r \leq p,\ c \leq a \quad \Rightarrow \quad r \perp c \tag{20}$$

$$a \perp b \quad \Rightarrow \quad b \perp a \tag{21}$$

$$a \perp b,\ c \leq a,\ d \leq b \quad \Rightarrow \quad c \perp d \tag{22}$$

$$a \perp b \quad \Rightarrow \quad a \wedge b = 0 \tag{23}$$

3 A SET OF AXIOMS

In this section we put forward a set of axioms and derive the consequences for the structure of the state property space of this set of axioms. We make an attempt to introduce all the axioms in a way which is as operational as possible.

3.1 The axiom of property determination

The first axiom expresses a relation between the states and the properties. We consider two properties $a, b \in \mathcal{L}$ of the entity S, and suppose that $\kappa(a) = \kappa(b)$. This means that each state which make property a actual also makes property b actual, and vice versa. It also means that we cannot distinguish between property a and property b by means of the states of the entity S. Hence, this means that for entity S, property a and property b are equivalent.

AXIOM 24 Property determination. We say that the axiom of property determination is satisfied for a state property space $(\Sigma, \mathcal{L}, \xi)$ iff for $a, b \in \mathcal{L}$ we have:

$$\kappa(a) = \kappa(b) \Rightarrow a = b \tag{24}$$

DEFINITION 25 Partial order relation. Suppose that we have a set Z. We say that $\leq$ is a partial order relation on Z iff $\leq$ is a pre-order relation for which equivalent elements are equal.

If axiom 24 is satisfied for a state property space $(\Sigma, \mathcal{L}, \xi)$, the pre-order relation on the set of properties $\mathcal{L}$ is then a partial order relation.

THEOREM 26. *Consider a state property space $(\Sigma, \mathcal{L}, \xi)$ for an entity S for which axiom 24 is satisfied. The 'property implication' on $\mathcal{L}$ is then a partial order relation on $\mathcal{L}$.*

Proof. Suppose that axiom 24 is satisfied for $(\Sigma, \mathcal{L}, \xi)$, and consider $a, b \in \mathcal{L}$ such that $a \leq b$ and $b \leq a$. Then we have $\kappa(a) \subseteq \kappa(b)$ and $\kappa(b) \subseteq \kappa(a)$, and hence $\kappa(a) = \kappa(b)$. As a consequence, because of axiom 24, we have $a = b$. This proves that $\leq$ is a partial order relation on $\mathcal{L}$. ■

The two archetypical examples we have introduced both satisfy the axiom of property determination. Indeed, consider the first example of classical mechanics. Since

κ is the identity, we have for $A, B \in \mathcal{P}(\Omega)$ that $\kappa(A) = \kappa(B)$ implies $A = B$. For the second example of quantum mechanics, consider $A, B \in \mathcal{L}(\mathcal{H})$, and suppose that $\kappa_{\mathcal{H}}(A) = \kappa_{\mathcal{H}}(B)$. Consider the vector $x \in \mathcal{H}$ such that $Ax = x$. Since $\kappa_{\mathcal{H}}(A) = \kappa_{\mathcal{H}}(B)$ this implies that $Bx = x$. This proves that $AB = A$. In an analoguous way we prove that $AB = B$, and hence $A = B$.

3.2 The axiom of completeness

We want to be able to distinguish between properties that are not necessarily of the type that they are meet properties, and between properties which are meet properties. In [Aerts, 1981; Aerts, 1982] we have introduced in this way a subset of properties $\mathcal{T} \subseteq \mathcal{L}$, and called it a 'generating set of properties' for the state property space $(\Sigma, \mathcal{L}, \xi)$.

AXIOM 27 Property completeness. We say that the axiom of property completeness is satisfied for a state property space $(\Sigma, \mathcal{L}, \xi)$ iff there exists a subset $\mathcal{T} \subseteq \mathcal{L}$ such that for each $(a_i)_i \subseteq \mathcal{T}$ there exists $a \in \mathcal{L}$ such that

$$\kappa(a) = \bigcap_i \kappa(a_i) \tag{25}$$

and, each property $a \in \mathcal{L}$ is of this form, i.e. for $a \in \mathcal{L}$ there exists a subset $(a_i)_i \subseteq T$ such that (25) is satisfied. We call $\mathcal{T} \subseteq \mathcal{L}$ a generating set of properties of the state property space $(\Sigma, \mathcal{L}, \xi)$, and call the property a of (25) a meet of the set of properties $(a_i)_i$, and denote it by

$$a = \bigwedge_i a_i \tag{26}$$

The following definition and proposition explain why we have chosen to call axiom 27 the axiom of completeness.

DEFINITION 28 Complete pre-ordered set. Suppose that $Z, \leq$ is a pre-ordered set. We say that Z is a complete pre-ordered set iff for each subset of elements of Z there exists an infimum and a supremum in Z.

PROPOSITION 29. *Consider a state property space $(\Sigma, \mathcal{L}, \xi)$ for which axiom 27 is satisfied. Then $\mathcal{L}, \leq$ is a complete pre-ordered set, and if for a subset $(a_i)_i \subseteq \mathcal{L}$ we denote an infimum of $(a_i)_i$ by $\wedge_i a_i$ we have*

$$\kappa(\bigwedge_i a_i) = \bigcap_i \kappa(a_i) \tag{27}$$

Proof. Consider an arbitrary set $(a_i)_i \subseteq \mathcal{L}$ of properties. We need to prove that there exists an infimum and a supremum in $\mathcal{L}$ for this set of properties $(a_i)_i$. From axiom 27 we know that for each a_i there is a set $(b^i_{j_i})_{j_i} \subseteq \mathcal{T}$, such that $a_i = \wedge_{j_i} b^i_{j_i}$, and $\kappa(a_i) = \cap_{j_i} \kappa(b^i_{j_i})$. From the same axiom 27 follows that for the subset

$(b^i_{j_i})^i_{j_i} \subseteq \mathcal{T}$ there exists a property $a \in \mathcal{L}$ such that $\kappa(a) = \cap_i \cap_{j_i} \kappa(b^i_{j_i}) = \cap_i \kappa(a_i)$. Let us prove that a is an infimum for the set $(a_i)_i \subseteq \mathcal{L}$. Since $\kappa(a) = \cap_i \kappa(a_i)$ we have $\kappa(a) \subseteq \kappa(a_j)\ \forall j$, and hence $a \leq a_j\ \forall\ j$, which proves that a is a lower bound for $(a_i)_i$. Consider $x \in \mathcal{L}$ such that $x \leq a_j\ \forall j$. This implies that $\kappa(x) \subseteq \kappa(a_j)\ \forall j$, and hence $\kappa(x) \subseteq \cap_i \kappa(a_i) = \kappa(a)$. From this follows that $x \leq a$, which proves that a is a greatest lower bound or infimum. It is a consequence that for each subset $(a_i)_i \subseteq \mathcal{L}$, there exists also a supremum in $\mathcal{L}$, let us denote it by $\vee_i a_i$. It is given by

$$\bigvee_i a_i = \bigwedge_{x \in \mathcal{L}, a_i \leq x \forall i} x \tag{28}$$

This proves that $\mathcal{L}, \leq$ is a complete pre-ordered set. ■

Remark that the supremum for elements of $\mathcal{L}$, although it exists, has no simple operational meaning.

DEFINITION 30 Complete lattice. Suppose that $Z, \leq$ is a partially ordered set. We say that $Z, \leq, \wedge, \vee$ is a complete lattice iff for each subset $(x_i)_i \subseteq Z$ of elements of Z there exists an infimum $\wedge_i x_i \in Z$ and a supremum $\vee_i x_i \in Z$ in Z. A complete lattice has a minimal element which we denote 0, and which is the infimum of all elements of Z, and a maximal element, which we denote I, and which is the supremum of all elements of Z.

THEOREM 31. *Consider a state property space $(\Sigma, \mathcal{L}, \xi)$ for which axioms 24 and 27 are satisfied. Then $\mathcal{L}, \leq, \wedge, \vee$ is a complete lattice. For I the maximum of $\mathcal{L}$, $(a_i)_i \subseteq \mathcal{L}$ and $p \in \Sigma$ we have:*

$$\kappa(I) = \Sigma \tag{29}$$

$$\bigcap_i \kappa(a_i) = \kappa(\bigwedge_i a_i) \tag{30}$$

$$\bigcup_i \kappa(a_i) \subseteq \kappa(\bigvee_i a_i) \tag{31}$$

$$a_i \in \xi(p)\ \forall i \Leftrightarrow \bigwedge_i a_i \in \xi(p) \tag{32}$$

Proof. From proposition 26 follows that $\mathcal{L}, \leq$ is a partially ordered set, and from proposition 29 follows that $\mathcal{L}, \leq, \wedge, \vee$ is a complete lattice. We have $\kappa(I) \subseteq \Sigma$. For an arbitrary $p \in \Sigma$ consider $\xi(p)$. Since $a \leq I\ \forall a \in \xi(p)$, we have $I \in \xi(p)$, and hence $p \in \kappa(I)$. This proves that $\Sigma \subseteq \kappa(I)$. As a consequence we have $\kappa(I) = \Sigma$. From (27) of proposition 29 follows (30). Let us prove (31). Since $\vee_i a_i$ is a supremum of $(a_i)_i$ we have $a_j \leq \vee_i a_i\ \forall\ j$. Hence $\kappa(a_j) \subseteq \kappa(\vee_i a_i)\ \forall\ j$. This proves that $\cup_i \kappa(a_i) \subseteq \kappa(\vee_i a_i)$. Suppose that $a_i \in \xi(p)\ \forall i$, then $p \in \kappa(a_i)\ \forall i$, and hence $p \in \cap_i \kappa(a_i) = \kappa(\wedge_i a_i)$. From this follows that $\wedge_i a_i \in \xi(p)$, and hence we have proven one of the implications of (32). Let us prove the other one, and hence suppose that $\wedge_i a_i \in \xi(p)$. From this follows that $p \in \kappa(\wedge_i a_i) = \cap_i \kappa(a_i)$. As a consequence we have $p \in \kappa(a_i)\ \forall i$, and hence $a_i \in \xi(p)\ \forall i$. ■

If axiom 24 and 27 are satisfied for a state property space $(\Sigma, \mathcal{L}, \xi)$, and hence the set of properties $\mathcal{L}$ is a complete lattice, we can represent the states by means of properties.

DEFINITION 32 Property state. Consider a state property space $(\Sigma, \mathcal{L}, \xi)$ for which axioms 24 and 27 are satisfied. For each state $p \in \Sigma$ we define the 'property state' corresponding to p as the property

$$s(p) = \bigwedge_{a \in \xi(p)} a \tag{33}$$

PROPOSITION 33. *Consider a state property space $(\Sigma, \mathcal{L}, \xi)$ for which axioms 24 and 27 are satisfied. For $p, q \in \Sigma$ and $a \in \mathcal{L}$ we have:*

$$s(p) \in \xi(p) \tag{34}$$

$$a \in \xi(p) \Leftrightarrow s(p) \leq a \tag{35}$$

$$p \leq q \Leftrightarrow s(p) \leq s(q) \tag{36}$$

$$\xi(p) = \{a \mid a \in \mathcal{L}, s(p) \leq a\} = [s(p), I] \tag{37}$$

Proof. That $s(p) \in \xi(p)$ follows directly from (32). Suppose $a \in \xi(p)$ then $\wedge_{a \in \xi(p)} a \leq a$ and hence $s(p) \leq a$. We have that $s(p) \in \xi(p)$, and if $s(p) \leq a$, from proposition 7 follows then that $a \in \xi(p)$. Suppose that $p \leq q$. Then we have $\xi(q) \subseteq \xi(p)$. From this follows that $s(p) = \wedge_{a \in \xi(p)} a \leq \wedge_{a \in \xi(q)} a = s(q)$. Suppose now that $s(p) \leq s(q)$. Take $a \in \xi(q)$, then we have $s(q) \leq a$. Hence also $s(p) \leq a$. But this implies that $a \in \xi(p)$. Hence this shows that $\xi(q) \subseteq \xi(p)$ and as a consequence we have $p \leq q$. Consider $b \in [s(p), I]$. This means that $s(p) \leq b$, and hence $b \in \xi(p)$. Consider now $b \in \xi(p)$. Then $s(p) \leq b$ and hence $b \in [s(p), I]$. ■

For a state property space satisfying axioms 24 and 27 we can prove that the set of property states is a full set for the complete lattice $\mathcal{L}$.

THEOREM 34 Full set of property states. *Consider a state property space $(\Sigma, \mathcal{L}, \xi)$ for which axioms 24 and 27 are satisfied. For $a \in \mathcal{L}$ we have*

$$\kappa(a) = \bigcup\nolimits_{a \in \xi(p)} \kappa(s(p)) \tag{38}$$

$$a = \bigvee\nolimits_{a \in \xi(p)} s(p) \tag{39}$$

Proof. From (35) follows that if $a \in \xi(p)$ we have $s(p) \leq a$ and hence $\kappa(s(p)) \subseteq \kappa(a)$. This proves that $\cup_{a \in \xi(p)} \kappa(s(p)) \subseteq \kappa(a)$. From (34) follows that for $p \in \Sigma$ we have $p \in \kappa(s(p))$ and hence $\{p\} \subseteq \kappa(s(p))$. This proves that $\kappa(a) = \cup_{p \in \kappa(a)} \{p\} \subseteq \cup_{p \in \kappa(a)} \kappa(s(p)) = \cup_{a \in \xi(p)} \kappa(s(p))$. From (35) we have $a \in \xi(p)$ then $s(p) \leq a$. This proves that $\vee_{a \in \xi(p)} s(p) \leq a$. Using (31) we have $\kappa(a) = \cup_{a \in \xi(p)} \kappa(s(p)) \subseteq \kappa(\vee_{a \in \xi(p)} s(p))$. This proves that $a \leq \vee_{a \in \xi(p)} s(p)$. Hence we have proven that $a = \vee_{a \in \xi(p)} s(p)$. ■

The two archetypical examples of classical mechanics and quantum mechanics satisfy the axiom of completeness. Consider the state property space $(\Omega, \mathcal{P}(\Omega), \xi_\Omega)$ of a classical mechanical physical system with state space Ω. Consider a set of properties $(A_i)_i \subseteq \mathcal{P}(\Omega)$ of the classical mechanical system. The property $A = \cap_i A_i$ makes axiom 27 to be satisfied. Indeed, consider an arbitrary state $p \in \Omega$. We have $\cap_i A_i \in \xi_\Omega(p) \Leftrightarrow p \in \cap_i A_i \Leftrightarrow p \in A_i \ \forall i \Leftrightarrow A_i \in \xi_\Omega(p) \ \forall i$. From (30) follows that axiom 27 is satisfied.

Next, consider the state property space $(\Sigma(\mathcal{H}), \mathcal{L}(\mathcal{H}), \xi_\mathcal{H})$ corresponding to a quantum mechanical physical system described by means of a complex Hilbert space $\mathcal{H}$. Consider a set of properties $(A_i)_i \subseteq \mathcal{L}(\mathcal{H})$ of the quantum mechanical physical system. The property $\cap_i A_i \in \mathcal{L}(\mathcal{H})$ makes axiom 27 to be satisfied. Indeed, consider an arbitrary state $x \in \Sigma(\mathcal{H})$. We have $\cap_i A_i \in \xi_\mathcal{H}(P) \Leftrightarrow (\cap_i A_i)x = x \Leftrightarrow A_i x = x \ \forall i \Leftrightarrow A_i \in \xi_\mathcal{H}(P) \ \forall i$. From (30) follows that axiom 27 is satisfied.

3.3 Ortho tests

We have come to the point where we will introduce the first operational element which is specifically quantum, in the sense that it does not necessarily correspond with our intuition about reality. We will suppose that a special type test exists, which we call an ortho test.

DEFINITION 35 Ortho test. A test α is called an ortho test if it is such that if the physical entity is in a state $p \perp a$, where a is a property tested by α, then $\widetilde{\alpha}$ is true, and if the physical entity is in state $q \perp b$, where b is a property tested by $\widetilde{\alpha}$, then α is true.

PROPOSITION 36. *Consider a test α. If α is an ortho test then $\widetilde{\alpha}$ is an ortho test.*

Proof. Follows directly from the definition. ■

We can see immediately that ortho tests are special types of test because of the next proposition, where we prove that a product test is never an ortho test, except when it is a trivial product test of equivalent tests.

PROPOSITION 37. *Consider a set of tests $(\alpha_i)_i$. The product test $\Pi_i \alpha_i$ is an ortho test iff α_j is an ortho test for each j, and $\alpha_j \approx \alpha_k$ for each j, k.*

Proof. Suppose that $\Pi_i \alpha_i$ is an ortho test, and let us call a_i a property tested by α_i, and hence $\wedge_i a_i$ a property tested by $\Pi_i \alpha_i$. Consider an arbitrary α_j of the set $(\alpha_i)_i$, and a state p such that $p \perp a_j$. From (20) follows that $p \perp \wedge_i a_i$. Since $\Pi_i \alpha_i$ is an ortho test, we have that whenever $p \perp \wedge_i a_i$ the test $\widetilde{\Pi_i \alpha_i} = \Pi_i \widetilde{\alpha}_i$ is true. This means that $\widetilde{\alpha}_k$ is true for all k. Hence $\widetilde{\alpha}_j$ is true. Hence, we have proven that if p is a state orthogonal to a_j, then $\widetilde{\alpha}_j$ is true, which is one of the necessary conditions for α_j to be an ortho test. Let us proceed proving the other. Suppose that $(b_i)_i$ is a set of properties such that each b_i is a property tested by

$\widetilde{\alpha}_i$, and hence $\wedge_i b_i$ is a property tested by $\Pi_i \widetilde{\alpha}_i$. Let us consider $\widetilde{\alpha}_j$ which tests b_j. Consider a state q such that $q \perp b_j$. From (20) follows that $q \perp \wedge_i b_i$, and since $\Pi_i \widetilde{\alpha}_i$ is an ortho test, we have that $\widetilde{\Pi_i \widetilde{\alpha}_i} = \Pi \alpha_i$ is true. This implies that α_i is true for all i, and hence α_j is true. This proves that α_j is an ortho test. Since we had chosen j arbitrary, this proves that all tests α_i are ortho tests. Let us next prove that all test are equivalent. Consider α_j and suppose that the entity is in state p such that α_j is true. From this follows that $p \perp b_j$ and hence $p \perp \wedge_i b_i$. Since $\Pi_i \widetilde{\alpha}_i$ is an ortho test, this implies that $\Pi_i \alpha_i$ is true, and hence α_k is true for all k. Hence we have proven that $\alpha_j \leq \alpha_k$ for all j and k, and as a consequence all the tests are equivalent. ■

Proposition 37 tells us something important. Ortho tests are the test which exist commonly in quantum mechanics. This proposition proves that such an ortho test cannot be generated in a non trivial way by the product test mechanism. This means that if a property can be tested by an ortho test, hence this ortho test is specific for this property, even if this property is a meet property. It is not a product of other tests, except in a trivial way, when these other tests are also ortho tests testing this same property. But ortho test and the corresponding ortho properties have other unsuspected features.

PROPOSITION 38. *Consider two tests α, β such that α is an ortho test. We have*

$$\alpha \leq \beta \Rightarrow \widetilde{\beta} \leq \widetilde{\alpha} \tag{40}$$

Proof. Suppose that $\alpha \leq \beta$ and that the state p of the entity is such that $\widetilde{\beta}$ is true. This means that $p \perp b$ where b is a property tested by β. Suppose that a is a property tested by α, then we have $a \leq b$, and hence $p \perp a$. Since α is an ortho test, it follows that $\widetilde{\alpha}$ is true. Hence we have proven that $\widetilde{\beta} \leq \widetilde{\alpha}$. ■

3.4 The axiom of orthocomplementation

There is a specific structure, namely the structure of an orthocomplementation, which has been identified mathematically in the formalism of quantum mechanics, and this structure has played an important role in the mathematical axiomatization, for example the one worked out in [Piron, 1964; Piron, 1976]. With the notion of ortho test we introduce an orthocomplementation in an operational way. Lets first define what an orthocomplementation on a partially ordered set with minimum is.

DEFINITION 39 Orthocomplementation. Suppose that we have a set Z with a partial-order $\leq$ and a smallest element 0. A bijective map $^\perp : Z \to Z$ is an orthocomplementation if for $x, y \in Z$ we have

$$(x^\perp)^\perp = x \tag{41}$$

$$x \leq y \Rightarrow y^\perp \leq x^\perp \tag{42}$$

$$0 \text{ is the infimum of } x \text{ and } x^\perp \tag{43}$$

The ortho test satisfies (41), (42) and (43). Of course, the set of tests corresponding to a physical entity is not a partially ordered set, but only a pre-ordered set. But the set of properties, if axiom 24 of property determination is satisfied, is a partially ordered set. Hence our aim is to make operational steps such that on the set of properties an orthocomplementation arises. This makes us introduce the following definition for an ortho property.

DEFINITION 40 Ortho property. Consider a state property space $(\Sigma, \mathcal{L}, \xi)$. We say that $a \in \mathcal{L}$ is an ortho property if there exists an ortho test testing a. If α is the ortho test testing a, we denote by $a^\perp$ the property tested by $\widetilde{\alpha}$.

Let us introduce the following definition.

DEFINITION 41 Orthogonal set. For a subset of states $A \subseteq \Sigma$ we define the orthogonal $A^\perp$ of this subset

$$A^\perp = \{p \mid p \perp q \ \forall \ q \in A\} \tag{44}$$

PROPOSITION 42. *Consider a state property space* $(\Sigma, \mathcal{L}, \xi)$. *If* $a \in \mathcal{L}$ *is an ortho property then we have*

$$\kappa(a)^\perp = \kappa(a^\perp) \tag{45}$$

Proof. Suppose that $a \in \mathcal{L}$ is an ortho property. This means that $p \perp a \Leftrightarrow a^\perp \in \xi(p)$. Hence $p \in \kappa(a)^\perp \Leftrightarrow p \in \kappa(a^\perp)$. And as a consequence we have $\kappa(a)^\perp = \kappa(a^\perp)$. ■

AXIOM 43 Orthocomplementation. Consider a state property space $(\Sigma, \mathcal{L}, \xi)$. For each property $a \in \mathcal{L}$ there exists an ortho test α testing this property.

THEOREM 44. *Consider a state property space* $(\Sigma, \mathcal{L}, \xi)$ *and suppose that axiom 24, axiom 27 and axiom 43 are satisfied. For* $a \in \mathcal{L}$ *and* α *an ortho test testing* a *let us denote the property tested by* $\widetilde{\alpha}$ *by* $a^\perp$. *For* $a, b \in \mathcal{L}$ *and* $p, q \in \Sigma$ *we then have*

$$(a^\perp)^\perp = a \tag{46}$$

$$a \leq b \Rightarrow b^\perp \leq a^\perp \tag{47}$$

$$a \wedge a^\perp = 0 \tag{48}$$

which proves that $^\perp : \mathcal{L} \to \mathcal{L}$ *is an orthocomplementation.*

Proof. First we remark that if α and β are ortho tests testing property a, and hence $\alpha \approx \beta$, we have that $\widetilde{\alpha} \approx \widetilde{\beta}$. This shows that $^\perp$ is a function. Consider now $a \in \mathcal{L}$ and α an ortho test testing a. Then $\widetilde{\alpha}$ is an ortho test testing $a^\perp$. We have that $(a^\perp)^\perp$ is the property tested by $\widetilde{\widetilde{\alpha}} = \alpha$. Hence $(a^\perp)^\perp = a$. Consider $a, b \in \mathcal{L}$ such that $a \leq b$, and α and β ortho tests testing respectively a and b. Hence $\alpha \leq \beta$. From this follows that $\widetilde{\beta} \leq \widetilde{\alpha}$, and hence $b^\perp \leq a^\perp$. Consider $a \in \mathcal{L}$ and α an ortho test testing a. Hence $\widetilde{\alpha}$ tests $a^\perp$. The infimum property $a \wedge a^\perp$ in $\mathcal{L}$ is tested by $\alpha \cdot \widetilde{\alpha}$ and hence $a \wedge a^\perp = 0$. ■

PROPOSITION 45. *Consider a state property space $(\Sigma, \mathcal{L}, \xi)$ for which axioms 24, 27 and 43 are satisfied. We have for $a, b \in \mathcal{L}$, $(a_i)_i \subseteq \mathcal{L}$ and $p, q \in \mathcal{L}$*

$$(\bigvee_i a_i)^\perp = \bigwedge_i a_i^\perp \tag{49}$$

$$(\bigwedge_i a_i)^\perp = \bigvee_i a_i^\perp \tag{50}$$

$$0^\perp = I \quad I^\perp = 0 \tag{51}$$

$$a \vee a^\perp = I \tag{52}$$

$$p \perp q \Leftrightarrow \exists\ c \in \mathcal{L} \text{ such that } c \in \xi(p) \text{ and } c^\perp \in \xi(q) \tag{53}$$

Proof. Let us prove (49) and (50). We have $\wedge_i a_i \leq a_j\ \forall j$, which implies that $a_j^\perp \leq (\wedge_i a_i)^\perp\ \forall j$, and hence $\vee_i a_i^\perp \leq (\wedge_i a_i)^\perp$. This also implies $\vee_i (a_i^\perp)^\perp = \vee_i a_i \leq (\wedge_i a_i^\perp)^\perp$. Hence $((\wedge_i a_i^\perp)^\perp)^\perp = \wedge_i a_i^\perp \leq (\vee_i a_i)^\perp$. We also have $a_j \leq \vee_i a_i\ \forall j$, which implies that $(\vee_i a_i)^\perp \leq a_j^\perp\ \forall j$. Hence $(\vee_i a_i)^\perp \leq \wedge_i a_i^\perp$. This also implies that $(\vee_i a_i^\perp)^\perp \leq \wedge_i (a_i^\perp)^\perp = \wedge_i a_i$. Hence $(\wedge_i a_i)^\perp \leq ((\vee_i a_i^\perp)^\perp)^\perp = \vee_i a_i^\perp$. Consider $a \in \mathcal{L}$, then $0 \leq a^\perp$, and hence $a \leq 0^\perp$. This proves that $0^\perp$ is a maximal element of $\mathcal{L}$, and hence $0^\perp = I$. In a analogous way we prove that $I^\perp = 0$. We have $I = 0^\perp = (a \wedge a^\perp)^\perp = a^\perp \vee a$ which proves (52). To prove (53) we remark that if $c \in \mathcal{L}$ is such that $c \in \xi(p)$ and $c^\perp \in \xi(q)$, we have $p \in \kappa(c)$ and $q \in \kappa(c^\perp)$. Since $\kappa(c^\perp) = \kappa(c)^\perp$ we have $p \perp q$. ■

In foregoing work on quantum axiomatics we have worked most of the time with state property systems [Aerts *et al.*, 1999; Aerts and van Steirteghem, 2000; Aerts *et al.*, 2000; Aerts and Deses, 2002; Aerts and van Valckenborgh, 2002; Aerts and Deses, 2005; Aerts *et al.*, in press; Aerts and Pulmannova, 2006].

DEFINITION 46 State property system. We say that $(\Sigma, \mathcal{L}, \xi)$ is a state-property system if $(\Sigma, \leq)$ is a pre-ordered set, $(\mathcal{L}, \leq, \wedge, \vee)$ is a complete lattice with the greatest element I and the smallest element 0, and ξ is a function

$$\xi : \Sigma \to \mathcal{P}(\mathcal{L}) \tag{54}$$

such that for $p \in \Sigma$ and $(a_i)_i \subseteq \mathcal{L}$, we have

$$I \in \xi(p), \tag{55}$$

$$0 \notin \xi(p), \tag{56}$$

$$a_i \in \xi(p)\ \forall i \Leftrightarrow \wedge_i a_i \in \xi(p)\ \text{(for an arbitrary set of indices)} \tag{57}$$

and for $p, q \in \Sigma$ and $a, b \in \mathcal{L}$ we have

$$p \leq q \Leftrightarrow \xi(q) \subseteq \xi(p) \tag{58}$$

$$a \leq b \Leftrightarrow \forall r \in \Sigma : a \in \xi(r) \Rightarrow b \in \xi(r) \tag{59}$$

Elements of Σ are called *states*, elements of $\mathcal{L}$ are called *properties*.

A state property space for which the three axioms which we have formulated are satisfied is a state property system.

THEOREM 47. *A state property space for which axioms 24, 27 and 43 are satisfied is a state property system.*

Proof. Consider a state property space $(\Sigma, \mathcal{L}, \xi)$ for which axioms 24, 27 and 43 are satisfied. From (7) follows that $(\Sigma, \leq)$ is a pre-ordered set. In theorem 31 we prove that $\mathcal{L}, \leq, \wedge, \vee$ is a complete lattice, and from (29) follows that $I \in \xi(p)\ \forall\ p \in \Sigma$. We have $a \wedge a^{\perp} = 0$ and hence $\kappa(0) = \kappa(a) \cap \kappa(a)^{\perp} = \emptyset$. This proves $0 \notin \xi(p)\ \forall\ p \in \Sigma$. From (25) and (26) of axiom 27 follows (57), and (58) and (59) follows respectively from (7) and (6). ■

4 MORPHISMS

We derive the notion of morphism from a covariance situation. Consider two state property spaces $(\Sigma, \mathcal{L}, \xi)$ and $(\Sigma', \mathcal{L}', \xi')$, describing respectively entities S and S'. We will arrive at the notion of morphism by analyzing the situation where the entity S is a sub-entity of the entity S'. In that case, the following three natural requirements should be satisfied:

i) If the entity S' is in a state p' then the state $m(p')$ of S is determined. This defines a function m from the set of states of S' to the set of states of S;

ii) If we consider a property a of the entity S, then to a corresponds a property $n(a)$ of the 'bigger' entity S'. This defines a function n from the set of properties of S to the set of properties of S';

iii) We want a and $n(a)$ to be two descriptions of the 'same' property of S, once considered as an entity on itself, once as a sub-entity of S'. In other words we want a and $n(a)$ to be actual at once. This means that for a state p' of S' (and a corresponding state $m(p')$ of S) we want the following 'covariance principle' to hold:

$$a \in \xi(m(p')) \Leftrightarrow n(a) \in \xi'(p') \tag{60}$$

We are now ready to present a formal definition of a morphism of state property spaces.

DEFINITION 48 Morphism. Consider two state property spaces $(\Sigma, \mathcal{L}, \xi)$ and $(\Sigma', \mathcal{L}', \xi')$. We say that

$$(m, n) : (\Sigma', \mathcal{L}', \xi') \longrightarrow (\Sigma, \mathcal{L}, \xi) \tag{61}$$

is a 'morphism' (of state property spaces) if m is a function:

$$m : \Sigma' \rightarrow \Sigma \tag{62}$$

and n is a function:

$$n : \mathcal{L} \to \mathcal{L}' \tag{63}$$

such that for $a \in \mathcal{L}$ and $p' \in \Sigma'$ the following holds:

$$a \in \xi(m(p')) \Leftrightarrow n(a) \in \xi'(p') \tag{64}$$

PROPOSITION 49. *Consider two state property spaces* $(\Sigma, \mathcal{L}, \xi)$ *and* $(\Sigma', \mathcal{L}', \xi')$, *and functions*

$$m : \Sigma' \to \Sigma \quad n : \mathcal{L} \to \mathcal{L}' \tag{65}$$

We have that

$$(m, n) : (\Sigma', \mathcal{L}', \xi') \longrightarrow (\Sigma, \mathcal{L}, \xi) \tag{66}$$

is a morphism iff for $a \in \mathcal{L}$, *and* $p' \in \Sigma'$

$$m(p') \in \kappa(a) \Leftrightarrow p' \in \kappa'(n(a)) \tag{67}$$

Proof. Let us prove (64) to show that (m, n) is a morphism. We have $a \in \xi(m(p')) \Leftrightarrow m(p') \in \kappa(a) \Leftrightarrow p' \in \kappa'(n(a)) \Leftrightarrow n(a) \in \xi'(p')$. ■

The next theorem gives some properties of morphisms.

THEOREM 50. *Consider two state property spaces* $(\Sigma, \mathcal{L}, \xi)$ *and* $(\Sigma', \mathcal{L}', \xi')$ *connected by a morphism* $(m, n) : (\Sigma', \mathcal{L}', \xi') \to (\Sigma, \mathcal{L}, \xi)$. *For* $p', q' \in \Sigma'$ *and* $a, b \in \mathcal{L}$ *we have:*

$$p' \leq q' \Rightarrow m(p') \leq m(q') \tag{68}$$

$$a \leq b \Rightarrow n(a) \leq n(b) \tag{69}$$

Proof. Suppose that $p' \leq q'$. We then have $\xi'(q') \subseteq \xi'(p')$. Consider $a \in \xi(m(q'))$, then (64) implies that $n(a) \in \xi'(q')$, and hence $n(a) \in \xi'(p')$, which means that $a \in \xi(m(p'))$. As a consequence we have $\xi(m(q')) \subseteq \xi(m(p'))$, whence $m(p') \leq m(q')$. Next consider $a \leq b$. We then have $\kappa(a) \subseteq \kappa(b)$. Let $r' \in \Sigma'$ be such that $n(a) \in \xi'(r')$. Then we have $a \in \xi(m(r'))$ and hence $m(r') \in \kappa(a) \subseteq \kappa(b)$. This yields $b \in \xi(m(r'))$. From this follows that $n(b) \in \xi'(r')$. So we have shown that $n(a) \leq n(b)$. ■

THEOREM 51. *Consider two state property spaces* $(\Sigma, \mathcal{L}, \xi)$ *and* $(\Sigma', \mathcal{L}', \xi')$ *connected by a morphism* $(m, n) : (\Sigma', \mathcal{L}', \xi') \to (\Sigma, \mathcal{L}, \xi)$ *for which the axioms 24, 27 are satisfied. For* I *and* I' *the maximum of respectively* $\mathcal{L}$ *and* $\mathcal{L}'$ *and* $(a_i)_i \subseteq \mathcal{L}$ *we have:*

$$n(I) = I' \tag{70}$$

$$n(\bigwedge_i a_i) = \bigwedge_i n(a_i) \tag{71}$$

Proof. We clearly have $n(I) \leq I'$. Hence remains to show that $I' \leq n(I)$. Consider $r' \in \Sigma' = \kappa'(I')$, then $m(r') \in \Sigma = \kappa(I)$. From (67) follows that $r' \in \kappa'(n(I))$. This proves that $\kappa'(I') \subseteq \kappa'(n(I))$, and hence $I' \leq n(I)$. Hence we have proven that $n(I) = I'$.
From $\wedge_i a_i \leq a_j \ \forall j$ we obtain $n(\wedge_i a_i) \leq n(a_j) \ \forall j$. This yields $n(\wedge_i a_i) \leq \wedge_i n(a_i)$. We still have to show that $\wedge_i n(a_i) \leq n(\wedge_i a_i)$. Let $r' \in \Sigma'$ be such that $r' \in \kappa'(\wedge_i n(a_i))$. Using (30) we have $r' \in \cap_i \kappa'(n(a_i))$, and hence $r' \in \kappa'(n(a_i)) \ \forall i$. From (67) follows that this implies that $m(r') \in \kappa(a_i) \ \forall i$, and hence $m(r') \in \cap_i \kappa(a_i) = \kappa(\wedge_i a_i)$ using again (30). From (67) this implies that $r' \in \kappa'(n(\wedge_i a_i))$. Hence we have shown that $\kappa'(\wedge_i n(a_i)) \subseteq \kappa'(n(\wedge_i a_i))$, and it follows that $\wedge_i n(a_i) \leq n(\wedge_i a_i)$. Hence we have proven that $n(\wedge_i a_i) = \wedge_i n(a_i)$. ■

DEFINITION 52 Orthomorphism. Consider two state property spaces $(\Sigma, \mathcal{L}, \xi)$ and $(\Sigma', \mathcal{L}', \xi')$ for which the axioms 24, 27, 43 are satisfied. A morphism $(m, n) : (\Sigma', \mathcal{L}', \xi') \rightarrow (\Sigma, \mathcal{L}, \xi)$ is an orthomorphism iff

$$m(p') \in \kappa(a^\perp) \Leftrightarrow p' \in \kappa'(n(a)^\perp) \tag{72}$$

THEOREM 53. *Consider two state property spaces $(\Sigma, \mathcal{L}, \xi)$ and $(\Sigma', \mathcal{L}', \xi')$ for which the axioms 24, 27, 43 are satisfied and connected by an orthomorphism $(m, n) : (\Sigma', \mathcal{L}', \xi') \rightarrow (\Sigma, \mathcal{L}, \xi)$. For 0 and 0′ being the minimal elements of $\mathcal{L}$ and $\mathcal{L}'$ respectively, $a \subseteq \mathcal{T}$ and $p', q' \in \Sigma'$ we have:*

$$n(a^\perp) = n(a)^\perp \tag{73}$$

$$n(0) = 0' \tag{74}$$

$$m(p') \perp m(q') \Rightarrow p' \perp q' \tag{75}$$

Proof.$p' \in \kappa'(n(a)^\perp) \Leftrightarrow m(p') \in \kappa(a^\perp) \Leftrightarrow p' \in \kappa'(n(a^\perp))$. This proves that $\kappa'(n(a)^\perp) = \kappa'(n(a^\perp))$, and hence $n(a)^\perp = n(a^\perp)$. We have $0 = a \wedge a^\perp$, and hence $n(0) = n(a \wedge a^\perp) = n(a) \wedge n(a^\perp) = n(a) \wedge n(a)^\perp = 0'$. Consider $p', q' \in \Sigma'$ such that $m(p') \perp m(q')$. This means that there exists $a \in \mathcal{L}$ such that $m(p') \in \kappa(a)$ and $m(q') \in \kappa(a^\perp)$. Hence $p' \in \kappa'(n(a))$ and $q' \in \kappa'(n(a^\perp)) = \kappa'(n(a)^\perp)$. This proves that $p' \perp q'$. ■

5 DECOMPOSITION OF A STATE PROPERTY SPACE

In this section we introduce the notion of classical test, classical property and classical state. This will lead us to elaborate a decomposition theorem for a state property space into non classical components over a classical state space.

5.1 The classical state property space

In this section we identify the classical part of an entity S. We start by introducing the notion of classical test. The basic idea for a classical test is that it is a test which does not contain any indeterminism. This means that for such a test the outcome 'yes' is certain or the outcome 'no' is certain for each state of the physical entity. Hence we put forward the following definition

DEFINITION 54 Classical test. A test α is a classical test if for any arbitrary state p of the physical entity we have α is true or $\widetilde{\alpha}$ is true.

For the product of classical tests we can prove the following

PROPOSITION 55. *If $(\alpha_i)_i$ is a set of tests, then $\Pi_i\alpha_i$ is a classical test iff each of the α_j are classical tests and $\alpha_j \approx \alpha_k$ for all j, k.*

Proof. Suppose that $\Pi_i\alpha_i$ is a classical test. Consider α_j and a state p such that α_j is not true if the entity is in state p. Then $\Pi_i\alpha$ is not true and since $\Pi_i\alpha_i$ is a classical test, we have that $\widetilde{\Pi_i\alpha_i} = \Pi\widetilde{\alpha}_i$ is true. But then $\widetilde{\alpha}_i$ is true for all i, and hence $\widetilde{\alpha}_j$ is true. This proves that α_j is a classical test. Since we had chosen j arbitrary, we can conclude that α_i is a classical test for all i. Suppose now that α_j is true. Then $\widetilde{\alpha}_j$ is not true, and hence $\Pi\widetilde{\alpha}_i = \widetilde{\Pi\alpha_i}$ is not true. But then, since $\widetilde{\Pi\alpha_i}$ is a classical test, we have that $\Pi_i\alpha_i$ is true, and hence α_k is true for all k. Hence we have proven that $\alpha_j \leq \alpha_k$ for all k. Hence $\alpha_j \approx \alpha_k$ for all j, k. ■

It is easy to see that a classical test is always an ortho test.

PROPOSITION 56. *If α is a classical test then α is an ortho test.*

Proof. Suppose that α is a classical test, and consider a state p such that $p \perp a$ where a is a property tested by α. This means that $a \notin \xi(p)$, and hence $\widetilde{\alpha}$ is true for the physical entity in state p. In an analogous way we prove that for $q \perp a^\perp$ when $a^\perp$ is a property tested by $\widetilde{\alpha}$ and the physical entity in state q we have α is true. This proves that α is an ortho test. ■

DEFINITION 57 Classical property. A classical property $a \in \mathcal{L}$ is a property such that there exists a set $(\alpha_i)_i$ of classical tests α_i such that $\Pi_i\alpha_i$ tests this property. We denote $\mathcal{C}$ the set of all classical properties. A basic classical property $a \in \mathcal{L}$ is a property such that there exists a classical test α testing this property. We denote $\mathcal{K}$ the set of basic classical properties.

DEFINITION 58 Classical elements. Consider a state property space for which axioms 24, 27 and 43 are satisfied. For $p \in \Sigma$, we introduce

$$\omega(p) = \bigwedge_{a \in \xi(p) \cap \mathcal{C}} a \tag{76}$$

$$\tag{77}$$

and call $\omega(p)$ the classical state of the entity S whenever S is in a state $p \in \Sigma$. The set of all classical states is denoted by Ω. We introduce

$$\xi_c : \Omega \rightarrow \mathcal{C} \quad \omega(p) \mapsto \xi(p) \cap \mathcal{C} \tag{78}$$

$$\kappa_c : \mathcal{C} \rightarrow \mathcal{P}(\Omega) \quad a \mapsto \{\omega(p) \ | a \in \xi(p)\} \tag{79}$$

and call κ_c the classical Cartan map of the state property space $(\Sigma, \mathcal{L}, \xi)$.

PROPOSITION 59. *Consider a state property space for which axioms 24 and 27 are satisfied. For classical states $\omega(p), \omega(q) \in \Omega$, classical property $a \in \mathcal{C}$, and states $p, q \in \Sigma$ we have*

$$a \in \xi(p) \Leftrightarrow \omega(p) \leq a \tag{80}$$
$$\omega(p) \leq \omega(q) \Leftrightarrow \xi_c(q) \subseteq \xi_c(p) \tag{81}$$
$$p \leq q \Rightarrow \omega(p) \leq \omega(q) \tag{82}$$

Proof. Suppose that $a \in \mathcal{C}$ and $a \in \xi(p)$. Since $\omega(p) = \wedge_{a \in \xi(p) \cap \mathcal{C}} a$ we have $\omega(p) \leq a$. Suppose now that $\omega(p) \leq a$. Since $\omega(p) \in \xi(p)$ we have $a \in \xi(p)$. Consider $a \in \xi_c(q) = \xi(q) \cap \mathcal{C}$ and $\omega(p) \leq \omega(q)$. This implies that $\omega(q) \leq a$ and hence $\omega(p) \leq a$. From this follows that $a \in \xi(p)$ and hence $a \in \xi(p) \cap \mathcal{C} = \xi_c(p)$. Hence we have proven that $\xi_c(q) \subseteq \xi_c(p)$. Suppose now that $\xi_c(q) \subseteq \xi_c(p)$, then $\omega(p) = \wedge_{a \in \xi_c(p)} \leq \wedge_{a \in \xi_c(q)} = \omega(q)$. Suppose that $p \leq q$ and hence $\xi(q) \subseteq \xi(p)$. We then have $\xi_c(q) = \xi(q) \cap \mathcal{C} \subseteq \xi(p) \cap \mathcal{C} = \xi_c(p)$, and hence $\omega(p) \leq \omega(q)$. ■

Let us consider our two physics examples, and see what the notion of classical property and classical state means in these cases. Consider first the state property space $(\Omega, \mathcal{P}(\Omega), \xi_\Omega)$ for a classical physical system, and consider a property $A \in \mathcal{P}(\Omega)$. Take $p \in \Omega$, then we have $p \in A$ or $p \in A^C$. This proves that any arbitrary property A is a classical property for the state property system $(\Omega, \mathcal{P}(\Omega), \xi_\Omega)$ corresponding to a classical physical system. Clearly, for such a state property system the states coincide with the classical states, which proves that any state is a classical state.

Consider now the state property system $(\Sigma(\mathcal{H}), \mathcal{L}(\mathcal{H}), \xi_\mathcal{H})$ corresponding to a quantum physical system, and consider a property $A \in \mathcal{L}(\mathcal{H})$ such that $A \neq 0$ and $A \neq \mathcal{H}$. In this case we have $A^{orth} \neq \mathcal{H}$ and $A^{orth} \neq 0$. Take $x \in A, x \neq 0$ and $y \in A^{orth}, y \neq 0$, and consider the vector $z = x + y$. Then $z \notin A$ and $z \notin A^{orth}$, and as a consequence $\bar{z} \notin A$ and $\bar{z} \notin A^{orth}$, which proves that A is not a classical property. We have proven that for the state property system corresponding to a quantum physical system the only classical properties are the minimal property and the maximal property. Moreover, the only classical state of the state property system corresponding to a quantum physical system is the classical state corresponding to $\mathcal{H}$ itself. This is the state describing the situation 'the entity is present'.

DEFINITION 60 Classical orthogonality relation. Consider a state property space describing a physical entity S for which axioms 24 and 27 are satisfied, and classical states $\omega(p), \omega(q) \in \Omega$ of this physical entity. We say that $\omega(p) \perp_c \omega(q)$ if there exists a classical test γ such that γ is true if $\omega(p)$ is actual, hence if the entity is in classical state $\omega(p)$, and $\widetilde{\gamma}$ is true if $\omega(q)$ is actual, hence if the entity is in classical state $\omega(q)$.

DEFINITION 61 Classical ortho test. Consider a state property space describing a physical entity S for which axioms 24 and 27 are satisfied. A classical test α is a classical ortho test if it is such that if the physical entity is in classical state $\omega(p) \perp_c a$, where a is the property tested by α, then $\widetilde{\alpha}$ is true, and if the physical entity is in state $\omega(q) \perp_c b$, where b is the property tested by $\widetilde{\alpha}$, then α is true.

PROPOSITION 62. *Consider a state property space for which axioms 24 and 27 are satisfied, and classical states $\omega(p), \omega(q) \in \Omega$ of this physical entity. We have*

$$\omega(p) \neq \omega(q) \Leftrightarrow \omega(p) \perp_c \omega(q) \Leftrightarrow \omega(p) \perp \omega(q) \Leftrightarrow \omega(p) \notin \xi(q) \Leftrightarrow q \perp \omega(p) \quad (83)$$

Proof. If $\omega(p) \perp_c \omega(q)$ then obviously $\omega(p) \neq \omega(q)$. Suppose now that $\omega(p) \neq \omega(q)$. Since $\omega(p)$ and $\omega(q)$ as classical states are also both classical properties there exist $(\alpha_i)_i$ and $(\beta_j)_j$ where α_i and β_j are classical tests for all i, j and such that $\Pi_i \alpha_i$ tests $\omega(p)$ and $\Pi_j \beta_j$ tests $\omega(q)$. If $\omega(p) \neq \omega(q)$ this can mean that $\omega(p) \not\leq \omega(q)$ or that $\omega(q) \not\leq \omega(p)$. Suppose we have that $\omega(p) \not\leq \omega(q)$, and suppose that $\omega(p)$ is actual. Since in this case $\omega(q)$ is not actual there is at least one β_j which is not true. But then $\widetilde{\beta_j}$ is true. If $\omega(q)$ is actual we have that β_j is true. Hence we have proven that $\omega(p) \perp_c \omega(q)$. Analogously we prove that $\omega(p) \perp_c \omega(q)$ if $\omega(q) \not\leq \omega(p)$. Suppose that $\omega(p) \perp_c \omega(q)$ and let γ be the test which is true if $\omega(p)$ is actual such that $\widetilde{\gamma}$ is true if $\omega(q)$ is actual. Consider states $r, s \in \Sigma$ such that $\omega(p) \in \xi(r)$ and $\omega(q) \in \xi(s)$. If c is the property tested by γ and d the property tested by $\widetilde{\gamma}$ we have $\omega(p) \leq c$ and $\omega(q) \leq d$. Hence $c \in \xi(r)$ and $d \in \xi(s)$. This proves that $r \perp s$, and hence $\omega(p) \perp \omega(q)$. Suppose now that $\omega(p) \perp \omega(q)$. Then certainly $\omega(p) \neq \omega(q)$ and hence $\omega(p) \perp_c \omega(q)$. Suppose that $\omega(p) \notin \xi(q)$. Then $\omega(q) \not\leq \omega(p)$, and as a consequence we have $\omega(q) \neq \omega(p)$. Hence $\omega(p) \perp \omega(q)$. If $\omega(p) \perp \omega(q)$ then $\omega(q) \not\leq \omega(p)$ and hence $\omega(p) \notin \xi(q)$. ■

PROPOSITION 63. *Consider a state property space for which axioms 24 and 27 are satisfied. A classical test α is a classical ortho test.*

Proof. Consider a classical test α such that a is the property tested by α and b is the property tested by $\widetilde{\alpha}$, and suppose we have $\omega(p) \perp_c a$. Consider $q \in \Sigma$ such that $a \in \xi(q)$. Then we have $\omega(q) \leq a$, and hence $\omega(p) \perp_c \omega(q)$. As a consequence we have $\omega(p) \perp q$ and hence $p \perp q$. This implies that $b \in \xi(p)$, and hence $\omega(p) \leq b$. In an analogous way we show that $\omega(q) \leq a$ if $\omega(q) \perp_c b$. This proves that α is a classical ortho test. ■

Suppose we consider a state property space $(\Sigma, \mathcal{L}, \xi)$ describing a physical entity S for which axioms 24, 27 and 43 are satisfied. We wonder whether $(\Omega, \mathcal{C}, \xi_c)$ is a state property space satisfying 24, 27 and 43. If this is the case we can consider

$(\Omega, \mathcal{C}, \xi_c)$ as the state property space describing the classical aspects of the physical entity S.

THEOREM 64. *Consider a state property space $(\Sigma, \mathcal{L}, \xi)$ describing a physical entity S for which axioms 24, 27 and 43 are satisfied. Consider $\omega(p), \omega(q) \in \Omega$ and $a, b \in \mathcal{C}$. We have*

$$a \in \xi(p) \Leftrightarrow \omega(p) \leq a \tag{84}$$

$$\omega(p) \leq \omega(q) \Leftrightarrow \xi_c(\omega(q)) \subseteq \xi_c(\omega(p)) \tag{85}$$

$$\omega(p) \in \kappa_c(a) \Leftrightarrow a \in \xi_c(\omega(p)) \tag{86}$$

$$a \leq b \Leftrightarrow \kappa_c(a) \subseteq \kappa_c(b) \tag{87}$$

$$\kappa_c(a) = \kappa_c(b) \Rightarrow a = b \tag{88}$$

$$\kappa_c(\wedge_i a_i) = \cap_i \kappa_c(a_i) \tag{89}$$

$$\kappa(a^\perp) = \Sigma \setminus \kappa(a) \tag{90}$$

$$\kappa_c(a^{\perp_c}) = \Omega \setminus \kappa_c(a) \tag{91}$$

$$\text{There exists a classical test } \alpha \text{ testing } a, \text{ hence } \mathcal{C} = \mathcal{K} \tag{92}$$

and $(\Omega, \mathcal{C}, \xi_c)$ is a state property space satisfying axioms 24, 27 and 43.

Proof. Suppose that $a \in \xi(p)$. Since $\omega(p) = \wedge_{a \in \xi(p) \cap \mathcal{C}} a$ we have $\omega(p) \leq a$. Suppose now that $\omega(p) \leq a$. Since $\omega(p) \in \xi(p)$ we have $a \in \xi(p)$. This proves (84). Suppose that $\omega(p) \leq \omega(q)$ and consider $a \in \xi_c(\omega(q))$ and hence $\omega(q) \leq a$. Then we have $\omega(p) \leq a$ and hence $a \in \xi_c(\omega(p))$. This proves that $\xi_c(\omega(q)) \subseteq \xi_c(\omega(p))$. Suppose now that $\xi_c(\omega(q)) \subseteq \xi_c(\omega(p))$, and hence $\xi(q) \cap \mathcal{C} \subseteq \xi(p) \cap \mathcal{C}$. Then we have $\omega(p) = \wedge_{a \in \xi(p) \cap \mathcal{C}} a \leq \wedge_{a \in \xi(q) \cap \mathcal{C}} a = \omega(q)$. This proves (85). Suppose that $\omega(p) \in \kappa_c(a)$, then $a \in \xi(p)$ and hence $\omega(p) \leq a$ which shows that $a \in \xi_c(\omega(p))$. Contrary, suppose that $a \in \xi_c(\omega(p)) = \xi(p) \cap \mathcal{C}$. Then we have $a \in \xi(p)$, and hence $\omega(p) \in \kappa_c(a)$. This proves (86). Suppose that $a \leq b$ and consider $\omega(p) \in \kappa_c(a)$. We then have $a \in \xi(p)$ and hence $b \in \xi(p)$. From this follows that $\omega(p) \in \kappa_c(b)$. Hence we have proven that $\kappa_c(a) \subseteq \kappa_c(b)$. Suppose now that $\kappa_c(a) \subseteq \kappa_c(b)$ and suppose we have $p \in \Sigma$ such that $a \in \xi(p)$. This means that $\omega(p) \in \kappa_c(a)$, and hence $\omega(p) \in \kappa_c(b)$, which implies that $b \in \xi(p)$. Hence we have proven that $a \leq b$. This proves (87). Suppose that $\kappa_c(a) = \kappa_c(b)$ and consider $p \in \kappa(a)$. Then we have $a \in \xi(p)$ and hence $\omega(p) \in \kappa_c(a)$. From this follows that $\omega(p) \in \kappa_c(b)$, and hence $b \in \xi(p)$. As a consequence we have $p \in \kappa(b)$. This means that we have proven that $\kappa(a) \subseteq \kappa(b)$. Analogously we prove that $\kappa(b) \subseteq \kappa(a)$. Since axiom 24 is satisfied for the state property space $(\Sigma, \mathcal{L}, \xi)$ we have $a = b$. This proves (88). Consider $(a_i)_i \subseteq \mathcal{K} \subset \mathcal{T}$. Since axiom 27 is satisfied for $(\Sigma, \mathcal{L}, \xi)$ we have a property $\wedge_i a_i \in \mathcal{L}$ such that $\kappa(\wedge_i a_i) = \wedge_i \kappa(a_i)$. We have $\wedge_i a_i \in \mathcal{C}$ since $a_i \in \mathcal{C} \; \forall \; i$. This means that we have the following $\omega(p) \in \kappa_c(\wedge_i a_i) \Leftrightarrow p \in \kappa(\wedge_i a_i) \Leftrightarrow p \in \cap_i \kappa(a_i) \Leftrightarrow p \in \kappa(a_i) \; \forall \; i \Leftrightarrow \omega(p) \in \kappa_c(a_i) \; \forall \; i \Leftrightarrow \omega(p) \in \cap_i \kappa_c(a_i)$. This means that $\kappa_c(\wedge_i a_i) = \cap_i \kappa_c(a_i)$, and hence we have proven (89). Consider $p \in \Sigma$ and $p \notin \kappa(a)$. Then we have $\omega(p) \not\leq a$ and hence $\omega(p) \neq \omega(q) \; \forall \; \omega(q) \leq a$. From this follows that $\omega(p) \perp_c \omega(q) \; \forall \; \omega(q) \leq a$, and hence $\omega(p) \perp \omega(q) \; \forall \; \omega(q) \leq a$.

As a consequence we have that $p \perp q \ \forall \ q \in \kappa(a)$, and hence $p \in \kappa(a^\perp)$. This proves that $\Sigma \setminus \kappa(a) \subseteq \kappa(a^\perp)$. We obviously have that $\kappa(a^\perp) \subseteq \Sigma \setminus \kappa(a)$. Hence we have proven (90). Consider $\omega(p) \in \Omega$ and $\omega(p) \notin \kappa_c(a)$. This means that $\omega(p) \neq \omega(q) \ \forall \ \omega(q) \leq \kappa_c(a)$, and hence $\omega(p) \perp_c \omega(q) \ \forall \ \omega(q) \leq \kappa_c(a)$. As a consequence we have $\omega(p) \leq \kappa_c(a)^{\perp_c}$. This proves that $\Omega \setminus \kappa_c(a) \subseteq \kappa_c(a)^{\perp_c}$. We also have that $\kappa_c(a)^{\perp_c} \subseteq \Omega \setminus \kappa_c(a)$. Hence we have proven (91). Remark that (85) shows that ξ_c defines the pre-order relation on the set of states Ω in a way which is necessary for $(\Omega, \mathcal{C}, \xi_c)$ to be a state property space, while (86) shows that the classical Cartan map is indeed the Galois inverse of the function ξ_c which has to be the case if $(\Omega, \mathcal{C}, \xi_c)$ is a state property space. With (87) we prove that the classical Cartan map indeed defines the pre-order relation on the set of properties. Hence we have proven that $(\Omega, \mathcal{C}, \xi_c)$ is a state property space. That axiom 24 is satisfied for the state property space $(\Omega, \mathcal{C}, \xi_c)$ is proven by (88). Consider now an arbitrary $a \in \mathcal{C}$. From the definition of $\mathcal{C}$ follows that there exists $(a_i)_i \subseteq \mathcal{K}$ such that for an arbitrary $p \in \Sigma$ we have $a_i \in \xi(p) \ \forall \ i \Leftrightarrow a \in \xi(p)$. Hence $p \in \kappa(a_i) \ \forall \ i \Leftrightarrow p \in \kappa(a)$. This means that $\kappa(a) = \cap_i \kappa(a_i)$. This proves that axiom 27 is satisfied for the state property space $(\Omega, \mathcal{C}, \xi_c)$. Let us prove now that axiom 43 is satisfied for $(\Omega, \mathcal{C}, \xi_c)$. Consider $a \in \mathcal{C}$. From (90) we know that $\kappa(a^\perp) = \Sigma \setminus \kappa(a)$. Since axiom 43 is satisfied for $(\Sigma, \mathcal{L}, \xi)$ there exists an ortho test α, such that α tests a and $\widetilde{\alpha}$ tests $a^\perp$. Since for $p \in \Sigma$ we have $p \in \kappa(a)$, and then α is true, or $p \in \kappa(a^\perp)$ and then $\widetilde{\alpha}$ is true, it follows that α is a classical test. This proves (92). Using proposition 63 it follows that α is a classical ortho test, which proves that axiom 43 is satisfied for $(\Omega, \mathcal{C}, \xi_c)$. ■

PROPOSITION 65. *Suppose that $(\Sigma, \mathcal{L}, \xi)$ is a state property space describing an entity S for which axioms 24, 27 and 43 are satisfied. For $p \in \Sigma$ and $a \in \mathcal{L}$ we have*

$$a \in \mathcal{C} \Leftrightarrow \kappa(a) \cup \kappa(a^\perp) = \Sigma \tag{93}$$

Proof. Suppose that a is a classical property. From (92) follows that a is a basic classical property, and hence there exists a classical test α such that α tests a. Consider the ortho test β that tests a. We have that $\alpha \approx \beta$ and hence $\widetilde{\alpha} \approx \widetilde{\beta}$. This means that $\widetilde{\alpha}$ tests $a^\perp$. Since α is true or $\widetilde{\alpha}$ is true we have $a \in \xi(p)$ or $a^\perp \in \xi(p)$ for an arbitrary $p \in \Sigma$, and hence $\Sigma = \kappa(a) \cup \kappa(a^\perp)$. Suppose now that $\Sigma = \kappa(a) \cup \kappa(a^\perp)$. This means that for an arbitrary $p \in \Sigma$ we have $a \in \xi(p)$ or $a^\perp \in \xi(p)$, and consider the ortho test α testing a such that $\widetilde{\alpha}$ tests $a^\perp$. For this ortho test we have that α is true or $\widetilde{\alpha}$ is true, which proves that α is a classical test. Hence $a \in \mathcal{K} = \mathcal{C}$. ■

We can prove that the classical state property system $(\Omega, \mathcal{C}, \xi_c)$ is isomorphic to the canonical state property system $(\Omega, \mathcal{P}(\Omega), Id)$.

THEOREM 66. *$\kappa_c : \mathcal{C} \to \mathcal{P}(\Omega)$ is an isomorphism.*

Proof. From (87) it follows that κ_c is an injective function. Let us prove that κ_c is a surjective function. Take an arbitrary element $A \in \mathcal{P}(\Omega)$. Consider the property

$$a = \bigwedge_{\kappa_c(\omega(p)) \subseteq \Omega \setminus A} \omega(p)^{\perp_c} \tag{94}$$

We have

$$\begin{aligned} \kappa_c(a) &= \kappa_c(\bigwedge_{\kappa_c(\omega(p)) \subseteq \Omega \setminus A} \omega(p)^{\perp_c}) = \bigcap_{\kappa_c(\omega(p)) \subseteq \Omega \setminus A} \kappa_c(\omega(p)^{\perp_c}) = \\ & \bigcap_{\kappa_c(\omega(p)) \subseteq \Omega \setminus A} \kappa_c(\Omega \setminus \omega(p)) \qquad (95) \\ &= \Omega \setminus \bigcup_{\kappa_c(\omega(p)) \subseteq \Omega \setminus A} \kappa_c(\omega(p)) = \Omega \setminus (\Omega \setminus A) = A \qquad (96) \end{aligned}$$

■

Let us consider again our two archetypical examples. For the state property system $(\Omega, \mathcal{P}(\Omega), Id)$ corresponding to a classical physical system, we have that classical state property system coincides with this state property system. This shows that in the case of a classical physical system our construction comes out as it should be, the classical state property system is the state property system of this classical physics system. For the state property system $(\Sigma(\mathcal{H}), \mathcal{P}(\mathcal{H}), \xi_{\mathcal{H}})$ of a quantum physical system, the classical state property system $(\Omega, \mathcal{C}, \kappa_{\mathcal{C}})$ is the following: $\Omega = \{\mathcal{H}\}$, $\mathcal{C} = \{0, \mathcal{H}\}$, $\kappa_{\mathcal{C}} : \mathcal{C} \rightarrow \mathcal{P}(\Omega)$, such that $\kappa_{\mathcal{C}}(0) = \emptyset$ and $\kappa_{\mathcal{C}}(\mathcal{H}) = \{\mathcal{H}\}$. This classical state property system describes the aspect of the quantum physical system which has to do with the properties 'the system is present' and 'the system is not present', properties that even for quantum systems are classical properties. In the next chapter we decompose an arbitrary state property system into its non-classical components and its classical state space. This structure shows us how we can describe a general situation.

5.2 The non classical components of a state property space

In this section we study the description of a physical entity whenever it is in a classical state. This leads to the existence of a non classical property space for each classical state describing the non classical elements of the entity. Consider a classical state $\omega \in \Omega$. Then $\omega \in \mathcal{C}$ is also a classical property. Hence there exists a classical test testing ω.

DEFINITION 67 ω-test. A classical test testing the classical state $\omega \in \Omega$ is called a ω-test, and we denote it α_ω.

DEFINITION 68 ω-inverse. Consider a test α and the product test $\alpha \cdot \alpha_\omega$. We define $\widetilde{\alpha \cdot \alpha_\omega}^{\omega} = \tilde{\alpha} \cdot \alpha_\omega$ and call $\widetilde{\alpha \cdot \alpha_\omega}^{\omega}$ the ω-inverse of $\alpha \cdot \alpha_\omega$.

PROPOSITION 69. *We have*

$$\widetilde{\widetilde{\alpha \cdot \alpha_\omega}^\omega}^\omega = \alpha \cdot \alpha_\omega \tag{97}$$

such that the operation is an inverse operation on the set of tests of the form $\alpha \cdot \alpha_\omega$.

Proof. We have $\widetilde{\widetilde{\alpha \cdot \alpha_\omega}^\omega}^\omega = \widetilde{\widetilde{\alpha} \cdot \alpha_\omega}^\omega = \widetilde{\widetilde{\alpha}} \cdot \alpha_\omega = \alpha \cdot \alpha_\omega$. ■

Let us explain the physical meaning of this. Suppose we consider a typical classical property ω in standard quantum mechanics, for example the property 'the neutron is there', in case the entity we are considering is a neutron. The test α_ω consists of verifying whether the neutron is there, for example by absorbing it on a detection screen. In general such a verification of the presence of the neutron destroys the neutron, which means that if we want to test another property, this time a non classical property of the neutron, we need to make recourse to the product test. And hence indeed, when we test the quantum test α, for example the spin of the neutron, then actually we perform the test $\alpha \cdot \alpha_\omega$. We test whether the neutron is there 'and' whether it has spin in a certain direction, by making sure that which ever of the two tests α_ω or α we perform, the outcome will be 'yes'. But we do not have to perform both tests together, it is sufficient to perform one 'or' the other. We are in a similar situation as the one with the piece of wood tested to burn well 'and' float on water, by performing one of the both tests.

DEFINITION 70 ω-orthogonality. Consider two states $p, q \in \Sigma$ such that $\omega \in \xi(p) \cap \xi(q)$ where $\omega \in \Omega$. We say that p and q are ω-orthogonal, and denote $p \perp_\omega q$, if there exists a test α such that $\alpha \cdot \alpha_\omega$ is true if the entity is in state p and $\widetilde{\alpha \cdot \alpha_\omega}^\omega$ is true if the entity is in state q.

PROPOSITION 71. *For $x \in \mathcal{L}$ and $a \in \mathcal{C}$ we have*

$$x = (x \wedge a) \vee (x \wedge a^\perp) \tag{98}$$

$$\kappa(x) = \kappa(x \wedge a) \cup \kappa(x \wedge a^\perp) \tag{99}$$

Proof. Since $x \wedge a \leq x$ and $x \wedge a^\perp \leq x$ we have $(x \wedge a) \vee (x \wedge a^\perp) \leq x$. Since $a \in \mathcal{C}$ we have $\kappa(a) \cup \kappa(a^\perp) = \Sigma$. This gives $\kappa(x) = \kappa(x) \cap (\kappa(a) \cup \kappa(a^\perp)) = (\kappa(x) \cap \kappa(a)) \cup (\kappa(x) \cap \kappa(a^\perp)) = \kappa(x \wedge a) \cup \kappa(x \wedge a^\perp) \subseteq \kappa((x \wedge a) \vee (x \wedge a^\perp))$. This proves (98) and (99). ■

PROPOSITION 72. *For $x, y \in \mathcal{L}$ and $a \in \mathcal{C}$ such that $x \leq a$ and $y \leq a^\perp$ we have*

$$(x \vee y)^\perp = (x^\perp \wedge a) \vee (y^\perp \wedge a^\perp) \tag{100}$$

$$(x \vee y) \wedge a = x \tag{101}$$

Proof. We have $a^\perp \leq x^\perp$ and $a \leq y^\perp$. From this it follows that $y^\perp \wedge a^\perp \leq x^\perp$ and $x^\perp \wedge a \leq y^\perp$. This implies that $x^\perp \wedge y^\perp \wedge a^\perp = y^\perp \wedge a^\perp$ and $x^\perp \wedge y^\perp \wedge a = x^\perp \wedge a$. Since $a \in \mathcal{C}$ we have $x^\perp \wedge y^\perp = (x^\perp \wedge y^\perp \wedge a) \vee (x^\perp \wedge y^\perp \wedge a^\perp)$. So $x^\perp \wedge y^\perp = (x^\perp \wedge a) \vee (y^\perp \wedge a^\perp)$. Hence $x \vee y = (x \vee a^\perp) \wedge (y \vee a)$. But then $(x \vee y) \wedge a = (x \vee a^\perp) \wedge a$. We know that $x^\perp = (x^\perp \wedge a) \vee (x^\perp \wedge a^\perp) = (x^\perp \wedge a) \vee a^\perp$. Hence $x = (x \vee a^\perp) \wedge a$. This proves that $(x \vee y) \wedge a = x$. ■

PROPOSITION 73. *For $x, x_i \in \mathcal{L}$ and $a \in \mathcal{C}$ we have*

$$a \wedge (\vee_i x_i) = \vee_i (a \wedge x_i) \tag{102}$$

$$a = (a \wedge x) \vee (a \wedge x^\perp) \tag{103}$$

Proof. We have $a \wedge (\vee_i x_i) = a \wedge (\vee_i ((x_i \wedge a) \vee (x_i \wedge a^\perp)) = a \wedge (\vee_i (x_i \wedge a) \vee \vee_i (x_i \wedge a^\perp)) = \vee_i (x_i \wedge a)$. We have $a = a \wedge (x \vee x^\perp)$. From (102) it follows that $a \wedge (x \vee x^\perp) = (a \wedge x) \vee (a \wedge x^\perp)$, which proves (103). ■

PROPOSITION 74. *For $a \in \mathcal{L}$ we have*

$$a = \bigvee_{\omega \in \Omega} (a \wedge \omega) \tag{104}$$

$$\kappa(a) = \bigcup_{\omega \in \Omega} \kappa(a \wedge \omega) \tag{105}$$

with

$$a \wedge \omega \perp a \wedge \omega' \quad \text{and} \quad \kappa(a \wedge \omega) \cap \kappa(a \wedge \omega') = \emptyset \quad \text{for} \quad \omega \neq \omega' \tag{106}$$

Proof. We have that $a \wedge \omega \leq a \ \forall \omega \in \Omega$, hence $\kappa(a \wedge \omega) \subseteq \kappa(a) \ \forall \ \omega \in \Omega$, and as a consequence $\cup_{\omega \in \Omega} \kappa(a \wedge \omega) \subseteq \kappa(a)$. Consider $p \in \kappa(a)$. We have $p \in \kappa(\omega(p))$, and hence $p \in \kappa(a) \cap \kappa(\omega(p)) = \kappa(a \wedge \omega(p)) \subseteq \cup_{\omega \in \Omega} \kappa(a \wedge \omega)$. So we have shown that $\kappa(a) \subseteq \cup_{\omega \in \Omega} \kappa(a \wedge \omega)$. This proves (105), namely $\kappa(a) = \cup_{\omega \in \Omega} \kappa(a \wedge \omega)$. We have that $a \wedge \omega \leq a \ \forall \omega \in \Omega$, hence $\vee_{\omega \in \Omega} (a \wedge \omega) \leq a$. Consider $p \in \kappa(a)$. We have $p \in \cup_{\omega \in \Omega} \kappa(a \wedge \omega) \subseteq \kappa(\vee_{\omega \in \Omega} (a \wedge \omega))$. So we have shown that $\kappa(a) \subseteq \kappa(\vee_{\omega \in \Omega} (a \wedge \omega))$. From this it follows that $a \leq \vee_{\omega \in \Omega} (a \wedge \omega)$, which proves (104), namely $a = \vee_{\omega \in \Omega} (a \wedge \omega)$. Consider $\omega \neq \omega'$, then we have $\omega \leq \omega'^\perp$. As a consequence $a \wedge \omega \leq \omega'^\perp \leq a^\perp \vee \omega'^\perp = (a \wedge \omega')^\perp$, which proves that $a \wedge \omega \perp a \wedge \omega'$. From this it follows that $\kappa(a \wedge \omega) \cap \kappa(a \wedge \omega') = \emptyset$. ■

COROLLARY 75. *We have*

$$\Sigma = \bigcup_{\omega \in \Omega} \kappa(\omega) \tag{107}$$

with

$$\kappa(\omega) \cap \kappa(\omega') = \emptyset \quad \text{for} \quad \omega \neq \omega' \tag{108}$$

PROPOSITION 76. *Consider a_ω such that $a_\omega \leq \omega \ \forall \ \omega \in \Omega$. We have*

$$\kappa(\bigvee_{\omega\in\Omega} a_\omega) = \bigcup_{\omega\in\Omega} \kappa(a_\omega) \tag{109}$$

with

$$\kappa(a_\omega) \cap \kappa(a_{\omega'}) = \emptyset \quad \text{for} \quad \omega \neq \omega' \tag{110}$$

Proof. We have $\kappa(\vee_{\omega\in\Omega} a_\omega) = \cup_{\omega'\in\Omega}\kappa((\vee_{\omega\in\Omega} a_\omega)\wedge\omega')$. From (101) it follows that $(\vee_{\omega\in\Omega} a_\omega)\wedge\omega' = a_{\omega'}$. Hence $\kappa(\vee_{\omega\in\Omega} a_\omega) = \cup_{\omega'\in\Omega}\kappa(a_{\omega'})$. This proves (109). ■

Let us now investigate the nonclassical parts of the state property system $(\Sigma, \mathcal{L}, \kappa)$.

DEFINITION 77 Nonclassical components. Suppose that $(\Sigma, \mathcal{L}, \xi)$ is the state property space of an entity satisfying axioms 24 and 27. For $\omega \in \Omega$ we introduce

$$\begin{aligned} \Sigma_\omega &= \{p \mid \omega \in \xi(p), p \in \Sigma\} & (111)\\ \mathcal{L}_\omega &= \{a \mid a \leq \omega, a \in \mathcal{L}\} & (112)\\ \xi_\omega(p) &= \xi(p) \cap \mathcal{L}_\omega & (113) \end{aligned}$$

and we call $(\Sigma_\omega, \mathcal{L}_\omega, \xi_\omega)$ the nonclassical components of $(\Sigma, \mathcal{L}, \xi)$ corresponding to ω. We also introduce Cartan map corresponding to ω

$$\kappa_\omega(a) = \{p \mid p \in \Sigma_\omega, a \in \xi_\omega(p)\} \tag{114}$$

THEOREM 78. *Consider $(\Sigma, \mathcal{L}, \xi)$ the state property space of an entity satisfying axioms 24, 27 and 43. For $a, b \in \mathcal{L}_\omega$, $(a_i)_i \subseteq \mathcal{L}_\omega$ and $p, q \in \Sigma_\omega$ we have*

$$\begin{aligned} a \in \xi_\omega(p) &\Leftrightarrow a \in \xi(p) & (115)\\ p \leq q &\Leftrightarrow \xi_\omega(q) \subseteq \xi_\omega(p) & (116)\\ p \in \kappa_\omega(a) &\Leftrightarrow p \in \kappa(a) & (117)\\ a \leq b &\Leftrightarrow \kappa_\omega(a) \subseteq \kappa_\omega(b) & (118)\\ \kappa_\omega(a) = \kappa_\omega(b) &\Rightarrow a = b & (119)\\ \kappa_\omega(\wedge_i a_i) &= \cap_i \kappa_\omega(a_i) & (120)\\ p \perp_\omega q &\Leftrightarrow p \perp q & (121)\\ \kappa_\omega(a^{\perp_\omega}) &= \kappa(a^\perp \wedge \omega) = \kappa_\omega(a)^{\perp_\omega} & (122) \end{aligned}$$

and $(\Sigma_\omega, \mathcal{L}_\omega, \xi_\omega)$ is a state property space that satisfies axioms 24, 27 and 43.

Proof. Suppose that $a \in \xi_\omega(p)$. This means that $a \in \xi(p) \cap \mathcal{L}_\omega$, and hence $a \in \xi(p)$. Suppose now that $a \in \xi(p)$. Since $a \in \mathcal{L}_\omega$ we have $a \in \xi(p) \cap \mathcal{L}_\omega = \xi_\omega(p)$. This proves (115). Suppose that $p \leq q$, then we have $\xi(q) \subseteq \xi(p)$, and hence

$\xi_\omega(q) = \xi(q) \cap \mathcal{L}_\omega \subseteq \xi(p) \cap \mathcal{L}_\omega = \xi_\omega(p)$. Suppose next that $\xi_\omega(q) \subseteq \xi_\omega(p)$, and consider $a \in \xi(q)$. Applying (115) this gives that $a \in \xi_\omega(q)$ and hence $a \in \xi_\omega(p)$. Applying again (115) this gives that $a \in \xi(p)$. Hence we have proven that $\xi(q) \subseteq \xi(p)$ and hence $p \leq q$. This proves (116). We have $p \in \kappa_\omega(a) \Leftrightarrow a \in \xi_\omega(a) \Leftrightarrow a \in \xi(p) \Leftrightarrow p \in \kappa(a)$. This proves (117). Suppose that $a \leq b$ and consider $p \in \kappa_\omega(a)$. Applying (117) this gives $p \in \kappa(a)$ and since $\kappa(a) \subseteq \kappa(b)$ this gives $p \in \kappa(b)$. Applying again (117) this gives $p \in \kappa_\omega(b)$. Hence we have proven that $\kappa_\omega(a) \subseteq \kappa_\omega(b)$. Suppose now that $\kappa_\omega(a) \subseteq \kappa_\omega(b)$, and consider $p \in \kappa(a)$. Applying (117) this gives $p \in \kappa_\omega(a)$ and hence $p \in \kappa_\omega(b)$. Applying again (117) this gives $p \in \kappa(b)$. So we have proven that $\kappa(a) \subseteq \kappa(b)$ and from this follows that $a \leq b$. This proves (118). Suppose that $\kappa_\omega(a) = \kappa_\omega(b)$ and consider $p \in \kappa(a)$. Then we have $p \in \kappa_\omega(a)$ and hence $p \in \kappa_\omega(b)$. From this follows that $p \in \kappa(b)$. This means that we have proven that $\kappa(a) \subseteq \kappa(b)$. Since axiom 24 is satisfied for the state property space $(\Sigma, \mathcal{L}, \xi)$ we have $a = b$. This proves (119). We have $p \in \kappa_\omega(\wedge_i a_i) \Leftrightarrow p \in \kappa(\wedge_i a_i) \Leftrightarrow p \in \cap_i \kappa(a_i) \Leftrightarrow p \in \kappa(a_i)\ \forall\ i \Leftrightarrow p \in \kappa_\omega(a_i)\ \forall\ i \Leftrightarrow p \in \cap_i \kappa_\omega(a_i)$. This proves (120). Suppose that $p \perp_\omega q$. This means that there exists a test α such that $\alpha \cdot \alpha_\omega$ is true if the entity is in state p and $\widetilde{\alpha \cdot \alpha_\omega}^\omega$ is true if the entity is in state q. We have $\widetilde{\alpha \cdot \alpha_\omega}^\omega = \widetilde{\alpha} \cdot \alpha_\omega$. This means that α is true if the entity is in state p and $\widetilde{\alpha}$ is true if the entity is in state q. Hence $p \perp q$. Suppose now that $p \perp q$. This means that there exists a test α such that α is true if the entity is in state p and $\widetilde{\alpha}$ is true if the entity is in state q. Since $p, q \in \Sigma_\omega$ we have that ω is actual and hence α_ω is true. Hence $\alpha \cdot \alpha_\omega$ is true if the entity is in state p and $\widetilde{\alpha} \cdot \alpha_\omega = \widetilde{\alpha \cdot \alpha_\omega}^\omega$ is true if the entity is in state q. This means that $p \perp_\omega q$. This proves (121). Suppose that $a \in \mathcal{L}_\omega$. In this case $a \in \mathcal{L}$ and hence there exists an ortho test α testing a. Consider the test $\alpha \cdot \alpha_\omega$ and a state q such that $q \perp_\omega a$ and $q \in \Sigma_\omega$. This means that $q \perp_\omega r\ \forall\ r$ such that $r \in \kappa_\omega(a)$. Hence $q \perp r\ \forall\ r$ such that $r \in \kappa(a)$. Hence $q \in \kappa(a)^\perp = \kappa(a^\perp)$. This means that $\widetilde{\alpha}$ is true if the entity is in state q. Since $q \in \Sigma_\omega$ we also have that α_ω is true if the entity is in state q. Hence $\widetilde{\alpha} \cdot \alpha_\omega = \widetilde{\alpha \cdot \alpha_\omega}^\omega$ is true if the entity is in state q. Note that property tested by $\widetilde{\alpha \cdot \alpha_\omega}^\omega$ is $a^\perp \wedge \omega$. Hence we have proven that $q \perp_\omega a \Rightarrow q \in \kappa_\omega(a^\perp \wedge \omega)$, or $\kappa(a)^{\perp_\omega} \subseteq \kappa_\omega(a^\perp \wedge \omega)$. Consider $r \in \Sigma_\omega$ such that $r \perp_\omega a^\perp \wedge \omega$. This means that $r \perp_\omega s\ \forall\ s$ such that $a^\perp \wedge \omega \in \xi(s)$. From this follows that $r \perp s\ \forall\ s$ such that $a^\perp \wedge \omega \in \xi(s)$. Hence $r \in \kappa(a^\perp \wedge \omega)^\perp$. Since $r \in \Sigma_\omega$ we have $r \perp \omega^\perp$, and hence $r \perp a^\perp \wedge \omega^\perp$, such that $r \in \kappa(a^\perp \wedge \omega^\perp)^\perp$. From (98) and (99) we know that $\kappa(a^\perp) = \kappa(a^\perp \wedge \omega) \cup \kappa(a^\perp \wedge \omega^\perp)$, and hence $\kappa(a) = \kappa(a^\perp)^\perp = \kappa(a^\perp \wedge \omega)^\perp \cap \kappa(a^\perp \wedge \omega^\perp)^\perp$. Hence we have $r \in \kappa(a)$. This proves that for the entity being in state r we have α is true, and hence $\alpha \cdot \alpha_\omega$ is true. This proves that $\alpha \cdot \alpha_\omega$ is an ortho test for the orthogonality relation $\perp_\omega$. We can now denote $a^\perp \wedge \omega = a^{\perp_\omega}$. And it follows that we have proven (122). Remark that from (119) follows that axiom 24 is satisfied for $(\Sigma_\omega, \mathcal{L}_\omega, \xi_\omega)$, and from (120) follows that axiom 27 is satisfied. From (122) follows that axiom 43 is satisfied for $(\Sigma_\omega, \mathcal{L}_\omega, \xi_\omega)$. ■

5.3 A decomposition theorem

To see in more detail in which way the classical and nonclassical parts are structured within the lattice $\mathcal{L}$, we need to introduce some additional structures.

DEFINITION 79 Direct union of state property spaces. Consider a set of state property spaces $(\Sigma_\omega, \mathcal{L}_\omega, \xi_\omega)$ that all satisfy axioms 24, 27 and 43. The direct union $\textcircled{\vee}_\omega(\Sigma_\omega, \mathcal{L}_\omega, \xi_\omega)$ of these state property spaces is the state property space $(\cup_\omega \Sigma_\omega, \textcircled{\vee}_\omega \mathcal{L}_\omega, \textcircled{\vee}_\omega \xi_\omega)$, where
(i) $\cup_\omega \Sigma_\omega$ is the disjoint union of the sets Σ_ω
(ii) $\textcircled{\vee}_\omega \mathcal{L}_\omega$ is the direct union of the lattices $\mathcal{L}_\omega$, which means the set of sequences $a = (a_\omega)_\omega$, such that

$$(a_\omega)_\omega \leq (b_\omega)_\omega \quad \Leftrightarrow \quad a_\omega \leq b_\omega \ \forall \omega \in \Omega \tag{123}$$
$$(a_\omega)_\omega \wedge (b_\omega)_\omega \quad = \quad (a_\omega \wedge b_\omega)_\omega \tag{124}$$
$$(a_\omega)_\omega \vee (b_\omega)_\omega \quad = \quad (a_\omega \vee b_\omega)_\omega \tag{125}$$
$$(a_\omega)_\omega^\perp \quad = \quad (a_\omega^{\perp_\omega})_\omega \tag{126}$$

(iii) $\textcircled{\vee}_\omega \xi_\omega$ is defined as follows:

$$\textcircled{\vee}_\omega \xi_\omega : \cup_\omega \Sigma_\omega \quad \rightarrow \quad \mathcal{P}(\textcircled{\vee}_\omega \mathcal{L}_\omega) \tag{127}$$
$$p_{\omega'} \quad \mapsto \quad \{(a_\omega)_\omega \mid a_{\omega'} \in \xi_{\omega'}(p_{\omega'}), a_\omega \in \mathcal{L}_\omega \ \forall \ \omega \neq \omega'\} \tag{128}$$

and hence the corresponding Cartan map is the following

$$\textcircled{\vee}_\omega \kappa_\omega : \textcircled{\vee}_\omega \mathcal{L}_\omega \quad \rightarrow \quad \cup_\omega \Sigma_\omega \tag{129}$$
$$(a_\omega)_\omega \quad \mapsto \quad \cup_{\omega \in \Omega} \kappa_\omega(a_\omega) \tag{130}$$

We remark that if $\mathcal{L}_\omega$ are complete orthocomplemented lattices, then also $\textcircled{\vee}_{\omega \in \Omega} \mathcal{L}_\omega$ is a complete orthocomplemented lattice. A fundamental decomposition theorem can now be proven.

THEOREM 80 Decomposition theorem. *Consider the state property space* $(\Sigma, \mathcal{L}, \xi)$, *and suppose that axioms 24, 27 and 43 are satisfied. Then*

$$(\Sigma, \mathcal{L}, \xi) \cong \textcircled{\vee}_{\omega \in \Omega}(\Sigma_\omega, \mathcal{L}_\omega, \xi_\omega) \tag{131}$$

where Ω *is the set of classical states of* $(\Sigma, \mathcal{L}, \xi)$, Σ_ω *is the set of states and* $\mathcal{L}_\omega$ *the lattice of properties of the nonclassical component state property space* $(\Sigma_\omega, \mathcal{L}_\omega, \xi_\omega)$.

Proof. We use the notion of orthomorphism of state property systems, and need to prove that there exists an isomorphism of ortho state property systems between $(\Sigma, \mathcal{L}, \xi)$ and $\textcircled{\vee}_{\omega \in \Omega}(\Sigma_\omega, \mathcal{L}_\omega, \xi_\omega)$. From (107) it follows that m can be defined in the following way:

$$m : \Sigma \quad \rightarrow \quad \cup_{\omega \in \Omega} \Sigma_\omega \tag{132}$$
$$p \quad \mapsto \quad p \tag{133}$$

The function n is defined in the following way:

$$n : \bigodot_{\omega\in\Omega}\mathcal{L}_\omega \rightarrow \mathcal{L} \tag{134}$$

$$(a_\omega)_\omega \mapsto \vee_{\omega\in\Omega}a_\omega \tag{135}$$

The function m is a bijection by definition. Consider $(a_\omega)_\omega, (b_\omega)_\omega \in \bigodot_{\omega\in\Omega}\mathcal{L}_\omega$ and suppose that $n((a_\omega)_\omega) = n((b_\omega)_\omega)$, hence $\vee_{\omega\in\Omega}a_\omega = \vee_{\omega\in\Omega}b_\omega$. Then $(\vee_{\omega\in\Omega}a_\omega) \wedge \omega' = (\vee_{\omega\in\Omega}b_\omega) \wedge \omega' \ \forall \ \omega' \in \Omega$. From (101) it follows that $(\vee_{\omega\in\Omega}a_\omega) \wedge \omega' = a_{\omega'}$ and $(\vee_{\omega\in\Omega}b_\omega) \wedge \omega' = b_{\omega'}$. Hence $a_{\omega'} = b_{\omega'} \ \forall \ \omega' \in \Omega$. As a consequence we have $(a_\omega)_\omega = (b_\omega)_\omega$. This proves that n is injective. Let us prove that n is surjective. Consider an arbitrary element $a \in \mathcal{L}$. From (104) it follows that $a = \vee_{\omega\in\Omega}(a \wedge \omega)$. Consider the element $(a \wedge \omega)_\omega \in \bigodot_{\omega\in\Omega}\mathcal{L}_\omega$. Then $n((a \wedge \omega)_\omega) = a$ which proves that n is surjective. Hence we have proven that m as well as n are bijections. Let us show that we have an orthomorphism. We need to prove (67) and (72) hence:

$$m(p) \in \bigodot_\omega \kappa_\omega((a_\omega)_\omega) \Leftrightarrow p \in \kappa(\vee_{\omega\in\Omega}a_\omega) \tag{136}$$

$$m(p) \in \bigodot_\omega \kappa_\omega((a_\omega)_\omega^\perp) \Leftrightarrow p \in \kappa((\vee_{\omega\in\Omega}a_\omega)^\perp) \tag{137}$$

Let us calculate $\bigodot_\omega \kappa_\omega((a_\omega)_\omega) = \cup_{\omega\in\Omega}\kappa_\omega(a_\omega) = \cup_{\omega\in\Omega}\kappa(a_\omega)$. On the other hand we have $\kappa(n((a_\omega)_\omega)) = \kappa(\vee_{\omega\in\Omega}a_\omega)$, and following (109), we have $\kappa(\vee_{\omega\in\Omega}a_\omega) = \cup_{\omega\in\Omega}\kappa(a_\omega)$. This means that (136) is satisfied. We have $\bigodot_\omega \kappa_\omega((a_\omega)_\omega^\perp) = \bigodot_\omega \kappa_\omega((a_\omega^{\perp_\omega})_\omega) = \cup_\omega \kappa_\omega(a_\omega^{\perp_\omega}) = \cup_\omega \kappa_\omega(a^\perp \wedge \omega) = \kappa(\vee_{\omega\in\Omega}(a^\perp \wedge \omega)) = \kappa(a^\perp) = \kappa((\vee_{\omega\in\Omega}a_\omega)^\perp)$. This proves (137). Hence we have proven that (m, n) is an isomorphism of ortho state property spaces. ■

6 ADDITIONAL AXIOMS

In the foregoing we have introduced three axioms. If these three axioms are satisfied we can decompose the state property space of an entity into its classical state property space such that for each classical state there is an underlying non classical property space describing the entity being in this classical state. We have proven that the classical state property space is isomorphic to the property space of classical physics. The underlying non classical state property spaces are however not necessarily isomorphic to the state property space of quantum mechanics. To make these underlying non classical state property spaces isomorphic to the state property space of quantum mechanics we need to introduce additional axioms.

6.1 The axiom of atomisticity

The axiom of property determination makes the pre-order relation on the set of properties $\mathcal{L}$ of a state property space into a partial order relation. The pre-order relation existing on the set of states is not necessarily a partial order relation. This means that it is possible for two states $p, q \in \Sigma$ to be different states even if $\xi(p) = \xi(q)$, which means that the properties which are actual if the entity is in state p are the same as the properties which are actual if the entity is in state q. The next axiom we introduce makes the pre-order relation on Σ into a trivial order, i.e. $p \leq q$ iff $p = q$.

AXIOM 81 Atomisticity. Consider a state property space $(\Sigma, \mathcal{L}, \xi)$ for which axioms 24, 27 and 43 are satisfied. We say that the axiom of atomisticity is satisfied if for $p, q \in \Sigma$ we have

$$\xi(q) \subseteq \xi(p) \Rightarrow p = q \tag{138}$$

DEFINITION 82. Consider a partially ordered set $Z, \leq$. We say that $s \in Z$ is an atom, if whenever $0 \leq a \leq s$ we have $a = 0$ ot $a = s$. A lattice $Z, \leq$, is atomistic, if there exists a set of atoms $\mathcal{A}$ which is ordering. This means that for $x, y \in Z$ we have $x \leq y \Leftrightarrow \{s \mid s \in \mathcal{A}, s \leq x\} \subseteq \{s \mid s \in \mathcal{A}, s \leq y\}$.

THEOREM 83. *Consider a state property space $(\Sigma, \mathcal{L}, \xi)$ for which axioms 24, 27, 43 and 81 are satisfied. For $p, q \in \Sigma$ and $a, b \in \mathcal{L}$ we have*

$$p \leq q \quad \Rightarrow \quad p = q \tag{139}$$

$$0 \leq a \leq s(p) \quad \Rightarrow \quad a = 0 \text{ or } a = s(p) \tag{140}$$

$$a \leq b \quad \Leftrightarrow \quad \{s(p) \mid s(p) \leq a\} \subseteq \{s(p) \mid s(p) \leq b\} \tag{141}$$

and $\mathcal{L}$ is an atomistic lattice with set of atoms the set of state properties $\mathcal{A} = \{s(p) \mid p \in \Sigma\}$.

Proof. Suppose $p \leq q$, then we have $\xi(q) \subseteq \xi(p)$, and hence $p = q$. This proves (139). Consider a such that $0 \leq a \leq s(p) = \wedge_{b \in \xi(p)} b$. If $a \neq 0$ there exists $q \in \Sigma$ such that $a \in \xi(q)$. Hence we have $s(p) \in \xi(q)$, and as a consequence we have $b \in \xi(q)\ \forall\ b \in \xi(p)$. Hence $\xi(p) \subseteq \xi(q)$, and hence $p = q$. This implies that $s(p) = s(q)$. Since $s(q) \leq a$ we have $a = s(p)$. So we have proven (140), which means that $s(p)$ is an atom of $\mathcal{L}$. If $a \leq b$ we obviously have that $\{s(p) \mid s(p) \leq a\} \subseteq \{s(p) \mid s(p) \leq b\}$. Suppose that $\{s(p) \mid s(p) \leq a\} \subseteq \{s(p) \mid s(p) \leq b\}$. From (39) follows that $a = \vee_{s(p) \leq a} s(p) \leq \vee_{s(p) \leq b} s(p) = b$, Hence we have proven (141), which means that $\mathcal{A}$ is an ordering set for $\mathcal{L}$, and hence $\mathcal{L}$ is a complete orthocomplemented atomistic lattice. ■

PROPOSITION 84. *Consider a state property space $(\Sigma, \mathcal{L}, \xi)$ for which axioms 24, 27 and 43 are satisfied. The classical state property space $(\Omega, \mathcal{C}, \xi_c)$ corresponding to $(\Sigma, \mathcal{L}, \xi)$ satisfies axiom 81. If $(\Sigma, \mathcal{L}, \xi)$ satisfies also axiom 81, then each non classical component $(\Sigma_\omega, \mathcal{L}_\omega, \xi_\omega)$ of the decomposition of $(\Sigma, \mathcal{L}, \xi)$ satisfies axiom 81.*

Proof. Consider $\omega(p), \omega(q) \in \Omega$ such that $\xi_c(\omega(q)) \subset \xi_c(\omega(p))$. From (85) follows that then $\omega(p) \leq \omega(q)$, and hence $\omega(p) \wedge \omega(q) = \omega(p)$. Suppose now that $\omega(p) \neq \omega(q)$, then from (83) follows that $\omega(p) \perp \omega(q)$. But then $\omega(p) \wedge \omega(q) = 0$, which would lead to $\omega(p) = 0$. This is not possible, and hence this proves that $\omega(p) =$

$\omega(q)$. Hence we have proven that $(\Omega, \mathcal{C}, \kappa_c)$ satisfies axiom 81. Suppose now that axiom 81 is satisfied for $(\Sigma, \mathcal{L}, \xi)$, and consider $p, q \in \Sigma_\omega$ such that $\xi_\omega(q) \subseteq \xi_\omega(p)$. From (116) follows then that $p \leq q$ and hence $\xi(q) \subseteq \xi(p)$. Since axiom 81 is satisfied for $(\Sigma, \mathcal{L}, \xi)$ we have that $p = q$. Hence we have proven that axiom 81 is satisfied for $(\Sigma_\omega, \mathcal{L}_\omega, \xi_\omega)$. ■

PROPOSITION 85. *Consider a state property space $(\Sigma, \mathcal{L}, \xi)$ for which axioms 24, 27, 43 and 81 are satisfied. A property $a \in \mathcal{L}$ is classical, hence $a \in \mathcal{C}$, iff a is a central element of the lattice $\mathcal{L}$, i.e. $x = (x \wedge a) \vee (x \wedge a^\perp)\ \forall\ x \in \mathcal{L}$. The lattice of properties $\mathcal{L}_\omega$ of a non classical component property space $(\Sigma_\omega, \mathcal{L}_\omega, \xi_\omega)$ is an irreducible lattice.*

Proof. From (98) follows that a classical property a is a central element of the lattice $\mathcal{L}$. Consider a central element a of the lattice $\mathcal{L}$, and an arbitrary state $p \in \Sigma$. Because axiom 81 is satisfied we have that $s(p)$ is an atom of $\mathcal{L}$. We have $s(p) \wedge a \leq s(p)$ and $s(p) \wedge a^\perp \leq s(p)$ and hence $s(p) \wedge a = s(p)$ or $s(p) \wedge a = 0$, and $s(p) \wedge a^\perp = s(p)$ or $s(p) \wedge a^\perp = 0$. Since a is a central element of $\mathcal{L}$ we have $s(p) = (s(p) \wedge a) \vee (s(p) \wedge a^\perp)$, and hence we cannot have $s(p) \wedge a = 0$ and $s(p) \wedge a^\perp = 0$, which means that at least one of $s(p) \wedge a = s(p)$ or $s(p) \wedge a^\perp = s(p)$ is true. From this follows that $s(p) \leq a$ or $s(p) \leq a^\perp$. Hence $a \in \xi(p)$ or $a^\perp \in \xi(p)$. Since axiom 43 is satisfied there exists an ortho test α testing a and hence $\widetilde{\alpha}$ testing $a^\perp$. From the foregoing follows that this ortho test is a classical test, and hence a is a classical property. Consider a central element $a \in \mathcal{L}_\omega$ of the lattice of properties of a non classical component $(\Sigma_\omega, \mathcal{L}_\omega, \xi_\omega)$. Since $a \in \mathcal{C}$, we have $a = 0$ or $a = \omega$ which proves that $\mathcal{L}_\omega$ is irreducible. ■

6.2 *The axiom of weak modularity*

If we consider a closed subspace $A \in \mathcal{L}(\mathcal{H})$ of a Hilbert space $\mathcal{H}$. The closed subspace A in itself is a Hilbert space. This means that we can consider the set $\mathcal{L}(A)$ of closed subspaces contained in A. One can prove that $\mathcal{L}(A)$ is a complete orthocomplemented lattice. The relative orthocomplementation $B^{\perp_A}$ for $B \in \mathcal{L}(A)$ is defined as follows: $B^{\perp_A} = B^\perp \cap A$. An important equality for $\perp_A$ to be an orthocomplementaion is the following: $(B^{\perp_A})^{\perp_A} = B$. This gives $B = (B^\perp \cap A)^\perp \cap A$ or $B = (B \vee A^\perp) \wedge A$. This is the way the requirement of 'weak modularity' is usually introduced, hence more specifically: for $B, A \in \mathcal{L}$ and $B \leq A$ we have $B = (B \vee A^\perp) \wedge A$. We however want to introduce 'weak modularity' in an operational way. To formulate the following axiom we first introduce the idea of relative state property space.

PROPOSITION 86 Relative state property space. *Suppose we have a state property space $(\Sigma, \mathcal{L}, \xi)$ and for $a \in \mathcal{L}$ we consider $(\Sigma, \mathcal{L}, \xi)_a = (\kappa(a), \mathcal{L}_a, \xi_a)$ where*

$$\mathcal{L}_a = \{b \mid b \leq a\} \tag{142}$$

$$\xi_a : \kappa(a) \rightarrow \mathcal{P}(\mathcal{L}_a) \quad p \mapsto \xi(p) \cap \mathcal{L}_a \tag{143}$$

then $(\Sigma, \mathcal{L}, \xi)_a$ *is a state property space and for* $b \in \mathcal{L}_a$ *we have*

$$\kappa_a(b) = \kappa(b) \tag{144}$$

We call $(\Sigma, \mathcal{L}, \xi)_a$ *the state property space relative to property* $a \in \mathcal{L}$.

Proof. Suppose the entity is in state $p \in \kappa(a)$ and the property $b \leq a$ is actual. This means that $b \in \xi(p)$. Since $b \leq a$ we have $b \in \xi(p) \cap \mathcal{L}_a = \xi_a(p)$. On the other hand, suppose we have $p \in \kappa(a)$ and $b \leq a$, and $b \in \xi_a(p)$. This means that $b \in \xi(p)$, and hence, if the entity is in state p the property b is actual. This proves that $(\Sigma, \mathcal{L}, \xi)_a$ is a state property space describing the same entity as the one described by the state property space $(\Sigma, \mathcal{L}, \xi)$. Suppose that $p \in \kappa_a(b)$, and hence $b \in \xi_a(p)$. This means that $b \in \xi(p)$, and hence $p \in \kappa(b)$. Hence we have $\kappa_a(b) \subseteq \kappa(b)$. Suppose now that $p \in \kappa(b)$ and hence $b \in \xi(p)$. Since we have $b \leq a$ we also have $b \in \mathcal{L}_a$, and hence $b \in \xi(p) \cap \mathcal{L}_a = \xi_a(p)$. From this follows that $p \in \kappa_a(b)$. Hence we have proven that $\kappa_a(b) \subseteq \kappa(b)$. As a consequence we have $\kappa_a(b) = \kappa(b)$. ∎

The operational meaning of the relative state property space is the following. We study the entity S in the special circumstance when we manage to keep the property a actual during the study. This means concretely that we can consider a test α testing a, and hence we consider only the states $\kappa(a)$ of the entity which make this test true.

PROPOSITION 87. *Consider a state property space* $(\Sigma, \mathcal{L}, \xi)$ *for which axioms 24 and 27 are satisfied, and* $a \in \mathcal{L}$. *Then the relative state property space* $(\Sigma, \mathcal{L}, \xi)_a$ *satisfies axioms 24 and 27.*

Proof. Consider $b, c \in \mathcal{L}_a$ such that $\kappa_a(b) = \kappa_a(c)$. From (144) follows that $\kappa(b) = \kappa(c)$, and since axiom 24 is satisfied for $(\Sigma, \mathcal{L}, \xi)$ we have $b = c$. This proves that axiom 24 is satisfied for $(\Sigma, \mathcal{L}, \xi)_a$. Consider $b \in \mathcal{L}_a$. Since axiom 27 is satisfied for $(\Sigma, \mathcal{L}, \xi)$ there exists $(b_i)_i \subseteq \mathcal{T}$ such that $\kappa(b) = \cap_i \kappa(b_i)$. Consider $\mathcal{T}_a = \{b \wedge a \mid b \in \mathcal{T}\}$. Since $b \leq a$ we have $b = b \wedge a$. Hence $\kappa(b) = \kappa(b) \cap \kappa(a)$ and as a consequence we have $\kappa(b) = \cap_i \kappa(b_i) \cap \kappa(a) = \cap_i(\kappa(b_i) \cap \kappa(a)) = \cap_i \kappa(b_i \wedge a)$. Consider now an arbitrary property $b \in \mathcal{L}_a$. Since $b \in \mathcal{L}$ there exists $(b_i)_i \subseteq \mathcal{T}$ such that $\kappa(b) = \cap_i \kappa(b_i)$. This gives that $\kappa(b) = \cap_i \kappa(b_i \wedge a)$ for $(b_i \wedge a)_i \subseteq \mathcal{T}_a$. Hence we have proven that $(\Sigma, \mathcal{L}, \xi)_a$ satisfies axiom 27. ∎

The axiom 43 of orthocomplementation is however not necessarily satisfied for a relative state property space. The next axiom, the axiom of weak modularity, is meant to make sure that also the axiom 43 of orthocomplementation is satisfied for an arbitrary relative state property space. Before we formulate the axiom of weak modularity, let us analyse why the axiom 43 of orthocomplementation is not necessarily satisfied for a relative state property space. We start by introducing the notion of relative inverse.

DEFINITION 88 Relative inverse. Consider a state property space $(\Sigma, \mathcal{L}, \xi)$ and for $a \in \mathcal{L}$ the relative state property space $(\Sigma, \mathcal{L}, \xi)_a$. For a test β we consider the test $\beta \cdot \alpha$ where α is the test testing the property a, and we introduce $\widetilde{\beta \cdot \alpha}^{\alpha} = \widetilde{\beta} \cdot \alpha$.

PROPOSITION 89. *$\widetilde{\beta \cdot \alpha}^{\alpha}$ is an inverse for tests testing properties of the relative state property space $(\Sigma, \mathcal{L}, \xi)_a$.*

Proof. Consider an arbitrary property $b \in \mathcal{L}_a$ and a test β testing b. Since $b \leq a$ we have that also $\beta \cdot \alpha$ tests property b. Obviously also $\widetilde{\beta \cdot \alpha}^{\alpha} = \widetilde{\beta} \cdot \alpha$ is a test testing a property of $\mathcal{L}_a$, and we have $\widetilde{\widetilde{\beta \cdot \alpha}^{\alpha}}^{\alpha} = \widetilde{\widetilde{\beta} \cdot \alpha}^{\alpha} = \widetilde{\widetilde{\beta}} \cdot \alpha = \beta \cdot \alpha$. This proves that $\widetilde{\beta \cdot \alpha}^{\alpha}$ defines an inverse on tests of properties of $\mathcal{L}_a$. ■

DEFINITION 90 Relative orthogonality. Consider a state property space $(\Sigma, \mathcal{L}, \xi)$, and for $a \in \mathcal{L}$ the relative state property space $(\Sigma, \mathcal{L}, \xi)_a$. Consider a test α testing a. For $p, q \in \kappa(a)$ we say that p is relatively orthogonal to q with respect to α, and denote $p \perp_\alpha q$, if there exists a test $\beta \cdot \alpha$ such that $\beta \cdot \alpha$ is true if the entity is in state p and $\widetilde{\beta \cdot \alpha}^{\alpha}$ is true if the entity is in state q.

PROPOSITION 91. *Consider a state property space $(\Sigma, \mathcal{L}, \xi)$, and for $a \in \mathcal{L}$ the relative state property space $(\Sigma, \mathcal{L}, \xi)_a$, and a test α testing a. For $p, q \in \kappa(a)$ we have*

$$p \perp_\alpha q \Leftrightarrow p \perp q \tag{145}$$

Proof. Suppose that $p \perp_\alpha q$. This means that there exists $\beta \cdot \alpha$ such that $\beta \cdot \alpha$ is true if the entity is in state p and $\widetilde{\beta} \cdot \alpha$ is true if the entity is in state q. This means that β is true if the entity is in state p and $\widetilde{\beta}$ is true if the entity is in state q. Hence $p \perp q$. Suppose that $p \perp q$. This means that there exists a test γ such that γ is true if the entity is in state p and $\widetilde{\gamma}$ is true if the entity is in state q. If we consider $\gamma \cdot \alpha$ and $\widetilde{\gamma} \cdot \alpha$, then, since $p, q \in \kappa(a)$ and α tests a, we have that $\gamma \cdot \alpha$ is true if the entity is in state p and $\widetilde{\gamma} \cdot \alpha$ is true if the entity is in state q. This proves that $p \perp_\alpha q$. ■

PROPOSITION 92. *Consider a state property space $(\Sigma, \mathcal{L}, \xi)$ for which 24, 27 and 43 are satisfied, and for $a \in \mathcal{L}$ the relative state property space $(\Sigma, \mathcal{L}, \xi)_a$. Axiom 43 is satisfied for the state property space $(\Sigma, \mathcal{L}, \xi)_a$ if and only if for $b \leq a$ we have $b = (b \vee a^\perp) \wedge a$.*

Proof. Consider an ortho test α testing property a and an ortho test β testing property b and $p \in \kappa(a)$ such that $p \perp_\alpha b$. From proposition 91 follows that $p \perp b$ and hence, since axiom 43 is satisfied for $(\Sigma, \mathcal{L}, \xi)$ we have $\widetilde{\beta}$ is true if the entity is in state p. Since $p \in \kappa(a)$ we have that $\widetilde{\beta} \cdot \alpha$ is true if the entity is in state

p. Hence $p \in \kappa(b^\perp \wedge a)$. On the other hand, consider $q \in \kappa(a)$ and $q \perp b^\perp \wedge a$. Since $b^\perp \wedge a \in \mathcal{L}$ there exists an ortho test γ, such that $\widetilde{\gamma}$ is true if the entity is in state q. This means that $q \in \kappa((b^\perp \wedge a)^\perp \wedge a))$. The test $\widetilde{\gamma} \cdot \alpha$ will test b, and hence be an ortho test for the relative inverse with respect to a, if and only if $\kappa(b) = \kappa((b^\perp \wedge a)^\perp \wedge a))$. This is equivalent to $b = (b^\perp \wedge a)^\perp \wedge a) = (b \vee a^\perp) \wedge a$. ■

AXIOM 93 Weak modularity. Consider a state property space $(\Sigma, \mathcal{L}, \xi)$ for which axioms 24, 27 and 43 are satisfied. We say that the axiom of weak modularity is satisfied if for $a, b \in \mathcal{L}$ we have

$$b \leq a \quad \Rightarrow \quad b = (b \vee a^\perp) \wedge a \tag{146}$$

Operationally this axiom means the following. Consider two properties $b \leq a$, an ortho test β testing b, and a test α testing a. Then $\widetilde{\beta} \cdot \alpha$ tests $b^\perp \wedge a$. Consider an ortho test $\gamma \approx \widetilde{\beta} \cdot \alpha$, then also γ tests $b^\perp \wedge a$. The test $\widetilde{\gamma} \cdot \alpha$ tests the property $(b^\perp \wedge a)^\perp \wedge a$. The axiom of weak modularity means that we want $\widetilde{\gamma} \cdot \alpha$ also to test the property b. Hence we want $\widetilde{\gamma} \cdot \alpha \approx \beta$.

PROPOSITION 94. *Consider a state property space $(\Sigma, \mathcal{L}, \xi)$ for which axioms 24, 27 and 43 are satisfied. The classical state property space $(\Omega, \mathcal{C}, \xi_c)$ corresponding to $(\Sigma, \mathcal{L}, \xi)$ satisfies axiom 93. If $(\Sigma, \mathcal{L}, \xi)$ satisfies also axiom 93, then each non classical component $(\Sigma_\omega, \mathcal{L}_\omega, \xi_\omega)$ of the decomposition of $(\Sigma, \mathcal{L}, \xi)$ satisfies axiom 93.*

Proof. Consider $a, b \in \mathcal{C}$ such that $b \leq a$. From (102) follows that $(b \vee a^\perp) \wedge a = (b \wedge a) \vee (a^\perp \wedge a) = b \vee 0 = b$. This proves that $(\Omega, \mathcal{C}, \xi_c)$ satisfies axiom 93. Suppose now that axiom 93 is satisfied for $(\Sigma, \mathcal{L}, \xi)$ and consider $a, b \in \mathcal{L}_\omega$ such that $b \leq a$. Using (122), and hence $b^{\perp_\omega} = b^\perp \wedge \omega$, we have $(b \vee a^{\perp_\omega}) \wedge a = (b^{\perp_\omega} \wedge a)^{\perp_\omega} \wedge a = (b^\perp \wedge \omega \wedge a)^{\perp_\omega} \wedge a = (b^\perp \wedge a)^{\perp_\omega} \wedge a = (b^\perp \wedge a)^\perp \wedge \omega \wedge a = (b^\perp \wedge a)^\perp \wedge a = (b \vee a^\perp) \wedge a = b$. This proves that axiom 93 is satisfied for $(\Sigma_\omega, \mathcal{L}_\omega, \xi_\omega)$. ■

6.3 The axiom 'the covering law'

The covering law is the root of the linear structure of quantum mechanics, which means that it is a very important axiom. In some sense it demands something similar to the axiom of atomisticity, but then for all parts of the lattice of properties.

AXIOM 95 The covering law. Consider a state property space $(\Sigma, \mathcal{L}, \xi)$ for which axioms 24, 27, 43 and 81 are satisfied. For $a, b \in \mathcal{L}$ and $p \in \Sigma$ we have

$$s(p) \wedge a = 0 \text{ and } a \leq b \leq a \vee s(p) \Rightarrow b = a \text{ or } b = a \vee s(p) \tag{147}$$

PROPOSITION 96. *Consider a state property space $(\Sigma, \mathcal{L}, \xi)$ for which axioms 24, 27 and 43 are satisfied. The classical state property space $(\Omega, \mathcal{C}, \kappa_c)$ corresponding* *to $(\Sigma, \mathcal{L}, \xi)$ satisfies axiom 95. If $(\Sigma, \mathcal{L}, \xi)$ satisfies also axiom 95, then each non classical component $(\Sigma_\omega, \mathcal{L}_\omega, \xi_\omega)$ of the decomposition of $(\Sigma, \mathcal{L}, \xi)$ satisfies axiom 95.*

Proof. Consider $a, b \in \mathcal{C}$ and $\omega(p) \in \Omega$, such that $a \wedge \omega(p) = 0$ and $a \leq b \leq a \vee \omega(p)$. Making use of theorem 66 we have $\kappa_c(a) \subseteq \kappa_c(b) = \kappa_c(a \vee \omega(p)) = \kappa_c(a) \cup \{\omega(p)\}$. Hence $\kappa_c(b) = \kappa_c(a)$ or $\kappa_c(b) = \kappa_c(a) \cup \{\omega(p)\}$. From (87) follows then that $b = a$ or $b = a \vee \omega(p)$. This proves that $(\Omega, \mathcal{C}, \kappa_c)$ satisfies axiom 95. Suppose now that axiom 95 is satisfied for $(\Sigma, \mathcal{L}, \xi)$ and consider $a, b \in \mathcal{L}_\omega$ and $p \in \Sigma_\omega$ such that $a \wedge s(p) = 0$ and $a \leq b \leq a \vee s(p)$. From axiom 95 being satisfied for $(\Sigma, \mathcal{L}, \xi)$ follows that $b = a$ or $b = a \vee s(p)$, which proves that axiom 95 is satisfied for $(\Sigma_\omega, \mathcal{L}_\omega, \xi_\omega)$. ■

6.4 The axiom of plane transitivity

The seventh axiom that brings us directly to the structure of one of the three standard Hilbert spaces is much more recent [Aerts and van Steirteghem, 2000].

AXIOM 97 Plane transitivity. Consider a state property space $(\Sigma, \mathcal{L}, \xi)$ for which axioms 24, 27 and 43 are satisfied. The state property space is plane transitive if for an arbitrary classical state $\omega \in \Omega$ and states $p, q \in \Sigma_\omega$ there exist two distinct atoms $s_1, s_2 \in \mathcal{L}_\omega$ and an automorphism (m, n) of $(\Sigma_\omega, \mathcal{L}_\omega, \xi_\omega)$ such that $n|_{[0, s_1 \vee s_2]}$ is the identity and $m(p) = q$.

Both classical entities and quantum entities can be described by a state property space where the set of properties is a complete atomistic orthocomplemented lattice that satisfies the covering law, is weakly modular and plane transitive. In section 8 we consider the converse, namely how this structure leads us to classical physics and to quantum physics. But first we want to look into one of the basic notions of quantum mechanics, namely the notion of superposition state.

7 SUPERPOSITION

One of the aspects which is often put forward as the most characteristic feature of quantum mechanics is the existence of 'superposition states'. In principle, the notion of superposition of states is intrinsically linked to the linearity of the Hilbert space. It is however possible to introduce it on a more fundamental level, which is what we will do in this section.

DEFINITION 98 Superposition. Consider a state property space $(\Sigma, \mathcal{L}, \xi)$ and states $p, q, r \in \Sigma$. We say that r is a superposition of p and q if $\xi(p) \cap \xi(q) \subseteq \xi(r)$. More generally, for a set of states $\Gamma \subseteq \Sigma$ we say that r is a superposition of Γ if $\cap_{p \in \Gamma} \xi(p) \subseteq \xi(r)$. We call

$$\bar{\Gamma} = \{r \mid r \in \Sigma, \cap_{p \in \Gamma} \xi(p) \subseteq \xi(r)\} \tag{148}$$

the superposition set corresponding to $\Gamma \subseteq \Sigma$.

PROPOSITION 99. *Consider a state property space* $(\Sigma, \mathcal{L}, \xi)$ *and* $\Gamma \subseteq \Sigma$. *For* $\Gamma, \Delta \subseteq \Sigma$ *we have*

$$\Gamma \subseteq \bar{\Gamma} \quad (149)$$

$$\Gamma \subseteq \Delta \Rightarrow \bar{\Gamma} \subseteq \bar{\Delta} \quad (150)$$

$$\bar{\bar{\Gamma}} = \bar{\Gamma} \quad (151)$$

which proves that $^-$ *is a closure operator.*

Proof. We have $\cap_{p \in \Gamma} \xi(p) \subseteq \xi(r) \ \forall \ r \in \Gamma$, and hence that $\Gamma \subseteq \bar{\Gamma}$, which proves (149). Suppose we have $\Gamma \subseteq \Delta \subseteq \Sigma$ and consider $r \in \bar{\Gamma}$. We have $\cap_{p \in \Delta} \xi(p) \subseteq \cap_{p \in \Gamma} \xi(p) \subseteq \xi(r)$ and hence $r \in \bar{\Delta}$. As a consequence we have $\bar{\Gamma} \subseteq \bar{\Delta}$, which proves (150). From (149) follows that $\bar{\Gamma} \subseteq \bar{\bar{\Gamma}}$. Consider $r \in \bar{\bar{\Gamma}}$, which means that $\cap_{p \in \bar{\Gamma}} \xi(p) \subseteq \xi(r)$. We have $\cap_{p \in \Gamma} \xi(p) \subseteq \cap_{p \in \bar{\Gamma}} \xi(p) \subseteq \xi(r)$, and hence $r \in \bar{\Gamma}$. Hence we have that $\bar{\bar{\Gamma}} \subseteq \bar{\Gamma}$. This means that we have proven (151). ■

PROPOSITION 100. *Consider a state property space* $(\Sigma, \mathcal{L}, \xi)$ *for which axioms 24 and 27 are satisfied, and* $\Gamma \subseteq \Sigma$. *We have*

$$r \in \bar{\Gamma} \Leftrightarrow s(r) \leq \vee_{p \in \Gamma} s(p) \quad (152)$$

Proof. Consider $r \in \bar{\Gamma}$, which means that $\cap_{p \in \Gamma} \xi(p) \subseteq \xi(r)$. We have $s(r) = \wedge_{a \in \xi(r)} a \leq \wedge_{a \in \cap_{p \in \Gamma} \xi(p)} a = \wedge_{a \in \xi(p) \forall p \in \Gamma} a = \wedge_{s(p) \leq a \forall p \in \Gamma} a = \vee_{p \in \Gamma} s(p)$. Suppose now that $s(r) \leq \vee_{p \in \Gamma} s(p)$. This means that $\wedge_{a \in \cap_{p \in \Gamma} \xi(p)} a \leq \wedge_{a \in \xi(r)} a$ and hence $\cap_{p \in \Gamma} \xi(p) \subseteq \xi(r)$, which proves that $r \in \bar{\Gamma}$. ■

DEFINITION 101 Superselection. Consider a state property space $(\Sigma, \mathcal{L}, \xi)$. We say that $p, q \in \Sigma$ are separated by a superselection rule, and we denote p ssr q, if the only superpositions of p and q are contained in p or in q. Hence, if for $r \in \Sigma$ such that $\xi(p) \cap \xi(q) \subseteq \xi(r)$ we have $\xi(p) \subseteq \xi(r)$ or $\xi(q) \subseteq \xi(r)$.

PROPOSITION 102. *Consider a state property space* $(\Sigma, \mathcal{L}, \xi)$ *for which axioms 24 and 27 are satisfied. For* $p, q \in \Sigma$ *we have*

$$p \text{ ssr } q \Leftrightarrow \kappa(s(p) \vee s(q)) = \{r \mid r \in \Sigma, r \leq p \text{ or } r \leq q\} \quad (153)$$

Proof. Suppose that p ssr q and consider $r \in \kappa(s(p) \vee s(q))$ and hence $s(r) \leq s(p) \vee s(q)$. From (152) follows that $r \in \{p, q\}^- = \{r \mid r \in \Sigma, r \leq p \text{ or } r \leq q\}$. Hence $r \leq p$ or $r \leq q$. Suppose that $\kappa(s(p) \vee s(q)) = \{r \mid r \in \Sigma, r \leq p \text{ or } r \leq q\}$ and consider $r \in \{p, q\}^-$. From (152) follows that $s(r) \leq s(p) \vee s(q)$ and hence $r \in \kappa(s(p) \vee s(q))$. As a consequence we have $r \leq p$ or $r \leq q$. This proves that p ssr q. ■

THEOREM 103. *Consider a state property space $(\Sigma, \mathcal{L}, \xi)$ for which axioms 24, 27, 43, 81 and 93 are satisfied. For $p, q \in \Sigma$ we have*

$$p \text{ ssr } q \Rightarrow p = q \text{ or } p \perp q \tag{154}$$

Proof. Suppose that p ssr q and $p \not\perp q$, and consider r such that $r \in \kappa((s(p) \vee s(q)) \wedge s(q)^\perp)$. This means that $r \in \kappa((s(p) \vee s(q))$ and $r \in \kappa(s(q)^\perp)$. From (153) we have that $r \leq p$ or $r \leq q$, but since axiom 81 is satisfied, this gives $r = p$ or $r = q$. However, since $r \in \kappa(s(q)^\perp)$ we cannot have $r = q$. Hence $r = p$. Hence we have proven that $\kappa((s(p) \vee s(q)) \wedge s(q)^\perp) = \kappa(s(p) \wedge s(q)^\perp)$. We have $s(p) \wedge s(q)^\perp \leq s(p)$ and since $s(p)$ is an atom of $\mathcal{L}$ we have $s(p) \wedge s(q)^\perp = s(p)$ or $s(p) \wedge s(q)^\perp = 0$. If $s(p) \wedge s(q)^\perp = s(p)$ we have $s(p) \leq s(q)^\perp$ and hence $p \perp q$, which is not true. This means that $s(p) \wedge s(q)^\perp = 0$, and hence $(s(p) \vee s(q)) \wedge s(q)^\perp = 0$. From this follows that $(s(p)^\perp \wedge s(q)^\perp) \vee s(q) = I$, and as a consequence we have $((s(p)^\perp \wedge s(q)^\perp) \vee s(q)) \wedge s(q)^\perp = s(q)^\perp$. Since axiom 93 is satisfied we have $((s(p)^\perp \wedge s(q)^\perp) \vee s(q)) \wedge s(q)^\perp = s(p)^\perp \wedge s(q)^\perp$. Hence we have $s(q)^\perp = s(p)^\perp \wedge s(q)^\perp$, and as a consequence $s(q)^\perp \leq s(p)^\perp$. From this follows that $s(p) \leq s(q)$, and since $s(q)$ is an atom of $\mathcal{L}$ we have $s(p) = s(q)$, and hence $p = q$. ■

THEOREM 104. *Consider a state property space $(\Sigma, \mathcal{L}, \xi)$ for which axioms 24, 27, 43, 81 and 95 are satisfied. For $p, q \in \Sigma$ we have*

$$p \text{ ssr } q \Rightarrow p = q \text{ or } p \perp q \tag{155}$$

Proof. In the same way as in the proof of theorem 103 we prove that $(s(p)^\perp \wedge s(q)^\perp) \vee s(q) = I$ if p ssr q and $p \not\perp q$. This means that I covers $s(p)^\perp \wedge s(q)^\perp$. Since $s(p)^\perp \wedge s(q)^\perp \leq s(q)^\perp \leq I$, and axiom 95 is satisfied, we have $s(p)^\perp \wedge s(q)^\perp = s(q)^\perp$ and hence $s(q)^\perp \leq s(p)^\perp$. From this follows that $s(p) \leq s(q)$, and since $s(q)$ is an atom, we have $s(p) = s(q)$ and hence $p = q$. ■

From these theorems follows that if axioms 24, 27, 43, 81 and 93 are satisfied or if axioms 24, 27, 43, 81 and 95 are satisfied, and two different states p and q are separated by a superselection rule then they are orthogonal. It also means that for two different states p and q that are not orthogonal there always exists a third state r which is a superposition of p and q.

8 HILBERT SPACE REPRESENTATIONS

In this section we make further steps to get closer to standard quantum mechanics in a complex Hilbert space. A first step is based on Piron's representation theorem for an irreducible complete orthocomplemented weakly modular lattice satisfying the covering law [Piron, 1964]. Piron proved that such a lattice can be represented as the set of closed subspaces of a generalized Hilbert space.

8.1 Representation in Generalized Hilbert spaces

Starting from the general decomposition theorem 80 we have proven in section 5.3, and using the extra axioms 81, 93, 95 introduced in section 6, we can prove the following theorem for each one of the non classical components of the decomposition.

THEOREM 105. *Consider a state property space* $(\Sigma, \mathcal{L}, \xi)$*, and suppose that axioms 24, 27, 43, 81, 93 and 95 are satisfied. Consider* $\textcircled{\vee}_{\omega \in \Omega}(\Sigma_\omega, \mathcal{L}_\omega, \xi_\omega)$ *the decomposition of* $(\Sigma, \mathcal{L}, \xi)$ *in its non classical components. For each nonclassical component* $(\Sigma_\omega, \mathcal{L}_\omega, \xi_\omega)$*, which has at least four orthogonal states, there exists a vector space* V_ω*, over a division ring* K_ω*, with an involution of* K_ω*, which means a function*

$$^* : K_\omega \rightarrow K_\omega \tag{156}$$

such that for $k, l \in K_\omega$ *we have:*

$$(k^*)^* = k \tag{157}$$

$$(k \cdot l)^* = l^* \cdot k^* \tag{158}$$

and an Hermitian product on V_ω*, which means a function*

$$\langle\ ,\ \rangle : V_\omega \times V_\omega \rightarrow K_\omega \tag{159}$$

such that for $x, y, z \in V_\omega$ *and* $k \in K_\omega$ *we have:*

$$\langle x + ky, z\rangle = \langle x, z\rangle + k\langle x, y\rangle \tag{160}$$

$$\langle x, y\rangle^* = \langle y, x\rangle \tag{161}$$

$$\langle x, x\rangle = 0 \Leftrightarrow x = 0 \tag{162}$$

and such that for $M \subset V_\omega$ *we have:*

$$M^\perp + (M^\perp)^\perp = V_\omega \tag{163}$$

where $M^\perp = \{y \mid y \in V_\omega, \langle y, x\rangle = 0, \forall x \in M\}$*. Such a vector space is called a generalized Hilbert space or an orthomodular vector space. And we have that:*

$$(\Sigma_\omega, \mathcal{L}_\omega, \xi_\omega) \cong (\mathcal{R}(V_\omega), \mathcal{L}(V_\omega), \nu_\omega) \tag{164}$$

where $\mathcal{R}(V_\omega)$ *is the set of rays of* V*,* $\mathcal{L}(V_\omega)$ *is the set of biorthogonally closed subspaces (subspaces that are equal to their biorthogonal) of* V_ω*, and* ν_ω *makes correspond with each ray the set of biorthogonally closed subspaces that contain this ray.*

Proof. If axioms 24, 27, 43, 81, 93 and 95 are satisfied for $(\Sigma, \mathcal{L}, \xi)$, then from theorem 78 and propositions 84, 94 and 96 follows that $(\Sigma_\omega, \mathcal{L}_\omega, \xi_\omega)$ satisfies axioms

24, 27, 43, 81, 93 and 95. Hence the lattice $\mathcal{L}_\omega$ is a complete orthocomplemented atomistic weakly modular lattice satisfying the covering law. Furthermore from proposition 85 follows that $\mathcal{L}_\omega$ is irreducible, and since $(\Sigma_\omega, \mathcal{L}_\omega, \xi_\omega)$ has at least four orthogonal states, it follows that $\mathcal{L}_\omega$ has at least four orthogonal atoms. This means that for $\mathcal{L}_\omega$ we can employ Piron's representation theorem [Piron, 1964; Piron, 1976; Maeda and Maeda, 1970; Faure and Frölicher, 1995], and hence infer that there exists a vector space V_ω, over a division ring K_ω, with an involution * of K_ω and an Hermitian product $\langle\ ,\ \rangle$ on V_ω and such that for $M \subset V_\omega$ we have:

$$M^\perp + (M^\perp)^\perp = V_\omega \tag{165}$$

where $M^\perp = \{y \mid y \in V_\omega, \langle y, x\rangle = 0, \forall x \in M\}$, and such that

$$\mathcal{L}_\omega \cong \mathcal{L}(V_\omega) \tag{166}$$

where $\mathcal{L}(V_\omega)$ is the set of all biorthogonal subspaces of V_ω, i.e.

$$\mathcal{L}(V_\omega) = \{M \mid M \subseteq V_\omega, (M^\perp)^\perp = M\} \tag{167}$$

Each atom $s(p)$ of the lattice $\mathcal{L}_\omega$ is represented by a ray, i.e. a one dimensional subspace of V_ω. This means that Σ_ω can be put equal to $\mathcal{R}(V_\omega)$ the set of rays of the vector space V_ω. If we define ν_ω as the function from $\mathcal{R}(V_\omega)$ to $\mathcal{P}(\mathcal{L}(V_\omega))$, that makes correspond with each ray the set of biorthogonally closed subspaces that contain this ray, we have proven that $(\mathcal{R}(V_\omega), \mathcal{L}(V_\omega), \nu_\omega)$ is isomorphic to $(\Sigma_\omega, \mathcal{L}_\omega, \xi_\omega)$. ■

8.2 *Representation in classical Hilbert spaces*

Maria Pia Solèr has proven that if V_ω contains an infinite orthonormal sequence, then $K = \mathbb{R}, \mathbb{C}$ or $\mathbb{H}$ and V_ω is the corresponding Hilbert space [Solèr, 1995]. Holland has shown that it is enough to demand the existence of a nonzero $\lambda \in K$ and an infinite orthogonal sequence $(e_n)_n \in V_\omega$ such that $< e_n, e_n >= \lambda$ for every n. To be precise, either $(V_\omega, K, < \cdot, \cdot >)$ or $(V_\omega, K, - < \cdot, \cdot >)$ is then a classical Hilbert space [Holland, 1995]. In [Aerts and van Steirteghem, 2000] we proved some alternatives to Solèr's result, by means of automorphisms of $\mathcal{L}(V_\omega)$.

PROPOSITION 106. *Let $(V, K, < \cdot, \cdot >)$ be an orthomodular space and let $\mathcal{L}(V)$ be the lattice of its closed subspaces. The following are equivalent:*
(1) $(V, K, < \cdot, \cdot >)$ is an infinite dimensional Hilbert space over $K = \mathbb{R}, \mathbb{C}$ or $\mathbb{H}$.
(2) V is infinite–dimensional and given two orthogonal atoms $\bar{x}, \bar{y}$ in $\mathcal{L}(V)$, there is a unitary operator U such that $U(\bar{x}) = \bar{y}$.
(3) There exist $a, b \in \mathcal{L}(V)$, where b is of dimension at least 2, and an ortholattice automorphism f of $\mathcal{L}(V)$ such that $f(a) \lneq a$ and $f|_{[0,b]}$ is the identical map.
(4) V is infinite dimensional and given two orthogonal atoms $\bar{x}, \bar{y}$ in $\mathcal{L}(V)$ there exist distinct atoms $\bar{x}_1, \bar{y}_2$ and an ortholattice automorphism f of $\mathcal{L}(V)$ such that $f|_{[0, \bar{x}_1 \vee \bar{y}_2}$ is the identity and $f(\bar{x}) = \bar{y}$.

Condition (2) is Holland's Ample Unitary Group axiom [Holland, 1995] and (3) is due to Mayet [Mayet, 1998]. Using the properties listed in section 2 of [Mayet, 1998], one can easily prove that (4) implies (2).

THEOREM 107. *Consider a state property space* $(\Sigma, \mathcal{L}, \xi)$, *and suppose that axioms 24, 27, 43, 81, 93, 95 and 97 are satisfied. Consider* $\textcircled{\vee}_{\omega \in \Omega}(\Sigma_\omega, \mathcal{L}_\omega, \xi_\omega)$ *the decomposition of* $(\Sigma, \mathcal{L}, \xi)$ *in its non classical components. Each nonclassical component* $(\Sigma_\omega, \mathcal{L}_\omega, \xi_\omega)$, *which has at least four orthogonal states, is isomorphic to the canonical state property space* $(\Sigma(\mathcal{H}_\omega), \mathcal{L}(\mathcal{H}_\omega), \xi_{\mathcal{H}_\omega})$ *where* $\mathcal{H}_\omega$ *are real, complex or quaternionic Hilbert spaces.*

Proof. An immediate consequence of proposition 106. ■

Theorem 107 proves that if axioms 24, 27, 43, 81, 93, 95 and 97 are satisfied, our theory reduces to standard quantum mechanics with super selection variables, and eventually on a quaternionic Hilbert space.

BIBLIOGRAPHY

[Aerts, 1981] D. Aerts. *The One and the Many: Towards a Unification of the Quantum and Classical Description of One and Many Physical Entities*, Doctoral dissertation, Brussels Free University, 1981.

[Aerts, 1982] D. Aerts. Description of many physical entities without the paradoxes encountered in quantum mechanics, *Foundations of Physics*, **12**, 1131-1170, 1982.

[Aerts, 1983] D. Aerts. Classical theories and non classical theories as a special case of a more general theory, *Journal of Mathematical Physics*, **24**, 2441-2453, 1983.

[Aerts, 1999a] D. Aerts. Foundations of quantum physics: a general realistic and operational approach, *International Journal of Theoretical Physics*, **38**, 289-358, 1999. Archive ref and link: quant-ph/0105109.

[Aerts, 1999b] D. Aerts. Quantum mechanics: structures, axioms and paradoxes, in *Quantum Mechanics and the Nature of Reality*, eds. Aerts, D. and Pykacz, J., Kluwer Academic, Dordrecht, 1999. Archive ref and link: quant-ph/0106132.

[Aerts, 2002] D. Aerts. Being and change: foundations of a realistic operational formalism, in *Probing the Structure of Quantum Mechanics: Nonlinearity, Nonlocality, Probability and Axiomatics*, eds. Aerts, D, Czachor, M. and Durt, T, World Scientific, Singapore, 2002.

[Aerts *et al.*, 1999] D. Aerts, E. Colebunders, A. Van der Voorde, and B. Van Steirteghem. State property systems and closure spaces: a study of categorical equivalence, *International Journal of Theoretical Physics*, **38**, 359-385, 1999. Archive ref and link: quant-ph/0105108.

[Aerts *et al.*, 2000] D. Aerts, E. Colebunders, A. Van der Voorde, and B. Van Steirteghem. On the amnestic modification of the category of state property systems. *Applied Categorical Structures*, **10**, 469-480, 2000.

[Aerts and van Steirteghem, 2000] D. Aerts and B. Van Steirteghem. Quantum axiomatics and a theorem of M. P. Solèr, *International Journal of Theoretical Physics*, **39**, 497-502, 2000. Archive ref and link: quant-ph/0105107.

[Aerts and Deses, 2002] D. Aerts and D. Deses. State property systems and closure spaces: extracting the classical and nonclassical parts. In D. Aerts, M. Czachor and T. Durt (Eds.), *Probing the Structure of Quantum Mechanics: Nonlinearity, Nonlocality, Probability and Axiomatics* (pp. 130-148). Singapore: World Scientific, 2002.

[Aerts and van Valckenborgh, 2002] D. Aerts and F. Valckenborgh. The linearity of quantum mechanics at stake: the description of separated quantum entities. In D. Aerts, M. Czachor and T. Durt (Eds.), *Probing the Structure of Quantum Mechanics: Nonlinearity, Nonlocality, Probability and Axiomatics* (pp. 20-46). Singapore: World Scientific, 2002.

[Aerts and Aerts, 2004] D. Aerts and S. Aerts. Towards a general operational and realistic framework for quantum mechanics and relativity theory. In A. C. Elitzur, S. Dolev and N. Kolenda (Eds.), *Quo Vadis Quantum Mechanics? Possible Developments in Quantum Theory in the 21st Century* (pp. 153-208). New York: Springer, 2004.

[Aerts and Deses, 2005] D. Aerts and D. Deses. State property systems and orthogonality. *International Journal of Theoretical Physics*, **44**, 919-929, 2005.

[Aerts and Pulmannova, 2006] D. Aerts and S. Pulmannova. Representation of state property systems. *Journal of Mathematical Physics*, **47**, 1-18, 2006.

[Aerts *et al.*, in press] D. Aerts, D. Deses, and A. Van der Voorde. Classicality and connectedness for state property systems and closure spaces. *International Journal of Theoretical Physics*, in press.

[Birkhoff and von Neumann, 1936] G. Birkhof and J. von Neumann. The logic of quantum mechanics, *Annals of Mathematics*, **37**, 823-843, 1936.

[Faure and Frölicher, 1995] Cl.-A. Faure and A. Frölicher. Dualities for infinite–dimensional projective geometries, *Geometriae Dedicata*, **56**, 225-236, 1995.

[Foulis *et al.*, 1983] D. Foulis, C. Piron, and C. Randall. Realism, operationalism, and quantum mechanics, *Foundations of Physics*, **13**, 813, 1983.

[Holland, 1995] S. S. Holland, Jr. Orthomodularity in infinite dimensions; a theorem of M. Solèr, *Bulletin of the American Mathematical Society*, **32**, 205-234, 1995.

[Keller, 1980] H. Keller. Ein nicht-klassischer Hilbertscher Raum, *Mathematische Zeitschrift*, **172**, 41-49, 1980.

[Mackey, 1963] G. Mackey. *Mathematical Foundations of Quantum Mechanics*, Benjamin, New York, 1963.

[Maeda and Maeda, 1970] F. Maeda and S. Maeda. *Theory of symmetric lattices*, Springer–Verlag, Berlin, 1970.

[Mayet, 1998] R. Mayet. Some characterizations of the underlying division ring of a Hilbert lattice by automorphisms, *International Journal of Theoretical Physics*, **37**, 109-114, 1998.

[Mittelstaedt, 1963] P. Mittelstaedt. *Philosophische Probleme der Modernen Physik*, Bibliographisches Institut, Manheim, 1963.

[Piron, 1964] C. Piron. Axiomatique quantique, *Helvetica Physica Acta*, **37**, 439-468, 1964.

[Piron, 1976] C. Piron. *Foundations of Quantum Physics*, Benjamin, Massachusetts, 1976.

[Randall and Foulis, 1983] C. Randall and D. Foulis. Properties and operational propositions in quantum mechanics, *Foundations of Physics*, **13**, 835, 1983.

[Solèr, 1995] M. P. Solèr. Characterization of Hilbert spaces by orthomodular spaces, *Communications in Algebra*, **23**, 219-243, 1995.

[von Neumann, 1932] J. von Neumann. *Mathematische Grundlagen der Quantenmechanik*, Springer, Berlin, 1932.

[Wigner, 1959] E. P. Wigner. *Group Theory and its Applications to Quantum Mechanics of Atomic Spectra*, Academic Press, New York, 1959.

Edited by K. Engesser, D. M. Gabbay and D. Lehmann

QUANTUM LOGIC AND NONCLASSICAL LOGICS

Gianpiero Cattaneo, Maria Luisa Dalla Chiara, Roberto Giuntini and Francesco Paoli

1 INTRODUCTION

Classical logic is sometimes described as "the logic of an omniscient mind in a deterministic universe". From an intuitive point of view the basic features of classical semantics can be summarized as follows:

1) any problem is *semantically decided*: for any sentence α, either α or its negation $\neg\alpha$ is true (*excluded middle principle*); at the same time, a sentence α and its negation $\neg\alpha$ cannot be both true (*noncontradiction principle*).

2) Meanings behave in a *compositional* way: the meaning of a *compound* expression is determined by the meanings of its *parts*.

3) Meanings are *sharp* and *unambiguous*.

Some (possibly all) of these principles have been put in question by different forms of nonclassical logic. In some significant cases, the *objective and descriptional* notion of truth (which is characteristic of classical logic) has been replaced by an *epistemic* conception. Accordingly, *truth* has been identified with *what is known* by non-omniscient minds, acting in a universe that may be either deterministic or indeterministic. The first choice is compatible with the intuitionistic approaches to logic and to mathematics, while the second choice represents the basic assumption of the quantum logical investigations. In both cases, the classical notion of truth has been replaced by the following relation:

an information i forces us to assert the truth of a sentence α.

One also briefly says that the information i *forces* (or *verifies*) the sentence α (and one usually writes: $i \models \alpha$). Should i represent a *noncontradictory and complete* information-system, our forcing relation would naturally collapse into the classical notion of truth. However, human information is generally incomplete and not necessarily consistent.

As expected, in the case of physical theories, significant pieces of information correspond to *what is known by an observer about the physical systems under investigation*. In this connection, one usually speaks of *physical states* (briefly, *states*). In the "happiest situations", a state may represent a *maximal knowledge* of the observer: a piece of information that cannot be consistently extended to a richer information, in the framework of the theory. Even a hypothetical *omniscient mind* could not know more about the system in question (if the theory is correct). States of this kind are usually called *pure states*, both in classical and in quantum physics. Pieces of information that are not maximal are generally represented by *mixtures of pure states* (also called *mixed states*). There is an important difference that concerns the logical behavior of classical and of quantum pure states. In classical mechanics, *maximality* implies *logical completeness*: any pure state *semantically decides* any *physical property* (or *event*) that may hold for the system under investigation (in other words, the state attributes to the system either the property or its negation).This is in accordance not only with classical logic, but also with a number of important nonclassical logics (like intuitionistic logic), where any *noncontradictory and maximal formal system* is *logically complete*. Quantum pure states, instead, give rise to a somewhat "mysterious" divergence between maximality and logical completeness, which represents the origin of most logical anomalies of the quantum world. Although representing a maximal information, a quantum pure state is never logically complete. This is a consequence of Heisenberg's uncertainty principle, according to which there are pairs of *complementary events* that cannot be simultaneously decided by any pure state.

Both in classical and in quantum mechanics, physical states are represented by special kinds of mathematical objects. In classical mechanics (CM), a pure state of a single particle can be represented by a sequence of six real numbers $(r_1, \ldots, r_6)$, where the first three numbers correspond to the position-coordinates, while the last ones are the momentum-components. The set $\mathbb{R}^6$ of all sextuples of real numbers represents the *phase-space* for the particle in question. Similarly for the case of compound systems, consisting of a finite number n of particles. Hence, any pure state of a classical particle-system is represented by a point of an appropriate phase space Σ. How to represent the physical events that may occur to a given particle? Following the standard ideas of classical (extensional) semantics, it is quite natural to assume that the such events are mathematically represented by suitable subsets of Σ. What about the structure of all events? As is well known, the power set of any set gives rise to a *Boolean algebra*. And also the set $\mathcal{F}(\Sigma)$ of all *measurable subsets* of Σ (which is more tractable than the full power set of Σ, from a measure-theoretic point of view) turns out to have a Boolean structure. Hence, we may refer to the following Boolean field of sets:

$$\mathcal{EV}^{\mathcal{C}} = \langle \mathcal{F}(\Sigma), \cap, \cup, {}^{c}, \emptyset, \Sigma \rangle ,$$

where the set-theoretic operations $\cap, \cup, {}^{c}$ represent respectively the *conjunction*, the *disjunction* and the *negation* of classical events.

As a consequence, the logic of CM turns out to be in perfect agreement with classical logic. Furthermore, pure states are logically complete: for any point p of the phase-space Σ and for any event E in $\mathcal{F}(\Sigma)$, either $p \in E$ or $p \in E^c$.

What happens in the case of quantum theory (QT)? As opposed to classical mechanics, QT is *essentially probabilistic*. A pure state generally assigns to a quantum event a probability-value (a real number in the interval $[0, 1]$). As a consequence, a quantum event may be semantically *indeterminate* for a given pure state, and the excluded middle principle is violated.

According to von Neumann's axiomatization of QT, the mathematical interpretation of any quantum system is a *complex separable Hilbert space* $\mathcal{H}$, which has, for QT, the role played by phase-spaces in CM.[1] Any pure state is mathematically represented by a unit vector $|\psi\rangle$ of $\mathcal{H}$, while mixed states are represented by special operators called *density operators*. What about quantum events? After Birkhoff and von Neumann's celebrated article "*The logic of quantum mechanics*" (which is considered the birth-date of quantum logic) the mathematical representatives of quantum events are identified with the *closed subspaces* of the Hilbert space $\mathcal{H}$ associated to the quantum system $\mathfrak{S}$ under investigation. Why are the *mere* subsets of $\mathcal{H}$ not adequate mathematical representatives for quantum events, as in the phase-space case? The reason depends on the *superposition principle*, which represents one of the basic dividing lines between the quantum and the classical case. As opposed to classical mechanics, in quantum mechanics, any unit vector, that is a linear combination of pure states, gives rise to a new pure state. Suppose two pure states $|\psi_1\rangle, |\psi_2\rangle$ are orthogonal vectors and suppose that a pure state $|\psi\rangle$ is a linear combination (a *superposition*) of $|\psi_1\rangle, |\psi_2\rangle$. In other words:

$$|\psi\rangle = c_1|\psi_1\rangle + c_2|\psi_2\rangle \text{ (where } |c_1|^2 + |c_2|^2 = 1).$$

According to one of the basic axioms of QT (the so called *Born rule*), this means that a quantum system in state $|\psi\rangle$ might verify with probability $|c_1|^2$ those events that are certain for state $|\psi_1\rangle$ (and are not certain for $|\psi\rangle$) and might verify with probability $|c_2|^2$ those events that are certain for state $|\psi_2\rangle$ (and are not certain for $|\psi\rangle$). A similar relation also holds for infinite superpositions $\sum_i c_i|\psi_i\rangle$. As a consequence, the mathematical representatives of events should be closed under finite and infinite linear combinations. The closed subspaces of $\mathcal{H}$ are just the mathematical objects that can realize such a role.

What will be the meaning of *negation*, *conjunction* and *disjunction* in the realm of quantum events? According to Birkhoff and von Neumann: "The mathematical representative of the *negative* of any experimental proposition is the *orthogonal complement* of the mathematical representative of the proposition itself". The orthogonal complement A' of a closed subspace A is defined as the set of all vectors that are orthogonal to all elements of A. In other words, $|\psi\rangle \in A'$ iff $|\psi\rangle \perp A$ iff, for any $|\varphi\rangle \in A$, the inner product $\langle\psi|\varphi\rangle$ is 0.[2] From the point of view of the physical interpretation, the orthogonal complement (called also *orthocomplement*) is particularly interesting, since it satisfies the following property: a pure state $|\psi\rangle$ assigns to an event A probability 1 (0, respectively) iff $|\psi\rangle$ assigns to the orthocomplement of A probability 0 (1, respectively). As a consequence, one is

[1] See Deff. 146, 148.

[2] See Def.142.

dealing with an operation that *inverts* the two extreme probability-values, which naturally correspond to the truth-values *Truth* and *Falsity* (as in the classical truth-table of negation).

As for conjunction, Birkhoff and von Neumann point out that this can be still represented by the set-theoretic intersection (as in the classical case). For, the intersection $A \cap B$ of two closed subspaces is again a closed subspace. Hence, we will obtain the usual truth-table for the connective *and*:

$$|\psi\rangle \text{ verifies } A \cap B \text{ iff } |\psi\rangle \text{ verifies both members.}$$

Disjunction, however, cannot be represented here as a set-theoretic union. For, generally, the union $A \cup B$ of two closed subspaces is not a closed subspace, except in special circumstances. In spite of this, we have at our disposal another good representative for the connective *or*: the *supremum* $A \vee B$ of two closed subspaces, that is the smallest closed subspace including both A and B. Of course, $A \vee B$ will include $A \cup B$.

As a consequence, we obtain the following structure:

$$\mathcal{EV}^{\mathcal{Q}} = \langle \mathcal{C}(\mathcal{H})\,,\,\wedge\,,\,\vee\,,\,'\,,\,\mathbf{0}\,,\mathbf{1}\rangle\,,$$

where $\wedge$ is the set-theoretic intersection; $\vee\,,\,'$ are defined as above; while $\mathbf{0}$ and $\mathbf{1}$ represent, respectively, the null subspace (the singleton consisting of the null vector, which is the smallest possible subspace) and the total space $\mathcal{H}$.

The quantum event structure $\mathcal{EV}^{\mathcal{Q}}$ turns out to simulate a "quasi-Boolean behavior"; however, it is not a Boolean algebra. Something very essential is missing. Conjunction and disjunction are not distributive. There are triplets of quantum events A, B, C such that:

$$A \wedge (B \vee C) \neq (A \wedge B) \vee (A \wedge C).$$

In fact, $\mathcal{EV}^{\mathcal{Q}}$ belongs to the variety of all *orthocomplemented orthomodular lattices*, that are not necessarily distributive.[3] Structures of this kind are also called *Hilbert lattices*. By the one-to-one correspondence between the set $\mathcal{C}(\mathcal{H})$ of all closed subspaces and the set $\Pi(\mathcal{H})$ of all projection-operators of $\mathcal{H}$, the lattice based on $\Pi(\mathcal{H})$ turns out to be isomorphic to the lattice based on $\mathcal{C}(\mathcal{H})$.[4] Hence, quantum events can be equivalently represented either by closed subspaces or by projection-operators of the Hilbert space associated to the quantum system under investigation.[5] For any quantum system $\mathfrak{S}$, the pair $(\mathcal{C}(\mathcal{H}), \mathcal{S})$ consisting of the set

[3]See Deff.126 and 129.

[4]See Deff.157 and 149.

[5]It is worth-while recalling that von Neumann was not completely satisfied with the basic idea of the "quantum logical approach", according to which the lattice of all closed subspaces of a Hilbert space should faithfully represent the structure of quantum events. The basic reason was the failure of *modularity* in the infinite dimensional case (see Def.130). In fact, von Neumann considered this property as an essential condition for a frequency-interpretation of quantum probabilities. An interesting analysis of von Neumann's doubts and critiques can be found in [Rédei, 1996].

$\mathcal{C}(\mathcal{H})$ of all closed subspaces of $\mathcal{H}$ and of the set $\mathcal{S}$ of all possible (pure and mixed) states for $\mathfrak{S}$ is usually called the *event-state system* for $\mathfrak{S}$ (which is equivalently represented also by the pair $(\Pi(\mathcal{H}), \mathcal{S})$). Given a state, represented by a density operator ρ ($\in \mathcal{S}$) and an event, represented by a projection-operator P ($\in \Pi(\mathcal{H})$), the probability $\rho(P)$ that the physical system $\mathfrak{S}$ in state ρ verifies the event P is determined by the *Born-rule*, according to which:

$$\rho(P) = \mathtt{tr}(\rho P),$$

(where $\mathtt{tr}$ is the *trace-functional*[6]).

A *similarity space* naturally arises in the framework of any event-state system $(\mathcal{C}(\mathcal{H}), \mathcal{S})$.

DEFINITION 1. Two states ρ and σ are called *similar* ($\rho \not\perp \sigma$) iff there is no event A that is certain for ρ and impossible for σ (in other words, $\rho(A) = 1 \curvearrowright \sigma(A) \neq 0$).

One can easily show that $\not\perp$ is a a genuine similarity relation (reflexive and symmetric, but generally not transitive). Hence, any pair $(\mathcal{S}, \not\perp)$ represents an example of a *similarity space*.

Any quantum similarity space $(\mathcal{S}, \not\perp)$ automatically determines a "twin space" $(\mathcal{S}, \perp)$ (also called *preclusivity space*), where $\perp$ is the negation of $\not\perp$ (in other words, $\rho \perp \sigma$ iff not $\rho \not\perp \sigma$).

Consider now the power set $\mathcal{P}(\mathcal{S})$ of the set of all states $\mathcal{S}$. The preclusivity relation $\perp$ permits one to define on $\mathcal{P}(\mathcal{S})$ a unary operation $^{\perp}$ (called the *preclusive complement*), which turns out to be a weak complement. For any set X of states:

$$X^{\perp} := \{x \in \mathcal{S} : \forall t \in X (s \perp t)\}.$$

The preclusive complement $^{\perp}$ satisfies the following properties for any sets X, Y of states:

- $X \subseteq X^{\perp\perp}$;
- $X \subseteq Y$ implies $Y^{\perp} \subseteq X^{\perp}$;
- $X \cap X^{\perp} = \emptyset$.

At the same time, the strong double negation principle ($X^{\perp\perp} \subseteq X$) and the excluded middle principle ($X \cup X^{\perp} = \mathcal{S}$) generally fails.

Consider now the map $^{\perp\perp} : \mathcal{P}(\mathcal{S}) \mapsto \mathcal{P}(\mathcal{S})$ such that:

$$X \rightarrowtail X^{\perp\perp}, \text{ for any } X \subseteq \mathcal{S}.$$

One can easily check that this map is a *closure operator*, satisfying the following conditions:

$$\emptyset^{\perp\perp} = \emptyset;\ X \subseteq X^{\perp\perp};\ X^{\perp\perp} = X^{\perp\perp\perp\perp};\ X \subseteq Y \curvearrowright X^{\perp\perp} \subseteq Y^{\perp\perp}.$$

[6]See Def.158.

Consider then the set $\mathbb{C}(\mathcal{P}(\mathcal{S}))$ of all *closed* elements of the power set of $\mathcal{S}$. By definition, we have:

$$X \in \mathbb{C}(\mathcal{P}(\mathcal{S})) \text{ iff } X = X^{\perp\perp}.$$

The elements of $\mathbb{C}(\mathcal{P}(\mathcal{S}))$ are called *closed* sets of states. As we will see, such sets play a very significant role for the semantics of quantum logic.

Some important properties of the closed sets of states are described by the following lemmas, which refer to any event-state system $(\mathcal{C}(\mathcal{H}), \mathcal{S})$.

THEOREM 2. *The structure*

$$\langle \mathbb{C}(\mathcal{P}(\mathcal{S})), \subseteq, ^{\perp}, \emptyset, \mathcal{S} \rangle$$

is a complete bounded ortholattice, where for any family $\{X_i\}_{i \in I} \subseteq \mathbb{C}(\mathcal{P}(\mathcal{S}))$*:*

- *the meet* $\bigwedge \{X_i\}_{i\in I}$ *exists and coincides with* $\bigcap \{X_i\}_{i\in I}$*;*
- *the join* $\bigvee \{X_i\}_{i\in I}$ *exists and coincides with* $(\bigcup \{X_i\}_{i\in I})^{\perp\perp}$*;*
- *the preclusive complement* $^{\perp}$ *is an orthocomplement.*[7]

The lattice $\mathbb{C}(\mathcal{P}(\mathcal{S}))$ fails to be distributive.

LEMMA 3. *The lattice* $\mathbb{C}(\mathcal{P}(\mathcal{S}))$ *is isomorphic to the lattice based on the set* $\mathcal{C}(\mathcal{H})$ *of all closed subspaces of* $\mathcal{H}$.

LEMMA 4. *Let* X *be any subset of* $\mathcal{S}$. *Then,* X *is closed iff* X *satisfies the following condition:*

$$\forall \rho [\rho \in X \textit{ iff } \forall \sigma \not\perp \rho \exists \tau \not\perp \sigma (\tau \in X)].$$

2 HOW QUANTUM LOGIC EMERGES FROM QUANTUM EVENT-STATE SYSTEMS

Quantum similarity spaces have naturally suggested an abstract *possible world semantics* for quantum logic (first proposed by Dishkant [Dishkant, 1972] and further developed by Goldblatt [Goldblatt, 1974]). Such semantics can be regarded as a "quantum variant" of the semantics proposed by Kripke for intuitionistic logic. Accordingly, one usually speaks of *Kripkean semantics* of quantum logic.

As is well known, Kripkean models for intuitionistic logic are based on sets of *possible worlds* possibly correlated by an *accessibility relation*, which is reflexive and transitive. According to a canonical interpretation, the possible worlds of an intuitionistic Kripkean model, can be regarded as *states of knowledge in progress*. When a world j is accessible to another world i, the state of knowledge corresponding to j is more informative with respect to the state of knowledge represented by i. In this framework, knowledge is *conservative*: when a state of knowledge i

[7]See Def. 117-126.

knows a given sentence, then all the states of knowledge that are accessible to i know the sentence in question.

The Kripkean characterization of quantum logic is based on a different intuitive idea. The possible worlds of a quantum logical model can be regarded as *pieces of information* that correspond to *states* of physical objects. What about the *accessibility relation*? This can be identified with the *similarity relation* that arises in a given event-state system. From an intuitive point of view, one can easily understand the reason why semantic models based on similarity spaces are physically significant. In fact, physical theories are not generally concerned with *possible evolutions of states of knowledge* with respect to a constant world (as happens in the case of intuitionistic logic), but rather with *sets of physical situations* that may be *similar*, where *states of knowledge* must single out some *invariants*.

We will now briefly sketch the basic concepts of an abstract possible world semantics for a weak form of quantum logic, that Dishkant had called *minimal quantum logic*, while Goldblatt's preferred denomination is *orthologic*. This logic fails to satisfy an important property of quantum event-structures: orthomodularity. Following Goldblatt's terminology, we will distinguish *orthologic* (**OL**) from *orthomodular quantum logic* (**OQL**), which is often simply called *quantum logic*. The sentential language of both logics consists of sentential letters and of the following primitive connectives: $\neg$ (not), $\curlywedge$ (and). The notion of *sentence* (or *formula*) is defined in the expected way. We will use the following metavariables: $\mathbf{p}$, $\mathbf{q}$, ... for atomic sentences and α, β, γ, ... for sentences. The disjunction $\curlyvee$ (or) is supposed to be defined via de Morgan law ($\alpha \curlyvee \beta := \neg(\neg\alpha \curlywedge \neg\beta)$).

Consider an abstract similarity space $(I, \not\perp)$, consisting of a non-empty set I of *possible worlds* and of a similarity relation $\not\perp$ (which is a reflexive and symmetric binary relation defined on I). In semantic contexts, similarity relations are dealt with as special cases of *accessibility relations*, while similarity spaces are also called *orthoframes*. Given an orthoframe $(I, \not\perp)$, we will use $i, j, k, \ldots$ as variables ranging over the set I of possible worlds.

Like in the concrete Hilbert-space case, any abstract similarity space has a "twin space" that is an abstract preclusivity space. The preclusivity relation, that represents the negation of the accessibility relation $\not\perp$, will be denoted by $\perp$. In other words:

$$i \perp j \text{ iff not } i \not\perp j.$$

Whenever $i \perp j$ we will say that j is *inaccessible* or *orthogonal* to i.

On this basis, one can define a *preclusive complement* $^{\perp}$ on the power set $\mathcal{P}(I)$ of I:

$$\forall X \subseteq I[X^{\perp} := \{i \in I : i \perp X\}].$$

The following conditions hold:

- the map $^{\perp\perp} : \mathcal{P}(I) \mapsto \mathcal{P}(I)$ is a closure operator;
- the structure $\langle \mathbb{C}(\mathcal{P}(I)), \subseteq, {}^{\perp}, \emptyset, I \rangle$ based on the set of all closed subsets of

I is an ortholattice.[8] Hence, in particular, $^{\perp}$ is an orthocomplement;

- X is a closed subset of I iff $\forall i[i \in X$ iff $\forall j \not\perp i \exists k \not\perp j(k \in X)]$.

In the framework of semantic applications, the closed subsets of I are usually called *propositions* of the orthoframe $(I, \not\perp)$.

The following Lemma sums up some basic properties of propositions:

LEMMA 5. *Let $(I, \not\perp)$ be an orthoframe.*

(i) I and $\emptyset$ are propositions;

(ii) If X is any set of worlds, then $X^{\perp}$ is a proposition;

(iii) If C is a family of propositions, then $\bigcap C$ is a proposition.

The notion of *Kripkean model* for **OL** can be now defined as follows:

DEFINITION 6. *Kripkean model* for **OL**
A *Kripkean model* for **OL** is a system $\mathcal{K} = \langle I, \not\perp, Pr, V\rangle$, where:

(i) $(I, \not\perp)$ is an orthoframe and Pr is a set of propositions of the frame that contains $\emptyset, I$ and is closed under the orthocomplement $^{\perp}$ and the set-theoretic intersection $\cap$;

(ii) V is a function that associates to any sentence α a proposition in Pr, satisfying the following conditions:
$V(\neg\beta) = V(\beta)^{\perp}$;
$V(\beta \curlywedge \gamma) = V(\beta) \cap V(\gamma)$.

Instead of $i \in V(\alpha)$, one usually writes $i \models \alpha$ and reads: "the information i forces α to be true" (or briefly, "α is true in the world i"). If T is a set of sentences, $i \models T$ will mean $i \models \beta$ for any $\beta \in T$.

THEOREM 7. *For any Kripkean model $\mathcal{K}$ and any sentence α:*

$$i \models \alpha \text{ iff } \forall j \not\perp i \exists k \not\perp j (k \models \alpha).$$

LEMMA 8. *In any Kripkean model $\mathcal{K}$:*

(i) $i \models \neg\beta$ *iff* $\forall j \not\perp i [j \not\models \beta]$;

(ii) $i \models \beta \curlywedge \gamma$ *iff* $i \models \beta$ *and* $i \models \gamma$.

On this basis, the notions of *truth*, *logical truth*, *consequence*, *logical consequence* are defined in the expected way.

DEFINITION 9. *Truth and logical truth*
A sentence α is *true* in a model $\mathcal{K} = \langle I, \not\perp, Pr, V\rangle$ (abbreviated $\models_{\mathcal{K}} \alpha$) iff $V(\alpha) = I$;

[8] See Def. 126.

α is a *logical truth* of **OL** ($\models_{\mathbf{OL}} \alpha$) iff $\models_{\mathcal{K}} \alpha$ for any model $\mathcal{K}$.

DEFINITION 10. *Consequence in a model and logical consequence*
Let T be a set of sentences and let $\mathcal{K}$ be a model. A sentence α is a *consequence in* $\mathcal{K}$ *of* T ($T \models_{\mathcal{K}} \alpha$) iff for any world i of $\mathcal{K}$, $i \models T \curvearrowright i \models \alpha$.
A sentence α is a *logical consequence* of T ($T \models_{\mathbf{OL}} \alpha$) iff for any model $\mathcal{K}$, $T \models_{\mathcal{K}} \alpha$.

The set of propositions of any Kripkean model for **OL** gives rise to an ortholattice. As a consequence, Kripkean models for **OL** can be canonically transformed into *algebraic models*, where the meaning of any sentence is identified with an element of an ortholattice, while the connectives are interpreted as the corresponding lattice-operation.

DEFINITION 11. *Algebraic model for* **OL**
An *algebraic model* for **OL** is a pair $\mathcal{A} = \langle \mathcal{B}, v \rangle$, consisting of an ortholattice $\mathcal{B} = \langle B, \leq, {}', \mathbf{0}, \mathbf{1} \rangle$ and a *valuation-function* v that associates to any sentence α of the language an element in B, satisfying the following conditions:

(i) $v(\neg\beta) = v(\beta)'$;

(ii) $v(\beta \curlywedge \gamma) = v(\beta) \wedge v(\gamma)$.

The notions of *truth*, *logical truth*, *consequence* and *logical consequence* are then defined in the expected way.

DEFINITION 12. *Truth and logical truth*
A sentence α is *true* in a model $\mathcal{A} = \langle \mathcal{B}, v \rangle$ (abbreviated as $\models_{\mathcal{A}} \alpha$) iff $v(\alpha) = \mathbf{1}$;
α is a *logical truth* of **OL** ($\models_{\mathbf{OL}} \alpha$) iff for any algebraic model $\mathcal{A} = \langle \mathcal{B}, v \rangle$, $\models_{\mathcal{A}} \alpha$.

When for any $\beta \in T, \models_{\mathcal{A}} \beta$, we will also write: $\models_{\mathcal{A}} T$.

DEFINITION 13. *Consequence in a model and logical consequence*
Let T be a set of sentences and let $\mathcal{A} = \langle \mathcal{B}, v \rangle$ be a model. A sentence α is a *consequence in* $\mathcal{A}$ of T ($T \models_{\mathcal{A}} \alpha$) iff for any element a of B:
if for any $\beta \in T$, $a \leq v(\beta)$ then $a \leq v(\alpha)$.
A sentence α is a *logical consequence* of T ($T \models_{\mathbf{OL}} \alpha$) iff for any algebraic model $\mathcal{A}$: $T \models_{\mathcal{A}} \alpha$.

One can prove that the Kripkean and the algebraic semantics characterize the same logic **OL**.[9]

In order to characterize *orthomodular quantum logic* (or *quantum logic*) one shall require a stronger condition both in the definition of Kripkean model and of algebraic model for **OL**:

DEFINITION 14. *Kripkean model for* **OQL**
A *Kripkean model* for **OQL** is a Kripkean model $\mathcal{K} = \langle I, \not\perp, Pr, V \rangle$ for **OL**, where the set of propositions Pr satisfies the *orthomodular property*: $X \subseteq Y \curvearrowright Y = X \vee (X^{\perp} \wedge Y)$.

[9]See [Dalla Chiara and Giuntini, 2002].

DEFINITION 15. *Algebraic model for* **OQL**
An *algebraic model* for **OQL** is an algebraic model $\mathcal{A} = \langle \mathcal{B}, v \rangle$ for **OL**, where $\mathcal{B}$ is an orthomodular lattice.[10]

We will indicate by **QL** either **OL** or **OQL**. Both logics are characterized by a deep asymmetry between conjunction and disjunction. By definition of Kripkean model, we have:

- $i \models \beta \curlywedge \gamma$ iff $i \models \beta$ and $i \models \gamma$;
- $i \models \beta \curlyvee \gamma$ iff $\forall j \not\perp i \, \exists k \not\perp j \, (k \models \beta$ or $k \models \gamma)$.

Hence, a disjunction may be true, even if both members are not true.
A consequence of this asymmetry is the failure of the distributivity principle:

$$\alpha \curlywedge (\beta \curlyvee \gamma) \not\models_{\mathbf{QL}} (\alpha \curlywedge \beta) \curlyvee (\alpha \curlywedge \gamma).$$

The semantic behavior of the quantum logical disjunction, which may appear *prima facie* somewhat strange, seems to reflect pretty well a number of concrete quantum situations. In quantum theory one is often dealing with alternatives that are semantically determined and true, while both members are, in principle, indeterminate. For instance, suppose we are referring to a spin one-half particle (say an electron) whose spin in a certain direction may assume only two possible values: either *up* or *down*. Now, according to one of the *uncertainty principles*, the spin in the x direction ($spin_x$) and the spin in the y direction ($spin_y$) represent two *incompatible* quantities that cannot be simultaneously measured. Suppose an electron in state $|\psi\rangle$ verifies the proposition "$spin_x$ is up". As a consequence of the uncertainty principle both propositions "$spin_y$ is up" and "$spin_y$ is down" shall be indeterminate. However the disjunction "either $spin_y$ is up or $spin_y$ is down" must be true.

As expected, the Kripkean models of **OQL** admit a quite natural realization in the framework of the Hilbert event-state systems. Consider a quantum system $\mathfrak{S}$ with associated Hilbert space $\mathcal{H}$. Let $(\mathcal{C}(\mathcal{H}), \mathcal{S}(\mathcal{H}))$ be the event-state system based on $\mathcal{H}$. Consider now a sentential language $\mathcal{L}^{\mathfrak{S}}$ for $\mathfrak{S}$, whose atomic sentences refer to quantum events in $\mathcal{C}(\mathcal{H})$. We can construct the following Kripkean model for $\mathfrak{S}$:

$$\mathcal{K}^{\mathfrak{S}} = \langle I \,,\, \not\perp \,,\, Pr \,,\, V \rangle, \text{ where:}$$

- I is the set $\mathcal{S}(\mathcal{H})$ of the states of $\mathfrak{S}$;
- $\not\perp$ is the similarity relation that is defined on $\mathcal{S}(\mathcal{H})$. In other words:

$$i \not\perp j \text{ iff not } \exists E \in \mathcal{C}(\mathcal{H})[i(E) = 1 \text{ and } j(E) = 0];$$

- $Pr = \mathbb{C}(\mathcal{P}(\mathcal{S}))$ (= the set of all closed subsets of $\mathcal{S}(\mathcal{H})$);

[10]See Def.129.

- for any atomic sentence $\mathbf{p}$,

$$V(\mathbf{p}) \text{ is the closed subspace which } \mathbf{p} \text{ refers to.}$$

One immediately realizes that $\mathcal{K}^{\mathfrak{S}}$ is a Kripkean model. For:

- $\not\perp$ is a similarity relation (reflexive and symmetric);
- Pr is a set of propositions, because every element X of $\mathbb{C}(\mathcal{P}(\mathcal{S}))$ is a closed set such that $X = X^{\perp\perp}$.

  Furthermore, Pr contains $\emptyset$ and I, and is closed under the operations $^{\perp}$ and $\cap$;
- for any $\mathbf{p}$, $V(\mathbf{p}) \in Pr$.

Interestingly enough, the accessibility relation turns out to have the following physical meaning: $i \not\perp j$ iff j is a state into which i can be transformed after the performance of a physical measurement that concerns an *observable* (i.e. a *physical quantity*) of the system (by application of von Neumann-Lüders axiom, the so called "collapse of the wave function").

Let us now return to our general definition of Kripkean model for **OQL**. Apparently, orthomodularity has not been characterized in terms of properties of the accessibility relation. Hence, the following important question arises:

> *is it possible to express the orthomodularity of the proposition-structure in an orthoframe* $(I, \not\perp)$ *as an elementary (first-order) property of the accessibility relation* $\not\perp$ *?*

In 1984, Goldblatt gave a negative answer to this question, proving that:

orthomodularity is not elementary.[11]

Goldblatt's theorem has revealed a kind of *metalogical intractability* of **OQL**. As a consequence of this negative result, properties like *decidability* and the *finite model property* (which are positively solved for **OL**) have stubbornly resisted to many attempts of solution in the case of **OQL**, and are still open problems.

At the same time, **OQL** seems to have some logical advantages that are not shared by the weaker **OL**. For instance, interestingly enough, a conditional connective $\rightarrow$ turns out to be definable in terms of the primitive connectives of the quantum logical language. The most natural definition (originally proposed by Finch [1970] and Mittelstaedt [1972] and further investigated by Hardegree [1976] and other authors) is the following:

$$\alpha \rightarrow \beta := \neg[\alpha \curlywedge \neg(\alpha \curlywedge \beta)]$$

(which is equivalent to $\neg\alpha \curlyvee (\alpha \curlywedge \beta)$).

[11]See [Goldblatt, 1984].

In the quantum logical literature, such a conditional connective is often called *Sasaki hook*. Of course, in classical logic (by distributivity), the Sasaki hook is equivalent to the standard *Philo's conditional* $\neg\alpha \curlyvee \beta$. Notice that this classical conditional could not represent a "good" conditional for quantum logic, because it does not generally satisfy *Modus Ponens*. One can easily show that there are worlds i of a Kripkean model $\mathcal{K}$ such that:

$$i \models \alpha; \; i \models \neg\alpha \curlyvee \beta; \; i \not\models \beta.$$

The Sasaki hook, instead, turns out to be well-behaved with respect to *Modus Ponens*, in the case of **OQL** (but not in the case of **OL**!).

Both **OL** and **OQL** are axiomatizable logics. Many axiomatizations have been proposed: in the Hilbert-Bernays style and in the Gentzen-style (*natural deduction* and *sequent-calculi*).[12]

We present here a calculus (in the natural deduction style) which is a slight modification of the version proposed by Goldblatt in 1974.

This calculus (which has no axioms) is determined as a set of *rules*. Let $T_1, \ldots, T_n$ be finite or infinite (possibly empty) sets of sentences. Any rule has the form

$$\frac{T_1 \vdash \alpha_1, \ldots, T_n \vdash \alpha_n}{T \vdash \alpha}$$

(if α_1 has been inferred from $T_1, \ldots, \alpha_n$ has been inferred from T_n, then α can be inferred from T).

We call any expression of the form $T \vdash \alpha$ a *configuration*. The configurations $T_1 \vdash \alpha_1, \ldots, T_n \vdash \alpha_n$ represent the *premisses* of the rule, while $T \vdash \alpha$ is the *conclusion*. As a limit case, we may have a rule in which the set of premisses is empty; in such a case we will speak of an *improper rule*. Instead of $\frac{\emptyset}{T \vdash \alpha}$ we will write $T \vdash \alpha$; instead of $\emptyset \vdash \alpha$, we will write $\vdash \alpha$.

Rules of **OL**

(OL1) $\quad T \cup \{\alpha\} \vdash \alpha$ (identity)

(OL2) $\quad \dfrac{T \vdash \alpha, \; T^* \cup \{\alpha\} \vdash \beta}{T \cup T^* \vdash \beta}$ (transitivity)

(OL3) $\quad T \cup \{\alpha \curlywedge \beta\} \vdash \alpha$ ($\curlywedge$-elimination)

(OL4) $\quad T \cup \{\alpha \curlywedge \beta\} \vdash \beta$ ($\curlywedge$-elimination)

[12] An axiomatization of **OQL** in the Hilbert-Bernays style see has been proposed by Hardegree in 1976 (see [Hardegree, 1976]). Sequent calculi for different forms of quantum logic have been investigated by Nishimura [1980] and by Battilotti and Sambin [1999]. See also [Battilotti and Faggian, 2002].

(OL5) $$\frac{T \models \alpha,\ T \models \beta}{T \models \alpha \curlywedge \beta}$$ ($\curlywedge$-introduction)

(OL6) $$\frac{T \cup \{\alpha, \beta\} \models \gamma}{T \cup \{\alpha \curlywedge \beta\} \models \gamma}$$ ($\curlywedge$-introduction)

(OL7) $$\frac{\{\alpha\} \models \beta,\ \{\alpha\} \models \neg\beta}{\neg\alpha}$$ (absurdity)

(OL8) $T \cup \{\alpha\} \models \neg\neg\alpha$ (weak double negation)

(OL9) $T \cup \{\neg\neg\alpha\} \models \alpha$ (strong double negation)

(OL10) $T \cup \{\alpha \curlywedge \neg\alpha\} \models \beta$ (Duns Scotus)

(OL11) $$\frac{\{\alpha\} \models \beta}{\{\neg\beta\} \models \neg\alpha}$$ (contraposition)

An axiomatization of **OQL** can be obtained by adding to the **OL**-calculus the following rule:

(OQL) $\alpha \curlywedge \neg(\alpha \curlywedge \neg(\alpha \curlywedge \beta)) \models \beta.$ (orthomodularity)

On this basis, all the standard syntactical notions (*derivation, derivability, logical theorem*) are defined in the expected way.

DEFINITION 16. *Derivation*
A *derivation* of **QL** is a finite sequence of configurations $T \models \alpha$, where any element of the sequence is either the conclusion of an improper rule or the conclusion of a proper rule whose premisses are previous elements of the sequence.

DEFINITION 17. *Derivability*
A sentence α is *derivable* from T ($T \models_{\mathbf{QL}} \alpha$) iff there is a derivation such that the configuration $T \models \alpha$ is the last element of the derivation.

Instead of $\{\alpha\} \models_{\mathbf{QL}} \beta$ we will write $\alpha \models_{\mathbf{QL}} \beta$.

DEFINITION 18. *Logical theorem*
A sentence α is a *logical theorem* of **QL** ($\models_{\mathbf{QL}} \alpha$) iff $\emptyset \models_{\mathbf{QL}} \alpha$.

A soundness and a completeness theorem have been proved for both logics with standard techniques (using the notion of *canonical model*)[13]:

THEOREM 19. Soundness theorem.

$$T \vdash_{\mathbf{QL}} \alpha \ \curvearrowright\ T \models_{\mathbf{QL}} \alpha.$$

THEOREM 20. Completeness theorem

$$T \models_{\mathbf{QL}} \alpha \ \curvearrowright\ T \vdash_{\mathbf{QL}} \alpha.$$

[13]See [Dalla Chiara and Giuntini, 2002].

To what extent does orthomodular quantum logic represent a completely faithful abstraction from QT? As we have seen, the prototypical models of **OQL** that are interesting from the physical point of view are based on the class $\mathbb{H}$ of all Hilbert lattices. Let us call *Hilbert quantum logic* (**HQL**) the logic that is semantically characterized by $\mathbb{H}$ (both in the Kripkean and in the algebraic semantics). An important problem that has been discussed at length is the following: do **OQL** and **HQL** represent one and the same logic? In 1981 Greechie gave a negative answer to this question: there is an ortholattice-theoretical equation, the so-called *orthoarguesian law*[14] that holds in $\mathbb{H}$, but fails in a particular orthomodular lattice. As a consequence, **OQL** does not represent a faithful logical abstraction from its quantum theoretical origin. The axiomatizability of **HQL** is still an open problem.

3 QUANTUM LOGIC *VS* POSITIVE, MINIMAL AND INTUITIONISTIC LOGIC

Quantum logic represents a "singular point" in the variety of nonclassical logics, giving rise to a number of *logical and metalogical anomalies*.

In order to understand some "strange" aspects of quantum logic a comparison with some logics that are at least as strong as *positive logic* (the positive fragment of intuitionistic logic) will be useful. Unlike **QL**, the language of positive logic (**PL**) contains as primitive connectives the conjunction $\curlywedge$, the disjunction $\curlyvee$ and the conditional $\rightarrow$. The notion of *Kripkean model* for **PL** is defined as follows:

DEFINITION 21. *Kripkean model* for **PL**
A *Kripkean model* for **PL** is a system $\mathcal{K} = \langle I\, , \mathcal{R}\, , Pr\, , V\rangle$, where:

(i) I is a nonempty set of possible worlds;

(ii) $\mathcal{R}$ is a reflexive and transitive relation on I;

(iii) Pr is the set of all *possible propositions*; where a *proposition* is a set X of possible worlds that is $\mathcal{R}$-closed. In other words:
$X \in Pr$ iff $X \subseteq I$ and $\forall i, j \in I[i \in X$ and $i\mathcal{R}j \curvearrowright j \in X]$;

(iv) V is a function that associates to any sentence α a proposition in Pr, satisfying the following conditions:

$$V(\beta \curlywedge \gamma) = V(\beta) \cap V(\gamma);$$
$$V(\beta \curlyvee \gamma) = V(\beta) \cup V(\gamma);$$
$$V(\beta \rightarrow \gamma) = \{i \in I : \forall j[i\mathcal{R}j \text{ and } j \in V(\beta) \curvearrowright j \in V(\gamma)]\}.$$

We will write $i \models \alpha$ instead of $i \in V(\alpha)$ (like in **QL**). The notions of *truth*, *logical truth,consequence* and *logical consequence* are then defined as in the case of **QL**.

[14]See [Greechie, 1981]. See also [Kalmbach, 1983].

As is well known, two important nonclassical logics that are stronger than **PL** are *minimal logic* (**ML**) and *intuitionistic logic* (**IL**). Minimal logic can be characterized by adding to the language of **LP** a privileged atomic sentence **f** representing the *Falsity*. The negation-connective is then defined as follows:

$$\neg\alpha := \alpha \to \mathbf{f}.$$

The Kripkean models of **ML** are defined like in the case of **PL**. In any minimal model $\mathcal{K}$, the proposition $J = V(\mathbf{f})$ (representing the meaning of the false sentence) is called the set of all *absurd worlds*. As a consequence, one immediately obtains:

$$i \models \neg\alpha \text{ iff } \forall j[i\mathcal{R}j \text{ and } j \models \alpha \curvearrowright j \in J].$$

Finally, intuitionistic logic is characterized by the class of all minimal models where the set J of all absurd worlds is empty. As a consequence, intuitionistic models turn out to satisfy the following condition:

$$i \models \neg\alpha \text{ iff } \forall j[i\mathcal{R}j \curvearrowright j \not\models \alpha].$$

We will now briefly focus upon the basic differences between the quantum logical Kripkean models and the Kripkean models of **PL**, **ML** and **IL** (briefly called *knowledge-in progress models*). As we will see, these differences are responsible for the main logical and metalogical anomalies of **QL**.

As we have seen, in any knowledge-in progress model, the accessibility relation is reflexive and transitive, and propositions are closed with respect to the accessibility relation. As a consequence, truth turns out to be conservative:

$$i \models \alpha \text{ and } i\mathcal{R}j \curvearrowright j \models \alpha.$$

Quantum logical models, instead, are based on similarity frames, where truth is not conservative and propositions X satisfy the following *stability condition*:

$$i \in X \text{ iff } \forall j[i \not\perp j \curvearrowright \exists k(j \not\perp k \text{ and } k \in X)].$$

Interestingly enough, such condition turns out to characterize propositions also in the case of knowledge-in progress models. In fact, one can prove that for any model $\mathcal{K} = \langle I\,,\,\mathcal{R}\,,\,Pr\,,\,V\rangle$ and for any set of worlds X, the two following conditions are equivalent:

I) X is $\mathcal{R}$-closed;

II) $i \in X$ iff $\forall j[i\mathcal{R}j \curvearrowright \exists k(j\mathcal{R}k \text{ and } k \in X)]$.

Of course, condition I) and II) are not equivalent in the case of **QL**.

Let us now focus on the different truth-conditions for the logical connectives.

Negation
The truth-condition for the connective $\neg$ turns out to be the same for **QL** and for **IL**:

$$i \models \neg\alpha \text{ iff for any world } j \text{ accessible to } i, j \not\models \alpha.$$

In both cases, one is dealing with a non-compositional situation: the truth of a negated sentence $\neg\alpha$ in a given world i essentially depends on the truth-status of the positive sentence α in all the worlds that are accessible to i.

Conjunction

Conjunction is the only (primitive) connective that has a truth-functional (compositional) behavior both in the quantum logical and in the knowledge-in progress models: a conjunction is true in a given world i iff both members are true in i (like in classical semantics).

Disjunction

The most important divergence between quantum logical and knowledge-in progress models concerns the behavior of disjunction. The quantum logical disjunction is essentially non-compositional: a world i may verify a disjunction $\alpha \curlyvee \beta$ even if both members of the disjunction (α, β) are indeterminate for i. The truth-status of $\alpha \curlyvee \beta$ in a given world depends on the truth-status of α and β in other worlds, according to the following condition:

$$i \models \alpha \curlyvee \beta \text{ iff } \forall j \not\perp i \exists k \not\perp j [k \models \alpha \text{ or } k \models \beta].$$

The positive (minimal and intuitionistic) disjunction is, instead, truth-functional (like in classical semantics):

$$i \models \alpha \curlyvee \beta \text{ iff } i \models \alpha \text{ or } i \models \beta.$$

It is worth-while recalling that intuitionistic disjunction is strongly *prime* even from the proof-theoretic point of view. For, $\alpha \curlyvee \beta$ is an intuitionistic logical theorem iff either α or β is an intuitionistic logical theorem.

As is well known (unlike classical and quantum disjunction) the intuitionistic disjunction is not definable in terms of negation and of conjunction (via de Morgan law). Of course, a *de Morgan disjunction* $\curlyvee_{dM}$ can be trivially defined also in the framework of **IL**:

$$\alpha \curlyvee_{dM} \beta := \neg(\neg\alpha \curlywedge \neg\beta).$$

We have: $\models_{\mathbf{IL}} \alpha \curlyvee \beta \to \alpha \curlyvee_{dM} \beta$, but not the other way around. Apparently, the truth of the intuitionistic de Morgan disjunction, does not imply the truth of at least one member of the disjunction (as happens in the case of **QL**). This failure of truth-functionality, however, is not sufficient to bring about a violation of distributivity. One can show that, in spite of their asymmetrical behavior, the intuitionistic connectives $\curlywedge$ and $\curlyvee_{dM}$ do satisfy the distributivity-relations. In fact, quantum logical non-distributivity seems to be essentially connected with the non-transitive character of the accessibility relation.

Conditional

The positive (minimal and intuitionistic) conditional is a primitive connective, whose behavior is governed by the following (non-truth functional) condition:

$$i \models \alpha \to \beta \text{ iff } \forall j \mathcal{R} i [j \models \alpha \curvearrowright j \models \beta].$$

The quantum logical conditional is, instead, defined in terms of $\neg$ and of $\curlywedge$:

$$\alpha \to \beta := \neg[\alpha \curlywedge \neg(\alpha \curlywedge \beta)].$$

As a consequence one obtains that:

$$i \models \alpha \to \beta \text{ iff } \forall j \not\perp i [j \models \alpha \curvearrowright \exists k \not\perp j (k \models \alpha \text{ and } k \models \beta)].$$

Such a condition (which is clearly weaker than the corresponding condition for the positive conditional) is responsible for most *anomalies* of the quantum logical conditional. For instance, the following *laws* that hold for *positive conditionals* are violated in **OQL**:

$\alpha \to (\beta \to \alpha)$;

$(\alpha \to (\beta \to \gamma)) \to ((\alpha \to \beta) \to (\alpha \to \gamma))$;

$(\alpha \to \beta) \to ((\beta \to \gamma) \to (\alpha \to \gamma))$;

$(\alpha \curlywedge \beta \to \gamma) \to (\alpha \to (\beta \to \gamma))$;

$(\alpha \to (\beta \to \gamma)) \to (\beta \to (\alpha \to \gamma))$.

Another interesting characteristic of the quantum logical conditional is a *weak nonmonotonic behavior*. For, we have:

$$\alpha \to \gamma \not\models_{\mathbf{QL}} \alpha \curlywedge \beta \to \gamma.$$

In 1975 Hardegree[15] has suggested that such anomalous aspects might be explained by conjecturing that the quantum logical conditional represents a kind of *counterfactual conditional.* This hypothesis seems to be confirmed by some significant physical examples. Let us consider again the Kripkean models that are associated to a quantum system $\mathfrak{S}$.

Following Hardegree, we restrict our attention to the case of pure states. As a consequence, we consider Kripkean models having the following form:

$$\mathcal{K}^{\mathfrak{S}} = \langle I \,,\, \not\perp \,,\, Pr \,,\, V \rangle \,, \text{ where :}$$

- I is the set of all pure states of $\mathfrak{S}$;
- $\not\perp$ is the nonorthogonality relation defined on I;

[15] See [Hardegree, 1975].

- Pr is the set of all *pure propositions* of the event-state system $(\mathcal{C}(\mathcal{H}), \mathcal{S}(\mathcal{H}))$. In other words: $Z \in Pr$ iff Z is a closed set of pure states (i.e., such that $Z = Z^{\perp\perp}$);
- $V(\mathbf{p})$ is the pure proposition consisting of all pure states that assign probability-value 1 to the quantum event which $\mathbf{p}$ refers to.

Hardegree has shown that, in such a case, the conditional $\rightarrow$ turns out to receive a quite natural counterfactual interpretation (in the sense of Stalnaker [16]). More precisely, one can define, for any sentence α of the language $\mathcal{L}^{\mathfrak{S}}$, a partial *Stalnaker-function* f_α in the following way:

$$f_\alpha : \mathrm{Dom}(f_\alpha) \mapsto I,$$

where:

$$\mathrm{Dom}(f_\alpha) := \{i \in I \,:\, i \not\perp V(\alpha)\}\,.$$

In other words, f_α is defined exactly for all the pure states that are not orthogonal to the proposition of α.

If $i \in \mathrm{Dom}(f_\alpha)$, then:

$$f_\alpha(i) := P^{V(\alpha)} i,$$

where $P^{V(\alpha)}$ is the projection determined by the closed subspace that is uniquely associated with the *pure proposition* $V(\alpha)$. The following condition holds:

$$i \models \alpha \rightarrow \beta \text{ iff either } \forall j \not\perp i (j \not\models \alpha) \text{ or } f_\alpha(i) \models \beta.$$

From an intuitive point of view, one can say that $f_\alpha(i)$ represents the "pure state nearest" to i, that verifies α, where "nearest" is here defined in terms of the metric of the Hilbert space $\mathcal{H}$. By definition and in virtue of von Neumann- Lüders axiom (*the collapse of the wave-function*), $f_\alpha(i)$ turns out to have the following physical meaning: it represents the transformation of state i after the performance of a measurement concerning the physical event expressed by α, provided the result was positive. As a consequence, one obtains: $\alpha \rightarrow \beta$ is true in a state i iff either α is impossible for i or the state into which i has been transformed after a positive α-test, verifies β.

As we have seen, the minimal and the intuitionistic negation can be defined (in terms of $\rightarrow$ and of $\mathbf{f}$). Is such a definition possible also in the case of **OQL**? Suppose that the language of **OQL** contains an atomic sentence $\mathbf{f}$ such that for any orthomodular model $\mathcal{K} = \langle I, \not\perp, Pr, V\rangle$ and any world i, $i \not\models \mathbf{f}$. One can easily show that the following equivalence holds (for any $\mathcal{K}$ and for any i):

$$i \models \neg\alpha \text{ iff } i \models \alpha \rightarrow \mathbf{f}.$$

Hence, an alternative description of **OQL** might assume as primitive logical constants the false sentence $\mathbf{f}$ and the positive connectives $\curlywedge$ and $\rightarrow$. Then the

[16] See [Stalnaker, 1981].

following conditions should be required for any orthomodular model $\mathcal{K}$ and for any world i of $\mathcal{K}$:

$i \not\models \mathbf{f}$;

$i \models \beta \curlywedge \gamma$ iff $i \models \beta$ and $i \models \gamma$;

$i \models \beta \rightarrow \gamma$ iff $\forall j \not\perp i[j \models \beta \curvearrowright \exists k \not\perp j(k \models \beta$ and $k \models \gamma)]$.

On this basis, both the quantum logical negation and the quantum logical disjunction can be dealt with as defined logical constants. Needless to stress, such a construction would not be possible in the case of orthologic, that does not admit a well-behaved conditional.

4 THE "LINDENBAUM-ANOMALY"

The consistency-property behaves differently in the framework of different logics. Let **L** represent either **QL** or **ML** or **IL**. The following concepts characterize different semantic aspects of the intuitive idea of consistency. Let T be a set of sentences of the logic **L**.

DEFINITION 22.

- T is *verifiable* in **L** iff there is an **L**-model $\mathcal{K}$ such that $\models_{\mathcal{K}} T$.
- T is *realizable* in **L** iff there is an **L**-model $\mathcal{K}$ and a world i of $\mathcal{K}$ such that $i \models T$.
- T is *semantically noncontradictory* in **L** iff for any sentence α, $T \not\models_{\mathbf{L}} \alpha \curlywedge \neg\alpha$.
- T is *semantically α-consistent* in **L** iff $T \not\models_{\mathbf{L}} \alpha$.
- T is *semantically consistent* in **L** iff there exists a sentence α such that T is semantically α-consistent.

As is well known, **IL** gives rise to a strong relation between these different aspects of the intuitive notion of consistency (relation that also holds in the case of **CL**).

THEOREM 23. *The following conditions are equivalent for* **IL**:

- *T is verifiable.*
- *T is realizable.*
- *T is semantically noncontradictory.*
- *T is semantically consistent.*
- *T is semantically* **f**-*consistent.*

The equivalence between semantic noncontradictory and semantic consistent depends on the fact that intuitionistic logic is a *scotian* logic, where "ex absurdo sequitur quodlibet" ($\alpha \curlywedge \neg\alpha \models_{\mathbf{IL}} \beta$ and $\mathbf{f} \models_{\mathbf{IL}} \beta$). Minimal logic, instead, is an important example of a *non-scotian* logic (for, $\alpha \curlywedge \neg\alpha \not\models_{\mathbf{ML}} \beta$ and $\mathbf{f} \not\models_{\mathbf{ML}} \beta$).

At the same time, the noncontradiction principle, which is an immediate consequence of *Modus Ponens* ($\alpha \curlywedge (\alpha \to \mathbf{f}) \to \mathbf{f}$) holds for **ML**. Furthermore, the sentence $\mathbf{f}$ turns out to be equivalent to the contradictory sentence $\mathbf{f} \curlywedge \neg\mathbf{f}$ ($:= \mathbf{f} \curlywedge (\mathbf{f} \to \mathbf{f})$) and to any contradiction expressed in a negative form ($\neg\alpha \curlywedge \neg\neg\alpha$).

As a consequence, the following weaker theorem holds for **ML**:

THEOREM 24.

- *T is verifiable iff T is realizable iff T is semantically noncontradictory iff T is $\mathbf{f}$-consistent.*
- *If T is semantically noncontradictory, then T is semantically consistent, but generally not the other way around.*

On this basis, one can prove that both **ML** and **IL** satisfy a strong metalogical condition, that we will call *pre-Lindenbaum*. Let **L** be either **ML** or **IL**.

LEMMA 25. *Pre-Lindenbaum Lemma*
*If T is semantically noncontradictory in **L** and $T \not\models_{\mathbf{L}} \neg\alpha$, then $T \cup \{\alpha\}$ is semantically noncontradictory in **L**.*

An important consequence of the pre-Lindenbaum Lemma is a strong relation between two relevant metalogical properties: *logical completeness* and *maximality*.

DEFINITION 26.

- T is *logically complete* in the logic **L** iff for any sentence α, either $T \models_{\mathbf{L}} \alpha$ or $T \models_{\mathbf{L}} \neg\alpha$.
- T is *maximal* in the logic **L** iff for any sentence α, the set $T \cup \{\alpha\}$ is semantically contradictory, if $T \not\models_{\mathbf{L}} \neg\alpha$.

Maximality and logical completeness turn out to be equivalent both in **ML** and in **IL**.

THEOREM 27. *A set T of sentences is maximal in **L** iff T is logically complete in **L**.*

The proof of the left-to-right implication essentially uses the pre-Lindenbaum Lemma.

On this basis, both a weak and a strong version of the Lindenbaum-theorem can be proved for minimal and intuitionistic logic. Let **L** be either **ML** or **IL**.

THEOREM 28. *Weak Lindenbaum*
*For any set of sentences T that is semantically noncontradictory in **L**, there exists a set of sentences T' such that:*

- *$T \subseteq T'$;*
- *T' is semantically noncontradictory in* **L***;*
- *T' is maximal in* **L**.

As is well known, the standard proof of the Lindenbaum-theorem refers to an enumeration of the sentences of the language $(\beta_1, \beta_2, \dots)$. On this basis, an infinite sequence of set of sentences is defined:

$T_0 = T$

$$T_{n+1} = \begin{cases} T_n \cup \{\beta_{n+1}\}, & \text{if } T_n \cup \{\beta_{n+1}\} \text{ is semantically noncontradictory;} \\ T_n, & \text{otherwise.} \end{cases}$$

By putting $T' = \bigcup_n T_n$, one can prove that T' is noncontradictory and maximal.[17]

By Lemma 25 one immediately obtains that also the strong version of the Lindenbaum-theorem holds for **L**.

THEOREM 29. *Strong Lindenbaum*
For any set of sentences T that is semantically noncontradictory in **L**, *there exists a set of sentences T' such that:*

- *$T \subseteq T'$;*
- *T' is semantically noncontradictory in* **L***;*
- *T' is logically complete in* **L**.

In other words, any noncontradictory set of sentences T can be extended to a set T', where any problem (expressed in the language) is *semantically decided*.

What happens in the case of **QL**? As we already know, **QL** is a scotian logic (where $\alpha \curlywedge \neg\alpha \models \beta$). Hence, semantically contradictory and semantically consistent collapse into one and the same concept (like in **IL** and in **CL**). At the same time (unlike **ML** and **IL**) verifiability and realizability split into two different concepts.

THEOREM 30. *In* **QL***:*

I) T is semantically noncontradictory iff T is realizable.

II) If T is verifiable, then T is realizable, but generally not the other way around.

Proof. The proof of I) is trivial. As to II), verifiability trivially implies realizability. A counterexample to the opposite implication is given by the following sentence γ (which represents the negation of the *a-fortiori principle*):

$$\gamma = \neg[\alpha \to (\beta \to \alpha)] = \neg\neg[\alpha \curlywedge \neg(\alpha \curlywedge \neg(\beta \curlywedge \neg(\beta \curlywedge \alpha)))].$$

---

[17] Of course, the proof makes essential use of the completeness theorem for the logic **L** (or, alternatively, of the semantic *compactness theorem*).

Consider a Kripkean model $\mathcal{K}^{\mathfrak{S}} = \langle I, \not\perp, Pr, V \rangle$, associated to a quantum system $\mathfrak{S}$ whose state-event system is $(\mathcal{S}(\mathcal{H}), \mathcal{C}(\mathcal{H}))$, where $\mathcal{H}$ is the two-dimensional Hilbert space $\mathbb{C}^2$. Suppose that α and β are two atomic sentences such that $V(\alpha)$ and $V(\beta)$ are propositions corresponding to two different non-orthogonal one-dimensional closed subspaces of $\mathcal{H}$. One immediately obtains that:

$$V(\gamma) = V(\alpha) \neq \emptyset.$$

Hence, there exists a pure state of $\mathfrak{S}$ that verifies γ, and consequently γ is realizable. At the same time, γ cannot be verifiable. For, one can easily show that the hypothesis that there exists a model $\mathcal{K} = \langle I, \not\perp, Pr, V \rangle$ such that $V(\gamma) = I$, leads to a contradiction. ■

As a consequence, one immediately obtains the failure of the strong Lindenbaum-property for **QL**. Suppose, by contradiction, that any semantically noncontradictory T admits a semantically noncontradictory and logically complete extension T'. One can easily show that any semantically noncontradictory and logically complete T' is verifiable by a *classical Kripkean model* $\mathcal{K} = \langle I, \not\perp, Pr, V \rangle$ such that:

- I is a singleton set $\{i\}$, where i is any object;
- $\not\perp$ is the identity relation on I;
- Pr is the power-set of I;
- $V(\mathbf{p}) = \begin{cases} \{i\}, & \text{if } T' \models_{\mathbf{QL}} \mathbf{p}; \\ \emptyset, & \text{otherwise.} \end{cases}$

Since T' is semantically noncontradictory and logically complete, $\mathcal{K}$ is well defined. Moreover we have:

$$i \models \alpha \text{ iff } \models_{\mathcal{K}} \alpha \text{ iff } T' \models_{\mathbf{QL}} \alpha.$$

Consequently, T' is verifiable.

Although the strong version of the Lindenbaum-theorem is violated in **QL**, the weak version of the theorem is valid. In other words, any semantically noncontradictory set of sentences T can be extended to a set T' that is semantically noncontradictory and maximal. The proof runs like in the case of **ML** and of **IL**. However (unlike **ML**, **IL** and **CL**) maximality does not imply logical completeness in **QL**.

THEOREM 31. *If T is logically complete (in* **QL***), then T is maximal (in* **QL***), but generally not the other way around.*

Proof. The proof that logical completeness implies maximality is trivial. Maximality, instead, cannot imply logical completeness. For, as we have seen, there are examples of semantically noncontradictory sentences γ that admit a maximal noncontradictory extension and do not admit any noncontradictory logically complete extension. ■

Also the pre-Lindenbaum property is violated in **QL**. A counterexample can be obtained as follows.

Put $T = \{\beta\}$ and $\alpha = \neg(\gamma \to \beta)$. We have $\beta \not\models_{\mathbf{QL}} \neg\alpha$, because $\beta \not\models_{\mathbf{QL}} \gamma \to \beta$, as shown by counterexamples in $\mathbb{C}^2$. At the same time, $T \cup \{\alpha\}$ is clearly semantically contradictory.

Only the following weak pre-Lindenbaum condition holds for **QL**.

THEOREM 32. *If T is semantically noncontradictory and $T \not\models_{\mathbf{QL}} \alpha$, then there exists a set of sentences T' such that:*

1) $T' \models_{\mathbf{QL}} \neg\alpha$;

2) T and T' are logically compatible. *In other words, for any sentence β, if $T \models_{\mathbf{QL}} \beta$, then $T' \not\models_{\mathbf{QL}} \neg\beta$.*[18]

5 QUANTUM LOGIC AND THE HIDDEN VARIABLE PROBLEM

The failure of the Lindenbaum property in **QL** has represented a powerful metalogical tool that has been used to prove the impossibility of *completing* QT via some *(non-contextual) hidden variable hypotheses*.[19]

The debate concerning the question whether QT can be considered a *physically complete* account of microphenomena has a long and deep history. A turning point in this discussion has been the celebrated Einstein-Bohr debate, with the ensuing charge of incompleteness raised by the Einstein-Podolsky-Rosen argument (EPR).

As we already know, in the framework of orthodox QT, physical systems can be prepared in *pure states* that have, in general, positive dispersion for most physical quantities. In the EPR argument, the attention is focused on the question whether the account of the microphysical phenomena provided by QT is to be regarded as an *exhaustive* description of the physical reality to which those phenomena are supposed to refer, a question to which Einstein himself answered in the negative.

There is a mathematical side of the completeness issue: the question becomes whether *states with positive dispersion* can be represented as a different, *dispersion-free*, kind of states in a way that is *consistent* with the mathematical constraints of the quantum theoretical formalism. In his book on the mathematical foundations of quantum mechanics, von Neumann proved a celebrated "No go theorem" asserting the logical incompatibility between the quantum formalism and the existence of dispersion-free states (satisfying some general conditions). Already in the preface, von Neumann anticipates the program and the conclusion concerning the possibility of 'neutralizing' the statistical character of QT:

> There will be a detailed discussion of the problem as to whether it is possible to trace the statistical character of quantum mechanics to an ambiguity (i.e., incompleteness) in our description of nature. Indeed,

[18] See [Dalla Chiara and Giuntini, 2002].

[19] See, for instance, [Giuntini, 1991b].

> such an interpretation would be a natural concomitant of the general principle that each probability statement arises from the incompleteness of our knowledge. This explanation "by hidden parameters" [...] has been proposed more than once. However, it will appear that this can scarcely succeed in a satisfactory way, or more precisely, such an explanation is incompatible with certain qualitative fundamental postulates of quantum mechanics.

According to the advocates of hidden variables, QT is a *physically incomplete theory*. The intuitive idea that represents the common background to almost all hidden variable theories can be described in the following way:

(I) the reason why a physical theory is *statistical* depends on the fact that the description provided by the states is *incomplete*.

(II) It is possible to add a set Ξ of parameters (*hidden variables*) in such a way that

- for every state s and for every $\omega \in \Xi$, there exists a dispersion-free (dichotomous) state s_ω which semantically decides every *property* (*event*) of the physical system at issue;
- the statistical predictions of the original theory should be recovered by averaging over these dichotomous states;
- the algebraic structures determined by the properties (events) of the system should be preserved in the hidden variable extension.

The hidden variable theories based on the assumptions (I) and (II) are usually called *non-contextual*, because they require the existence of a single space Ξ of hidden variables determining dispersion-free states. A weaker position is represented by the *contextual hidden variable theories*, according to which the choice of the hidden variable space depends on the physical quantity to be dealt with. As pointed out by Beltrametti and Cassinelli [1981]:

> Despite the absence of mathematical obstacles against contextual hidden variable theories, it must be stressed that their calling for completed states that are probability measures not on the whole proposition [event] lattice $\mathcal{E}$ but only on a subset of $\mathcal{E}$ is rather far from intuitive physical ideas of what a state of a physical system should be. Thus, contextual hidden variable theorists, in their search for the restoration of some classical deterministic aspects, have to pay, on other sides, in quite radical departures from properties of classical states.

Von Neumann's proof of his "No go theorem" was based on a general assumption that has been, later, considered too strong. The condition asserts the following:

Let s_ω be a dispersion-free state and let A, B be two (possibly noncompatible) observables. Then, $Exp(A + B, s_\omega) = Exp(A, s_\omega) + Exp(B, s_\omega)$.

In other words, the *expectation functional Exp* determined by the completed state s_ω is linear.

In the late Sixties, Kochen and Specker published a series of articles, developing a purely logical argument for a "No go theorem," such that von Neumann's strong assumption can be relaxed.[20]

Kochen and Specker's proof is based on a variant of quantum logic, that has been called *partial classical logic* (**PCL**). The basic semantic idea is the following: unlike orthologic and orthomodular quantum logic (which are *total logics*, because the meaning of any sentence is always defined), molecular sentences of **PCL** can be semantically undefined. From the semantic point of view, the crucial relation is represented by a *compatibility* relation, that may hold between the meanings of two sentences. As expected, the intended physical interpretation of the compatibility relation is the following: two sentences α and β have *compatible meanings* iff α and β can be simultaneously tested. Models of **PCL** are special kinds of algebraic models based on *partial Boolean algebras* (weaker versions of Boolean algebras where the meet and the join are only defined for pairs of compatible elements).

All these investigations have revealed that there is a *deep logical connection* between the two following questions:

- does a quantum system $\mathfrak{S}$ admit a non-contextual hidden variable theory?
- Does **PCL** satisfy a version of the Lindenbaum property with respect to the algebraic models concerning the events that may occur to the system $\mathfrak{S}$?

6 THE UNSHARP APPROACHES TO QT

The essential indeterminism of QT gives rise to a kind of *ambiguity* of the quantum world. Such ambiguity can be investigated at different levels. The first level concerns the characteristic features of quantum pure states, which represent pieces of information that are at the same time *maximal* and *logically incomplete*. As we have seen, the divergence between maximality and logical completeness is the origin of most logical anomalies of the quantum phenomena.

A second level of ambiguity is connected with a possibly *fuzzy* character of the physical events that are investigated. We can try and illustrate the difference between two "fuzziness-levels" by referring to a nonscientific example. Let us consider the two following sentences, which apparently have no definite truth-value:

I) Hamlet is 1.70 meters tall;

II) Brutus is an honourable man.

The semantic uncertainty involved in the first example seems to depend on the logical incompleteness of the *individual concept* associated to the name "Hamlet."

[20] See Kochen and Specker [1965a; 1965b; 1967].

In other words, the property "being 1.70 meters tall" is a *sharp* property. However, our concept of Hamlet is not able to *decide* whether such a property is satisfied or not. Unlike real persons, literary characters have a number of indeterminate properties. On the contrary, the semantic uncertainty involved in the second example, is mainly caused by the ambiguity of the concept "honourable." What does it mean "being honourable?" One need only recall how the ambiguity of the adjective "honourable" plays an important role in the famous Mark Anthony's monologue in Shakespeare's "Julius Caesar." Now, orthodox QT generally takes into consideration examples of the first kind (our first level of fuzziness): events are sharp, while all semantic uncertainties are due to the *logical incompleteness* of the individual concepts, that correspond to pure states of quantum objects. This is the reason why orthodox QT is sometimes called *sharp* QT, in contrast with *unsharp* QT, which also investigates examples of the second kind (second level of fuzziness).

Strangely enough, the abstract researches on fuzzy logics and on quantum structures have undergone quite independent developments for many decades during the 20-th century. Only after the Eighties, there emerged an interesting convergence between the investigations about fuzzy and quantum structures, in the framework of the so called *unsharp approach to quantum theory*. In this connection a significant conjecture has been proposed: perhaps some apparent *mysteries* of the quantum world should be described as special cases of some more general *fuzzy phenomena*, whose behavior has not yet been fully understood.

In 1983 the German physicist G. Ludwig published the book *Foundations of Quantum Mechanics*, which has been later regarded as the birth of the unsharp approach to QT. Paradoxically enough, Ludwig has always been an "enemy" of quantum logic. In spite of this, his ideas have greatly contributed to the revival of the quantum logical investigations during the last two decades. Ludwig's pioneering work has been further developed by many scholars (Kraus, Davies, Mittelstaedt, Busch, Lahti, Bugajski, Beltrametti, Nisticò, Foulis, Bennett, Gudder, Greechie, Pulmannová, Dvurečenskij, Riečan, Riečanova, Schroeck and many others including the authors of this chapter).

The starting point of the unsharp approach is deeply connected with a general problem that naturally arises in the framework of Hilbert space QT. Let us consider the event-state system $(\Pi(\mathcal{H}), \mathcal{S}(\mathcal{H}))$ of a quantum system $\mathfrak{S}$, where $\Pi(\mathcal{H})$ is the set of projections, while $\mathcal{S}(\mathcal{H})$ is the set of density operators of the Hilbert space $\mathcal{H}$ (associated to $\mathfrak{S}$). One can ask the following question: do the sets $\Pi(\mathcal{H})$ and $\mathcal{S}(\mathcal{H})$ correspond to an *optimal* possible choice of adequate mathematical representatives for the intuitive notions of *event* and of *state*, respectively?

Consider first the notion of *state*. Once $\Pi(\mathcal{H})$ is fixed, Gleason's Theorem[21] guarantees that any probability measure defined on $\Pi(\mathcal{H})$ is determined by a density operator of $\mathcal{H}$ (provided the dimension of $\mathcal{H}$ is greater than or equal to 3). Hence, $\mathcal{S}(\mathcal{H})$ corresponds to an *optimal* notion of state.

Let us discuss then the notion of *event* and let us ask whether $\Pi(\mathcal{H})$ represents

[21] See [Gleason, 1957] and [Dvurečenskij, 1993].

the largest set of operators assigned a probability-value, according to the Born rule. The answer to this question is negative.

One can easily recognize the existence of bounded linear operators E that are not projections and that satisfy the following condition:

$$\text{for any density operator } \rho, \ \mathtt{tr}(\rho E) \in [0,1].$$[22]

Recalling the Born rule, this means that such operators E "behave as possible events," because any state assigns to them a probability value.

An interesting example of this kind is represented by the operator $\frac{1}{2}\mathbb{I}$ (where $\mathbb{I}$ is the identity operator). One immediately realizes that $\frac{1}{2}\mathbb{I}$ is a linear bounded operator that is *not* a projection, because:

$$\frac{1}{2}\mathbb{I}\frac{1}{2}\mathbb{I} = \frac{1}{4}\mathbb{I} \neq \frac{1}{2}\mathbb{I}$$

(hence $\frac{1}{2}\mathbb{I}$ fails to be idempotent).

At the same time, for any density operator ρ we have:

$$\mathtt{tr}(\rho\frac{1}{2}\mathbb{I}) = \frac{1}{2}.$$

Thus, $\frac{1}{2}\mathbb{I}$ seems to represent a totally *indeterminate* event, to which each state assigns probability $\frac{1}{2}$. Apparently, the event $\frac{1}{2}\mathbb{I}$ plays the role that, in fuzzy set theory, is played by the *semitransparent* fuzzy set $\frac{1}{2}\mathbf{1}$ such that for any object x of the universe:

$$\frac{1}{2}\mathbf{1}(x) = \frac{1}{2}.$$

This situation suggests that we liberalize the notion of *quantum event* and extend the set $\Pi(\mathcal{H})$ to a new set of operators. Following Ludwig, the elements of this new set have been called *effects*. The precise mathematical definition of effect is the following:

DEFINITION 33. *Effects*

An *effect* of $\mathcal{H}$ is a bounded linear operator E that satisfies the following condition, for any density operator ρ:

$$\mathtt{tr}(\rho E) \in [0,1].$$

We denote by $\mathcal{E}(\mathcal{H})$ the set of all effects of $\mathcal{H}$.

Clearly, $\mathcal{E}(\mathcal{H})$ properly includes $\Pi(\mathcal{H})$. Because:

- any projection satisfies the definition of effect;
- there are examples of effects that are not projections (for instance the effect $\frac{1}{2}\mathbb{I}$, that is usually called *the semitransparent effect*).

[22] See Deff. 152, 153, 158.

By definition, effects turn out to represent a kind of maximal mathematical representative for the notion of quantum event, in agreement with the basic statistical rule of QT (the *Born rule*).

Unlike projections, effects represent quite general mathematical objects that describe at the same time events and states. Let E be any effect in $\mathcal{E}(\mathcal{H})$. The following conditions hold:

- E represents a sharp event ($\in \Pi(\mathcal{H})$) iff E is idempotent ($EE = E$);
- E is a density operator (representing a state) iff $\mathtt{tr}(E) = 1$;
- E represents a pure state iff E is at the same time a projection and a density operator.

6.1 *Algebraic effect-structures*

There are different algebraic structures that can be induced on the set of all effects in a Hilbert space.

One immediately realizes that the set $\mathcal{E}(\mathcal{H})$ can be naturally structured as a *regular involution bounded poset*[23]:

$$\langle \mathcal{E}(\mathcal{H}), \leq, ', \mathbf{0}, \mathbf{1} \rangle ,$$

where

(i) $\leq$ is the *natural* order determined by the set of all density operators. In other words:

$$E \leq F \text{ iff for any density operator } \rho \in D(\mathcal{H}), \mathtt{tr}(\rho E) \leq \mathtt{tr}(\rho F).$$

(i.e., any state assigns to E a probability-value that is less or equal than the probability-value assigned to F);

(ii) $E' = \mathbf{1} - E$ (where $-$ is the standard operator difference);

(iii) $\mathbf{0}$, $\mathbf{1}$ are the null projection ($\mathbb{O}$) and the identity projection ($\mathbb{I}$), respectively.

One can easily check that:

- $\leq$ is a partial order;
- $'$ is an involution;
- $\mathbf{0}$ and $\mathbf{1}$ are respectively the minimum and the maximum with respect to $\leq$;
- the *regularity* condition holds. In other words:

$$E \leq E' \text{ and } F \leq F' \text{ implies } E \leq F'.$$

[23]See Def. 117- 124.

The effect poset $\mathcal{E}(\mathcal{H})$ turns out to be properly fuzzy. The noncontradiction principle is violated: for instance the semitransparent effect $\frac{1}{2}\mathbb{I}$ satisfies the following condition:

$$\frac{1}{2}\mathbb{I} \wedge (\frac{1}{2}\mathbb{I})' = \frac{1}{2}\mathbb{I} \wedge \frac{1}{2}\mathbb{I} = \frac{1}{2}\mathbb{I} \neq \mathbf{0}.$$

This is one of the reasons why proper effects (those that are not projections) may be regarded as representing *unsharp physical events*. Accordingly, we will also call the involution operation of an effect-structure a *fuzzy complement*.

At the same time, the effect-poset fails to be a lattice. As proved by Greechie and Gudder in 1996, some pairs of effects have no meet.[24]

The effect poset $\mathcal{E}(\mathcal{H})$ can be expanded to a richer structure, equipped with a new complement $^{\sim}$, that has an intuitionistic-like behavior. Such operation $^{\sim}$ has been called the *Brouwer complement*.[25]

DEFINITION 34. *The Brouwer complement*

$$\forall E \in \mathcal{E}(\mathcal{H}) : E^{\sim} = P_{Ker(E)}.$$

In other words, the Brouwer complement of E is the projection operator $P_{Ker(E)}$ whose range is $Ker(E)$, the kernel of E.[26]

By definition, the Brouwer complement of an effect is always a projection. In the particular case, when E is a projection, it turns out that $E' = E^{\sim}$, in other words, the fuzzy and the intuitionistic complement collapse into one and the same operation.

The structure $\langle \mathcal{E}(\mathcal{H}), \leq, ', ^{\sim}, \mathbf{0}, \mathbf{1} \rangle$ turns out to be a particular example of a kind of abstract structure that has been termed *Brouwer Zadeh poset*.[27] The abstract definition of *Brouwer Zadeh posets* is the following:

DEFINITION 35. *Brouwer Zadeh poset*
A *Brouwer Zadeh poset* (or *BZ-poset*) is a structure

$$\langle B, \leq, ', ^{\sim}, \mathbf{0}, \mathbf{1} \rangle,$$

where

(i) $\langle B, \leq, ', \mathbf{0}, \mathbf{1} \rangle$ is a regular poset;

(ii) $^{\sim}$ is a unary operation that behaves like an intuitionistic complement:

- (iia) $a \wedge a^{\sim} = \mathbf{0}$;
- (iib) $a \leq a^{\sim\sim}$;
- (iic) $a \leq b$ implies $b^{\sim} \leq a^{\sim}$.

---

[24]See [Gudder and Greechie, 1996].
[25]See [Cattaneo and Nisticó, 1989].
[26]The *kernel* of E is the set of all vectors of $\mathcal{H}$ that are transformed by E into the null vector.
[27]See [Cattaneo and Nisticó, 1989].

(iii) The following relation connects the fuzzy and the intuitionistic complement:

$$a^{\sim\prime} = a^{\sim\sim}.$$

Of course, any BZ-poset $\langle B\,,\,\leq\,,\,'\,,\,^{\sim}\,,\,\mathbf{0}\,,\mathbf{1}\rangle$ where the two complements $'$ and $^{\sim}$ coincide, turns out to be an orthoposet (i.e. a bounded involution poset, where the involution $'$ satisfies the noncontradiction and the excluded middle principles).

One can prove that the concrete effect-structure

$$\langle \mathcal{E}(\mathcal{H})\,,\,\leq\,,\,'\,,\,^{\sim}\,,\,\mathbf{0}\,,\mathbf{1}\rangle$$

is a Brouwer Zadeh poset, that is not an orthoposet.

An interesting feature of the Brouwer Zadeh structures is the possibility to define two unary operations ν and μ, which turn out to behave as the modal operators *necessarily* and *possibly*, respectively.

DEFINITION 36. *The modal operators*
Let $\langle B\,,\,\leq\,,\,'\,,\,^{\sim}\,,\,\mathbf{0}\,,\mathbf{1}\rangle$ be a Brouwer Zadeh poset.

$$\nu(a) := a^{\prime\sim}; \quad \mu(a) := a^{\sim\prime}.$$

In other words, *necessity* is identified with the intuitionistic negation of the fuzzy negation, while *possibility* is identified with the fuzzy negation of the intuitionistic negation.

The modal operators ν and μ turn out to have a typical $\mathbf{S_5}$-like behavior. For, the following conditions are satisfied:

- $\nu(a) \leq a$

  Necessarily a implies a.

- If $a \leq b$, then $\nu(a) \leq \nu(b)$

  If a implies b, then the necessity of a implies the necessity of b.

- $a \leq \nu(\mu(a))$

  a implies the necessity of its possibility.

- $\nu(\nu(a)) = \nu(a)$

  Necessity is equivalent to the necessity of the necessity.

- $\nu(\mu(a)) = \mu(a)$

  The necessity of a possibility is equivalent to the possibility.

Of course, in any BZ-poset $\langle B\,,\,\leq\,,\,'\,,\,^{\sim}\,,\,\mathbf{0}\,,\mathbf{1}\rangle$ where the two complements $'$ and $^{\sim}$ coincide, we obtain a collapse of the modalities. In other terms, $\nu(a) = a = \mu(a)$.

Let us now return to concrete (Hilbert-space) Brouwer Zadeh posets

$$\langle \mathcal{E}(\mathcal{H})\,,\,\leq\,,\,'\,,\,^{\sim}\,,\,\mathbf{0}\,,\mathbf{1}\rangle,$$

and consider the necessity $\nu(E)$ of a given effect E (which may be either sharp or unsharp). One can easily prove the following lemma.

LEMMA 37.

(i) *E is a projection iff* $E = \nu(E) = E'^{\sim} = P_{Ker(E')}$.

(ii) *Let P be any projection.*

$$P \leq E \textit{ implies } P \leq \nu(E).$$

As a consequence, we can say that $\nu(E)$ represents a kind of "best sharp lower approximation of E."

Brouwer Zadeh posets do not represent the only interesting way of structuring the set of all concrete effects. Other important structures that have naturally emerged from effect-systems are *effect algebras* and *quantum MV algebras*. Such structures (introduced in the late Eighties and in the Nineties) have represented a privileged object of research for the logico-algebraic approach to QT at the turn of the century.

We will first sketch the definition of effect algebra (also called *unsharp orthoalgebras*)[28]. One is dealing with a particular kind of *partial structure*, equipped with a basic operation $\boxplus$ that is only defined for special pairs of elements. From an intuitive point of view, such an operation can be regarded as an *exclusive disjunction* (*aut*), defined for events that are *logically incompatible*.

The abstract definition of *effect algebra* is the following.

DEFINITION 38. *Effect algebra*
An *effect algebra* is a partial structure

$$\mathcal{A} = \langle A\,, \boxplus\,, \mathbf{0}\,, \mathbf{1}\rangle\,,$$

where $\boxplus$ is a partial binary operation on A, and $\mathbf{0}$ and $\mathbf{1}$ are special distinct elements of A. When $\boxplus$ is defined for a pair $a\,, b \in A$, we will write $\exists(a \boxplus b)$.

The following conditions hold:

(i) *Weak commutativity*
$\exists(a \boxplus b)$ implies $\exists(b \boxplus a)$ and $a \boxplus b = b \boxplus a$;

(ii) *Weak associativity*
$\exists(b \boxplus c)$ and $\exists(a \boxplus (b \boxplus c))$ implies $\exists(a \boxplus b)$ and $\exists((a \boxplus b) \boxplus c)$ and $a \boxplus (b \boxplus c) = (a \boxplus b) \boxplus c$;

(iii) *Strong excluded middle*
For any a, there exists a unique x such that $a \boxplus x = \mathbf{1}$;

(iv) *Weak consistency*
$\exists(a \boxplus \mathbf{1})$ implies $a = \mathbf{0}$.

[28]See [Giuntini and Greuling, 1989; Foulis and Bennett, 1994; Dalla Chiara and Giuntini, 1994; Dvurečenskij, 2000; Dalla Chiara *et al.*, 2004].

An orthogonality relation $\perp$, a partial order relation $\leq$ and a generalized complement $'$ (which generally behaves as a fuzzy complement) can be defined in any effect algebra.

DEFINITION 39. Let $\langle A, \boxplus, \mathbf{0}, \mathbf{1}\rangle$ be an effect algebra and let $a, b \in A$.

(i) $a \perp b$ iff $a \boxplus b$ is defined in A.

(ii) $a \leq b$ iff $\exists c \in A$ such that $a \perp c$ and $b = a \boxplus c$.

(iii) The *generalized complement* of a is the unique element a' such that $a \boxplus a' = \mathbf{1}$.

One can show that any effect algebra $\langle A, \boxplus, \mathbf{0}, \mathbf{1}\rangle$ gives rise to a bounded involution poset $\langle A, \leq, ', \mathbf{0}, \mathbf{1}\rangle$, where $\leq$ and $'$ are defined according to Definition 39.

(iii) The category of all effect algebras turns out to be (categorically) equivalent to the category of all *difference posets*, which have been first studied by Kôpka and Chovanec and further investigated by Pulmannová and others.[29]

Effect algebras represent weak examples of *orthoalgebras*, a category of partial structures that Foulis and Randall had introduced in 1981.[30] Roughly, orthoalgebras are effect algebras that satisfy the noncontradiction principle. In such algebras, the involution $'$ becomes an orthocomplementation.

The precise mathematical definition is the following:

DEFINITION 40. *Orthoalgebras*
An *orthoalgebra* is an effect algebra $\langle A, \boxplus, \mathbf{0}, \mathbf{1}\rangle$ such that the following condition is satisfied:

$$\exists(a \boxplus a) \text{ implies } a = \mathbf{0} \quad (\textit{Strong consistency}).$$

In other words: $\mathbf{0}$ is the only element that is orthogonal to itself.

One can easily realize that orthoalgebras always determine an orthoposet. Let $\mathcal{A} = \langle A, \boxplus, \mathbf{0}, \mathbf{1}\rangle$ be an orthoalgebra. The structure

$$\langle A, \leq, ', \mathbf{0}, \mathbf{1}\rangle$$

(where $\leq$ and $'$ are the partial order and the generalized complement of $\mathcal{A}$) is an *orthoposet*. For, given any $a \in A$, the infimum $a \wedge a'$ exists and is equal to $\mathbf{0}$; equivalently, the supremum $a \vee a'$ exists and is equal to $\mathbf{1}$.

THEOREM 41. *Any orthoalgebra $\mathcal{A} = \langle A, \boxplus, \mathbf{0}, \mathbf{1}\rangle$ satisfies the following condition: if $a, b \in A$ and $a \perp b$, then $a \boxplus b$ is a minimal upper bound for a and b in $\mathcal{A}$.*

[29] See [Kôpka and Chovanec, 1994],[Pulmannová, 1995].
[30] See [Foulis and Randall, 1981].

COROLLARY 42. *Any orthoalgebra* $\mathcal{A} = \langle A\,,\boxplus\,,\mathbf{0}\,,\mathbf{1}\rangle$ *satisfies the following condition: for any* $a, b \in A$ *such that* $a \perp b$*, if the supremum* $a \vee b$ *exists, then* $a \vee b = a \boxplus b$.

Orthoalgebras and orthomodular posets turn out to be deeply connected. Any orthomodular poset $\langle A\,,\leq\,,'\,,\mathbf{0}\,,\mathbf{1}\rangle$ determines an orthoalgebra

$$\langle A\,,\boxplus\,,\mathbf{0}\,,\mathbf{1}\rangle\,,$$

where: $a \boxplus b$ is defined iff $a \leq b'$. Furthermore, when defined, $a \boxplus b = a \vee b$. At the same time, not every orthoalgebra is an orthomodular poset (as shown by Wright in 1990[31]).

Genuine examples of effect algebras (which are not generally orthoalgebras) can be naturally obtained in the domain of fuzzy set systems.

EXAMPLE 43. *Effect algebras of fuzzy sets*
Let B be the set of all fuzzy subsets of a universe U (in other words, B is the set of all functions assigning to any element of U a value in the real interval $[0, 1]$).

A partial operation $\boxplus$ can be defined on B. For any $f, g \in B$:

$$\exists(f \boxplus g) \text{ iff } \forall x \in U:\ f(x) + g(x) \leq 1,$$

where $+$ is the usual sum of real numbers. Furthermore:

$$\text{if } \exists(f \boxplus g), \text{ then } \quad f \boxplus g := f + g,$$

where:

$$\forall x \in U\,\{(f+g)(x) := f(x) + g(x)\}\,.$$

Let $\mathbf{1}$ be the classical characteristic function of the total set U, while $\mathbf{0}$ is the classical characteristic function of the empty set $\emptyset$.

The structure $\langle B\,,\boxplus\,,\mathbf{0}\,,\mathbf{1}\rangle$ is an effect algebra.
It turns out that the effect-algebra generalized complement $'$ coincides with the fuzzy complement. In other words:

$$\forall x \in U : f'(x) = 1 - f(x).$$

Furthermore, the effect-algebra partial order relation coincides with the natural partial order of B. In other words:

$$\forall x \in U\,[f(x) \leq g(x)] \text{ iff } \exists h \in B[f \perp h \text{ and } g = f \boxplus h].$$

The effect algebra $\langle B\,,\boxplus\,,\mathbf{0}\,,\mathbf{1}\rangle$ is not an orthoalgebra, because the strong consistency condition is violated by some genuine fuzzy sets (such as the *semitransparent* fuzzy set $\frac{1}{2}\mathbf{1}$ that assigns to any object x value $\frac{1}{2}$).

---

[31] See [Wright, 1990].

How can we induce the structure of an effect algebra on the set $\mathcal{E}(\mathcal{H})$ of all effects of the Hilbert space $\mathcal{H}$? As in the fuzzy-set case, it is sufficient to define the partial sum $\boxplus$ as follows:

$$\exists(E \boxplus F) \text{ iff } E + F \in \mathcal{E}(\mathcal{H}),$$

where $+$ is the usual sum-operator. Furthermore:

$$E \boxplus F := E + F, \text{ if } \exists(E \boxplus F).$$

It turns out that the structure $\langle \mathcal{E}(\mathcal{H}), \boxplus, \mathbb{O}, \mathbb{I} \rangle$ is an effect algebra (called *standard effect algebra* or *Hilbert effect algebra*), where the generalized complement of any effect E is just $\mathbb{I} - E$. Furthermore, the effect-algebra order relation coincides with the *natural* order defined on $\mathcal{E}(\mathcal{H})$. In other words:

$$\forall \rho \in D(\mathcal{H})[\mathtt{tr}(\rho E) \leq \mathtt{tr}(\rho F)] \text{ iff } \exists G \in \mathcal{E}(\mathcal{H})[E \perp G \text{ and } F = E \boxplus G].$$

At the same time, this structure fails to be an orthoalgebra. For instance, the semitransparent effect $\frac{1}{2}\mathbb{I}$ gives rise to a counterexample to the strong consistency condition:

$$\frac{1}{2}\mathbb{I} \neq \mathbb{O} \text{ and } \frac{1}{2}\mathbb{I} \boxplus \frac{1}{2}\mathbb{I} = \frac{1}{2}\mathbb{I} + (\frac{1}{2}\mathbb{I})' = \mathbb{I}.$$

Let us now turn to the other kind of structure that naturally emerges from concrete effect systems. One is dealing with *quantum MV algebras* (QMV algebras): they are *weak variants* of MV algebras (which represent privileged abstractions from classical fuzzy set structures).[32]

Before introducing the notion of QMV algebra, it will be useful to sum up some basic properties of MV algebras. As is well known, the set of all fuzzy subsets of a given set X gives rise to a de Morgan lattice, where the noncontradiction and the excluded middle principles are possibly violated. In this framework, the lattice operations (the meet $\wedge$, the join $\vee$ and the fuzzy complement $'$) do not represent the only interesting fuzzy operations that can be defined. An important role is played by a new kind of conjunction and disjunction, which have been first investigated in the framework of Łukasiewicz' approach to many valued logics. These operations are usually called *Łukasiewicz operations*.

The definition of Łukasiewicz conjunction and disjunction in the framework of fuzzy set structures turns out to be quite natural. Fuzzy sets are nothing but generalized characteristic functions whose range is the real interval $[0, 1]$. Of course, $[0, 1]$ is not closed under the ordinary real sum $+$ (we may have $x, y \in [0, 1]$ and $x + y \notin [0, 1]$). However, one can introduce a new operation $\oplus$, which is called *truncated sum*:

$$\forall x, y \in [0, 1] \, \{x \oplus y := \min(1, x + y)\} .$$

In other words, $x \oplus y$ is the ordinary sum $x + y$, whenever this sum belongs to the interval; otherwise $x \oplus y$ collapses into the maximum element 1.

[32]See [Giuntini, 1996].

One immediately realizes that $[0,1]$ is closed under the operation $\oplus$. Now, we can use the truncated sum in order to define the Łukasiewicz disjunction between fuzzy sets (since no confusion is possible, it will be expedient to use the same symbol $\oplus$ both for the truncated sum and for the Łukasiewicz disjunction).

Let f, g be fuzzy subsets of a set X. The Łukasiewicz disjunction $\oplus$ is defined as follows:

$$\forall x \in X \, \{(f \oplus g)(x) := f(x) \oplus g(x) = \min(1, f(x) + g(x))\}.$$

On this basis, the Łukasiewicz conjunction $\odot$ can be defined, via de Morgan, in terms of $\oplus$ and $'$:

$$\forall x \in X \, \{(f \odot g)(x) := (f' \oplus g')'(x)\}.$$

As a consequence, one obtains:

$$(f \odot g)(x) = \max(0, f(x) + g(x) - 1).$$

From an intuitive point of view, the Łukasiewicz operations and the lattice operations represent different notions of conjunction and disjunction that can be used in a fuzzy situation. Consider two fuzzy sets f and g; they can be intuitively regarded as two *ambiguous properties*. The number $f(x)$ represents the "degree of certainty" according to which the object x satisfies the property f. A similar comment holds for g and $g(x)$. What does it mean that the object x satisfies the disjunctive property "f *or* g" with a given degree of certainty? If we interpret "or" as the lattice join, we assume the following choice: an object satisfies a disjunction to a degree that corresponds to the maximum between the degrees of the two members. If we, instead, interpret "or" as the Łukasiewicz disjunction, we assume the following choice: the degrees of the members of the disjunction have to be summed in such a way that one never goes beyond the absolute certainty (the value 1). Of course, in the limit-case represented by *crisp sets* (i.e., classical characteristic functions) the Łukasiewicz disjunction and the lattice join will coincide. Suppose $x, y \in \{0,1\}$, then $x \oplus y = \max(x, y)$.

From the definitions, one immediately obtains that the Łukasiewicz operations are not generally idempotent. It may happen:

$$a \oplus a \neq a \text{ and } a \odot a \neq a.$$

As noticed by Mundici[33], this is a typical semantic situation that seems to be governed by the principle "repetita iuvant!" (repetitions are useful!). Of course repetitions are really useful in all physical circumstances that are accompanied by a certain *noise*. As a consequence, $\oplus$ and $\odot$ do not give rise to a lattice structure. At the same time, as with the lattice operations, they turn out to satisfy commutativity and associativity:

$$f \oplus g = g \oplus f; \quad f \odot g = g \odot f;$$

[33]See [Mundici, 1992].

$$f \oplus (g \oplus h) = (f \oplus g) \oplus h; \quad f \odot (g \odot h) = (f \odot g) \odot h.$$

Unlike the fuzzy lattice operations, the Łukasiewicz conjunction and disjunction do satisfy both the excluded middle and the noncontradiction principle:

$$f \oplus f' = \mathbf{1}; \quad f \odot f' = \mathbf{0}.$$

Another important difference concerns the distributivity property. As opposed to the case of $\wedge$ and $\vee$ (which satisfy distributivity in the fuzzy set environment), it may happen:

$$f \odot (g \oplus h) \neq (f \odot g) \oplus (f \odot h);$$
$$f \oplus (g \odot h) \neq (f \oplus g) \odot (f \oplus h).$$

What can be said about the relationships between the Łukasiewicz operations and the lattice operations? Interestingly enough, the lattice operations turn out to be definable in terms of the fuzzy complement and of the Łukasiewicz operations. For, we have:

$$f \wedge g := (f \oplus g') \odot g;$$
$$f \vee g := (f \odot g') \oplus g.$$

In this framework, the conjunction $\wedge$ will be also called *et*, while the disjunction $\vee$ will be called *vel*.

An interesting algebraic abstraction from fuzzy set structures can be obtained if we restrict our attention to the fuzzy complement, the lattice operations and the Łukasiewicz operations.

This gives rise to the abstract notion of an *MV algebra* (*multi-valued algebra*), that Chang introduced in 1958 in order to provide an adequate semantic characterization for Łukasiewicz' many-valued logics.[34] MV algebras represent a weakening of Boolean algebras, where the notion of *conjunction* (*disjunction*) is split into two different operations. The first kind of operation behaves like a Łukasiewicz conjunction (disjunction) and is generally nonidempotent; the second kind of operation is a lattice-meet (join). These algebras are also equipped with a generalized complement. In this framework, the lattice operations turn out to be defined in terms of the generalized complement and of the Łukasiewicz operations. Whenever the two conjunctions (resp., disjunctions) collapse into one and the same operation, one obtains a Boolean algebra.

Let us now recall the formal definition of MV algebra.

DEFINITION 44. *MV algebra*[35]
An *MV algebra* is a structure

$$\mathcal{M} = \langle M \,, \oplus \,, ' \,, \mathbf{0} \,, \mathbf{1} \rangle \,,$$

where $\oplus$ is a binary operation, $'$ is a unary operation and $\mathbf{0}$, $\mathbf{1}$ are special distinct elements satisfying the following conditions:

[34] See [Chang, 1958; Chang, 1959].
[35] See [Mangani, 1973; Cignoli and D'Ottaviano, 2000].

(MV1) $a \oplus b = b \oplus a$;

(MV2) $a \oplus (b \oplus c) = (a \oplus b) \oplus c$;

(MV3) $a \oplus a' = \mathbf{1}$;

(MV4) $a \oplus \mathbf{0} = a$;

(MV5) $a \oplus \mathbf{1} = \mathbf{1}$;

(MV6) $a'' = a$;

(MV7) $(a' \oplus b)' \oplus b = (b' \oplus a)' \oplus a$.

In any MV algebra $\mathcal{M} = \langle M\,, \oplus\,, '\,, \mathbf{0}\,, \mathbf{1}\rangle$, the Łukasiewicz conjunction $\odot$, the lattice operations $\wedge$ and $\vee$, the Łukasiewicz implication $\to_L$, the partial order relation $\leq$ can be defined as follows:

- $a \odot b := (a' \oplus b')'$;
- $a \wedge b := (a \oplus b') \odot b$;
- $a \vee b := (a \odot b') \oplus b$;
- $a \to_L b := a' \oplus b$;
- $a \leq b$ iff $a \wedge b = a$.

It is not difficult to see that $\forall a, b \in M$: $a \leq b$ iff $a \to_L b = a' \oplus b = \mathbf{1}$. Hence, the operation $\to_L$ represents a well behaved conditional.[36]

LEMMA 45. *Let $\mathcal{M} = \langle M\,, \oplus\,, '\,, \mathbf{0}\,, \mathbf{1}\rangle$ be an MV algebra. Consider the structure*

$$\langle M\,, \leq\,, '\,, \mathbf{0}\,, \mathbf{1}\rangle\,,$$

where $\leq$ is the partial order defined on $\mathcal{M}$. Such structure is a distributive bounded involution lattice, where $\wedge$ and $\vee$ represent the infimum and the supremum, respectively. The noncontradiction principle ($a \wedge a' = \mathbf{0}$) and the excluded middle ($a \vee a' = \mathbf{1}$) are possibly violated.[37]

A privileged example of MV algebra can be defined by assuming as support the real interval $[0, 1]$.

DEFINITION 46. *The $[0,1]$-MV algebra*
The $[0,1]$-*MV algebra* is the structure

$$\mathcal{M}^{[0,1]} = \langle [0,1]\,, \oplus\,, '\,, \mathbf{0}\,, \mathbf{1}\rangle\,,$$

where

[36]Generally, a binary operation $\to$ of a structure (which is at least a bounded poset) is considered a *well behaved conditional*, when: $a \leq b$ iff $a \to b = 1$, for any elements a and b. By assuming a natural logical interpretation, this means that the conditional $a \to b$ is "true" iff the "implication-relation" $a \leq b$ holds.

[37]See, for instance, [Cignoli and D'Ottaviano, 2000].

- $\oplus$ is the truncated sum. In other words:
  $\forall x, y \in [0,1] \, \{x \oplus y = \min(1, x+y)\}$;
- $\forall x \in [0,1] \, \{x' = 1 - x\}$;
- $\mathbf{0} = 0$;
- $\mathbf{1} = 1$.

One can easily realize that $\mathcal{M}^{[0,1]}$ is a special example of MV algebra where:

- the partial order $\leq$ is a total order (coinciding with the natural real order);
- $x \wedge y = \min(x, y)$;
- $x \vee y = \max(x, y)$.

Let us now return to the concrete effect-structure $\langle \mathcal{E}(\mathcal{H}) \, , \boxplus \, , \mathbf{0} \, , \mathbf{1} \rangle$. The partial operation $\boxplus$ can be naturally extended to a total operation $\oplus$ that behaves similarly to a *truncated sum*. For any $E, F \in \mathcal{E}(\mathcal{H})$,

$$E \oplus F := \begin{cases} E + F & \text{if } \exists (E \boxplus F), \\ \mathbf{1} & \text{otherwise.} \end{cases}$$

Furthermore, let us define:

$$E' := \mathbb{I} - E.$$

The structure $\langle \mathcal{E}(\mathcal{H}) \, , \oplus \, , ' \, , \mathbf{0} \, , \mathbf{1} \rangle$ turns out to be "very close" to an MV algebra. However, something is missing: $\mathcal{E}(\mathcal{H})$ satisfies the first six axioms of the definition of an MV algebra; at the same time one can easily check that the final axiom (usually called "Łukasiewicz axiom") is violated. For instance, consider two nontrivial projections P, Q such that P is not orthogonal to Q' and Q is not orthogonal to P'. Then, by the definition of $\oplus$ given immediately above, we have that $P \oplus Q' = \mathbb{I}$ and $Q \oplus P' = \mathbb{I}$. Hence, $(P' \oplus Q)' \oplus Q = Q \neq P = (Q' \oplus P)' \oplus P$.

As a consequence, the Łukasiewicz axiom must be conveniently weakened to obtain an adequate description of concrete effect structures. This can be done by means of the notion of *quantum MV algebra* (QMV algebra).[38]

As with MV algebras, QMV algebras are total structures having the following form: $\mathcal{M} = \langle M \, , \oplus \, , ' \, , \mathbf{0} \, , \mathbf{1} \rangle$, where:

(i) $\mathbf{0} \, , \mathbf{1}$ represent the impossible and the certain object, respectively;

(ii) $'$ is the negation-operation;

(iii) $\oplus$ represents a disjunction (*or*) which is generally nonidempotent ($a \oplus a \neq a$).

[38] See [Giuntini, 1996].

A (generally nonidempotent) conjunction (*and*) is then defined via the de Morgan law:

$$a \odot b := (a' \oplus b')'.$$

On this basis, a pair consisting of an idempotent conjunction *et* and of an idempotent disjunction *vel* is then defined. As we have already discussed, in any MV algebra such idempotent operations behave as a lattice-meet and lattice-join, respectively. However, this is not the case for QMV algebras. As a consequence, in such a more general situation, we will denote the *et* operation by the symbol $\Cap$, while the *vel* will be indicated by $\Cup$. The definition of *et* and *vel* is as in the MV-case:

$$a \Cap b := (a \oplus b') \odot b$$

$$a \Cup b := (a \odot b') \oplus b.$$

DEFINITION 47. *QMV algebra*
A *quantum MV algebra* (*QMV algebra*) (QMV) is a structure

$$\mathcal{M} = \langle M\,,\,\oplus\,,\,'\,,\,\mathbf{0}\,,\mathbf{1}\rangle\,,$$

where $\oplus$ is a binary operation, $'$ is a unary operation, and $\mathbf{0}, \mathbf{1}$ are special distinct elements of M. For any $a, b \in M$: $a \odot b := (a' \oplus b')'$, $a \Cap b := (a \oplus b') \odot b$, $a \Cup b := (a \odot b') \oplus b$. Assume that the following conditions hold:

(QMV1) $a \oplus b = b \oplus a$;

(QMV2) $a \oplus (b \oplus c) = (a \oplus b) \oplus c$;

(QMV3) $a \oplus a' = \mathbf{1}$;

(QMV4) $a \oplus \mathbf{0} = a$;

(QMV5) $a \oplus \mathbf{1} = \mathbf{1}$;

(QMV6) $a'' = a$;

(QMV7) $a \oplus [(a' \Cap b) \Cap (c \Cap a')] = (a \oplus b) \Cap (a \oplus c)$.

By Axioms (QMV3), (QMV1) and (QMV4), one immediately obtains that

$$\mathbf{0}' = \mathbf{1}.$$

The operations $\Cap$ and $\Cup$ of a QMV algebra $\mathcal{M}$ are generally noncommutative. As a consequence, they do not represent lattice-operations. It is not difficult to prove that $\Cap$ is commutative iff $\Cup$ is commutative iff (MV7) of Definition 44 holds. From this it easily follows that a QMV algebra $\mathcal{M}$ is an MV algebra iff $\Cap$ or $\Cup$ is commutative.

At the same time (as in the MV-case), we can define in any QMV algebra $\langle M\,,\,\oplus\,,\,'\,,\,\mathbf{0}\,,\mathbf{1}\rangle$ the following relation:

$$a \leq b \text{ iff } a \Cap b = a.$$

The structure

$$\langle M\,,\,\leq\,,\,'\,,\,\mathbf{0}\,,\,\mathbf{1}\rangle$$

turns out to be a bounded involution poset.

One can prove that the concrete effect structure $\langle \mathcal{E}(\mathcal{H})\,,\,\oplus\,,\,'\,,\,\mathbf{0}\,,\,\mathbf{1}\rangle$ is a QMV algebra (which is not an MV algebra).

7 UNSHARP QUANTUM LOGICS

Orthologic, orthomodular quantum logic and partial classical logic are all examples of *sharp logics*. Both the *logical* and the *semantic* version of the noncontradiction principle hold:

- any contradiction $\alpha \curlywedge \neg\alpha$ is always false;[39]
- a sentence α and its negation $\neg\alpha$ cannot both be true.

Some *unsharp* forms of quantum logic have been proposed (in the late Eighties and in the Nineties) as natural logical abstractions from the *effect-state systems*.[40] The most obvious unsharp weakening of orthologic is represented by a logic that has been called *paraconsistent quantum logic* (briefly, **PQL**).[41] In the algebraic semantics, this logic is characterized by the class of all models based on a bounded involution lattice, where the noncontradiction principle ($a \wedge a' = \mathbf{0}$) is possibly violated. In the Kripkean semantics, instead, **PQL** is characterized by the class of all models $\mathcal{K} = \langle I\,,\,\mathcal{R}\,,\,Pr\,,\,V\rangle$, where the accessibility relation $\mathcal{R}$ is symmetric (but not necessarily reflexive), while Pr behaves as in the **OL** case (i.e., Pr is a set of propositions that contains I, $\emptyset$ and is closed under the operations $\cap$ and $'$). Any pair $\langle I\,,\,\mathcal{R}\rangle$, where $\mathcal{R}$ is a symmetric relation on I, is called a *symmetric frame*. All the other semantic definitions are given as in the case of **OL**, *mutatis mutandis*. On this basis, one can show that our algebraic and Kripkean semantics characterize the same logic.

Unlike **OL** and **OQL**, a world i of a **PQL**-Kripkean model may verify a contradiction. Since $\mathcal{R}$ is generally not reflexive, it may happen that $i \in V(\alpha)$ and $i \perp V(\alpha)$. Hence, $i \models \alpha \curlywedge \neg\alpha$. In spite of this, a contradiction cannot be verified by all worlds of a model $\mathcal{K}$. It is worth-while noticing that, unlike **OQL**, the connective negation could not be defined in terms of $\mathbf{f}$ and of $\rightarrow$ in the framework of **PQL**. In fact, should $\rightarrow$ satisfy *Modus Ponens*, we would immediately obtain for any i: $i \models \neg(\alpha \curlywedge \neg\alpha)(:= \alpha \curlywedge (\alpha \rightarrow \mathbf{f}) \rightarrow \mathbf{f})$.

Hilbert-space models for **PQL** can be constructed in a natural way. In the Kripkean semantics, consider the models based on the following frame

$$\langle \mathcal{E}(\mathcal{H}) - \{\mathbf{0}\}\,,\,\not\perp\rangle\,,$$

[39] Of course, in the case of partial classical logic, contradictions are false only if defined.

[40] See [Dalla Chiara *et al.*, 2004].

[41] See [Dalla Chiara and Giuntini, 1989].

where $\not\perp$ represents the nonorthogonality relation between effects ($E \not\perp F$ iff $E \not\leq F'$). Unlike the corresponding case involving projections, in this situation the accessibility relation is symmetric but generally nonreflexive. For instance, the semi-transparent effect $\frac{1}{2}\mathbb{I}$ (representing the prototypical ambiguous property) is a fixed point of the generalized complement $'$. Hence,

$$\frac{1}{2}\mathbb{I} \perp \frac{1}{2}\mathbb{I} \text{ and } (\frac{1}{2}\mathbb{I})' \perp (\frac{1}{2}\mathbb{I})'.$$

From the physical point of view, possible worlds are here identified with possible pieces of information about the physical system under investigation. Any information may correspond to:

- a pure state (a maximal information);
- a proper mixture (a non-maximal information);
- a projection (a sharp property);
- a proper effect (an unsharp property).

Thus, unlike the sharp models of orthomodular quantum logic, here possible worlds do not always correspond to states of the quantum system under investigation. As expected, violations of the noncontradiction principle will be determined by unsharp (ambiguous) pieces of knowledge.

PQL represents a somewhat rough logical abstraction from the class of all effect-structures. As we already know, a characteristic condition that holds in all effect structures is the *regularity property* (which may fail in a generic **PQL**-model).

DEFINITION 48. An algebraic **PQL**-model $\langle \mathcal{B}, v \rangle$ is called *regular* iff the bounded involution lattice $\mathcal{B}$ is regular.[42]

The regularity property can be naturally formulated also in the framework of the Kripkean semantics:

DEFINITION 49. A **PQL**-Kripkean model $\langle I, \mathcal{R}, Pr, V \rangle$ is *regular* iff its frame $\langle I, \mathcal{R} \rangle$ is *regular*. In other words, $\forall i, j \in I$: $i \perp i$ and $j \perp j \curvearrowright i \perp j$.

One can prove that a symmetric frame $\langle I, \mathcal{R} \rangle$ is regular iff the involutive bounded lattice of all propositions of $\langle I, \mathcal{R} \rangle$ is regular. As a consequence, an algebraic model is regular iff its Kripkean transformation is regular and viceversa (where the Kripkean [algebraic] transformation of an algebraic [Kripkean] model is defined like in **OL**).

On this basis one can introduce a proper extension of **PQL**: *regular paraconsistent quantum logic* (**RPQL**). Semantically **RPQL** is characterized by the class of all regular models (both in the algebraic and in the Kripkean semantics).

An axiomatization of **PQL** can be obtained by dropping the *absurdity rule* and the *Duns Scotus rule* in the **OL** calculus. As with **OL**, the logic **PQL** satisfies the finite model property and is consequently decidable.

---

[42] See Deff. 124, 125.

The calculus for **RPQL** is obtained by adding to the **PQL**-calculus the following rule:

$$\alpha \curlywedge \neg\alpha \vdash \beta \curlyvee \neg\beta \qquad \textit{Kleene rule}$$

A completeness theorem for both **PQL** and **RPQL** can be proved, similarly to the case of **OL**.

From the logical point of view, an interesting feature of **PQL** is represented by the fact that this logic is a common sublogic in a wide class of important logics. In particular, **PQL** is a sublogic of Girard's linear logic, of Łukasiewicz' infinitely many-valued logic and of some relevant logics.[43]

An interesting question concerns the relation between **PQL** and the orthomodular property.

Let $\mathcal{B} = \langle B, \leq, ', \mathbf{0}, \mathbf{1} \rangle$ be an ortholattice. It is well known that the following three conditions (expressing possible definitions of the orthomodular property) turn out to be equivalent:

(i) $\forall a, b \in B$: $a \leq b \curvearrowright b = a \vee (a' \wedge b)$;

(ii) $\forall a, b \in B$: $a \leq b$ and $a' \wedge b = \mathbf{0} \curvearrowright a = b$;

(iii) $\forall a, b \in B$: $a \wedge (a' \vee (a \wedge b)) \leq b$.

However, this equivalence breaks down in the case of bounded involution lattices. One can only prove:

LEMMA 50. *Let $\mathcal{B}$ be a bounded involution lattice. If $\mathcal{B}$ satisfies condition (i), then $\mathcal{B}$ satisfies conditions (ii) and (iii).*

LEMMA 51. *Any bounded involution lattice $\mathcal{B}$ that satisfies condition (iii) is an ortholattice.*

As a consequence, we can conclude that there exists no proper orthomodular paraconsistent quantum logic when orthomodularity is understood in the sense (i) or (iii). A residual possibility for a proper paraconsistent quantum logic to be orthomodular is orthomodularity in the sense (ii). In fact, there are examples of lattices that are orthomodular (ii) but not orthomodular (i).[44]

Hilbert space models for orthomodular paraconsistent quantum logic can be constructed in the algebraic semantics by taking as support the following proper subset of the set of all effects:

$$\mathcal{E}_c(\mathcal{H}) := \{a\mathbb{I} \,:\, a \in [0,1]\} \cup \Pi(\mathcal{H}).$$

In other words, a possible meaning of a sentence is either a sharp event (projection) or an unsharp effect that can be represented as a multiple of the universal event ($\mathbb{I}$). Hence all proper unsharp effects are supposed to have a very special form. We will call the elements of $\mathcal{E}_c(\mathcal{H})$ *eccentric effects*.

---

[43]See also Section 9.

[44]See [Giuntini, 1990].

Unlike the case of $\mathcal{E}(\mathcal{H})$ (which is not a lattice), the set $\mathcal{E}_c(\mathcal{H})$ of all eccentric effects turns out to be closed under $\wedge$ and $\vee$. As a consequence, $\mathcal{E}_c(\mathcal{H})$ determines an orthomodular regular bounded involution lattice, where the partial order is the partial order of $\mathcal{E}(\mathcal{H})$ restricted to $\mathcal{E}_c(\mathcal{H})$, while the fuzzy complement is defined like in the class of all effects ($E' := \mathbb{I} - E$).

As we have seen, **PQL** is expressed in the same language of orthologic and of orthomodular quantum logic, representing a weakening thereof. The Brouwer Zadeh structures (emerging from the concrete effect-state systems) have suggested a stronger example of unsharp quantum logic, called *Brouwer Zadeh logic* (also *fuzzy intuitionistic logic*).

As expected, a characteristic property of Brouwer Zadeh logic is a splitting of the connective "not" into two forms of negation: a *fuzzy-like* negation, that gives rise to a *paraconsistent* behaviour, and an *intuitionistic-like* negation. The fuzzy "not" represents a weak negation, that inverts the two extreme truth-values (truth and falsity), satisfies the double negation principle but generally violates the noncontradiction principle. The second "not" is a stronger negation, a kind of necessitation of the fuzzy "not".

The language of *Brouwer Zadeh logic* (**BZL**) is an extension of the language of **QL**. The primitive connectives are: the *conjunction* ($\curlywedge$), the *fuzzy* negation ($\neg$), the *intuitionistic* negation ($\sim$).

Disjunction is metatheoretically defined in terms of conjunction and of the fuzzy negation:

$$\alpha \curlyvee \beta := \neg(\neg\alpha \curlywedge \neg\beta)\,.$$

A *necessity* operator is defined in terms of the intuitionistic and of the fuzzy negation:

$$L\alpha := \sim \neg\alpha\,.$$

A *possibility* operator is defined in terms of the necessity operator and of the fuzzy negation:

$$M\alpha := \neg L \neg\alpha\,.$$

As happens to **OL**, **OQL** and **PQL**, also **BZL** can be characterized by an algebraic and by a Kripkean semantics.

DEFINITION 52. *Algebraic model for* **BZL**
An *algebraic model* of **BZL** is a pair $\langle \mathcal{B}\,, v\rangle$, consisting of a BZ-lattice $\langle B\,, \leq\,, {}'\,, {}^{\sim}\,, \mathbf{0}\,, \mathbf{1}\rangle$ and a valuation-function v that associates to any sentence α an element in B, satisfying the following conditions:

(i) $v(\neg\beta) = v(\beta)'$

(ii) $v(\sim\beta) = v(\beta)^{\sim}$

(iii) $v(\beta \curlywedge \gamma) = v(\beta) \wedge v(\gamma)$.

The definitions of truth, consequence in an algebraic realization for **BZL**, logical truth and logical consequence are given similarly to the case of **OL**.

A characteristic feature of the Kripkean semantics for **BZL** is the use of frames with two accessibility relations.

DEFINITION 53. *Kripkean model for* **BZL**
A *Kripkean model* for **BZL** is a system $\mathcal{K} = \langle I\,, \not\perp\,, \not\stackrel{\sim}{\perp}\,, Pr\,, V\rangle$ where:

(i) $\langle I\,, \not\perp\,, \not\stackrel{\sim}{\perp}\rangle$ is a frame with a non empty set I of possible worlds and two accessibility relations: $\not\perp$ (the *fuzzy accessibility* relation) and $\not\stackrel{\sim}{\perp}$ (the *intuitionistic accessibility* relation).

Two worlds $i\,, j$ are called *fuzzy-accessible* iff $i \not\perp j$. They are called *intuitionistically-accessible* iff $i \not\stackrel{\sim}{\perp} j$. Instead of not $(i \not\perp j)$ and not $(i \not\stackrel{\sim}{\perp} j)$, we will write $i \perp j$ and $i \stackrel{\sim}{\perp} j$, respectively.

The following conditions are required for the two accessibility relations:

(ia) $\langle I, \not\perp\rangle$ is a regular symmetric frame;

(ib) any world is fuzzy-accessible to at least one world:

$$\forall i\, \exists j (i \not\perp j)\,.$$

(ic) $\langle I, \not\stackrel{\sim}{\perp}\rangle$ is an orthoframe;

(id) Fuzzy accessibility implies intuitionistic accessibility:

$$i \not\perp j \quad \curvearrowright \quad i \not\stackrel{\sim}{\perp} j.$$

(ie) Any world i has a kind of "twin-world" j such that for any world k:

(a) $i \not\stackrel{\sim}{\perp} k$ iff $j \not\stackrel{\sim}{\perp} k$

(b) $i \not\stackrel{\sim}{\perp} k \quad \curvearrowright \quad j \not\perp k$.

For any set X of worlds, the *fuzzy orthogonal* set $X^{\perp}$ is defined like in **OL**:

$$X^{\perp} = \{i \in I\, :\, \forall j \in X\, (\, i \perp j)\}\,.$$

Similarly, the *intuitionistic orthogonal* set $X^{\sim}$ is defined as follows:

$$X^{\sim} = \{i \in I \mid \forall j \in X\, (i \stackrel{\sim}{\perp} j)\}\,.$$

The notion of *proposition* is defined like in **OL**: a set of worlds X is a proposition iff $X = X^{\perp\perp}$.

One can prove that for any set of worlds X, both $X^{\perp}$ and $X^{\sim}$ are propositions. Further, like in **OL**, $X \wedge Y$ (the greatest proposition included in the propositions X and Y) is $X \cap Y$, while $X \vee Y$ (the smallest proposition including X and Y) is $(X \cup Y)^{\perp\perp}$.

(ii) Pr is a set of propositions that contains I, and is closed under $^{\perp}$, $^{\sim}$, $\wedge$.

(iii) V associates to any sentence a proposition in Pr according to the following conditions:

$V(\neg\beta) = V(\beta)^{\perp}$;

$$V(\sim \beta) = V(\beta)^{\sim};$$
$$V(\beta \curlywedge \gamma) = V(\beta) \wedge V(\gamma).$$

THEOREM 54. *Let* $\langle I\,, \not\perp\,, \not\perp\!\!\!\!\!\perp \rangle$ *be a BZ-frame (i.e. a frame satisfying the conditions of Definition 53) and let* Pr^0 *be the set of all propositions of the frame. Then, the structure* $\left\langle Pr^0\,, \subseteq\,, {}^{\perp}\,, {}^{\sim}\,, \emptyset\,, I \right\rangle$ *is a complete BZ-lattice such that for any set* $\Gamma \subseteq Pr^0$*:*

$$\bigwedge \Gamma = \bigcap \Gamma \quad \textit{and} \quad \bigvee \Gamma = \left(\bigcup \Gamma\right)^{\perp\perp}.$$

As a consequence, the proposition-structure $\left\langle Pr\,, \subseteq\,, {}^{\perp}\,, {}^{\sim}\,, \emptyset\,, I \right\rangle$ of a **BZL**-model, turns out to be a BZ-lattice.

The definitions of truth, consequence in a Kripkean model, logical truth and logical consequence, are given similarly to the case of **OL**.

One can prove, with standard techniques, that the algebraic and the Kripkean semantics for **BZL** characterize the same logic.

We will now introduce a calculus that represents an adequate axiomatization for the logic **BZL**. The most intuitive way to formulate our calculus is to present it as a *modal extension* of the axiomatic version of regular paraconsistent quantum logic **RPQL**. (Recall that the modal operators of **BZL** are defined as follows: $L\alpha := \sim \neg\alpha$; $M\alpha := \neg L \neg \alpha$).

Rules of **BZL.**

The BZL-calculus includes, besides the rules of **RPQL** the following modal rules:

(BZ1) $L\alpha \vdash \alpha$

(BZ2) $L\alpha \vdash LL\alpha$

(BZ3) $ML\alpha \vdash L\alpha$

(BZ4) $\dfrac{\alpha \vdash \beta}{L\alpha \vdash L\beta}$

(BZ5) $L\alpha \curlywedge L\beta \vdash L(\alpha \curlywedge \beta)$

(BZ6) $\emptyset \vdash \neg(L\alpha \curlywedge \neg L\alpha)$

The rules (BZ1)–(BZ5) give rise to a $\mathbf{S}_5$–like modal behaviour. The rule (BZ6) (the noncontradiction principle for necessitated sentences) is, of course, trivial in any classical modal system.

One can [Dalla Chiara and Giuntini, 2002] prove a soundness and completeness Theorem with respect to the Kripkean semantics (by an appropriate modification of the corresponding proofs for **OL**).

THEOREM 55. *Soundness theorem*

$$T \vdash_{\mathbf{BZL}} \alpha \curvearrowright T \models_{\mathbf{BZL}} \alpha.$$

THEOREM 56. *Completeness theorem*

$$T \models_{\mathbf{BZL}} \alpha \curvearrowright T \vdash_{\mathbf{BZL}} \alpha.$$

Characteristic logical properties of **BZL** are the following:

(a) like in **PQL**, the distributive principles, Duns Scotus, the non-contradiction and the excluded middle principles break down for the fuzzy negation;

(b) like in intuitionistic logic:

$$\models_{\mathbf{BZL}} \sim (\alpha \curlywedge \sim \alpha); \quad \not\models_{\mathbf{BZL}} \alpha \curlyvee \sim \alpha; \quad \alpha \models_{\mathbf{BZL}} \sim\sim \alpha; \quad \sim\sim \alpha \not\models_{\mathbf{BZL}} \alpha;$$

$$\sim\sim\sim \alpha \models_{\mathbf{BZL}} \sim \alpha; \; \alpha \models_{\mathbf{BZL}} \beta \curvearrowright \sim \beta \models_{\mathbf{BZL}} \sim \alpha;$$

(c) furthermore:

$$\sim \alpha \models_{\mathbf{BZL}} \neg\alpha; \quad \neg\alpha \not\models_{\mathbf{BZL}} \sim \alpha; \quad \neg \sim \alpha \models_{\mathbf{BZL}} \sim\sim \alpha;$$

We have seen that concrete effect-systems give also rise to examples of partial algebraic structures, where the basic operations are not always defined. How to give a semantic characterization for a logic that corresponds to the class of all effect algebras? Such a logic has been called *unsharp partial quantum logic* (**UPaQL**).

The language of **UPaQL** consists of a set of atomic sentences and of two primitive connectives: the *negation* $\neg$ and the *exclusive disjunction* $\veebar$ (aut). The set of sentences is defined in the usual way. A *conjunction* is metalinguistically defined, via de Morgan law:

$$\alpha \barwedge \beta := \neg(\neg\alpha \veebar \neg\beta).$$

The intuitive idea underlying the semantics for **UPaQL** is the following: disjunctions and conjunctions are always considered "legitimate" from a mere linguistic point of view. However, semantically, a disjunction $\alpha \veebar \beta$ will have the intended meaning only in the "appropriate cases:" where the values of α and β are orthogonal in the corresponding effect algebra. Otherwise, $\alpha \veebar \beta$ will have any meaning whatsoever (generally not connected with the meanings of α and β). As is well known, a similar semantic "trick" is used in some classical treatments of the description operator ι ("the unique individual satisfying a given property"; for instance, "the present king of Italy"). Apparently, one is dealing with a different idea with respect to the semantics of partial classical logic (**PaCL**), where the meaning of a sentence is not necessarily defined. One has proved that **UPaQL** is an axiomatizable logic.[45]

[45] See [Dalla Chiara and Giuntini, 2002].

Also the theory of QMV algebras has naturally suggested the semantic characterization of another form of quantum logic (called *Łukasiewicz quantum logic* (**ŁQL**)), which generalizes both **OQL** and $\mathbf{L}_{\aleph}$ (Łukasiewicz' infinite many valued logic). The language of **ŁQL** contains the same primitive connectives as **UPaQL** ($\forall\!\!\!\forall$, $\neg$). The conjunction ($\curlywedge$) is defined via the de Morgan law (as with **UPaQL**). Furthermore, a new pair of conjunction ($\Cap$) and disjunction ($\Cup$) connectives are defined as follows:

$$\alpha \Cap \beta := (\alpha \forall\!\!\!\forall \neg\beta) \curlywedge \beta$$

$$\alpha \Cup \beta := \neg(\neg\alpha \Cap \neg\beta)$$

ŁQL can be easily axiomatized by means of a calculus that simply mimics the axioms of QMV algebras.[46]

8 QUANTUM LOGIC AND PARACONSISTENT LOGIC

As we have seen, the orthodox quantum logics whose investigation can be traced back to the work by Birkhoff and von Neumann, such as orthologic or orthomodular quantum logic, are not paraconsistent: both logics validate the principle of *ex absurdo quodlibet* (any sentence β follows from a contradiction $\alpha \curlywedge \neg\alpha$) and the law of noncontradiction $\neg(\alpha \curlywedge \neg\alpha)$, although the acceptance of the latter is clearly compatible with paraconsistency[47]. Thus, any inconsistent theory in either **OL** or **OQL** is bound to be trivial. We have learnt, however, that other quantum logics — in particular, unsharp logics like **PQL** — have a marked paraconsistent character. As a consequence, a question naturally arises: should a "good" quantum logic be paraconsistent? At least two different stances can be adopted in this connection:

- On the one hand, we might believe that orthodox quantum logics are inadequate to account for some aspects of quantum theory which are better captured through recourse to a paraconsistent logic.[48] For example, if we liberalize the notion of quantum event in such a way as to allow for proper effects as mathematical counterparts of unsharp properties, we end up adopting one of the paraconsistent quantum logics (either **RPQL** or **BZL** or **ŁQL** or **UPaQL**).

- On the other hand, we might claim that orthodox quantum logics — whether or not they adequately capture every single aspect of quantum theory — are, in a way, already paraconsistent in themselves. For example, it is remarked in [Restall, 2002] that both **OL** and **OQL** (unlike, for example,

[46]See [Dalla Chiara *et al.*, 2004].

[47]Recall the famous anecdote of the judge who had been given the task of settling a controversy between two litigants asserting opposite claims. After hearing the former's pleading, he decreed he was in the right; yet, when the latter set out his reasons, he conceded that he also was in the right. As his assistant protested that no two parties asserting mutually contradictory statements can both be in the right, he exclaimed: "You're in the right, too!". Similarly, it is possible for a logic to uphold the noncontradiction principle while allowing for nontrivial inconsistent theories.

[48]See [Dalla Chiara *et al.*, 2004], [da Costa *et al.*, 2006].

propositional intuitionistic logic) tolerate classical inconsistencies, because the *falsum*-sentence **f** does not follow therein from the classically inconsistent formula

$$\alpha \curlywedge (\beta \curlyvee \gamma) \curlywedge \neg((\alpha \curlywedge \beta) \curlyvee (\alpha \curlywedge \gamma)).$$

Having discussed at some length in the preceding sections the interaction between the unsharp approach to quantum logic and paraconsistency, we will focus here on another application of paraconsistent logic to quantum theory, viz. the logical treatment of the *complementarity* problem. According to Bohr's interpretation, quantum theory contains pairs of *complementary* sentences which, although not mutually contradictory as such, jointly entail a contradiction. Examples could be "p is a corpuscle" and "p is a wave", where p refers to a designated photon — the former sentence is not the negation of the latter, but it *entails* the negation of the latter. If the logic underlying our physical theory is classical logic, which does not support nontrivial inconsistent theories, the triviality of quantum theory readily follows.

A first attempt to cope with this problem was done as early as in 1937 by P. Destouches-Février, who (informally) introduced a three-valued logic by adjoining to the classical values *True* and *False* a third one (*Absolutely False*), meant to act as the truth value of the conjunctions of complementary sentences [Destouches-Février, 1937]. Here, "absolutely false" should not be understood as "necessarily false" or "definitely false", but rather as "meaningless" or "ungrammatical", i.e. as expressing the fact that strictly speaking complementary sentences could not be conjoined. A paraconsistent version of this logic — where not only the value *True*, but also the value *Absolutely False* is designated — can be found in [da Costa and Krause, 2005]; the authors claim this modification of Février's logic can meet some of the objections levelled against it in the literature.

The same authors also devised another possible way out of the logical problem aroused by the complementarity phenomenon [da Costa *et al.*, 2006]: they introduced a paraconsistent logical consequence relation $\vdash_{\mathbf{P}}$, termed *paraclassical* and specified in the following guise:

DEFINITION 57. *Paraclassical consequence*
Let $\mathcal{L}$ be the language of **CL** (classical sentential logic), containing the connectives $\neg, \curlywedge, \curlyvee, \rightarrow$ and let T be a set of sentences of $\mathcal{L}$. A sentence α (of $\mathcal{L}$) is called a **P**-*consequence* of T (in symbols, $T \vdash_{\mathbf{P}} \alpha$) iff:

P1 $\alpha \in T$, or

P2 α is a classical tautology, or

P3 there exists a classically consistent subset $T' \subseteq T$ such that $T' \vdash_{\mathbf{CL}} \alpha$ (where $\vdash_{\mathbf{CL}}$ is the classical consequence relation).

The notions of **P**-*inconsistency* and **P**-*triviality* are introduced in the expected way.

DEFINITION 58. A set T of sentences of $\mathcal{L}$ is **P**-*inconsistent* iff there exists a sentence α s.t. $T \vdash_{\mathbf{P}} \alpha$ and $T \vdash_{\mathbf{P}} \neg\alpha$; it is **P**-*trivial* iff $T \vdash_{\mathbf{P}} \alpha$ for every formula α.[49]

P-inconsistency and **P**-triviality are distinct concepts: if $\mathbf{p}$ is an atomic sentence, the set $\{\mathbf{p}, \neg\mathbf{p}\}$ is **P**-inconsistent, but not **P**-trivial. Also, **P**-triviality and **P**-inconsistency imply, respectively, classical triviality and classical inconsistency, but the converse relations do not generally hold.

Complementary sentences easily find a home within this framework. First of all, one defines a C-*theory* in the language $\mathcal{L}$ as a set of $\mathcal{L}$-sentences closed under the relation $\vdash_{\mathbf{P}}$, and checks that there exist classically inconsistent C-theories which are not **P**-trivial. With these ingredients at our disposal, we can introduce a proper definition of *complementary sentences*:

DEFINITION 59. Let T be a C-theory in the language $\mathcal{L}$. Two sentences α and β of $\mathcal{L}$ are said to be T-*complementary* iff: 1) $T \vdash_{\mathbf{P}} \alpha$ and $T \vdash_{\mathbf{P}} \beta$; 2) there is a sentence γ of $\mathcal{L}$ such that: $T, \alpha \vdash_{\mathbf{P}} \gamma$ and $T, \beta \vdash_{\mathbf{P}} \neg\gamma$.

It follows from the definitions that C-theories containing complementary theorems are classically inconsistent, but not **P**-trivial: if α and β are complementary theorems and γ is such that $T, \alpha \vdash_{\mathbf{P}} \gamma$ and $T, \beta \vdash_{\mathbf{P}} \neg\gamma$, then in general $\gamma \curlywedge \neg\gamma$ does not follow from T. Da Costa et al. conclude by remarking:

> (Complementary) theories are closer to those theories scientists *actually* use in their everyday activity than those theories with the classical concept of deduction. In other words, paraclassical logic (and paraconsistent logics in general) seems to fit more accurately the way scientists reason when stating their theories [da Costa *et al.*, 2006].

9 QUANTUM LOGIC AND LINEAR LOGIC

9.1 A brief survey of linear logic

Since Heyting, followers of the constructive approach to logic have singled out the notion of *proof* as the fundamental concept of their discipline, stressing at the same time that what really matters is not *whether* a given formula is provable in a certain logical calculus, but *how* it can be proved therein. In other words, the emphasis is not on the outcome, but on the process, the idea being that we must sharply distinguish between different proofs of one and the same formula. Given these basic tenets, it does no harm to identify a formula with the set of its proofs, so that a proof of α from the assumptions $\alpha_1, ..., \alpha_n$ — seen as a method for converting any given proofs of $\alpha_1, ..., \alpha_n$ into a proof of α — boils down to a function $f(x_1, ..., x_n)$ which associates to elements $a_i \in \alpha_i$ a result $f(a_1, ..., a_n) \in \alpha$.

[49]A terminological remark: here "trivial" corresponds to "semantically inconsistent" (in the sense of Def. 22), while "inconsistent" corresponds to "semantically contradictory".

This much is already implicitly contained in the general lines of the so-called BHK (Brouwer-Heyting-Kolmogorov) interpretation of intuitionistic logic. In the 1960's, W.A. Howard [Howard, 1980] added to it the identification of intuitionistic natural deduction proofs with terms of typed lambda calculus. In a nutshell, a proof of the formula α is associated with a term of type α, whence it becomes possible to spell out the computational content of the inference rules in the $\{\curlywedge, \rightarrow\}$-fragment of the intuitionistic natural deduction calculus:

- if t and s are terms having respective types α and β, then $\langle t, s\rangle$ (the *pairing* of t and s) is a term of type $\alpha \curlywedge \beta$;
- if t is a term of type $\alpha \curlywedge \beta$, then $\pi_1(t)$ and $\pi_2(t)$ (the first and second *projections* of t) are terms having respective types α and β;
- if x is a variable of type α and t is a term of type β, then $\lambda x.t$ (the *abstraction* of t w.r.t. x) is a term of type $\alpha \rightarrow \beta$;
- if t and s are terms having respective types $\alpha \rightarrow \beta$ and α, then ts (the *application* of t to s) is a term of type β.

The ensuing correspondence between intuitionistic natural deduction proofs and terms in the lambda calculus with projection and pairing functors can be seen as a full-fledged isomorphism (and it is indeed referred to as the *Curry-Howard isomorphism*) in that there is a perfect match between the notions of *conversion*, *normality* and *reduction* introduced in the two frameworks.

In the light of the Curry-Howard isomorphism, it was readily acknowledged that the problem of finding a "semantics of proofs" for a given constructive logic and the problem of providing lambda calculus (or, for that matter, functional programming) with a semantical interpretation were two sides of the same coin. In Dana Scott's *domain theory*, a first attempt to accomplish this task, a type α was interpreted by means of a particular topological space. In the mid 1980's, on the other hand, Jean-Yves Girard [Girard, 1987] introduced for this purpose the notion of *coherent space* over a set X. In the spirit of the BHK tradition, Girard identifies each formula with the set of its proofs, which are in turn represented as "coherent" sets of information bits. More precisely:

- the points in X are supposed to represent atomic *information tokens*;
- a *coherence* relation between information tokens is defined;
- sets of pairwise coherent tokens correspond to *proofs*;
- a coherent space represents a *formula*, viewed as the set of its proofs.

DEFINITION 60. *Polar*
Let X be a set and let $\alpha \subseteq \mathcal{P}(X)$. The *polar* of α (indicated by $\neg\alpha$) is defined as follows:

$$\neg\alpha := \{Y \subseteq X : \forall Z \in \alpha(card(Y \cap Z) \leq 1)\}.$$

In other words, given a family α of subsets of X, $\neg\alpha$ is the family of all subsets of X whose intersection with all elements of α is either the empty set or a singleton.

DEFINITION 61. *Coherent space*
A *coherent space* over the set X is a family α of subsets of X s.t. $\alpha = \neg\neg\alpha$.

EXAMPLE 62. Let $X = \{a, b, c\}$.

- $\alpha = \{\emptyset, \{a\}, \{b\}, \{c\}, \{a, b\}\}$ is a coherent space over X;
- $\beta = \{\emptyset, \{b, c\}, \{a, b, c\}\}$ is not a coherent space over X.

A *coherence* relation $\not\perp_\alpha$ can be now defined in any coherent space α.

DEFINITION 63. If α is a coherent space over X and $a, b \in X$, we set:

$$a \not\perp_\alpha b \text{ iff } \{a, b\} \in \alpha.$$

Interestingly enough, coherent spaces turn out to be deeply connected with orthoframes.

LEMMA 64. *Let X be a set. There is a 1-1 correspondence between coherent spaces over X and orthoframes with universe X.*

Proof. If α is a coherent space over X, we set, in accordance with Definition 63:

$$\mathcal{F}^\alpha = (X, \not\perp_\alpha).$$

Conversely, let $\mathcal{F} = (X, \not\perp)$ be an orthoframe with universe X; we define:

$$\alpha^{\mathcal{F}} = \{Y \subseteq X : a \not\perp b \text{ for any } a, b \in Y\}.$$

The system $\mathcal{F}^\alpha$ is an orthoframe. For, the induced relation $\not\perp_\alpha$ is clearly symmetric and is likewise reflexive in that $\alpha = \neg\neg\alpha$ contains the singletons of all tokens. A moment's reflection also shows that $\alpha^{\mathcal{F}}$ is a coherent space over X: it is a family of subsets of X, and it equals its own double polar. It therefore suffices to show that: $\mathcal{F} = \mathcal{F}^{\alpha^{\mathcal{F}}}$ and $\alpha = \alpha^{\mathcal{F}^\alpha}$. As to the former claim, just remark that $a \not\perp_{\alpha^{\mathcal{F}}} b$ iff $\{a, b\} \in \alpha^{\mathcal{F}}$ iff $a \not\perp b$. As regards the latter,

$$\begin{aligned}
\alpha^{\mathcal{F}^\alpha} &= \{Y \subseteq X : a \not\perp_\alpha b \text{ for any } a, b \in Y\} \\
&= \{Y \subseteq X : \{a, b\} \in \alpha \text{ for any } a, b \in Y\} \\
&= \neg\neg\alpha = \alpha.
\end{aligned}$$

■

Definition 60 already provides us with a semantics of proofs for the negation connective. What about the other logical connectives? Let us consider conjunction first. Here, Girard departs from the intuitionistic tradition inasmuch as he distinguishes two non-equivalent proof conditions for a conjunctive sentence:

- in a sense, I have a proof of a conjunctive sentence when I can assemble together two different proofs, one for each conjunct;
- yet, in a different sense, I can prove a conjunctive sentence when I have just one proof, which however suffices to yield *any one* of the two conjuncts.

From a constructive viewpoint, different proof conditions correspond to different logical constants; thus, classical (or, for that matter, intuitionistic) conjunction "splits up" into a *multiplicative* conjunction $\otimes$ and an *additive* conjunction $\sqcap$. Translating the above into the formal jargon of coherent spaces, we are led to the following

DEFINITION 65. Let α and β be two coherent spaces, respectively over X and over Y (which we assume without loss of generality to be disjoint sets[50]). We define:

$$\begin{aligned}&\alpha \otimes \beta := \{Z \subseteq X \times Y : \text{there exist } W \in \alpha, V \in \beta \text{ s.t. } Z \subseteq V \times W\};\\ &\alpha \sqcap \beta := \{Z \cup W : Z \in \alpha, W \in \beta\}.\end{aligned}$$

One can prove that $\alpha \otimes \beta$ is a coherent space over $X \times Y$, while $\alpha \sqcap \beta$ is a coherent space over $X \cup Y$.

Let us now come to implication, which can be defined out of negation and multiplicative conjunction as $\neg(\alpha \otimes \neg\beta)$. This does not quite seem to tally with the usual picture of a proof for an implicative sentence, i.e. of a method (function) for extracting a proof for the consequent from a proof for the antecedent. The following definition and fact, however, provide the required bridge:

DEFINITION 66. Let α, β be coherent spaces. A function f from α to β is called *linear* whenever it preserves disjoint unions.

LEMMA 67. *If α, β are coherent spaces, respectively over X and Y, $Z \in \alpha$ and $W \in \alpha \to \beta$, then*

$$\widetilde{W}(Z) = \{b \in Y : \langle a, b\rangle \in W \textit{ for some } a \in Z\}$$

is a member of β, $\widetilde{W}$ is a linear function from α to β, and

$$\alpha \to \beta = \left\{W : \widetilde{W}(Z) \in \beta \textit{ for every } Z \in \alpha\right\}.$$

The resulting implication is accordingly termed *linear* implication. In addition, Girard considers a coherent space $\alpha \twoheadrightarrow \beta$ corresponding to intuitionistic implication, which is decomposed in terms of linear implication and a necessity-like unary operator, !. More precisely, the orthoframe associated to the coherent space $\alpha \twoheadrightarrow \beta$ is isomorphic (as a first order structure) to the orthoframe associated to $!\alpha \to \beta$.

Having seen how implication can be dealt with, we observe that disjunction, like conjunction, is affected by ambiguity. In fact:

[50] If they are not, it suffices to consider disjoint bijective copies of X and Y, built in the standard way.

- I have a proof of a disjunctive sentence when I have a method for converting a proof for the negation of any disjunct into a proof for the other;
- yet, in a different sense, I can prove a disjunctive sentence when I have a proof for one of the two disjuncts.

Thus, once again, we have to distinguish between a *multiplicative* disjunction $\oplus$ and an *additive* disjunction $\sqcup$:

DEFINITION 68. Let α and β be two coherent spaces, respectively over X and over Y (which we assume to be disjoint sets). We define:

$$\alpha \oplus \beta := \neg\alpha \to \beta;$$
$$\alpha \sqcup \beta := \alpha \cup \beta.$$

One can prove that $\alpha \oplus \beta$ is a coherent space over $X \times Y$, while $\alpha \sqcup \beta$ is a coherent space over $X \cup Y$.

How can we make intuitive sense of this logic? A possible option is viewing formulas as concrete resources that, once they are consumed in a deduction to get some conclusion, cannot be recycled or reused. Formulas of the form $!\alpha$, on the other hand, represent "ideal" resources that can be reused at will. Thus, while the availability of an intuitionistic implication $\alpha \twoheadrightarrow \beta$ means that using as many α's as I might need I can get one β, the availability of a linear implication $\alpha \to \beta$ expresses the fact that, using just one α, I can get one β — something that perfectly squares with the coherent space isomorphism pointed out above. We can also view the other compound formulas of linear logic as concrete resources: for example, $\alpha \otimes \beta$ expresses the availability of both resource α and resource β, while $\alpha \sqcap \beta$ expresses the availability of *any* one between such resources — i.e. we can have just one between α and β and not both, but we are in a position to choose which[51].

In his 1987 seminal paper, Girard sets up a *sequent calculus* for his newly discovered logic, henceforth called **LL**.[52]

Unlike the calculi we have considered in the previous sections, a *sequent calculus* for a logic **L** is based on *axioms* and *rules* that govern the behavior of *sequents*. Any sequent has the form

$$\Gamma \Rightarrow \Delta,$$

where Γ and Δ are finite (possibly empty) *multisets* of formulas.[53] Axioms are

[51] A standard example goes like this: suppose that a cup of coffee and a newspaper cost 1 Euro each. Thus, for 1 Euro I can get a cup of coffee and ($\sqcap$) a newspaper, but for the same amount I cannot get a cup of coffee and ($\otimes$) a newspaper.

[52] One *caveat*, though: our notation for logical constants does not exactly match Girard's original one. For example, he resorts to the symbol $\oplus$ to denote *additive*, not multiplicative, disjunction.

[53] A *multiset* is a set of pairs such that the first element of every pair denotes an object, while the second element denotes the multiplicity of the occurrences of our object. Two multisets are equal iff all their pairs are equal, that is all their objects together with their multiplicities are equal.

particular sequents. Any rule has the form:

$$\frac{\Gamma_1 \Rightarrow \Delta_1, \ldots, \Gamma_n \Rightarrow \Delta_n}{\Gamma \Rightarrow \Delta}$$

where $\Gamma_1 \Rightarrow \Delta_1, \ldots \Gamma_n \Rightarrow \Delta_n$ are the *premisses* of the rule while $\Gamma \Rightarrow \Delta$ is the *conclusion*. Rules can be either *structural* or *operational*. Operational rules introduce a new connective either on the left or on the right side of a sequent. Accordingly, one usually speaks of *left* and of *right* introduction rule. For example, consider Gentzen's calculus **LK** for classical logic. The left introduction rule for the connective $\curlywedge$ can be written as follows:

$$\frac{\alpha, \Gamma \Rightarrow \Delta}{\alpha \curlywedge \beta, \Gamma \Rightarrow \Delta} \qquad \frac{\beta, \Gamma \Rightarrow \Delta}{\alpha \curlywedge \beta, \Gamma \Rightarrow \Delta} \ (\curlywedge \mathrm{L})$$

Structural rules, instead, only deal with the structure of sequents (orders, repetitions, etc.). Two important examples of structural rules are *weakening* and *contraction*:

$$\frac{\Gamma \Rightarrow \Delta}{\Gamma, \Pi \Rightarrow \Sigma, \Delta} \ (\text{weakening}); \qquad \frac{\Gamma, \Pi, \Pi \Rightarrow \Delta, \Delta, \Sigma}{\Gamma, \Pi \Rightarrow \Delta, \Sigma,} \ (\text{contraction})$$

A *derivation* (in the logic **L**) is a sequence of sequents where any element is either an axiom or the conclusion of a rule whose premisses are previous elements of the sequence. A sequent $\Gamma \Rightarrow \Delta$ is said to be *derivable* in the logic **L** (abbreviated $\vdash_{\mathbf{L}} \Gamma \Rightarrow \Delta$) iff $\Gamma \Rightarrow \Delta$ is the last element of a derivation of **L**.[54]

The completeness (and soundness) theorem for Gentzen's **LK** can be formulated as follows:

$$\vdash_{\mathbf{LK}} \{\alpha_1, \ldots, \alpha_n\} \Rightarrow \{\beta_1, \ldots, \beta_m\} \ \text{iff}$$

$$\alpha_1 \curlywedge \ldots \curlywedge \alpha_n \rightarrow \beta_1 \curlyvee \ldots \curlyvee \beta_m \ \text{ is a logical truth of classical logic.}$$

Girard's **LL** differs from **LK** in that:

- **LL** has no weakening or contraction rules — so that its language contains additive and multiplicative versions of both conjunction and disjunction, as well as of the *verum* and *falsum* constants. The absence of weakening and contraction is readily explained in terms of the resource interpretation: for example, contraction would say that if you can get a resource β by using n copies of the resource α, you can still get β by using just one copy of α, which is just wishful thinking.

- The language of **LL** includes the modalities *of course!* (!) and *why not?* (?). The addition of modalities reintroduces weakening and contraction for special classes of formulas; more precisely, a formula preceded by an exclamation mark can be introduced through unrestricted *left* weakening and

[54] Since a rule has generally many premisses, derivations can be also conveniently represented as special tree-configurations.

> contraction inferences, while a formula preceded by a question mark can be introduced through unrestricted *right* weakening and contraction inferences. The modalities abide at the same time by structural rules and modal rules: if we have a look at the sequent calculi for modal logics, in fact, it is easy to see that ! behaves as an **S4** necessity operator, and ? as an **S4** possibility operator.

The importance of the modalities is fully realized if we take into account the fact that Girard is admittedly not interested in setting up a logic which is weaker than classical or intuitionistic logic: he rather aims for a logic which permits a better analysis of proofs through a stricter control of structural rules. Modalities are there precisely to recapture the deductive power of weakening and contraction, an aim which is attained — in a sense — by showing that both classical logic and intuitionistic logic can be embedded into linear logic. Confining ourselves to classical logic, Girard provides in fact translations of multisets of classical formulas into multisets of linear formulas in such a way that $\Gamma \Rightarrow \Delta$ is provable in **LK** iff a sequent compounded out of an appropriate translation of its formulas is provable in **LL**. Beside this desirable aspect, modalities have their down side too: they are to blame for the *undecidability* of propositional linear logic [Lincoln *et al.*, 1992].

The language $\mathcal{L}_1$ of **LL** contains the connectives $\otimes, \oplus, \rightarrow, \sqcap, \sqcup$ (binary), $\neg, !, ?$ (unary), $\top, \bot, \mathbf{f}, \mathbf{t}$ (nullary). The notations $!\Gamma$ and $?\Gamma$ mean that the modality at issue must be prefixed to each element of the given sequence; $\mathcal{L}_0$ will denote the language obtained by barring from $\mathcal{L}_1$ the modalities ! and ?.

9.2 Quantum logic vs *linear logic: proof theory*

A prominent feature shared by quantum logics and linear logic alike is the failure of *lattice distributivity*. This means that any adequate proof system for these logics — for the sake of definiteness, say any sequent calculus — should clamp in some way or another the mechanism which allows the proof of $\alpha \curlywedge (\beta \curlyvee \gamma) \Rightarrow (\alpha \curlywedge \beta) \curlyvee (\alpha \curlywedge \gamma)$ in **LK**:[55]

$$\dfrac{\dfrac{\dfrac{\dfrac{\dfrac{\alpha \Rightarrow \alpha}{\alpha, \beta \Rightarrow \alpha, \alpha \curlywedge \gamma} \quad \dfrac{\beta \Rightarrow \beta}{\alpha, \beta \Rightarrow \beta, \alpha \curlywedge \gamma}}{\alpha, \beta \Rightarrow \alpha \curlywedge \beta, \alpha \curlywedge \gamma} \quad \dfrac{\dfrac{\alpha \Rightarrow \alpha}{\alpha, \gamma \Rightarrow \alpha \curlywedge \beta, \alpha} \quad \dfrac{\gamma \Rightarrow \gamma}{\alpha, \gamma \Rightarrow \alpha \curlywedge \beta, \gamma}}{\alpha, \gamma \Rightarrow \alpha \curlywedge \beta, \alpha \curlywedge \gamma}}{\alpha, \beta \curlyvee \gamma \Rightarrow \alpha \curlywedge \beta, \alpha \curlywedge \gamma}}{\alpha \curlywedge (\beta \curlyvee \gamma), \alpha \curlywedge (\beta \curlyvee \gamma) \Rightarrow (\alpha \curlywedge \beta) \curlyvee (\alpha \curlywedge \gamma), (\alpha \curlywedge \beta) \curlyvee (\alpha \curlywedge \gamma)}}{\alpha \curlywedge (\beta \curlyvee \gamma) \Rightarrow (\alpha \curlywedge \beta) \curlyvee (\alpha \curlywedge \gamma)}$$

As we have seen, in **LL** — whose operational rules for the additives are the same as we used in the above **LK** proof — we attain such an aim by doing away with the structural rules of weakening and contraction outright. In the various sequent systems which have been suggested for quantum logics, on the other hand, one

[55]The converse half corresponds to an inequality which is valid in the class of general lattices and, therefore, holds both in linear logic and in all mainstream quantum logics.

Axioms and rules of **LL**

$\alpha \Rightarrow \alpha$ (ID)

$\bot, \Gamma \Rightarrow \Delta$ ($\bot$L) $\qquad$ $\Gamma \Rightarrow \Delta, \top$ ($\top$R)

$\mathbf{f} \Rightarrow$ (**f**L) $\qquad$ $\dfrac{\Gamma \Rightarrow \Delta}{\Gamma \Rightarrow \Delta, \mathbf{f}}$ (**f**R)

$\dfrac{\Gamma \Rightarrow \Delta}{\mathbf{t}, \Gamma \Rightarrow \Delta}$ (**t**L) $\qquad$ $\Rightarrow \mathbf{t}$ (**t**R)

$$\frac{\alpha, \beta, \Gamma \Rightarrow \Delta}{\alpha \otimes \beta, \Gamma \Rightarrow \Delta}\ (\otimes \mathrm{L}) \qquad \frac{\Gamma \Rightarrow \Delta, \alpha \qquad \Pi \Rightarrow \Sigma, \beta}{\Gamma, \Pi \Rightarrow \Delta, \Sigma, \alpha \otimes \beta}\ (\otimes \mathrm{R})$$

$$\frac{\alpha, \Gamma \Rightarrow \Delta \qquad \beta, \Pi \Rightarrow \Sigma}{\alpha \oplus \beta, \Gamma, \Pi \Rightarrow \Delta, \Sigma}\ (\oplus \mathrm{L}) \qquad \frac{\Gamma \Rightarrow \Delta, \alpha, \beta}{\Gamma \Rightarrow \Delta, \alpha \oplus \beta}\ (\oplus \mathrm{R})$$

$$\frac{\alpha, \Gamma \Rightarrow \Delta}{\alpha \sqcap \beta, \Gamma \Rightarrow \Delta} \quad \frac{\beta, \Gamma \Rightarrow \Delta}{\alpha \sqcap \beta, \Gamma \Rightarrow \Delta}\ (\sqcap \mathrm{L}) \qquad \frac{\Gamma \Rightarrow \Delta, \alpha \qquad \Gamma \Rightarrow \Delta, \beta}{\Gamma \Rightarrow \Delta, \alpha \sqcap \beta}\ (\sqcap \mathrm{R})$$

$$\frac{\alpha, \Gamma \Rightarrow \Delta \qquad \beta, \Gamma \Rightarrow \Delta}{\alpha \sqcup \beta, \Gamma \Rightarrow \Delta}\ (\sqcup \mathrm{L}) \qquad \frac{\Gamma \Rightarrow \Delta, \alpha}{\Gamma \Rightarrow \Delta, \alpha \sqcup \beta} \quad \frac{\Gamma \Rightarrow \Delta, \beta}{\Gamma \Rightarrow \Delta, \alpha \sqcup \beta}\ (\sqcup \mathrm{R})$$

$$\frac{\Gamma \Rightarrow \Delta, \alpha \qquad \beta, \Pi \Rightarrow \Sigma}{\alpha \to \beta, \Gamma, \Pi \Rightarrow \Delta, \Sigma}\ (\to \mathrm{L}) \qquad \frac{\alpha, \Gamma \Rightarrow \Delta, \beta}{\Gamma \Rightarrow \Delta, \alpha \to \beta}\ (\to \mathrm{R})$$

$$\frac{\Gamma \Rightarrow \Delta, \alpha}{\neg\alpha, \Gamma \Rightarrow \Delta}\ (\neg \mathrm{L}) \qquad \frac{\alpha, \Gamma \Rightarrow \Delta}{\Gamma \Rightarrow \Delta, \neg\alpha}\ (\neg \mathrm{R})$$

$$\frac{\alpha, \Gamma \Rightarrow \Delta}{!\alpha, \Gamma \Rightarrow \Delta}\ (!\mathrm{L}) \qquad \frac{!\Gamma \Rightarrow ?\Delta, \alpha}{!\Gamma \Rightarrow ?\Delta, !\alpha}\ (!\mathrm{R})$$

$$\frac{\Gamma \Rightarrow \Delta}{!\alpha, \Gamma \Rightarrow \Delta}\ (!\mathrm{W}) \qquad \frac{!\alpha, !\alpha, \Gamma \Rightarrow \Delta}{!\alpha, \Gamma \Rightarrow \Delta}\ (!\mathrm{C})$$

$$\frac{\alpha, !\Gamma \Rightarrow ?\Delta}{?\alpha, !\Gamma \Rightarrow ?\Delta}\ (?\mathrm{L}) \qquad \frac{\Gamma \Rightarrow \Delta, \alpha}{\Gamma \Rightarrow \Delta, ?\alpha}\ (?\mathrm{R})$$

$$\frac{\Gamma \Rightarrow \Delta}{\Gamma \Rightarrow \Delta, ?\alpha}\ (?\mathrm{W}) \qquad \frac{\Gamma \Rightarrow \Delta, ?\alpha, ?\alpha}{\Gamma \Rightarrow \Delta, ?\alpha}\ (?\mathrm{C})$$

usually manages to avoid distribution by means of appropriate restrictions on *side formulas*.[56] For example, the calculus for orthologic implicit in [Schütte Monting, 1981][57] is exactly like **LK** formulated in the $\{\neg, \curlywedge, \curlyvee\}$-language, except for the fact that sequents can contain *at most two* formulas. Weakening and contraction rules are there indeed, but many applications thereof are blocked because they "take up too much room". Several steps in the displayed proof of distribution, for example, turn out to be incorrect by such a standard. If we tighten even further the mentioned constraint, by allowing only sequents which contain *exactly two* formulas, we obtain a calculus for an unbounded version of paraconsistent quantum logic (where *all* weakening and contraction inferences clearly become unfeasible, leading to the unprovability of sequents like $\alpha \curlywedge \neg\alpha \Rightarrow \beta$ or $\alpha \Rightarrow \beta \curlyvee \neg\beta$).[58]

There is, however, a different standpoint we could take while comparing the proof theories of linear logic and quantum logics. One could, indeed, search for a *common abstraction* of such logics, i.e. try to formulate a more general logic from which they can be obtained through the addition of special postulates. This perspective has been suggested in [Sambin *et al.*, 2000] and [Battilotti and Faggian, 2002], where the resulting logic is called *basic logic*[59].

Three general assumptions lay the groundwork for the development of basic logic: the principles of *reflection, symmetry,* and *visibility.* The principle of reflection amounts, roughly speaking, to the assumption that in the framework of sequent calculi each propositional connective reflects at the level of the object language a *link* between assertions formulated in an appropriate metalanguage. Such assertions are compounded out of basic assertions of the form α *is* by means of the links *and* and *yields*, which are sufficient to express all the metalinguistic assertions occurring in a sequent calculus:

[56]In any operational rule, the formula in the conclusion that contains the connective introduced by the rule in question is called the *principal formula*; the formulas in the premisses that are the components of the formula introduced by the rule are called the *auxiliary formulas*. All other formulas occurring in the rule are called *side formulas* (or also *context-formulas*).

[57]More precisely, it is a calculus which generates the involution lattice inequalities valid in all ortholattices.

[58]In the algebraic semantics, this logic is characterized by the class of all models based on involution lattices that are not necessarily bounded. Hence, *truth* and *logical truth* cannot be defined in this framework; at the same time, *consequence* and *logical consequence* are defined in the usual way.

[59]This label, as a matter of fact, is all too overworked in the field of non-classical logics. To the best of our knowledge, there are three more established logics with this same name (a fuzzy logic, introduced by Hajek; a relevant logic, introduced by Routley; and a constructive logic, introduced by Visser).

the sequent calculus configuration	abbreviates	the complex assertion
$\alpha_1, ..., \alpha_n$	$\hookrightarrow$	$(\alpha_1\ is)\ and...and\ (\alpha_n\ is)$
$\Gamma \Rightarrow \Delta$	$\hookrightarrow$	$\Gamma\ yields\ \Delta$
$\dfrac{\Gamma \Rightarrow \Delta \quad \Pi \Rightarrow \Sigma}{\Lambda \Rightarrow \Theta}$	$\hookrightarrow$	$(\Gamma \Rightarrow \Delta)\ and\ (\Pi \Rightarrow \Sigma)$ $yields\ (\Lambda \Rightarrow \Theta)$

The truth conditions for a given connective are given by its *definitional equation*, a metalinguistic biconditional which relates a definiendum, containing the connective, and a definiens, not containing it. Definitional equations provide the justification for the inference rules of the calculus, obtained by "solving" such equations according to a general method[60]. More precisely, every connective has a *formation rule*, derived from the direction of the definitional equation which gives sufficient conditions for asserting a formula containing the connective at issue, and a *reflection rule*, which licenses deductions from an already available formula containing the connective. For example, here are the definitional equations for the multiplicative ($\otimes$) and the additive ($\sqcap$) conjunction of basic logic:

(DE$\otimes$) $(\alpha \otimes \beta\ is)\ yields\ \Gamma$ iff $(\alpha\ is\ and\ \beta\ is)\ yields\ \Gamma$;

(DE$\sqcap$) $\Gamma\ yields\ (\alpha \sqcap \beta\ is)$ iff $(\Gamma\ yields\ (\alpha\ is))\ and\ (\Gamma\ yields\ (\beta\ is))$.

Remark that multiplicative conjunction reflects an *and* link within the scope of a *yields* link, while additive conjunction reflects a principal *and* link. Solving such definitional equations one gets the rules

$$\frac{\alpha, \beta \Rightarrow \Gamma}{\alpha \otimes \beta \Rightarrow \Gamma}\ (\otimes \mathrm{F}) \qquad \frac{\Gamma \Rightarrow \alpha \qquad \Delta \Rightarrow \beta}{\Gamma, \Delta \Rightarrow \alpha \otimes \beta}\ (\otimes \mathrm{R})$$

$$\frac{\Gamma \Rightarrow \alpha \qquad \Gamma \Rightarrow \beta}{\Gamma \Rightarrow \alpha \sqcap \beta}\ (\sqcap \mathrm{F}) \qquad \frac{\alpha \Rightarrow \Gamma}{\alpha \sqcap \beta \Rightarrow \Gamma} \qquad \frac{\beta \Rightarrow \Gamma}{\alpha \sqcap \beta \Rightarrow \Gamma}\ (\sqcap \mathrm{R})$$

One can observe how the formation rule for multiplicative conjunction operates on the antecedent of the sequent, whereas the reflection rule operates on the succedent; the rules for additive conjunction, instead, follow the opposite pattern. All the other connectives of basic logic can be likewise divided into *left* connectives and *right* connectives, according as they behave like $\otimes$ or like $\sqcap$ under the mentioned respect. Every left connective is matched by a *symmetric* right connective: for example, $\otimes$ is matched by a right multiplicative disjunction $\oplus$, while $\sqcap$ is matched by a left additive disjunction $\sqcup$. Furthermore, one can notice that all the rules of basic logic satisfy, like in the example cited above, a *visibility* requirement: there are no side formulas on the same side of either the principal, or the auxiliary formulas. Said otherwise, basic logic adds to the control of structural rules typical

[60]for the details see e.g. [Sambin *et al.*, 2000].

of linear logic a control of side formulas, both on the left and on the right of the arrow.

How can we get back from here to where we started, namely to linear and quantum logics? Well, we can extend basic logic in at least three ways: relaxing the visibility constraints, either on the left (**L**) or on the right (**R**) of the arrow, or restoring the structural rules of weakening and contraction (**S**). According to the chosen path, we get:

- **BL**, intuitionistic linear logic without exponentials[61];
- **BR**, "dual intuitionistic" linear logic without exponentials;
- **BS**, basic orthologic, an expansion of paraconsistent quantum logic;
- **BLR**, linear logic without exponentials;
- **BLS**, intuitionistic logic;
- **BRS**, "dual intuitionistic" logic;
- **BLRS**, classical logic.

In particular, a calculus for **PQL** can be obtained by: (i) confining ourselves to the $\{\sqcap, \sqcup\}$ fragment of **BS**; (ii) introducing negation definitionally: in other words, we start from *positive literals* (atomic sentences) and *negative literals* (the negations of such) and we inductively stipulate that:

$$\neg\neg\mathbf{p} := \mathbf{p};$$
$$\neg(\alpha \sqcap \beta) := \neg\alpha \sqcup \neg\beta;$$
$$\neg(\alpha \sqcup \beta) := \neg\alpha \sqcap \neg\beta.$$

This calculus is open to further simplification once we consider that weakening and contraction rules are redundant in it: the calculus admits elimination of contraction, while left and right weakening are respectively simulated by ($\sqcap$R) and ($\sqcup$R). In addition, we can turn it into a sequent system for orthologic if we restore the weakening and contraction rules and we adjoin two rules of *transfer* — a means to displace whole multisets of formulae across the arrow:

$$\frac{\Gamma \Rightarrow \Delta}{\Gamma, \neg\Delta \Rightarrow}\ (tr1) \qquad\qquad \frac{\Gamma \Rightarrow \Delta}{\Rightarrow \neg\Gamma, \Delta}\ (tr2)$$

The visibility and symmetry properties enable to overcome in a simple way two problems that usually make cut elimination for orthologic so complicated: respectively, the constraints on contexts and negation. For a presentation of the cut elimination proof, see e.g. [Battilotti and Faggian, 2002].

Another significant connection between the proof theories of linear logic and quantum logics can be found in the formalism of *proofnets*. Girard [Girard, 1987] introduced proofnets in order to amend three flaws of sequent calculi:

[61]To be precise, what we obtain is an expansion of this logic by a non-associative multiplicative disjunction.

- Some "informal" proofs of linear logical sequents are represented in **LL** by more than one proof; formally speaking, they count as distinct proofs only because the order of application of the rules is different. For example:

$$\mathcal{D}: \dfrac{\dfrac{\alpha \Rightarrow \alpha \qquad \beta \Rightarrow \beta}{\alpha, \beta \Rightarrow \alpha \otimes \beta}}{\beta \Rightarrow \neg\alpha, \alpha \otimes \beta} \qquad \mathcal{D}': \dfrac{\dfrac{\alpha \Rightarrow \alpha}{\Rightarrow \alpha, \neg\alpha} \qquad \beta \Rightarrow \beta}{\beta \Rightarrow \neg\alpha, \alpha \otimes \beta}$$

  We might therefore want to find a calculus where these "bureaucratical variants" correspond to one and the same proof, as it would seem appropriate.

- Sequent proofs are extremely redundant, because side formulas keep being copied again and again with each inference of the proof itself.

- The cut elimination procedure is highly nondeterministic. For example, consider the proof:

$$\dfrac{\dfrac{\begin{matrix}\vdots\\ \Gamma \Rightarrow \alpha, \beta, \sigma\end{matrix}}{\Gamma \Rightarrow \alpha \oplus \beta, \sigma} \qquad \dfrac{\begin{matrix}\vdots\\ \sigma, \gamma, \delta \Rightarrow \Delta\end{matrix}}{\sigma, \gamma \otimes \delta \Rightarrow \Delta}}{\gamma \otimes \delta, \Gamma \Rightarrow \Delta, \alpha \oplus \beta}$$

  The indicated cut can be pushed upwards and replaced by a cut whose premisses are $\Gamma \Rightarrow \alpha, \beta, \sigma$ and $\sigma, \gamma, \delta \Rightarrow \Delta$. But what comes next? We can either go on with a ($\otimes$L) inference and then apply the ($\oplus$R) rule, or proceed the other way around. Both choices are equally legitimate.

The calculus of proofnets is free of these shortcomings and works perfectly well at least for the constant-free multiplicative fragment of linear logic. Put briefly, one associates sequent proofs to special undirected graphs, in such a way that a graph actually represents a sequent proof if and only if it satisfies a simple, purely geometrical, criterion (namely, if and only if all the graphs one obtains from it by omitting edges according to a specified procedure are acyclic and connected: [Danos and Reignier, 1989]). This is all well and good so far as we remain confined within the multiplicative fragment; however, a completely satisfactory way to fit *additives* into the picture has not yet been found, despite several attempts by [Girard, 1996], [Tortora De Falco, 2003], [Hughes and van Glabbeek, 2003]. Restall and Paoli [Restall and Paoli, 2005] reversed, so to speak, the perspective and provided a simple calculus of proofnets for the linear logical additives *alone*, i.e. for the logic of general (possibly nondistributive) lattices, also giving a purely geometrical correctness criterion. Even though the possibility to extend such a calculus to the multiplicative fragment remains to be explored, this approach yielded a bonus — viz., a proofnet formulation for a number of quantum logics expressed in the language of involution lattices, such as an unbounded version of paraconsistent quantum logic and orthologic.

9.3 *Quantum logic* vs *linear logic: semantics of provability*

Although Girard repeatedly pointed out that, in his opinion, the only meaningful semantics for a logical system is its *semantics of proofs*, in his 1987 paper he does provide linear logic with a more traditional, Kripke-style semantics of provability, which he dubs *phase semantics.*

This circumstance suggests to highlight another noteworthy similarity between linear and quantum logics. As we already know, both orthologic (or weakenings thereof) and linear logic share the failure of lattice distributivity. In particular, the fragment of linear logic that includes just negation and the *additive* connectives is nothing but a version of the paraconsistent quantum logic **PQL**, whose algebraic counterpart, as we have seen, is the class of bounded involution lattices, and whose proper Kripkean counterpart is the class of *symmetric frames.* In linear logic, however, we have to model a wider range of connectives, including the multiplicatives and the modalities. True to form, adequate Kripkean models for full linear logic can be obtained by tinkering with symmetric frames in an appropriate way. Let us now fill the details, which we partly draw from [Gallier, 1991]. For the sake of simplicity, we will leave the modalities aside; throughout this section, therefore, **LL** will not denote the whole calculus, but only its fragment corresponding to the language $\mathcal{L}_0$.

DEFINITION 69. *Monoidal symmetric frame*
A *monoidal symmetric frame* is a first order structure $\mathbf{I} = (I, \bullet, \mathbf{1}, \not\perp)$ such that $(I, \not\perp)$ is a symmetric frame, $(I, \bullet, \mathbf{1})$ is an Abelian monoid[62] and, for every $a, b, c \in I$,

$$a \not\perp b \bullet c \text{ iff } a \bullet b \not\perp c$$

Intuitively, the elements of I are information tokens which may or may not verify a given formula; the monoidal operation can be understood as *composition* of information bits; $\mathbf{1}$ represents a "true" piece of information; finally, that a is in the relation $\not\perp$ with b should be taken to mean that the information in a does not conflict with the information in b.

The additional monoidal operation turns out to be the right key to access linear logical multiplicative connectives. Recalling the notion of closure of a subset in a symmetric frame[63], and recalling the notion of *generalized subset product* in a monoid:

$$XY = \{a \bullet b : a \in X, b \in Y\}$$

we can define the following operations between subsets of a monoidal symmetric frame $\mathbf{I}$:

$$X \boxtimes Y = (XY)^{\perp\perp};$$
$$X \boxplus Y = (X^{\perp}Y^{\perp})^{\perp};$$
$$X \rightsquigarrow Y = (X \boxtimes Y^{\perp})^{\perp}.$$

---

[62]See Def.137.

[63]Given a symmetric frame, define $X^{\perp} := \{a \in X : a \perp X\}$. The operation ${}^{\perp\perp}$ turns out to be a closure operator. See Section 1.

It is immediate to verify that $X \boxplus Y$ and $X \rightsquigarrow Y$ are closed subsets of I, that $X \boxtimes Y = (X^{\perp} \boxplus Y^{\perp})^{\perp}$, and that $X \rightsquigarrow Y = (XY^{\perp})^{\perp}$. Furthermore, the definition of implication makes sense in the light of the following

LEMMA 70. *Let X, Y be closed subsets of I and $a \in I$. We have:* *$a \in X \rightsquigarrow Y$ iff for every $b \in I$, if $b \in X$ then $a \bullet b \in Y$.*

Monoidal symmetric frames can be used to interpret the formulas and sequents of linear logic. Basically, we associate to each formula the set of information bits which verify it — formally speaking, a closed subset in a monoidal symmetric frame.

DEFINITION 71. *Monoidal symmetric model*
A *monoidal symmetric model* is a system $\mathcal{M} = \langle I, \bullet, \mathbf{1}, \not\perp, v \rangle$, where

(i) $\langle I, \bullet, \mathbf{1}, \not\perp \rangle$ is a monoidal symmetric frame;

(ii) v is a function that associates to any sentence α a closed subset of I, satisfying the following conditions:

$$\begin{array}{ll} v(\neg\alpha) = v(\alpha)^{\perp}; & v(\alpha \otimes \beta) = v(\alpha) \boxtimes v(\beta); \\ v(\alpha \oplus \beta) = v(\alpha) \boxplus v(\beta); & v(\alpha \rightarrow \beta) = v(\alpha) \rightsquigarrow v(\beta); \\ v(\alpha \sqcap \beta) = v(\alpha) \cap v(\beta); & v(\alpha \sqcup \beta) = (v(\alpha) \cup v(\beta))^{\perp\perp}; \\ v(\mathbf{t}) = \{\mathbf{1}\}^{\perp\perp}; & v(\mathbf{f}) = \{a \in I : a \perp \mathbf{1}\}; \\ v(\top) = I; & v(\bot) = I^{\perp}. \end{array}$$

DEFINITION 72. *Truth and logical truth*
A formula α is *true* in a model $\mathcal{M} = \langle I, \bullet, \mathbf{1}, \not\perp, v \rangle$ (abbreviated $\models_{\mathcal{M}} \alpha$) iff $\mathbf{1} \in v(\alpha)$;
α is a *logical truth* (or a *valid formula*) of **LL** ($\models_{\mathbf{LL}} \alpha$) iff for any model $\mathcal{M}$, $\models_{\mathcal{M}} \alpha$.

DEFINITION 73. Let Γ be a finite, possibly empty multiset of formulas in $\mathcal{L}_0$; we define:

$$t^-(\Gamma) = \begin{cases} \alpha_1 \otimes \ldots \otimes \alpha_n, \text{ if } \Gamma = \{\alpha_1, \ldots, \alpha_n\}; \\ \mathbf{t}, \text{ otherwise.} \end{cases}$$

$$t^+(\Gamma) = \begin{cases} \alpha_1 \oplus \ldots \oplus \alpha_n, \text{ if } \Gamma = \{\alpha_1, \ldots, \alpha_n\}; \\ \mathbf{f}, \text{ otherwise.} \end{cases}$$

LEMMA 74. $\vdash_{\mathbf{LL}} \Gamma \Rightarrow \Delta$ *iff* $\vdash_{\mathbf{LL}} \Rightarrow t^-(\Gamma) \rightarrow t^+(\Delta)$.

We have the following completeness theorem:

THEOREM 75. $\vdash_{\mathbf{LL}} \Gamma \Rightarrow \Delta$ *iff* $t^-(\Gamma) \rightarrow t^+(\Delta)$ *is valid.*[64]

In Girard's paper, phase semantics is not introduced in terms of monoidal symmetric frames, but rather in terms of a slightly different class of structures, called *phase structures*. Although we preferred the alternative presentation in that it simplifies comparison with the relational semantics for quantum logics, both perspectives are readily seen to be absolutely equivalent.

---

[64] See [Girard, 1987].

9.4 *Quantum logic* vs *linear logic: semantics of proofs*

Girard's attitude towards quantum logic is utterly disparaging. Rather than abstracting away from the structures of quantum mechanics in order to get a logic — a process which is bound to yield not a "theory of reasoning", but only a game of algebra written in a different guise — he claims that one should go the other way around and use quantum structures to interpret logical proofs, in particular linear logical proofs [Girard, 1999], [Girard, 2004a], [Girard, 2004b]. Once we strip away from it similar anathemas — which are invariably, we think, the most infelicitous component of Girard's contributions — this particular variant of denotational semantics may be worth a mention. Thus, we now turn to expound its main features.

Some linear logical proofs have a probabilistic flavour. Consider e.g. a possible proof for the sequent $\Rightarrow \alpha \sqcup \beta$. How could we have obtained it? Well, for sure we applied the ($\sqcup$R) rule to get to it, but we can only guess as to the form of its premiss: the best we can say is that there is a 50% chance for $\Rightarrow \alpha$ and a 50% chance for $\Rightarrow \beta$ (the 50-50 assignment being a default one, of course). Thus, we cannot maintain with certainty that the information tokens constituting a proof of $\Rightarrow \alpha$ belong to the envisaged proof of $\Rightarrow \alpha \sqcup \beta$; we can only assign a probability to their so doing. This leads us to revise our previous concept of a proof as a "crisp" set of information tokens belonging to a set X — or, in other words, as a (characteristic) function from X to $\{0, 1\}$. A more realistic proof model will rather be a function from X to the *whole real interval* $[0, 1]$, intuitively representing a probability assignment which outputs, for each information token, a "likelihood measure" of its membership in the given proof. Formally:

DEFINITION 76. Let X be a finite non empty set and let $\alpha \subseteq \mathbb{R}^{+X}$ (where $\mathbb{R}^{+X}$ is the set of all functions from X to the positive reals). The *polar* of α (denoted by $\neg\alpha$), is defined as follows:

$$\neg\alpha = \{f \in \mathbb{R}^{+X} : \text{ for every } g \in \alpha, \sum_{x \in X} f(x) \cdot g(x) \leq 1\}.$$

The rationale for this definition is clear enough: if f, g are the characteristic functions, respectively, of the sets Y, Z, then $\sum_{x \in X} f(x) \cdot g(x)$ is nothing but the cardinality of $Y \cap Z$. Definition 76 is therefore a natural generalization of Definition 60, meaning that two proofs are "orthogonal" whenever, allowing for mutual compensation effects, there is a maximal probability that they share at most one piece of information.

DEFINITION 77. A *probabilistic coherent space* over the set X is a set $\alpha \subseteq \mathbb{R}^{+X}$ s.t. $\alpha = \neg\neg\alpha$.

Let us now examine the treatment of additive connectives within this framework. Since there is nothing probabilistic about additive conjunction, we may simply generalize the classical definition to our real-valued function setting; to cope with additive disjunction, on the other hand, we need to identify each one of its proofs with a probability assignment that weighs up the respective chances of its stemming from a proof of the former or from a proof of the latter disjunct. We are thus led to the following definitions:

DEFINITION 78. If $f \in \mathbb{R}^{+X}$ and $g \in \mathbb{R}^{+Y}$, where X, Y are disjoint sets, then $f \cup g \in \mathbb{R}^{+X \cup Y}$ is so defined:

$$(f \cup g)(a) = \begin{cases} f(a), \text{ if } a \in X; \\ g(a), \text{ if } a \in Y. \end{cases}$$

DEFINITION 79. Let α and β be two probabilistic coherent spaces, respectively over X and over Y (which we assume to be disjoint sets). We define:

$$\begin{aligned} \alpha \sqcap \beta &= \{f \cup g : f \in \alpha, g \in \beta\}; \\ \alpha \sqcup \beta &= \{\lambda f \cup (1-\lambda) g : f \in \alpha, g \in \beta, 0 \leq \lambda \leq 1\}. \end{aligned}$$

This definition of additive conjunction squares with the standard coherent space framework in that one can prove that $\alpha \sqcup \beta = \neg\neg(\alpha \cup \beta)$, as expected.

The standard coherent space perspective dictates that a proof for an implicative sentence be a function which extracts a proof for the consequent from a proof for the antecedent; here, however, proofs are themselves functions — it follows that the objects which correspond to proofs of implications will be *functionals*. Yet, this is not exactly what we want: to parallel our development of coherent spaces in the previous section, we need to simulate each linear functional from $\mathbb{R}^{+X}$ to $\mathbb{R}^{+Y}$ by means of a function in $\mathbb{R}^{+X \times Y}$. In other words, we want an implication whose antecedent is a probabilistic space α over X and whose consequent is a probabilistic space β over Y to be a space $\alpha \to \beta$ over $X \times Y$ whose members are "linear traces" of functionals from α to β.

DEFINITION 80. If $h : X \times Y \mapsto \mathbb{R}^{+}$, then $\widetilde{h} : \mathbb{R}^{+X} \mapsto \mathbb{R}^{+Y}$ is defined in such a way that, for every $f \in \mathbb{R}^{+X}$ and for every $a \in X$,

$$\left(\widetilde{h}(f)\right)(a) = \sum_{x \in X} h((x, a)) \cdot f(x)$$

(Remenber that X is a finite set!) The next lemma guarantees the adequacy of the previous definition:

LEMMA 81. *The map sending h to $\widetilde{h}$ is a bijection from $\mathbb{R}^{+X \times Y}$ onto the set of linear functions from $\mathbb{R}^{+X}$ to $\mathbb{R}^{+Y}$.*

A fairly straightforward adaptation of the classical case thus gives:

DEFINITION 82. Let α, β be probabilistic coherent spaces, respectively over X and Y. We define $\alpha \to \beta$ over $X \times Y$ as follows:

$$\alpha \to \beta = \left\{h : \widetilde{h}(f) \in \beta \text{ for every } f \in \alpha\right\}.$$

One then proceeds to show that $\alpha \to \beta$ is indeed a probabilistic coherent space, a fact that carries along the obvious definitions for the multiplicatives:

DEFINITION 83. Let α and β be two probabilistic coherent spaces, respectively over X and over Y (which we assume to be disjoint sets). We define:

$$\alpha \otimes \beta = \neg(\alpha \to \neg\beta);$$
$$\alpha \oplus \beta = \neg\alpha \to \beta.$$

It is not infrequent to find in the linear logical literature the following analogy: formulas are to proofs as states (of a system) are to transitions between states. This perspective suggests a rather natural generalization of the above, obtained by formalizing linear logical proofs not through real-valued functions, but by means of self-adjoint operators on a complex finite-dimensional Hilbert space — the idea being that each state is identified with the set of all transitions leading to it.

DEFINITION 84. Let $\mathcal{H}_{\mathbf{X}}, \mathcal{H}_{\mathbf{Y}}$ be complex finite-dimensional Hilbert spaces. We denote by $\mathcal{L}(\mathcal{H}_{\mathbf{X}}, \mathcal{H}_{\mathbf{Y}})$ the set of all linear maps from $\mathcal{H}_{\mathbf{X}}$ to $\mathcal{H}_{\mathbf{Y}}$, abbreviating $\mathcal{L}(\mathcal{H}_{\mathbf{X}}, \mathcal{H}_{\mathbf{X}})$ by $\mathcal{L}(\mathcal{H}_{\mathbf{X}})$. By $\mathcal{H}^{\mathcal{L}(\mathcal{H}_{\mathbf{X}})}$ we mean the set of all self-adjoint operators on $\mathcal{H}_{\mathbf{X}}$.

The set $\mathcal{H}^{\mathcal{L}(\mathcal{H}_{\mathbf{X}})}$ can be turned in the usual way into a vector space (over $\mathbb{R}$), endowed with an inner product *à la* Hilbert-Schmidt:

$$\langle U \mid V \rangle = \mathtt{tr}(UV).$$

DEFINITION 85. Let $\alpha \subseteq \mathcal{H}^{\mathcal{L}(\mathcal{H}_{\mathbf{X}})}$. The *polar* of α (denoted by $\neg\alpha$), is defined as follows:

$$\neg\alpha = \{U : \text{ for every } V \in \alpha, 0 \leq \langle U \mid V \rangle \leq 1\}$$

As far as we stay within the domain of positive reals, the inner product of two operators is a measure of their orthogonality: the lower the product, "the more orthogonal" they are.

DEFINITION 86. A *quantum coherent space* over the space $\mathcal{H}_{\mathbf{X}}$ is a set $\alpha \subseteq \mathcal{H}^{\mathcal{L}(\mathcal{H}_{\mathbf{X}})}$ s.t. $\alpha = \neg\neg\alpha$.

Recall that, in standard coherent space semantics, additive (respectively, multiplicative) coherent spaces were defined over the disjoint union (respectively, the cartesian product) of the sets underlying the constituent spaces. Here, disjoint union is replaced by direct sum of vector spaces, while tensor product plays the role of cartesian product [65]. We begin with the additives:

DEFINITION 87. Let α and β be two quantum coherent spaces over $\mathcal{H}_{\mathbf{X}}$ and $\mathcal{H}_{\mathbf{Y}}$, respectively. $\alpha \sqcap \beta$ and $\alpha \sqcup \beta$ are spaces over the direct sum $\mathcal{H}_{\mathbf{X}} \oplus \mathcal{H}_{\mathbf{Y}}$, defined as

[65] See Deff. 161 and 162

follows [66]:

$$\alpha \sqcap \beta = \{U : P_{\mathbf{X}} U P_{\mathbf{X}} = V \oplus \mathbb{O} \text{ and } P_{\mathbf{Y}} U P_{\mathbf{Y}} = \mathbb{O} \oplus W, \text{ for } V \in \alpha, W \in \beta\};$$
$$\alpha \sqcup \beta = \{\lambda(V \oplus \mathbb{O}) + (1-\lambda)(\mathbb{O} \oplus W) : V \in \alpha, W \in \beta, 0 \leq \lambda \leq 1\}.$$

While it is fairly clear why disjunction is treated like this (cp. Definition 79), the definition of conjunction deserves a bit of elucidation. Intuitively, $P_{\mathbf{X}} U P_{\mathbf{X}}$ and $P_{\mathbf{Y}} U P_{\mathbf{Y}}$ can be thought of as the "restrictions" of the operator U to the subspaces $\mathcal{H}_{\mathbf{X}}$ and $\mathcal{H}_{\mathbf{Y}}$, respectively. Indeed, if U is a self-adjoint operator on $\mathcal{H}_{\mathbf{X}} \oplus \mathcal{H}_{\mathbf{Y}}$ of the form $U_1 \oplus U_2$, then, for any $|\psi\rangle \in \mathcal{H}_{\mathbf{X}}$ and for any $|\varphi\rangle \in \mathcal{H}_{\mathbf{Y}}$:

$$P_{\mathbf{X}} U P_{\mathbf{X}}(|\psi\rangle \oplus |\varphi\rangle) = U_1(|\psi\rangle) \oplus \mathbb{O}$$

and

$$P_{\mathbf{Y}} U P_{\mathbf{Y}}(|\psi\rangle \oplus |\varphi\rangle) = \mathbb{O} \oplus U_2(|\varphi\rangle).$$

Accordingly, if we define

$$P_{\mathbf{X}} U P_{\mathbf{X}} \cup P_{\mathbf{Y}} U P_{\mathbf{Y}} = P_{\mathbf{X}} U P_{\mathbf{X}} + P_{\mathbf{Y}} U P_{\mathbf{Y}},$$

we obtain in this case:

$$(P_{\mathbf{X}} U P_{\mathbf{X}} \cup P_{\mathbf{Y}} U P_{\mathbf{Y}})(|\psi\rangle \oplus |\varphi\rangle) = U_1(|\psi\rangle) \oplus U_2(|\varphi\rangle).$$

Notice, however, that, in general, not every self-adjoint operator on $\mathcal{H}_{\mathbf{X}} \oplus \mathcal{H}_{\mathbf{Y}}$ has the form $U_1 \oplus U_2$.

As regards the multiplicatives, our starting point is the observation that, if $\mathcal{H}_{\mathbf{X}}$ and $\mathcal{H}_{\mathbf{Y}}$ are finite-dimensional Hilbert spaces, the space of the linear maps from the space of the operators over $\mathcal{H}_{\mathbf{X}}$ to the space of the operators over $\mathcal{H}_{\mathbf{Y}}$ is isomorphic to the space of the operators over the tensor product $\mathcal{H}_{\mathbf{X}} \otimes \mathcal{H}_{\mathbf{Y}}$.

DEFINITION 88. If $U \in \mathcal{L}(\mathcal{H}_{\mathbf{X}} \otimes \mathcal{H}_{\mathbf{Y}})$, then $\widetilde{U}$ is a linear map belonging to $\mathcal{L}(\mathcal{L}(\mathcal{H}_{\mathbf{X}}), \mathcal{L}(\mathcal{H}_{\mathbf{Y}}))$, defined in such a way that, $\forall V \in \mathcal{L}(\mathcal{H}_{\mathbf{X}})$, $\forall |\psi\rangle, |\varphi\rangle \in \mathcal{H}_{\mathbf{Y}}$:

$$\left\langle \left(\widetilde{U}(V)\right)(|\psi\rangle) | \varphi \right\rangle = \mathtt{tr}(U \cdot (V \otimes W_{|\psi\rangle,|\varphi\rangle})),$$

where, $\forall |\chi\rangle \in \mathcal{H}_{\mathbf{Y}}$: $W_{|\psi\rangle,|\varphi\rangle}(|\chi\rangle) = \langle \chi|\psi\rangle|\varphi\rangle$.

Restricting ourselves to self-adjoint operators we get, in full analogy with the probabilistic case:

LEMMA 89. *The map sending U to $\widetilde{U}$ is a bijection from $\mathcal{H}^{\mathcal{L}(\mathcal{H}_{\mathbf{X}} \otimes \mathcal{H}_{\mathbf{Y}})}$ onto $\mathcal{L}\left(\mathcal{H}^{\mathcal{L}(\mathcal{H}_{\mathbf{X}})}, \mathcal{H}^{\mathcal{L}(\mathcal{H}_{\mathbf{Y}})}\right)$.*

---

[66] $P_{\mathbf{X}}$ and $P_{\mathbf{Y}}$ denote the projection operators onto the spaces $\mathcal{H}_{\mathbf{X}}$ and $\mathcal{H}_{\mathbf{Y}}$, respectively. In other terms, for any $|\psi\rangle \in \mathcal{H}_{\mathbf{X}}$ and for any $|\varphi\rangle \in \mathcal{H}_{\mathbf{Y}}$: $P_{\mathbf{X}}(|\psi\rangle \oplus |\varphi\rangle) = |\psi\rangle \oplus \underline{0}$ and $P_{\mathbf{Y}}(|\psi\rangle \oplus |\varphi\rangle) = \underline{0} \oplus |\varphi\rangle$.

If U and V are self-adjoint operators on $\mathcal{H}_{\mathbf{X}}$ and $\mathcal{H}_{\mathbf{Y}}$, respectively, and $\lambda \in \mathbb{R}$, then $U \oplus V$ and $\lambda(U \oplus V)$ are (self-adjoint) operators on $\mathcal{H}_{\mathbf{X}} \oplus \mathcal{H}_{\mathbf{Y}}$ such that for any $|\psi\rangle \in \mathcal{H}_{\mathbf{X}}$ and for any $|\varphi\rangle \in \mathcal{H}_{\mathbf{Y}}$: $(U \oplus V)(|\psi\rangle \oplus |\varphi\rangle) = U(|\psi\rangle) \oplus V(|\varphi\rangle)$ and $\lambda(U \oplus V)(|\psi\rangle \oplus |\varphi\rangle) = \lambda U(|\psi\rangle) \oplus \lambda V(|\varphi\rangle)$.

Thus, as expected,

DEFINITION 90. Let α, β be quantum coherent spaces, respectively over $\mathcal{H}_{\mathbf{X}}$ and $\mathcal{H}_{\mathbf{Y}}$. We define $\alpha \rightarrow \beta$ over $\mathcal{H}_{\mathbf{X}} \otimes \mathcal{H}_{\mathbf{Y}}$ as follows:

$$\alpha \rightarrow \beta = \left\{ U : \widetilde{U}\,(V) \in \beta \text{ for every } V \in \alpha \right\}.$$

One then goes on to show that $\alpha \rightarrow \beta$ is indeed a quantum coherent space, a fact that leads once again to the appropriate definitions for the multiplicatives:

DEFINITION 91. Let α and β be two quantum coherent spaces, respectively over $\mathcal{H}_{\mathbf{X}}$ and over $\mathcal{H}_{\mathbf{Y}}$. We define:

$$\alpha \otimes \beta = \neg(\alpha \rightarrow \neg\beta);$$
$$\alpha \oplus \beta = \neg\alpha \rightarrow \beta.$$

10 QUANTUM LOGIC AND QUANTUM COMPUTATION

Quantum computation has suggested new forms of quantum logic that have been called *quantum computational logics*.[67] The main difference between these new logics and traditional (sharp and unsharp) quantum logics concerns a basic semantic question: how to represent the *meanings* of the sentences of a given language? As we have learnt, the answer given by "orthodox" quantum logic was the following: the meanings of the elementary experimental sentences of quantum theory have to be regarded as determined by appropriate sets of *states* of quantum objects (mathematically represented by closed subspaces of a Hilbert space). The answer given in the framework of quantum computational logics is quite different. The *meaning* of a sentence is identified with a quantum information quantity: a system of *qubits* or, more generally, a *mixture* of systems of qubits.[68]

Two kinds of *quantum computational semantics* have been investigated:

- A *compositional* semantics, where (like in classical logic) the *meaning* of a compound sentence is determined by the meanings of its parts.
- A *holistic* semantics, which makes essential use of the characteristic "holistic" features of the quantum-theoretic formalism. Hence, in this framework, the meaning of a compound sentence generally determines the meanings of its parts, but not the other way around.

Let us first recall some basic definitions of quantum computation. Consider the two-dimensional Hilbert space $\mathbb{C}^2$ (where any vector $|\psi\rangle$ is represented by a pair of complex numbers). Let $\mathcal{B}^{(1)} = \{|0\rangle, |1\rangle\}$ be the canonical orthonormal basis for $\mathbb{C}^2$, where $|0\rangle = (1, 0)$ and $|1\rangle = (0, 1)$.

---

[67] See [Dalla Chiara *et al.*, 2003].

[68] Other logical approaches have investigated some interesting relations between quantum computation and linear logic. See, for instance, [Pratt, 1993],[Selinger, 2004],[van Tonder, 2004].

DEFINITION 92. *Qubit*
A *qubit* is a unit vector $|\psi\rangle$ of the Hilbert space $\mathbb{C}^2$.

From an intuitive point of view, a qubit can be regarded as a quantum variant of the classical notion of bit: a kind of "quantum perhaps". In this framework, the two basis-elements $|0\rangle$ and $|1\rangle$ represent the two classical bits 0 and 1, respectively. From a physical point of view, a qubit represents a *state* of a single particle, carrying an atomic piece of quantum information. In order to carry the information stocked by n qubits, we need of course a compound system, consisting of n particles.

DEFINITION 93. *Quregister*
An n-qubit system (also called n-*quregister*) is a unit vector in the n-fold tensor product Hilbert space $\otimes^n\mathbb{C}^2 := \underbrace{\mathbb{C}^2 \otimes \ldots \otimes \mathbb{C}^2}_{n-times}$ (where $\otimes^1\mathbb{C}^2 := \mathbb{C}^2$).[69]

We will use $x, y, \ldots$ as variables ranging over the set $\{0,1\}$. At the same time, $|x\rangle, |y\rangle, \ldots$ will range over the basis $\mathcal{B}^{(1)}$. Any factorized unit vector $|x_1\rangle\otimes\ldots\otimes|x_n\rangle$ of the space $\otimes^n\mathbb{C}^2$ will be called a *classical register*. Instead of $|x_1\rangle \otimes \ldots \otimes |x_n\rangle$ we will simply write $|x_1, \ldots, x_n\rangle$. The set $\mathcal{B}^{(n)}$ of all classical registers is an orthonormal basis for the space $\otimes^n\mathbb{C}^2$.

Quregisters are *pure states*: maximal pieces of information about the particles under consideration. In quantum computation (as well as in quantum theory), one cannot help referring also to *mixed states* (pieces of information that are not maximal and might be enriched). In the framework of quantum computation, mixed states (mathematically represented by density operators of an appropriate Hilbert space) are also called *qumixes*.

DEFINITION 94. *Qumix*
A *qumix* is a density operator of $\otimes^n\mathbb{C}^2$ (where $n \geq 1$).[70]

Needless to say, quregisters correspond to particular qumixes that are *pure states* (i.e. projections onto one-dimensional closed subspaces of a given $\otimes^n\mathbb{C}^n$). We will indicate by $\mathfrak{D}(\otimes^n\mathbb{C}^2)$ the set of all density operators of $\otimes^n\mathbb{C}^2$. Hence the set $\mathfrak{D} = \bigcup_{n=1}^{\infty} \mathfrak{D}(\otimes^n\mathbb{C}^2)$ will represent the set of all possible qumixes.

A classical register $|x_1, \ldots, x_n\rangle$ is called *true*, when $x_n = 1$; *false*, otherwise. The idea is that any classical register corresponds to a classical truth-value that is determined by its last element. Hence, in particular, the bit $|1\rangle$ corresponds to the truth-value *Truth*, while the bit $|0\rangle$ corresponds to the truth-value *Falsity*.

On this basis, we can identify, in any space $\otimes^n\mathbb{C}^2$, two special projection-operators ($P_1^{(n)}$ and $P_0^{(n)}$) that represent, in this framework, the *Truth-property* and the *Falsity-property*, respectively. The projection $P_1^{(n)}$ is determined by the closed subspace spanned by the set of all true registers, while $P_0^{(n)}$ is determined by the closed subspace spanned by the set of all false registers. As we already know, in quantum theory, projections have the role of *mathematical representatives* of

[69]See Def. 161.
[70]See Def. 159.

possible *events* (or *physical properties*) of the quantum objects under investigation. Hence, it turns out that *Truth* and *Falsity* behave here as special cases of quantum events.

As a consequence, one can naturally apply the *Born rule* that determines *the probability-value that a quantum system in a given state satisfies a physical property*. Consider any qumix ρ, which represents a possible state of a quantum system in the space $\otimes^n \mathbb{C}^2$. By applying the Born rule, we obtain that the probability-value that a physical system in state ρ satisfies the *Truth-property* $P_1^{(n)}$ is the number $\mathtt{tr}(\rho P_1^{(n)})$ (where $\mathtt{tr}$ is the *trace functional*).[71] This suggests the following natural definition of the notion of *probability* of a given qumix.

DEFINITION 95. *Probability of a qumix*
For any qumix $\rho \in \mathfrak{D}(\otimes^n \mathbb{C}^2)$:

$$\mathtt{p}(\rho) = \mathtt{tr}(\rho P_1^{(n)}).$$

From an intuitive point of view, $\mathtt{p}(\rho)$ represents the probability that the information stocked by the qumix ρ is true. In the particular case where ρ corresponds to the qubit

$$|\psi\rangle = c_0|0\rangle + c_1|1\rangle,$$

we obtain that $\mathtt{p}(\rho) = |c_1|^2$.

Given a quregister $|\psi\rangle$, we will also write $\mathtt{p}(|\psi\rangle)$ instead of $\mathtt{p}(P_{|\psi\rangle})$, where $P_{|\psi\rangle}$ is the density operator represented by the projection onto the one-dimensional subspace spanned by the vector $|\psi\rangle$.

10.1 Quantum gates

In quantum computation, *quantum logical gates* (briefly, *gates*) are unitary operators that transform quregisters into quregisters.[72] Being unitary, gates represent characteristic *reversible transformations*. The canonical gates (which are studied in the literature) can be naturally generalized to qumixes. Generally, gates correspond to some basic *logical operations* that admit a reversible behavior. We will consider here the following gates: the *negation*, the *Petri-Toffoli gate* and the *square root of the negation*.

Let us first describe these gates in the framework of quregisters.

DEFINITION 96. *The negation*
For any $n \geq 1$, the negation on $\otimes^n \mathbb{C}^2$ is the linear operator $\mathtt{Not}^{(n)}$ such that for every element $|x_1, \ldots, x_n\rangle$ of the basis $\mathcal{B}^{(n)}$:

$$\mathtt{Not}^{(n)}(|x_1, \ldots, x_n\rangle) := |x_1, \ldots, x_{n-1}\rangle \otimes |1 - x_n\rangle.$$

In other words, $\mathtt{Not}^{(n)}$ inverts the value of the last element of any basis-vector of $\otimes^n \mathbb{C}^2$.

---

[71] See Def. 158.
[72] See Def. 160.

DEFINITION 97. *The Petri-Toffoli gate*
For any $n \geq 1$ and any $m \geq 1$ the Petri-Toffoli gate is the linear operator $\mathtt{T}^{(n,m,1)}$ defined on $\otimes^{n+m+1}\mathbb{C}^2$ such that for every element $|x_1, \ldots, x_n\rangle \otimes |y_1, \ldots, y_m\rangle \otimes |z\rangle$ of the basis $\mathcal{B}^{(n+m+1)}$:

$$\mathtt{T}^{(n,m,1)}(|x_1, \ldots, x_n\rangle \otimes |y_1, \ldots, y_m\rangle \otimes |z\rangle) := |x_1, \ldots, x_n\rangle \otimes |y_1, \ldots, y_m\rangle \otimes |x_n y_m \oplus z\rangle,$$

where $\oplus$ represents the sum modulo 2.

One can easily show that both $\mathtt{Not}^{(n)}$ and $\mathtt{T}^{(n,m,1)}$ are unitary operators. Consider now the set $\mathfrak{R}$ of all quregisters $|\psi\rangle$. The gates $\mathtt{Not}$ and $\mathtt{T}$ can be uniformly defined on this set in the expected way:

$$\begin{aligned} &\mathtt{Not}(|\psi\rangle) := \mathtt{Not}^{(n)}(|\psi\rangle), \qquad \text{if } |\psi\rangle \in \otimes^n\mathbb{C}^2 \\ &\mathtt{T}(|\psi\rangle \otimes |\varphi\rangle \otimes |\chi\rangle) := \mathtt{T}^{(n,m,1)}(|\psi\rangle \otimes |\varphi\rangle \otimes |\chi\rangle), \\ &\qquad\qquad\qquad\qquad \text{if } |\psi\rangle \in \otimes^n\mathbb{C}^2,\ |\varphi\rangle \in \otimes^m\mathbb{C}^2 \text{ and } |\chi\rangle \in \mathbb{C}^2. \end{aligned}$$

On this basis, a conjunction $\mathtt{And}$, a disjunction $\mathtt{Or}$, can be defined for any pair of quregisters $|\psi\rangle$ and $|\varphi\rangle$:

$$\mathtt{And}(|\psi\rangle, |\varphi\rangle) := \mathtt{T}(|\psi\rangle \otimes |\varphi\rangle \otimes |0\rangle);$$

$$\mathtt{Or}(|\psi\rangle, |\varphi\rangle) := \mathtt{Not}(\mathtt{And}(\mathtt{Not}(|\psi\rangle), \mathtt{Not}(|\varphi\rangle))).$$

Clearly, $|0\rangle$ represents an "ancilla" in the definition of $\mathtt{And}$.

The quantum logical gates we have considered so far are, in a sense, "semiclassical". A quantum logical behavior only emerges in the case where our gates are applied to superpositions. When restricted to classical registers, such operators turn out to behave as classical (reversible) truth-functions. We will now consider an important example of a *genuine quantum gate* that transforms classical registers (elements of $\mathcal{B}^{(n)}$) into quregisters that are superpositions. This gate is the *square root of the negation.*

DEFINITION 98. *The square root of the negation*
For any $n \geq 1$, the square root of the negation on $\otimes^n\mathbb{C}^2$ is the linear operator $\sqrt{\mathtt{Not}}^{(n)}$ such that for every element $|x_1, \ldots, x_n\rangle$ of the basis $\mathcal{B}^{(n)}$:

$$\sqrt{\mathtt{Not}}^{(n)}(|x_1, \ldots, x_n\rangle) := |x_1, \ldots, x_{n-1}\rangle \otimes \frac{1}{2}((1+i)|x_n\rangle + (1-i)|1-x_n\rangle),$$

where $i := \sqrt{-1}$.

One can easily show that $\sqrt{\mathtt{Not}}^{(n)}$ is a unitary operator. The basic property of $\sqrt{\mathtt{Not}}^{(n)}$ is the following:

$$\text{for any } |\psi\rangle \in \otimes^n\mathbb{C}^2,\ \sqrt{\mathtt{Not}}^{(n)}(\sqrt{\mathtt{Not}}^{(n)}(|\psi\rangle)) = \mathtt{Not}^{(n)}(|\psi\rangle).$$

In other words, applying twice the square root of the negation means negating.

From a logical point of view, $\sqrt{\mathtt{Not}}^{(n)}$ can be regarded as a "tentative partial negation" (a kind of "half negation") that transforms *precise pieces of information* into *maximally uncertain* ones. For, we have:

$$\mathtt{p}(\sqrt{\mathtt{Not}}^{(1)}(|1\rangle)) = \frac{1}{2} = \mathtt{p}(\sqrt{\mathtt{Not}}^{(1)}(|0\rangle)).$$

As expected, also $\sqrt{\mathtt{Not}}$ can be uniformly defined on the set $\mathfrak{R}$ of all quregisters.

Interestingly enough, the gate $\sqrt{\mathtt{Not}}$ seems to represent a typically *quantum logical operation* that does not admit any counterpart either in classical logic or in standard fuzzy logics.

THEOREM 99. *[Dalla Chiara and Leporini, 2005]*

1. *There is no function* $f : \{0,1\} \mapsto \{0,1\}$ *such that for any* $x \in \{0,1\}$: $f(f(x)) = 1 - x$.
2. *There is no continuous function* $f : [0,1] \mapsto [0,1]$ *such that for any* $x \in [0,1] : f(f(x)) = 1 - x$.

The gates considered so far can be naturally generalized to qumixes [Gudder, 2003]. When our gates will be applied to density operators, we will write: $\mathtt{NOT}$, $\sqrt{\mathtt{NOT}}$, $\mathbb{T}$, $\mathtt{AND}$, $\mathtt{OR}$ (instead of $\mathtt{Not}$, $\sqrt{\mathtt{Not}}$, $\mathtt{T}$, $\mathtt{And}$, $\mathtt{Or}$).

DEFINITION 100. *The negation*
For any qumix $\rho \in \mathfrak{D}(\otimes^n \mathbb{C}^2)$,

$$\mathtt{NOT}^{(n)}(\rho) := \mathtt{Not}^{(n)}\, \rho\, \mathtt{Not}^{(n)}.$$

DEFINITION 101. *The square root of the negation*
For any qumix $\rho \in \mathfrak{D}(\otimes^n \mathbb{C}^2)$,

$$\sqrt{\mathtt{NOT}}^{(n)}(\rho) := \sqrt{\mathtt{Not}}^{(n)}\, \rho\, \sqrt{\mathtt{Not}}^{(n)*},$$

where $\sqrt{\mathtt{Not}}^{(n)*}$ is the adjoint of $\sqrt{\mathtt{Not}}^{(n)}$.[73]

It is easy to see that for any $n \in \mathbb{N}^+$, both $\mathtt{NOT}^{(n)}(\rho)$ and $\sqrt{\mathtt{NOT}}^{(n)}(\rho)$ are qumixes of $\mathfrak{D}(\otimes^n \mathbb{C}^2)$.

DEFINITION 102. *The conjunction*
Let $\rho \in \mathfrak{D}(\otimes^n \mathbb{C}^2)$ and $\sigma \in \mathfrak{D}(\otimes^m \mathbb{C}^2)$.

$$\mathtt{AND}^{(n,m,1)}(\rho, \sigma) = \mathbb{T}^{(n,m,1)}(\rho, \sigma, P_0^{(1)}) := \mathtt{T}^{(n,m,1)}(\rho \otimes \sigma \otimes P_0^{(1)})\mathtt{T}^{(n,m,1)}.$$

Like in the quregister-case, the gates $\mathtt{NOT}$, $\sqrt{\mathtt{NOT}}$, $\mathbb{T}$, $\mathtt{AND}$, $\mathtt{OR}$ can be uniformly defined on the set $\mathfrak{D}$ of all qumixes.

An interesting preorder relation can be defined on the set of all qumixes.

DEFINITION 103. *Preorder*
$\rho \preceq \sigma$ iff the following conditions hold:

[73] See Def. 155.

(i) $\mathtt{p}(\rho) \leq \mathtt{p}(\sigma)$;

(ii) $\mathtt{p}(\sqrt{\mathtt{NOT}}(\sigma)) \leq \mathtt{p}(\sqrt{\mathtt{NOT}}(\rho))$.

One immediately shows that $\preceq$ is reflexive and transitive, but not antisymmetric. Counterexamples can be easily found in $\mathfrak{D}(\mathbb{C}^2)$.

10.2 The compositional quantum computational semantics

Both the compositional and the holistic semantics are based on the following intuitive idea: any *sentence* α of the language is interpreted as an appropriate qumix, that generally depends on the logical form of α; at the same time, the logical connectives are interpreted as special operations defined in terms of *gates*. We will consider a *minimal (sentential) quantum computational language* $\mathcal{L}$ that contains a privileged atomic sentence $\mathbf{f}$ (whose intended interpretation is the truth-value *Falsity*) and the following primitive connectives: the *negation* ($\neg$), the *square root of the negation* ($\sqrt{\neg}$), a ternary *conjunction* $\bigwedge$ (which corresponds to the Petri-Toffoli gate). For any sentences α and β, the expression $\bigwedge(\alpha, \beta, \mathbf{f})$ is a sentence of $\mathcal{L}$. In this framework, the usual conjunction $\alpha \curlywedge \beta$ is dealt with as metalinguistic abbreviation for the ternary conjunction $\bigwedge(\alpha, \beta, \mathbf{f})$. The disjunction connective ($\curlyvee$) is supposed to be defined via de Morgan ($\alpha \curlyvee \beta := \neg(\neg\alpha \curlywedge \neg\beta)$). This minimal quantum computational language can be extended to richer languages containing other primitive connectives.

We will first introduce the notion of *compositional quantum computational model* (briefly, *compositional QC-model* or *qumix-model*).

DEFINITION 104. *Compositional QC-model*
A *compositional QC-model* of $\mathcal{L}$ is a map $\mathtt{Qum}$ that associates a qumix to any sentence α of $\mathcal{L}$, satisfying the following conditions:

$$\mathtt{Qum}(\alpha) := \begin{cases} \text{a density operator of } \mathfrak{D}(\mathbb{C}^2) & \text{if } \alpha \text{ is an atomic sentence;} \\ P_0 & \text{if } \alpha = \mathbf{f}; \\ \mathtt{NOT}(\mathtt{Qum}(\beta)) & \text{if } \alpha = \neg\beta; \\ \sqrt{\mathtt{NOT}}(\mathtt{Qum}(\beta)) & \text{if } \alpha = \sqrt{\neg}\beta; \\ \mathbb{T}(\mathtt{Qum}(\beta), \mathtt{Qum}(\gamma), \mathtt{Qum}(\mathbf{f})) & \text{if } \alpha = \bigwedge(\beta, \gamma, \mathbf{f}). \end{cases}$$

The concept of compositional QC-model seems to have a "quasi intensional" feature: the meaning $\mathtt{Qum}(\alpha)$ of the sentence α partially reflects the logical form of α. In fact, the dimension of the Hilbert space where $\mathtt{Qum}(\alpha)$ "lives" depends on the number of occurrences of atomic sentences in α.

DEFINITION 105. *The atomic complexity of* α

$$At(\alpha) = \begin{cases} 1 & \text{if } \alpha \text{ is an atomic sentence;} \\ At(\beta) & \text{if } \alpha = \neg\beta \text{ or } \alpha = \sqrt{\neg}\beta; \\ At(\beta) + At(\gamma) + 1 & \text{if } \alpha = \bigwedge(\beta, \gamma, \mathbf{f}). \end{cases}$$

LEMMA 106. *If* $At(\alpha) = n$, *then* $\mathtt{Qum}(\alpha) \in \mathfrak{D}(\otimes^n \mathbb{C}^2)$.

We can say that the space $\otimes^{At(\alpha)}\mathbb{C}^2$ represents the *semantic space* where all possible meanings of α should "live". Accordingly we will also write $\mathcal{H}^\alpha$ instead of $\otimes^{At(\alpha)}\mathbb{C}^2$.

Given a model $\mathtt{Qum}$, any sentence α has a natural probability-value, which can be also regarded as its *extensional meaning* with respect to $\mathtt{Qum}$.

DEFINITION 107. *The probability-value of α in a model* $\mathtt{Qum}$

$$\mathtt{p}_{\mathtt{Qum}}(\alpha) := \mathtt{p}(\mathtt{Qum}(\alpha)).$$

As we have learnt, qumixes are preordered by the relation $\preceq$. This suggests a natural definition of a *logical consequence relation.*

DEFINITION 108. *Consequence in a model* $\mathtt{Qum}$
A sentence β is a *consequence in a model* $\mathtt{Qum}$ of a sentence α ($\alpha \models_{\mathtt{Qum}} \beta$) iff $\mathtt{Qum}(\alpha) \preceq \mathtt{Qum}(\beta)$.

DEFINITION 109. *Logical consequence*
A sentence β is a *logical consequence of a sentence* α ($\alpha \models \beta$) iff for any model $\mathtt{Qum}$, $\alpha \models_{\mathtt{Qum}} \beta$.

We call *quantum computational logic* (**QCL**) the logic that is semantically characterized by the logical consequence relation we have just defined. Hence, β is a logical consequence of α in the logic **QCL** ($\alpha \models_{\mathbf{QCL}} \beta$) iff β is a consequence of α in any model $\mathtt{Qum}$.

So far we have considered models, where the meaning of any sentence is represented by a qumix. A natural question arises: do density operators have an essential role in characterizing the logic **QCL**? This question has a negative answer. In fact, one can prove that quregisters are sufficient for our logical aims in the case of the *minimal* quantum computational language $\mathcal{L}$.

Let us first introduce the notion of *qubit-model.*

DEFINITION 110. *Qubit-model*
A *qubit-model* of $\mathcal{L}$ is a function $\mathtt{Qub}$ that associates a quregister to any sentence α of $\mathcal{L}$, satisfying the following conditions:

$$\mathtt{Qub}(\alpha) := \begin{cases} \text{a qubit in } \mathbb{C}^2 & \text{if } \alpha \text{ is an atomic sentence;} \\ |0\rangle & \text{if } \alpha = \mathbf{f}; \\ \mathtt{Not}(\mathtt{Qub}(\beta)) & \text{if } \alpha = \neg\beta; \\ \sqrt{\mathtt{Not}}(\mathtt{Qub}(\beta)) & \text{if } \alpha = \sqrt{\neg}\beta; \\ \mathtt{T}(\mathtt{Qub}(\beta), \mathtt{Qub}(\gamma), \mathtt{Qub}(\mathbf{f})) & \text{if } \alpha = \bigwedge(\beta, \gamma, \mathbf{f}). \end{cases}$$

The notions of *consequence and logical consequence* are defined like in the case of qumix-models, *mutatis mutandis.* One can prove that the qubit-semantics and the qumix-semantics characterize the same logic.[74]

Quantum computational logic turns out to be a non-standard form of quantum logic. Conjunction and disjunction do not correspond to lattice operations, because

[74]See [Dalla Chiara *et al.*, 2003].

they are not generally idempotent ($\alpha \nvDash_{\mathbf{QCL}} \alpha \curlywedge \alpha$, $\alpha \curlyvee \alpha \nvDash_{\mathbf{QCL}} \alpha$). Unlike the usual (sharp and unsharp) versions of quantum logic, the weak distributivity principle breaks down ($(\alpha \curlywedge \beta) \curlyvee (\alpha \curlywedge \gamma) \nvDash_{\mathbf{QCL}} \alpha \curlywedge (\beta \curlyvee \gamma)$). At the same time, the strong distributivity, that is violated in orthodox quantum logic, is here valid ($\alpha \curlywedge (\beta \curlyvee \gamma) \vDash_{\mathbf{QCL}} (\alpha \curlywedge \beta) \curlyvee (\alpha \curlywedge \gamma)$). Both the excluded middle and the noncontradiction principles are violated. As a consequence, one can say that the logic arising from quantum computation represents, in a sense, a new example of *fuzzy logic.*

The axiomatizability of **QCL** is an open problem.

10.3 Quantum trees

The *meaning* and the probability-value of any sentence α can be naturally described (and calculated) by means of a special configuration called *quantum tree,* that illustrates a kind of reversible transformation of the atomic subformulas of α. The notion of quantum tree can be dealt with either in the framework of the qubit-semantics or in the framework of the qumix-semantics. In the first case quantum trees will be called *qubit trees,* while in the second case we will speak of *qumix trees.*

Any sentence α can be naturally decomposed into its parts, giving rise to a special configuration called the *syntactical tree* of α (indicated by $STree^{\alpha}$).

Roughly,[75] $STree^{\alpha}$ can be represented as a sequence of *levels*:

$$Level_k(\alpha)$$

$$\cdots$$

$$Level_1(\alpha),$$

where:

- each $Level_i(\alpha)$ (with $1 \leq i \leq k$) is a sequence of subformulas of α;
- the *bottom level* ($Level_1(\alpha)$) consists of α;
- the *top level* ($Level_k(\alpha)$) is the sequence of all atomic occurrences of α;
- for any i (with $1 \leq i < k$), $Level_{i+1}(\alpha)$ is the sequence obtained by dropping the *principal connective*[76] in all molecular formulas occurring at $Level_i(\alpha)$, and by repeating all the atomic sentences that possibly occur at $Level_i(\alpha)$.

As an example, consider the following sentence: $\alpha = \mathbf{q} \curlywedge \neg\mathbf{q} = \bigwedge(\mathbf{q}, \neg\mathbf{q}, \mathbf{f})$. The syntactical tree of α is the following configuration:

[75] A formal definition of *syntactical tree* can be found in [Dalla Chiara and Leporini, 2005].

[76] The principal connective of α is $\begin{cases} \neg, & \text{if } \alpha = \neg\beta; \\ \sqrt{\neg}, & \text{if } \alpha = \sqrt{\neg}\beta; \\ \bigwedge, & \text{if } \alpha = \bigwedge(\beta, \gamma, \mathbf{f}). \end{cases}$

$$Level_3(\alpha) = (\mathbf{q}, \mathbf{q}, \mathbf{f});$$
$$Level_2(\alpha) = (\mathbf{q}, \neg\mathbf{q}, \mathbf{f});$$
$$Level_1(\alpha) = (\bigwedge(\mathbf{q}, \neg\mathbf{q}, \mathbf{f})).$$

By *Height* of α (indicated by $Height(\alpha)$ we mean the number of levels of the syntactical tree of α. For instance, $Height(\bigwedge(\mathbf{q}, \neg\mathbf{q}, \mathbf{f})) = 3$.

The syntactical tree of α (which represents a purely syntactical object) uniquely determines a sequence of gates that are all defined on the semantic space of α. We will call this gate-sequence the *qubit tree* of α.

Consider a sentence α such that $At(\alpha) = t$ and $Height(\alpha) = k$. Let $Level_i^j(\alpha)$ represent the j-th *node* of $Level_i(\alpha)$. Each $Level_i^j(\alpha)$ (where $1 \leq i < Height(\alpha)$) can be naturally associated to a unitary operator Op_i^j, according to the following *operator-rule*:

$$Op_i^j := \begin{cases} \mathbb{I}^{(1)} & \text{if } Level_i^j(\alpha) \text{ is an atomic sentence;} \\ \mathtt{Not}^{(r)} & \text{if } Level_i^j(\alpha) = \neg\beta \text{ and } At(\beta) = r; \\ \sqrt{\mathtt{Not}}^{(r)} & \text{if } Level_i^j(\alpha) = \sqrt{\neg}\beta \text{ and } At(\beta) = r; \\ \mathtt{T}^{(r,s,1)} & \text{if } Level_i^j(\alpha) = \bigwedge(\beta, \gamma, \mathbf{f}),\ At(\beta) = r \text{ and } At(\gamma) = s, \end{cases}$$

where $\mathbb{I}^{(1)}$ is the identity operator of $\mathbb{C}^2$.

On this basis, one can associate a gate G_i^α to each $Level_i(\alpha)$ (such that $1 \leq i < Height(\alpha)$):

$$G_i^\alpha := \bigotimes_{j=1}^{|Level_i(\alpha)|} Op_i^j,$$

where $|Level_i(\alpha)|$ is the length of the sequence $Level_i(\alpha)$.

Being the tensor product of unitary operators, every G_i^α turns out to be a unitary operator. One can easily show that all G_i^α are defined on the same space, $\mathcal{H}^\alpha$.

DEFINITION 111. *The qubit tree of* α
The *qubit tree* of α (denoted by $QTree^\alpha$) is the gate-sequence

$$(G_1^\alpha, \ldots, G_{Height(\alpha)-1}^\alpha)$$

that is uniquely determined by the syntactical tree of α.

As an example, consider again the sentence: $\alpha = \bigwedge(\mathbf{q}, \neg\mathbf{q}, \mathbf{f})$.

In order to construct the qubit tree of α, let us first determine the operators Op_i^j corresponding to each node of $Stree^\alpha$. We will obtain:

- $Op_1^1 = \mathtt{T}^{(1,1,1)}$, because $\bigwedge(\mathbf{q}, \neg\mathbf{q}, \mathbf{f})$ is connected with $(\mathbf{q}, \neg\mathbf{q}, \mathbf{f})$ (at $Level_2(\alpha)$);

- $Op_2^1 = \mathbb{I}^{(1)}$, because $\mathbf{q}$ is connected with $\mathbf{q}$ (at $Level_3(\alpha)$);
- $Op_2^2 = \mathtt{Not}^{(1)}$, because $\neg\mathbf{q}$ is connected with $\mathbf{q}$ (at $Level_3(\alpha)$);
- $Op_2^3 = \mathbb{I}^{(1)}$, because $\mathbf{f}$ is connected with $\mathbf{f}$ (at $Level_3(\alpha)$).

The qubit tree of α is represented by the gate-sequence (G_1^α, G_2^α), where:

$$G_1^\alpha = Op_1^1 = \mathtt{T}^{(1,1,1)};$$
$$G_2^\alpha = Op_2^1 \otimes Op_2^2 \otimes Op_2^3 = \mathbb{I}^{(1)} \otimes \mathtt{Not}^{(1)} \otimes \mathbb{I}^{(1)}.$$

As we have seen, qubit trees consist of unitary operators (which can be applied to quregisters). The notion of qubit tree can be naturally generalized to qumixes. In such a case we will speak of *qumix trees*, and we will call *quantum tree* either a qubit tree or a qumix tree. Let $(G_1^\alpha, \ldots, G_{k-1}^\alpha)$ be the qubit tree of α. We can define the following sequence of functions on the set $\mathfrak{D}(\mathcal{H}^\alpha)$:

$$^{\mathcal{D}}G_1^\alpha(\rho) = G_1^\alpha \, \rho \, G_1^{\alpha *}$$

$$\ldots$$

$$^{\mathcal{D}}G_{k-1}^\alpha(\rho) = G_{k-1}^\alpha \, \rho \, G_{k-1}^{\alpha *}.$$

One can easily prove that, for any $\rho \in \mathfrak{D}(\mathcal{H}^\alpha)$ and for any i $(1 \leq i \leq k-1)$, $^{\mathcal{D}}G_i^\alpha(\rho)$ is a density operator of $\mathfrak{D}(\mathcal{H}^\alpha)$. The sequence

$$QumTree^\alpha = (^{\mathcal{D}}G_1^\alpha, \ldots, {}^{\mathcal{D}}G_{k-1}^\alpha)$$

is called the *qumix tree* of α.

Consider now a sentence α and let $(^{\mathcal{D}}G_1^\alpha, \ldots, {}^{\mathcal{D}}G_{k-1}^\alpha)$ be the qumix tree of α. Any choice of a qumix ρ in $\mathcal{H}^\alpha$ determines a sequence $(\rho_k, \ldots, \rho_1)$ of qumixes of $\mathcal{H}^\alpha$, where:

$$\rho_k = \rho$$
$$\rho_{k-1} = {}^{\mathcal{D}}G_{k-1}^\alpha(\rho_k)$$
$$\ldots$$
$$\rho_1 = {}^{\mathcal{D}}G_1^\alpha(\rho_2)$$

The qumix ρ_k can be regarded as a possible *input-information* concerning the atomic parts of α, while ρ_1 represents the *output-information* about α, given the input-information ρ_k. Each ρ_i corresponds to the *information* about $Level_i(\alpha)$, given the input-information ρ_k.

How to determine an information about the parts of α under a given input? It is natural to apply the standard quantum-theoretic rule that determines the *states of the parts of a compound system.*

Suppose that:

$$Level_i(\alpha) = (\beta_{i_1}, \ldots \beta_{i_r})$$

We have:

$$\mathcal{H}^{\alpha} = \mathcal{H}^{\beta_{i_1}} \otimes \ldots \otimes \mathcal{H}^{\beta_{i_r}}$$

We know that $QumTree^{\alpha}$ and the choice of an input ρ_k (in $\mathcal{H}^{\alpha}$) determine a sequence of qumixes:

$$\rho_k \leftrightsquigarrow Level_k(\alpha) = (\mathbf{q}_1, \ldots \mathbf{q}_t)$$

$$\ldots$$

$$\rho_i \leftrightsquigarrow Level_i(\alpha) = (\beta_{i_1}, \ldots, \beta_{i_r})$$

$$\ldots$$

$$\rho_1 \leftrightsquigarrow Level_1(\alpha) = (\alpha)$$

Consider $red^j(\rho_i)$, the *reduced state of ρ_i with respect to the j-th subsystem* (where $1 \le j \le r$).[77] From a semantic point of view, this state can be regarded as a *contextual information* about β_{i_j} (the subformula of α occurring at the j-th position at $Level_i(\alpha)$) under the input ψ_k. Apparently, a contextual information about a subformula is generally a mixture.

An interesting situation arises when the qumix ρ_k, representing a global information about the atomic parts of α, is an *entangled* pure state.[78]

As an example, consider the sentence $\alpha = \neg \bigwedge(\mathbf{q}, \neg\mathbf{q}, \mathbf{f})$ (which represents an example of the *noncontradiction principle* formalized in the quantum computational language). The input-information might be the following entangled state:

$$|\psi_4\rangle = \frac{1}{\sqrt{2}}|110\rangle + \frac{1}{\sqrt{2}}|000\rangle \leftrightsquigarrow Level_4(\alpha) = (\mathbf{q}, \mathbf{q}, \mathbf{f})$$

The reduced states of $|\psi_4\rangle$ turn out to be the following:
$red^1(\frac{1}{\sqrt{2}}|110\rangle + \frac{1}{\sqrt{2}}|000\rangle) = \frac{1}{2}P_0^{(1)} + \frac{1}{2}P_1^{(1)}$
$red^2(\frac{1}{\sqrt{2}}|110\rangle + \frac{1}{\sqrt{2}}|000\rangle) = \frac{1}{2}P_0^{(1)} + \frac{1}{2}P_1^{(1)}$
$red^3(\frac{1}{\sqrt{2}}|110\rangle + \frac{1}{\sqrt{2}}|000\rangle) = P_0^{(1)}$

Hence, the contextual information about both occurrences of $\mathbf{q}$ is the (proper) mixture

$$\frac{1}{2}P_0^{(1)} + \frac{1}{2}P_1^{(1)}.$$

---

[77] We recall that $red^j(\rho_i)$ is the unique density operator that satisfies the following condition: for any self-adjoint operator A^j of $\mathcal{H}^{\beta_{i_j}}$,

$$\mathtt{tr}(red^j(\rho_i)A^j) = \mathtt{tr}(\rho_i(\mathbb{I}^1 \otimes \ldots \otimes \mathbb{I}^{j-1} \otimes A^j \otimes \mathbb{I}^{j+1} \otimes \ldots \otimes \mathbb{I}^r)),$$

(where $\mathbb{I}^h$ is the identity operator of $\mathcal{H}^{\beta_{i_h}}$). As a consequence, ρ_i and $red^j(\rho_i)$ are statistically equivalent with respect to the j-th subsystem of the compound system described by ρ_i.

[78] As is well known, the basic features of an *entangled state* $|\psi\rangle$ are the following: 1) $|\psi\rangle$ is a maximal information (a pure state) that describes a compound physical system S; 2) the pieces of information determined by $|\psi\rangle$ about the parts of S are, generally, non-maximal (proper mixtures). Hence, the information about the *whole* is more precise than the information about the *parts*.

At the same time, the contextual information about **f** is the false projection $P_0^{(1)}$.

Quantum trees can be naturally regarded as examples of quantum circuits that compute outputs under given inputs. Since both qubit trees and qumix trees are determined by the syntactical tree of a given sentence, one can also say that any sentence of the quantum computational language plays the role of an intuitive and "economical" description of a quantum circuit.

10.4 Holistic semantics

As we have seen, in the *compositional quantum computational semantics*, the meaning of a molecular sentence is determined by the meanings of its parts (like in classical logic). In this framework, the input-information about the top level of the syntactical tree of a sentence α is always associated to a factorized state $\rho_1 \otimes \ldots \otimes \rho_t$, where t is the atomic complexity of α and $\rho_1, \ldots, \rho_t$ are qumixes of $\mathbb{C}^2$. As a consequence, the meaning of a molecular α cannot be a pure state, if the meanings of some atomic parts of α are proper mixtures.

The *holistic quantum compositional semantics*[79] is based on a more "liberal" assumption: the input information about the top-level of the syntactical tree of α can be represented by any qumix "living" in the semantic space of α. As a consequence, the meanings of all levels of $STree^\alpha$ are not, generally, factorized states.

Suppose that:

$$Level_i(\alpha) = (\beta_1, \ldots, \beta_r).$$

As we already know, the space $\mathcal{H}^\alpha$ can be naturally regarded as the Hilbert space of a compound physical system consisting of r parts (mathematically represented by the spaces $\mathcal{H}^{\beta_1}, \ldots, \mathcal{H}^{\beta_r}$), where each part may be compound. On this basis, for any qumix ρ_i (associated to $Level_i(\alpha)$) and for any node $Level_i^j(\alpha)$, we can consider the *reduced state* $red^j(\rho_i)$ with respect to the j-th subsystem of the system described by ρ_i. From an intuitive point of view, $red^j(\rho_i)$ describes the j-th subsystem on the basis of the *global* information ρ_i. Since $Level_i(\alpha) = (\beta_1, \ldots, \beta_r)$, the qumix $red^j(\rho_i)$ (which is a density operator of the space $\mathcal{H}^{\beta_j}$) represents a *possible meaning* of the sentence β_j.

We can now introduce the basic definitions of the holistic semantics. Unlike compositional models, a *holistic quantum computational model* is a function `Hol` that assigns to any sentence α of the quantum computational language a *global meaning* that cannot be generally inferred from the meanings of the parts of α. Of course, the function `Hol` shall respect the logical form of α.

In order to define the concept of *holistic quantum computational model*, we will first introduce the notions of *atomic holistic model* and of *tree holistic model*.

DEFINITION 112. *Atomic holistic model*
An *atomic holistic model* is a map $\mathtt{Hol}^{At}$ that associates a qumix to any sentence α of $\mathcal{L}$, satisfying the following conditions:

[79]In [Dalla Chiara and Leporini, 2005] we have presented a weaker version of holistic semantics.

(1) $\mathtt{Hol}^{At}(\alpha) \in \mathfrak{D}(\mathcal{H}^{\alpha})$;

(2) Let $At(\alpha) = n$ and $Level_{Heigth(\alpha)} = \mathbf{q}_1, \ldots, \mathbf{q}_n$. Then,

(2.1) if $\mathbf{q}_j = \mathbf{f}$, then $red^j(\mathtt{Hol}^{At}(\alpha)) = P_0$;

(2.2) if $\mathbf{q}_j$ and $\mathbf{q}_h$ are two occurrences in α of the same atomic sentence, then $red^j(\mathtt{Hol}^{At}(\alpha)) = red^h(\mathtt{Hol}^{At}(\alpha))$.

Apparently, $\mathtt{Hol}^{At}(\alpha)$ represents a *global interpretation* of the atomic sentences occurring in α. At the same time, $red^j(\mathtt{Hol}^{At}(\alpha))$, the *reduced state* of the compound system (described by $\mathtt{Hol}^{At}(\alpha)$) with respect to the j-th subsystem, represents a *contextual meaning* of $\mathbf{q}_j$ with respect to the *global meaning* $\mathtt{Hol}^{At}(\alpha)$. Conditions (2.1) and (2.2) guarantee that $\mathtt{Hol}^{At}(\alpha)$ is well behaved. For, the contextual meaning of $\mathbf{f}$ is always the *Falsity*, while two different occurrences (in α) of the same atomic sentence have the same contextual meaning.

The map $\mathtt{Hol}^{At}$ (which assigns a meaning to the top-level of the syntactical tree of any sentence α) can be naturally extended to a map $\mathtt{Hol}^{Tree}$ that assigns a meaning to each level of the syntactical tree of any α, following the prescriptions of the qumix tree of α.

Consider a sentence α such that:

$$QumTree^{\alpha} = ({}^{\mathcal{D}}G_1^{\alpha}, \ldots, {}^{\mathcal{D}}G_{Heigth(\alpha)-1}^{\alpha}).$$

The map $\mathtt{Hol}^{Tree}$ is defined as follows:

$$\mathtt{Hol}^{Tree}(Level_{Heigth(\alpha)}) = \mathtt{Hol}^{At}(\alpha)$$

$$\mathtt{Hol}^{Tree}(Level_i(\alpha)) = {}^{\mathcal{D}}G_i^{\alpha}(\mathtt{Hol}^{Tree}(Level_{i+1}(\alpha))$$

(where $Heigth(\alpha) > i \geq 1$).

On this basis, one can naturally define the notion of *holistic (quantum computational) model* of $\mathcal{L}$.

DEFINITION 113. *Holistic model*
A map $\mathtt{Hol}$ that assigns to any sentence α a qumix of the space $\mathcal{H}^{\alpha}$ is called a *holistic (quantum computational) model* of $\mathcal{L}$ iff there exists an atomic holistic model $\mathtt{Hol}^{At}$ s.t.:

$$\mathtt{Hol}(\alpha) = \mathtt{Hol}^{Tree}(Level_1(\alpha)),$$

where $\mathtt{Hol}^{Tree}$ is the extension of $\mathtt{Hol}^{At}$.

Given a sentence γ, $\mathtt{Hol}$ determines the *contextual meaning*, with respect to the context $\mathtt{Hol}(\gamma)$, of any occurrence of a subformula β in γ.

DEFINITION 114. *Contextual meaning of a node*
Let β be a subformula of γ occurring at the $j - th$ position of the $i - th$ level of the syntactical tree of γ. We indicate by $\beta[^i_j]$ the node of $STree^{\gamma}$ corresponding

to such occurrence. The contextual meaning of $\beta[^i_j]$ with respect to the context $\texttt{Hol}(\gamma)$ is defined as follows:

$$\texttt{Hol}^\gamma(\beta[^i_j]) := red^j(\texttt{Hol}^{Tree}(Level_i(\gamma))).$$

Hence, we have:

$$\texttt{Hol}^\gamma(\gamma) = \texttt{Hol}^{Tree}(Level_1(\gamma)) = \texttt{Hol}(\gamma).$$

Suppose that $\beta[^i_j]$ and $\beta[^h_k]$ are two nodes of the syntactical tree of γ, representing two occurrences of the same subformula β. One can show that:

$$\texttt{Hol}^\gamma(\beta[^i_j]) = \texttt{Hol}^\gamma(\beta[^h_k]).$$

In other words, two different occurrences of one and the same subformula in a sentence γ receive the same contextual meaning with respect to the context $\texttt{Hol}(\gamma)$.

On this basis, one can define the *contextual meaning* of a subformula β of γ, with respect to the context $\texttt{Hol}(\gamma)$:

$$\texttt{Hol}^\gamma(\beta) := \texttt{Hol}^\gamma(\beta[^i_j]),$$

where $\beta[^i_j]$ is any occurrence of β at a node of $STree^\gamma$.

Suppose now that β is a subformula of two different formulas γ and δ. Generally, we have:

$$\texttt{Hol}^\gamma(\beta) \neq \texttt{Hol}^\delta(\beta).$$

In other words, sentences may receive different contextual meanings in different contexts!

Apparently, $\texttt{Hol}^\gamma$ is a partial function that only assigns meanings to the subformulas of γ. Given a sentence γ, we will call the partial function $\texttt{Hol}^\gamma$ a *contextual holistic model* of the language.

As expected, compositional models turn out to be limit-cases of holistic models. One can easily prove that $\texttt{Hol}$ represents a compositional model iff the following condition is satisfied for any sentence α:

$$\texttt{Hol}^{At}(\alpha) = \texttt{Hol}(\mathbf{q}_1) \otimes \ldots \otimes \texttt{Hol}(\mathbf{q}_t),$$

where $\mathbf{q_1}, \ldots, \mathbf{q_t}$ are the atomic sentences occurring in α.

Unlike holistic models, compositional models are, of course, context-independent. Suppose that β is a subformula of two different formulas γ and δ, and let $\texttt{Hol}$ represent a compositional model. We have:

$$\texttt{Hol}^\gamma(\beta) = \texttt{Hol}^\delta(\beta) = \texttt{Hol}(\beta).$$

The notion of logical consequence in the framework of the holistic quantum computational semantics can be now defined in a natural way.

Let us first define the notion of *consequence in a given contextual model.*

DEFINITION 115. *Consequence in a given contextual model* $\mathtt{Hol}^\gamma$
A sentence β is a consequence of a sentence α in a given contextual model $\mathtt{Hol}^\gamma$ ($\alpha \models_{\mathtt{Hol}^\gamma} \beta$) iff

1. α and β are subformulas of γ;
2. $\mathtt{Hol}^\gamma(\alpha) \preceq \mathtt{Hol}^\gamma(\beta)$ (where $\preceq$ is the preorder relation defined in Def. 103).

DEFINITION 116. *Logical consequence (in the holistic semantics)*
A sentence β is a consequence of a sentence α (in the holistic semantics) iff for any sentence γ such that α and β are subformulas of γ and for any $\mathtt{Hol}$,

$$\alpha \models_{\mathtt{Hol}^\gamma} \beta.$$

We call **HQCL** the logic that is semantically characterized by the logical consequence relation we have just defined. Hence, $\alpha \models_{\mathbf{HQCL}} \beta$ iff for any sentence γ such that α and β are subformulas of γ and for any $\mathtt{Hol}$,

$$\alpha \models_{\mathtt{Hol}^\gamma} \beta.$$

Although the basic ideas of the holistic and of the compositional quantum computational semantics are quite different, one can prove that **HQCL** and **QCL** are the same logic.[80] In other words, for any sentences α and β,

$$\alpha \models_{\mathbf{HQCL}} \beta \text{ iff } \alpha \models_{\mathbf{QCL}} \beta.$$

This means that the logics (formalized in our "poor" sentential languages) are not able to capture the difference between an analytical and a holistic semantic procedure.

The holistic quantum computational semantics provides a formalism that might represent a useful abstract tool for describing *gestaltic semantic patterns*, which arise in a number of different rational and perceptual activities. As we have seen, an important role in this game is played by the notion of *tensor product*, which is mainly responsible for most holistic quantum phenomena. The compositional and analytical features of classical semantics (and of many other non-classical approaches) are, instead, generally based on *cartesian products*. In this connection, an interesting question arises: to what extent is it possible (and reasonable) to try and generalize the tensor-product formalism to some abstract semantic situations, that might be quite independent of the notion of Hilbert space?

11 MATHEMATICAL APPENDIX

We give here a survey of the definitions of some basic mathematical concepts that play a fundamental role in quantum logic.

[80] See [Dalla Chiara *et al.*, 2006].

11.1 Algebraic structures

DEFINITION 117. *Poset*
A *partially ordered set* (called also *poset*) is a structure

$$\mathcal{B} = \langle B\,,\,\leq\rangle\,,$$

where: B (the *support* of the structure) is a nonempty set and $\leq$ is a *partial order relation* on B. In other words, $\leq$ satisfies the following conditions for all $a, b, c \in B$:

(i) $a \leq a$ (reflexivity);
(ii) $a \leq b$ and $b \leq a$ implies $a = b$ (antisymmetry);
(iii) $a \leq b$ and $b \leq c$ implies $a \leq c$ (transitivity).

DEFINITION 118. *Chain*
Let $\mathcal{B} = \langle B\,,\,\leq\rangle$ be a poset. A *chain* in $\mathcal{B}$ is a subset $C \subseteq B$ such that $\forall a, b \in C$: $a \leq b$ or $b \leq a$.

DEFINITION 119. *Bounded poset*
A *bounded poset* is a structure

$$\mathcal{B} = \langle B\,,\,\leq\,,\,\mathbf{0}\,,\mathbf{1}\rangle\,,$$

where:

(i) $\langle B\,,\,\leq\rangle$ is a poset;

(ii) $\mathbf{0}$ and $\mathbf{1}$ are special elements of B: the *minimum* and the *maximum* with respect to $\leq$. In other words, for all $b \in B$:

$$\mathbf{0} \leq b \text{ and } b \leq \mathbf{1}.$$

DEFINITION 120. *Lattice*
A *lattice* is a poset $\mathcal{B} = \langle B\,,\,\leq\rangle$ in which any pair of elements a, b has a *meet* $a \wedge b$ (also called *infimum*) and a *join* $a \vee b$ (also called *supremum*) such that:

(i) $a \wedge b \leq a, b$, and $\forall c \in B$: $c \leq a, b$ implies $c \leq a \wedge b$;
(ii) $a, b \leq a \vee b$, and $\forall c \in B$: $a, b \leq c$ implies $a \vee b \leq c$.

In any lattice the following condition holds:

$$a \leq b \text{ iff } a \wedge b = a \text{ iff } a \vee b = b.$$

DEFINITION 121. *Complemented lattice*
A *complemented* lattice is a bounded lattice $\mathcal{B}$ where: $\forall a \in B\ \exists b \in B$ such that $a \wedge b = \mathbf{0}$ and $a \vee b = \mathbf{1}$.

Let X be any set of elements of a lattice $\mathcal{B}$. If existing, the *infimum* $\bigwedge X$ and the *supremum* $\bigvee X$ are the elements of B that satisfy the following conditions:

(ia) $\forall a \in X : \bigwedge X \leq a$;

(ib) $\forall c \in B : \forall a \in X[c \leq a]$ implies $c \leq \bigwedge X$;

(iia) $\forall a \in X : a \leq \bigvee X$;

(iib) $\forall c \in B : \forall a \in X[a \leq c]$ implies $\bigvee X \leq c$.

On can show that, when they exist the infimum and the supremum are unique. A lattice is *complete* iff for any set of elements X the infimum $\bigwedge X$ and the supremum $\bigvee X$ exist. A lattice is *σ-complete* iff for any countable set of elements X the infimum $\bigwedge X$ and the supremum $\bigvee X$ exist.

In many situations, a poset (or a lattice) is closed under a unary operation that represents a weak form of logical *negation*. Such a finer structure is represented by a *bounded involution poset*.

DEFINITION 122. *Bounded involution poset*
A *bounded involution poset* is a structure $\mathcal{B} = \langle B\,, \leq\,, '\,, \mathbf{0}\,, \mathbf{1}\rangle$ where:

(i) $\langle B\,, \leq\,, \mathbf{0}\,, \mathbf{1}\rangle$ is a bounded poset;

(ii) $'$ is a unary operation (called *involution* or *generalized complement*) that satisfies the following conditions:

(a) $a = a''$ (double negation);

(b) $a \leq b$ implies $b' \leq a'$ (contraposition).

The presence of a negation-operation permits us to define an *orthogonality relation* $\perp$, that may hold between two elements of a bounded involution poset.

DEFINITION 123. *Orthogonality*
Let a and b belong to a bounded involution poset. The object a is *orthogonal* to the object b (indicated by $a \perp b$) iff $a \leq b'$. A set of elements S is called a *pairwise orthogonal* set iff $\forall a, b \in S$ such that $a \neq b$, $a \perp b$.
A *maximal* set of pairwise orthogonal elements is a set of pairwise orthogonal elements that is not a proper subset of any set of pairwise orthogonal elements.

When a is not orthogonal to b we write: $a \not\perp b$. The orthogonality relation $\perp$ is sometimes also called *preclusivity*; while its negation $\not\perp$ is also called *accessibility*.

Since, by definition of bounded involution poset, $a \leq b$ implies $b' \leq a'$ (contraposition) and $a = a''$ (double negation), one immediately obtains that $\perp$ is a symmetric relation.

Notice that $\mathbf{0} \perp \mathbf{0}$ and that $\perp$ is not necessarily irreflexive. It may happen that an object a (different from the null object $\mathbf{0}$) is orthogonal to itself:

$$a \perp a \text{ (because } a \leq a'\text{)}.$$

Objects of this kind are called *self-inconsistent*. Suppose we have two self-inconsistent objects a and b, and let us ask whether in such a case a is necessarily orthogonal to b. Generally, the answer to this question is negative. There are examples of bounded involution posets such that for some objects a and b:

$$a \perp a \text{ and } b \perp b \text{ and } a \not\perp b.$$

DEFINITION 124. *Kleene poset*
A bounded involution poset is a *Kleene poset* (or also a *regular poset*) iff it satisfies the *Kleene condition* for any pair of elements a and b:

$$a \perp a \text{ and } b \perp b \text{ implies } a \perp b.$$

DEFINITION 125. *Bounded involution lattice*
A *bounded involution lattice* is a bounded involution poset that is also a lattice.

A *Kleene lattice* (or *regular lattice*) is a Kleene poset that is also a lattice. One can prove that a bounded involution lattice is regular iff $a \wedge a' \leq b \wedge b'$ (for any pair of elements a and b).

Generally, bounded involution lattices and Kleene lattices may violate both the noncontradiction principle and the excluded middle. In other words, it may happen that:

$$a \wedge a' \neq \mathbf{0} \quad \text{and} \quad a \vee a' \neq \mathbf{1}.$$

DEFINITION 126. *Orthoposet and ortholattice*
An *orthoposet* is a bounded involution poset $\mathcal{B} = \langle B\,, \leq\,, '\,, \mathbf{0}\,, \mathbf{1}\rangle$ that satisfies the conditions:

(i) $a \wedge a' = \mathbf{0}$ (noncontradiction principle);

(ii) $a \vee a' = \mathbf{1}$ (excluded middle principle).

An *ortholattice* is an orthoposet that is also a lattice.

The involution operation $'$ of an orthoposet (ortholattice) is also called *orthocomplementation* (or shortly *orthocomplement*).

A *σ-orthocomplete orthoposet* (*σ-orthocomplete ortholattice*) is an orthoposet (ortholattice) $\mathcal{B}$ such that for any countable set $\{a_i\}_{i \in I}$ of pairwise orthogonal elements the supremum $\bigvee \{a_i\}_{i \in I}$ exists in $\mathcal{B}$.

DEFINITION 127. *Distributive lattice*
A lattice $\mathcal{B} = \langle B\,, \wedge\,, \vee\rangle$ is *distributive* iff the meet $\wedge$ is distributed over the join $\vee$ and vice versa. In other words:

(i) $a \wedge (b \vee c) = (a \wedge b) \vee (a \wedge c)$;

(ii) $a \vee (b \wedge c) = (a \vee b) \wedge (a \vee c)$.

Distributive involution lattices are also called *de Morgan lattices*.

In this framework, Boolean algebras can be then defined as particular examples of de Morgan lattices.

DEFINITION 128. *Boolean algebra*
A *Boolean algebra* is a structure

$$\mathcal{B} = \langle B\,, \wedge\,, \vee\,, '\,, \mathbf{0}\,, \mathbf{1}\rangle$$

that is at the same time an ortholattice and a de Morgan lattice.

In other words, Boolean algebras are distributive ortholattices.

DEFINITION 129. *Orthomodular poset and orthomodular lattice*
An *orthomodular poset* is an orthoposet

$$\mathcal{B} = \langle B\,,\,\leq\,,\,'\,,\,\mathbf{0}\,,\mathbf{1}\rangle$$

that satisfies the following conditions:

(i) $\forall a, b \in B,\ \ a \perp b$ implies $a \vee b \in B$;

(ii) $\forall a, b \in B,\ \ a \leq b$ implies $b = a \vee (a \vee b')'$.

An *orthomodular lattice* is an orthomodular poset that is also a lattice.

Clearly, any distributive ortholattice (i.e., any Boolean algebra), is orthomodular.

DEFINITION 130. *Modularity*
A lattice $\mathcal{B}$ is called *modular* iff $\forall a, b \in B$,

$$a \leq b \text{ implies } \forall c \in B[a \vee (b \wedge c) = (a \vee b) \wedge (a \vee c)].$$

Every modular ortholattice is orthomodular, but not the other way around. Furthermore, any distributive lattice is modular.

A bounded poset (lattice) $\mathcal{B}$ may contain some special elements, called *atoms*.

DEFINITION 131. *Atom*
An element b of B is called an *atom* of $\mathcal{B}$ iff b *covers* $\mathbf{0}$. In other words, $b \neq \mathbf{0}$ and $\forall c \in B$: $c \leq b$ implies $c = \mathbf{0}$ or $c = b$.

Apparently, atoms are nonzero elements such that no other element lies between them and the lattice-minimum.

DEFINITION 132. *Atomicity*
A bounded poset $\mathcal{B}$ is *atomic* iff $\forall a \in B - \{\mathbf{0}\}$ there exists an atom b such that $b \leq a$.

Of course, any finite bounded poset is atomic. At the same time, there are examples of infinite bounded posets that are *atomless* (and hence nonatomic), the real interval $[0, 1]$ being the most familiar example.

It turns out that any atomic orthomodular lattice $\mathcal{B}$ is *atomistic* in the sense that any element can be represented as the supremum of a set of atoms, i.e., for any element a there exists a set $\{b_i\}_{i \in I}$ of atoms such that $a = \bigvee \{b_i\}_{i \in I}$.

DEFINITION 133. *Covering property*
A lattice $\mathcal{B}$ satisfies the *covering property* iff $\forall a, b \in B$: if a covers $a \wedge b$, then $a \vee b$ covers b.

It turns out that an atomic lattice $\mathcal{B}$ has the covering property iff for every atom a of $\mathcal{B}$ and for every element $b \in B$ such that $a \wedge b = \mathbf{0}$, the element $a \vee b$ covers b.

One of the most significant quantum relations, *compatibility*, admits a purely algebraic definition.

DEFINITION 134. *Compatibility*
Let $\mathcal{B}$ be an orthomodular lattice and let a and b be elements of B. The element a is called *compatible* with the element b iff

$$a = (a \wedge b') \vee (a \wedge b).$$

One can show that the compatibility relation is symmetric. The proof uses the orthomodular property in an essential way.

Clearly, if $\mathcal{B}$ is a Boolean algebra, then any element is compatible with any other element by distributivity.

One can prove that a, b are compatible in the orthomodular lattice $\mathcal{B}$ iff the subalgebra of $\mathcal{B}$ generated by $\{a, b\}$ is Boolean.

DEFINITION 135. *Irreducibility*
Let $\mathcal{B}$ be an orthomodular lattice. $\mathcal{B}$ is said to be *irreducible* iff

$$\{a \in B : \forall b \in B\, (a \text{ is compatible with } b)\} = \{\mathbf{0}, \mathbf{1}\}.$$

If $\mathcal{B}$ is not irreducible, it is called *reducible*.

DEFINITION 136. *Separability*
An orthomodular lattice $\mathcal{B}$ is called *separable* iff every set of pairwise orthogonal elements of B is countable.

DEFINITION 137. *Group*
A *group* is a structure $\mathcal{G} = \langle G, +, -, 0\rangle$, where $+$ is a binary operation, $-$ is a unary operation, 0 is a special element. The following conditions hold:

(i) $\langle G, +, 0\rangle$ is a *monoid*. In other words,

  (a) the operation $+$ is associative:
  $a + (b + c) = (a + b) + c$;

  (b) 0 is the *neutral element*:
  $a + 0 = a$;

(ii) $\forall a \in G$, $-a$ is the *inverse* of a:
$a + (-a) = 0$.

An *Abelian monoid* (*group*) is a monoid (group) in which the operation $+$ is commutative: $a + b = b + a$.

DEFINITION 138. *Ring*
A *ring* is a structure $\mathbf{D} = \langle D, +, \cdot, -, 0\rangle$ that satisfies the following conditions:

(i) $\langle D, +, 0 \rangle$ is an Abelian group;

(ii) the operation $\cdot$ is associative:

$a \cdot (b \cdot c) = (a \cdot b) \cdot c$;

(iii) the operation $\cdot$ distributes over $+$ on both sides, i.e., $\forall a, b, c \in D$:

(a) $a \cdot (b + c) = (a \cdot b) + (a \cdot c)$;

(b) $(a + b) \cdot c = (a \cdot c) + (b \cdot c)$.

If there is an element 1 in D that is neutral for $\cdot$ (i.e., if $\langle D, \cdot, 1 \rangle$ is a monoid), then the ring is called a *ring with unity*.

A ring is *trivial* in case it has only one element, otherwise it is *nontrivial*. It is easy to see that a ring with unity is nontrivial iff $0 \neq 1$.

A *commutative ring* is a ring in which the operation $\cdot$ is commutative.

DEFINITION 139. *Division ring*
A *division ring* is a nontrivial ring **D** with unity such that any nonzero element is *invertible*; in other words, for any $a \in D$ $(a \neq 0)$, there is an element $b \in D$ such that $a \cdot b = b \cdot a = 1$.

DEFINITION 140. *Field*
A *field* is a commutative division ring.

Both the real numbers ($\mathbb{R}$) and the complex numbers ($\mathbb{C}$) give rise to a field. An example of a genuine division ring (where $\cdot$ is not commutative) is given by the quaternions ($\mathbb{Q}$).

11.2 Hilbert spaces

DEFINITION 141. *Vector space*
A *Vector space* over a division ring **D** is a structure $\mathcal{V} = \langle V, +, -, \cdot, \underline{0} \rangle$ that satisfies the following conditions:

(i) $\langle V, +, -, \underline{0} \rangle$ (the vector structure) is an Abelian group, where $\underline{0}$ (the *null vector*) is the neutral element;

(ii) for any element a of the division ring **D** and any vector $|\varphi\rangle$ of V, $a|\varphi\rangle$ (the *scalar product* of a and $|\varphi\rangle$) is a vector in V. The following conditions hold for any $a, b \in D$ and for any $|\varphi\rangle, |\psi\rangle \in V$:

(a) $a(|\varphi\rangle + |\psi\rangle) = (a|\varphi\rangle) + (a|\psi\rangle)$;

(b) $(a + b)|\varphi\rangle = (a|\varphi\rangle) + (b|\varphi\rangle)$;

(c) $a(b|\varphi\rangle) = (a \cdot b)|\varphi\rangle$;

(d) $1|\varphi\rangle = |\varphi\rangle$.

The elements (vectors) of a vector space $\mathcal{V}$ are indicated by $|\varphi\rangle, |\psi\rangle, |\chi\rangle, \ldots$, while $a, b, c, \ldots$ represent elements (scalars) of the division ring $\mathbf{D}$.
Any finite sum of vectors $|\psi_1\rangle, \ldots, |\psi_n\rangle$ is indicated by $|\psi_1\rangle + \ldots + |\psi_n\rangle$ (or $\sum_{i \in K} |\psi_i\rangle$, when $K = \{1, \ldots, n\}$.)

On this basis, one can introduce the notion of *pre-Hilbert space*. Hilbert spaces are then defined as special cases of pre-Hilbert spaces. We will only consider pre-Hilbert spaces (and Hilbert spaces) whose division ring is either $\mathbb{R}$ or $\mathbb{C}$.

DEFINITION 142. *Pre-Hilbert space*
Let $\mathbf{D}$ be the field of the real or the complex numbers. A *pre-Hilbert space* over $\mathbf{D}$ is a vector space $\mathcal{V}$ over $\mathbf{D}$, equipped with an inner product $\langle .|. \rangle$ that associates to any pair of vectors $|\varphi\rangle, |\psi\rangle \in V$ an element $\langle \varphi|\psi\rangle \in D$. The following conditions are satisfied for any $|\varphi\rangle, |\psi\rangle, |\chi\rangle \in V$ and any $a \in D$:

(i) $\langle \varphi|\varphi\rangle \geq 0$;

(ii) $\langle \varphi|\varphi\rangle = 0$ iff $|\varphi\rangle = \underline{0}$;

(iii) $\langle \psi|a\varphi\rangle = a\langle \psi|\varphi\rangle$;

(iv) $\langle \varphi|\psi + \chi\rangle = \langle \varphi|\psi\rangle + \langle \varphi|\chi\rangle$;

(v) $\langle \varphi|\psi\rangle = \langle \psi|\varphi\rangle^*$, where * is the identity if $D = \mathbb{R}$, and the complex conjugation if $D = \mathbb{C}$.

The inner product $\langle .|. \rangle$ permits one to generalize some geometrical notions of ordinary 3-dimensional spaces.

DEFINITION 143. *Norm of a vector*
The *norm* $\| |\varphi\rangle \|$ of a vector $|\varphi\rangle$ is the number $\langle \varphi|\varphi\rangle^{1/2}$.

A *unit* (or *normalized*) *vector* is a vector $|\psi\rangle$ such that $\| |\psi\rangle \| = 1$.
Two vectors $|\varphi\rangle, |\psi\rangle$ are called *orthogonal* iff $\langle \varphi|\psi\rangle = 0$.

DEFINITION 144. *Orthonormal set of vectors*
A set $\{|\psi_i\rangle\}_{i \in I}$ of vectors is called *orthonormal* iff its elements are pairwise *orthogonal* unit vectors. In other words:

(i) $\forall i, j \in I (i \neq j) : \langle \psi_i|\psi_j\rangle = 0$;
(ii) $\forall i \in I : \| |\psi_i\rangle \| = 1$.

The norm $\|.\|$ induces a *metric* d on the pre-Hilbert space $\mathcal{V}$:

$$d(|\psi\rangle, |\varphi\rangle) := \| |\psi\rangle - |\varphi\rangle \|.$$

We say that a sequence $\{|\psi_i\rangle\}_{i \in \mathbb{N}}$ of vectors in V *converges in norm* (or simply *converges*) to a vector $|\varphi\rangle$ of V iff $\lim_{i \to \infty} d(|\psi_i\rangle, |\varphi\rangle) = 0$. In other words, $\forall \varepsilon > 0 \, \exists n \in \mathbb{N} \, \forall k > n : \; d(|\psi_k\rangle, |\varphi\rangle) < \varepsilon$.

A *Cauchy sequence* is a sequence $\{|\psi_i\rangle\}_{i \in \mathbb{N}}$ of vectors in V such that $\forall \varepsilon > 0 \, \exists n \in \mathbb{N} \, \forall h > n \, \forall k > n : \; d(|\psi_h\rangle, |\psi_k\rangle) < \varepsilon$.

It is easy to see that whenever a sequence $\{|\psi_i\rangle\}_{i\in\mathbb{N}}$ of vectors in V converges to a vector $|\varphi\rangle$ of V, then $\{|\psi_i\rangle\}_{i\in\mathbb{N}}$ is a Cauchy sequence. The crucial question is the converse one: which are the pre-Hilbert spaces in which every Cauchy sequence converges to *an element in the space*?

DEFINITION 145. *Metrically complete pre-Hilbert space*
A pre-Hilbert space $\mathcal{V}$ with inner product $\langle .|.\rangle$ is *metrically complete* with respect to the metric d induced by $\langle .|.\rangle$ iff every Cauchy sequence of vectors in V converges to a vector of V.

DEFINITION 146. *Hilbert space*
A *Hilbert space* is a metrically complete pre-Hilbert space.

A *real* (*complex*) *Hilbert space* is a Hilbert space whose division ring is $\mathbb{R}$ ($\mathbb{C}$). The notion of pre-Hilbert space (Hilbert space) can be generalized to the case where the division ring is represented by $\mathbb{Q}$ (the division ring of all quaternions).

Consider a Hilbert space $\mathcal{H}$ over a division ring $\mathbf{D}$.

DEFINITION 147. (*Hilbert*) *linear combination*
Let $\{|\psi_i\rangle\}_{i\in I}$ be a set of vectors of $\mathcal{H}$ and let $\{a_i\}_{i\in I} \subseteq D$. A vector $|\psi\rangle$ is called a (*Hilbert*) *linear combination* (or *superposition*) of $\{|\psi_i\rangle\}_{i\in I}$ (with scalars $\{a_i\}_{i\in I}$) iff $\forall\varepsilon \in \mathbb{R}^+$ there is a finite set $J \subseteq I$ such that for any finite subset K of I including J:

$$\||\psi\rangle - \sum_{i\in K} a_i|\psi_i\rangle\| \leq \varepsilon.$$

Apparently, when existing, the linear combination of $\{|\varphi_i\rangle\}_{i\in I}$ (with scalars $\{a_i\}_{i\in I}$) is unique. We denote it by $\sum_{i\in I} a_i|\psi_i\rangle$. When no confusion is possible, the index set I will be omitted.

DEFINITION 148. *Orthonormal basis*
An *orthonormal basis* of $\mathcal{H}$ is a *maximal orthonormal set* $\{|\psi_i\rangle\}_{i\in I}$ of $\mathcal{H}$. In other words, $\{|\psi_i\rangle\}_{i\in I}$ is an orthonormal set such that no orthonormal set includes $\{|\psi_i\rangle\}_{i\in I}$ as a proper subset.

One can prove that every Hilbert space $\mathcal{H}$ has an orthonormal basis and that all orthonormal bases of $\mathcal{H}$ have the same cardinality. The *dimension* of $\mathcal{H}$ is then defined as the cardinal number of any basis of $\mathcal{H}$.

Let $\{|\psi_i\rangle\}_{i\in I}$ be any orthonormal basis of $\mathcal{H}$. One can prove that every vector $|\varphi\rangle$ of $\mathcal{H}$ can be expressed in the following form:

$$|\varphi\rangle = \sum_{i\in I} \langle\psi_i|\varphi\rangle|\psi_i\rangle.$$

Hence, $|\varphi\rangle$ is a linear combination of $\{|\psi_i\rangle\}_{i\in I}$ with scalars $\langle\psi_i|\varphi\rangle$ (the scalars $\langle\psi_i|\varphi\rangle$ are also called *Fourier coefficients*.)
A Hilbert space $\mathcal{H}$ is called *separable* iff $\mathcal{H}$ has a countable orthonormal basis. In the following, we will always refer to separable Hilbert spaces.

DEFINITION 149. *Closed subspace*

A *closed subspace* of $\mathcal{H}$ is a set X of vectors that satisfies the following conditions:

(i) X is a *subspace* of $\mathcal{H}$. In other words, X is closed under finite linear combinations. Hence,

$$|\psi\rangle, |\varphi\rangle \in X \text{ implies } a|\psi\rangle + b|\varphi\rangle \in X;$$

(ii) X is closed under limits of Cauchy sequences. In other words: if each element of a Cauchy sequence of vectors belongs to X, then also the limit of the sequence belongs to X.

The set of all closed subspaces of $\mathcal{H}$ is indicated by $\mathcal{C}(\mathcal{H})$. For any vector $|\psi\rangle$, we indicate by $[|\psi\rangle]$ the unique 1-dimensional closed subspace that contains $|\psi\rangle$.

DEFINITION 150. *Operator*
An operator of $\mathcal{H}$ is a map

$$A : Dom(\mathcal{H}) \mapsto \mathcal{H},$$

where $Dom(A)$ (the *domain* of A) is a subset of $\mathcal{H}$.

DEFINITION 151. *Densely defined operator*
A *densely defined operator* of $\mathcal{H}$ is an operator A that satisfies the following condition: $\forall \varepsilon \in \mathbb{R}^+ \, \forall |\psi\rangle \in \mathcal{H} \, \exists |\varphi\rangle \in Dom(A) \, [d(|\psi\rangle, |\varphi\rangle) < \varepsilon]$, where d represents the metric induced by $\langle . | . \rangle$.

DEFINITION 152. *Linear operator*
A *linear operator* on $\mathcal{H}$ is an operator A that satisfies the following conditions:

(i) $Dom(A)$ is a closed subspace of $\mathcal{H}$;

(ii) $\forall |\psi\rangle, |\varphi\rangle \in Dom(A) \, \forall a, b \in D : A(a|\psi\rangle + b|\varphi\rangle) = aA|\psi\rangle + bA|\varphi\rangle$.

In other words, a characteristic of linear operators is preserving the linear combinations.

DEFINITION 153. *Bounded operator*
A linear operator A is called *bounded* iff there exists a positive real number a such that $\forall |\psi\rangle \in \mathcal{H} : \| A|\psi\rangle \| \leq a \| |\psi\rangle \|$.

The set $\mathcal{B}(\mathcal{H})$ of all bounded operators of $\mathcal{H}$ turns out to be closed under the operator sum, the operator product and the scalar product. In other words, if $A \in \mathcal{B}(\mathcal{H})$ and $B \in \mathcal{B}(\mathcal{H})$, then $A + B \in \mathcal{B}(\mathcal{H})$ and $A.B \in \mathcal{B}(\mathcal{H})$; for any scalar a, if $B \in \mathcal{B}(\mathcal{H})$, then $aB \in \mathcal{B}(\mathcal{H})$.

DEFINITION 154. *Positive operator*
A bounded operator A is called *positive* iff $\forall |\psi\rangle \in \mathcal{H} : \langle \psi | A\psi \rangle \geq 0$.

DEFINITION 155. *The adjoint operator*

Let A be a densely defined linear operator of $\mathcal{H}$. The *adjoint* of A is the unique operator A^* such that

$$\forall|\psi\rangle \in Dom(A)\ \forall|\varphi\rangle \in Dom(A^*) : \langle A\psi|\varphi\rangle = \langle\psi|A^*\varphi\rangle .$$

DEFINITION 156. *Self-adjoint operator*
A *self-adjoint operator* is a densely defined linear operator A such that $A = A^*$.

If A is self-adjoint, then $\forall|\psi\rangle, |\varphi\rangle \in Dom(A) : \langle A\psi|\varphi\rangle = \langle\psi|A\varphi\rangle$.
If A is self-adjoint and everywhere defined (i.e., $Dom(A) = \mathcal{H}$), then A is bounded.

DEFINITION 157. *Projection operator*
A *projection operator* is an everywhere defined self-adjoint operator P that satisfies the idempotence property: $\forall|\psi\rangle \in \mathcal{H} : P|\psi\rangle = PP|\psi\rangle$.
There are two special projections $\mathbb{O}$ and $\mathbb{I}$ called *the zero* (or *null projection*) and *the identity projection* which are defined as follows: $\forall|\psi\rangle \in \mathcal{H}$,

$$\mathbb{O}|\psi\rangle = \underline{0} \text{ and } \mathbb{I}|\psi\rangle = |\psi\rangle.$$

Any projection other than $\mathbb{O}$ and $\mathbb{I}$ is called a *nontrivial projection.*

Thus, P is a projection operator if $Dom(P) = \mathcal{H}$ and $P = P^2 = P^*$. The set of all projection operators will be indicated by $\Pi(\mathcal{H})$.

One can prove that the set $\mathcal{C}(\mathcal{H})$ of all closed subspaces and the set $\Pi(\mathcal{H})$ of all projections of $\mathcal{H}$ are in one-to-one correspondence.

Let X be a closed subspace of $\mathcal{H}$. By the *projection theorem* every vector $|\psi\rangle \in \mathcal{H}$ can be uniquely expressed as a linear combination $|\psi_1\rangle + |\psi_2\rangle$, where $|\psi_1\rangle \in X$ and $|\psi_2\rangle$ is orthogonal to any vector of X. Accordingly, we can define an operator P_X on $\mathcal{H}$ such that

$$\forall|\psi\rangle \in \mathcal{H} : P_X|\psi\rangle = |\psi_1\rangle$$

(in other words, P_X transforms any vector $|\psi\rangle$ into the "X-component" of $|\psi\rangle$)

It turns out that P_X is a projection operator of $\mathcal{H}$.

Conversely, we can associate to any projection P its range,

$$X_P = \{|\psi\rangle : \exists|\varphi\rangle(P\varphi = |\psi\rangle)\} ,$$

which turns out to be a closed subspace of $\mathcal{H}$.

For any closed subspace X and for any projection P, the following conditions hold:

$$X_{(P_X)} = X; \quad P_{(X_P)} = P.$$

DEFINITION 158. *The trace functional*
Let $\{|\psi_i\rangle\}_{i\in I}$ be any orthonormal basis for $\mathcal{H}$ and let A be a positive operator. The *trace* of A (indicated by $\mathtt{tr}(A)$) is defined as follows:

$$\mathtt{tr}(A) := \sum_i \langle\psi_i|A\psi_i\rangle .$$

One can prove that the definition of $\mathtt{tr}$ is independent of the choice of the basis.

For any positive operator A, there exists a unique positive operator B such that: $B^2 = A$. If A is a (not necessarily positive) bounded operator, then A^*A is positive. Let $|A|$ be the unique positive operator such that $|A|^2 = A^*A$. A bounded operator A is called a *trace-class operator* iff $\mathtt{tr}(|A|) < \infty$.

DEFINITION 159. *Density operator*
A density operator is a positive, self-adjoint, trace-class operator ρ such that $\mathtt{tr}(\rho) = 1$.

It is easy to see that, for any vector $|\psi\rangle$, the projection $P_{[|\psi\rangle]}$ onto the 1-dimensional closed subspace $[|\psi\rangle]$ is a density operator.

DEFINITION 160. *Unitary operator*
A *unitary operator* is a linear operator U such that:

- $Dom(U) = \mathcal{H}$;
- $UU^* = U^*U = \mathbb{I}$.

One can show that the unitary operators U are precisely the operators that preserve the inner product. In other words, for any $|\psi\rangle, |\varphi\rangle \in \mathcal{H}$:

$$\langle \psi | \varphi \rangle = \langle U\psi | U\varphi \rangle .$$

Any pair of Hilbert spaces $\mathcal{H}_1, \mathcal{H}_2$ gives rise to two new Hilbert spaces $\mathcal{H}_1 \otimes \mathcal{H}_2$ and $\mathcal{H}_1 \oplus \mathcal{H}_2$ that represent the *tensor product* of $\mathcal{H}_1$ and $\mathcal{H}_2$ and the *direct sum* of $\mathcal{H}_1$ and $\mathcal{H}_2$, respectively.

Tensor products play an important role for the mathematical representation of compound quantum systems. They are also systematically used in the mathematical formalism of quantum computation.

DEFINITION 161. *Tensor product Hilbert space*
Let $\mathcal{H}_1$ and $\mathcal{H}_2$ be two Hilbert spaces over the same field $\mathbf{D}$ (the real or the complex numbers). A Hilbert space $\mathcal{H}$ is the *tensor product* of $\mathcal{H}_1$ and $\mathcal{H}_2$ iff the following conditions are satisfied:

(i) there exists a map (called *tensor product*) from the cartesian product $\mathcal{H}_1 \times \mathcal{H}_2$ into $\mathcal{H}$ that satisfies the following conditions:

  (a) the tensor product $\otimes$ is linear in each "slot"; in other words, $\forall |\psi\rangle, |\varphi\rangle \in \mathcal{H}_1 \, \forall |\chi\rangle, |\delta\rangle \in \mathcal{H}_2 \, \forall a, b \in \mathbf{D}$:
  (a1) $(a|\psi\rangle + b|\varphi\rangle) \otimes |\chi\rangle = (a|\psi\rangle) \otimes |\chi\rangle + (b|\varphi\rangle) \otimes |\chi\rangle$;
  (a2) $|\psi\rangle \otimes (a|\chi\rangle + b|\delta\rangle) = |\psi\rangle \otimes (a|\chi\rangle) + |\psi\rangle \otimes (b|\delta\rangle)$;

  (b) the external product with a scalar carries across the tensor product; in other words, $\forall |\psi\rangle \in \mathcal{H}_1 \, \forall |\varphi\rangle \in \mathcal{H}_2 \, \forall a \in \mathbf{D}$:
  $a(|\psi\rangle \otimes |\varphi\rangle) = (a|\psi\rangle) \otimes |\varphi\rangle = |\psi\rangle \otimes (a|\varphi\rangle)$.

(ii) every vector of $\mathcal{H}$ can be expressed as a linear combination of vectors of the set $\{|\varphi\rangle \otimes |\psi\rangle \,:\, |\varphi\rangle \in \mathcal{H}_1\,, |\psi\rangle \in \mathcal{H}_2\}$.

One can show that the tensor product is unique up to isomorphism.

As required by condition (ii), every vector of $\mathcal{H}_1 \otimes \mathcal{H}_2$ can be expressed as a linear combination of vectors of the form $|\psi\rangle \otimes |\varphi\rangle$ (where $|\psi\rangle \in \mathcal{H}_1$, $|\varphi\rangle \in \mathcal{H}_2$). At the same time, there are vectors of $\mathcal{H}_1 \otimes \mathcal{H}_2$ that cannot be written as a single product $|\psi\rangle \otimes |\varphi\rangle$ for any $|\psi\rangle \in \mathcal{H}_1$, $|\varphi\rangle \in \mathcal{H}_2$. These vectors are called *non factorized.*

If $\{|\psi_i\rangle\}_{i\in I}$ and $\{|\varphi_j\rangle\}_{j\in J}$ are orthonormal bases for $\mathcal{H}_1$ and $\mathcal{H}_2$, respectively, then the set $\{|\varphi_i\rangle \otimes |\psi_j\rangle \,:\, i \in I\,, j \in J\}$ is an orthonormal basis of the tensor product Hilbert space. In particular, if $\{|\psi_i\rangle, \ldots, |\psi_n\rangle\}$ and $\{|\varphi_1\rangle, \ldots |\varphi_m\rangle\}$ are orthonormal bases of the finite dimensional Hilbert spaces $\mathcal{H}_1$, $\mathcal{H}_2$, then every vector $|\psi\rangle \in \mathcal{H}_1 \otimes \mathcal{H}_2$ can be written as

$$|\psi\rangle = \sum_{i=1}^{n}\sum_{j=1}^{m} a_{ij}|\psi_i\rangle \otimes |\varphi_j\rangle.$$

This shows that the dimension of the tensor product Hilbert space $\mathcal{H}_1 \otimes \mathcal{H}_2$ is the product of the dimensions of $\mathcal{H}_1$ and $\mathcal{H}_2$.

DEFINITION 162. *Direct sum Hilbert space*
Let $\mathcal{H}_1$ and $\mathcal{H}_2$ be two Hilbert spaces over the same field $\mathbf{D}$ (the real or the complex numbers), with inner products $\langle\cdot|\cdot\rangle_1$ and $\langle\cdot|\cdot\rangle_2$, respectively. The *direct sum* of $\mathcal{H}_1$ and $\mathcal{H}_2$ (denoted by $\mathcal{H}_1 \oplus \mathcal{H}_2$) is the vector space based on the cartesian product $\mathcal{H}_1 \times \mathcal{H}_2$, where the vector operations are defined as follows $\forall\,|\psi_1\rangle, |\varphi_1\rangle \in \mathcal{H}_1$, $\forall|\psi_2\rangle, |\varphi_2\rangle \in \mathcal{H}_2$, $\forall a \in \mathbf{D}$:

(i) $(|\psi_1\rangle, |\psi_2\rangle) + (|\varphi_1\rangle, |\varphi_2\rangle) = (|\psi_1\rangle + |\varphi_1\rangle\,,\, |\psi_2\rangle + |\varphi_2\rangle)$;

(ii) $a(|\psi_1\rangle, |\psi_2\rangle) = (a|\psi_1\rangle, a|\psi_2\rangle)$.

The inner product $\langle\cdot|\cdot\rangle$ of $\mathcal{H}_1 \oplus \mathcal{H}_2$ is defined in the following way:

(iii) $\langle(|\psi_1\rangle, |\psi_2\rangle)|(|\varphi_1\rangle, |\varphi_2\rangle)\rangle = \langle\psi_1|\varphi_1\rangle_1 + \langle\psi_2|\varphi_2\rangle_2$.

One can easily show that $\mathcal{H}_1 \oplus \mathcal{H}_2$ is a Hilbert space.

Instead of $(|\psi_1\rangle, |\psi_2\rangle)$ we will write $|\psi_1\rangle \oplus |\psi_2\rangle$. Clearly, the subspaces $\mathcal{H}_1 \times \{\underline{0}\}$ and $\{\underline{0}\} \times \mathcal{H}_2$ of $\mathcal{H}_1 \oplus \mathcal{H}_2$ are isomorphic to $\mathcal{H}_1$ and $\mathcal{H}_2$, respectively.

ACKNOWLEDGEMENTS

We warmly thank Giuseppe Sergioli for his precious suggestions concerning Section 9.

BIBLIOGRAPHY

[Aerts, 2000] D. Aerts and B. Van Steirteghem, *Quantum axiomatics and a theorem of M.P. Solér*, International Journal of Theoretical Physics **39** (2000), 497–502.

[Altenkirch and Grattage, 2005] T. Altenkirch and J. Grattage, *A functional quantum programming language*, Proceedings of the XX Annual Symposium on Logic and Computer Science, IEEE, 2005.

[Arrighi and Dowek, 2004] P. Arrighi and G. Dowek, *Operational semantics for formal tensorial calculus,* Languages (P. Selinger, ed.), Turku Centre for Computer Science, 2004.

[Battilotti, 1998] G. Battilotti, *Embedding classical logic into basic orthologic with a primitive modality*, Logic Journal of the IGPL, **6** (1998), 383–402.

[Battilotti and Faggian, 2002] G. Battilotti and C. Faggian, *Quantum logic and the cube of logics*, Handbook of Philosophical Logic (D. M. Gabbay and F. Guenthner, eds.), vol. 6, Kluwer Academic Publishers, Dordrecht, 2002, pp. 213–226, section 17 in [Dalla Chiara and Giuntini, 2002].

[Battilotti and Sambin, 1999] G. Battilotti and G. Sambin, *Basic logic and the cube of its extensions*, Logic and Foundations of Mathematics (A. Cantini, E. Casari, and P. Minari, eds.), Kluwer Academic Publishers, Dordrecht, 1999, pp. 165–186.

[Beltrametti and Bugajski, 1995] E. Beltrametti and S. Bugajski, *A classical extension of quantum mechanics*, Journal of Physics A: Mathematical and General **28** (1995), 247–261.

[Beltrametti and Bugajski, 1997] E. Beltrametti and S. Bugajski. *Effect algebras and statistical physical theories*, Journal of Mathematical Physics **38** (1997), 3020–3030.

[Beltrametti and Cassinelli, 1981] E. Beltrametti and G. Cassinelli, *The logic of quantum mechanics*, Encyclopedia of Mathematics and its Applications, vol. 15, Addison-Wesley, Reading, 1981.

[Bennett and Foulis, 1997] M. K. Bennett and D. J. Foulis, *Interval and scale effect algebras*, Advances in Mathematics **19** (1997), 200–215.

[Birkhoff, 1940] G. Birkhoff, *Lattice Theory*, 3^{rd}(new) ed., Colloquium Publications, vol. 25, American Mathematical Society, Providence, 1967.

[Birkhoff and von Neumann, 1936] G. Birkhoff and J. von Neumann, *The logic of quantum mechanics*, Annals of Mathematics **37** (1936), 823-843, in [von Neumann, 1961b].

[Bruns *et al.*, 1990] G. Bruns, R. J. Greechie, J. Harding, and M. Roddy, *Completions of orthomodular lattices*, Order **7** (1990), 789–807.

[Bub, 1999] J. Bub, *Interpreting the quantum world*, Cambridge University Press, Cambridge, 1999.

[Bugajski, 1993] S. Bugajski, *Delinearization of quantum logic*, International Journal of Theoretical Physics **32** (1993), 389–398.

[Busch, 1985] P. Busch, *Elements of unsharp reality in the EPR experiment*, Symposium on the Foundations of Modern Physics (P. Lahti and P. Mittelstaedt, eds.), World Scientific, Singapore, 1985, pp. 343–357.

[Busch *et al.*, 1995] P. Busch, M. Grabowski, and P. Lahti, *Operational quantum mechanics*, Lectures Notes in Physics, no. m31, Springer, Berlin, 1995.

[Busch *et al.*, 1991] P. Busch, P. Lahti, and P. Mittelstaedt, *The quantum theory of measurement*, Lectures Notes in Physics, no. m2, Springer, Berlin, 1991.

[Cattaneo, 1993] G. Cattaneo, *Fuzzy quantum logic II: the logics of unsharp quantum mechanics*, International Journal of Theoretical Physics **32** (1993), 1709–1734.

[Cattaneo, 1997] G. Cattaneo. *A unified framework for the algebra of unsharp quantum mechanics*, International Journal of Theoretical Physics **36** (1997), 3085–3117.

[Cattaneo *et al.*, 1999] G. Cattaneo, M. L. Dalla Chiara, and R. Giuntini, *How many notions of 'sharp'?*, International Journal of Theoretical Physics **38** (1999), 3153–3161.

[Cattaneo *et al.*, 1989] G. Cattaneo, C. Garola, and G. Nisticó, *Preparation-effect versus question-proposition structures*, Physics Essays **2** (1989), 197–216.

[Cattaneo and Giuntini, 1995] G. Cattaneo and R. Giuntini, *Some results on BZ structures from hilbertian unsharp quantum physics*, Foundations of Physics **25** (1995), 1147–1182.

[Cattaneo and Gudder, 1999] G. Cattaneo and S. P. Gudder, *Algebraic structures arising in axiomatic unsharp quantum physics*, Foundations of Physics **29** (1999), 1607–1637.

[Cattaneo and Laudisa, 1994] G. Cattaneo and F. Laudisa, *Axiomatic unsharp quantum theory (from Mackey to Ludwig)*, Foundations of Physics **24** (1994), 631–683.

[Cattaneo and Nisticó, 1989] G. Cattaneo and G. Nisticó, *Brouwer-Zadeh posets and three-valued Łukasiewicz posets*, Fuzzy Sets and Systems **33** (1986), 165–190.

[Chang, 1958] C. C. Chang, *Algebraic analysis of many valued logics*, Transactions of the American Mathematical Society **88** (1958), 74–80.

[Chang, 1959] C. C. Chang. *A new proof of the completeness of Łukasiewicz axioms*, Transactions of the American Mathematical Society **93** (1959), 467–490.

[Cignoli and D'Ottaviano, 2000] R. Cignoli, I. M. L. D'Ottaviano, and D. Mundici, *Algebraic foundations of many-valued reasoning*, Trends in Logic, vol. 7, Kluwer Academic Publishers, Dordrecht, 2000.

[Coecke *et al.*, 2000] B. Coecke, D. Moore and A. Wilce, *Operational quantum logic: An overview*, Current Research in Operational Quantum Logic (B. Coecke, D. Moore and A. Wilce, eds.), Kluwer, Dordrecht, 2000, pp. 1-36.

[Cutland and Gibbins, 1982] N. J. Cutland and P .F. Gibbins, *A regular sequent calculus for quantum logic in which* $\wedge$ *and* $\vee$ *are dual*, Logique et Analyse — Nouvelle Serie - **25** (1982), no. 45, 221–248.

[da Costa and Krause, 2005] N.C.A. da Costa, D. Krause, *Remarks on the applications of paraconsistent logic to physics,* Filosofia, Ciencia e Historia (M. Pietrocola and O. Freire, eds.), Discurso Editorial, S. Paulo, 2005, pp. 337-359.

[da Costa *et al.*, 2006] N.C.A. da Costa, D. Krause and O. Bueno, *Paraconsistent logic and paraconsistency: Technical and philosophical developments,* Handbook of Philosophy of Science (D. Jacquette, ed.), Elsevier, Amsterdam, 2006.

[Dalla Chiara, 1981] M. L. Dalla Chiara, *Some metalogical pathologies of quantum logic*, Current Issues in Quantum Logic (E. Beltrametti and B. van Fraassen, eds.), Ettore Majorana International Science Series, vol. 8, Plenum, New York, 1981, pp. 147–159.

[Dalla Chiara and Giuntini, 1989] M. L. Dalla Chiara and R. Giuntini, *Paraconsistent quantum logics*, Foundations of Physics **19** (1989), 891–904.

[Dalla Chiara and Giuntini, 1994] M. L. Dalla Chiara and R. Giuntini. *Unsharp quantum logics*, Foundations of Physics **24** (1994), 1161–1177.

[Dalla Chiara and Giuntini, 1995] M. L. Dalla Chiara and R. Giuntini. *The logics of orthoalgebras*, Studia Logica **55** (1995), 3–22.

[Dalla Chiara and Giuntini, 1999] M. L. Dalla Chiara and R. Giuntini, *Łukasiewicz theory of truth, from the quantum logical point of view*, Alfred Tarski and the Vienna Circle (J. Woléński and E. Köhler, eds.), Kluwer, Dordrecht, 1999, pp. 127–134.

[Dalla Chiara and Giuntini, 2002] M. L. Dalla Chiara and R. Giuntini. *Quantum logics*, Handbook of Philosophical Logic (D.M. Gabbay and F. Guenthner, eds.), vol. 6, Kluwer Academic Publishers, Dordrecht, 2002, pp. 129–228.

[Dalla Chiara *et al.*, 2004] M. L. Dalla Chiara, R. Giuntini, R. Greechie, *Reasoning in quantum theory. Sharp and unsharp quantum logics*, Kluwer Academic Publishers, Dordrecht, 2004.

[Dalla Chiara *et al.*, 2003] M. L. Dalla Chiara, R. Giuntini, and R. Leporini, *Quantum Computational Logics. A Survey*, Studia Logica, **21** (2003), 213–255.

[Dalla Chiara *et al.*, 2004a] M. L. Dalla Chiara, R. Giuntini and R. Leporini, *Quantum computational logics and Fock space semantics*, International Journal of Quantum Information, **2** (2004), 1–8.

[Dalla Chiara and Leporini, 2005] M. L. Dalla Chiara, R. Giuntini and R. Leporini, *Logics from quantum computation*, International Journal of Quantum Information, **3** (2005), 293–337.

[Dalla Chiara *et al.*, 2006] M. L. Dalla Chiara, R. Giuntini and R. Leporini, *Compositional and Holistic Quantum Computational Semantics*, to appear in Natural Computing, 2006.

[Danos and Reignier, 1989] V. Danos and L. Reignier, *The structure of multiplicatives*, Archive for Mathematical Logic **28** (1989), 181-203.

[Davies, 1976] E. B. Davies, *Quantum theory of open systems*, Academic, New York, 1976.

[Destouches-Février, 1937] P. Destouches-Février, *Les relations d'incertitude de Heisenberg et la logique*, Comptes Rendus de l'Académie de Sciences **204** (1937), 481-483.

[Dishkant, 1972] H. Dishkant, *Semantics of the minimal logic of quantum mechanics*, Studia Logica **30** (1972), 17–29.

[Dunn-Hardegree, 2001] J. M. Dunn and G. M. Hardegree, *Algebraic methods in philosophical logic*, Clarendon Press, Oxford, 2001.

[Dvurečenskij, 1993] A. Dvurečenskij, *Gleason's theorem and its applications*, Mathematics and its Applications, no. 60, Kluwer, Dordrecht, 1993.

[Dvurečenskij, 1997] A. Dvurečenskij, *Measures and ⊥-decomposable measures of effects of a Hilbert space*, Atti del Seminario Matematico e Fisico dell'Università di Modena **45** (1997), 259–288.

[Dvurečenskij, 2000] A. Dvurečenskij, *New trends in quantum structures*, Mathematics and Its Applications, vol. 516, Kluwer Academic Publishers, Dordrecht, 2000.

[Einstein *et al.*, 1035] A. Einstein, B. Podolsky, and N. Rosen, *Can quantum-mechanical description of reality be considered complete?*, Physical Review **47** (1935), 777–780.

[Engesser and Gabbay, 2002] K. Engesser and D. Gabbay, *Quantum logic, Hilbert space, revision theory*, Artificial Intelligence **136** (2002), 61–100.

[Faggian, 1998] C. Faggian, *Classical proofs via basic logic*, Computer Science Logic 11th International Workshop, CSL'97 (M. Nielson and W. Thomas, eds.), Lecture Notes in Computer Science, vol. 1414, Springer Verlag, 1998, pp. 203–219.

[Faggian and Sambin, 1997] C. Faggian and G. Sambin, *From basic logic to quantum logics with cut-elimination*, International Journal of Theoretical Physics **12** (1997), 31–37.

[Finch, 1970] P. D. Finch, *Quantum logic as an implication algebra*, Bulletin of the Australian Mathematical Society **2** (1970), 101–106.

[Fitting, 1969] M. Fitting, *Intuitionistic logic, model theory and forcing*, North-Holland, Amsterdam, 1969.

[Foulis, 1999] D. J. Foulis, *A half-century of quantum logic, what have we learned?*, Quantum Structures and the Nature of Reality (D. Aerts and J. Pykacz, eds.), vol. 7, Kluwer Academic Publishers, Dordrecht, 1999, pp. 1–36.

[Foulis, 2000] D. J. Foulis, *MV and Heyting effect algebras*, Foundations of Physics **30** (2000), 1687–1706.

[Foulis and Bennett, 1994] D. J. Foulis and M. K. Bennett, *Effect algebras and unsharp quantum logics*, Foundations of Physics **24** (1994), 1325–1346.

[Foulis and Greechie, 2000] D. J. Foulis and R. J. Greechie, *Specification of finite effect algebras*, International Journal of Theoretical Physics **39** (2000), 665–676.

[Foulis and Randall, 1981] D. J. Foulis and C. H. Randall, *Empirical logic and tensor product*, Interpretation and Foundations of Quantum Mechanics, Grundlagen der exakten Naturwissenschaften, vol. 5, Bibliographisches Institut, Mannheim, 1981, pp. 9–20.

[Foulis and Randall, 1983] D. J. Foulis and C. H. Randall, *Properties and operational propositions in quantum mechanics*, Foundations of Physics **13** (1983), 843–857.

[Foulis *et al.*, 1996] D.J. Foulis, R.J. Greechie, M. L. Dalla Chiara, and R. Giuntini, *Quantum Logic*, Encyclopedia of Applied Physics (G. Trigg, ed.), vol. 15, VCH Publishers, 1996, pp. 229–255.

[Gallier, 1991] J. Gallier, *Constructive Logics. II: Linear Logic and Proof Nets*, Research Report no. 9, Digital PRL, 1991.

[Garola, 1980] C. Garola, *Propositions and orthocomplementation in quantum logic*, International Journal of Theoretical Physics **19** (1980), 369–378.

[Garola, 1985] C. Garola, *Embedding of posets into lattices in quantum logic*, International Journal of Theoretical Physics **24** (1985), 423–433.

[Gibbins, 1985] P. F. Gibbins, *A user-friendly quantum logic*, Logique-et-Analyse.-Nouvelle-Serie **28** (1985), 353–362.

[Gibbins, 1987] P. F. Gibbins, *Particles and paradoxes — the limits of quantum logic*, Cambridge University Press, Cambridge, 1987.

[Girard, 1987] J. Y. Girard, *Linear logic*, Theoretical Computer Science **50** (1987), 1–102.

[Girard, 1996] J.-Y. Girard, *Proofnets: The parallel syntax for proof theory*, Logic and Algebra (A. Ursini and P. Aglianó, eds.), Dekker, New York, 1996, pp. 97-124.

[Girard, 1999] J.-Y. Girard, *Coherent Banach spaces: A continuous denotational semantics*, Theoretical Computer Science **227** (1999), 275-297.

[Girard, 2004a] J.-Y. Girard, *Between logic and quantic: A tract*, Linear Logic in Computer Science (P. Ruet, J. Ehrhard, J.-Y. Girard and P. Scott, eds.), Cambridge University Press, Cambridge, 2004, pp. 346-390.

[Girard, 2004b] J.-Y. Girard, *Logique à Rome*, electronic notes, Ch. 17, 2004.

[Giuntini, 1990] R. Giuntini, *Brouwer-Zadeh logic and the operational approach to quantum mechanics*, Foundations of Physics **20** (1990), 701–714.

[Giuntini, 1991a] R. Giuntini, *Quantum logic and hidden variables*, Grundlagen der exakten Naturwissenschaften, no. 8, Bibliographisches Institut, Mannheim, 1991.

[Giuntini, 1991b] R. Giuntini, *A semantical investigation on Brouwer-Zadeh logic*, Journal of Philosophical Logic **20** (1991), 411–433.

[Giuntini, 1992] R. Giuntini, *Brouwer-Zadeh logic, decidability and bimodal systems*, Studia Logica **51** (1992), 97–112.

[Giuntini, 1993] R. Giuntini, *Three-valued Brouwer-Zadeh logic*, International Journal of Theoretical Physics **32** (1993), 1875–1887.

[Giuntini, 1995a] R. Giuntini, *Quasilinear QMV algebras*, International Journal of Theoretical Physics **34** (1995), 1397–1407.

[Giuntini, 1995b] R. Giuntini, *Unsharp orthoalgebras and quantum MV algebras*, The Foundations of Quantum Mechanics — Historical Analysis and Open Questions (C. Garola and A. Rossi, eds.), Kluwer, Dordrecht, 1995, pp. 325–337.

[Giuntini, 1996] R. Giuntini, *Quantum MV algebras*, Studia Logica **56** (1996), 393–417.

[Giuntini, 2000] R. Giuntini, *An independent axiomatization of QMV algebras*, The Foundations of Quantum Mechanics (C. Garola and A. Rossi, eds.), World Scientific, Singapore, 2000.

[Giuntini, 2002] R. Giuntini, *Weakly linear QMV algebras*, Algebra Universalis, 53, 1, 2005, 45-72.

[Giuntini and Greuling, 1989] R. Giuntini and H. Greuling, *Toward an unsharp language for unsharp properties*, Foundations of Physics **19** (1989), 931–945.

[Giuntini and Pulmannová, 2000] R. Giuntini and S. Pulmannová, *Ideals and congruences in QMV algebras*, Communications in Algebra **28** (2000), 1567–1592.

[Gleason, 1957] A. M. Gleason, *Measures on the closed subspaces of a Hilbert space*, Journal of Mathematics and Mechanics **6** (1957), 885–893.

[Goldblatt, 1974] R. Goldblatt, *Semantical analysis of orthologic*, Journal of Philosophical Logic **3** (1974), 19–35.

[Goldblatt, 1984] R. Goldblatt, *Orthomodularity is not elementary*, The Journal of Symbolic Logic **49** (1984), 401–404.

[Greechie, 1968] R. J. Greechie, *On the structure of orthomodular lattices satisfying the chain condition*, Journal of Combinatorial Theory **4** (1968), 210–218.

[Greechie, 1974] R. J. Greechie, *Some results from the combinatorial approach to quantum logic*, Synthese **29** (1974), 113–127.

[Greechie, 1981] R. J. Greechie, *A non-standard quantum logic with a strong set of states*, Current Issues in Quantum Logic (E. Beltrametti and B. van Fraassen, eds.), Ettore Majorana International Science Series, vol. 8, Plenum, New York, 1981, pp. 375–380.

[Greechie and Foulis, 1995] R. J. Greechie and D. J. Foulis, *The transition to effect algebras*, International Journal of Theoretical Physics **34** (1995), 1369–1382.

[Gudder, 1979] S. P. Gudder, *A survey of axiomatic quantum mechanics*, The Logico-Algebraic Approach to Quantum Mechanics (C. A. Hooker, ed.), vol. II, Reidel, Dordrecht, 1979, pp. 323–363.

[Gudder, 1995] S. P. Gudder, *Total extensions of effect algebras*, Foundations of Physics Letters **8** (1995), 243–252.

[Gudder, 1998] S. P. Gudder, *Sharply dominating effect algebras*, Tatra Mountains Mathematical Publications **15** (1998), 23–30.

[Gudder, 2003] S. Gudder, Quantum computational logic, *International Journal of Theoretical Physics* **42** (2003), 39–47.

[Gudder and Greechie, 1996] S. P. Gudder and R. J. Greechie, *Effect algebra counterexamples*, Mathematica Slovaca **46** (1996), 317–325.

[Hájek, 1998] P. Hájek, *Metamathematics of fuzzy logic*, Trends in Logic, vol. 4, Kluwer Academic Publishers, Dordrecht, 1998.

[Halmos, 1951] P. R. Halmos, *Introduction to Hilbert space and the theory of spectral multiplicity*, Chelsea, New York, 1951.

[Hardegree, 1975] G. M. Hardegree, *Stalnaker conditionals and quantum logic*, Journal of Philsophical Logic **4** (1975), 399–421.

[Hardegree, 1976] G. M. Hardegree, *The conditional in quantum logic*, Logic and Probability in Quantum Mechanics (P. Suppes, ed.), Reidel, Dordrecht, 1976, pp. 55–72.

[Hardegree, 1981] G. M. Hardegree, *An axiom system for orthomodular quantum logic*, Studia Logica **40** (1981), 1–12.

[Holland, 1995] S. S. Holland, *Orthomodularity in infinite dimensions: a theorem of M. Solèr*, Bulletin of the American Mathematical Society **32** (1995), 205–232.

[Howard, 1980] W.A. Howard, *The formulae-as-types notion of construction,* To H.B. Curry: Essays on Combinatory Logic, Lambda Calculus, and Formalism (J.R. Hindley and J. Seldin, eds.), Academic Press, New York, 1980, pp. 479-490.

[Hughes, 1985] R. I. G. Hughes, *Semantic alternatives in partial Boolean quantum logic*, Journal of Philosophical Logic **14** (1985), 411–446.

[Hughes, 1987] R. I. G. Hughes, *The structure and interpretation of quantum mechanics*, Cambridge University Press, Cambridge, 1987.

[Hughes and van Glabbeek, 2003] D.J.D. Hughes, R.J. van Glabbeek, *Proof nets for unit free multiplicative additive linear logic* (extended abstract), Proceedings of the XVIII Annual Symposium on Logic and Computer Science, IEEE, 2003, pp. 1-10.

[Jammer, 1974] M. Jammer, *The philosophy of quantum mechanics*, Wiley-Interscience, New York, 1974.

[Jauch, 1968] J. M. Jauch, *Foundations of quantum mechanics*, Addison-Wesley, London, 1968.

[Kalmbach, 1983] G. Kalmbach, *Orthomodular lattices*, Academic Press, New York, 1983.

[Keller, 1980] H. A. Keller, *Ein nichtklassischer hilbertscher Raum*, Mathematische Zeitschrift **172** (1980), 41–49.

[Kochen and Specker, 1965a] S. Kochen and E. P. Specker, *The calculus of partial propositional functions*, Proceedings of the 1964 International Congress for Logic, Methodology and Philosophy of Science (Y. Bar-Hillel, ed.), North-Holland, Amsterdam, 1965, pp. 45–57.

[Kochen and Specker, 1965b] S. Kochen and E. P. Specker, *Logical structures arising in quantum theory*, The Theory of Models (J. Addison, L. Henkin, and A. Tarski, eds.), North-Holland, Amsterdam, 1965, pp. 177–189.

[Kochen and Specker, 1967] S. Kochen and E. P. Specker, *The problem of hidden variables in quantum mechanics*, Journal of Mathematics and Mechanics **17** (1967), 59–87.

[Kôpka and Chovanec, 1994] F. Kôpka and F. Chovanec, *D-posets*, Mathematica Slovaca **44** (1994), 21–34.

[Kraus, 1983] K. Kraus, *States, effects and operations*, Lecture Notes in Physics, vol. 190, Springer, Berlin, 1983.

[Lincoln *et al.*, 1992] P.D. Lincoln, J.C. Mitchell, A. Scedrov, N. Shankar, *Decision problems for propositional linear logic*, Annals of Pure and Applied Logic **56** (1992), 239-311.

[Ludwig, 1983] G. Ludwig, *Foundations of quantum mechanics*, vol. 1, Springer, Berlin, 1983.

[Łukasiewicz, 1936] J. Łukasiewicz, *Logistic and philosophy*, Selected Work (L. Borkowski, ed.), North-Holland, Amsterdam, 1970, pp. 218–235.

[Łukasiewicz, 1946] J. Łukasiewicz, *On determinism*, Selected Work (L. Borkowski, ed.), North-Holland, Amsterdam, 1970, pp. 110–128.

[Łukasiewicz, 1970] J. Łukasiewicz, *On three-valued logic*, Selected Work (L. Borkowski, ed.), North-Holland, Amsterdam, 1970.

[Mackey, 1957] G. Mackey, *The Mathematical Foundations of Quantum Mechanics*, Benjamin, New York, 1957.

[Mangani, 1973] P. Mangani, *Su certe algebre connesse con logiche a più valori*, Bollettino Unione Matematica Italiana **8** (1973), 68–78.

[Minari, 1987] P. Minari, *On the algebraic and kripkean logical consequence relation for orthomodular quantum logic*, Reports on Mathematical Logic **21** (1987), 47–54.

[Mittelstaedt, 1972] P. Mittelstaedt, *On the interpretation of the lattice of subspaces of Hilbert space as a propositional calculus*, Zeitschrift für Naturforschung, Dordrecht, textbf27a (1972), 1358–1362.

[Mittelstaedt, 1978] P. Mittelstaedt, *Quantum logic*, Reidel, Dordrecht, 1978.

[Mittelstaedt, 1985] P. Mittelstaedt (ed.), *Recent developments in quantum logic*, Grundlagen der exakten Naturwissenschaften, no. 6, Bibliographisches Institut, Mannheim, 1985.

[Mundici, 1992] D. Mundici, *The logic of Ulam's game with lies*, Knowledge, Belief and Strategic Interaction (C. Bicchieri and M. L. Dalla Chiara, eds.), Cambridge University Press, Cambridge, 1992.

[Navara, 1999] M. Navara, *Two descriptions of state spaces of orthomodular structures*, International Journal of Theoretical Physics **38** (1999), 3163–3178.

[Neubrunn and Riečan, 1997] T. Neubrunn and B. Riečan, *Integral, measure and ordering*, Kluwer Academic Publishers, Dordrecht, 1997.

[Nielsen and Chuang, 2000] M.A. Nielsen and I.L. Chuang, *Quantum computation and quantum information*, Cambridge University Press, Cambridge, 2000.

[Nishimura, 1980] H. Nishimura, *Sequential method in quantum logic*, Journal of Symbolic Logic **45** (1980), 339–352.

[Nishimura, 1994] H. Nishimura, *Proof theory for minimal quantum logic I and II*, International Journal of Theoretical Physics **33** (1994), 102–113, 1427–1443.

[Paoli, 2002] F. Paoli, *Substructural logics: A primer*, Trends in Logic, vol. 13, Kluwer Academic Publishers, 2002.

[Peres, 1995] A. Peres, *Quantum theory: Concepts and methods*, Kluwer Academic Publishers, Dordrecht, 1995.

[Piron, 1976] C. Piron, *Foundations of quantum physics*, W. A. Benjamin, Reading, 1976.

[Pitowsky, 1989] I. Pitowsky, *Quantum probability — quantum logic*, Lectures Notes in Physics, no. 321, Springer, Berlin, 1989.

[Pratt, 1993] V. Pratt, *Linear logic for generalized quantum mechanics*, Workshop on Physics and Computation (PhysComp'92) (Dallas), IEEE, 1993, pp. 166–180.

[Pták and Pulmannová, 1991] P. Pták and S. Pulmannová, *Orthomodular structures as quantum logics*, Fundamental Theories of Physics, no. 44, Kluwer, Dordrecht, 1991.

[Pulmannová, 1995] S. Pulmannová, *Representation of D-posets*, International Journal of Theoretical Physics **34** (1995), 1689–1696.

[Putnam, 1969] H. Putnam, *Is logic empirical?*, Boston Studies in the Philosophy of Science (R. S. Cohen and M. W. Wartofsky, eds.), vol. 5, Reidel, Dordrecht, 1969, pp. 216–241.

[Pykacz, 2000] J. Pykacz, *Łukasiewicz operations in fuzzy set theories and many-valued representations of quantum logics*, Foundations of Physics, **30** (2000), 1503–1524.

[Rédei, 2001] M. Rédei, *Von Neumann's concept of quantum logic and quantum probability*, in [Rédei and Stoeltzner, 2001].

[Rédei, 1999] M. Rédei, *"Unsolved problems in mathematics" J. von Neumann's address to the International Congress of Mathematicians Amsterdam, September 2-9, 1954*, The Mathematical Intelligencer **21** (1999), 7-12.

[Rédei, 1998] M. Rédei, *Quantum logic in algebraic approach*, Kluwer Academic Publishers, Dordrecht, Holland, 1998.

[Rédei, 1996] M. Rédei, *Why John von Neumann did not like the Hilbert space formalism of quantum mechanics (and what he liked instead)*, Studies in the History and Philosophy of Modern Physics **27** (1996), 493-510.

[Rédei, forthcoming] M. Rédei: *The birth of quantum logic*, History and Philosophy of Logic, forthcoming.

[Rédei and Stoeltzner, 2001] M. Rédei and M. Stöltzner, *John von Neumann and the Foundations of Quantum Physics*, M. Stöltzner, M. Rédei (eds.), Kluwer Academic Publishers, Dordrecht, Boston, London 2001.

[Redhead, 1987] M. Redhead, *Incompleteness, nonlocality and realism — a prolegomenon to the philosophy of quantum mechanics*, Clarendon Press, Oxford, 1987.

[Reed and Simon, 1972] M. Reed and B. Simon, *Methods of modern mathematical physics*, vol. I, Academic Press, New York, 1972.

[Restall, 2002] G. Restall, *Paraconsistency everywhere*, Notre Dame Journal of Formal Logic **43** (2002), 147-156.

[Restall and Paoli, 2005] G. Restall, F. Paoli, *The geometry of nondistributive logics*, Journal of Symbolic Logic **70** (2005), 1108-1126.

[Rosenthal, 1990] K. I. Rosenthal, *Quantales and their applications*, Longman, New York, 1990.

[Sambin *et al.*, 2000] G. Sambin, G. Battilotti, and C. Faggian, *Basic logic: reflection, symmetry, visibility*, The Journal of Symbolic Logic **65** (2000), 979–1013.

[Schroeck, 1996] F. E. Schroeck, *Quantum mechanics on phase space*, Fundamental Theories of Physics, vol. 74, Kluwer Academic Publishers, Dordrecht, 1996.

[Schütte Monting, 1981] J. Schülte Monting, *Cut elimination and word problem for varieties of lattices*, Algebra Universalis **12** (1981), 290-321.

[Selinger, 2004] P. Selinger, *Towards a quantum programming language*, Mathematical Structures in Computer Science **14** (2004), 527-586.

[Smets, 2001] S. Smets, *The logic of physical properties in static and dynamic perspective*, PhD Thesis, Vrije Universiteit Brussels, 2001.

[Solèr, 1995] M. P. Solèr, *Characterization of Hilbert spaces by orthomodular spaces*, Communications in Algebra **23** (1995), 219–243.

[Stalnaker, 1981] R. Stalnaker, *A theory of conditionals*, Ifs. Conditionals, Belief, Decision, Chance, and Time (W. Harper, G. Pearce, and R. Stalnaker, eds.), Reidel, Dordrecht, 1981, pp. 41–55.

[Svozil, 1998] K. Svozil, *Quantum logic*, Springer, Singapore, 1998.

[Takeuti, 1981] G. Takeuti, *Quantum set theory*, Current Issues in Quantum Logic (E. G. Beltrametti and B. C. van Fraassen, eds.), Ettore Majorana International Science Series, vol. 8, Plenum, New York, 1981, pp. 303–322.

[Tamura, 1988] S. Tamura, *A Gentzen formulation without the cut rule for ortholattices*, Kobe Journal of Mathematics **5** (1988), 133–150.

[Tortora De Falco, 2003] L. Tortora De Falco, *The additive multiboxes*, Annals of Pure and Applied Logic **120** (2003), 489-524.

[van Fraassen, 1991] B. van Fraassen, *Quantum mechanics. An empiricist view*, Clarendon Press, Oxford, 1991.

[Varadarajan, 1985] V. S. Varadarajan, *Geometry of quantum theory*, 2 ed., Springer, Berlin, 1985.

[van Tonder, 2004] A. van Tonder, *A lambda calculus for quantum computation*, SIAM Journal of Computing **33** (2004), 1109-1135.

[von Neumann, 1927a] J. von Neumann, *Mathematische Begründung der Quantenmechanik*, Göttinger Nachrichten (1927), 1-57, in [von Neumann, 1962], pp. 151-207.

[von Neumann, 1927b] J. von Neumann, *Wahrscheinlichkeitstheoretischer Aufbau der Quantenmechanik*, Göttinger Nachrichten (1927), 245-272, in [von Neumann, 1962], pp. 208-235.

[von Neumann, 1927c] J. von Neumann, *Thermodynamik quantenmechanischer Gesamtheiten*, Göttinger Nachrichten (1927), 245-272, in [von Neumann, 1962],pp. 236-254.

[von Neumann, 1932] J. von Neumann, *Mathematische Grundlagen der Quantenmechanik*, Dover Publications, New York, 1943 (first American Edition; first edition: Springer Verlag, Heidelberg, 1932).

[von Neumann, 1962] J. von Neumann, *Collected Works Vol. I. Logic, Theory of Sets and Quantum Mechanics* , A.H. Taub (ed.), Pergamon Press, 1962.

[von Neumann, 1961] J. von Neumann, *Collected Works Vol. III. Rings of Operators*, A.H. Taub (ed.), Pergamon Press, 1961.

[von Neumann, 1961b] J. von Neumann, *Collected Works Vol. IV. Continuous Geometry and Other Topics*, A.H. Taub (ed.), Pergamon Press, 1961.

[von Neumann, 1981] J. von Neumann,*Continuous Geometries with Transition Probability*, Memoirs of the American Mathematical Society **34** No. 252 (1981) 1-210.

[Wright, 1990] R. Wright, *Generalized urn models*, Foundations of Physics **20** (1990), 881–903.

[Zadeh, 1965] L. Zadeh, *Fuzzy sets*, Information and Control **8** (1965), 338–353.

[Zierler, 1961] N. Zierler, *Axioms for non-relativistic quantum mechanics*, Pacific Journal of Mathematics **11** (1961), 1151–1169.

Edited by K. Engesser, D. M. Gabbay and D. Lehmann

GENTZEN METHODS IN QUANTUM LOGIC

Hirokazu Nishimura

1 INTRODUCTION

Since Birkhoff and von Neumann [Birkhoff and von Neumann, 1936] a new area of logical investigation has grown up under the name of quantum logic. During its early days emphasis was put exclusively on its algebraic aspects. A new impetus came from Dishkant and Goldblatt's ([Dishkant, 1972], [Dishkant, 1977] and [Goldblatt, 1974]) remarkable discovery on the relationship between ortholattices and the Brouwerian modal logic **B** in the 1970's, which is comparable to Mckinsey and Tarski's [McKinsey and Tarski, 1948] translation of intuitionistic logic into the modal logic **S**4. As the semantics of possible worlds has been one of the main tools in modal logic since Kripke [Kripke, 1963], the discovery naturally admitted to a Kripkian relational semantics of minimal quantum logic. Since it was then well known that there is a close relationship between Gentzen-style formulations of modal logics and their Kripkian relational semantics (cf. [Nishimura, 1983] and [Sato, 1977]), Nishimura [Nishimura, 1980] was driven on closing days of the 1970's to a Gentzen-style formulation of minimal quantum logic, which regrettably failed to enjoy the cut-elimination theorem. A more natural Gentzen-style formulation of minimal quantum logic with closer inspection on its relationship to the relational semantics was given by Cutland and Gibbins [Cutland and Gibbins, 1982], but it still failed to acquiesce in the cut-elimination property. The first cut-free Gentzen-style formulation of minimal quantum logic was presented by Tamura [Tamura, 1988], though it suffered from unnecessary clumsiness, which made his system appear more esoteric than it really was. A final step was taken again by Nishimura ([Nishimura, 1994a] and [Nishimura, 1994b]), which was followed by Takano's [Takano, 1995] significant remark that the inference rule from a sequent to its contraposition is redundant. The first stage of the story has thus ended, and the principal objective in this paper is to present its fruits to a novice thoroughly.

In Section 2 we will present our cut-free Gentzen-style sequential system **GMQL**. We will remark, following [Cutland and Gibbins, 1982], that admitting unrestricted (cut) as an inference rule would force our system **GMQL** to degenerated into classical logic. In Section 3 we will show, following [Cutland and Gibbins, 1982], that the inference rule from a sequent to its contraposition is admissible in **GMQL**. In Section 4 we will establish the fundamental fact that the negation $'$ is involutive with respect to its proof-theoretical behaviors. In Section 5 the desired cut-elimination theorem is to be demonstrated. The final section is devoted to the completeness theorem with respect to the relational semantics of Dishkant and Goldblatt.

The reader may wonder what is to be the second stage of the story. We will give two suggestions. The modal logics **S**4 and **B** stand to the modal logic **S**5 in opposite directions, but they are complementary against **S**5, as may be illustrated in the following figure:

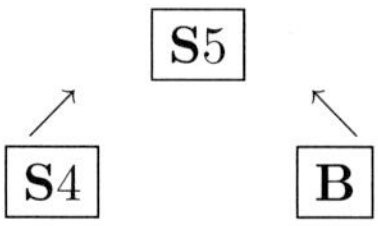

The complementarity of the modal logics **S**4 and **B** corresponds to the following complementarity of intuitionistic logic and minimal quantum logic against classical logic, as may be illustrated in the following figure:

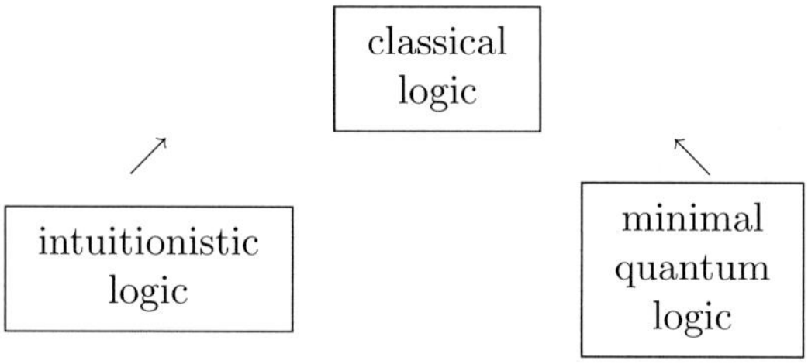

Logics between classical logic and intuitionistic logic have been studied vigorously under the name of intermediate logics. It would be interesting to investigate logics between classical logic and minimal quantum logic, among which you can find quantum logic.

The other intriguing topic for future study is a semantical proof of the cut-elimination theorem of **GMQL**. In other words, it would be interesting to give a proof of the completeness theorem with respect to the Kripkian relational semantics without any recourse to the cut-elimination theorem, which would surely open a new area of research.

2 MINIMAL QUANTUM LOGIC IN GENTZEN STYLE

The sequential system **GMQL** that we have enunciated for minimal quantum logic in our [Nishimura, 1994a] and that has then been elaborated by Takano in [Takano, 1995] consists of the following inference rules:

$$\frac{\Gamma \rightarrow \Delta}{\pi, \Gamma \rightarrow \Delta, \Sigma} \quad \text{(extension)}$$

$$\frac{\alpha, \Gamma \rightarrow \Delta}{\alpha \wedge \beta, \Gamma \rightarrow \Delta} \quad \frac{\beta, \Gamma \rightarrow \Delta}{\alpha \wedge \beta, \Gamma \rightarrow \Delta} \quad (\wedge \rightarrow)$$

$$\frac{\Gamma \rightarrow \Delta, \alpha}{\Gamma \rightarrow \Delta, \alpha \vee \beta} \quad \frac{\Gamma \rightarrow \Delta, \beta}{\Gamma \rightarrow \Delta, \alpha \vee \beta} \quad (\rightarrow \vee)$$

$$\frac{\alpha \to \Delta \qquad \beta \to \Delta}{\alpha \vee \beta \to \Delta} \quad (\vee \to)$$

$$\frac{\Gamma \to \alpha \qquad \Gamma \to \beta}{\Gamma \to \alpha \wedge \beta} \quad (\to \wedge)$$

$$\frac{\Gamma \to \Delta}{\Delta', \Gamma \to} \quad ('\to)$$

$$\frac{\Gamma \to \Delta}{\to \Delta, \Gamma'} \quad (\to')$$

$$\frac{\alpha, \Gamma \to \Delta}{\alpha'', \Gamma \to \Delta} \quad (''\to)$$

$$\frac{\Gamma \to \Delta, \alpha}{\Gamma \to \Delta, \alpha''} \quad (\to'')$$

$$\frac{\alpha', \Gamma \to \Delta}{(\alpha \vee \beta)', \Gamma \to \Delta} \quad \frac{\beta', \Gamma \to \Delta}{(\alpha \vee \beta)', \Gamma \to \Delta} \quad (\vee' \to)$$

$$\frac{\Gamma \to \Delta, \alpha'}{\Gamma \to \Delta, (\alpha \wedge \beta)'} \quad \frac{\Gamma \to \Delta, \beta'}{\Gamma \to \Delta, (\alpha \wedge \beta)'} \quad (\to \wedge')$$

$$\frac{\alpha' \to \Delta \quad \beta' \to \Delta}{(\alpha \wedge \beta)' \to \Delta,} \quad (\wedge' \to)$$

$$\frac{\Gamma \to \alpha' \quad \Gamma \to \beta'}{\Gamma \to (\alpha \vee \beta)'} \quad (\to \vee')$$

$$\frac{\Gamma \to \alpha' \quad \Gamma \to \beta'}{\alpha \vee \beta, \Gamma \to} \quad (\vee \to')$$

$$\frac{\alpha' \to \Delta \quad \beta' \to \Delta}{\to \Delta, \alpha \wedge \beta} \quad ('\to \wedge)$$

Now some notational and terminological comments are in order. In this paper we adopt $'$ (negation), $\wedge$ (conjunction), and $\vee$ (disjunction) as primitive logical symbols. Propositional variables are denoted by $p, q, \ldots$, while wffs (well-formed formulas), also called formulas, are denoted by $\alpha, \beta, \ldots$. The *grade of* a wff α, denoted by $\mathcal{G}(\alpha)$, is defined inductively as follows:

1. $\mathcal{G}(p) = 0$ for any propositional variable p.

2. $\mathcal{G}(\alpha') = \mathcal{G}(\alpha) + 1$.

3. $\mathcal{G}(\alpha \wedge \beta) = \mathcal{G}(\alpha \vee \beta) = \mathcal{G}(\alpha) + \mathcal{G}(\beta) + 2$.

Finite (possibly empty) sets of wffs are denoted by $\Gamma, \Delta, \Pi, \ldots$. Given a finite set Γ of wffs, Γ' denotes the set $\{\alpha' | \alpha \in \Gamma\}$. A *sequent* $\Gamma \to \Delta$ means the ordered pair (Γ, Δ) of finite sets Γ and Δ of wffs, while the sets Γ and Δ are called the *antecedent* and the *succedent of* the sequent $\Gamma \to \Delta$, respectively. Such self-explanatory notations as $\Pi, \Gamma \to \Delta, \Sigma$ for $\Pi \cup \Gamma \to \Delta \cup \Sigma$ are used freely. A sequent of the form $\alpha \to \alpha$ is called an *axiom sequent.* Given a sequent $\Gamma \to \Delta$, the sequent $\Delta' \to \Gamma'$ is called the *contraposition* of $\Gamma \to \Delta$.

The notion of a *proof* P of a sequent $\Gamma \to \Delta$ with *length* n is defined inductively as follows:

1. Any axiom sequent $\alpha \to \alpha$ is a proof of itself with length 0.

2. If P is a proof of a sequent $\Gamma \to \Delta$ with length n and

$$\frac{\Gamma \to \Delta}{\Pi \to \Sigma}$$

is an instance of an inference rule of **GMQL**, then

$$\frac{P}{\Pi \to \Sigma}$$

is a proof of the sequent $\Pi \to \Sigma$ with length $n + 1$.

3. If P_i is a proof of a sequent $\Gamma_i \to \Delta_i$ with length n_i $(i = 1, 2)$ and

$$\frac{\Gamma_1 \to \Delta_1 \quad \Gamma_2 \to \Delta_2}{\Pi \to \Sigma}$$

is an instance of an inference rule of **GMQL**, then

$$\frac{P_1 \quad P_2}{\Pi \to \Sigma}$$

is a proof of the sequent $\Pi \to \Sigma$ with length $\max\{n_1, n_2\} + 1$.

The length of a proof P is denoted by $l(P)$. A sequent $\Gamma \to \Delta$ is said to be *provable* if it has a proof. Otherwise it is called *consistent.*

Although our cut-free sequential system **GMQL** does not satisfy the so-called subformula property in its strict sense, it gives a decision procedure for the word problem of free ortholattices once the completeness theorem is established, for which it suffices to note that $\mathcal{G}(\alpha') < \mathcal{G}((\alpha \wedge \beta)')$ and $\mathcal{G}(\beta') < \mathcal{G}((\alpha \wedge \beta)')$ for the rule $(\wedge' \to)$ by way of example. For algebraic and semantical decision procedures, the reader is referred to [Bruns, 1976], [Goldblatt, 1974] and [Goldblatt, 1975]. Fortunately, minimal quantum logic enjoys these three kinds of decision procedures. However, algebraic and semantical approaches to the decision problem of quantum logic have not succeeded so far. This is why we should try the third one.

Generally speaking, (cut) is the inference rule of the following form:

$$\frac{\Gamma_1 \to \Delta_1, \alpha \quad \alpha, \Gamma_2 \to \Delta_2}{\Gamma_1, \Gamma_2 \to \Delta_1, \Delta_2} \ (\text{cut})$$

However, following [Cutland and Gibbins, 1982], we should remark that the inference rule (cut) in such an unrestricted form forces our system **GMQL** to degenerate into classical logic. In other words, we have

PROPOSITION 1. *If we add the inference rule (cut) to the system* ***GMQL****, then we obtain classical logic. Schematically, we have*

$$\textit{\textbf{GMQL}}\textit{+(cut)} = \textit{classical logic}$$

Proof. It suffices to show that the following three rules are admissible in **GMQL**+(cut):

$$\frac{\Gamma \to \Delta, \alpha}{\alpha', \Gamma \to \Delta} \ ('\to)_c$$

$$\frac{\alpha, \Gamma \to \Delta}{\Gamma \to \Delta, \alpha'} \ (\to')_c$$

$$\frac{\Gamma \to \Delta, \alpha \quad \Gamma \to \Delta, \beta}{\Gamma \to \Delta, \alpha \wedge \beta} \ (\to \wedge)_c$$

Since we have

$$\frac{\dfrac{}{\Gamma \to \Delta, \alpha} \quad \dfrac{\alpha \to \alpha}{\alpha, \alpha' \to}\ ('\to)}{\alpha', \Gamma \to \Delta} \ (\text{cut})$$

the inference rule $('\to)_c$ is admissible in **GMQL**+(cut). Similarly, since we have

$$\frac{\dfrac{\alpha \to \alpha}{\to \alpha', \alpha}\ (\to') \qquad \alpha, \Gamma \to \Delta}{\Gamma \to \Delta, \alpha'} \ (\text{cut})$$

the inference rule $(\to')_c$ is admissible in **GMQL**+(cut). Now we deal with the last inference rule $(\to \wedge)_c$. The sequents $\Gamma, \Delta' \to \alpha$ and $\Gamma, \Delta' \to \beta$ follow from the sequents $\Gamma \to \Delta, \alpha$ and $\Gamma \to \Delta, \beta$ respectively by a finite number of applications of the inference rule $('\to)_c$. Now we have

$$\frac{\Gamma, \Delta' \to \alpha \quad \Gamma, \Delta' \to \beta}{\Gamma, \Delta' \to \alpha \wedge \beta} \ (\to \wedge)$$

The sequent $\Gamma \to \Delta'', \alpha \wedge \beta$ follows from the sequent $\Gamma, \Delta' \to \alpha \wedge \beta$ by a finite number of applications of the inference rule $(\to')_c$. Since we have

$$\frac{\dfrac{\gamma \to \gamma}{\to \gamma, \gamma'}\ (\to')}{\gamma'' \to \gamma} \ ('\to)_c$$

we are sure that the sequent $\gamma'' \rightarrow \gamma$ is provable in **GMQL**+(cut) for any $\gamma \in \Delta$. Therefore the desired sequent $\Gamma \rightarrow \Delta, \alpha \wedge \beta$ follows from the sequent $\Gamma \rightarrow \Delta'', \alpha \wedge \beta$ by a finite number of applications of the inference rule (cut). ■

This is the reason why Cutland and Gibbins [Cutland and Gibbins, 1982] should have proposed (cut) in the following restricted form:

$$\frac{\Gamma \rightarrow \Delta_1, \alpha \quad \alpha \rightarrow \Delta_2}{\Gamma \rightarrow \Delta_1, \Delta_2} \quad \text{(cut-1)}$$

$$\frac{\Gamma_1 \rightarrow \alpha \quad \alpha, \Gamma_2 \rightarrow \Delta}{\Gamma_1, \Gamma_2 \rightarrow \Delta} \quad \text{(cut-2)}$$

The wff α in (cut-1) and (cut-2) is called the *cut formula*. Both (cut-1) and (cut-2) are called $(\text{cut})_q$ as a whole. Roughly speaking, if we deprive our system **GMQL** of the inference rules $(\vee' \rightarrow)$, $(\rightarrow \wedge')$, $(\wedge' \rightarrow)$, $(\rightarrow \vee')$, $(\vee \rightarrow')$, and $('\rightarrow \wedge)$ and we agree to admit the inference rules (cut-1) and (cut-2), then we obtain the system of Cutland and Gibbins [Cutland and Gibbins, 1982]. We will prove in Section 5 that the inference rules (cut-1) and (cut-2) are admissible in **GMQL**.

Tamura [Tamura, 1988] gave a cut-free system by exploiting the legacy of Cutland and Gibbins [Cutland and Gibbins, 1982] but incorporating their inference rules surely except (cut-1) and (cut-2) into his system in an unnecessarily restricted manner. This unreasonable restriction forced him in the proof of the cut-elimination theorem to combine wffs in the antecedent of a sequent by conjunction and wffs in its succedent by disjunction, and then to dissolve such unnatural combinations. Such a proof is not compatible with Gentzen's [Gentzen, 1935] original philosophy and aesthetics, and is to be avoided if possible. Furthermore, the conceptual significance of Lemma 4 in Tamura's [Tamura, 1988] paper remained vague at best there. This is distilled into the duality theorem in Section 4, which is followed by the so-called cut-elimination theorem in Section 5.

Our original **GMQL**, proposed in [Nishimura, 1994a], contains the following inference rule besides the above ones:

$$\frac{\Gamma \rightarrow \Delta}{\Delta' \rightarrow \Gamma'} \quad ('\rightarrow')$$

It was pointed out by Takano [Takano, 1995] that the rule is redundant, which is the topic of the succeeding section.

3 THE CONTRAPOSITION THEOREM

The principal objective in this section is to show the following theorem on the lines of Takano [Takano, 1995].

THEOREM 2. *The following inference rule is admissible in* ***GMQL***.

$$\frac{\Gamma \rightarrow \Delta}{\Delta' \rightarrow \Gamma'} \quad ('\rightarrow')$$

To establish the above theorem, we introduce an auxiliary formal system to be denoted by **GMQL**$^{\#}$ and to be obtained from **GMQL** by admitting not only sequents $\alpha \to \alpha$ but also sequents $\alpha, \alpha' \to$ and $\to \alpha, \alpha'$ as axiom sequents while deleting the inference rules $(' \to)$ and $(\to ')$ and adding the following two inference rules:

$$\frac{\Gamma \to \alpha \quad \Gamma \to \beta}{(\alpha \wedge \beta)', \Gamma \to} \ (\wedge' \to)^{\#}$$

$$\frac{\alpha \to \Delta \quad \beta \to \Delta}{\to \Delta, (\alpha \vee \beta)'} (\to \vee')^{\#}$$

We need three lemmas so as to establish the equivalence of **GMQL** and **GMQL**$^{\#}$.

LEMMA 3.

1. *If a sequent* $\alpha'', \Gamma \to \Delta$ *is provable in* ***GMQL***$^{\#}$, *then so is* $\alpha, \Gamma \to \Delta$.
2. *If a sequent* $\Gamma \to \Delta, \alpha''$ *is provable in* ***GMQL***$^{\#}$, *then so is* $\Gamma \to \Delta, \alpha$.

Proof. We prove only the first statement by induction on the length $l(P)$ of a proof P of the sequent $\alpha'', \Gamma \to \Delta$, while leaving a similar treatment of the second statement to the reader. Our treatment is divided into several cases, some of which are again divided into several subcases.

1. The case that the sequent $\alpha'', \Gamma \to \Delta$ is an axiom sequent: We divide this case into three subcases.

   (a) The subcase that the sequent $\alpha'', \Gamma \to \Delta$ is $\alpha'' \to \alpha''$: Since we have

   $$\frac{\alpha \to \alpha}{\alpha \to \alpha''} \ (\to '')$$

   the sequent $\alpha \to \alpha''$ is also provable.

   (b) The subcase that the sequent $\alpha'', \Gamma \to \Delta$ is $\alpha'', \alpha' \to$: The sequent $\alpha, \alpha' \to$ is an axiom, and so is provable.

   (c) The subcase that the sequent $\alpha'', \Gamma \to \Delta$ is $\alpha'', \alpha''' \to$: Since we have

   $$\frac{\alpha, \alpha' \to}{\alpha, \alpha''' \to} \ ('' \to)$$

   the sequent $\alpha, \alpha''' \to$ is also provable.

2. The case that the last inference is (extension): The last inference has one of the following two forms:

   $$\frac{\Gamma_1 \to \Delta_1}{\alpha'', \Gamma_2, \Gamma_1 \to \Delta_1, \Delta_2} \ \text{(extension)}$$

$$\frac{\alpha'', \Gamma_1 \to \Delta_1}{\alpha'', \Gamma_2, \Gamma_1 \to \Delta_1, \Delta_2} \text{ (extension)}$$

In the former case, since we have

$$\frac{\Gamma_1 \to \Delta_1}{\alpha, \Gamma_2, \Gamma_1 \to \Delta_1, \Delta_2} \text{ (extension)}$$

the sequent $\alpha, \Gamma_2, \Gamma_1 \to \Delta_1, \Delta_2$ is also provable. In the latter case, since the sequent $\alpha, \Gamma_1 \to \Delta_1$ is provable by induction hypothesis and we have

$$\frac{\alpha, \Gamma_1 \to \Delta_1}{\alpha, \Gamma_2, \Gamma_1 \to \Delta_1, \Delta_2} \text{ (extension)}$$

the sequent $\alpha, \Gamma_2, \Gamma_1 \to \Delta_1, \Delta_2$ is also provable.

3. The case that the last inference is $(''\to)$: We divide this case into two subcases according as the principal formula of the last inference is α'' or not.

   (a) The subcase that the principal formula of the last inference is α'': The last inference has one of the following two forms:

   $$\frac{\alpha, \Gamma \to \Delta}{\alpha'', \Gamma \to \Delta} \ (''\to)$$

   $$\frac{\alpha'', \alpha, \Gamma \to \Delta}{\alpha'', \Gamma \to \Delta} \ (''\to)$$

   In the former case the sequent $\alpha, \Gamma \to \Delta$ is palpably provable, while in the latter case it should be provable by induction hypothesis.

   (b) The subcase that the principal formula of the last inference is not α'': The last inference has the form

   $$\frac{\alpha'', \beta, \Gamma_1 \to \Delta}{\alpha'', \beta'', \Gamma_1 \to \Delta} \ (''\to)$$

   Since the sequent $\alpha, \beta, \Gamma_1 \to \Delta$ is provable by induction hypothesis and we have

   $$\frac{\alpha, \beta, \Gamma_1 \to \Delta}{\alpha, \beta'', \Gamma_1 \to \Delta} \ (''\to)$$

   the sequent $\alpha, \beta'', \Gamma_1 \to \Delta$ is also provable.

4. The case that the last inference is neither (extension) nor $(''\to)$:Similar to the subcase (3-b).

■

LEMMA 4. *The inference rule* $('\to')$ *is admissible in* $\boldsymbol{GMQL}^{\#}$.

Proof. We will prove that if a sequent $\Gamma \to \Delta$ is provable in $\mathbf{GMQL}^{\#}$, then its contraposition $\Delta' \to \Gamma'$ is also provable in $\mathbf{GMQL}^{\#}$. The proof is carried out by induction on the length of a proof P of the given sequent $\Gamma \to \Delta$. Our treatment is divided into several cases.

1. The case that the given sequent $\Gamma \to \Delta$ is an axiom sequent: The sequent $\Gamma \to \Delta$ has one of the following three forms $\alpha \to \alpha$, $\alpha', \alpha \to$ and $\to \alpha, \alpha'$, whose contrapositions are also axioms $\alpha' \to \alpha'$, $\to \alpha', \alpha''$ and $\alpha'', \alpha' \to$.

2. The case that the last inference in P is (extension), $(\wedge \to)$, $(\to \wedge)$, $(\vee \to)$, $(\to \vee)$, $(''\!\to)$ or $(\to'')$: All these cases can be dealt with similarly, so we deal only with the case that the last inference is $(\to \wedge)$ as follows:

$$\frac{\Gamma \to \alpha \qquad \Gamma \to \beta}{\Gamma \to \alpha \wedge \beta}\ (\to \wedge)$$

Since the sequents $\alpha' \to \Gamma'$ and $\beta' \to \Gamma'$ are provable by induction hypothesis and we have

$$\frac{\alpha' \to \Gamma' \qquad \beta' \to \Gamma'}{(\alpha \wedge \beta)' \to \Gamma'}\ (\wedge' \to)$$

the sequent $(\alpha \wedge \beta)' \to \Gamma'$ is also provable.

3. The case that the last inference in P is either $(\wedge' \to)^{\#}$ or $(\to \vee')^{\#}$: Here we deal only with the former case, leaving a similar treatment of the latter case to the reader. So we suppose that the last inference in P is

$$\frac{\Gamma_1 \to \alpha \quad \Gamma_1 \to \beta}{(\alpha \wedge \beta)', \Gamma_1 \to}\ (\wedge' \to)^{\#}$$

Since the sequents $\alpha' \to \Gamma_1'$ and $\beta' \to \Gamma_1'$ are provable by induction hypothesis and we have

$$\frac{\dfrac{\alpha' \to \Gamma_1' \quad \beta' \to \Gamma_1'}{\to \Gamma_1', \alpha \wedge \beta}\ ({}' \to \wedge)}{\to \Gamma_1', (\alpha \wedge \beta)''}\ (\to'')$$

we are sure that the sequent $\to \Gamma_1', (\alpha \wedge \beta)''$ is also provable.

4. The case that the last inference in P is either $({}'\!\to \wedge)$ or $(\vee \to')$: Here we deal only with the former case, leaving a similar treatment of the latter case to the reader. So we suppose that the last inference in P is

$$\frac{\alpha' \to \Delta_1 \qquad \beta' \to \Delta_1}{\to \Delta_1, \alpha \wedge \beta}\ ({}'\!\to \wedge)$$

The sequents $\Delta_1' \to \alpha''$ and $\Delta_1' \to \beta''$ are provable by induction hypothesis, which imply by Lemma 3 that the sequents $\Delta_1' \to \alpha$ and $\Delta_1' \to \beta$ are also provable. Since we have

$$\frac{\Delta_1' \to \alpha \quad \Delta_1' \to \beta}{(\alpha \wedge \beta)', \Delta_1' \to}\ (\wedge' \to)^{\#}$$

we are sure that the sequent $(\alpha \wedge \beta)', \Delta_1' \rightarrow$ is also provable.

5. The case that the last inference in P is $(\wedge' \rightarrow)$, $(\rightarrow \wedge')$, $(\vee' \rightarrow)$ or $(\rightarrow \vee')$: Here we deal only with the first case, leaving similar treatments of the remaining three cases to the reader. So we suppose that the last inference in P is
$$\frac{\alpha' \rightarrow \Delta_1 \quad \beta' \rightarrow \Delta_1}{(\alpha \wedge \beta)' \rightarrow \Delta_1} \ (\wedge' \rightarrow)$$
The sequents $\Delta_1' \rightarrow \alpha''$ and $\Delta_1' \rightarrow \beta''$ are provable by induction hypothesis, which imply by Lemma 3 that the sequents $\Delta_1' \rightarrow \alpha$ and $\Delta_1' \rightarrow \beta$ are also provable. Since we have
$$\frac{\dfrac{\Delta_1' \rightarrow \alpha \quad \Delta_1' \rightarrow \beta}{\Delta_1' \rightarrow \alpha \wedge \beta} \ (\rightarrow \wedge)}{\Delta_1' \rightarrow (\alpha \wedge \beta)''} \ (\rightarrow'')$$
we are sure that the sequent $\Delta_1' \rightarrow (\alpha \wedge \beta)''$ is also provable.

■

LEMMA 5.

1. *If a sequent $\Gamma \rightarrow \Delta$ is provable in **GMQL#**, then so is $\Delta', \Gamma \rightarrow$.*
2. *If a sequent $\Gamma \rightarrow \Delta$ is provable in **GMQL#**, then so is $\rightarrow \Delta, \Gamma'$.*

Proof. The proof is by induction on the length of a proof P of the given sequent $\Gamma \rightarrow \Delta$. We deal only with the first statement, leaving a similar treatment of the second treatment to the reader. Our treatment is divided into several cases.

1. The case that the given sequent $\Gamma \rightarrow \Delta$ is an axiom sequent: The sequent $\Gamma \rightarrow \Delta$ is one of the three forms $\alpha \rightarrow \alpha$, $\alpha', \alpha \rightarrow$ and $\rightarrow \alpha, \alpha'$. Then the sequent $\Delta', \Gamma \rightarrow$ is one of the two forms $\alpha', \alpha \rightarrow$ and $\alpha'', \alpha' \rightarrow$, both of which are axioms.

2. The case that the last inference in P is (extension), $(\wedge \rightarrow)$, $(\rightarrow \vee)$, $(''\rightarrow)$, $(\rightarrow'')$ or $(\vee' \rightarrow)$: Here we deal only with the third case, leaving similar treatments of the remaining five cases to the reader. Thus the last inference of P is of the following form:
$$\frac{\Gamma \rightarrow \Delta_1, \alpha}{\Gamma \rightarrow \Delta_1, \alpha \vee \beta} \ (\rightarrow \vee)$$
Since the sequent $\alpha', \Delta_1', \Gamma \rightarrow$ is provable by induction hypothesis and we have
$$\frac{\alpha', \Delta_1', \Gamma \rightarrow}{(\alpha \vee \beta)', \Delta_1', \Gamma \rightarrow} \ (\vee' \rightarrow)$$
we are sure that the sequent $(\alpha \vee \beta)', \Delta_1', \Gamma \rightarrow$ is also provable.

3. The case that the last inference in P is $(\rightarrow \wedge')$: The last inference in P is in the following form:
$$\frac{\Gamma \rightarrow \Delta_1, \alpha'}{\Gamma \rightarrow \Delta_1, (\alpha \wedge \beta)'} \ (\rightarrow \wedge')$$
The sequent $\alpha'', \Delta_1', \Gamma \rightarrow$ is provable by induction hypothesis, which implies by Lemma 3 that the sequent $\alpha, \Delta_1', \Gamma \rightarrow$ is also provable. Since we have
$$\frac{\dfrac{\alpha, \Delta_1', \Gamma \rightarrow}{\alpha \wedge \beta, \Delta_1', \Gamma \rightarrow} \ (\wedge \rightarrow)}{(\alpha \wedge \beta)'', \Delta_1', \Gamma \rightarrow} \ ('' \rightarrow)$$
we are sure that the sequent $(\alpha \wedge \beta)'', \Delta_1', \Gamma \rightarrow$ is also provable.

4. The case that the last inference in P is $(\rightarrow \wedge)$: The last inference in P is of the following form:
$$\frac{\Gamma \rightarrow \alpha \quad \Gamma \rightarrow \beta}{\Gamma \rightarrow \alpha \wedge \beta} \ (\rightarrow \wedge)$$
Since we have
$$\frac{\Gamma \rightarrow \alpha \quad \Gamma \rightarrow \beta}{(\alpha \wedge \beta)', \Gamma \rightarrow} \ (\wedge' \rightarrow)^{\#}$$
we are sure that the sequent $(\alpha \wedge \beta)', \Gamma \rightarrow$ is also provable.

5. The case that the last inference in P is $(\rightarrow \vee')$: The last inference in P is of the following form:
$$\frac{\Gamma \rightarrow \alpha' \quad \Gamma \rightarrow \beta'}{\Gamma \rightarrow (\alpha \vee \beta)'} \ (\rightarrow \vee')$$
Since we have
$$\frac{\dfrac{\Gamma \rightarrow \alpha' \quad \Gamma \rightarrow \beta'}{\alpha \vee \beta, \Gamma \rightarrow} \ (\vee \rightarrow')}{(\alpha \vee \beta)'', \Gamma \rightarrow} \ ('' \rightarrow)$$
we are sure that the sequent $(\alpha \vee \beta)'', \Gamma \rightarrow$ is also provable.

6. The case that the last inference in P is either $(' \rightarrow \wedge)$ or $(\wedge' \rightarrow)$: The last inference in P is one of the following two forms:
$$\frac{\alpha' \rightarrow \Delta_1 \quad \beta' \rightarrow \Delta_1}{\rightarrow \Delta_1, \alpha \wedge \beta} \ (' \rightarrow \wedge)$$
$$\frac{\alpha' \rightarrow \Delta_1 \quad \beta' \rightarrow \Delta_1}{(\alpha \wedge \beta)' \rightarrow \Delta_1} \ (\wedge' \rightarrow)$$
In both cases, the sequents $\Delta_1' \rightarrow \alpha''$ and $\Delta_1' \rightarrow \beta''$ are provable by Lemma 4, which implies by dint of Lemma 3 that the sequents $\Delta_1' \rightarrow \alpha$ and $\Delta_1' \rightarrow \beta$ are also provable. Since we have
$$\frac{\Delta_1' \rightarrow \alpha \quad \Delta_1' \rightarrow \beta}{(\alpha \wedge \beta)', \Delta_1' \rightarrow} \ (\wedge' \rightarrow)^{\#}$$
we are sure that the sequent $(\alpha \vee \beta)'', \Gamma \rightarrow$ is also provable.

7. The case that the last inference in P is $(\vee \rightarrow)$: The last inference in P is of the following form:

$$\frac{\alpha \rightarrow \Delta \quad \beta \rightarrow \Delta}{\alpha \vee \beta \rightarrow \Delta} \ (\vee \rightarrow)$$

The sequents $\Delta' \rightarrow \alpha'$ and $\Delta' \rightarrow \beta'$ are provable by Lemma 4. Since we have

$$\frac{\Delta' \rightarrow \alpha' \quad \Delta' \rightarrow \beta'}{\Delta', \alpha \vee \beta \rightarrow} \ (\vee \rightarrow')$$

we are sure that the sequent $\Delta', \alpha \vee \beta \rightarrow$ is also provable.

8. The case that the last inference in P is $(\rightarrow \vee')^{\#}$: The last inference in P is of the following form:

$$\frac{\alpha \rightarrow \Delta \quad \beta \rightarrow \Delta}{\rightarrow \Delta, (\alpha \vee \beta)'} (\rightarrow \vee')^{\#}$$

The sequents $\Delta' \rightarrow \alpha'$ and $\Delta' \rightarrow \beta'$ are provable by Lemma 4. Since we have

$$\frac{\dfrac{\Delta' \rightarrow \alpha' \quad \Delta' \rightarrow \beta'}{\Delta', \alpha \vee \beta \rightarrow} \ (\vee \rightarrow')}{\Delta', (\alpha \vee \beta)'' \rightarrow} \ ('' \rightarrow)$$

9. The case that the last inference in P is either $(\vee \rightarrow')$ or $(\wedge' \rightarrow)^{\#}$: There is nothing to prove, for the succedent Δ of the given sequent $\Gamma \rightarrow \Delta$ is empty.

■

Now we are ready to present a proof of the main theorem.

THEOREM 6. *A sequent* $\Gamma \rightarrow \Delta$ *is provable in* $\boldsymbol{GMQL}^{\#}$ *iff it is provable in* ***GMQL***.

Proof.

1. First we deal with the only-if part. Since

$$\frac{\alpha \rightarrow \alpha}{\alpha', \alpha \rightarrow} \ (' \rightarrow)$$

and

$$\frac{\alpha \rightarrow \alpha}{\rightarrow \alpha, \alpha'} \ (\rightarrow ')$$

sequents $\alpha, \alpha' \rightarrow$ and $\rightarrow \alpha, \alpha'$ are provable in **GMQL**. Since

$$\frac{\dfrac{\Gamma \rightarrow \alpha \qquad \Gamma \rightarrow \beta}{\Gamma \rightarrow \alpha \wedge \beta} \ (\rightarrow \wedge)}{(\alpha \wedge \beta)', \Gamma \rightarrow} \ (' \rightarrow)$$

and

$$\dfrac{\dfrac{\alpha \to \Delta \qquad \beta \to \Delta}{\alpha \wedge \beta \to \Delta}\ (\vee \to)}{\to \Delta, (\alpha \vee \beta)'}\ (\to')$$

the inferene rules $(\wedge' \to)^{\#}$ and $(\to \vee')^{\#}$ are admissible in **GMQL**. Thus the only-if part has been established

2. The if part follows directly from Lemmas 4 and 5

■

Our desired theorem at the beginning of this section follows at once from the above theorem. Since the inference rule $('\to')$ is admissible in **GMQL**, we will often take it as a basic inference rule of the system **GMQL**.

4 THE DUALITY THEOREM

Two wffs α and β are said to be *provably equivalent*, in notation $\alpha \simeq \beta$, if for any finite sets Γ and Δ of wffs we have that

1. the sequent $\alpha, \Gamma \to \Delta$ is provable iff the sequent $\beta, \Gamma \to \Delta$ is provable; and
2. the sequent $\Gamma \to \Delta, \alpha$ is provable iff the sequent $\Gamma \to \Delta, \beta$ is provable.

It is easy to see that this is indeed an equivalence relation among wffs. We will show that it is even a congruence relation.

THEOREM 7. *(The fundamental theorem of provability equivalence). If* $\alpha_1 \simeq \beta_1$ *and* $\alpha_2 \simeq \beta_2$*, then* $\alpha_1' \sim \beta_1', \alpha_1 \wedge \alpha_2 \simeq \beta_1 \wedge \beta_2$*, and* $\alpha_1 \vee \alpha_2 \simeq \beta_1 \vee \beta_2$.

Proof. If $\gamma, \delta_1, ..., \delta_n$ are wffs and $p_1, ..., p_n$ are distinct propositional variables, we write $\gamma[\delta_1/p_1, ..., \delta_n/p_n]$ for the wff obtained from γ by replacing every occurrence of p_i by δ_i $(1 \leq i \leq n)$. Whenever we use this notation, it will always be assumed that the propositional variables at issue are distinct. The theorem follows readily from the following two statements:

1. If $\delta_1 \simeq \sigma_1, ..., \delta_n \simeq \sigma_n$ and a sequent $\gamma[\delta_1/p_1, ..., \delta_n/p_n], \Gamma \to \Delta$ has a proof P with $l(P) \leq m$, then the sequent $\gamma[\sigma_1/p_1, ..., \sigma_n/p_n], \Gamma \to \Delta$ is also provable.
2. If $\delta_1 \simeq \sigma_1, ..., \delta_n \simeq \sigma_n$ and a sequent $\Gamma \to \Delta, \gamma[\delta_1/p_1, ..., \delta_n/p_n]$ has a proof P with $l(P) \leq m$, then the sequent $\Gamma \to \Delta, \gamma[\sigma_1/p_1, ..., \sigma_n/p_n]$ is also provable.

These two statements are proved simultaneously by double induction principally on $\mathcal{G}(\gamma)$ and secondly on m. The proof is divided into cases according to which inference rule is used as the last inference in P. The details are safely left to the reader. ■

THEOREM 8. *(The first duality theorem). If $\alpha \simeq \beta$, then $\alpha \simeq \beta''$.*

Proof. It suffices to show the following claim:

CLAIM 9.

1. If a sequent $\alpha, \Gamma \to \Delta$ is provable, then the sequent $\beta'', \Gamma \to \Delta$ is also provable.

2. If a sequent $\Gamma \to \Delta, \alpha$ is provable, then the sequent $\Gamma \to \Delta, \beta''$ is also provable.

3. If a sequent $\alpha'', \Gamma \to \Delta$ is provable, then the sequent $\beta, \Gamma \to \Delta$ is also provable.

4. If a sequent $\Gamma \to \Delta, \alpha''$ is provable, then the sequent $\Gamma \to \Delta, \beta$ is also provable.

It is easy to see that the first and second statements of the above claim follow at once from a simple application of the inference rules $('' \to)$ and $(\to '')$, respectively, while 3 and 4 of the above claim follow at once from the following, ostensibly more general statement.

CLAIM 10. If $\alpha_1 \simeq \beta_1, ..., \alpha_n \simeq \beta_n, \alpha_{n+1} \simeq \beta_{n+1}, ..., \alpha_{n+m} \simeq \beta_{n+m}$ and a sequent $\alpha_1'', ..., \alpha_n'', \Gamma \to \Delta, \alpha_{n+1}'', ..., \alpha_{n+m}''$ has a proof P with $l(P) \leq k$ then the sequent $\beta_1, ..., \beta_n, \Gamma \to \Delta, \beta_{n+1}, ..., \beta_{n+m}$ is also provable.

We will prove Claim 10 by induction on k. The proof is divided into cases according to which inference rule is used in the last step of P. To make the notation simpler, we proceed as if $n = 1$ and $m = 0$, leaving safely easy but due modifications to the reader. In dealing with the rules $(\wedge \to), (\to \vee), (\vee' \to)$ and $(\to \wedge')$, each of which consists of two forms, we treat only one of them.

1. The case that the sequent $\alpha_1'', \Gamma \to \Delta$ is an axiom sequent: It must be that $\alpha_1'' \to \alpha_1''$. Since $\beta_1 \to \beta_1$ is an axiom sequent and $\alpha_1 \simeq \beta_1$ by assumption, the sequent $\beta_1 \to \alpha_1$ is provable, which implies that the sequent $\beta_1 \to \alpha_1''$ is also provable as follows:

$$\frac{\beta_1 \to \alpha_1}{\beta_1 \to \alpha_1''} \quad (\to '')$$

2. The case that the last inference of the proof of the sequent $\alpha_1'', \Gamma \to \Delta$ is (extension), $(\wedge \to), (\to \vee), (\wedge \to), (\to ''), (\vee' \to), (\to \wedge'), (\to \vee')$, or $(\vee \to')$: All the cases can be dealt with similarly, so here we deal only with the case in which the last inference of the proof is $(\to \wedge)$ as follows:

$$\frac{\alpha_1'', \Gamma \to \beta \qquad \alpha_1'', \Gamma \to \gamma}{\alpha_1'', \Gamma \to \beta \wedge \gamma} \quad (\to \wedge)$$

By the induction hypothesis the sequents $\beta_1, \Gamma \to \beta$ and $\beta_1, \Gamma \to \gamma$ are provable, which gives the desired result as follows:

$$\frac{\beta_1, \Gamma \to \beta \qquad \beta_1, \Gamma \to \gamma}{\beta_1, \Gamma \to \beta \wedge \gamma} \quad (\to \wedge)$$

3. The case that the last inference of the proof of $\alpha_1'', \Gamma \to \Delta$ is $(''\!\to)$: Then the last inference is one of the following two forms.

$$\frac{\alpha_1, \Gamma \to \Delta}{\alpha_1'', \Gamma \to \Delta} \quad (''\!\to \wedge) \qquad\qquad \frac{\alpha_1'', \beta'', \Gamma_1 \to \Delta}{\alpha_1'', \beta'', \Gamma_1 \to \Delta} \quad (''\!\to)$$

In the former case the sequent $\beta_1, \Gamma \to \Delta$ is provable for $\alpha_1 \simeq \beta_1$ and the sequent $\alpha_1, \Gamma \to \Delta$ is provable by assumption. In the latter case the sequent $\beta_1, \beta, \Gamma_1 \to \Delta$ is provable by the induction hypothesis, which implies that the sequent $\beta_1, \beta'', \Gamma_1 \to \Delta$ is provable as follows:

$$\frac{\beta_1, \beta, \Gamma_1 \to \Delta}{\beta_1, \beta'', \Gamma_1 \to \Delta} \quad (''\!\to)$$

4. The case that the last inference of the proof of the sequent $\alpha_1'', \Gamma \to \Delta$ is $('\!\to)$: This case is divided into several subcases according to how the upper sequent of $('\!\to)$ is obtained.

   (a) The case that the upper sequent of $('\!\to)$ is an axiom sequent: In this case the axiom sequent must be $\alpha_1' \to \alpha_1'$, so the proof that we must consider is as follows:

   $$\frac{\alpha_1' \to \alpha_1'}{\alpha_1'', \alpha_1' \to} \quad ('\!\to)$$

   Since the sequent $\alpha_1 \to \alpha_1$ is an axiom sequent and $\alpha_1 \simeq \beta_1$ by assumption, the sequent $\beta_1 \to \alpha_1$ is provable, which implies that the desired sequent $\beta_1, \alpha_1' \to$ is also provable as follows:

   $$\frac{\beta_1, \alpha_1 \to}{\alpha_1', \beta_1 \to} \quad ('\!\to)$$

   (b) The case that the upper sequent of $('\!\to)$ is obtained as the lower sequent of (extension), $(\wedge \to)$, $(''\!\to)$, or $(\vee' \to)$: All these cases can be dealt with similarly, so here we consider only the case of $(''\!\to)$, in which the last two steps of the proof go as follows:

   $$\cfrac{\cfrac{\beta, \Gamma_2 \to \alpha_1', \Gamma_1}{\beta'', \Gamma_2 \to \alpha_1', \Gamma_1}\ (''\!\to)}{\alpha_1'', \Gamma_1', \beta'', \Gamma_2 \to} \quad ('\!\to)$$

The sequent $\alpha_1'', \Gamma_1', \beta, \Gamma_2 \rightarrow$ has a shorter proof than the sequent $\alpha_1'', \Gamma_1', \beta'', \Gamma_2 \rightarrow$, as follows:

$$\frac{\beta, \Gamma_2, \rightarrow \alpha_1', \Gamma_1'}{\alpha_1'', \Gamma_1', \beta, \Gamma_2 \rightarrow} \quad ('\rightarrow)$$

Therefore the sequent $\beta_1, \Gamma_1', \beta, \Gamma_2 \rightarrow$ is provable by the induction hypothesis, which implies that the desired sequent $\beta_1, \Gamma_1', \beta'', \Gamma_2 \rightarrow$ is also provable as follows:

$$\frac{\beta_1, \Gamma_1', \beta, \Gamma_2 \rightarrow}{\beta_1, \Gamma_1', \beta'', \Gamma_2 \rightarrow} \quad (''\rightarrow)$$

(c) The case that the upper sequent of $('\rightarrow)$ is obtained as the lower sequent of $(\rightarrow'')$: The last two steps of the proof that we must consider can be supposed to be one of the following two forms:

$$\frac{\dfrac{\Gamma_2 \rightarrow \beta, \Gamma_1}{\Gamma_2 \rightarrow \beta'', \Gamma_1} \quad (\rightarrow'')}{\beta'', \Gamma_1, \Gamma_2 \rightarrow} \quad ('\rightarrow)$$

$$\frac{\dfrac{\Gamma_2 \rightarrow \alpha_1', \beta, \Gamma_1}{\Gamma_2 \rightarrow \alpha_1', \beta'', \Gamma_1} \quad (\rightarrow'')}{\alpha_1'', \beta''', \Gamma_1, \Gamma_2 \rightarrow} \quad ('\rightarrow)$$

In the former case α_1 is supposed to be β', Since the latter case can be dealt with in a similar manner to the case (2), here we deal with the former case, in which the sequent $\alpha_1, \Gamma_1', \Gamma_2 \rightarrow$ is provable with a shorter proof than that of the sequent $\alpha_1'', \Gamma_1', \Gamma_2 \rightarrow$ as follows:

$$\frac{\Gamma_2, \rightarrow \beta, \Gamma_1}{\beta', \Gamma_1', \Gamma_2 \rightarrow} \quad ('\rightarrow)$$

Thus the desired sequent $\beta_1, \Gamma_2', \Gamma_2 \rightarrow$ is also provable by hypothesis,

(d) The case that the upper sequent of $('\rightarrow)$ is obtained as the lower sequent of $(\rightarrow \vee)$: The last two steps of the proof go as follows:

$$\frac{\dfrac{\Gamma_2 \rightarrow \alpha_1', \beta, \Gamma_1}{\Gamma_2 \rightarrow \alpha_1', \beta \vee \gamma, \Gamma_1} \quad (\rightarrow \vee)}{\alpha_1'', (\beta \vee \gamma)', \Gamma_1', \Gamma_2 \rightarrow} \quad ('\rightarrow)$$

The sequent $\alpha_1'', \beta', \Gamma_1', \Gamma_2 \rightarrow$ has a shorter proof than the sequent $\alpha_1'', (\beta \vee \gamma)', \Gamma_1', \Gamma_2 \rightarrow$ as follows:

$$\frac{\Gamma_2, \rightarrow \alpha_1', \beta, \Gamma_1}{\alpha_1'', \beta', \Gamma_1', \Gamma_2 \rightarrow} \quad ('\rightarrow)$$

Therefore the sequent $\beta_1, \beta', \Gamma_1', \Gamma_2 \rightarrow$ is provable by the induction hypothesis, which implies that the desired sequent $\beta_1, (\beta \vee \gamma)', \Gamma_1', \Gamma_2 \rightarrow$ is provable as follows:

$$\frac{\beta_1, \beta', \Gamma_1', \Gamma_2 \rightarrow}{\beta_1, (\beta \vee \gamma)', \Gamma_1', \Gamma_2 \rightarrow} \ (\vee' \rightarrow)$$

(e) The case that the upper sequent of $(' \rightarrow)$ is obtained as the lower sequent of $(\rightarrow \wedge')$; The last two steps of the proof are of one of the following two forms:

$$\frac{\dfrac{\Gamma_2 \rightarrow \alpha_1', \beta', \Gamma_1}{\Gamma_2 \rightarrow \alpha_1', (\beta \wedge \gamma)', \Gamma_1} \ (\rightarrow \wedge')}{\alpha_1'', (\beta \wedge \gamma)'', \Gamma_1', \Gamma_2 \rightarrow} \ (' \rightarrow)$$

$$\frac{\dfrac{\Gamma_2 \rightarrow \beta', \Gamma_1}{\Gamma_2 \rightarrow (\beta \wedge \gamma)', \Gamma_1} \ (\rightarrow \wedge')}{(\beta \wedge \gamma)'', \Gamma_1', \Gamma_2 \rightarrow} \ (' \rightarrow)$$

In the latter case α_1 is assumed to be $\beta \wedge \gamma$. Here we deal only with the former case, leaving a similar treatment of the latter case to the reader. The sequent $\alpha_1'', \beta'', \Gamma_1', \Gamma_2 \rightarrow$ has a shorter proof than the sequent $\alpha_1'', (\beta \wedge \gamma)'', \Gamma_1', \Gamma_2 \rightarrow$ as follows:

$$\frac{\Gamma_2, \rightarrow \alpha_1', \beta', \Gamma_1}{\alpha_1'', \beta'', \Gamma_1', \Gamma_2 \rightarrow} \ (' \rightarrow)$$

This implies by the induction hypothesis that the sequent $\beta_1, \beta, \Gamma_1', \Gamma_2 \rightarrow$ is also provable. Thus the desired sequent $\beta_1, (\beta \wedge \gamma)'', \Gamma_1', \Gamma_2 \rightarrow$ is also provable, as follows:

$$\frac{\dfrac{\beta_1, \beta, \Gamma_1', \Gamma_2 \rightarrow}{\beta_1, \beta \wedge \gamma, \Gamma_1', \Gamma_2 \rightarrow} \ (\wedge \rightarrow)}{\beta_1, (\beta \wedge \gamma)'', \Gamma_1', \Gamma_2 \rightarrow} \ ('' \rightarrow)$$

(f) The case that the upper sequent of $(' \rightarrow)$ is obtained as the lower sequent of $(\vee \rightarrow)$: The last two steps of the proof that we must consider go as follows:

$$\frac{\dfrac{\beta \rightarrow \alpha_1', \Gamma_1 \qquad \gamma \rightarrow \alpha_1', \Gamma_1}{\beta \vee \gamma \rightarrow \alpha_1', \Gamma_1} \ (\vee \rightarrow)}{\alpha_1'', \Gamma_1', \beta \vee \gamma \rightarrow} \ (' \rightarrow)$$

The sequents $\alpha_1'', \Gamma_1' \rightarrow \beta'$ and $\alpha_1'', \Gamma_1' \rightarrow \gamma'$ are provable with shorter proofs than that of $\alpha_1'', \Gamma_1', \beta \vee \gamma \rightarrow$ as follows

$$\frac{\beta \rightarrow \alpha_1', \Gamma_1}{\alpha_1'', \Gamma_1' \rightarrow \beta'} \ (' \rightarrow ')$$

$$\frac{\gamma \rightarrow \alpha_1', \Gamma_1}{\alpha_1'', \Gamma_1' \rightarrow \gamma'} \quad ('\!\rightarrow')$$

Therefore the sequents $\beta_1, \Gamma_1' \rightarrow \beta'$ and $\beta_1, \Gamma_1' \rightarrow \gamma'$ are provable by the induction hypothesis, which implies that the desired sequent $\beta \vee \gamma, \beta_1, \Gamma_1' \rightarrow$ is also provable as follows:

$$\frac{\beta_1, \Gamma_1' \rightarrow \beta' \qquad \beta_1, \Gamma_1' \rightarrow \gamma'}{\beta \vee \gamma, \beta_1, \Gamma_1' \rightarrow} \quad (\vee \rightarrow')$$

(g) The case that the upper sequent of $('\!\rightarrow)$ is obtained as the lower sequent of $(\wedge' \rightarrow)$: The last two steps of the proof that we must consider go as follows:

$$\frac{\dfrac{\beta' \rightarrow \alpha_1', \Gamma_1 \qquad \gamma' \rightarrow \alpha_1', \Gamma_1}{(\beta \wedge \gamma)' \rightarrow \alpha_1, \Gamma_1} \; (\wedge' \rightarrow)}{\alpha_1'', \Gamma_1', (\beta \vee \gamma)' \rightarrow} \quad ('\!\rightarrow)$$

The sequents $\alpha_1'', \Gamma_1' \rightarrow \beta''$ and $\alpha_1'', \Gamma_1' \rightarrow \gamma''$ are provable with shorter proofs than that of the sequent $\alpha_1'', \Gamma_1', (\beta \vee \gamma)' \rightarrow$ as follows:

$$\frac{\beta' \rightarrow \alpha_1', \Gamma_1}{\alpha_1'', \Gamma_1' \rightarrow \beta''} \quad ('\!\rightarrow')$$

$$\frac{\gamma' \rightarrow \alpha_1', \Gamma_1}{\alpha_1'', \Gamma_1' \rightarrow \gamma''} \quad ('\!\rightarrow')$$

Thus the sequents $\beta_1, \Gamma_1' \rightarrow \beta$ and $\beta_1, \Gamma_1' \rightarrow \gamma$ are provable by the induction hypothesis, which implies that the desired sequent $(\beta \wedge \gamma)', \beta_1, \Gamma_1' \rightarrow$ is also provable as follows:

$$\frac{\dfrac{\beta_1, \Gamma_1' \rightarrow \beta \qquad \beta_1, \Gamma_1' \rightarrow \gamma}{\beta_1, \Gamma_1' \rightarrow \beta \wedge \gamma,} \; (\rightarrow \wedge)}{(\beta \wedge \gamma)', \beta_1, \Gamma_1' \rightarrow} \quad ('\!\rightarrow)$$

(h) The case that the upper sequent of $('\!\rightarrow)$ is obtained as the lower sequent of $(\rightarrow \vee')$: The last two steps of the proof that we must consider go as follows:

$$\frac{\dfrac{\Gamma_1' \rightarrow \beta' \qquad \Gamma_1 \rightarrow \gamma'}{\Gamma_1' \rightarrow (\beta \vee \gamma)'} \; (\rightarrow \vee')}{(\beta \vee \gamma)'', \Gamma_1 \rightarrow} \quad ('\!\rightarrow)$$

Here α_1 is supposed to be $\beta \vee \gamma$. The sequent $\beta \vee \gamma, \Gamma_1 \rightarrow$ is provable as follows:

$$\frac{\Gamma_1 \rightarrow \beta' \qquad \Gamma_1 \rightarrow \gamma'}{\beta \vee \gamma, \Gamma_1 \rightarrow} \quad (\vee \rightarrow')$$

Since $\beta_1 \simeq \alpha_1$ by assumption, the desired sequent $\beta_1, \Gamma_1 \rightarrow$ is also provable.

(i) The case that the upper sequent of $('\!\rightarrow)$ is obtained as the lower sequent of $('\!\rightarrow \wedge)$: The last two steps of the proof that we must consider go as follows:

$$\dfrac{\dfrac{\beta' \rightarrow \alpha_1', \Gamma_1' \qquad \gamma' \rightarrow \alpha_1', \Gamma_1'}{\rightarrow \beta \wedge \gamma, \alpha_1', \Gamma_1'}\ ('\!\rightarrow \wedge)}{(\beta \wedge \gamma)', \alpha_1'', \Gamma_1' \rightarrow}\ ('\!\rightarrow)$$

The sequents $\alpha_1', \Gamma_1' \rightarrow \beta$ and $\alpha_1'', \Gamma_1' \rightarrow \gamma''$ are provable with shorter proofs than that of the sequent $(\beta \wedge \gamma)', \alpha_1'', \Gamma_1' \rightarrow$ as follows:

$$\dfrac{\beta' \rightarrow \alpha_1', \Gamma_1}{\alpha_1'', \Gamma_1' \rightarrow \beta''}\ ('\!\rightarrow')$$

$$\dfrac{\gamma' \rightarrow \alpha_1', \Gamma_1'}{\alpha_1'', \Gamma_1' \rightarrow \gamma''}\ ('\!\rightarrow')$$

Thus the sequents $\beta_1, \Gamma_1' \rightarrow \beta$ and $\beta_1, \Gamma_1' \rightarrow \gamma$ are provable by the induction hypothesis, which implies that the desired sequent $(\beta\wedge\gamma)', \beta_1, \Gamma_1' \rightarrow$ is also provable as follows:

$$\dfrac{\dfrac{\beta_1, \Gamma_1' \rightarrow \beta \qquad \beta_1, \Gamma_1' \rightarrow \gamma}{\beta_1, \Gamma_1' \rightarrow \beta \wedge \gamma}\ (\rightarrow \wedge)}{(\beta \wedge \gamma)', \beta_1, \Gamma_1' \rightarrow}\ ('\!\rightarrow)$$

(j) The case that the upper sequent of $('\!\rightarrow)$ is obtained as the lower sequent of $('\!\rightarrow')$: The last two steps of the proof that we must consider go as follows:

$$\dfrac{\dfrac{\alpha_1, \Gamma_1 \rightarrow \Gamma_2}{\Gamma_2' \rightarrow \alpha_1', \Gamma_1'}\ ('\!\rightarrow')}{\alpha_1'', \Gamma_1'', \Gamma_2' \rightarrow}\ ('\!\rightarrow)$$

Since the sequent $\alpha_1, \Gamma_1 \rightarrow \Gamma_2$ is provable and $\alpha_1 \simeq \beta_1$ by assumption, $\beta_1, \Gamma_1 \rightarrow \Gamma_2$ is also provable, which implies that the desired sequent $\beta_1, \Gamma_1'', \Gamma_2' \rightarrow$ is provable, as follows:

$$\dfrac{\dfrac{\beta_1, \Gamma_1 \rightarrow \Gamma_2}{\beta_1, \Gamma_1, \Gamma_2' \rightarrow}\ ('\!\rightarrow)}{\beta_1, \Gamma_1'', \Gamma_2' \rightarrow}\ (''\!\rightarrow)$$

5. The case that the upper sequent of $('\!\rightarrow)$ is obtained as the lower sequent of $(\rightarrow')$: We can proceed similarly to (4-j). The case that the last inference of the proof of the sequent $\alpha_1'', \Gamma \rightarrow \Delta$ is $('\!\rightarrow')$: This case is divided into several subcases according to how the upper sequent of $('\!\rightarrow')$ is obtained.

   (a) The case that the upper sequent of $('\!\rightarrow')$ is an axiom sequent: The treatment of this case is similar to (4-a) and is safely left to the reader.

(b) The case that the upper sequent of $('\rightarrow')$ is obtained as the lower sequent of (extension): This case can safely be left to the reader.

(c) The case that the upper sequent of $('\rightarrow')$ is obtained as the lower sequent of $(\wedge \rightarrow)$: The last two steps of the proof that we must consider go as follows:

$$\dfrac{\dfrac{\alpha, \Delta_1 \rightarrow \alpha_1', \Gamma_1}{\alpha \wedge \beta, \Delta_1 \rightarrow \alpha_1', \Gamma_1}\ (\wedge \rightarrow)}{\alpha_1'', \Gamma_1' \rightarrow (\alpha \wedge \beta)', \Delta_1'}\ ('\rightarrow')$$

The sequent $\alpha_1'', \Gamma_1' \rightarrow \alpha', \Delta_1'$ is provable with a shorter proof than that of $\alpha_1'', \Gamma_1' \rightarrow (\alpha \wedge \beta)', \Delta_1'$ as follows:

$$\dfrac{\alpha, \Delta_1 \rightarrow \alpha_1', \Delta_1'}{\alpha_1'', \Gamma_1 \rightarrow \alpha', \Delta_1'}\ ('\rightarrow')$$

Thus the sequent $\beta_1, \Gamma_1' \rightarrow \alpha', \Delta_1'$ is provable by the induction hypothesis, which implies that the desired sequent $\beta_1, \Gamma_1' \rightarrow (\alpha \wedge \beta)', \Delta_1'$ is also provable as follows:

$$\dfrac{\beta_1, \Gamma_1' \rightarrow \alpha', \Delta_1'}{\beta_1, \Gamma_1' \rightarrow (\alpha \wedge \beta)', \Delta_1'}\ (\rightarrow \wedge')$$

(d) The case that the upper sequent of $('\rightarrow')$ is obtained as the lower sequent of $(\rightarrow \vee)$: The treatment is similar to (5-c) and is safely left to the reader.

(e) The case that the upper sequent of $('\rightarrow')$ is obtained as the lower sequent of $(\vee \rightarrow)$: The last two steps of the proof that we have to consider go as follows:

$$\dfrac{\dfrac{\alpha \rightarrow \alpha_1', \Gamma_1 \qquad \beta \rightarrow \alpha_1', \Gamma_1}{\alpha \vee \beta \rightarrow \alpha_1', \Gamma_1}\ (\vee \rightarrow)}{\alpha_1'', \Gamma_1' \rightarrow (\alpha \vee \beta)'}\ ('\rightarrow')$$

The sequents $\alpha_1'', \Gamma_1 \rightarrow \alpha'$ and $\alpha_1'', \Gamma_1' \rightarrow \beta'$ are provable with shorter proofs than that of $\alpha_1'', \Gamma_1' \rightarrow (\alpha \vee \beta)'$ as follows:

$$\dfrac{\alpha \rightarrow \alpha_1', \Gamma_1}{\alpha_1'', \Gamma_1' \rightarrow \alpha'}\ ('\rightarrow') \qquad \dfrac{\beta \rightarrow \alpha_1', \Gamma_1}{\alpha_1'', \Gamma_1' \rightarrow \beta'}\ ('\rightarrow')$$

Thus the sequents $\beta_1, \Gamma_1' \rightarrow \alpha'$ and $\beta_1, \Gamma_1' \rightarrow \beta'$ are provable by the induction hypothesis, which implies that the desired sequent $\alpha_1'', \Gamma_1' \rightarrow (\alpha \vee \beta)'$ is provable, as follows:

$$\dfrac{\beta_1, \Gamma_1' \rightarrow \alpha' \qquad \beta_1, \Gamma_1' \rightarrow \beta'}{\beta_1, \Gamma_1' \rightarrow (\alpha \vee \beta)'}\ (\rightarrow \vee')$$

(f) The case that the upper sequent of $({}'{\rightarrow}')$ is the lower sequent of $({\rightarrow}')$: The last two steps of the proof that we should consider can be supposed to be one of the following two forms:

$$\cfrac{\cfrac{\alpha_1, \Gamma_1 \rightarrow \Gamma_2}{\rightarrow \alpha_1', \Gamma_1', \Gamma_2}\ ({\rightarrow}')}{\alpha_1'', \Gamma_1'', \Gamma_2' \rightarrow}\ ({}'{\rightarrow}')$$

$$\cfrac{\cfrac{\Gamma_1 \rightarrow \alpha_1', \Gamma_2}{\rightarrow \alpha_1', \Gamma_1', \Gamma_2}\ ({\rightarrow}')}{\alpha_1'', \Gamma_1'', \Gamma_2' \rightarrow}\ ({}'{\rightarrow}')$$

In the former case the sequent $\alpha_1, \Gamma_1'', \Gamma_2' \rightarrow$ is provable as follows:

$$\cfrac{\cfrac{\alpha_1, \Gamma_1 \rightarrow \Gamma_2}{\alpha_1, \Gamma_1, \Gamma_2' \rightarrow}\ ({}'{\rightarrow})}{\alpha_1, \Gamma_1'', \Gamma_2' \rightarrow}\ ({}''{\rightarrow})$$

Since $\alpha_1 \simeq \beta_1$ by assumption, the desired sequent $\beta_1, \Gamma_1'', \Gamma_2' \rightarrow$ is also provable. As for the latter case, the sequent $\alpha_1'', \Gamma_2' \rightarrow \Gamma_1'$ is provable with a shorter proof than that of the sequent $\alpha_1'', \Gamma_1'', \Gamma_2 \rightarrow$, as follows:

$$\cfrac{\Gamma_1 \rightarrow \alpha_1', \Gamma_2}{\alpha_1'', \Gamma_2' \rightarrow \Gamma_1'}\ ({}'{\rightarrow}')$$

By the induction hypothesis the sequent $\beta_1, \Gamma_2' \rightarrow \Gamma_1'$ is also provable, which implies that the desired sequent $\beta_1, \Gamma_1'', \Gamma_2' \rightarrow$ is provable, as follows:

$$\cfrac{\beta_1, \Gamma_2' \rightarrow \Gamma_1'}{\beta_1, \Gamma_1'', \Gamma_2' \rightarrow}\ ({}'{\rightarrow})$$

(g) The case that the upper sequent of $({}'{\rightarrow}')$ is obtained as the lower sequent of $({}''{\rightarrow})$ or $({\rightarrow}'')$: The treatment is similar to (4-c) and is safely left to the reader.

(h) The case that the upper sequent of $({}'{\rightarrow}')$ is the lower sequent of another $({}'{\rightarrow}')$: The last two steps of the proof that we have to consider go as follows:

$$\cfrac{\cfrac{\alpha_1, \Gamma_1 \rightarrow \Delta_1}{\Delta_1' \rightarrow \alpha_1', \Gamma_1'}\ ({}'{\rightarrow}')}{\alpha_1'', \Gamma_1'' \rightarrow \Delta_1''}\ ({}'{\rightarrow}')$$

Since the sequent $\alpha_1, \Gamma_1 \rightarrow \Delta_1$ has a shorter proof than the sequent $\alpha_1'', \Gamma_1'' \rightarrow \Delta_1''$, the sequent $\beta_1, \Gamma_1 \rightarrow \Delta_1$ is also provable by the induction hypothesis, which implies that the desired sequent $\beta_1, \Gamma_1'' \rightarrow \Delta_1''$ is also provable, as follows

$$\cfrac{\cfrac{\beta_1, \Gamma_1 \rightarrow \Delta_1}{\beta_1, \Gamma_1'' \rightarrow \Delta_1}\ ({}''{\rightarrow})}{\beta_1, \Gamma_1'' \rightarrow \Delta_1''}\ ({\rightarrow}'')$$

(i) The case that the upper sequent of $('\!\rightarrow')$ is obtained as the lower sequent of $(\vee' \rightarrow), (\rightarrow \wedge'), (\wedge' \rightarrow)$, or $(\rightarrow \vee')$: These four cases can be . dealt with similarly, so here we deal only with the case of $(\rightarrow \vee')$, in which the last two steps of the proof that we must consider go as follows:

$$\dfrac{\dfrac{\Delta_1 \rightarrow \alpha' \qquad \Delta_1 \rightarrow \beta'}{\Delta_1 \rightarrow (\alpha \vee \beta)'}\ (\rightarrow \vee')}{(\alpha \vee \beta)'' \rightarrow \Delta_1'}\ ('\!\rightarrow')$$

Here α_1 is supposed to be $\alpha \vee \beta$. The sequents $\alpha'' \rightarrow \Delta_1'$ and $\beta'' \rightarrow \Delta_1'$ are provable with shorter proofs than that of $(\alpha \vee \beta)'' \rightarrow \Delta_1'$ as follows:

$$\dfrac{\Delta_1 \rightarrow \alpha'}{\alpha'' \rightarrow \Delta_1'}\ ('\!\rightarrow') \qquad \dfrac{\Delta_1' \rightarrow \beta''}{\beta'' \rightarrow \Delta_1'}\ ('\!\rightarrow')$$

Therefore the sequents $\alpha \rightarrow \Delta_1'$ and $\beta \rightarrow \Delta_1'$ are provable by the induction hypothesis, which implies that the sequent $\alpha \vee \beta \rightarrow \Delta_1'$ is also provable, as follows:

$$\dfrac{\alpha \rightarrow \Delta_1 \qquad \beta \rightarrow \Delta_1}{\alpha \vee \beta \rightarrow \Delta_1}\ (\vee \rightarrow)$$

Since $\beta_1 \simeq \alpha_1 = \alpha \vee \beta$ by assumption, the desired sequent $\beta_1 \rightarrow \Delta_1$ is provable.

(j) The case that the upper sequent of $('\!\rightarrow')$ is obtained as the lower sequent of $('\!\rightarrow \wedge)$: The last two steps of the proof that we must consider go as follows:

$$\dfrac{\dfrac{\alpha' \rightarrow \alpha_1', \Gamma_1 \qquad \beta' \rightarrow \alpha_1', \Gamma_1}{\rightarrow \alpha_1', \Gamma_1', \alpha \wedge \beta}\ ('\!\rightarrow \wedge)}{\alpha_1'', \Gamma_1', (\alpha \wedge \beta)' \rightarrow}\ ('\!\rightarrow')$$

The sequents $\Gamma_1', \alpha_1'' \rightarrow \alpha''$ and $\Gamma_1', \alpha_1'' \rightarrow \beta''$ are provable with shorter proofs than that of $\alpha_1', \Gamma_1', (\alpha \wedge \beta)'' \rightarrow$ as follows:

$$\dfrac{\alpha' \rightarrow \alpha_1', \Gamma_1}{\alpha_1'', \Gamma_1' \rightarrow \alpha''}\ ('\!\rightarrow') \qquad \dfrac{\beta' \rightarrow \alpha_1', \Gamma_1}{\alpha_1'', \Gamma_1' \rightarrow \beta''}\ ('\!\rightarrow')$$

By the induction hypothesis the sequents $\beta_1, \Gamma_1' \rightarrow \alpha$ and $\beta_1, \Gamma_1' \rightarrow \beta$ are provable, which implies that the desired sequent $\beta_1, \Gamma_1', (\alpha \wedge \beta)' \rightarrow$ is also provable, as follows:

$$\dfrac{\dfrac{\beta_1, \Gamma_1' \rightarrow \alpha \qquad \beta_1, \Gamma_1' \rightarrow \beta}{\beta_1, \Gamma_1' \rightarrow \alpha \wedge \beta}\ (\rightarrow \wedge)}{\beta_1, \Gamma_1, (\alpha \wedge \beta)' \rightarrow}\ ('\!\rightarrow)$$

■

COROLLARY 11.

1. *If a sequent* $\Gamma, \Pi' \to \Delta, \Sigma'$ *is provable, then the sequent* $\Delta', \Sigma \to \Gamma', \Pi$ *is also provable.*

2. *If a sequent* $\Gamma \to \Delta'$ *is provable, then the sequent* $\Gamma, \Delta \to$ *is provable.*

3. *If a sequent* $\Gamma' \to \Delta$ *is provable, then the sequent* $\to \Gamma, \Delta$ *is also provable.*

Proof. By Theorem 8 it suffices only to take into account the rules $('\!\to')$, $('\!\to)$, and $(\to')$. ∎

THEOREM 12. *(The second duality theorem). If* $\alpha_1 \simeq \beta_1$ *and* $\alpha_2 \simeq \beta_2$*, then* $\alpha_1 \wedge \alpha_2 \simeq (\beta_1' \vee \beta_2')'$ *and* $\alpha_1 \vee \alpha_2 \simeq (\beta_1' \wedge \beta_2')'$*.*

Proof. First we show the following claim:

CLAIM 13.

1. If a sequent $\alpha_1 \wedge \alpha_2, \Gamma \to \Delta$ is provable, then the sequent $(\beta_1' \vee \beta_2')', \Gamma \to \Delta$ is also provable.

2. If a sequent $\Gamma \to \Delta, \alpha_1 \wedge \alpha_2$ is provable, then the sequent $\Gamma \to \Delta, (\beta_1' \vee \beta_2')'$ is also provable.

3. If a sequent $\alpha_1 \vee \alpha_2, \Gamma \to \Delta$ is provable, then the sequent $(\beta_1' \wedge \beta_2'), \Gamma \to \Delta$ is also provable.

4. If a sequent $\Gamma \to \Delta, \alpha_1 \vee \alpha_2$ is provable, then the sequent $\Gamma \to \Delta, (\beta_1' \wedge \beta_2')'$ is also provable.

Here we deal only with the second statement in the above claim, leaving the remaining three statements to the reader. The proof is carried out by induction on the construction of a proof P of the sequent $\Gamma \to \Delta, \alpha_1 \wedge \alpha_2$. Here we deal only with the critical case in which the last inference is $(\to \wedge)$ as follows:

$$\frac{\Gamma \to \alpha_1 \qquad \Gamma \to \alpha_2}{\Gamma \to \alpha_1 \wedge \alpha_2}\ (\to \wedge)$$

Since $\alpha_1 \simeq \beta_1$ and $\alpha_2 \simeq \beta_2$ by assumption, the sequents $\Gamma \to \beta_1$ and $\Gamma \to \beta_2$ are provable, which implies that the sequent $\Gamma'' \to (\beta_1' \vee \beta_2')$ is is provable, as follows:

$$\cfrac{\cfrac{\cfrac{\Gamma \to \beta_1}{\beta_1' \to \Gamma'}('\!\to') \qquad \cfrac{\Gamma \to \beta_2}{\beta_2' \to \Gamma'}('\!\to')}{\beta_1' \vee \beta_2' \to \Gamma'}(\vee \to)}{\Gamma'' \to (\beta_1' \vee \beta_2')'}('\!\to')$$

Therefore the sequent $\Gamma \to (\beta_1' \vee \beta_2')'$ s provable by Theorem 8. To establish the remaining half of the theorem smoothly, we introduce a useful notion weaker than provability equivalence. A wff β is said to *be provably dominated by* a wff α, in notation $\alpha \succeq \beta$, if we have that for any finite sets Γ rand Δ of wffs:

1. Whenever the sequent $\alpha, \Gamma \to \delta$ is provable, the sequent $\beta, \Gamma \to \Delta$ is also provable.

2. Whenever the sequent $\Gamma \to \Delta, \alpha$ is provable, the sequent $\Gamma \to \Delta, \beta$ is also provable.

We notice that what we have really proved in Claim 13 is that if $\gamma_1 \widetilde{\rightarrow} \delta_1$ and $\gamma_2 \widetilde{\rightarrow} \delta_2$, then $\gamma_1 \wedge \gamma_2 \widetilde{\rightarrow} (\delta_1' \vee \delta_2')'$ and $\gamma_1 \vee \gamma_2 \widetilde{\rightarrow} (\delta_1' \wedge \delta_2')'$. Similarly, what we have really proved in the proof of Theorem 7 is that if $\gamma_1 \widetilde{\rightarrow} \delta_1$ and $\gamma_2 \widetilde{\rightarrow} \delta_2$, then $\gamma_1' \widetilde{\rightarrow} \delta_1'$, $\gamma_1 \wedge \gamma_2 \widetilde{\rightarrow} \delta_1 \wedge \delta_2$ nd $\gamma_1 \vee \gamma_2 \widetilde{\rightarrow} \delta_1 \vee \delta_2$, while what we have really proved in the proof of Theorem 8 is that if $\alpha \widetilde{\rightarrow} \beta$, , then $\alpha'' \widetilde{\rightarrow} \beta$. It is easy to see that two wffs α and β are provably equivalent iff each of them is provably dominated by the other. Thus, to conclude the proof of the theorem, it suffices to notice that

$$\alpha_1 \wedge \alpha_2 \widetilde{\rightarrow} (\beta_1' \vee \beta_2')' \widetilde{\rightarrow} (\alpha_1'' \wedge \alpha_2'') \widetilde{\rightarrow} \alpha_1 \wedge \alpha_2$$

■

COROLLARY 14.

1. *If* $\alpha_1 \simeq \beta_1$ *and* $\alpha_2 \simeq \beta_2$, *then* $\alpha_1' \wedge \alpha_2' \simeq (\beta_1 \vee \beta_2)'$ *and* $\alpha_1' \vee \alpha_2' \simeq (\beta_1 \wedge \beta_2)'$.

Proof. By Theorems 7, 8, and 12, we have that $\alpha_1' \wedge \alpha_2' \simeq (\alpha_1'' \vee \alpha_2'')' \simeq (\beta_1 \vee \beta_2)'$ and $\alpha_1' \vee \alpha_2' \simeq (\alpha_1'' \wedge \alpha_2'')' \simeq (\beta_1 \wedge \beta_2)'$. ■

5 THE CUT-ELIMINATION THEOREM

THEOREM 15. *A sequent* $\alpha, \beta, \Gamma \to \Delta$ *is provable iff the sequent* $\alpha \wedge \beta, \Gamma \to \Delta$ *is provable. Similarly, a sequent* $\Pi \to \Sigma, \gamma, \delta$ *is provable iff the sequent* $\Pi \to \Sigma, \gamma \vee \delta$ *is provable.*

Proof. For both statements, the only-if part follows readily from $(\wedge \to)$ or $(\to \vee)$. The if part can be established by induction on the construction of a proof of $\alpha \wedge \beta, \Gamma \to \Delta$ or $\Gamma \to \Delta, \alpha \vee \beta$. ■

COROLLARY 16. *A sequent* $\alpha', \beta', \Gamma \to \Delta$ *is provable iff the sequent* $(\alpha \vee \beta)', \Gamma \to \Delta$ *is provable. Similarly, a sequent* $\Pi \to \Sigma, \gamma', \delta'$ *is provable iff the sequent* $\Pi \to \Sigma, (\gamma \wedge \delta)'$ *is provable.*

Proof. Follows from Corollary 1 and Theorem 15. ■

THEOREM 17. *If a sequent* $\alpha \vee \beta, \Gamma \to \Delta$ *is provable, then the sequents* $\alpha, \Gamma \to \Delta$ *and* $\beta, \Gamma \to \Delta$ *are provable. Similarly, if a sequent* $\Pi \to \Sigma, \gamma \wedge \delta$ *is provable, then the sequents* $\Pi \to \Sigma, \gamma$ *and* $\Pi \to \Sigma, \delta$ *are provable.*

Proof. By induction on the construction of a proof of $\alpha \vee \beta, \Gamma \to \Delta$ or $\Pi \to \Sigma, \gamma \wedge \delta$. Here we deal only with the case that the last step of a proof of a sequent $\alpha \vee \beta, \Gamma \to \Delta$ is $(\vee \to')$. So it must be one of the following two forms.

$$\frac{\Gamma \to \alpha' \qquad \Gamma \to \beta'}{\alpha \vee \beta, \Gamma \to} \ (\vee \to')$$

$$\frac{\alpha \vee \beta, \Gamma_1 \to \sigma' \qquad \alpha \vee \beta, \Gamma_1 \to \rho'}{\alpha \vee \beta, \sigma \vee \rho, \Gamma_1 \to} \ (\vee \to')$$

In the former case the sequents $\alpha'', \Gamma \to$ and $\beta'', \Gamma \to$ are provable by $('\to)$. So the desired sequents $\alpha, \Gamma \to$ and $\beta, \Gamma \to$ are provable by Theorem 8. In the latter case the sequents $\alpha, \Gamma_1 \to \sigma', \beta, \Gamma_1 \to \sigma'$, and $\beta, \Gamma_1 \to \rho'$, are provable by the induction hypothesis. So the desired sequents $\alpha, \sigma \vee \rho$, $\Gamma_1 \to$ and $\beta, \sigma \wedge \rho, \Gamma_1 \to$ are provable as follows:

$$\frac{\alpha, \Gamma_1 \to \sigma' \qquad \alpha, \Gamma_1 \to \rho'}{\alpha, \sigma \vee \rho, \Gamma_1 \to} \ (\vee \to')$$

$$\frac{\beta, \Gamma_1 \to \sigma' \qquad \beta, \Gamma_1 \to \rho'}{\beta, \sigma \vee \rho, \Gamma_1 \to} \ (\vee \to')$$

■

COROLLARY 18. *If a sequent $(\alpha \wedge \beta)', \Gamma \to \Delta$ is provable, then the sequents $\alpha', \Gamma \to \Delta$ and $\beta', \Gamma \to \Delta$ are provable. Similarly, if a sequent $\Pi \to \Sigma, (\gamma \vee \delta)'$ is provable, then the sequents $\Pi \to \Sigma, \gamma'$ and $\Pi \to \Sigma, \delta'$ are provable.*

Proof. This follows from Theorem 17 and Corollary 1. ■

THEOREM 19. *(The cut-elimination theorem). If sequents $\Gamma_1 \to \Delta_1, \alpha$ and $\alpha, \Gamma_2 \to \Delta_2$ are provable with either $\Delta_1 = \varnothing$ or $\Gamma_2 = \varnothing$, then the sequent $\Gamma_1, \Gamma_2 \to \Sigma_1, \Delta_2$ is also provable. In other words, $(cut)_q$ is permissible in **GMQL**.*

Proof. Suppose that the sequents $\Gamma_1 \to \Delta_1, \alpha$ and $\alpha, \Gamma_2 \to \Delta_2$ have proofs P_1 and P_2, respectively. We prove the theorem by double induction principally on $\mathcal{G}(\alpha)$ and secondarily on $l(P_1) + l(P_2)$. By Theorem 12 we can assume that there is no occurrence of the disjunction symbol $\vee$ in P_1 or P_2. As in the proof of Theorem 8, whenever we are forced to deal with the rules $(\wedge \to)$ or $(\to \wedge')$, each of which consists of two forms, only one of them is treated. Our proof is divided into several cases according to which inference rule is used in the last step of P_1 or P_2 as follows:

1. The case that one of the sequents $\Gamma_1 \to \Delta_1, \alpha$ and $\alpha, \Gamma_2 \to \Delta_2$ is an axiom sequent: There is nothing to prove.

2. The case that one of the sequents $\Gamma_1 \to \Delta_1, \alpha$ and $\alpha, \Gamma_2 \to \Delta_2$ is obtained as the lower sequent of (extension): Here we deal only with the case that the former sequent $\Gamma_1 \to \Delta_1, \alpha$ is obtained as the lower sequent of (extension), leaving the dual case to the reader. Then the last step of the proof P_1 is in one of the following two forms:

$$\frac{\Gamma_{11} \to \Delta_{11}, \alpha}{\Gamma_{11}, \Gamma_{12} \to \Delta_{11}, \Delta_{12}, \alpha} \text{(extension)}$$

$$\frac{\Gamma_{11} \to \Delta_{11}}{\Gamma_{11}, \Gamma_{12} \to \Delta_{11}, \Delta_{12}, \alpha} \text{(extension)}$$

In the former case the desired sequent $\Gamma_{11}, \Gamma_{12}, \Gamma_2 \to \Delta_{11}, \Delta_{12}, \Delta_2$ is provable by induction hypothesis as follows.

$$\frac{\dfrac{\Gamma_{11} \to \Delta_{11}, \alpha \qquad \alpha, \Gamma_2 \to \Delta_2}{\Gamma_{11}, \Gamma_2 \to \Delta_{11}, \Delta_2} \text{(cut)}_q}{\Gamma_{11}, \Gamma_{12}, \Gamma_2 \to \Delta_{11}, \Delta_{12}, \Delta_2} \text{(extension)}$$

In the latter case the desired sequent $\Gamma_{11}, \Gamma_{12}, \Gamma_2 \to \Delta_{11}, \Delta_{12}, \Delta_2$ is obtained as follows.

$$\frac{\Gamma_{11} \to \Delta_{11}}{\Gamma_{11}, \Gamma_{12}, \Gamma_2 \to \Delta_{11}, \Delta_{12}, \Delta_2} \text{(extension)}$$

3. The case that either the sequent $\Gamma_1 \to \Delta_1, \alpha$ is obtained as the lower sequent of one of the inference rules $('' \to)$ and $(\wedge \to)$ or the sequent $\alpha, \Gamma_2 \to \Delta_2$ is obtained as the lower sequent of one of the inference rules $(\to '')$ and $(\to \wedge')$: Here we deal only with the case that the sequent $\alpha, \Gamma_2 \to \Delta_2$ is obtained as the lower sequent of $(\to \wedge')$, leaving the remaining three cases to the reader. So the last step of P_2 is of the following form:

$$\frac{\alpha, \Gamma_2 \to \Sigma_2, \beta'}{\alpha, \Gamma_2 \to \Sigma_2, (\beta \vee \gamma)'} (\to \wedge')$$

The desired sequent $\Gamma_1, \Gamma_2 \to \Delta_1, \Sigma_2, (\beta \wedge \gamma)'$ is provable by induction hypothesis as follows:

$$\frac{\dfrac{\Gamma_1 \to \Delta_1, \alpha \qquad \alpha, \Gamma_2 \to \Sigma_2, \beta'}{\Gamma_1, \Gamma_2 \to \Delta_1, \Sigma_2, \beta'} \text{(cut)}_q}{\Gamma_1, \Gamma_2 \to \Delta_1, \Sigma_2, (\beta \wedge \gamma)'} (\to \wedge')$$

4. The case that either the sequent $\Gamma_1 \to \Delta_1, \alpha$ is obtained as the lower sequent of $(\wedge' \to)$ or the sequent $\alpha, \Gamma_2 \to \Delta_2$ is obtained as the lower sequent of $(\to \wedge)$: Here we deal only with the former case, leaving a similar treatment of the latter case to the reader. So the last step of P_1 goes as follows:

$$\frac{\beta' \to \Delta_1, \alpha \qquad \gamma' \to \Delta_1, \alpha}{(\beta \wedge \gamma)' \to \Delta_1, \alpha} \ (\wedge' \to)$$

If $\gamma_2 = \varnothing$, then the desired sequent $(\beta \wedge \gamma)' \to \Delta_1, \Delta_2$ is provable by the induction hypothesis as follows:

$$\frac{\dfrac{\beta' \to \Delta_1, \alpha \qquad \alpha \to \Delta_2}{\beta_1' \to \Delta_1, \Delta_2}\ (\text{cut})_q \qquad \dfrac{\gamma' \to \Delta_1, \alpha \qquad \alpha \to \Delta_2}{\gamma_2' \to \Delta_1, \Delta_2}\ (\text{cut})_q}{(\beta \wedge \gamma)' \to \Delta_1, \Delta_2} \ (\wedge' \to)$$

Unless $\Gamma_2 = \varnothing$, the situation can be classified into cases according to which inference rule is used in the last step of P_2. If $\Gamma_2 \neq \varnothing$ and it is not the case that the last inference of P_2 is $(\to \wedge)$, the situation is subsumed under the cases that have been or will be dealt with. If $\Gamma_2 \neq \varnothing$ and the last inference of P_2 is $(\to \wedge)$, then surely $\Gamma_1 \neq \varnothing$, so that the situation can be handled dually to the case that $\Gamma_2 = \varnothing$.

5. The case that either the sequent $\Gamma_1 \to \Delta_1, \alpha$ is obtained as the lower sequent of one of the inference rules $(\to'')$ and $(\to \wedge')$ or the sequent $\alpha, \Gamma_2 \to \Delta_2$ is obtained as the lower sequent of one of the inference rules $(''\!\to)$ and $(\wedge \to)$: Here we deal only with the case that the sequent $\Gamma_1 \to \Delta_1, \alpha$ is obtained as the lower sequent of $(\to \wedge')$, leaving the remaining three cases to the reader. So the last step of P_1 is in one of the following two forms:

$$\frac{\Gamma_1 \to \Sigma, \beta', \alpha}{\Gamma_1 \to \Sigma, (\beta \wedge \gamma)', \alpha} \ (\to \wedge')$$

$$\frac{\Gamma_1 \to \Delta_1, \beta'}{\Gamma_1 \to \Delta_1, (\beta \wedge \gamma)'} \ (\to \wedge')$$

In the latter case α is supposed to be $(\beta \wedge \gamma)'$. In the former case the $(\text{cut})_q$ at issue is an instance of (cut-1), so that $\Gamma_2 = \varnothing$, and the desired sequent $\Gamma_1 \to \Sigma, (\beta \wedge \gamma)', \Delta_2$ is provable by induction hypothesis, as follows:

$$\frac{\dfrac{\Gamma_1 \to \Sigma, \beta', \alpha \qquad \alpha \to \Delta_2}{\Gamma_1 \to \Sigma, \beta'}\ (\text{cut})_q}{\Gamma_1 \to \Sigma, (\beta \wedge \gamma)'} \ (\to \wedge')$$

As for the latter case, the cut formula is $(\beta \wedge \gamma)'$, and the sequent, $\beta', \Gamma_2 \to \Delta_2$ is provable by Corollary 18. Thus the desired sequent $\Gamma_1, \Gamma_2 \to \Delta_1, \Delta_2$ is provable by the induction hypothesis, as follows:

$$\frac{\Gamma_1 \to \Delta_1, \beta' \qquad \beta', \Gamma_2 \to \Delta_2}{\Gamma_1, \Gamma_2 \to \Delta_1, \Delta_2} \ (\text{cut})_q$$

6. The case that either the sequent $\Gamma_1 \to \Delta_1, \alpha$ is obtained as the lower sequent of $(\to \wedge)$ or the sequent $\alpha, \Gamma_2 \to \Delta_2$ is obtained as the lower sequent . of $(\wedge' \to)$: Here we deal only with the latter case, leaving a similar treatment of the latter to the reader. So the last step of P_2 goes as follows:

$$\frac{\beta' \to \Delta_2 \qquad \gamma' \to \Delta_2}{(\beta \wedge \gamma)' \to \Delta_2}\ (\wedge' \to)$$

Here α is supposed to be $(\beta\wedge\gamma)'$, and the $(\mathrm{cut})_q$ at issue is an instance of (cut-1) with the cut formula $(\beta \wedge \gamma)'$. By Corollary 8 the sequent $\Gamma_1 \to \Delta_1, \beta', \gamma'$ is provable, so that the desired sequent $\Gamma_1 \to \Delta_1, \Delta_2$ is also provable, as follows:

$$\frac{\dfrac{\Gamma_1 \to \Delta_1, \beta', \gamma' \qquad \beta' \to \Delta_2}{\Gamma_1 \to \Delta_1, \Delta_2, \gamma'}\ (\mathrm{cut})_q \qquad \gamma' \to \Delta_2}{\Gamma_1 \to \Delta_1, \Delta_2}\ (\mathrm{cut})_q$$

7. The case that the sequent $\Gamma_1 \to \Delta_1, \alpha$ is obtained as the lower sequent of $('\!\to \wedge)$: The last step of P_1 is in one of the following two forms:

$$\frac{\beta' \to \Sigma, \alpha \qquad \gamma' \to \Sigma, \alpha}{\to \Sigma, \beta \wedge \gamma, \alpha}\ ('\!\to \wedge)$$

$$\frac{\beta' \to \Delta_1 \qquad \gamma' \to \Delta_1}{\to \Delta_1, \beta \wedge \gamma}\ ('\!\to \wedge)$$

In the latter case α is assumed to be $\beta \wedge \gamma$. First we deal with the former case, in which the $(\mathrm{cut})_q$ at issue is (cut-1) so that $\Gamma_2 = \varnothing$. Then the desired sequent $\to \Sigma, \beta \wedge \gamma, \Delta_2$ is provable by the induction hypothesis as follows:

$$\frac{\dfrac{\beta' \to \Sigma, \alpha \qquad \alpha \to \Delta_2}{\beta_1' \to \Sigma, \Delta_2}\ (\mathrm{cut})_q \qquad \dfrac{\gamma' \to \Sigma, \alpha \qquad \alpha \to \Delta_2}{\gamma' \to \Sigma, \Delta_2}\ (\mathrm{cut})_q}{\to \Sigma, \beta \wedge \gamma, \Delta_2}\ ('\!\to \wedge)$$

As for the latter case, suppose first that $\Delta_1 \neq \varnothing$, so that $\Gamma_2 = \varnothing$. Then the sequents $\Delta_1' \to \beta$ and $\Delta_1' \to \gamma$ are provable by Corollary 11, while the sequent $\beta, \gamma \to \Delta_1'', \Delta_2$ is provable by Theorem 15. Thus the sequent $\to \Delta_1'', \Delta_2$ is provable by the induction hypothesis, as follows:

$$\frac{\dfrac{\Delta_1' \to \gamma \qquad \dfrac{\Delta_1' \to \beta \qquad \beta, \gamma \to \Delta_2}{\gamma, \Delta_1' \to \Delta_2}\ (\mathrm{cut})_q}{\Delta_1' \to \Delta_2}\ (\mathrm{cut})_q}{\to \Delta_1'', \Delta_2}\ (\to')$$

Thus the desired sequent $\to \Delta_1, \Delta_2$ is provable by Theorem 8. If $\Delta_1 = \varnothing$, then the sequents $\to \beta$ and $\to \gamma$ are provable by Corollary 11, while the sequent $\beta, \gamma, \Gamma_2 \to \Delta_2$ is provable by Theorem 15. Thus the desired sequent $\Gamma_2 \to \Delta_2$ is provable by the induction hypothesis as follows:

$$\dfrac{\dfrac{\to \beta \qquad \beta, \gamma, \Gamma_2 \to \Delta_2}{\to \gamma \qquad \gamma, \Gamma_2 \to \Delta_2}\ (\text{cut})_q}{\Gamma_2 \to \Delta_2}\ (\text{cut})_q$$

8. The case that one of the sequents $\Gamma_1 \to \Delta_1, \alpha$ and $\alpha, \Gamma_2 \to \Delta_2$ is obtained as the lower sequent of $(\to')$: Here we deal only with the case that the sequent $\Gamma_1 \to \Delta_1, \alpha$ is obtained as the lower sequent of $(\to')$, leaving the dual case to the reader. So the last step of the proof P_1 is in one of the following two forms:

$$\dfrac{\Delta'_{12} \to \Delta_{11}, \alpha}{\to \Delta_{11}, \Delta'_{12}, \alpha}\ (\to')$$

$$\dfrac{\Delta'_{12}, \beta \to \Delta_{11}}{\to \Delta_{11}, \Delta'_{12}, \beta'}\ (\to')$$

In the latter case α is supposed to be β'. First we deal with the former case. If $\Gamma_2 = \varnothing$, then the desired sequent $\to \Delta_{11}, \Delta'_{12}, \Delta_2$ is provable by the induction hypothesis, as follows:

$$\dfrac{\dfrac{\Delta_{12} \to \Delta_{11}, \alpha \qquad \alpha \to \Delta_2}{\Delta_{12} \to \Delta_{11}, \Delta_2}\ (\text{cut})_q}{\to \Delta_{11}, \Delta'_{12}, \Delta_2}\ (\to')$$

If $\Gamma_2 \neq \varnothing$, then α is of the form γ' and the sequent $\Delta_{12} \to \Delta_{11}, \alpha$ is $\gamma \to \gamma'$. The sequents $\gamma \to$ and $\Delta'_2 \to \gamma$ are provable by Corollary 11, which implies that the sequent $\Delta'_2 \to$ is also provable by the induction hypothesis, as follows:

$$\dfrac{\Delta'_2 \to \gamma \qquad \gamma \to}{\Delta'_2 \to}\ (\text{cut})_q$$

By Corollary 11 the sequent $\to \Delta_2$ is provable, which implies that the desired sequent $\to \Delta_1, \Delta_2$ is provable as follows:

$$\dfrac{\to \Delta_2}{\to \Delta_1, \Delta_2}\ (\text{extension})$$

Now we deal with the latter case. If $\Gamma_2 = \varnothing$, then the sequent $\Delta'_2 \to \beta$ is provable by Corollary 11, and the sequent $\to \Delta_{11}, \Delta'_{12}, \Delta''_2$ is also provable by the induction hypothesis as follows: .

$$\dfrac{\dfrac{\Delta'_2 \to \beta \qquad \Delta_{12}, \beta \to \Delta_{11}}{\Delta_{12}, \Delta'_2 \to \Delta_{11}}\ (\text{cut})_q}{\to \Delta_{11}, \Delta'_{12}, \Delta''_2}\ (\to')$$

Thus the desired sequent $\rightarrow \Delta_{11}, \Delta_{12}, \Delta_2$ is provable by Theorem 8. If $\Gamma_2 \neq \varnothing$, then the sequent $\Delta_{12}, \beta \rightarrow \Delta_{11}$ must be $\beta \rightarrow$ or $\beta \rightarrow \beta'$, the latter of which implies by Corollary 11 that the sequent $\beta \rightarrow$ is provable. Thus in any case the sequent $\beta \rightarrow$ is provable. Since the sequent $\Delta_2' \rightarrow \Gamma_2', \beta$ is provable by Corollary 11, the sequent $\Delta_2' \rightarrow \Gamma_2'$ is provable by induction hypothesis as follows:

$$\frac{\Delta_2' \rightarrow \Gamma_2', \beta \qquad \beta \rightarrow}{\Delta_2' \rightarrow \Gamma_2'} \ (\mathrm{cut})_q$$

Therefore the sequent $\Gamma_2 \rightarrow \Delta_2$ is provable by Corollary 11, which implies that the desired sequent $\Gamma_2 \rightarrow \Delta_1, \Delta_2$ is provable as follows:

$$\frac{\Gamma_2 \rightarrow \Delta_2}{\Gamma_2 \rightarrow \Delta_1, \Delta_2} \ (\mathit{extension})$$

9. The case that both the sequent $\Gamma_1 \rightarrow \Delta_1, \alpha$ and the sequent $\alpha, \Gamma_2 \rightarrow \Delta_2$ are obtained as the lower sequent of $('\!\rightarrow')$: The last steps of the proofs P_1 and P_2 go as follows:

$$\frac{\Sigma_1, \beta \rightarrow \Pi_1}{\Pi_1' \rightarrow \Sigma_1', \beta} \ ('\!\rightarrow')$$

$$\frac{\Sigma_2, \rightarrow \beta, \Pi_2}{\beta', \Pi_2' \rightarrow \Sigma_2'} \ ('\!\rightarrow')$$

In the above α is supposed to be β'. The desired sequent $\Pi_1, \Pi_2 \rightarrow \Sigma_1, \Sigma_2$ is provable by the induction hypothesis as follows:

$$\dfrac{\dfrac{\Sigma_2 \rightarrow \Pi_2, \beta \qquad \beta, \Sigma_1 \rightarrow \Pi_1}{\Sigma_1, \Sigma_2 \rightarrow \Pi_1, \Pi_2} \ (\mathit{cut})}{\Pi_1', \Pi_2' \rightarrow \Sigma_1', \Sigma_2'} \ ('\!\rightarrow')$$

■

6 THE COMPLETENESS THEOREM

An *O-frame* is a pair $(X, \perp)$ of a nonempty set X and an orthogonality relation (i.e., an irreflexive and symmetric binary relation) on X. Given $Y \subseteq X$, we write $Y^{\perp}$ for the set $\{x \in X | x \perp y \text{ for any } y \in Y\}$. A subset Y of X is said to be $\perp$-*closed* if $Y = Y^{\perp\perp}$.

An *O-model* is a triple $(X, \perp, D)$, where $(X, \perp)$ is an *O-frame* and D assigns to each propositional variable p a $\perp$-closed subset $D(p)$ of X. The notation $\|\alpha\|$ for a wff α is defined inductively as follows:

1. $\|p\| = D(p)$ for any propositional variable p.

2. $\|\alpha \wedge \beta\| = \|\alpha\| \cap \|\beta\|$.
3. $\|\alpha'\| = \|\alpha\|^{\perp}$.
4. $\|\alpha \vee \beta\| = \|\alpha' \wedge \beta'\| = (\|\alpha\|^{\perp} \cap \|\beta\|^{\perp})^{\perp}$.

Given $x \in X$ and a wff α, we write $V(\alpha; x) = 1$ if $x \in \|\alpha\|$ and $V(\alpha; x) = 0$ if $x \notin \|\alpha\|$. Given $x \in X$ and a sequent $\Gamma \to \Delta$, we write $V(\Gamma \to \Delta; x) = 1$ if $x \in \bigcap\{\|\alpha\| \mid \alpha \in \Gamma\}$ and $x \notin (\cup\{\|\beta\|^{\perp} | \beta \in \Delta\})^{\perp}$, and $V(\Gamma \to \Delta; x) = 0$ otherwise.

A sequent $\Gamma \to \Delta$ is said to be *realizable* if there exists an *O-model* $(X, \perp, D)$ and some $x \in X$ such that $V(\Gamma \to \Delta; x) = 1$. The sequent $\Gamma \to \Delta$ is called *valid* otherwise.

THEOREM 20. *(The soundness theorem). If a sequent* $\Gamma \to \Delta$ *is provable, then it is valid.*

Proof. By induction on the construction of a proof of the sequent $\Gamma \to \Delta$. ■

A set Ω of wffs is said to be *admissible* if it satisfies the following conditions:

1. If p is a propositional variable and $p \in \Omega$, then $p' \in \Omega$.
2. If $\alpha \in \Omega$ and β is a subformula of α, then $\beta \in \Omega$.
3. If $(\alpha \vee \beta) \in \Omega$, then $(\alpha' \wedge \beta')' \in \Omega$.

A finite set Γ of wffs is said to be *inconsistent* if for some wff α, both of the sequents $\Gamma \to \alpha$ and $\Gamma \to \alpha'$ are provable. Otherwise the set Γ is said to be *consistent.*

LEMMA 21. *A finite set* Γ *of wffs is inconsistent iff the sequent* $\Gamma \to$ *is provable.*

Proof. The if part is obvious. The only-if part can be shown easily as follows:

$$\dfrac{\dfrac{\Gamma \to \alpha \qquad \Gamma \to \alpha'}{\Gamma \to \alpha \wedge \alpha'}\,(\to \wedge) \qquad \dfrac{\dfrac{\alpha \to \alpha}{\alpha, \alpha' \to}\,('\!\to)}{\alpha \wedge \alpha' \to}\,(\wedge \to)}{\Gamma \to}\,(\mathrm{cut})_q$$

■

Given an admissible set Ω of wffs, the Ω-*canonical* O-model $\mathcal{M}(\Omega) = (X_\Omega, \perp_\Omega, D_\Omega)$ is defined as follows:

1. X_Ω is the set of all the consistent subsets of Ω.
2. For any $\Gamma_1, \Gamma_2 \in X_\Omega, \Gamma_1 \perp_\Omega \Gamma_2$ iff for some $\alpha' \in \Omega$, either: (a) both of the sequents $\Gamma_1 \to \alpha$ and $\Gamma_1 \to \alpha'$ are provable, or (b) both of the sequents $\Gamma_1 \to \alpha'$ and $\Gamma_2 \to \alpha$ are provable.

3. If $p \notin \Omega$, then $D_\Omega(p) = \varnothing$, while if $p \in \Omega$, then $D_\Omega(p)$ consists of all the consistent subsets Γ of Ω such that the sequent $\Gamma \to p$ is provable.

THEOREM 22. *$\mathcal{M}(\Omega)$ is an O-model.*

Proof. Obviously the relation $\perp_\Omega$ is symmetric. That the relation $\perp_\Omega$ is irreflexive follows from our assumption that every element of X_Ω is a consistent set of wffs. Now it remains to show that $D_\Omega(p)$ is $\perp_\Omega$-closed for any propositional variable p. Unless $p \in \Omega$, there is nothing to prove. So let $p \in \Omega$. Let Γ be an element of X_Ω such that the sequent $\Gamma \to p$ is not provable. Suppose for the sake of contradiction that the set $\{p'\}$ is inconsistent, which implies by Lemma 21 that the sequent $p' \to$ is provable. By Corollary 11 the sequent $\to p$ is provable, which implies by (extension) that the sequent $\Gamma \to p$ is provable. This is a contradiction. So $\{p'\} \in X_\Omega$. Suppose, for the sake of contradiction, that for some $\alpha' \in \Omega$, either both of the sequents $\Gamma \to \alpha'$ and $p' \to \alpha$ are provable or both of the sequents $\Gamma \to \alpha'$ and $p' \to \alpha$ are provable. Here we deal only with the former case, leaving a similar treatment of the latter to the reader. By Corollary 11 the sequent $\alpha \to p$ is provable, which implies by (cut) that the sequent $\Gamma \to p$ is provable. This is a contradiction. Thus it cannot be the case that $\Gamma \perp_\Omega \{p'\}$, while for any $\Delta \in X_\Omega$ such that the sequent $\Delta \to p$ is provable, $\Delta \perp_\Omega \{p\}$. This implies that the set of all $\Delta \in X_\Omega$ such that the sequent $\Delta \to p$ is provable is $\perp_\Omega$-closed. ■

The *disjunction grade* of a wff α, denoted by $\mathcal{G}_\vee(\alpha)$, is defined inductively as follows:

1. $\mathcal{G}_\vee(p) = o$ for any propositional variable p.
2. $\mathcal{G}_\vee(\alpha') = \mathcal{G}_\vee(\alpha)$.
3. $\mathcal{G}_\vee(\alpha \wedge \beta) = \mathcal{G}_\vee(\alpha) + \mathcal{G}_\vee(\beta)$.
4. $\mathcal{G}_\vee(\alpha \vee \beta) = \mathcal{G}_\vee(\alpha) + \mathcal{G}_\vee(\beta) + 1$.

THEOREM 23. *(The fundamental theorem for $\mathcal{M}(\Omega)$). For any $\alpha \in \Omega$ and any $\Gamma \in X_\Omega$, the sequent $\Gamma \to \alpha$ is provable iff $\Gamma \in \|\alpha\|$ in $\mathcal{M}(\Omega)$.*

Proof. The proof is carried out by double induction principally on $\mathcal{G}_\vee(\alpha)$ and secondarily on $\mathcal{G}(\alpha)$. The proof is divided into several cases.

1. In the case that α is a propositional variable: It follows from the definition of D_Ω.

2. In the case that $\alpha = \beta'$ for some wff β: If $\Gamma \to \beta'$ is provable, then $\Gamma \perp_\Omega \|\beta\|$ by induction hypothesis, which implies that $\Gamma \in \|\beta'\|$. Suppose, for the sake of contradiction, that the set $\{\beta\}$ is inconsistent, which implies by Lemma 21 that the sequent $\beta \to$ is provable. Thus the sequent $\Gamma \to \beta'$ is provable as follows:

$$\dfrac{\dfrac{\beta \rightarrow}{\rightarrow \beta'}\ (\rightarrow')}{\Gamma \rightarrow \beta'}\ \text{(extension)}$$

This is a contradiction. So it must be the case that $\{\beta\} \in X_\Omega$. Suppose, for the sake of contradiction, that for some $\gamma' \in \Omega$, either both of the sequents $\Gamma \rightarrow \gamma'$ and $\beta \rightarrow \gamma$ are provable or both of the sequents $\Gamma \rightarrow \gamma$ and $\beta \rightarrow \gamma'$ are provable. Here we deal only with the former case, leaving safely a similar treatment of the latter to the reader. The desired contradiction is obtained as follows:

$$\dfrac{\Gamma \rightarrow \gamma' \qquad \dfrac{\beta \rightarrow \gamma}{\gamma' \rightarrow \beta'}\ ('\!\rightarrow')}{\Gamma \rightarrow \beta}\ (\text{cut})_q$$

Thus it cannot be the case that $\Gamma \perp_\Omega \{\beta\}$. Since $\{\beta\} \in \|\beta\|$ by induction hypothesis, this means that $\Gamma \notin \|\beta\| \perp \Omega = \|\beta'\|$.

3. In the case that α is of the form $\beta \wedge \gamma$ for some wffs β and γ: If the sequent $\Gamma \rightarrow \alpha$ is provable, then both of the sequents $\Gamma \rightarrow \beta$ and $\Gamma \rightarrow \gamma$ are provable by Theorem 17, which implies by induction hypothesis that $\Gamma \in \|\beta\|$ and $\Gamma \in \|\gamma\|$. So $\Gamma \in \|\beta\| \cap \|\gamma\| = \|\beta \wedge \gamma\|$. Unless the sequent $\Gamma \rightarrow \alpha$ is provable, suppose, for the sake of contradiction, that both of the sequents $\Gamma \rightarrow \beta$ and $\Gamma \rightarrow \gamma$ are provable. The desired conclusion is obtained as follows:

$$\dfrac{\Gamma \rightarrow \beta \qquad \Gamma \rightarrow \gamma}{\Gamma \rightarrow \beta \wedge \gamma}\ (\rightarrow \wedge)$$

Thus one of the sequents $\Gamma \rightarrow \beta$ and $\Gamma \rightarrow \gamma$ is consistent, which implies by induction hypothesis that $\Gamma \notin \|\beta\|$ or $\Gamma \notin \|\gamma\|$. So $\Gamma \notin \|\beta \wedge \gamma\| = \|\beta\| \cap \|\gamma\|$.

4. In the case that α is of the form $\beta \vee \gamma$ for some wffs β and γ: Use Theorem 12.

■

THEOREM 24. *(The completeness theorem). A sequent* $\Gamma \rightarrow \Delta$ *is realizable iff it is consistent.*

Proof. The only-if part is the soundness theorem already established. To see the if part, take an admissible set Ω such that $\Gamma \cup \{\beta_1 \vee ... \vee \beta_n\} \subseteq \Omega$, where $\Delta = \{\beta_1, ..., \beta_n\}$. By Theorem 15 the sequent $\Gamma \rightarrow \Delta$ is consistent iff the sequent $\Gamma \rightarrow \beta_1 \vee ... \vee \beta_n$ is consistent. The desired conclusion follows readily from Theorem 23. ■

We remark in passing that in the proof of Theorem 24 it does not matter how to insert parentheses in $\beta_1 \vee ... \vee \beta_n$.

BIBLIOGRAPHY

[Birkhoff and von Neumann, 1936] G. Birkhoff and J. von Neumann. The logic of quantum mechanics, *Annals of Mathematics*, **37**, 823-843, 1936.

[Bruns, 1976] G. Bruns. Free ortholattices, *Canadian Journal of Mathematics*, 28, 977-985, 1976.

[Cutland and Gibbins, 1982] N. J. Cutland and P. F. Gibbins. A regular sequent calculus for quantum logic in which $\wedge$ and $\vee$ are dual, *Logique et Analyse*, 99, 221-248, 1982.

[Dalla Chiara, 1986] M. L. Dalla Chiara. Quantum logic, in *Handbook of Philosophical Logic*, D. Gabbay and F. Guenthner, eds., Reidel, Dordrecht, Volume III, pp. 427-469, 1986.

[Dishkant, 1972] H. Dishkant. Semantics for the minimal logic of quantum mechanics, *Studia Logica*, 30, 23-36, 1972.

[Dishkant, 1977] H. Dishkant. Imbedding of the quantum logic in the modal system of Brower, *Journal of Symbolic Logic*, 42, 321-328, 1977.

[Foulis and Randall, 1971] D. J. Foulis and C. H. Randall. Lexicographic orthogonality, *Journal of Combinatorial Theory*, **11**, 157-162, 1971.

[Gentzen, 1935] G. Gentzen. Untersuchungen fiber das logische Schliessen, I and II, *Mathematische Zeitschrift*, 39, 176-210, 405-431, 1935.

[Goldblatt, 1974] R. I. Goldblatt. Semantical analysis of orthologic, *Journal of Philosophical Logic*, 3, 19-36, 1974.

[Goldblatt, 1975] R. I. Goldblatt. The Stone space of an ortholattice, *Bulletin of the London Mathematical Society*, 7, 45-48. 1975.

[Goldblatt, 1984] R. I. Goldblatt. Orthomodularity is not elementary. *Journal of Symbolic Logic*, 49, 401-404, 1984.

[Jammer, 1974] M. Jammer. *The Philosophy of Quantum Mechanics*, Wiley, New York, 1974.

[Kripke, 1963] S. Kripke. Semantical analysis of modal logic, I, normal modal propositional calculi, *Z. Math. Logik Grundlagen Math.*, **9**, 67-96, 1963.

[Maeda, 1980] S. Maeda. *Lattice Theory and Quantum Logic*, Maki, Tokyo [in Japanese], 1980.

[McKinsey and Tarski, 1948] J. C. McKinsey and A. Tarski. Some theorems about the sentential calculi of Lewis and Heyting, *Journal of Symbolic Logic*, **13**, 1-15, 1948.

[Nishimura, 1980] H. Nishimura. Sequential method in quantum logic, *Journal of Symbolic Logic*, 45, 339-352, 1980.

[Nishimura, 1983] H. Nishimura. A cut-free sequential system for the propositional modal logic of finite chains, *Publications of RIMS, Kyoto University*, **19**, 305-316, 1983.

[Nishimura, 1994a] H. Nishimura. Proof theory for minimal quantum logic I, *International Journal of Theoretical Physics*, **33**, 103-113, 1994.

[Nishimura, 1994b] H. Nishimura. Proof theory for minimal quantum logic I, *International Journal of Theoretical Physics*, **33**, 1427-1443, 1994.

[Randall and Foulis, 1970] C. H. Randall and D. J. Foulis. An approach to empirical logic, *American Mathematical Monthly*, **77**, 363-374, 1970.

[Sato, 1977] M. Sato. A study of Kripke-type models for some modal logics by Gentzen's sequential methods, *Publications of RIMS, Kyoto University*, **13**, 381-468, 1977.

[Takano, 1995] M. Takano. Proof theory for minimal quantum logic:a remark, *International Journal of Theoretical Physics*, **34**, 649-654, 1995.

[Takeuti, 1975] G. Takeuti. *Proof Theory*, North-Holland, Amsterdam, 1975.

[Tamura, 1988] S. Tamura. A Gentzen formulation without the cut rule for ortholattices, *Kobe Journal of Mathematics*, 5, 133-150, 1988.

HANDBOOK OF QUANTUM LOGIC AND QUANTUM STRUCTURES: QUANTUM LOGIC
Edited by K. Engesser, D. M. Gabbay and D. Lehmann

CATEGORICAL QUANTUM MECHANICS

Samson Abramsky and Bob Coecke

1 INTRODUCTION

Our aim is to revisit the mathematical foundations of quantum mechanics from a novel point of view. The standard axiomatic presentation of quantum mechanics in terms of Hilbert spaces, essentially due to von Neumann [1932], has provided the mathematical bedrock of the subject for over 70 years. Why, then, might it be worthwhile to revisit it now?

First and foremost, the advent of *quantum information and computation* (QIC) as a major field of study has breathed new life into basic quantum mechanics, asking new kinds of questions and making new demands on the theory, and at the same time reawakening interest in the foundations of quantum mechanics.

As one key example, consider the changing perceptions of *quantum entanglement* and its consequences. The initial realization that this phenomenon, so disturbing from the perspective of classical physics, was implicit in the quantum-mechanical formalism came with the EPR *Gedanken*-experiment of the 1930's [Einstein *et al.*, 1935], in the guise of a "paradox". By the 1960's, the paradox had become a *theorem* — Bell's theorem [Bell, 1964], demonstrating that non-locality was an essential feature of quantum mechanics, and opening entanglement to experimental confirmation. By the 1990's, entanglement had become a *feature*, used in quantum teleportation [Bennett *et al.*, 1993], in protocols for quantum key distribution [Ekert, 1991], and, more generally, understood as a computational and informatic *resource* [Bouwmeester *et al.*, 2001].

1.1 *The Need for High-Level Methods*

The current tools available for developing quantum algorithms and protocols, and more broadly the whole field of quantum information and computation, are *deficient* in two main respects.

Firstly, they are too *low-level*. One finds a plethora of ad hoc calculations with 'bras' and 'kets', normalizing constants, matrices etc. The arguments for the benefits of a high-level, conceptual approach to designing and reasoning about quantum computational systems are just as compelling as for classical computation. In particular, we have in mind the hard-learned lessons from Computer Science of the importance of *compositionality*, *types*, *abstraction*, and the use of tools from algebra and logic in the design and analysis of complex informatic processes.

At a more fundamental level, the standard mathematical framework for quantum mechanics is actually *insufficiently comprehensive* for informatic purposes. In describing a protocol such as teleportation, or any quantum process in which *the outcome of a measurement is used to determine subsequent actions*, the von Neumann formalism leaves feedback of information from the classical or macroscopic level back to the quantum *implicit* and *informal*, and hence not subject to rigorous analysis and proof. As quantum protocols and computations grow more elaborate and complex, this point is likely to prove of increasing importance.

Furthermore, there are many fundamental issues in QIC which remain very much open. The current low-level methods seem unlikely to provide an adequate basis for addressing them. For example:

- What are the precise structural relationships between superposition, entanglement and mixedness as quantum informatic resources? Or, more generally,
- Which features of quantum mechanics account for differences in computational and informatic power as compared to classical computation?
- How do quantum and classical information interact with each other, and with a spatio-temporal causal structure?
- Which quantum control features (e.g. iteration) are possible and what additional computational power can they provide?
- What is the precise logical status and axiomatics of No-Cloning and No-Deleting, and more generally, of the quantum mechanical formalism as a whole?

These questions gain additional force from the fact that a variety of different quantum computational architectures and information-processing scenarios are beginning to emerge. While at first it seemed that the notions of Quantum Turing Machine [Deutsch, 1985] and the quantum circuit model [Deutsch, 1989] could supply canonical analogues of the classical computational models, recently some very different models for quantum computation have emerged, e.g. Raussendorf and Briegel's *one-way quantum computing* model [Raussendorf and Briegel, 2001; Raussendorf *et al.*, 2003] and *measurement based quantum computing* in general [Jozsa, 2005], *adiabatic quantum computing* [Farhi *et al.*, 2000], *topological quantum computing* [Freedman *et al.*, 2004], etc. These new models have features which are both theoretically and experimentally of great interest, and the methods developed to date for the circuit model of quantum computation do not carry over straightforwardly to them. In this situation, we can have no confidence that a comprehensive paradigm has yet been found. It is more than likely that we have overlooked many new ways of letting a quantum system compute.

Thus there is a need to design structures and develop methods and tools which apply to these *non-standard quantum computational models* . We must also address the question of how the various models compare — can they be interpreted in each other, and which computational and physical properties are preserved by such interpretations?

1.2 High-Level Methods for Quantum Foundations

Although our initial motivation came from quantum information and computation, in our view the development of high-level methods is potentially of great significance for the development of the foundations of quantum mechanics, and of fundamental physical theories in general. We shall not enter into an extended discussion of this here, but simply mention some of the main points:

- By identifying the fundamental mathematical structures at work, at a more general and abstract level than that afforded by Hilbert spaces, we can hope to gain new structural insights, and new ideas for how various physical features can be related and combined.
- We get a new perspective on the logical structure of quantum mechanics, radically different to the traditional approaches to quantum logic.
- We get a new perspective on "No-Go" theorems, and new tools for formulating general results applying to whole classes of physical theories.
- Our structural tools yield an *effective calculational formalism* based on a diagrammatic calculus, for which automated software tool-support is currently being developed. This is not only useful for quantum information and computation, it may also yield new ways of probing key foundational issues. Again, this mirrors what has become the common experience in Computer Science. In the age of QIC, *Gedanken*-experiments turn into programs!

We shall take up some of these issues again in the concluding sections.

1.3 Outline of the Approach

We shall use *category theory* as the mathematical setting for our approach. This should be no surprise. Category theory is the language of modern structural mathematics, and the fact that it is not more widely used in current foundational studies is a regrettable consequence of the sociology of knowledge and the encumbrances of tradition. Computer Science, once again, leads the way in the applications of category theory; abstract ideas can be very practical!

We shall assume a modest familiarity with basic notions of category theory, including symmetric monoidal categories. Apart from standard references such as [MacLane, 1998], a number of introductions and tutorials specifically on the use of monoidal categories in physics are now available [Abramsky and Tzevelekos, 2008; Baez and Stay, 2008; Coecke and Paquette, 2008]. More advanced textbooks in the area are [Kock, 2003; Street, 2007].

We shall give an axiomatic presentation of quantum mechanics at the abstract level of *strongly compact closed categories with biproducts* — of which the standard von Neumann presentation in terms of Hilbert spaces is but one example. Remarkably enough, all the essential features of modern quantum protocols such as *quantum teleportation* [Bennett *et al.*, 1993], *logic-gate teleportation* [Gottesman and Chuang, 1999], and *entanglement swapping* [Żukowski *et al.*, 1993] —which exploit quantum mechanical effects in an essential way— find natural counterparts at this abstract level. More specifically:

- The basic structure of a symmetric monoidal category allows *compound systems* to be described in a resource-sensitive fashion (cf. the 'no cloning' [Dieks, 1982; Wootters and Zurek, 1982] and 'no deleting' [Pati and Braunstein, 2000] theorems of quantum mechanics).
- The compact closed structure allows *preparations and measurements of entangled states* to be described, and their key properties to be proved.
- The strong compact closed structure brings in the central notions of adjoint, unitarity and sesquilinear inner product —allowing an involution such as complex conjugation to play a role— and it gives rise to a two-dimensional generalization of Dirac's *bra-ket* calculus [Dirac, 1947], in which the structure of compound systems is fully articulated, rather than merely implicitly encoded by labelling of basis states.
- Biproducts allow *probabilistic branching* due to measurements, *classical communication* and *superpositions* to be captured. Moreover, from the combination of the—apparently purely qualitative—structures of strong compact closure and biproducts there emerge *scalars* and a *Born rule.*

We are then able to use this abstract setting to give precise formulations of quantum teleportation, logic gate teleportation, and entanglement swapping, and to prove correctness of these protocols — for example, proving correctness of teleportation means showing that the final state of Bob's qubit equals the initial state of Alice's qubit.

1.4 Development of the Ideas

A first step in the development of these ideas was taken in [Abramsky and Coecke, 2003], where it was recognized that compact-closed structure could be expressed in terms of bipartite projectors in Hilbert space, thus in principle enabling the structural description of information flows in entangled quantum systems. In [Coecke, 2003] an extensive analysis of a range of quantum protocols was carried out concretely, in terms of Hilbert spaces, with a highly suggestive but informal graphical notation of information-flow paths through networks of projectors. The decisive step in the development of the categorical approach was taken in [Abramsky and Coecke, 2004], with [Abramsky and Coecke, 2005] as a supplement improving the definition of strongly compact closed category. The present article is essentially an extended and revised version of [Abramsky and Coecke, 2004]. There have been numerous subsequent developments in the programme of categorical quantum mechanics since [Abramsky and Coecke, 2004]. We shall provide an overview of the main developments in Section 7, but the underlying programme as set out in [Abramsky and Coecke, 2004] still stands, and we hope that the present article will serve as a useful record of this approach in its original conception.

1.5 Related Work

To set our approach in context, we compare and contrast it with some related approaches.

Quantum Logic

Firstly, we discuss the relationship with quantum logic as traditionally conceived, *i.e.* the study of lattices abstracted from the lattice of closed linear subspaces of Hilbert space [Birkhoff and von Neumann, 1936].

We shall not emphasize the connections to logic in the present article, but in fact our categorical axiomatics can be seen as the algebraic or semantic counterpart to a *logical type theory* for quantum processes. This type theory has a resource-sensitive character, in the same sense as Linear logic [Girard, 1987] — and this is directly motivated by the no-cloning and no-deleting principles of quantum information. The correspondence of our formalism to a logical system, in which a notion of *proof-net* (a graphical representation of multiple-conclusion proofs) gives a diagrammatics for morphisms in the free strongly compact closed category with biproducts, and simplification of diagrams corresponds to *cut-elimination*, is developed in detail in [Abramsky and Duncan, 2006].

This kind of connection with logic belongs to the proof-theory side of logic, and more specifically to the Curry-Howard correspondence, and the three-way connection between logic, computation and categories which has been a staple of categorical logic, and of logical methods in computer science, for the past three decades [Lambek and Scott, 1986; Abramsky and Tzevelekos, 2008].

The key point is that we are concerned with the direct mathematical representation of *quantum processes*. By contrast, traditional quantum logic is concerned with *quantum propositions*, which express *properties* of quantum systems. There are many other differences. For example, compound systems and the tensor product are central to our approach, while quantum logic has struggled to accommodate these key features of quantum mechanics in a mathematically satisfactory fashion. However, connections between our approach and the traditional setting of orthomodular posets and lattices have been made by John Harding [2007; 2008].

Categories in Physics

There are by now several approaches to using category theory in physics. For comparison, we mention the following:

- [Baez and Dolan, 1995; Crane, 2006]. Higher-dimensional categories, TQFT's, categorification, etc.

- [Isham and Butterfield, 1998; Doëring and Isham, 2007]. The topos-theoretic approach.

Comparison with the topos approach The topos approach aims ambitiously at providing a general framework for the formulation of physical theories. It is still in an early stage of development. Nevertheless, we can make some clear comparisons.

Our approach		Topos approach
monoidal	*vs.*	cartesian
linear	*vs.*	intuitionistic
processes	*vs.*	propositions
geometry of proofs	*vs.*	geometric logic

Rather as in our comparison with quantum logic, the topos approach is primarily concerned with quantum propositions, whereas we are concerned directly with the representation of quantum processes. Our underlying logical setting is linear, theirs is cartesian, supporting the intuitionistic logic of toposes. It is an interesting topic for future work to relate, and perhaps even usefully combine, these approaches.

Comparison with the n-categories approach The n-categories approach is mainly motivated by the quest for quantum gravity. In our approach, we emphasize the following key features which are essentially absent from the n-categories work:

- operational aspects

- the interplay of quantum and classical

- compositionality

- open *vs.* closed systems.

These are important for applications to quantum informatics, but also of foundational significance.

There are nevertheless some intriguing similarities and possible connections, notably in the rôle played by *Frobenius algebras*, which we will mention briefly in the context of our approach in Section 7.

1.6 *Outline of the Article*

In Section 2, we shall give a rapid review of quantum mechanics and some quantum protocols such as teleportation. In Sections 3, 4 and 5, we shall present the main ingredients of the formalism: compact and strongly compact categories, and biproducts. In Section 6, we shall show how quantum mechanics can be axiomatized in this setting, and how the formalism can be applied to the complete specification and verification of a number of important quantum protocols. In Section 7 we shall review some of the main developments and advances made within the categorical quantum mechanics programme since [Abramsky and Coecke, 2004], thus giving a picture of the current state of the art.

2 REVIEW OF QUANTUM MECHANICS AND TELEPORTATION

In this paper, we shall only consider *finitary* quantum mechanics, in which all Hilbert spaces are finite-dimensional. This is standard in most current discussions of quantum computation and information [Nielsen and Chuang, 2000], and corresponds physically to considering only observables with finite spectra, such as *spin*. (We refer briefly to the extension of our approach to the infinite-dimensional case in the Conclusion.)

Finitary quantum theory has the following basic ingredients (for more details, consult standard texts such as [Isham, 1995]).

1. The *state space* of the system is represented as a finite-dimensional Hilbert space $\mathcal{H}$, *i.e.* a finite-dimensional complex vector space with a 'sesquilinear' inner-product written $\langle \phi \mid \psi \rangle$, which is conjugate-linear in the first argument and linear in the second. A *state* of a quantum system corresponds to a one-dimensional subspace $\mathcal{A}$ of $\mathcal{H}$, and is standardly represented by a vector $\psi \in \mathcal{A}$ of unit norm.

2. For informatic purposes, the basic type is that of *qubits*, namely 2-dimensional Hilbert space, equipped with a *computational basis* $\{|0\rangle, |1\rangle\}$.

3. *Compound systems* are described by tensor products of the component systems. It is here that the key phenomenon of *entanglement* arises, since the general form of a vector in $\mathcal{H}_1 \otimes \mathcal{H}_2$ is
$$\sum_{i=1}^{n} \alpha_i \cdot \phi_i \otimes \psi_i$$
Such a vector may encode *correlations* between the first and second components of the system, and cannot simply be resolved into a pair of vectors in the component spaces.

The *adjoint* to a linear map $f : \mathcal{H}_1 \to \mathcal{H}_2$ is the linear map $f^\dagger : \mathcal{H}_2 \to \mathcal{H}_1$ such that, for all $\phi \in \mathcal{H}_2$ and $\psi \in \mathcal{H}_1$,
$$\langle \phi \mid f(\psi) \rangle_{\mathcal{H}_2} = \langle f^\dagger(\phi) \mid \psi \rangle_{\mathcal{H}_1} .$$

Unitary transformations are linear isomorphisms $U : \mathcal{H}_1 \to \mathcal{H}_2$ such that

$$U^{-1} = U^\dagger : \mathcal{H}_2 \to \mathcal{H}_1 \,.$$

Note that all such transformations *preserve the inner product* since, for all $\phi, \psi \in \mathcal{H}_1$,

$$\langle U(\phi) \mid U(\psi)\rangle_{\mathcal{H}_2} = \langle (U^\dagger U)(\phi) \mid \psi\rangle_{\mathcal{H}_1} = \langle \phi \mid \psi\rangle_{\mathcal{H}_1} \,.$$

Self-adjoint operators are linear transformations $M : \mathcal{H} \to \mathcal{H}$ such that $M = M^\dagger$.

4. The *basic data transformations* are represented by unitary transformations. Note that all such data transformations are necessarily *reversible*.

5. The *measurements* which can be performed on the system are represented by self-adjoint operators.

The act of measurement itself consists of two parts:

5a. The observer is informed about the measurement outcome, which is a value x_i in the spectrum $\sigma(M)$ of the corresponding self-adjoint operator M. For convenience we assume $\sigma(M)$ to be *non-degenerate* (linearly independent eigenvectors have distinct eigenvalues).

5b. The state of the system undergoes a change, represented by the action of the *projector* P_i arising from the *spectral decomposition*

$$M = x_1 \cdot \mathrm{P}_1 + \ldots + x_n \cdot \mathrm{P}_n$$

In this spectral decomposition the projectors $\mathrm{P}_i : \mathcal{H} \to \mathcal{H}$ are idempotent, self-adjoint, and mutually orthogonal

$$\mathrm{P}_i \circ \mathrm{P}_i = \mathrm{P}_i \qquad \mathrm{P}_i = \mathrm{P}_i^\dagger \qquad \mathrm{P}_i \circ \mathrm{P}_j = 0, \quad i \neq j.$$

This spectral decomposition always exists and is unique by the *spectral theorem* for self-adjoint operators. By our assumption that $\sigma(M)$ was non-degenerate each projector P_i has a one-dimensional subspace of $\mathcal{H}$ as its fixpoint set (which equals its image).

The probability of $x_i \in \sigma(M)$ being the actual outcome is given by the *Born rule* which does not depend on the value of x_i but on P_i and the system state ψ, explicitly

$$\mathsf{Prob}(\mathrm{P}_i, \psi) = \langle \psi \mid \mathrm{P}_i(\psi)\rangle \,.$$

The status of the Born rule within our abstract setting will emerge in Section 8. The derivable notions of *mixed states* and *non-projective measurements* will not play a significant rôle in this paper.

The values $x_1, \ldots, x_n$ are in effect merely labels distinguishing the projectors $\mathrm{P}_1, \ldots, \mathrm{P}_n$ in the above sum. Hence we can abstract over them and think of a measurement as a list of n mutually orthogonal projectors $(\mathrm{P}_1, \ldots, \mathrm{P}_n)$ where n is the dimension of the Hilbert space.

Although real-life experiments in many cases destroy the system (e.g. any measurement of a photon's location destroys it) measurements always have the same shape in the quantum formalism. When distinguishing between 'measurements which preserve the system' and 'measurements which destroy the system' it would make sense to decompose a measurement explicitly in two components:

- *Observation* consists of receiving the information on the outcome of the measurement, to be thought of as specification of the index i of the outcome-projector P_i in the above list. Measurements which destroy the system can be seen as 'observation only'.

- *Preparation* consists of producing the state $\mathrm{P}_i(\psi)$.

In our abstract setting these arise naturally as the two 'building blocks' which are used to construct projectors and measurements.

We now discuss some important quantum protocols which we chose because of the key rôle entanglement plays in them — they involve both initially entangled states, and measurements against a basis of entangled states.

2.1 Quantum teleportation

The quantum teleportation protocol [Bennett *et al.*, 1993] (see also [Coecke, 2003, §2.3&§3.3]) involves three qubits a, b and c and two spatial regions A (for "Alice") and B (for "Bob").

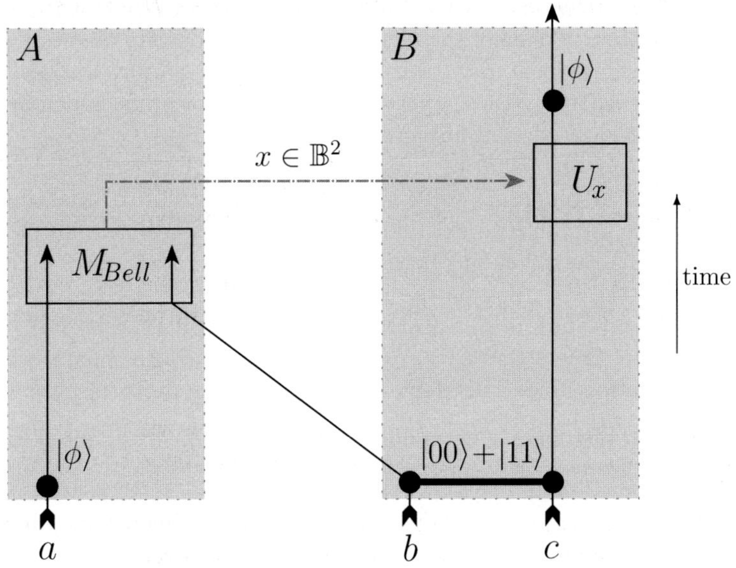

Qubit a is in a state $|\phi\rangle$ and located in A. Qubits b and c form an 'EPR-pair', that is, their joint state is $|00\rangle + |11\rangle$. We assume that these qubits are initially in B e.g. Bob created them. After *spatial relocation* so that a and b are located in A, while c is positioned in B, or in other words, "Bob sends qubit b to *Alice*", we can start the actual teleportation of qubit a. Alice performs a *Bell-base measurement* on a and b at A, that is, a measurement such that each P_i projects on one of the one-dimensional subspaces spanned by a vector in the *Bell basis*:

$$b_1 := \frac{|00\rangle + |11\rangle}{\sqrt{2}} \quad b_2 := \frac{|01\rangle + |10\rangle}{\sqrt{2}} \quad b_3 := \frac{|00\rangle - |11\rangle}{\sqrt{2}} \quad b_4 := \frac{|01\rangle - |10\rangle}{\sqrt{2}}.$$

This measurement can be of the type 'observation only'. Alice observes the outcome of the measurement and "sends these two classical bits ($x \in \mathbb{B}^2$) to Bob". Depending on which classical bits he receives Bob then performs one of the unitary transformations

$$\beta_1 := \begin{pmatrix} 1 & 0 \\ 0 & 1 \end{pmatrix} \quad \beta_2 := \begin{pmatrix} 0 & 1 \\ 1 & 0 \end{pmatrix} \quad \beta_3 := \begin{pmatrix} 1 & 0 \\ 0 & -1 \end{pmatrix} \quad \beta_4 := \begin{pmatrix} 0 & -1 \\ 1 & 0 \end{pmatrix}$$

on c — $\beta_1, \beta_2, \beta_3$ are all self-inverse while $\beta_4^{-1} = -\beta_4$. The final state of c proves to be $|\phi\rangle$ as well. (Because of the measurement, a no longer has this state — the information in the source has been 'destroyed' in transferring it to the target). Note that the state of a constitutes continuous data —an arbitrary pair of complex numbers (α, β) satisfying $|\alpha|^2 + |\beta|^2 = 1$— while the actual physical data transmission only involved two classical bits. We will be able to derive this fact in our abstract setting. Teleportation is simply the most basic of a family of quantum protocols, and already illustrates the basic ideas, in particular the use of *preparations of entangled states* as carriers for information flow, performing *measurements* to propagate information, using *classical information* to control branching behaviour to ensure the required behaviour despite quantum indeterminacy, and performing local data transformations using *unitary operations*. (Local here means that we apply these operations only at A or at B, which are assumed to be spatially separated, and not simultaneously at both).

Since in quantum teleportation a continuous variable has been transmitted while the actual *classical communication* involved only two bits, besides this *classical information flow* there has to exist some kind of *quantum information flow*. The nature of this quantum flow has been analyzed by one of the authors in [Coecke, 2003; Coecke, 2004], building on the joint work in [Abramsky and Coecke, 2003]. We recover those results in our abstract setting (see Subsection 3.5), which also reveals additional 'fine structure'. To identify it we have to separate it from the classical information flow. Therefore we decompose the protocol into:

1. a *tree* with the operations as nodes, and with *branching* caused by the indeterminism of measurements;

2. a *network* of the operations in terms of the order they are applied and the subsystem to which they apply.

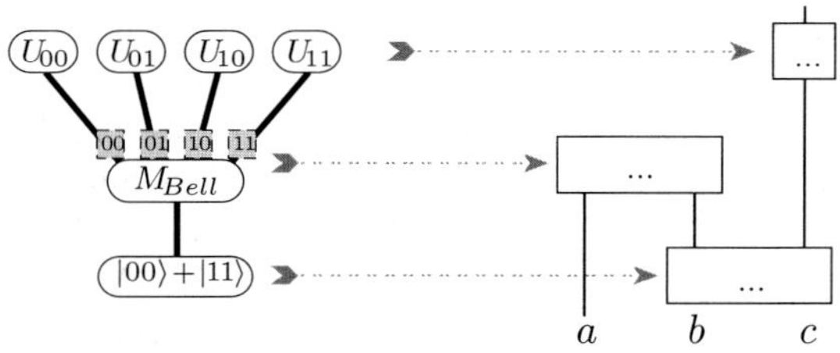

The nodes in the tree are connected to the boxes in the network by their temporal coincidence. Classical communication is encoded in the tree as the dependency of operations on the branch they are in. For each path from the root of the tree to a leaf, by 'filling in the operations on the included nodes in the corresponding boxes of the network', we obtain an *entanglement network*, that is, a network

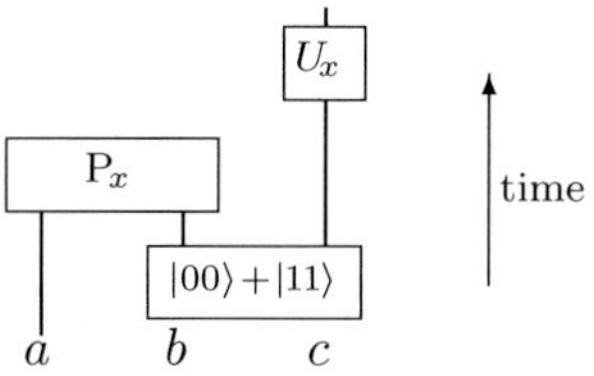

for each of the four values x takes. A component P_x of an observation will be referred to as an *observational branch*. It will be these networks, from which we have removed the classical information flow, that we will study in Subsection 3.5. (There is a clear analogy with the idea of unfolding a Petri net into its set of 'processes' [Petri, 1977]). The classical information flow will be reintroduced in Section 5.

2.2 Logic gate teleportation

Logic gate teleportation [Gottesman and Chuang, 1999] (see also [Coecke, 2003]§3.3) generalizes the above protocol in that b and c are initially not necessarily an EPR-pair but may be in some other (not arbitrary) entangled state $|\Psi\rangle$. Due to this modification the final state of c is not $|\phi\rangle$ but $|f_\Psi(\phi)\rangle$ where f_Ψ is a linear map which depends on Ψ. As shown in [Gottesman and Chuang, 1999], when this construction is applied to the situation where a, b and c are each a pair of qubits rather than a single qubit, it provides a universal quantum computational primitive which is moreover fault-tolerant [Shor, 1996] and enables the construction of a quantum computer based on single qubit unitary operations, Bell-base measurements and only one kind of prepared state (so-called GHZ states). The connection between Ψ, f_Ψ and the unitary corrections $U_{\Psi,x}$ will emerge straightforwardly in our abstract setting.

2.3 Entanglement swapping

Entanglement swapping [Żukowski *et al.*, 1993] (see also [Coecke, 2003]§6.2) is another modification of the teleportation protocol where a is not in a state $|\phi\rangle$ but is a qubit in an EPR-pair together with an ancillary qubit d. The result is that after the protocol c forms an EPR-pair with d. If the measurement on a and b is non-destructive, we can also perform a unitary operation on a, resulting in a and b also constituting an EPR-pair. Hence we have 'swapped' entanglement:

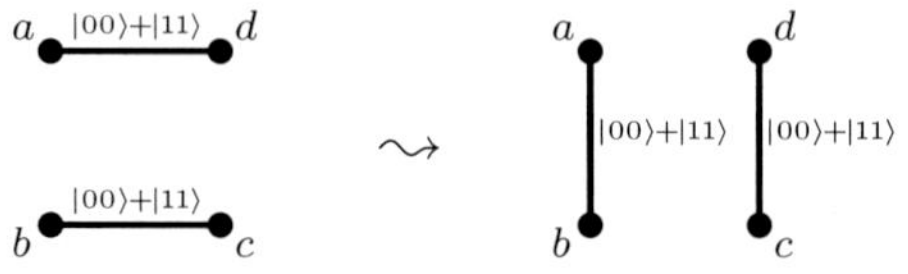

In this case the entanglement networks have the shape:

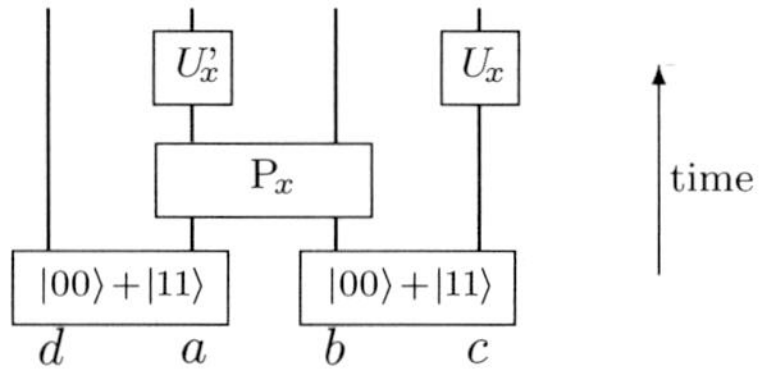

Why this protocol works will again emerge straightforwardly from our abstract setting, as will generalizations of this protocol which have a much more sophisticated compositional content (see Subsection 3.5).

3 COMPACT CLOSED CATEGORIES AND THE LOGIC OF ENTANGLEMENT

3.1 Monoidal Categories

Recall that a *symmetric monoidal category* consists of a category $\mathbf{C}$, a bifunctorial *tensor*

$$- \otimes - : \mathbf{C} \times \mathbf{C} \to \mathbf{C},$$

a *unit* object I, and natural isomorphisms

$$\lambda_A : A \simeq \mathrm{I} \otimes A \qquad\qquad \rho_A : A \simeq A \otimes \mathrm{I}$$

$$\alpha_{A,B,C} : A \otimes (B \otimes C) \simeq (A \otimes B) \otimes C$$

$$\sigma_{A,B} : A \otimes B \simeq B \otimes A$$

which satisfy certain coherence conditions [MacLane, 1998].

Examples The following two examples are of particular importance and will recur through this section.

1. The category $\mathbf{FdVec}_{\mathbb{K}}$, of finite-dimensional vector spaces over a field $\mathbb{K}$ and linear maps. The tensor product is the usual construction on vector spaces. The unit of the tensor is $\mathbb{K}$, considered as a one-dimensional vector space over itself.

2. The category **Rel** of sets and relations, with cartesian product as the 'tensor', and a one-element set as the unit. Note that cartesian product is *not* the categorical product in **Rel**.

The Logic of Tensor Product Tensor can express *independent* or *concurrent* actions (mathematically: bifunctoriality):

$$\begin{array}{ccc} A_1 \otimes A_2 & \xrightarrow{f_1 \otimes 1} & B_1 \otimes A_2 \\ {\scriptstyle 1 \otimes f_2}\downarrow & & \downarrow{\scriptstyle 1 \otimes f_2} \\ A_1 \otimes B_2 & \xrightarrow[f_1 \otimes 1]{} & B_1 \otimes B_2 \end{array}$$

But tensor is *not* a categorical product, in the sense that *we cannot reconstruct an 'element' of the tensor from its components.*

This turns out to comprise the *absence* of *diagonals* and *projections*:

$$\begin{array}{ccc} & A \xrightarrow{\Delta} A \otimes A & A_1 \otimes A_2 \xrightarrow{\pi_i} A_i \\ \text{Cf.} & A \vdash A \wedge A & A_1 \wedge A_2 \vdash A_i \end{array}$$

Hence monoidal categories provide a setting for *resource-sensitive* logics such as Linear Logic [Girard, 1987]. No-Cloning and No-Deleting are built in! *Any* symmetric monoidal category can be viewed as *a setting for describing processes in a resource sensitive way, closed under sequential and parallel composition*

3.2 The 'miracle' of scalars

A key step in the development of the categorical axiomatics for Quantum Mechanics was the recognition that the notion of *scalar* is meaningful in great generality — in fact, in any monoidal (not necessarily symmetric) category.

Let $(\mathbf{C}, \otimes, \mathrm{I}, \lambda, \alpha, \sigma)$ be a monoidal category . We define a *scalar* in $\mathbf{C}$ to be a morphism $s : \mathrm{I} \to \mathrm{I}$, *i.e.* an endomorphism of the tensor unit.

EXAMPLE 1. In $\mathbf{FdVec}_{\mathbb{K}}$, linear maps $\mathbb{K} \to \mathbb{K}$ are uniquely determined by the image of 1, and hence correspond biuniquely to elements of $\mathbb{K}$; composition corresponds to multiplication of scalars. In **Rel**, there are just two scalars, corresponding to the Boolean values 0, 1.

The (multiplicative) monoid of scalars is then just the endomorphism monoid $\mathbf{C}(\mathrm{I},\mathrm{I})$. The first key point is the elementary but beautiful observation by Kelly and Laplaza [Kelly and Laplaza, 1980] that this monoid is always commutative.

LEMMA 2. $\mathbf{C}(\mathrm{I},\mathrm{I})$ *is a commutative monoid*

Proof:

$$
\begin{array}{ccccccc}
\mathrm{I} & \xrightarrow{\rho_{\mathrm{I}}} & \mathrm{I}\otimes\mathrm{I} & = & \mathrm{I}\otimes\mathrm{I} & \xrightarrow{\lambda_{\mathrm{I}}^{-1}} & \mathrm{I} \\
{\scriptstyle s}\uparrow & & \uparrow{\scriptstyle s\otimes 1} & & \downarrow{\scriptstyle 1\otimes t} & & \downarrow{\scriptstyle t} \\
\mathrm{I} & \xrightarrow{\rho_{\mathrm{I}}} & \mathrm{I}\otimes\mathrm{I} & \xrightarrow{s\otimes t} & \mathrm{I}\otimes\mathrm{I} & \xrightarrow{\lambda_{\mathrm{I}}^{-1}} & \mathrm{I} \\
{\scriptstyle t}\downarrow & & \downarrow{\scriptstyle 1\otimes t} & & \uparrow{\scriptstyle s\otimes 1} & & \uparrow{\scriptstyle s} \\
\mathrm{I} & \xrightarrow[\lambda_{\mathrm{I}}]{} & \mathrm{I}\otimes\mathrm{I} & = & \mathrm{I}\otimes\mathrm{I} & \xrightarrow[\rho_{\mathrm{I}}^{-1}]{} & \mathrm{I}
\end{array}
$$

using the coherence equation $\lambda_{\mathrm{I}} = \rho_{\mathrm{I}}$. ■

The second point is that a good notion of *scalar multiplication* exists at this level of generality. That is, each scalar $s : \mathrm{I} \to \mathrm{I}$ induces a natural transformation

$$s_A : A \xrightarrow{\simeq} \mathrm{I}\otimes A \xrightarrow{s\otimes 1_A} \mathrm{I}\otimes A \xrightarrow{\simeq} A\,.$$

with the naturality square

$$
\begin{array}{ccc}
A & \xrightarrow{s_A} & A \\
{\scriptstyle f}\downarrow & & \downarrow{\scriptstyle f} \\
B & \xrightarrow[s_B]{} & B
\end{array}
$$

We write $s \bullet f$ for $f \circ s_A = s_B \circ f$. Note that

$$
\begin{array}{lcl}
1 \bullet f & = & f \\
s \bullet (t \bullet f) & = & (s \circ t) \bullet f \\
(s \bullet g) \circ (t \bullet f) & = & (s \circ t) \bullet (g \circ f) \\
(s \bullet f) \otimes (t \bullet g) & = & (s \circ t) \bullet (f \otimes g)
\end{array}
$$

which exactly generalizes the multiplicative part of the usual properties of scalar multiplication. Thus scalars act globally on the whole category.

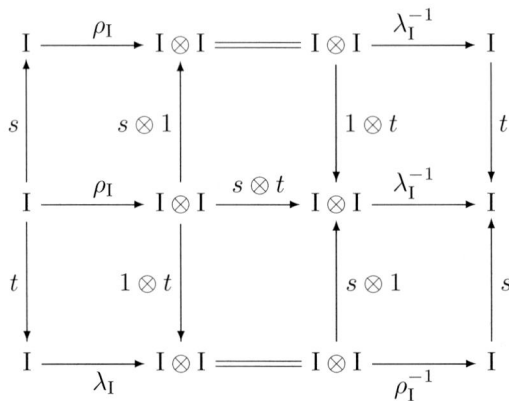

3.3 Compact Closure

A category **C** is $*$-*autonomous* [Barr, 1979] if it is symmetric monoidal, and comes equipped with a full and faithful functor

$$(\ )^* : \mathbf{C}^{op} \to \mathbf{C}$$

such that a bijection

$$\mathbf{C}(A \otimes B, C^*) \simeq \mathbf{C}(A, (B \otimes C)^*)$$

exists which is natural in all variables. Hence a $*$-autonomous category is closed, with

$$A \multimap B := (A \otimes B^*)^* .$$

These $*$-autonomous categories provide a categorical semantics for the multiplicative fragment of linear logic [Seely, 1989].

A *compact closed category* [Kelly, 1972] is a $*$-autonomous category with a self-dual tensor, i.e. with natural isomorphisms

$$u_{A,B} : (A \otimes B)^* \simeq A^* \otimes B^* \qquad u_{\mathrm{I}} : \mathrm{I}^* \simeq \mathrm{I} .$$

It follows that

$$A \multimap B \simeq A^* \otimes B .$$

A very different definition arises when one considers a symmetric monoidal category as a one-object bicategory. In this context, compact closure simply means that every object A, qua 1-cell of the bicategory, has a specified adjoint [Kelly and Laplaza, 1980].

DEFINITION 3 Kelly-Laplaza. *A* compact closed category *is a symmetric monoidal category in which to each object A a* dual object A^*, *a* unit

$$\eta_A : \mathrm{I} \to A^* \otimes A$$

and a counit

$$\epsilon_A : A \otimes A^* \to \mathrm{I}$$

are assigned, in such a way that the diagram

$$\begin{array}{ccccc}
A & \xrightarrow{\rho_A} & A \otimes \mathrm{I} & \xrightarrow{1_A \otimes \eta_A} & A \otimes (A^* \otimes A) \\
{\scriptstyle 1_A}\downarrow & & & & \downarrow{\scriptstyle \alpha_{A,A^*,A}} \\
A & \xleftarrow[\lambda_A^{-1}]{} & \mathrm{I} \otimes A & \xleftarrow[\epsilon_A \otimes 1_A]{} & (A \otimes A^*) \otimes A
\end{array}$$

and the dual one for A^* both commute.

Examples The symmetric monoidal categories $(\mathbf{Rel}, \times)$ of sets, relations and cartesian product and $(\mathbf{FdVec}_{\mathbb{K}}, \otimes)$ of finite-dimensional vector spaces over a field $\mathbb{K}$, linear maps and tensor product are both compact closed. In $(\mathbf{Rel}, \times)$, we simply set $X^* = X$. Taking a one-point set $\{*\}$ as the unit for $\times$, and writing $R^{\cup}$ for the converse of a relation R:

$$\eta_X = \epsilon_X^{\cup} = \{(*, (x, x)) \mid x \in X\}\,.$$

For $(\mathbf{FdVec}_{\mathbb{K}}, \otimes)$, we take V^* to be the dual space of linear functionals on V. The unit and counit in $(\mathbf{FdVec}_{\mathbb{K}}, \otimes)$ are

$$\eta_V : \mathbb{K} \to V^* \otimes V :: 1 \mapsto \sum_{i=1}^{i=n} \bar{e}_i \otimes e_i \qquad \text{and} \qquad \epsilon_V : V \otimes V^* \to \mathbb{K} :: e_i \otimes \bar{e}_j \mapsto \bar{e}_j(e_i)$$

where n is the dimension of V, $\{e_i\}_{i=1}^{i=n}$ is a basis of V and $\bar{e}_i$ is the linear functional in V^* determined by $\bar{e}_j(e_i) = \delta_{ij}$.

DEFINITION 4. *The* name $\ulcorner f \urcorner$ *and the* coname $\llcorner f \lrcorner$ *of a morphism* $f : A \to B$ *in a compact closed category are*

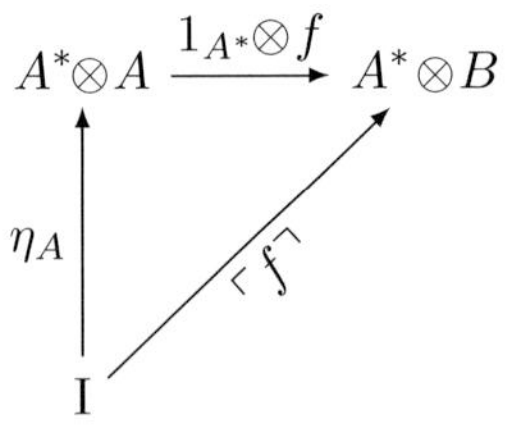

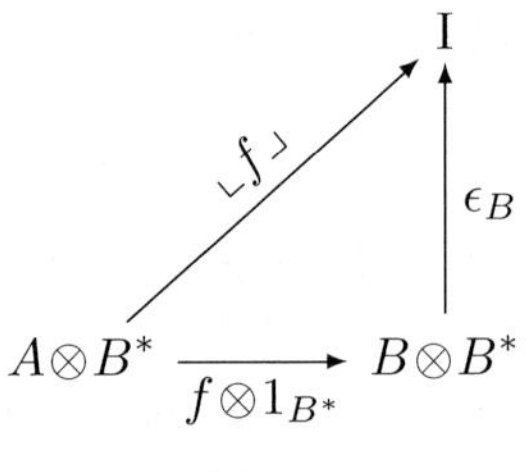

For $R \in \mathbf{Rel}(X, Y)$ we have

$$\ulcorner R \urcorner = \{(*, (x, y)) \mid xRy, x \in X, y \in Y\} \quad \text{and} \quad \llcorner R \lrcorner = \{((x, y), *) \mid xRy, x \in X, y \in Y\}$$

and for $f \in \mathbf{FdVec}_{\mathbb{K}}(V, W)$ with (m_{ij}) the matrix of f in bases $\{e_i^V\}_{i=1}^{i=n}$ and $\{e_j^W\}_{j=1}^{j=m}$ of V and W respectively

$$\ulcorner f \urcorner : \mathbb{K} \to V^* \otimes W :: 1 \mapsto \sum_{i,j=1}^{i,j=n,m} m_{ij} \cdot \bar{e}_i^V \otimes e_j^W$$

and

$$\llcorner f \lrcorner : V \otimes W^* \to \mathbb{K} :: e_i^V \otimes \bar{e}_j^W \mapsto m_{ij}.$$

Given $f : A \to B$ in any compact closed category $\mathbf{C}$ we can define $f^* : B^* \to A^*$

as

$$\begin{array}{ccccc}
B^* & \xrightarrow{\lambda_{B^*}} & \mathrm{I} \otimes B^* & \xrightarrow{\eta_A \otimes 1_{B^*}} & A^* \otimes A \otimes B^* \\
{\scriptstyle f^*}\downarrow & & & & \downarrow{\scriptstyle 1_{A^*} \otimes f \otimes 1_{B^*}} \\
A^* & \xleftarrow[\rho_{A^*}^{-1}]{} & A^* \otimes \mathrm{I} & \xleftarrow[1_{A^*} \otimes \epsilon_B]{} & A^* \otimes B \otimes B^*
\end{array}$$

This operation $(\ )^*$ is functorial and makes Definition 3 coincide with the one given at the beginning of this section. It then follows by

$$\mathbf{C}(A \otimes B^*, \mathrm{I}) \simeq \mathbf{C}(A, B) \simeq \mathbf{C}(I, A^* \otimes B)$$

that every morphism of type $\mathrm{I} \to A^* \otimes B$ is the name of some morphism of type $A \to B$ and every morphism of type $A \otimes B^* \to \mathrm{I}$ is the coname of some morphism of type $A \to B$. In the case of the unit and the counit we have

$$\eta_A = \ulcorner 1_A \urcorner \qquad \text{and} \qquad \epsilon_A = \llcorner 1_A \lrcorner \,.$$

For $R \in \mathbf{Rel}(X, Y)$ the dual is the converse, $R^* = R^{\cup} \in \mathbf{Rel}(Y, X)$, and for $f \in \mathbf{FdVec}_{\mathbb{K}}(V, W)$, the dual is

$$f^* : W^* \to V^* :: \phi \mapsto \phi \circ f \,.$$

The following holds by general properties of adjoints and symmetry of the tensor [Kelly and Laplaza, 1980]§6.

PROPOSITION 5. *In a compact closed category* $\mathbf{C}$ *there is a natural isomorphism* $d_A : A^{**} \simeq A$ *and the diagrams*

$$\begin{array}{ccc}
A^* \otimes A & \xrightarrow{\sigma_{A^*,A}} & A \otimes A^* \\
{\scriptstyle 1_{A^*} \otimes d_A^{-1}}\downarrow & & \downarrow{\scriptstyle \epsilon_A} \\
A^* \otimes A^{**} & \xrightarrow[\epsilon_{A^*}]{} & \mathrm{I}
\end{array}
\qquad\qquad
\begin{array}{ccc}
\mathrm{I} & \xrightarrow{\eta_{A^*}} & A^{**} \otimes A^* \\
{\scriptstyle \eta_A}\downarrow & & \downarrow{\scriptstyle d_A \otimes 1_{A^*}} \\
A^* \otimes A & \xrightarrow[\sigma_{A^*,A}]{} & A \otimes A^*
\end{array}$$

commute for all objects A *of* $\mathbf{C}$.

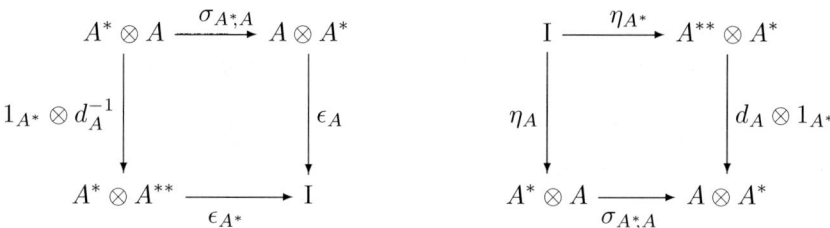

Graphical representation. Complex algebraic expressions for morphisms in symmetric monoidal categories can rapidly become hard to read. Graphical representations exploit two-dimensionality, with the vertical dimension corresponding to composition and the horizontal to the monoidal tensor, and provide more intuitive presentations of morphisms. We depict objects by wires, morphisms by boxes with input and output wires, composition by connecting outputs to inputs, and the monoidal tensor by locating boxes side-by-side. We distinguish between an object and its dual in terms of directions of the wires. In particular, $g \circ f$, $f \otimes g$, $\ulcorner f \urcorner$ and $\llcorner f \lrcorner$ will respectively be depicted by

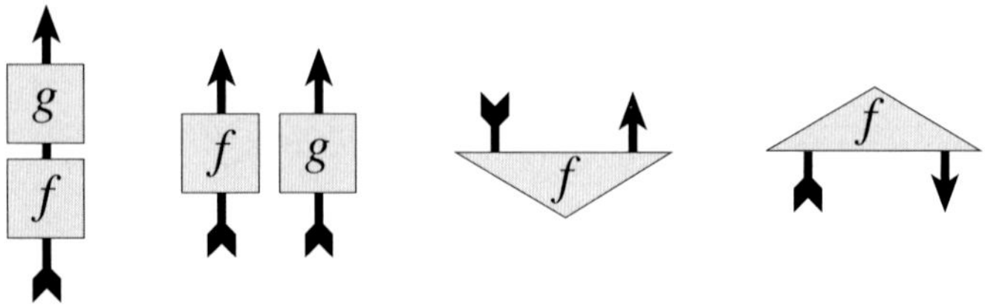

Implicit in the use of this graphical notation is that we assume we are working in a *strict* monoidal category, in which the unit and associativity isomorphisms are identities. We can always do this because of the coherence theorem for monoidal categories [MacLane, 1998]. Similarly, strictness is assumed for the duality in compact closed categories:

$$A^{**} = A, \qquad (A \otimes B)^* = A^* \otimes B^*, \qquad \mathrm{I}^* = \mathrm{I}\,.$$

Pointers to references on diagrammatic representations and corresponding calculi are in Section 7.8.

3.4 Key lemmas

The following Lemmas constitute the core of our interpretation of entanglement in compact closed categories. It was however observed by Radha Jagadeesan [2004] that they can be shown in arbitrary $*$-autonomous categories using some of the results in [Cockett and Seely, 1997].

LEMMA 6 absorption. *For* $A \xrightarrow{f} B \xrightarrow{g} C$ *we have that*

$$(1_{A^*} \otimes g) \circ \ulcorner f \urcorner = \ulcorner g \circ f \urcorner.$$

Proof: Straightforward by Definition 4. ■

In a picture,

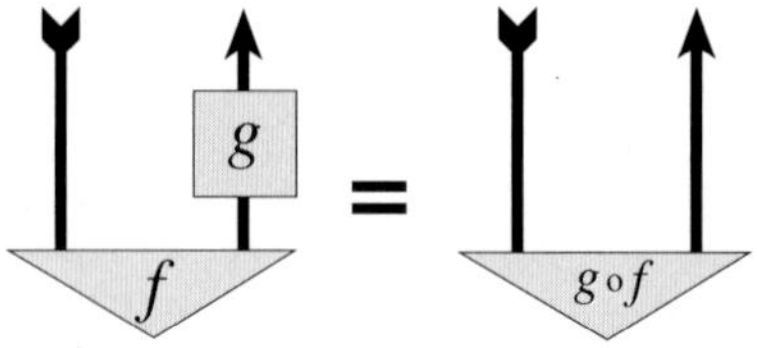

LEMMA 7 Compositionality. *For* $A \xrightarrow{f} B \xrightarrow{g} C$ *we have that*

$$\lambda_C^{-1} \circ (\llcorner f \lrcorner \otimes 1_C) \circ (1_A \otimes \ulcorner g \urcorner) \circ \rho_A = g \circ f\,.$$

Proof:

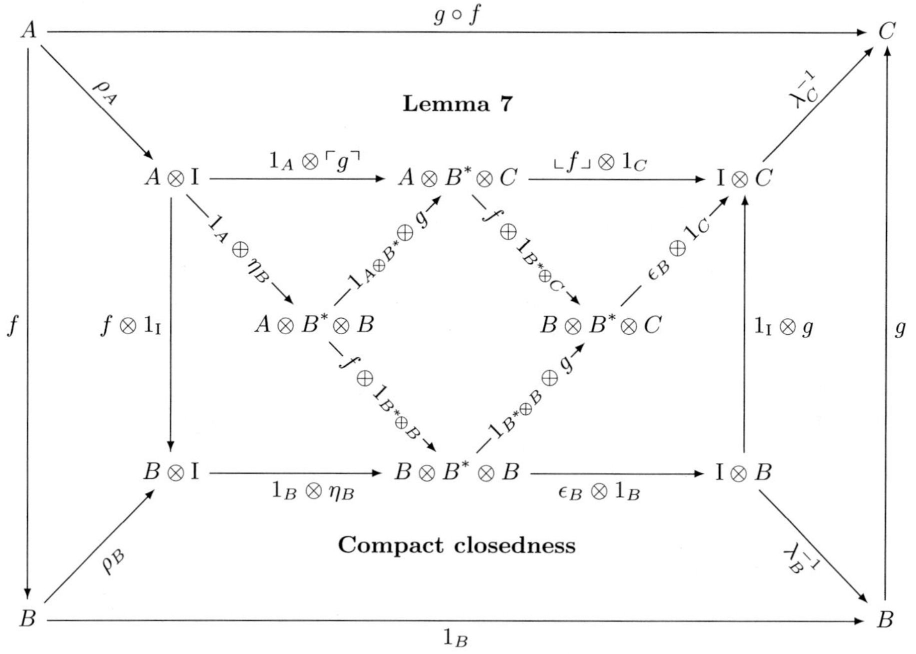

The top trapezoid is the statement of the Lemma. The diagram uses bifunctoriality and naturality of ρ and λ. ■

In a picture,

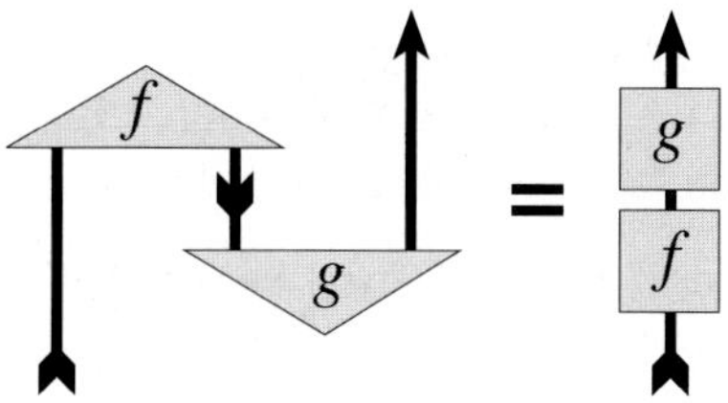

LEMMA 8 Compositional CUT. *For* $A \xrightarrow{f} B \xrightarrow{g} C \xrightarrow{h} D$ *we have that*

$$(\rho_A^{-1} \otimes 1_{D^*}) \circ (1_{A^*} \otimes \llcorner g \lrcorner \otimes 1_D) \circ (\ulcorner f \urcorner \otimes \ulcorner h \urcorner) \circ \rho_I = \ulcorner h \circ g \circ f \urcorner.$$

Proof:

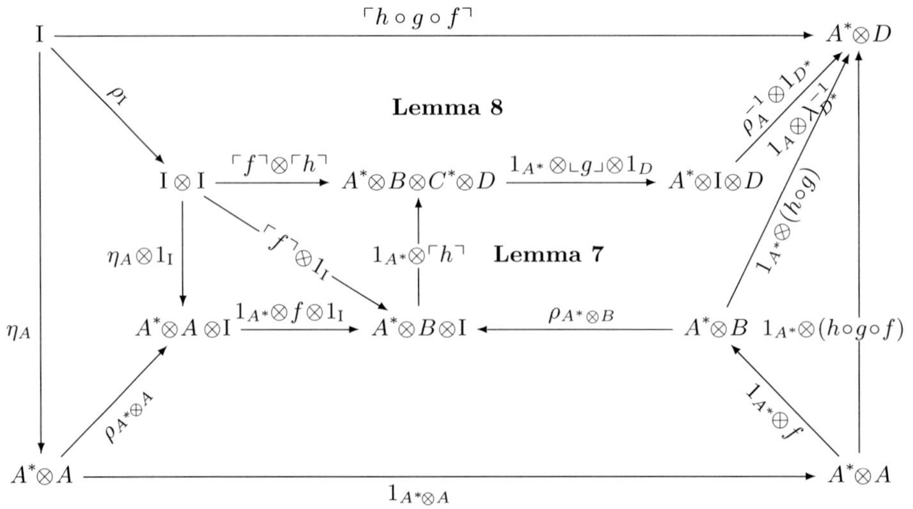

The top trapezoid is the statement of the Lemma. The diagram uses Lemma 7 and naturality of ρ and λ. ■

In a picture,

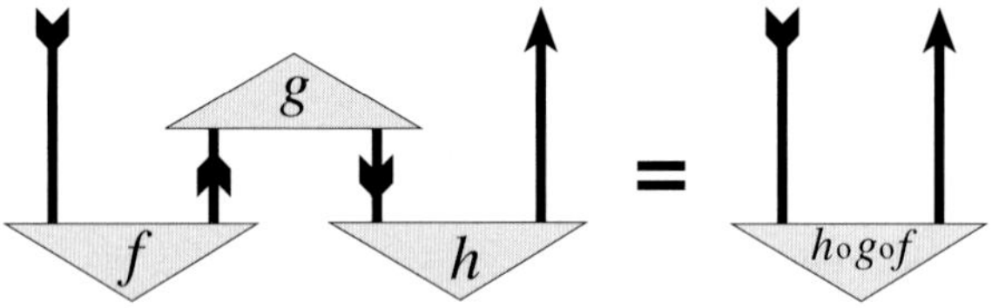

Discussion. On the right hand side of Lemma 7 we have $g \circ f$, that is, we first apply f and then g, while on the left hand side we first apply the coname of g, and then the coname of f. In Lemma 8 there is a similar, seemingly 'acausal' inversion of the order of application, as g gets inserted between h and f.

For completeness we add the following 'backward' absorption lemma, which again involves a reversal of the composition order.

LEMMA 9 backward absorption. *For* $C \xrightarrow{g} A \xrightarrow{f} B$ *we have that*

$$(g^* \otimes 1_{A^*}) \circ \ulcorner f \urcorner = \ulcorner f \circ g \urcorner.$$

Proof: This follows by unfolding the definition of g^*, then using naturality of λ_{A^*}, $\lambda_{\mathrm{I}} = \rho_{\mathrm{I}}$, and finally Lemma 8. ■

In a picture,

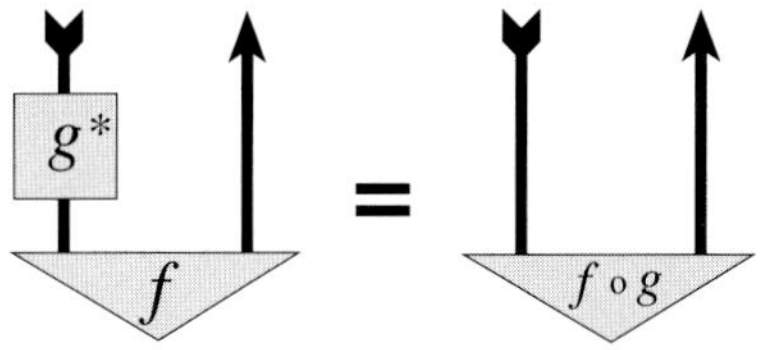

The obvious analogues of Lemma 6 and 9 for conames also hold.

3.5 Quantum information flow in entanglement networks

We claim that Lemmas 6, 7 and 8 capture the quantum information flow in the (logic-gate) teleportation and entanglement swapping protocols. We shall provide a full interpretation of finitary quantum mechanics in Section 6 but for now the following rule suffices:

- We interpret *preparation* of an entangled state as a *name* and an *observational branch* as a *coname.*

For an entanglement network of teleportation-type shape, applying Lemma 7 yields

$$U \circ \left(\lambda_C^{-1} \circ (\llcorner f \lrcorner \otimes 1)\right) \circ ((1 \otimes \ulcorner g \urcorner) \circ \rho_A) = U \circ g \circ f\,.$$

In a picture,

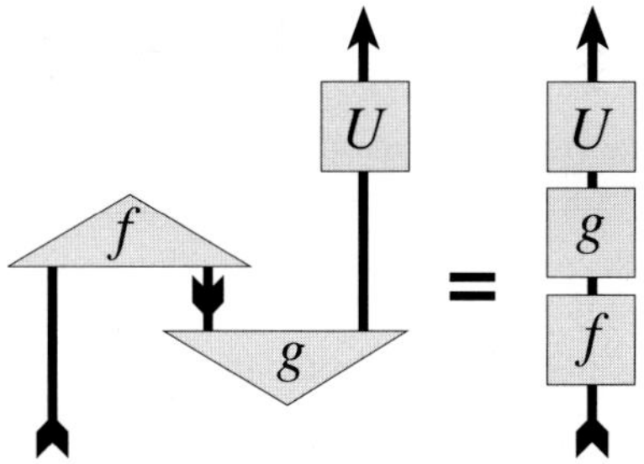

We make the information flow more explicit in the following version of the same picture:

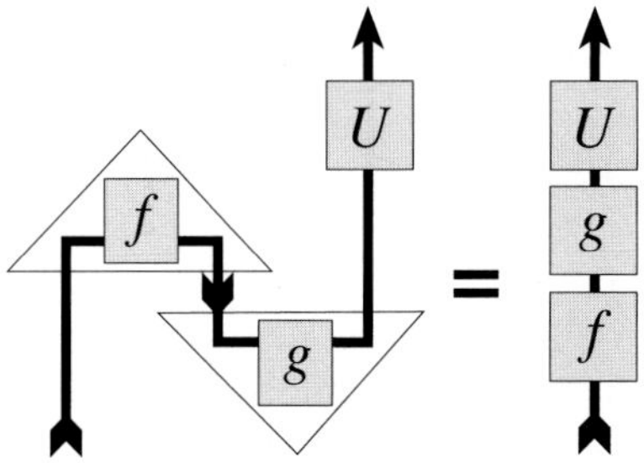

Note that the quantum information seems to flow 'following the line' while being acted on by the functions whose name or coname labels the boxes (and this fact remains valid for much more complex networks [Coecke, 2003]).

Teleporting the input requires $U \circ g \circ f = 1_A$ — we assume all functions have type $A \to A$. Logic-gate teleportation of $h : A \to B$ requires $U \circ g \circ f = h$.

We calculate this explicitly in **Rel**. For initial state $x \in X$ after preparing $\ulcorner S \urcorner \subseteq \{*\} \times (Y \times Z)$ we obtain $\{x\} \times \{(y,z) \mid * \ulcorner S \urcorner (y,z)\}$ as the state of the system. For observational branch $\llcorner R \lrcorner \subseteq (X \times Y) \times \{*\}$ we have that $z \in Z$ is the output iff $\llcorner R \lrcorner \times 1_Z$ receives an input $(x,y,z) \in X \times Y \times Z$ such that $(x,y) \llcorner R \lrcorner *$. Since

$$* \ulcorner S \urcorner (y,z) \Leftrightarrow ySz \qquad \text{and} \qquad (x,y) \llcorner R \lrcorner * \Leftrightarrow xRy$$

we indeed obtain $x(R;S)z$. This illustrates that the compositionality is due to a mechanism of imposing constraints between the components of the tuples.

In $\mathbf{FdVec}_{\mathbb{C}}$ the vector space of all linear maps of type $V \to W$ is $V \multimap W$ and hence by $V^* \otimes W \simeq V \multimap W$ we have a bijective correspondence between linear maps $f : V \to W$ and vectors $\Psi \in V^* \otimes W$ (see also [Coecke, 2003; Coecke, 2004]):

$$\Psi_f = \frac{1}{\sqrt{2}} \cdot \ulcorner f \urcorner (1) \qquad \text{and} \qquad \llcorner f \lrcorner = \langle \sqrt{2} \cdot \Psi_f | - \rangle \,.$$

In particular we have for the Bell base:

$$b_i = \frac{1}{\sqrt{2}} \cdot \ulcorner \beta_i \urcorner (1) \qquad \text{and} \qquad \llcorner \beta_i \lrcorner = \langle \sqrt{2} \cdot b_i | - \rangle \,.$$

Setting $g := \beta_1 = 1_V$, $f := \beta_i$ and $U := \beta_i^{-1}$ indeed yields $\beta_i^{-1} \circ 1_A \circ \beta_i = 1_A$, which expresses the correctness of the teleportation protocol along each branch.

Setting $g := h$ and $f := \beta_i$ for logic-gate teleportation requires U_i to satisfy $U_i \circ h \circ \beta_i = h$ that is $h \circ \beta_i = U^\dagger \circ h$ (since U has to be unitary). Hence we have derived the laws of logic-gate teleportation — one should compare this calculation to the size of the calculation in Hilbert space.

Deriving the swapping protocol using Lemma 6 and Lemma 8 proceeds analogously to the derivation of the teleportation protocol.

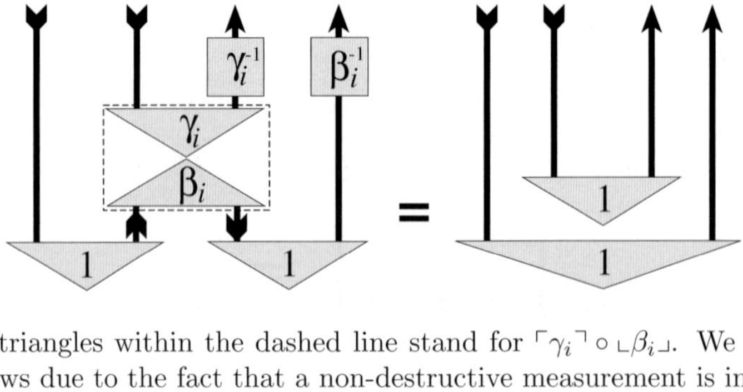

– the two triangles within the dashed line stand for $\ulcorner \gamma_i \urcorner \circ \llcorner \beta_i \lrcorner$. We obtain two distinct flows due to the fact that a non-destructive measurement is involved.

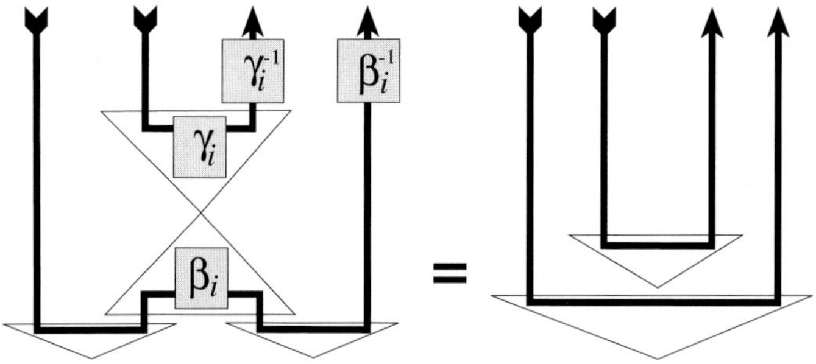

How γ_i has to relate to β_i such that they make up a true projector will be discussed in Section 6.

For a general entanglement network of the swapping-type (without unitary correction and observational branching) by Lemma 8 we obtain the following 'reduction':

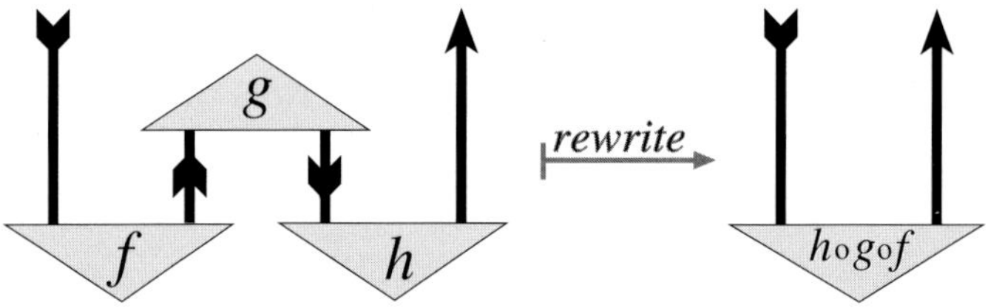

This picture, and the underlying algebraic property expressed by Lemma 3.5, is in fact directly related to *Cut-Elimination* in the logic corresponding to compact-closed categories. If one turns the above picture upside-down, and interprets names as Axiom-links and conames as Cut-links, then one has a normalization rule for proof-nets. This perspective is developed in [Abramsky and Duncan, 2006].

4 STRONGLY COMPACT CLOSED CATEGORIES AND 2-DIMENSIONAL DIRAC NOTATION

The key example In Section 3 we analysed the compact closed structure of $\mathbf{FdVec}_{\mathbb{K}}$, where we took the dual of a vector space V to be the vector space of its linear functionals V^*. In the case that V is equipped with an inner product we can refine this analysis. We discuss this for the key example of *finite-dimensional Hilbert spaces*, *i.e.* finite-dimensional complex vector spaces with a *sesquilinear* inner product: the inner product is linear in the second argument, and

$$\langle \phi \mid \psi \rangle = \overline{\langle \psi \mid \phi \rangle}$$

which implies that it is *conjugate-linear* rather than linear in its first argument.

We organize these spaces into a category **FdHilb**, where the morphisms are linear maps. Note that we do *not* require morphisms to preserve the inner product.

This category provides the basic setting for finite-dimensional quantum mechanics and for quantum information and computation.[1]

In the setting of Hilbert spaces, we can replace the dual space by a more elementary construction. In a Hilbert space, each linear functional $\bar{\psi} : \mathcal{H} \to \mathbb{C}$ is witnessed by some $\psi \in \mathcal{H}$ such that $\bar{\psi} = \langle \psi \mid \cdot \rangle$. This however does *not* induce an isomorphism between $\mathcal{H}$ and $\mathcal{H}^*$, due to the conjugate-linearity of the inner product in its first argument. This leads us to introduce the *conjugate space* $\bar{\mathcal{H}}$ of a Hilbert space $\mathcal{H}$: this has the same additive group of vectors as $\mathcal{H}$, while the scalar multiplication and inner product are "twisted" by complex conjugation:

$$\alpha \bullet_{\bar{\mathcal{H}}} \phi := \bar{\alpha} \bullet_{\mathcal{H}} \phi \qquad \langle \phi \mid \psi \rangle_{\bar{\mathcal{H}}} := \langle \psi \mid \phi \rangle_{\mathcal{H}}$$

We can define $\mathcal{H}^* = \bar{\mathcal{H}}$, since $\mathcal{H}$ and $\bar{\mathcal{H}}$ have the same orthornormal bases, and we can define the counit by

$$\epsilon_{\mathcal{H}} : \mathcal{H} \otimes \bar{\mathcal{H}} \to \mathbb{C} :: \phi \otimes \psi \mapsto \langle \psi \mid \phi \rangle_{\mathcal{H}}$$

which is indeed (bi)linear rather than sesquilinear! Note that

$$\bar{\bar{\mathcal{H}}} = \mathcal{H}, \qquad \overline{A \otimes B} = \bar{A} \otimes \bar{B}\,.$$

4.1 Why compact closure does not suffice

Note that the categories **FdHilb** and $\mathbf{FdVec}_{\mathbb{C}}$ are equivalent! This immediately suggests that some additional categorical structure must be identified to reflect the rôle of the inner product.

A further reason for seeking additional categorical structure is to reflect the centrally important notion of *adjoint* in Hilbert spaces:

$$\frac{A \xrightarrow{f} B}{A \xleftarrow{f^\dagger} B} \qquad \langle f\phi \mid \psi \rangle_B = \langle \phi \mid f^\dagger \psi \rangle_A$$

This is *not* the same as the dual — the types are different! In "degenerate" CCC's in which $A^* = A$, e.g. **Rel** or real inner-product spaces, we have $f^* = f^\dagger$. In Hilbert spaces, the isomorphism $A \simeq A^*$ is not linear, but *conjugate linear*:

$$\langle \lambda \bullet \phi \mid - \rangle \;\; = \;\; \bar{\lambda} \bullet \langle \phi \mid - \rangle$$

and hence does not live in the category **Hilb** at all!

[1]Much of quantum information is concerned with *completely positive maps* acting on *density matrices*. An account of this extended setting in terms of a general categorical construction within our framework is discussed in Section 7.

4.2 Solution: Strong Compact Closure

A key observation is this: the assignment $\mathcal{H} \mapsto \mathcal{H}^*$ on objects has a *covariant* functorial extension $f \mapsto f_*$, which is essentially identity on morphisms; and then we can *define*

$$f^\dagger = (f^*)_* = (f_*)^* .$$

Concretely, in terms of matrices $()^*$ is *transpose*, $()_*$ is *complex conjugation*, and the adjoint is the *conjugate transpose.* Each of these three operations can be expressed in terms of the other two. For example, $f^* = (f^\dagger)_*$. All three of these operations are important in articulating the foundational structure of quantum mechanics. All three can be presented at the abstract level as functors, as we shall now show.

4.3 Axiomatization of Strong Compact Closure

We shall adopt the most concise and elegant axiomatization of strongly compact closed categories, which takes the adjoint as primitive, following [Abramsky and Coecke, 2005].

It is convenient to build the definition up in several stages, as in [Selinger, 2007].

DEFINITION 10. A *dagger category* is a category $\mathbf{C}$ equipped with an identity-on-objects, contravariant, strictly involutive functor $f \mapsto f^\dagger$:

$$1^\dagger = 1, \qquad (g \circ f)^\dagger = f^\dagger \circ g^\dagger, \qquad f^{\dagger\dagger} = f .$$

We define an arrow $f : A \to B$ in a dagger category to be *unitary* if it is an isomorphism such that $f^{-1} = f^\dagger$. An endomorphism $f : A \to A$ is *self-adjoint* if $f = f^\dagger$.

DEFINITION 11. A *dagger symmetric monoidal category* $(\mathbf{C}, \otimes, \mathrm{I}, \lambda, \rho, \alpha, \sigma, \dagger)$ combines dagger and symmetric monoidal structure, with the requirement that the natural isomorphisms λ, ρ, α, σ are componentwise unitary, and moreover that $\dagger$ is a strong monoidal functor:

$$(f \otimes g)^\dagger = f^\dagger \otimes g^\dagger .$$

Finally we come to the main definition.

DEFINITION 12. A *strongly compact closed category* is a dagger symmetric monoidal category which is compact closed, and such that the following diagram commutes:

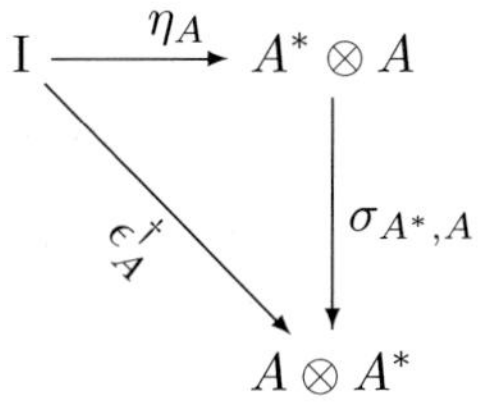

This implies that the counit is *definable* from the unit and the adjoint:

$$\epsilon_A = \eta_A^\dagger \circ \sigma_{A,A^*}$$

and similarly the unit can be defined from the counit and the adjoint. Furthermore, it is in fact possible to replace the two commuting diagrams required in the definition of compact closure by one. We refer to [Abramsky and Coecke, 2005] for the details.

DEFINITION 13. In any strongly compact closed category **C**, we can define a covariant monoidal functor

$$A \mapsto A^*, \qquad f : A \to B \mapsto f_* = (f^\dagger)^* : A^* \to B^* .$$

Examples Our central example is of course **FdHilb**. Any compact closed category such as **Rel**, in which $(\ )^*$ is the identity on objects, is trivially strongly compact closed (we just take $f^\dagger = f^*$). Note that in this case $f_* = f^{**} = f$. Thus in **Rel** the adjoint is relational converse. The category of finite-dimensional real inner product spaces and linear maps, with $A = A^*$, offers another example of this situation. A construction of free strongly compact closed categories over dagger categories is given in [Abramsky, 2005].

Scalars Self-adjoint scalars $s = s^\dagger$ in strongly compact closed categories are of special interest. In the case of **FdHilb**, these are the *positive reals* $\mathbb{R}^+$. The passage from s to $ss^\dagger$, which is self-adjoint, will track the passage in quantum mechanics from *amplitudes* to *probabilities*.

4.4 Inner Products and Dirac Notation

With the adjoint available, it is straightforward to interpret Dirac notation — the indispensable everyday notation of quantum mechanics and quantum information. A *ket* is simply an arrow $\psi : \mathrm{I} \to A$, which we can write as $|\psi\rangle$ for emphasis. We think of kets as *states*, of a given type of system A. The corresponding *bra* will then be $\psi^\dagger : A \to \mathrm{I}$, which we can think of as a *costate*.

Example In **FdHilb**, a linear map $f : \mathbb{C} \to \mathcal{H}$ can be identified with the vector $f(1) = \psi \in \mathcal{H}$: by linearity, all other values of f are determined by ψ. Even better, we can identify f with its image, which is the *ray* or one-dimensional subspace of $\mathcal{H}$ spanned by ψ — the proper notion of (pure) state of a quantum system.

DEFINITION 14. Given $\psi, \phi : \mathrm{I} \to A$ we define their *abstract inner product* $\langle \psi \mid \phi \rangle$ as

$$\psi^\dagger \circ \phi : \mathrm{I} \longrightarrow \mathrm{I} .$$

Note that this is a scalar, as it should be. In **FdHilb**, this definition coincides with the usual inner product. In **Rel**, for $x, y \subseteq \{*\} \times X$:

$$\langle x \mid y \rangle = 1_{\mathrm{I}}, \quad x \cap y \neq \varnothing \qquad \text{and} \qquad \langle x \mid y \rangle = 0_{\mathrm{I}}, \quad x \cap y = \varnothing.$$

We now show that two of the basic properties of adjoints in **FdHilb** hold in generality in the abstract setting.

PROPOSITION 15. *For $\psi : \mathrm{I} \to A$, $\phi : \mathrm{I} \to B$ and $f : B \to A$ we have*

$$\langle f^\dagger \circ \psi \mid \phi \rangle_B = \langle \psi \mid f \circ \phi \rangle_A .$$

Proof: $\langle f^\dagger \circ \psi \mid \phi \rangle = (f^\dagger \circ \psi)^\dagger \circ \phi = \psi^\dagger \circ f \circ \phi = \langle \psi \mid f \circ \phi \rangle$. ■

PROPOSITION 16. *Unitary morphisms $U : A \to B$ preserve the inner product, that is for all $\psi, \phi : \mathrm{I} \to A$ we have*

$$\langle U \circ \psi \mid U \circ \phi \rangle_B = \langle \psi \mid \phi \rangle_A .$$

Proof: By Proposition 15, $\langle U \circ \psi \mid U \circ \phi \rangle_B = \langle U^\dagger \circ U \circ \psi \mid \phi \rangle_A = \langle \psi \mid \phi \rangle_A$. ■

Finally, we show how the inner product can be defined in terms of the 'complex conjugate' functor $()_*$.

PROPOSITION 17. *For $\psi, \phi : \mathrm{I} \to A$ we have:*

$$\langle \psi \mid \phi \rangle_A = \mathrm{I} \xrightarrow{\rho_{\mathrm{I}}} \mathrm{I} \otimes \mathrm{I} \xrightarrow{1_{\mathrm{I}} \otimes u_{\mathrm{I}}} \mathrm{I} \otimes \mathrm{I}^* \xrightarrow{\phi \otimes \psi_*} A \otimes A^* \xrightarrow{\epsilon_A} \mathrm{I}.$$

Proof: Since $u_{\mathrm{I}} = \rho_{\mathrm{I}^*}^{-1} \circ \eta_{\mathrm{I}}$ by naturality of ρ we have

$$\eta_{\mathrm{I}} = \rho_{\mathrm{I}^*} \circ \rho_{\mathrm{I}^*}^{-1} \circ \eta_{\mathrm{I}} = \rho_{\mathrm{I}^*} \circ u_{\mathrm{I}} = (u_{\mathrm{I}} \otimes 1_{\mathrm{I}}) \circ \rho_{\mathrm{I}}$$

where we use $\rho^{-1} = \rho^\dagger$ and similarly we obtain $\epsilon_{\mathrm{I}} = \rho_{\mathrm{I}}^\dagger \circ (1_{\mathrm{I}} \otimes u_{\mathrm{I}}^\dagger)$. Hence by $1_{\mathrm{I}} = u_{\mathrm{I}}^\dagger \circ u_{\mathrm{I}}$ and the analogues to Lemmas 6 and 9 for the counit we obtain

$$\begin{aligned}
\psi^\dagger \circ \phi = \rho_{\mathrm{I}}^\dagger \circ ((\psi^\dagger \circ \phi) \otimes 1_{\mathrm{I}}) \circ \rho_{\mathrm{I}} &= \epsilon_{\mathrm{I}} \circ (\psi^\dagger \otimes 1_{\mathrm{I}^*}) \circ (\phi \otimes 1_{\mathrm{I}^*}) \circ \epsilon_{\mathrm{I}}^\dagger \\
&= \llcorner \psi^\dagger \lrcorner \circ (\phi \otimes 1_{\mathrm{I}^*}) \circ \epsilon_{\mathrm{I}}^\dagger \\
&= \epsilon_{\mathrm{I}} \circ (1_{\mathrm{I}} \otimes \psi_*) \circ (\phi \otimes 1_{\mathrm{I}^*}) \circ \epsilon_{\mathrm{I}}^\dagger
\end{aligned}$$

which is equal to $\epsilon_{\mathrm{I}} \circ (\phi \otimes \psi_*) \circ (1_{\mathrm{I}} \otimes u_{\mathrm{I}}) \circ \rho_{\mathrm{I}}$. ■

4.5 Dissection of the bipartite projector

Projectors are a basic building block in the von Neumann-style foundations of quantum mechanics, and in standard approaches to quantum logic. It is a notable feature of our approach that we are able, at the abstract level of strongly compact closed categories, to delineate a *fine-structure* of bipartite projectors, which can be applied directly to the analysis of information flow in quantum protocols.

We define a *projector* on an object A in a strongly compact closed category to be an arrow $\mathrm{P} : A \to A$ which is idempotent and self-adjoint:

$$\mathrm{P}^2 = \mathrm{P}, \qquad \mathrm{P} = \mathrm{P}^\dagger .$$

PROPOSITION 18. *Suppose we have a state* $\psi : \mathrm{I} \to A$ *which is* normalized, *meaning* $\langle \psi \mid \psi \rangle = 1_\mathrm{I}$. *Then the 'ket-bra'* $|\psi\rangle\langle\psi| = \psi \circ \psi^\dagger : A \to A$ *is a projector.*

Proof: Self-adjointness is clear. For idempotence:

$$\psi \circ \psi^\dagger \circ \psi \circ \psi^\dagger = \langle \psi \mid \psi \rangle \bullet \psi \circ \psi^\dagger = \psi \circ \psi^\dagger .$$

■

We now want to apply this idea in a more refined form to a state $\psi : \mathrm{I} \to A^* \otimes B$ of a compound system. Note that, by Map-State duality:

$$\mathbf{C}(\mathrm{I}, A^* \otimes B) \equiv \mathbf{C}(A, B)$$

any such state ψ corresponds biuniquely to the name of a map $f : A \to B$, *i.e.* $\psi = \ulcorner f \urcorner$. This arrow witnesses an information flow from A to B, and we will use this to expose the information flow inherent in the corresponding projector.

Explicitly, we define

$$\mathrm{P}_f := \ulcorner f \urcorner \circ (\ulcorner f \urcorner)^\dagger = \ulcorner f \urcorner \circ \llcorner f_* \lrcorner : A^* \otimes B \to A^* \otimes B ,$$

that is, we have an assignment

$$\mathrm{P}_\_ : \mathbf{C}(\mathrm{I}, A^* \otimes B) \longrightarrow \mathbf{C}(A^* \otimes B, A^* \otimes B) :: \Psi \mapsto \Psi \circ \Psi^\dagger$$

from bipartite elements to bipartite projectors. Note that the strong compact closed structure is essential in order to define P_f as an endomorphism.

We can *normalize* these projectors P_f by considering $s_f \bullet \mathrm{P}_f$ for $s_f := (\llcorner f_* \lrcorner \circ \ulcorner f \urcorner)^{-1}$ (provided this inverse exists in $\mathbf{C}(\mathrm{I}, \mathrm{I})$), yielding

$$(s_f \bullet \mathrm{P}_f) \circ (s_f \bullet \mathrm{P}_f) = s_f \bullet (\ulcorner f \urcorner \circ (s_f \bullet (\llcorner f_* \lrcorner \circ \ulcorner f \urcorner)) \circ \llcorner f_* \lrcorner) = s_f \bullet \mathrm{P}_f ,$$

and also

$$(s_f \bullet \mathrm{P}_f) \circ \ulcorner f \urcorner = \ulcorner f \urcorner \qquad \text{and} \qquad \llcorner f_* \lrcorner \circ (s_f \bullet \mathrm{P}_f) = \llcorner f_* \lrcorner .$$

A picture corresponding to this decomposed bipartite projector is:

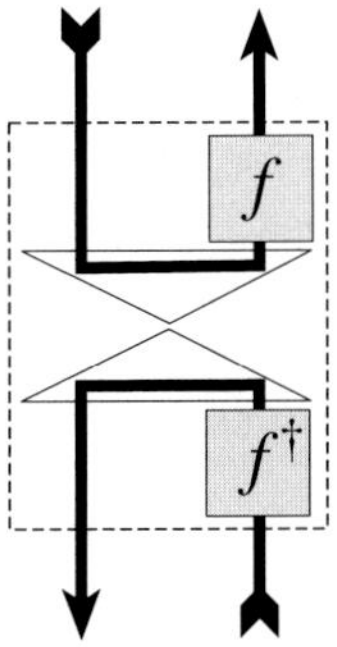

4.6 Trace

Another essential mathematical instrument in quantum mechanics is the *trace* of a linear map. In quantum information, extensive use is made of the more general notion of *partial trace*, which is used to trace out a subsystem of a compound system.

A general categorical axiomatization of the notion of partial trace has been given by Joyal, Street and Verity [Joyal, Street and Verity, 1996]. A trace in a symmetric monoidal category **C** is a family of functions

$$\mathsf{Tr}^U_{A,B} : \mathbf{C}(A \otimes U, B \otimes U) \longrightarrow \mathbf{C}(A, B)$$

for objects A, B, U of **C**, satisfying a number of axioms, for which we refer to [Joyal, Street and Verity, 1996]. This specializes to yield the total trace for endomorphisms by taking $A = B = \mathrm{I}$. In this case, $\mathsf{Tr}(f) = \mathsf{Tr}^U_{\mathrm{I,I}}(f) : \mathrm{I} \to \mathrm{I}$ is a scalar. Expected properties such as the invariance of the trace under cyclic permutations

$$\mathsf{Tr}(g \circ f) = \mathsf{Tr}(f \circ g)$$

follow from the general axioms.

Any compact closed category carries a canonical (in fact, a unique) trace. The definition can be given slightly more elegantly in the strongly compact closed case. For an endomorphism $f : A \to A$, the total trace is defined by

$$\mathsf{Tr}(f) = \epsilon_A \circ (f \otimes 1_{A^*}) \circ \epsilon_A^\dagger .$$

More generally, if $f : A \otimes C \to B \otimes C$, $\mathsf{Tr}^C_{A,B}(f) : A \to B$ is defined to be:

$$A \xrightarrow{\ \rho_A\ } A\otimes\mathrm{I} \xrightarrow{1\otimes\epsilon_C^\dagger} A\otimes C\otimes C^* \xrightarrow{f\otimes 1_{C^*}} B\otimes C\otimes C^* \xrightarrow{1\otimes\epsilon_C} B\otimes\mathrm{I} \xrightarrow{\ \rho_B^{-1}\ } B .$$

These definitions give rise to the standard notions of trace and partial trace in **FdHilb**.

5 BIPRODUCTS, BRANCHING AND MEASUREMENTS

As we have seen, many of the basic ingredients of quantum mechanics are present in strongly compact closed categories. What is lacking is the ability to express the probabilistic branching arising from measurements, and the information flows from quantum to classical and back. We shall find this final piece of expressive power in a rather standard piece of additional categorical structure, namely *biproducts*.

5.1 Biproducts

Biproducts have been studied as part of the structure of Abelian categories. For further details, and proofs of the general results we shall cite in this sub-section, see e.g. [Mitchell, 1965; MacLane, 1998].

Recall that a *zero object* in a category is one which is both initial and terminal. If $\mathbf{0}$ is a zero object, there is an arrow

$$0_{A,B} : A \longrightarrow \mathbf{0} \longrightarrow B$$

between any pair of objects A and B. Let $\mathbf{C}$ be a category with a zero object and binary products and coproducts. Any arrow

$$A_1 \coprod A_2 \xrightarrow{\ f\ } A_1 \prod A_2$$

with injections $q_i : A_i \to A_1 \coprod A_2$ and projections $p_j : A_1 \prod A_2 \to A_j$ can be written uniquely as a matrix

$$\begin{pmatrix} f_{11} & f_{21} \\ f_{12} & f_{22} \end{pmatrix}$$

where $f_{ij} := p_j \circ f \circ q_i : A_i \to A_j$. If the arrow

$$\begin{pmatrix} 1 & 0 \\ 0 & 1 \end{pmatrix}$$

is an isomorphism for all A_1, A_2, then we say that $\mathbf{C}$ has *biproducts*, and write $A \oplus B$ for the biproduct of A and B.

PROPOSITION 19 Semi-additivity. *If* $\mathbf{C}$ *has biproducts, then we can define an operation of addition on each hom-set* $\mathbf{C}(A, B)$ *by*

$$\begin{array}{ccc} A & \xrightarrow{\ f+g\ } & B \\ {\scriptstyle \Delta}\downarrow & & \uparrow{\scriptstyle \nabla} \\ A \oplus A & \xrightarrow[\ f \oplus g\ ]{} & B \oplus B \end{array}$$

for $f, g : A \to B$, where $\Delta = \langle 1_A, 1_A \rangle$ and $\nabla = [1_B, 1_B]$ are respectively the diagonal and codiagonal. This operation is associative and commutative, with 0_{AB} as an identity. Moreover, composition is bilinear with respect to this additive structure. Thus **C** *is enriched over abelian monoids.*

Because of this automatic enrichment of categories with biproducts over abelian monoids, we say that such a category is *semi-additive.*

PROPOSITION 20. *If* **C** *has biproducts, we can choose projections $p_1, \ldots, p_n$ and injections $q_1, \ldots, q_n$ for each $\bigoplus_{k=1}^{k=n} A_k$ satisfying*

$$p_j \circ q_i = \delta_{ij} \qquad \text{and} \qquad \sum_{k=1}^{k=n} q_k \circ p_k = 1_{\bigoplus_k A_k}$$

where $\delta_{ii} = 1_{A_i}$, and $\delta_{ij} = 0_{A_i, A_j}$, $i \neq j$.

5.2 Strongly compact closed categories with biproducts

We now come to the full mathematical structure we shall use as a setting for finitary quantum mechanics: namely *strongly compact closed categories with biproducts.*

A first point is that, because of the strongly self-dual nature of compact closed categories, weaker assumptions suffice in order to guarantee the presence of biproducts. The following elegant result is due to Robin Houston [2006], and was in fact directly motivated by [Abramsky and Coecke, 2004], the precursor to the present article.

THEOREM 21. *Let* **C** *be a monoidal category with finite products and coproducts, and suppose that for every object A of* **C***, the functor $A \otimes -$ preserves products and the functor $- \otimes A$ preserves coproducts. Then C has finite biproducts.*

Because a compact closed category is closed and self-dual, the existence of products implies that of coproducts, and vice versa, and the functor $- \otimes A$ is a left adjoint and hence preserves coproducts. Moreover, since $A^* \multimap B \simeq A^{**} \otimes B \simeq A \otimes B$, the functor $A \otimes -$ is a right adjoint and preserves products. Hence this result specializes to the following:

PROPOSITION 22. *If* **C** *is a compact closed category with either products or coproducts, then it has biproducts, and hence is semiadditive.*

Examples There are many examples of compact closed categories with biproducts: the category of relations for a regular category with stable disjoint coproducts; the category of finitely generated projective modules over a commutative ring; the category of finitely generated free semimodules over a commutative semiring; and the category of free semimodules over a complete commutative semiring are all semi-additive compact closed categories. Examples have also arisen in a Computer Science context in the first author's work on Interaction Categories [Abramsky,

Gay and Nagarajan, 1995]. Compact closed categories with biproducts with additional assumptions, in particular that the category is abelian, have been studied in the mathematical literature on *Tannakian categories* [Deligne, 1990].

In the case of strongly compact closed categories, we need a coherence condition between the dagger and the biproduct structure. We say that a category is *strongly compact closed with biproducts* if we can choose biproduct structures p_i, q_i as in Proposition 20 such that $p_i^\dagger = q_i$ for $i = 1, \ldots, n$.

PROPOSITION 23. *If* $\mathbf{C}$ *is strongly compact closed with biproducts, then*

$$\sum_{k=1}^{k=n} p_k^\dagger \circ p_k = \sum_{k=1}^{k=n} q_k \circ q_k^\dagger = 1_{\bigoplus_k A_k} .$$

Moreover, there are natural isomorphisms

$$\nu_{A,B} : (A \oplus B)^* \simeq A^* \oplus B^* \qquad \text{and} \qquad \nu_{\mathrm{I}} : \mathbf{0}^* \simeq \mathbf{0} ,$$

and $(\ )^\dagger$ *preserves biproducts and hence is additive:*

$$(f \oplus g)^\dagger = f^\dagger \oplus g^\dagger , \qquad (f+g)^\dagger = f^\dagger + g^\dagger \qquad \text{and} \qquad 0_{A,B}^\dagger = 0_{B,A} .$$

Examples Examples of semi-additive strongly compact closed categories are the category $(\mathbf{Rel}, \times, +)$, where the biproduct is the disjoint union, and the category $(\mathbf{FdHilb}, \otimes, \oplus)$, where the biproduct is the direct sum.

Distributivity and classical information flow As we have already seen, in a compact closed category with biproducts, tensor distributes over the biproduct. This abstract-seeming observation in fact plays a crucial rôle in the representation of *classical information flow*. To understand this, consider a quantum system $A \otimes B$, composed from subsystems A(lice) and B(ob). Now suppose that Alice performs a local measurement, which we will represent as resolving her part of the system into say $A_1 \oplus A_2$. Here the biproduct is used to represent the different *branches* of the measurement. At this point, by the functorial properties of $\oplus$, Alice can perform actions $f_1 \oplus f_2$, which *depend on which branch of the measurement has been taken.* The global state of the system is $(A_1 \oplus A_2) \otimes B$, and as things stand Bob has no access to this measurement outcome. However, under distributivity we have

$$(A_1 \oplus A_2) \otimes B \simeq (A_1 \otimes B) \oplus (A_2 \otimes B)$$

which corresponds to propagating the classical information as to the measurement outcome 'outwards', so that it is now accessible to Bob, who can perform an action depending on this outcome, of the form $1_A \otimes (g_1 \oplus g_2)$.

We shall record distributivity in an explicit form for future use.

PROPOSITION 24 Distributivity of $\otimes$ over $\oplus$. *In any monoidal closed category there is a right distributivity natural isomorphism* $\tau_{A,B,C} : A \otimes (B \oplus C) \simeq (A \otimes B) \oplus (A \otimes C)$*, which is explicitly defined as*

$$\tau_{A,\cdot,\cdot} := \langle 1_A \otimes p_1, 1_A \otimes p_2 \rangle \qquad \text{and} \qquad \tau^{-1}_{A,\cdot,\cdot} := [1_A \otimes q_1, 1_A \otimes q_2] .$$

A left distributivity isomorphism $\upsilon_{A,B,C} : (A \oplus B) \otimes C \simeq (A \otimes C) \oplus (A \otimes C)$ *can be defined similarly.*

Semiring of scalars. In a strongly compact closed category with biproducts, the scalars form a commutative semiring. Moreover, scalar multiplication satisfies the usual additive properties

$$(s_1 + s_2) \bullet f = s_1 \bullet f + s_2 \bullet f, \qquad 0 \bullet f = 0$$

as well as the multiplicative ones. For Hilbert spaces, this commutative semiring is the field of complex numbers. In **Rel** the commutative semiring of scalars is the Boolean semiring $\{0, 1\}$, with disjunction as sum.

Matrix calculus. We can write any arrow of the form $f : A \oplus B \to C \oplus D$ as a matrix

$$M_f := \begin{pmatrix} p_1^{C,D} \circ f \circ q_1^{A,B} & p_1^{C,D} \circ f \circ q_2^{A,B} \\ p_2^{C,D} \circ f \circ q_1^{A,B} & p_2^{C,D} \circ f \circ q_2^{A,B} \end{pmatrix}.$$

The sum $f + g$ of such morphisms corresponds to the matrix sum $M_f + M_g$ and composition $g \circ f$ corresponds to matrix multiplication $M_g \cdot M_f$. Hence categories with biproducts admit a matrix calculus.

5.3 *Spectral Decompositions*

We define a *spectral decomposition* of an object A to be a unitary isomorphism

$$U : A \to \bigoplus_{i=1}^{i=n} A_i .$$

(Here the 'spectrum' is just the set of indices $1, \ldots, n$). Given a spectral decomposition U, we define morphisms

$$\psi_j := U^\dagger \circ q_j : A_j \to A \qquad \text{and} \qquad \pi_j := \psi_j^\dagger = p_j \circ U : A \to A_j ,$$

diagramatically

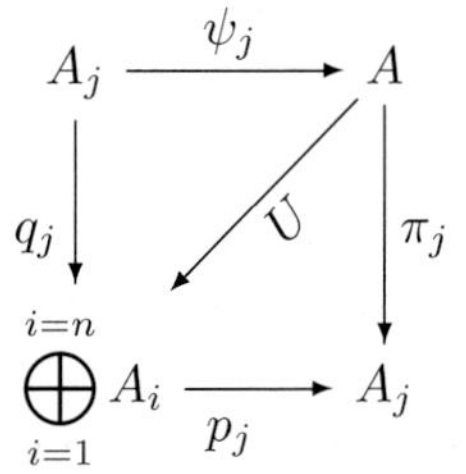

and finally *projectors*

$$\mathrm{P}_j := \psi_j \circ \pi_j : A \to A\,.$$

These projectors are *self-adjoint*

$$\mathrm{P}_j^\dagger = (\psi_j \circ \pi_j)^\dagger = \pi_j^\dagger \circ \psi_j^\dagger = \psi_j \circ \pi_j = \mathrm{P}_j$$

idempotent and *orthogonal*

$$\mathrm{P}_i \circ \mathrm{P}_j = \psi_i \circ \pi_i \circ \psi_j \circ \pi_j = \psi_i \circ \delta_{ij} \circ \pi_j = \delta_{ij}^A \circ \mathrm{P}_i.$$

Moreover, they yield a *resolution of the identity*:

$$\begin{aligned}\sum_{i=1}^{i=n} \mathrm{P}_i &= \sum_{i=1}^{i=n} \psi_i \circ \pi_i = \sum_{i=1}^{i=n} U^\dagger \circ q_i \circ p_i \circ U \\ &= U^\dagger \circ (\sum_{i=1}^{i=n} q_i \circ p_i) \circ U = U^{-1} \circ 1_{\bigoplus_i A_i} \circ U = 1_A\,.\end{aligned}$$

5.4 Bases and dimension

Writing $n \cdot X$ for types of the shape $\bigoplus_{i=1}^{i=n} X$ it follows by self-duality of the tensor unit I that

$$\nu_{\mathrm{I},\ldots,\mathrm{I}}^{-1} \circ (n \cdot u_\mathrm{I}) \ : n \cdot \mathrm{I} \simeq (n \cdot \mathrm{I})^*\,.$$

A *basis* for an object A is a unitary isomorphism

$$\mathsf{base} : n \cdot \mathrm{I} \to A\,.$$

Given bases base_A and base_B for objects A and B respectively we can define the matrix (m_{ij}) of any morphism $f : A \to B$ in those two bases as the matrix of

$$\mathsf{base}_B^\dagger \circ f \circ \mathsf{base}_A : n_A \cdot \mathrm{I} \to n_B \cdot \mathrm{I}\,.$$

PROPOSITION 25. *Given* $f : A \to B$, $\mathsf{base}_A : n_A \cdot \mathrm{I} \to A$ *and* $\mathsf{base}_B : n_B \cdot \mathrm{I} \to A$ *the matrix* (m'_{ij}) *of* $f^\dagger$ *in these bases is the conjugate transpose of the matrix* (m_{ij}) *of* f.

Proof: $m'_{ij} = p_i \circ \mathsf{base}_A^\dagger \circ f^\dagger \circ \mathsf{base}_B \circ q_j = (p_j \circ \mathsf{base}_B^\dagger \circ f \circ \mathsf{base}_A \circ q_i)^\dagger = m_{ji}^\dagger$. ■

If in addition to the assumptions of Proposition 15 and Proposition 16 there exist bases for A and B, we can prove converses to both of them.

PROPOSITION 26. *If there exist bases for* A *and* B *then* $f : A \to B$ *is the adjoint to* $g : B \to A$ *if and only if*

$$\langle f \circ \psi \mid \phi \rangle_B = \langle \psi \mid g \circ \phi \rangle_A$$

for all $\psi : \mathrm{I} \to A$ *and* $\phi : \mathrm{I} \to B$.

Proof: Let (m_{ij}) be the matrix of $f^\dagger$ and (m'_{ij}) the matrix of g in the given bases.

$$\begin{aligned}
m_{ij} &= p_i \circ \mathsf{base}_A^\dagger \circ f^\dagger \circ \mathsf{base}_B \circ q_j \\
&= \langle f \circ \mathsf{base}_A \circ q_i \mid \mathsf{base}_B \circ q_j \rangle_B \\
&= \langle f \circ \psi \mid \phi \rangle_B \\
&= \langle \psi \mid g \circ \phi \rangle_A \\
&= \langle \mathsf{base}_A \circ q_i \mid g \circ \mathsf{base}_B \circ q_j \rangle_A \\
&= p_i \circ \mathsf{base}_A^\dagger \circ g \circ \mathsf{base}_B \circ q_j \\
&= m'_{ij} .
\end{aligned}$$

Hence the matrix elements of g and $f^\dagger$ coincide so g and $f^\dagger$ are equal. The converse is Proposition 15. ■

PROPOSITION 27. *If there exist bases for A and B then a morphism $U : A \to B$ is unitary if and only if it preserves the inner product, that is for all $\psi, \phi : \mathrm{I} \to A$ we have*

$$\langle U \circ \psi \mid U \circ \phi \rangle_B = \langle \psi \mid \phi \rangle_A .$$

Proof: We have $\langle U^{-1} \circ \psi \mid \phi \rangle_A = \langle U \circ U^{-1} \circ \psi \mid U \circ \phi \rangle_B = \langle \psi \mid U \circ \phi \rangle_B$ and hence by Proposition 26, $U^\dagger = U^{-1}$. The converse is given by Proposition 16. ■

Note also that when a basis is available we can assign to $\psi^\dagger : A \to \mathrm{I}$ and $\phi : \mathrm{I} \to A$ matrices

$$\begin{pmatrix} \psi_1^\dagger & \cdots & \psi_n^\dagger \end{pmatrix} \qquad \text{and} \qquad \begin{pmatrix} \phi_1 \\ \vdots \\ \phi_n \end{pmatrix}$$

respectively, and the inner product becomes

$$\langle \psi \mid \phi \rangle = \begin{pmatrix} \psi_1^\dagger & \cdots & \psi_n^\dagger \end{pmatrix} \begin{pmatrix} \phi_1 \\ \vdots \\ \phi_n \end{pmatrix} = \sum_{i=1}^{i=n} \psi_i^\dagger \circ \phi_i .$$

Dimension Interestingly, two different notions of dimension arise in our setting. We assign an *integer dimension* $\mathsf{dim}(A) \in \mathbb{N}$ to an object A provided there exists a base

$$\mathsf{base} : \mathsf{dim}(A) \cdot \mathrm{I} \to A .$$

Alternatively, we introduce the *scalar dimension* as

$$\mathsf{dim}_s(A) := \mathsf{Tr}(1_A) = \epsilon_A \circ \epsilon_A^\dagger \in \mathbf{C}(\mathrm{I}, \mathrm{I}).$$

We also have:

$$\mathsf{dim}_s(\mathrm{I}) = 1_\mathrm{I} \qquad \mathsf{dim}_s(A^*) = \mathsf{dim}_s(A) \qquad \mathsf{dim}_s(A \otimes B) = \mathsf{dim}_s(A)\mathsf{dim}_s(B)$$

In $\mathbf{FdVec}_{\mathbb{K}}$ these notions of dimension coincide, in the sense that $\mathsf{dim}_s(V)$ is multiplication with the scalar $\mathsf{dim}(V)$. In **Rel** the integer dimension corresponds to the cardinality of the set, and is only well-defined for finite sets, while $\mathsf{dim}_s(X)$ always exists; however, $\mathsf{dim}_s(X)$ can only take two values, 0_{I} and 1_{I}, and the two notions of dimension diverge for sets of cardinality greater than 1.

5.5 Towards a representation theorem

As the results in this section have shown, under the assumption of biproducts we can replicate many of the familiar linear-algebraic calculations in Hilbert spaces. One may wonder how far we really are from Hilbert spaces.

The deep results by Deligne [1990] and Doplicher-Roberts [1989] on Tannakian categories, the latter directly motivated by algebraic quantum field theory, show that under additional assumptions, in particular that the category is abelian as well as compact closed, we obtain a representation into finite-dimensional modules over the ring of scalars. One would like to see a similar result in the case of strongly compact closed categories with biproducts, with the conclusion being a representation into inner-product spaces.

6 ABSTRACT QUANTUM MECHANICS: AXIOMATICS AND QUANTUM PROTOCOLS

We can identify the basic ingredients of finitary quantum mechanics in any semi-additive strongly compact closed category.

1. A *state space* is represented by an object A.

2. A *basic variable* ('type of qubits') is a state space Q with a given unitary isomorphism

$$\mathsf{base}_Q : \mathrm{I} \oplus \mathrm{I} \to Q$$

which we call the *computational basis* of Q. By using the isomorphism $n \cdot \mathrm{I} \simeq (n \cdot \mathrm{I})^*$ described in Section 5, we also obtain a computational basis for Q^*.

3. A *compound system* for which the subsystems are described by A and B respectively is described by $A \otimes B$. If we have computational bases base_A and base_B, then we define

$$\mathsf{base}_{A\otimes B} := (\mathsf{base}_A \otimes \mathsf{base}_B) \circ d_{nm}^{-1}$$

where

$$d_{nm} : n \cdot \mathrm{I} \otimes m \cdot \mathrm{I} \simeq (nm) \cdot \mathrm{I}$$

is the canonical isomorphism constructed using first the left distributivity isomorphism υ, and then the right distributivity isomorphism τ, to give the usual lexicographically-ordered computational basis for the tensor product.

4. Basic data transformations are unitary isomorphisms.

5a. A *preparation* in a state space A is a morphism $\psi : \mathrm{I} \to A$ for which there exists a unitary $U : \mathrm{I} \oplus B \to A$ such that

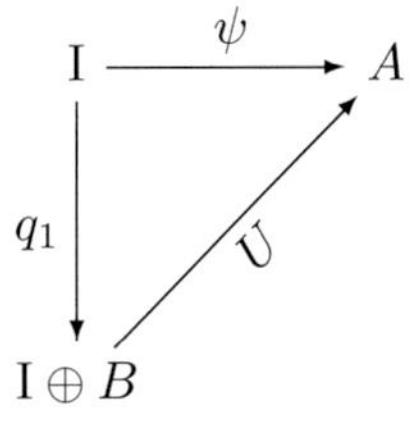

commutes.

5b. Consider a spectral decomposition

$$U : A \to \bigoplus_{i=1}^{i=n} A_i$$

with associated projectors P_j. This gives rise to the *non-destructive measurement*

$$\langle \mathrm{P}_i \rangle_{i=1}^{i=n} : A \to n \cdot A.$$

The projectors $\mathrm{P}_i : A \to A$ for $i = 1, \ldots, n$ are called the *measurement branches.* This measurement is *non-degenerate* if $A_i = \mathrm{I}$ for all $i = 1, \ldots, n$. In this case we refer to U itself as a *destructive measurement* or *observation.* The morphisms $\pi_i = p_i \circ U : A \to \mathrm{I}$ for $i = 1, \ldots, n$ are called *observation branches.*

Note that the type of a non-destructive measurement makes it explicit that it is an operation which involves a non-deterministic transition (by contrast with the standard Hilbert space quantum mechanical formalism).

6a. Explicit biproducts represent the *branching* arising from the indeterminacy of measurement outcomes.

Hence an operation f acting on an explicit biproduct $A \oplus B$ should itself be an explicit biproduct, *i.e.* we want

$$f = f_1 \oplus f_2 : A \oplus B \to C \oplus D\,,$$

for $f_1 : A \to C$ and $f_2 : B \to D$. The dependency of f_i on the branch it is in captures *local* classical communication. The full force of non-local classical communication is enabled by Proposition 24.

6b. Distributivity isomorphisms represent *non-local classical communication.*

To see this, suppose e.g. that we have a compound system $Q \otimes A$, and we (non-destructively) measure the qubit in the first component, obtaining a new system state described by $(Q \oplus Q) \otimes A$. At this point, we know 'locally', *i.e.* at the site of the first component, what the measurement outcome is, but we have not propagated this information to the rest of the system A. However, after applying the distributivity isomorphism

$$(Q \oplus Q) \otimes A \simeq (Q \otimes A) \oplus (Q \otimes A)$$

the information about the outcome of the measurement on the first qubit has been propagated globally throughout the system, and we can perform operations on A depending on the measurement outcome, e.g. $(1_Q \otimes U_0) \oplus (1_Q \otimes U_1)$ where U_0, U_1 are the operations we wish to perform on A in the event that the outcome of the measurement we performed on Q was 0 or 1 respectively.

6.1 The Born rule

We now show how the *Born rule*, which is the key quantitative feature of quantum mechanics, emerges automatically from our abstract setting.

For a preparation $\psi : \mathrm{I} \to A$ and spectral decomposition $U : A \to \bigoplus_{i=1}^{i=n} A_i$, with corresponding non-destructive measurement

$$\langle \mathrm{P}_i \rangle_{i=1}^{i=n} : A \to n \cdot A \,,$$

we can consider the protocol

$$\mathrm{I} \xrightarrow{\ \psi\ } A \xrightarrow{\langle \mathrm{P}_i \rangle_{i=1}^{i=n}} n \cdot A \,.$$

We define scalars

$$\mathsf{Prob}(\mathrm{P}_i, \psi) := \langle \psi \mid \mathrm{P}_i \mid \psi \rangle = \psi^\dagger \circ \mathrm{P}_i \circ \psi \,.$$

PROPOSITION 28. *With notation as above,*

$$\mathsf{Prob}(\mathrm{P}_i, \psi) = (\mathsf{Prob}(\mathrm{P}_i, \psi))^\dagger$$

and

$$\sum_{i=1}^{i=n} \mathsf{Prob}(\mathrm{P}_i, \psi) = 1 \,.$$

Hence we think of the scalar $\mathsf{Prob}(\mathrm{P}_j, \psi)$ *as 'the probability of obtaining the* j*'th outcome of the measurement* $\langle \mathrm{P}_i \rangle_{i=1}^{i=n}$ *on the state* ψ*'.*

Proof: From the definitions of preparation and the projectors, there are unitaries U, V such that

$$\mathsf{Prob}(\mathrm{P}_i, \psi) = (V \circ q_1)^\dagger \circ U^\dagger \circ q_i \circ p_i \circ U \circ V \circ q_1$$

for each i. Hence

$$\begin{aligned}\sum_{i=1}^{i=n} \mathsf{Prob}(\mathrm{P}_i, \psi) &= \sum_{i=1}^{i=n} p_1 \circ V^\dagger \circ U^\dagger \circ q_i \circ p_i \circ U \circ V \circ q_1 \\ &= p_1 \circ V^\dagger \circ U^\dagger \circ \Big(\sum_{i=1}^{n} q_i \circ p_i\Big) \circ U \circ V \circ q_1 \\ &= p_1 \circ V^{-1} \circ U^{-1} \circ 1_{n \cdot \mathrm{I}} \circ U \circ V \circ q_1 = p_1 \circ q_1 = 1_{\mathrm{I}}\,.\end{aligned}$$

■

Moreover, since by definition $\mathrm{P}_j = \pi_j^\dagger \circ \pi_j$, we can rewrite the Born rule expression as

$$\mathsf{Prob}(\mathrm{P}_j, \psi) = \psi^\dagger \circ \mathrm{P}_j \circ \psi = \psi^\dagger \circ \pi_j^\dagger \circ \pi_j \circ \psi = (\pi_j \circ \psi)^\dagger \circ \pi_j \circ \psi = s_j^\dagger \circ s_j$$

for some scalar $s_j \in \mathbf{C}(\mathrm{I}, \mathrm{I})$. Thus s_j can be thought of as the 'probability amplitude' giving rise to the probability $s_j^\dagger \circ s_j$, which is of course self-adjoint. If we consider the protocol

$$\mathrm{I} \xrightarrow{\ \psi\ } A \xrightarrow{\langle \pi_i \rangle_{i=1}^{i=n}} n \cdot \mathrm{I}\,.$$

which involves an observation $\langle \pi_i \rangle_{i=1}^{i=n}$, then these scalars s_j correspond to the branches

$$\mathrm{I} \xrightarrow{\ \psi\ } A \xrightarrow{\ \pi_j\ } \mathrm{I}\,.$$

We now turn to the description of the quantum protocols previously discussed in Section 2 within our framework. In each case, we shall give a complete description of the protocol, including the quantum-to-classical information flows arising from measurements, and the subsequent classical-to-quantum flows corresponding to the classical communications and the actions depending on these performed as steps in the protocols. We shall in each case verify the correctness of the protocol, by proving that a certain diagram commutes. Thus these case studies provide evidence for the expressiveness and effectiveness of the framework.

Our general axiomatic development allows for considerable generality. The standard von Neumann axiomatization fits Quantum Mechanics perfectly, with no room to spare. Our basic setting of strongly compact closed categories with biproducts is general enough to allow very different models such as **Rel**, the category of sets and relations. When we consider specific protocols such as teleportation, a kind of 'Reverse Arithmetic' (by analogy with Reverse Mathematics [Simpson, 1999]) arises. That is, we can characterize what requirements are placed on the semiring of scalars $\mathbf{C}(\mathrm{I}, \mathrm{I})$ (where I is the tensor unit) in order for the protocol to be realized. This is often much less than requiring that this be the field of complex numbers, but in the specific cases which we shall consider, the requirements are sufficient to exclude **Rel**.

6.2 *Quantum teleportation*

DEFINITION 29. *A* teleportation base *is a scalar s together with a morphism*

$$\mathsf{prebase}_{\mathrm{T}} : 4 \cdot \mathrm{I} \to Q^* \otimes Q$$

such that:

- $\mathsf{base}_{\mathrm{T}} := s \bullet \mathsf{prebase}_{\mathrm{T}}$ *is unitary.*
- *the four maps* $\beta_j : Q \to Q$*, where* β_j *is defined by* $\ulcorner \beta_j \urcorner := \mathsf{prebase}_{\mathrm{T}} \circ q_j$*, are unitary.*
- $2s^\dagger s = 1$.

The morphisms $s \bullet \ulcorner \beta_j \urcorner$ *are the* base vectors *of the teleportation base. A teleportation base is a* Bell base *when the* Bell base maps $\beta_1, \beta_2, \beta_3, \beta_4 : Q \to Q$ *satisfy*[2]

$$\beta_1 = 1_Q \qquad \beta_2 = \sigma_Q^{\oplus} \qquad \beta_3 = \beta_3^{\dagger} \qquad \beta_4 = \sigma_Q^{\oplus} \circ \beta_3$$

where

$$\sigma_Q^{\oplus} := \mathsf{base}_Q \circ \sigma_{\mathrm{I},\mathrm{I}}^{\oplus} \circ \mathsf{base}_Q^{-1} .$$

A teleportation base defines a teleportation observation

$$\langle s^\dagger \bullet \llcorner \beta_i \lrcorner \rangle_{i=1}^{i=4} : Q \otimes Q^* \to 4 \cdot \mathrm{I} .$$

To emphasize the identity of the individual qubits we label the three copies of Q we shall consider as Q_a, Q_b, Q_c. We also use labelled identities, e.g. $1_{bc} : Q_b \to Q_c$, and labelled Bell bases. Finally, we introduce

$$\Delta_{ac}^4 := \langle s^\dagger s \bullet 1_{ac} \rangle_{i=1}^{i=4} : Q_a \to 4 \cdot Q_c$$

as the *labelled, weighted diagonal.* This expresses the intended behaviour of teleportation, namely that the input qubit is propagated to the output along each branch of the protocol, with 'weight' $s^\dagger s$, corresponding to the probability amplitude for that branch. Note that the sum of the corresponding probabilities is

$$4(s^\dagger s)^\dagger s^\dagger s = (2s^\dagger s)(2s^\dagger s) = 1 .$$

[2]This choice of axioms is sufficient for our purposes. One might prefer to axiomatize a notion of Bell base such that the corresponding Bell base maps are exactly the Pauli matrices — note that this would introduce a coefficient i in β_4.

THEOREM 30. *The following diagram commutes.*

$$
\begin{array}{ccccl}
Q_a & = & Q_a & & \\
\Big| & & \Big\downarrow \rho_a & & \textbf{import unknown state} \\
\Big| & & Q_a \otimes \mathrm{I} & & \\
\Big| & & \Big\downarrow 1_a \otimes (s \bullet \ulcorner 1_{bc} \urcorner) & & \textbf{produce EPR-pair} \\
\Big| & & Q_a \otimes (Q_b^* \otimes Q_c) & & \\
\Big| & & \Big\downarrow \alpha_{a,b,c} & & \textbf{spatial delocation} \\
\Delta_{ac}^4 \Big| & & (Q_a \otimes Q_b^*) \otimes Q_c & & \\
\Big| & & \Big\downarrow \langle s^\dagger \bullet \llcorner \beta_i^{ab} \lrcorner \rangle_{i=1}^{i=4} \otimes 1_c & & \textbf{teleportation observation} \\
\Big| & & (4 \cdot \mathrm{I}) \otimes Q_c & & \\
\Big| & & \Big\downarrow (4 \cdot \lambda_c^{-1}) \circ \upsilon_c & & \textbf{classical communication} \\
\Big| & & 4 \cdot Q_c & & \\
\Big\downarrow & & \Big\downarrow \bigoplus_{i=1}^{i=4} (\beta_i^c)^{-1} & & \textbf{unitary correction} \\
4 \cdot Q_c & = & 4 \cdot Q_c & &
\end{array}
$$

The right-hand-side of the above diagram is our formal description of the teleportation protocol; the commutativity of the diagram expresses the correctness of the protocol. Hence any strongly compact closed category with biproducts admits quantum teleportation provided it contains a teleportation base. If we do a Bell-base observation then the corresponding unitary corrections are

$$\beta_i^{-1} = \beta_i \quad \textit{for} \quad i \in \{1, 2, 3\} \qquad \textit{and} \qquad \beta_4^{-1} = \beta_3 \circ \sigma_Q^{\oplus} .$$

Proof: For each $j \in \{1, 2, 3, 4\}$ we have a commutative diagram of the form below. The top trapezoid is the statement of the Theorem. We ignore the scalars – which cancel out against each other – in this proof.

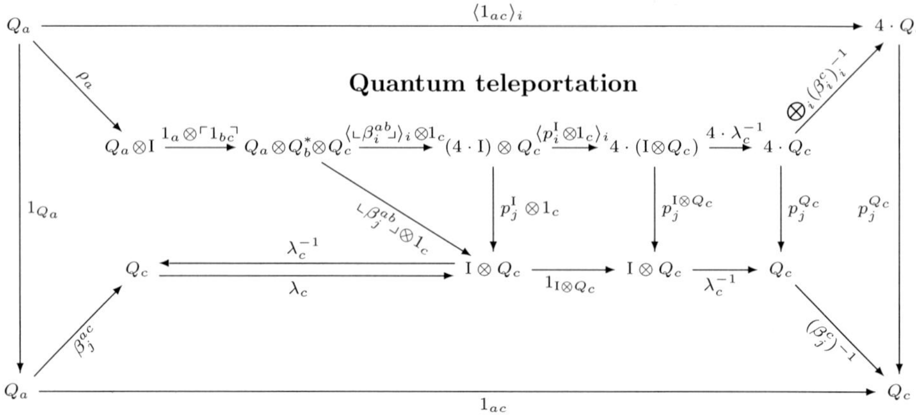

We use the universal property of the product, naturality of λ and the explicit form of the natural isomorphism $v_c := \langle p_i^{\mathrm{I}} \otimes 1_c \rangle_{i=1}^{i=4}$. In the specific case of a Bell-base observation we use $1_Q^\dagger = 1_Q$, $(\sigma_Q^\oplus)^\dagger = \sigma_Q^\oplus$ and $(\sigma_Q^\oplus \circ \beta_3)^\dagger = \beta_3^\dagger \circ (\sigma_Q^\oplus)^\dagger = \beta_3 \circ \sigma_Q^\oplus$.
■

Although in **Rel** teleportation works for 'individual observational branches' it fails to admit the full teleportation protocol since there are only two automorphisms of Q (which is just a two-element set, *i.e.* the type of 'classical bits'), and hence there is no teleportation base.

We now consider sufficient conditions on the ambient category **C** for a teleportation base to exist. We remark firstly that if $\mathbf{C}(\mathrm{I},\mathrm{I})$ contains an additive inverse for 1, then it is a ring, and moreover all additive inverses exist in each hom-set $\mathbf{C}(A,B)$, so **C** is enriched over Abelian groups. Suppose then that $\mathbf{C}(\mathrm{I},\mathrm{I})$ is a ring with $1 \neq -1$. We can define a morphism

$$\mathsf{prebase}_{\mathrm{T}} = \mathsf{base}_{Q^*\otimes Q} \circ M : 4\cdot \mathrm{I} \to Q^* \otimes Q$$

where M is the endomorphism of $4 \cdot \mathrm{I}$ determined by the matrix

$$\begin{pmatrix} 1 & 0 & 1 & 0 \\ 0 & 1 & 0 & 1 \\ 0 & 1 & 0 & -1 \\ 1 & 0 & -1 & 0 \end{pmatrix}$$

The corresponding morphisms β_j will have 2×2 matrices determined by the columns of this 4×4 matrix, and will be unitary. If $\mathbf{C}(\mathrm{I},\mathrm{I})$ furthermore contains a scalar s satisfying $2s^\dagger s = 1$, then $s \bullet \mathsf{prebase}_{\mathrm{T}}$ is unitary, and the conditions for a teleportation base are fulfilled. Suppose we start with a ring R containing an element s satisfying $2s^2 = 1$. (Examples are plentiful, e.g. any subring of $\mathbb{C}$, or of $\mathbb{Q}(\sqrt{2})$, containing $\frac{1}{\sqrt{2}}$). The category of finitely generated free R-modules and R-linear maps is strongly compact closed with biproducts, and admits a teleportation base (in which s will appear as a scalar with $s = s^\dagger$), hence realizes teleportation.

6.3 *Logic-gate teleportation*

Logic gate teleportation of qubits requires only a minor modification as compared to the teleportation protocol.

THEOREM 31. *Let unitary morphism $f : Q \to Q$ be such that for each $i \in \{1, 2, 3, 4\}$ a morphism $\varphi_i(f) : Q \to Q$ satisfying $f \circ \beta_i = \varphi_i(f) \circ f$ exists. The diagram of Theorem 30 with the modifications made below commutes.*

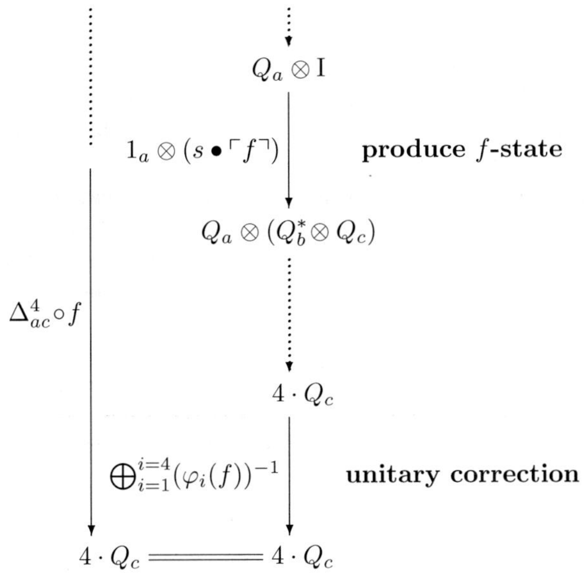

The right-hand-side of the diagram is our formal description of logic-gate teleportation of $f : Q \to Q$; the commutativity of the diagram under the stated conditions expresses the correctness of logic-gate teleportation for qubits.

Proof: The top trapezoid is the statement of the Theorem. The a, b and c-labels are the same as in the proof of teleportation. For each $j \in \{1, 2, 3, 4\}$ we have a diagram of the form below. Again we ignore the scalars in this proof.

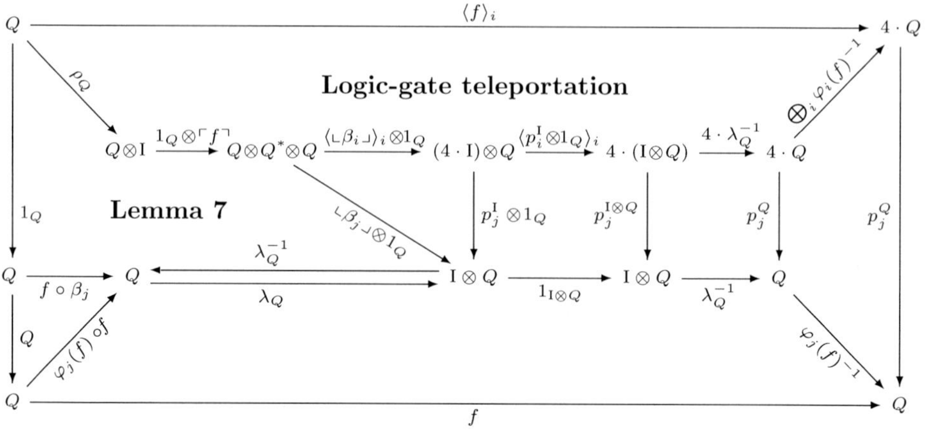

■

This two-dimensional case does not yet provide a universal computational primitive, which requires teleportation of $Q \otimes Q$-gates [Gottesman and Chuang, 1999]. We present the example of teleportation of a CNOT gate [Gottesman and Chuang, 1999] (see also [Coecke, 2003] Section 3.3).

Given a Bell base we define a CNOT gate as one which acts as follows on tensors of the Bell base maps:

$$\mathrm{CNOT} \circ (\sigma_Q^{\oplus} \otimes 1_Q) = (\sigma_Q^{\oplus} \otimes \sigma_Q^{\oplus}) \circ \mathrm{CNOT} \qquad \mathrm{CNOT} \circ (1_Q \otimes \sigma_Q^{\oplus}) = (1_Q \otimes \sigma_Q^{\oplus}) \circ \mathrm{CNOT}$$

$$\mathrm{CNOT} \circ (\beta_3 \otimes 1_Q) = (\beta_3 \otimes 1_Q) \circ \mathrm{CNOT} \qquad \mathrm{CNOT} \circ (1_Q \otimes \beta_3) = (\beta_3 \otimes \beta_3) \circ \mathrm{CNOT}$$

It follows from this that

$$\mathrm{CNOT} \circ (\beta_4 \otimes 1_Q) = (\beta_4 \otimes \sigma_Q^{\oplus}) \circ \mathrm{CNOT} \qquad \mathrm{CNOT} \circ (1_Q \otimes \beta_4) = (\beta_3 \otimes \beta_4) \circ \mathrm{CNOT}$$

from which in turn it follows by bifunctoriality of the tensor that the required unitary corrections factor into single qubit actions, for which we introduce a notation by setting

$$\mathrm{CNOT} \circ (\beta_i \otimes 1_Q) = \varphi_1(\beta_i) \circ \mathrm{CNOT} \qquad \mathrm{CNOT} \circ (1_Q \otimes \beta_i) = \varphi_2(\beta_i) \circ \mathrm{CNOT}$$

The reader can verify that for

$$4^2 \cdot (Q_{c_1} \otimes Q_{c_2}) := 4 \cdot (4 \cdot (Q_{c_1} \otimes Q_{c_2}))$$

and

$$\Delta_{ac}^{4^2} := \langle s^\dagger s \bullet \langle s^\dagger s \bullet 1_{ac} \rangle_{i=1}^{i=4} \rangle_{i=1}^{i=4} : Q_{a_1} \otimes Q_{a_2} \to 4^2 \cdot (Q_{c_1} \otimes Q_{c_2})$$

the following diagram commutes.

$$
\begin{array}{ccccl}
Q_{a_1}\otimes Q_{a_2} & = & Q_{a_1}\otimes Q_{a_2} & & \\
 & & \downarrow \rho_a & & \textbf{import unknown state} \\
 & & (Q_{a_1}\otimes Q_{a_2})\otimes \mathrm{I} & & \\
 & & \downarrow 1_a \otimes (s^2 \bullet \ulcorner \mathrm{CNOT} \urcorner) & & \textbf{produce CNOT-state} \\
 & & (Q_{a_1}\otimes Q_{a_2})\otimes ((Q_{b_1}\otimes Q_{b_2})^* \otimes (Q_{c_1}\otimes Q_{c_2})) & & \\
 & & \downarrow (\alpha,\sigma)\circ(1_a\otimes(u_b\otimes 1_c)) & & \textbf{spatial delocation} \\
 & & ((Q_{a_1}\otimes Q_{b_1}^*)\otimes(Q_{c_1}\otimes Q_{c_2}))\otimes(Q_{a_2}\otimes Q_{b_2}^*) & & \\
 & & \downarrow ((\langle s^\dagger \bullet \llcorner \beta_i^{a_1 b_1} \lrcorner\rangle_{i=1}^{i=4}\otimes 1_c)\otimes 1_2 & & \textbf{1st observation} \\
 & & ((4\cdot \mathrm{I})\otimes(Q_{c_1}\otimes Q_{c_2}))\otimes(Q_{a_2}\otimes Q_{b_2}^*) & & \\
\downarrow \Delta_{ac}^{4^2}\circ \mathrm{CNOT} & & \downarrow ((4\cdot\lambda_c^{-1})\circ \upsilon_c)\otimes 1_2 & & \textbf{1st communication} \\
 & & (4\cdot(Q_{c_1}\otimes Q_{c_2}))\otimes(Q_{a_2}\otimes Q_{b_2}^*) & & \\
 & & \downarrow (\bigoplus_{i=1}^{i=4}(\varphi_1^c(\beta_i))^{-1})\otimes 1_2 & & \textbf{1st correction} \\
 & & (4\cdot(Q_{c_1}\otimes Q_{c_2}))\otimes(Q_{a_2}\otimes Q_{b_2}^*) & & \\
 & & \downarrow (4\cdot 1_c)\otimes\langle s^\dagger \bullet \llcorner \beta_i^{a_2 b_2} \lrcorner\rangle_{i=1}^{i=4} & & \textbf{2nd observation} \\
 & & (4\cdot(Q_{c_1}\otimes Q_{c_2}))\otimes(4\cdot \mathrm{I}) & & \\
 & & \downarrow (4\cdot\rho_{4c}^{-1})\circ\tau_{4c} & & \textbf{2nd communication} \\
 & & (4\cdot(4\cdot(Q_{c_1}\otimes Q_{c_2}))) & & \\
 & & \downarrow \bigoplus_{i=1}^{i=4}(4\cdot\varphi_2^c(\beta_i))^{-1} & & \textbf{2nd correction} \\
4^2\cdot(Q_{c_1}\otimes Q_{c_2}) & = & 4^2\cdot(Q_{c_1}\otimes Q_{c_2}) & &
\end{array}
$$

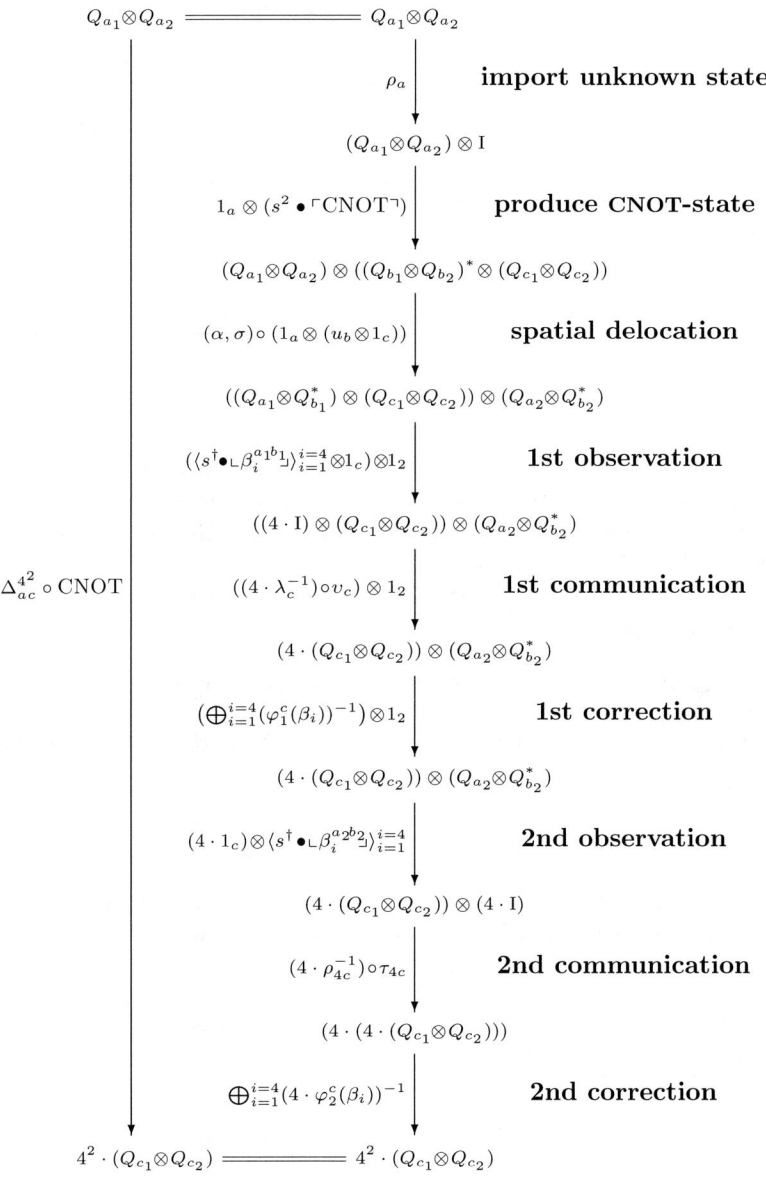

6.4 Entanglement swapping

THEOREM 32. *Setting*

$$\begin{aligned}
\gamma_i &:= (\beta_i)_* \\
\mathrm{P}_i &:= s^\dagger s \bullet (\ulcorner \gamma_i \urcorner \circ \llcorner \beta_i \lrcorner) \\
\zeta_i^{ac} &:= \bigoplus_{i=1}^{i=4} \left((1_b^* \otimes \gamma_i^{-1}) \otimes (1_d^* \otimes \beta_i^{-1}) \right) \\
\Theta_{ab} &:= 1_d^* \otimes \langle \mathrm{P}_i \rangle_{i=1}^{i=4} \otimes 1_c \\
\Omega_{ab} &:= \langle s^\dagger s^3 \bullet (\ulcorner 1_{ba} \urcorner \otimes \ulcorner 1_{dc} \urcorner) \rangle_{i=1}^{i=4}
\end{aligned}$$

the following diagram commutes.

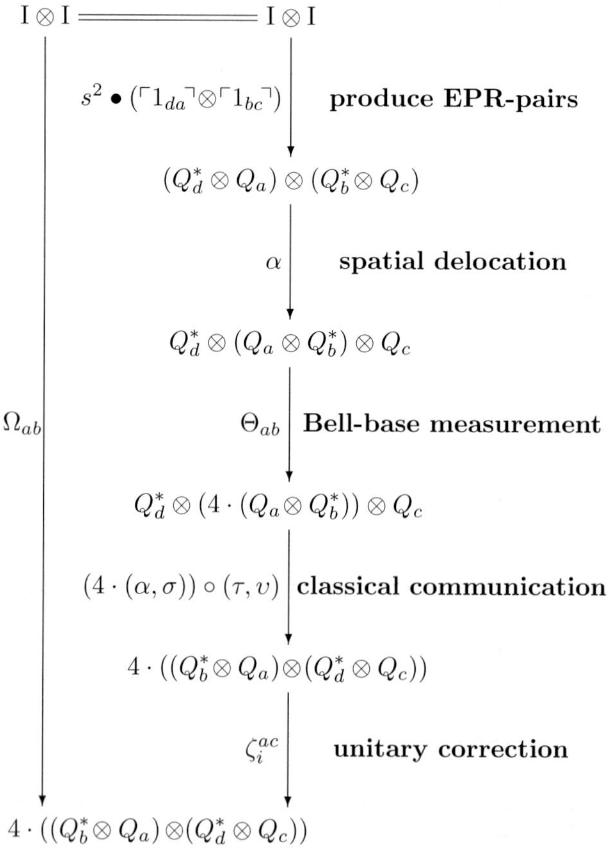

The right-hand-side of the above diagram is our formal description of the entanglement swapping protocol.

Proof: The top trapezoid is the statement of the Theorem. We have a diagram of the form below for each $j \in \{1, 2, 3, 4\}$. To simplify the notation of the types we set (a^*, b, c^*, d) for $Q_a^* \otimes Q_b \otimes Q_c^* \otimes Q_d$ etc. Again we ignore the scalars in this proof.

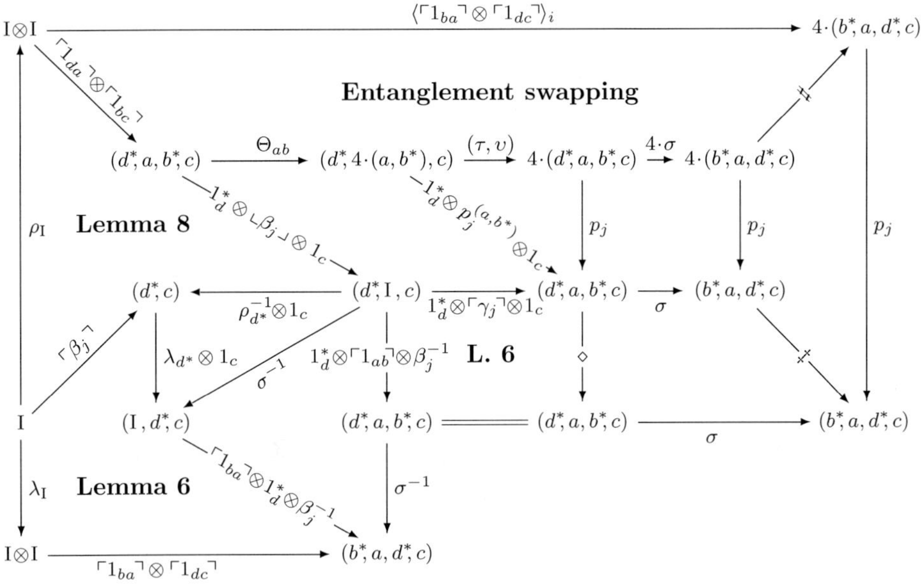

where

$$\sharp := \bigoplus_i (1_b^* \otimes \gamma_i^{-1} \otimes 1_d^* \otimes \beta_i^{-1}) \quad , \quad \ddagger := 1_b^* \otimes \gamma_j^{-1} \otimes 1_d^* \otimes \beta_j^{-1} \quad \text{and} \quad \diamond := 1_d^* \otimes \gamma_j^{-1} \otimes 1_b^* \otimes \beta_j^{-1}$$

■

We use $\gamma_i = (\beta_i)_*$ rather than β_i to make P_i an endomorphism and hence a projector. The general definition of a 'bipartite entanglement projector' is

$$\mathrm{P}_f := \ulcorner f \urcorner \circ \llcorner f_* \lrcorner = \ulcorner f \urcorner \circ \llcorner f^\dagger \lrcorner \circ \sigma_{A^*,B} : A^* \otimes B \to A^* \otimes B$$

for $f : A \to B$, so in fact $\mathrm{P}_i = \mathrm{P}_{(\beta_i)_*}$.

7 EXTENSIONS AND FURTHER DEVELOPMENTS

Since its first publication in 2004, a number of elaborations on the categorical quantum axiomatics described above have been proposed, by ourselves in collaboration with members of our group, Ross Duncan, Dusko Pavlovic, Eric Oliver Paquette, Simon Perdrix and Bill Edwards, and also by others elsewhere, most notably Peter Selinger and Jamie Vicary. We shall present some of the main developments.

7.1 Projective structure

We shall discuss our first topic at considerably greater length than the others we shall cover in this survey. The main reason for this is that it concerns the passage to a projective point of view, which makes for an evident comparison with the standard approaches to quantum logic going back to [Birkhoff and von Neumann, 1936]. Thus it seems appropriate to go into some detail in our coverage of this topic, in the context of the Handbook in which this article will appear.

The axiomatics we have given corresponds to the pure state picture of quantum mechanics. The very fact that we can faithfully carry out linear-algebraic calculations using the semi-additive structure provided by biproducts means that states will typically carry redundant global phases, as is the case for vectors in Hilbert spaces. Eliminating these means 'going projective'. The *quantum logic* tradition provides one way of doing so [Birkhoff and von Neumann, 1936]. Given a Hilbert space one eliminates global scalars by passing to the projection lattice. The non-Boolean nature of the resulting lattice is then taken to be characteristic for quantum behaviour. This leads one then to consider certain classes of non-distributive lattices as 'quantum structures'.

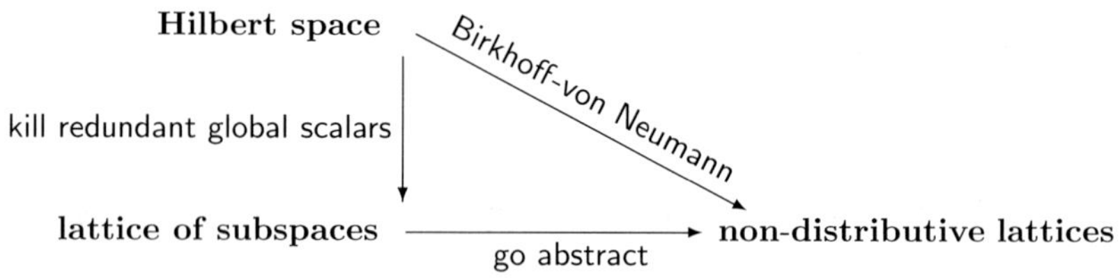

It is well-known that there is no obvious counterpart for the Hilbert space tensor product when passing to these more general classes of lattices. This is one reason why Birkhoff-von Neumann style quantum logic never penetrated the mainstream physics community, and is particularly unfortunate in the light of the important role that the tensor product plays in quantum information and computation.

But one can also start from the whole category of finite dimensional Hilbert spaces and linear maps **FdHilb**. Then we can consider 'strongly compact closed categories + some additive structure' as its appropriate abstraction, and hope to find some abstractly valid counterpart to 'elimination of redundant global scalars'.

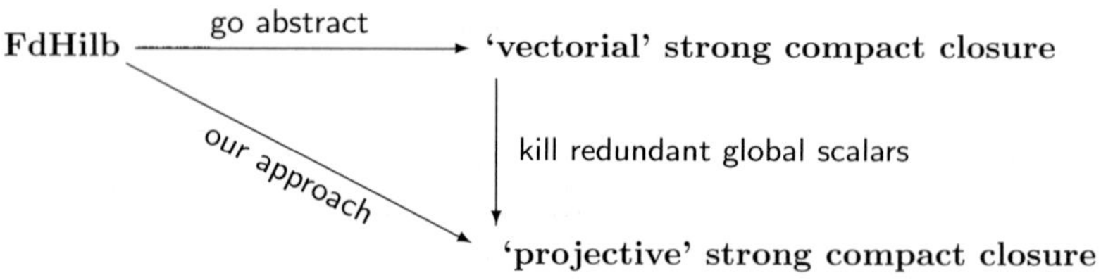

The major advantage which such a construction has is that the tensor product is now part of the mathematical object under consideration, and hence will not be lost in the passage from vectorial spaces to projective ones.

This passage was realised in [Coecke, 2007a] as follows. For morphisms in **FdHilb**, i.e. linear maps, if we have that $f = e^{i\theta} \cdot g$ with $\theta \in [0, 2\pi[$ for $f, g : \mathcal{H}_1 \to \mathcal{H}_2$, then

$$f \otimes f^\dagger = e^{i\theta} \cdot g \otimes (e^{i\theta} \cdot g)^\dagger = e^{i\theta} \cdot g \otimes e^{-i\theta} \cdot g^\dagger = g \otimes g^\dagger \, .$$

Now in abstract generality, given a strongly compact closed category $\mathbf{C}$, we can define a new category $\mathsf{WProj}(\mathbf{C})$ with the same objects as those of $\mathbf{C}$, but with

$$\mathsf{WProj}(\mathbf{C})(A, B) := \{ f \otimes f^\dagger \mid f \in \mathbf{C}(A, B) \}$$

as hom-sets and in which composition is given by

$$(f \otimes f^\dagger) \,\bar{\circ}\, (g \otimes g^\dagger) := (f \circ g) \otimes (f \circ g)^\dagger.$$

One easily shows that $\mathsf{WProj}(\mathbf{C})$ is again a strongly compact closed category. The abstract counterpart to elimination of global phases is expressed by the following propositions.

PROPOSITION 33. [Coecke, 2007a] *For morphisms f, g and scalars s, t in a strongly compact closed category, we have*

$$s \bullet f = t \bullet g \ \wedge \ s \circ s^\dagger = t \circ t^\dagger = 1_{\mathrm{I}} \quad \Longrightarrow \quad f \otimes f^\dagger = g \otimes g^\dagger .$$

PROPOSITION 34. [Coecke, 2007a] *For morphisms f and g in a strongly compact closed category with scalars S we have*

$$f \otimes f^\dagger = g \otimes g^\dagger \quad \Longrightarrow \quad \exists s, t \in S.\, s \bullet f = t \bullet g \ \wedge \ s \circ s^\dagger = t \circ t^\dagger .$$

In particular we can set

$$s := (\ulcorner f \urcorner)^\dagger \circ \ulcorner f \urcorner \qquad \text{and} \qquad t := (\ulcorner g \urcorner)^\dagger \circ \ulcorner f \urcorner .$$

While the first proposition is straightforward, the second one is somewhat more surprising. It admits a simple graphical proof. We represent units by dark triangles and their adjoints by the same triangle but depicted upside down. Other morphisms are depicted by square boxes as before, with the exception of scalars which are depicted by 'diamonds'. The scalar $s := (\ulcorner f \urcorner)^\dagger \circ \ulcorner f \urcorner$ is depicted as

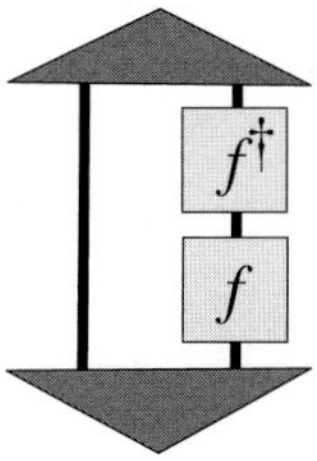

Bifunctoriality means that we can move these boxes upward and downward, and naturality provides additional modes of movement, e.g. scalars admit arbitrary movements. Now, given that $f \otimes f^\dagger = g \otimes g^\dagger$, that is, in a picture,

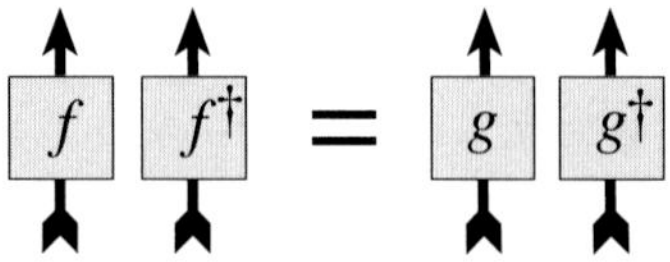

we need to show that $s \bullet f = t \bullet g$ and $s \circ s^\dagger = t \circ t^\dagger$ for some choice of scalars s and t, that is, in a picture,

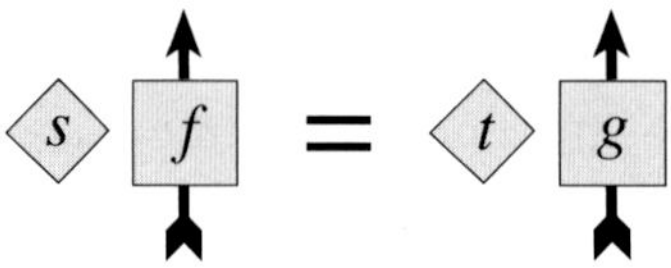

The choice that we will make for s and t is

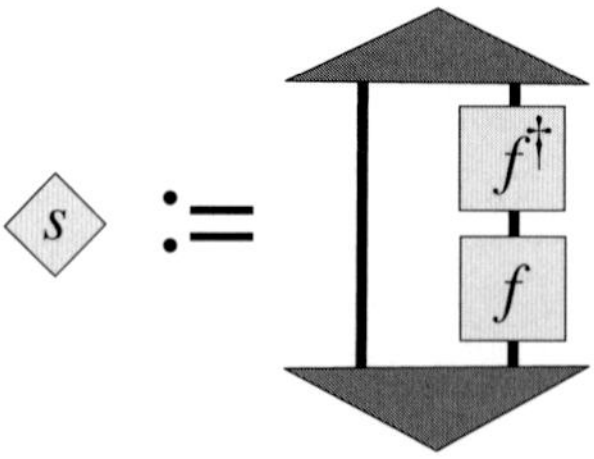

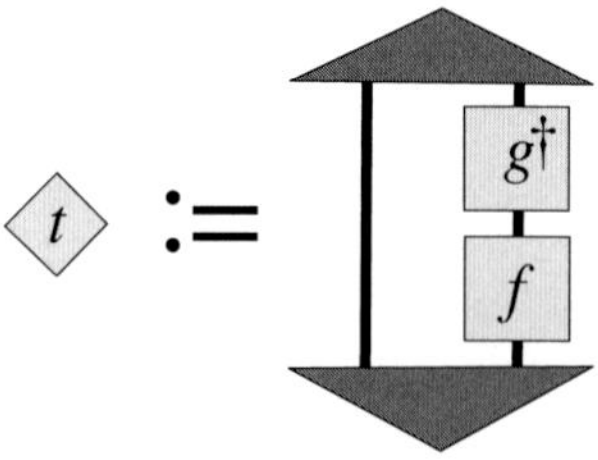

Then we indeed have $s \bullet f = t \bullet g$ since in

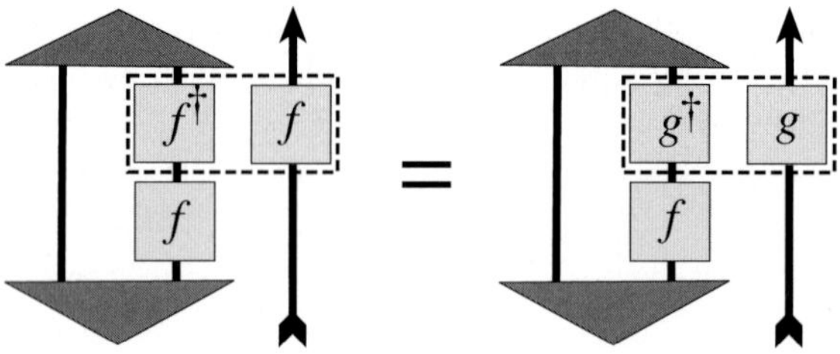

the areas within the dotted line are equal by assumption. We also have that $s \circ s^\dagger = t \circ t^\dagger$ since

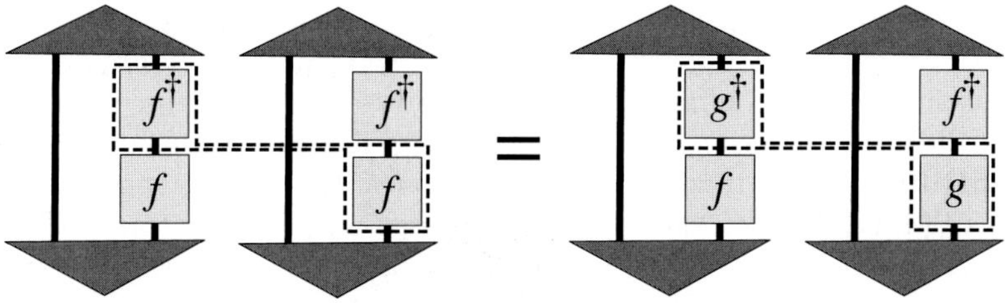

which completes the proof.

As expected, biproducts do not survive the passage from $\mathbf{C}$ to $\mathsf{WProj}(\mathbf{C})$ but the weaker structure which results still suffices for a comprehensive description of the protocols we have discussed in this article. In particular, the distributivity natural isomorphisms

$$\mathrm{dist}_{0,l} : A \otimes 0 \simeq 0 \qquad \mathrm{dist}_l : A \otimes (B \oplus C) \simeq (A \otimes B) \oplus (A \otimes C)$$

$$\mathrm{dist}_{0,r} : 0 \otimes A \simeq 0 \qquad \mathrm{dist}_r : (B \oplus C) \otimes A \simeq (B \otimes A) \oplus (C \otimes A) .$$

carry over to $\mathsf{WProj}(\mathbf{C})$. Details can found in [Coecke, 2007a].

Our framework also allows a precise general statement of the incompatibility of biproducts with projective structure.

Call a strongly compact closed category *projective* iff equality of projections implies equality of the corresponding states, that is,

(1) $\forall\, \psi, \phi : \mathrm{I} \to A . \quad \psi \circ \psi^\dagger = \phi \circ \phi^\dagger \implies \psi = \phi .$

PROPOSITION 35. [Coecke, 2007a] *If a strongly compact closed category with biproducts is projective and the semiring of scalars admits negatives, i.e. is a ring, then we have* $1 = -1$*, that is, there are no non-trivial negatives.*

Having no non-trivial negatives of course obstructs the description of interference.

7.2 Mixed states and Completely Positive Maps

The categorical axiomatics set out in this article primarily refers to the pure-state picture of quantum mechanics. However, for many purposes, in particular those of quantum information, it is *mixed states*, acted on by *completely positive maps*, which provide the most appropriate setting. Peter Selinger [2007] proposed a general categorical construction, directly in the framework of the categorical axiomatics of [Abramsky and Coecke, 2004] which has been described in this article, to capture the passage from the pure states to the mixed states picture.

The construction proceeds as follows. Given any strongly compact closed category $\mathbf{C}$ we define a new category $\mathsf{CPM}(\mathbf{C})$ with the same objects as $\mathbf{C}$ but with morphisms given by

$\mathsf{CPM}(\mathbf{C})(A, B) :=$

$$\left\{(1_B \otimes \epsilon_C \otimes 1_{B^*}) \circ (1_{B\otimes C} \otimes \sigma_{B^*,C^*}) \circ (f \otimes f_*) \mid f \in \mathbf{C}(A, B \otimes C)\right\}$$

where for simplicity we assume that the monoidal structure is strict. Composition in $\mathsf{CPM}(\mathbf{C})$ is inherited pointwise from $\mathbf{C}$. The morphisms of the category $\mathsf{CPM}(\mathbf{FdHilb})$ are exactly the *completely positive maps*, and the morphisms in the hom-set $\mathsf{CPM}(\mathbf{FdHilb})(\mathbb{C}, \mathcal{H})$ are exactly the self-adjoint operators with positive trace on $\mathcal{H}$. The category $\mathsf{WProj}(\mathbf{C})$ faithfully embeds in $\mathsf{CPM}(\mathbf{C})$ by setting

$$f \otimes f^\dagger \mapsto f \otimes f_*.$$

Metaphorically, we have

$$\frac{\mathsf{CPM}(\mathbf{C})}{\mathsf{WProj}(\mathbf{C})} = \frac{\text{density operators}}{\text{projectors}}.$$

For more details on the CPM-construction we refer the reader to [Selinger, 2007].

Recently it was shown that the CPM-construction does not require strong compact closure, but only dagger symmetric monoidal structure. Details are in [Coecke, 2007]. An axiomatic presentation of categories of completely positive maps is given in [Coecke, 2008].

7.3 Generalised No-Cloning and No-Deleting theorems

The No-Cloning theorem [Dieks, 1982; Wootters and Zurek, 1982] is a basic limitative result for quantum mechanics, with particular significance for quantum information. It says that there is no unitary operation which makes perfect copies of an unknown (pure) quantum state. A stronger form of this result is the No-Broadcasting theorem [Barnum *et al.*, 1996], which applies to mixed states. There is also a No-Deleting theorem [Pati and Braunstein, 2000].

The categorical and logical framework which we have described provides new possibilities for exploring the structure, scope and limits of of quantum information processing, and the features which distinguish it from its classical counterpart. One area where some striking progress has already been made is the axiomatics of No-Cloning and No-Deleting. It is possible to delimit the classical-quantum boundary here in quite a subtle way. On the one hand, we have the strongly compact closed structure which is present in the usual Hilbert space setting for QIC, and which we have shown accounts in generality for the phenomena of entanglement. Suppose we were to assume that *either* copying *or* deleting were available in a strongly compact closed category as *uniform operations*. Mathematically, a uniform copying operation means a *natural diagonal*

$$\Delta_A : A \to A \otimes A$$

i.e. a monoidal natural transformation, which moreover is co-associative and co-

commutative:

$$
\begin{array}{ccccc}
A & \xrightarrow{\Delta} & A \otimes A & \xrightarrow{1 \otimes \Delta} & A \otimes (A \otimes A) \\
\Vert & & & & \downarrow{\scriptstyle \alpha_{A,A,A}} \\
A & \xrightarrow[\Delta]{} & A \otimes A & \xrightarrow[\Delta \otimes 1]{} & (A \otimes A) \otimes A
\end{array}
\qquad\qquad
\begin{array}{ccc}
A & \xrightarrow{\Delta} & A \otimes A \\
& {\scriptstyle \Delta} \searrow & \downarrow{\scriptstyle \sigma_{A,A}} \\
& & A \otimes A
\end{array}
$$

Thinking of the diagonal associated with the usual cartesian product, one sees immediately that co-commutativity and co-associativity are basic requirements for a copying operation: if I have two copies of the same thing, it does not matter which order they come in, and if I produce three copies by iterating the copying operation, which copy I choose to perform the second copying operation on is immaterial. Naturality, on the other hand, corresponds essentially to *basis-independence* in the Hilbert space setting; it says that the operation exists 'for logical reasons', in a representation-independent form.

We have shown recently that under these assumptions *the category trivializes*; in other words, that this combination of quantum and classical features is *inconsistent*, leading to a collapse of the structure. The precise form of the result is that under these hypotheses every endomorphism in the category is a scalar multiple of the identity.

Similar generalizations of the No-Deleting theorem [Pati and Braunstein, 2000] and the No-Broadcasting theorem [Barnum *et al.*, 1996] also hold. Papers on these results are in preparation.

One striking feature of these results is that they are visibly in the same genre as a well-known result by Joyal in categorical logic [Lambek and Scott, 1986] showing that a 'Boolean cartesian closed category' trivializes, which provides a major roadblock to the computational interpretation of classical logic. In fact, they strengthen Joyal's result, insofar as the assumption of a full categorical product (diagonals *and* projections) in the presence of a classical duality is weakened. This shows a heretofore unsuspected connection between limitative results in proof theory and No-Go theorems in quantum mechanics.

Another interesting point is the way that this result is delicately poised. The basis structures to be discussed in the next sub-section do assume commutative comonoid structures existing in strongly compact closed categories—indeed with considerable additional properties, such as the Frobenius identity. Not only is this consistent, such structures correspond to a major feature of Hilbert spaces, namely orthonormal bases. The point is that there are many such bases for a given Hilbert space, and none are canonical. Indeed, the *choice* of basis corresponds to the *choice* of measurement set-up, to be made by a 'classical observer'. The key ingredient which leads to inconsistency, and which basis structures lack, is *naturality*, which, as we have already suggested, stands as an abstract proxy for basis-independence.

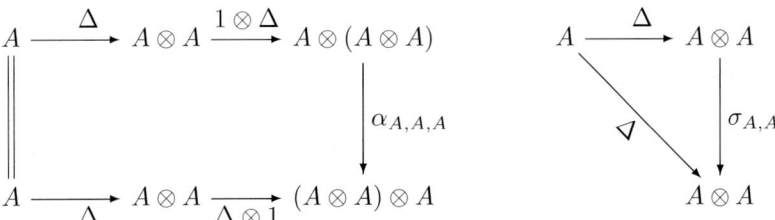

7.4 Basis Structures and Classical Information

In this article, an approach to measurements and classical information has been developed based on biproducts. This emphasizes the *branching structure* of measurements due to their probabilistic outcomes.

One may distinguish the 'multiplicative' from the 'additive' levels of our axiomatization (using the terminology of Linear logic [Girard, 1987]). The multiplicative, purely tensorial level of strongly compact closed categories shows, among other things, how a remarkable amount of multilinear algebra, encompassing much of the structure needed for quantum mechanics and quantum information, can be done without any substrate of linear algebra. Moreover, this level of the axiomatization carries a very nice diagrammatic calculus, which we have sampled informally. In general, the return on structural insights gained from the axiomatization seems very good. The additive level of biproducts reinstates a linear (or 'semilinear') level of structure, albeit with fairly weak assumptions, and there is more of a sense of recapitulating familiar definitions and calculations. While a diagrammatic calculus is still available here (see [Abramsky and Duncan, 2006]), it is subject to a combinatorial unwieldiness familiar from process algebra in Computer Science [Milner, 1989] (cf. the 'Expansion Theorem').

An alternative approach to measurements and classical information has been developed in a series of papers [Coecke and Pavlovic, 2007; Coecke and Paquette, 2006; Coecke, Pavlovic and Vicary, 2008a; Coecke, Paquette and Perdrix, 2008] under various names, the best of which is probably 'basis structure'. Starting from the standard idea that a measurement set-up corresponds to a choice of orthonormal basis, the aim is to achieve an axiomatization of the notion of basis as an additional structure. Of course, the notion of basis developed in Section 5 has all the right properties, but it is defined in terms of biproducts, while the aim here is to achieve an axiomatization purely at the multiplicative level.

This is done in an interesting way, bringing the informatic perspective to the fore. One can see the choice of a basis as determining a notion of 'classical data', namely the basis vectors. These vectors are subject to the classical operations of *copying* and *deleting*, so in a sense classical data, defined with respect to a particular choice of basis, stands as a contrapositive to the No-Cloning and No-Deleting theorems. Concretely, having chosen a basis $\{|i\rangle\}$ on a Hilbert space $\mathcal{H}$, we can define linear maps

$$\mathcal{H} \longrightarrow \mathcal{H} \otimes \mathcal{H} :: |i\rangle \mapsto |ii\rangle, \qquad \mathcal{H} \longrightarrow \mathbb{C} :: |i\rangle \mapsto 1$$

which do correctly copy and delete the basis vectors (the 'classical data'), although not of course the other vectors.

These considerations lead to the following definition. A *basis structure* on an object A in a strongly compact closed category is a commutative comonoid structure on A

$$Copy : A \to A \otimes A, \qquad Delete : A \to \mathrm{I}$$

subject to a number of additional axioms, the most notable of which is the *Frobenius identity* [Carboni and Walters, 1987]. In **FdHilb** these structures exactly correspond to orthonormal bases [Coecke, Pavlovic and Vicary, 2008a], which justifies their name and interpretation.

Quantum measurements can be defined relative to these structures, as self-adjoint Eilenberg-Moore coalgebras for the comonads induced by the above comonoids [Coecke and Pavlovic, 2007]. In **FdHilb** these indeed correspond exactly to projective spectra. The Eilenberg-Moore coalgebra square

$$\begin{array}{ccc} A & \xrightarrow{\quad Measure \quad} & X \otimes A \\ {\scriptstyle Measure}\downarrow & & \downarrow{\scriptstyle 1_X \otimes Measure} \\ X \otimes A & \xrightarrow[\quad Copy \otimes 1_A \quad]{} & X \otimes X \otimes A \end{array}$$

can be seen as an operational expression of von Neumann's projection postulate in a resource-sensitive setting: measuring twice is the same as measuring once and then copying the measurement outcome. This abstract notion of measurement admits generalisation to POVMs and PMVMs, for which a generalised Naimark dilatation theorem can be proved at the abstract level [Coecke and Paquette, 2006].

Within $\mathsf{CPM}(\mathbf{C})$ the *decoherence* aspect of quantum measurement, which, concretely in **FdHilb**, is the completely positive map which eliminates non-diagonal elements relative to the measurement basis, arises as

$$Copy \circ Copy^{\dagger} : X \otimes X \to X \otimes X$$

where X is now taken to be self-dual, that is, $X = X^*$.

These basis structures not only allow for classical data, measurement and control operations to be described but also provide useful expressiveness when discussing multipartite states and unitaries. For example, they capture GHZ-states in a canonical fashion [Coecke and Pavlovic, 2007], and enable an elegant description of the state-transfer protocol [Coecke, Paquette and Perdrix, 2008].

Vicary showed that if one drops the co-commutativity requirement of basis structures in **FdHilb**, then, rather than all orthonormal bases, one finds exactly all finite dimensional C*-algebras [Vicary, 2008].

The fact that these multiplicative basis structures allow measurements to be expressed without any explicit account of branching may be compared to the way that the pure λ-calculus can be used to encode booleans and conditionals [Barendregt, 1984]. From this perspective, explicit branching can be seen to have its merits, while *model-checking* [Clarke *et al.*, 1999] has done much to ameliorate the combinatorial unwieldiness mentioned above. It is likely that further insights will be gained by a deeper understanding of the relationships between the additive and multiplicative levels.

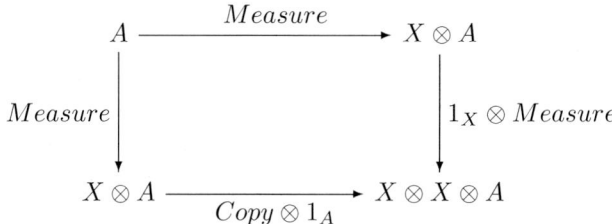

7.5 Complementary observables and phases

A further step is taken in [Coecke and Duncan, 2008], where abstract counterparts to (relative) phases are defined. Given a basis structure on X and a point $\psi : \mathrm{I} \to X$ its action on X is defined to be the morphism

$$\Delta(\psi) := Copy^{\dagger} \circ (\psi \otimes 1_X) : X \to X.$$

From the axioms of basis structures it follows that the set of all these actions on the hom-set $\mathbf{C}(\mathrm{I}, X)$ is a commutative monoid. Defining *unbiassed points* as those $\psi \in \mathbf{C}(\mathrm{I}, X)$ for which $\Delta(\psi)$ is unitary, the corresponding set of *unbiassed actions* is always an abelian group, which we call the *phase group*. In the case of the qubit in **FdHilb** the phase group corresponds to the equator of the Bloch sphere, that is, indeed, to relative phase data.

Also in [Coecke and Duncan, 2008] an axiomatics is proposed for *complementary observables*. It is shown that for all known constructions of complementary quantum observables, the corresponding basis structures obey a 'scaled' variant of the bialgebra laws. This scaled bialgebra structure together with the phase group is sufficiently expressive to describe all linear maps, hence all mutipartite states and unitary operators, in **FdHilb**. It provides an abstract means to reason about quantum circuits and to translate between quantum computational models, such as the circuit model and the measurement-based model.

As an application, a description is given of the quantum Fourier transform, the key ingredient of Shor's factoring algorithm [Shor, 1994], the best-known example of a quantum algorithm.

7.6 The quantum harmonic oscillator

Jamie Vicary [2007] gave a purely categorical treatment of the quantum harmonic oscillator, directly in the setting described in this article, of strongly compact closed categories with biproducts. In Linear logic terminology, he introduced an 'exponential level' of structure, corresponding to Fock space. This provides a monoidal adjunction that encodes the raising and lowering operators into a co-commutative comonoid. Generalised coherent states arise through the hom-set isomorphisms defining the adjunction, and it is shown that they are eigenstates of the lowering operators. Similar results were independently obtained in [Fiore, 2007] in an abstract 'formal power series' context, with a motivation stemming from Joyal's theory of species.

7.7 Automated quantum reasoning

The structures uncovered by the research programme we have described provide a basis for the design of software tools for automated reasoning about quantum phenomena, protocols and algorithms. Several MSc students at Oxford University

Computing Laboratory have designed and implemented such tools for their Masters Thesis projects. An ongoing high-level comprehensive approach has recently be initiated by Lucas Dixon and Ross Duncan [2008].

7.8 Diagrammatic reasoning

We have used a diagrammatic notation for tensor categories in an informal fashion. In fact, this diagrammatic notation, which can be traced back at least to Penrose [1971], was made fully formal by Joyal and Street [1991]; topological applications can be found in [Turaev, 1994].

The various structures which have arisen in the above discussion, such as strong compact closure, biproducts, dagger Frobenius comonoids, phase groups, scaled bialgebras, and the exponential structures used in the description of the quantum harmonic oscilator, all admit intuitive diagrammatic presentations in this tradition. References on these include [Abramsky and Duncan, 2006; Coecke and Paquette, 2006; Coecke and Duncan, 2008; Vicary, 2007]. Tutorial introductions to these diagrammatic calculi are given in [Coecke and Paquette, 2008; Selinger, 2008a]

These diagrammatic calculi provide very effective tools for the communication of the structural ideas. The software tools mentioned in the previous sub-section all support the presentation and manipulation of such diagrams as their interface to the user.

7.9 Free constructions

In [Abramsky, 2005] a number of free constructions are described in a simple, synthetic and conceptual manner, including the free strongly compact closed category over a dagger category, and the free traced monoidal category. The Kelly-Lapalaza [1980] construction of the free compact closed category is recovered in a structured and conceptual fashion.

These descriptions of free categories in simple combinatorial terms provide a basis for the use of diagrammatic calculi as discussed in the previous sub-section.

7.10 Temperley-Lieb algebra and connections to knot theory and topological quantum field theory

Our basic categorical setting has been that of symmetric monoidal categories. If we weaken the assumption of symmetry, to *braided* or *pivotal* categories, we come into immediate contact with a wide swathe of developments relating to knot theory, topology, topological quantum field theories, quantum groups, etc. We refer to [Freyd and Yetter, 1989; Kaufmann, 1991; Turaev, 1994; Yetter, 2001; Kock, 2003; Street, 2007] for a panorama of some of the related literature.

In [Abramsky, 2007], connections are made between the categorical axiomatics for quantum mechanics developed in this article, and the Temperley-Lieb algebra, which plays a central rôle in the Jones polynomial and ensuing developments. For illustration, we show the defining relations of the Temperley-Lieb algebra, in the diagrammatic form introduced by Kauffman:

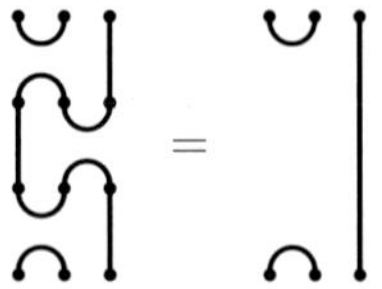

$U_1U_2U_1 = U_1$

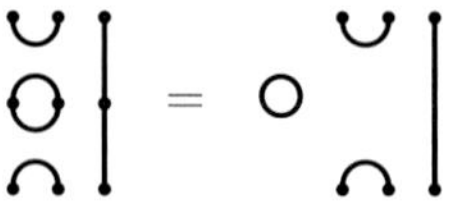

$U_1^2 = \delta U_1$

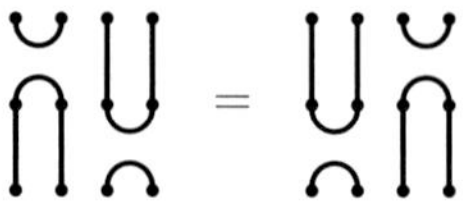

$U_1U_3 = U_3U_1$

The relationship with the diagrammatic notation we have been using should be reasonably clear, The 'cups' and 'caps' in the above diagrams correspond to the triangles we have used to depict units and counits.

An important mediating rôle is played by the *geometry of interaction* [Girard, 1989; Abramsky, 1996], which provides a mathematical model of information flow in logic (cut-elimination of proofs) and computation (normalization of λ-terms).

The Temperley-Lieb algebra is essentially the (free) *planar* version of our quantum setting; and new connections are made between logic and geometry in [Abramsky, 2007]. For example, a simple, direct description of the Temperley-Lieb algebra, with no use of quotients, is given in [Abramsky, 2007]. This leads in turn to full completeness results for various non-commutative logics. Moreover, planarity is shown to be an invariant of the information flow analysis of cut elimination.

This leads to a number of interesting new kinds of questions:

- It seems in practice that few naturally occurring quantum protocols require the use of the symmetry maps. (For example, none of those described in this paper do). How much of Quantum Informatics can be done 'in the plane'? What is the significance of this constraint?

- Beyond the planar world we have *braiding*, which carries 3-dimensional geometric information. Does this information have some computational significance? Some ideas in this direction have been explored by Kauffman and Lomonaco [2002], but no clear understanding has yet been achieved.

- Beyond this, we have the general setting of Topological Quantum Field Theories [Witten, 1988; Atiyah, 1998] and related notions. This may be relevant to Quantum Informatic concerns in (at least) two ways:

    1. A novel and promising paradigm of *Topological Quantum Computing* has recently been proposed [Freedman *et al.*, 2004].

    2. The issues arising from *distributed quantum computing*, *quantum security protocols* etc. mean that the interactions between quantum informatics and spatio-temporal structure will need to be considered.

7.11 Logical syntax

In [Abramsky and Duncan, 2006] a strongly normalising proof-net calculus corresponding to the logic of strongly compact closed categories with biproducts is presented. The calculus is a full and faithful representation of the free strongly compact closed category with biproducts on a given category with an involution. This syntax can be used to represent and reason about quantum processes.

In [Duncan, 2006] this is extended to a description of the free strongly compact category generated by a monoidal category. This is applied to the description of the measurement calculus of [Danos *et al.*, 2007].

7.12 Completeness

In [Selinger, 2008] Selinger showed that finite-dimensional Hilbert spaces are equationally complete for strongly compact closed categories. This result shows that if we want to verify an equation expressed purely in the language of strongly compact closed categories, then it suffices to verify that it holds for Hilbert spaces.

7.13 Toy quantum categories

In [Coecke and Edwards, 2008] it is shown that Spekkens' well-known 'toy model' of quantum mechanics described in [Spekkens, 2007] can be regarded as an instance of the categorical quantum axiomatics. The category **Spek** is defined to be the dagger symmetric monoidal subcategory of **Rel** generated by those objects whose cardinality is a power of 4, the symmetry group on 4 elements, and a well-chosen copying-deleting pair for the 4 element set.

BIBLIOGRAPHY

[Abramsky, 1996] S. Abramsky. *Retracing some paths in process algebra.* In: Proceedings of CONCUR 96, Lectures Notes in Computer Science Volume 1119, pp. 1–17. Springer-Verlag, 1996.

[Abramsky, 2004] S. Abramsky. *High-level methods for quantum computation and information.* In: Proceedings of the 19th Annual IEEE Symposium on Logic in Computer Science, IEEE Computer Science Press, 2004.

[Abramsky, 2005] S. Abramsky. *Abstract scalars, loops, and free traced and strongly compact closed categories.* In: Proceedings of CALCO 2005, pages 1–31. Springer Lecture Notes in Computer Science **3629**, 2005.

[Abramsky, 2007] S. Abramsky. *Temperley-Lieb algebra: From knot theory to logic and computation via quantum mechanics.* In: Mathematics of Quantum Computing and Technology, G. Chen, L. Kauffman and S. Lamonaco (eds), pages 415–458. Taylor and Francis, 2007.

[Abramsky, Blute and Panangaden, 1999] S. Abramsky, R. Blute and P. Panangaden. *Nuclear and trace ideals in tensored $*$-categories.* Journal of Pure and Applied Algebra **143**, 3–47, 1999.

[Abramsky and Coecke, 2003] S. Abramsky and B. Coecke. *Physical traces: Quantum vs. classical information processing.* Electronic Notes in Theoretical Computer Science **69**. arXiv:cs/0207057, 2003.

[Abramsky and Coecke, 2004] S. Abramsky and B. Coecke. *A categorical semantics of quantum protocols.* In: Proceedings of 19th IEEE conference on Logic in Computer Science, pages 415–425. IEEE Press. arXiv:quant-ph/0402130, 2004.

[Abramsky and Coecke, 2005] S. Abramsky and B. Coecke. *Abstract physical traces.* Theory and Applications of Categories **14**, 111–124. http://www.tac.mta.ca/tac/volumes/14/6/14-06abs.html, 2005.

[Abramsky and Duncan, 2006] S. Abramsky and R. W. Duncan. *A categorical quantum logic.* Mathematical Structures in Computer Science **16**, 469–489. arXiv:quant-ph/0512114, 2006

[Abramsky, Gay and Nagarajan, 1995] S. Abramsky, S. J. Gay and R. Nagarajan. *Interaction categories and the foundations of typed concurrent programming.* In: Deductive Program Design: Proceedings of the 1994 Marktoberdorf International Summer School, NATO Advanced Science Institutes Series F, pages 35–113. Springer-Verlag, 1995.

[Abramsky and Tzevelekos, 2008] S. Abramsky and N. Tzevelekos. *Introduction to categories and categorical logic.* In: New Structures for Physics, B. Coecke, Ed, Springer lecture Notes in Physics, 2008.

[Atiyah, 1998] M. F. Atiyah. *Topological quantum field theory.* Publications Mathématiques de l'IHES **68**, 175–186, 1998.

[Baez, 2006] J. C. Baez. *Quantum quandaries: a category-theoretic perspective.* In: The Structural Foundations of Quantum Gravity, D. Rickles, S. French and J. T. Saatsi (Eds), pages 240–266. Oxford University Press. arXiv:quant-ph/0404040, 2006

[Baez and Dolan, 1995] J. C. Baez and J. Dolan. *Higher-dimensional algebra and topological quantum field theory.* Journal of Mathematical Physics **36**, 60736105. arXiv:q-alg/9503002, 1995

[Baez and Stay, 2008] J. C. Baez and M. Stay (2008) *Physics, topology, logic and computation: A Rosetta Stone.* In: New Structures for Physics, B. Coecke, Ed, Springer lecture Notes in Physics, 2008.

[Barendregt, 1984] H. Barendregt. *The Lambda Calculus: Revised Edition.* Elsevier, 1984.

[Barnum *et al.*, 1996] H. Barnum, C. M. Caves, C. A. Fuchs, R. Jozsa, and B. Schumacher. *Noncommuting mixed states cannot be broadcast.* Physical Review Letters, **76**, 2818–2821, 1996.

[Barr, 1979] M. Barr. *$*$-Autonomous Categories.* Lecture Notes in Mathematics **752**, Springer, 1979.

[Bell, 1964] J. S. Bell. *On the Einstein Podolsky Rosen paradox.* Physics **1**, 195, 1964.

[Bennett *et al.*, 1993] C. H. Bennett, G. Brassard, C. Crépeau, R. Jozsa, A. Peres and W. K. Wooters. *Teleporting an unknown quantum state via dual classical and Einstein-Podolsky-Rosen channels.* Physical Review Letters **70**, 1895–1899, 1993.

[Birkhoff and von Neumann, 1936] G. Birkhoff and J. von Neumann. *The logic of quantum mechanics.* Annals of Mathematics **37**, 823–843, 1936.

[Bouwmeester *et al.*, 2001] D. Bouwmeester, A. Ekert and A. Zeilinger, eds. *The Physics of Quantum Information.* Springer-Verlag, 2001.

[Carboni and Walters, 1987] A. Carboni and R. F. C. Walters. *Cartesian bicategories I.* Journal of Pure and Applied Algebra **49**, 11–32, 1987.

[Clarke *et al.*, 1999] E. M. Clarke, O. Grumberg and D. Peled. *Model Checking.* Springer, 1999.

[Cockett and Seely, 1997] R. Cockett and R. A. G. Seely. *Weakly distributive categories.* Journal of Pure and Applied Algebra **114**, 85–131, 1997.

[Coecke, 2003] B. Coecke. *The logic of entanglement. An invitation.* Technical Report PRG-RR-03-12. web.comlab.ox.ac.uk/oucl/publications/tr/rr-03-12.html, 2003.

[Coecke, 2004] B. Coecke. *The logic of entanglement.* arXiv:quant-ph/0402014, 2004.

[Coecke, 2007] B. Coecke. *Complete positivity without positivity and without compactness.* Oxford University Computing Laboratory Research Report PRG-RR-07-05, 2007. web.comlab.ox.ac.uk/ oucl/publications/tr/rr-07-05.html

[Coecke, 2007a] B. Coecke. *De-linearizing linearity: projective quantum axiomatics from strong compact closure.* Electronic Notes in Theoretical Computer Science **170**, 47–72, 2007. arXiv:quant-ph/0506134

[Coecke, 2008] B. Coecke. *Axiomatic description of mixed states from Selinger's CPM-construction.* Proceedings of QPL'06. Electronic Notes in Theoretical Computer Science (to appear, 2008).

[Coecke and Duncan, 2008] B. Coecke and R. W. Duncan. *Interacting quantum observables.* In: Proceedings of the 35th International Colloquium on Automata, Languages and Programming (ICALP), pages 298–310, Lecture Notes in Computer Science **5126**, Springer-Verlag, 2008.

[Coecke and Edwards, 2008] B. Coecke and W. A. Edwards. *Toy quantum categories.* In: Proc. Quantum Physics and Logic/Development of Computational Models (QPL-DCM). Electronic Notes in Theoretical Computer Science, to appear 2008.

[Coecke and Paquette, 2006] B. Coecke and E. O. Paquette. *POVMs and Naimark's theorem without sums.* Electronic Notes in Theoretical Computer Science (To appear). arXiv:quant-ph/0608072, 2006.

[Coecke and Paquette, 2008] B. Coecke and E. O. Paquette. *Categories for the practising physicist.* In: New Structures for Physics, B. Coecke, Ed, Springer lecture Notes in Physics, 2008.

[Coecke, Paquette and Perdrix, 2008] B. Coecke, E. O. Paquette and S. Perdrix. *Bases in diagramatic quantum protocols.* In: Proc. Mathematical Foundations for Programming Semantics XXIV, to appear, 2008.

[Coecke and Pavlovic, 2007] B. Coecke and D. Pavlovic. *Quantum measurements without sums.* In: Mathematics of Quantum Computing and Technology, G. Chen, L. Kauffman and S. Lamonaco (eds), pages 567–604. Taylor and Francis, 2007. arXiv:quant-ph/0608035.

[Coecke, Pavlovic and Vicary, 2008a] B. Coecke, D. Pavlovic and J. Vicary. *Commutative †-Frobenius algebras in* **FdHilb** *are bases.* In: New Structures for Physics, B. Coecke, ed, Springer Lecture Notes in Physics, 2008.

[Crane, 2006] L Crane. *Categorical geometry and the mathematical foundations of quantum general relativity*, 2006. arXiv:gr-qc/0602120v

[Danos *et al.*, 2007] V. Danos, E. Kashefi and P. Panangaden. *The Measurement Calculus.* Journal of the ACM **54**, 2007. arXiv:quant-ph/0412135

[Deligne, 1990] P. Deligne. *Catégories tannakiennes.* In: The Grothendieck Festschrift Volume II, Progress in Mathematics **87**, pages 111–196. Birkhäuser, 1990.

[Deutsch, 1985] D. Deutsch. *Quantum theory, the Church-Turing principle and the universal quantum computer.* Proceedings of the Royal Society of London. Series A, Mathematical and Physical Sciences **400**, 97–117, 1985.

[Deutsch, 1989] D. Deutsch. *Quantum computational networks.* Proceedings of the Royal Society of London, Series A, Mathematical and Physical Sciences **425**, 73–90, 1989.

[Dieks, 1982] D. G. B. J. Dieks. *Communication by EPR devices.* Physics Letters A **92**, 271–272, 1982.

[Dirac, 1947] P. A. M. Dirac, (1947) *The Principles of Quantum Mechanics* (third edition). Oxford University Press, 1947.

[Dixon and Duncan, 2008] L. Dixon and R. W. Duncan. *Extending graphical representations for compact closed categories with applications to symbolic quantum computation.* Proceedings of AISC'08, Lecture Notes in Artificial Intelligence, Springer, 2008.

[Doëring and Isham, 2007] A. Doëring and C. J. Isham. *A topos foundation for theories of physics: I. Formal languages for physics*, 2007. arXiv:quant-ph/0703060

[Doplicher and Roberts, 1989] S. Doplicher and J. E. Roberts. *A new duality theory for compact groups.* Inventiones Mathematicae **98**, 157, 1989.

[Duncan, 2003] R. W. Duncan. *Quantum entanglement and multiplicative linear logic.* Transfer Report Oxford University, November 2003, 2003.

[Duncan, 2006] R. W. Duncan. *Types for Quantum Computing.* D.Phil. thesis. University of Oxford, 2006.

[Einstein *et al.*, 1935] A. Einstein, B. Podolsky and N. Rosen. *Can quantum-mechanical description of physical reality be considered complete?* Physical Review **47**, 777, 1935.

[Ekert, 1991] A. K. Ekert. *Quantum cryptography based on Bell's theorem.* Physical review letters **67**, 661–663, 1991.

[Farhi *et al.*, 2000] E. Farhi, J. Goldstone, S. Gutmann and M. Sipser (2000) *Quantum computation by adiabatic evolution*, 2000. arXiv:quant-ph/0001106

[Fiore, 2007] M. Fiore. *An axiomatics and a combinatorial model for creation/annihilation operators and differential structure.* Talk given at the workshop on Categorical Quantum Logic, Oxford University, Jul. 2007. Slides available at http://se10.comlab.ox.ac.uk:8080/FOCS/CQL program en.html

[Freedman *et al.*, 2004] M. H. Freedman, A. Kitaev, M. J. Larsen and Z. Wang. *Topological quantum computation*, 2004. arXiv:quant-ph/0101025

[Freyd and Yetter, 1989] P. Freyd and D. Yetter. *Braided compact closed categories with applications to low-dimensional topology.* Advances in Mathematics **77**, 156–182, 1989.

[Girard, 1987] J.-Y. Girard. *Linear logic.* Theoretical Computer Science **50**, 1–102, 1987.

[Girard, 1989] J.-Y. Girard. *Geometry of Interaction I: Interpretation of System F.* In: Logic Colloquium '88, R. Ferro et al., pp. 221–260. North-Holland, 1989.

[Gottesman and Chuang, 1999] D. Gottesman and I. L. Chuang. *Quantum teleportation is a universal computational primitive.* Nature **402**, 390–393, 1999. arXiv:quant-ph/9908010

[Harding, 2007] J. Harding, *Some quantum logic and a few categories.* Talk given at the workshop on Categorical Quantum Logic, Oxford University, Jul. 2007. Slides available at http://se10.comlab.ox.ac.uk:8080/FOCS/CQL program en.html

[Harding, 2008] J. Harding. *Orthomodularity in dagger biproduct categories.* Preprint, 2008.

[Houston, 2006] R. Houston. *Finite products are biproducts in a compact closed category.* Journal of Pure and Applied Algebra **212**, 2006. arXiv:math/0604542

[Isham, 1995] C. J. Isham/ *Lectures on Quantum Theory.* Imperial College Press, 1995.

[Isham and Butterfield, 1998] C. J. Isham and J. Butterfield. *Topos perspective on the Kochen-Specker theorem: I. Quantum states as generalized valuations.* International Journal of Theoretical Physics **37**, 2669–2733, 1998. arXiv:quant-ph/9803055

[Jagadeesan, 2004] R. Jagadeesan. Private communication, 2004.

[Joyal and Street, 1991] A. Joyal and R. Street. *The geometry of tensor calculus I.* Advances in Mathematics **88**, 55–112, 1991.

[Joyal, Street and Verity, 1996] A. Joyal, R. Street and D. Verity. *Traced monoidal categories.* Proceedings of the Cambridge Philosophical Society **119**, 447–468, 1996.

[Jozsa, 2005] R. Jozsa. *An introduction to measurement based quantum computation*, 2005. arXiv:quant-ph/0508124.

[Kaufmann, 1991] L. H. Kauffman. *Knots and Physics.* World Scientific, 1991.

[Kauffman and Lomonaco, 2002] L. H. Kauffman and S. J. Lomonaco Jr. *Quantum entanglement and topological entanglement.* New Journal of Physics **4**, 1–18, 2002.

[Kauffman, 2005] L. H. Kauffman. *Teleportation topology.* Optics and Spectroscopy **99**, 227–232, 2005. arXiv:quant-ph/0407224

[Kelly, 1972] G. M. Kelly. *Many-variable functorial calculus.* In: Coherence in Categories, G. M. Kelly, M. L. Laplaza, G. Lewis and S. Mac Lane (eds), pages 66–105. Lecture notes in Mathematics **281**, Springer-Verlag, 1972.

[Kelly and Laplaza, 1980] G. M. Kelly and M. L. Laplaza. *Coherence for compact closed categories.* Journal of Pure and Applied Algebra **19**, 193–213, 1980.

[Kock, 2003] J. Kock. *Frobenius Algebras and 2D Topological Quantum Field Theories.* Cambridge University Press, 2003.

[Lambek and Scott, 1986] J. Lambek and P. J. Scott. *Introduction to Higher-Order Categorical Logic.* Cambridge University Press, 1986.

[MacLane, 1998] S. MacLane. *Categories for the Working Mathematician. 2nd edition.* Springer, 1998.

[Mackey, 1963] G. M. Mackey. *The mathematical foundations of quantum mechanics.* W. A. Benjamin, 1963.

[Milner, 1989] R. Milner. *Communication and Concurrency.* Prentice Hall, 1989.

[Mitchell, 1965] B. Mitchell. *Theory of Categories.* Academic Press, 1965.

[Nielsen and Chuang, 2000] M. A. Nielsen and L. Chuang. *Quantum computation and quantum information.* Cambridge University Press, 2000.

[Pati and Braunstein, 2000] A. K. Pati and S. L. Braunstein. *Impossibility of deleting an unknown quantum state.* Nature **404**, 164–165, 2000. arXiv:quant-ph/9911090

[Penrose, 1971] R. Penrose. *Applications of negative dimensional tensors.* In: Combinatorial Mathematics and its Applications, pages 221–244. Academic Press, 1971.

[Petri, 1977] C. A. Petri. *Non-sequential processes.* Technical Report ISF-77-5, GMD, St-Augustin, 1977.

[Piron, 1976] C. Piron. *Foundations of Quantum Physics.* W. A. Benjamin, 1976.

[Raussendorf and Briegel, 2001] R. Raussendorf and H.-J. Briegel. *A one-way quantum computer.* Physical Review Letters **86**, 5188, 2001.

[Raussendorf *et al.*, 2003] R. Raussendorf, D. E. Browne and H.-J. Briegel. *Measurement-based quantum computation on cluster states.* Physical Review A **68**, 022312, 2003. arXiv:quant-ph/0301052.

[Seely, 1989] R. A. G. Seely. *Linear logic, $*$-autonomous categories and cofree algebras.* In: Categories in Computer Science and Logic, Contemporary Mathematics **92**, 1989.

[Selinger, 2004] P. Selinger. *Towards a quantum programming language.* Mathematical Structures in Computer Science **14**, 527–586, 2004.

[Selinger, 2007] P. Selinger. *Dagger compact categories and completely positive maps.* Electronic Notes in Theoretical Computer Science **170**, 139–163. Extended version to appear in New Structures for Physics, B. Coecke (ed). Springer Lecture Notes in Physics, 2007.

[Selinger, 2008] P. Selinger. *Finite dimensional Hilbert spaces are complete for dagger compact closed categories.* In: Proc. Quantum Physics and Logic/Development of Computational Models (QPL-DCM). Electronic Notes in Theoretical Computer Science, to appear, 2008.

[Selinger, 2008a] P. Selinger. *TBA.* In: New Structures for Physics, B. Coecke, Ed, Springer lecture Notes in Physics, 2008.

[Shor, 1994] P. W. Shor. *Algorithms for quantum computation: discrete logarithms and factoring.* Proceedings of the 35th Annual IEEE Symposium on Foundations of Computer Science, IEEE Computer Science Press, 1994.

[Shor, 1996] P. W. Shor. *Fault-tolerant quantum computation.* In: Proceedings of the 37nd Annual Symposium on Foundations of Computer Science, pages 56–65. IEEE Computer Society Press, 1996.

[Simpson, 1999] S. G. Simpson. *Subsystems of Second-Order Arithmetic.* Springer-Verlag, 1999.

[Spekkens, 2007] R. Spekkens. *Evidence for the epistemic view of quantum states: A toy theory.* Physical Review A **75**, 032110, 2007. arXiv:quant-ph/0401052

[Street, 2007] R. Street. *Quantum Groups: A Path to Current Algebra.* Cambridge UP, 207.

[Turaev, 1994] V. Turaev. *Quantum Invariants of Knots and 3-Manifolds*, de Gruyter, 1994.

[van Tonder, 2003] A. van Tonder. *Quantum computation, categorical semantics, and linear logic*, 2003. arXiv:quant-ph/0312174

[Vicary, 2007] J. Vicary. *A categorical framework for the quantum harmonic oscillator.* International Journal of Theoretical Physics, to appear, 2007. arXiv:0706.0711

[Vicary, 2008] J. Vicary (2008) *Categorical formulation of C^*-algebras.* In: Proc. Quantum Physics and Logic/Development of Computational Models (QPL-DCM). Electronic Notes in Theoretical Computer Science, to appear, 2008.

[von Neumann, 1932] J. von Neumann. *Mathematische grundlagen der quantenmechanik.* Springer-Verlag, 1932.

[Witten, 1988] E. Witten. *Topological quantum field theory.* Communications in Mathematical Physics **117**, 353–386, 1988.

[Wootters and Zurek, 1982] W. K. Wootters and W. Zurek. *A single quantum cannot be cloned. Nature* **299**, 802–803, 1982.

[Yetter, 2001] D. N. Yetter. *Functorial Knot Theory. Categories of Tangles, Coherence, Categorical Deformations, and Topological Invariants.* World Scientific, 2001.

[Żukowski *et al.*, 1993] M. Żukowski, A. Zeilinger, M. A. Horne and A. K. Ekert. *Event-ready-detectors' Bell experiment via entanglement swapping.* Physical Review Letters **71**, 4287–4290, 1993.

HANDBOOK OF QUANTUM LOGIC AND QUANTUM STRUCTURES: QUANTUM LOGIC
Edited by K. Engesser, D. M. Gabbay and D. Lehmann

EXTENDING CLASSICAL LOGIC FOR REASONING ABOUT QUANTUM SYSTEMS

Rohit Chadha, Paulo Mateus, Amílcar Sernadas, and Cristina Sernadas

1 INTRODUCTION

A new logic EQPL (exogenous quantum propositional logic) was proposed in [Mateus and Sernadas, 2004a; Mateus and Sernadas, 2004b; Mateus and Sernadas, 2006] for modeling and reasoning about quantum systems, embodying all that is stated in the relevant Postulates of quantum physics (as presented, for instance, in [Cohen-Tannoudji *et al.*, 1977; Nielsen and Chuang, 2000]). The logic was designed from the semantics upwards, starting with the key idea of adopting superpositions of classical models as the models of the proposed quantum logic.

This novel approach to quantum reasoning is different from the mainstream approach [Foulis, 1999; Chiara *et al.*, 2004]. The latter, as initially proposed by Birkhoff and von Neumann [Birkhoff and von Neumann, 1936], focuses on the lattice of closed subspaces of a Hilbert space and replaces the classical connectives by new connectives representing the lattice-theoretic operations. The former adopts superpositions of classical models as the models of the quantum logic, leading to a natural extension of the classical language containing the classical connectives (just as modal languages are extensions of the classical language). Furthermore, EQPL allows quantitative reasoning about amplitudes and probabilities, being in this respect much closer to the possible worlds logics for probability reasoning than to the mainstream quantum logics. Finally, EQPL is designed to reason about finite collections of qubits and, therefore, it is suitable for applications in quantum computation and information. The models of EQPL are superpositions of classical valuations that correspond to unit vectors expressed in the computational basis of the Hilbert space resulting from the tensor product of the independent qubit systems.

Therefore, in EQPL we can express a wide range of properties of states of such a finite collection of qubits. For example, we can impose that some qubits are independent of (that is, not entangled with) other qubits; we can prescribe the amplitudes of a specific quantum state; we can assert the probability of a classical outcome after a projective measurement over the computational basis; and, we can also impose classical constraints on the admissible quantum states.

Herein, we concentrate on presenting a decidable fragment of EQPL by suitably relaxing the semantic structures of EQPL. Instead of considering Hilbert spaces we work with inner product spaces over an arbitrary real closed field and its algebraic closure. The decidability results from the fact that the first order theory of such fields is decidable [Tarski, 1948; Hodges, 1993; Basu *et al.*, 2003]. This technique was inspired by related work on probabilistic logic [Abadi and Halpern, 1994]. Furthermore, the decidable fragment of EQPL so established turns out to be strongly complete although we concentrate on weak completeness. The price we have to pay for decidability is a weak arithmetic language — we loose the analytic aspects of complex numbers.

The exogenous approach to extending a given logic is discussed and illustrated in Section 2. Section 3 presents dEQPL step by step: design options, models, language and its interpretation, sound axiomatization, and some useful metatheorems. In Section 4 we show that dEQPL is weakly complete and decidable. The proof of weak completeness can easily be adapted to a proof of strong completeness but we refrained to do so since our primary interest is in applications involving finitely presented theories. We illustrate the use of dEQPL with two worked examples in Section 5. First we reason about a Bell state. Afterwards, we reason about the quantum teleportation protocol proposed in [Bennett *et al.*, 1993]. Finally, in Section 6 we assess what was achieved and provide an outlook of further developments of the proposed approach to quantum reasoning.

2 EXOGENOUS APPROACH

The exogenous semantics approach to enriching a given logic roughly consists of taking as models of the new logic sets of models of the original logic, possibly together with some additional structure. This general mechanism for building new logics is described in detail in [Mateus *et al.*, 2005; Caleiro *et al.*, 2006]. The first example of the approach appeared in the context of probabilistic logics [Nilsson, 1986; Nilsson, 1993], although by then not yet recognized as a general construction.

The adjective "exogenous" is used as a counterpoint to "endogenous". For instance, in order to enrich some given logic with probabilistic reasoning it may be convenient to tinker with the models of the original logic. This endogenous approach has been used extensively. For example, the domains of first-order structures are endowed with probability measures in [Halpern, 1990]. Other examples include labeling the accessibility pairs with probabilities in the case of Kripke structures [Hansson and Jonsson, 1995] for reasoning about probabilistic transition systems.

By not tinkering with the original models and only adding some additional structure on collections of those models as they are, the exogenous approach has the potential for providing general mechanisms for enriching a given logic with some additional reasoning dimension. As we shall see, in our case the exogenous approach has the advantage of closely guiding the design of the language around the underlying concepts of quantum physics while keeping the classical connectives.

The exogenous approach of collecting the original models as proposed in [Mateus and Sernadas, 2004a; Mateus and Sernadas, 2004b] is inspired by the possible worlds semantics of modal logic [Kripke, 1963]. It is also akin to the society semantics for many-valued logic [Carnielli and Lima-Marques, 1999] and to the possible translations semantics for paraconsistent logic [Carnielli, 2000]. The possible worlds approach also plays a role in probabilistic logic [Nilsson, 1986; Nilsson, 1993; Bacchus, 1990b; Bacchus, 1990a; Fagin *et al.*, 1990; Abadi and Halpern, 1994; Chadha *et al.*, 2007].

As an introductory example of the exogenous approach, we briefly explain how a probabilistic logic can be obtained from classical propositional logic, following closely [Mateus *et al.*, 2005]. Since quantum reasoning subsumes probabilistic reasoning, this example will also be useful for our purposes. However, before we proceed to explain the probabilistic logic, we first concentrate on a fragment of the probabilistic logic called *global propositional logic.* Global logic is also a fragment of the quantum logic proposed in this paper.

We start by taking a set Π of propositional symbols. From a semantic point of view, the models of global logic are sets of valuations over Π. The language of global logic consists of:

- Classical propositional formulas constructed from Π using the classical connectives $\bot$ and $\Rightarrow$.
- Global formulas constructed from the classical propositional formulas by the global connectives $\perp\!\!\!\perp$ and $\sqsupset$. The global connectives mimic the classical connectives in a sense which we will make precise shortly.

The satisfaction relation between the semantic models and the formulas is as follows. A model V (V is some set of "classical" valuations) of the global logic satisfies a classical propositional formula α if every classical valuation $v \in V$ satisfies α. Therefore, any classical tautology is a global tautology.

Analogous to the case of classical logic, a global valuation V satisfies the global formula $\gamma_1 \sqsupset \gamma_2$ if either V satisfies γ_2 or V does not satisfy γ_1. The global connective $\perp\!\!\!\perp$ is never satisfied. Clearly this is a copy of the classical propositional logic and indeed, if we replace the classical connectives in a classical tautology by their global counterparts we will get a global tautology.

As we just saw, there are two copies of the classical propositional logic in the global logic. A natural question to ask is whether the two copies are necessarily distinct. The answer is yes and while the connectives $\bot$ and $\perp\!\!\!\perp$ collapse, it is not the case with the two implications. However, there is a relation between those two and if V satisfies $\alpha_1 \Rightarrow \alpha_2$ then V also satisfies $\alpha_1 \sqsupset \alpha_2$. The reverse does not hold in general.

There is a sound and strongly complete axiomatization for global logic which contains five axiom schemas and an inference rule. One axiom schema says that every classical tautology is a global tautology while the other says that replacing classical connectives by their global counterparts results in a global tautology. One axiom identifies $\perp$ and $\perp\!\!\!\perp$, a second one axiomatizes the relation between the two implications that we mentioned above, and the last one says that the classical and global conjunctions (global conjunction is introduced as usual) collapse. The inference rule is the global counterpart of modus ponens.

Global logic is the first step towards creating the exogenous probabilistic logic. The probabilistic logic is obtained "exogenously" by assigning probabilities to each of the classical valuations in a global valuation V. This allows us to reason about the probability that a classical propositional formula is true in V: the probability of φ is the sum of the probabilities of the valuations that satisfy φ. Given a set Π of propositional symbols, the language of the probabilistic logic consists of:

- Classical propositional formulas constructed from Π using the classical connectives $\perp$ and $\Rightarrow$.
- A set of terms that include:
  - real-valued variables and real computable numbers;
  - probability terms denoting probabilities of classical formulas; and
  - sum and product of terms.
- Comparison formulas of the form $t_1 \leq t_2$ where t_1 and t_2 are terms.
- Formulas constructed from classical propositional formulas and comparison formulas using the global connectives $\perp\!\!\!\perp$ and $\sqsupset$.

A model for the probabilistic logic, that is a *probabilistic valuation*, contains a global valuation along with a probability measure which assigns to each classical valuation a real value between 0 and 1. As explained, this gives us an interpretation of the probability terms in the language. The satisfaction of classical formulas is the same as in the global logic. Observe that if V satisfies a classical formula α then the probability of α being true is 1 regardless of the probability measure on V. Hence, the probability of a classical tautology in any model is always 1.

In order to interpret the variables, the model also contains an assignment of variables to real numbers. This helps to interpret the terms and the comparison formulas in the natural way. The interpretation of the global connectives is the same as before.

An axiomatization of probabilistic logic is obtained by extending the axiomatization for global logic as follows. The connection between the classical connectives and probability terms is obtained by three axioms:

1. The probability of any classical tautology is 1.
2. If the probability of the classical formula $\alpha_1 \wedge \alpha_2$ is 0 then the probability of $\alpha_1 \vee \alpha_2$ is the sum of the probabilities of α_1 and α_2. This is the finite additivity of probability measures.

3. If the probability of the classical formula $\alpha_1 \Rightarrow \alpha_2$ is 1 then the probability of α_1 is less than the probability of α_2. This is the monotonicity property of probability measures.

For the comparison formulas, an oracle is used which gives the valid comparison formulas. The axiomatization is sound and *weakly* complete modulo the oracle. However, even with the oracle strong completeness fails as the logic is not compact.

The development of the exogenous quantum logic herein follows the same lines as the development of the probabilistic one. Instead of assigning probabilities, we assign amplitudes to the classical valuations in a global valuation. The classical valuations themselves represent the computational basis of the qubits in a quantum system. In fact, we are only interested in quantum systems composed of a finite number of qubits since applications in quantum computation and information only deal with such systems. A superposition of these classical valuations will then give the state of the quantum system. We will explicitly have terms in the language to interpret these amplitudes and they will be at the core of the design of our language. We postpone the detailed discussion of the language and the logic to Section 3. The resulting quantum logic is a decidable fragment of the logic in [Mateus and Sernadas, 2006].

These quantum logics obtained using the exogenous approach are philosophically closer to some probabilistic logics (like [Fagin *et al.*, 1990; Abadi and Halpern, 1994]) than to the mainstream quantum logics in the tradition of Birkhoff and von Neumann [Birkhoff and von Neumann, 1936; Foulis, 1999; Chiara *et al.*, 2004]. Both types of quantum logic are motivated by semantic considerations, albeit very different ones. The mainstream quantum logics are based on the idea of replacing the Boolean algebras of truth values by the more relaxed notion of orthomodular lattices. Thus, they end up with non classical connectives reflecting the properties of meets and joins of those lattices. The exogenous quantum logics are based on the idea of replacing classical valuations by superpositions of classical valuations while preserving the classical connectives. On the other hand, in both types of quantum logic a formula and a propositional symbol in particular denotes a subspace of the Hilbert space at hand. However, in the exogenous quantum logics a quantum system is assumed to be composed of n qubits and, hence, the underlying Hilbert space has dimension 2^n.

Our semantics of quantum logic, although inspired by modal logic, is also completely different from the alternative Kripke semantics given to mainstream quantum logics (as first proposed in [Dishkant, 1972]). That Kripke semantics is based on orthomodular lattices.

The quantum logic proposed in [van der Meyden and Patra, 2003b; van der Meyden and Patra, 2003a; Patra, 2008] is also inspired by probabilistic logics [Fagin *et al.*, 1990] and capitalizes on some techniques first proposed for those logics, but it has aspects of both mainstream quantum logics and exogenous quantum logics. In short, it is a classical logic of probabilistic measurements over a quantum system where quantum formulas denote projectors, quantum negation stands for orthogonal complement and quantum conjunction stands for composition.

Note also that no amplitude terms appear in [van der Meyden and Patra, 2003b; van der Meyden and Patra, 2003a] contrarily to exogenous quantum logics where amplitudes replace probabilities as the central concept.

The tensor product plays a key role in the exogenous quantum logics as it does in the categorical semantics proposed in [Abramsky and Coecke, 2004; Abramsky and Duncan, 2006]. However, in our logics we still use the concrete characterization of tensor product of qubits (represented in our language by the propositional symbols).

3 DECIDABLE FRAGMENT OF EQPL

We start by discussing design issues, and then proceed to introduce the logic.

3.1 Design issues

In this section, we shall discuss how the Postulates of quantum mechanics [Cohen-Tannoudji *et al.*, 1977] guided the design of the proposed logic, and give a brief introduction to the relevant concepts and results. The first Postulate of quantum mechanics states:

Postulate 1: *Every isolated quantum system is described by a Hilbert space. The states of the quantum system are the unit vectors of the corresponding Hilbert space.*

Please recall that a Hilbert space is a complete inner product space over $\mathbb{C}$ (the field of complex numbers). In quantum computation and information the quantum systems are composed of qubits. For example, the states of an isolated qubit are vectors of the form $z_0|0\rangle + z_1|1\rangle$ where $z_0, z_1 \in \mathbb{C}$ and $|z_0|^2 + |z_1|^2 = 1$. In other words, they are unit vectors in the (unique up to isomorphism) Hilbert space of dimension two. As pointed out in the introduction, instead of working with a Hilbert space we shall consider a "generalized" inner product space over the algebraic closure of an arbitrary *real closed field.* This design decision has the advantage that the resulting logic is decidable. It is possible to work with Hilbert spaces and still get a weakly-complete calculus as was the case in EQPL [Mateus and Sernadas, 2006], a previous version of the logic developed herein. Indeed, the logic defined here identifies a decidable fragment of EQPL, and hence we shall call it dEQPL. In addition to being decidable, dEQPL turns out to be strongly complete and, therefore, compact. In fact, the source of the non compactness of EQPL mentioned in [Mateus and Sernadas, 2006] was in its arithmetic component.

We shall now briefly review some definitions and results concerning real closed fields and their algebraic closures.

DEFINITION 1 Real closed fields.

An ordered field $\mathcal{K} = (K, +, ., 1, 0, \leq)$ is said to be a real closed field if the following hold:

- Every non-negative element of the K has a square root in K.
- Any polynomial of odd degree with coefficients in K has at least one solution in K.

We shall use $\mathcal{K}_1, \mathcal{K}_2, \ldots$ to range over real closed fields and $k_1, k_2, \ldots$ to range over the elements of a real closed field. The set of real numbers with the usual multiplication, addition and order constitute a real closed field. The set of computable real numbers with the same operations is another example of a real closed field.

The *algebraic closure* of a real closed field $\mathcal{K} = (K, +, \times, 1, 0, \leq)$ is obtained by adjoining an element δ to $\mathcal{K}$ such $\delta^2 + 1 = 0$. The algebraic closure, denoted by $\mathcal{K}(\delta)$, is a two-dimensional vector space over $\mathcal{K}$. Each element in $\mathcal{K}(\delta)$ is of the form $k_1 + k_2\delta$ where $k_1, k_2 \in K$. The addition and multiplication are defined as:

$$\begin{array}{lcl} (k_1 + k_2\,\delta) + (k_1' + k_2'\,\delta) & = & (k_1 + k_1'\,\delta) + (k_2 + k_2'\,\delta) \\ (k_1 + k_2\,\delta).(k_1' + k_2'\,\delta) & = & (k_1.k_1' - k_2.k_2') + (k_1.k_2' + k_1'.k_2\delta) \\ & & \text{where } -k_2.k_2' \text{ is the additive inverse of } k_2.k_2' \end{array}$$

We shall use $c_1, c_2, \ldots$ to range over the elements of $\mathcal{K}(\delta)$. For example, the field of complex numbers is the algebraic closure of the set of real numbers with $\delta = i$. The standard notion of conjugation, absolute value and real and imaginary parts from complex numbers can be generalized to $\mathcal{K}(\delta)$ as follows:

$$\begin{array}{lcl} Re(k_1 + k_2\,\delta) & = & k_1 \\ Im(k_1 + k_2\,\delta) & = & k_2 \\ |k_1 + k_2\,\delta| & = & k_1^2 + k_2^2 \\ \overline{k_1 + k_2\delta} & = & k_1 + (-k_2)\delta \text{ where } -k_2 \text{ is the additive inverse of } k_2 \end{array}$$

The conjugation allows us to generalize the notion of inner product and normed vector space over $\mathbb{C}$ to an arbitrary $\mathcal{K}(\delta)$ as follows:

DEFINITION 2 $\mathcal{K}(\delta)$-inner product space. A $\mathcal{K}(\delta)$-*inner product space* is a vector space W over the field $\mathcal{K}(\delta)$ together with a map

$$\langle \cdot, \cdot \rangle : W \times W \to \mathcal{K}(\delta)$$

such that for all $w, w_1, w_2 \in V$ and $k \in \mathcal{K}(\delta)$, the following hold:

1. $\langle w, w_1 + w_2 \rangle = \langle w, w_1 \rangle + \langle w, w_2 \rangle$.
2. $\langle w, w \rangle \in \mathcal{K}$ and $\langle w, w \rangle \geq 0$.
3. $\langle w, w \rangle = 0$ if and only if $w = 0$.
4. $\langle w_1, w_2 \rangle = \overline{\langle w_2, w_1 \rangle}$.

5. $\langle w_1, cw_2 \rangle = c\langle w_1, w_2 \rangle$.

DEFINITION 3 $\mathcal{K}(\delta)$-normed vector space. A $\mathcal{K}(\delta)$-*normed space* is a vector space W over the field $\mathcal{K}(\delta)$ together with a map

$$||.|| : W \times W \to \mathcal{K}$$

such that for all $w, w_1, w_2 \in V$ and $k \in \mathcal{K}$, the following hold

1. $||w|| \geq 0$.

2. $||w|| = 0$ if and only if $w = 0$.

3. $||cw|| = |c|||w||$ where $|c|$ is the absolute value of c.

4. $||w_1 + w_2|| \leq ||w_1|| + ||w_2||$.

We shall say that a vector w is a *unit vector* if $||w|| = 1$.

As in the case of inner product spaces over complex numbers, a $\mathcal{K}(\delta)-$inner product space $(W, \langle \cdot, \cdot \rangle)$ gives rise to a norm by letting:

$$||w|| = \sqrt{\langle w, w \rangle}.$$

For example, the field $\mathcal{K}(\delta)$ together with the map: $\langle c_1, c_2 \rangle = c_1.\overline{c_2}$ is itself a $\mathcal{K}(\delta)$-inner product space. In this case, the resulting norm ($||c|| = \sqrt{c.\overline{c}}$) is the absolute value function.

Any Hilbert space is a C-inner product space. However, we shall model quantum systems as $\mathcal{K}(\delta)-$inner product spaces instead of Hilbert spaces, and the field $\mathcal{K}(\delta)$ will be a part of our semantic structure. Therefore, any theorem we prove in the logic would remain valid if we had just used Hilbert spaces.

It is also worthwhile to point out that, unlike Hilbert spaces, $\mathcal{K}(\delta)-$inner product spaces in general may not have an analytical structure. So, we will not be able to express properties that necessarily depend upon the analytical structure[1].

Moreover, as the logic is intended to be applied for quantum computation and information, we shall work only with a special kind of $\mathcal{K}(\delta)-$inner product spaces that are defined by free construction from finite sets:

DEFINITION 4 Free $\mathcal{K}(\delta)$-inner product space. Given an arbitrary finite set $\mathcal{B}$, we can construct the free $\mathcal{K}(\delta)$-inner product space $\mathcal{H}_{\mathcal{K}(\delta)}(\mathcal{B})$ as:

- Each element of $\mathcal{H}_{\mathcal{K}(\delta)}(\mathcal{B})$ is a map $|\psi\rangle : \mathcal{B} \to \mathcal{K}(\delta)$.

- $|\psi_1\rangle + |\psi_2\rangle$ is pointwise addition, *i.e.*,

$$(|\psi_1\rangle + |\psi_2\rangle)(b) = |\psi_1\rangle(b) + |\psi_2\rangle(b).$$

[1]For example, we cannot define the exponential function on an arbitrary $\mathcal{K}(\delta)$.

- $c|\psi\rangle$ is pointwise scalar multiplication, *i.e.*,

$$(c|\psi\rangle)(b) = c\,(|\psi\rangle(b)).$$

- The inner product is given by[2]

$$\langle\psi_1|\psi_2\rangle = \sum_{b\in\mathcal{B}} \overline{|\psi_1\rangle(b)}\,|\psi_2\rangle(b).$$

The dimension of the vector space $\mathcal{H}_{\mathcal{K}(\delta)}(\mathcal{B})$ is the cardinality of the set $\mathcal{B}$. Given $b \in \mathcal{B}$, let $|b\rangle \in \mathcal{H}_{\mathcal{K}(\delta)}(\mathcal{B})$ be the vector defined as

$$|b\rangle(b) = 1 \text{ and } |b\rangle(b_1) = 0 \text{ for every } b_1 \neq b.$$

It can be easily checked that the set $\{|b\rangle : b \in \mathcal{B}\}$ forms a basis of the vector space $\mathcal{H}_{\mathcal{K}(\delta)}(\mathcal{B})$. Furthermore, it is the case that $\langle b|b\rangle = 1$ and $\langle b|b_1\rangle = 0$ for every $b \neq b_1$. For obvious reasons, we say that $\{|b\rangle : b \in \mathcal{B}\}$ is an *orthonormal basis* of $\mathcal{H}_{\mathcal{K}(\delta)}(\mathcal{B})$. This basis plays an important role in the semantics of dEQPL and for this reason we will henceforth refer to it as being the *canonical basis* of $\mathcal{H}_{\mathcal{K}(\delta)}$.

A natural question that arises in this context is how do we choose $\mathcal{B}$. The answer lies in our interest in quantum systems composed of qubits. As mentioned before, the states of an isolated qubit are vectors of the form $z_0|0\rangle + z_1|1\rangle$ where $z_0, z_1 \in \mathbb{C}$ and $|z_0|^2 + |z_1|^2 = 1$. The set of states can be identified with (upto isomorphism) the unit vectors in the free $\mathbb{C}$-inner product $\mathcal{H}_{\mathbb{C}}(\mathcal{B})$ where $\mathcal{B}$ is an set of 2 elements. Keeping this is mind, it is natural to represent a qubit by a propositional symbol (henceforth called a qubit symbol) and take $\mathcal{B}$ in this case to be the set of two possible classical valuations of the qubit symbol: 0 that assigns false to the qubit symbol and 1 that assigns true to it.

Similarly, the states of a isolated *pair* of qubits are of the form $z_{00}|00\rangle + z_{01}|01\rangle + z_{10}|10\rangle + z_{11}|11\rangle$, where $z_{00}, z_{10}, z_{01}, z_{11} \in \mathbb{C}$ and $|z_{00}|^2 + |z_{01}|^2 + |z_{10}|^2 + |z_{11}|^2 = 1$. The set of states in this case can be identified with the unit vectors in the free $\mathbb{C}$-inner product $\mathcal{H}_{\mathbb{C}}(\mathcal{B})$ where $\mathcal{B}$ is the set of the four classical valuations over the pair of qubit symbols representing the two qubits at hand.

The pattern becomes clear, and in general, we will fix a *finite* set of qubit symbols [3]:

$$\mathsf{qB} = \{\mathsf{qb}_k : 0 < k \leq n\}.$$

These will represent the n qubits in our system. As we need to work with the algebraic closure of arbitrary real closed fields, the states in our systems will be unit vectors in the free $\mathcal{K}(\delta)$-inner product space $\mathcal{H}_{\mathcal{K}(\delta)}(2^{\mathsf{qB}})$, where 2^{qB} is the set

[2] We adopt here the Dirac notation, given its widespread use by the community of quantum physics and computation.

[3] In [Mateus and Sernadas, 2006], the set of qubits was infinite. However, the set was restricted when judgments were considered.

of 2^n possible *classical valuations* of the n qubit symbols. We shall call these unit vectors $\mathcal{K}(\delta)$-*quantum valuations* over the set qB.

Another characteristic of quantum systems that we are likely to encounter in applications in computation and information is that they will be built from independent sub-systems. We shall model the sub-systems by partitioning the set qB, and a semantic structure will contain this partition. Each member of the partition, henceforth called a *component*, will then model the qubits of an independent sub-system.

If $A \subseteq \mathsf{qB}$ is a component, then the states of the A sub-system will be quantum valuations over A, *i.e.*, unit vectors in $\mathcal{H}_{\mathcal{K}(\delta)}(2^A)$. If S is the partition, then the semantic structure also includes a collection $\{|\varphi\rangle_A : A \in S\}$, where $|\varphi\rangle_A$ is a quantum valuation over A. These represent the states of the sub-systems.

In addition to reasoning about component sub-systems, we also need to reason about bigger sub-systems. The sets of qubits of bigger sub-systems are given by union of qubits of the component sub-systems. Therefore, given a partition S of qB, we define $\mathrm{Alg(S)} = \{\cup_i A_i : A_i \in S\}$. A member $F \in \mathrm{Alg(S)}$ models the qubits of the component systems. It is easy to see that Alg(S) satisfies the following properties[4]:

1. $\emptyset$, $\mathsf{qB} \in S$.
2. $G \in S$ implies that $\mathsf{qB} \setminus G \in S$.
3. $G_1, G_2 \in S$ implies that $G_1 \cup G_2 \in S$

We also need a way to construct the states of sub-systems from smaller ones. For this, we take recourse to the second Postulate of quantum mechanics:

Postulate 2: *The Hilbert space of a quantum system composed of a finite number of independent components is the tensor product of the component Hilbert spaces.*

Therefore, for instance, the state of a sub-system composed of two independent sub-systems is the "tensor product" of the states of the sub-system. Of course, we remember that we are not working with Hilbert spaces. Therefore, we need a definition of a $\mathcal{K}(\delta)$-tensor product. For this, we will assume that the reader is familiar with tensor products of vector spaces. Given two vector $(K)(\delta)$ vector spaces W_1 and W_2, we shall denote the tensor product by $W_1 \otimes W_2$. Please recall that the vector space $W_1 \otimes W_2$ is generated by vectors of form $w_1 \otimes w_2$ where $w_1 \in W_1$ and $w_2 \in W_2$. We are ready to define $\mathcal{K}(\delta)$-tensor products:

DEFINITION 5 $\mathcal{K}(\delta)$-tensor product. The tensor product of two $\mathcal{K}(\delta)$-in-ner product spaces $(W_1, \langle\cdot,\cdot\rangle_1)$ and $(W_2, \langle\cdot,\cdot\rangle_2)$, is the pair $(W_1 \otimes W_2, \langle\cdot,\cdot\rangle)$, where $\langle\cdot,\cdot\rangle$ is defined as:

$$\langle \sum_i a_i\, v_i \otimes w_i \,,\, \sum_j b_j\, v'_j \otimes w'_j\rangle \;=\; \sum_{i,j} a_i\overline{b_j}\,\langle v_i, v'_j\rangle\langle w_i, w'_j\rangle$$

[4]These properties define a structure often called an algebra in probability theory.

Observe also that given $w \in W_1 \otimes W_2$ it is not always possible to find $w_1 \in W_1$ and $w_2 \in W_2$ such that $w = w_1 \otimes w_2$. Furthermore, when that factorization is possible it is not necessarily unique.

Please also observe that in our case, the $\mathcal{K}$-vector spaces over the set of qubits A are generated by vectors $|v\rangle$ where v is a classical valuation over A. Therefore, if S is the partition of qB in our model and $A_1, A_2 \in S$ then the sub-system composed of A_1 and A_2 will be generated by vectors of the form $|v_1\rangle \otimes |v_2\rangle$ where $v_i \in \mathcal{H}(2^{A_i})$. We will identify $|v_1\rangle \otimes |v_2\rangle$ with the vector $|v_1 v_2\rangle \in \mathcal{H}(2^{A_1 \cup A_2})$ where $v_1 v_2$ is the unique valuation that extends v_1 and v_2. Furthermore, the state of sub-system composed of A_1 and A_2 is the tensor product $\psi_{A_1} \otimes \psi_{A_2}$. (Please note that the tensor product of two unit vectors is again a unit vector.)

When given a quantum state $|\psi\rangle \in \mathcal{H}_{\mathcal{K}(\delta)}(2^{\mathsf{qB}})$ and non empty $G \subsetneq \mathsf{qB}$, we say that the qubits in G are *not entangled* with the other qubits if there are $|\psi_1\rangle \in \mathcal{H}_{\mathcal{K}(\delta)}(2^G)$ and $|\psi_2\rangle \in \mathcal{H}_{\mathcal{K}(\delta)}(2^{\mathsf{qB} \setminus G})$ such that $|\psi\rangle = |\psi_1\rangle \otimes |\psi_2\rangle$.

Therefore, given any $G \in \mathrm{Alg(S)}$, the qubits in G are not entangled with the other qubits, thanks to the way we build the whole state of the system from the states of the components. Hence, qubits taken from any two independent components of the system are not entangled in every possible quantum state.

Please note also that (contrarily to what was adopted in [Mateus and Sernadas, 2006]) we do not require that each component state be non factorisable. This relaxation of the notion of quantum structure had no impact on the entailment relation.

Another key concept in the design of our logic is the concept of logical amplitudes. Given a $\mathcal{K}(\delta)$-quantum valuation $|\psi\rangle$ and a classical valuation v, the inner product $\langle v|\psi\rangle$ is said to be the *logic amplitude* of $|\psi\rangle$ for v. As we shall see, these logical amplitudes are at the core of dEQPL. These amplitudes appear in two ways in the structure which we discuss below.

It is also sometimes convenient to work with $V \subsetneq 2^{\mathsf{qb}}$, as we may want to impose classical constraints on the quantum valuations. For example, we may want to impose $(\mathsf{qb}_1 \vee \mathsf{qb}_2)$ requiring states to have (logical) amplitude zero for every classical valuation not satisfying this classical formula. In our semantics structures, we shall therefore explicitly have a set $V \subseteq 2^{\mathsf{qB}}$ and we shall call V the *set of admissible classical valuations.* Furthermore, for any $v \notin V$, we will require that the amplitude $\langle v|\psi\rangle = 0$ where ψ is the quantum state of the full system.

Note also that every subset A of qB can be identified with a classical valuation v over qB: v assigns true to qb if and only if $\mathsf{qb} \in A$. This, of course, can be generalized. Any set $A \subset G \subset \mathsf{qB}$ can be identified with a classical valuation v_A^G over G: v_A^G assigns true to all elements of A and false to all elements of $G \setminus A$.

Finally, we also have a collection of $\mathcal{K}(\delta)$ values $\nu = \{\nu_{GA}\}_{G \subseteq \mathsf{qB},\, A \subseteq G}$ in the semantic structure. We impose that if $G \subset \mathrm{Alg(S)}$ then $\nu_{GA} = \langle v_A^G|\psi\rangle_G$ where $|\psi\rangle_G$ is the state of the sub-system composed of qubits modeled by G. In other words, they are logic amplitudes when the qubits in G constitute an independent sub-system.

It should be stressed that these values are not always physically meaningful. A term ν_{GA} is meaningful only if $G \in Alg(S)$. The others are nevertheless useful for our purposes and help to avoid partial denotation maps. We are now ready to assemble the different pieces of our semantic structure:

DEFINITION 6 Quantum structure. A *quantum structure over* qB is a tuple

$$\mathbf{w} = (\mathcal{K}, \delta, V, \mathcal{S}, |\psi\rangle, \nu)$$

where:

- $\mathcal{K}$ is a real closed field and $\mathcal{K}(\delta)$ is its algebraic closure;
- V is a nonempty subset of 2^{qB};
- $\mathcal{S}$ is a partition of qB;
- $|\psi\rangle = \{|\psi\rangle_S\}_{S\in\mathcal{S}}$ where each $|\psi\rangle_S$ is a unit vector of $\mathcal{H}_S$. We extend $|\psi\rangle$ to Alg($\mathcal{S}$) as follows:
  1. $|\psi\rangle_\emptyset = 1$;
  2. $|\psi\rangle_{S_1\cup\cdots\cup S_n} = |\psi\rangle_{S_1} \otimes \cdots \otimes |\psi\rangle_{S_n}$;
- $\langle v|\psi\rangle_{\mathsf{qB}} = 0$ if $v \notin V$;
- $\nu : \{\nu_{GA}\}_{G\subseteq\mathsf{qB}, A\subseteq G}$ where $\nu_{GA} = \langle v_A^G|\psi\rangle_G$ if $G \in \text{Alg}(\mathcal{S})$. In particular, $\nu_{\emptyset\emptyset} = 1$.

The proposed quantum logic will be interpreted over these quantum structures. Obviously, we have some redundancy in the notion of quantum structure, namely, $|\psi\rangle$ can be reconstructed from ν. However, this redundancy pays off in ease of use and in clarifying the connection to quantum physics.

The first two Postulates were sufficient to guide us in the task of setting up the notion of quantum structures over which we shall be able to define the semantics of dEQPL. Now, we turn our attention to the Postulates concerning measurements of physical quantities.

Postulate 3: *Every measurable physical quantity of an isolated quantum system is described by an observable acting on its Hilbert space.*

Please recall that an observable is a Hermitian operator such that the direct sum of its eigensubspaces coincides with the underlying Hilbert space. Also recall that the *spectrum* Ω of a Hermitian operator (set of its eigenvalues) is a subset of the set of real numbers, $\mathbb{R}$. For each $e \in \Omega$, we denote the corresponding eigensubspace by H_e, and the projector onto the subspace E_e by P_e.

It might seem at first that we need to extend the definition of Hermitian operators to an arbitrary $\mathcal{K}(\delta)$ as Hermitian operators are usually defined over Hilbert spaces. However, as we shall see shortly, fortunately that is not required. This is because we do not have constructs in the language for denoting such measurement operators. In order to use Postulate 3, we need to consider Postulate 4.

Postulate 4: *The possible outcomes of the measurement of a physical quantity are the eigenvalues of the corresponding observable. When the physical quantity is measured using observable A on a system in a state $|\psi\rangle$, the resulting outcomes are ruled by the probability space $\mathcal{P}^A_{|\psi\rangle} = (\Omega, \mathcal{E}|_\Omega, \mu^A_{|\psi\rangle})$ where (in the case A has a countable spectrum)*

- *Ω is the spectrum of the observable A,*
- *$\mathcal{E}|_\Omega$ is $\wp\Omega$ the power-set of Ω , and*
- *$\mu^A_{|\psi\rangle} : \mathcal{E}|_\Omega \to \mathbb{R}$ is the probability measure defined as*

$$\mu^A_{|\psi\rangle}(E) = \sum_{e \in E} ||P_e|\psi\rangle||^2 .$$

For the applications in quantum computation and information that we have in mind, only *logic projective measurements* are relevant. Given a quantum system with the set of qubits qB and a set of classical valuations V, these are measurements A such that:

- The spectrum of A is *equipotent*[5] to V, *i.e.*, there is a bijection between the spectrum of A and V.
- If we identify V with the spectrum of A then for each $v \in V$, the corresponding eigenspace H_v is generated by the vector $|v\rangle$. The projector P_v is the operator $|v\rangle\langle v|$, *i.e.*, $P_v|\psi\rangle = \langle v, \psi\rangle\ |v\rangle$ for each vector $\psi \in \mathcal{H}_\mathbb{C}(2^{\mathsf{qB}})$.

Postulate 4 then tells us that the stochastic result of making a logic projective measurement A given a quantum structure $\mathbf{w} = (\mathcal{K}, \delta, V, \mathcal{S}, |\psi\rangle, \nu)$ is described by the finite probability space $\mathcal{P}_\mathbf{w} = (V, \wp V, \mu_\mathbf{w})$ where for each $U \subseteq V$:

$$(1) \quad \mu_\mathbf{w}(U) = \sum_{v \in U} |\langle v|\psi\rangle|^2 .$$

For example, if the quantum system is in the particular state

$$\alpha_{00\omega_1}|00\omega_1\rangle + \alpha_{01\omega_2}|01\omega_2\rangle + \alpha_{01\omega_3}|01\omega_3\rangle + \alpha_{10\omega_4}|10\omega_4\rangle$$

[5]The chosen bijection depends on how the qubits are physically implemented. For example, when implementing a qubit using the spin of an electron, we may impose that spin $+\frac{1}{2}$ corresponds to true and spin $-\frac{1}{2}$ corresponds to false. But, as we shall see, the semantics of EQPL does not depend on the choice of the bijection, as long as one exists. The same happens in the case of classical logic — its semantics does not depend on how bits are implemented. The details of which voltages correspond to which truth values are irrelevant.

then the probability of observing the first two qubits $\mathsf{qb}_0, \mathsf{qb}_1$ in the classical valuation 01 (here we take V as $\{00\omega_1, 00\omega_2, 00\omega_3, 00\omega_4\}$) is given by $|\alpha_{01\omega_2}|^2 + |\alpha_{01\omega_3}|^2$.

We have probability terms in the language of the proposed logic and Equation 1 is all that we need from Postulates 3 and 4 for interpreting them as we shall see in Section 3.2.

Once again, we recall that we are working with an arbitrary real closed field. Given a quantum structure $\mathbf{w} = (\mathcal{K}, \delta, V, \mathcal{S}, |\psi\rangle, \nu)$, we define the *probability map* $\mu_{\mathbf{w}} : \wp(V) \to \mathcal{K}$ as:

$$(2) \quad \mu_{\mathbf{w}}(U) = \sum_{v \in U} |\langle v|\psi\rangle|^2 .$$

The essential difference between Equation 1 and 2 is that summands in the former are real numbers while the summands of the latter one are elements of a real closed field given by the quantum structure. It is easy to check that μ_V defined in Equation 2 satisfies the "usual" finite probability axioms:

1. $\mu_V(\emptyset) = 0$ and $\mu_V(V) = 1$, and
2. $\mu_V(U_1 \cup U_2) = \mu_V(U_1) + \mu_V(U_2)$ if U_1 and U_2 are disjoint sets.

Therefore, given a quantum structure $\mathbf{w}$, we have the means for interpreting dEQPL terms of the form $(\int \alpha)$ that denote probabilities.

Finally, although irrelevant to the design of dEQPL, we mention *en passant* Postulate 5 that rules how quantum systems evolve beyond measurements:

Postulate 5: *Excluding measurements, the evolution of a quantum system is described by unitary transformations.*

This last Postulate becomes relevant only when designing a dynamical extension of the logic (see for instance [Mateus and Sernadas, 2004b]).

3.2 Language and semantics

There are two kinds of terms in dEQPL, one denoting elements of real closed field in the quantum structure and the other denoting elements in its algebraic closure. The formulas of dEQPL, henceforth called *quantum formulas*, are constructed from classical propositional formulas, formulas denoting sub-system and comparison formulas (comparing terms denoting elements of real closed fields) using global connectives introduced in Section 2. We present language of dEQPL in Table 1 using an abstract version of BNF notation [Naur, 1963] for a compact presentation of inductive definitions. We discuss the language in detail below.

The first syntactic category is classical formulas. Please recall that we fixed a finite set of qubit symbols qB. Classical formulas are built from qubit symbols in qB using the classical disjunctive connectives, falsum $\bot$ and implication $\Rightarrow$. As usual,

Classical formulas

$$\alpha \quad := \quad \bot \;[\!]\; \mathsf{qb} \;[\!]\; (\alpha \Rightarrow \alpha)$$

Term language (with the proviso $A \subseteq G \subseteq \mathsf{qB}$)

$$t \quad := \quad x \;[\!]\; 0 \;[\!]\; 1 \;[\!]\; (t+t) \;[\!]\; (t\,t) \;[\!]\; \mathrm{Re}(u) \;[\!]\; \mathrm{Im}(u) \;[\!]\; |u| \;[\!]\; (\textstyle\int \alpha)$$

$$u \quad := \quad z \;[\!]\; |\top\rangle_{GA} \;[\!]\; t + it \;[\!]\; \overline{u} \;[\!]\; (u+u) \;[\!]\; (u\,u) \;[\!]\; (\alpha \rhd u;\, u)$$

Quantum formulas (with the proviso $F \subseteq \mathsf{qB}$):

$$\gamma \quad := \quad \alpha \;[\!]\; (t \leq t) \;[\!]\; [F] \;[\!]\; \perp\!\!\!\perp \;[\!]\; (\gamma \sqsupset \gamma)$$

Table 1. Language of dEQPL

other classical connectives like $\neg, \wedge, \vee, \Leftrightarrow$ and $\top$ are introduced as abbreviations. We denote the set of qubit symbols occurring in α by $\mathsf{qB}(\alpha)$, and say that a classical formula α is over a set S of qubit symbols if $\mathsf{qB}(\alpha) \subseteq S$.

For the term language, we pick two disjoint denumerable sets of variables. The first set of variables $X = \{x_k : k \in \mathbb{N}\}$ is interpreted in the real closed field of the quantum structure, and the second set $Z = \{z_k : k \in \mathbb{N}\}$ is interpreted in the closure of the real closed fields. As we shall see in Section 5, variables are often useful for applications that we have in mind. There are two syntactic categories t and u for terms, which are mutually defined. The syntactic category t denotes the elements of a real closed field and u denotes the elements of its closure respectively. We will often abuse the notation by saying that t is a real term and u is a complex term.

Most of the term constructs are self-explanatory and already motivated in the previous section. The term $|\top\rangle_{GA}$ denotes the logical amplitude ν_{GA} in the quantum structure, and henceforth will be called an *amplitude term*. The term $(\int \alpha)$ denotes the probability that classical formula α holds for an outcome of a logical projective measurement, and will be called a *probability term*. The denotation of the *alternative term* $(\alpha \rhd u_1;\, u_2)$ will be the value denoted by u_1 if α is true, and the value denoted by u_2 otherwise.

As usual, we may define the notion of occurrence of a term t_1 in a term t, and the notion of replacing zero or more occurrences of terms t_1 in t by t_2. If $\vec{x}$, $\vec{t}$, $\vec{z}$ and $\vec{u}$ are sequences of real variables, real terms, complex variables and complex terms respectively, we will write $t\{\!|\vec{x}/\vec{t}, \vec{z}/\vec{u}|\!\}$ to mean the real term obtained by substituting <u>all</u> occurrences of x_i by t_i and all occurrences of z_j by u_j. The complex term $u\{\!|\vec{x}/\vec{t}, \vec{z}/\vec{u}|\!\}$ is similarly defined.

The quantum formulas are built from classical formulas α, *sub-system formulas* $[F]$ and *comparison formulas* $(t \leq t)$ using the connectives $\perp\!\!\!\perp$ and $\sqsupset$. The formulas consisting of just the classical formulas, sub-system and comparison formulas are called *quantum atoms*, and the set of quantum atoms shall henceforth be called qAtom. We shall use δ, δ' to range over elements of qAtom. Please note that quantum bottom $\perp\!\!\!\perp$ and quantum implication $\sqsupset$ are global connectives and should not be confused with their classical (local) counterparts.

The notion of occurrence of a term t in a quantum formula γ can be easily defined. However, we have to be careful while defining the notion of occurrence of a quantum formula γ in the quantum formula γ_1. This is because we want γ to occur as a *quantum sub-formula* of γ_1 and rule out situations where γ occurs as classical sub-formula. More precisely, we define γ_1 *q-occurs* in γ inductively as:

- if γ is a classical formula, a comparison formula, a sub-system formula, or $\perp\!\!\!\perp$, then γ_1 q-occurs in γ if and only if γ_1 is γ and;
- if γ is $\gamma' \sqsupset \gamma''$ then γ_1 q-occurs in γ if and only if one of the following holds:
  - γ_1 is γ, or
  - γ_1 q-occurs in γ', or
  - γ_1 q-occurs in γ''.

The notion of replacing zero or more q-occurrences of a quantum formula γ_1 in γ by γ' can now be suitably defined.

For example, the classical formula qb q-occurs in $(\mathsf{qb} \sqsupset \mathsf{qb}_1)$ and replacing one q-occurrence of qb by qb_2 will yield the quantum formula $(\mathsf{qb}_2 \sqsupset \mathsf{qb}_1)$. On the other hand qb does not q-occur in $(\mathsf{qb} \Rightarrow \mathsf{qb}_1)$ (qb is a classical sub-formula and not quantum sub-formula). The replacement qb by qb_2 in $(\mathsf{qb}\Rightarrow\mathsf{qb}_1)$ has no effect. Similarly, qb does not q-occur in $[\{\mathsf{qb}\}]$.

For clarity sake, we shall often drop parenthesis in formulas and terms if it does not lead to ambiguity. As expected, other quantum connectives will be introduced as abbreviations. However, before introducing a whole set of useful abbreviations, we present the semantics of the language.

The language is interpreted in a quantum structure as defined in Section 3.1. Given a quantum structure $\mathbf{w} = (\mathcal{K}, \delta, V, \mathcal{S}, |\psi\rangle, \nu)$, recall that $\mathcal{K}$ is a real closed field with $\mathcal{K}(\delta)$ as its algebraic closure, V is a set of valuations over qB, $\mathcal{S}$ is a partition of qB, $|\psi\rangle$ is a collection of $\mathcal{K}(\delta)$-quantum states, and ν is a collection of amplitude terms. We shall assume the semantics of classical propositional logic, and say that $v \Vdash_c \alpha$ if the classical valuation v satisfies the classical formula α.

For interpreting the probability terms, we shall use the *probability map* $\mu_{\mathbf{w}} : \wp(V) \to \mathcal{K}$ defined in Section 3.1 as:

$$\mu_{\mathbf{w}}(U) = \sum_{v \in U} ||\langle v|\psi\rangle||^2 .$$

For the probability terms, we shall also need the *extent* at a set V of classical formulas over S defined as:

$$|\alpha|_V = \{v \in V : v \Vdash_c \alpha\}.$$

For interpreting the variables, we need the concept of an assignment. Given a real closed field $\mathcal{K}$, a *$\mathcal{K}$-assignment* ρ is a map such that $\rho(x) \in \mathcal{K}$ for each $x \in X$ and $\rho(z) \in \mathcal{K}(\delta)$ for each $z \in Z$. Please note that when $\mathcal{K}$ is clear from the context, we shall drop $\mathcal{K}$.

Given a quantum structure $\mathbf{w} = (\mathcal{K}, \delta, V, \mathcal{S}, |\psi\rangle, \nu)$ and a $\mathcal{K}$-assignment ρ. The *denotation of terms* and *satisfaction of quantum formulas* at $\mathbf{w}$ and ρ and is inductively defined in Table 2 (omitting the obvious ones).

Denotation of terms

$$
\begin{array}{lcl}
[\![x]\!]_{\mathbf{w}\rho} & = & \rho(x) \\
[\![t_1 + it_2]\!]_{\mathbf{w}\rho} & = & [\![t_1]\!]_{\mathbf{w}\rho} + \delta[\![t_2]\!]_{\mathbf{w}\rho} \\
[\![(\textstyle\int \alpha)]\!]_{\mathbf{w}\rho} & = & \mu_{\mathbf{w}}(|\alpha|_V) \\
[\![z]\!]_{\mathbf{w}\rho} & = & \rho(z) \\
[\![|\top\rangle_{GA}]\!]_{\mathbf{w}\rho} & = & \nu_{GA} \\
[\![(\alpha \rhd u_1;\, u_2)]\!]_{\mathbf{w}\rho} & = & \begin{cases} [\![u_1]\!]_{\mathbf{w}\rho} & \text{if } |\alpha|_V = V \\ [\![u_2]\!]_{\mathbf{w}\rho} & \text{otherwise} \end{cases}
\end{array}
$$

Satisfaction of quantum formulas

$$
\begin{array}{lll}
\mathbf{w}\rho \Vdash \alpha & \text{iff} & |\alpha|_V = V \\
\mathbf{w}\rho \Vdash (t_1 \leq t_2) & \text{iff} & [\![t_1]\!]^{\mathbf{w}}_{\rho} \leq [\![t_2]\!]^{\mathbf{w}}_{\rho} \\
\mathbf{w}\rho \Vdash [A] & \text{iff} & A \in \mathrm{Alg}(\mathcal{S}) \\
\mathbf{w}\rho \nVdash \perp\!\!\!\perp & & \\
\mathbf{w}\rho \Vdash (\gamma_1 \sqsupset \gamma_2) & \text{iff} & \mathbf{w}\rho \nVdash \gamma_1 \text{ or } \mathbf{w}\rho \Vdash \gamma_2
\end{array}
$$

Table 2. Semantics of dEQPL

Please observe that the set V is sufficient to interpret the classical formulas, and the partition $\mathcal{S}$ is sufficient to interpret the sub-system formulas. The $\mathcal{K}$-assignment ρ is sufficient to interpret a useful sub-language of the formulas defined as:

$$
\begin{array}{lcl}
\kappa & := & (a \leq a) \,[\!]\, \perp\!\!\!\perp \,[\!]\, (\kappa \sqsupset \kappa) \\
a & := & x \,[\!]\, 0 \,[\!]\, 1 \,[\!]\, (a + a) \,[\!]\, (a\,a) \,[\!]\, \mathrm{Re}(b) \,[\!]\, \mathrm{Im}(b) \,[\!]\, |b| \\
b & := & z \,[\!]\, a + ia \,[\!]\, \overline{b} \,[\!]\, (b + b) \,[\!]\, (b\,b)
\end{array}
$$

Henceforth, the terms of this sub-language will be called *arithmetical terms* and the formulas will be called *arithmetical formulas.*

We may use the satisfaction relation to define *entailment* as expected: we say that a set of quantum formulas Γ entails a quantum formula η, written $\Gamma \vDash \eta$, if $\mathbf{w}\rho \Vdash \eta$ for every $\mathbf{w}$ and ρ satisfying every element of Γ. We say a quantum formula η is valid when it is entailed by the empty set of quantum formulas. Please note also that the metatheorem of entailment holds: $\Gamma, \eta_1 \vDash \eta_2$ iff $\Gamma \vDash (\eta_1 \sqsupset \eta_2)$. That is, quantum implication internalizes the notion of quantum entailment. The following are some examples of entailment:

$$\begin{array}{rcl} & \vDash & (\neg\,\alpha) \sqsupset (\boxminus\,\alpha) \\ & \vDash & (\alpha_1 \wedge \alpha_2) \equiv (\alpha_1 \sqcap \alpha_2) \\ [G_1],\ [G_2] & \vDash & [G_1 \cap G_2] \\ \alpha & \vDash & ((\int \alpha) = 1) \\ [G] & \vDash & ((\sum_{A \subseteq G} ||\top\rangle_{GA}|^2) = 1) \end{array}$$

We shall now present some useful abbreviations, and give some small examples.

3.3 *Abbreviations and examples*

As anticipated, the proposed quantum language with the semantics above is rich enough to express interesting properties of quantum systems. To this end, it is quite useful to introduce other operations, connectives and modalities through abbreviations. We start with some additional quantum connectives:

- quantum negation: $(\boxminus\,\gamma)$ for $(\gamma \sqsupset \perp\!\!\!\perp)$;
- quantum disjunction: $(\gamma_1 \sqcup \gamma_2)$ for $((\boxminus\,\gamma_1) \sqsupset \gamma_2)$;
- quantum conjunction: $(\gamma_1 \sqcap \gamma_2)$ for $(\boxminus((\boxminus\,\gamma_1) \sqcup (\boxminus\,\gamma_2)))$;
- quantum equivalence: $(\gamma_1 \equiv \gamma_2)$ for $((\gamma_1 \sqsupset \gamma_2) \sqcap (\gamma_2 \sqsupset \gamma_1))$.

It is also useful to introduce some additional comparison formulas:

- $(t_1 < t_2)$ for $((t_1 \leq t_2) \sqcap (\boxminus(t_2 \leq t_1)))$;
- $(t_1 = t_2)$ for $((t_1 \leq t_2) \sqcap (t_2 \leq t_1))$;
- $(u_1 = u_2)$ for $((\mathrm{Re}(u_1) = \mathrm{Re}(u_2)) \sqcap (\mathrm{Im}(u_1) = \mathrm{Im}(u_2)))$

Please note that the only constants in our term language are 0 and 1. As every real closed field $\mathcal{K}$ has characteristic 0, we can embed a copy of rationals in $\mathcal{K}$. It is also possible to take square roots of positive numbers. Hence, it will be useful to use the following abbreviations (with the proviso $n > 0$):

- $(t = n)$ for $t = \underbrace{((1 + (1 + \ldots\ldots)))}_{n \text{ times}}$;
- $(t = \frac{m}{n})$ for $((m.t) = n)$;
- $(t_1 = \sqrt{t_2})$ for $((t_2 \geq 0) \sqcap ({t_1}^2 = t_2))$.

Given $A \subseteq G \subseteq \mathsf{qB}$, the following classical formula will also be useful:

- $(\wedge_G A)$ is $((\wedge_{\mathsf{qb}_k \in A} \mathsf{qb}_k) \wedge (\wedge_{\mathsf{qb}_k \in G \setminus A} (\neg\,\mathsf{qb}_k))$.

The classical formula $(\wedge_G A)$ specifies the unique classical valuation that satisfies all the qubit symbols in A and does not satisfy the qubit symbols in $G \setminus A$. We will often need this classical formula in the case the set G is the full set of qubit symbols qB. Therefore, we will often use the following abbreviation

- $(\wedge A)$ for $(\wedge_{\mathsf{qB}} A)$.

The logical amplitude terms, $|\top\rangle_{GA}$, are easily extendible to any classical formula as (with the provisos $\mathsf{qB}(\alpha) \subseteq G$ and $A \subseteq G \subseteq \mathsf{qB}$):

- $|\alpha\rangle_{GA}$ for $(((\wedge_G A) \Rightarrow \alpha) \rhd |\top\rangle_{GA}; 0)$.

Intuitively, the amplitude term $|\alpha\rangle_{GA}$ coincides with $|\top\rangle_{GA}$ when the valuation $\wedge_G A$ satisfies with α and is 0 otherwise. We will often use this term in the case G is the full set of qubit symbols qB. Therefore, the following abbreviation will also be useful:

- $|\alpha\rangle_A$ for $|\alpha\rangle_{\mathsf{qB}A}$.

We introduce a couple of probability modalities as abbreviations:

- $(\Diamond\alpha)$ for $(0 < (\int\alpha))$;
- $(\Box\alpha)$ for $(1 = (\int\alpha))$.

Finally, we can also define a quantum modality as an abbreviation:

- $([G]\Diamond\ \alpha : u)$ for $([G] \sqcap (|u| > 0) \sqcap (\sqcup_{A \subseteq G}(|\alpha\rangle_{GA} = u)))$.

Intuitively $([G]\Diamond\ \alpha : u)$ is true iff G is a sub-system, there is a subset A of G such that the classical valuation $\wedge_G A$ satisfies α and the logical amplitude $|\top\rangle_{GA}$ takes the non-zero value u.

We discuss a small example where we demonstrate the usefulness of dEQPL to specify properties of a quantum system. We postpone the discussion of more involved examples to Section 5. Consider the following variant of Schrödinger's cat. The attributes of the cat that we consider are: being inside or outside the box, alive or dead, and moving or still. We choose three qubit symbols $\mathsf{qb}_0, \mathsf{qb}_1, \mathsf{qb}_2$ to represent these attributes. For the sake of readability, we use **cat-in-box**, **cat-alive** and **cat-moving** instead of the symbols $\mathsf{qb}_0, \mathsf{qb}_1$ and qb_2 respectively. The following dEQPL formulas constrain the state of the cat at different levels of detail:

1. $[\textbf{cat-in-box}, \textbf{cat-alive}, \textbf{cat-moving}]$;
2. $(\textbf{cat-moving} \Rightarrow \textbf{cat-alive})$;
3. $((\Diamond\,\textbf{cat-alive}) \sqcap (\Diamond\,(\neg\,\textbf{cat-alive})))$;
4. $(\boxminus[\textbf{cat-alive}])$;
5. $((\int \textbf{cat-alive}) = \frac{1}{3})$;

6. $([\mathbf{cat\text{-}alive}, \mathbf{cat\text{-}moving}] \sqcap ((\textstyle\int \mathbf{cat\text{-}alive} \wedge \mathbf{cat\text{-}moving}) = \frac{1}{6})$
$\sqcap ((\textstyle\int \mathbf{cat\text{-}alive} \wedge (\neg\, \mathbf{cat\text{-}moving})) = \frac{1}{6})$
$\sqcap ((\textstyle\int (\neg\, \mathbf{cat\text{-}alive}) \wedge (\neg\, \mathbf{cat\text{-}moving})) = \frac{2}{3}))$.

Please observe that all the above assertions are consistent with each other. Intuitively, the first assertion states that the qubits **cat-in-box**, **cat-alive** and **cat-moving** form a sub-system and therefore, are not entangled with the other qubits of the cat system. The second is a classical constraint on the set of admissible valuations: if the cat is moving then it is alive. The third assertion is a consequence of the famous paradox: the cat can be in a state where it is possible that the cat is alive and it is possible that the cat is dead. The fourth assertion states that the qubit **cat-alive** is necessarily entangled with other qubits. The fifth assertion states that the cat is in a state where the probability of observing it alive (after collapsing the wave function) is $\frac{1}{3}$. Finally, the sixth assertion states that the qubits **cat-alive**, **cat-moving** are not entangled with other qubits, and that the cat is in quantum state where: the probability of observing it alive and moving is $\frac{1}{6}$, the probability of observing it alive and not moving is $\frac{1}{6}$, and the probability of observing it dead (and, thus also not moving by second assertion) is $\frac{2}{3}$.

3.4 The axiomatization

We shall present a Hilbert-style axiomatization of the dEQPL. We need two new concepts for the axiomatization, one of quantum tautology and the second of a valid arithmetical formula.

Let P be a countable set of propositional symbols disjoint from qB. Given a classical formula β over P, let β_q be the syntactic entity obtained by replacing all occurrences of $\bot$ by $\perp\!\!\!\perp$ and $\Rightarrow$ by $\sqsupset$. A quantum formula σ is said to be a *quantum tautology* if there is a classical tautology β over P and a map $\sigma : \mathrm{P} \to \mathsf{qAtom}$ such that σ coincides with $\beta_q\sigma$ where $\beta_q\sigma$ is the quantum formula obtained from β_q by replacing each $p \in \mathrm{P}$ by $\sigma(p)$. For instance, the quantum formula $((x_1 \leq x_2) \sqsupset (x_1 \leq x_2))$ is tautological (obtained, for example, from the classical tautology $p \Rightarrow p$).

Please recall that an arithmetical formula in the dEQPL is any formula that does not have probability terms, amplitude terms, alternative terms, classical formulas and sub-system formulas. As noted in Section 3.2, given an quantum structure with $\mathcal{K}_0$ as the underlying real closed field, a $\mathcal{K}_0$-assignment is enough to interpret all arithmetical formulas. We say that an arithmetical formula κ is a *valid arithmetical formula* if it holds for any assignment that maps variables into an arbitrary real closed field $\mathcal{K}$. Clearly, a valid arithmetical formula holds for all semantic structures of dEQPL. It is a well-known fact from the theory of quantifier elimination [Hodges, 1993; Basu *et al.*, 2003] that the set of valid arithmetical formulas so defined is decidable[6]. However, we shall not go into details of this

[6]For the arithmetical sub-language, we may treat the global connectives as classical connectives.

result as we want to focus our attention on reasoning about quantum aspects only.

The axioms and inference rules of dEQPL are listed in Table 3. In total, we have two inference rules and sixteen axioms. The two inference rules are modus ponens for classical implication **CMP** and *modus ponens* for quantum implication **QMP**[7]. The axioms are better understood in the following groups.

Axioms

[CTaut]	$\vdash$	α for each classical tautology α
[QTaut]	$\vdash$	γ for each quantum tautology γ
[Lift⇒]	$\vdash$	$((\alpha_1 \Rightarrow \alpha_2) \sqsupset (\alpha_1 \sqsupset \alpha_2))$
[Eqv⊥]	$\vdash$	$(\bot \equiv \bot\!\!\!\bot)$
[Ref⊓]	$\vdash$	$((\alpha_1 \sqcap \alpha_2) \sqsupset (\alpha_1 \wedge \alpha_2))$
[Sub∅]	$\vdash$	$[\emptyset]$
[Sub∪]	$\vdash$	$([G_1] \sqsupset ([G_2] \sqsupset [G_1 \cup G_2]))$
[Sub\\]	$\vdash$	$([G] \equiv [\mathsf{qB} \setminus G])$
[RCF]	$\vdash$	$\kappa\{\!\!\{\vec{x}/\vec{t}, \vec{z}/\vec{u}\}\!\!\}$ where κ is a valid arithmetical formula, $\vec{x}$, $\vec{z}$, $\vec{t}$ and $\vec{u}$ are sequences of real variables, complex variables, real terms and complex terms respectively
[If⊤]	$\vdash$	$(\alpha \sqsupset ((\alpha \rhd u_1; u_2) = u_1))$
[If⊥]	$\vdash$	$((\boxminus \alpha) \sqsupset ((\alpha \rhd u_1; u_2) = u_2))$
[Empty]	$\vdash$	$(\vert\top\rangle_{\emptyset\emptyset} = 1)$
[NAdm]	$\vdash$	$((\neg(\wedge A)) \sqsupset (\vert\top\rangle_{\mathsf{qB}A} = 0))$
[Unit]	$\vdash$	$([G] \sqsupset ((\sum_{A \subseteq G} \vert\vert\top\rangle_{GA}\vert^2) = 1))$
[Mul]	$\vdash$	$(([G_1] \sqcap [G_2]) \sqsupset (\vert\top\rangle_{G_1 \cup G_2 A_1 \cup A_2} = \vert\top\rangle_{G_1 A_1} \vert\top\rangle_{G_2 A_2}))$ where $G_1 \cap G_2 = \emptyset$, $A_1 \subseteq G_1$ and $A_2 \subseteq G_2$
[Prob]	$\vdash$	$((\int \alpha) = (\sum_A \vert\vert\alpha\rangle_A\vert^2))$

Inference rules

[CMP]	$\alpha_1, (\alpha_1 \Rightarrow \alpha_2) \vdash \alpha_2$
[QMP]	$\gamma_1, (\gamma_1 \sqsupset \gamma_2) \vdash \gamma_2$

Table 3. Axioms for dEQPL

[7]Actually, **CMP** can be derived from **QMP** and **Lift⇒**.

We have as axioms the classical tautologies and the quantum tautologies (**CTaut** and **QTaut**, respectively). Since the set of classical tautologies and the set of quantum tautologies are both recursive, there is no need to spell out the details of tautological reasoning.

The axioms **Lift**$\Rightarrow$, **Eqv**$\perp$ and **Ref**$\sqcap$ are sufficient to relate (local) classical reasoning and (global) quantum tautological reasoning. These are exactly the axioms that relate classical connectives and global connectives in global logic (see Section 2). We refer to [Mateus *et al.*, 2005] for more details.

The axioms **Sub**, **Sub**$\cup$, and **Sub**$\setminus$ are enough to reason about sub-systems. Together, they impose that sub-systems are closed under set-theoretic operations (closure under intersection and set difference appear as theorems).

The axiom **RCF** says that if κ is a valid arithmetical formula, then any formula obtained by replacing variables with the terms of dEQPL is a tautology. Since the set of valid arithmetical formulas is recursive, we refrain from spelling out the details.

The axioms **If**$\top$ and **If**$\perp$ are self-explanatory, and will be used in the completeness proof to remove alternative terms.

The axioms **Empty**, **NAdm**, **Unit** and **Mul** rule logical amplitudes. Each of them closely reflects a property of our semantic structures. The axiom empty says that the logical amplitude $|\top\rangle_{\emptyset\emptyset}$ is always 1. The axiom **Unit** says that the state of each sub-system is a unit vector. The axiom **NAdm** says that the amplitude of a non-admissible classical valuation is 0. The axiom **Mul** says that the state of a system composed of two subs-systems is a tensor product of the two sub-systems.

Finally, the axiom **Prob** relates probabilities and amplitudes, closely following Postulate 4 of quantum mechanics.

As expected, we say that a formula γ is a *theorem*, written $\vdash \gamma$, if we can build a derivation of γ from the axioms using the inference rules. We say that a (possibly infinite) set of formulas Γ *derives* γ, written $\Gamma \vdash \gamma$, if we can build a derivation of γ from axioms and the inference rules using formulas in Γ as hypothesis. As an illustration of the axiomatization, we establish the following theorems:

PROPOSITION 7. For any classical formulas α_1, α_2, we have

[**Lift**$\wedge$] $\vdash (\alpha_1 \wedge \alpha_2) \sqsupset (\alpha_1 \sqcap \alpha_2)$.
[**PUnit**] $\vdash ((\int \top) = 1)$.

Proof. Derivation of [**Lift**$\wedge$]:

1	$(\alpha_1 \wedge \alpha_2) \Rightarrow \alpha_1$	**CTaut**
2	$((\alpha_1 \wedge \alpha_2) \Rightarrow \alpha_1) \sqsupset ((\alpha_1 \wedge \alpha_2) \sqsupset \alpha_1)$	**Lift⇒**
3	$(\alpha_1 \wedge \alpha_2) \sqsupset \alpha_1$	**QMP**:1,2
4	$(\alpha_1 \wedge \alpha_2) \Rightarrow \alpha_2$	**CTaut**
5	$((\alpha_1 \wedge \alpha_2) \Rightarrow \alpha_2) \sqsupset ((\alpha_1 \wedge \alpha_2) \sqsupset \alpha_2)$	**Lift⇒**
6	$(\alpha_1 \wedge \alpha_2) \sqsupset \alpha_2$	**QMP**:4,5
7	$((\alpha_1 \wedge \alpha_2) \sqsupset \alpha_1) \sqsupset (((\alpha_1 \wedge \alpha_2) \sqsupset \alpha_2) \sqsupset ((\alpha_1 \wedge \alpha_2) \sqsupset (\alpha_1 \sqcap \alpha_2)))$	**Qtaut**
8	$((\alpha_1 \wedge \alpha_2) \sqsupset \alpha_2) \sqsupset ((\alpha_1 \wedge \alpha_2) \sqsupset (\alpha_1 \sqcap \alpha_2))$	**QMP**:3,8
9	$(\alpha_1 \wedge \alpha_2) \sqsupset (\alpha_1 \sqcap \alpha_2)$	**QMP**:6,8

Derivation of [**PUnit**]

1	$[\emptyset]$	**Sub∅**
2	$[\emptyset] \sqsupset [\mathsf{qB}]$	**Sub**
3	$[\mathsf{qB}]$	**QMP**:1,2
4	$([\mathsf{qB}] \sqsupset ((\sum_{A \subseteq \mathsf{qB}} \lvert \lvert \top \rangle_{\mathsf{qB}A} \rvert^2) = 1))$	**Unit**
5	$((\sum_{A \subseteq \mathsf{qB}} \lvert \lvert \top \rangle_{\mathsf{qB}A} \rvert^2) = 1)$	**QMP**:3,4
6	$((\textstyle\int \top) = (\sum_{A \subseteq \mathsf{qB}} \lvert \lvert \top \rangle_{\mathsf{qB}A} \rvert^2))$	**Prob**
7	$(((\textstyle\int \top) = (\sum_{A \subseteq \mathsf{qB}} \lvert \lvert \top \rangle_{\mathsf{qB}A} \rvert^2)) \sqsupset (((\sum_{A \subseteq \mathsf{qB}} \lvert \lvert \top \rangle_{\mathsf{qB}A} \rvert^2) = 1) \sqsupset ((\textstyle\int \top) = 1)))$	**RCF**
8	$(((\sum_{A \subseteq \mathsf{qB}} \lvert \lvert \top \rangle_{\mathsf{qB}A} \rvert^2) = 1) \sqsupset ((\textstyle\int \top) = 1))$	**QMP**:6,7
9	$((\textstyle\int \top) = 1)$	**QMP**:5,8

■

We finish this section with a list of interesting theorems. The first three shall be proved in Section 3.6 using the metatheorems of the logic. The first two relate local equivalence and negation with their global counterparts, while the third one says sub-systems are closed under set intersection and the fourth one says that sub-systems are closed under set difference.

[**Lift⇔**]	$\vdash (\alpha_1 \Leftrightarrow \alpha_2) \sqsupset (\alpha_1 \equiv \alpha_2)$.
[**Lift¬**]	$\vdash \neg\, \alpha \sqsupset \boxminus \alpha$.
[**Sub∩**]	$\vdash_F ([G_1] \sqsupset ([G_2] \sqsupset [G_1 \cap G_2]))$.
[**SubDiff**]	$\vdash_F ([G_1] \sqsupset ([G_2] \sqsupset [G_1 \setminus G_2]))$.

The following theorems give some insight on the major properties of logical amplitudes.

[**AAdd**] $\vdash ((|(\alpha_1 \vee \alpha_2)\rangle_G + |(\alpha_1 \wedge \alpha_2)\rangle_G) = (|\alpha_1\rangle_G + |\alpha_2\rangle_G))$.
[**AMon**] $\vdash ((\alpha_1 \Rightarrow \alpha_2) \sqsupset (||\alpha_1\rangle_G| \leq ||\alpha_2\rangle_G|))$.
[**ASoE**] $\vdash ((\alpha_1 \Leftrightarrow \alpha_2) \sqsupset (|\alpha_1\rangle_G = |\alpha_2\rangle_G))$.
[**ANec**] $\vdash (\alpha \sqsupset (|\alpha\rangle_G = |\top\rangle_G))$.
[**AMExc**] $\vdash ((|\alpha\rangle_G + |(\neg\alpha)\rangle_G) = |\top\rangle_G)$.

The first of the following theorems about probability after measurements just states finite additivity. The second relates logical reasoning with probability reasoning (monotonicity). These two theorems and the theorem **PUnit** illustrated in Proposition 7 are axioms in the exogenous probabilistic logic discussed in Section 2.

[**PAdd**] $\vdash (((\textstyle\int(\alpha_1 \vee \alpha_2)) + (\textstyle\int(\alpha_1 \wedge \alpha_2))) = ((\textstyle\int\alpha_1) + (\textstyle\int\alpha_2)))$.
[**PMon**] $\vdash ((\alpha_1 \Rightarrow \alpha_2) \sqsupset ((\textstyle\int\alpha_1) \leq (\textstyle\int\alpha_2)))$.

The following theorems show that probability modalities behave as normal modalities.

[**PNec**] $\vdash (\alpha \sqsupset (\Box\alpha))$.
[**PNorm**] $\vdash ((\Box(\alpha \Rightarrow \alpha')) \sqsupset ((\Box\alpha) \sqsupset (\Box\alpha')))$.

The quantum modalities also behave as normal modalities.

[**QNorm**] $\vdash (([G]\Diamond\ (\alpha \vee \alpha') : u) \equiv (([G]\Diamond\ \alpha : u) \sqcup ([G]\Diamond\ \alpha' : u)))$.
[**QMon**] $\vdash ((\alpha \Rightarrow \alpha') \sqsupset (([G]\Diamond\ \alpha : u) \sqsupset ([G]\Diamond\ \alpha' : u)))$.
[**QCong**] $\vdash ((u = u') \sqsupset (([G]\Diamond\ \alpha : u) \sqsupset ([G]\Diamond\ \alpha : u')))$.

3.5 Soundness

We now show that the calculus is strongly sound, *i.e.*, if $\Gamma \vdash \gamma$ then $\Gamma \vDash \gamma$. It suffices to show that each of the axioms is valid, *i.e.*, if $\vdash \gamma_1$ is an axiom, then every semantic structure satisfies γ_1.

LEMMA 8. The axiom **QTaut** is valid.

Proof. Assume that β is a classical tautology over the set of propositional symbols P and let $\sigma : \mathrm{P} \to \mathsf{qAtom}$ be a map from P into quantum atoms. We show that $\beta_q\sigma$ is valid in all models of dEQPL.

Take an arbitrary quantum structure $\mathbf{w} = (K, \delta, V, \mathcal{S}, |\psi\rangle, \nu)$, and consider the classical valuation v' over P such that

$$v'(p) = \begin{cases} 1 & \text{if } \mathbf{w}\rho \Vdash \sigma(p) \\ 0 & \text{otherwise} \end{cases} .$$

We show that for any classical formula β' over P

$$v' \Vdash \beta' \text{ iff } \mathbf{w}\rho \Vdash \beta'_q\sigma$$

by induction on the structure of β' as follows.

If β' is a propositional symbol then it follows from the definition of v'. The case where β' is the connective $\bot$ is immediate.

If β' is $(\beta_1 \Rightarrow \beta_2)$, then v' satisfies β_2 or v' does not satisfy β_1. If v' satisfies β_2 then by induction hypothesis $\mathbf{w}\rho \Vdash (\beta_2)_q$. If v' does not satisfy β_1, then by induction hypothesis once again, $v' \not\Vdash (\beta_1)_q$. Therefore, in either case, $\mathbf{w}\rho \Vdash (\beta_1)_q \sqsupset (\beta_2)_q$. Now, note that β'_q is $(\beta_1)_q \sqsupset (\beta_2)_q$.

The lemma now follows by observing that $v' \Vdash \beta$. ■

LEMMA 9. The axioms are valid.

Proof. The axioms **CTaut**, **Eqv**$\bot$, **RCF**, **If**$\top$, **If**$\bot$, **Empty**, **Sub**$\emptyset$, **Sub**$\cup$ and **Sub**$\setminus$ are easy to show. For the rest, let $\mathbf{w} = (K, \delta, V, \mathcal{S}, |\psi\rangle, \nu)$ be a quantum structure and ρ a $\mathcal{K}$-assignment. We consider the other axioms one by one:

- **Lift**$\Rightarrow$. Assume that $\mathbf{w}\rho \Vdash (\alpha_1 \Rightarrow \alpha_2)$. Then, by definition, all classical valuations in V must satisfy $(\alpha_1 \Rightarrow \alpha_2)$. Therefore, if all classical valuations in V satisfy α_1 they must satisfy α_2 also. Hence, either $|\alpha_1|_V \neq V$ or $|\alpha_2|_V = V$. We conclude, by definition, $\mathbf{w} \Vdash \alpha_1 \sqsupset \alpha_2$.

- **Ref**$\sqcap$. Similar to the axiom **Lift**$\Rightarrow$.

- **NAdm**. Assume that $\mathbf{w}\rho \Vdash (\neg(\wedge A))$. This means that the classical valuation v_A^{qB} that assigns 1 to the qubit symbols in A and 0 to all other qubits is not an element of V. Therefore, $\nu_{\mathsf{qB}A} = \langle v_A^{\mathsf{qB}} | \psi \rangle_{\mathsf{qB}} = 0$ and hence $\mathbf{w}\rho \Vdash |\top\rangle_{\mathsf{qB}A} = 0$.

- **Prob**. Using the definition $[\![(\int \alpha)]\!]_{\mathbf{w}\rho} = \mu_{\mathbf{w}}(|\alpha|_V) = \sum_{v \in |\alpha|_V} |\langle v|\psi\rangle_{\mathsf{qB}}|^2$, it suffices to show that

$$\sum_{v \in |\alpha|_V} |\langle v|\psi\rangle_{\mathsf{qB}}|^2 = [\![\sum_A ||\alpha\rangle_A|^2]\!]_{\mathbf{w}\rho}.$$

Please note that $[\![|\alpha\rangle_A]\!]_{\mathbf{w}\rho} = \begin{cases} \nu_{\mathsf{qB}A} & \text{if } v_A^{\mathsf{qB}} \in |\alpha|_V \\ 0 & \text{otherwise} \end{cases}$.

Also, by definition, $\nu_{\mathsf{qB}A} = \langle v_A^{\mathsf{qB}} | \psi \rangle_{\mathsf{qB}}$. Therefore,

$$[\![\sum_A ||\alpha\rangle_A|^2]\!]_{\mathbf{w}\rho} = \sum_{v_A^{\mathsf{qB}} \in |\alpha|_V} |\langle v_A^{\mathsf{qB}} | \psi \rangle_{\mathsf{qB}}|^2$$

We conclude by observing that every v is a v_A^{qB} for some unique $A \subseteq \mathsf{qB}$.

- **Unit**. Assume that $\mathbf{w}\rho \Vdash [G]$. Then $G \in \mathrm{Alg}(\mathcal{S})$. Please note that $\{|v_A^G\rangle : A \subseteq G\}$ forms an orthonormal basis of $H(2^G)$. Hence,

$$|\psi\rangle_G = \sum_{A \subseteq G} \langle v_A^G|\psi\rangle_G \ |v_A^G\rangle.$$

Again, by definition, $\langle v_A^G|\psi\rangle_G = \nu_{GA}$ and so

$$|\psi\rangle_G = \sum_{A \subseteq G} \nu_{GA} \ |v_A^G\rangle.$$

Since $|\psi\rangle_G$ is a unit vector, we get

$$\sum_{A \subseteq G} |\nu_{GA}|^2 = 1.$$

We conclude by noting that $[\![|\top\rangle_{GA}]\!]_{\mathbf{w}\rho} = \nu_{GA}$ by definition.

- **Mul.** Assume that $\mathbf{w}\rho \Vdash [G_1] \sqcap [G_2]$ where $G_1 \cap G_2 = \emptyset$. Then $G_1, G_2 \in \mathrm{Alg}(\mathcal{S})$. The definition of quantum structure says that $|\psi\rangle_{G_1 \cup G_2} = |\psi\rangle_{G_1} \otimes |\psi\rangle_{G_2}$.

  The definition of tensor product says that $|v_{A_1 \cup A_2}^{G_1 \cup G_2}\rangle = |v_{A_1}^{G_1}\rangle \otimes |v_{A_2}^{G_2}\rangle$.
  The definition of quantum structure gives

$$\nu_{G_1 \cup G_2 A_1 \cup A_2} = \langle v_{A_1 \cup A_2}^{G_1 \cup G_2} \ |\psi\rangle_{G_1 \cup G_2}.$$

The definition of tensor product then gives,

$$\nu_{G_1 \cup G_2 A_1 \cup A_2} = \langle v_{A_1}^{G_1} \otimes v_{A_2}^{G_2} \mid \psi_{G_1} \otimes \psi_{G_2}\rangle_{G_1 \cup G_2} = \langle v_{A_1}^{G_1}|\psi\rangle_{G_1} \langle v_{A_2}^{G_2}|\psi\rangle_{G_2}.$$

We conclude by observing that $\nu_{G_1 A_1}$ is $\langle v_{A_1}^{G_1}|\psi\rangle_{G_1}$ and $\nu_{G_2 A_2}$ is $\langle v_{A_2}^{G_2}|\psi\rangle_{G_2}$.

■

THEOREM 10 **Soundness.** The proof system of dEQPL is sound.

Proof. The proof now follows by induction on the number of steps in the derivation. ■

3.6 Metatheorems

We now prove some useful metatheorems for dEQPL. We start by showing that the inference rule Hypothetical Syllogism holds for dEQPL.

LEMMA 11 Hypothetical Syllogism. Let $\gamma_1, \gamma_2, \gamma_3$ be quantum formulas. Then,

$$[\textbf{HypSyl}] \ \Gamma \vdash \gamma_1 \sqsupset \gamma_2 \text{ and } \Gamma \vdash \gamma_2 \sqsupset \gamma_3 \text{ imply } \Gamma \vdash \gamma_1 \sqsupset \gamma_3.$$

Proof. Observe that by **QTaut**,

$$\vdash (\gamma_1 \sqsupset \gamma_2) \sqsupset ((\gamma_2 \sqsupset \gamma_3) \sqsupset (\gamma_1 \sqsupset \gamma_3)).$$

The proposition follows by using two instances of **QMP**. ■

The inference rule **HypSyl** is a useful rule as illustrated in the derivation of the theorem **Lift¬** below:

PROPOSITION 12. For any classical formula α, we have

$$[\textbf{Lift}\neg] \quad \vdash \neg\,\alpha \sqsupset \boxminus\,\alpha.$$

Proof.

1	$((\bot \sqsupset \bot\!\!\bot) \sqcap (\bot\!\!\bot \sqsupset \bot))$	**Eqv⊥**
2	$((\bot \sqsupset \bot\!\!\bot) \sqcap (\bot\!\!\bot \sqsupset \bot)) \sqsupset (\bot \sqsupset \bot\!\!\bot)$	**QTaut**
3	$(\bot \sqsupset \bot\!\!\bot)$	**QMP**: 1,2
4	$(\bot \sqsupset \bot\!\!\bot) \sqsupset ((\alpha \sqsupset \bot) \sqsupset (\alpha \sqsupset \bot\!\!\bot))$	**QTaut**
5	$(\alpha \sqsupset \bot) \sqsupset (\alpha \sqsupset \bot\!\!\bot)$	**QMP**: 3,4
6	$(\alpha \Rightarrow \bot) \sqsupset (\alpha \sqsupset \bot)$	**Lift⇒**
7	$(\alpha \Rightarrow \bot) \sqsupset (\alpha \sqsupset \bot\!\!\bot)$	**HypSyl**: 5,6

■

The axiomatization also enjoys the metatheorem of deduction:

THEOREM 13 Metatheorem of deduction. Let Γ be a set of quantum formulas and γ_1, γ_2 be quantum formulas. Then,

$$\Gamma \cup \{\gamma_1\} \vdash \gamma_2 \text{ iff } \Gamma \vdash \gamma_1 \sqsupset \gamma_2.$$

Proof. ($\leftarrow$) Assume that $\Gamma \vdash \gamma_1 \sqsupset \gamma_2$. Let Π be a proof of the derivation $\Gamma \vdash \gamma_1 \sqsupset \gamma_2$ and assume that the length of Π is n. We can extend Π to obtain $\Gamma \cup \{\gamma_1\} \vdash \gamma_2$ as follows:

n	$\gamma_1 \sqsupset \gamma_2$	Π
n+1	γ_1	**Hyp**
n+2	γ_2	**QMP**: n,n+1

($\rightarrow$) Assume that $\Gamma \cup \{\gamma_1\} \vdash \gamma_2$. We will prove $\Gamma \vdash \gamma_1 \sqsupset \gamma_2$ by induction on n, the length of proof of $\Gamma \cup \{\gamma_1\} \vdash \gamma_2$. The base step $n = 1$ will be subsumed by the inductive step. In the inductive step, we consider the last rule applied. There are three cases:

- γ_2 is either an hypothesis or an axiom. In this case:

1	γ_2	axiom or hypothesis
2	$\gamma_2 \sqsupset (\gamma_1 \sqsupset \gamma_2)$	**QTaut**
3	$\gamma_1 \sqsupset \gamma_2$	**QMP**: 1,2

- γ_2 is obtained from γ and $\gamma \sqsupset \gamma_2$ by **QMP** where γ and $\gamma \sqsupset \gamma_2$ are also derived from $\Gamma \cup \{\gamma_1\}$. Then, by the induction hypothesis,

  - $\Gamma \vdash \gamma_1 \sqsupset \gamma$;
  - $\Gamma \vdash \gamma_1 \sqsupset (\gamma \sqsupset \gamma_2)$

  Let Π_1 and Π_2 be the proofs of $\Gamma \vdash \gamma_1 \sqsupset \gamma$ and $\Gamma \vdash \gamma_1 \sqsupset (\gamma \sqsupset \gamma_2)$ of lengths m_1 and m_2, respectively. Let m_3 be $m_1 + m_2$. The proof of $\Gamma \vdash \gamma_1 \sqsupset \gamma_2$ is as follows:

m_1.	$\gamma_1 \sqsupset \gamma$	Π_1
m_3.	$\gamma_1 \sqsupset (\gamma \sqsupset \gamma_2)$	Π_2
$m_3 + 1$.	$(\gamma_1 \sqsupset (\gamma \sqsupset \gamma_2)) \sqsupset ((\gamma_1 \sqsupset \gamma) \sqsupset (\gamma_1 \sqsupset \gamma_2))$	**QTaut**
$m_3 + 2$.	$(\gamma_1 \sqsupset \gamma) \sqsupset (\gamma_1 \sqsupset \gamma_2)$	**QMP**: m_3,$m_3 + 1$
$m_3 + 3$.	$\gamma_1 \sqsupset \gamma_2$	**QMP**: m_1,$m_3 + 2$

- γ_2 is obtained from γ and $\gamma \Rightarrow \gamma_2$ by **CMP** where γ and $\gamma \Rightarrow \gamma_2$ are also derived from $\Gamma \cup \{\gamma_1\}$. Then, by the induction hypothesis,

  - $\Gamma \vdash \gamma_1 \sqsupset \gamma$;
  - $\Gamma \vdash \gamma_1 \sqsupset (\gamma \Rightarrow \gamma_2)$.

  By the axiom **Lift⇒** we also have $\Gamma \vdash (\gamma \Rightarrow \gamma_2) \sqsupset (\gamma \sqsupset \gamma_2)$.

By hypothetical syllogism (Lemma 11) we also have $\Gamma \vdash \gamma_1 \sqsupset (\gamma \sqsupset \gamma_2)$. The proof now proceeds as in the previous case. ■

We get as a corollary:

COROLLARY 14 Metatheorem of reductio ad absurdum. Let Γ be a set of quantum formulas and γ be a quantum formula. Then,

$$\text{If } \Gamma \cup \{\gamma\} \vdash \perp\!\!\!\perp \text{ then } \Gamma \vdash \boxminus \gamma.$$

We use the metatheorem of equivalence to derive the following theorems:

PROPOSITION 15. For every classical formulas α_1 and α_2 and subsets $G_1, G_2 \in$ qB, we have

[**Lift≡**]	$\vdash (\alpha_1 \Leftrightarrow \alpha_2) \sqsupset (\alpha_1 \equiv \alpha_2)$.
[**Sub∩**]	$\vdash_F ([G_1] \sqsupset ([G_2] \sqsupset [G_1 \cap G_2]))$.

Proof. We shall use metatheorem of deduction to show each of the theorems:
Lift$\equiv$. It suffices to show that $(\alpha_1 \Leftrightarrow \alpha_2) \vdash (\alpha_1 \equiv \alpha_2)$

1	$(\alpha_1 \Rightarrow \alpha_2) \wedge (\alpha_2 \Rightarrow \alpha_1)$	Hyp
2	$((\alpha_1 \Rightarrow \alpha_2) \wedge (\alpha_2 \Rightarrow \alpha_1)) \Rightarrow (\alpha_1 \Rightarrow \alpha_2)$	**CTaut**
3	$(\alpha_1 \Rightarrow \alpha_2)$	**CMP**: 1,2
4	$(\alpha_1 \Rightarrow \alpha_2) \sqsupset (\alpha_1 \sqsupset \alpha_2)$	**Lift$\Rightarrow$**
5	$(\alpha_1 \sqsupset \alpha_2)$	**QMP**: 3, 4
6	$((\alpha_1 \Rightarrow \alpha_2) \wedge (\alpha_2 \Rightarrow \alpha_1)) \Rightarrow (\alpha_2 \Rightarrow \alpha_1)$	**CTaut**
7	$(\alpha_2 \Rightarrow \alpha_1)$	**CMP**: 1,2
8	$(\alpha_2 \Rightarrow \alpha_1) \sqsupset (\alpha_2 \sqsupset \alpha_1)$	**Lift$\Rightarrow$**
9	$(\alpha_2 \sqsupset \alpha_1)$	**QMP**: 7, 8
10	$(\alpha_1 \sqsupset \alpha_2) \sqsupset ((\alpha_2 \sqsupset \alpha_1) \sqsupset (\alpha_1 \equiv \alpha_2))$	**QTaut**
11	$(\alpha_2 \sqsupset \alpha_1) \sqsupset (\alpha_1 \equiv \alpha_2)$	**QMP**: 5,10
12	$(\alpha_1 \equiv \alpha_2)$	**QMP**: 9,11

Sub$\cap$. It suffices to show that $[G_1], [G_2] \vdash [G_1 \cap G_2]$

1	$[G_1]$	Hyp
2	$[G_1] \sqsupset [\mathsf{qB} \setminus G_1]$	**Sub$\setminus$**
3	$[\mathsf{qB} \setminus G_1]$	**QMP**: 1,2
4	$[G_2]$	Hyp
5	$[G_2] \sqsupset [\mathsf{qB} \setminus G_2]$	**Sub$\setminus$**
6	$[\mathsf{qB} \setminus G_2]$	**QMP**: 4,5
7	$[\mathsf{qB} \setminus G_1] \sqsupset ([\mathsf{qB} \setminus G_2] \sqsupset [\mathsf{qb} \setminus (G_1 \cap G_2)])$	**Sub$\cup$**
8	$[\mathsf{qB} \setminus G_2] \sqsupset [\mathsf{qb} \setminus (G_1 \cap G_2)]$	**QMP**: 3,7
9	$[\mathsf{qb} \setminus (G_1 \cap G_2)]$	**QMP**: 6,8
10	$[\mathsf{qb} \setminus (G_1 \cap G_2)] \sqsupset [G_1 \cap G_2])$	**Sub$\setminus$**
11	$[G_1 \cap G_2]$	**QMP**: 9,10

■

We also have the principles of substitution of equal terms and equivalent formulas.

THEOREM 16 Principle of substitution of equal terms. Given a quantum formula γ, two real terms t_1 and t_2, let γ' be a quantum formula obtained from γ by replacing zero or more occurrences of t_1 in γ_1 by t_2. Then,

$$\vdash t_1 = t_2 \sqsupset (\gamma \equiv \gamma').$$

Proof. The proof is by a straightforward induction on the structure of γ. We note that in the case where γ is $t \leq t'$, we use the axiom **RCF**. The other cases are immediate. ■

Substitution of equivalent terms preserves quantum equivalence:

THEOREM 17 Principle of substitution of equivalent formulas. Given three quantum formulas γ, γ_1 and γ_2, let γ' be obtained from γ by replacing zero or more q-occurrences of γ_1 in γ by γ_2. Then,

$$\vdash (\gamma_1 \equiv \gamma_2) \sqsupset (\gamma \equiv \gamma').$$

Proof. The case γ_1 does not q-occur in γ is trivial. We just consider the case in which γ_1 has at least one q-occurrence in γ and γ' is obtained by replacement of at least one such q-occurrence. The proof is carried out by induction on the structure of γ. There are two cases:

1. γ is a quantum atom or $\perp\!\!\!\perp$. Then γ_1 is γ, γ_1 q-occurs in γ exactly once, and replacement of q-occurrence of γ_1 in γ by γ_2 yields γ_2. Hence, in that case γ' is γ_2. So the theorem holds trivially by the following assertion (justified by the axiom **QTaut**):

$$\vdash (\gamma_1 \equiv \gamma_2) \sqsupset (\gamma_1 \equiv \gamma_2).$$

2. γ is $\gamma_a \sqsupset \gamma_b$. Then there are two cases.

   - γ_1 is γ. Then the theorem follows as in the previous case.
   - γ_1 q-occurs in γ_a or γ_b (it may occur in both). Let γ' be $\gamma'_a \sqsupset \gamma'_b$ where γ'_a and γ'_b are obtained by replacing zero or more occurrences of γ_a and γ_b respectively. Then, by the induction hypothesis we have

$$(\gamma_1 \equiv \gamma_2) \vdash (\gamma_a \equiv \gamma'_a)$$

   and

$$(\gamma_1 \equiv \gamma_2) \vdash (\gamma_b \equiv \gamma'_b).$$

   We show that

$$(\gamma_1 \equiv \gamma_2), \gamma \vdash \gamma'$$

   as follows.

1	$\gamma_1 \equiv \gamma_2$	Hyp
2	$\gamma_a \sqsupset \gamma_b$	Hyp
3	$\gamma_b \sqsupset \gamma'_b$	1, Induction Hypothesis
4	$\gamma_a \sqsupset \gamma'_b$	**HypSyl**: 2, 3
5	$\gamma'_a \sqsupset \gamma_a$	1, Induction Hypothesis
6	$\gamma'_a \sqsupset \gamma'_b$	**HypSyl**: 4, 5

We can show similarly that

$$(\gamma_1 \equiv \gamma_2), \gamma' \vdash \gamma.$$

The theorem now follows from metatheorem of deduction.

■

We get as a corollary that substitution of classically equivalent formulas preserves quantum equivalence:

COROLLARY 18. Given a quantum formula γ, two classical formulas α_1, α_2, let γ' be obtained from γ by replacing zero or more q-occurrences of α_1 in γ by α_2. Then

$$\vdash (\alpha_1 \Leftrightarrow \alpha_2) \sqsupset (\gamma \equiv \gamma').$$

Proof. We observe that by **Lift**$\Leftrightarrow$, we have $\vdash (\alpha_1 \Leftrightarrow \alpha_2) \sqsupset (\alpha_1 \equiv \alpha_2)$. The result then follows from principle of substitution of equivalent formulas and hypothetical syllogism. ■

Please note that we are only concerned with occurrence of classical formulas only as quantum sub-formulas and not as classical formulas. Indeed, replacement of a classical formula by a quantum formula may not always yield valid a quantum formula. Even in the case it yields a valid quantum formula, the principle of substitution does not hold. For example, let α_1 be qb_1, α_2 be qb_2 and γ be qb_3. Now, consider the quantum formula:

$$(\mathsf{qb}_1 \equiv \mathsf{qb}_2) \sqsupset ((\mathsf{qb}_1 \Rightarrow \mathsf{qb}_3) \equiv (\mathsf{qb}_2 \Rightarrow \mathsf{qb}_3)).$$

Let V be the set of two valuations v_1, v_2 such that:

- $v_1(\mathsf{qb}_1) = v_1(\mathsf{qb}_3) = 0$, $v_1(\mathsf{qb}_2) = 1$;
- $v_2(\mathsf{qb}_1) = v_2(\mathsf{qb}_3) = 1$, $v_2(\mathsf{qb}_2) = 0$.

Any quantum structure with V as the set of valuations would then invalidate the above quantum formula.

4 COMPLETENESS AND DECIDABILITY

We shall prove weak completeness of dEQPL — if Γ is a finite set of quantum formulas, then $\Gamma \vDash \gamma$ implies that $\Gamma \vdash \gamma$. As our proof system enjoys principle of deduction, it suffices to demonstrate weak completeness when the set Γ is empty. The proof of weak completeness will go hand-in-hand with the proof of decidability, and can be adapted to a proof of strong completeness as we will sketch later.

The proof of weak completeness essentially follows the proof in [Mateus and Sernadas, 2006], which in turn was inspired by the Fagin-Megiddo-Halpern technique for probabilistic logic [Fagin *et al.*, 1990]. The main difference is in the way the sub-system formulas are treated here. The other difference is that the proof is carried out in a manner so as to facilitate the proof of decidability.

The central result in the proof is the Model Existence Lemma, namely, if γ is consistent then there is a quantum structure $\mathbf{w}$ and an assignment ρ such that $\mathbf{w}\rho \Vdash \gamma$. A quantum formula γ is said to be *consistent* if $\nvdash (\boxminus \gamma)$. It will suffice to show that the model existence lemma holds for specials kinds of quantum formula, namely *quantum molecular formulas.* A quantum molecular formula is a quantum disjunction of *quantum literals* (a quantum literal is either a quantum atom or the quantum negation of a quantum atom). Please recall that quantum atoms are classical formulas, comparison terms and sub-system assertions.

The first steps in the proof of the Model Existence Lemma are to remove the probability and alternative terms using the axioms **Prob**, **If**$\top$ and **If**$\bot$. Next, we use the weak completeness of classical propositional logic to construct the set of valuations V in the envisaged quantum structure. The partition S is constructed by considering the sub-system literals in the quantum molecule, and the construction is guided by the fact that sub-systems are closed under set operations (axioms **Sub**$\emptyset$, **Sub**$\cup$ and **Sub**$\setminus$). The logical amplitudes ν_{GA} are constructed by first adding all consistent equations using the axioms **NAdm**, **Unit**, **Empty** and **Mul**, and then "solving" for the (in)equations in the quantum molecule using **RCF**.

Before proceeding with carrying out the above outline, we start with a few abbreviations and notations. We introduce the following abbreviation where $Q \subset$ qAtom and $D \subseteq Q$:

- $(\bigsqcap_Q D)$ for $((\bigsqcap_{\mu \in D} \mu) \sqcap (\bigsqcap_{\mu \in (Q \setminus D)} (\boxminus \mu)))$.

We shall say that D is the positive part of the quantum molecule $(\bigsqcap_Q D)$ and that $Q \setminus D$ is its negative part. Given a molecule η, we denote by η^+ and η^- the positive and negative parts respectively. We denote by η_c the conjunction of the classical literals in η. In a similar way we define $\eta_\leq$ and η_s.

As is the case with classical propositional logic, every dEQPL formula has a *quantum disjunctive normal form.* A quantum formula is said to be in quantum disjunctive normal form if it is a disjunction of quantum molecules.

PROPOSITION 19. Every quantum formula is equivalent to a quantum disjunctive normal form. Furthermore, there is an algorithm that computes the quantum disjunctive normal form.

Proof. It is easier to prove a stronger result. That is, we show that any quantum formula η has both a quantum disjunctive normal form and a quantum conjunctive normal form. We say that η is in quantum conjunctive normal form if it is a quantum conjunction of quantum disjunctions of literals. The proof is constructive and follows by induction on the structure of the quantum formula as in the case of classical logic. The construction also gives the algorithm for computing the normal forms. ■

From now on we will assume that every quantum formula is in quantum disjunctive normal form. The following proposition will ensure that to decide consistency of a quantum formula we only need to check if one of its molecules is consistent.

PROPOSITION 20. A quantum formula is consistent iff one of its molecules is consistent.

Proof. ($\Rightarrow$) It suffices to show that the quantum disjunction of two inconsistent quantum formulas γ_1 and γ_2 is inconsistent. If γ_1 and γ_2 are inconsistent then $\vdash (\boxminus \gamma_1)$ and $\vdash (\boxminus \gamma_2)$. We can easily show that in this case $\vdash \boxminus(\gamma_1 \sqcup \gamma_2)$ as follows:

1	$(\boxminus\gamma_1) \sqsupset ((\boxminus\gamma_2) \sqsupset \boxminus(\gamma_1 \sqcup \gamma_2))$	**QTaut**
2	$(\boxminus\gamma_1)$	Hyp
3	$(\boxminus\gamma_2)$	Hyp
4	$(\boxminus\gamma_2) \sqsupset \boxminus(\gamma_1 \sqcup \gamma_2)$	**QMP**: 1,2
5	$\boxminus(\gamma_1 \sqcup \gamma_2)$	**QMP**: 3,4

Hence the formula, $(\gamma_1 \sqcup \gamma_2)$ is inconsistent.

($\Leftarrow$) Assume that η is inconsistent. Then $\vdash (\boxminus \eta)$. Let η be $\eta_1 \sqcup \ldots \sqcup \eta_n$. By **QTaut**, $\vdash (\boxminus \eta) \equiv (\boxminus \eta_1 \sqcap \ldots \sqcap \boxminus \eta_n)$. Using **QMP** and **QTaut** we can easily show that η_i is inconsistent for $i = 1, \ldots, n$. ■

The first step in the proof is to remove the probability terms.

PROPOSITION 21. Given a quantum molecule η, there is a η' such that η' has no probability terms and $\vdash \eta \equiv \eta'$. Furthermore, there is an algorithm that computes η'.

Proof. Let η be a molecule. For every probability term of the form $(\int \alpha)$ replace it by $(\sum_A ||\alpha\rangle_A|^2))$. Then by axiom **Prob** and the principle of substitution of equal terms, the resulting formula is equivalent to η. ■

The following proposition allows us to remove alternative terms in quantum molecules.

PROPOSITION 22. A quantum molecule η is consistent iff there is a consistent quantum molecule η' such that η' has no alternative terms and $\vdash (\eta' \sqsupset \eta)$. Moreover, if there is an algorithm for deciding the consistency of quantum molecules without alternative terms then there is an algorithm for deciding the consistency of quantum molecules.

Proof. The existence of a consistent η' such that $\vdash (\eta' \sqsupset \eta)$ clearly implies the consistency of η. For the other direction, consider an ordering $\alpha_0, \ldots, \alpha_m$ of the guards of alternative terms occurring in η. Let α_i^0 be α_i and α_i^1 be $\boxminus \alpha_i$ for $i = 0, \ldots, m$.

Given $b_0 \ldots b_m \in \{0,1\}^m$, let

$$\eta_{b_0 \ldots b_m} := \eta \sqcap \alpha_0^{b_0} \sqcap \ldots \sqcap \alpha_m^{b_m}.$$

Using **QTaut** we get,

$$\vdash \eta \equiv \bigsqcup_{b_0 \ldots b_m \in \{0,1\}^m} \eta_{b_0 \ldots b_m}.$$

Observe that, using the axioms **If**$\top$ and **If**$\bot$ and the principle of substitution of equal terms, each $\eta_{b_0 \ldots b_m}$ is equivalent to a formula in which the alternative $(\alpha_i \rhd u_i^0; u_i^1)$ is replaced by $u_i^{b_i}$. Let $\overline{\eta}_{b_0 \ldots b_m}$ be the resulting formula. Therefore,

$$\vdash \eta \equiv \bigsqcup_{b_0 \ldots b_m \in \{0,1\}^m} \overline{\eta}_{b_0 \ldots b_m}.$$

With a reasoning similar to the one in Proposition 20, we conclude that η is consistent iff $\overline{\eta}_{b_0 \ldots b_m}$ is consistent for some $b_0 \ldots b_m \in \{0,1\}^m$. Please note that $\vdash \overline{\eta}_{b_0 \ldots b_m} \sqsupset \eta$ for each $b_0 \ldots b_m \in \{0,1\}^m$.

Finally, as the construction of each $\overline{\eta}_{b_0 \ldots b_m}$ can be defined by an algorithm, we get the proposition. ■

We shall now build the set of classical valuations V. Given a classical formula α and a non-empty set of valuations V, we write $V \Vdash_c \alpha$ if every element of V classically satisfies α. We say that $V \Vdash_c \eta$ if $V \Vdash_c \alpha$ for every $\alpha \in \eta_c^+$ and $V \nVdash_c \beta$ for every $\beta \in \eta_c^-$.

We will consider only a special kind of molecular formulas which will allow us to deal with the restrictions imposed by the axiom **NAdm**. Please recall that given $A \subseteq \mathsf{qB}$, v_A is the valuation that assigns true to qubit symbols in A and false to qubit symbols in $\mathsf{qb} \setminus A$. A molecular formula η is said to be *maximal with respect to admissible classical valuations* if for every subset A of qB and set of valuations V such that $V \Vdash_c \eta$, we have:

$$v_A \notin V \text{ iff } (\neg(\wedge A)) \in \eta_c^+.$$

The following proposition ensures that it suffices to consider molecular formulas maximally consistent with classical valuations.

PROPOSITION 23. A molecule η is consistent iff there is a consistent molecule η' such that η' is maximal with respect to admissible classical valuations and $\vdash \eta' \sqsupset \eta$. Moreover, if there is an algorithm for deciding the consistency of quantum molecules maximal with respect to admissible valuations then there is an algorithm for deciding consistency of quantum molecules.

Proof. Let $A_1, \ldots, A_m$ be an ordering of the subsets of qB. Let A_i^0 be $(\neg(\wedge A_i))$ and A_i^1 be $\boxminus(\neg(\wedge A_i))$ for $i = 0, \ldots, m$. Given $b_0 \ldots b_m \in \{0,1\}^m$, let

$$\eta_{b_0 \ldots b_m} := \eta \sqcap A_0^{b_0} \sqcap \ldots \sqcap A_m^{b_m}.$$

Using **QTaut**,

$$\vdash \eta \equiv \bigsqcup_{b_0 \ldots b_m \in \{0,1\}^m} \eta_{b_0 \ldots b_m}.$$

With a reasoning similar to the one in Proposition 20, we can conclude that η is consistent iff $\eta_{b_0 \ldots b_m}$ is consistent for some $b_0 \ldots b_m \in \{0,1\}^m$. Please note that $\vdash \eta_{b_0 \ldots b_m} \sqsupset \eta$ for each $b_0 \ldots b_m \in \{0,1\}^m$. We claim that each $\eta_{b_0 \ldots b_m}$ is maximal with respect to admissible valuations. Fix one $\eta_{b_0 \ldots b_m}$.

Let V be a set of valuations such that $V \Vdash_c \eta_{b_0 \ldots b_m}$. We will show that

$$v_{A_i} \notin V \text{ iff } (\neg(\wedge A_i)) \in (\eta_{b_0 \ldots b_m})_c^+.$$

Clearly if $(\neg(\wedge A_i)) \in (\eta_{b_0 \ldots b_m})_c^+$ then $v_{A_i} \notin V$.

For the other part, if $v_{A_i} \notin V$ it suffices to show that $b_i = 0$. Suppose that $b_i = 1$. Then $V \not\Vdash_c (\neg(\wedge A_i))$. That means there is $v \in V$ such that $v \not\Vdash_c (\neg(\wedge A_i))$. This means that $v \Vdash_c \wedge A_i$ which in turn implies that v is equal to v_{A_i}. Therefore $v_{A_i} \in V$ contradicting the assumption $b_i = 1$.

As the construction of $\eta_{b_0 \ldots b_m}$ can be defined by an algorithm, we get the proposition. ∎

We will say that η is *g-satisfiable* if there is a set of valuations V such that $V \Vdash_c \eta$. Given a consistent molecule η, we now construct V such that $V \Vdash_c \eta$ as follows.

LEMMA 24 *g*-satisfiability. If η is consistent then η is g-satisfiable. Furthermore, there is an algorithm to decide if η is g-satisfiable.

Proof. Let V be the set of valuations v such that $v \Vdash_c \alpha$ for every $\alpha \in \eta_c^+$. This set can be computed since the set of qubit symbols is finite.

If V is empty then η is not g-satisfiable. If V is not empty, then η is g-satisfiable iff $V \not\Vdash_c \beta$ for every $\beta \in \eta_c^-$. As V and η_c^- are finite sets, this gives us an algorithm to check if η is g-satisfiable.

Assume that η is a consistent formula. Please note that using the theorem **Lift**$\wedge$ and the principle of substitution, it is easy to show that if η is consistent then $(\wedge_{\alpha \in \eta_c^+} \alpha)$ is consistent as a classical propositional formula.

We show that η is g-satisfiable. As $(\wedge_{\alpha\in\eta_c^+}\alpha)$ is consistent (in propositional logic), there is a classical valuation v that satisfies every α. As above, let V be the set of valuations v such that $v \Vdash_c \alpha$ for every $\alpha \in \eta_c^+$. It suffices to show that $V \nVdash_c \beta$ for every $\beta \in \eta_c^-$.

We proceed by contradiction. Assume that there is $\beta \in \eta_c^-$ such that $V \Vdash_c \beta$. Fix one such β say β_0. Therefore, by construction of V, we get:

$$\Vdash_c \left(\left(\bigwedge_{\alpha\in\eta_c^+} \alpha\right) \Rightarrow \beta_0\right).$$

So, by **CTaut** we get:

$$\vdash \left(\left(\bigwedge_{\alpha\in\eta_c^+} \alpha\right) \Rightarrow \beta_0\right).$$

Thus, by **Lift⇒**, we obtain

$$\vdash \left(\left(\bigwedge_{\alpha\in\eta_c^+} \alpha\right) \sqsupset \beta_0\right).$$

Thus, by **Ref⊓** and **QTaut** (transitivity of $\sqsupset$) we get

$$\vdash \left(\left(\bigsqcap_{\alpha\in\eta_c^+} \alpha\right) \sqsupset \beta_0\right).$$

Therefore, by **QTaut** (right weakening of $\sqsupset$)

$$\vdash \left(\left(\bigsqcap_{\alpha\in\eta_c^+} \alpha\right) \sqsupset \left(\bigsqcup_{\beta\in\eta_c^-} \beta\right)\right)$$

leading to

$$\vdash \left(\boxminus\left(\left(\bigsqcap_{\alpha\in\eta_c^+} \alpha\right) \sqcap \left(\bigsqcap_{\beta\in\eta_c^-} (\boxminus\beta)\right)\right)\right)$$

by several obvious tautological steps. That is, we have $\vdash (\boxminus\eta)$, contradicting the consistency of η. ■

Please observe that if η has neither probability nor alternative terms then η' as constructed in the above proof also does not have probability and alternative terms.

Given a sub-system formula $[G]$ and a partition $\mathcal{S}$ of the set of qubits, we write $\mathcal{S} \Vdash_s [G]$ if $G \in \mathrm{Alg}(\mathcal{S})$. We say that $\mathcal{S} \Vdash_s \eta$ if $\mathcal{S} \Vdash_s [G]$ for every $[G] \in \eta_s^+$

and $\mathcal{S} \not\Vdash_s [G]$ for every $[G] \in \eta_s^-$. We will say that η is *s-satisfiable* if there is a partition $\mathcal{S}$ such that $\mathcal{S} \Vdash_s \eta$. We construct the partition S in the proof of Model Existence Lemma as follows.

LEMMA 25 s-satisfiability. If η is consistent then η is s-satisfiable. There is an algorithm to decide if η is s-satisfiable.

Proof.

Assume that η is consistent. We will show that η is s-satisfiable. Please recall that an algebra of sets on a domain X is a non-empty collection of subsets of X closed under complements and unions. Let $\mathrm{Alg}(\eta_s^+)$ be the smallest algebra on qB containing η_s^+.

Find the minimal elements for $\mathrm{Alg}(\eta_s^+)$: a set $G \in \mathrm{Alg}(\eta_s^+)$ is minimal if $G' \subseteq G$ and $G \in \mathrm{Alg}(\eta_s^+)$ implies that G' is either the empty set or G itself. Take $\mathcal{S}$ to be the set of minimal elements of $\mathrm{Alg}(\eta_s^+)$ (it can be easily shown that they form a partition). Therefore, by construction, $\mathcal{S} \Vdash_s [G]$ for every $[G] \in \eta_s^+$.

If $[H] \in \eta_s^-$ then we need to show $[H] \notin \mathrm{Alg}(\eta_s^+)$. We proceed by contradiction and assume $H \in \mathrm{Alg}(\eta_s^+)$. Then $H = H_1 \cup \ldots \cup H_m$, where either $H_i \in \eta_s^+$ or $\mathsf{qB} \setminus H_i \in \eta_s^+$ for each $1 \leq i \leq m$. Using the axioms **QTaut**, **Sub$\setminus$** and **Sub$\cup$**, we can show that

$$\vdash \eta \equiv \eta \sqcap [H].$$

Now as $[H] \in \eta_s^-$, we get

$$\vdash \eta \equiv \eta \sqcap [H] \equiv \eta \sqcap [H] \sqcap \boxminus[H].$$

Now, by **QTaut**,

$$\vdash (\eta \sqcap [H] \sqcap \boxminus[H]) \sqsupset \perp\!\!\!\perp .$$

Then, by principle of substitution of equivalent formulas, we get

$$\vdash (\boxminus \eta).$$

This contradicts the consistency of η.

The algorithm for checking the s-consistency is as follows. Take η_s^+ and generate the algebra $\mathrm{Alg}(\eta_s^+)$ with them. This algebra can be computed since the set of qubit symbols is finite. The formula η is s-satisfiable iff $G \notin \mathrm{Alg}(\eta_s^+)$ for every $[G] \in \eta_s^-$. This can be checked by an algorithm again as the set of qubits is finite. ∎

We are now ready to construct the model (the amplitudes ν_{GA} will be constructed in the proof). We need some auxiliary definitions. Recall that assignments are enough to interpret the arithmetical formulas. Let κ be a quantum conjunction of comparison literals. Let $\mathcal{K}$ be a real closed field with algebraic closure $\mathcal{K}(\delta)$ and ρ be a $\mathcal{K}$-assignment. We say that $\mathcal{K}(\delta)$, $\rho \Vdash_i \kappa$ if

- $[\![s]\!]_\rho \leq [\![t]\!]_\rho$ if $s \leq t \in \kappa^+$;

- $[\![s]\!]_\rho \not\leq [\![t]\!]_\rho$ if $s \leq t \in \kappa^-$.

We say that ρ is a *solution* of κ in $\mathcal{K}(\delta)$. We say that κ is $\leq$-*consistent* if there is a real closed field $\mathcal{K}$ with algebraic closure $\mathcal{K}(\delta)$, and a $\mathcal{K}$-assignment ρ such that $\mathcal{K}(\delta),\, \rho \Vdash_i \kappa$. Please note that the theory of elimination of quantifiers ensures that there is an algorithm to decide the $\leq$-consistency [Hodges, 1993; Basu *et al.*, 2003].

THEOREM 26 Model Existence Theorem. If the molecule η is consistent then there is a quantum structure $\mathbf{w} = (\mathcal{K}, \delta, V, \mathcal{S}, |\psi\rangle, \nu)$ and a $\mathcal{K}$-assignment ρ such that $\mathbf{w}\rho \Vdash \eta$.

Proof. As a result of Propositions 21 and 22, we can assume that η does not have any probability and alternative terms and is maximally consistent with respect to admissible valuations.

Using Lemma 24 and Lemma 25, we find V and $\mathcal{S}$ such that $V \Vdash_c \eta$ and $\mathcal{S} \Vdash_s \eta$. We can show that $\vdash \eta \equiv (\eta \sqcap \prod_{[G]\in \mathrm{Alg}(\mathcal{S})} [G])$ using axioms **Sub**$\emptyset$, **Sub**$\cup$ and **Sub**$\backslash$.

Please observe that the axiom **Unit** allows us to establish for every $[G] \in \mathrm{Alg}(\mathcal{S})$:

$$\vdash \eta \sqsupset (\sum_{A\subseteq G} ||\top\rangle_{GA}|^2 = 1).$$

Let η_1 be the formula

$$\eta \sqcap \prod_{G\in \mathrm{Alg}(\mathcal{S})} (\sum_{A\subseteq G} ||\top\rangle_{GA}|^2 = 1).$$

As a result we get that $\vdash (\eta_1 \equiv \eta)$.

We also get as a result of the axiom **NAdm**, for every $(\neg(\wedge A))$ occurring in η:

$$\vdash \eta_1 \sqsupset (|\top\rangle_{\mathsf{qB}A} = 0).$$

Let η_2 be the formula

$$\eta_1 \sqcap \prod_{(\neg(\wedge A)) \text{ in } \eta_c^+} (|\top\rangle_{\mathsf{qB}A} = 0).$$

As a result of axiom **Mul**, for every G_1, G_2, A_1, A_2 such that $G_1, G_2 \in \mathrm{Alg}(\mathcal{S})$, $A_1 \subseteq G_1$ and $A_2 \subseteq G_2$, we get

$$\vdash \eta_2 \sqsupset (|\top\rangle_{G_1\cup G_2 A_1\cup A_2} = |\top\rangle_{G_1 A_1} |\top\rangle_{G_2 A_2}).$$

Let η_3 be the formula

$$\eta_2 \sqcap \prod_{\substack{G_1, G_2 \in \mathrm{Alg}(\mathcal{S}) \\ A_1 \subseteq G_1, A_2 \subseteq G_2}} (|\top\rangle_{G_1\cup G_2 A_1\cup A_2} = |\top\rangle_{G_1 A_1} |\top\rangle_{G_2 A_2}).$$

The axiom **Empty** gives us

$$\vdash \eta_3 \sqsupset (|\top\rangle_{\emptyset\emptyset} = 1).$$

Let $\eta^\bullet$ be the formula

$$\eta_3 \sqcap (|\top\rangle_{\emptyset\emptyset} = 1).$$

Observe that we can show:

$$\vdash (\eta \equiv \eta^\bullet).$$

Please recall that $\eta^\bullet{}_\leq$ is the conjunction of the (in)equations in $\eta^\bullet$. Let η_R be the formula obtained from $\eta^\bullet$ by replacing each term of the form $|\top\rangle_{GA}$ by a fresh variable $z_{|\top\rangle_{GA}}$. Please observe that $\eta^\bullet$ is $\eta_R\{\!|z_{|\top\rangle_{GA}}/\ |\top\rangle_{GA}|\!\}$.

Now, either there is a real closed field $\mathcal{K}$ with $\mathcal{K}(\delta)$ as its algebraic closure, and a $\mathcal{K}$-assignment ρ such that $K(\delta)$, $\rho \Vdash_i (\eta_R)_\leq$ or not. If there is no such $\mathcal{K}$ and ρ then it must be the case that $\boxminus(\eta_R)_\leq$ is a valid arithmetic formula. So, by axiom **RCF**,

$$\vdash (\boxminus(\eta_R)_\leq)\{\!|z_{|\top\rangle_{GA}}/\ |\top\rangle_{GA}|\!\}.$$

However, the formula $(\boxminus(\eta_R)_\leq)\{\!|z_{|\top\rangle_{GA}}/\ |\top\rangle_{GA}|\!\}$ is $(\boxminus\,\eta_\leq)$ and this will imply that η is inconsistent.

Therefore there are $\mathcal{K}(\delta)$ and ρ such that $\mathcal{K}(\delta)$, $\rho \Vdash_i (\eta_R)_\leq$. We fix such a $\mathcal{K}$, $\mathcal{K}(\delta)$ and ρ.

We now construct $|\psi\rangle = \{|\psi\rangle_S\}_{S\in\mathcal{S}}$ as follows:

- $|\psi\rangle_{[\emptyset]} = 1$;
- Let $\nu_{SA} = \rho(z_{|\top\rangle_{SA}})$ for every $S \in \mathcal{S}$ and $A \subseteq S$. Then,

$$|\psi\rangle_{[S]} = \sum_{A\subseteq S} \nu_{SA}|v_A^S\rangle.$$

We construct $\nu = \{\nu_{GA}\}_{G\subset \mathsf{qB}, A\subseteq G}$ as follows:

$$\nu_{GA} = \begin{cases} \rho(z_{|\top\rangle_{GA}}) & \text{if } z_{|\top\rangle_{GA}} \text{ is a variable in } \eta_R \\ 0 & \text{otherwise} \end{cases}.$$

Please note that, by construction $\nu_{GA} = \langle v_A^G|\psi\rangle_{[G]}$ if $G \in \mathrm{Alg}(\mathcal{S})$. Let $\mathbf{w}$ be $(\mathcal{K}, \delta, V, \mathcal{S}, |\psi\rangle, \nu)$. We can easily show that $\mathbf{w}$ is a quantum structure and $\mathbf{w}\rho \Vdash \eta$. ■

The decidability of consistency of molecular formulas follows as a corollary to the proof of the Model Existence Lemma.

COROLLARY 27. There is an algorithm to decide if a quantum molecule η is consistent.

Proof. As a result of Propositions 21 and 22, we can assume that η does not have any probability and alternative terms and is maximally consistent with respect to admissible valuations. Now as a result of the model existence lemma, all we need to do is to check if there is a quantum structure $\mathbf{w}$ such that $\mathbf{w} \Vdash \eta$. We refer to the proof of model existence lemma.

We first check if η is g-satisfiable and s-satisfiable which is algorithmic by Lemmas 24 and 25. If not then η is not consistent. Otherwise, let V and $\mathcal{S}$ be as in the proof of the model existence lemma.

Now, we construct η_R as in the same proof. Note that the construction is algorithmic. We check if $(\eta_R)_\leq$ is $\leq$-consistent or not. If it is not the case then η is not consistent. If $(\eta_R)_\leq$ is $\leq$-consistent then we can construct $\mathbf{w}$ as in that proof such that $\mathbf{w} \Vdash \eta$. Therefore η will be consistent if $(\eta_R)_\leq$ is $\leq$-consistent. ■

Please note any formula γ is equivalent to a disjunction of quantum molecular formulas. Furthermore, if γ is consistent, so is one of its molecules, say η. Theorem 26 gives a quantum structure $\mathbf{w}$ and an assignment ρ such that $\mathbf{w}\rho \models \eta$. As η is a quantum molecule of Γ we get easily $\mathbf{w}\rho \models \gamma$. Hence, if any quantum formula γ is consistent then γ has a model. We can now deduce the weak completeness of dEQPL in the standard way.

THEOREM 28 Completeness. The proof system of dEQPL is weakly complete, *i.e.*, $\models \gamma$ implies $\vdash \gamma$.

Proof.
We prove completeness by contradiction. Assume that $\not\vdash \gamma$. So by **Qtaut** and **QMP**, we have $\not\vdash (\boxminus(\boxminus \gamma))$. Therefore, $\boxminus \gamma$ is consistent, and hence there is a quantum structure $\mathbf{w}$ and an assignment ρ such that $\mathbf{w}\rho \models \boxminus \gamma$. Therefore, $\mathbf{w}\rho \not\models \gamma$. ■

Finally, we get the decidability of dEQPL.

THEOREM 29 Decidability. The set of theorems is decidable.

Proof. As a result of soundness and completeness we have, $\vdash \eta$ iff $\boxminus \eta$ is inconsistent. We can decide consistency of a formula by Corollary 27, Proposition 19 and Proposition 20. ■

We finish this section by observing two things. The first observation is that the proof of weak completeness can be adapted to a proof of strong completeness as follows. The key in the proof is again the Model Existence Lemma. Given a possibly infinite consistent set of quantum formulas Γ, we construct a maximally consistent set (the usual Henkin-Lindenbaum construction). Next, by looking at the classical formulas in Γ, we construct V using the strong completeness of propositional logic. The construction of the partition S is by considering the subsystem literals in Γ and is similar to the one in the above proof. Finally, just as in the proof above, we replace the amplitude terms in comparison-literals by fresh variables and "solve" the resulting equations using the strong completeness of first-order logic (note that as Γ is maximal all the maximally consistent information about logical amplitudes is already in Γ).

The second observation is that in our semantic structures, if G is the set of qubits of a sub-system then the qubits in G are necessarily not entangled with the rest. That is, the following is a theorem in dEQPL:

$$\vdash [G] \sqsupset \prod_{A_1 \subseteq G\,,\, A_2 \subseteq \mathsf{qB} \setminus G} (|\top\rangle_{\mathsf{qB}(A_1 \cup A_2)} = |\top\rangle_{GA_1} |\top\rangle_{(\mathsf{qB}\setminus G)A_2}).$$

In EQPL [Mateus and Sernadas, 2006], the reverse implication was also true. That is in [Mateus and Sernadas, 2006], it was the case that G is a sub-system if and only if the qubits in G are not entangled with the rest. We can extend our results to such semantic structures by considering the (finite) set of formulas $\Gamma = \{\gamma_G \,|\, G \subseteq \mathsf{qB}\}$ where

$$\gamma_G := ([G] \equiv (\prod_{A_1 \subseteq G\,,\, A_2 \subseteq \mathsf{qB} \setminus G} (|\top\rangle_{\mathsf{qB}(A_1 \cup A_2)} = |\top\rangle_{GA_1} |\top\rangle_{(\mathsf{qB}\setminus G)A_2}))).$$

Clearly $\Gamma \vDash \gamma$ if and only γ holds in all the semantic structures where every set of qubits not entangled with the rest forms a sub-system. If were to augment our axiom system with elements of Γ, then γ is a theorem in the augmented axiomatization if and only if $\Gamma \vdash \gamma$. The weak completeness and decidability in the augmented system then follow from the results of this section.

5 APPLICATION EXAMPLES

As it is, dEQPL is appropriate for reasoning about quantum states only. For reasoning about the evolution of quantum systems through the application of measurements and unitary transformations we will need to extend it towards a dynamic logic, as already sketched in [Mateus and Sernadas, 2004a; Mateus and Sernadas, 2004b].

Herein, we first illustrate how dEQPL can be used to reason about a Bell state. Afterwards, we turn our attention to quantum teleportation and outline there some of the relevant constructs of the envisaged dynamic logic. In the following examples, we write $|F\rangle$ as an abbreviation for the vector $(|\top\rangle_{FA})_{A \subseteq F}$ assuming the lexicographic ordering of the subsets of F. We may also abbreviate $\{\mathsf{qb}_{k_1}, \ldots, \mathsf{qb}_{k_m}\}$ by $\mathsf{qb}_{k_1,\ldots,k_m}$ in amplitude terms.

5.1 Reasoning about Bell states

Bell states were first discussed by Einstein, Podolsky and Rosen [Einstein *et al.*, 1935] and have been very useful in designing quantum protocols. An independent sub-system composed of a pair of qubits is said to be in a Bell state if they are maximally entangled. For instance,

$$|\psi\rangle = \frac{1}{\sqrt{2}}(|10\rangle - |01\rangle)$$

is a Bell state.

In order to represent this pair in our logic, we choose two qubit symbols, say qb_0 and qb_1. The fact that these qubits are independent from other qubits can be written as

$$\gamma_{\text{ind}} := [\mathsf{qb}_0, \mathsf{qb}_1].$$

We can express the state as the following formula

$$\gamma_{\text{EPR}} := (|\mathsf{qb}_{01}\rangle = \frac{1}{\sqrt{2}}(0, -1, 1, 0)).$$

We can use our logic to derive that these qubits are necessarily entangled, that is, neither qb_0 nor qb_1 form an independent sub-system. In other words we will show that

$$\gamma_{\text{ind}}, \gamma_{\text{EPR}} \vdash \boxminus[\mathsf{qb}_0] \sqcap \boxminus[\mathsf{qb}_1].$$

The proof will follow by applying the metatheorem theorem of deduction. In particular, we show

$$\gamma_{\text{ind}}, \gamma_{\text{EPR}}, [\mathsf{qb}_0] \vdash \perp\!\!\!\perp,$$

as follows:

1	$[\mathsf{qb}_0, \mathsf{qb}_1]$	Hyp
2	$[\mathsf{qb}_0]$	Hyp
3	$([\mathsf{qb}_0, \mathsf{qb}_1] \sqsupset ([\mathsf{qb}_0] \sqsupset [\mathsf{qb}_1]))$	**SubDiff**
4	$([\mathsf{qb}_0] \sqsupset [\mathsf{qb}_1]))$	**QMP**: 1,3
5	$[\mathsf{qb}_1]$	**QMP** :2,4
6	$(\vert\top\rangle_{\mathsf{qb}_{01}\emptyset} = 0) \sqcap ((\vert\top\rangle_{\mathsf{qb}_{01}\mathsf{qb}_0} = -\frac{1}{\sqrt{2}}) \sqcap (\vert\top\rangle_{\mathsf{qb}_{01}\mathsf{qb}_1} = \frac{1}{\sqrt{2}}) \sqcap (\vert\top\rangle_{\mathsf{qb}_{01}\mathsf{qb}_{01}} = 0))$	Hyp
7	$(\gamma_1 \sqcap \gamma_2) \sqsupset \gamma_2$	**QTaut**
8	$(\vert\top\rangle_{\mathsf{qb}_{01}\mathsf{qb}_0} = -\frac{1}{\sqrt{2}}) \sqcap (\vert\top\rangle_{\mathsf{qb}_{01}\mathsf{qb}_1} = \frac{1}{\sqrt{2}}) \sqcap (\vert\top\rangle_{\mathsf{qb}_{01}\mathsf{qb}_{01}} = 0)$	**QMP**: 6,7
9	$\vert\top\rangle_{\mathsf{qb}_{01}\mathsf{qb}_0} = \vert\top\rangle_{\mathsf{qb}_0\mathsf{qb}_0}\vert\top\rangle_{\mathsf{qb}_1\emptyset}$	**Mul**: 2,5
10	$\vert\top\rangle_{\mathsf{qb}_{01}\mathsf{qb}_1} = \vert\top\rangle_{\mathsf{qb}_0\emptyset}\vert\top\rangle_{\mathsf{qb}_1\mathsf{qb}_1}$	**Mul**: 2,5
11	$\vert\top\rangle_{\mathsf{qb}_{01}\mathsf{qb}_1} = \vert\top\rangle_{\mathsf{qb}_0\mathsf{qb}_0}\vert\top\rangle_{\mathsf{qb}_1\mathsf{qb}_1}$	**Mul**: 2,5
12	$\perp$	**RCF**: 8–11
13	$\perp\!\!\!\perp$	**Eqv**$\perp$: 12

Therefore, by metatheorem of deduction, we get

$$\gamma_{\rm ind}, \gamma_{\rm EPR} \vdash \boxminus[\mathsf{qb}_0].$$

In a similar way, we can derive

$$\gamma_{\rm ind}, \gamma_{\rm EPR} \vdash \boxminus[\mathsf{qb}_1],$$

and consequently, we get

$$\gamma_{\rm ind}, \gamma_{\rm EPR} \vdash \boxminus[\mathsf{qb}_0] \sqcap \boxminus[\mathsf{qb}_1].$$

In the next section, we consider a protocol which uses this Bell state to achieve teleportation.

5.2 Reasoning about quantum teleportation

A protocol for quantum teleportation was first proposed in [Bennett *et al.*, 1993]. The idea is to move a qubit from one agent to another who share an entangled pair of qubits while exchanging only classical information.

Before describing and verifying the protocol we need to extend dEQPL with some features from dynamic logic. Namely, we shall use formulas, called *Hoare triples* for historical reasons [Hoare, 1969], of the form

$$\{\gamma_1\}\, P\, \{\gamma_2\}$$

where γ_1 and γ_2 are dEQPL formulas and P is a quantum program denoting some composition of unitary transformations and measurements. It is often useful to reserve some qubits that are always in a classical state. Let us call them classical bits and use the symbols $\mathsf{cb}_1, \ldots, \mathsf{cb}_m$ to range over them. We shall avoid going into the details of the quantum program language and semantics, better left to a specific paper on a dynamic extension of dEQPL. However, we shall provide the needed intuitions. Namely, the Hoare triple above means that if the system is in a quantum state satisfying γ_1 then after running P it reaches a state satisfying γ_2.

The protocol in [Bennett *et al.*, 1993] uses three qubits, say qb_0, qb_1 and qb_2 plus two classical bits cb_0 and cb_2. The purpose is to transfer the quantum state of qb_0 to qb_1, using qb_2 and the classical bits as auxiliary variables. Initially, qb_1 and qb_2 will be prepared in a Bell state not entangled with qb_0. Afterwards, a measurement of qb_0 and qb_2 is made (by Alice). Note that this measurement will also affect qubit 1 because it is entangled with qubit 2. The classical bits are used to store the result of measuring the corresponding qubits. The classical information to be exchanged is precisely the contents of the classical bits after the measurement. Finally, this information is used (by Bob) to decide which unitary transformation to apply on qb_1 in order to achieve the required state. In short, the protocol QTP is as follows:

$$
\begin{array}{lllll}
M_{\mathsf{qb}_{02}}; & & & & \\
IF & & & & \\
 & |\mathsf{cb}_{02}\rangle = (1,0,0,0) & \rightarrow & -I_{\mathsf{qb}_1} \\
 & |\mathsf{cb}_{02}\rangle = (0,1,0,0) & \rightarrow & -Z_{\mathsf{qb}_1} \\
 & |\mathsf{cb}_{02}\rangle = (0,0,1,0) & \rightarrow & X_{\mathsf{qb}_1} \\
 & |\mathsf{cb}_{02}\rangle = (0,0,0,1) & \rightarrow & -X_{\mathsf{qb}_1} Z_{\mathsf{qb}_1} \\
FI & & & &
\end{array}
$$

where I is the identity operator and X and Z are the standard Pauli operators (not and phase flip, respectively).

The initial state of the system (after preparing the qubits 1 and 2) is assumed to comply with:

$$\gamma_{\text{init}} := [\mathsf{qb}_0] \sqcap (|\mathsf{qb}_{12}\rangle = \frac{1}{\sqrt{2}}(0,1,-1,0)) \sqcap (|\mathsf{qb}_0\rangle = (z_0, z_1)) \, .$$

Observe that we are not constraining the state of qubit 0. We just need to refer to it which we achieve by using the (rigid) variables z_0 and z_1. Note also that in such a state the qubits 1 and 2 are entangled. Actually, they are in a Bell state as discussed in the previous example.

We want the final state of the system (after running the protocol) to comply with:

$$\gamma_{\text{fin}} := [\mathsf{qb}_1] \sqcap (|\mathsf{qb}_1\rangle = (z_0, z_1)) \, .$$

In other words, we want to establish:

$$\text{Spec} := \{\gamma_{\text{init}}\} \, \text{QTP} \, \{\gamma_{\text{fin}}\} \, .$$

To this end, it is enough to assume that the measurement operator $M_{\mathsf{qb}_{02}}$ complies with the following non probabilistic specification:

$$\{\gamma_{\text{init}}\} \, M_{\mathsf{qb}_{02}} \, \{\sqcup_{k=1}^{4} \gamma_k\}$$

where

$$
\begin{aligned}
\gamma_1 &:= (|\mathsf{cb}_{02}\rangle = (1,0,0,0)) \sqcap (|\mathsf{qb}_{02}\rangle = \tfrac{1}{\sqrt{2}}(1,0,0,1)) \sqcap (|\mathsf{qb}_1\rangle = -(z_0, z_1)); \\
\gamma_2 &:= (|\mathsf{cb}_{02}\rangle = (0,1,0,0)) \sqcap (|\mathsf{qb}_{02}\rangle = \tfrac{1}{\sqrt{2}}(-1,0,0,1)) \sqcap (|\mathsf{qb}_1\rangle = (-z_0, z_1)); \\
\gamma_3 &:= (|\mathsf{cb}_{02}\rangle = (0,0,1,0)) \sqcap (|\mathsf{qb}_{02}\rangle = \tfrac{1}{\sqrt{2}}(0,1,1,0)) \sqcap (|\mathsf{qb}_1\rangle = (z_1, z_0)); \\
\gamma_4 &:= (|\mathsf{cb}_{02}\rangle = (0,0,0,1)) \sqcap (|\mathsf{qb}_{02}\rangle = \tfrac{1}{\sqrt{2}}(0,-1,1,0)) \sqcap (|\mathsf{qb}_1\rangle = (z_1, -z_0)).
\end{aligned}
$$

Observe that the IF part of the protocol QTP complies with:

$$\{\gamma_k\} \, \text{IF} \, \{\gamma_{\text{fin}}\}$$

for $k = 1, \ldots, 4$. Therefore, we can derive Spec using the traditional composition rules of dynamic logic.

6 CONCLUDING REMARKS

A decidable quantum logic allowing us to reason about amplitudes of quantum states and probabilities of classical outcomes was obtained as a fragment of EQPL. Decidability was achieved by relaxing the semantics, replacing Hilbert spaces by inner product spaces over arbitrary real closed fields and their algebraic closures. The proof of decidability was carried out hand in hand with the proof of weak completeness and follows the Fagin-Halpern-Megiddo technique (originally proposed for probabilistic logics [Fagin *et al.*, 1990; Abadi and Halpern, 1994]).

We envision to use this decidable quantum logic in the specification and verification of quantum procedures and protocols, either via model checking or theorem proving. To this end, the hardness of the proposed decision algorithm needs to be analyzed. We also intend to enrich this decidable quantum logic with Hoare triples as outlined in Section 5 and in [Mateus and Sernadas, 2004a; Mateus and Sernadas, 2004b]. Temporal extensions of dEQPL should also be explored to reason about liveness and progress properties of quantum computations. Another interesting line of research would be to develop a first-order quantum logic based on the exogenous semantics approach.

Both EQPL and dEQPL allow us to express amplitudes of pure quantum states of collections of qubits, so these logics are not insensitive to the global phase of the quantum state. One may argue that it should be insensitive since no physical measurement will ever be able to distinguish two quantum states that are equivalent up to global phase. We decided to leave dEQPL as it is (that is, sensitive to global phase) for two reasons. In practice, physicists and quantum computer scientists need to work with both levels of abstraction. Sometimes they want to work with states as unit vectors and other times they want to abstract away the global phase. So, a calculus supporting the former level of abstraction is also useful. The second reason is a consequence of the fact that forgetting global phase requires a major semantic shift. Indeed, it is better solved by identifying a quantum state with a density operator working on the underlying inner product space, that is, working with probabilistic ensembles or mixed quantum states in general.

Such a shift toward a semantics based on density operators will lead to a quite different quantum logic (but still extending classical logic by applying the exogenous approach) that will also be useful for reasoning about quantum systems evolving under partial tracing, besides unitary transformations and measurements. Clearly, this is yet another line of research that will deserve attention.

The relationship between the exogenous quantum logics and the more traditional quantum logics (based on the original Birkhoff and von Neumann proposal) should be further explored. At the preliminary stage of work in this direction, it seems that most of the qualitative assertions possible in the latter can be made in the former and that the latter can be easily extended with quantitative aspects of the former. In other words, it seems feasible to combine the two quantum logics into a single logic by using fibring techniques [Gabbay, 1996; Caleiro *et al.*, 2005].

ACKNOWLEDGMENTS

The authors wish to express their gratitude to the regular participants in the QCI Seminar at SQIG-IT (formerly at CLC), and also to Dave Marker and Anand Pillay for their help on real closed fields and their algebraic completion. This work was partially supported by FCT and FEDER through POCTI, namely via the QuantLog POCTI/MAT/55796/2004 (Quantum Logic), KLog PTDC/MAT/68723/2006 (Kleistic Logic) and QSec PTDC/EIA/67661/2006 (Quantum Security) projects.

BIBLIOGRAPHY

[Abadi and Halpern, 1994] M. Abadi and J. Y. Halpern. Decidability and expressiveness for first-order logics of probability. *Information and Computation*, 112(1):1–36, 1994.

[Abramsky and Coecke, 2004] S. Abramsky and B. Coecke. A categorical semantics of quantum protocols. In *Proceedings of the 19th Annual IEEE Symposium on Logic in Computer Science (LICS 2004)*, pages 415–425. IEEE Computer Science Press, 2004. Extended version at arXiv:quant-ph/0402130).

[Abramsky and Duncan, 2006] S. Abramsky and R. Duncan. A categorical quantum logic. *Mathematical Structures in Computer Science*, 16(3):469–489, 2006.

[Bacchus, 1990a] F. Bacchus. On probability distributions over possible worlds. In *Uncertainty in Artificial Intelligence, 4*, volume 9 of *Machine Intelligence and Pattern Recognition*, pages 217–226. North-Holland, 1990.

[Bacchus, 1990b] F. Bacchus. *Representing and Reasoning with Probabilistic Knowledge.* MIT Press Series in Artificial Intelligence. MIT Press, 1990.

[Basu *et al.*, 2003] S. Basu, R. Pollack, and R. Marie-Françoise. *Algorithms in Real Algebraic Geometry.* Springer, 2003.

[Bennett *et al.*, 1993] C. H. Bennett, G. Brassard, C. Crépeau, R. Jozsa, A. Peres, and W. K. Wootters. Teleporting an unkown quantum state via dual classical and einstein-podolsky-rosen channels. *Physical Review Letters*, 70(13):1895–1899, 1993.

[Birkhoff and von Neumann, 1936] G. Birkhoff and J. von Neumann. The logic of quantum mechanics. *Annals of Mathematics*, 37(4):823–843, 1936.

[Caleiro *et al.*, 2005] C. Caleiro, A. Sernadas, and C. Sernadas. Fibring logics: Past, present and future. In S. Artemov, H. Barringer, A. S. d'Avila Garcez, L. C. Lamb, and J. Woods, editors, *We Will Show Them: Essays in Honour of Dov Gabbay, Volume One*, pages 363–388. College Publications, 2005.

[Caleiro *et al.*, 2006] C. Caleiro, P. Mateus, A. Sernadas, and C. Sernadas. Quantum institutions. In K. Futatsugi, J.-P. Jouannaud, and J. Meseguer, editors, *Algebra, Meaning, and Computation — Essays Dedicated to Joseph A. Goguen on the Occasion of His 65th Birthday*, volume 4060 of *Lecture Notes in Computer Science*, pages 50–64. Springer-Verlag, 2006.

[Carnielli and Lima-Marques, 1999] W. A. Carnielli and M. Lima-Marques. Society semantics and multiple-valued logics. In *Advances in Contemporary Logic and Computer Science (Salvador, 1996)*, volume 235 of *Contemporary Mathematics*, pages 33–52. AMS, 1999.

[Carnielli, 2000] W. A. Carnielli. Possible-translations semantics for paraconsistent logics. In *Frontiers of Paraconsistent Logic (Ghent, 1997)*, volume 8 of *Studies in Logic and Computation*, pages 149–163. Research Studies Press, 2000.

[Chadha *et al.*, 2007] R. Chadha, L. Cruz-Filipe, P. Mateus, and A. Sernadas. Reasoning about probabilistic sequential programs. *Theoretical Computer Science*, 379(1-2):142–165, 2007.

[Chiara *et al.*, 2004] M. L. D. Chiara, R. Giuntini, and R. Greechie. *Reasoning in Quantum Theory.* Kluwer Academic Publishers, 2004.

[Cohen-Tannoudji *et al.*, 1977] C. Cohen-Tannoudji, B. Diu, and F. Laloë. *Quantum Mechanics.* John Wiley, 1977.

[Dishkant, 1972] H. Dishkant. Semantics of the minimal logic of quantum mechanics. *Studia Logica*, 30:23–32, 1972.

[Einstein *et al.*, 1935] A. Einstein, B. Podolsky, and N. Rosen. Can quantum-mechanical description of physical reality be considered complete? *Physical Review*, 47:777–780, 1935.

[Fagin *et al.*, 1990] R. Fagin, J. Y. Halpern, and N. Megiddo. A logic for reasoning about probabilities. *Information and Computation*, 87(1-2):78–128, 1990.

[Foulis, 1999] D. J. Foulis. A half-century of quantum logic. What have we learned? In *Quantum Structures and the Nature of Reality*, volume 7 of *Einstein Meets Magritte*, pages 1–36. Kluwer Acad. Publ., 1999.

[Gabbay, 1996] D. M. Gabbay. Fibred semantics and the weaving of logics: Part 1. *Journal of Symbolic Logic*, 61(4):1057–1120, 1996.

[Halpern, 1990] J. Y. Halpern. An analysis of first-order logics of probability. *Artificial Intelligence*, 46:311–350, 1990.

[Hansson and Jonsson, 1995] H. Hansson and B. Jonsson. A logic for reasoning about time and reliability. *Formal Aspects of Computing*, 6:512–535, 1995.

[Hoare, 1969] C. Hoare. An axiomatic basis for computer programming. *Communications of the ACM*, 12:576–583, 1969.

[Hodges, 1993] W. Hodges. *Model Theory*. Cambridge University Press, 1993.

[Kripke, 1963] S. A. Kripke. Semantical analysis of modal logic. I. Normal modal propositional calculi. *Zeitschrift für Mathematische Logik und Grundlagen der Mathematik*, 9:67–96, 1963.

[Mateus and Sernadas, 2004a] P. Mateus and A. Sernadas. Exogenous quantum logic. In W. A. Carnielli, F. M. Dionísio, and P. Mateus, editors, *Proceedings of CombLog'04, Workshop on Combination of Logics: Theory and Applications*, pages 141–149, 1049-001 Lisboa, Portugal, 2004. Departamento de Matemática, Instituto Superior Técnico. Extended abstract.

[Mateus and Sernadas, 2004b] P. Mateus and A. Sernadas. Reasoning about quantum systems. In J. Alferes and J. Leite, editors, *Logics in Artificial Intelligence, Ninth European Conference, JELIA'04*, volume 3229 of *Lecture Notes in Artificial Intelligence*, pages 239–251. Springer-Verlag, 2004.

[Mateus and Sernadas, 2006] P. Mateus and A. Sernadas. Weakly complete axiomatization of exogenous quantum propositional logic. *Information and Computation*, 204(5):771–794, 2006.

[Mateus *et al.*, 2005] P. Mateus, A. Sernadas, and C. Sernadas. Exogenous semantics approach to enriching logics. In G. Sica, editor, *Essays on the Foundations of Mathematics and Logic*, volume 1 of *Advanced Studies in Mathematics and Logic*, pages 165–194. Polimetrica, 2005.

[Naur, 1963] P. Naur. Revised report on the algorithmic language Algol 60. *The Computer Journal*, 5:349–367, 1963.

[Nielsen and Chuang, 2000] M. A. Nielsen and I. L. Chuang. *Quantum Computation and Quantum Information*. Cambridge University Press, 2000.

[Nilsson, 1986] N. J. Nilsson. Probabilistic logic. *Artificial Intelligence*, 28(1):71–87, 1986.

[Nilsson, 1993] N. J. Nilsson. Probabilistic logic revisited. *Artificial Intelligence*, 59(1-2):39–42, 1993.

[Patra, 2008] M. Patra. A logic for quantum computation and classical simulation of quantum algorithms. *International Journal of Quantum Information*, 6(2), 2008. In print.

[Tarski, 1948] A. Tarski. A decision method for elementary algebra and geometry. Manuscript. Santa Monica, CA: RAND Corp., 1948. Republished as *A Decision Method for Elementary Algebra and Geometry*, 2nd ed. Berkeley, CA: University of California Press, 1951.

[van der Meyden and Patra, 2003a] R. van der Meyden and M. Patra. Knowledge in quantum systems. In M. Tennenholtz, editor, *Theoretical Aspects of Rationality and Knowledge*, pages 104–117. ACM, 2003.

[van der Meyden and Patra, 2003b] R. van der Meyden and M. Patra. A logic for probability in quantum systems. In M. Baaz and J. A. Makowsky, editors, *Computer Science Logic*, volume 2803 of *Lecture Notes in Computer Science*, pages 427–440. Springer-Verlag, 2003.

Edited by K. Engesser, D. M. Gabbay and D. Lehmann

SOLÈR'S THEOREM

Alexander Prestel

Solèr's Theorem gives an axiomatic characterization (in algebraic terms) of infinite dimensional Hilbert spaces over the reals, the complex, and the quaternions. At the same time it gives a characterization of the lattices that are isomorphic to the lattice of closed subspaces of the just mentioned Hilbert spaces. These lattices play an important role in quantum logic (see [Holland, 1995]) and, more generally, in Hilbert space logic (see [Engesser and Gabbay, 2002]). More about the history of Solèr's Theorem and its consequences in areas like Baer *-rings, infinite dimensional projective geometry, orthomodular lattices, and the logic of quantum mechanics can be found in S. Holland's exposition [Holland, 1995].

The aim of this paper is to provide a complete proof of Solèr's Theorem for the reader of the 'Handbook of Quantum Logic and Quantum Structures'.[1]

Solèr's Theorem deals with infinite dimensional hermitian spaces $(E, <>)$ over an arbitrary skew field K which are *orthomodular*, i.e., every subspace X of E satisfies

(1) $$X = (X^\perp)^\perp \Rightarrow E = X \oplus X^\perp.$$

The theorem can be stated as follows (for definitions see below):

Solèr's Theorem *Let $(K,^*)$ be a skew field together with an (anti-)involution *, and let $(E, <>)$ be an infinite dimensional hermitian vector space over $(K,^*)$. If $(E, <>)$ is orthomodular and contains an infinite orthonormal sequence $(e_n)_{n\in\mathbb{N}}$, then*

(i) *$K = \mathbb{R}, \mathbb{C}$, or $\mathbb{H}$ where in the first case * is the identity and in the case of $\mathbb{C}$ and the quaternions $\mathbb{H}$, * is the canonical conjugation,*

(ii) *$(E, <>)$ is a Hilbert space over $\mathbb{R}, \mathbb{C}$ or $\mathbb{H}$, resp.*

The main part of this theorem actually is (i). Once we know that K is the field of real or complex numbers, or the skew field $\mathbb{H}$ of the quaternions, it is not difficult to prove (ii) (cf. the end of Section 2). While the (skew) fields $\mathbb{R}, \mathbb{C}$ and $\mathbb{H}$ carry a canonical metric that makes them complete, there is no mention of any metric on

[1]The proof presented here is identical with that of [Prestel, 1995].

K in the assumption of the theorem. Not even any topology is mentioned. Nevertheless, the seemingly 'algebraic' conditions of orthomodularity and the existence of an infinite orthonormal sequence will lead to the surprising fact that K is $\mathbb{R}, \mathbb{C}$ or $\mathbb{H}$.

It has been a long standing open problem whether orthomodularity of $(E, <>)$ would already force K to be one of $\mathbb{R}, \mathbb{C}$ or $\mathbb{H}$. In 1980, finally, H. Keller (see [Keller, 1980]) constructed "non-classical Hilbert spaces", i.e., orthomodular hermitian spaces, not isomorphic to one of the classical Hilbert spaces. Only in 1995, M.P. Solèr showed in her Ph.D. thesis (cf [Solèr, 1995]) that adding the existence of an infinite orthonormal sequences expelles all the non-classical Hilbert spaces.

In sections 2 to 4 below we shall first treat the commutative case, i.e., we let K be a commutative field. Under this additional assumption we give a complete proof of Solèr's Theorem. In Section 4 we deal with the general case, i.e., we let K be an arbitrary skew field. The proof in this case needs a few refinements of the earlier ones which we shall explain. The structure of the proof in the non-commutative case, however, is essentially the same as that in the commutative case.

1 PRELIMINARIES AND STRUCTURE OF THE PROOF

Let K be a (commutative) field and $^* : K \to K$ an *involution* on K, i.e. * satisfies

$$(\alpha+\beta)^* = \alpha^* + \beta^*, \quad (\alpha\beta)^* = \alpha^*\beta^*, \quad \alpha^{**} = \alpha$$

for all $\alpha, \beta \in K$. The identity on $\mathbb{R}$ and the complex conjugation $^-$ on $\mathbb{C}$ are our standard examples for such an involution.

Furthermore, let E be an infinite dimensional K-vector space and $<>: E \times E \to K$ a *hermitian form* on E, i.e. $<>$ satisfies

$$\begin{aligned} < \alpha x + \beta y, z > &= \alpha < x, z > + \beta < y, z > \\ < z, \alpha x + \beta y > &= \alpha^* < z, x > + \beta^* < z, y > \\ < x, z > &= < z, x >^* \end{aligned}$$

for all $\alpha, \beta \in K$ and $x, y, z \in E$, The pair $(E, <>)$ is called a *hermitian space* if $<>$ is a hermitian form on E. The hermitian form $<>$ is called *anisotropic* if for $x \in E$,

$$< x, x > = 0 \text{ implies } x = 0.$$

As usual, we define *orthogonality* $x \perp y$ by $< x, y >= 0$ for vectors $x, y \in E$. The *orthogonal space* $U^\perp$ to a subset U of E is defined by

$$U^\perp = \{x \in E | \; x \perp u \text{ for all } u \in U\}.$$

We simply write $U^{\perp\perp}$ for $(U^\perp)^\perp$. The space $U^{\perp\perp}$ is called the *closure* of U, and U is called *closed* if $U^{\perp\perp} = U$. In case $(E, <>)$ is anisotropic, every finite dimensional subspace of E is closed. Since $U^{\perp\perp\perp} = U^{\perp}$, $U^{\perp}$ is also closed. Thus, in particular, orthomodularity of $(E, <>)$ may be equivalently expressed by

(1) $$E = U^{\perp} \oplus U^{\perp\perp} \text{ for all subspaces } U \text{ of } E.$$

Two important consequences of orthomodularity of a hermitian space are

(2) if U and V are orthogonal subspaces of E then $(U+V)^{\perp\perp} = U^{\perp\perp} + V^{\perp\perp}$ and $U^{\perp\perp} \perp V^{\perp\perp}$,

(3) if U is a closed subspace of E, then U together with the restriction of $<>$ to $U \times U$ is an orthomodular space, too.

Applying (1) to $U = \{x\}$ we see that orthomodular spaces are anisotropic. A finite dimensional hermitian space is orthomodular if and only if it is anisotropic. It is also well-known that any $\mathbb{R}$-Hilbert space and any $\mathbb{C}$-Hilbert space is orthomodular (w.r.t. its defining inner product). The content of the above theorem is that there are no other examples of hermitian spaces $(E, <>)$ which are orthomodular and contain an infinite sequence $(e_n)_{n \in \mathbb{N}}$ which is *orthonormal*, i.e. for all $n, m \in \mathbb{N}$ we have:

(4) $$< e_n, e_n >= 1 \text{ and } e_n \perp e_m \text{ for } n \neq m.$$

As H. Keller has shown in [Keller, 1980], infinite dimensional orthomodular spaces exist which do not contain any orthonormal sequence, hence cannot be real or complex Hilbert spaces.

Now let us explain the three steps into which we will divide the proof of the theorem.

In *Step 1* we will show that $<>$ is 'positive definite', i.e. we will show that the *fixed field*

$$F = \{\alpha \in K | \ \alpha^* = \alpha\}$$

of 'symmetric' elements admits an ordering $\leq$ such that $< x, x > \geq 0$ for all $x \in E$. (Note that $< x, x >$ is symmetric.) In fact, the ordering on F will be given by

(5) $$\alpha \leq \beta \text{ iff } \beta - \alpha \in P,$$

where $P = \{< x, x > | \ x \in E\}$ is the set of 'lengths' of vectors $x \in E$.

In *Step 2* we will show first that $\leq$ is *archimedean* on F, i.e.

(6) $$\text{to every } \alpha \in P \text{ there exists } n \in \mathbb{N} \text{ such that } \alpha \leq n.$$

At this point we make use of the fact that, in order to prove that (5) defines an ordering, it suffices to know here that $\leq$ is a *semi-ordering* (as introduced in

[Prestel, 1984] by the author), i.e. $\leq$ linearly orders F such that in addition we have for $\alpha, \beta \in F$:

$$\begin{array}{rcl} 0 \leq \alpha,\ 0 \leq \beta & \Longrightarrow & 0 \leq \alpha + \beta \\ 0 \leq \alpha & \Longrightarrow & 0 \leq \alpha\beta^2 \\ 0 \leq 1 & & \end{array} \tag{7}$$

In fact, an archimedean semi-ordering is already an *ordering*, i.e., it satisfies in addition

$$0 \leq \alpha,\ 0 \leq \beta \Longrightarrow 0 \leq \alpha \cdot \beta$$

for all $\alpha, \beta \in F$ ([Prestel, 1984], Theorem 1.20). Thus, in Step 1 it therefore suffices to show that (5) defines a semi-ordering on F. The linearity of $\leq$ is obtained from Solèr's main Lemma 5, which also gives the archimedeanity (Lemma 6). Actually, the linearity of $\leq$, i.e. $P \cup -P = F$, is not important in the commutative case, since it may be simply obtained by maximalizing a subset P of F satisfying $P + P \subset P$, $PF^2 \subset P$, $P \cap -P = \{0\}$ and $1 \in P$ by Zorn's Lemma to some P_0, w.r.t. to these properties. The maximal object P_0 then satisfies $P_0 \cup -P_0 = F$ (see [Prestel, 1984], Lemma 1.13). In the non-commutative case, however, this method may not work (see the remarks at the end of Section 6).

As it is well-known, an archimedean ordered field $(F, \leq)$ contains (an isomorphic) copy of the rational number field $\mathbb{Q}$ as a dense subfield. Thus, it suffices to prove that every Dedekind cut is realized in $(F, \leq)$ in order to find that F is isomorphic to $\mathbb{R}$. This is done in Lemma 7. Since F is the fixed field of the involution *, we have $[K : F] \leq 2$. Hence $K = \mathbb{R}$ and *=id or $K = \mathbb{C}$ and * is the complex conjugation.

Now we know that K is $\mathbb{R}$ or $\mathbb{C}$ and $<>$ is positive definite, i.e., $(E, <>)$ is a pre-Hilbert space. Thus, in *Step 3* it remains to prove that every orthomodular pre-Hilbert space is complete. This, however, can already be found in the literature (see [Maeda and Maeda, 1970], Theorem 34.9). The argument runs as follows.

Let $\hat{E}$ be the completion of E w.r.t. the metric induced by $<>$. Given $a \in \hat{E}$ we have to show that a already belongs to E. Choosing $c \in E$ suitably such that $< a, c > \neq 0$, we find $b \in \hat{E}$ such that $a \perp b$ and $c = a + b$. By standard arguments we then find sequences $(a_n)_{n\in\mathbb{N}}$ and $(b_n)_{n\in\mathbb{N}}$ in E converging to a and b resp. such that $a_n \perp b_m$ for all $n, m \in \mathbb{N}$. If we then take

$$A = \{b_n |\ n \in \mathbb{N}\}^{\perp}$$

we see that $A^{\perp\perp} = A$ in E and thus the modularity (1) of E gives us a decomposition $c = c_1 + c_2$ with $c_1 \in A$ and $c_2 \in A^{\perp}$.

Denoting by $\overline{A}$ the topological closure of A in $\hat{E}$ we trivially have $c_1 \in \overline{A}$ and by continuity of $<>$ also $a \in \overline{A}$. Again by continuity arguments we see that b and c_2 belong to $(\overline{A})^{\perp}$ in $\hat{E}$. Since $\hat{E}$ is a Hilbert space, the closed subspace $\overline{A}$ yields a decomposition

$$\hat{E} = \overline{A} \oplus (\overline{A})^{\perp}.$$

Thus, it follows from $a + b = c = c_1 + c_2$ that $a = c_1 \in E$.

2 CONSTRUCTION OF THE SEMI-ORDERING ON F

Let $(E, <>)$ be an orthomodular hermitian space and $(f_i)_{i\in I}$ be some *orthogonal* sequence in E, i.e., $f_i \perp f_j$ for $i \neq j$. Then the map

$$x \longmapsto (< x, f_i >)_{i\in I}$$

is K-linear and injective on the subspace $U = {(f_i)_{i\in I}}^{\perp\perp}$ of E. In fact, if $< x, f_i >= 0$ for all $i \in I$, then $x \in {(f_i)_{i\in I}}^{\perp}$. Thus, for $x \in U$ we get $x \perp x$ which implies $x = 0$, since $<>$ is anisotropic. Hence, every vector x from U is uniquely determined by the sequence $(< x, f_i >)_{i\in I}$. We therefore write

$$(1) \qquad x = \sum_{i\in I} \alpha_i f_i \text{ with } \alpha_i := \frac{< x, f_i >}{< f_i, f_i >}$$

and call α_i the 'Fourier-coefficient' of x w.r.t. $(f_i)_{i\in I}$. It should be clear, however, that $\sum$ in (1) is not an infinite sum converging w.r.t. some metric. It is only a *formal notation*, expressing nothing else than: x is the unique vector in U satisfying $< x, f_i >= \alpha_i < f_i, f_i >$ for all $i \in I$.

In case I is a finite set, $\sum$ in (1) is actually a finite sum and the identity in (1) is familiar.

As an ordered field, K clearly will have characteristic zero. This, however, is already clear from the existence of an infinite orthonormal sequence $(e_n)_{n\in\mathbb{N}}$ and the anisotropy of $<>$. In fact, for any prime p

$$0 \neq < e_1 + ... + e_p, e_1 + ... + e_p >= p.$$

The next proposition contains one main tool for constructing the desired ordering on F.

PROPOSITION 1. *Let $(f_n)_{n\in\mathbb{N}}$ and $(g_n)_{n\in\mathbb{N}}$ be two orthonormal sequences such that $f_n \perp g_m$ for all $n, m \in \mathbb{N}$. Then to every vector $x = \sum_n \alpha_n f_n \in {(f_n)_{n\in\mathbb{N}}}^{\perp\perp}$ a vector $y = \sum_n \alpha_n g_n \in {(g_n)_{n\in\mathbb{N}}}^{\perp\perp}$ exists such that $< x, x >=< y, y >$.*

Proof. Since char $K \neq 2$, the systems $(f_n)_{n\in\mathbb{N}} \cup (g_n)_{n\in\mathbb{N}}$ and $(f_n + g_n)_{n\in\mathbb{N}} \cup (f_n - g_n)_{n\in\mathbb{N}}$ generate the same linear subspace H of E. From ${(f_n)_{n\in\mathbb{N}}}^{\perp\perp} \subset H^{\perp\perp}$ and (2) we get $x = x_1 + x_2$ with

$$\begin{aligned} x_1 &= \sum_n \alpha_n'(f_n + g_n) \in {(f_n + g_n)_{n\in\mathbb{N}}}^{\perp\perp} \\ x_2 &= \sum_n \alpha_n''(f_n - g_n) \in {(f_n - g_n)_{n\in\mathbb{N}}}^{\perp\perp} \end{aligned}$$

(Note that $(f_n + g_n) \perp (f_m - g_m)$ for all $n, m \in \mathbb{N}$.) For α'_n we find

$$\alpha'_n = \frac{< x_1, f_n + g_n >}{< f_n + g_n, f_n + g_n >} = \frac{< x, f_n + g_n >}{2} - \frac{< x_2, f_n + g_n >}{2} = \frac{1}{2}\alpha_n$$

Similarly, we get $\alpha''_n = \frac{1}{2}\alpha_n$.

From $(f_n + g_n)_{n\in\mathbb{N}}{}^{\perp\perp} \subset H^{\perp\perp}$ and (2) we next get $2x_1 = \sum_n \alpha_n(f_n+g_n) = y_1+y_2$ with

$$\begin{aligned} y_1 &= \sum_n \beta'_n f_n \in (f_n)_{n\in\mathbb{N}}{}^{\perp\perp} \\ y_2 &= \sum_n \beta''_n g_n \in (g_n)_{n\in\mathbb{N}}{}^{\perp\perp}. \end{aligned}$$

For β'_n we find

$$\beta'_n =< y_1, f_n >=< 2x_1, f_n >=< x_1, (f_n+g_n)+(f_n-g_n) >=< x_1, f_n+g_n >= \alpha_n$$

and similarly

$$\beta''_n =< y_2, g_n >=< 2x_1, g_n >=< x_1, (f_n + g_n) - (f_n - g_n) >= \alpha_n.$$

Thus, we first see that $y_1 = x$, and, taking

$$y := y_2 = \sum_n \alpha_n g_n \in (g_n)_{n\in\mathbb{N}}{}^{\perp\perp},$$

we then find $< y, y >=< x, x >$ from $y = 2x_1 - x$ and $< x, x_1 >=< x_1, x_1 >=$ $< x_1, x >$. ∎

In order to apply this proposition we will make the

ASSUMPTION 2. $E = (e_n)_{n\in\mathbb{N}}{}^{\perp\perp}$ *is orthomodular.*

By (3) this is without restriction since by Step 3 of Section 2 it suffices to prove that $F \cong \mathbb{R}$.

LEMMA 3. *Let* $(E, <>)$ *satisfy Assumption 2 and let* $(f_n)_{n\in\mathbb{N}}, (g_n)_{n\in\mathbb{N}}$ *be two othonormal sequences in* E. *Then the map assigning to every* $x = \sum_n \alpha_n f_n \in (f_n)_{n\in\mathbb{N}}{}^{\perp\perp}$ *the vector* $y = \sum_n \alpha_n g_n \in (g_n)_{n\in\mathbb{N}}{}^{\perp\perp}$ *defines an isometry of the subspaces* $(f_n)_{n\in\mathbb{N}}{}^{\perp\perp}$ *and* $(g_n)_{n\in\mathbb{N}}{}^{\perp\perp}$ *of* E.

Proof. The only difference of this lemma to Proposition 1 is that the sequences $(f_n)_{n\in\mathbb{N}}$ and $(g_n)_{n\in\mathbb{N}}$ may not be orthogonal to each other.

Let $E_i = (e_{4n+i})_{n\in\mathbb{N}}{}^{\perp\perp}$ with $i \in \{0, 1, 2, 3\}$. Then

(2) $$E = E' \oplus E'' \text{ with } E' = E_1 \oplus E_2 \text{ and } E'' = E_3 \oplus E_4.$$

From Proposition 1 we immediately get the isometries

$$E' \cong E'',\ E'' \cong E_1,\ E'' \cong E_2.$$

Together we therefore also obtain $E = E' \oplus E'' \cong E_1 \oplus E_2 = E'$ and $E \cong E''$.

Now we are able to deduce the lemma from Proposition 1. In fact, we first move the sequence $(f_n)_{n\in\mathbb{N}}$ from E to E' and the sequence $(g_n)_{n\in\mathbb{N}}$ from E to E''. Then we apply Proposition 1. ■

From this lemma we get two easy consequences which will be used in the next sections.

COROLLARY 4. *Let $(E, <>)$ satisfy Assumption 2. Then:*

(a) E contains a subspace U isomorphic to the infinite orthogonal sum $E \oplus E \oplus E \oplus ...$

(b) There is no vector $x \in E$ having Fourier-coefficient $< x, e_n >= 1$ for all $n \in \mathbb{N}$.

Proof. (a) Let $\mathbb{N} = \bigcup_{m\in\mathbb{N}} I_m$ be a disjoint partition of $\mathbb{N}$ into infinite sets I_n. Then by Lemma 3, every closure $(I_m)^{\perp\perp}$ is isometric to E. Now let

$$U = I_1^{\perp\perp} + I_2^{\perp\perp} + I_3^{\perp\perp} + ...$$

(b) Assume $x \in E$ has $< x, e_n >= 1$ for all $n \in \mathbb{N}$. Observing that $E = (e_{2n})_{n\in\mathbb{N}}{}^{\perp\perp} + (e_{2n+1})_{n\in\mathbb{N}}{}^{\perp\perp}$ we get $x = x_1 + x_2$ with $x_1 \in (e_{2n})_{n\in\mathbb{N}}{}^{\perp\perp}$ and $x_2 \in (e_{2n+1})_{n\in\mathbb{N}}{}^{\perp\perp}$. By Lemma 3 we get $< x, x >=< x_1, x_1 >=< x_2, x_2 >$. Now $x_1 \perp x_2$ yields the contradiction

$$< x, x >=< x_1, x_1 > + < x_2, x_2 >= 2 < x, x > .$$

■

We are now able to prove that the set of lengths

$$P = \{< x, x > \mid x \in E\}$$

satisfies the following properties:

(3)
$$\begin{array}{ll} (i) & P + P \subset P \\ (ii) & PF^2 \subset P \\ (iii) & P \cap -P = \{0\} \\ (iv) & 1 \in P \end{array}$$

Obviously, (ii) and (iv) are trivial. To see (i), let $\alpha =< x,x >$ and $\beta = < y,y >$. By (2) choose $x' \in E'$ and $y' \in E''$ such that $< x',x' >=< x,x >$ and $< y',y' >=< y,y >$. Since E' and E'' are orthogonal, we get $< x'+y',x'+y' >= \alpha + \beta$. For proving (iii), let $\alpha =< x,x >$ and $-\alpha =< y,y >$. Now we find $< x'+y',x'+y' >= \alpha - \alpha = 0$. Since $<>$ is anisotropic, we have $x'+y' = 0$. Thus, $x' = y' = 0$ and $\alpha = 0$.

From (3) we see that

$$\alpha \leq \beta \Longleftrightarrow \beta - \alpha \in P$$

defines a partial order on F which satisfies (7). Thus, $\leq$ is be a semi-ordering on F, if we can prove its linearity, i.e. $\alpha \leq \beta$ or $\beta \leq \alpha$ for all $\alpha, \beta \in F$. This clearly is equivalent to $P \cup -P = F$.

The next lemma will easily imply linearity of $\leq$. This lemma actually is the heart of Solèr's Thesis.

LEMMA 5. *Let $(E, <>)$ satisfy Assumption 2 and let $\alpha \in F\backslash\{0, \pm 1\}$. Then either the vector $a = \sum_n \alpha^n e_n$ exists in E with length $< a,a >= (1-\alpha^2)^{-1}$ or the $a' = \sum_n \alpha^{-n} e_n$ exists in E with length $< a',a' >= (1-\alpha^{-2})^{-1}$.*

Proof. We define for all $n \in \mathbb{N}$:

$$x_n = \sum_{i=0}^{n} \alpha^i e_i \quad , \quad y_n = x_n - \frac{1}{1-\alpha^2} e_0$$

$$a_n = x_{2n+1} - \alpha^2 x_{2n} \ , \quad b_n = y_{2n+2} - \alpha^2 y_{2n+1}$$

From these definitions we find

$$\begin{aligned} \frac{< x_n, y_n >}{< x_{n-1}, y_{n-1} >} &= 1 + \alpha^{2n} < x_{n-1}, y_{n-1} >^{-1} \\ &= 1 + \alpha^{2n} (1 - \frac{1}{1-\alpha^2} + \alpha^2 + \alpha^4 + ... + \alpha^{2(n-1)})^{-1} = \alpha^2 \end{aligned}$$

An easy computation shows

$$a_n \perp b_m \text{ for all } n, m \in \mathbb{N}.$$

The subspace $A = (b_n)_{n \in \mathbb{N}}{}^{\perp}$ is closed. Hence orthomodularity of E yields

$$E = A \oplus A^{\perp}.$$

Thus, we get $a \in A$ and $b \in A^{\perp}$ such that

$$\frac{1}{1-\alpha^2} e_0 = a - b.$$

Clearly, $a \perp a_n$ and $b \perp b_n$ for all $n \in \mathbb{N}$.

We now set

$$a = \sum_n (\alpha^n + \varepsilon_n) e_n$$

where ε_n is suitably chosen from K. The following computations then yield recursive conditions on the ε_n:

$$\begin{aligned}
0 &= < b, a_n >=< b, x_{2n+1} - \alpha^2 x_{2n} > \\
&= < a, x_{2n+1} - \alpha^2 x_{2n} > -(1-\alpha^2)^{-1} < e_0, x_{2n+1} - \alpha^2 x_{2n} > \\
&= < a, y_{2n+2} - \alpha^2 y_{2n+1} - \alpha^{2n+2} e_{2n+2} + \alpha^2 \alpha^{2n+1} e_{2n+1} + e_0) - (1-\alpha^2)^{-1}(1-\alpha) \\
&= -(\alpha^{2n+2} + \varepsilon_{2n+2})\alpha^{2n+2} + (\alpha^{2n+1} + \varepsilon_{2n+1})\alpha^{2n+3} + (1+\varepsilon_0) - 1 \\
&= -\varepsilon_{2n+2}\alpha^{2n+2} + \varepsilon_{2n+1}\alpha^{2n+3} + \varepsilon_0
\end{aligned}$$

Hence

(5) $$\varepsilon_{2n+2} = \varepsilon_{2n+1}\alpha + \varepsilon_0 \alpha^{-(2n+2)}$$

Similarly we get

$$\begin{aligned}
0 &= < a, y_{2n+2} - \alpha^2 y_{2n+1} > \\
&= < b, x_{2n+3} - \alpha^2 x_{2n+2} - \alpha^{2n+3} e_{2n+3} + \alpha^{2n+4} e_{2n+2} - e_0 > +(1 - (1-\alpha^2)^{-1}) \\
&= -\varepsilon_{2n+3}\alpha^{2n+3} + \varepsilon_{2n+2}\alpha^{2n+4} - \varepsilon_0
\end{aligned}$$

and thus

(6) $$\varepsilon_{2n+3} = \varepsilon_{2n+2}\alpha - \varepsilon_0 \alpha^{-(2n+3)}$$

Finally

$$0 =< b, a_0 >=< a - \frac{1}{1-\alpha^2} e_0 \, , \, (1-\alpha^2)e_0 + \alpha e_1 >= \varepsilon_0(1-\alpha^2) + \varepsilon_1 \alpha$$

gives

(7) $$\varepsilon_1 = -\varepsilon_0 \alpha^{-1}(1-\alpha^2)$$

If now $\varepsilon_0 = 0$, the recursive conditions (5), (6) and (7) imply $\varepsilon_n = 0$ for all $n \in \mathbb{N}$. Hence, in this case we get

$$a = \sum_n \alpha^n e_n.$$

In case $\varepsilon_0 \neq 0$, by Lemma 3 and the recursion formulas (5) and (6) the following vector exists in E:

$$\begin{aligned}
a'' &= \sum_{n \geq 2} (\alpha^n + \varepsilon_n) e_n - \alpha \sum_{n \geq 2} (\alpha^{n-1} + \varepsilon_{n-1}) e_n \\
&= \sum_{n \geq 2} (\alpha^n + \alpha\varepsilon_{n-1} + (-1)^n \alpha^{-n} \varepsilon_0) e_n - \alpha \sum_{n \geq 2} (\alpha^{n-1} + \varepsilon_{n-1}) e_n \\
&= \sum_{n \geq 2} ((-1)^n \alpha^{-n} \varepsilon_0) e_n
\end{aligned}$$

Thus, again by Lemma 3 the vector

$$a' = \sum_n \alpha^{-n} e_n$$

exists in E.

This proves the difficult part of the Lemma. The computation of $< a, a >$ or $< a', a' >$ resp. is easy (note that we shall use Lemma 3 in the last step):

$$\begin{aligned} < a, a > &= 1 + \langle \sum_{n \geq 1} \alpha^n e_n, \sum_{n \geq 1} \alpha^n e_n \rangle = 1 + \alpha^2 \langle \sum_{n \geq 1} \alpha^{n-1} e_n, \sum_{n \geq 1} \alpha^{n-1} e_n \rangle \\ &= 1 + \alpha^2 \langle \sum_n \alpha^n e_{n+1}, \sum_n \alpha^n e_{n+1} \rangle = 1 + \alpha^2 \langle \sum_n \alpha^n e_n, \sum_n \alpha^n e_n \rangle \end{aligned}$$

This clearly gives

$$< a, a > = \frac{1}{1 - \alpha^2}$$

and similarly

$$< a', a' > = \frac{1}{1 - \alpha^{-2}}$$

■

From this lemma we get the linearity of $\leq$ as follows. For every $\alpha \in F \backslash \{0, \pm 1\}$ we have

$$\frac{1}{1 - \alpha^2} \in P \quad \text{or} \quad \frac{1}{1 - \alpha^{-2}} \in P.$$

Using (3)(ii) twice we find for every $\alpha \in F$:

$$1 - \alpha^2 \in P \quad \text{or} \quad \alpha^2 - 1 \in P.$$

Now let $\beta \in F \backslash \{-1\}$ be given and take $\alpha = \frac{2}{\beta + 1} - 1$. This clearly yields

$$\frac{4\beta}{(\beta + 1)^2} \in P \quad \text{or} \quad -\frac{4\beta}{(\beta + 1)^2} \in P$$

Again by (3)(ii) we finally find

$$\beta \in P \quad \text{or} \quad -\beta \in P.$$

Thus linearity of $\leq$ is proved.

3 ARCHIMEDEANITY AND COMPLETENESS OF $(F, \leq)$

The next lemma shows that $\leq$ is archimedean. Thus by [Prestel, 1984], Theorem 1.20, $\leq$ is even an ordering and therefore F may be considered as a subfield of $\mathbb{R}$ with $\leq$ induced by the ordinary ordering of $\mathbb{R}$.

LEMMA 6. *Let $(E, <>)$ satisfy Assumption 2 and let $\leq$ be a semi-ordering on the fixed field F such that $< x, x > \geq 0$ for all $x \in E$. Then $\leq$ is archimedean.*

Proof. Assume that $\leq$ is not archimedean. Then $\gamma \in F$ exists such that $n < \gamma$ for all $n \in \mathbb{N}$. Then by the laws of semi-orderings (cf.[Prestel, 1984], Lemma 1.18), we have $0 < 2^n\delta < 1$ for all $n \in \mathbb{N}$ in case we take $\delta = \gamma^{-1}$. Thus by Lemma 5 a vector h'_n exists such that $< h'_n, h'_n >= (1-(2^n\delta)^2)^{-1}$. Scaling h'_n by $1-(2^n\delta)^2$ we obtain some vector h_n with $< h_n, h_n >= 1 - (2^n\delta)^2$ for each $n \in \mathbb{N}$. By Corollary 4(a) we may also assume that $h_n \perp h_m$ for all $n \neq m$. Moreover we may assume that there exists an orthonormal sequence $(e'_n)_{n\in\mathbb{N}}$ orthogonal to all vectors h_n. Then the vectors $f_n = h_n + 2^n\delta e'_n$ obviously have length $< f_n, f_n >= 1$ and are pairwise orthogonal. Thus $(f_n)_{n\in\mathbb{N}}$ is an orthonormal sequence in E.

Now by Lemma 5 together with Lemma 3 the vector

$$x' = \sum_n 2^{-n} e'_n$$

exists in ${(e'_n)_{n\in\mathbb{N}}}^{\perp\perp}$. From the orthomodularity of E we find by (2.1)

$$E = {(f_n)_{n\in\mathbb{N}}}^{\perp\perp} + {(f_n)_{n\in\mathbb{N}}}^{\perp}$$

and thus $x' = f + g$ with $f \in {(f_n)_{n\in\mathbb{N}}}^{\perp\perp}$ and $g \in {(f_n)_{n\in\mathbb{N}}}^{\perp}$. Computing the Fourier coefficients of f w.r.t. the orthonormal system $(f_n)_{n\in\mathbb{N}}$, we find

$$\begin{aligned} < f, f_n > &= < x', f_n >=< x', h_n + (2^n\delta)e'_n > \\ &= < x', 2^n\delta e'_n >= 2^n\delta < x', e'_n >= \delta \,. \end{aligned}$$

Thus $\delta^{-1} f$ would have Fourier coefficients 1 w.r.t. $(f_n)_{n\in\mathbb{N}}$ which contradicts Corollary 4 (b) (together with Lemma 3).

■

LEMMA 7. *Let $(E, <>)$ satisfy Assumption 2. Then every Dedekind cut is realized in the archimedean ordered field $(F, \leq)$.*

Proof. It clearly suffices that every Dedekind cut in the interval from 0 to $\frac{16}{5}$ is realized.

Taking $\alpha = \frac{3}{4}$ in Lemma 5 we see that the vector $x = \sum_n \alpha^n e_n$ exists in E and has length

$$< x, x >= \frac{16}{5} = \frac{1}{1-\alpha^2} = \sum_n \alpha^{2n}.$$

Now let β be any real number from the interval $[0, \frac{16}{5}]$. It is not difficult to see that a subset M of $\mathbb{N}$ exists such that

(1) $$\beta = \sum_{n \in M} \alpha^{2n}$$

From (2) we get

$$E = (e_n)_{n \in M}{}^{\perp\!\!\perp} + (e_n)_{n \notin M}{}^{\perp\!\!\perp}.$$

Thus x admits a decomposition $x = y + z$ with $y \in (e_n)_{n \in M}{}^{\perp\!\!\perp}$ and $z \in (e_n)_{n \notin M}{}^{\perp\!\!\perp}$. We claim that $< y, y >$ realizes the Dedekind cut determined by the real number β of (1) in the interval $[0, \frac{16}{5}]$. In fact, if I is any finite subset of M we find again from (2) that

$$(e_n)_{n \in M}{}^{\perp\!\!\perp} = (e_n)_{n \in I}{}^{\perp\!\!\perp} + (e_n)_{n \in M \setminus I}{}^{\perp\!\!\perp}.$$

Thus y admits a representation $y = y' + y''$ with $y' \in (e_n)_{n \in I}{}^{\perp\!\!\perp}$ and $y'' \in (e_n)_{n \in M \setminus I}{}^{\perp\!\!\perp}$. Since $(e_n)_{n \in I}$ is a finite orthonormal system, we get

$$< y, y >= \sum_{n \in I} \alpha^{2n} + < y'', y'' >$$

and, hence, in particular

(2) $$\sum_{n \in I} \alpha^{2n} \leq < y, y > .$$

Arguing similarly for the vector z which has length $< z, z >= \sum_{n \notin M} \alpha^{2n} = \frac{16}{5} - \beta$, we find for every finite subset J of $\mathbb{N} \setminus M$

(3) $$\sum_{n \in J} \alpha^{2n} \leq < z, z > .$$

From (2) and (3) we finally obtain

(4) $$\sum_{n \in I} \alpha^{2n} \leq < y, y > \leq \frac{16}{5} - \sum_{n \in J} \alpha^{2n}.$$

Since in (4) the left hand and the right hand side both are rational numbers converging to β in $\mathbb{R}$ for increasing sets I and J, and since $\mathbb{Q}$ is dense in F (by Lemma 6), the length $< y, y >$ realizes in F the Dedekind cut determined by β. ∎

4 THE NON-COMMUTATIVE CASE

As alraedy mentioned at the beginning of this paper, M.P. Solèr's Thesis also covers Hilbert spaces over the quaterions $\mathbb{H}$. We will now explain this and also show how some slight changes in the above proof of the commutative case give the theorem in full generality.

Let now K be a skew field (commutative or not) and $^* : K \to K$ an *involution*[2], i.e. for all $\alpha, \beta \in K$ we have

$$(\alpha + \beta)^* = \alpha^* + \beta^*, \quad (\alpha\beta)^* = \beta^* \alpha^*, \quad \alpha^{**} = \alpha$$

The main non-commutative example is the skew field $K = \mathbb{H} = \mathbb{R} + \mathbb{R}i + \mathbb{R}j\mathbb{R}k$ of quaterions with $i^2 = j^2 = -1$, $ij = k = -ji$, and

$$(\alpha + \beta i + \gamma j + \delta k)^* = \alpha - \beta i - \gamma j - \delta k$$

for all $\alpha, \beta, \gamma, \delta \in \mathbb{R}$.

Now the set

$$F = \{\alpha \in K | \alpha^* = \alpha\}$$

of *symmetric elements* of K need no longer be a sub-skew-field of K, in fact, F need not be closed under multiplication.

Next let E be an infinite dimensional K-vector space and $<>: E \times E \to K$ a *hermitian form* on E which now has the properties

$$\begin{aligned} < \alpha x + \beta y, z > &= \alpha < x, z > + \beta < y, z > \\ < z, \alpha x + \beta y > &= < z, x > \alpha^* + < z, y > \beta^* \\ < x, z > &= < z, x >^* \end{aligned}$$

for all $\alpha, \beta \in K$ and $x, y, z \in E$. All the other notions like orthogonality, anisotropy, orthomodularity and orthonormal sequences are defined as in the commutative case.

The Theorem of Solèr then says that the only infinite dimensional hermitian K-vector spaces $(E, <>)$ which are orthomodular and admit an orthonormal sequence $(e_n)_{n \in \mathbb{N}}$ are the usual Hilbert spaces over $\mathbb{R}, \mathbb{C}$, and $\mathbb{H}$.

The proof of this theorem can be obtained from the one above by observing the following modifications.

We first observe that Lemma 3 and Corollary 4 also hold in the generalized situation.[3] The proofs are literally the same. Also Solèr's main lemma (Lemma 5) now holds. The proof, however, is more involved (see [Solèr, 1995], Lemma 4,

[2]Sometimes also called an *anti-involution*

[3]By an 'isometry' of subspaces U and V we here mean a K-linear isomorphism f satisfying $< f(x), f(x) > = < x, x >$ for all $x \in U$. In the commutative case this clearly implies $< f(x), f(y) > = < x, y >$ for all $x, y \in U$.

and its proof). Nevertheless, it only uses Lemma 3 and Corollary 4, with a slight generalization in (b):

(b') *Let* $\alpha \neq 0$ *and* $(f_n)_{n\in\mathbb{N}}$ *be a sequence of pairwise orthogonal vectors such that*
$< f_n, f_n >= \alpha$ *for all* $n \in \mathbb{N}$. *Then there is no vector* $f \in (f_n)_{n\in\mathbb{N}}^{\perp\perp}$ *having all Fourier coefficients equal to 1 w.r.t.* $(f_n)_{n\in\mathbb{N}}$.

The proof of (b') is just the same as that of (b) using (3.2).

From Lemma 3 and Corollary 4 we obtain that the set $P = \{< x, x > |x \in E\}$ of lengths defines a *Baer-ordering* on K, i.e.

$$\alpha \leq \beta \text{ iff } \beta - \alpha \in P,$$

linearly orders F and satisfies

$$(1) \qquad \begin{array}{rcl} 0 \leq \alpha,\ 0 \leq \beta & \Rightarrow & 0 \leq \alpha + \beta \\ 0 \leq \alpha & \Rightarrow & 0 \leq \gamma\alpha\gamma^* \\ & 0 \leq 1 & \end{array}$$

for all $\alpha, \beta \in F$ and $\gamma \in K$. All these properties are proved as in Section 3. Concerning linearity, i.e. $0 \leq \alpha$ or $\alpha \leq 0$ for all $\alpha \in F$, one should be aware of the fact that $L = \mathbb{Q}(\alpha)$ is a commutative sub-skew-field of K, and that $\leq$ restricted to L therefore yields a semi-ordering.

This very fact also allows us to use the proof of Lemma 6 in order to see that $\leq$ is an *archimedean* Baer-ordering of K, i.e. $0 \leq \lambda < \frac{1}{n}$ for all $n \in \mathbb{N}\backslash\{0\}$ can only hold for $\lambda = 0$.

At this point we make use of S. Holland's result (see [Holland, 1977], Theorem 2 and Corollary 3) that a skew field K with involution * which admits an archimedean Baer-ordering is isomorphic to a sub-skew-field of $\mathbb{R}, \mathbb{C}$, or $\mathbb{H}$ and is actually equal to $\mathbb{R}, \mathbb{C}$, or $\mathbb{H}$ in case every Dedekind cut is realized in F. This last fact, however, follows literally as in Lemma 7.

In [Holland, 1995] Holland actually shows that a non-commutative skew field K with involution * can not be generated as a ring by the set F of symmetric elements of K. But then a theorem from Dieudonné (cf. [Holland, 1995], Lemma 1) states that K is a generalized quaterion algebra $K = Z(\alpha\beta)$ over its center Z, that is

$$K = Z + Zi + Zj + Zk$$

with $i^2 = \alpha, \quad j^2 = \beta, \quad ij = -ji = k$, and

$$(\alpha_0 + \alpha_1 i + \alpha_2 j + \alpha_3 k)^* = \alpha_0 - \alpha_1 i - \alpha_2 j - \alpha_3 k$$

for all $\alpha_0, \ldots, \alpha_3 \in Z$.

It now follows that $F = Z$ is a commutative subfield of $K$and thus by the above arguments has to be isomorphic to $\mathbb{R}$. This then clearly implies that $K = \mathbb{H}$.

Now that we know that the field K of scalars is $\mathbb{R}, \mathbb{C}$, or $\mathbb{H}$, and that $<>$ is positive definite, the completeness of $(E, <>)$ follows as in Step 3 of Section 2.

Remark. It should be pointed out that the linearity of the Baer-ordering $\leq$ defined by P was essential in the above arguments. If we did not have $P \cup -P = F$, i.e. $\leq$ is only a partial ordering of F satisfying (1), it is not known whether P could be extended to some Baer-ordering on K or not.

BIBLIOGRAPHY

[Engesser and Gabbay, 2002] K. Engesser and D. M. Gabbay. Quantum logic, Hilbert space, revision theory. *Artificial Intelligence*, 136 (2002), 61-100.

[Holland, 1977] S. S. Holland, Jr. Orderings and square roots in *-fields. *J. of Algebra*, 46(1977), 207-219.

[Holland, 1995] S. S. Holland, Jr. Orthomodularity in infinite dimensions; a theorem of M. Solèr. *Bull. AMS*, 32 (1995), 205-234.

[Keller, 1980] H. A. Keller. Ein nicht klassischer Hilbertscher Raum. *Math.Z.*, 172(1980), 41-49.

[Keller *et al.*, 1998] H. A. Keller, U. M. Künzi, and M. P. Solèr. Orthomodular spaces. In: Orthogonal geometry in infinite dimensional vector spaces. *Bayreuther Mathematische Schriften*, 53 (1998), 171-250.

[Maeda and Maeda, 1970] F. Maeda and S. Maeda. *Theory of symmetric lattices.* Springer-Verlag, Heidelberg, 1970.

[Prestel, 1984] A. Prestel. *Lectures on formally real fields.* Lecture Notes in Math. 1093 (Springer), 1984.

[Prestel, 1995] A. Prestel. On Solèr's characterization of Hilbert spaces. *Manuscripta Math.*, 86 (1995), 225-238.

[Solèr, 1995] M. P. Solèr. Characterization of Hilbert spaces by orthomodular spaces. *Communications in Algebra*, 23 (1995), 219-243.

Edited by K. Engesser, D. M. Gabbay and D. Lehmann

OPERATIONAL QUANTUM LOGIC: A SURVEY AND ANALYSIS

David J. Moore and Frank Valckenborgh

1 INTRODUCTION

One of the basic assumptions of standard quantum physics is that the collection of experimentally verifiable propositions that are associated with a quantum system has a logico-algebraic structure isomorphic with the lattice of closed linear subspaces of some complex Hilbert space or, dually, with the lattice of orthogonal projections on these subspaces. The quantum logic approach to the mathematical and conceptual foundations of quantum mechanics attempts to provide an advanced perspective on the rather high-level languages of standard text-book classical and quantum theory, in the hope that the resulting theory leads to additional simplifications in the conceptual framework of conventional quantum theory. For example, for a slightly idiosyncratic basic text-book on elementary quantum mechanics with an operational twist, the intrepid reader can consult the blue book by Constantin Piron 1998).

The goal of the particular branch of *operational quantum logic* — the subject of this work — consists first and foremost in developing a suitable global framework in which the general discussion of physical processes can take place. The driving principle here is the conviction that the primitive terms of a candidate mathematical framework in which at a second stage more concrete models for physical phenomena can be developed, should be based, as much as possible, on *concrete* physical notions and operations. In other words, the abstract language should be developed in terms of operationally well-defined entities, closely related to the actual practice of the experimental physicist, rather than on mathematically possibly more convenient or elegant but experimentally less motivated terms. That is, in the development of a suitable framework we have to adopt the more concrete line of thinking of the experimental physicist, where the characterisation of a system depends on how we can act upon that system. One then exploits explicitly the empirical fact that there exists a certain class of systems, so-called measuring instruments, that are capable of undergoing macroscopically observational changes triggered by their interaction with single micro-systems (Kraus, 1983). By way of contrast, in the construction of a mathematical theory the choice of primitive terms is usually motivated internally (how else?) by the elegance and simplicity of the resulting theory; in physics this choice then should ideally be justified externally.

Instead of defining properties of a physical system by reference to its putative *internal* structure, one thus proceeds by analyzing its *external* relationships with other systems; this information is then used to attribute properties to the system under investigation. We will argue that an operational approach — at least from a methodological point of view — on fundamental physical concepts such as system and particle, in particular composite particle, evolution and dynamics, space and spacetime, and so on, possibly also leads to additional clarifications and maybe even a reformulation of the foundations of physical theory. More concretely, we will indicate how the resulting language encompasses many of the essentials of both classical physics and quantum physics, and so is obviously of considerable independent interest.

From a philosophical perspective, it is important to remark that the word *operational* in this context has to be considered as a pragmatic attitude — in which we privilege mathematical terms possessing a definite physical heuristic in the development of a framework theory overarching the plethora of specific models — rather than a doctrine. Specifically, we do not commit ourselves to a definite metaphysical position when we insist on developing a general mathematical language by abstracting from concrete physical operations. On the contrary, one of the important underlying motivations to do so is based upon the conviction that many of the interpretational issues that haunt standard quantum mechanics, in particular those associated with the wave-mechanical formalism, will present themselves in a different light when the primitive terms of the formalism have a clear physical interpretation. For that matter, the Q in OQL seems to have more historical roots, when the necessity for a deeper analysis of the descriptional framework of physical systems became more acute with the investigation of micro-systems. Finally, trying to analyze the meaning of the L would almost certainly stir up an academic hornet's nest, so we happily refrain from doing so, positions of members of the quantum logic research community covering the whole available spectrum from a purely algebraic perspective on the usage of the word 'logic' to much stronger philosophical commitments. We have to emphasise, however, that OQL has an explicit physical origin, contrary to the metaphysical basis of empirical quantum logic *sensu* Finkelstein (1968) and Putnam (1968).

It is also from this perspective that at least some part of the general operational methodology and philosophy of the researchers active in the thriving domains of quantum information science and quantum computation is actually quite close in spirit to the conceptual methodology used by OQL, the behaviour of individual physical systems rather than statistical ensembles of such objects moving more into the focus of attention. On the other hand, it is only fair to mention that there are also important differences. Specifically, we will argue that at least the original incarnations of OQL have some difficulty when confronted with the notions of composite systems on the one hand, and with the explicit sequential composition of quantum processes on the other. Indeed, the explicit incorporation of multistage quantum processes in more general terms seems to require some form of typing, which reflects a conditional aspect in performing consecutive operations, due to the probabilistic character of measurements on quantum systems (at least from the perspective of the system itself). In other words, both the serial composition and parallel composition of physical operations are usually not explicitly accounted for. In this context, we will also have another look at the important work of Pool (1968a, b).

Distinct approaches to formulating an adequate framework for a better understanding of physical phenomena attribute basic physical significance to different concepts, reflecting the fact that various notions play an important conceptual role in physical thinking. Consequently, it is clear that *a priori* various choices for the primitive terms can be made, and this handbook reflects some parts of the corresponding spectrum of possibilities. Most authors ascribe a fundamental importance to the two concepts of *state* and *observable*; the physical interpretation of these notions, however, varies considerably from one author to another. Let us briefly discuss these notions from an operationally oriented perspective, more details about the particular incarnations in our perception of OQL being referred to a later section.

The abstract notion of a *state* that can be assigned to a physical system is a very natural one from the more static perspective that focuses on sufficiently stable physical objects in the here and now, but it is not necessarily restricted to this setting. States come in two conceptually distinct flavours. In principle, knowledge of the state allows one to predict the experimental behaviour to the best of one's knowledge, but often only in probabilistic terms. Indeed, it is one of the essential features of quantum physics that for each state — even states that apparently are not reducible to more fundamental states — there exist many observables with highly non-trivial probability distributions. Contrary to the orthodox ensemble-based interpretation of quantum mechanics, we claim that *pure* states can be construed — and that it is a conceptual advantage to do so — as abstract names that encode the possible singular realisations of a given particular physical system. In other words, pure states are admittedly somewhat idealised objects that adopt the somewhat privileged role of being able to refer to *individual* physical systems in an ensemble, at least under mildly favourable conditions. As a consequence, there will always exist some kind of duality between the set of pure states and the collection of properties attributed to a physical system, reflecting the well-established practice among physicists to describe a general system either by its properties or by its states. In fact, on the one hand knowledge of the pure state of the individual system allows one to predict its behaviour, or its properties, in the best possible way, and on the other hand we expect that the knowledge of all its actual properties determines the pure state completely. Often, it is sufficient to know a particular subset of its actual properties to identify the state; such a collection of properties is then state-determining.

Alternatively, *mixed* states are often and meaningfully regarded as a convenient heuristic that summarises the probabilistic behaviour of an ensemble of systems that can be subjected to various experimental protocols. We can always interpret these, with some care, as a probabilistic mixture of pure states; the mixed state then also incorporates the maximal information we can assign *a priori* to the individual objects in the ensemble. Mathematically speaking, for a given physical system we then have to assume the existence of an abstract collection of states, say $\mathfrak{M}$, in this generalised sense, that forms a convex set, the pure states corresponding with the extreme points, say $\partial_e\mathfrak{M}$. In practice, pure or mixed states can often be assigned to physical systems by a physically well-defined *preparation* procedure. Sometimes however, this is not the case, and we are forced to attribute a state to an object that is beyond our immediate control, based on our experience with objects that can be prepared in a certain state.

Observables on the other hand can be construed as encoding the abstract relation between some physically relevant collection of properties of a measurement device and corresponding properties assigned to our physical system. For example, after performing an experiment — an interaction between a physical system and the appropriate device — the position of a pointer of the measurement device should better reflect a possible property of the system under investigation. Operationally speaking, *properties* assigned to a system will then be construed as candidate elements of reality corresponding to definite experimental projects that are defined for that particular physical system. From a mathematical perspective, the collection of physically useful properties associated with the measurement device can be represented by a boolean algebra of some sort, and this structure should then be faithfully represented also at the level of the description of the physical system. Often, such a correspondence will be subject to additional constraints at the level of the measurement devices, due to symmetry considerations, and such constraints should then be appropriately incorporated, also with regard to the system. Standard examples are given by the position and momentum observables with respect to a given reference frame, coordinates and intervals of space and time being defined in a physically standardised way (Audoin & Guinot, 2001). In many approaches, a distinguished role is played by the subset of two-valued observables. In conventional quantum mechanics, such observables will correspond with pairs of orthogonal projections in the complex Hilbert space.

In our view, a general operational attitude — methodologically speaking — towards the formulation of physical theories has other advantages. In fact, one is inclined to critically examine the various concepts that otherwise would be taken for granted. For example, the world-view of many working physicists is still (dangerously?) close to the classical Descartian one that pictures physical particles as point-like particles that move — possibly under the influence of fields of some sort — in an otherwise independent spatio-temporal containing arena (Piron, 2002). We think that one of the lessons that we have to learn from the progress of technology is to take non-locality seriously, although it is admittedly very difficult for us macroscopic beings to think in terms of individual microscopic objects that have an extension but no parts, acting as a single whole. Specifically, neutron interferometric experiments performed during the last decennia of the previous century indicate that our current notion of space-time is more subtle than anticipated. Indeed, individual neutrons can be manipulated in spatially separated regions, hence at least for a micro-system the property of being localised in a certain region of space cannot be decided with certainty from the knowledge of its state. As some sort of an afterthought, we can also add that it may be useful to regard the notion of a physical system as being 'composed' of more fundamental entities in an operational light.

For the sake of completeness, for the convenience of the reader, and to make the paper as self-contained as possible, we have included many of the proofs of at least the simpler statements, so that the reader is not only spared the interruption of the train of thought by feeling the need to show a particular result before proceeding with the text, but also the task of having to look up the widely dispersed and — with the progress of time — not necessarily easily available original sources, both papers and monographs, if she or he wishes to do so.

2 SOME HISTORICAL REMARKS

The subject of quantum logic started essentially with the monumental treatise of John von Neumann (1932, 1955), in which the complex Hilbert space model of quantum physics was developed and its structural, logical and physical implications explored. One can argue that the intellectual heritage of this work was the essential dominance of the complex Hilbert space model for the mathematical description and the physical and philosophical analysis of all physical systems, *a priori* without any additional restrictions. It is, however, rarely appreciated that von Neumann himself started expressing his dissatisfaction with the resulting Hilbert space framework not long after (Rédei, 1996). One manifestation of this eventually led to the pioneering paper — co-authored by Garrett Birkhoff — that was appropriately titled "The Logic of Quantum Mechanics", published in 1936 in the Annals of Mathematics (Birkhoff & von Neumann, 1936). In this paper, the authors take a more structural view, based upon a deeper analysis of the "logical" structure one may expect to find in general physical theories, and in particular quantum mechanics, for the description of physical systems for which the collection of experimentally verifiable propositions does not seem to conform to the laws of classical logical discourse. In this paper, the idea was presented that the primitive mathematical building blocks (and the relations connecting them) that are used in the abstract mathematical description of physical systems, should take into account the basic structure of some sort of more concrete and admittedly slightly idealised operational calculus about physical systems, in which one tries to isolate key aspects of the general behaviour of such systems in an experimental context. In particular, notions of "experimental proposition", "observation space", "experimental implication" and the like gained some ground in developing the structural foundations of a physical theory.

This perspective led almost automatically to a shift of the burden of the conceptual weight to the *projective* structure associated with the traditional complex Hilbert space model, the linear structure becoming a secondary — though technically extremely useful — construction, in view of the first fundamental theorem of projective geometry. A more detailed analysis of this part of the story is given in this handbook by the paper by Isar Stubbe & Bart Van Steirteghem (2007).

In 1957, George W. Mackey, in an influential article in the American Mathematical Monthly that was later expanded in an important monograph (Mackey, 1957, 1963), sketched a probabilistic framework for the mathematical foundations of quantum (and classical) mechanics. In his set-up, he starts with an abstract set of observables $\mathcal{O}$ that can be effectuated on a physical system; this set is closed under the action of real Borel functions, which represent measurements of the same observable followed by a simple computation of f. States attributed to this system are conceived as functions $\mu : \mathcal{O} \longrightarrow \mathcal{M}$, where the latter set denotes the collection of probability measures on the Borel or Lebesgue sets in $\mathbb{R}$. An important role is played by Mackey's *questions*, which correspond with the two-valued observables that take their values in the outcome set $\{0, 1\}$. Each observable A generates a family of questions of the form $\chi_E(A)$, where χ_E is the characteristic function of some Borel set $E \subseteq \mathbb{R}$. The whole mathematical structure is then determined by the restriction of the states to the set of questions. Mackey then observes that one can *partially order* the set of questions by the prescription

$$(1) \quad Q_1 \leq Q_2 \quad \text{iff} \quad \forall \mu : \mu(Q_1)(\{1\}) \leq \mu(Q_2)(\{1\})$$

and define a relation of orthogonality on this set by setting $Q_1 \perp Q_2$ iff $Q_1 \leq 1 - Q_2$. Notice the reliance on probabilistic arguments to define this order relation. In his 1963 monograph, this framework was refined and extended into a system of nine axioms for the mathematical foundations of quantum mechanics. His Axiom VII is somewhat remarkable and *ad hoc* in the sense that it explicitly demands that the partially ordered set of all questions associated with a quantum system should be isomorphic with the lattice of all closed subspaces of a separable, infinite-dimensional complex Hilbert space. Parts of this work were generalised in an important paper by Varadarajan (1962) into a more general probabilistic formalism, that considerable extended the axiomatic model that was proposed by Kolmogorov (1956) to take into account the phenomenology associated with quantum systems. More explicitly, the author proceeds in developing an experimentally motivated probability calculus on orthomodular σ-orthoposets. In particular, he developed and thoroughly analysed the notion of *observables* — generalised random variables with respect to his framework — and their mathematical characterisation in terms of Boolean subalgebras. In addition, he generalised the functional calculus of the more conventional quantum mechanical observables in this setting, and the related problem of *simultaneous measurability* for such observables.

Some of the central tenets of the axiomatic structure of quantum theory as it was developed by the so-called Geneva school from the early sixties or so consist in the affirmation that it is possible to erect the framework of quantum theory on the basis of the notion of certainty instead of probability, and in the assumption that and the physical motivation why the collection of experimentally verifiable propositions has the structure of a *lattice* and not merely a partially ordered set of some sort (Piron, 1964; Jauch, 1968; Jauch & Piron, 1969; Piron, 1976). Remarkably, the axiomatic framework that was developed by this group of researchers is explicitly non-probabilistic; in the words of Jauch & Piron (1969):

> "If one introduces probability at this stage of the axiomatics one has difficulties of avoiding the criticism of Einstein that a state is not an attribute of an individual system but merely the statistical property of a homogeneous ensemble of similarly prepared identical systems."

This *desideratum* requires that any probabilistic characteristics associated with the system itself have to be eliminated right from the start, and so the restriction to *pure* states only as part of the primitive objects of the framework is essential in this program. Specifically, the aims of Piron were at least twofold: (1) Develop a general theory valid for both classical and quantal systems; (2) Justify the usual formalisms for these two extremal cases by an intrinsic characterisation of the structure of their sets of observables. This program culminates in Piron's celebrated representation theorem, about which more will be said in the next section (Piron, 1964, 1976). At the end of the day, one then likes to reconstruct explicitly the usual concrete particle models of classical Hamiltonian mechanics and Hilbert space quantum physics in this light. Hereto, one has to appeal to more advanced techniques of group representation theory. The by now classical work of Mackey on the induced representations of locally compact groups has proven instrumental in this respect; for a rather leisurely overview, see Mackey (1968).

The operational statistics approach that was developed in an extensive series of papers by Charles Randall and Dave Foulis in Amherst is in many respects a parallel branch in the evolutionary tree of foundational approaches (see Foulis & Randall, 1972; Randall & Foulis, 1973). For a more elaborate analysis of this work and its more recent development, see the contribution on test spaces by Alex Wilce (2008) in this handbook. For the relation between the Geneva school approach and the Amherst approach, see Foulis, Piron & Randall (1983) and Coecke, Moore & Wilce (2000) for a more general discussion.

For the sake of completeness, let us also mention that a third highly operationally motivated sibling was developed by the school of Ludwig, starting from the early 1960's (see, for example Ludwig, 1983). For another account of parts of this approach, the reader can consult the monograph by Kraus (1983).

In the light of the recent explosion of the domains of quantum computation and quantum information science, new axiom systems have been proposed that emphasise various concepts that stem from the more specialised perspectives of this type of research. Here is a quick anthology of some more recent attempts: Fuchs (2001); Hardy (2001); D'Ariano (2007a, b), and there are undoubtedly many others. Without going into details, we only remark that — to avoid some of the conceptual and methodological pitfalls of the past on the one hand, and to highlight the successes achieved and the weaknesses exposed in the operational quantum logic approach to the foundations of the physical sciences on the other — we consider it useful and appropriate also from this more specialised perspective to give a reasonably detailed annotated exposition of what has been achieved so far by this part of the quantum logic community.

3 QUANTUM OBJECTS

The empirical basis for the statistical laws of quantum physics consists in the reproducibility of the relative frequencies of the results one obtains when subjecting micro-systems, prepared according to a given experimental protocol, to various measurement procedures. It is indeed an empirically well-established fact that there exist macroscopic devices capable of undergoing macroscopically observable changes when interacting with such micro-systems (Kraus, 1983). Experience also tells us that the outcome of a single experiment is not determined completely by the specifications of the preparation procedure and the macroscopic change that occurred in the measuring instrument after the interaction between the system and the device.

As we have already indicated, there exist various established schools in the quantum logic enterprise and they attribute basic physical significance to different concepts as the primitive building blocks. *A priori*, various choices can be made: in general, a physical system can be described by a class of events, propositions, properties, operations, etc., and a class of states. Sometimes the events alone are seen as fundamental and states are considered as derived entities, sometimes both events and states are considered as primitive, and sometimes the collection of states plays a central role. In operational quantum logic, the Geneva school approach occupies a central position. The central observation of this approach is that the abstract sets $\partial_e\mathfrak{M}$ of pure states and $\mathcal{L}$ of properties each admit an abstract mathematical structure, induced by concrete physical considerations on the concrete primitive notions of *particular physical system* and *definite experimental project*. It is important to remark that these properties can be introduced in a non-probabilistic way. In the following development, we will first work in the apparently slightly more general probabilistically oriented framework, and subsequently add the refinements that are particular to the Geneva approach.

The highly idealised concept of a *physical system* is central in physics, and usually taken for granted, without much further comments. As such, the potential difficulties it embodies tend to escape attention and scrutiny. From the more object-oriented viewpoint that we adopt in this section, however, it may be wise to add a few comments on this potentially problematic notion. Indeed, it is important to clearly express the ambiguities inherent in any precisification of a physical concept rather than trying to ignore or minimise their cognitive importance and consequences. A physical system is usually conceived as a sufficiently well-circumscribed part of reality external to the physicist, in the sense that its interaction with its surroundings can either be ignored or modeled in an effective way. An extreme case corresponds with the notion of a *closed* system, where the interaction with the external world can be totally neglected. In practice however, it is hard, if not impossible, to isolate physical systems completely from their environment, and the classical thermodynamic concept of an *open* system is perhaps a more appropriate term in the other extreme case where the very identity of the system becomes somewhat blurred. For example, we can choose to investigate the properties of a certain part of space-time, identified with respect to some reference frame that attributes coordinates of space and time to potential events, and then electromagnetic radiation generated by the environment is often an issue. Another yet not unrelated reason why we have to be careful in this case, is the experimental reality of non-locality, that manifests itself in some well-known cases (see for example, Rauch *et al.*, 1975). The typical adventures of the paradigmatic couple Alice and Bob dramatically illustrates some of these aspects. Of course, it is part of conventional scientific methodology that some idealisation is inescapable in the elaboration of our physical models, considered as abstract and hence partial reflections of concrete world situations. In this text, we will take the point of view that the state attributed to an individual member of a closed or almost closed physical system is always a pure state. This then can also be taken as a characterization of the type of physical systems that we favour in this development. On the other hand, if we happen to encounter a physical system where this would not be the case, we may be inclined to embark on a more comprehensive investigation of that particular phenomenon.

Physical systems can be investigated by their potential interaction with a privileged class of macroscopic probing devices. Notice that the inherently active nature of experimentation, which implies that probing a system will in general perturb and sometimes even destroy it entirely, is an essential part of modern scientific methodology. A *definite experimental project* — also called *question* or sometimes *test* — then consists of a complete experimental protocol: a measurement device and instructions on how to properly use it, together with a rule for interpreting its possible results in terms of two alternatives only: either the positive result, identified with a particular predefined set of configurations of the measurement device before the actual measurement takes place, would be obtained (*yes*), or not (*no*). We emphasise that this notion explicitly refers to the empirical existence of independent macroscopic arrangements that can interact with single individual samples of the physical system under investigation, and that can leave an objective macroscopic effect due to such a singular interaction; this effect is then interpreted as the occurrence of the positive result for the measurement. In other words, this approach is explicitly relational in that systems are characterised by their interaction — hypothetical or *de facto* — with other systems. The class of definite experimental projects relative to the system will be denoted by $\mathfrak{Q}$. To avoid any confusion and possible conflation with Mackey's notion of a question, we will avoid the use of the term "question" in the context of the Geneva approach. Observe that the notion of a definite experimental project is sufficiently general, since each multi-outcome experiment can be chopped up into a collection of two-valued tests.

We say that a definite experimental project $\alpha \in \mathfrak{Q}$ is *certain* for a given singular realisation of a particular physical system (with state $\mu \in \mathfrak{M}$) when the positive result would be obtained with certainty, should the experiment be (properly) executed. For notational brevity, we shall encode this relationship between the set of all possible states $\mathfrak{M}$ and the class of definite experimental projects $\mathfrak{Q}$ succinctly as $\mu \vDash \alpha$, or sometimes also $T(\mu, \alpha)$; alternatively, we use the notation $F(\mu, \alpha)$ whenever the negative result would obtain with certainty. Note the counterfactual locution in this statement! It implies that the notion of certainty has a sense before, after, and even in the absence of an experiment. The crucial point is that definite experimental projects are *hypothetical*, so that their certainty or otherwise can be regarded as an objective feature of the particular physical system. In practice, we often convince ourselves that a definite experimental project is certain for some preparation of a physical system by running the experiment on a number of samples of the physical system; if the positive result is always obtained, we have the right to claim that similarly prepared new samples will also yield a positive result, if we would actually run the experiment (Aerts, 1983). The emphasis of the founders of the Geneva approach on the notion of certainty, in contrast to probability, at this early level of the formalisation circumvents Einstein's criticism on conventional quantum mechanics that a state is not an attribute of an individual system but rather of a homogeneous ensemble of similarly prepared systems (Jauch & Piron, 1969).

The class $\mathfrak{Q}$ can then be endowed with an elementary mathematical structure by exploiting these concrete epistemological considerations. Specifically, there exists a physically natural pre-order relation — encoding phenomenological implication — where we say that $\alpha \preceq \beta$ when α certain implies the certainty of β, for any preparation or realisation of the particular physical system under investigation. Symbolically,

(2) $\alpha \preceq \beta \quad \text{iff} \quad \forall\, \mu \in \mathfrak{M} : \mu \vDash \alpha \text{ implies } \mu \vDash \beta$

The advantage of this counterfactual notion of certainty is that it allows one to give a sensible justification for claiming that more than one definite experimental project is certain for a given single sample of a particular physical system, even when the various experimental conditions are incompatible. From an operational perspective then, the problem is the following: Given a family of not necessarily compatible experimental protocols, is it possible to construct a definite experimental project for the conjunction of these properties? In fact, the counterfactual notion of certainty, given a bunch of definite experimental projects $\{\alpha_j \mid j \in \mathcal{J}\}$, allows one to define a new project, denoted by $\prod\{\alpha_j \mid j \in \mathcal{J}\}$, that tests the certainty of all these questions. For example, pick any $j \in \mathcal{J}$ at random and perform the corresponding experiment on an individual sample of the physical system under investigation; this then is an explicit observational procedure for the conjunction of a collection of properties. Formally,

(3) $\mu \vDash \prod\Big\{\alpha_j \mid j \in \mathcal{J}\Big\} \quad \text{iff} \quad \mu \vDash \alpha_j \text{ for all } j \in \mathcal{J}$

Mathematically speaking, the thin category $\mathfrak{Q}$ is then finitely complete, σ-complete, or complete, depending upon whether one allows only finite products, countable products, or all products, depending somewhat upon one's philosophical attitude. In the Geneva approach proper, one always takes $\mathfrak{Q}$ to be complete.

This definition of the notion of certainty immediately leads to a second definition, the notion of two definite experimental projects being *equivalent*:

$$(4) \quad \alpha \sim \beta \quad \text{iff} \quad \alpha \preceq \beta \text{ and } \beta \preceq \alpha$$

In other words, $\alpha \sim \beta$ iff for all $\mu \in \mathfrak{M}$, $\mu \vDash \alpha \Leftrightarrow \mu \vDash \beta$.

There is another natural elementary operation on $\mathfrak{Q}$, which consists in simply interchanging the positive and the negative labels for a given test α. This is a new test $\alpha^{\sim}$ — the *inverse* test — although it is associated with the same piece of experimental equipment as α. Clearly, $\mu \vDash \alpha^{\sim}$ iff $F(\mu, \alpha)$; notice that $\alpha^{\sim\sim} = \alpha$. In addition, one can always add two ideal elements 0 and 1 to $\mathfrak{Q}$, where $1^{\sim} = 0$. A definite experimental project that tests 1 could for example be described by using some given experimental protocol and always assign the positive result. A remarkable consequence of these heuristic definitions that has led to considerable confusion is the following:

$$(5) \quad \left(\prod\left\{\alpha_j \mid j \in \mathcal{J}\right\}\right)^{\sim} \quad \sim \quad \prod\left\{\alpha_j^{\sim} \mid j \in \mathcal{J}\right\}$$

In particular, given any $\alpha \in \mathfrak{Q}$, it is always the case that

$$(6) \quad \prod\{\alpha, 1\} \sim \alpha \qquad \text{and} \qquad \left(\prod\{\alpha, 1\}\right)^{\sim} \sim 0$$

In order to build a well defined mathematical theory one must pass from the concrete primitive notions of particular physical system and definite experimental project to abstract concepts susceptible to formal symbolical analysis. That is, we must provide a physical relation between particular physical systems and definite experimental projects which lifts to a mathematical relationship between states and properties. Two definite experimental projects whose positive responses are determined with certainty for exactly the same singular realisations of a given particular physical system have the same epistemic content and so should refer to the same property attributed to the system. Consequently, the abstract property assigned to the system should correspond with an equivalence class in $\mathfrak{Q}$ induced by the equivalence relation $\sim$ and one writes $\mathcal{L} := \mathfrak{Q}/\sim$. It is easy to see that $\mathcal{L}$ inherits the pre-order relation and the completeness from $\mathfrak{Q}$. More precisely, the pre-order becomes a partial order, hence $\mathcal{L}$ is a complete meet-semilattice. Since elements of $\mathcal{L}$ are sometimes construed as potential properties of the physical system under investigation, it is often called the *property lattice* associated with the physical system, although strictly speaking there is a cognitive distinction between a property assigned to a system and an equivalence class of definite experimental projects, and so one should not conflate the two concepts; this distinction has also led to some confusion in the past. Alternatively, its elements are sometimes called *propositions*, certainly in the older literature on the subject. In any case, one envisages a one-to-one correspondence between properties attributed to the system, and propositions about the system. According to one of our previous remarks, any equivalence class of definite experimental projects contains a test whose inverse is in the absurd equivalence class.

One should note that the meet in the complete meet-semilattice $\mathcal{L}$ is then operationally well-defined, whereas the join, defined by the prescription

$$(7)\quad \bigvee\left\{a_j \mid j \in \mathcal{J}\right\} := \bigwedge\left\{c \in \mathcal{L} \mid \forall j \in \mathcal{J} : a_j \leq c\right\}$$

admits no direct physical meaning. Consequently, physical arguments must proceed from the meet as encoded by the product operation rather than the join whose existence is induced by completeness of the lattice. Also, the attentive reader should not fail to recall that $\alpha \sim \beta$ does not imply $\alpha^{\sim} \sim \beta^{\sim}$ in general; that is, the operation of taking inverses is not compatible with the equivalence relation induced by the pre-order $\preceq$.

A property is said to be *actual* if any — hence all — of the definite experimental projects in the corresponding equivalence class is certain; in this case, the system is also said to have an *element of reality* corresponding to this test, in a sense derived from the terminology used in the famous paper of Einstein, Podolsky & Rosen (1935). It is worth noting that the derived notion of actuality is also counterfactually defined, and that the clever device of product questions then allows one to attribute more than one property to a given singular realisation of a particular physical system. Consequently, by the usual abuse of language, we shall sometimes use the term "property" also for these equivalence classes. To make the terminology complete, a property that is not actual is said to be *potential*.

Just as properties should not be conflated with equivalence classes of definite experimental projects, we insist that the notion of state is an independent primitive concept whose empirical realisation at the level of properties is something to be established and not merely posited. In its most general form, the state — pure or mixed — attributed to a general physical system should be representable as some sort of probability measure on $\mathfrak{Q}$, since empirically speaking physical experiments often yield only probabilistic information. In more detail, a very general formalisation of these ideas would then lead to the consideration of test-state triples $(\mathfrak{Q}, \mathfrak{M}, \mathbb{P})$, where $\mathbb{P} : \mathfrak{Q} \times \mathfrak{M} \longrightarrow [0,1]$ encodes the probability $\mathbb{P}(\alpha, \mu)$ to obtain a positive result for the test α for a physical system in an initial state μ, should the corresponding experiment be performed, and proceed by developing an axiomatic framework for these objects that reduces the generality and increases the applicability. In this way, we indeed obtain a correspondence

$$(8)\quad \mathfrak{M} \longrightarrow [0,1]^{\mathfrak{Q}} : \mu \mapsto \mathbb{P}(-, \mu)$$

We have already indicated that some approaches treat the notion of state as a derived concept, defined in terms of the definite experimental projects. Here, it is one of our aims to characterise the independent notion of state, in particular pure states, in terms of the algebraic structure on $\mathfrak{Q}$, and as independent of the probabilistic structure as possible. This can be done by associating to a given state μ the collection of all tests that are certain, or properties that are actual, for a system in this state μ. In general, one should not expect this correspondence to be one-to-one, since it is conceivable that distinct mixtures of the same set of pure states may lead to the same collection of certain tests.

From an operational perspective then, it may be useful to make the distinction between the *a priori* primitive concrete notion of a preparation, represented by some sort of probability measure on $\mathcal{Q}$, and the more ideal elements that represent pure states that can be attributed to single samples of the system, more explicit. This cognitive distinction may be particularly relevant from the perspective of evolutions, interactions and possibly also composite systems. In more detail, each preparation procedure μ — possibly corresponding with a mixed state — defines the subsets $\mathrm{S}_1(\mu) \subseteq \mathfrak{Q}$ and $\mathrm{S}_0(\mu) \subseteq \mathfrak{Q}$ of all definite experimental projects that are certain, respectively impossible for that particular preparation μ, and in this sense determines the maximal *a priori* information the physicist has about the actual properties of individual single samples of the system prepared according to μ. By the very definition of the pre-order on $\mathfrak{Q}$, it is clear that $\mathrm{S}_1(\mu)$ is always a filter in $\mathfrak{Q}$. It is a basic assumption of a realistic philosophical attitude, at least in combination with a careful notion of physical system, that these specifications should be caught by certain ideal elements — the pure states — that are mathematically realised by *principal ultrafilters* in $\mathfrak{Q}$. This *desideratum* reflects the requirement that the actualisation of some potential properties for a given system is impossible without the concomitant disappearance of some currently actual properties in the realm of potentiality. In other words, pure states correspond exactly with principal ultrafilters in $\mathfrak{Q}$, while for each mixed state the corresponding filter should always be contained in at least one principal ultrafilter. Metaphorically, in the words of one of the authors: The web of potentialities captures the particularity of a given physical system, whereas the filter of actualities captures the singularity of a given realisation. An immediate consequence of this requirement is that we can then reformulate all the previous definitions only in terms of pure states:

$$(9) \quad \alpha \preceq \beta \quad \text{iff} \quad \forall\, \mu \in \partial_e\mathfrak{M} : \mu \vDash \alpha \text{ implies } \mu \vDash \beta$$

and this prescription then corresponds with the fundamental definition of the Geneva approach.

As a consequence, we automatically obtain some sort of operational duality at the level of the representing objects, between the set of (pure) states and the class of definite experimental projects associated with a given physical system. This duality can be expressed by the mapping

$$(10) \quad \kappa_Q : \mathfrak{Q} \longrightarrow \mathcal{P}(\partial_e\mathfrak{M}) : \alpha \mapsto \left\{\mu \in \partial_e\mathfrak{M} \mid \mu \vDash \alpha\right\}$$

By the definition (4) of equivalence in $\mathfrak{Q}$, we immediately infer that there exists a unique injection

$$(11) \quad \kappa_L : \mathcal{L} \longrightarrow \mathcal{P}(\partial_e\mathfrak{M})$$

as expressed in the diagram

$$(12)\quad \begin{array}{ccc} \mathfrak{Q} & \xrightarrow{\quad q \quad} & \mathcal{L} \\ & {\scriptstyle \kappa_Q} \searrow & \vdots\, {\scriptstyle \kappa_L} \\ & & \mathcal{P}(\partial_e \mathfrak{M}) \end{array}$$

where q is the canonical quotient. We will sometimes use the derived notation $\lhd \subseteq \mathfrak{M} \times \mathcal{L}$, where $\mu \lhd a$ iff $\mu \models \alpha$ for any (hence all) $\alpha \in a$. Observe that κ_Q and κ_L are order-preserving in the appropriate sense. In addition, our specifications indicate that $\partial_e \mathfrak{M}$ should be strongly order-determining, in the sense that two non-equivalent tests can always be distinguished by some (pure) state μ. The function κ_L is sometimes called the *Cartan map*, because it was Elie Cartan who introduced the notion of a state space, as opposed to the phase space, in mathematical physics in the early 1920's (Cartan, 1971).

Summarising, in the Geneva approach $\mathfrak{Q}$ is taken to be complete as a thin category, and the primitive set of (pure) states are explicitly non-statistical, referring to a singular realisation of a physical system and not to an ensemble; that is, pure states are construed as attributes of individual samples of physical systems; general mixed states then being a derived concept. In addition, there exist two mappings

$$(13)\quad \mathcal{E}_Q : \mathfrak{M} \longrightarrow \mathcal{P}(\mathfrak{Q}) : \mu \mapsto S_1(\mu) = \left\{ \alpha \in \mathfrak{Q} \mid \mu \models \alpha \right\}$$

and its associated

$$(14)\quad \mathcal{E}_L = q^{\rightarrow} \circ \mathcal{E}_Q : \mathfrak{M} \longrightarrow \mathcal{P}(\mathcal{L}) : \mu \mapsto \left\{ a \in \mathcal{L} \mid \mu \lhd a \right\}$$

where $q^{\rightarrow}$ denotes the direct-image function induced by q, and the range of these mappings is included in the class of principal filters on $\mathfrak{Q}$ and $\mathcal{L}$ respectively, due to the completeness of both structures. Now any state, in particular a pure state, that is assigned to a physical system completely determines its associated filter of certain tests or actual properties, but for physical reasons a partial converse should also be true: Knowing the collection of all certain tests or actual properties for a singular realisation of a physical system should determine its *pure* state uniquely, and so it is natural to require that the restrictions to pure states of both mappings are one-to-one. Again, during the evolution of a physical system some properties become actual and some potential; it is an old and venerable principle that a given system does not obtain new actual properties *ex nihilo*. Mathematically, this requirement translates in the condition that the (necessarily non-trivial) filter associated with a *pure* state should be *maximal*, and so corresponds with an atom in $\mathfrak{Q}$ and $\mathcal{L}$, respectively. Next, we also have to require for the class of physical systems under consideration that the collection of *pure* states is strongly order-determining on $\mathfrak{Q}$, in the sense that we can always find a *pure* state $\mu \in \partial_e \mathfrak{M}$ such that $\mu \models \alpha$ and $\mu \not\models \beta$ whenever $\alpha \not\prec \beta$. Collecting all these requirements in a single statement, we obtain

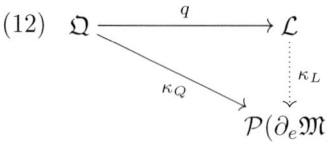

THEOREM 1. *Suppose that the restriction $\mathcal{E}_Q|_{\partial_e\mathfrak{M}}$ is one-to-one and its range consists of maximal ultrafilters in $\mathfrak{Q}$. In addition, let $\partial_e\mathfrak{M}$ be strongly order-determining on $\mathfrak{Q}$. Then $\mathfrak{Q}$ and $\mathcal{L}$ are atomistic, atoms corresponding with pure states.*

Proof. The completeness of $\mathfrak{Q}$ and the interpretation of the product test force all filters in the range of $\mathcal{E}_Q$ and $\mathcal{E}_L$ to be principal, and the maximality condition implies that each filter associated with some $\mu \in \partial_e\mathfrak{M}$ is represented by an atom p_μ in $\mathfrak{Q}$. If α and β are non-trivial tests and $\alpha \not\prec \beta$, we can find a pure state $\mu \vDash \alpha$, $\mu \not\vDash \beta$. In other words, $\partial_e\mathfrak{M}$ corresponds with the order-generating set $\mathrm{A}(\mathfrak{Q})$ of all atoms in $\mathfrak{Q}$, and this property is equivalent with atomisticity. ■

In conventional quantum theory, the operational duality between states and properties is manifested by the one-to-one correspondence between the complete lattices of projection operators on the one hand and closed subspaces of the complex Hilbert space on the other.

For many physical systems, there exist non-trivial definite experimental projects that are always either certain or false. In other words, for each $\mu \in \partial_e\mathfrak{M}$, we have either $\mu \vDash \alpha$ or $\mu \vDash \alpha^\sim$. Such tests are called *classical tests*, and the properties that they determine *classical properties.* Minimal non-trivial classical properties are also called *macroscopic states*, since they correspond with the atoms in the Boolean sublattice of all classical properties (Piron, 1998); we come back to this situation in the next section. In the same vein, two pure states μ_1 and μ_2, represented by atoms p_{μ_1} and p_{μ_2} are said to be separated by a *superselection rule*, in a usage derived from that of Wick, Wightman & Wigner (1952), if there exist no superposition states relative to μ_1 and μ_2, in the sense that the only atoms under $p_{\mu_1} \vee p_{\mu_2}$ are p_{μ_1} and p_{μ_2}.

Readers who are interested in a more formal presentation of the Geneva approach along logical lines may consult the papers by Cattaneo *et al.* (1988) and Cattaneo & Nisticó (1991). In this work, the global (relative) coherence of the formalism — in the usual model-theoretic sense — is explicitly shown by presenting a concrete Hilbert space model, where $\mathfrak{Q}$ consists of the effect algebra associated with this space, properties correspond to the projections, (pure) preparations are given by non-zero vectors, and pure states become equivalent with one-dimensional subspaces; two effects E_1 and E_2 being equivalent whenever $\ker(1-E_1) = \ker(1-E_2)$. Recall that a Hilbert space effect is a non-negative operator E satisfying $0 \leq E \leq 1$. In this model, the inverse effect is given by $E^\sim := 1 - E$, and the product of a collection $\{ E_j \mid j \in \mathcal{J} \}$ of effects is defined as

$$(15) \quad \prod\Big\{E_j \mid j \in \mathcal{J}\Big\} := \frac{1}{2}\Big(P_{M_1(\mathcal{J})} + P_{M_0(\mathcal{J})^\perp}\Big)$$

where $M_1(\mathcal{J}) = \bigcap_{j\in\mathcal{J}} \ker(1 - E_j)$ and $M_0(\mathcal{J}) = \bigcap_{j\in\mathcal{J}} \ker(E_j)$. In particular, the product of the sharp effect given by a projection P and the maximal effect 1 is explicitly given by the effect

$$(16)\quad \prod\{P,1\} = \frac{1}{2}\left(1+P\right) \quad ; \quad \left(\prod\{P,1\}\right)^{\sim} = \frac{1}{2}\left(1-P\right)$$

Since $\ker \frac{1}{2}\left(1-P\right) = \ker\left(1-P\right)$, we see that indeed $\prod\{P,1\} \sim P$. On the other hand, $\ker \frac{1}{2}\left(I1+P\right) = \{0\}$, and so indeed $\left(\prod\{P,1\}\right)^{\sim} \sim 0$.

Another important operational ingredient, the relation of *orthogonality*, has its most natural formulation on the set of all states. In the Geneva approach, orthogonality of states is usually defined at the level of $\partial_e\mathfrak{M}$, using the structural properties of $\mathfrak{Q}$ (Aerts, 1982):

$$(17)\quad \mu_1 \perp \mu_2 \quad \text{iff} \quad \exists\, \alpha \in \mathfrak{Q} : \mu_1 \in \kappa_Q(\alpha),\ \mu_2 \in \kappa_Q(\alpha^{\sim})$$

which turns the pair $(\partial_e\mathfrak{M}, \perp)$ into an *orthogonality space*; specifically, $\perp$ defines an anti-reflexive and symmetric relation on $\partial_e\mathfrak{M}$ that under appropriate conditions also separates points in the following sense: If $\mu_1 \neq \mu_2$, there exists $\nu \in \partial_e\mathfrak{M}$ such that $\nu \perp \mu_1$, $\nu \not\perp \mu_2$ (Moore, 1995). The first two properties are trivial, the third follows from the fact that $\partial_e\mathfrak{M}$ is strongly order-determining on $\mathfrak{Q}$, and so also on $\mathcal{L}$. Physically, two orthogonal states can be separated by the execution of a *single* measurement. A standard argument then constructs the complete atomistic ortholattice of all biorthogonally closed subsets $\mathcal{F}(\partial_e\mathfrak{M}, \perp)$ from this space. Of course, there exists also a converse construction, which yields an orthogonality space from a complete atomistic ortholattice $\mathcal{L}$. Explicitly, the space is given by the set of all atoms $A(\mathcal{L})$ of the lattice, and the relation of orthogonality becomes $p \perp q$ iff $p \leq q^{\perp}$. Actually, this correspondence between mathematical objects can be extended into a dual equivalence of the categories of complete atomistic lattices and join-preserving mappings that map atoms to atoms or 0 on the one hand, and T_1-closure spaces and continuous partially defined functions on the other. For more details, we refer to Moore (1995) and Faure & Frölicher (2000); the intrepid reader should find it fairly easy to fill in the details. We will come back to this point in the next section, after discussing appropriate classes of structure-preserving mappings. Observe also that if $a = [\alpha]$ and $b = [\beta]$ are disjoint classical properties, in the sense that $a \wedge b = 0$, and $\mu_1 \vDash \alpha$, $\mu_2 \vDash \beta$, then $\mu_1 \perp \mu_2$. Indeed, in this case necessarily $\mu_2 \vDash \alpha^{\sim}$, from which the assertion follows.

At this point of structural detail, there is no need for the complete atomistic ortholattice of biorthogonally closed subsets of $\partial_e\mathfrak{M}$ and the complete atomistic property lattice $\mathcal{L}$ to be related to each other. Additional axioms make this correspondence more precise. First, Piron (1976) postulates the existence of a so-called *compatible* complement $a^{\perp}$ for each property a (his axiom C), $a^{\perp}$ being a complement in the lattice-theoretic sense ($a \wedge a^{\perp} = 0$, $a \vee a^{\perp} = 1$). This is the only axiom in the traditional development of the Geneva approach in which the structural properties of $\mathfrak{Q}$ and $\mathcal{L}$ are required to interlock at a deeper level than the equivalence relation; the point is that one requires the existence of a privileged test $\alpha \in a$ such that $\alpha^{\sim} \in a^{\perp}$. Second, Piron (1976) formulates his axiom P, which postulates that the sublattice generated by $\{a, a^{\perp}, b, b^{\perp}\}$ should be distributive whenever $a \leq b$. If both conditions are satisfied, compatible complements are unique and $\mathcal{L}$ acquires the structure of a complete atomistic orthocomplemented lattice.[1] Operationally speaking, the existence of an orthocomplementation on $\mathcal{L}$ is not completely trivial. For example, Aerts (1984) gives an example of a combinatorial structure obtained from two complete atomistic ortholattices where the orthocomplementation fails. A more elaborate discussion of the interplay of and the possible confusions between the mappings $\alpha \mapsto \alpha^{\sim}$ and $a \mapsto a^{\perp}$ is given in the paper by Cattaneo and Nisticó (1991).

At the level of $\mathfrak{Q}$, the orthocomplementation seems to be related to some sort of symmetry between the certainly-true and certainly-false domains of a test and its compatible complement. In fact, recall that $T(\mu, \alpha^{\sim})$ iff $F(\mu, \alpha)$. We have remarked previously that the equivalence of two tests does not imply the equivalence of their inverses, due to the asymmetry in the definition of the pre-order relation $\preceq$ with respect to the certainly-true and certainly-false domains. We may, however, require the existence of some sort of *sharp* test $\Box\alpha$ for each property $[\alpha]$ that maximises the certainly-false domain, in the sense of

(18) $\beta \sim \Box\alpha \quad \text{only if} \quad F(mu, \beta) \Rightarrow F(p, \Box\alpha)$

for all $\mu \in \partial_e\mathfrak{M}$. These considerations lead us to formulate an axiom that is similar in spirit to the axiom CC proposed by Cattaneo and Nisticó (1991):

AXIOM 2 C3. There exists a mapping $\mathfrak{Q} \longrightarrow \mathfrak{Q} : \alpha \mapsto \Box\alpha$ that satisfies the conditions

(1) $\alpha \sim \Box\alpha$

(2) $\Box\alpha \preceq \beta$ only if $\forall\, \mu \in \partial_e\mathfrak{M} : F(\mu, \beta)$ implies $F(\mu, \Box\alpha)$

(3) $\Box\alpha \sim (\Box(\Box\alpha)^{\sim})^{\sim}$

In this case, two states μ and ν will be orthogonal iff $T(\mu, \Box\alpha)$ and $T(\nu, (\Box\alpha)^{\sim})$ for some $\alpha \in \mathfrak{Q}$:

(19) $\mu \perp \nu \quad \text{iff} \quad T(\mu, \Box\alpha),\ T(\nu, (\Box\alpha)^{\sim})$ for some $\alpha \in \mathfrak{Q}$

Indeed, if $T(\mu, \alpha)$ and $T(\nu, \alpha^{\sim})$, then obviously $T(\mu, \Box\alpha)$. On the other hand, $F(\nu, \alpha)$ implies $F(\nu, \Box\alpha)$, and so $T(\nu, (\Box\alpha)^{\sim})$.

THEOREM 3. *If $\mathfrak{Q}$ satisfies Axiom C3, the mapping*

(20) $q^{\sim} : \mathfrak{Q} \longrightarrow \mathcal{L} : \alpha \mapsto [(\Box\alpha)^{\sim}]$

is well-defined, and there exists a unique mapping $\mathcal{L} \longrightarrow \mathcal{L} : a \mapsto a^{\perp}$ such that the following diagram commutes:

(21)
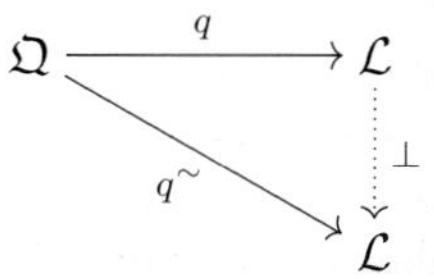

[1]An orthocomplementation on a lattice is an order-reversing involution $a \mapsto a'$ such that $a \wedge a' = 0$ and $a \vee a' = 1$.

This mapping defines an orthocomplementation on $\mathcal{L}$. In addition, two states μ and ν are orthogonal iff $\mu \in \kappa_L(a)$ and $\nu \in \kappa_L(a^\perp)$ for some $a \in \mathcal{L}$.

Proof. If $\square\alpha \preceq \square\beta$, then $T(\mu, (\square\beta)^\sim)$ iff $F(\mu, \square\beta)$ only if $F(\mu, \square\alpha)$ iff $T(\mu, (\square\alpha)^\sim)$, and so $(\square\beta)^\sim \preceq (\square\alpha)^\sim$. It follows also that

(22) $\square\alpha \sim \square\beta \quad \text{iff} \quad (\square\beta)^\sim \sim (\square\alpha)^\sim$

Consequently, the mapping $q^\sim$ is well-defined, and correspondingly induces a unique mapping

(23) $\perp : \mathcal{L} \longrightarrow \mathcal{L} : [\alpha] \mapsto [\alpha]^\perp := [(\square\alpha)^\sim]$

such that $\perp \circ q = q^\sim$. It is clear that $a^{\perp\perp} = a$ and that $a \le b$ implies $b^\perp \le a^\perp$. Indeed, if $\alpha \in a$ and $\beta \in b$, there exist $\square\alpha \in a$ and $\square\beta \in b$, and we can recycle the argument above at the start of the proof. Similarly, $a \wedge a^\perp \ni \square\alpha \cdot (\square\alpha)^\sim$, and so $a \wedge a^\perp = 0$. Next, since $\square 0 \sim 0$, we always have $F(\mu, \square 0)$, and so $T(\mu, (\square 0)^\sim)$, hence $(\square 0)^\sim \sim 1$, hence $0^\perp = 1$. Finally, if both $a \le c$ and $a^\perp \le c$, we obtain $c^\perp \le a \le c$, and so $c^\perp = c \wedge c^\perp = 0$, from which $c = 1$. The last result follows from the argument immediately before the formulation of the proposition. ■

Two elements a and b in an orthocomplemented lattice are said to be *orthogonal* if $a \le b^\perp$; one denotes this situation also as $a \perp b$.

If a is a classical property determined by a classical test α_c, we expect that α_c is sharp. This is indeed the case, because if $\alpha_c \preceq \beta$ then $\neg F(\mu, \alpha_c)$ iff $T(\mu, \alpha_c)$ only if $T(\mu, \beta)$ only if $\neg F(\mu, \beta)$. Consequently, $a^\perp = [\alpha_c^\sim]$ is also a classical property.

If $\mathfrak{Q}$ is represented by the effect algebra associated with a complex Hilbert space, this axiom is satisfied. Specifically, the mapping $F \mapsto P_{\ker(1-F)}$ — where P_M denotes the projection operator on the subspace M — clearly satisfies the first requirement. As for the second condition, suppose that P is a projection with $P \le F$; that is, $\ker(1-P) \subseteq \ker(1-F)$. We have to show that $\ker(F) \subseteq \ker(P)$; if $\phi \in \ker(F)$ and $\psi \in \ker(P)^\perp = \ker(1-P)$, we have

$$\langle \phi, \psi \rangle = \langle \phi, P\psi \rangle = \langle \phi, F\psi \rangle = \langle F\phi, \psi \rangle = 0$$

from which our assertion follows. The third condition is easily verified.

Additional mathematical structure can then be progressively introduced at the level of the sets of pure states $\partial_e\mathfrak{M}$, the class of tests $\mathfrak{Q}$, or the property lattice $\mathcal{L}$. Unfortunately, or fortunately, there is no shortage of physically motivated axioms, but they can apply to different levels, and this often leads to considerable confusion. So far, we have introduced some general axioms indicated by physical reflection on the nature of orthogonality and products. Now we have to look for more specific axioms for which the justification based on physical or operational demonstration is only partial. In fact, we may have to restrict our attention to a particular subclass of physical systems — the class to which the additional requirements apply — and try to strike a balance between the competing *desiderata* of cognitive accessibility and methodological power. In such cases, we should at the very least aim for a justification of the additional structure in retrospect, with the benefit of hindsight.

In algebraic quantum logic, orthomodular structures play a predominant role. From an operational perspective, it is the opinion of the authors that the physical meaning of the orthomodular law in the property lattice $\mathcal{L}$, in particular its relation to the class of definite experimental projects $\mathfrak{Q}$, needs additional clarification. The *orthomodular property* states that

(24) $a \leq b \quad \text{implies} \quad b = a \vee (b \wedge a^{\perp})$

This statement is valid for the lattice of closed subspaces of a complex Hilbert space, and has important physical consequences. From a more pragmatic perspective, it is fair to say that it is more convenient to do calculations in orthomodular structures. On the other hand, the orthomodular law for example fails for the complete atomistic orthocomplemented lattice of all subspaces of a pre-Hilbert space that is not complete in the topological sense.

Here is a selection of some equivalent characterisations of the orthomodular property:

PROPOSITION 4. *Let $\mathcal{L}$ be an orthocomplemented poset. The following two properties are equivalent:*

(i) $\mathcal{L}$ is orthomodular.

(ii) For all $a, b \in \mathcal{L}$, if $b \leq a^{\perp}$, $a \wedge b = 0$ and $a \vee b = 1$, then $b = a^{\perp}$.

Proof. First, suppose that $\mathcal{L}$ is orthomodular, and let a, b satisfy the assumptions in (ii). Orthomodularity yields

$$a^{\perp} = b \vee (a^{\perp} \wedge b^{\perp}) = b \vee (a \vee b)^{\perp} = b$$

Conversely, let $a \leq b$, hence also $a \vee (a^{\perp} \wedge b) \leq b$. We then have

$$\begin{aligned} (a \vee (a^{\perp} \wedge b)) \vee b^{\perp} &= (a^{\perp} \wedge b)^{\perp} \vee (a^{\perp} \wedge b) = 1 \\ (a \vee (a^{\perp} \wedge b)) \wedge b^{\perp} &\leq b \wedge b^{\perp} = 0 \end{aligned}$$

and we infer that $b = (a \vee (a^{\perp} \wedge b))$. ■

In other words, orthogonal complements are always unique in an orthomodular lattice.

For future reference, we also state the following auxilliary result.

LEMMA 5. *If $\mathcal{L}$ is an orthomodular lattice, $a \leq b$ and $a \leq c^{\perp}$, then*

$$a \vee (b \wedge c) = b \wedge (c \vee a)$$

Proof. Two successive applications of the orthomodular property yield

$$\begin{aligned} b \wedge (c \vee a) &= a \vee (b \wedge (c \vee a) \wedge a^{\perp}) \\ &= a \vee (b \wedge c) \end{aligned}$$

■

As we will see later in this section, this result can be succinctly expressed under the more condensed form

(25) $c \perp a \quad \Rightarrow \quad (c, a)\, M^*$

the expression at the right meaning that (c, a) forms a dual modular pair in the sense of Maeda & Maeda (1970).

There is another physically fairly transparent property which automatically leads to the orthomodular property. If $\mathfrak{M}$ is a collection of probability measures on an orthocomplemented lattice $\mathcal{L}$, then it is said to be *separating* if $\mathfrak{M}$ can distinguish the elements of the lattice, in the sense that $\mu(a) = \mu(b)$ for all $\mu \in \mathfrak{M}$ only if $a = b$. Recall that a probability measure on an orthocomplemented lattice is a mapping that should behave as a classical probability measure when restricted to Boolean subalgebras of $\mathcal{L}$. To be precise, we will use the next

DEFINITION 6. A probability measure on a σ-complete orthocomplemented lattice $\mathcal{L}$ is a map $\mu : \mathcal{L} \longrightarrow [\,0,1\,]$ that satisfies the following requirements:

(1) $\mu(0) = 0 \quad ; \quad \mu(1) = 1$

(2) For any pairwise orthogonal countable family $\left\{a_n \mid n \in \mathbb{N}\right\}$ in $\mathcal{L}$:

$$\mu(\bigvee\left\{a_n \mid n \in \mathbb{N}\right\}) = \sum_{n \in \mathbb{N}} \mu(a_n)$$

(3) If $a, b \in \mathcal{L}$ satisfy $\mu(a) = \mu(b) = 1$, then also $\mu(a \wedge b) = 1$

The last property is the so-called *Jauch-Piron property*. We add it explicitly to the list of requirements in the definition, in view of the interpretation of the infimum in property lattices. The definition can be relaxed slightly by requiring that (2) only holds for finite families. In this case, the definition also makes sense when $\mathcal{L}$ is an orthocomplemented lattice, and we obtain the next

PROPOSITION 7. *If an orthocomplemented lattice $\mathcal{L}$ admits a separating set $\mathfrak{M}$ of probability measures, then $\mathcal{L}$ is orthomodular.*

Proof. Again, suppose that $a \leq b$, and let $\mu \in \mathfrak{M}$. We compute

$$\begin{aligned} \mu(a \vee (b \wedge a^{\perp})) &= \mu(a) + \mu(b \wedge a^{\perp}) \\ &= \mu(a) + \mu(b^{\perp} \vee a)^{\perp}) \\ &= \mu(a) + 1 - \mu(b^{\perp} \vee a) \\ &= \mu(a) + 1 - \mu(b^{\perp}) - \mu(a) \\ &= \mu(b) \end{aligned}$$

Since μ was arbitrary, we deduce that $a \vee (b \wedge a^{\perp}) = b$. ■

According to the following physically interesting result, only orthogonal atoms in a complete atomistic orthomodular lattice can be separated by a superselection rule (Pulmannová, 1985).

PROPOSITION 8. *If $\mathcal{L}$ is a complete atomistic orthomodular lattice and p and q are two distinct atoms separated by a superselection rule, then $p \leq q^{\perp}$.*

Proof. First, observe that

$$\begin{aligned} (p \vee q) \wedge q^{\perp} &= \bigvee \left\{ r \in \mathrm{A}(\mathcal{L}) \mid r \leq (p \vee q) \wedge q^{\perp} \right\} \\ &= p \wedge q^{\perp} \end{aligned}$$

Suppose that $p \wedge q^{\perp} = 0$, then orthomodularity yields

$$p \vee q = q \vee ((p \vee q) \wedge q^{\perp}) = q \vee (p \wedge q^{\perp}) = q$$

and this is impossible. Consequently, $p \wedge q^{\perp} = p$. ■

Last but not least, there is the *covering property*, which states that for each atom $p \in \mathrm{A}(\mathcal{L})$ and $a \in \mathcal{L}$ we should have

(26) $p \not\leq a$ implies $a \lessdot a \vee p$

where $b \lessdot c$ means that $b \leq x \leq c$ implies $x = b$ or $x = c$; one says that c covers b. Also here, the immediate operational significance of this requirement is unclear, but it is satisfied by the collection of closed subspaces of a complex Hilbert space. As will become apparent later, the covering property, in combination with orthomodularity, is important from a more dynamic perspective, in view of the following well-known results. Here, the *exchange property* states that, given atoms $p, q \in \mathcal{L}$ and an $x \in \mathcal{L}$, the two conditions $p \not\leq x$ and $p \leq q \vee x$ imply that $q \leq p \vee x$, and so p and q can be interchanged.

PROPOSITION 9. *Let $\mathcal{L}$ be an atomistic orthomodular lattice. The following properties are equivalent:*

(1) $\mathcal{L}$ has the covering property.

(2) $\mathcal{L}$ has the exchange property.

(3) For all $p \in A(\mathcal{L})$ and $a \in \mathcal{L}$, $(p \vee a^{\perp}) \wedge a \in A(\mathcal{L})$ whenever $p \not\leq a^{\perp}$.

(4) For all $p \in A(\mathcal{L})$ and $a, b \in \mathcal{L}$ such that $a \perp b$, if $p \leq a \vee b$, $p \not\leq a$, $p \not\leq b$, there exist unique $p_1, p_2 \in A(\mathcal{L})$ such that $p_1 \leq a$, $p_2 \leq b$, and $p \leq p_1 \vee p_2$.

Proof. (1) implies (2): Take $p, q \in \mathrm{A}(\mathcal{L})$, with $p \not\leq a$ and $p \leq a \vee q$. On the one hand, we have $a < a \vee p \leq a \vee q$; on the other hand, since also $q \not\leq a$, the covering property yields $a \lessdot a \vee q$. Consequently, $a \vee p = a \vee q$, hence $q \leq a \vee p$.
(2) implies (3): Let $p \in \mathrm{A}(\mathcal{L})$, $p \not\leq a^{\perp}$. Since

$$a^{\perp} \vee ((p \vee a^{\perp}) \wedge a) \quad = \quad p \vee a^{\perp} \quad > \quad a^{\perp}$$

it follows that $(p \vee a^{\perp}) \wedge a > 0$. By the atomistic property, there is an atom r under $(p \vee a^{\perp}) \wedge a$. Consequently,

$$r \quad \leq \quad a^{\perp} \vee ((p \vee a^{\perp}) \wedge a) \quad = \quad a^{\perp} \vee p$$

and the exchange property implies that $r \vee a^{\perp} = p \vee a^{\perp}$, and so

$$(p \vee a^{\perp}) \wedge a \quad = \quad (r \vee a^{\perp}) \wedge a \quad = \quad r \quad \in \quad \mathrm{A}(\mathcal{L})$$

where we have used orthomodularity twice.
(3) implies (4): By our assumptions, both $p_1 := (p \vee a^{\perp}) \wedge a$ and $p_2 := (p \vee b^{\perp}) \wedge b$ are atoms, since $p \not\leq a^{\perp}, b^{\perp}$. Indeed, $p \leq a^{\perp}$ would imply $p \leq a^{\perp} \wedge (a \vee b) = b$ by orthomodularity, and similarly $p \not\leq b^{\perp}$. Next, observe that

$$\begin{aligned} p \quad &\leq \quad (p \vee a) \wedge (p \vee a^{\perp}) \\ &\leq \quad (p \vee a) \wedge (((p \vee a^{\perp}) \wedge a) \vee a^{\perp}) \\ &\leq \quad ((p \vee a^{\perp}) \wedge a) \vee ((p \vee a) \wedge a^{\perp}) \end{aligned}$$

where we have used lemma 5 in the last step. Next, observe that

$$(p \vee a) \wedge a^{\perp} \quad \leq \quad (a \vee b) \wedge a^{\perp} \quad = \quad b$$

and so

$$(p \vee a) \wedge a^{\perp} \quad \leq \quad (p \vee b^{\perp}) \wedge b$$

Since $p \not\leq a$ and $p \not\leq b^{\perp}$, both expressions are atoms in $\mathcal{L}$, hence equality holds. This proves existence. As for uniqueness, suppose that $p \leq p_1 \vee p_2$, with $p_1 \leq a$ and $p_2 \leq b$; then

$$\begin{aligned} (p \vee a^{\perp}) \wedge a \quad &\leq \quad (p_1 \vee p_2 \vee a^{\perp}) \wedge a \\ &\leq \quad (p_1 \vee a^{\perp}) \wedge a \quad = \quad p_1 \end{aligned}$$

and so $p_1 = (p \vee a^{\perp}) \wedge a$. Similarly, $p_2 = (p \vee b^{\perp}) \wedge b$.
(4) implies (3): Suppose that $p \not\leq a^{\perp}$, hence $a^{\perp} \vee p > a^{\perp}$. By the orthomodular identity,

$$p \vee a^{\perp} = ((p \vee a^{\perp}) \wedge a) \vee a^{\perp}$$

and so $(p \vee a^{\perp}) \wedge a \neq 0$. If $p \perp a^{\perp}$, then $p \leq a$, and so $(p \vee a^{\perp}) \wedge a = p$. If not, then we obtain two atoms r_1, r_2 with $r_1 \leq a$, $r_2 \leq a^{\perp}$ such that $p \leq r_1 \vee r_2$. We then have

$$\begin{aligned} (p \vee a^{\perp}) \wedge a \quad &\leq \quad (r_1 \vee r_2 \vee a^{\perp}) \wedge a \\ &\leq \quad (r_1 \vee a^{\perp}) \wedge a \quad = \quad r_1 \end{aligned}$$

and this shows that $(p \vee a^{\perp}) \wedge a \in \mathrm{A}(\mathcal{L})$.
(3) implies (1): Suppose that p is an atom, with $p \not\leq a$, and let $a \leq x \leq a \vee p$. We then have also $0 \leq x \wedge a^{\perp} \leq (a \vee p) \wedge a^{\perp}$, which is an atom in $\mathcal{L}$. Either $x \wedge a^{\perp} = 0$, and so $x = a \vee (x \wedge a^{\perp}) = a$, invoking orthomodularity; or $x \wedge a^{\perp} = (a \vee p) \wedge a^{\perp}$, and then

$$x \quad = \quad a \vee (x \wedge a^{\perp}) \quad = \quad a \vee ((a \vee p) \wedge a^{\perp}) \quad = \quad a \vee p$$

The proof is complete. ■

According to this result, in an orthomodular lattice with the covering property the orthogonal projections of a given atom on a pair of orthogonal subspaces are necessarily unique, a property with a nice physical interpretation. The mappings

(27) $\phi_a : \mathcal{L} \longrightarrow \mathcal{L} : x \mapsto (x \vee a^{\perp}) \wedge a$

are usually called the Sasaki projections, and are important in the theory of measurement.

Here are two physically useful examples of complete atomistic orthomodular lattices that satisfy the covering property, at extreme ends of the spectrum of possibilities. The first one consists of the complete lattice of all subsets of a set Σ; verification of the axioms is trivial. The second model corresponds with the usual "quantum logic" of conventional quantum physics, given by the collection of all closed linear subspaces of a complex separable Hilbert space; in this case, establishing the properties required by the axioms is definitely less trivial. This example also leads to another perspective on the physical meaning of the infimum of a collection of not necessarily compatible properties, in analogy with von Neumann's alternating projections theorem, which states that the infimum of a pair of projection operators E and F in the projection lattice determined by a complex Hilbert space is given by the following expression which involves a strong limit of a sequence of finite alternating products of the operators (von Neumann, 1950):

(28) $E \wedge F \quad = \quad \operatorname*{s\text{-}lim}_{n \to +\infty} (E\,F)^n$

Specifically, Jauch (1968) interprets the property corresponding to this expression to be actual for a particular quantum system when it passes with certainty through the filter that consists in an alternating sequence of any chosen length of passive elementary filters associated with the individual projections.

By far, the most important result in the Geneva school approach — according to some more critical physicists even the only significant result of the whole quantum logic enterprise — is the celebrated representation theorem proved by Constantin Piron in the early 1960's. For a more detailed analysis, we refer to the original source (Piron, 1964, 1976), to Valckenborgh (2000), and, in the first volume of this handbook, to Stubbe & Van Steirteghem (2007). In a nutshell, given a complete atomistic orthomodular lattice with the covering property $\mathcal{L}$ and with chains of length at least 4, this result is established by the construction of a family of orthomodular spaces $\mathcal{H}(\mathcal{L})_{\omega}$, indexed by the minimal non-trivial classical properties

$\omega \in \Omega$, such that $\mathcal{L}$ is ortho-isomorphic with the product lattice, each component consisting of the set of closed subspaces of $\mathcal{H}(\mathcal{L})_\omega$.[23] It is important to realise that one allows the existence of *continuous* superselection rules in the formalism, and so it is a proper generalisation of conventional quantum theory. From the perspective of the system, it is immaterial whether such apparent constraints are induced by the environment or arise for intrinsic reasons. Operationally, superselection rules correspond with the existence of classical tests in $\mathfrak{Q}$. Conceptually, contrary to the orthodox formalism, the superselection rules don't need to be introduced *a posteriori* in the usual rather *ad hoc* fashion, but live in the structural framework right from the start. For example, this type of superselection rule may play a role in the explanation of stable chiral states, so important in chemistry and molecular biology (Amann, 1988).

The first steps of establishing the representation of a property lattice in terms of an underlying vector space usually consist in the observation that the collection of atoms in the property lattice has a natural projective geometric structure. Recall that a *projective geometry* $\mathfrak{G}$ can be regarded either as a set G with a ternary "collinearity" relation ℓ, where $\ell(p,q,r)$ expresses the fact that the three points p,q,r lie on a line; or alternatively, as a set G endowed with an operation $\star : G \times G \longrightarrow \mathcal{P}(G)$, where $p \star q$ is the line incident with both p and q.[4] The subsets of G that contain all lines through its points are the projective subspaces, and one can show without too much difficulties that the collection $\mathscr{L}(\mathfrak{G})$ of all projective subspaces always has the structure of a *projective lattice*: a complete atomistic meet-continuous modular lattice. See Faure & Frölicher (2000) and Stubbe & Van Steirteghem (2007) for additional details.

The crucial property in establishing that a given lattice defines a projective geometry is the so-called *intersection property.* The intersection property holds in a lattice $\mathcal{L}$ if for each pair p,q of distinct atoms and for any $a \in \mathcal{L}$ such that $p \leq a \vee q$, there exists an atom r with $r \leq (p \vee q) \wedge a$.

THEOREM 10. *The atom space $A(\mathcal{L})$ of a lattice $\mathcal{L}$ with the intersection property, endowed with the ternary operation $\ell(p,q,r)$ iff $p \leq q \vee r$ or $q = r$ — or $p \star q = \left\{r \in A(\mathcal{L}) \mid r \leq p \vee q\right\}$ — forms a projective geometry $\mathfrak{G}(\mathcal{L})$. If $\mathcal{L}$ is also atomistic, the mapping*

$$(29)\quad \eta_L : \mathcal{L} \longrightarrow \mathscr{L}(\mathfrak{G}(\mathcal{L})) : a \mapsto \left\{p \in A(\mathcal{L}) \mid p \leq a\right\}$$

is meet-preserving and injective.

---

[2] Recall that the length of a finite chain $C \subseteq \mathcal{L}$ is given by $l(C) = \#C - 1$.

[3] An orthomodular space $(V, \mathbb{F})$, or generalised Hilbert space, is a (left) vector space over an involutive division ring $\mathbb{F}$ endowed with a positive definite Hermitian form.

[4] Specifically, the axioms for a projective geometry (G,ℓ) in terms of the ternary collinearity relation ℓ are: (1) $\ell(a,b,a)$ for all $a,b \in G$; (2) $\ell(a,p,q), \ell(b,p,q), p \neq q$ only if $\ell(a,b,p)$; (3) $\ell(p,a,b), \ell(p,c,d)$ implies $\ell(q,a,c), \ell(q,b,d)$ for some $q \in G$. Alternatively, the axioms for $(G,\star)$ in terms of the operator $\star : G \times G \longrightarrow \mathcal{P}(G)$ are: (1) $a \star a = \{a\}$; (2) $a \in b \star a$; (3) $a \in b \star p$ and $p \in c \star d$ and $a \neq c$ imply $(a \star c) \cap (b \star d) \neq \emptyset$, for all $a,b,c,d,p \in G$.

Proof. First, we verify that the relation $\ell \subseteq \mathrm{A}(\mathcal{L}) \times \mathrm{A}(\mathcal{L}) \times \mathrm{A}(\mathcal{L})$ satisfies the requirements for a projective geometry. Since $p \leq q \vee p$, it is always the case that $\ell(p,q,p)$. Second, suppose that $\ell(a,p,q)$ and $\ell(b,p,q)$ for $p \neq q$. If $p = b$, then trivially $\ell(a,b,p)$; if not, we have $b \leq p \vee q$, and the intersection property implies the existence of an atom $r \leq (b \vee p) \wedge q$, and so $r = q$, hence $q \leq (b \vee p)$, hence $p \vee q \leq b \vee p$, and we infer that also $a \leq b \vee p$, and so $\ell(a,b,p)$. Third, suppose that $\ell(p,a,b)$ and $\ell(p,c,d)$; we have to show that there exists an atom q for which both $\ell(q,a,c)$ and $\ell(q,b,d)$. Without loss of generality, we can assume that a,b,c,d are all distinct, and so either $p \neq a$ or $p \neq b$; let's suppose that $p \neq b$. The intersection property applied to $p \leq a \vee b$ implies that $a \leq p \vee b$, hence $a \leq b \vee c \vee d$, and since $a \neq c$, a second application of the intersection property yields an atom $q \leq (a \vee c) \wedge (b \vee d)$. Next, we show that the mapping η_L is well-defined and preserves all existing meets. In fact, $\eta_L(a)$ is always a projective subspace, since $p \leq a$ and $q \leq a$ implies $p \vee q \leq a$, and so $\eta_L(a)$ contains all lines incident with points in $\eta_L(a)$. If $\mathcal{L}$ is atomistic and $a \neq b$ and $a \neq 0$, there exists an atom $p \in \mathrm{A}(\mathcal{L})$ such that $p \leq a$, $p \not\leq b$, and so $\eta_L(a) \neq \eta_L(b)$. Finally, it follows from the definition that

$$(30)\quad \eta_L\Big(\bigwedge\Big\{a_i \mid i \in I\Big\}\Big) \quad = \quad \bigcap\Big\{\eta_L(a_i) \mid i \in I\Big\}$$

and so η_L preserves all existing infima. ■

It follows trivially that η_L is order-preserving. An atomistic lattice that has the covering property is sometimes called an AC-lattice (Maeda & Maeda, 1970).

THEOREM 11. *Each orthocomplemented AC-lattice has the intersection property.*

Proof. Take any $p,q \in \mathrm{A}(\mathcal{L})$ such that $p \neq q$, and let $a \in \mathcal{L}$, $p \leq a \vee q$. We have to show that there exists an atom $r \leq (p \vee q) \wedge a$. By the atomicity of $\mathcal{L}$, it is sufficient to show that $(p \vee q) \wedge a \neq 0$. If $p \leq a$ we are done; if not, because of the covering property, $a^\perp \wedge p^\perp \lessdot a^\perp$, and so there is an atom s such that

$$a^\perp = (a \vee p)^\perp \vee s = (a \vee p \vee q)^\perp \vee s = (a^\perp \wedge (p \vee q)^\perp) \vee s$$

where we have used the covering law again in the second step. Consequently, we also have

$$a^\perp \vee (p \vee q)^\perp = (p \vee q)^\perp \vee s$$

and so $(p \vee q)$ either covers or equals $a \wedge (p \vee q)$. Since $p \neq q$, the assertion follows. ■

In a projective geometry, the join of two projective subspaces satisfies the *projective law*: If M and N are subspaces of a projective geometry, then

$$(31)\quad M \vee N = \bigcup\Big\{p \star q \mid p \in M, q \in N\Big\}$$

This expression follows essentially from the simpler case

(32) $M \vee \{q\} = \bigcup\left\{q \star y \mid y \in M\right\}$

where q is a point of the projective geometry. If $\mathfrak{G}(\mathcal{L})$ is the projective geometry associated with a lattice that has the intersection property, the latter equality is easy to prove. In fact, we can do better. If $p \leq q \vee b$ and $q \not\leq b$, we can assume that $p \neq q$, in which case the intersection property yields an atom $r \leq (p \vee q) \wedge b$. Since $r \neq q$, a second application gives $p \leq q \vee r$, where $r \leq b$. The other inclusion is obvious, and so

$$(33)\quad \eta_L(q \vee b) \quad = \quad \bigcup\left\{\eta_L(q \vee r) \mid r \in \mathrm{A}(\mathcal{L}), r \leq b\right\} = \eta_L(q) \vee \eta_L(b)$$

since the union is clearly contained in $\eta_L(q) \vee \eta_L(b)$, and it obviously also the case that $\eta_L(q) \vee \eta_L(b) \subseteq \eta_L(q \vee b)$. The former expression can be proved in a similar fashion, by reiterating the same process several times and considering the various cases that arise. In this context, the projective law then states that

$$(34)\quad \eta_L(a) \vee \eta_L(b) \quad = \quad \bigcup\left\{\eta_L(q \vee r) \mid q, r \in \mathrm{A}(\mathcal{L}), q \leq a, r \leq b\right\}$$

In a general lattice a pair $a, b \in \mathcal{L}$ is said to form a *modular pair*, denoted by $(a, b)\, M$ iff for all $x \leq b$, $(x \vee a) \wedge b = x \vee (a \wedge b)$; a *dual modular pair*, denoted by $(a, b)\, M^*$ iff for all $y \geq b$, we have $(a \vee b) \wedge y = (a \wedge y) \vee b$ (Maeda & Maeda, 1970). In an orthocomplemented lattice, it is easy to verify that $(a, b)\, M$ iff $(a^\perp, b^\perp)\, M^*$. Lemma 5 then states that in an orthomodular lattice, orthogonal pairs always form dual modular pairs. The following related result can be found in Aerts & Piron (1979).

THEOREM 12. *Let $\mathcal{L}$ be an orthocomplemented AC-lattice. Then*

$$(35)\quad (a, b)M^* \quad \Leftrightarrow \quad \eta_L(a \vee b) = \eta_L(a) \vee \eta_L(b)$$

Proof. First, suppose that $\eta_L(a \vee b) = \eta_L(a) \vee \eta_L(b)$, and let $y \geq b$. Since $\mathcal{L}$ is atomistic, it is sufficient to show that each atom dominated by $(a \vee b) \wedge y$ is also under $(a \wedge y) \vee b$. Let q be such an atom; since $q \in \eta_L(a \vee b) = \eta_L(a) \vee \eta_L(b)$, the projective law implies that $q \leq p \vee r$ for some $p \in \eta_L(a)$ and $r \in \eta_L(b)$. If $q = r$, we are done; if not, since also $q \leq y$ and $r \leq y$, the intersection property implies that $p \leq q \vee r \leq y$, and so $p \leq a \wedge y$. Consequently, $q \leq (a \wedge y) \vee b$. Conversely, suppose that $\eta_L(a \vee b) > \eta_L(a) \vee \eta_L(b)$. In that case, there is an atom q such that $q \leq a \vee b$ but $q \notin \eta_L(a) \vee \eta_L(b)$. In particular, $q \not\leq b$. We show that $(b \vee q) \wedge a = b \wedge a$. In fact, if r is an atom for which $r \leq (b \vee q) \wedge a$, then $r \leq a$ and $r \leq \beta \vee q$ for some atom $\beta \leq b$. If $r = \beta$, then $r \leq a \wedge b$; if not, the intersection property implies that $q \leq r \vee \beta$, and so $q \in \eta_L(a) \vee \eta_L(b)$, which is impossible. Therefore, $r = \beta$ and so $r \leq a \wedge b$. It follows that

$$((b \vee q) \wedge a) \vee b = (b \wedge a) \vee b = b$$

$$(b \vee q) \wedge (a \vee b) = b \vee q$$

and so $(a, b) \notin M^*$. ∎

An easy consequence of this theorem is the fact that the relation M^* — and so also the relation M — is symmetric in an orthocomplemented AC-lattice. In other words, such lattices are always *semimodular*. Indeed, it is easy to see that in an orthocomplemented lattice, M is symmetric iff M^* is symmetric. The combination of this result with the projective law then states that

$$(36)\quad \eta_L(a \vee b) = \bigcup \left\{ \eta_L(q \vee r) \mid q, r \in \mathrm{A}(\mathcal{L}), q \leq a, r \leq b \right\}$$

in an orthocomplemented AC-lattice whenever $a \perp b$. This also corresponds with statement (4) in proposition 9.

PROPOSITION 13. *Let $\mathcal{L}$ be an atomic orthomodular lattice. Then $\mathcal{L}$ has the covering property iff it is semimodular.*

Proof. First, we assert that each atomic orthomodular lattice is actually atomistic. Indeed, suppose that $x \geq p$ for all $p \in \eta_L(a)$. To prove that $a = \bigvee \eta_L(a)$, we have to show that $x \geq a$. Suppose not, then $a \wedge x < a$, and so orthomodularity yields

$$a = (a \wedge x) \vee (a \wedge (a \wedge x)^{\perp})$$

hence $a \wedge (a \wedge x)^{\perp} \neq 0$. By the atomicity of $\mathcal{L}$, there exists an atom q under $a \wedge (a \wedge x)^{\perp} \leq (a \wedge x)^{\perp}$. Since $q \leq a$, we have $q \leq x$, and so $q \leq a \wedge x$, hence $q = 0$ which is impossible. Therefore, $a \wedge x = a$, and so $a \leq x$. This shows that our first assertion holds. Second, if the covering property holds it follows directly from this observation and the previous proposition that $\mathcal{L}$ is semimodular. Conversely, semimodularity of $\mathcal{L}$ implies the covering property. First, one verifies easily that $(a, p)M$ whenever $p \in \mathrm{A}(\mathcal{L})$ and $a \in \mathcal{L}$. Semimodularity then implies that $(p, a)M$ for all $a \in \mathcal{L}$. It then follows also that $(p, a)M^*$ for all $p \in \mathrm{A}(\mathcal{L})$ and $a \in \mathcal{L}$, and so the conditions $p \not\leq a$, $a \leq x \leq a \vee p$ and $a \vee (p \wedge x) = (a \vee p) \wedge x$ together imply that $x = a \vee (p \wedge x)$, and so $x = a$ or $x = a \vee p$, depending on whether $p \wedge x$ equals 0 or p respectively. ■

In the final paragraphs of this section, we briefly sketch some additional developments and potentially interesting digressions, without going into too much detail, because of lack of space. We start with the observation that for the next step in the transition from the abstract framework in terms of property lattices to the conventional models of classical and quantum theory, it is necessary to implement additional structural constraints on the framework. Indeed, the nature of the underlying involution division ring of the orthomodular space that is determined by a property lattice with the appropriate technical conditions is left unspecified by the standard axioms. It is however known that the projective geometries associated with two (left) vector spaces $(V, \mathbb{F})$ and $(W, \mathbb{G})$ of dimension at least three are isomorphic iff they have the same dimension as vector spaces and $\mathbb{F}$ and $\mathbb{G}$ are isomorphic division rings (Baer, 1952). The situation at this stage is not unlike the relation between classical probability theory on the one hand and the impressive fortress of classical mechanics in terms of symplectic spaces on the other, where important physical observables subject to additional symmetry constraints of a group-theoretic nature are fully integrated in the basic framework of measurable spaces and accordingly bring them to life.

We then note that also in the general case the properties of a physical system often have a more concrete interpretation as the properties defined by a convenient set of physical observables, such as the measurement of a position or momentum for instance. In this way, a whole collection of properties in the property lattice are related when they refer to the same observable, giving a certain additional coherence to the property lattice as a whole. We will use the following slightly idealised mathematical translation of this idea, in the spirit of the conventional probability calculus:

DEFINITION 14. A weak observable is a mapping F from the Borel sets $\mathfrak{B}(S)$ associated with a second countable locally compact Hausdorff space S to the property lattice $\mathcal{L}$ that satisfies the following properties:

(i) $F(\emptyset) = 0$; $F(S) = 1$;

(ii) If $A \cap B = \emptyset$, then $F(A) \perp F(B)$;

(iii) For any sequence $n \mapsto A_n$ of Borel sets in S, we have

$$F\Big(\bigcap_{n\in\mathbb{N}} A_n\Big) = \bigwedge_{n\in\mathbb{N}} F(A_n)$$

In case we also have

(iv) $F(A) \vee F(A^c) = 1$

then F is said to be an observable.

In the most common situation, the appropriate topological space is of course given by $\mathbb{R}$ with its natural topology, or a suitable subspace. Not every experimental physical procedure yields an observable in the strong sense. For example, Jauch & Piron (1967) argue that the property of (weak) localisation for elementary particles of zero mass corresponds with a weak observable, but not with the stronger notion of an observable. Notice that for an observable F each pure state in $\mathcal{L}$ can be written uniquely as a superposition of a pure state under $F(A)$ and a pure state under $F(A^c)$ under the appropriate conditions, according to proposition 9. The latter property does not need to hold in the more general case of a weak observable. The potential existence of observables in the strong sense for a physical system is particularly interesting, in view of the following result:

PROPOSITION 15. *If F is an observable and $\mathcal{L}$ is orthomodular, then also*

$$F\Big(\bigcup_{n\in\mathbb{N}} A_n\Big) = \bigvee_{n\in\mathbb{N}} F(A_n)$$

Proof. Properties (i) and (iii) imply that $F(A) \wedge F(A^c) = 0$, and (ii) yields $F(A) \perp F(A^c)$. Taking into account (iv), it follows that $F(A^c) = F(A)^\perp$ by the uniqueness of the orthogonal complement in an orthomodular lattice. To prove that F preserves countable joins, we compute

$$\begin{aligned} F\Big(\bigcup_{n\in\mathbb{N}} A_n\Big) &= F\Big(\bigcap_{n\in\mathbb{N}} A_n^c\Big)^\perp \\ &= \Big(\bigwedge_{n\in\mathbb{N}} F(A_n^c)\Big)^\perp = \bigvee_{n\in\mathbb{N}} F(A_n) \end{aligned}$$

■

Consequently, the range of an observable is always a Boolean σ-algebra in $\mathcal{L}$. Conversely, a Boolean σ-algebra in $\mathcal{L}$ can always be regarded as some sort of abstract observable, with domain a concrete σ-algebra of sets, after invoking Loomis' representation theorem (Loomis, 1947). In particular, observables corresponding with maximal Boolean σ-algebras will potentially play an important role in the representation theory of property lattices, giving some sort of global coherence to the whole structure.

For the sake of completeness, we notice that it is also possible to define observables in a similar way but requiring that the domain Boolean algebra of the observable is actually complete instead, the observable then preserving all meets. Without going into details, we can remark that this eventually leads to some rather subtle set-theoretic questions. See Piron (1998) for a brief discussion.

Next, various more or less equivalent definitions that try to capture the compatibility of two experimental procedures have been proposed in the quantum logic literature, but not always with an immediate operational significance. For example, two elements a, b in an ortholattice $\mathcal{L}$ are often said to be compatible whenever there exist three mutually orthogonal elements a_0, b_0 and c such that $a = a_0 \vee c$, $b = b_0 \vee c$. An alternative definition may be formulated in terms of the requirement that the sub-ortholattice generated by a and b in $\mathcal{L}$ should be distributive. In orthomodular lattices, the two notions are equivalent (Beltrametti & Cassinelli, 1981). Since we are constrained by operational considerations in this work, we have to look for a more operationally inclined notion of compatibility. The definition of Varadarajan (1962) appears the most useful from our perspective: $C(a, b)$ iff there exists an observable such that both a and b are in its range. This definition then allows the latter author to develop the theory of simultaneous observability of families of observables in his setting, and he shows that a countable family $n \mapsto F_n$ of observables is simultaneously observable iff there exists a global observable ξ such that all observables in the initial family are functionally related to ξ, in the sense that there exists a family of Borel functions $n \mapsto u_n$ such that $F_n(B) = \xi(u_n^\leftarrow(B))$ for all Borel sets B, where $u_n^\leftarrow$ is the inverse-image function induced by u_n. This classic paper is warmly recommended to all prospective students of quantum logic.

An important class of definite experimental projects corresponds with the collection of so-called ideal experiments of the first kind, a notion which is due to Pauli (1958) and for which a more sophisticated theory of measurement exists. With an appropriate notion of compatibility, recall that a definite experimental project α is said to be

(i) *ideal* iff $C(\beta, \alpha)$ and $\mu \in \kappa_Q(\beta)$ imply $\phi_\alpha(\mu) \in \kappa_Q(\beta)$;

(ii) *first-kind* iff $\phi_\alpha(\mu) \in \kappa_Q(\alpha)$.

where $\phi_\alpha(\mu)$ is the state of the physical system immediately after the measurement has been executed on a system in an initial state $\mu \in \partial_e\mathfrak{M}$ in case the result of the experiment was positive.[5] Observe that this prescription requires that ϕ_α is a partially defined mapping that maps pure states to pure states.

The notion of an ideal experiment of the first kind is intimately linked with the covering property in an orthomodular lattice. Indeed, we have shown earlier that the covering property is valid iff the following condition is essentially valid:

$$(37)\quad \forall\ \mu \in \partial_e\mathfrak{M} : \mu \notin \kappa_Q((\Box\alpha)^\sim) \quad \text{implies} \quad \phi_{\Box\alpha}(\mu) \in \partial_e\mathfrak{M}$$

A definite experimental project α is said to be *perfect* if both α and $\alpha^\sim$ are ideal first-kind. One can argue that this abstract notion of an ideal experiment of the first kind may well reflect some sort of optimal balance of the vague idea that if one acts upon nature, nature acts back; in other words, the fact that nature appears to be sensitive to our touch requires an adequate theory of measurement in terms of reasonably well-behaved measurement protocols.

It is amusing to observe that the theory of orthomodular lattices — and the theory of order structures in general — can be presented in fully category-theoretic terms, and this presentation makes the connection with Heyting algebras more explicit (Khatcherian, 1991). In a nutshell, one can regard a lattice $\mathcal{L}$ as a thin category, and a *projectale* as a lattice (with 0) equipped with a class of endofunctors $a \sqcap - : \mathcal{L} \longrightarrow \mathcal{L}$, one for each $a \in \mathcal{L}$, with the additional properties

$$\begin{array}{ll} (1) & \forall x \in \mathcal{L} : a \sqcap x \leq x \\ (2) & \forall x \in \mathcal{L} : x \leq a \Rightarrow a \sqcap x = x \\ (3) & \forall x \in \mathcal{L} : a \sqcap x = 0 \Leftrightarrow x \sqcap a = 0 \end{array}$$

and such that for each $a \in \mathcal{L}$, $a\sqcap-$ has a right adjoint $a \Downarrow -$. Notice the correspondence between some of these properties and the physical properties required for a first kind experiment. It turns out that a projectale is a Heyting algebra iff it is commutative, i.e. $a \sqcap x = x \sqcap a$ for all $a, x \in \mathcal{L}$; in this case, $a\sqcap-$ corresponds with

[5] By way of contrast, an experiment is said to be of the second kind if the state of the system immediately after the measurement is changed in such a way that the repetition of the measurement will not give the same result, but the change of state happened in a controllable fashion and an unambiguous conclusion can be drawn regarding the quantity that was measured for the system before the measurement (Pauli, 1958).

$a \wedge -$. Orthomodular lattices are related to general projectales in much the same way as Boolean algebras are related to Heyting algebras. In particular, in analogy with the Heyting algebra case — for which $(a \wedge -) \dashv (a \Rightarrow -)$ — the opposite of a is defined as $a^\perp = a \Downarrow 0$, and it turns out that this operation similarly defines a contravariant endofunctor on $\mathcal{L}$ that makes $\mathcal{L}$ into an orthomodular lattice under the additional assumption of regularity, i.e. $a^{\perp\perp} = a$, or equivalently $a \vee a^\perp = 1$, where $1 = 0^\perp$. For more details, see the reference above.

4 MORPHISMS

The quest for an appropriate class of morphisms that relates the mathematical representants for physical systems as presented in the previous section, has led to the identification of various classes of structure-preserving mappings. In the conventional model of quantum theory — complex Hilbert spaces — the problem is arguably obscured by the fact that distinct physical concepts can be represented by similar mathematical constructs. For example, physical observables correspond with spectral measures on the Hilbert space, and the spectral theorem leads to a duality between such measures and self-adjoint operators defined on the space. Smooth evolutions are represented by groups or semigroups of unitary operators, and ideal measurements of the first kind by projections, which also represent the properties one can attribute to the given system. Probability measures correspond, via Gleason's theorem, with trace class operators of trace one (Gleason, 1957).

From an operational perspective, the meet on the property lattice and the orthogonality relation on the state space have a physically transparent meaning, and so it makes sense to use this operation respectively binary relation as the primitive ingredients that should be preserved by a privileged class of morphisms. Various such classes of morphisms have been investigated by Coecke & Moore (2002), among others. On the other hand, the standard representation theorems in terms of underlying linear or semilinear functions on the representing vector spaces specifically refer to join-preserving mappings. Fortunately, there is a deep connection between meet-preserving and join-preserving mappings, which is briefly sketched in this section. In essence, the theory of Galois adjunctions is the theory of adjoint functors specialised to the context of lattices, since each partially ordered set can be regarded as a thin category. In other words, we can apply the tools of category theory to convert categories of arrows with a deeper physical significance into categories for which the arrows are representable in terms of more convenient mathematical structures.

Suppose then that $g : \mathcal{L}_2 \longrightarrow \mathcal{L}_1$ is a function between lattices with 0 and 1 that preserves all existing meets. It is important that g should also preserve empty meets, that is $g(1) = 1$. The adjoint functor theorem then yields a unique function, say $g_* : \mathcal{L}_1 \longrightarrow \mathcal{L}_2$ such that

$$(38) \quad a \leq g(b) \quad \Leftrightarrow \quad g_*(a) \leq b$$

for all $a \in \mathcal{L}_1$, $b \in \mathcal{L}_2$. More generally, the equation (38) can also be regarded as the definition of an adjunction between partially ordered sets. From an order-theoretic perspective, the function g has then the property that the inverse images of principal up-sets are *principal* up-sets, and dually g_* has the property that the inverse images of principal down-sets are again *principal* down-sets. Therefore, if g has this property, one can define a new function g_* that gives the minimal element of the principal up-set $g^{\leftarrow}(\uparrow a)$: $g^{\leftarrow}(\uparrow a) = \uparrow g_*(a)$. Categorically speaking, this function g_* is the *left adjoint* of g, and one writes $g_* \dashv g$. It is easy to see that g_* preserves all existing suprema. In fact,

$$\begin{aligned} g_*(\bigvee_{i\in I} a_i) \leq b \quad &\Leftrightarrow \quad \bigvee_{i\in I} a_i \leq g(b) \\ &\Leftrightarrow \quad \forall i \in I : a_i \leq g(b) \\ &\Leftrightarrow \quad \forall i \in I : g_*(a_i) \leq b \\ &\Leftrightarrow \quad \bigvee_{i\in I} g_*(a_i) \leq b \end{aligned}$$

In particular, $g_*(0) = 0$. Dually, each join-preserving mapping $f : \mathcal{L}_1 \longrightarrow \mathcal{L}_2$ has a unique meet-preserving right adjoint $f^* : \mathcal{L}_2 \longrightarrow \mathcal{L}_1$, $f \dashv f^*$. This follows also from the fact that each meet-preserving mapping can be regarded as a join-preserving mapping between the opposite lattices, *viz.* the lattices with the order operation reversed. Join-preserving mappings between lattices are often called *residuated mappings* and sometimes hemimorphisms, their right adjoints then referred to as *residual* mappings. One can easily verify that $(g_2 \circ g_1)_* = (g_1)_* \circ (g_2)_*$, and so functional composition defines an endo-operation in the set of residuated mappings. We then obtain a unital semigroup of residuated endomappings on $\mathcal{L}$, since the identity function is always residuated.

The lattices of the form $\mathcal{P}(E)$ for a set E are complete join-semilattices, the join corresponding with set-theoretic union. Any relation $R \subseteq A \times B$ defines a function

$$(39)\quad \xi_R : \mathcal{P}(A) \longrightarrow \mathcal{P}(B) : X \mapsto \Big\{ y \in B \mid \exists x \in X : (x, y) \in R \Big\}$$

and it is easy to verify that ξ_R is residuated, since it preserves unions. Conversely, it is not too difficult to show that each residuated mapping $\xi : \mathcal{P}(A) \longrightarrow \mathcal{P}(B)$ can be regarded as the residuated mapping induced by a relation $R \subseteq A \times B$. Specifically, define a relation $R_\xi \subseteq A \times B$ by setting $(x, y) \in R_\xi$ iff $y \in \xi(\{x\})$, and verify that ξ and ξ_{R_ξ} agree on atoms. Observe also that the relation R reduces to a genuine function $A \longrightarrow B$ whenever ξ_R maps atoms in $\mathcal{P}(A)$ to atoms in $\mathcal{P}(B)$; to a partial function whenever ξ_R maps atoms in $\mathcal{P}(A)$ to atoms or to the empty set in $\mathcal{P}(B)$. According to our definition, a second important example is given by an observable, in the strong sense, from a σ-algebra to a complete ortholattice $\mathcal{L}$ under the appropriate conditions.

If the lattices involved are complete, the left and right adjoints can be explicitly expressed in terms of their respective adjoints. If $g : \mathcal{L}_2 \longrightarrow \mathcal{L}_1$ preserves all existing infima, then

$$(40)\quad g_*(a) = \bigwedge_{i\in I}\left\{b \in \mathcal{L}_2 \mid a \leq g(b)\right\}$$

since in this case the minimal element of the principal up-set always exists, with a similar expression and argument for the right adjoint of a join-preserving mapping $f : \mathcal{L}_1 \longrightarrow \mathcal{L}_2$:

$$(41)\quad f^*(b) = \bigvee_{i\in I}\left\{a \in \mathcal{L}_1 \mid f(a) \leq b\right\}$$

In the terminology of operational quantum logic, the adjunction relation has a direct physical interpretation: $g_*(a)$ corresponds with the strongest property in $\mathcal{L}_2$ for which the actuality of the image (by g) is guaranteed by $a \in \mathcal{L}_1$, and $f^*(b)$ is the weakest property in $\mathcal{L}_1$ for which the image (by f) is guaranteeing the actuality of $b \in \mathcal{L}_2$.

These considerations naturally lead to a dual equivalence between the category JCLattt with objects complete lattices and arrows join-preserving maps and the category MCLatt with the same objects and meet-preserving mappings respectively. The first category is the prototypic example of a so-called *quantaloid, viz.* a category enriched in join-semilattices.

An important special case is given by the situation where the direct image $g_*^{\rightarrow}(A(\mathcal{L}_1)) \subseteq A(\mathcal{L}_2) \cup \{0\}$. In this case, g_* restricts and corestricts to a partial function $A(\mathcal{L}_1) - \rightarrow A(\mathcal{L}_2)$. If the domain and codomain both have the structure of a projective geometry, the tools of this discipline become available, and the second fundamental theorem usually yields a family of semilinear functions such that each restriction of g_* to an irreducible component of the projective geometry corresponds with a semilinear function defined on the underlying vector spaces. In more detail, two points p and q in $A(\mathcal{L}_1)$ are said to be in the same irreducible component if there exists a third distinct point $r \leq p \vee q$. In this case, $g_*(r) \leq g_*(p) \vee g_*(q)$, $g_*(p) \leq g_*(q) \vee g_*(r)$ and $g_*(q) \leq g_*(p) \vee g_*(r)$, since g_* preserves joins, and we deduce that $g_*(p)$ and $g_*(q)$ also belong to the same component. In other words, g_* restricts and corestricts to a family of irreducible components, and under mild technical assumptions on its range each restriction can be represented by a semilinear function on the underlying vector spaces generated by the irreducible projective geometries. See Faure & Frölicher (2000), and also Stubbe & Van Steirteghem (2007) in the first volume of this handbook.

If all lattices involved are orthocomplemented and $f : \mathcal{L}_1 \longrightarrow \mathcal{L}_2$ is residuated, with $f \dashv f^*$, one can define a new mapping

$$(42)\quad f^\dagger : \mathcal{L}_2 \longrightarrow \mathcal{L}_1 : b \mapsto f^*(b^\perp)^\perp = \left(\bigvee\left\{a \in \mathcal{L}_1 \mid f(a) \leq b^\perp\right\}\right)^\perp$$

Notice that $f^\dagger$ goes in the reverse direction. It is easy to see that $f^\dagger$ is also residuated, since f^* preserves infima. In addition, $f^\dagger$ is the unique mapping that satisfies the relation

$$\begin{aligned} f^\dagger(a) \leq b^\perp \quad &\Leftrightarrow \quad b \leq f^*(a^\perp) \\ &\Leftrightarrow \quad f(b) \leq a^\perp \end{aligned}$$

from which it easily follows that $(f^\dagger)^\dagger = f$ and $(f_2 \circ f_1)^\dagger = f_1^\dagger \circ f_2^\dagger$. The mapping $f^\dagger$ is sometimes called the *ortho-adjoint* of f.

In our earlier example, the lattices of the form $\mathcal{P}(E)$ are complete, atomistic and orthocomplemented, and so the previous construction applies. If $\xi_R : \mathcal{P}(A) \longrightarrow \mathcal{P}(B)$ is the residuated mapping associated with a relation $R \subseteq A \times B$, then one verifies that $(\xi_R)^\dagger = \xi_{R^d} : \mathcal{P}(B) \longrightarrow \mathcal{P}(A)$ is the residuated mapping associated with the converse relation.

If f is residuated and maps atoms to atoms or 0, and the same holds for $f^\dagger$, then the tools of projective geometry apply to both f and $f^\dagger$ under appropriate conditions, and the similarities with the concepts of standard operator theory become even more conspicuous. For example, f is said to be unitary iff $f^\dagger$ is inverse to f. In this case, $f(a) \perp f(b)$ whenever $a \perp b$ and $f(a^\perp) = f(a)^\perp$.

The physical duality that exists at the object-level between states and properties was encoded by the Cartan map $\kappa_L : \mathcal{L} \longrightarrow \mathcal{P}(\partial_e \mathfrak{M})$ and it is easy to see that κ_L preserves all infima. We then automatically obtain a join-preserving left adjoint, which was called the *operational resolution* by Amira, Coecke and Stubbe (1998) and is explicitly given by

$$(43)\quad \rho : \mathcal{P}(\partial_e \mathfrak{M}) \longrightarrow \mathcal{L} : A \mapsto \bigvee \left\{ p_\mu \mid \mu \in A \right\}$$

This duality can then be extended into a categorical equivalence, and this reformulation gives access to the tools of category theory (Moore, 1995, 1999). Indeed, the fact that the standard representation theorems are based on the join requires a categorical formalism enabling adequate translations from one level to the other. Specifically, we can then translate physically meaningful statements about the meet into mathematically convenient statements about the join, since the powerful mathematical representation theorems from the field of projective geometry live mainly at the level of the second category. Consequently, we are confronted with the highly non-trivial problem of identifying a *physically relevant* notion of structure-preserving mapping between state spaces as encoded by orthogonality spaces on the one hand, and property lattices as represented by complete atomistic orthocomplemented lattices on the other.

As indicated previously, one class to contemplate consists of the set of all meet-preserving mappings between two property lattices, in view of the physical meaning attributed to the infimum of a collection of properties. On the other hand, observe that this class does not contain the physically important class of Sasaki projections:

$$(44)\quad \phi_a : \mathcal{L} \longrightarrow \mathcal{L} : x \mapsto (x \vee a^\perp) \wedge a$$

since $\phi_a(1) = a \neq 1$. In addition, Sasaki projections do not preserve the orthogonality relation when restricted to $A(\mathcal{L})$. On the other hand, for a physicist this is not a surprise, because of the rather drastic evolution represented by a Sasaki projection, $\phi_a(p_\mu)$ representing the state of the system after a measurement of an ideal experiment of the first kind has been performed on the system in an initial state μ and the positive result is obtained. Following Moore (1995), we construct the category Prop with objects complete atomistic orthocomplemented lattices and arrows residuated mappings that map atoms to atoms or 0, which is dual to the category Prop* with same objects and arrows given by meet-preserving mappings $g : \mathcal{L}_1 \longrightarrow \mathcal{L}_2$ that satisfy the additional property

(45) $\forall\, p_2 \in \mathrm{A}(\mathcal{L}_2) : \exists\, p_1 \in \mathrm{A}(\mathcal{L}_1) : p_2 \leq g(p_1)$

and the category State with objects orthogonality spaces and morphisms partial functions $f : \Sigma_1 - \rightarrow \Sigma_2$ such that $\mathcal{K}_1 \cup f^{\leftarrow}(F_2)$, $\mathcal{K}_1$ being the kernel of f, is a biorthogonal subset of Σ_1 whenever F_2 is a biorthogonal subset of Σ_2.[6] It is then straightforward but somewhat tedious to show that Prop and State are equivalent in the precise sense of category theory (Moore, 1995).

This categorical reformulation of the basic ideas of the Geneva approach also makes the decomposition of property lattices into its irreducible components mathematically more precise. Given a complete atomistic orthocomplemented lattice $\mathcal{L}$, first construct the subset $\mathscr{C}$ of its classical properties. If $\{c_j \mid j \in \mathscr{J}\}$ is a collection of classical properties and $\mu \in \partial_e \mathfrak{M}$, then $p_\mu \not\leq \bigwedge\{c_j \mid j \in \mathscr{J}\}$ implies that $p_\mu \not\leq c_k$ for some $k \in \mathscr{J}$, hence $p_\mu \leq c_k^\perp$ for some $k \in \mathscr{J}$, and so $p_\mu \leq \bigvee\{c_j^\perp \mid j \in \mathscr{J}\}$; a similar argument holds for the join of a collection of classical properties. Next, if c is a classical property, then obviously $c^\perp$ is. In addition, $\mathscr{C}$ is atomistic, with atoms given by the elements $c_p = \bigwedge\{c \in \mathscr{C} \mid p \leq c\}$, where p runs through the atoms of $\mathcal{L}$. Finally, if $c_p \leq \bigvee\{c_j \mid j \in \mathscr{J}\}$, then necessarily $c_p \leq c_k$ for some $k \in \mathscr{J}$, which implies that $\mathscr{J}$ is distributive. Altogether, $\mathscr{C}$ is a distributive complete atomistic orthocomplemented sublattice of $\mathcal{L}$. Consider then the canonical projections

(46) $\mathrm{pr}_{c_p} : \mathcal{L} \longrightarrow [0, c_p] : a \mapsto a \wedge c_p$

It is easy to see that all pr_{c_p} preserve infima; moreover, for each $p \in \mathrm{A}(\mathcal{L})$, $p \leq c_p$, it is obvious that $p \leq p = \mathrm{pr}_{c_p}(p)$, and so the projections are arrows in Prop*. Since for $a \in [0, c_p]$ we also have

(47) $a \leq c_p \wedge b \quad \text{iff} \quad a \leq b$

and so we obtain the adjunction $\mathrm{copr}_{c_p} \dashv \mathrm{pr}_{c_p}$, with the canonical coprojection $\mathrm{copr}_{c_p} : [0, c_p] \longrightarrow \mathcal{L}$ the left adjoint of pr_{c_p}, hence an arrow in Prop. It is then not difficult to show that the source given by

(48) $\left(\mathrm{pr}_{c_p} : \mathcal{L} \longrightarrow [0, c_p]\right)_{c_p \in \mathrm{A}(\mathscr{C})}$

defines a product in Prop*, hence the sink

(49) $\left(\mathrm{copr}_{c_p} : [0, c_p] \longrightarrow \mathcal{L}\right)_{c_p \in \mathrm{A}(\mathscr{C})}$

[6] Recall that the kernel $\mathcal{K}$ of a partially defined function is the subset of the domain where f is not defined.

defines a coproduct in Prop by the dual equivalence of both categories. In addition, the state spaces in State accordingly form the underlying objects of a coproduct at their level of the description of a physical system.

If $\mu_1, \mu_2 \in \partial_e\mathfrak{M}$ are two states such that $\mu_1 \vDash c_p$ and $\mu_2 \vDash c_q$ for $p, q \in \mathrm{A}(\mathcal{L})$, $c_p \neq c_q$, then μ_1 and μ_2 are orthogonal, as we have seen previously, but they are also separated by a superselection rule. In fact, first observe that for each $a \in \mathcal{L}$ we have

$$(50) \quad a = \bigvee\Big\{a \wedge c_p \mid p \in \mathrm{A}(\mathcal{L})\Big\}$$

which follows easily from the fact that $\mathcal{L}$ is atomistic, and each atom $r \in \mathrm{A}(\mathcal{L})$, $r \leq a$ is under (exactly) one of the c_q, for some $q \in \mathrm{A}(\mathcal{L})$. Second, we also have, for any pair of atoms $p, q \in \mathrm{A}(\mathcal{L})$ for which $c_p \neq c_q$

$$(51) \quad (p \vee q) \wedge c_p = p \qquad ; \qquad (p \vee q) \wedge c_q = q$$

In fact, we only have to show that $p^\perp \leq (p^\perp \wedge q^\perp) \vee c_p^\perp$. Now if $r \perp p$, $r \in \mathrm{A}(\mathcal{L})$, then either $r \leq c_r$ for some $c_r \neq c_p$, and then $r \leq c_p^\perp$ (since all atoms in c_p are then automatically orthogonal to r), or $r \leq p^\perp \wedge c_p$, in which case $r \leq p^\perp \wedge q^\perp$, for essentially the same reason. If the pure states μ_1 and μ_2 are represented by the atoms p_{μ_1} and p_{μ_2} respectively, the initial assertion follows immediately from equation (51).

The dagger operation defines an involution on the unital semigroup of residuated endomappings on $\mathcal{L}$. Recall that involution semigroups $(S, \circ, *)$ have an intrinsic notion of projection: these are idempotent elements e for which $e = e^*$. More generally, at the category-theoretic level the category JCOLatt with objects complete orthocomplemented lattices and arrows join-preserving mappings has the properties of a *dagger category*, a notion that was introduced by Peter Selinger (2007) by abstraction from some of the categorical structures that arise in the pioneering paper by Abramsky & Coecke (2004).

If $\mathcal{L}$ happens to be orthomodular, the unital involution semigroup of residuated endomappings becomes a *Foulis semigroup*, a particular kind of Baer $*$-semigroup, the latter notion being due to Foulis (1960). In this case, the right annihilator

$$(52) \quad R(x) = \Big\{y \in S \mid x \circ y = 0\Big\}$$

of each element $x \in S$ is required to be a right principal ideal $x'\,S$ generated by a projection $x' \in S$. One can show that the set of projections that satisfy this condition play a privileged role and they are referred to as *closed* projections. The collection of closed projections of a Baer $*$-semigroup always forms an orthomodular lattice (Foulis, 1960).

The prototypic Baer $*$-semigroups are given by the collection of bounded linear operators on a complex Hilbert space, with involution $A \mapsto A^\dagger$ and Baer $*$-projection $A \mapsto A' = \mathbb{P}_{\ker A}$, and by its projective counterpart — the collection of equivalence classes of such operators under the equivalence relation

(53) $A \sim B$ iff $\exists\, \lambda \in \mathbb{C} : \lambda \neq 0$ and $A = \lambda B$

The collection of residuated endomappings on a complete orthocomplemented lattice $\mathcal{L}$ has another related mathematical structure that has surfaced in more recent literature (Mulvey & Pelletier, 1992). Specifically, it has the structure of a so-called *Gelfand quantale*, and in particular a *Hilbert quantale*. Recall that a quantale is a complete join-semilattice $\mathcal{L}$ endowed with a second binary operation $\& : \mathcal{L} \times \mathcal{L} \longrightarrow \mathcal{L}$ that distributes at the left and at the right with all joins. If there is a unit element u for &, the quantale is unitary. If it has an involution that respects joins and such that $(a\&b)^* = b^*\&a^*$, the quantale is involutive. A Gelfand quantale is an involutive unital quantale that satisfies the additional property $a\&a^*\&a = a$ for all $a \in \mathcal{L}$. Finally, a Hilbert quantale is a Gelfand quantale that is isomorphic — in the sense of unital involutive quantales — to the Gelfand quantale of residuated mappings for some orthocomplemented complete lattice. In this case, the binary operation "&" is given by functional composition.

Finally, we remark that Aerts & Piron (1979) investigated a category of complete orthomodular lattices for which the morphisms were taken to satisfy the additional requirement of preserving dual modular pairs, i.e. join-preserving mappings f that map orthogonal elements to orthogonal elements and $(a, b)\, M^*$ only if $(f(a), f(b))\, M^*$.

5 DYNAMICAL EVOLUTION

To illustrate the operational ideas and the fruitful interplay between physical reasoning and mathematical deduction that is so characteristic for OQL, we will consider some models of important physical phenomena in some detail. In particular, we will consider the deterministic evolution — mathematically represented by so-called Schrödinger flows — and the indeterministic case associated with the action of an ideal experiment of the first kind on a physical system (Faure, Moore and Piron, 1995; Piron, 1998). In this section, we will investigate some structural consequences of explicitly allowing for a deterministic evolution in the conceptual framework, a discussion of the second model being referred to the next section.

To incorporate the notion of an evolution at this basic level of the framework requires first and foremost that the observer is equipped with an appropriate physical clock, so that the potential (consecutive) performance of experiments can be timed. Operationally speaking, the second is currently defined as the duration of 9 192 631 770 periods of the radiation corresponding to the transition between the two hyperfine levels of the ground state of the Cs_{55}^{133} atom (Audoin and Guinot, 2001).

In this way, the class of experimental projects can be regarded as a trivial bundle over $T \subseteq \mathbb{R}$: for each $t \in T$ we have a complete sub-prelattice $\mathcal{Q}_t$, where T is related to the stability of the system under investigation. The central idea consists in the observation that a given evolution can be regarded as part of an experimental project (Daniel, 1989). In fact, for each pair $(t_0, t_1) \in T^2$, with

$t_0 \leq t_1$, we can define a mapping

(54) $\phi(t_1, t_0) : \mathcal{Q}_{t_1} \longrightarrow \mathcal{Q}_{t_0}$

Specifically, the experimental project $\phi(t_1, t_0)(\alpha) \in \mathcal{Q}_{t_0}$ has the same operational prescriptions as the project α, modulo a waiting period of $t_1 - t_0$ time units; $\phi(t_1, t_0)(\alpha)$ is certain (at t_0) iff the execution of the experimental project associated with α (at time t_1) would yield the positive result with certainty should this experiment be performed at time t_1. In this way, the *desideratum* of describing a given evolution explicitly defines a whole bunch of experimental projects, and to be consistent we have to incorporate these projects into $\mathcal{Q}$. Of course, the system under investigation will naturally evolve during this time period, so we need to distinguish between various possible physical scenarios, depending on the possible interactions of the system with its environment or otherwise.

If $\phi(t_1, t_0)(\alpha) \not\sim \phi(t_1, t_0)(\beta)$, the system can evolve into a state such that only one of α or β is certain, and so $\alpha \not\sim \beta$; indeed, the evolution of the system should not depend on the particular experimental project we may or may not decide to perform in the future. Consequently, we have a natural factorisation as displayed in the picture

(55)
$$\begin{array}{ccc} \mathcal{Q}_{t_1} & \xrightarrow{\quad q_1 \quad} & \mathcal{L}_{t_1} \\ {\scriptstyle \phi(t_1,t_0)} \downarrow & & \vdots\ {\scriptstyle \Phi(t_1,t_0)} \\ \mathcal{Q}_{t_0} & \xrightarrow[\quad q_0 \quad]{} & \mathcal{L}_{t_0} \end{array}$$

where the q_j are the usual quotient maps. Observe that

(56) $\phi(t_1, t_0)(\alpha^\sim) = \phi(t_1, t_0)(\alpha)^\sim$

In addition, by definition of product questions in the Geneva approach, the mappings ϕ will preserve products, and so the corresponding mappings $\Phi(t_1, t_0)$ are meet-preserving, hence also order-preserving. In principle, it can happen that the system vanishes for some initial states, and so we cannot assume in general that $\Phi(t_1, t_0)(1) = 1$. However, we can always consider the appropriate corestrictions of the various mappings, conveniently denoted by the same symbol

(57) $\Phi(t_1, t_0) : \mathcal{L}_{t_1} \longrightarrow [\,0, \Phi(t_1, t_0)(1)\,]$

and these mappings will preserve all infima, including the empty one, and so the theory of Galois adjunctions (or the adjoint functor theorem) provides a left adjoint $\Psi(t_0, t_1) \dashv \Phi(t_1, t_0)$, where

(58) $\Psi(t_0, t_1) : [\,0, \Phi(t_1, t_0)(1)\,] \longrightarrow \mathcal{L}_{t_1}$

preserves all suprema, and is explicitly given by the expression

(59) $\Psi(t_0, t_1)(a) = \bigwedge \Big\{ b \in \mathcal{L}_{t_1} \mid a \leq \Phi(t_1, t_0)(b) \Big\}$

In words, $\Psi(t_0, t_1)(a)$ is the strongest property of which the actuality at t_1 is guaranteed by the actuality of a at t_0. Physically speaking, there then exists some sort of duality between propagation and causation.

From now on, we will assume that $\mathcal{Q}_t \cong \mathcal{Q}_{t_0} =: \mathcal{Q}$ for all $t \in T$. In other words, the physical system is sufficiently stable for the experimental projects one can perform at some initial time t_0 to make sense at the appropriate later times. If $t_0 \leq t_1 \leq t_2$, it is clear by the prescriptions of the corresponding experimental projects that $\phi(t_1, t_0) \circ \phi(t_2, t_1) = \phi(t_2, t_0)$, and so also $\Phi(t_1, t_0) \circ \Phi(t_2, t_1) = \Phi(t_2, t_0)$ and

(60) $\Psi(t_1, t_2) \circ \Psi(t_0, t_1) = \Psi(t_0, t_2)$

by the standard properties of left adjoints. In other words, we obtain a unital semigroup of mappings that describes the evolution.

In particular, for $p \in \mathrm{A}(\mathcal{L})$, $\Psi(t_0, t_1)(p)$ is the strongest property of which the actuality at time t_1 is guaranteed by the actuality of p at time t_0. It is then natural to define a given evolution as *deterministic* relative to an initial pure state when its associated mapping $\Psi(t_0, t_1)$ maps the corresponding atom to an atom; *maximally deterministic* if $\Psi(t_0, t_1)$ maps all atoms under $\Phi(t_1, t_0)(1)$ to atoms (Faure, Moore & Piron, 1995). If the atom space of $\mathcal{L}$ — hence also of the sublattice $[\,0, \Phi(t_1, t_0)(1)\,]$ — has the structure of a projective geometry, the second fundamental theorem of projective geometry applies, and the mapping $\Psi(t_0, t_1)$ can be represented as a collection of semilinear mappings defined on the underlying vector spaces. For additional details, see the previous reference.

Notice that maximally deterministic evolutions reverse the orthogonality relation on the atom space associated with their domain. In fact, if $\Psi(t_0, t_1)(p)$ is orthogonal to $\Psi(t_0, t_1)(q)$, there exists an experimental project $\alpha \in \mathcal{Q}$ for which

$$\Psi(t_0, t_1)(p) \in \kappa_Q(\alpha) \quad \text{and} \quad \Psi(t_0, t_1)(q) \in \kappa_Q(\alpha^{\sim})$$

According to our previous specifications, this means that

$$p \in \kappa_Q(\phi(t_1, t_0)(\alpha)) \quad \text{and} \quad q \in \kappa_Q(\phi(t_1, t_0)(\alpha^{\sim})) = \kappa_Q(\phi(t_1, t_0)(\alpha)^{\sim})$$

which proves our assertion. In particular, unitary evolutions on complex Hilbert spaces become an important example of this part of the formalism.

For the concrete example of applying this formalism to the evolution of so-called separated physical systems, see Ishi (2000, 2001).

6 FROM BEING TO BECOMING

The postulates of the quantum logic approach are usually introduced in a way that is more or less independent of the general theory of measurement. If one tries to incorporate explicitly the potential concatenation of successive measurements *explicitly* into the standard framework, one needs to take into account the potential effects at the end of a single observation procedure on the initial state assigned to a singular realisation of a particular physical system. It is in this context that the role of notions of typing, of conditional probability, and of partially defined operations, become more prominent.

In classical probability theory, this problem does not occur, since the actual observation of a random variable in general does not need to take into account a potential change of state of the physical system. On the other hand, it is amusing to note that a probabilistic model may have to refer explicitly to the operational procedure that is used to obtain the results, as is vividly illustrated by Bertrand's random-chord paradox, for example. In quantum theory, this is no longer the case in general, and the potential order of successive operations is indeed an important part of the conceptual framework. In other words, measuring an observable B immediately before or after a measurement of A in general yields different results. This experimental fact is usually interpreted as resulting from a change of state of the system after the interaction with a measurement device associated with the observable A. In other words, in quantum physics there appears an irreducible element in the interaction between a measurement device and a particular physical system.

In quantum logic, this problem was — to the best of our knowledge — first tackled in a systematic way in the late 1960's in the work of Pool (1968a, b). Some years earlier, Foulis (1960) had established an intimate but purely abstract connection between orthomodular lattices on the one hand and a particular type of involution semigroups on the other, the Baer $*$-semigroups. Given this duality, and in view of the central role of orthomodular structures in algebraic quantum logic, it was a natural idea to investigate the potential phenomenological role and interpretation of these semigroups in fundamental physical theory. It turns out that the semigroup perspective adds a more active twist as compared with the essentially passive picture provided by the object-focused state space and property lattice perspective.

Pool starts with assigning an *event-state structure* to a physical system. This is a triple $(\mathfrak{E}, \mathcal{S}, P)$ where $\mathfrak{E}$ denotes the collection of events, $\mathcal{S}$ the collection of (mixed) states, and $P : \mathfrak{E} \times \mathcal{S} \longrightarrow [0,1]$ a probability function; although events in his sense explicitly refer to observational procedures, they are more similar to the concept of property that arises in the Geneva school approach, since also in this framework events e_1 and e_2 are actually identified when $\mathcal{S}_1(e_1) = \mathcal{S}_1(e_2)$, where $\mathcal{S}_1(e)$ consists of the collection of all states μ for which $P(e, \mu) = 1$. In this way, he obtains the structure of a partially ordered set for $\mathfrak{E}$, the partial order reflecting the same phenomenological implication as in the Geneva school approach:

(61) $e_1 \leq e_2 \quad \text{iff} \quad \mathcal{S}_1(e_1) \subseteq \mathcal{S}_1(e_2)$

The usual progression of axioms then guarantees that $\mathfrak{E}$ is an orthomodular σ-orthoposet, and $\mathcal{S}$ a strongly order-determining σ-convex set, which can be identified with the collection of generalised probability measures defined on $\mathfrak{E}$.

A central role in the work of Pool is played by the notion of an *operation* associated with a definite experimental project — a concept that is inspired by the work of Haag and Kastler (1964) in the setting of algebraic quantum field theory — and which tries to formalise the potential state transition suffered by the system, after it has been tortured by the execution of a measurement and the positive response is obtained. Actually, Pool associates a simple operation Ω_e not with each definite experimental project, but with each event e, which may alternatively be regarded as the choice from a privileged set of definite experimental projects associated with the individual events, in the spirit of our axiom C3. Notice that such a procedure only makes sense when the probability for a positive response is indeed non-zero. In other words, there is a partially defined operation Ω_e associated with the event e that can be formalised as a *partial* endo-function of the state space, defined on those states μ for which $P(e, \mu) > 0$ — and the type of the system doesn't change — with restricted domain

$$(62)\quad \mathcal{D}(\Omega_e) = \Big\{\mu \in \mathcal{S} \mid P(e, \mu) \neq 0\Big\}$$

The state Ω_e is of couse conceived as the state of the system immediately after the privileged experimental project associated with the event e has been effectuated and a positive result is obtained. Alternatively, from a more probabilistic perspective one can also regard the state transition Ω_e as a procedure that provides the new state as conditioned on the event e.

A general operation is then regarded as a succession of simple operations, with the domain defined in the standard way when one deals with the composition of partial functions:

$$(63)\quad \mathcal{D}(\Omega_{e_n} \circ \cdots \circ \Omega_{e_1}) = \Big\{\mu \in \mathcal{S} \mid (\Omega_{e_k} \circ \cdots \circ \Omega_{e_1})(\mu) \in \mathcal{D}(\Omega_{e_{k+1}}), 1 \leq k \leq n-1\Big\}$$

In this way, he obtains a semigroup of operations, which has a zero element and a unit element. Incorporating the operations as primitive objects into the fundamental framework, Pool defines an *event-state-operation structure* as a quadruple $(\mathfrak{E}, \mathcal{S}, P, \Omega)$, where $(\mathfrak{E}, \mathcal{S}, P)$ is an event-state structure and

$$(64)\quad \Omega : \mathfrak{E} \rightarrow \underline{\mathrm{Pfn}}(\mathcal{S}, \mathcal{S})$$

generates a subsemigroup of the set $\underline{\mathrm{Pfn}}(\mathcal{S}, \mathcal{S})$ of all partial endofunctions on $\mathcal{S}$ by composition of partial functions, subject to various additional conditions. In particular, he requires that $P(e, \mu) = 1$ only if $\Omega_e(\mu) = \mu$, and $P(e, \Omega_e(\mu)) = 1$ when this expression is defined; these two conditions are reminiscent of the notion of an ideal first-kind experiment. In fact, if we demand that the rules of classical probability theory should hold on Boolean subalgebras of $\mathfrak{E}$, the state transitions should be consistent with the usual requirements of conditional probability, and so we naturally require

$$(65)\quad P(b, \Omega_a(\mu)) = \frac{P(a \wedge b, \mu)}{P(a, \mu)}$$

whenever $C(a, b)$ in an appropriate sense, and $\mu \in \mathcal{D}(\Omega_a)$. If $C(a, b)$ and $P(b, \mu) = 1$, we obtain $P(b, \Omega_a \mu) = 1$, and so the operation Ω_a corresponds with an ideal measurement. Specifically, we also have

$$\begin{aligned} P(a \wedge b, \mu) &= P(b, \mu)\, P(a, \Omega_b(\mu)) \\ &= P(a, \Omega_b(\mu)) \\ &= P(a, \mu) \end{aligned}$$

since $\Omega_b(\mu) = \mu$ follows from $P(b, \mu) = 1$.

An involution on this semigroup is then defined by reversing a succession of operations:

$$(66)\quad (\Omega_{e_n} \circ \cdots \circ \Omega_{e_1})^* = \Omega_{e_1} \circ \cdots \circ \Omega_{e_n}$$

For this definition to make sense, it is necessary to require, together with Pool, that

$$(67)\quad \Omega_{e_n} \circ \cdots \circ \Omega_{e_1} = \Omega_{f_m} \circ \cdots \circ \Omega_{f_1} \quad \Longrightarrow \quad \Omega_{e_1} \circ \cdots \circ \Omega_{e_n} = \Omega_{f_1} \circ \cdots \circ \Omega_{f_m}$$

Again, we require the existence of a privileged collection of definite experimental projects such that this reversal makes sense and is well-defined. The final axiom asserts that for each succession of elementary operations $X = \Omega_{e_n} \circ \cdots \circ \Omega_{e_1}$ there is an event e_X for which

$$(68)\quad \mathcal{S}_1(e_X) = \left\{\mu \in \mathcal{S} \mid \mu \notin \mathcal{D}(X)\right\}$$

The heuristic motivation for this axiom is the following: Given a particular state, one can always experimentally verify whether or not this state is in $\mathcal{D}(X)$ by the following procedure: $\mu \notin \mathcal{D}(X)$ iff $\mu \notin \mathcal{D}(\Omega_{e_1})$ or $\mu \in \mathcal{D}(\Omega_{e_k} \circ \cdots \circ \Omega_{e_1})$ and $\mu \notin \mathcal{D}(\Omega_{e_{k+1}} \circ \cdots \circ \Omega_{e_1})$ for some $1 \leq k < n$. We then require the existence of an event e_X associated with this experimental procedure. In this way, the involution semigroup generated by the elementary operations becomes a Baer $*$-semigroup, the closed projections being the elementary operations of the form Ω_e, and the correspondence $e \mapsto \Omega_e$ is an isomorphism of partially ordered sets. The latter poset is an orthomodular lattice, and so the set of events also obtains a *lattice* structure in this way, which is *a priori* not obvious in this setting. Both the conventional probabilistic model *sensu* Kolmogorov (1956) and von Neumann's Hilbert space model are instances of the general notion of an event-state-operation structure.

On the other hand, if $(\mathfrak{E}, \mathcal{S}, P, \Omega)$ is an event-state-operation structure with $\mathfrak{E}$ an orthomodular lattice, one can consider the associated Baer $*$-semigroup $(S(\mathfrak{E}), \circ, \dagger)$ of residuated endomappings on $\mathfrak{E}$, and investigate the relation between this semigroup and the semigroup of operations induced by the mapping Ω; the latter semigroup consists of partial endofunctions on $\mathcal{S}$, while the first one corresponds with endomappings on $\mathfrak{E}$. The bridge is formed by the notion of *support* of a state μ. This is conceived as an event s_μ such that for all events e, $P(e, \mu) = 0$ iff $e \perp s_\mu$; in other words, it is an event s_μ such that the filter of all certain events in $\mathfrak{E}$ induced by the state μ is the principal filter $\uparrow s_\mu$. Physically speaking, the support associated with a general state corresponds with the strongest actual property we actually know about the single individual samples of a system prepared according to this state. If $\mathfrak{E}$ is complete, supports for states will always exist; otherwise, the existence has to be assumed. Since

$$\begin{aligned} P(s_\mu^\perp \wedge e, \mu) &\leq P(s_\mu^\perp, \mu) \\ &\leq 1 - P(s_\mu, \mu) = 0 \end{aligned}$$

and so

$$\begin{aligned} P(s_\mu \vee e^\perp, \Omega_e(\mu)) &= 1 - P(s_\mu^\perp \wedge e, \Omega_e(\mu)) \\ &= 1 - \frac{P(s_\mu^\perp \wedge e, \mu)}{P(e, \mu)} = 1 \end{aligned}$$

we infer that $s_{\Omega_e(\mu)} \leq s_\mu \vee e^\perp$. In addition, $P(e, \Omega_e(\mu)) = 1$ for $\mu \in \mathcal{D}(\Omega_e)$, and so $s_{\Omega_e(\mu)} \leq e$. Consequently, it is always the case that

(69) $s_{\Omega_e \mu} \leq (s_\mu \vee e^\perp) \wedge e = \phi_e(s_\mu)$

where ϕ_e corresponds again with a suitable Sasaki projection associated with the event e, this time defined on the collection of supports of states. Pool (1968b) then postulates that equality actually holds in all cases, and shows that one can construct a Baer $*$-semigroup morphism Θ — an involutive semigroup morphism also preserving the unary operation $'$ — from the semigroup of operations into the semigroup of residuated mappings. This mapping is defined on generators as

(70) $\Theta(\Omega_e) = \phi_e$

and extended by functional composition. It is then also true that

(71) $(\Theta(\Omega_e))(s_\mu) = s_{\Omega_e(\mu)} = \phi_e(s_\mu)$

and so the (partial) state transforming map Ω_e is converted into the corresponding support transforming map ϕ_e. Finally, the covering law reduces to the statement that the restriction of Ω_e to the set of pure states only also corestricts to pure states.

Conversely, one can show that each event-state structure that consists of an orthomodular lattice $\mathcal{L}$ equipped with a strongly separating set of (generalised) probability measures $\mathcal{M}$ admits a set of (partial) endomappings on $\mathcal{M}$ that can be interpreted as ideal first-kind measurements (Cassinelli & Beltrametti, 1975).

More recent work that focusses on the quantale and quantaloid properties of the mathematical objects, and a discussion of how this leads to a more dynamic perspective on operational quantum logic, can be found in Coecke & Stubbe (2000) and Coecke, Moore & Stubbe (2001).

7 COMPOSITE QUANTUM OBJECTS

Let us start with a brief discussion of and some of the problems associated with an elementary model in standard quantum mechanics for a typical "2-particle", the hydrogen atom. For a more extensive discussion, see Piron's blue book (Piron, 1998). In standard Hilbert space quantum mechanics, the behaviour of protons and electrons is conventionally modeled with the aid of the complex Hilbert space $\mathrm{L}^2(\mathbb{R}^3, \lambda) \otimes \mathbb{C}^2$, where λ denotes Lebesgue measure, and $\mathbb{R}^3$ refers, for example, to the spatial coordinates that diagonalise the respective position operators. The hydrogen atom is conceived as consisting of a proton and an electron interacting with one another. Taking into account the spins, according to the prescriptions of standard textbook quantum mechanics the appropriate Hilbert space becomes

$$(72)\quad \left(\mathrm{L}^2(\mathbb{R}^3, \lambda) \otimes \mathbb{C}^2\right) \otimes \left(\mathrm{L}^2(\mathbb{R}^3, \lambda) \otimes \mathbb{C}^2\right) \cong \bigoplus_{j=1}^{4} \mathrm{L}^2(\mathbb{R}^6, \lambda)$$

For the sake of the discussion, we will disregard the spin-orbit coupling and consider the Hamiltonian operator

$$(73)\quad H = -\frac{\hbar^2}{2m_p}\sum_{i=1}^{3}\frac{\partial^2}{\partial {x_p^i}^2} - \frac{\hbar^2}{2m_e}\sum_{j=1}^{3}\frac{\partial^2}{\partial {x_e^j}^2} - \frac{1}{4\pi\epsilon_0}\frac{e^2}{\|\vec{r_p} - \vec{r_e}\|}$$

acting on each of the four components of the wave function in the space above. The usual transformations to the centre-of-mass coordinate system

$$(74)\quad \begin{array}{ll} \vec{R} = \frac{1}{m_p+m_e}\left(m_p\vec{r_p} + m_e\vec{r_e} \right) & \vec{P} = \vec{p_p} + \vec{p_e} \\ \vec{r} = \vec{r_e} - \vec{r_p} & \vec{p} = \frac{1}{m_p+m_e}\left(m_p\vec{r_e} - m_e\vec{r_p} \right) \end{array}$$

lead us to the transformed Hamiltonian

$$(75)\quad H' = -\frac{\hbar^2}{2M}\sum_{i=1}^{3}\frac{\partial^2}{\partial {X^i}^2} - \frac{\hbar^2}{2\mu}\sum_{j=1}^{3}\frac{\partial^2}{\partial {x^j}^2} - \frac{1}{4\pi\epsilon_0}\frac{e^2}{\|\vec{r}\|}\mathbb{1}$$

where

$$(76)\quad M = m_p + m_e \quad ; \quad \frac{1}{\mu} = \frac{1}{m_p} + \frac{1}{m_e}$$

and this PDE can be separated into a free part and a part that is familiar from the related one-particle problem. The free part, which has a purely continuous spectrum $(0, +\infty)$, and therefore leads to a *purely continuous* spectrum and hence also the absence of stationary states for the total Hamiltonian (75), is usually dismissed by physical hand-waving arguments, because the second equation conforms to the experimentally observed discrete spectrum, at least at this level of precision. In addition, the solution associated with the centre-of-mass part will quickly expand in space (Piron, 1998).

From the perspective of OQL, there is a corresponding treatment of the problem, using a model with continuous superselection rules instead (Piron, 1965; 1998). Specifically, the observables associated with $\vec{Q}$, $\vec{P}$ and t are regarded as determining classical properties in the property lattice, and can be represented by functions on $\mathbb{R}^7$. The appropriate Hilbert space then becomes the trivial Hilbert bundle

$$(77)\quad (\vec{Q},\vec{P},t) \mapsto \mathrm{L}^2(\mathbb{R}^3,\lambda)\otimes\mathbb{C}^2\otimes\mathbb{C}^2 \cong \bigoplus_{j=1}^{4}\mathrm{L}^2(\mathbb{R}^3,\lambda)$$

In other words, pure states of this system are completely determined by the specification of a point $(\vec{Q},\vec{P},t)\in\mathbb{R}^7$ and a ray in the direct sum Hilbert space. The Schrödinger operator (73) becomes, after transforming to CM-coordinates,

$$(78)\quad H'' = \frac{P^2}{2M}\mathbb{1} - \frac{\hbar^2}{2\mu}\sum_{j=1}^{3}\frac{\partial^2}{\partial {x^j}^2} - \frac{1}{4\pi\epsilon_0}\frac{e^2}{\|\vec{r}\|}\mathbb{1}$$

acting on each of the four components in the direct sum. In this model, the first term becomes a rather innocent constant that does not affect the calculation of energy differences, at least not at this level of precision. Consequently, the spectrum of H'' has a discrete part, stationary states exist, and all is well. Observe that this is a model which uses a continuous superselection rule. From the endo-perspective of the system, the potential origin of this apparent "decoherence" is immaterial. It is there because of its potential interaction with the appropriate class of definite experimental projects, and it is usually considered as part of current scientific practice that the scientist investigates a reasonably circumscribed part of the external world.

So what do we mean then in more general terms with a *composite* physical system? Which criteria do we have to use to assign the label "composite" to a given physical system? From the perspective of OQL, it is arguably desirable that such criteria are purely operational. Before continuing the discussion, we want to point out that the very notion of a composite physical system creates a potential conceptual conflict between the two archetypical approaches — top-bottom and bottom-top — towards the description of a physical system that is conceived as composite, or at least moves the potential schizophrenia that is involved in these two possible extreme viewpoints into sharper focus. More precisely, the first type of approach describes a physical system by investigating and abstracting its interaction as a single entity with all sorts of external devices, leading to the notion of an "n-particle"; the second type of approach will rather try to describe a composite system by attributing some appropriate intrinsic substructure *ab initio*, using independent empiric or other information and/or considerations of a theoretical nature, presumably leading to some notion of n interacting particles. To make matters worse, there is also a dynamical component involved. For example, we are inclined to construe the singlet state as referring to a composite system because we can interact with the system in such a way that we obtain two individual components, but the interaction involved is rather disruptive.

In the second case, we are led to expect that the behaviour of a physical system that is being conceived as composite can be deduced from the properties attributed to its putative constituents. We then have to require that and investigate how this independent knowledge can be incorporated in the description. It is in this context that a typical *subsystem recognition* problem appears: How do we recognise the constituents in the global system? At this stronger level, we may then actively look for the operational recognition of potential subsystems in a given system. In other words, construct an explicit embedding of the properties attributed to the putative subsystem in the collection of properties assigned to the larger composite system. For a more elaborate discussion of some of these aspects, see Valckenborgh (2000).

Some of these difficulties with forming an appropriate notion of a composite physical system are highlighted in the early work of Aerts (1982). Conceptually speaking, one expects that the simplest type of compound physical system corresponds with a system of two objects that are separated from one another. Using the operational principles and prescriptions of the Geneva approach, Aerts constructed explicitly a mathematical model for the description of several operationally identified and separated systems as one global system. It turned out that two of the axioms formulated in Piron (1976) — weak modularity and the covering property — in general fail in this case, unless all but at most one of the constituents have only classical properties. More precisely, it turns out that the property lattice that results from the combination of two *operationally separated* pure quantum systems cannot be embedded in a natural way in the collection of closed subspaces of an appropriate complex Hilbert space. Even worse, the collection of pure states attributed to the composite entity no longer forms a projective geometry, even in the extended sense of Faure and Frölicher (2000). Since these ingredients are necessary for the representation of the property lattice as the collection of closed subspaces of a suitable orthomodular space, these results highlight some of the tensions that exist between quantum physics and the more classical notions of a composite system and separation of systems. Although the physical justification takes some work, from a mathematical perspective the property lattice, denoted by $\mathcal{L}_1 \circledwedge \mathcal{L}_2$, is easy to construct, given the (pure) state spaces $\partial_e\mathfrak{M}_1$ and $\partial_e\mathfrak{M}_2$ associated with the two components. Specifically, $\mathcal{L}_1 \circledwedge \mathcal{L}_2$ consists of the collection of all biorthogonally closed subsets of the set $\partial_e\mathfrak{M}_1 \times \partial_e\mathfrak{M}_2$, with Aerts' orthogonality relation given by

$$(79)\quad (\mu_1, \mu_2) \perp (\nu_1, \nu_2) \quad \text{iff} \quad \mu_1 \perp_1 \nu_1 \text{ or } \mu_2 \perp_2 \nu_2$$

For the sake of argument, suppose that $\mathcal{L}_1$ and $\mathcal{L}_2$ are both non-trivial irreducible complete atomistic orthomodular lattices that satisfy the covering property. Take three distinct pure states $p_1, q_1, r_1 \in \partial_e\mathfrak{M}_1$ with $r_1 \in \{p_1, q_1\}^{\perp\perp}$, and do the same for $\partial_e\mathfrak{M}_2$; in other words, r_j is on the line generated by p_j and q_j for $j = 1, 2$. Using the standard results

$$(80)\quad \left(\bigcup\{m \mid m \in M\}\right)^{\perp} = \bigcap\{m^{\perp} \mid m \in M\}$$

which holds in any orthogonality space, and the (79)-specific result

(81) $\{(p_1, p_2)\}^\perp = \left(\{p_1\}^\perp \times \partial_e\mathfrak{M}_2\right) \cup \left(\partial_e\mathfrak{M}_1 \times \{p_2\}^\perp\right)$

it is then not difficult to verify that the orthogonality relation (79) implies that

(82) $\{(p_1, p_2), (r_1, r_2)\}^{\perp\perp} = \{(p_1, p_2), (r_1, r_2)\}$

(83) $\{(q_1, p_2), (r_1, q_2)\}^{\perp\perp} = \{(q_1, p_2), (r_1, q_2)\}$

and so

(84) $\{(p_1, p_2), (r_1, r_2)\}^{\perp\perp} \cap \{(q_1, p_2), (r_1, q_2)\}^{\perp\perp} = \emptyset$

although $(r_1, p_2) \in \{(p_1, p_2), (q_1, p_2)\}^{\perp\perp} \cap \{(r_1, r_2), (r_1, q_2)\}^{\perp\perp}$. In other words, we have a pair of intersecting lines and two points on each line for which the new pair of lines generated by these points is not intersecting, and so axiom (3) of a projective geometry in terms of the ternary collinearity relation fails to hold for the pair $(\partial_e\mathfrak{M}_1 \times \partial_e\mathfrak{M}_2, \ell_A)$, where the ternary relation ℓ_A satisfies

(85) $\ell_A(a, b, c)$ iff $b = c$ or $a \leq \{b, c\}^{\perp\perp}$

Similarly, properties of the form $\{(p_1, p_2), (q_1, q_2)\}$, where we require that $p_1 \neq q_1, p_1 \not\perp q_1$ and $p_2 \neq q_2, p_2 \not\perp q_2$, lie at the basis of the failure of the orthomodular law. Specifically, we obviously have

(86) $(p_1, p_2) \in \{(p_1, p_2), (q_1, q_2)\}$

(87) $\{(p_1, p_2)\}^\perp \cap \{(p_1, p_2), (q_1, q_2)\} = \emptyset$

and so we have found two properties a and b such that $a < b$ and $a^\perp \wedge b = 0$. Consequently

(88) $a \vee (b \wedge a^\perp) = a < b$

and the orthomodular law fails.

In the same spirit, if μ is a generalised probability measure on an ortholattice $\mathcal{L}$ in the sense of definition 6, then $a \leq b$ and $a^\perp \wedge b = 0$ automatically imply that $\mu(a) = \mu(b)$. Indeed,

$$\begin{aligned} \mu(a^\perp \wedge b) &= \mu((a \vee b^\perp)^\perp) \\ &= 1 - \mu(a \vee b^\perp) \\ &= 1 - \mu(a) - \mu(b^\perp) \\ &= \mu(b) - \mu(a) \end{aligned}$$

where the third equality holds because of the orthogonality of a and $b^\perp$. Consequently, the collection of all probability measures — in the sense of definition 6

— on $\mathcal{L}_1 \circledwedge \mathcal{L}_2$ is *not* separating, although the collection of atoms, corresponding with the set of pure states attributed to the separated physical system, obviously does. In other words, whereas one physically expects that each pure state induces a probability measure in some general sense on $\mathcal{L}_1 \circledwedge \mathcal{L}_2$, these probabilities will behave erratically with respect to Boolean algebras in $\mathcal{L}_1 \circledwedge \mathcal{L}_2$.

Finally, separated physical systems also exhibit a very anomalous behaviour with respect to superselection sectors. More precisely, the particularities of the orthogonality relation (79) imply that there exist superselection rules for some *non-orthogonal* states, a situation that does not occur in conventional quantum theory, because of proposition 8. For example, the states (p_1, p_2) and (q_1, q_2) used in the previous argument are separated by a superselection rule, because the line they generate contains no other points. As a consequence, proposition 8 then implies again that the orthomodular law fails for the separated product of two non-classical complete atomistic ortholattices. We happily refer the reader to Aerts & Valckenborgh (2002) for additional discussion, and the application of these ideas to the physically more concrete example of two separated spin-1/2 systems.

The fact that a given physical system can be regarded as composite, for example due to its possible behaviour under certain experimental conditions, seems to compel us to explicitly try to incorporate this additional information in one way or another. From the more operational top-bottom perspective, it is conceivable that such *a priori* information would be only statistical, due to the potential disruptive aspects of an explicit interaction of the initial system with the measurement device that may break it apart, not unlike the situation that arises in the measurement problem. If we have access to only one of the parts of a system that has been prepared in this state, a situation which seems to make sense from the perspective of the enigmatic couple Alice and Bob, we can only use the class of definite experimental projects that applies to the perceived subsystem. With respect to the property lattice that is so constructed, the initial state attributed to one of the components necessarily becomes the mixed state that arises from the partial tracing of the pure state of the global system.

In orthodox quantum mechanics in Hilbert space, composite physical systems are mathematically described by means of the Hilbert tensor product $\mathscr{H}_1 \otimes \mathscr{H}_2$ of the Hilbert spaces $\mathscr{H}_1$ and $\mathscr{H}_2$ that represent the individual subsystems, or an appropriate subspace thereof, if there are additional externally imposed symmetry constraints in the interaction, at least before the potential disruption takes place. This weaker type of predicting the behaviour of the components is achieved in this case by the mathematical device of partial tracing, while the stronger notion would correspond with the actual explicit existence of an embedding

(89) $j_1 : \mathfrak{P}(\mathscr{H}_1) \longrightarrow \mathfrak{P}(\mathscr{H}_1 \otimes \mathscr{H}_2)$

at the level of the projective geometries associated with the Hilbert spaces, and similarly for j_2. Aerts & Daubechies (1978) have used such *desiderata* to operationally justify the use of the Hilbert space tensor product in quantum physics. In more detail, they show that when the following three conditions are imposed on mappings $h_1 : \mathcal{L}(\mathscr{H}_1) \longrightarrow \mathcal{L}(\mathscr{H})$ and $h_2 : \mathcal{L}(\mathscr{H}_2) \longrightarrow \mathcal{L}(\mathscr{H})$, where $\mathcal{L}(\mathscr{H})$ denotes the lattice of closed subspaces of a Hilbert space $\mathscr{H}$

(1) h_1 and h_2 are unital, preserve infima and the orthogonality relation

(2) $C(h_1(a_1), h_2(a_2))$ for all $a_1 \in \mathcal{L}(\mathscr{H}_1)$ and $a_2 \in \mathcal{L}(\mathscr{H}_2)$

(3) $h_1(p_1) \wedge h_2(p_2) \in \mathrm{A}(\mathcal{L}(\mathscr{H}))$ for all $p_1 \in \mathrm{A}(\mathcal{L}(\mathscr{H}_1))$, $p_2 \in \mathrm{A}(\mathcal{L}(\mathscr{H}_2))$

then $\mathcal{L}(\mathscr{H})$ will be canonically isomorphic with the lattice of closed subspaces of either the Hilbert space $\mathscr{H}_1 \otimes \mathscr{H}_2$ or the space $\mathscr{H}_1 \otimes \mathscr{H}_2^\dagger$, where $\mathscr{H}^\dagger$ denotes the dual space of $\mathscr{H}$. Some structural subtleties that arise are discussed in Valckenborgh (2000).

8 QUANTUM PROCESSES

We want to end our exposition with some musing on the more dynamical ideas on quantum processes and indulge in some speculations about work that, in our opinion, remains to be done.

In mathematical terms, the categorical view on mathematical structures characterises objects not intrinsically, but rather by the way how they relate to other objects of the same kind, by means of the network of arrows into and out of the object (Mac Lane, 1998). In other words, it is the relational pattern that matters, and not the potential intrinsic properties of a particular object. Operationally speaking, applying such principles in the physical context points to a more operational and dynamical attitude towards physical systems: the way the system behaves with respect to the external world is given physical preference above the way the system is supposed to be.

There exists a general type of category that allows the expression of a coherent calculus of sequential *and* parallel operations; these are the so-called *monoidal categories.* Interestingly, many categories that are of interest in current physical thinking are examples of monoidal categories, the monoidal structure reflecting the conventional representations of composite physical systems. For example, the category of finite-dimensional Hilbert spaces and linear transformations is a symmetric monoidal category with respect to the usual Hilbert space tensor product, as is the category of sets and functions when we pick the cartesian product for the monoidal product. Here, category theory itself becomes an appropriate language to express the formalism, in contrast to its more organising role in the setting of the previous sections. Specifically, *symmetric monoidal categories* can be regarded as an appropriate mathematical vehicle in which general processes are conveniently expressed. From this perspective, a quantum process can almost literally be regarded as a path in a symmetric monoidal category, and parallel processes are formed by invoking the properties of the monoidal bifunctor $\otimes$. This visualisation is very intuitive, and can be formally justified by the existence of a powerful graphical calculus for symmetric monoidal categories (Joyal & Street, 1991). For more details on the development of such a categorical framework for the discussion of abstract and various concrete processes that are relevant for quantum computation and information processing, see Abramsky & Coecke (2004). For an overview of the fascinating interplay between the domains of physics, topology, logic and computation to which this unifying framework leads, see the recent paper by Baez & Stay (2008).

A *dagger category* is a pair consisting of a category and an involutive, identity-on-objects endofunctor (Selinger, 2007). Therefore, the set of all endomorphisms of an object in a dagger category has always the structure of an involutive monoid, and so each object comes with a partially ordered set of projections. The relation between the role of involution semigroups in the more object-oriented quantum logic approach and the more dynamical afterthoughts in this section becomes more obvious if we restrict ourselves to the more gentle experimental procedures that do not change the type of the physical system under investigation (Amira, Coecke & Stubbe, 1998). Indeed, in many important cases, the phenomenon under investigation is sufficiently stable and this observation should allow us to recover the more traditional static picture from this dynamic setting, and some recent work suggests how this can be done (Coecke, Paquette & Pavlovic, 2007). Another domain where one expects this approach to become important, is in quantum scattering theory, where a change in the types of objects after a dynamical process has occurred is the rule rather than the exception. It is possible that the structural axioms *sensu* Abramsky & Coecke (2004) have to be extended in this setting. More precisely, in that paper preference is given to a particular type of symmetric monoidal category, the *compact closed* categories.

One can argue that physics is, first and foremost, the study of detailed models of specific situations. Consequently, to narrow the remaining gap between OQL and standard quantum theory, additional empirical structure seems necessary. In particular, the characterisation of the underlying division ring D — an invariant of the projective geometry — remains unclear. By requiring the existence of a certain type of observable, in combination with some additional technical assumptions, Gudder & Piron (1971) were able to show that D contains at least a copy of $\mathbb{R}$. One possible strategy would consist in studying the structural consequences when one explicitly requires the representability of symmetry groups of a certain type — in particular second countable locally compact Hausdorff groups, a collection that encompasses all groups that are currently conceived as physically important — in the existing framework. In particular, the incorporation of the physically important notion of *localisability* at a much earlier stage of the discussion may lead to important constraints on the admissible lattice-theoretic and related constructions. We know, for example, that the mathematical property of modularity is incompatible with localisability (Piron, 1964; Jauch & Piron, 1967). In the categorical framework that was discussed, these ideas lead to the investigation of the structural properties of appropriate subcategories of categories of representations of a symmetry group G, in particular of the categories $\underline{\text{Prop}}^G$, $\underline{\text{State}}^G$ and $\underline{\text{Proj}}^G$. More specifically, for groups that have the additional internal algebraic and topological structure that are important in physics.

ACKNOWLEDGEMENTS

Part of this work was carried out during the employment of one of the authors (FV) as a Macquarie University Research Fellow, in local collaboration with John Corbett. We also want to thank our long term colleague Bob Coecke and other members of the former Brussels community for the many useful — though often beer-induced — discussions on the limitations and successes of the quantum logic approach, and its foundational relation to the more dynamic viewpoint that is currently explored.

BIBLIOGRAPHY

[Abramsky and Coecke, 2004] S. Abramsky and B. Coecke. A categorical semantics of quantum protocols. Proceedings of the 19th Annual IEEE Symposium on Logic in Computer Science (LiCS'04), IEEE Computer Science Press, 2004.

[Aerts, 1982] D. Aerts. Description of many separated physical entities without the paradoxes encountered in quantum mechanics. *Found. Phys.* **12**: 1131–1170, 1982.

[Aerts, 1983] D. Aerts. Classical theories and nonclassical theories as special cases of a more general theory. J. Math. Phys. **24**: 2441–2453, 1983.

[Aerts, 1984] D. Aerts. 1984. Construction of the tensor product for the lattices of properties of physical entities. *J. Math. Phys.* **25**: 1434–1441, 1984.

[Aerts and Piron, 1979] D. Aerts and C. Piron. The role of the modular pairs in the category of complete orthomodular lattice. *Lett. Math. Phys.* **3**: 1–10, 1979.

[Aerts and Valckenborgh, 2002] D. Aerts and F. Valckenborgh. Linearity and compound physical systems: the case of two separated spin 1/2 entities. In: D. Aerts, M. Czachor & T. Durt. *Probing the Structure of Quantum Mechanics: Nonlinearity, Nonlocality, Computation and Axiomatics.* World Scientific, Singapore, 2002.

[Amann, 1988] A. Amann. Chirality as a classical observable in algebraic quantum mechanics. In: A. Amann, L. Cederbaum & W. Gans. *Fractals, Quasicrystals, Chaos, Knots and Algebraic Quantum Mechanics.* Kluwer, Dordrecht, 1988.

[Amira *et al.*, 1998] H. Amira, B. Coecke and I. Stubbe. How Quantales Emerge by Introducing Induction within the Operational Approach. *Helv. Phys. Acta* **71**: 554–572, 1988.

[Audoin and Guinot, 2001] C. Audoin and B. Guinot. *The Measurement of Time.* Cambridge University Press, 2001.

[Baer, 1952] R. Baer. *Linear Algebra and Projective Geometry.* Academic Press, New York, 1952.

[Baez and Stay, 2008] J. Baez and M. Stay. Physics, Topology, Logic and Computation: A Rosetta Stone. Preprint. To appear in: B. Coecke. *New Structures of Physics*, 2008.

[Beltrametti and Cassinelli, 1981] E. G. Beltrametti and G. Cassinelli. *The Logic of quantum Mechanics.* Addison-Wesley, Reading, Massachusetts, 1981.

[Birkhoff and von Neumann, 1936] G. Birkhoff and J. von Neumann. The logic of quantum mechanics. *Annals of Math.* **37**: 823–843, 1936.

[Cartan, 1971] E. Cartan. *Leçons sur les invariants intégraux.* Hermann, Paris VI, 1971.

[Cassinelli and Beltrametti, 1975] G. Cassinelli and E. G. Beltrametti. Ideal, First-kind Measurements in a Proposition-State Structure. *Commun. math. Phys.* **40**: 7–13, 1975.

[Cattaneo *et al.*, 1988] G. Cattaneo, C. Dalla Pozza, C. Garola and G. Nisticó. On the Logical Foundations of the Jauch-Piron Approach to Quantum Physics. *Int. J. Theor. Phy.* **27**: 1313–1349, 1988.

[Cattaneo and Nisticó, 1991] G. Cattaneo and G. Nisticó. Axiomatic Foundations of Quantum Physics: Critiques and Misunderstandings. Piron's Question-Proposition System. *Int. J. Theor. Phys.* **30**: 1293–1336, 1991.

[Coecke and Moore, 2000] B. Coecke and D. Moore. Operational Galois adjunctions. In: B. Coecke, D. J. Moore and A. Wilce (eds.). *Current Research in Operational Quantum Logic: Algebras, Categories, Languages.*. Kluwer Academic Publishers. Dordrecht, 2000.

[Coecke *et al.*, 2001] B. Coecke, D. Moore and I. Stubbe. 2001. Quantaloids describing causation and propagation of physical properties. *Found. Phys. Lett.* **14**: 133–145, 2001.

[Coecke *et al.*, 2000] B. Coecke, D. Moore and A. Wilce. Operational quantum logic: An overview. In: B. Coecke, D. J. Moore and A. Wilce (eds.). *Current Research in Operational Quantum Logic: Algebras, Categories, Languages.* Kluwer Academic Publishers. Dordrecht, 2000.

[Coecke *et al.*, 2007] B. Coecke, E. O. Paquette and D. Pavlovic. Classical and quantum structures. Preprint, 2007.

[Coecke and Stubbe, 2000] B. Coecke and I. Stubbe. State Transitions as Morphisms for Complete Lattices. *Int. J. Theor. Phys.* **39**: 605–614, 2000.

[Daniel, 1989] W. Daniel. 1989. Axiomatic description of irreversible and reversible evolution of a physical system. *Helv. Phys. Acta* **62**: 941–968, 1989.

[D'Ariano, 2007a] G. M. D'Ariano. Operational Axioms for Quantum Mechanics. arXiv:quant-ph/0611094v3, 2007.

[D'Ariano, 2007b] G. M. D'Ariano. Operational axioms for a C*-algebraic formulation of quantum mechanics. arXiv:quant-ph/0710.1448v1, 2007.

[Einstein *et al.*, 1935] A. Einstein, B. Podolsky and N. Rosen. Can quantum mechanical description of physical reality be considered complete? *Phys. Rev.* **47**: 777–780, 1935.

[Faure and Frólicher, 2000] C. Faure and A. Frólicher. *Modern Projective Geometry.* Kluwer, Dordrecht, 2000.

[Faure *et al.*, 1995] C. Faure, D. J. Moore and C. Piron. Deterministic Evolutions and Schrödinger Flows. *Helv. Phys. Acta* **68**: 150–157, 1995.

[Finkelstein, 1968] D. Finkelstein. Matter, space and logic. Boston Studies in the Philosophy of Science V, D. Reidel, Dordrecht, 1968.

[Fuchs, 2001] C. Fuchs. 2001. Quantum Foundations in the Light of Quantum Information. In: A. Gonis and P. E. A. Turchi. *Decoherence and its Implications in Quantum Computation and Information Transfer: Proceedings of the NATO Advanced Research Workshop, Mykonos Greece, June 25–30, 2000.* IOS Press, Amsterdam. arXiv:quant-ph/0106166, 2001

[Foulis, 1960] D. J. Foulis. Baer *-semigroups. *Proc. Am. Math. Soc.* **11**: 648–654, 1960.

[Foulis *et al.*, 1983] D. J. Foulis, C. Piron and C. H. Randall. Realism, operationalism, and quantum mechanics. *Found. Phys.* **13**: 813–841, 1983.

[Foulis and Randall, 1972] D. J. Foulis and C. H. Randall. Operational Statistics. I. Basic concepts. *J. Math. Phys.* **13**: 1667–1675, 1972.

[Gleason, 1957] A. M. Gleason. 1957. Measures on the closed subspaces of a Hilbert space. *J. Math. Mech.* **6**: 885–893, 1957.

[Gudder and Piron, 1971] S. Gudder and C. Piron. Observables and the Field in Quantum Mechanics. *J. Math. Phys.* **12**: 1583–1588, 1971.

[Haag and Kastler, 1964] R. Haag and D. Kastler. An Algebraic Approach to Quantum Field Theory. *J. Math. Phys.* **5**: 848–861, 1964.

[Hardy, 2001] L. Hardy. Quantum Theory From Five Reasonable Axioms. arXiv:quant-ph/0101012, 2001.

[Jauch, 1968] J. Jauch. *Foundations of Quantum Mechanics.* Addison-Wesley, 1968.

[Jauch and Piron, 1967] J. Jauch and C. Piron. Generalized Localizability. *Helv. Phys. Acta* **40**: 559–570, 1967.

[Jauch and Piron, 1969] J. Jauch and C. Piron. On the stucture of quantal proposition systems. *Helv. Phys. Acta* **42**: 842–848, 1969.

[Joyal and Street, 1991] A. Joyal and R. Street. The Geometry of Tensor Calculus I. *Adv. Math.* **88**: 55–112, 1991.

[Kolmogorov, 1956] A. N. Kolmogorov. *Foundations of the Theory of Probability.* Chelsea Publishing Company, New York, 1956.

[Khatcherian, 1991] G. Khatcherian. Projectales. *J. Pure and Applied Alg.* **74**: 177–195, 1991.

[Kraus, 1983] K. Kraus. *States, Effects, and Operations.* Lecture Notes in Physics 190. Springer, Berlin, 1983.

[Loomis, 1947] L. H. Loomis. On the representation of σ-complete Boolean algebras. *Bull. Amer. Math. Soc.* **53**: 757–760, 1947.

[Ludwig, 1983] G. Ludwig. *Foundations of Quantum Mechanics I.* Springer, New York, 1983.

[Mackey, 1957] G. W. Mackey. Quantum Mechanics and Hilbert Space. *Amer. Math. Monthly* **64**: 45-57, 1957.
[Mackey, 1963] G. W. Mackey. *Mathematical Foundations of Quantum Mechanics.* Benjamin, New York, 1963.
[Mackey, 1968] G. W. Mackey. *Induced Representations of Groups and Quantum Mechanics.* W. A. Benjamin, New York, 1968.
[Mac Lane, 1998] S. Mac Lane. *Categories for the Working Mathematician.* Springer, New York, 1998.
[Maeda and Maeda, 1970] F. Maeda and S. Maeda. *Theory of Symmetric Lattices.* Springer, New York, 1970.
[Moor, 1995] D. J. Moore. Categories of representations of physical systems. *Helv. Phys. Acta* **68**: 658–678, 1995.
[Moore, 1999] D. J. Moore. On State Spaces and Property Lattices. *Stud. Hist. Phil. Mod. Phys.* **30**: 61–83, 1999.
[Mulvey and Pelletier, 1992] C. J. Mulvey and J. W. Pelletier. Category Theory 1991, CMS Conference Proceedings 13, Amer. Math. Soc, pp. 345-360, 1992.
[Pauli, 1958] W. Pauli. *Die allgemeinen Prinzipien der Wellenmechanik.* Handbuch der Physik 5, part I. Springer-Verlag, Berlin, 1958.
[Piron, 1964] C. Piron. Axiomatique quantique. *Helv. Phys. Acta* **37**: 439–468, 1964.
[Piron, 1965] C. Piron. Sur la quantification du système de deux particules. *Helv. Phys. Acta* **38**: 104–108, 1965.
[Piron, 1976] C. Piron. *Foundations of Quantum Physics.* W.A. Benjamin, Reading, Massachusetts, 1976.
[Piron, 1998] C. Piron. *Mécanique quantique. Bases et applications.* Presses polytechniques et universitaires romandes, Lausanne, 1998.
[Piron, 2002] C. Piron. Quantum Theory without Quantification. arXiv:math-ph/0204046 v1, 2002.
[Pool, 1968a] J. C. T. Pool. Baer*-Semigroups and the Logic of Quantum Mechanics. *Commun. math. Phys.* **9**: 118–141, 1968.
[Pool, 1968b] J. C. T. Pool. Semimodularity and the logic of quantum mechanics, 1968.
[Pulmannová, 1985] S. Pulmannová. Tensor products of quantum logics. *J. Math. Phys.* **26**: 1–5, 1985.
[Putnam, 1968] H. Putnam. Is logic empirical? In: R.S. Cohen and M.W. Wartofsky. Boston Studies in the Philosophy of Science V. D. Reidel, Dordrecht, 1968.
[Randall and Foulis, 1973] C. Randall and D. J. Foulis. Operational statistics. II. Manuals of operations and their logics. *J. Math. Phys.* **14**: 1472–1480, 1973.
[Rauch *et al.*, 1975] H. Rauch, A. Zeilinger, G. Badurek, A. Wilfing, W. Bauspies and U. Bonse. *Verification of coherent spinor rotation of fermions.* Phys. Lett. **54 A**: 425, 1975.
[Rédei, 1996] M. Rédei. Why John von Neumann did not Like the Hilbert Space Formalism of Quantum Mechanics (and What he Liked Instead). *Stud. Hist. Phil. Mod. Phys.* **27** : 493–510, 1996.
[Selinger, 2007] P. Selinger. Dagger Compact Closed Categories and Completely Positive Maps. *Electronic Notes in Theoretical Computer Science* **170** : 139–163, 2007.
[Stubbe and Van Steirteghem, 2007] I. Stubbe and B. Van Steirteghem. Propositional systems, Hilbert lattices and generalized Hilbert spaces. In: Engesser, Gabbay and Lehmann. *Handbook of Quantum Logic and Quantum Structures: Quantum Structures.* Elsevier, 2007.
[Valckenborgh, 2000] F. Valckenborgh. Operational axiomatics and compound systems. In: B. Coecke, D. J. Moore and A. Wilce (eds.). *Current Research in Operational Quantum Logic: Algebras, Categories, Languages.* Kluwer Academic Publishers. Dordrecht, 2000.
[Varadarajan, 1962] V. S. Varadarajan. Probability in Physics and a Theorem on Simultaneous Observability. *Comm. Pure Appl. Math.* **15**: 189–217, 1962.
[von Neumann, 1950] J. von Neumann. *Functional Operators Volume II: The Geometry of Orthogonal Spaces.* Annals of Mathematics Studies **22**, 1950.
[von Neuman, 1932] J. von Neumann. *Mathematische Grundlagen der Quantenmechanik.* Springer-Verlag, Berlin, 1932.
[von Neumann, 1955] J. von Neumann. *Mathematical Foundations of Quantum Mechanics.* Princeton University Press, 1955.
[Wick *et al.*, 1952] G. C. Wick, A. S. Wightman and E. P. Wigner. The Intrinsic Parity of Elementary Particles. *Phys. Rev.* **88**: 101–105, 1952.

Edited by K. Engesser, D. M. Gabbay and D. Lehmann

TEST SPACES

Alexander Wilce

0 INTRODUCTION AND OVERVIEW

As is well known, the mathematical framework for quantum physics can be reduced to a generalization of classical probability theory, in which boolean algebras are replaced by what one might broadly call projective geometries — more technically, the projection lattices of von Neumann algebras.

In the simplest cases, e.g., that of a small number of non-relativistic particles, a quantum-mechanical system is represented by a Hilbert space $\mathbf{H}$, in such a way that every physical state of the system is encoded by a density operator W on $\mathbf{H}$, and each real-valued observable, by a self- adjoint operator A on $\mathbf{H}$. The probability that the observable A takes a value in a Borel set $B \subseteq \mathbb{R}$, when the system is in state W, is given by $\mathrm{Tr}\,(\pi_A(B)W)$, where π_A is the spectral measure of A. Evidently, we can regard the projection $\pi_A(B)$ as encoding the *proposition* that the observable A will take its value in the set B. Two such propositions are simultaneously testable iff they are in the range of the spectral projection of a single observable — in other words, iff the corresponding projections commute.

The lattice $L(\mathbf{H})$ of all projection operators $\mathbf{H}$ can thus be regarded as an algebraic model for the "logic" of all *measurement propositions*, i.e, propositions asserting that an observable takes a value in a particular range.[1] Remarkably, once we have decided so to represent this logic, all of the remaining structure of quantum mechanics, up to the choice of a particular Hamiltonian, is simply determined. Although $L(\mathbf{H})$ is not a Boolean algebra, it retains enough "locally" Boolean structure to support a natural definition of probability measures; Gleason's theorem identifies these with the density operators on $\mathbf{H}$. Reversible temporal evolution would naturally be modeled by a pointwise-continuous one-parameter group of symmetries of $L(\mathbf{H})$; by Wigner's Theorem, such a one-parameter group corresponds to a strongly continuous one-parameter group of unitaries on $\mathbf{H}$. By Stone's Theorem, this in turn has the form $t \mapsto e^{-itH}$ for a self-adjoint operator H, which we identify with the system's Hamiltonian observable. In the representation in which H is diagonal, this reduces to the usual Schrödinger equation. [2]

---

[1] This point of view was emphasized by von Neumann in his monograph of 1932 [von Neumann, 1932], and again in his joint paper with Birkhoff [Birkhoff and von Neumann, 1936] in 1936.

[2] For a more detailed sketch of the reduction of quantum mechanics to probability theory on $L(\mathbf{H})$, see the book [Beltrametti and Cassinelli, 1981] of Beltrametti and Cassinelli. For the details, see the book [Varadarajan, 1985] of V. S. Varadarjan. The story is somewhat more

Quantum Logics. That so much structure should follow from one simple postulate — that the logic of measurement propositions is the projection lattice of a Hilbert space — is entrancing. However, once we decide that the logic of measurement should allow for non-simultaneously testable propositions, we open the door to a host of formal possibilities, most of them far less tame than a Hilbert space projection lattice. Why, then, should Nature (or we) hew to the latter? In order even to approach such a question, we need to be able to step outside of quantum probability theory. That is, we require a conceptual and mathematical framework for a generalized probability theory that allows for incommensurable random quantities ("observables"), but that is as far as possible independent of any special physical assumptions. Within this, we may hope to characterize quantum probability theory in a way that casts some light on the question posed above. Broadly speaking, *any* formal structure of "propositions" (or the like) and probabilities, serving as such a framework, counts as a "quantum logic".

There have been various proposals. The most prominent are those of Mackey [Mackey, 1963], Kochen and Specker [Kochen and Specker, 1967], and Piron [Piron, 1964]. For Mackey, a quantum logic is (essentially) an orthomodular poset with a strong order-determining sets of states; for Kochen and Specker, it is a partial boolean algebra; and for Piron, a complete, atomistic, irreducible orthomodular lattice satisfying the covering law — which, as Piron's celebrated representation theorem shows, is representable as the lattice of orthogonally closed subspaces of an inner product space over an involutive division ring. However, all three approaches are problematic. First, all make, either explicitly or tacitly, technical assumptions that are difficult to motivate. For instance, both Mackey, and Kochen and Specker assume that pairwise compatible propositions should be jointly compatible. Piron's approach, too, contains *ad hoc* elements, notably the conditions required to make his lattice of propositions orthocomplemented. Secondly, as was pointed out by Aerts (in connection with Piron lattices) and by Foulis and Randall (in connection with orthomodular posets), these structures do not admit natural natural tensor products. Finally, all three approaches are poorly equipped to deal with situations involving repeated measurements, where, in quantum mechanics, one needs to take account of relative phase information.

Test Spaces. A different approach, free from these difficulties, was pursued by D. J. Foulis and C. H. Randall in a series of more than two dozen papers (e.g., [Foulis *et al.*, 1985; Foulis and Randall, 1971; Foulis and Randall, 1981a; Foulis and Randall, 1981b; Randall and Foulis, 1970; Randall and Foulis, 1976; Randall and Foulis, 1978; Randall and Foulis, 1983a; Randall and Foulis, 1983b; Randall *et al.*, 1973]), beginning in the early 1970s and continuing until Randall's

complicated for quantum field theory, where $L(\mathbf{H})$ must be replaced by the projection lattice of a von Neumann algebra. However, the key result, Gleason's Theorem, as extended by Christensen [Christensen, 1982] and Yeadon [Yeadon, 1983], applies here too: if $\mathcal{A}$ is a von Neumann algebra having no type I_2 factor, then every countably-additive probability measure on the projection lattice of $\mathcal{A}$ extends uniquely to a state on $\mathcal{A}$.

untimely death in 1987. The Foulis-Randall theory begins with the simple idea of a *test space*. Mathematically, this is just a hypergraph, i.e., a collection $\mathfrak{A}$ of sets; but the intended interpretation is that each set $E \in \mathfrak{A}$ represents the outcome set of some discrete, classical statistical experiment.[3] Accordingly, each $E \in \mathfrak{A}$ is called a *test*; subsets of tests are called *events*, and elements of tests — that is, points $x \in X$ — are called *outcomes*. The simplest example, of course, is that in which $\mathfrak{A}$ consists of just one test. More generally, in classical probability theory, it is assumed that the various sets $E \in \mathfrak{A}$ can be represented as partitions of a common set M, which may then be regarded as the outcome-set of a single, idealized measurement, of which each test in $\mathfrak{A}$ is a "coarse-grained" version; one further assumes that any two tests $E, F \in \mathfrak{A}$ have a common refinement that again belongs in $\mathfrak{A}$. In contrast, the simplest quantum test space takes for $\mathfrak{A}$ the set of orthonormal bases for a Hilbert space. In this case, each test $E \in \mathfrak{A}$ is maximally informative — there exists *no* common refinement for two distinct tests.

In both of the preceding examples, distinct tests are permitted to overlap, that is, to share outcomes. In fact, from the combinatorial structure of $\mathfrak{A}$ alone one can recover, in the classical case, the boolean algebra $\mathcal{B}$ of measurable sets, and, in the quantum case, the lattice $L(\mathbf{H})$ of closed subspaces of $\mathbf{H}$. More generally, the combinatorial structure of an arbitrary test space $\mathfrak{A}$ allows one to associate to it various order-theoretic and partial-algebraic invariants that can serve as "logics" of measurement propositions. However, these logics are regarded, in the Foulis-Randall approach, only as useful mathematical invariants of the underlying test space, which is the real object of interest.

As has already been mentioned, test spaces avoid all of the problems described above in connection with order-theoretic approaches to quantum logic. In particular, test spaces admit a very natural tensor product, and can be composed in a way that faithfully mirrors the construction of iterated experiments in orthodox quantum mechanics, phases and all. But perhaps the single greatest advantage test spaces enjoy, as a framework for quantum logic, is that they are conceptually very simple, and hence, transparent.

The theory of test spaces commands, by now, a literature consisting of scores of papers, scattered through a variety of journals, as well as a good number of unpublished doctoral dissertations. However, it has never enjoyed a monographic treatment, and so remains in some degree inaccessible, and hence, less well known than it ought to be. I hope the present paper will help to correct this. I have tried to write with a general mathematical audience in mind, assuming little background beyond the essential functional-analytic machinery of quantum theory (in particular, elementary spectral theory), some acquaintance with ordered sets and lattices, and basic abstract algebra. I have tried, also, to produce something that a non-expert might find agreeable to read. Accordingly, I have in many cases included fully detailed proofs of even rather elementary propositions, where I thought they might help the reader to absorb the flavor of the subject, while omitting some

[3]By which one means, a little circularly, some process, such as a measurement, that *has* a well-defined outcome-set.

longer, and more technically involved, arguments in the interest of maintaining the flow of the discussion.

SYNOPSIS

Background on Quantum Logic. In order to make these notes reasonably self-contained, I have collected in Section 1 some essential background material on quantum logics, beginning with a sketch of the elementary theory of orthomodular lattices and posets, orthoalgebras, and effect algebras, and their characterization in terms of partial abelian semigroups. Included also is a précis of Mackey's axioms for quantum mechanics, a brief discussion of the Piron representation theorem, the characterization of complete orthomodular lattices in terms of orthogonality spaces due to J. R. Dacey, and the celebrated "Loop Lemma" of R. Greechie.

Test Space Basics. Section 2 develops the elementary theory of test spaces. After introducing the concept of a test space, and illustrating it with many examples, I discuss certain basic constructions with, and mappings between, test spaces. This section concludes with a brief discussion of spaces of vector-valued weights associated with test spaces, which may be regarded as optional (as it is referred to again only at the end of Section 5).

Logics of Test Spaces. Associated to any test space is an involutive poset called its *logic*. This is the subject of Section 3. In all but pathological cases, the logic of a test space is orthocomplemented, but little more can be said unless some additional conditions are imposed. In particular, if the test space satisfies a simple combinatorial condition called *algebraicity* (in older terminology, if it is a *manual*), then its logic has the structure of an orthoalgebra; conversely, every orthoalgebra arises as the logic of a canonical algebraic test space. Among other results proved in this section is a generalization of the Loop Lemma, due to Foulis, Greechie and Rüttimann, giving a sufficient condition for an algebraic test space to have a lattice-ordered logic.

Supports, Entities and Property Lattices. Section 4 concerns might be called the *FPR formalism.* This derives from the joint paper [Foulis *et al.*, 1985] of Foulis, Piron and Randall, in which a test space is equipped with a family Σ of *supports* — roughly, sets of outcomes that are possible in various states — and from this is constructed a complete lattice $\mathcal{L}(\mathfrak{A}, \Sigma)$, called the *property lattice* of the entity. We present a basic theorem, due to Foulis and Randall, giving sufficient conditions for the logic Π and the property lattice $\mathcal{L}$ to be isomorphic — in which case, each is a complete OML. We then discuss how one can formulate many parts of the standard machinery of algebraic quantum logic – in particular, centers, Sasaki projections, and the covering law — in terms of supports.

Sections 1 - 4, which are designed to be read in sequence, present what I would regard as the essential core of the Foulis-Randall theory. In contrast, sections 5,

6, and 7 are more nearly independent of one another, and represent three areas of still active development:

Tensor Products. Section 5 concerns tensor products of test spaces, orthoalgebras, and entities. I reproduce the basic example, due to Foulis and Randall [Foulis and Randall, 1981a], showing that no suitable tensor product construction is available for orthomodular posets or lattices, and then discuss the bilateral product of test spaces, and associated tensor products of orthoalgebras and property lattices. I also discuss, briefly, the representation of states on the bilateral tensor product of frame manuals as bilinear forms, and the resulting "unentangled Gleason Theorem".

Symmetric Test Spaces. The basic classical test space, consisting of a single discrete outcome-set, and the basic quantum-mechanical test space, consisting of frames of a Hilbert space, both exhibit strong symmetry properties. In Section 6, I summarize recent work [Wilce, 2005c] on test spaces subject to similarly strong symmetry requirements. In particular, I show that a minimum of group-theoretic data suffices to fix the structure of such a test space, and give a group-theoretic characterization of algebraicity in this context. I also discuss how one can extend a test space to accommodate a larger symmetry group.

Topological Test Spaces. Section 7 deals with test spaces in which the underlying set of outcomes carries a topology. This is a natural direction in which to develop this theory, since the basic quantum-mechanical example enjoys a rich and very relevant topological structure. I outline some of the basic general theory of topological test spaces and their logics, following [Wilce, 2005a; Wilce, 2005b]. Among other results, I show that a very large class of topological test spaces always have dense semi-classical sub-test spaces. This shows that the Meyer/Clifton/Kent "non-contextual hidden variables" theorems [Meyer, 1999; Clifton-Kent, 2000] are fairly generic.

What has been omitted. These notes deal primarily on logico-algebraic and combinatorial aspects of the theory of test spaces. In the interest of producing a work that is short enough and focussed enough to be of use to the general quantum logician (a narrow enough audience as it is!), I have omitted a great deal of analytic machinery supporting a generalized probability theory — notably, the generalized Bayesian probability theory of, e.g., [Randall and Foulis, 1976] and [Gaudard, 1977]. Omitted, too, is any discussion of the way in which the Foulis-Randall machinery can be extended to accommodate effect algebras — for that, see [Pulmannová and Wilce, 1995; Foulis *et al.*, 1996]. I should also mention that there exists a certain body of work applying category-theoretic methods to the theory of test spaces including, e.g., [Lock, P., 1981; Nishimura, 1995], which is not dealt with here.

All three of these topics bear on a particularly interesting issue, namely, the degree to which test spaces can serve as a foundation for a general non-classical information theory, extending (and, one would hope, clarifying) what is currently known about quantum information theory. This, however, is a project just barely begun (see, e.g., [Barnum *et al.*, 2005]).

1 BACKGROUND ON QUANTUM LOGICS

Before introducing test spaces, it seems prudent to begin with a detailed review of quantum logic in its more familiar, order-theoretic formulation. I begin by sketching the basic algebraic theory of orthomodular lattices and posets, from both an order-theoretic and a partial algebraic point of view, and then discuss the coordinatization theorems of Piron, Amemiya and Araki, and Solér, characterizing lattice-theoretically the projection lattices of Hilbert spaces. Readers well familiar with this material may wish to proceed directly to sub-section 1.4, which develops the representation theory for complete ortholattices in terms of so-called orthogonality spaces. Here, I discuss Dacey's Theorem, characterizing complete orthomodular lattices in these terms, as well as Greechie's celebrated Loop Lemma. This section closes with a critique of traditional apporaches to quantum logic, the centerpiece of which is the Foulis-Randall example showing that there is no satisfactory tensor product of orthomodular lattices or posets.

1.1 Orthomodular Lattices and Posets

Recall that a *lattice* is a partially ordered set $(L, \leq)$ in which every two elements, say a and b, have both a greatest lower bound, or *meet*, $a \wedge b$, and a least upper bound or *join*, $a \wedge b$. Note that $a \leq b$ iff $a = a \wedge b$ iff $a \vee b = b$. We say that L is *complete* iff every subset A of L has a meet and join, denoted $\bigwedge A$ and $\bigvee A$, respectively. A lattice is *bounded* iff it has a least element, usually denoted 0, and a greatest element, usually denoted $\mathbf{1}$. Two elements, a and b, in a bounded lattice L are said to be *complements* of one another iff $a \wedge b = 0$ and $a \vee b = \mathbf{1}$. If every element has a complement, then L is *complemented.* A lattice L is said to be *distributive* iff the two distributive laws

$$a \wedge (b \vee c) = (a \wedge b) \vee c \text{ and } a \vee (b \wedge c) = (a \vee b) \wedge (a \vee c)$$

hold for all $a, b, c \in L$. We have the following very basic

LEMMA 1. *An element of a bounded distributive lattice can have at most one complement.*

Proof. Suppose that $b, c \in L$ are complements for $a \in L$. Then

$$b = b \wedge \mathbf{1} = b \wedge (a \vee c) = (b \wedge a) \vee (b \wedge c) = 0 \vee (b \wedge c) = b \wedge c.$$

Thus, $b \leq c$. By the same token, $c \leq b$. ■

DEFINITION 2. A *Boolean algebra* is a complemented, distributive lattice.

By Lemma 1, every element a in a Boolean algebra has a *unique* complement a'. Thus, a lattice in which an element has two or more complements is *ipso facto* not distributive. A familiar example is the lattice of subspaces of a finite-dimensional vector space. Indeed, if $\mathbf{V}$ is a vector space (real or complex, say) of dimension greater than 1, then every proper subspace $\mathbf{M}$ of $\mathbf{V}$ has infinitely many distinct complements. On the other hand, in the presence of an inner product on $\mathbf{V}$, there is a *preferred* complement, namely, the subspace $\mathbf{M}^{\perp} = \{y \in \mathbf{V} | \langle x, y \rangle = 0 \forall x \in \mathbf{M}\}$ orthogonal to $\mathbf{M}$. The assignment $\mathbf{M} \mapsto \mathbf{M}^{\perp}$ enjoys a number of formal properties reminiscent of complementation in a Boolean algebra.

DEFINITION 3. Let L be a bounded poset (that is, a poset with least element 0 and greatest element $\mathbf{1}$). An *orthocomplementation* on L is a map $' : L \to L$ such that for all $a, b \in L$,

(i) $a \leq b \Rightarrow b' \leq a'$

(ii) $a = a''$

(iii) $a \wedge a' = 0$.

Note that from this we have $\mathbf{1}' = \mathbf{1} \wedge \mathbf{1}' = 0$, whence $\mathbf{1}' = 0$. An *orthoposet* is a pair $(L,')$ where L is a poset and $'$ is a distinguished orthocomplementation on L. An orthoposet that is a lattice is called an *ortholattice*.

For later reference: a mapping $' : L \to L$ satisfying conditions (i) and (ii), and with $\mathbf{1}' = 0$, but not necessarily satisfying condition (iii), is called an *involution* on L.

Familiar examples of ortholattices include boolean algebras and lattices of subspaces of finite-dimensional inner-product spaces. Note that any subset of an ortholattice that contains both 0 and 1, and is closed under the taking of orthocomplements, is an orthoposet. Note, too, that a version of deMorgan's laws hold in any orthoposet. Here, we denote the meet and join of a family $\{a_i | i \in I\}$ of elements of L by $\bigvee_i a_i$ and $\bigwedge_i a_i$, anti-respectively:

LEMMA 4. *Let L be any orthoposet. Then for any family of elements $\{a_i\} \subseteq L$,*

$$\left(\bigvee_i p_i\right)' = \bigwedge_i p_i' \quad \textit{and} \quad \left(\bigwedge_i p_i\right)' = \bigvee_i p_i'$$

(in the sense that, if either side of the above equation is defined, so is the other, and the two are equal).

Proof. From (i) in Definition 3, we have $(\bigvee_i a_i)' \leq a_j'$ for every $j \in I$. If $b \leq a_j'$ for every $j \in I$, then $a_j = a_j'' \leq b'$ for every j, whence, $\bigvee_i a_i \leq b'$, whence, $b = b'' \leq (\bigvee_i a_i)'$. Thus, $(\bigvee_i a_i)' = \bigwedge_i a_i$. The other identity now follows by taking orthocomplements. ■

An immediate consequence is that the dual of condition (iii) in the definition of an orthocomplement holds, namely, $a \vee a' = \mathbf{1}$ for all $a \in L$.

It is convenient to introduce the following usages and notations:

DEFINITION 5. If L is an orthoposet and $a, b \in L$, then

(i) if $a \leq b$, we write $b - a$ for $b \wedge a'$, *provided this exists*, and speak of it as the *relative difference* between b and a.

(ii) We say that elements $a, b \in L$ are *orthogonal* iff $a \leq b'$, and indicate this relation by writing $a \perp b$.

Note that if $a \leq b$ and $b - a$ exists, then $a \perp (b - a)$, since

$$(a - b)' = (b \wedge a')' = b' \vee a \geq a.$$

The principal focus of this section is the class of orthoposets in which the relative difference operation is reasonably well-behaved.

DEFINITION 6. An *orthomodular poset* (OMP) is an orthoposet P satisfying

(i) $a \vee b$ exists whenever $a \perp b$.

(ii) $a \leq b \Rightarrow (b - a) \vee a = b$.

Condition (ii) is often called the *orthomodular law*.[4] An *orthomodular lattice* (OML) is simply an orthomodular poset that happens to be a lattice (in which case, note, condition (i) above is vaccuously satisfied). Orthomodularity has a particularly nice characterization in this case. I leave the proof of the following to the reader:

LEMMA 7. *Let* $(L, \leq, ')$ *be an ortholattice. The following are equivalent:*

(a) L *is orthomodular;*

(b) For all $a, b \in L$, *if* $a \leq b$ *and* $b - a = 0$, *then* $a = b$

EXAMPLES 8.

(i) Clearly, any boolean algebra is an OML.

(ii) Let $L(\mathbf{H})$ denote the lattice of closed subspaces of a Hilbert space $\mathbf{H}$. This is orthocomplemented by the map $\mathbf{M} \mapsto \mathbf{M}^{\perp}$. If $\mathbf{N} \in L(\mathbf{H})$, let $p_{\mathbf{N}}$ denote the orthogonal projection onto $\mathbf{N}$. If $\mathbf{N} \subseteq M$ and $x \in \mathbf{M} \setminus \mathbf{N}$, then $x - p_{\mathbf{N}}(x) \in \mathbf{M} \cap N^{\perp}$, where $p_{\mathbf{N}}(x)$ is the orthogonal projection of x onto $\mathbf{M}$. Thus, $\mathbf{M} \cap \mathbf{N}^{\perp} = \{0\}$ implies $\mathbf{M} = \mathbf{N}$. By Lemma 7, then, $L(\mathbf{H})$ is orthomodular.

[4]The existence of the quantity $(b - a) \vee a$ requires some comment. Notice that if $a \leq b$, then $a \perp b'$, so $(a \vee b')' = b - a$ exists, by (i). By the preceding Lemma, $(b - a) \vee a$ therefore also exists. Conversely, if $a \leq b \Rightarrow b - a$ exists, then $a \leq b' \Rightarrow b' \wedge a' = (b \vee a)'$ exists, whence, so does $a \vee b$.

Note that $L(\mathbf{H})$ can equivalently be described as the lattice of projection operators on $\mathbf{H}$, under the obvious isomorphism $\mathbf{M} \mapsto p_{\mathbf{M}}$. We'll use the notation $L(\mathbf{H})$ for both lattices, usually leaving it to context to make it clear which we intend.

(iii) A simple example of a *non*-orthomodular ortholattice is the so-called *benzene ring*, the Hasse diagram of which appears below:

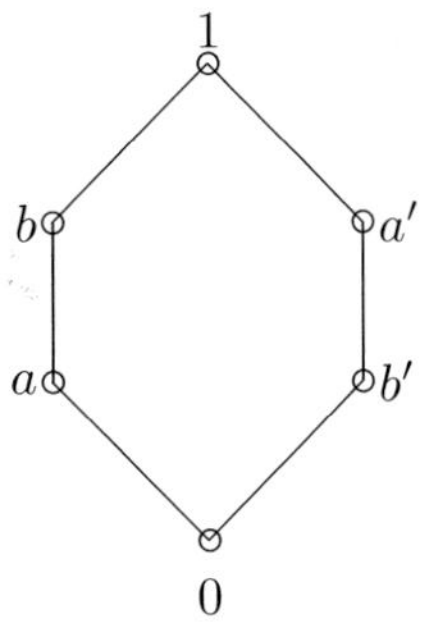

Here $a \leq b$, but $b \wedge a' = 0$, so that $(b \wedge a') \vee a = a$, rather than b as required by the orthomodular law. Indeed, it is easy to see that a lattice L is orthomodular if and only if it contains no sub-ortholattice isomorphic to the benzine ring.

(iv) Here is the Hasse diagram of a small non-boolean OML:

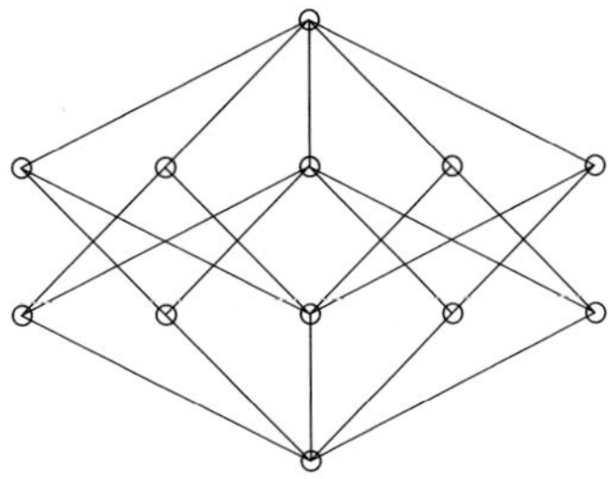

Notice that this lattice consists of two two Boolean algebras "pasted" together at $0, c, c'$ and 1.

(v) Here is an example of an OMP that is not a lattice. Let X be a finite set of even cardinality six or greater, and let L be the collection of subsets of X having even cardinality, ordered by set-inclusion. Then, for all $a, b \in L$, $a \perp b$ iff a and b are disjoint, and in this case $a \cup b$ belongs to L, where it serves as $a \vee b$. Also note that if $a, b \in L$ with $a \leq b$, then $b' \wedge a$ exists in L, being just the set-theoretic difference $b \setminus a \in L$. It follows that if $a \leq b$, then $(b' \wedge a) \vee a = (b \setminus a) \cup a = b$, so L is orthomodular. On the other hand, L is not a lattice. Indeed, let u, v, x, y and z be five distinct elements of X, and let $a = \{x, y\}$ and $b = \{y, z\}$. Then $\{x, y, z, u\}$ and $\{x, y, z, v\}$ are distinct minimal upper bounds for a and b in L.

9 THE CENTER OF AN OML. If L is an OML and $a, b \in L$, one says that a and b *commute*, writing aCb, iff $a = (a \wedge b) \vee (a \wedge b')$. One can show that a and b commute iff they belong to a common boolean sub-OML of L. If $L = L(\mathbf{H})$ for a Hilbert space $\mathbf{H}$, projections $a, b \in L(\mathbf{H})$ commute in the above sense iff $ab = ba$. The *center* of L, i.e., the set $\mathbf{C}(L)$ of all elements of L commuting with every element of L, is a boolean sub-OML of L. It can be shown that $a \in \mathbf{C}(L)$, then $L \simeq [0, a] \times [0, a']$. Thus, the possible direct-product decompositions of L are parametrized by $\mathbf{C}(L)$. In particular, L is *irreducible* — that is, has no direct-product decompositions — iff $\mathbf{C}(L) = \{0, 1\}$. For details, see Kalmbach [Kalmbach, 1983].

10 SASAKI PROJECTIONS. On any ortholattice L, we may define a binary operation $\phi : L \times L \to L$ by $\phi(a, b) := a \wedge (a' \vee b)$. If L is Boolean, of course, $\phi(a, b)$ is simply $a \wedge b$. If L is the lattice of closed subspaces of a Hilbert space and $\mathbf{M}, \mathbf{N} \in L(\mathbf{H})$, $\phi(\mathbf{M}, \mathbf{N}) = \mathbf{M} \cap (\mathbf{M}^{\perp} \vee \mathbf{N})$ is the subspace obtained by projecting $\mathbf{N}$ orthogonally onto $\mathbf{M}$. It is not difficult to show that an ortholattice L is orthomodular iff, for all $a, b \in L$, $b \leq a \Rightarrow \phi(a, b) = b$. In this context, ϕ is usually called the *Sasaki map*, and the mapping $\phi_a : L \to L$ taking b to $\phi(a, b)$ is the *Sasaki projection* associated with $a \in L$.

11 ATOMS. An *atom* in an ordered set L with least element 0, is a minimal non-zero element. A poset L is said to be *atomic* iff there exists at least one atom under each non-zero element, and *atomistic* in case every element of L is the join of a set of atoms. It is not hard to see that if L is any complete atomic OML, then it is automatically atomistic. Indeed, let $a \in L$, and let A denote the set of atoms under a. Let $b = \bigvee A$; then $b \leq a$; hence, $a \ominus b$ exists in L. If $a \ominus b$ is non-zero, then, as L is atomic, there exists an atom x under $a \ominus b$. But then $x \not\leq b$, which is absurd, as b is the joint of all atoms under a.

STATES ON OMPS

DEFINITION 12. Let L be an OMP. A *state* on L is a map $\mu : L \to [0, 1]$ such that

(i) $\mu(1) = 1$

(ii) $a \perp b \Rightarrow \mu(a \oplus b) = \mu(a) + \mu(b)$.

If L is a Boolean algebra, then a state is just a (finitely additive) probability measure on L. Accordingly, one also speaks of states on arbitrary orthomodular posets or, more generally, effect algebras, as probability measures. This suggests the following

DEFINITION 13. A state μ on an orthomodular poset L is σ*-additive* iff, for every pair-wise orthogonal sequence (a_n) in L for which $\bigvee_n a_n$ exists, $\mu(\bigvee_n a_n) =$

$\sum_n \mu(a_n)$. An OMP in which the join of a countable pair-wise orthogonal set always exists is called a *σ-OMP*.

The following celebrated result characterizes the σ-additive states on $L(\mathbf{H})$:

THEOREM 14 [Gleason, 1957]. *Every countably additive state* μ *on* $L(\mathbf{H})$, *for a separable Hilbert space* $\mathbf{H}$ *of dimension* ≥ 3, *extends uniquely to a positive linear functional* $\phi : \mathcal{B}(\mathbf{H}) \to \mathbb{R}$ *with* $\phi(\mathbf{1}) = 1$, *and hence, has the form* $\mu(p) = Tr(Wp)$ *where* W *is a density operator — that is, a positive, trace-class operator with trace* 1 *— on* $\mathbf{H}$.

Gleason's theorem has since been extended to non-separable Hilbert spaces [Eilers and Horst, 1975], to finitely-additive measures, and to the projection lattices of von Neumann algebras not containing type-I_2 summands [Christensen, 1982; Yeadon, 1983].

The connection between an orthomodular poset or lattice L and its state-space can be very loose. A theorem of Schultz [Schultz, 1974] shows that any compact convex set can be represented as the state-space of an orthomodular lattice. On the other hand, there exist orthomodular lattices admitting *no* states whatever; other examples exist of orthomodular lattices admitting just a single state. Indeed, results of Pták [Pták, 1983], Navara [Navara, 1992] and Harding and Navara [Harding and Navara, 2000] show that one can manufacture orthomodular lattices having any desired center, automorphism group, and (compact) state-space. This notwithstanding, a good deal of classical measure theory can be made to work with OMPs in place of boolean algebras. This includes versions of the Brookes-Jewett and Nikodym convergence theorems [D'Andrea and De Lucia, 1991; Habil, 1995], and the Yosida-Hewitt and Lebsgue decompositions [De Simone and Navara, 2001; De Simone and Navara, 2002].

1.2 Orthomodular Posets as Partial Semigroups

It is possible, and in some ways advantageous, to view orthomodular posets as a special class of partial semigroups. In any orthoposet, we may define a partial binary operation $\oplus$ — the *orthogonal sum* — by setting $a \oplus b = a \vee b$ iff $a \leq b'$ and the join exists. Thus, condition (i) of Definition 6 says that, in an orthomodular poset, $a \oplus b$ is defined iff $a \perp b$. Using the notation $b \ominus a$ for $b \wedge a'$, then the orthomodular law reads $a \leq b \Rightarrow (b \ominus a) \oplus a = b$.

Let us agree to call a partial binary operation $*$ on a set S *commutative* iff, for all $a, b \in S$, the existence of $a * b$ implies that of $b * a$, and the two agree. Similarly, we say that $*$ is *associative* iff, for all $a, b, c \in S$, whenever either of $a * (b * c)$ and $(a * b) * c$ is defined, the other is also defined, and the two agree. Finally, let us call $*$ *cancellative* iff $a * b = a * c \Rightarrow b = c$ for all $a, b, c \in L$.

Note that, in an arbitrary poset P, the partial operation $\vee$ is associative only in the weak sense that if $a \vee (b \vee c)$ and $(a \vee b) \vee c$ *both* exist, then they agree — one can easily construct examples in which one of these exists but the other does not. Note, too, that $a \vee b = a \vee c$ does not generally entail that $b = c$; so the join operation is not cancellative.

LEMMA 15. *Let $(L, \leq, ')$ be an orthoposet in which $a \oplus b := a \vee b$ is defined for every orthogonal pair a, b. Then*

(a) $\oplus$ is associative;

(b) L is orthomodular iff $\oplus$ is cancellative.

Proof. (a) To see that $\oplus$ is associative, suppose $a, b, c \in L$ with $a \oplus (b \oplus c)$ defined. Then $b \leq c'$ and $a \leq (b \oplus c)' = (b \vee c)' = b' \wedge c'$. Hence, $a \leq b'$, whence $a \oplus b = a \vee b$ exists. Since $a \leq c'$ and $b \leq c'$, $a \oplus b \leq c'$; whence, $(a \oplus b) \oplus c$ also exists. By the weak associativity of existing joins, the two sums are the same.

(b) Suppose L is orthomodular. To see that orthogonal addition is cancellative, we exploit the orthomodular identity: If $a \perp b$, $a \perp c$, and $a \oplus b = a \oplus c$, then $c = (a \oplus c) - a = (a \oplus b) - a = b$. Conversely, suppose $\oplus$ is cancellative. If $a \leq b$, then $a \perp (b - a)$ by Lemma 4. As $b \wedge a' = (b' \vee a)'$, we have

$$(b \wedge a') \vee (b' \vee a) = \mathbf{1}.$$

Both of the joins appearing above are *orthogonal* joins, so we have

$$\mathbf{1} = ((b - a) \oplus a) \oplus b'$$

by the associativity (and evident commutativity) of $\oplus$. Since 1 is also equal to $b \vee b'$, cancellativity yields $(b - a) \oplus a = b$. But this is just the orthomodular law. ■

A pair $(L, \oplus)$ consisting of a set L and a commutative, associative partial operation $\oplus$ on L, is called a *partial abelian semigroup* or PAS [Wilce, 1998]. A *zero element* for $(L, \oplus)$ is an element $0 \in L$ satisfying $0 \oplus a = a \oplus 0 = a$ for every $a \in L$. Such an element, if it exists, is unique. Evidently, one can formally adjoin a zero to any PAS; accordingly, we assume in what follows that every PAS has a zero.[5]

According to Lemma 15, every OMP $(L, \leq)$ gives rise to a PAS $(L, \oplus)$. One can ask for an abstract characterization of OMPs in terms of partial abelian semigroups. It is easy to see that the PAS obtained from an OMP satisfies all of the following conditions:

DEFINITION 16. A partial abelian semigroup $(L, \oplus)$ is said to be

(i) *positive* iff, for all $a, b \in L$, $a \oplus b = \mathbf{0}$ implies $a = b = 0$;

(ii) *cancellative* iff, for all $a, b, c \in L$, $a \oplus b = a \oplus c$ implies $b = c$;

(iii) *unital* iff it contains an element $\mathbf{1}$ such that for every $a \in L$, there exists an element $b \in L$ with $a \oplus b = \mathbf{1}$. Such an element is termed a *unit* for L; and

[5]Arguably, a PAS with zero element should be called a "partial abelian monoid"; but I think considerations of euphony dictate otherwise.

(iv) *orthocoherent* iff, for all $a, b, c \in L$, if $a \oplus b, b \oplus c$ and $c \oplus a$ all exist, then so does $a \oplus b \oplus c$.

It is straightforward to show that if a PAS L is positive and cancellative, then it can be partially ordered by setting $a \leq b$ iff $b = a \oplus c$ for some $c \in L$ [Wilce, 1998]. In this context, the unit element $\mathbf{1}$ of condition (iii) is unique, and maximal in the poset $(L, \leq)$; moreover, for each element $a \in L$, there is a unique element $a' \in L$ with $a \oplus a' = \mathbf{1}$. The mapping $a \mapsto a'$ is an involution, and $a \oplus b$ exists iff $a \leq b'$.

The following is not hard to prove directly, but it will emerge even more easily as a consequence of material discussed below.

LEMMA 17. *Let $(L, \oplus)$ be a positive, cancellative, unital PAS. Then the associated involutive poset $(L, \leq, ')$ is an OMP iff L is orthocoherent. In this case, $a \oplus b = a \vee b$ for all $a \leq b'$.*

MACKEY'S AXIOMS

In an influential paper [Mackey, 1957] and subsequent monograph [Mackey, 1963], George Mackey proposed a set of axioms for a generalized probability calculus, within which he hoped to characterize quantum probability. Mackey's axioms were framed in terms of a structure $(\mathfrak{O}, S, p)$, where $\mathfrak{O}$ is a set of "observables", S is a set of "states", and p is a function associating to each $a \in \mathfrak{O}$ and $s \in S$, a Borel probability measure on the real line, intended to represent the distribution of possible values of a when measured in the state s. Rather than reproducing Mackey's axiomatic in detail, I'll paraphrase it as follows. Let S be any set, and let $\mathbb{R}^S$ be the set of all real-valued functions $f : S \to \mathbb{R}$, ordered pointwise. Let $\mathbf{0}$ and $\mathbf{1}$ denote the constant functions on S with value 0 and 1, respectively. By the *unit interval* in $\mathbb{R}^S$, I mean the set $[\mathbf{0}, \mathbf{1}] = \{f \in \mathbb{R}^S | 0 \leq f(s) \leq 1| \text{ for all } s \in \S\}$.

DEFINITION 18. A *Mackey system* over S is a set $L \subseteq [\mathbf{0}, \mathbf{1}]$ such that

(i) $\mathbf{1} \in L$;

(ii) $\forall f \in L, \mathbf{1} - f \in L$;

(iii) If $f_1, f_2, f_3 \in L$ and $f_i + f_j \leq 1$ for all $i \neq j$, $i, j = 1, 2, 3$, then $f_1 + f_2 + f_3 \in L$.

Remark: From (i) and (ii), we see that $\mathbf{0} \in L$. It follows from (iii) (with $f_3 = \mathbf{0}$) that for any $f_1, f_2 \in L$, $f_1 + f_2 \leq \mathbf{1} \Rightarrow f_1 + f_2 \in L$. Notice that (iii) is a strong property. In particular, it is not satisfied by $[\mathbf{0}, \mathbf{1}]$.

The intended interpretation of a Mackey system over S is that every $s \in S$ represents a possible *state* of a physical system, every $f \in L$ represents a binary observable on that system, and $f(s)$ gives the probability that f has value 1 when measured in state s. The following simple result[6] captures in a compact way much of the content of Mackey's construction.

[6]Attributed to Maczynski in [Beltrametti and Cassinelli, 1981]

PROPOSITION 19. *Let L be a Mackey system over S, ordered pointwise. For each $f \in L$, let $f' := 1 - f$. Then $(L, \leq, ')$ is an orthomodular poset. Moreover, each $s \in S$ defines a state $\hat{s}$ on L via $\hat{s}(f) = f(s)$.*

Proof. If $f, g \in L$, define $f \oplus g = f + g$, provided that this lies in L. It is straightforward to check that $(L, \oplus)$ is a cancellative, positive PAS. Conditions (i) and (ii) in the definition of a Mackey system guarantee that L has a unit, and condition (iii) is essentially orthocoherence. Thus, by Lemma 17, $(L, \leq)$ is an OMP. ■

DEFINITION 20. A set Δ of states on an OMP L is said to be *order determining* iff, for all $a, b \in L$,

$$\mu(a) \leq \mu(b) \ \forall \mu \in \Delta \ \Rightarrow \ a \leq b.$$

One says that Δ is *strongly order-determining* iff

$$a \not\leq b \ \Rightarrow \ \exists \mu \in \Delta \ \mu(a) = 1 \ \& \ \mu(b) < 1.$$

It is not difficult to see that the states $\hat{s}$, $s \in S$, constitute a strong, order-determining set of states for a Mackey system L over S. Conversely, if Δ is any strong, order-determining set of states for an orthomodular poset L, then one can embed L in $[0,1]^S$ via the evaluation mapping $a \mapsto \hat{a}$ with $\hat{a}(\mu) = \mu(a)$ for every $\mu \in S$. Then $\{ \hat{a} \mid a \in L \}$ is a Mackey system over S, and, for every $s \in S, a \in L$, $\hat{s}(\hat{a}) = s(a)$. Thus, the models of Mackey's axioms are pairs (L, S) where L is an orthomodular poset and S is a strong, order-determining, convex set of states on L. In much of the literature from the mid 1960s onwards, a *quantum logic* is taken to be, by definition, just such a pair.[7]

Effect Algebras and Orthoalgebras

A cancellative, positive unital partial abelian is more usually called an *effect algebra* [Bennett and Foulis, 1994] . In this language, then, Lemma 17 says that an OMP is effectively the same thing as an orthocoherent effect algebra. Examples of non-orthocoherent effect algebras abound: if G is any ordered abelian group, the interval $[0, e]$ of positive elements below any particular positive element e, is an effect algebra under the operation $a \oplus b = a + b$, provided $a + b \leq e$. One can show that all cancellative, unital partial abelian semigroups arise as extensions of effect algebras by abelian groups [Feldman and Wilce, 1998].

If a is an element of an effect algebra L, the unique element $a' \in L$ such that $a \oplus a' = \mathbf{1}$ is called the *orthosupplement* of a. As mentioned above, the mapping $a \mapsto a'$ is an involution. A particularly natural class of effect algebras is that in which the orthosupplement behaves as an ortho*complement*.

LEMMA 21. *Let $(L, \oplus)$ be an effect algebra, ordered in the usual way. The following are equivalent:*

[7] To be scrupulously correct, I should note that Mackey required that L be a σ-OMP and S, a convex set of σ-additive state.

(a) *For all* $a \in L$, $a \wedge a' = 0$;

(b) *The mapping* $a \mapsto a'$ *is an orthocomplementation on* L;

(c) $a \oplus a$ *is defined only if* $a = 0$;

(d) *For any orthogonal elements* $a, b \in L$, $a \oplus b$ *is a* minimal *upper bound for* a *and* b.

Proof. Since $a \mapsto a'$ is in any event an involution, condition (a) is all that is required to make it an orthocomplementation. If it *is* an orthocomplementation, and $a \oplus a$ is defined, then $a \leq a'$, whence, $a = a \wedge a' = 0$. This establishes that (a) implies (b) and (b) implies (c). Now suppose that condition (c) holds, and that a and b are orthogonal elements of L. If $c \in L$ satisfies $a, b < c \leq a \oplus b$, then there exist elements $x, y, z \in L$ with $c = a \oplus x = b \oplus y$ and $a \oplus b = c \oplus z$. But then, by associativity, we have

$$a \oplus b = (a \oplus x) \oplus z = a \oplus (x \oplus z).$$

By cancellativity, $x \oplus z = b$. Similarly, $y \oplus z = a$. But then, as $a \oplus b = (x \oplus z) \oplus (y \oplus z)$ is defined, we can invoke commutativity and associativity to conclude that $(x \oplus y) \oplus (z \oplus z)$, and hence, in particular, $z \oplus z$, is defined — whence, $z = 0$, and hence $c = a \oplus b$. Thus, (c) implies (d). Finally, suppose (d) holds. Let $c \leq a$ and $c \leq a'$. We must show that $c = 0$. We have $a = c \oplus x$ and $a' = c \oplus y$ for some $x, y \in L$. Again invoking commutativity and associativity, we may write

$$\mathbf{1} = a \oplus a' = (c \oplus x) \oplus (c \oplus y) = c \oplus (x \oplus c \oplus y).$$

In particular, the element $u := (x \oplus c \oplus y)$ is defined. Plainly, $u = (x \oplus c) \oplus y = a \oplus y \geq a$ and likewise $u = x \oplus (c \oplus y) = x \oplus a' \geq a'$, so u is an upper bound for a and a'. But by (d), $\mathbf{1}$ is a minimal upper bound for a and a', so $u = \mathbf{1}$. On the other hand, $c \oplus u = \mathbf{1}$, so $c \leq u' = \mathbf{1}' = 0$. This shows that (d) implies (a), and completes the proof. ■

DEFINITION 22. An effect algebra satisfying the equivalent conditions of Lemma 21 is called an *orthoalgebra.*

In view of condition (d), any orthomodular poset is an (orthocoherent) orthoalgebra. On the other hand, suppose L is an orthocoherent effect algebra: then $(L, \leq, ')$ is an orthoposet, for which the orthogonal sum $a \oplus b = a \vee b$ is defined whenever $a \leq b'$; hence, by Lemma 15, $(L, \leq, ')$ is an OMP. This supplies the proof of Lemma 17. Thus, we can, and shall, identify OMPs with orthocoherent orthoalgebras. As will emerge in Section 2, there exist plentiful examples of non-orthocoherent orthoalgebras. We shall see that orthoalgebras provide in many ways a more natural framework for quantum logic than do orthomodular posets or lattices.

23 JOINT ORTHOGONALITY. A finite pair-wise orthogonal set $F = \{a_1, ..., a_n\}$ in an orthoalgebra L is said to be *jointly orthogonal* iff it is pairwise orthogonal, and the sum $\bigoplus F := a_1 \oplus \cdots \oplus a_n$ (as defined, say, by the obvious recursion) exists. An infinite set $A \subseteq L$ is jointly orthogonal iff every finite subset $F \subseteq A$ is jointly orthogonal. If the join of all partial sums $\bigoplus F$, F ranging over finite subsets of A, exists, we denote it by $\bigoplus A$, and say that A is *orthosummable.* We shall say that L is *orthocomplete* (respectively, σ- orthocomplete) if every jointly orthogonal subset (respectively, every countable jointly orthogonal subset) of L has a sum in this sense. An orthoalgebra is *atomistic* iff every element of L can be expressed as the sum of a jointly orthogonal set of atoms.

24 BOOLEAN ORTHOALGEBRAS AND COMPATIBILITY. An orthoalgebra $(L, \oplus)$ is said to be *Boolean* iff the corresponding orthoposet $(L, \leq, ', 0, 1)$ is a Boolean lattice. A subset of L is said to be *compatible* iff it is contained in a Boolean sub-orthoalgebra of L. One can show that two elements $a, b \in L$ are compatible iff there exist elements a_1, b_1 and c with $a = a_1 \oplus c$, $b = c \oplus b_1$, and $a \perp b_1$, so that $a_1 \oplus c \oplus b_1$ exists [5]. Equivalently, a and b are compatible iff there exists an element $c \leq a, b$ with $a \perp (b \ominus c)$. The triple $(a_1, c, b_1) = (a \ominus c, c, b \ominus c)$ is then called a *Mackey decomposition* for a and b. If L is Boolean, then every pair of elements $a, b \in L$ has a *unique* Mackey decomposition, namely, $(a \ominus b, a \wedge b, b \ominus a)$. It is possible, even in an OMP, for a pairwise compatible set of elements not to be compatible. [8] An orthoalgebra in which pairwise compatible sets *are* compatible is said to be *regular.* Note that such an orthoalgebra is automatically orthocoherent, i.e., an OMP.

25 THE CENTER OF AN ORTHOALGEBRA. If L_1 and L_2 are orthoalgebras, then $L_1 \times L_2$ is an orthoalgebra under the orthogonal sum defined by

$$(a, b) \oplus (c, d) = (a \oplus c, b \oplus d)$$

provided that $a \perp c$ and $b \perp d$. The unit element is given by $\mathbf{1} = (\mathbf{1}_1, \mathbf{1}_2)$ where $\mathbf{1}_i$ is the unit element of L_i. For any element a of an orthoalgebra L, the interval $[0, a] = \{x \in L | 0 \leq x \leq a\}$ is itself an orthoalgebra, the orthogonal sum of $x, y \leq a$ being given by $x \oplus y$, provided this exists in L *and* is again below a. There is a natural mapping $[0, a] \times [0, a'] \to L$ given by $(x, y) \mapsto x \oplus y$. If this is an isomorphism, a is said to be *central.* The *center* of L is the set $\mathbf{C}(L)$ of all central elements of L. It can be shown [8] that $\mathbf{C}(L)$ is a Boolean sub-orthoalgebra of L. In particular, L is Boolean iff $L = \mathbf{C}(L)$. L is *irreducible* iff $\mathbf{C}(L) = \{0, 1\}$, i.e., L admits no non-trivial decomposition as a direct product.

1.3 Lattices of Subspaces

Mackey's axioms lead only to a σ-complete orthomodular poset. This is a far cry from the lattice $L(\mathbf{H})$ of closed subspaces of a Hilbert space, upon which quantum

[8] Indeed, there exist non-Boolean OAs, called *centeria,* in which *every* pair of elements is compatible [Obeid, 1990].

probability rests. In [Mackey, 1963], Mackey asks for additional axioms that will bridge this gulf. In considering what such axioms might look like, it is natural to begin by trying to find an ortho-lattice characterization of the subspace lattices of Hilbert spaces. For *finite-dimensional* Hilbert spaces, such a characterization is a chapter of classical projective geometry. It is worth a few paragraphs to review this.

MODULAR ORTHOLATTICES
A lattice L is *modular* iff, for all $a, b, c \in L$,

$$c \leq a \;\Rightarrow\; a \wedge (b \vee c) = (a \wedge b) \vee c$$

for all $a, b, c \in L$.[9] Besides distributive lattices, examples of modular lattices include the lattice of subgroups of an arbitrary group, and, more significantly for our purposes, the lattice $\mathcal{L}(\mathbf{V})$ of all submodules of a module over an arbitrary ring R. There is a partial converse. A lattice L has *finite height* iff there is a finite upper bound on the length of a chain in L. In this case, the *height* of L is $n-1$, where n is the maximal length of a chain in L. The classical coordinatization theorem of projective geometry asserts that every irreducible, complemented modular lattice of finite height four or greater, is isomorphic to the lattice $\mathcal{L}(\mathbf{V})$ for a finite-dimensional vector space $\mathbf{V}$ over a division ring.

Of course, the lattice of subspaces of a vector space is not naturally *ortho*-complemented: our interest here is rather in lattices of subspaces of inner product spaces. To be a bit more general, let $\mathbf{V}$ be a (left) vector space over a division ring $\mathbb{D}$. An *involution* on $\mathbb{D}$ is a mapping $* : \mathbb{D} \to \mathbb{D}$ such that, for all $a, b \in \mathbb{D}$, $(ab)^* = b^*a^*$, $(a+b)^* = a^* + b^*$, and $a^{**} = a$. If $\mathbb{D}$ is equipped with a particular involution, we call it an *involutive division ring.* An *inner product* on a vector space $\mathbf{V}$ over an involutive division ring $\mathbb{D}$ is a mapping $q : \mathbf{V} \times \mathbf{V} \to \mathbb{D}$ that is linear in its first variable, satisfies $q(x,y) = q(y,x)^*$ for all $x, y \in \mathbf{V}$, and is *non-degenerate* in the sense that $q(x,x) = 0$ implies $x = 0$ for all $x \in \mathbf{V}$. We call vectors in $\mathbf{V}$ *orthogonal* relative to q iff $q(x,y) = 0$, writing $x \perp y$. If $A \subseteq \mathbf{V}$, then $A^\perp := \{x \in \mathbf{V} | x \perp y \forall y \in A\}$ is a subspace of $\mathbf{V}$. If $\mathbf{V}$ is finite-dimensional, then $\mathbf{V} = \mathbf{M} \oplus \mathbf{M}^\perp$ for all subspaces $\mathbf{M}$ of $\mathbf{V}$; in this case, $\mathbf{M} \mapsto \mathbf{M}^\perp$ is an orthocomplementation on $L(\mathbf{V})$.

The greater part of von Neumann and Birkhoff's paper [Birkhoff and von Neumann, 1936] on quantum logic is devoted to the following result (the proof of which is laid out in detail in Chapter IV of Varadarajan's book [Varadarajan, 1985]):

THEOREM 26 [Birkhoff-von Neumann, 1937]. *Let $\mathbf{V}$ be a vector space of finite dimension $n \geq 3$ over a division ring $\mathbb{D}$. If $'$: $\mathcal{L}(\mathbf{V}) \to \mathcal{L}(\mathbf{V})$ is any orthocomplementation on the lattice $\mathcal{L}(\mathbf{V})$ of subspaces of $\mathbf{V}$, then there exists an involution on $\mathbb{D}$ and, relative to this, an inner product on $\mathbf{V}$ such that $\mathbf{M}' = \mathbf{M}^\perp$.*

[9] More generally, a pair of elements a and b for which the foregoing holds for all $c \leq a$ is said to be a *modular pair.* It is not difficult to show that a lattice is orthomodular iff every orthogonal pair is a modular pair — indeed, this is the origin of the term.

Thus, *irreducible modular ortholattices of finite height* ≥ 4 are exactly the subspace lattices of finite-dimensional inner product spaces over involutive division rings.

PIRON'S THEOREM

If $\mathbf{H}$ is *infinite* dimensional, then, while the lattice of *all* subspaces of $\mathbf{H}$ is modular, the lattice of *closed* subspaces is not.[10] In light of this, it is natural to seek a coordinatization theorem characterizing Hilbert-space projection lattices in terms of non-modular orthomodular lattices. The major step towards such a coordinatization theorem for infinite-dimensional Hilbert space lattices was achieved by C. Piron in [Piron, 1964]. Before stating Piron's theorem, we need some preliminary definitions. A *quadratic space* over an involutive division ring $\mathbb{D}$ is a pair $(\mathbf{V}, q)$, where $\mathbf{V}$ is a vector space over $\mathbb{D}$ and q is a non-degenerate quadratic form on $\mathbf{V}$. A subspace $\mathbf{M} \subseteq \mathbf{V}$ is said to be *ortho-closed* iff $\mathbf{M} = \mathbf{M}^{\perp\perp}$. The collection of all ortho-closed subspaces of $\mathbf{V}$ is denoted by $L(\mathbf{V})$. If $\mathbf{V}$ is finite-dimensional, then it is easy to show that every subspace is ortho-closed; otherwise, $L(\mathbf{V})$ is a proper subset of $\mathcal{L}(\mathbf{V})$. It is easily verified that $L(\mathbf{V})$ is a complete lattice, with $\bigwedge_i \mathbf{M}_i = \bigcap_i \mathbf{M}_i$ and $\bigvee_i \mathbf{M}_i = (\bigcup_i \mathbf{M}_i)^{\perp\perp}$.

LEMMA 27. *Let $(\mathbf{V}, q)$ be a quadratic space. The following are equivalent:*

(a) $L(\mathbf{V})$ is an orthomodular lattice;

(b) $\mathbf{V} = \mathbf{M} \oplus \mathbf{M}^{\perp}$ for every $\mathbf{M} \in L(\mathbf{V})$.

Proof. To see that (a) implies (b), just note that (b) makes $(L(\mathbf{V}), \oplus)$ an orthoalgebra. To see that (b) implies (a), let $\mathbf{M}, \mathbf{N} \in L(\mathbf{V})$ with $\mathbf{M} \subseteq \mathbf{N}$. As $\mathbf{V} = \mathbf{M} \oplus \mathbf{M}^{\perp}$, each $n \in \mathbf{N}$ has a unique expression of the form $n = a(n) + b(n)$ where $a(n) \in \mathbf{M}$ and $b(n) \in \mathbf{M}^{\perp}$. Since $\mathbf{M} \subseteq \mathbf{N}$, $a(n) \in \mathbf{N}$, from which it follows that $b(n) = a - a(n) \in \mathbf{M}$ as well. Thus, $b(n) \in \mathbf{N} \cap \mathbf{N}^{\perp}$. Hence, $\mathbf{N} \subseteq \mathbf{M} \vee (\mathbf{N} \cap \mathbf{M}^{\perp})$, which suffices to show that L is orthomodular. ■

A quadratic space $(\mathbf{V}, q)$ satisfying the equivalent conditions of Lemma 27 is called an *orthomodular space*, or a *generalized Hilbert space*. Beyond being a complete OML, the lattice $L(\mathbf{V})$ of subspaces of an orthomodular space has some additional structural features that stand out. For one thing, it is atomic. For another, it satisfies the so-called *covering law*. Recall that, in any poset L, one element y is said to *cover* another, x, iff $x < y$ and there is no element of L properly between x and y.

[10] If $\mathbf{H}$ is infinite dimensional, one can find two closed subspaces $\mathbf{M}$ and $\mathbf{N}$ of $\mathbf{H}$ such that $\mathbf{M} + \mathbf{N}$ is not closed (cf [Halmos, 1957], p. 20). Let $x \in (\mathbf{M} \vee \mathbf{N}) \ominus (\mathbf{M} + \mathbf{N})$. Let $\mathbf{V} = \mathbf{M} \vee (x)$, i.e., the span of $\mathbf{M}$ and x. Equivalently, $\mathbf{V} = \{a + tx | t \in \mathbb{C}\}$. Clearly, $\mathbf{V}$ is closed, and $\mathbf{M} \leq \mathbf{V}$. Note that $x \in \mathbf{V} \cap (\mathbf{M} \vee \mathbf{N})$. Now, consider any vector $y \in \mathbf{V} \cap \mathbf{N}$: We have $y = a + tx$ for some scalar t and some $a \in \mathbf{M}$. It follows that $tx = b - a \in \mathbf{N} + \mathbf{M}$ — which, by the choice of x, is impossible unless $t = 0$, in which case, $y = a \in \mathbf{M}$. It follows that $\mathbf{V} \cap \mathbf{N} \subseteq \mathbf{M}$. Thus, $(\mathbf{V} \cap \mathbf{N}) \vee \mathbf{M} = \mathbf{M}$ — which definitely does not contain x. Thus, $\mathbf{V} \cap (\mathbf{M} \vee \mathbf{N}) \not\subseteq (\mathbf{V} \cap \mathbf{N}) \vee \mathbf{M}$. Modularity fails.

DEFINITION 28. An atomistic lattice L satisfies the *covering law* iff, for every atom x of L and every element $a \in L$ with $x \not\leq a$, $a \vee x$ covers a.

The covering law can be regarded as a weakening of modularity — to which it reduces, for ortholattices of finite height.

We are now ready to state Piron's result:

THEOREM 29 [Piron, 1964]. *Let L be a complete, irreducible atomistic OML satisfying the covering law. If L has height 4 or more, then there exists an involutive division ring $\mathbb{D}$, a vector space (left module) $\mathbf{V}$ over $\mathbb{D}$, and a non-degenerate sesquilinear form on $\mathbf{V}$, such that L is isomorphic to the ortholattice $L(\mathbf{V})$ of ortho-closed subspaces of $\mathbf{V}$.*

The question arises: Are there any "non-classical" — that is, non Hilbert — examples of orthomodular quadratic spaces? If the involutive ring $\mathbb{D}$ of $\mathbf{V}$ is one of the classical division rings — $\mathbb{R}$, $\mathbb{C}$ or $\mathbb{H}$ then the answer is *no*. For a proof of the following, see the original paper [Amemiya and Araki, 1967] of Araki and Amemiya:

THEOREM 30 [Amemiya-Araki, 1967]. *If $\mathbf{V}$ is an orthomodular space over the $\mathbb{D} = \mathbb{R}, \mathbb{C}$ or $\mathbb{H}$ (the quaternions), then $\mathbf{V}$ is complete, i.e., a Hilbert space.*

For a time, it was thought that perhaps the *only* orthomodular quadratic spaces were in fact Hilbert spaces. However, in the late 1970s, Keller produced examples of orthomodular spaces over more general division rings [Keller, 1980]. Efforts to find sharp characterizations of the classical examples culminated in the work of Solèr [Solèr, 1995], which shows that if an orthomodular quadratic space contains any infinite orthonormal sequence, then $\mathbb{D} = \mathbb{R}, \mathbb{C}$, or $\mathbb{H}$. This has an equivalent, purely lattice-theoretic formulation in terms of a cetain angle-bisection axiom.[11]

Remark: Pitowsky [Pitowsky, 2005] has recently argued that the hypotheses of Solér's angle-bisection axiom can be interpreted as natural rationality constraints on gambles, similar in spirit to such constraints in classical Bayesian probability theory. This might suggest that project of deducing quantum theory from a priori principles is essentially complete; but later sections of this paper may temper such optimism.

31 PIRON'S AXIOMS. In order to motivate the hypotheses in the representation theorem, Piron presented a series of axioms for quantum logic in which the primitive notion is, just as for Mackey, the idea of a yes-no question. This is understood to mean a realizable experiment having two possible outcomes, designated *yes* and *no*. A question α is deemed *stronger* than another question β iff β is certain to yield the affirmative outcome in any state in which α is certain so to do. This defines a pre-order on the class of questions. By a *property* of the sys-

[11]Prestel [Prestel, 1995] gives a nice treatment of the case in which D is commutative. Further discussion can be found in Holland [Holland, 1995]. See [Mayet, 1998] for an interesting variant on Solér's Theorem.

tem, Piron means an equivalence class of questions with respect to the equivalence relation $\alpha \approx \beta$ iff $\alpha \leq \beta \leq \alpha$ — that is, $\alpha \approx \beta$ iff α and β are certain to yield affirmative outcomes in all and only the same states. The set $\mathcal{L}$ of all properties is then partially ordered by $[\alpha] \leq [\beta]$ iff $\alpha \leq \beta$. The various axioms adduced by Piron constrain $\mathcal{L}$ to have exactly the structure of an irreducible, complete, atomistic OML with the covering law. I will not reproduce Piron's axioms here; see [Coecke *et al.*, 2000] for a detailed exposition and commentary. Suffice it to say that, while some of Piron's axioms are fairly innocent, it is at least arguable that the orthocomplementation on $\mathcal{L}$ is introduced by force.

1.4 Orthogonality Spaces

In the mid-1960s, it was discovered by Foulis and his students that one can construct examples of OMPs and OMLs from very simple combinatorial objects called *orthogonality spaces*. The motivating example is the unit sphere X of a Hilbert space $\mathbf{H}$. The relation $\perp$ of orthogonality — $x \perp y$ meaning $\langle x, y\rangle = 0$ — entirely determines the structure of $\mathbf{H}$ (since maximal pairwise orthogonal subsets of X are precisely the orthonormal bases of $\mathbf{H}$).

DEFINITION 32. A symmetric, irreflexive binary relation on a set X is called an *orthogonality relation* on X. An *orthogonality space* is a pair $(X, \perp)$ consisting of a set X and a fixed orthogonality relation $\perp$ on X.

If $(X, \perp)$ is an orthogonality space, we shall call a subset of X *orthogonal* iff it is pairwise so. Clearly, the union of a chain of orthogonal subsets of X is still pairwise orthogonal, so, by Zorn's Lemma, every orthogonal subset of X is contained in a maximal orthogonal set. We shall denote the collection of all such maximal orthogonal subsets of X by $\mathcal{O}(X, \perp)$.

EXAMPLES 33.

(i) For any set X, $(X, \neq)$ is an orthogonality space having only one maximal orthogonal subset — X itself.

(ii) Any collection $\mathcal{E}$ of subsets of a set X is an orthogonality space under the orthogonality relation $a \perp b \Leftrightarrow a \cap b = \emptyset$. Note that any partition of X into cells belonging to $\mathcal{E}$ gives a maximal orthogonal subset of $\mathcal{E}$, but that there may well be additional maximal orthogonal sets not of this form.

(iii) If $\mathbf{V}$ is an inner product space, the unit sphere $X = \{x \in \mathbf{V} | \|x\| = 1\}$ is an orthogonality space under the relation $x \perp y \Leftrightarrow \langle x, y\rangle = 0$. In particular, if $\mathbf{V}$ is a Hilbert space, then the elements of $\mathcal{O}(X, \perp)$ are precisely the (unordered) orthonormal bases for $\mathbf{V}$.

(iv) If L is an orthomodular poset, let $X = L \setminus \{0\}$, and, for $a, b \in L$, let $a \perp b$ iff $a \leq b'$. Then any orthopartition in L — that is, any finite set E of non-zero elements with $\bigvee E = \mathbf{1}$ — is a maximal orthogonal set. More generally, if E is pairwise orthogonal and $\bigvee E$ exists, then E is a maximal pairwise-orthogonal set iff $\bigvee E = 1$.

The preceding examples — particularly the last one — suggest the following general interpretation for an orthogonality space $(X, \perp)$: the set X consists of the possible *outcomes* of various measurements, and two outcomes $x, y \in X$ are orthogonal iff they are mutually exclusive and jointly testable, i.e., if they can be regarded as distinct outcomes of one and the same measurement. If, with Mackey, we take the view that all pairwise co-testable outcomes can be regarded as *jointly* co-testable, i.e., outcomes of some single measurement, then the collection $\mathcal{O}(X, \perp)$ of maximal pairwise-orthogonal subsets of $(X, \perp)$ serves as a model for the set of (outcome-sets of) the possible measurements. As we shall see below, this point of view runs into difficulties.

DEFINITION 34. Let $(X, \perp)$ be an orthogonality space. For $A \subseteq X$, let $A^{\perp} = \{x \in X | \forall a \in A\ x \perp a\}$. Write $x^{\perp}$ for $\{x\}^{\perp}$. A set $A \subseteq X$ is $\perp$-*orthoclosed* iff $A = A^{\perp\perp}$. The set of $\perp$-orthoclosed subsets of X is denoted by $\mathcal{C}(X, \perp)$.

It is easy to see that, for any sets $A, B \subseteq X$, $A \subseteq B$ implies $B^{\perp} \subseteq A^{\perp}$ and $A \subseteq A^{\perp\perp}$. From this, it follows that $A^{\perp} = A^{\perp\perp\perp}$. Using these observations, it is difficult to show that $\mathcal{C}(X, \perp)$ is a complete lattice, with the meet and join of a family $\{A_i\}_{i \in I}$ of orthoclosed sets given by

$$\bigwedge_i A_i = \bigcap_i A_i \text{ and } \bigvee_i A_i = \left(\bigcup_i A_i\right)^{\perp\perp}.$$

Moreover, the mapping $A \mapsto A^{\perp}$ provides an orthocomplementation on $\mathcal{C}(X, \perp)$.

If X is the unit sphere of a Hilbert space $\mathbf{H}$, then a set $A \subseteq X$ is orthoclosed iff $A = \mathbf{M} \cap X$ for a closed subspace $\mathbf{M}$. Accordingly, $\mathcal{C}(X, \perp)$ is isomorphic to the orthomodular lattice $L(\mathbf{H})$ of closed subspaces of $\mathbf{H}$. On the other hand, if $\mathbf{V}$ is an incomplete inner product space over $\mathbb{R}, \mathbb{C}$ or $\mathbb{H}$, the Araki-Amemiya Theorem (Theorem 30) tells us that $\mathcal{C}(X, \perp)$ is not orthomodular. It is natural, then, to ask whether one can identify conditions on an abstract orthogonality space that will guarantee that $\mathcal{C}(X, \perp)$ is an OML. This was accomplished by J. R. Dacey in his dissertation [Dacey, 1968]:

THEOREM 35 [J. R. Dacey, 1968]. *Let $(X, \perp)$ be an orthogonality space. The following are equivalent:*

(a) $\mathcal{C}(X, \perp)$ is an orthomodular lattice.

(b) For all orthogonal sets $D \subseteq X$ and all $x \in X$,

$$x \notin D^{\perp\perp} \Rightarrow D^{\perp} \cap (x^{\perp} \cap D^{\perp})^{\perp} \neq \emptyset.$$

(c) If D is a maximal orthogonal subset of $A \in \mathcal{C}(X, \perp)$, then $A = D^{\perp\perp}$.

Proof. $(a) \Rightarrow (b)$: Suppose $D^{\perp} \cap (x^{\perp} \cap D^{\perp})^{\perp} = \emptyset$. Since $x^{\perp} \cap D^{\perp} \subseteq D^{\perp}$, orthomodularity of $(X, \perp)$ yields $D^{\perp} = x^{\perp} \cap D^{\perp}$. Hence $D^{\perp} \subseteq x^{\perp}$, whence $x^{\perp\perp} \subseteq D^{\perp\perp}$. Since $x \in x^{\perp\perp}$, $x \in D^{\perp\perp}$, a contradiction.

$(b) \Rightarrow (c)$: Let D be a maximal orthogonal subset of $A \subseteq A^{\perp\perp}$. Suppose $x \in A \setminus D^{\perp\perp}$. By (b), there exists some $y \in D^{\perp} \cap (x^{\perp} \cap D^{\perp})^{\perp}$. Then $y \in x^{\perp\perp} \cup D^{\perp\perp} \subseteq A^{\perp\perp} = A$, which contradicts the maximality of D.

$(c) \Rightarrow (a)$: Let $A, B \in \mathcal{C}(X, \perp)$ with $A \subseteq B$ and $B \cap A^{\perp} = \emptyset$. Let $D \subseteq A$ be maximal among orthogonal subsets of A. By (c), $A = D^{\perp\perp}$. Hence, $A^{\perp} = D^{\perp}$, and it follows that D is maximal among orthogonal subsets of B as well. But then $B = D^{\perp\perp} = A$, and $\mathcal{C}(X, \perp)$ is orthomodular. ■

DEFINITION 36. An orthogonality space satisfying the conditions of Theorem 35 is called a *Dacey space.*[12]

Remark: One can ask whether or not there exists a *first-order* characterization of orthomodular spaces — that is, one that refers only to points, rather than subsets, of X. The answer, as shown by Goldblatt [Goldblatt, 1984], is *no*: orthomodularity is (as the Amemiya-Araki theorem already suggests) an inherently second-order property of $(X, \perp)$.

If L is any complete OML, the set $X \setminus \{0\}$, with the usual orthogonality relation $p \perp q \Leftrightarrow p \leq q'$, is an orthomodular space, with $\mathcal{C}(X, \perp) \simeq L$. It remains an important open question whether or not every OML can be embedded in a complete OML, i.e., in one of the form $\mathcal{C}(X, \perp)$. A discussion of this problem can be found in the paper [Bruns and Harding, 2000].

In [Greechie, 1966; Greechie, 1968], R. J. Greechie described a method for construcing orthomodular lattices and posets by "pasting" Boolean algebras together in various configurations.

DEFINITION 37. A maximal orthogonal subset of an orthogonality space is called a *block*. A *Greechie space* is an orthogonality space in which every block has at least three points, and in which two blocks intersect, if at all, in a single point. A *Greechie diagram* of such an orthogonality space displays the points of each block along a distinct line or other smooth curve.

EXAMPLE 38. Let X_5 be the orthogonality space, the graph of whose orthogonality relation is pictured below. This is plainly a Greechie space with five three-element blocks. The corresponding Greechie diagram appears at right.

[12]Dacey christened those orthogonality spaces satisfying the equivalent conditions of his theorem *orthomodular spaces*. Since that time, the term has come to be more commonly applied to orthomodular *quadratic* spaces — whose unit spheres, needless to say, are orthomodular spaces in Dacey's sense.

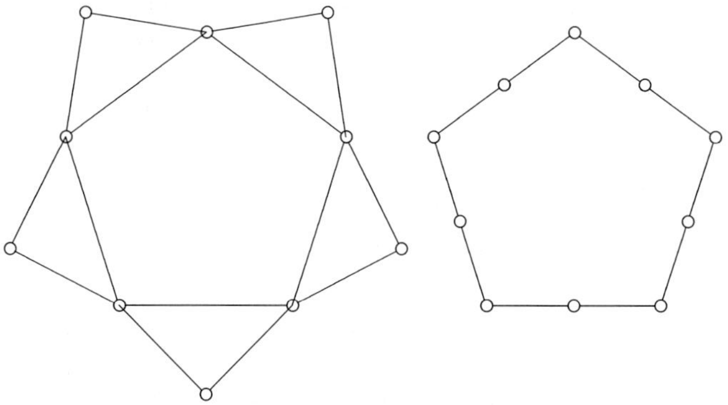

DEFINITION 39. A *loop* of length n in a Greechie space is a finite sequence $E_0, ..., E_{n-1}$ of blocks such that the consecutive intersections $E_1 \cap E_2$, ... $E_{n-1} \cap E_n$ and $E_n \cap E_1$ are all non-empty, and all distinct.

A simple version of Greechie's result takes the following form:

LEMMA 40 The Loop Lemma [Greechie, 1966]. *Let $(X, \perp)$ be a Greechie space. Then,*

(a) If X contains no loop of order less than four, $\mathcal{C}(X, \perp)$ is an OMP;

(b) If X contains no loop of order less than five, then $\mathcal{C}(X, \perp)$ is an OML.

I'll not pause to prove this here, as a more general result is discussed in section 2.4. For an example of the Loop Lemma, note that the Greechie space X_5 of the preceding example (which consists of a single loop of order five) has no loops of order less than 5. Accordingly, $\mathcal{C}(X_5, \perp)$ is an orthomodular lattice. It is not difficult to show that it hosts a strongly order-determining set of dispersion-free states, and hence, is embeddable in a Boolean algebra [Wright, 1978b]. In other words, $\mathcal{C}(X_5, \perp)$ seems a very respectable quantum logic.

ORTHOMONOIDS

If X is any set, let X^* denote the free monoid over X. Thus, X^* consists of finite strings $x = x_1....x_n$ of elements of X, multiplication in X^* is simply juxtaposition, and the unit element is the empty string, which we denote by 1. If $A, B \subseteq X^*$, we write AB for the set of all strings of the form ab with $a \in A$, $b \in B$.

DEFINITION 41. The *free orthomonoid* over an orthogonality space $(X, \perp)$ is the orthogonality space $(X^*, \#)$, where $\#$ is defined lexicographically: $\forall a, b \in X^*$, $a\#b \Leftrightarrow \exists x, y \in X$ and $c, d, e \in X^*$ with $a = cxd, b = cye$, and $x \perp y$.

To the extent that we regard $(X, \perp)$ as a model for the set of outcomes of a collection of possible measurements, it would seem reasonable to regard $(X^*, \#)$ as a model for the set of outcomes of possible *compound* measurements.

Remarks. We treat X as a subset of X^*. If $A \subseteq X^*$, we denote by $\pi(A)$ the set of first terms of strings in A. (When dealing with singletons, we write, e.g., $\pi(a) = x$ rather than $\pi(\{a\}) = \{x\}$. Note, though, that $\pi(1) = \emptyset$.) Notice that for any $a, b \in X^*$, $(ab)^\perp = a^\perp \cup ab^\perp$. Also observe that for $A \subseteq X$, $(AX^*)^\perp = A^\# X^*$.

The (non-trivial) proof of the following can be found in [Foulis and Randall, 1971]:

THEOREM 42. *Let $(X, \#)$ be a Dacey space. Then the free orthomonoid $(X^*, \perp)$ over X is also a Dacey space.*

The structure of the lattice $\mathcal{C}(X^*, \perp)$ of ortho-closed subsets of a free orthomonoid can be quite complex. In particular, $\mathcal{C}(X, \perp)$ is totally non-atomic [Foulis and Randall, 1971].

THE PROBLEM OF TENSOR PRODUCTS

A compound quantum-mechanical system having two component subsystems, represented by Hilbert spaces $\mathbf{H}_1$ and $\mathbf{H}_2$, is represented by the tensor product $\mathbf{H}_1 \otimes \mathbf{H}_2$. Naturally, any satisfactory generalized probability theory will have to include some analogous device for combining models of systems in a manner consistent with the Hilbert space tensor product. In particular, given two orthomodular posets L_1 and L_2, we should like to construct a orthomodular poset L that serves as their "tensor product". Some minimal requirements might be that

(a) There exists an injection $L_1 \times L_2 \subseteq L$ sending $(a, b) \in L_1 \times L_2$ to an element $ab \in L$;

(b) For all states α on L_1 and β on L_2, there exists a state $\alpha \otimes \beta$ on Z with $(\alpha \otimes \beta)(ab) = \alpha(a)\beta(b)$; and

(c) If $a_1 \perp a_2$ in L_1, then $a_1 b \perp a_2 b$ for all $b \in L_2$; similarly if $b_1 \perp b_2$ in L_2, then $ab_1 \perp ab_2$ for all $a \in L_1$.

Suppose that $L_1 = \mathcal{C}(X_1, \perp)$ and $L_2 = \mathcal{C}(X_2, \perp)$ for two Dacey spaces $(X_1, \perp)$ and $(X_2, \perp)$. A natural candidate for such a tensor product would then be $C(X_1 \times X_2, \#)$ where $\#$ is *defined* by condition (c). As it turns out, however, this is in general no longer Dacey. Indeed, as Foulis and Randall showed [Foulis and Randall, 1981a; Foulis and Randall, 1981b], in general, even for quite well-behaved quantum logics, *no such tensor product exists.* A proof of the following — which would be slightly tedious to write down using only the materials we have to hand — is given in Section 5:

EXAMPLE 43 [Foulis-Randall, 1979]. Let $L = \mathcal{C}(X_5, \perp)$, where X_5 is the Greechie space of Example. There exists no orthomodular poset L satisfying conditions (a), (b) and (c) for $L_1 = L_2 = \mathcal{C}(X_5, \perp)$, where X_5 is the Greechie space of Example 38.

As mentioned above, the lattice $\mathcal{C}(X_5, \perp)$ is very well-behaved, even to the point of being embeddable in a Boolean algebra. Thus, the Foulis-Randall example suggests very strongly that Mackey-style quantum logics — that is, orthomodular posets with ample state spaces — are simply too special to supply an adequate generalized probability theory. The crucial problem, as we'll see more clearly in later sections, is the ortho-coherence assumption, that any maximal pairwise orthogonal set should count as a test — a strong assumption in any case, with little heuristic motivation. Section 2 develops the theory of test spaces, which generalizes Mackey's theory in several ways, but most crucially, in relaxing this assumption. In sections 3 and 4, we'll see how one can associate two quite different "quantum logics" to a test spaces, in a manner that generalizes Mackey's construction. One of these turns out to be orthocomplemented, but rarely a lattice; the other turns out to be a complete lattice, but only rarely orthocomplemented. In section 5, we'll return to the problem of tensor products, and show that this framework, unlike that of Mackey and Piron, is stable under the formation of a tensor product.

2 TEST SPACE BASICS

As we've seen, the approach to quantum logic that begins with orthomodular lattices and posets is beset by serious conceptual and technical difficulties. In this chapter, we develop the concept of a *test space*, which offers a much less problematic foundation for quantum logic. In brief, a test space is simply the collection, or catalogue, of outcome-sets (that is, sample spaces) associated with some collection of discrete classical measurements. As we'll see, such objects are easy to manipulate, and, in particular, lend themselves to free constructions modelling sequential experiments. Our focus in this section is on test spaces and their state spaces exclusively. More traditional models of quantum logics such as orthocomplemented posets and lattices will emerge again in Chapter 3, as useful (but not complete) mathematical invariants associated with test spaces.

2.1 *Examples of Test Spaces*

In its simplest formulation, classical probability theory concerns a set E of possible *outcomes* — as of some process, experiment, measurement, operation, selection or *test*. It is understood that, on each realization of this test, exactly one outcome $x \in E$ will *occur*.[13] A *probability weight* on E is a mapping $\omega : E \to [0,1]$ assigning, to each outcome $x \in E$, a number $0 \leq \omega(x) \leq 1$, to be construed as the probability of the occurrence of that outcome, and summing to unity. The theory of test spaces generalizes this simple apparatus in the most direct imaginable way:

DEFINITION 44. A *test space* is a collection $\mathfrak{A}$ of non-empty subsets $E, F, \ldots$. Elements of $\mathfrak{A}$ are called *tests*; elements of tests are called *outcomes*, and subsets

[13] I want to stress here that *any* phenomenon associated with a definite and exhaustive set of mutually alternatives, to which a modality of *occurrence* is appropriate, may count as a test. In particular, tests need not involve any notion of *agency*.

of tests are called *events*. The *outcome set* of $\mathfrak{A}$ is the collection $X = \bigcup \mathfrak{A}$ of all outcomes of all tests $E \in \mathfrak{A}$. A *probability weight*, or *state*, on $\mathfrak{A}$ is a mapping $\omega : X \to [0,1]$ such that $\sum_{x \in E} \omega(x) = 1$ for every test $E \in \mathfrak{A}$.[14] The set of all states on $\mathfrak{A}$, denoted by $\Omega(\mathfrak{A})$, is called the *state space* of $\mathfrak{A}$. (Note that this is a convex subset of $\mathbb{R}^X$.)

Remark. It will sometimes be convenient to denominate a test space $\mathfrak{A}$ with outcome-set X as a pair $(X, \mathfrak{A})$. In this case, we write $\mathcal{E}(X, \mathfrak{A})$ for the set of events, $\Omega(X, \mathfrak{A})$ for the set of states, *etc.*

A test space $\mathfrak{A}$ is *locally finite* iff every test $E \in \mathfrak{A}$ is finite. The *rank* of $\mathfrak{A}$ is the supremum of the cardinalities of its tests. If the outcome set X is finite, then $\mathfrak{A}$ is *totally finite*.[15] It is usually assumed that $\mathfrak{A}$ is *irredundant* — that is, that no test properly contains another. In some contexts, it is convenient to relax this assumption, but I'll generally adhere to it in this paper. Thus, **unless otherwise noted, all test spaces are irredundant.** However, there is no requirement that the sets comprising $\mathfrak{A}$ be disjoint. In other words, distinct tests are permitted to share outcomes. This certainly occurs in practice, as we shall see below. The definition of a state requires that, where an outcome is shared by two or more tests, its probability in a given state not depend upon which test is used to secure it. (To use the current term of art, these probabilities are *non-contextual.*)

DEFINITION 45. A test space in which distinct tests *do not* intersect is said to be *semi-classical.* A test space consisting of but a single test is *classical.*

As the following examples illustrate, *non*-semiclassical test spaces occur abundantly in practice.

EXAMPLE 46 Borel Test Spaces. Let $(M, \mathcal{F})$ be a measurable space. The corresponding *Borel test space*, or *Borel manual*[16], is the collection $\mathfrak{B} = \mathfrak{B}(M, \mathcal{F})$ of all countable $\mathcal{F}$- measurable partitions of M. States on $\mathfrak{B}$ are in an obvious one-to-one correspondence with probability measures on $(M, \mathcal{F})$. Thus, classical probability theory is a special case of the theory of states on test spaces.

EXAMPLE 47 Frame Manuals. Let $\mathbf{H}$ be a Hilbert space. Let X denote the unit sphere of $\mathbf{H}$, and let $\mathfrak{F} = \mathfrak{F}(\mathbf{H})$ be the collection of all *frames*, i.e., unordered orthonormal bases, for $\mathbf{H}$. Note that this is locally finite iff $\mathbf{H}$ is finite dimensional, and that then the rank of $\mathfrak{F}$ is the dimenion of $\mathbf{H}$. We can regard $\mathfrak{F}$ as a test space with outcome set X; and indeed, in orthodox non-relativistic quantum mechanics this interpretation is taken quite literally. If ϕ is a unit vector of $\mathbf{H}$ and $E \in \mathfrak{F}$ is an orthonormal basis, the completeness of $\mathbf{H}$ gives us $\sum_{x \in E} |\langle x, \phi \rangle|^2 = \|\phi\|^2 = 1$.

[14] If E is infinite, this is to be understood as an unordered sum, i.e., the limit of the net of finite partial sums.

[15] In this case, $\mathfrak{A}$ is what combinatorists call a *hypergraph* — that is, nothing but a finite set of finite sets. The language introduced above reflects the specific probabilistic interpretation we have in mind.

[16] In the older literature, a test space was often referred to as a *manual.* The term now survives mostly in connection with particular test spaces. See the remarks following Theorem 116

Hence,the unit vector ϕ defines a state $\omega_\phi(x) = |\langle\phi, x\rangle|^2$ on $\mathfrak{F}$, which we call a *vector state*. Gleason's theorem tells us that every state on $\mathfrak{F}$ is a σ-convex combination of such vector states. Thus, the standard probabilistic machinery of quantum mechanics is a special case of the theory of states on test spaces. The test space $\mathfrak{F}(\mathbf{H})$ is called the *frame manual* of $\mathbf{H}$.

EXAMPLE 48 Orthopartitions. Let L be an orthoalgebra. A set of non-zero elements of L is said to be an *orthopartition* (or *partition of unity*) iff it is jointly orthogonal, in the sense of def..., and sums to the unit element of L. Let $\mathfrak{A}_L$ be the set of all finite partitions of unity in L. Then $\mathfrak{A}_L$ is is naturally regarded as a test space. States on $\mathfrak{A}_L$ correspond to states on L in an obvious manner.

Many variations are possible on this example. One can, for example, consider the collection of all countable partitions of unity, or the set of partitions of unity by atoms, in the case in which L is atomistic. Notice, too, that since any Boolean algebra is an orthoalgebra, Example 49 subsumes Example 46.

EXAMPLE 49. Let $\mathfrak{A} = \{E, F\}$ be a semi-classical test space consisting of two disjoint, two-outcome tests $E = \{a, b\}$ and $F = \{c, d\}$. A *two-stage test over* $\mathfrak{A}$ is a test in which one a predetermined one of the two tests E and F is executed first, and then another, depending in a pre-determined way upon the outcome of the first. The ordered pair xy of outcomes thus secured is recorded as the outcome of the two-stage test. Thus, the outcome-set of the test, "execute E; upon obtaining a, execute E again; upon obtaining b, execute F" would be $\{aa, ab, bc, bd\}$; that of "execute E twice in succession", which would have outcome-set $\{aa, ab, ba, bb\}$. Let $\mathfrak{A}^2$ denote the collection of all eight such outcome-sets for two-stage tests over $\mathfrak{A}$: this has a rich combinatorial structure, and is certainly *not* semiclassical.

EXAMPLE 50. Let $\mathfrak{A}$ consist of the rows and columns of the array $\{1, 2, 3\}^2$. The states on $\mathfrak{A}$ are exactly the 3×3 doubly-stochastic matrices. Here is a "physical" model for this test space. Suppose two urns each contain three balls, labelled 1, 2 and 3. Consider the following experiment: a permutation σ of the set $\{1, 2, 3\}$ is selected, and then one of the urns is chosen at random, and a ball is selected. If ball i is drawn from urn 1, then the pair $(i, \sigma(i))$ is written down as the outcome of the experiment. If ball i is drawn from urn 2, the outcome is recorded as $(\sigma^{-1}(i), i)$. The possible outcomes for this experiment are the pairs (i, j) belonging to the graph of σ. It is easy to check that the graphs of the six possible permutations of $\{1, 2, 3\}$ have exactly the structure of the rows and columns of a 3×3 array.

51 Test Spaces vs Orthogonality Spaces. If $(X, \perp)$ is any orthogonality space, we can regard the space $\mathcal{O}(X, \perp)$ of maximal pairwise-orthogonal subsets of X as a test space. Both the Borel and Frame manuals discussed above are of this form. Conversely, if $\mathfrak{A}$ is a test space with outcome-set X, then we can define an orthogonality relation on X by setting $x \perp y$ iff x and y are distinct outcomes of a single test. In this case, every test will be pairwise-orthogonal; however, there is no guarantee that these will be maximal, nor is it necessarily the case that every maximal orthogonal set will be a test. We'll see some examples below.

52 Greechie Diagrams. One can adapt Greechie diagrams, introduced in section 1 in connection with orthogonality spaces, to represent small finite test spaces. In such a diagram, the outcome set of the test space is represented by a set of points or nodes, and each test is represented by a smooth arc running through the points corresponding its outcomes. The test space of Example 50, for instance, might be represented by the diagram

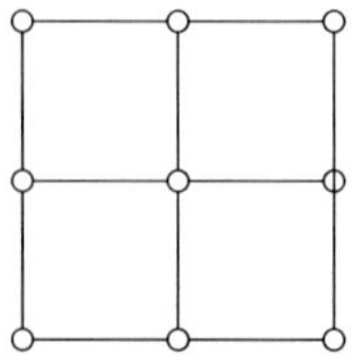

Some further examples of Greechie diagrams are the following:

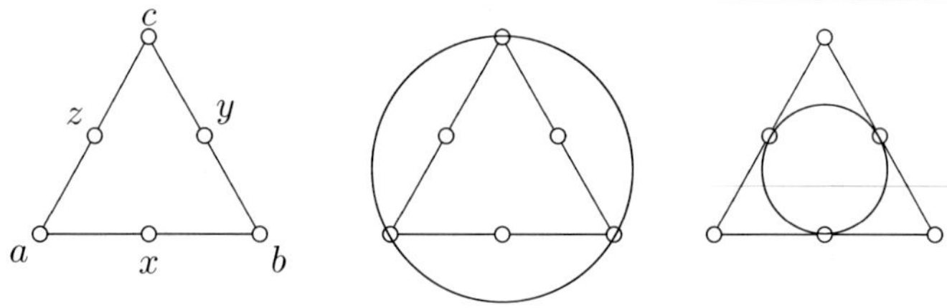

(i) (ii) (iii)

In figure (i), we have a Greechie diagram for a test space

$$\mathfrak{W} = \{\{a, x, b\}, \{b, y, c\}, \{c, z, a\}\}$$

consisting of three, three-outcome tests pasted together in a loop. Figure (ii) represents the test space obtained by adjoining the set $\{a, b, c\}$ to $\mathfrak{W}$ as a fourth test; similarly, Figure (iii) represents the test space obtained by adjoining the set $\{x, y, z\}$ to $\mathfrak{W}$.

Remark. Notice that only in figure (ii) is every maximal pairwise-orthogonal set a test! For a more striking illustration of the distinction between a test space and an orthogonality space, consider the test space having the following Greechie diagram:

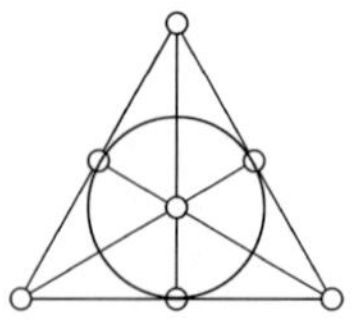

Here the tests are the lines of the Fano plane. In this example, every outcome is orthogonal to every other, so that the only maximal orthogonal set is X itself. In particular, *no* test is a maximal pairwise orthogonal set.

Greechie diagrams are very useful in the construction of finite test spaces having special properties. For a very simple example, the test space represented by the Greechie diagram

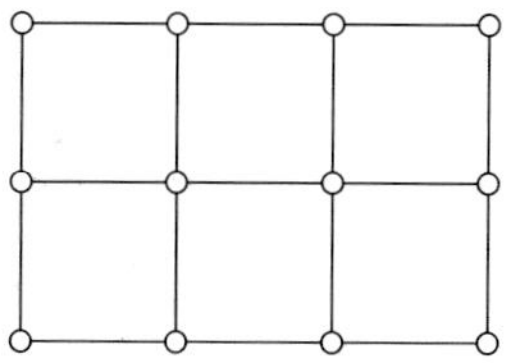

has no states at all (since $3 \neq 4$).

Where distinct tests overlap in more than a single outcome, Greechie diagrams become a bit less agreeable, but one can still manage. For instance, the test space

$$\mathfrak{A} = \{\{a,b,x,y\},\{b,c,y,z\},\{x,y,u,v\},\{y,z,v,w\}\}$$

can be represented by a Greechie diagram in which each test is indicated by four points lying on one of four intersecting circles. A test space in which distinct tests intersect, if at all, in single outcomes, is said to be *Greechie*. For instance, the projection manual of a three-dimensional Hilbert space is Greechie, but that of a higher dimensional Hilbert space is not.

TEST SPACES OF PARTITIONS

Many interesting test spaces arise as spaces of partitions of an underlying set. We've seen one example already, the Borel manual $\mathfrak{B}(M,\mathcal{F})$ of a measurable space $(M,\mathcal{F})$.

DEFINITION 53. A state ω on a test space $\mathfrak{A}$ is *dispersion-free* iff it takes only the values 0 and 1. A set $\Delta \subseteq \Omega(\mathfrak{A})$ of states on $\mathfrak{A}$ is *unital* iff $\forall x \in X$. A test space $\mathfrak{A}$ carrying a unital set of dispersion-free states is said to be *UDF*.

LEMMA 54. *Let $\mathfrak{A}$ be a test space with outcome-set X. The following are equivalent:*

(a) $\mathfrak{A}$ is UDF.

(b) There exists a set S and a mapping $\phi : X \to \mathcal{P}(S)$ sending each test $E \in \mathfrak{A}$ to a partition of S.

Proof. If (a) holds, we may take for S the set of all dispersion-free states on $(X,\mathfrak{A})$, and for ϕ, the mapping $\phi(x) = \{\omega \in S | \omega(x) = 1\}$. Conversely, if (b) holds, then for every point $s \in S$, we have a dispersion-free state ω_s on $(X,\mathfrak{A})$ defined by $\omega_s(x) = 1$ iff $s \in \phi(x)$, and $\omega_s(x) = 0$ otherwise. Evidently, $\omega_s(x) = 1$ for any $s \in \phi(x)$. ■

We shall make frequent reference in the sequel to the following simple but illuminating

EXAMPLE 55 The "Wright Triangle". Imagine a covered, three-sided box in which is trapped a firefly. Imagine, further, that the three walls of the box are transparent, and that the interior of the box is divided into three chambers, which communicate in such a way that the fire-fly can move freely among the chambers, but we can view only two of the chambers (through one of the side-windows) at a time.

Label the three chambers a, b and c. Consider the experiment of viewing chambers a and b: The possible outcomes are that we see the firefly in chamber a, or in chamber b, or that we see no light (either because the fire-fly is not lit, or because it occupies chamber c). Call this last outcome x. Thus, the outcome-set for this experiment is the set $E = \{a, x, b\}$. We have two similar experiments corresponding to the other two windows, with outcome-sets $F = \{b, y, c\}$ and $G = \{c, z, a\}$ (y and z being the "no light" outcomes for these experiments).

The test space $\mathfrak{W} = \{E, F, G\}$ is usually called the *Wright Triangle*[17]. It may be represented by the Greechie diagram of Figure 2.1.8 (i) above — which I'll reproduce for convenience:

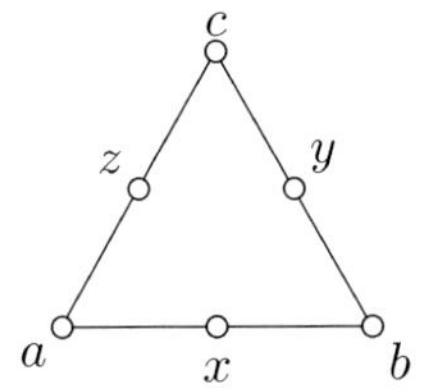

The Wright Triangle has a separating set of dispersion-free states, hence, can be embedded in a Borel test space. Indeed, let S denote the set of dispersion-free states on $\mathfrak{A}$, corresponding to the firefly's being in one of the three chambers, with its light on, or being anywhere, with its light off. Denote these states by $\{\alpha, \beta, \gamma, \delta\}$, where δ represents the "light off" state. Then we can assign each outcome to a subset of S: $a \mapsto \{\alpha\}$, $b \mapsto \{\beta\}$, $c \mapsto \{\gamma\}$, $x \mapsto \{\gamma, \delta\}$, and so on. In this way, each test corresponds to a partition of S.

However, not every state on $\mathfrak{W}$ survives this embedding. Consider the state $\omega(a) = \omega(b) = \omega(c) = 1/2$, $\omega(x) = \omega(y) = \omega(z) = 0$. This might represent a gregarious state of the firefly, in which the insect always contrives to occupy (lit, and with equal probability) one of the two rooms under observation. This state is clearly not a convex combination of dispersion-free states, and hence, does not arise from a probability distribution on S.

Remark. It is easy to construct variants of the Wright triangle involving "firefly boxes" with any number of chambers. In particular, the pentagonal Greechie diagram for the orthogonality space X_5 of Example 38 can be interpreted in terms of a five-sided firefly-box.

56 The Semiclassical Cover. A semiclassical test space is obviously UDF. Moreover, any state on any locally countable semiclassical test space arises as an

[17]So named for R. Wright, who made extensive use of it. The example is in fact due to Foulis.

average over dispersion free states.[18] In this respect, semiclassical test spaces are quite close to being classical. Now, *any* test space $\mathfrak{A}$ may be understood as arising from a semiclassical test space by the conflation of certain outcomes of distinct tests — perhaps to reflect some conviction that the conflated outcomes "mean" the same thing. Indeed, given any test space $\mathfrak{A}$, we can construct what we may call the *semiclassical cover*, of $\mathfrak{A}$, as follows: let $\widetilde{X}$ be the graph of the incidence (that is, element-hood) relation between outcomes and tests:

$$\widetilde{X} = \{(x, E) | x \in E \in \mathfrak{A}\}.$$

For each $E \in \mathfrak{A}$, let $\widetilde{E} = \{(x, E) | x \in E\} \subseteq \widetilde{X}$, and let

$$\widetilde{\mathfrak{A}} \ := \ \{\widetilde{E} | E \in \mathfrak{A}\}.$$[19]

Clearly, any state ω on $\mathfrak{A}$ lifts to a state on $\widetilde{\mathfrak{A}}$, given by $\widetilde{\omega}(x, E) = \omega(x)$; and this, in turn, may be interpreted as an average over dispersion-free states on $\widetilde{\mathfrak{A}}$. This seems to be the essential content of various "contextual" hidden variables schemes discussed in the literature.

2.2 Combinations of Test Spaces

Since they are such simple objects, test spaces are easy to manipulate. In particular, there are many, many ways in which test spaces can meaningfully and usefully be combined. In this section, we consider several of these. Throughout this section, $\mathfrak{A}$ and $\mathfrak{B}$ are test spaces with outcome sets X and Y, respectively.

HORIZONTAL AND DIRECT SUMS OF TEST SPACES

If A and B are any two sets, let's write $A \oplus B$ for their coproduct, or disjointified union (identifying this with $A \cup B$ if A and B happen to be disjoint.)[20] Certainly the simplest combination of $\mathfrak{A}$ and $\mathfrak{B}$ is the one that simply lays them side by side, as it were:

DEFINITION 57. The *horizontal sum* of $\mathfrak{A}$ and $\mathfrak{B}$ is the test space whose outcome set is $X \oplus Y$ and whose set of tests, $\mathfrak{A} \dot{+} \mathfrak{B}$, consists of (copies of) tests $E \in \mathfrak{A}$, understood as disjoint from Y, and (copies of) tests $F \in \mathfrak{B}$, understood as disjoint from X.

Note that the state space of $\mathfrak{A} \dot{+} \mathfrak{B}$ is essentially just the Cartesian product of the states spaces of $\mathfrak{A}$ and $\mathfrak{B}$.

DEFINITION 58. The *direct sum* of $\mathfrak{A}$ and $\mathfrak{B}$ is the test space on $X \oplus Y$ having tests $E \oplus F$, where $E \in \mathfrak{A}$ and $F \in \mathfrak{B}$. By abuse of notation, we write the set of such tests as $\mathfrak{A} \oplus \mathfrak{B}$.

---

[18] This is an easy consequence of the Bishop-deLeuw theorem; see [Wilce, 2002].

[19] This represents the extremity of "de-Occamization", to use Michael Redhead's nice phrase.

[20] That is, $A \oplus B = (A \times \{1\}) \cup (B \times \{2\})$. I'll follow the usual relaxed convention that identifies $a \in A$ and $b \in B$ with $(a, 1)$ and $(b, 2) \in A \oplus B$, respectively, that identifies $A \oplus B$ with $A \cup B$ in case A and B happen to be disjoint, and that treats $A \oplus B$ and $B \oplus A$ as the same.

A test in $\mathfrak{A} \oplus \mathfrak{B}$ is executed by choosing either of two tests, $E \in \mathfrak{A}$ or $F \in \mathfrak{B}$ (say, by flipping a coin), executing that test, and recording the result as the outcome of $E \oplus F$.

If $\mu \in \Omega(\mathfrak{A})$ and $\nu \in \Omega(\mathfrak{B})$, then we can regard both μ and ω as functions on $X \oplus Y$ by setting $\mu(y) = \nu(x) = 0$ for all $x \in X$ and all $y \in Y$. We can then meaningfully form convex combinations of μ and ν, e.g., $t\mu + (1-t)\nu$; it is easy to see that this will be a state on $\mathfrak{A} \oplus \mathfrak{B}$, and that every state on $\mathfrak{A} \oplus \mathfrak{B}$ has a unique representation as such a convex combination.

PRODUCT AND COMPOUND TEST SPACES

Given two test spaces $\mathfrak{A}$ and $\mathfrak{B}$, it is easy to construct a test space modelling the situation in which tests from $\mathfrak{A}$ and from $\mathfrak{B}$ are executed in parallel. It will be convenient to use the juxtapositive notation xy for an ordered pair (x, y) in $X \times Y$.

DEFINITION 59. The *cartesian product* of two test spaces $(X, \mathfrak{A})$ and $(Y, \mathfrak{B})$ is the test space $(XY, \mathfrak{A} \times \mathfrak{B})$ where

$$\mathfrak{A} \times \mathfrak{B} = \{EF | E \in \mathfrak{A}, F \in \mathfrak{B}\}.$$

A test $EF \in \mathfrak{A} \times \mathfrak{B}$ is called a *product test.* Such a test may be performed by separately performing the tests $E \in \mathfrak{A}$ and $F \in \mathfrak{B}$, and then collating the results. The following construction models the situation in which one executes a test $E \in \mathfrak{A}$ *first,* and *then* executes a test $F_x \in \mathfrak{B}$, *depending upon the outcome* $x \in E$ *that was secured.*

DEFINITION 60. Let $\mathfrak{A}$ and $\mathfrak{B}$ be test spaces. The *forward product* of $\mathfrak{A}$ and $\mathfrak{B}$ is the test space $\overrightarrow{\mathfrak{A}\mathfrak{B}}$ consisting of two-stage tests of the form $\bigcup_{x \in E} xF_x$, where $E \in \mathfrak{A}$ and, for each outcome $x \in E$, $F_x \in \mathfrak{B}$. Note that $\overrightarrow{\mathfrak{A}\mathfrak{B}}$ is semi-classical if, and only if, both $\mathfrak{A}$ and $\mathfrak{B}$ are classical.

The states on the forward product are easy to describe. Let ω be a state on the initial test space $\mathfrak{A}$, and, for each outcome $x \in X = \bigcup \mathfrak{A}$, let ω_x be a state $\mathfrak{B}$. Then we may define a state ω on $\overrightarrow{\mathfrak{A}\mathfrak{B}}$ by $\omega(xy) = \mu(x)\nu_x(y)$. It is easy to see that every state on $\overrightarrow{\mathfrak{A}\mathfrak{B}}$ has this form. In particular, then, the state at the second stage will in general depend, and may depend in an arbitrary way, upon which *outcome* was secured at the previous stage. We'll examine the forward product of test spaces in more detail in section 5, in connection with tensor products.

Obviously, one can construct compound tests involving any number of stages. The collection of all such multi-stage tests that can be built up from the tests belonging to a fixed test space $\mathfrak{A}$ gives us a new test space, called the *compounding* of $\mathfrak{A}$. This can be defined as follows:

DEFINITION 61. The *compounding* of a test space $\mathfrak{A}$ is the test space $\mathfrak{A}^c := \bigcup_{n \in \mathbb{N}} \mathfrak{A}^n$, where $\mathfrak{A}^n$ is defined recursively by

(i) $\mathfrak{A}^0 = \mathfrak{A}$;

(ii) $\mathfrak{A}^{n+1} = \overrightarrow{\mathfrak{A}^n\mathfrak{A}}$.

Outcomes of $\mathfrak{A}^c$ are, in effect, finite strings over the alphabet X. One can represent a test in $\mathfrak{A}^c$ by a rooted tree whose nodes are labelled by elements of $\mathfrak{A}$ and in which the edges leaving a node E are indexed by the outcomes $x \in E$. Clearly, even if $\mathfrak{A}$ is quite simple, $\mathfrak{A}^c$ will typically be very complex. Note that $\bigcup \mathfrak{A}^c$ is the free semi-group on X, and that the orthogonality relation induced by $\mathfrak{A}^c$ is the lexicographic one associated with the free orthomonoid over $(X, \perp)$, as discussed in section 1.4.

Remark. It should be mentioned that the construction of $\overrightarrow{\mathfrak{A}\mathfrak{B}}$ can be generalized to allow the tests at the second stage to come from a *family* of test spaces $\mathfrak{B}_x$ indexed by outcomes of $\mathfrak{A}$. The resulting test space is called the *Dacey sum* of the $\mathfrak{B}_x$ over $\mathfrak{A}$.

TEST SPACES OF BIJECTIONS

The following construction, discussed in more detail in [Wilce, 1997b], generalizes Example 50.

DEFINITION 62. Let E and F be two sets of the same cardinality, regarded as the outcome-sets of two tests. Let $B(E, F)$ denote the set of all bijections $f : E \to F$. Identifying f with its graph, we can regard $B(E, F)$ as a test space with outcome-set $E \times F$.

This has the following interpretation: to execute the test (corresponding to) $f \in B(E, F)$, choose and execute one of the two tests E and F. If E is executed and the outcome $x \in E$ is obtained, record $(x, f(x))$ as the outcome of f; if F is executed and the outcome $y \in F$ is obtained, record $(f^{-1}(y), y)$ as the outcome of f. If $|E| = |F| = 2$, then the structure of $B(E, F)$ is quite simple: it is a semi-classical test space consisting of two, two-outcome tests. On the other hand, if the cardinality of E and F is three or greater, the structure of $B(E, F)$ is quite complex. Indeed, in the next-simplest case, where $E = F = \{1, 2, 3\}$, we find that $B(E, E)$ is the 3×3 "array" test space of Example 50, whose states are the doubly-stochastic three-by-three matrices. Obviously, we can generalize this construction:

DEFINITION 63. Let $\mathfrak{A}$ and $\mathfrak{B}$ are test spaces, both of uniform rank n (finite or otherwise). Then $B(\mathfrak{A}, \mathfrak{B})$ is the union of the test spaces $B(E, F)$ where $E \in \mathfrak{A}$ and $F \in \mathfrak{B}$.

If $\mathfrak{A}$ and $\mathfrak{B}$ are both uniform and of the same rank n, then $B(\mathfrak{A}, \mathfrak{B})$ is also of uniform rank n, and in this case, $\bigcup B(\mathfrak{A}, \mathfrak{B}) = XY$. One can show that $B(\mathfrak{A}, \mathfrak{B})$ is effective as the direct product of $\mathfrak{A}$ and $\mathfrak{B}$ in a suitable category of uniform test spaces of a fixed rank.

2.3 *Orthogonality and Perspectivity of Events*

Above (in 51), we defined an orthogonality relation on the outcomes of a test space: two outcomes x and y are *orthogonal* iff they belong to a common test —

equivalently, iff $\{x,y\}$ is a two-element event. There is a natural orthogonality relation on the set of events of a test space, extending that on outcomes:

DEFINITION 64. Let $\mathfrak{A}$ be a test space. Events A and B of $\mathfrak{A}$ are *compatible* iff they are contained in a common test — equivalently, if their union is again an event. Events A and B are *orthogonal*, written $A \perp B$, iff they are compatible and disjoint.

If $A \perp B$ in $\mathcal{E}(\mathfrak{A})$, then $A \subseteq B^{\perp}$; but the converse is in general false. Consider the Wright triangle $\mathfrak{W} = \{\{a,x,b\},\{b,y,c\},\{c,z,a\}\}$ (Example 55): Then$\{a\} \subseteq \{b,c\}^{\perp}$, as $a \perp b$ and $a \perp c$; but $\{a\} \not\perp \{b,c\}$, since $\{a,b,c\}$ is not an event.

A large part of the theory of test spaces involves on the following notion of *perspectivity* between events:

DEFINITION 65. We say that two events A and B, of a test space $\mathfrak{A}$ are *complementary*, or that A is a *complement* for B, and write $A\mathsf{co}B$, iff $A \perp B$ and $A \cup B \in \mathfrak{A}$. Equivalently, A and B are complementary iff they partition a test. If events A and B share a complement, we say that they are *perspective*, and write $A \sim B$.

It is sometimes useful to illustrate the perspectivity of A and B via axis C by means of the following sort of diagram:

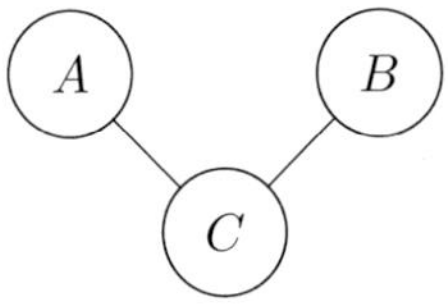

This is much like a Greechie diagram, but with the lines indicating complementarity rather than mere orthogonality. Call a diagram of this kind a co-*diagram* . We say here that C is an *axis of perspectivity* for A and B.

Note that, for any pair of tests $E, F \in \mathfrak{A}$, we have $E \sim F$ with axis $\emptyset$ and $E \backslash F \sim F \backslash E$ with axis $C = E \cap F$. Notice, too, that if A and C are complementary events, and ω is a state, then $\omega(A) = 1 - \omega(C)$. Hence, if $A \sim B$, $\omega(A) = \omega(B)$ for every state ω. The following observation is sometimes useful:

LEMMA 66. *Let $\mathfrak{A}$ be any test space. Then for all $A, B \in \mathcal{E}(\mathfrak{A})$,*

$$A \sim B \subseteq A \;\Rightarrow\; A = B.$$

Proof. If $A \sim B$, then for some event C, $A\mathsf{co}C\mathsf{co}B$. In particular, both $C \cup A$ and $C \cup B$ belong to $\mathfrak{A}$. If $B \subseteq A$, we have $C \cup B \subseteq C \cup A$. Since $\mathfrak{A}$ is irredundant, $C \cup A \subseteq C \cup B$. Since C is disjoint from A and B, it follows that $A \subseteq B$, whence, $A = B$. ■

EXAMPLE 67. If $\mathfrak{B} = \mathfrak{B}(M, \mathcal{F})$ is the Borel manual of all countable measurable partitions of a measurable space $(M, \mathcal{F})$, then events of $\mathfrak{B}$ are merely disjoint families of sets in $\mathcal{F}$. It follows easily that two such families are perspective iff they have the same union.

EXAMPLE 68. In the case of a quantum test space $\mathfrak{A} = \mathfrak{F}(\mathbf{H})$, events are pairwise orthogonal sets of unit vectors, and it is a simple exercise to show that two such sets A and B are orthogonal iff their spans are orthogonal. Note that, in this case $A \subseteq B^{\perp}$ implies $A \perp B$. Events are complementary iff their spans are complementary subspaces of $\mathbf{H}$, and hence, perspective iff they have the same closed span.

The preceding examples may suggest that perspectivity is a kind of "physical equivalence". This idea is reinforced by the observation, made above, that perspective events have the same probability in every state. However, the following examples should temper one's enthusiasm for this interpretation.

EXAMPLE 69. Consider — yet again! — the "Wright Triangle" test space

$$\mathfrak{W} = \{\{a, x, b\}, \{b, y, c\}, \{c, z, a\}\},$$

discussed above in Example 55. The events $\{a, x\}$ and $\{y, c\}$ are perspective with axis $\{b\}$. These two events represent two possible situations in which b might be observed, but is not. However, as the discussion in Example 46 shows, these two events are not, strictly speaking, physically equivalent, if we regard the choice of test as influencing the state of the "firefly".

Another obstacle to viewing perspectivity as an absolute physical equivalence is the fact that perspective events behave differently in two-stage experiments.

EXAMPLE 70. Consider a compound test space $\mathfrak{A}\mathfrak{B}$ where the first test space consists of two tests $E = \{a, c\}$ and $F = \{c, b\}$, and where $\mathfrak{B}$ is a classical test space consisting of the single test $G = \{x, y\}$. Then the outcomes a and b are perspective in $\mathfrak{A}$, but the outcomes ax and bx are *not* perspective in the forward product

$$\overrightarrow{\mathfrak{A}\mathfrak{B}} = \{\{ax, ay, cx, cy\}, \{cx, cy, bx, by\}\}.$$

Notice here that different states on $\overrightarrow{\mathfrak{A}\mathfrak{B}}$ will generally assign different probabilities to ax and bx.

As these last two example show, one should be careful not to take perspectivity too seriously as representing a strict notion of "physical equivalence". This caveat notwithstanding, however, the concept of perspectivity is the central structural tool in the theory of test spaces. It will be exploited at every turn in the sequel.

2.4 Mappings of Test Spaces

If we wish to employ test spaces with any fluency, we need to single out appropriate morphisms between them. It will be convenient, in what follows, to treat a relation between two sets X and Y as a set-valued mapping $\phi : X \to \mathcal{P}(Y)$. In this case the relational image of a set $A \subseteq X$ is given by $\phi(A) = \bigcup_{a \in A} \phi(a)$. The composition of two set-valued mappings $\phi : X \to \mathcal{P}(Y)$ and $\psi : Y \to \mathcal{P}(Z)$ is given by $(\psi \circ \phi)(x) = \bigcup_{y \in \phi(x)} \psi(y)$.

DEFINITION 71. Let $\mathfrak{A}$ and $\mathfrak{B}$ be test spaces with outcome-sets X and Y respectively. A *homomorphism* from $\mathfrak{A}$ to $\mathfrak{B}$ is an event-valued mapping $\phi : X \to \mathcal{E}(\mathfrak{B})$ taking events to events, and preserving both orthogonality and perspectivity — that is,

(i) $\forall A \in \mathcal{E}(\mathfrak{A}),\ \phi(A) \in \mathcal{E}(\mathfrak{B})$;

(ii) $\forall\ x, y \in X,\ x \perp y\ \Rightarrow\ \phi(x) \perp \phi(y)$; and

(iii) $\forall\ A, B \in \mathcal{E}(X, \mathfrak{A}),\ A \sim B\ \Rightarrow\ \phi(A) \sim \phi(B)$.

If ϕ also preserves *tests* — that is, if $\phi(E) \in \mathfrak{B}$ for every $E \in \mathfrak{A}$ — then we call ϕ an *interpretation.*

I'll abuse notation to the extent of writing $\phi : \mathfrak{A} \to \mathfrak{B}$ (rather than $\phi : X \to \mathcal{E}(\mathfrak{B})$) to indicate that ϕ is an homomorphism from $\mathfrak{A}$ to $\mathfrak{B}$. The set of all homomorphisms and the set of all interpretations from $\mathfrak{A}$ to $\mathfrak{B}$ will be denoted by $\mathrm{Hom}(\mathfrak{A}, \mathfrak{B})$ and $\mathrm{Int}(\mathfrak{A}, \mathfrak{B})$, respectively. Evidently, the composition of two homomorphisms is another homomorphism, and likewise for interpretations.

Remark. Any event-valued mapping $\phi : X \to \mathcal{E}(\mathfrak{B})$ satisfying the condition $\phi(x) \perp \phi(y)$ for $x \perp y$ in X and taking tests to tests, automatically preserves perspectivity, and hence, is a homomorphism, and, in particular, an interpretation. Thus, an interpretation simply identifies (interprets!) each test $E \in \mathfrak{A}$ as a partition — a *coarse-grained* version — of a test $\phi(E)$ in $\mathfrak{B}$. This is surely a very natural way of linking two test spaces. In contrast, the more general notion of a homomorphism may seem somewhat arbitrary. However, as we'll see below, there is a natural sense in which a homomorphism is simply an interpretation onto a certain fragment of its codomain, which we may regard as its image.

DEFINITION 72. Notice that the image of an outcome under a homomorphism may be empty. The *support* of a homomorphism $\phi : \mathfrak{A} \to \mathfrak{B}$ is the set

$$S_\phi = \{\, x \in X \mid \phi(x) \neq \emptyset \,\}.$$

We say that ϕ is *positive* iff $S_\phi = X$. We say that ϕ is *single-valued*, or *outcome-preserving*, iff $\forall x \in S_\phi$, $|\phi(x)| = 1$. In this case, we generally write $\phi(x) = y$ rather than $\phi(x) = \{y\}$.

Given any set $S \subseteq X$, we can define a set-valued characteristic function $\chi_S : X \to 2^S$ by setting

$$\chi_S(x) \;=\; \begin{cases} \{x\} & x \in S \\ \emptyset & x \notin S \end{cases}$$

for each $x \in X$. Notice that $\chi_S(A) = \emptyset$ iff $A \cap S$ is empty.

LEMMA 73. *Let $\mathfrak{A}$ be a test space with outcome set X, and let $S \subseteq X$. Let $\mathfrak{A}_S := \{E \cap S | E \in \mathfrak{A}\}$. The following are equivalent:*

(a) $S = S_\phi$ for some homomorphism $\phi : \mathfrak{A} \to \mathfrak{B}$;

(b) $A \sim B$ implies $S \cap A = \emptyset$ iff $S \cap B = \emptyset$, for all events $A, B \in \mathcal{E}$.

(c) $\mathfrak{A}_S$ is irredundant, hence, a test space in its own right.

(d) $\chi_S : X \to 2^S$ defines a positive interpretation from $\mathfrak{A}$ to $\mathfrak{A}_S$.

Proof. Obviously, (d) implies (a). To see that (a) implies (b), suppose that $S = S_\phi$ for a homomorphism $\phi : \mathfrak{A} \to \mathfrak{B}$. For any event A in $\mathcal{E}(\mathfrak{A})$, we have $A \cap S = \emptyset$ iff $\phi(A) = \emptyset$. If $A \sim B$, then $\phi(A) \sim \phi(B)$; hence, if $\phi(A) = \emptyset$, then $\phi(B) \sim \emptyset$. It follows from the irredundancy of $\mathfrak{B}$ that then $\phi(B) = \emptyset$, whence, $B \cap S = \emptyset$ as well. Thus, (a) implies (b). Suppose now that (b) holds. If $E, F \in \mathfrak{A}$ and $E \cap S \subseteq F \cap S$, then $S \cap (F \setminus E) = \emptyset$, whence, as $(F \setminus E) \sim (E \setminus F)$, $S \cap (E \setminus F) = \emptyset$ — that is, $S \cap F \subseteq S \cap E$. Thus, $\mathfrak{A}_S$ is irredundant. Finally, suppose that $\mathfrak{A}_S$ is irredundant, i.e., a test space in its own right. It is easy to check that the "characteristic function" given by

$$\chi_S(x) = \begin{cases} \{x\} & x \in S \\ \emptyset & x \notin S \end{cases}$$

defines a homomorphism — indeed, an interpretation — from $\mathfrak{A}$ to $\mathfrak{A}_S$. ■

We call a set $S \subseteq X$ satisfying the equivalent conditions of Lemma 73, a *support*. We call the test space $\mathfrak{A}_S$ the *restriction* of $\mathfrak{A}$ to S, and the interpretation $\mathfrak{A} \to \mathfrak{A}_S$ given by χ_S, the *restriction interpretation* associated to S.

EXAMPLES 74. (i) The mapping $x \mapsto \{x\}$ sending each outcome $x \in X = \bigcup \mathfrak{A}$ to the corresponding one-outcome event is the *identity interpretation* on the test space $\mathfrak{A}$.

(ii) Let E and F be any two sets, and let $f \subseteq F \times E$. Then $f^{-1} : E \to \mathcal{P}(F)$ is a morphism iff f is a partially-defined function, an interpretation iff f is totally defined, and outcome-preserving iff f is injective.

(iii) The *coarsening* of a test space $\mathfrak{A}$ is the test space $\mathfrak{A}^{\#}$ consisting of of finite partitions of $\mathfrak{A}$-tests into non-empty events. Note that each non-empty event of $\mathfrak{A}$ corresponds to an *outcome* of $\mathfrak{A}^{\#}$. There are natural outcome-preserving interpretations $\iota : \mathfrak{A} \to \mathfrak{A}^{\#}$, $\delta : \mathfrak{A}^{\#} \to \mathfrak{A}$. The former takes each outcome $x \in X$ to the corresponding outcome $\{\{x\}\}$, while the latter takes each *outcome* $A \in \bigcup \mathfrak{A}^{\#}$ to the corresponding *event* $A \in \mathcal{E}(\mathfrak{A})$.

(iv) Let L_1 and L_2 be two orthomodular lattices. Let $\mathfrak{A}_i$ be the test space of (finite) partitions of unity in L_i, $i = 1, 2$. Then any homomorphism $\phi : L_1 \to L_2$ defines a homomorphism $\mathfrak{A}_1 \to \mathfrak{A}_2$. This is an interpretation iff ϕ maps the unit of L_1 to that of L_2. As a special case, let $\mathfrak{B}_i = \mathfrak{B}(M_i, \mathcal{F}_i)$, $i = 1, 2$, be two Borel test spaces associated with measurable spaces $(M_1, \mathcal{F}_2)$ and $(M_2, \mathcal{F}_2)$ respectively. If f :

$M_2 \to M_1$ is any measurable function, the mapping $f^{-1} : \mathcal{F}_1 \to \mathcal{F}_2$ is a complete Boolean homomorphism, and hence maps partitions to partitions. We obtain an outcome-preserving interpretation $\phi : \mathfrak{B}_1 \to \mathfrak{B}_2$ given by $\phi(a) = \{f^{-1}(a)\}$ for all $a \in \mathcal{F}_1$.

(v) Let $\mathfrak{A}(\mathbf{H})$ be the quantum test spaces associated with the Hilbert space $\mathbf{H}$, and let $U : \mathbf{H} \to \mathbf{H}$ be unitary or anti- unitary mapping from $\mathbf{H}$ to itself. Since U maps orthonormal bases to orthonormal bases, we may define an interpretation $\phi : \mathfrak{A}(\mathbf{H}) \to \mathfrak{A}(\mathbf{H})$ by $\phi(x) = \{Ux\}$. Note that this interpretation is invertible. More generally, let c be a completely arbitrary mapping assigning to each unit vector x a complex number $c(x)$ with $|c(x)| = 1$. Then $\phi : x \mapsto \{c(x)U(x)\}$ is again an outcome- preserving interpretation. It follows from Wigner's theorem that every invertible interpretation $\phi : \mathfrak{A}(\mathbf{H}) \to \mathfrak{A}(\mathbf{H})$ arises in this manner. (Arbitrary interpretations between quantum test spaces can be characterized using a powerful generalization of Wigner's theorem due to R. Wright [Wright, 1978a].)

We now establish that arbitrary morphisms are essentially just interpretations in disguise. Evidently, if $\phi : \mathfrak{A} \to \mathfrak{B}$ is a homomorphism, then $\phi(\mathfrak{A}) := \{\phi(E)|E \in \mathfrak{A}\}$ is a set of mutually perspective events in $\mathfrak{B}$. The following gives us a converse:

LEMMA 75. *Let $\mathfrak{B} \subseteq \mathcal{E}(\mathfrak{A})$ be any collection of mutually perspective events. Let $Y = \bigcup \mathfrak{B}$. If we view $(Y, \mathfrak{B})$ as a test space, the natural injection $i : Y \to \mathcal{P}(X)$ given by $i(x) = \{x\}$ is a homomorphism.*

Proof. To see this, it suffices to check that if A and B are perspective as $\mathfrak{B}$-events, then they are perspective as $\mathfrak{A}$-events as well. But if $A \sim B$ in $\mathcal{E}(\mathfrak{B})$, there exists some event C such that $A \cap C = C \cap B = \emptyset$ and $A \cup C, C \cup B \in \mathfrak{B}$. But then, as the members of $\mathfrak{B}$ are mutually perspective in $\mathcal{E}(\mathfrak{A})$, there exists some event D of $\mathfrak{A}$ such that $(A \cup C) \cup D = D \cap (C \cup B) = \emptyset$ and $A \cup (C \cup D), (D \cup C) \cup B \in \mathcal{E}(\mathfrak{A})$ — whence, A and B are perspective in $\mathfrak{A}$. ■

We shall call $i_{\mathfrak{B}}$ the *inclusion homomorphism* associated with $\mathfrak{B}$. Evidently, any homomorphism $\phi : \mathfrak{A} \to \mathfrak{B}$ with support S factors as $\phi = i_{\phi(\mathfrak{A})} \circ \overline{\phi} \circ \chi_S$, where χ_S is the restriction interpretation associated with S, and where $\overline{\phi}$ is a positive interpretation. It seems reasonable to call the test space $\phi(\mathfrak{A}) = \{\phi(E)|E \in \mathfrak{A}\}$ the *image* of the homomorphism ϕ. Thus, in particular, every homomorphism defines an interpretation onto its image.

Categories of Test Spaces

It is easy to see that the composition of two homomorphisms, or of two interpretations, is again a homomorphism or interpretation. Accordingly, we may speak of the *category* of test spaces and homomorphisms, and the sub-category of test spaces and interpretations. I'll denote these, respectively, by TES and TES1. Note that TES comes equipped with two obvious functors, namely, the outcome-space functor and the forgetful functor that regards $\mathfrak{A}$ merely as a set. The former takes TES to REL (the category of sets and relations) and the latter, to SET (the category of sets and mappings).

Let $\mathfrak{A}$ and $\mathfrak{B}$ be test spaces with outcome sets X and Y, respectively, and let $\phi \in \mathrm{Int}(\mathfrak{A}, \mathfrak{B})$. It is not hard to show that ϕ is invertible iff there exists a bijective mapping $f : X \to Y$, taking the tests of $\mathfrak{A}$ bijectively to the those of $\mathfrak{B}$, such that $\phi(x) = \{f(x)\}$ for every $x \in X$. The characterization of epimorphisms and monomorphisms in TESP1 is much less straightforward. For example, let $\mathfrak{A} = \{\{a, b\}, \{c, d\}\}$ and let $\mathfrak{B} = \{\{x, y\}, \{y, z\}, \{x, z\}\}$. It is easy to check that the mapping $\phi : \mathfrak{A} \to \mathfrak{B}$ given by

$$\phi(a) = x, \ \phi(b) = \phi(c) = y, \ \phi(d) = z$$

is both an epimorphism and a monomorphism, even though it is neither injective on outcomes, nor surjective on tests.

For any test spaces $\mathfrak{A}_1$ and $\mathfrak{A}_2$, there are natural inclusion interpretations $\phi_i : \mathfrak{A}_i \to \mathfrak{A}_1 \oplus \mathfrak{A}_2$ given by $\phi_i(x) = \{x\}$ for $x \in X_i$, $i = 1, 2$. It is easy to show that these make $\mathfrak{A}_1 \oplus \mathfrak{A}_2$ the direct product of $\mathfrak{A}_1$ and $\mathfrak{A}_2$ in TES.

We have thus far been regarding interpretations as set-valued mappings, but, as remarked above, we can equally well regard them as relations. If $\phi : \mathfrak{A} \to \mathfrak{B}$ is an interpretation, its *graph* is the corresponding relation $R_\phi \subseteq X \times Y$ given by

$$R_\phi := \{(x, y) \in X \times Y | y \in \phi(x)\}$$

The set $\mathrm{Int}(\mathfrak{A}, \mathfrak{B})$ of all graphs of interpretations is an irredundant set of subsets of $X \times Y$, hence, can itself be regarded as a test space. The following is due to P. Lock [Lock, P., 1981]:

THEOREM 76. *Let $\mathfrak{A}, \mathfrak{B}$ and $\mathfrak{C}$ be test spaces. Then*

$$Int(\mathfrak{A} \times \mathfrak{B}, \mathfrak{C}) \simeq Int(\mathfrak{A}, Int(\mathfrak{B}, \mathfrak{C})).$$

For any fixed test space $\mathfrak{A}$, we have a functor $\mathrm{Int}(\mathfrak{A}, -)$ from TES to SET. Lock's theorem tells us that the endo-functor $- \times \mathfrak{B} : \mathfrak{A} \mapsto \mathfrak{A} \times \mathfrak{B}$ is left-adjoint to the endo-functor $\mathrm{Int}(\mathfrak{B}, -) : \mathfrak{C} \mapsto \mathrm{Int}(\mathfrak{B}, \mathfrak{C})$.[21]

2.5 Spaces of Weights on Test Spaces[22]

If a state on a test space $\mathfrak{A}$ is a generalization of a probability measure, it is natural to attempt to generalize the notion of an arbitrary bounded measure, or for that matter, a vector-valued measure, in a similar spirit. What follows is drawn mainly from [Cook, 1985; Wilce, 1995].

DEFINITION 77. Call a function $\omega : X(\mathfrak{A}) \to \mathbb{R}$ a *weight* on $\mathfrak{A}$ iff

[21] This divergence between the direct product and the adjoint of the Hom functor is equally a feature of the category of vector spaces and linear mappings, where the adjoint of $\mathrm{Int}(-, V)$ is the functor $- \otimes V$. Indeed, if $\mathfrak{A}$ and $\mathfrak{B}$ are test spaces consisting of bases for vector spaces V and W, then $\mathfrak{A} \times \mathfrak{B}$ may be regarded as a collection of bases for $V \otimes W$.

[22] This material will be used only in section 5.5, where we discuss tensor products of frame manuals.

(i) ω has uniformly bounded variation over elements of $\mathfrak{A}$, and

(ii) $\sum_{x \in E} \omega(x)$ — which exists, by (i) — is independent of $E \in \mathfrak{A}$.

Clearly, a weight on the Borel test space of a measurable space $(M, \mathcal{F})$ corresponds to a bounded measure on $\mathcal{F}$. It is an easy extension of Gleason's theorem that a weight on the test space of frames of a Hilbert space $\mathbf{H}$ is representable by a self-adjoint operator on $\mathbf{H}$.

It is not difficult to show that a linear combination of two weights on $\mathfrak{A}$ is another weight on $\mathfrak{A}$. Thus, the set of all weights on $\mathfrak{A}$ is a vector space, which we denote by $W(\mathfrak{A})$. This space was first studied by Cook in [Cook, 1985].

DEFINITION 78. Let $\mathfrak{A}$ be a test space, and $\mathbf{V}$, any (real) normed linear space. A function ω from $X = \bigcup \mathfrak{A}$ to $\mathbf{V}$ is $\mathfrak{A}$-*summable* iff

$$\|\omega\| := \sup_{E \in \mathfrak{A}} \sum_{x \in E} \|\omega(x)\| < \infty.$$

We refer to $\|\omega\|$ as the *variation* of ω. We will denote the space of all such functions by $\Lambda^1(\mathfrak{A}, \mathbf{V})$, abbreviating $\Lambda^1(\mathfrak{A}, \mathbb{R})$ to $\Lambda^1(\mathfrak{A})$.

LEMMA 79. *Let* $\mathfrak{A}$ *and* $\mathbf{V}$ *be as above. Then*

(a) For every event A *of* $\mathfrak{A}$ *and every* $\omega \in \Lambda_1(\mathfrak{A}, \mathbf{V})$, $\omega(A) := \sum_{x \in A} \omega(x)$ *exists;*

(b) If $\mathbf{V} = \mathbb{R}$ *or* $\mathbb{C}$, *then* $\|\cdot\|$ *is equivalent to the supremum norm given by*

$$\|\omega\|_s = \sup_{A \in \mathcal{E}(\mathfrak{A})} |\omega(A)|.$$

Proof. (a) Straightforward. (b) Let $\omega \in \Lambda_1(\mathfrak{A})$ and $E \in \mathfrak{A}$. If $\mathbf{V}$ is $\mathbb{R}$, let $A = \{x \in E \mid \omega(x) > 0\}$ and let $B = E \setminus A$. Then $\sum_{x \in E} |\omega(x)| = \omega(A) - \omega(B) \leq 2 \sup_{A \in \mathcal{E}} |\omega(A)|$. If $\mathbf{V}$ is $\mathbb{C}$, note that ω's real and imaginary parts are real-valued weights, and apply the foregoing argument to each of these. On the other hand, if $A \subset E$, E a test, then $|\omega(A)| \leq \sum_{x \in E} |\omega(x)|$. ■

THEOREM 80. *For any test space* $\mathfrak{A}$ *and for any Banach space* $\mathbf{V}$, *the space* $(\Lambda_1(\mathfrak{A}, \mathbf{V}), \|\cdot\|)$ *is complete.*

Proof. For any set S and any Banach space $\mathbf{V}$, let $\ell_\infty(S, \mathbf{V})$ denote the space of bounded $\mathbf{X}$-valued functions on S, with the supremum-norm. This is clearly complete. Let $\ell_1(S, \mathbf{V})$ denote the space of summable $\mathbf{V}$-valued functions on S, i.e., those with

$$\|f\|_1 := \sum_{s \in S} \|f(s)\| < \infty.$$

Note that $\|\cdot\|_1$ is a norm on $\ell_1(S, \mathbf{V})$, and that the latter is complete in this norm (using, e.g., the fact that a normed space is complete iff every norm-absolutely summable series is summable). Now, given any $\omega \in \Lambda_1(\mathfrak{A}, \mathbf{V})$, define a map $\hat{\omega} : \mathfrak{A} \to \ell_1(X, \mathbf{V})$, where $X = \bigcup \mathfrak{A}$, by $\hat{\omega}(E)(x) = \delta_{x,E}\omega(x)$, where $\delta_{x,E}$ is 1 if $x \in E$ and 0 otherwise. Then

$$\|\hat{\omega}\|_\infty = \sup_{E \in \mathfrak{A}} \|\hat{\omega}(E)\|_1 = \|\omega\|.$$

Thus, the map $\omega \mapsto \hat{\omega}$ provides an isometric embedding of $\Lambda_1(\mathfrak{A}, \mathbf{V})$ into the space $\ell_\infty(\mathfrak{A}, \ell_1(X, , \mathbf{V}))$.

We now show that the image of the map $\hat{\cdot}$ is closed: If $\phi \in \ell_\infty(\mathfrak{A}, \ell_1(X(\mathfrak{A}), \mathbf{V}))$ and $E \in \mathfrak{A}$, define $f_{E,x}(\phi) = \phi(E)(x)$. Clearly, $f_{E,x}$ is a bounded linear functional for each pair (E, x). It is equally clear that ϕ belongs to the image of $\hat{\cdot}$ iff $\phi(E)(x) = \phi(F)(x)$ for all $x \in E \cap F$, where E and F range over $\mathfrak{A}$. Thus,

$$\operatorname{ran}(\hat{\cdot}) = \bigcap_{x \in E \cap F} \ker(f_{E,x} - f_{F,x}),$$

a closed subspace of $\ell_\infty(\mathfrak{A}, \ell_1(X, \mathbf{V}))$. ■

DEFINITION 81. Let $\mathfrak{A}$ be a test space and $\mathbf{V}$, a Banach space. An element ω of $\Lambda_1(\mathfrak{A}, \mathbf{V})$ is called an $\mathbf{V}$-*valued weight* on $\mathfrak{A}$ iff for all $E, F \in \mathfrak{A}$,

$$\sum_{x \in E} \omega(x) = \sum_{y \in F} \omega(y).$$

The space of all $\mathbf{V}$-valued weights on $\mathfrak{A}$ will be denoted $W(\mathfrak{A}, \mathbf{V})$. We denote the space of scalar-valued weights by $W(\mathfrak{A})$.

COROLLARY 82. *$W(\mathfrak{A}, \mathbf{V})$ is closed as a subspace of $\Lambda_1(\mathfrak{A}, \mathbf{V})$, and hence, complete.*

Proof. For each test E of $\mathfrak{A}$, let T_E be the bounded linear map from $\Lambda_1(\mathfrak{A}, \mathbf{V})$ to $\mathbf{V}$ given by $T_E(\omega) = \sum_{x \in E} \omega(x)$. Then

$$W(\mathfrak{A}, \mathbf{V}) = \bigcap_{E, F \in \mathfrak{A}} \ker(f_E - f_F).$$

■

Any state on $\mathfrak{A}$ is a weight on $\mathfrak{A}$; indeed, $\Omega(\mathfrak{A})$ is a base for the positive cone W_+ of $W(\mathfrak{A})$. The space $W_+ - W_+$ spanned by $\Omega(\mathfrak{A})$ is usually denoted by $V(\mathfrak{A})$. It can be shown ([3]) that this space is complete in the base-norm induced by $\Omega(\mathfrak{A})$. In general, $V(\mathfrak{A})$ is a proper subspace of $W(\mathfrak{A})$ — that is, no analogue of the Jordan decomposition theorem obtains in this general setting [9]. However, we do have the following easy

LEMMA 83. *Let $\mathfrak{A}$ be a test space of finite rank n. Then every weight on $\mathfrak{A}$ is a difference of positive weights.*

Proof. Since every test in $\mathfrak{A}$ has, say, cardinality n, the constant function $\eta(x) \equiv 1$ is a weight on $\mathfrak{A}$. If ω is any real-valued weight, then $\mu := \omega + 2\|\omega\|\eta$ is a positive weight, with $\mu > \omega$; hence, $\omega = \mu - (\mu - \omega)$ is Jordan. ■

3 LOGICS OF TEST SPACE

In section 2.3, we acquired some scruples about interpreting perspective events as "physically equivalent". On the other hand, the fact that two perspective events have the same probability in every state suggests that it may be mathematically worthwhile to identify perspective events. Pursuing this idea, we arrive at a useful order-theoretic invariant of a test space, called its *logic*. This always carries a natural involution, and is often orthocomplemented.

3.1 An Implication Relation on Events

In general, perspectivity is not even transitive, much less an equivalence relation on the set of events of a test space. Thus, if we wish to identify perspective events with one another, we need first to find a suitable equivalence relation extending perspectivity.

DEFINITION 84. Let $\mathfrak{A}$ be any test space. For any events $A, B \in \mathcal{E}(\mathfrak{A})$, say that A *weakly implies* B iff there exists a chain $A_1, ..., A_n$ of events with $A = A_1$, $A_n = B$, and, for each $i = 1, ..., n-1$, either $A_i \sim A_{i+1}$ or $A_i \subseteq A_{i+1}$. In this case, write $A \leq B$. Evidently, $\leq$ is a pre-order on $\mathcal{E}$. Let $\equiv$ be the associated equivalence relation, i.e., given two events A and B, set $A \equiv B$ iff $A \leq B$ and $B \leq A$. Write $[A]$ for the equivalence class of A with respect to $\equiv$, and $\Pi(X, \mathfrak{A})$ for the ordered set $\mathcal{E}/\equiv$, with $[A] \leq [B]$ iff $A \leq B$. This is called the *logic* of $(X, \mathfrak{A})$.

EXAMPLE 85. Let $\mathfrak{B} = \mathfrak{B}(M, \mathcal{F})$ be the Borel manual of a measurable space $(M, \mathcal{F})$, that is, the test space of countable partitions of M by sets $a \in \mathcal{F}$. Events are thus countable, pairwise-disjoint families $A = \{a_i\}$ of non-empty sets $a_i \in M$. As discussed above, two events are perspective iff they have the same union. If $A \subseteq B$ then $a := \bigcup A \subseteq b := \bigcup B$. Hence, if $A \leq B$, then again $a \subseteq b$. Conversely, if $\bigcup A \subseteq \bigcup B$, then $B \sim A_1$ where $A_1 = A \cup \{b \setminus a\}$; accordingly, $A \subseteq A_1 \sim B$, so $A \leq B$. Consequently, the logic $\Pi(\mathfrak{B})$ is order-isomorphic to $\mathcal{F}$.

EXAMPLE 86. Let $\mathfrak{F}$ be the frame manual of a hilbert space $\mathbf{H}$, that is, the test space of unordered orthonormal bases for $\mathbf{H}$. Then events of $\mathfrak{F}$ are orthonormal subsets of $\mathbf{H}$; as discussed above, two events A and B are perspective iff they have the same closed span, i.e., iff $A^{\perp\perp} = B^{\perp\perp}$. Hence, if $A \leq B$, then $A^{\perp\perp} \subseteq B^{\perp\perp}$. Conversely, if $A^{\perp\perp} = \mathbf{M}$ and $B^{\perp\perp} = \mathbf{N}$, and $\mathbf{M} \subseteq \mathbf{N}$, then we can enlarge A to an orthonormal basis A_1 for $\mathbf{N}$; then $A \subseteq A_1 \sim B$, so $A \leq B$. Hence, $\Pi(\mathfrak{F})$ is order-isomorphic to the lattice $L(\mathbf{H})$ of closed subspaces of $\mathbf{H}$.

LEMMA 87. *Let A, B, C and D be events in $\mathcal{E}(\mathfrak{A})$ with A co C and B co D. Then $A \leq B \Rightarrow D \leq C$.*

Proof. This is clear if $A \sim B$, so it suffices to show that $A \subseteq B \Rightarrow D \leq C$. But $A \subseteq B$ co $D \Rightarrow A$ co $D \cup (B \setminus A)$. Hence, $D \cup (B \setminus A) \sim C$. Since $D \subseteq D \cup (B \setminus A)$, we have $D \leq C$. ■

By Lemma 87, the relation $\equiv$ respects complementarity of events. Thus, we may define, for any $A \in \mathcal{E}(A)$, $p(A)' := p(E \setminus A)$ where E is an arbitrary element of $\mathfrak{A}$ containing A. The map $(\cdot)'$ is evidently an involution on $\Pi(\mathfrak{A})$, but need not be an orthcomplementation — consider, for instance, the test space $\{\{a,b\},\{b,c\},\{c,a\}\}$:

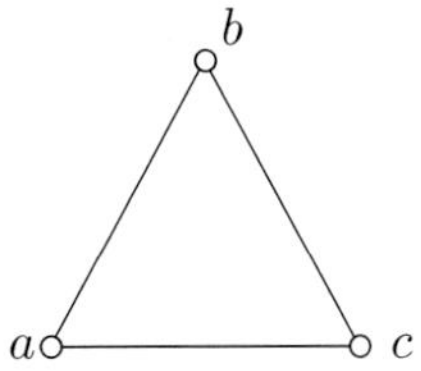

Here, all three outcomes (treated as singleton events) are perspective. Thus, the logic is the three-element chain $\{\mathbf{0}, p, \mathbf{1}\}$, where $p = p(a) = p(b) = p(c)$. This is not even complemented, and hence, supports no orthocomplementation.

DEFINITION 88. We say that $\mathfrak{A}$ is *consistent* iff its logic Π is orthocomplemented by $(\cdot)'$ — equivalently, iff $\forall p \in \Pi$, $p \leq p' \Rightarrow p = \mathbf{0}$.[23]

REMARK 89. If ω is a state on $\mathfrak{A}$, then $A \leq B \Rightarrow \omega(A) \leq \omega(B)$ for all events A and B. Hence, ω lifts to a function $\overline{\omega} : \Pi(\mathfrak{A}) \to [0,1]$ given by $\overline{\omega}(p(A)) = \omega(A)$. This function is monotone, takes values in $[0,1]$, and satisfies $\omega(p) + \omega(p') = 1$ for all $p \in \Pi(\mathfrak{A})$. A test space $\mathfrak{A}$ is said to be *semi-unital* iff $\forall x \in X$, $\exists \omega \in \Omega(\mathfrak{A})$ with $\omega(x) > 1/2$. It is immediate from what has just been said that if $\mathfrak{A}$ is semi-unital, then $\mathfrak{A}$ is consistent.

DEFINITION 90. Events A and B of $\mathfrak{A}$ are *weakly perspective*, and we write $A \overset{w}{\sim} B$, iff there is a chain of events $A = D_1 \sim D_2 \sim \ldots \sim D_n = B$.

Note that weakly equivalent events are equivalent, in the sense of Definition 84.

LEMMA 91. *In a consistent test space, equivalent events are weakly perspective.*

Proof. Suppose that, for $C, D \in \mathcal{E}(\mathfrak{A})$, $C \cap D = \emptyset$ and

$$A \leq D \subseteq D \cup C \leq B \leq A.$$

Let $E \in \mathfrak{A}$ be an test with $D \cup C \subseteq E$. Then $D \subseteq E \setminus C$ co C. Hence $C \subseteq D \cup C \leq A \leq D \subseteq E \setminus C$ — whence $p(C) \leq p(C)'$. If $\mathfrak{A}$ is consistent, it follows that $C = \emptyset$. Hence given a chain D_i with $A = D_1$, $B = D_n$ and $D_i \subseteq D_{i+1}$ or $D_i \sim D_{i+1}$, there can be no *proper* inclusions $D_i \subset D_{i+1}$. It follows that $D_i \sim D_{i+1}$ for all $i = 1, ..., n-1$, whence $A \overset{w}{\sim} B$. ■

THE LOGIC AS A FUNCTOR

Let $\mathfrak{A}$ and $\mathfrak{B}$ be test spaces, and $\phi : \mathfrak{A} \to \mathfrak{B}$, an homomorphism. If A and B are events of $\mathfrak{A}$ with $A \sim B$, then $\phi(A) \sim \phi(B)$; also, if $A \subseteq B$, then $\phi(A) \subseteq$

[23]It is easy to see that an involution is an orthocomplementation iff it satisfies $p \leq p' \Rightarrow p = 0$.

$\phi(B)$. Hence, $A \leq B$ in $\mathcal{E}(\mathfrak{A})$ implies that $\phi(A) \leq \phi(B)$ in $\mathcal{E}(\mathfrak{B})$. It follows that ϕ descends to a well-defined order-preserving mapping $\Pi\phi : \Pi(\mathfrak{A}) \to \Pi(\mathfrak{B})$, given by $(\Pi\phi)(p(A)) = p(\phi(A))$. It is easy to verify that $\Pi\phi(1) = 1$ iff ϕ is an interpretation. In this case, since an interpretation preserves complementarity of events, the mapping $\Pi\phi : \Pi(\mathfrak{A}) \to \Pi(\mathfrak{B})$ also preserves the involution, i.e., $\phi(p') = \phi(p)'$. It is also straightforward that $\Pi(\phi \circ \psi) = (\Pi\phi) \circ (\Pi\psi)$ whenever the composite $\phi \circ \psi$ is defined; thus, we can regard Π as a functor from the category of test spaces and interpretations to that of involutive posets and homomorphisms of these (that is, order- and involution-preserving mappings).

In general there will exist unital homomorphisms $f : \Pi(\mathfrak{A}) \to \Pi(\mathfrak{B})$ that do not have the form $\Pi\phi$ for any interpretations $\phi : \mathfrak{A} \to \mathfrak{B}$. In fact, this is the case even if $\mathfrak{A}$ and $\mathfrak{B}$ are classical. Let $\mathfrak{A} = \{E\}$ where E is any infinite set, and let $\mathfrak{B} = \{\{a\}\}$. Then $\Pi(\mathfrak{A}) \simeq 2^E$ and $\Pi(\mathfrak{B}) \simeq 2$. Any non- principal ultrafilter on E yields a homomorphism $f : 2^E \to 2$ such that $f(x) = 0$ for every $x \in E$; accordingly, f can not be obtained from any interpretation from $\mathfrak{A}$ to $\mathfrak{B}$.

If L_1 and L_2 are involutive posets, their cartesian product $L_1 \times L_2$ is again an involutive poset with respect to the natural slot-wise order involution. The following observation will be of some use later. I leave the straightworward proof to the reader.

LEMMA 92. *Let $\mathfrak{A}$ and $\mathfrak{B}$ be test spaces. Then* $\Pi(\mathfrak{A} \oplus \mathfrak{B}) \simeq \Pi(\mathfrak{A}) \times \Pi(\mathfrak{B})$.

3.2 Regular and Dacey Test Spaces

As noted in section 2.1, any test space $(X, \mathfrak{A})$ gives rise to an orthogonality space $(X, \perp)$. This in turn gives rise to the complete ortholattice $\mathfrak{C}(X, \perp)$ of ortho-closed subsets of X, as discussed in section 1.4. In certain cases there is an intimate connection between $\mathfrak{C}(X, \perp)$ and the logic $\Pi(\mathfrak{A})$ just constructed. For instance, if $\mathfrak{A}$ consists of all ortho-partitions of unity in a complete OML L, we have, for any events A and B, $A \sim B \;\Leftrightarrow\; A^{\perp\perp} = B^{\perp\perp}$. In this case, then, $\Pi(\mathfrak{A}) \simeq \mathcal{C}(X, \perp) \simeq L$. In general, however, even if $\mathfrak{A}$ is consistent, $\Pi(\mathfrak{A})$ need bear little relation to $\mathfrak{C}(X, \perp)$. There is, to be sure, an obvious map $\gamma : \mathcal{E}(\mathfrak{A}) \to \mathcal{C}(X, \perp)$ given by $\gamma(A) := A^{\perp\perp}$; but this map does not necessarily respect perspectivity, and hence, does not necessarily descend to a well-defined map $\Pi(\mathfrak{A}) \to \mathcal{C}(X, \perp)$.

DEFINITION 93. A test space $\mathfrak{A}$ is

(i) *regular* iff $A \sim B \;\Rightarrow\; A^{\perp} = B^{\perp}$ for all $A, B \in \mathcal{E}(\mathfrak{A})$, and

(ii) *Dacey* iff $A\text{co}B \;\Rightarrow\; A^{\perp} = B^{\perp\perp}$ for all $A, B \in \mathcal{E}(\mathfrak{A})$

In other words, $\mathfrak{A}$ is regular iff $\gamma(A) = \gamma(B)$ for all perspective events A and B, and Dacey iff $\gamma(A) = \gamma(B)'$ whenever A and B are complementary. Clearly, every Dacey test space is regular. The converse is false:

EXAMPLE 94. Let $\mathfrak{A}$ be the Greechie test space diagrammed below, consisting of four, three-outcome tests, together with the four-outcome test $\{a, b, c, d\}$. It is not difficult to check that $\mathfrak{A}$ is regular (indeed, any two perspective events A and B have a unique axis of perspectivity, consisting of a singleton event $\{v\}$, and $A^{\perp} = B^{\perp} = \{v\}$.) On the other hand, $\mathfrak{A}$ is not Dacey. For, consider the two complementary events $A = \{a, c\}$ and $B = \{b, d\}$: we have $A^{\perp} = \{b, d, x\}$ and $B^{\perp} = \{a, c, y\}$. But then $B^{\perp\perp} = \{b, d\} \neq A^{\perp}$. .

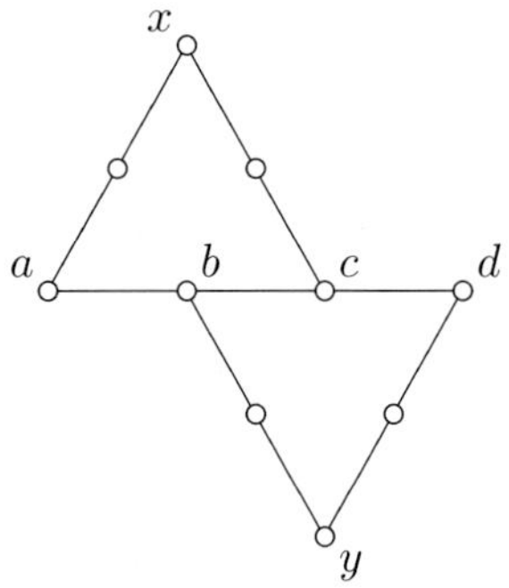

Virtually all test spaces that occur in practice are Dacey; however, this condition does not seem easy to motivate independently. Regularity, on the other hand, seems at least somewhat natural. In fact, any test space that admits the following abstract version of the projection postulate, must be regular:

LEMMA 95. *Suppose that, for every outcome $x \in X$, there exits a state ω_x on $\mathfrak{A}$ with $\omega_x(y) = 0$ iff $x \perp y$. Then $\mathfrak{A}$ is regular.*

Proof. Let $A \perp B$ and let $x \in A^{\perp}$. Then $\omega_x(A) = 0$, so $\omega_x(B) = 0$. In particular, $\omega_x(b) = 0$ for all $b \in B$, so $x \perp b$ — that is, $x \in B^{\perp}$. ∎

LEMMA 96. *If $\mathfrak{A}$ is regular, then*

(a) $A \leq B \Rightarrow B^{\perp} \subseteq A^{\perp}$ for all events A and B

(b) $E^{\perp} = \emptyset$ for all $E \in \mathfrak{A}$.

(c) $\mathfrak{A}$ is consistent.

Proof. (a) Certainly if $A \subseteq B$, $B^{\perp} \subseteq A^{\perp}$; regularity supplies the remaining case. (b) Suppose $x \in E^{\perp}$. Let $x \in F$. Then $E \sim F$, so $E^{\perp} = F^{\perp}$, whence $x \in F^{\perp}$, which is impossible. (c) Suppose $A \leq E \setminus A$. Then $(E \setminus A)^{\perp} \subseteq A^{\perp}$ by (a), whence $A \subseteq A^{\perp}$ and A is empty. Thus, $\mathfrak{A}$ is consistent. ∎

The following characterization of Dacey-ness is often useful:

LEMMA 97. *$\mathfrak{A}$ is Dacey iff, for all $E \in \mathfrak{A}$ and $x, y \in X$,*

$$E \subseteq x^{\perp} \cup y^{\perp} \;\Rightarrow\; x \perp y.$$

As an illustration, in the non-Dacey test space of Example 94, the test $E = \{a, b, c, d\}$ is contained in $x^{\perp} \cup y^{\perp}$, yet $x \not\perp y$.

Proof. Suppose $\mathfrak{A}$ is Dacey and $E \subseteq x^\perp \cup y^\perp$. Let $E_x = \{z \in E | z \perp x\}$ and $E_y = E \setminus E_x$. Then $E_x \text{co} E_y$, so $E_x^\perp = E_y^{\perp\perp}$. But $x \in E_x^\perp$ and $y \in E_y^\perp$. For the converse, suppose $\mathfrak{A}$ satisfies the given condition and $A \text{co} B$ in $\mathcal{E}(\mathfrak{A})$. Then certainly $A \subseteq B^\perp$, whence $B^{\perp\perp} \subseteq A^\perp$. Let $x \in A^\perp$. For any $y \in B^\perp$, $A \subseteq x^\perp$ and $B \subseteq x^\perp$, whence $E = A \cup B \subseteq x^\perp \cup y^\perp$, so $x \perp y$. Thus, $A^\perp \subseteq B^{\perp\perp}$. ■

Recall that, for any orthogonality space $(X, \perp)$, the set of maximal pairwise orthogonal subsets of X is denoted by $\mathcal{O}(X, \perp)$.

LEMMA 98. *If $\mathfrak{A}$ is Dacey, then $\mathfrak{A} \subseteq \mathcal{O}(X, \perp)$.*

Proof. If $\mathfrak{A}$ is Dacey and $x \in E^\perp$ for some $x \in X, E \in \mathfrak{A}$, then by the preceding Lemma, $E \subseteq x^\perp \cup x^\perp$, whence $x \perp x$, a contradiction. ■

Of course, $\mathcal{O}(X, \perp)$ can always be regarded as a test space in its own right. The following is a corollary to Dacey's Theorem (Theorem 35):

THEOREM 99. *If $\mathcal{C}(X, \perp)$ is an orthomodular lattice, then $\mathcal{O}(X, \perp)$ is Dacey.*

Proof. Let $A, B \in \mathcal{E}(\mathcal{O}(X, \perp))$ with $A \text{co} B$. Then $(A \cup B)^\perp = A^\perp \cap B^\perp = \emptyset$ and $A \subseteq B^\perp$. Thus, $A^{\perp\perp} \subseteq B^\perp$ and $B^\perp \cap A^{\perp\perp\perp} = B^\perp \cap A^\perp = \emptyset$. Since both $B^\perp$ and $A^{\perp\perp}$ are closed and $\mathcal{C}(X, \perp)$ is orthomodular, $B^\perp = A^{\perp\perp}$. Thus, $\mathcal{O}(X, \perp)$ is Dacey. ■

The converse is false:

EXAMPLE 100 Janowitz. Consider the test space the test space

$$\mathfrak{A} = \{\{a, x, b\}, \{b, y, c\}, \{c, z, d\}, \{d, w, a\}\} :$$

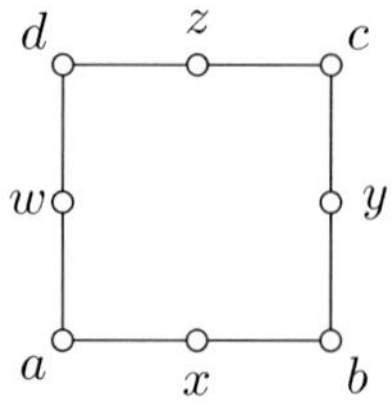

It is easily verified that $\mathfrak{A}$ is Dacey; but $\mathcal{C}(X, \perp)$ is not orthomodular. To see this, let $M = \{a, c\}$. Then $M^\perp = \{b, d\}$ and $M^{\perp\perp} = M$. Notice that $\{a\}$ is a maximal orthogonal subset of M, but $\{a\}^{\perp\perp} = \{a\} \neq M$.

3.3 Algebraic Test Spaces

We next consider a particularly important class of test spaces, the logics of which are orthoalgebras.

DEFINITION 101. A test space $\mathfrak{A}$ is *algebraic* iff perspective events share exactly the same complements — that is,

$$A \sim B \text{ co } C \Rightarrow A \text{ co } C$$

for all events $A, B, C \in \mathcal{E}(\mathfrak{A})$. In terms of the co-diagrams introduced in Section 2.3, this means that every co-diagram of the form $A\text{co}C\text{co}B\text{co}D$ "closes" with $D\text{co}A$:

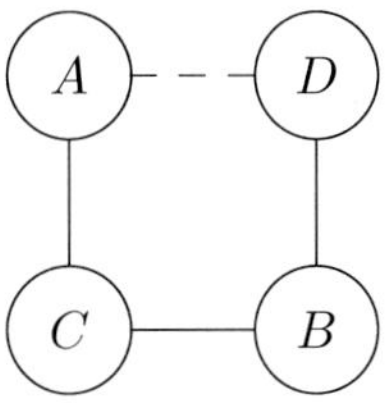

Most of the test spaces that arise in practice are either algebraic, or embeddable in one that is. For example, both the Borel test spaces of measurable spaces and the frame manuals of Hilbert spaces are algebraic. Any test space in which, for all E, F, $|E \cap F| < |E \setminus F|$ for all tests E and F, is algebraic by default. Thus, in particular, Greechie test spaces are algebraic. As we shall see below, any test space having a reasonably rich supply of states is embeddable in an algebraic test space having the same outcomes and the same state space.

It follows immediately from the definition of algebraicity that, if $\mathfrak{A}$ is algebraic, then perspectivity is an equivalence relation on $\mathcal{E}(\mathfrak{A})$. In fact, as we'll now see, the logic $\Pi(\mathfrak{A})$ is simply the set of perspectivity-classes of events.

LEMMA 102. *Let $\mathfrak{A}$ be algebraic. Then for all $\mathfrak{A}$-events A, B and C,*

(a) If $A \sim B \subseteq C$, then there is some event B' with $A \subseteq B' \sim C$;

(b) $A \leq B$ iff there exists an event B' with $A \subseteq B' \sim B$.

(c) $\mathfrak{A}$ is consistent

(d) $A \equiv B$ iff $A \sim B$.

Proof. (a) If $A \sim B \subseteq C$, let $C_1 = C \setminus B$, and let $D\text{co}C$. Then $A \sim B\text{co}(C_1 \cup D)$. Since $\mathfrak{A}$ is algebraic, $A\text{co}(C_1 \cup D)$. Set $B' = A \cup C_1$. Then $A \subseteq B'$ and $B'\text{co}D\text{co}C$, i.e., $B' \sim C$.

(b) If $A \leq B$, then there exists a chain of events $A_1, ..., A_n$ with $A = A_1, A_n = B$, and for each $i = 1, ..., n-1$, either $A_i \sim A_{i+1}$ or $A_i \subseteq A_{i+1}$. Since both $\subseteq$ and $\sim$ are transitive (the latter, since $\mathfrak{A}$ is algebraic), we can assume without loss of generality that $A_i \subseteq A_{i+1} \Rightarrow A_{i+1} \sim A_{i+2}$ and $A_i \sim A_{i+1} \Rightarrow A_{i+1} \subseteq A_{i+2}$ for $i = 1, ..., n-2$. Suppose that the length, n, or such a chain is at least 4. We are then faced with two possibilities:

$$\text{Case I: } A_1 \subseteq A_2 \sim A_3 \subseteq A_4 \sim \cdots \text{ or Case II: } A_1 \sim A_2 \subseteq A_3 \sim A_4 \subseteq \cdots$$

In Case I, part (a) gives us an event A_3' with $A_1 \subseteq A_2 \subseteq A_3 \sim A_4$; in case II, we have an event A_2' with $A_1 \subseteq A_2' \sim A_3 \sim A_4$. Thus, in either Case, we can replace the given chain by one of length $n-1$, namely,

$$A_1 \subseteq A_3' \subseteq A_4 \sim \cdots \quad \text{or} \quad A_1 \subseteq A_2' \sim A_4 \subseteq \cdots.$$

Evidently, then, any chain of length $n \geq 4$ can be reduced to one of length 3, for which, again, part (a) supplies the desired conclusion.

(c) It is sufficient to show that if $p(A) \leq p(A)'$ in $\Pi(\mathfrak{A})$, then $p(A) = 0$. But if $p(A) \leq p(A)'$, then $A \leq C$ for some $C\mathsf{co}A$. By part (b), there is some B' with $A \subseteq B'\mathsf{co}A$ — which is absurd unless $A = \emptyset$, i.e., $p(A) = 0$.

(d) As $\mathfrak{A}$ is consistent, equivalent events are weakly perspective, by Lemma 91. But since $\sim$ is already an equivalence relation on $\mathcal{E}$, weakly perspective events are in fact perspective. ■

Recall from Section 1.2 that an *orthoalgebra* is a positive, cancellative, unital partial abelian semigroup (an effect algebra) $(L, \oplus)$ in which $a \oplus a$ exists only if $a = 0$. As we'll now see, the logic of an algebraic test space is an orthoalgebra.

LEMMA 103 Additivity Lemma. *Let $\mathfrak{A}$ be algebraic, and let $A \sim A'$ and $B \sim B'$ in $\mathcal{E}(\mathfrak{A})$. Then*

$$A \perp B \ \Rightarrow A' \perp B' \text{ and } A \cup B \sim A' \cup B'.$$

Proof. Since $A \perp B$, we can find a test $E \in \mathfrak{A}$ with $A \cup B \in E \in \mathfrak{A}$. Let $C = E \setminus A \cup B$. Then $B \cup C\mathsf{co}A$, whence, as $A' \sim A$ and $\mathfrak{A}$ is algebraic, $B \cup C\mathsf{co}A'$. Now $B\mathsf{co}(C \cup A')$, whence, as $B' \sim B$, $B'\mathsf{co}(C \cup A')$ also. Hence, $B' \perp A'$ and $(A' \cup B')\mathsf{co}C\mathsf{co}(A \cup B)$, i.e., $A \cup B \sim A' \cup B'$. ■

It follows from Lemma 103 that, for all $p(A)$, $p(B)$ in $\Pi(\mathfrak{A})$, the relation

$$p(A) \perp p(B) \ \Leftrightarrow \ A \perp B$$

is well defined, as is the partial operation

$$p(A) \perp p(B) \ \Rightarrow \ p(A) \oplus p(B) := p(A \cup B).$$

PROPOSITION 104. *Let $\mathfrak{A}$ be an algebraic test space with logic Π. Then $(\Pi, \oplus)$ is an orthoalgebra.*

Proof. To see that $\oplus$ is associative, it suffices to note that, for $a = p(A), b = p(B)$ and $c = p(C)$ in Π, where $A, B, C \in \mathcal{E}$, $(a \oplus b) \oplus c$ is defined iff $A \perp B$ and $(A \cup B) \perp C$ — whence, $(a \oplus b) \oplus c = p(A \cup B \cup C) = a \oplus (b \oplus c)$. The rest of the proof is similarly straightforward. ■

The following representation theorem is proved in several places, e.g., [Gudder, 1988].

PROPOSITION 105. *If $(L, \oplus)$ is an abstract orthoalgebra, the set $\mathfrak{A}_L$ of all finite sets $E = \{a_1, ..., a_n\}$ of non-zero elements of L with $a_1 \oplus \cdots \oplus a_n = 1$, is an algebraic test space with $\Pi(\mathfrak{A}_L) \simeq L$ via the bijection $A \mapsto \bigoplus A$, where A is an event of $\mathcal{E}(\mathfrak{A}_L)$.*

Thus, orthoalgebras are the same things as logics of algebraic test spaces. Naturally, one wants to identify conditions on an algebraic test space $\mathfrak{A}$ that will force its logic to be an orthomodular poset, an orthomodular lattice, etc.

DEFINITION 106. A test space is *ortho-coherent* iff every pairwise orthogonal triple of events is jointly orthogonal.

If $\mathfrak{A}$ is algebraic, orthocoherence is equivalent to the condition that the orthoalgebra $\Pi(\mathfrak{A})$ be orthocoherent, i.e., an OMP (cf Lemma 21 and remarks following). Both Borel and frame manuals are manifestly orthocoherent. However, orthocoherence is difficult to motivate on purely "operational" grounds. Indeed, we have already seen a simple and plausible toy example in which it fails, namely, the "Wright triangle" of Example 55:

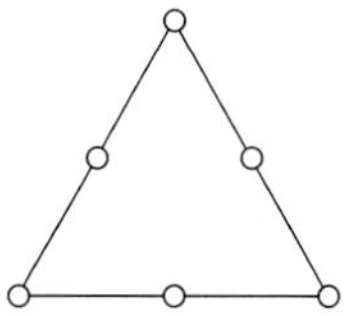

The three corner outcomes (corresponding, in the "firefly box" interpretation, to the appearance of a light in one of the three chambers) are pairwise, but not jointly, orthogonal.

A further reason to be suspicious of orthocoherence as a fundamental principle is that it is not stable under the formation of tensor products [Foulis and Randall, 1981a; Foulis and Randall, 1981b] — something we'll take up in Section 5.

PRE-ALGEBRAIC TEST SPACES

We now show that, under very weak conditions, a test space can be "filled out" to an algebraic test space without change to either its outcome set or its state space. Call a set Γ of states on a test space $\mathfrak{A}$ *semi-unital* iff, for every $x \in X = \bigcup \mathfrak{A}$, there exists a state $\omega \in \Gamma$ with $\omega(x) > 1/2$.

LEMMA 107. *'Let $\mathfrak{A}$ be a test space with a semi-unital set of states. Then there exists an algebraic test space $\mathfrak{A}' \supseteq \mathfrak{A}$, having the same outcome set, and the same states as $\mathfrak{A}$.*

Proof. Let $\mathfrak{A}$ consist of all subsets of X over which all states sum to 1. It is routine to verify that this test space is algebraic, and has exactly the same states as $\mathfrak{A}$. ■

One can go further, to show that any test space that can be embedded in an algebraic test space, has a canonical *minimal* such embedding. Let us call a test space $\mathfrak{A}$ *pre-algebraic* iff there exists an algebraic test space $\mathfrak{B}$ with $\mathfrak{A} \subseteq \mathfrak{B}$. (Note that we do not require that $\bigcup \mathfrak{B} = \bigcup \mathfrak{A}$.)

LEMMA 108. *Let $\mathfrak{A}$ be a pre-algebraic test space with outcome-set X. Suppose $\mathfrak{B}$ is any algebraic test space with $\mathfrak{A} \subseteq \mathfrak{B}$. Define*

$$\langle \mathfrak{A} \rangle = \bigcap \{ \mathfrak{C} \subseteq \mathfrak{B} \mid \mathfrak{A} \subseteq \mathfrak{C} \ \& \ \mathfrak{C} \text{ is algebraic} \}.$$

Then

(a) $\langle\mathfrak{A}\rangle$ *is algebraic.*

(b) $\bigcup\langle\mathfrak{A}\rangle = X$.

(c) $\langle\mathfrak{A}\rangle$ *does not depend upon* $\mathfrak{B}$.

Proof. (a) is trivial. To verify (b), let $\mathfrak{A} \subseteq \mathfrak{C} \subseteq \mathfrak{B}$, where C is algebraic. Let $y \in \bigcup\mathfrak{C} \setminus X$. Let $\widehat{\mathfrak{C}} = \{E \in \mathfrak{C} | y \notin E\}$ and note that $\widehat{C}$ is an algebraic test space containing $\mathfrak{A}$. Statement (c) now follows. ∎

The test space $\langle\mathfrak{A}\rangle$ is called the *algebraic closure* of $\mathfrak{A}$. Notice if $\mathfrak{A}$ is contained in any algebraic test space $\mathfrak{B}$, then $\langle\mathfrak{A}\rangle \subseteq \mathfrak{B}$. There is a constructive description of $\langle\mathfrak{A}\rangle$ due to P. Lock [Lock, P., 1981]. If $\mathfrak{A}$ is any test space, the *derived test space* $\mathfrak{A}'$ has the same outcomes, and tests given by

$$\mathfrak{A}' \;=\; \{\; A \cup D \mid \exists B \in \mathcal{E}(\mathfrak{A}) \; A \sim B \text{ co } D \;\}.$$

Define $\mathfrak{A}^{(n)}$ recursively: $\mathfrak{A}^{(0)} = \mathfrak{A}$, $\mathfrak{A}^{(n+1)} = (\mathfrak{A}^{(n)})'$.

THEOREM 109 [P. Lock, 1981]. *$\mathfrak{A}$ is pre-algebraic if and only if $\mathfrak{A}^{(n)}$ is consistent for all n. In this case, $\langle\mathfrak{A}\rangle = \bigcup_n \mathfrak{A}^{(n)}$.*

Proof. We show first that $\mathfrak{A}$ is algebraic iff $\mathfrak{A}$ is consistent and $\mathfrak{A} = \mathfrak{A}'$. Note that $\mathfrak{A} \subseteq \mathfrak{A}'$ (take $E = F = G$ in the definition). Let $\mathfrak{A}' \subseteq \mathfrak{A}$ and suppose A, C, B, D form an N-configuration. Then $A \cup D \in \mathfrak{A}$. If $\mathfrak{A}$ is consistent, then $A \cap D = \emptyset$; consequently A co D, i.e., the N-configuration closes. Thus $\mathfrak{A}$ is algebraic.

More generally, if $\mathfrak{A}$ is pre-algebraic, then $\mathfrak{A}' \subseteq \langle\mathfrak{A}\rangle$. Now let $\mathfrak{B} = \bigcup_n \mathfrak{A}^{(n)}$. Certainly $\mathfrak{A} \subseteq \mathfrak{B}$ and $\mathfrak{B}' = \mathfrak{B}$. Thus, our remaining task is to show that $\mathfrak{B}$ is consistent iff every $\mathfrak{A}^{(n)}$ is consistent. One implication is trivial, and the other, nearly so: If $\mathfrak{B}$ is the union of an ascending chain of test spaces $\mathfrak{A}_n$, then $\mathcal{E}(\mathfrak{A})$ is the ascending union of the sets $\mathcal{E}(\mathfrak{A}_n)$. Since the relation $\leq$ involves only a finite chain of comparable and complementary pairs of events, $A \leq B$ holds in $\mathcal{E}(\mathfrak{B})$ iff for some n, $A, B \in \mathcal{E}(\mathfrak{A}_n)$. and $A \leq B$ holds in $\mathfrak{A}_n$. Consequently, the union of an ascending chain of consistent test spaces is consistent. ∎

COROLLARY 110. *Let $\mathfrak{A}$ be a pre-algebraic test space. Then*

(a) $\Omega(\langle\mathfrak{A}\rangle) = \Omega(\mathfrak{A})$;

(b) For any algebraic test space $\mathfrak{B}$, Int$(\mathfrak{A}, \mathfrak{B}) =$ Int$(\langle\mathfrak{A}\rangle, \mathfrak{B})$.

Proof. (a) In light of Theorem 109, it suffices to show that every $\omega \in \Omega(\mathfrak{A})$ is a state on $\mathfrak{A}'$. Let $\omega \in \Omega(\mathcal{A})$. If $A \sim B$co D then $\omega(A) = \omega(B)$ and $\omega(D) = 1 - \omega(B)$, so $\omega(A) + \omega(D) = 1$. Since $\mathfrak{A}$ is consistent, $A \cap D = \emptyset$. Hence, $\omega(A \cup D) = 1$. The proof of (b) is similar. ∎

3.4 Coherent Test Spaces

As we discussed in Section 1, traditional approaches to quantum logic make, either explicitly or implicitly, some assumption to the effect that pairwise co-testable events are jointly testable. A strong form of this requirement is the following:

DEFINITION 111. A test space $\mathfrak{A}$ is said to be *coherent* iff for all events A and B in $\mathcal{E}(\mathfrak{A})$,

$$A \subseteq B^{\perp} \;\Rightarrow\; A \perp B.$$

Coherence is a stronger condition than ortho-coherence: If A, B and C are pairwise orthogonal events, then $C \subseteq (A \cup B)^{\perp}$, whence, if $\mathfrak{A}$ is coherent, $C \perp (A \cup B)$. If $\mathfrak{A}$ is locally finite, the two conditions coincide, and amount to the requirement that pair-wise orthogonal sets of *outcomes* be events.

LEMMA 112. *If $(X, \mathfrak{A})$ is coherent, then $\mathfrak{A} \subseteq \mathcal{O}(X, \perp)$.*

Proof. If $E \in \mathfrak{A}$ and $x \in E^{\perp}$, then by coherence, $x \perp E$, whence, $E \cup \{x\}$ is an event properly larger than E. Since $\mathfrak{A}$ is irredundant, this is impossible. Thus, $E^{\perp}$ is empty, i.e., E is a maximal orthogonal set in X. ■

In the presence of coherence, the three classes of test spaces introduced earlier coalesce:

LEMMA 113. *Let $(X, \mathfrak{A})$ be coherent. Then the following are equivalent:*

(a) $\mathfrak{A}$ is Dacey.

(b) $\mathfrak{A}$ is regular;

(c) $\mathfrak{A}$ is algebraic;

Proof. Suppose $\mathfrak{A}$ is coherent. If $\mathfrak{A}$ is Dacey, it is regular in any case. If $\mathfrak{A}$ is regular and $A \sim B \perp C$, then $C \subseteq B^{\perp}$ and $B^{\perp} = A^{\perp}$ by regularity, so $C \perp A$ by coherence; thus $\mathfrak{A}$ is algebraic. It remains to show that a coherent algebraic test space is Dacey. Suppose $E \subseteq x^{\perp} \cup y^{\perp}$. Let $A = E \cap x^{\perp}$ and $C = E \setminus A$. Then $A \subseteq x^{\perp}$ and $C \subseteq y^{\perp}$, so $A \perp \{x\}$ and $C \perp \{y\}$ by coherence. It follows that $p(x) \leq p(A)' \leq p(C) \leq p(y)'$; if $\mathfrak{A}$ is algebraic, it then follows that $x \perp y$. Thus $\mathfrak{A}$ is Dacey, by Lemma 97. ■

AN ISOMORPHISM THEOREM

Recall that, for any test space $\mathfrak{A}$ with outcome-set X, there is a natural mapping $\gamma : \mathcal{E}(\mathfrak{A}) \to \mathcal{C}(X, \perp)$ given by $\gamma A \mapsto A^{\perp\perp}$. If $\mathfrak{A}$ is regular, this descends to an injection $\gamma : \Pi(\mathfrak{A}) \to \mathcal{C}(X, \perp)$, given by $\gamma : p(A) \mapsto A^{\perp\perp}$. If $\mathfrak{A}$ is also Dacey, then $\gamma : p(C) \mapsto p(A)'$ where C is any event complementary to A.

THEOREM 114. *Let $\mathfrak{A}$ be a coherent, algebraic test space, and let $A, B \in \mathcal{E}(\mathfrak{A})$. Then $A \leq B$ iff $A^{\perp\perp} \subseteq B^{\perp\perp}$.*

Proof. If $A \subseteq C \sim B$, then since $\mathfrak{A}$ is regular, we have $B^{\perp} = C^{\perp} \subseteq A^{\perp}$, whence $A^{\perp\perp} \subseteq C^{\perp\perp} = B^{\perp\perp}$. Conversely, suppose $A^{\perp\perp} \subseteq B^{\perp\perp}$. Then $B^{\perp} \subseteq A^{\perp}$. Hence, if C is any event complementary to B, $C \subseteq A^{\perp}$, whence, $C \perp A$. Thus, there is some $A' \in \mathcal{E}$ with $A \subseteq A' \mathsf{co} C$, whence, $A \subseteq A' \sim B$ — i.e., $A \leq B$. ■

It follows that, for a coherent algebraic test space, the mapping $\gamma : \mathcal{E} \to \mathcal{C}(X, \perp)$ induces an ortho-embedding $\gamma : \Pi(\mathfrak{A}) \to \mathcal{C}(X, \perp)$, given by $\gamma : p(A) \mapsto A^{\perp\perp}$. As we shall now show, if Π is a complete lattice, then γ is in fact an ortho-isomorphism.

LEMMA 115. *Let $\mathfrak{A}$ be algebraic, with logic $\Pi = \Pi(\mathfrak{A})$. Then*

(a) If $\mathfrak{A}$ is regular, then for every event $A \in \mathcal{E}(\mathfrak{A})$, $p(A) = \bigvee_{a \in A} p(a)$, provided that the latter join exits in Π.

(b) $\mathfrak{A}$ is coherent iff, for every event $A \in \mathcal{E}(\mathfrak{A})$, $p(A) = \bigvee_{a \in A} p(a)$ in $\Pi(\mathfrak{A})$.

Proof. (a) Let $\bigvee_{a \in A} p(a)$ exist. Then there exists some event A_o in $\mathcal{E}(\mathfrak{A})$ with $\bigvee_{a \in A} p(a) = p(A_o)$. Since $p(a) \leq p(A)$ for every $a \in A$, we have $p(A_o) \leq p(A)$. Thus, by part (b) of Lemma 102, there exists some $A_1 \in \mathcal{E}(\mathfrak{A})$ with $A_o \subseteq A_1 \sim A$. Let $x \in A_1 \setminus A_o$ (supposing this set to be non-empty). Then $x \perp A_o$, whence, $p(A_o) \leq p(x)'$. It follows that $p(a) \leq p(x)'$ for every $a \in A$. Since $\mathfrak{A}$ is algebraic, $a \perp x$ for every $a \in A$, i.e., $x \in A^{\perp}$. Since $\mathfrak{A}$ is regular, and $A_1 \sim A$, it follows that $x \in A_1 \cap A_1^{\perp}$, a contradiction. Thus, $A_1 \setminus A_o = \emptyset$, i.e., $A_o = A_1$, whence, $p(A_o) = p(A)$.

(b) Suppose first that $p(A) = \bigvee_{a \in A} p(a)$ for every event A of $\mathfrak{A}$. Let $A \subseteq B^{\perp}$. Then $p(a) \leq p(B)'$ for every $a \in A$, whence, $p(A) = \bigvee_{a \in A} p(a) \leq p(B)'$, whence, as $\mathfrak{A}$ is algebraic, $B \perp A$. Thus, $\mathfrak{A}$ is coherent. Conversely, let $\mathfrak{A}$ be coherent, and let $A \in \mathcal{E}(\mathfrak{A})$. Clearly, $p(a) \leq p(A)$ for every $a \in A$. Suppose that $p(a) \leq p(B)$ for all $a \in A$, and let $C \mathsf{co} B$. Then $p(C) = p(B)' \leq p(a)'$, whence, $C \perp a$ for each a. By coherence, then, $C \perp A$. It follows that $p(A) \leq p(C)' = p(B)$. ■

Combining the preceding lemmas, we arrive at the following isomorphism theorem, due to Foulis and Randall [Randall and Foulis, 1983a]:

THEOREM 116. *Let $\mathfrak{A}$ be a regular algebraic test space with outcome-space X. If $\Pi(\mathfrak{A})$ is a complete lattice, then $\mathfrak{A}$ is coherent, and the mapping $\gamma : \Pi(\mathfrak{A}) \to \mathcal{C}(X, \perp)$ is an ortho- isomorphism. In particular, $\mathcal{C}(X, \perp)$ is a complete orthomodular lattice.*

Proof. Since Π is a complete lattice, $\bigvee_{a \in A} p(a)$ exists for every event $A \in \mathcal{E}(\mathfrak{A})$. Therefore, by Lemma 115 (a) $p(A) = \bigvee_{a \in A} p(a)$ for every event A. By Lemma 115 (b), then, $\mathfrak{A}$ is coherent. Thus, by the remarks following Theorem 114, γ is an ortho-embedding. It now suffices to show that γ is surjective. Suppose $M = M^{\perp\perp} \in \mathcal{C}(X, \perp)$: we wish to show that $M = A^{\perp\perp}$ for some $A \in \mathcal{E}$. Since Π is a complete lattice, the join $\bigvee_{m \in M} p(m)$ exists in Π; therefore, there exists some event $A \in \mathcal{E}$ with $p(A) = \bigvee_{m \in M} p(m)$. Let $x \in A^{\perp}$. Then, by coherence, $x \perp A$.

Hence, $p(x) \leq p(A)'$, i.e., $p(A) = \bigvee_{m \in M} p(m) \leq p(x)'$. Accordingly, $p(m) \leq p(x)'$ for every $m \in M$. Since $\mathfrak{A}$ is algebraic, it follows that $m \perp x$, for every $m \in M$, i.e., $x \in M^{\perp}$. Thus, $A^{\perp} \subseteq M^{\perp}$, whence, $M = M^{\perp\perp} \subseteq A^{\perp\perp}$. To see that $A^{\perp\perp} \subseteq M$, it suffices to show that $M^{\perp} \subseteq A^{\perp}$. Let $x \in M^{\perp}$: then $p(m) \leq p(x)'$ for every $m \in M$, so $\bigvee_{m \in M} p(m) = p(A) \leq p(x)'$. Since $\mathfrak{A}$ is algebraic, it follows that $A \perp x$, whence, $x \in A^{\perp}$. ■

Historical Remark: Prior to about 1979, the term *manual* was used to describe coherent test spaces, with most attention being devoted to Dacey manuals — equivalently, coherent algebraic test spaces, in the current terminology. Throughout the 1980s, algebraic test spaces (coherent or not) were termed manuals; since the early 1990s, the term manual has largely been dropped. In these notes, I retain the term *manual* for certain special test spaces — e.g., Borel and Frame manuals — that are both coherent and algebraic.

COHERENCE AND COARSE-GRAINING

There is a sense in which the property of being a coherent *algebraic* test space is simply a special form of regularity.

Recall (Example 74 (iii)) that the *coarsening* of a test space $\mathfrak{A}$ is the collection $\mathfrak{A}^{\#}$ of partitions of tests $E \in \mathfrak{A}$ into non-empty events. Note that an outcome of $\mathfrak{A}^{\#}$ is a non-empty event of $\mathfrak{A}$, while an event of $\mathfrak{A}^{\#}$ is a jointly orthogonal family of events of $\mathfrak{A}$. We regard tests in $\mathfrak{A}^{\#}$ as "coarse-grained" versions of tests in $\mathfrak{A}$.

THEOREM 117. *Let $(X, \mathfrak{A})$ be a test space. Then $\mathfrak{A}$ is a coherent algebraic test space iff $\mathfrak{A}^{\#}$ is regular.*

Proof. We shall identify $\mathfrak{A}$ with its canonical image in $\mathfrak{A}^{\#}$, i.e., we do not distinguish between x and $\{x\}$ for an outcome $x \in X$. Thus, we may speak of $x \in X$ and $A \in \mathfrak{A}$ as outcomes and events of $\mathfrak{A}^{\#}$ as well. To avoid a proliferation of curly brackets, let's agree to denote by x_A the outcome of $\mathfrak{A}^{\#}$ corresponding to the event $A \in \mathcal{E}(\mathfrak{A})$ — that is, $x_A = \{A\}$. Notice that if $A \subseteq E \in \mathfrak{A}$, then $\{x_A\} \cup (E \setminus A) \in \mathfrak{A}^{\#}$; so $\{x_A\} \sim A$. Thus, if $\mathfrak{A}^{\#}$ is regular, we have $x_A^{\perp} = A^{\perp}$. Suppose now that A and B are events of $\mathfrak{A}$ with $A \subseteq B^{\perp} = x_B^{\perp}$. For each $x \in A$, then, $x \perp x_B$; hence, $x_B \in A^{\perp} = x_A^{\perp}$. It follows that $x_B \perp x_A$. In view of the definition of $\mathfrak{A}^{\#}$, this is only possible if $A \perp B$. It follows that $\mathfrak{A}$ is coherent. It is not difficult to show that the regularity of $\mathfrak{A}^{\#}$ entails that of $\mathfrak{A}$; hence, by Lemma 113, $\mathfrak{A}$ is algebraic as well.

For the converse, suppose $\mathfrak{A}$ is both coherent and algebraic. Let $\{A_i\}$ and $\{B_j\}$ be two events in $\mathcal{E}(\mathfrak{A}^{\#})$. Thus, $\{A_i\}$ is a partition of an event $A = \bigcup_i A_i$ of $\mathfrak{A}$, and likewise $\{B_j\}$ is a partition of an event $B = \bigcup_j B_j$ of $\mathfrak{A}$. Note that $\{A_i\} \sim \{B_j\}$ if and only if $A \sim B$. Suppose this is the case, and suppose that C is an outcome of $\mathfrak{A}^{\#}$ — that is, an event of $\mathfrak{A}$ — with $C \in \{A_i\}^{\perp}$. Then $C \perp A_i$ for each i, whence, $C \in (\bigcup_i A_i)^{\perp} = A^{\perp}$. Since $\mathfrak{A}$ is coherent, $C \perp A$. Since $\mathfrak{A}$ is algebraic and $A \sim B$, we have $C \perp B$ as well. But then $C \in \{B_j\}^{\perp}$. It follows that $\mathfrak{A}^{\#}$ is regular. ■

Remark: It is not hard to show that, similarly, a test space $\mathfrak{A}$ is *ortho*-coherent iff the test space of *finite* coarse-grainings of $\mathfrak{A}$-tests, is regular.

The Loop Lemma

The following result, generalizes the "Loop Lemma" of Greechie [Greechie, 1971] (our Theorem 42). A *square* in a test space $(X, \mathfrak{A})$ is a sequence of four (distinct) outcomes a, b, c and d with $a \perp b, b \perp c$, $c \perp d$ and $d \perp a$, but $a \not\perp d$ and $b \not\perp c$. Call a test space $(X, \mathfrak{A})$ *square free* iff contains no squares.

THEOREM 118 The Generalized Loop Lemma [Foulis *et al.*, 1993]. *Let $(X, \mathfrak{A})$ be a coherent, square-free algebraic test space. Suppose, further, that there exists some integer n such that $|E \cap F| < n$ for all $E, F \in \mathfrak{A}$. Then $\Pi(X, \mathfrak{A})$ is an orthomodular lattice.*

Notice that if $(X, \mathfrak{A})$ has finite rank, the hypotheses reduce to the absence of three or four loops, hence the name,"loop lemma".

LEMMA 119. *Let $\mathfrak{A}$ be a coherent, algebraic test space, and let $A, B \in \mathfrak{A}$. If $D = A^{\perp\perp} \cap B^{\perp\perp}$ is an event, then $p(D) = p(A) \wedge p(B)$.*

Proof. If D is an event, then $D \subseteq A^{\perp\perp}$ implies $D \leq A$, by Theorem 114 above; likewise, $D \leq B$. Now if $C \leq A$ and $C \leq B$, then $C^{\perp\perp} \subseteq A^{\perp\perp}$ and $C \subseteq B^{\perp\perp}$, so $C \subseteq A^{\perp\perp} \cap B^{\perp\perp} = D$. Thus, $p(C) \leq p(D)$. ■

Proof of Theorem 117. Our approach will be to show that, if Π is not a lattice, then there is no upper bound to $|E \cap F|$ as E, F range over distinct tests. Suppose, then, that A and B are events of $\mathfrak{A}$ such that $p(A)$ and $p(B)$ have no meet in Π. Let $D = A^{\perp\perp} \cap B^{\perp\perp}$. We claim that D is pairwise orthogonal, but not an event. In particular, D is infinite.

To begin with, note that if D were an event, then we should have $p(D) = p(A) \wedge p(B)$, by Lemma 115. Now suppose that D is not pairwise orthogonal. Then we can find two elements $c, d \in D$ with $c \not\perp d$. Now $d \in A^{\perp\perp}$ implies $p(d) \leq p(A)$, by coherence, so $A \sim N \cup \{d\}$ for some event N. Likewise, there is some event M with $A \sim M \cup \{d\}$. Replacing A and B by $N \cup \{d\}$ and $M \cup \{d\}$ if necessary, we may now suppose that $d \in A \cap B$. Let A', B' be local complements for A and B. Since $p(A) \wedge p(B)$ does not exist, $p(A)$ and $p(B)$ can lie in no common block; hence, the same is true of $p(A)' = p(A')$ and $p(B)' = p(B')$. In particular, $A' \cup B'$ is not an event.

We now invoke coherence again to obtain a square in X. Since $A' \cup B'$ is not an event, it can not be pairwise orthogonal. Pick distinct elements $x \in A'$ and $y \in B'$ with $x \not\perp y$. Since $A' \subseteq A^{\perp}$ and $B' \subseteq B^{\perp}$, and $d \in A^{\perp} \cap B^{\perp}$, we now have $d \perp x$ and $d \perp y$. But now (d, c, x, y) is a square, contrary to our assumption that $(X, \mathfrak{A})$ is square-free.

This completes the proof that D is pairwise orthogonal. As D is not an event, it follows that D is infinite. Let C be any finite subset of D. By coherence, C is an event. As in the proof of the Lemma, we may replace A and B by equivalent events so that $C \subseteq A \cap B$ (this having no effect on $A^{\perp\perp}$ or $B^{\perp\perp}$, hence, no effect on D). Let E and F be tests with $A \subseteq E$ and $B \subseteq F$. Since $p(A) \wedge p(B)$ does not exist, $E \neq F$. Now $C \subseteq E \cap F$ implies that $|E \cap F| \geq |C|$. As C is an arbitrary finite subset of the infinite set D, the cardinalities of intersections of tests is unbounded, as claimed. ■

Planar Test Spaces

As an application of the Loop Lemma, we shall now characterize orthocomplemented projective planes in terms of test spaces.

DEFINITION 120. A Greechie test space $\mathfrak{A}$ is *cubic* iff it is of uniform rank 3. A test space $\mathfrak{A}$ has the *plane property* iff, for all $\mathfrak{A}$-outcomes x and y, $x^{\perp} \cap y^{\perp} \neq \emptyset$.

The following is due to Greechie [Greechie, 1974]:

PROPOSITION 121. *Let $(X, \mathfrak{A})$ be a cubic Greechie test space having the plane property and no 3-loops or 4-loops. Let $\mathcal{L} = \{x^{\perp} | x \in X\}$. Then $(X, \mathcal{L})$ is an orthocomplemented projective plane.*

Proof. By assumption, every pair of lines meets. If x and y are distinct points in X, let $u, v \in x^{\perp} \cap y^{\perp}$. Then we have $x \perp u \perp y \perp v \perp x$. Since L is lattice-ordered, $(X, \mathfrak{A})$ is square-free; hence, either $x \perp y$, $u \perp v$, $x = y$, or $u = v$. If $x \perp y$, then $\{x, u, y\}$ is a pairwise orthogonal triple. Since $(X, \mathfrak{A})$ is orthocoherent, this is a test. Likewise, $\{x, v, y\}$ is pairwise orthogonal, hence, a test. By assumption, no two (distinct) tests meet in two outcomes; hence $u = v$. Thus, every pair of lines meets in exactly one point. To see that every pair of points belongs to a unique line, again let x and y be distinct. Let $z \in x^{\perp} \cap y^{\perp}$: then $x, y \in z^{\perp}$, so x and y are co-linear. Since the intersection of two lines is unique, x and y lie on no other line. This establishes that $(X, \mathcal{L})$ is a projective plane. It is orthocomplemented by the mapping $x \mapsto x^{\perp}$, which interchanges lines and planes. ■

4 SUPPORTS AND ENTITIES

Recall that a *filter* in a Boolean algebra L is a set $\Phi \subseteq L$ such that

(i) $\forall a \in \Phi, \forall b \in L \; a \leq b \Rightarrow b \in \Phi$, and

(ii) $\forall a, b \in \Phi$, $a \wedge b \in \Phi$.

If we interpret elements of L as propositions about the outcomes of measurements, then a filter Φ may naturally be interpreted as a set of propositions that are *certain* to occur. In [von Neumann, 1932], von Neumann interpreted the projection lattice $L(\mathbf{H})$ of a Hilbert space $\mathbf{H}$ as a lattice of physical *properties* of the quantum-mechanical system modelled by $\mathbf{H}$. It has widely been supposed that physical properties can be individuated by keeping track of those outcomes of experiments that are certain to occur when the various properties are present, or *actual*. At the very least, where a preparation yields systems for which a given set of outcomes — and only those outcomes — are possible, then those systems do indeed share a physical property. This echoes the famous remark of Einstein, Podolsky and Rosen [Einstein *et al.*, 1935]:

> If, without in any way disturbing the system, we can predict with certainty ... the value of a physical quantity, then there exists an element of physical reality corresponding to this physical quantity.

This notion of property can be formalized very naturally in terms of test spaces. The main ideas of this section first appeared in the joint paper [Foulis *et al.*, 1985] of Foulis, Piron and Randall, and were further developed in [Randall and Foulis, 1983a; Foulis *et al.*, 1992].

4.1 Supports and Local Filters

Suppose we are given a physical system which is modelled (at least partially) by a test space $\mathfrak{A}$. Suppose P is a property of this system. Let $\mathfrak{f}(P)$ denote the set of all events that are *certain* to occur when the property P is actual (that is, when the system's state is such that P obtains). We would expect $\mathfrak{f}(P)$ to satisfy the conditions of the following definition.

DEFINITION 122. A *local filter* on a test space $\mathfrak{A}$ is a set of events $\mathfrak{f} \subseteq \mathcal{E}(\mathfrak{A})$

(i) $A \in \mathfrak{f}$, $A \subseteq B$ imply $B \in \mathfrak{f}$

(ii) $A_i \subseteq E \in \mathfrak{A}$ and $A_i \in \mathfrak{f}$ imply that $\bigcap A_i \in \mathfrak{f}$.

In computational practice, it is often easier to work, not with $\mathfrak{f}$, but with the associated set of *possible outcomes*. Notice that if $\mathfrak{f}$ is a local filter of $\mathfrak{A}$ and $E \in \mathfrak{A}$, then $\mathfrak{f} \cap 2^E$ is a complete filter on the Boolean algebra 2^E, and hence, is generated by a minimal element $A = \bigcap\{B \subseteq E | B \in \mathfrak{f}\}$. This event must be construed as the set of all outcomes of E that are *possible* of occurrence when the property P is actual. Noticing that an event $A \subseteq E$ is minimal in $\mathfrak{f} \cap 2^E$ iff it is minimal in $\mathfrak{f}$, we are led to the following

DEFINITION 123. If $\mathfrak{f}$ is a local filter, its *support* is the set

$$S_{\mathfrak{f}} = \bigcup\{A \in \mathfrak{f} | A \text{ is minimal in } \mathfrak{f}\}.$$

If $\mathfrak{f}$ represents the set of events that are certain when some property is actual, then $S_{\mathfrak{f}}$ represents the set of outcomes that are possible. Notice that the set $S_{\mathfrak{f}}$ completely determines $\mathfrak{f}$ — indeed,

$$\mathfrak{f} = \{ A \in \mathcal{E}(\mathfrak{f}) \mid \exists E \in \mathfrak{A} \; S_{\mathfrak{f}} \cap E \subseteq A \subseteq E \}.$$

Sets of the form $S_{\mathfrak{f}}$ can be characterized abstractly in a variety of ways, and thus are fairly easy to recognize. The proof of the following is straightforward:

LEMMA 124. *Let $\mathfrak{A}$ be a test space with outcome set X. For a subset $S \subseteq X$, the following are equivalent:*

(a) $S = S_{\mathfrak{f}}$ for some local filter $\mathfrak{f}$ on $\mathcal{E}(\mathfrak{A})$,

(b) *For every pair of tests* $E, F \in \mathfrak{A}$, $S \cap E \subseteq F \;\Rightarrow\; S \cap F \subseteq E$.

(c) *The collection* $\mathfrak{A}_S := \{E \cap S | E \in \mathfrak{A}\}$ *is irredundant, i.e., a test space.*

(d) *For every pair of events* A, B *of* $\mathfrak{A}$,

$$A \sim B \;\&\; S \cap A = \emptyset \Rightarrow S \cap B = \emptyset.$$

(e) *The collection of events*

$$\mathfrak{f}_S := \{A \in \mathcal{E} | \exists E \in \mathfrak{A} \; S \cap E \subseteq A \subseteq E\}$$

is a local filter.

In other words, supports of filters are the same things as the supports of interpretations, as discussed in Section 2.4. (see Lemma 73).

DEFINITION 125. A set S satisfying the equivalent conditions of Lemma 124 is said to be a *support* of $\mathfrak{A}$.

We denote the collection of all supports of $\mathfrak{A}$ by $\mathcal{S}(\mathfrak{A})$. Using Lemma 124, it is easy to see that the union of any collection of supports is again a support. Hence, $\mathcal{S}(\mathfrak{A})$ is a complete lattice under set-inclusion.

EXAMPLES 126. (i) Let $\mathfrak{A}$ be any test space. If ω is a state on $\mathfrak{A}$, let $S_\omega = \{x \in X | \omega(x) > 0\}$, i.e., the support of ω in the usual analytic sense. This is easily seen to be a support. Any support having this form is said to be *stochastic.* It is easy to construct finite test spaces having non-stochastic supports, i.e., supports that support no probability weights. For a simple example, consider the three-by-three "window" test space (Example 50), whose states are three-by-three doubly stochastic matrices. As we've seen, this has a simple physical interpretation. Omit one test, to obtain the test space with the Greechie diagram

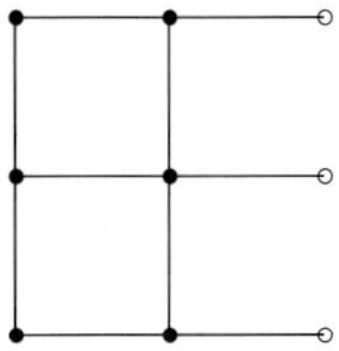

The set S of outcomes corresponding to the nodes shaded in black is a support, as is easily verified. However, $\mathfrak{A}_S$, a 2×3 window, admits no states at all.

(ii) Let $\mathfrak{B}_0(M, \mathcal{F})$ be the Borel test space of *finite* partitions of a measurable space $(M, \mathcal{F})$ by measurable sets. There is a one-to-one correspondence between the supports of $\mathfrak{B}_0$ and the filters of the Boolean algebra $\mathcal{F}$. Indeed, if Φ is any filter in the Boolean algebra $\mathcal{F}$, let $\mathfrak{f} = \{A \in \mathcal{E}(\mathfrak{B}) | \bigcup A \in \Phi\}$. It is not difficult to show that this is a local filter with corresponding support $S = \{a \in \mathcal{F} | \forall b \in \Phi \; a \cap b \neq \emptyset\}$.

(iii) If $\mathfrak{A} = \mathfrak{A}(\mathbf{H})$ is the quantum test space associated with a Hilbert space $\mathbf{H}$, then for every $x \in X$ (i.e., for every unit vector of $\mathbf{H}$), $X \setminus x^{\perp}$ is a support, and indeed, minimal in the lattice of all supports, as shown by Cohen and Svetlichny [Cohen and Svetlichny, 1987].

REGULARITY VIA SUPPORTS

There is a useful — and independently interesting — characterization of regular test spaces in terms of supports. Recall that the *forward product* of two test spaces $\mathfrak{A}$ and $\mathfrak{B}$ is the test space $\overrightarrow{\mathfrak{A}\mathfrak{B}}$ consisting of all two-stage tests of the form $\cup_{x\in E} xF_x$, where $E \in \mathfrak{A}$ and $F_x \in \mathfrak{B}$ for each $x \in E$.

LEMMA 127. *A set $V \subseteq XY$ is a support of $\overrightarrow{\mathfrak{A}\mathfrak{B}}$ iff $V = \bigcup_{x\in S} x\ S_x$ where $S \in \mathcal{S}(\mathfrak{A})$ and $S_x \in \mathcal{S}(\mathfrak{B})$ for each $x \in S$.*

Proof. Any subset V of XY can be written uniquely in the form $\bigcup_{x\in S} xS_x$: Let $S = \{x \in X | \exists\ y \in Y\ xy \in V\}$ and set $S_x = xY \cap V$. Then $V = \bigcup_{x\in S} x\ S_x$.

$$(\bigcup_{x\in S} x\ S_x) \cap (\bigcup_{x\in E} x\ F_x) = \bigcup_{x\in E\cap S} x\ (F_x \cap S_x).$$

It follows at once that if S and S_x are supports, so is V. Conversely, if V is a support, let $F \in \mathcal{B}$ be arbitrary and suppose $E, E' \in \mathcal{A}$ with $S \cap E \subseteq E'$: Then $V \cap EF \subseteq E'F$, whence $V \cap E'F \subseteq EF$, whence $S \cap E' \subseteq E$. Thus, S is a support. Suppose for some $x \in S$, $F_x, F'_x \in \mathcal{B}$ with $S_x \cap F_x \subseteq F'_x$. Fix $E \in \mathcal{A}$ containing x and $F \in \mathcal{B}$ arbitrarily, and set $F_y = F'_y = F$ for all $y \in E \setminus x$. Let $G = \bigcup_{y\in E} y\ F_y$ and $G' = \bigcup_{y\in E} y\ F'_y$. Then $V \cap G \subseteq G'$, whence $V \cap G' \subseteq G$, whence $S_x \cap F'_x \subseteq F_x$, and S_x is also a support. ■

THEOREM 128. *Let $\mathfrak{A}$ be a test space with outcome set X. Then the following are equivalent:*

(a) $\mathfrak{A}$ is regular

(b) $X \setminus x^{\perp}$ is a support of $\mathfrak{A}$ for every $x \in X$

(c) $X^2 \setminus \perp$ is a support of $\overrightarrow{\mathfrak{A}\mathfrak{A}}$.

Proof. (a) ⇒ (b): Suppose that $\mathfrak{A}$ is regular, and let $S = X \setminus x^{\perp}$ for an outcome $x \in X$. For any event A, $S \cap A = A \setminus x^{\perp} = \emptyset$ iff $A \subseteq x^{\perp}$ iff $x \in A^{\perp}$. Hence, if $A \sim B$, we have $S \cap A = \emptyset$ iff $x \in A^{\perp} = B^{\perp}$ iff $S \cap B = \emptyset$. Hence, by part (c) of Lemma 124, S is a support.

(b) ⇒ (c): Now suppose that $X \setminus x^{\perp}$ is a support for every $x \in X$. Then $X^2 \setminus \perp = \bigcup_{x\in X} x(X \setminus x^{\perp})$. By Lemma 127, this is a support of $\overrightarrow{\mathfrak{A}\mathfrak{B}}$.

(c) ⇒ (a): Suppose that $X \setminus x^{\perp}$ is a support of $\overrightarrow{\mathfrak{A}\mathfrak{A}}$. Let $A, B \in \mathfrak{A}$. For any $x \in X$, $xA \sim xB$ in $\overrightarrow{\mathfrak{A}\mathfrak{A}}$. Note that $xA \cap (X \setminus \perp)$ is empty iff $x \in A^{\perp}$, and similarly, $xB \cap (X \setminus \perp)$ is empty iff $x \in B^{\perp}$. Thus, if $X \setminus \perp$ is a support, we have $x \in A^{\perp}$ iff $x \in B^{\perp}$, i.e., $A^{\perp} = B^{\perp}$. ■

SUPPORTS, IDEALS, AND FILTERS IN ORTHOALGEBRAS
The supports — equivalently, the local filters — on an *algebraic* test space $\mathfrak{A}$ descend in a natural way to the orthoalgebra $L = \Pi(\mathfrak{A})$.

DEFINITION 129. An *ideal* in an orthoalgebra L is a set $I \subseteq L$ such that, for all $a, b \in L$ with $a \perp b$, $a \oplus b \in I \iff a, b \in I$. A *filter* in L is a set $\Phi \subseteq L$ such that $\{a' | a \in \Phi\}$ is an ideal. A *support* in L is the complement of an ideal.

A straightforward application of Lemma 124 yields the following

LEMMA 130. *Let* $L = \Pi(\mathfrak{A})$ *for an algebraic test space* $\mathfrak{A}$. *If* $\mathfrak{f}$ *is a local filter on* $\mathfrak{A}$, *then* $\Phi := \{p(A) | A \in \mathfrak{f}\}$, *is a filter on* L. *If* $\mathfrak{A}$ *is locally finite, then conversely, if* Φ *is a filter on* L, *the set* $\{A \in \mathcal{E}(\mathfrak{A}) | p(A) \in \Phi\}$ *is a local filter on* $\mathfrak{A}$.

The following result, due to Foulis (cf [Foulis *et al.*, 1992]) generalizes the compactness theorem of first order logic — equivalently, the prime ideal theorem for Boolean algebras — to the setting of an arbitrary orthoalgebra.

THEOREM 131 The Foulis Compactness Theorem. *Let* L *be an orthoalgebra, and let* M *be a subset of* L *such that* $S \cap M \neq \emptyset$ *for every non-empty support* $S \in \mathcal{S}(L)$. *Then some finite subset of* M *also has this property.*

Proof. Let T be the set of points $a \in L$ such that, for every finite set $N \subseteq M$, some proper minimal support S with $a \in S$, is disjoint from N. We shall show, first, that T is a support, and then, that $T = \emptyset$. From this, the result follows: since $1 \notin T$, there must exist some finite set $N \subseteq M$ meeting every support that contains 1. But every support contains **1**.

We proceed to show that T is a support. Let $a \in T$ and $a \leq b \in L$. Then any support containing a contains b, so $b \in T$ as well. Thus, T is an order filter. Now suppose that $p \oplus q \in T$. If neither a nor b belonged to T, there would exist finite sets $N_a, N_b \subseteq M$ such that for every support S containing a, $S \cap N_a \neq \emptyset$, and likewise $b \in S \Rightarrow S \cap N_b \neq \emptyset$. Thus, if $a \oplus b \in S$, we have either $a \in S$ or $b \in S$, whence, now we have $S \cap (N_a \cup N_b) \neq \emptyset$, contradicting $a \oplus b \in T$. Thus, T is a support. If T is non-empty, it must contain a minimal support S_o. By assumption, M meets every non-empty support, so in particular, M meets S_o. Let $c \in M \cap S_o$. Take $N = \{c\}$: then c is contained in some support S' with $S' \cap N = \emptyset$, which is absurd. Thus, $T = \emptyset$, whence, $\mathbf{1} \notin T$. But then there exists some finite $N \subseteq M$ such that, for any support S containing **1**, $S \cap N \neq \emptyset$. But every support contains **1**, so the proof is done. ■

Theorem 131 seems likely to be important in the future development of the theory of orthoalgebras; so far as I know, however, it has yet to be exploited.

4.2 Entities

Suppose the test space $\mathfrak{A}$ represents the set of measurements or observations by means of which we acquire knowledge of some particular physical system. As argued above, every detectable physical property of the system is represented by, a support of $\mathfrak{A}$. It does not follow, however, that every support corresponds to a physically meaningful property. Thus, we are led to the following definition [Foulis *et al.*, 1985]:

DEFINITION 132. An *entity* is a pair $(\mathfrak{A}, \Sigma)$ consisting of a test space $\mathfrak{A}$ and a distinguished collection of supports Σ, called *states*. The collection Σ is called the *state space* of the entity.

There is, of course, a conflict between this use of the term *state* and the usage of Section 2, in which states are probability weights. In general, context will make it clear which is meant; but if there is any danger of ambiguity, I'll refer to states *qua* probability weights, as *statistical states*, and to states *qua* supports as *realistic states* (to emphasize their interpretation as Einstein-Podolsky-Rosen "elements of reality"). Of course, any statistical state ω is associated with a realistic state, namely, its support S_ω; however, as many different probability weights may share the same support, S_ω typically carries less information than ω. On the other hand, as remarked above, many test spaces possess non-stochastic supports, which carry information not encoded in *any* statistical state.

Classically, any subset of the state-space Σ defines a categorical *property* of the entity $(\mathfrak{A}, \Sigma)$. However, it will not in general be possible to detect or discriminate between such properties using the tests comprising $\mathfrak{A}$. Indeed, if $\Lambda \subseteq \Sigma$, let

$$\overline{\Lambda} := \{S \in \Sigma | S \subseteq \bigcup \Lambda\},$$

and notice that $\bigcup \Lambda = \bigcup \overline{\Lambda}$. In other words, the properties Λ and $\overline{\Lambda}$ are associated with exactly the same sets of possible outcomes, and thus cannot be distinguished by tests in $\mathfrak{A}$. [24] This suggests the following

DEFINITION 133. A subset $\Lambda \subseteq \Sigma$ of the state-space of an entity $\mathfrak{A}$ is *detectable* iff $\Lambda = \overline{\Lambda}$. We say that such a property Λ is *actual* in a state $S \in \Sigma$ iff $S \in \Lambda$.

Note that the mapping $\Lambda \mapsto \overline{\Lambda}$ is a closure on 2^Σ. Notice also that there is a one-to-one correspondence between closed subsets of Σ and the collection of supports of the form $\bigcup \Lambda$, $\Lambda \subseteq X$. Thus, we can identify detectable properties of the entity $(\mathfrak{A}, \Sigma)$ with certain supports of $\mathfrak{A}$, as follows:

DEFINITION 134. The *property lattice* of an entity $(\mathfrak{A}, \Sigma)$ is the complete sublattice $\mathcal{L} = \mathcal{L}(\mathfrak{A}, \Sigma)$ of $\mathcal{S}(\mathfrak{A})$ generated by Σ — that is, $\mathcal{L}$ consists of all supports of the form $\bigcup \Lambda$ where $\Lambda \subseteq \Sigma$.

If A is an event of $\mathfrak{A}$, let Σ_A denote the set of all states making A certain to occur if tested — that is, the set of all supports $S \in \Sigma$ such that for all $E \in \mathfrak{A}$, $A \subseteq E \ \Rightarrow\ S \cap E \subseteq A$. The *principal property* generated by A is

$$[A] := \bigcup \Sigma_A \in \mathcal{L}$$

[24] States in $\overline{\Lambda}$ are, in an abstract sense, *superpositions* of states in Λ [Bennett and Foulis, 1990]. Indeed, if $\mathfrak{A} = \mathfrak{F}(\mathbf{H})$ is the test space of frames of a Hilbert space $\mathbf{H}$ and $\Sigma = \{X \setminus x^\perp | x \in X\}$ (X being the unit sphere of $\mathbf{H}$), then for $\Lambda = \{X \setminus x^\perp | x \in A\}$, $A \subseteq X$, one finds that $\overline{\Lambda} = \{X \setminus x^\perp | x \in \overline{\text{span}}(A) \cap X\}$.

Thus, $[A]$ represents the largest (detectable) property of the entity making the event A certain to occur if tested.

MAPPINGS OF ENTITIES

Let $\mathfrak{A}$ and $\mathfrak{B}$ be test spaces, with outcome sets X and Y, respectively, and let $\phi : \mathfrak{A} \to \mathfrak{B}$ be a homomorphism. If T is any subset of Y, define

$$\phi^+(T) = \{x \in X | \phi(x) \cap T \neq \emptyset\}.$$

(If we think of ϕ as representing a relation $R \subseteq X \times Y$, then $\phi^+(T)$ is simply the relational pre-image of T under R.) Notice that, for any set $A \subseteq X$,

$$\phi^+(T) \cap A = \emptyset \ \Leftrightarrow \ T \cap \phi(A) = \emptyset.$$

It follows easily (from 124 (d)) that, if T is a support of $\mathfrak{B}$, then $\phi^+(T)$ is a support of $\mathfrak{A}$. Observe also that $\phi^+(\bigcup_i T_i) = \bigcup_i \phi^+(T_i)$, for any family $\{T_i | i \in I\}$ of $\mathfrak{B}$-supports. Thus, ϕ^+ is a complete lattice-homomorphism from $\mathcal{S}(\mathfrak{B})$ to $\mathcal{S}(\mathfrak{A})$. This suggests the following

DEFINITION 135. Let $(\mathfrak{A}, \Sigma)$ and $(\mathfrak{B}, \Gamma)$ be entities. A *mapping of entities* from $(\mathfrak{A}, \Sigma)$ to $(\mathfrak{B}, \Gamma)$ is a homomorphism $\phi : \mathfrak{A} \to \mathfrak{B}$ such that $\phi^+(T) \in \Sigma$ for every $T \in \Gamma$. In this case, we write $\phi : (\mathfrak{A}, \Sigma) \to (\mathfrak{B}, \Gamma)$.

Obviously, mappings of entities compose to yield the same. Thus, we can speak of the *category* of entities. One can construct products and co-products et.c (refs); but we shall not pursue this here. Notice that if $\phi : (\mathfrak{A}, \Sigma) \to (\mathfrak{B}, \Gamma)$ is a mapping of entities, then the mapping $\phi^+ : \mathcal{S}(\mathfrak{B}) \to \mathcal{S}(\mathfrak{A})$ restricts to a complete lattice-homomorphism $\phi^+ : \mathcal{L}(\mathfrak{B}, \Gamma) \to \mathcal{L}(\mathfrak{A}, \Sigma)$. Thus, we have a natural contravariant functor $\mathcal{L}$ from the category of entities to the category of complete lattices.

STANDARD ENTITIES

Suppose $\mathfrak{A}$ is a regular test space with outcome space X. By Lemma 127, $X \setminus x^\perp$ is a support for every $x \in X$. Let $(\mathfrak{A}, \Sigma)$ be the entity obtained by taking $\Sigma = \{X \setminus x^\perp | x \in X\}$. Then the property lattice $\mathcal{L} = \mathcal{L}(X, \Sigma)$ consists of sets

$$S_Z := \bigcup_{x \in Z} (X \setminus x^\perp) = X \setminus Z^\perp$$

where $Z \subseteq X$ is arbitrary. Notice that $S_Z = S_{Z^{\perp\perp}}$. Thus, we have an order-preserving bijection from $\mathcal{C}$ to $\mathcal{L}$, sending each set $Z = Z^{\perp\perp}$ in the former to the set $S_Z = X \setminus Z^\perp$ in the latter. The inverse mapping is clearly also order-preserving, so $Z \mapsto S_Z$ is an order-isomorphism. Since both $\mathcal{L}$ and $\mathcal{C}$ are complete lattices, the mapping $S_{(\ .\ )}$ is in fact a lattice isomorphism. We can transfer the orthocomplementation from $\mathcal{C}$ to $\mathcal{L}$ by defining $S'_Z := S_{Z^\perp}$.

EXAMPLE 136. If $\mathfrak{A} = \mathfrak{A}(\mathbf{H})$, the quantum test space associated with a Hilbert space $\mathbf{H}$, then $\mathcal{C}(X, \perp) \simeq \mathcal{L} \simeq L(\mathbf{H})$. In the case of the Borel test space $\mathfrak{B}(M, \mathcal{F})$ of a measurable space $(M, \mathcal{F})$, $\mathcal{C} \simeq \mathcal{L} \simeq \mathcal{F}$.

The mapping $[\cdot] : \mathcal{E}(\mathfrak{A}) \to \mathcal{L}$ is order- preserving. Moreover, its image in $\mathcal{L}$ is meet-dense — indeed, for an arbitrary $P \in \mathcal{L}$, we have [Foulis *et al.*, 1985]

$$P = \bigwedge_{E \in \mathfrak{A}} [P \cap E].$$

In some special cases, a good deal more can be said. Recall that if $(X, \mathfrak{A})$ is a regular test space, then, for every outcome $x \in X$, $X \setminus x^{\perp}$ is a support.

DEFINITION 137. Let us call an entity $(\mathfrak{A}, \Sigma)$ *standard*[25] iff $\mathfrak{A}$ is regular and

$$\Sigma = \{ X \setminus x^{\perp} \mid x \in X \}$$

LEMMA 138. *Let $(\mathfrak{A}, \Sigma)$ be a standard entity. Then for every event $A \in \mathcal{E}(\mathfrak{A})$, $[A] = X \setminus C^{\perp\perp}$, where C is any event complementary to A.*

Proof. If $x \in X$, $X \setminus x^{\perp} \in \Sigma_A$ iff, for every test $E \supseteq A$, $E \setminus x^{\perp} \subseteq A$, or, equivalently, iff $E \setminus A \subseteq x^{\perp}$. Thus, $X \setminus x^{\perp} \in \Sigma_A$ iff $C \subseteq x^{\perp}$ iff $x \in C^{\perp}$, for every event $C \text{co} A$. However, as $\mathfrak{A}$ is regular and all events complementary to A are perspective, if $C_1, C_2 \text{co} A$, $C_1^{\perp} = C_2^{\perp}$. Hence, if C is any fixed event complementary to A, we have

$$\begin{aligned} [A] &= \bigcup \Sigma_A \\ &= \bigcup \{X \setminus x^{\perp} | x \in C^{\perp}\} \\ &= X \setminus \bigcap \{x^{\perp} | x \in C^{\perp}\} \\ &= X \setminus C^{\perp\perp} \end{aligned}$$

■

Thus, for a standard entity, we have canonical mappings linking $\mathcal{E}$, $\mathcal{L}$, and $\mathcal{C}$, as indicated the following diagram. Recall here that $\gamma : \mathcal{E} \to \mathcal{C}(X, \perp)$ is the mapping $A \mapsto A^{\perp\perp}$.

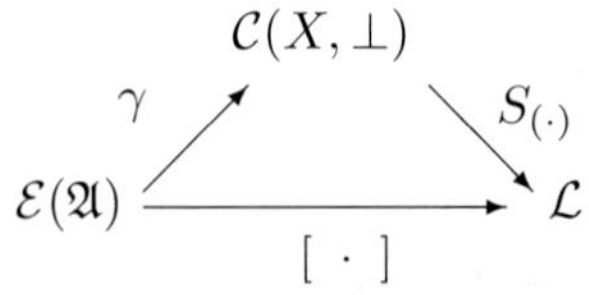

The question arises of when this diagram commutes. Evidently, what is required is that $[A] = X \setminus A^{\perp}$ for all events A, or, equivalently, that

$$A \text{co} C \Rightarrow A^{\perp} = C^{\perp\perp}$$

[25]or *biregular*, in [Randall and Foulis, 1983a]

for all $A, C \in \mathcal{E}(\mathfrak{A})$ – that is, $\mathfrak{A}$ must be Dacey. There is an interesting alternative characterization of Dacey-ness in terms of an orthogonality relation on $\mathcal{L}$:

DEFINITION 139. Let us call two supports S and T of $\mathfrak{A}$ *orthogonal* [26] iff there exists a test $E \in \mathfrak{A}$ such that $S \cap T \cap E = \emptyset$. In this case, we shall write $S \perp T$.

LEMMA 140. *Let $(\mathfrak{A}, \Sigma)$ be a standard entity. Then the following are equivalent:*

(a) $\mathfrak{A}$ is Dacey

(b) $[x] \perp [y] \;\Rightarrow x \perp y$ for all outcomes $x, y \in X$.

Proof. If $\mathfrak{A}$ is Dacey, then $[x] \perp [y]$ implies that there exists some $E \in \mathfrak{A}$ with $E \setminus (x^\perp \cup y^\perp) = \emptyset$, i.e., $E \subseteq x^\perp \cup y^\perp$. Let $A = E \cap x^\perp$ and $B = C \setminus A$. Note that $x \in A^\perp$ and $y \in C^\perp$. Since $A\text{co}C$, $A^\perp = C^{\perp\perp}$, so $x \perp y$. For the converse, suppose $[x] \perp [y] \Rightarrow x \perp y$. If $A\text{co}C$, then clearly $C^{\perp\perp} \subseteq A^\perp$. Let $x \in A^\perp$. For any $y \in C^\perp$, we have $A \subseteq x^\perp$ and $C \subseteq y^\perp$, so $E := A \cup C \subseteq x^\perp \cup y^\perp$. Equivalently, $E \setminus x^\perp = E \cap [x]$ and $E \setminus y^\perp = E \cap [y]$ are disjoint, whence, $[x] \perp [y]$. Hence, $x \perp y$. As x and y were arbitrary elements of $A^\perp$ and $C^\perp$, we have $A^\perp \subseteq C^{\perp\perp}$, concluding the proof. ■

If $\mathfrak{A}$ is algebraic, the mapping $[\cdot] : \mathcal{E}(\mathfrak{A}) \to \mathcal{L}$ descends (by Lemma 73 (B)) to a canonical, well-defined order-preserving mapping $[\cdot] : \Pi(\mathfrak{A}) \to \mathcal{L}$, given by $[p(A)] = [A]$ for all $p(A) \in \Pi(\mathfrak{A})$. The following theorem, the proof of which is omitted, is a slightly watered- down version of the main result (Theorem 16) of [Foulis *et al.*, 1985].[27]

THEOREM 141 [Foulis-Piron-Randall, 1985]. *Suppose $(\mathfrak{A}, \Sigma)$ is a standard entity, with $\mathfrak{A}$ algebraic and locally finite. Then the following are equivalent:*

(a) $\Pi(\mathfrak{A})$ is a complete lattice

(b) The canonical mapping $[\;\cdot\;]$ is an isomorphism

(c) $\mathfrak{A}$ is coherent, and the canonical mapping $[\;\cdot\;]$ is surjective.

What happens if the logic $\Pi(\mathfrak{A})$ is a lattice, but not complete? In that case (still assuming we are dealing with a regular algebraic test space $\mathfrak{A}$ and its standard entity $(\mathfrak{A}, \Sigma)$), both $\mathcal{L}$ and $\mathcal{C}$ are isomorphic to the McNeille completion of $\Pi(\mathfrak{A})$ — which, as is well known (cf [Bruns and Harding, 2000]), is seldom orthomodular.

[26] In [Foulis *et al.*, 1985], the term used is *uniformly* orthogonal.

[27] The hypotheses that the entity be standard and that $\mathfrak{A}$ be locally finite can be avoided, at the cost of complicating condition (c).

4.3 Central Supports

Recall from section 1.2 that the *center* of an orthoalgebra L is the set $\mathbf{C}(L)$ of all elements p such that both p and p' are principal — that is, $p \in C(L)$ the sum of two orthogonal elements of L lying below p is again below p, and likewise for p' — and such that for every $a \in L$, $a = x \oplus y$ where $x \leq p$ and $y \leq p'$. $\mathbf{C}(L)$ is a Boolean sub-orthoalgebra of L, and if $p \in C(L)$, then $L \simeq [0, p] \times [0, p']$. In this section, I shall discuss how the idea of centrality plays out in the setting of test spaces.

A starting point is the following observation: an element p in the logic $\Pi = \Pi(\mathfrak{A})$ of a test space is, in literal fact, a set of events. Moreover, by Lemma 66, p is irredundant (since we never have $A \sim B \subsetneq A$), and so, can be viewed as a test space in its own right. Let's agree to write $\mathfrak{A}_p$ when we wish to view p in this way, rather than as a point in an abstract structure Π.

LEMMA 142. *Let $\mathfrak{A}$ be algebraic. Then for every $p \in \Pi$,*

(a) $\mathfrak{A}_p$ is algebraic

(b) For all $B, C \in \mathcal{E}(\mathfrak{A}_p)$, $A \sim B$ in $\mathfrak{A}_p$ iff $A \sim B$ in $\mathfrak{A}$

(c) $\Pi(\mathfrak{A}_p) \simeq [0, p] \subseteq L$.

Proof. (a) Notice that $\mathcal{E}(\mathfrak{A}_p)$ is the set of all events A of $\mathfrak{A}$ such that $p(A) \leq p$. Let $A, B \in \mathcal{E}(\mathfrak{A}_p)$ be perspective in $\mathfrak{A}_p$; then there is some $C \in \mathcal{E}(\mathfrak{A}_p)$ such that $A \cap C = B \cap C = \emptyset$ and $A \cup C, B \cup C \in p$. Hence $p(A) \perp p(C)$, $p(B) \perp p(C)$ and $p(A) \oplus p(C) = p(B) \oplus p(C) = p$. Thus, $p(A) = p(B)$. If C' is any event of p complementary to A in $\mathcal{E}(\mathfrak{A}_p)$, then $p = p(A) \oplus p(C') = p(B) \oplus p(C')$, whence C' is also complementary to B. Parts (b) and (c) are straightforward. ■

THEOREM 143. *An algebraic test space $\mathfrak{A}$ is a direct sum iff there exists a support S such that $\mathfrak{A}_S = \mathfrak{A}_p$ for some $p \in \Pi(\mathfrak{A})$. In this case, $\mathfrak{A} \simeq \mathfrak{A}_p \oplus \mathfrak{A}_{p'}$.*

Proof. If $\mathfrak{C} = \mathfrak{A} \oplus \mathfrak{B}$, let $X = \bigcup \mathfrak{A}$ and $Y = \bigcup \mathfrak{B}$; then $Z := \bigcup \mathfrak{C} = X \oplus Y$, and X and Y are evidently supports of $\mathfrak{C}$ with $\mathfrak{A} = \mathfrak{C}_X$ and $\mathfrak{B} = \mathfrak{C}_Y$. Moreover, if $E, F \in \mathfrak{C}$, then

$$E \cap X \sim F \cap X \quad \text{and} \quad E \cap (Z \setminus X) \sim F \cap (Z \setminus X);$$

conversely, if $C \in \mathcal{E}(\mathfrak{C})$ is equivalent to $E \cap S$, then $C \in \mathfrak{C}_S$. Thus, the support S is the union of $\mathfrak{A}_p$ for some $p \in \Pi(\mathfrak{C})$, and $Z \setminus S = \bigcup \mathfrak{A}_{p'}$. ■

DEFINITION 144. Let $\mathfrak{A}$ be algebraic. A set $S \subseteq X$ is a *central support* iff

$$E, F \in \mathfrak{A} \Rightarrow S \cap E \sim S \cap F.$$

Equivalently, S is central iff $\mathfrak{A}_S = \mathfrak{A}_p$ for some $p \in \Pi(\mathfrak{A})$. The collection of all central supports is the *center* of $\mathfrak{A}$, denoted $\mathcal{Z}(\mathcal{A})$.

Note that a central support *is*, in fact, a support. For suppose $E, F \in \mathfrak{A}$ and $S \cap E \subseteq S \cap F$. Since $E \sim F$ and S is central, $S \cap E \sim S \cap F$. It follows easily (using Lemma 66) that $S \cap E = S \cap F$.

LEMMA 145. *$\mathcal{Z}(\mathfrak{A})$ is a field of sets.*

Proof. It is clear that the complement of a central support is again a central support. Thus, it suffices to show that the intersection of two central supports, say S and T, is again a central support. Let $E, F \in \mathfrak{A}$. Let $A = S \cap E, A' = S \cap F$, $B = T \cap E, B' = T \cap F$. We need to show that $A \cup B \sim A' \cup B'$. Let $C = E \setminus (A \cup B)$ and $C' = F \setminus (A' \cup B')$. Since $A \sim A'$, we have $A \cup (F \setminus A) = A \cup (B' \setminus A') \cup C' \in \mathfrak{A}$; similarly, since $B \sim B'$, we have $(A \setminus B) \cup C \cup B' \in \mathfrak{A}$. As S is a central support, we have

$$(A \setminus B) \cup (B' \cap A') = S \cap [(A \setminus B) \cup B' \cup C \ \sim \ S \cup F = A.'$$

It follows that

$$(A \setminus B) \cup (B' \cap A') \cup (B' \setminus A') \cup D \in \mathfrak{A}.$$

But now $(A \setminus B) \sim (A' \setminus B')$, so the additivity lemma (Lemma **??**) gives us

$$A \cup B = (A \setminus B) \cup B \sim (A' \setminus B') \cup B' = A' \cup B'.$$

■

If $\mathfrak{A} = \{E\}$ is a classical test space, then $\mathcal{Z}(\mathfrak{A}) = 2^E$. If $\mathfrak{A}$ is a *semi-classical* test space, i.e., if $E \cap F = \emptyset$ for any two distinct tests in $\mathfrak{A}$, then $\mathcal{Z}(\mathfrak{A}) = \{\emptyset, X\}$. This last example illustrates, by the way, that not every complemented support is central. Indeed, suppose that $\mathfrak{A}$ is semiclassical, then any set $S \subseteq X = \bigcup \mathfrak{A}$ such that, for every $E \in \mathfrak{A}$, $S \cap E$ is neither empty nor all of E, is a support, the complement of which is again a support.

LEMMA 146. *If $\mathfrak{A}$ is algebraic, the mapping $S \mapsto p(S \cap E)$ (where $E \in \mathfrak{A}$ is arbitrary) takes $\mathcal{Z}(\mathfrak{A})$ defines a Boolean embedding of $\mathcal{Z}(\mathfrak{A})$ into* $\mathbf{C}(\Pi(\mathfrak{A}))$.

Proof. By Lemma 92, $\Pi(\mathfrak{A}_{p(S)}) \times \Pi(\mathfrak{A}_{p(S)'}) = \Pi(\mathfrak{A}_S \oplus \mathfrak{A}_{S'}) = \Pi(\mathfrak{A})$; hence, $p(S) \in C(\Pi(\mathfrak{A}))$. The rest of the claim is clear. ■

The embedding of Lemma 146 need not be surjective, even when the logic in question is Boolean. Indeed, let $\mathfrak{A} = \{\{a, x, b\}, \{b, b'\}\}$. It is not difficult to show that $\mathcal{Z}(\mathfrak{A}) = \{\emptyset, \{a, x, b'\}, \{b\}, X\} \simeq 2^2$, while $\Pi(\mathfrak{A}) = 2^3$.[28]

On the other hand, it is not hard to establish the following. Call an orthoalgebra L *atomistic* iff every $p \in L$ has the form $p = \bigoplus_i a_i$ for a finite jointly orthogonal set of atoms a_i. If L is atomistic, let $\mathfrak{A}^0_L$ denote the collection of all finite jointly orthogonal sets E of atoms with $\bigoplus E = 1$. It is straightforward to show that ${\mathfrak{A}^0}_L$ is algebraic, with $\Pi(\mathfrak{A}^0_L) \simeq L$.

PROPOSITION 147. *If L is atomistic, then $C(L) \simeq \mathcal{Z}(\mathfrak{A}^o_L)$.*

[28]This reflects the fact that, in general, $\mathfrak{A}^{\#} \oplus \mathfrak{B}^{\#} \subsetneq (\mathfrak{A} \oplus \mathfrak{B})^{\#}$.

Proof. Let L be atomistic, and let X denote the set of atoms of L. Let $p \in C(L)$. Since p is principal, $(p) = \{q \in L | q \leq p\}$ is an ideal. If $a \in X$, then $a = x \oplus y$ for some $x \leq p$ and $y \leq p'$. As a is an atom, $x = 0$ and $a = y \leq p'$ or $y = 0$ and $a = x \leq p$. Thus, $S_p := X \cap (p)$ and and $S_{p'} := X \cap (p')$ are disjoint supports of $\mathfrak{A}^0_L$ partitioning X. Moreover, if $E \in \mathfrak{A}^L_0$ and $A = S_p \cap E$, then we have $q := \oplus A \leq p$ (since every $a \in A$ lies below p and p is principle). If $q \neq p$ then there exists $r = p - q \leq p$, whence, for any atom $x \leq r$, we have $x \perp A$. But since S_p is a support, $A \subseteq A \cup \{x\}$ implies $x \in A$, which is impossible since then $x \perp x$. Hence, $\bigoplus(S_p \cap E) = p$, whence, $S_p \cap E \sim S_p \cap F$ for all $E, F \in \mathfrak{A}^0_L$. Thus, S_p is central. That the mapping $p \mapsto S_p$ is a boolean isomorphism is straightforward. ■

4.4 Generalized Sasaki Projections

As we saw in Section 1.1, an ortholattice L is orthomodular iff the *Sasaki projection* $\phi_a : L \times L \to L$, given by $\phi_a(b) = a \wedge (a' \vee b)$, satisfies the condition $b \leq a \Rightarrow \phi_a(b) = b$. Recall also that, if L is an atomic OML, the covering law amounts to the statement that if x and y are non-orthogonal atoms of L, then $\phi_x(y)$ is another atom. In this section, following [Bennett and Foulis, 1998; Wilce, 2000], I'll describe a generalization of Sasaki projections to the context of an arbitrary entity. As we shall see, this greatly clarifies the interpretation of Sasaki projections in terms of conditioning.

In what follows, L is an orthoalgebra and $\mathfrak{A}$ is an algebraic test space having L as its logic. As in section 4.3, if $p \in L$ we write $\mathfrak{A}_p$ for the set of events comprising p when this is to be regarded as an algebraic test space in its own right. In this case $\Pi(\mathfrak{A}_p) \simeq [0, p]$. We also write X_p for $\bigcup \mathfrak{A}_p$, the set of outcomes $x \in X$ with $p(x) \leq p$. By Lemma 124 (d), for any support S of $\mathfrak{A}$, $S \cap X_p$ is a support of $\mathfrak{A}_p$. This suggests the following construction.

DEFINITION 148. Let $(\mathfrak{A}, \Sigma)$ be an entity with logic $L = \Pi(\mathfrak{A})$ and property lattice $\mathcal{L}$. For $p \in L$ and $S \in \mathcal{L}$, let $\mathcal{L}_{p,S}$ denote the collection of all properties $T \in \mathcal{L}$ such that

(i) $T \subseteq [p]$

(ii) $T \cap X_p \subseteq S$.

We define the *conditioning map* $\gamma_p : \mathcal{L} \to \mathcal{L}$ by $\gamma_p(S) := \bigcup \mathcal{L}_{p,S}$.

To motivate this, let $\Sigma_{p,S} = \Sigma \cap \mathcal{L}_{p,S}$. This is the set of states in which it is certain that p will be confirmed, and impossible that S will be refuted, by a test of p. Then $\mathcal{L}_{p,S}$ is the complete sub-lattice of the interval $[0, [p]]$ in $\mathcal{L}$ generated by $\Sigma_{p,S}$, and that $\gamma_p(S) = \bigcup \Sigma_{p,S}$. The maps $p, S \mapsto \Sigma_{p,S}$ and $p, S \mapsto \gamma_{p,S}$ represent a simple form of *conditioning*. If are given data from a large number of tests of $p \in L$, all confirming p, and if the actual state of the entity for all of these tests was S, then all our data lies in $X_p \cap S$. We will be inclined to infer not only that p is certain, but that the state of the entity belongs to $\Sigma_{p,S}$, and that the property $\gamma_p(S)$ is actual.

EXAMPLE 149. Let $X = \{a, x, b, y, c, z\}$ and $\mathfrak{A} = \{\{a, x, b\}, \{b, y, c\}, \{c, z, a\}\}$ be the "Wright triangle" of Example 55. (For convenience, a Greechie diagram is given below.) We shall compute $\gamma_p(S)$ for $p = b'$ and $S = [z] = \{b, z, y, z\}$ where Σ consists of all supports of X. Note that $X_{b'} = \{a, x, y, c\}$ and $[b'] = \{a, x, y, z, c\}$. Hence, $S \cap X_{b'} = \{x, y\}$. The largest support contained in $[b']$ having this same intersection with $X_{b'}$ is the support $\{x, y, z\}$. Hence, $\gamma_{b'}([z]) = \{x, y, z\}$.

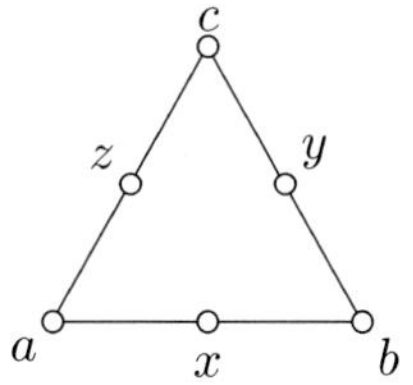

EXAMPLE 150. To illustrate the dependence of γ_a on Σ, let $\mathfrak{A}$ be as above but suppose that Σ consists only of the principal properties $[p] = X - p^{\perp}$ where p is an atom of L. Again, $S = [z]$, $S \cap X_{b'} = \{x, y\}$ and $[b'] = \{a, x, y, z, c\}$. However, in this case the only elements of Σ below $[b']$ are $[x] = \{x, y, z, c\}$, $[y] = \{a, x, y, z\}$, $[a] = \{a, y\}$ and $[c] = \{x, c\}$. None of these has intesection with $X_{b'} = \{a, x, y, c\}$ contained in $\{x, y\}$; hence, in this setting, $\gamma_{b'}([z]) = 0$.

THEOREM 151. *For any $p \in L$ and $S \in \mathcal{L}$, $\gamma_p(S) = \bigwedge_{A \in p}[S \cap A]$.*

Proof. Suppose that $T \in \mathcal{L}_{p,S}$. Then for every event $A \in p$, $T \cap A \subseteq S \cap A$ (since $T \cap X_p \subseteq S \cap X_p$ and $A \subseteq X_p$). Since $T \subseteq [p]$, we have for every test E containing A that $T \cap E \subseteq A$, so $T \cap E \subseteq T \cap A \subseteq S \cap A$. Hence, $T \in \Sigma_{S \cap A}$, whence, $T \subseteq [S \cap A]$. It follows that $\gamma_p(S) = \bigcup \mathcal{L}_{p,S} \subseteq [S \cap A]$. Now suppose $T \subseteq [S \cap A]$ for every $A \in p$. Then in particular, since $[S \cap A] \subseteq [A] = [p]$, we have $T \subseteq [p]$. Now, noting that for any test E containing A we have $T \cap A = T \cap E \subseteq S \cap A$, it follows that

$$T \cap X_p = \bigcup_{A \in p} T \cap A \subseteq \bigcup_{A \in p} S \cap A = S \cap X_p.$$

Hence, $T \in \mathcal{L}_{p,S}$, so $T \subseteq \gamma_p(S)$. ■

As observed above, the principal properties $[A]$, $A \in \mathcal{E}(\mathfrak{A})$, are meet-dense in $\mathcal{L}$. Hence, we can extend γ to a mapping $\gamma : \mathcal{L} \times \mathcal{L} \to \mathcal{L}$ given by

$$\gamma_Q(S) := \bigwedge_{Q \leq [p]} \gamma_p(S).$$

For the proof of the following, see [Wilce, 2000].

THEOREM 152. *Let L be an OML and let $\mathfrak{A}$ be the test space of orthopartitions of the unit in L. Let Σ consist of all supports of L. Then $\forall a, b \in L$,*

$$\gamma_a([\ b\ ]) = [\phi(a, b)].$$

In [Bennett and Foulis, 1998], Bennett and Foulis introduce (for any effect algebra) the quantity

$$\nabla(a,b) := \{x \leq a \mid b \leq x \oplus a'\}.$$

They then define a generalized Sasaki projection $\Phi(a,b)$ to be the set of all minimal elements of $\nabla(a,b)$ (if any). If L is an OML, there is a unique minimal element, namely $\phi(a,b)$. It can be shown that if L is an OMP satisfying the descending chain condition, then the two definitions are essentially the same [Wilce, 2000, Theorem 3]; however, they generally diverge for non-orthocoherent orthoalgebras.

4.5 Minimal Supports and the Covering Law

We now revisit the covering law from the point of view of supports. The main result (Theorem 157 below), due to D. Cohen and G. Svetlichny [Cohen and Svetlichny, 1987], shows that for an OML of finite rank, the covering law is equivalent to a certain minimality condition on supports of the form $X \setminus x^{\perp}$.

DEFINITION 153. A regular algebraic test space $\mathfrak{A}$ satisfies the *minimal support condition* (MSC) iff, for every $x \in X$ and every $p \in \Pi(\mathfrak{A})$ with rank $\mathfrak{A}_p > 2$, $X_p \setminus x^{\perp}$ is a minimal support of $\mathfrak{A}_p$. We shall call a regular algebraic test space satisfying the MSC an *MSC test space*, for short.

EXAMPLES 154. (i) If $\mathfrak{A}$ has rank 2, then it satisfies the MSC by default. It is easy to check that, for instance, the Wright Triangle also enjoys the MSC.

(ii) If $\mathfrak{A}$ satisfies the MSC, then so does $\mathfrak{A}_p$ for every $p \in \Pi(\mathfrak{A})$.

(iii) Using a clever geometric argument, which I omit here, Cohen and Svetlichny establish that, for a Hilbert space $\mathbf{H}$ of dimension at least three, the frame manual $\mathfrak{F}(\mathbf{H})$ is MSC.

Call a test space $\mathfrak{A}$ *chain-connected* iff, for any tests $E, F \in \mathfrak{A}$, there exists a finite sequence $E_1, ..., E_n$ of tests in $\mathfrak{A}$ with $E = E_1$, $E_n = F$, and $E_i \cap E_{i+1} \neq \emptyset$ for all $i = 1, ..., n-1$. It is easy to show that $\mathfrak{A}$ is chain-connected iff it can not be represented non-trivially as a horizontal sum.

LEMMA 155. *Any MSC test space is chain-connected.*

Proof. If $\mathfrak{A} = \mathfrak{A}_1 \dot{+} \mathfrak{A}_2$, and neither $\mathfrak{A}_1$ nor $\mathfrak{A}_2$ is empty, then the regularity of $\mathfrak{A}_1$ implies that of both $\mathfrak{A}_1$ and $\mathfrak{A}_2$. Choose any $x \in X_1 := \bigcup \mathfrak{A}_1$ and any $y \in X_2 := \bigcup \mathfrak{A}_2$. Then $(X_1 \setminus x^{\perp}) \cup (X_2 \setminus y^{\perp})$ is a support of $\mathfrak{A}$. But this is properly contained in $X \setminus x^{\perp} = (X_1 \setminus x^{\perp}) \cup X_2$, contradicting the MSC. ■

We can regard the minimal support condition as asserting that supports of the form $X \setminus x^{\perp}$ are maximally informative, and remain maximally informative when we condition upon various propositions in Π. This enforces a strong degree of uniformity on a test space.

LEMMA 156. *Let $\mathfrak{A}$ be an MSC test space. If $\mathfrak{A}$ contains a test of cardinality n, but no test of cardinality less than n, then it has uniform rank n.*

Proof. By induction on n. Notice that the conclusion is trivial if $n = 1$ or $n = 2$. Therefore, suppose $n \geq 3$, and that the result is known for all smaller values of n. Let $E, F \in \mathfrak{A}$ with $|E| = |F| = n$. We claim that $E \cap F \neq \emptyset$. For suppose $a \in E \cap F$. Then $\mathfrak{A}_{a'}$ has a test of size $n - 1$, but no smaller test; moreover, by the remarks following the Definition, it again has the MSC. Thus, it is $n - 1$ uniform. But then $F \setminus \{a\} \in \mathfrak{A}_{a'}$ also has cardinality n. Now let $\mathfrak{A}_n$ be the set of tests in $\mathfrak{A}$ having cardinality exactly n, and let $\mathfrak{B} = \mathfrak{A} \setminus \mathfrak{A}_n$. By the foregoing argument, $\mathfrak{A} = \mathfrak{A}_n \dot{+} \mathfrak{B}$. But then, by the Lemma above, $\mathfrak{B}$ is empty, finishing the proof. ■

Let L be a complete atomistic OML. Denote by X_L the set of atoms of L, and by $\mathfrak{A}_L$, the set of maximal pairwise orthogonal sets of atoms joining to $\mathbf{1}$. As discussed in Section 4.2, $(X_L, \mathfrak{A}_L)$ is a coherent algebraic test space (hence, regular) with $\Pi(\mathfrak{A}_L) \simeq L$. Note that for every $p \in L$, $\mathfrak{A}_p = \mathfrak{A}_{[0,p]}$.

THEOREM 157 [Cohen-Svetlichny, 1987]. *Let L be an OML of finite rank. Then L satisfies the covering law iff $\mathfrak{A}_L$ satisfies the minimal support condition.*

I refer the reader to the original paper of Cohen and Svetlichny for the somewhat involved inductive proof.

5 TENSOR PRODUCTS OF QUANTUM LOGICS

We now revisit the question, touched upon briefly in section 1, of how one ought to define the tensor product of two "quantum logics". Section 5.1 examines in more detail the example of Foulis and Randall, exhibited in section 1.4, that shows there can be no reasonable (and reasonably general) tensor product for orthomodular posets or lattices. In section 5.2, we'll see that there *does* exist a perfectly good tensor product for test spaces, and also for *unital* algebraic test spaces. This last allows one to define a tensor product of orthoalgebras having unital sets of states. In particular, we'll see that the tensor product of the offending orthomodular lattices of the Foulis-Randall example *does* exist — but as a non-orthocoherent orthoalgebra.

Sections 5.3 and 5.4 discusses work of R. Lock [Lock, R., 1981] on tensor products of UDF test spaces, and Golfin [Golfin, 1987] on tensor products of property lattices, respectively. To round out the discussion, section 5.5 characterizes the state-space of the tensor product of two frame manuals, following [Kläy *et al.*, 1987; Wilce, 1990; Wilce, 1992].

5.1 *The Problem of Tensor Products,* bis

If $\mathbf{H}_1$ and $\mathbf{H}_2$ are Hilbert spaces, understood as representing a pair of particles, one understands $\mathbf{H}_1 \otimes \mathbf{H}_2$ as a model for the system consisting of the two particles *together*. Now, if we are dealing with a pair of bosons, it is customary to use, not the entire tensor product, but the *symmetric* tensor product $\mathbf{H}_1 \otimes_s \mathbf{H}_2$, i.e., the null-space of the operator $A : \mathbf{H}_1 \otimes \mathbf{H}_2 \to \mathbf{H}_2 \otimes \mathbf{H}_1$ given by $A(x \otimes y) = y \otimes x$. If our particles are fermions, we use the *antisymmetric* tensor product, $\mathbf{H}_1 \otimes_a \mathbf{H}_2$, i.e., the orthogonal complement of $\mathbf{H}_1 \otimes_s \mathbf{H}_2$ in $\mathbf{H}_1 \otimes \mathbf{H}_2$. If we wish, we can view these two subspaces as representing two possible (and complementary) *properties* of the entity represented by $\mathbf{H}_1 \otimes \mathbf{H}_2$. Indeed, we are virtually forced to do so by the following considerations. Each orthonormal basis $E \in \mathfrak{F}(\mathbf{H}_1)$ represents (according to the usual interpretation) a possible experiment performable on the first particle — realizable, say, by some suitably involved collection of Stern- Gerlach experiments. Similarly, each $F \in \mathfrak{F}(\mathbf{H}_2)$ represents an experiment performable on the system represented by $\mathbf{H}_2$. Clearly, if the two systems are indeed separated (e.g., the apparatus for measuring E is in Copenhagen while that for measuring F is in Chicago) we can perform the experiment EF on the composite system: Measure the first component using E, the second using F; record the results as the pair xy, $x \in E$ and $y \in F$. Thus, we must allow test spaces corresponding to our composite system to contain *at least* the product test $EF \in \mathfrak{F}(\mathbf{H}_1) \times \mathfrak{F}(\mathbf{H}_2)$. If we insist on representing the composite entity by a frame manual $\mathfrak{F}(\mathbf{K})$ for some Hilbert space $\mathbf{K}$, it quickly follows that $\mathbf{K} = \mathbf{H}_1 \otimes \mathbf{H}_2$. The symmetric and anti-symmetric tensor products enter into the description of the possible *states* of a fermionic or bosonic system.

Recall that a quantum logic is traditionally defined as a pair (L, Δ) where L is at least an OMP and Δ is a strong order-determining set of states for L. We wish to consider the question: What is meant by the tensor product of two quantum logics? What is required, presumably, is some device $\otimes$ whereby two quantum logics (M, Δ) and (N, Γ) can be coupled to yields a quantum logic $(L, \Omega) = (M, \Delta) \otimes (N, \Gamma)$. As discussed in Section 1.4, such a device ought to satisfy at least the following requirements:

(1) There exists a map $\otimes : M \times N \to L$ such that $\mathbf{1}_M \otimes \mathbf{1}_N = 1_L$ and

$$(p \otimes q) \perp (p' \otimes q) \quad \& \quad (p \otimes q) \oplus (p' \otimes q) = 1 \otimes q$$

for all $p \in M, q \in N$; and similarly for $p \otimes q$ and $p \otimes q'$;

(2) For every $\mu \in \Delta$ and $\nu \in \Gamma$, there exists a state $\omega \in \Omega$ with $\omega(p \otimes q) = \mu(p)\nu(q)$.

We are now in a position to give an easy exposition of the Foulis-Randall counter-example (Example 43)

EXAMPLE 158. Let $\mathfrak{A}$ be the "pentagon" test space consisting of five three-element operations pasted in a loop:

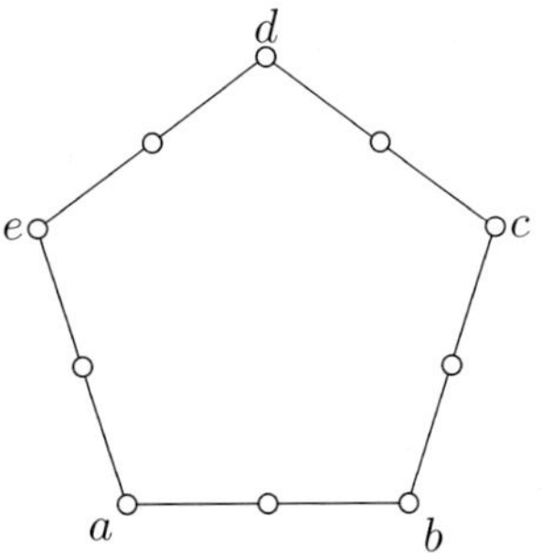

Let $M = N = \Pi(\mathfrak{A})$. Then M is an orthomodular poset. Indeed, it is an orthomodular lattice (by the Loop Lemma) and has a strongly order-determining state space. Suppose now that there exists an OMP L and a map $\otimes : M \times M \to L$ satisfying (1) and (2) as above. Let μ be the state on $\mathfrak{A}$ assigning value 1/2 to each of the five outcomes a, b, c, d, e in the figure above, and 0 to the rest. By assumption (2), there exists a state ω on $M \otimes M$ such that $\omega(a \otimes b) = \mu(a)\mu(b)$ for all $p, q \in M$. By assumption (1), $A := \{a \otimes a, b \otimes c, c \otimes e, e \otimes d, d \otimes b\}$ is pairwise orthogonal. Since L is an OMP, A is jointly orthogonal. But ω takes the value 1/4 on each element of this set, whence $\sum_{x \in A} \omega(x) = 5/4 > 1$, a contradiction. It follows that no such state ω — and hence, no tensor product L satisfying (1) and (2) — exists.

There are two ways in which we might react to such an example. One is to regard it as a pathology, to be eliminated by adopting a more restrictive definition of quantum logics. However, the pentagon test space is quite well-behaved: it is coherent, algebraic, UDF, and even embeddable in a Boolean algebra [Wright, 1978b; Foulis and Randall, 1981b]. Alternatively, we may conclude that the category of OMPs (or OMLs, or quantum logics) is too narrow to admit tensor products, and seek to enlarge the category. This is the strategy we pursue in the sequel. As we shall see, the category of *unital orthoalgebras* (and, more broadly, the category of unital test spaces) *does* support a perfectly sensible notion of a tensor product, satisfying the Foulis-Randall desiderata.

5.2 Bilateral and Tensor Products of Test Spaces

Recall from Section 2.2 that the *cartesian product* of two test spaces $\mathfrak{A}$ and $\mathfrak{B}$ (with outcome sets X and Y respectively), is $\mathfrak{A} \times \mathfrak{B} = \{EF | E \in \mathfrak{A}, F \in \mathfrak{B}\}$. Recall also that the *forward product* of $\mathfrak{A}$ and $\mathfrak{B}$ is

$$\overrightarrow{\mathfrak{A}\mathfrak{B}} = \{ \bigcup_{x \in E} xF_x \mid E \in \mathfrak{A} \ \& \ F_x \in \mathfrak{B} \ \forall \ x \in E \}.$$

Since $\bigcup(\overrightarrow{\mathfrak{A}\mathfrak{B}}) = XY = \bigcup \mathfrak{A} \times \mathfrak{B}$, every state on the former is a state on the latter.

We have the following characterization of states on $\overrightarrow{\mathfrak{A}\mathfrak{B}}$:

LEMMA 159. *A state* ω *on* $\mathfrak{A} \times \mathfrak{B}$ *belongs to* $\Omega(\overrightarrow{\mathfrak{A}\mathfrak{B}})$ *iff* $\omega(xF_1) = \omega(xF_2)$ *for all* $x \in X$ *and all* $F_1, F_2 \in \mathfrak{B}$.

Proof. For any $x \in X$, and for any $F_1, F_2 \in \mathfrak{B}$, $xF_1 \sim xF_2$ in $\overrightarrow{\mathfrak{A}\mathfrak{B}}$; therefore, for any state $\omega \in \Omega(\overrightarrow{\mathfrak{A}\mathfrak{B}})$, $\omega(xF_1) = \omega(xF_2)$. Conversely, if $\omega(xF)$ is independent of F, and $G = \bigcup_{x \in E} x\ F_x \in \overrightarrow{\mathfrak{A}\mathfrak{B}}$, then selecting any $F \in \mathfrak{A}$, we have

$$\sum_{xy \in G} \omega(xy) = \sum_{x \in E} \omega(xF_x) = \sum_{x \in E} \omega(xF) = \omega(EF) = 1$$

(where the convergence of $\omega(EF)$ guarantees that of $\sum_{xy \in G} \omega(xy)$). ■

We can interpret Lemma 159 as follows: $\omega(xF)$ is the probability to obtain x upon executing a test for x, given that F is executed on the second system. This ought not depend on the choice of F unless the system represented by $\mathfrak{B}$ has some influence on the system represented by $\mathfrak{A}$. The idea is that the execution of F might involve some perturbation of the one system, leading in turn to some statistically observable perturbation of the other, if such influence is possible. (This is necessarily a bit vague: we are, in effect, laying down a *definition* of statistical influence.) In this language, then, the states on $\overrightarrow{\mathfrak{A}\mathfrak{B}}$ are exactly those states on $\mathfrak{A} \times \mathfrak{B}$ that display *no influence* of $\mathfrak{B}$ on $\mathfrak{A}$.

Reversing the roles of $\mathfrak{A}$ and $\mathfrak{B}$, we define $\overleftarrow{\mathfrak{A}\mathfrak{B}}$ to be $\pi(\overrightarrow{\mathfrak{B}\mathfrak{A}})$ where $\pi : YX \to XY$ is the map $\pi(yx) = xy$. That is, $\overleftarrow{\mathfrak{A}\mathfrak{B}}$ consists of operations of the form $\bigcup_{y \in F} E_y\ y$. We call this the *backward product* of $\mathfrak{A}$ and $\mathfrak{B}$. Of course, the states on $\overleftarrow{\mathfrak{A}\mathfrak{B}}$ are just those states on $\mathfrak{A} \times \mathfrak{B}$ that display no influence of $\mathfrak{A}$ on $\mathfrak{B}$.

DEFINITION 160. The *bilateral product*[29] of two test spaces $\mathfrak{A}$ and $\mathfrak{B}$ is

$$\mathfrak{A}\mathfrak{B} := \overrightarrow{\mathfrak{A}\mathfrak{B}} \cup \overleftarrow{\mathfrak{A}\mathfrak{B}}\ .$$

Clearly, the states on $\mathfrak{A}\mathfrak{B}$ are precisely those states on $\mathfrak{A} \times \mathfrak{B}$ that exhibit no influence in either direction between $\mathfrak{A}$ and $\mathfrak{B}$. A situation in which we would certainly like to enforce this condition would be one in which $\mathfrak{A}$ and $\mathfrak{B}$ represent localized, space-like separated physical systems. For a slightly fanciful illustration, consider a pair of observatories, one on earth, the other on a planet orbiting a nearby star. Suppose the laboratories are in communication with one another (over a purely classical channel). One observatory takes measurements of some remote astronomical event, and relays instructions to the other to make specific specific follow-up measurements, depending upon what has been seen. The set of

[29] The term originally used by Foulis and Randall was *pre-tensor product.*

such compound experiments initiated by either observatory corresponds exactly to the bilateral product of their respective obervational repertoires.

It is worth noting that the bilateral product satisfies, almost trivially, a version of the Foulis-Randall desiderata for a model of coupled systems. Obviously, if $\mu \in \Omega(\mathfrak{A})$ and $\nu \in \Omega(\mathfrak{B})$, the product state $\mu \otimes \nu$ given by

$$(\mu \otimes \nu)(xy) = \mu(x)\nu(y)$$

is a state on $\mathfrak{A}\mathfrak{B}$. Also, if $A_1 \perp A_2 \in \mathcal{E}(\mathfrak{A})$, and $B \in \mathcal{E}(\mathfrak{B})$, then $A_1B \perp A_2B$ in $\mathfrak{A}\mathfrak{B}$, and similarly $AB_1 \perp AB_2$ if $A \in \mathcal{E}(\mathfrak{A})$ and $B_1 \perp B_2$ in $\mathcal{E}(\mathfrak{B})$.

TENSOR PRODUCTS OF ALGEBRAIC TEST SPACES

In general, $\mathfrak{A}\mathfrak{B}$ is not algebraic, or even pre-algebraic, even if $\mathfrak{A}$ and $\mathfrak{B}$ are both algebraic. For instance, it can be shown [Lock, R., 1981] that if $\mathfrak{A}$ the test space having the Greechie diagram

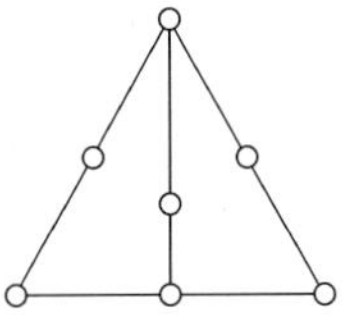

then $\mathfrak{A}\mathfrak{A}$ is not pre-algebraic. However, the following is easily verified:

LEMMA 161. *If $\mathfrak{A}$ and $\mathfrak{B}$ are unital, so is $\mathfrak{A}\mathfrak{B}$.*

Proof. If $xy \in XY$, let $\mu \in \Omega(\mathfrak{A})$ with $\mu(x) = 1$ and $\nu \in \Omega(\mathfrak{B})$ with $\nu(y) = 1$; then $(\mu \otimes \nu)(xy) = 1$. ■

DEFINITION 162. Two algebraic test spaces $\mathfrak{A}$ and $\mathfrak{B}$ *admit a tensor product* iff $\mathfrak{A}\mathfrak{B}$ is pre-algebraic. In that case, the *tensor product* of $\mathfrak{A}$ and $\mathfrak{B}$ is

$$\mathfrak{A} \otimes \mathfrak{B} := \langle \mathfrak{A}\mathfrak{B} \rangle.$$

Since a unital test space is pre-algebraic, the preceding Lemma guarantees that the tensor product of unital algebraic test spaces always exists.

Having defined a tensor product for algebraic test spaces, we have a natural definition of the tensor product of two orthoalgebras:

DEFINITION 163. Let L and M be orthoalgebras. We say that L and M *admit a tensor product* iff their respective test spaces $\mathfrak{A}_L$ and $\mathfrak{A}_M$ of orthopartitions of unity admit a tensor product, in which case we define

$$L \otimes M := \Pi(\mathfrak{A}_L \otimes \mathfrak{A}_M).$$

EXAMPLE 164. Let $\mathfrak{A}$ be the "pentagon" test space used in the example of section 1, and let $M = \Pi(\mathfrak{A})$. Note that $\mathfrak{A}$ is unital. Thus, there exists a tensor product

$M \otimes M$ in the category of (unital) orthoalgebras, though not in that of unital OMPs.

Let $\mathfrak{A}, \mathfrak{B}$ and $\mathfrak{C}$ be algebraic test spaces. A *bi-interpretation* from $\mathfrak{A} \times \mathfrak{B}$ to $\mathfrak{C}$ is a map $\Phi : X(\mathfrak{A}) \times X(\mathfrak{B}) \to \mathfrak{E}(\mathfrak{C})$ such that for every fixed $a \in X(\mathfrak{A})$ and $b \in X(\mathfrak{B})$ the maps $\Phi(a, \ \cdot \ ), \Phi(\ \cdot \ , b)$ satisfy

(i) $x \perp y \Rightarrow \Phi(x,b) \perp \Phi(y,b)$ for all $x, y \in X(\mathfrak{A})$ and $u \perp v \Rightarrow \Phi(a,u) \perp \Phi(a,v)$;

(ii) $\Phi(E,b) = \bigcup_{x \in E} \Phi(x,b) \in \mathcal{E}(\mathfrak{C})$ and $\Phi(a,F) = \bigcup_{y \in F} \Phi(a,y) \in \mathcal{E}(\mathfrak{C})$;

(iii) $\Phi(E,F) = \bigcup_{b \in F} \Phi(E,b) = \bigcup_{a \in E} \Phi(a,F) \in \mathfrak{C}$.

THEOREM 165. *Let $\mathfrak{A}$ and $\mathfrak{B}$ be locally finite algebraic test spaces admitting a tensor product. For any bi- interpretation $\Phi : \mathfrak{A} \times \mathfrak{B} \to \mathfrak{C}$, there exists a unique interpretation $\phi : \mathfrak{A} \otimes \mathfrak{B} \to \mathfrak{C}$ such that $\Phi(a,b) = \phi(a \otimes b)$ for all $a \in X(\mathfrak{A})$ and $b \in X(\mathfrak{B})$.*

Proof. Let $L = \Pi(\mathfrak{A})$, $M = \Pi(\mathfrak{B})$, and $N = \Pi(\mathfrak{C})$. It is straightforward that every bi-interpretation $\Phi : \mathfrak{A} \times \mathfrak{B} \to \mathfrak{C}$ determines a unique bi-morphism $L \times M \to N$, hence, extends to a unique unital homorphism $L \otimes M \to N$, which in turn yields an interpretation $\mathfrak{A}_{L \otimes M} \to \mathfrak{A}_N$. By restriction, we have an interpretation $\mathfrak{A} \otimes \mathfrak{B} \to \mathfrak{A}_N$. ■

COROLLARY 166. *If L and M are orthoalgebras admitting a tensor product, then $\mathfrak{A}_{L \otimes M} = \mathfrak{A}_L \otimes \mathfrak{A}_M$.*

Remark: Theorem 165 implies that if $\mathfrak{A}$ and $\mathfrak{B}$ are *locally finite* algebraic test spaces admitting a tensor product, then $\Pi(\mathfrak{A} \otimes \mathfrak{B})$ depends only upon $\Pi(\mathfrak{A})$ and $\Pi(\mathfrak{B})$. This need not be true, however, for more general test spaces. (For an example, see R. Lock [Lock, R., 1981], §7.3.)

5.3 Tensor Products of UDF test spaces

In this section, we consider bilateral and tensor products of test spaces of partitions, i.e., UDF test spaces. In particular, we present a result of R. Lock that gives a sharp characterization of the tensor product of two semiclassical test spaces. Among other things, Lock's result shows that the bilateral product of semiclassical test spaces is already algebraic, so that, in this case, the bilateral and tensor products coincide.

In general, a UDF test space can be represented in many inequivalent ways as a test space of partitions of a set M. However, there is always a smallest candidate for the set M, namely, the set of dispersion-free states on X. This gives rise to a useful duality.

DEFINITION 167. A *transversal* of a test space $\mathfrak{A}$ is a subset T of $X = \bigcup \mathfrak{A}$ such that for all $E \in \mathfrak{A}$, $\#(T \cap E) = 1$. The *Lock Dual* of a test space $\mathfrak{A}$ is the set $\mathfrak{A}^*$ of all transversals of $\mathfrak{A}$.

Remarks: Note that $\bigcup\mathfrak{A}^* \subseteq X = \bigcup\mathfrak{A}$, with equality iff $\mathfrak{A}$ is UDF. It is easy to see that $\mathfrak{A} \subseteq \mathfrak{A}^{**}$, and that if $\mathfrak{A}$ and $\mathfrak{B}$ are UDF test spaces having the same outcome set, then $\mathfrak{A} \subseteq \mathfrak{B}$ implies $\mathfrak{B}^* \subseteq \mathfrak{A}^*$. It follows that if $\mathfrak{A}$ is UDF (so that $\mathfrak{A}$ and $\mathfrak{A}^*$ have the same outcome set), $\mathfrak{A}^{***} = \mathfrak{A}^*$. A test space $\mathfrak{A}$ is *reflexive* iff $\mathfrak{A} = \mathfrak{A}^{**}$. Evidently, a test space is reflexive iff it is the Lock dual of a test space.

As long as it is non-empty, the Lock dual $\mathfrak{A}^*$ of a test space $\mathfrak{A}$ may be viewed as a test space in its own right.

LEMMA 168. *The lock dual of any test space is either algebraic, or empty.*

Proof. Let A, B, C and D be events of $\mathfrak{A}^*$ with $A\text{co}C\text{co}B\text{co}D$. Thus, $A \cup C$, $C \cup B$, and $B \cup D$ are transversals of $\mathfrak{A}$, with $A \cap C = C \cap B = B \cap D = \emptyset$. Let $E \in \mathfrak{A}$, and suppose that $A \cap E$ is non-empty. Then $A \cap E = \{x\}$ for some $x \in E$, and $C \cap E = \emptyset$. It follows that $B \cap E$ is non-empty, and, from this, that $D \cap E = \emptyset$. Reversing the argument, if $D \cap E$ is not empty, then it consists of a single point, and $A \cap E$ is empty. Hence $A \cap D = \emptyset$, and $A \cup D$ meets every test of $\mathfrak{A}$ in a single outcome, i.e., is a transversal. Thus $A\text{co}D$. ■

LEMMA 169. *Let $\mathfrak{A}$ and $\mathfrak{B}$ be UDF test spaces. Then* $\mathfrak{A}^* \times \mathfrak{B}^* \subseteq \overrightarrow{\mathfrak{A}\mathfrak{B}}^*, \overleftarrow{\mathfrak{A}\mathfrak{A}}^*$. $\overrightarrow{\mathfrak{A}\mathfrak{B}}, \overleftarrow{\mathfrak{A}\mathfrak{B}} \subseteq (\mathfrak{A}^* \times \mathfrak{B}^*)^*$.

Proof. Let $G = \bigcup_{x\in E} xF_x \in \overrightarrow{\mathfrak{A}\mathfrak{B}}$, where $E \in \mathfrak{A}$ and, for every $x \in E$, $F_x \in \mathfrak{B}$. Let $S \in \mathfrak{A}^*$ and $T \in \mathfrak{B}^*$. Let $S \cap E = \{x_o\}$ and $T \cap F_{x_o} = \{y_o\}$. Then

$$(ST) \cap G = \bigcup_{x\in E} xF_x \cap ST = \{x_o y_o\}.$$

Hence, $ST \in \overrightarrow{\mathfrak{A}\mathfrak{B}}^*$. A symmetrical argument shows that $ST \in \overleftarrow{\mathfrak{A}\mathfrak{B}}^*$. ■

It follows, upon taking duals on both sides, that $\overrightarrow{\mathfrak{A}\mathfrak{B}}, \overleftarrow{\mathfrak{A}\mathfrak{B}} \subseteq (\mathfrak{A}^* \times \mathfrak{B}^*)^*$. Since a non-empty Lock dual is always algebraic, we also have the following

COROLLARY 170. *Let $\mathfrak{A}$ and $\mathfrak{B}$ be algebraic UDF test spaces. Then* $\langle\mathfrak{A}\mathfrak{B}\rangle \subseteq (\mathfrak{A}^* \times \mathfrak{B}^*)^*$.

The following is due to Robin Lock [Lock, R., 1981].

THEOREM 171. *Let $\mathfrak{A}$ and $\mathfrak{B}$ be semiclassical test spaces. Then*

$$\mathfrak{A}\mathfrak{B} = (\mathfrak{A}^* \times \mathfrak{B}^*)^*.$$

Consequently, $\mathfrak{A}\mathfrak{B}$ is algebraic.

Remark: It is conjectured in [Lock, R., 1981] that the conclusion of Theorem 171 holds for all reflexive UDF test spaces. So far as I am aware, this remains unsettled.

Suppose that $\mathfrak{B}_1$ and $\mathfrak{B}_2$ are the partition manuals of two Boolean algebras (for instance, the Borel manuals associated with two measurable spaces). We should like to understand the structure of $\mathfrak{B}_1 \otimes \mathfrak{B}_2$ in terms of some natural product of the Boolean algebras The following is proved in [Lock, R., 1981, Theorem 7.18]:

THEOREM 172. *Let B_1 and B_2 be Boolean algebras, and let $\mathfrak{B}_1$ and $\mathfrak{B}_2$ denote their corresponding manuals of* finite *orthopartitions of unity. Let S_1 and S_2 be the Stone spaces of B_1 and B_2, respectively. Then the logic $\Pi(\mathfrak{B}_1 \otimes \mathfrak{B}_2)$ is isomorphic to the boolean algebra of clopen subsets of $S_1 \times S_2$.*

5.4 Tensor products of property lattices

We now characterize the supports on a bilateral product. This leads us to consider two possible definitions for the tensor product of a pair of entities, in the sense of section 4. The results in this section are due to A. Golfin [Golfin, 1987].

Recall (from Lemma 127) that a set $R \subseteq XY$ is a support of $\overrightarrow{\mathfrak{A}\mathfrak{B}}$ iff $R = \bigcup_{x \in S} x\ S_x$ where $S \in \mathfrak{S}(\mathfrak{A})$ and $S_x \in \mathcal{S}(\mathfrak{B})$ for each $x \in S$. This yields immediately the following

PROPOSITION 173. *$R \subseteq XY$ is a support of $\mathfrak{A}\mathfrak{B}$ iff*

$$R = \bigcup_{x \in S} x\ S_x = \bigcup_{y \in T} T_y\ y$$

where $S \in \mathcal{S}(\mathfrak{A}), T \in \mathcal{S}(\mathfrak{B})$, and for all $x \in S$ and $y \in T$, $S_x \in \mathcal{S}(\mathfrak{A})$ and $T_y \in \mathcal{S}(\mathfrak{B})$.

In particular, note that $\mathcal{S}(\mathcal{A}) \times \mathcal{S}(\mathcal{B}) \subseteq \mathcal{S}(\mathcal{A}\mathcal{B})$.

THEOREM 174. $\mathcal{S}(\mathfrak{A}\mathfrak{B}) = \mathcal{S}(\overrightarrow{\mathfrak{A}\mathfrak{B}}) \cap \mathcal{S}(\overleftarrow{\mathfrak{A}\mathfrak{B}})$.

Proof. One inclusion is obvious. For the other, let $R \in \mathcal{S}(\overrightarrow{\mathfrak{A}\mathfrak{B}}) \cap \mathcal{S}(\overrightarrow{\mathfrak{B}\mathfrak{A}})$. We wish to prove that $R \in \mathcal{S}(\mathfrak{A}\mathfrak{B})$. It will be sufficient to show that if $G \in \overrightarrow{\mathfrak{A}\mathfrak{B}}$ and $H \in \overleftarrow{\mathfrak{A}\mathfrak{B}}$, then $R \cap (G \setminus H) \neq \emptyset$ implies $R \cap (H \setminus G) \neq \emptyset$. Thus, let $G = \bigcup_{x \in E} xE_x$ and $H = \bigcup_{y \in F} F_y y$ (where $E \in \mathfrak{A}$ and $E_x \in \mathfrak{B}$ for all $x \in E$, and where $F \in \mathfrak{B}$ and $F_y \in \mathfrak{A}$ for all $y \in F$). Since R is a support of both the forward and backwards products of $\mathfrak{A}$ and $\mathfrak{B}$, by Lemma 127, there exist supports $S \in \mathcal{S}(\mathfrak{A})$, $T \in \mathcal{S}(\mathfrak{B})$, and, for all $x \in S$, $y \in T$, supports $S_x \in \mathcal{S}(\mathfrak{B})$ and $T_y \in \mathcal{S}(\mathfrak{A})$, such that

$$R = \bigcup_{x \in S} xS_x = \bigcup_{y \in T} T_y y.$$

Now suppose that $xy \in S \cap (G \setminus H)$. We wish to show that $S \cap (H \setminus G)$ is non-empty as well. We proceed by cases. First, suppose that $y \in F$. Then $x \notin F_y$, since otherwise $xy \in H$. Since $y \in T$ and $x \in T_y$, we have $x \in T_y \cap (E \setminus F_y)$. Since T_y is a support of $\mathfrak{A}$, there exists some $x' \in T_y \cap (F_y \setminus E)$. Then $x'y \in R \cap (H \setminus G)$.

Now suppose that $y \notin F$. Since $y \in E_x \cap S_x$, we have $y \in S_x \cap (E_x \setminus F)$. As S_x is a support of $\mathfrak{B}$, there is some $y' \in S_x \cap (F \setminus E_x)$. Note that $xy' \in R$. If $x \in F_{y'}$, then we have $xy' \in H$, but (as $y' \notin E_x$), $xy' \notin G$, whence, $xy' \in R \cap (H \setminus G)$. On the other hand, if $x \notin F_{y'}$, then, since $xy' \in R$, we have $y' \in T$ and $xy' \in T_{y'} \cap (E \setminus F_{y'})$, whence, there exists some $x'' \in T_{y'} \cap (F_{y'} \setminus E)$, whence, $x''y' \in R \cap (H \setminus G)$. ■

There are at least two ways in which one might define a tensor product of two property lattices $\mathcal{L}(X, \Sigma)$ and $\mathcal{L}(Y, \Gamma)$. The smallest reasonable candidate would be $\mathcal{L}(XY, \Sigma \times \Gamma)$, where $\Sigma \times \Gamma = \{ST | S \in \Sigma, T \in \Gamma\}$. The largest would be the following:

DEFINITION 175. Let Σ_1 and Σ_2 be two families of subsets of sets X_1 and X_2, respectively. The *square product* of two property lattices $\mathcal{L}_1 = \mathcal{L}(X_1, \Sigma_1)$ and $\mathcal{L}_2 = \mathcal{L}(X_2, \Sigma_2)$ is the lattice $\mathcal{L}_1 \boxtimes \mathcal{L}_2$ consisting of all sets $V \subseteq XY$ having marginals $\pi_x(V) = V \cap xX_2$ and $\pi^y(V) = X_1 y \cap V$ in $\mathcal{L}_1$ and $\mathcal{L}_2$, respectively for all $x \in X_1, y \in X_2$.

Using Theorem 174, one can characterize $\mathcal{L}_1 \boxtimes \mathcal{L}_2$ directly in terms of $\mathcal{L}_1$ and $\mathcal{L}_2$. Recall that a *Galois connection* between two lattices L and M is a pair (f, g) of order-preserving mappings $f : L \to M$ and $g : M \to L$, such that for all $a \in L, b \in M$, $a \leq g(b)$ iff $f(a) \leq b$. The set of all Galois connections between L and M is a lattice, and a complete lattice, if L and M are complete.

For a proof of the following, see [Golfin, 1987]

THEOREM 176. *$\mathcal{L}_1 \boxtimes \mathcal{L}_2$ is order-isomorphic to the lattice of Galois connections $\mathcal{L}_1 \to \mathcal{L}_2{}^{op}$.*

It is worth remarking that, in [Shmuely, 1974], Z. Shmuely had already introduced the lattice of anti-Galois connections as a categorical tensor product of two complete lattices.

It is not difficult to verify that the pentagon test space of section 1 is coherent and regular, and thus supports a standard bi-regular entity. By simply surveying the possible properties, one verifies that the canonical map is surjective for this entity, and therefore, by Theorem 141, an isomorphism. Thus the considerations of section 1 also show that there is no tensor product (satisfying the condition (1) and the obvious analogue of condition (2) in terms of supports) for standard entities that preserves the equivalent conditions of Theorem 141.

Remark: A complementary series of results by Aerts [Aerts, 1982], and more recently, Ischi [Ischi, 2002; Ischi, 2005], show that there is no reasonable tensor product for complete atomistic lattices that preserves both orthocomplementation and the covering law.

5.5 *The tensor product of frame manuals*

We now briefly sketch how the Foulis-Randall tensor product works for a pair of frame manuals, focussing on the finite-dimensional case. For further details, see [Barnum *et al.*, 2005] and [Kläy *et al.*, 1987], and, for an extension to the infinite-dimensional case, [Wilce, 1990; Wilce, 1992].

We begin by noticing that the signed weight space on a bilateral product has a natural representation as a space of linear mappings. Let $\mathfrak{A}$ and $\mathfrak{B}$ be test spaces. For a given state ω on $\mathfrak{A} \times \mathfrak{B}$, the function $y \mapsto \omega(Ey)$ is independent of $E \in \mathfrak{A}$ if and only if, for every fixed $y \in Y$, the map $\omega_y : x \mapsto \omega(xy)$ is a (non-normalized, but positive) *weight* on $\mathfrak{A}$. If it happens that $x \mapsto \omega(xF)$ is independent of $F \in \mathfrak{B}$ — in which case, let us call ω *influence-free* — then the map $y \mapsto \omega_y$ can be interpreted as a vector-valued weight on $\mathfrak{B}$ with values in the space $V(\mathfrak{A})$ of signed weights on $\mathfrak{A}$, as described in section 2.5.

If the spaces $V(\mathfrak{A})$ and $V(\mathfrak{B})$ of signed weights on $\mathfrak{A}$ and $\mathfrak{B}$ are finite-dimensional, we can immediately dualize the foregoing picture: each positive influence-free weight ω on $\mathfrak{A} \times \mathfrak{B}$ corresponds to a positive linear operator $\widehat{\omega} : V^*(\mathfrak{B}) \to V(\mathfrak{A})$ with the property that $\omega(1) \in \Omega(\mathfrak{A})$ where 1 denotes the constant function with that value on $\Omega(\mathfrak{B})$. Any such map ϕ, conversely, determines a influence-free weight ω via $\phi(f_x) = \omega(x)$ for all $x \in X$. Thus, we have

THEOREM 177. *The mapping $\omega \mapsto \widehat{\omega}$ is affine isomorphism between the cone of positive influence-free weights on $\mathfrak{A} \times \mathfrak{B}$ and the cone of positive linear maps from $V^*(\mathfrak{B})$ to $V(\mathfrak{A})$, sending the state space $\Omega(\mathfrak{A}\mathfrak{B})$ to the set of positive linear maps ω such that $\omega(\mathbf{1}) \in \Omega(\mathfrak{A})$. Thus, these two spaces are isomorphic as base-normed spaces.*

COROLLARY 178. *With $\mathfrak{A}$ and $\mathfrak{B}$ as above, there is a linear isomorphism*

$$V(\mathfrak{A}\mathfrak{B}) \simeq V(\mathfrak{A}) \otimes V(\mathfrak{B}).$$

EXAMPLE 179. Applied to the frame manuals of two finite-dimensional Hilbert spaces $\mathbf{H}$ and $\mathbf{K}$, this gives us, for every influence-free state on $\mathfrak{F}(\mathbf{H}) \times \mathfrak{F}(\mathbf{K})$, a positive linear map $\widehat{\omega} : \mathcal{B}_{sa}(\mathbf{H}) \to \mathcal{B}_{sa}(\mathbf{K})$ (where $\mathcal{B}_{sa}(\mathbf{H})$ is the space of bounded self-adjoint operators on $\mathbf{H}$) satisfying $\mathrm{Tr}\,(\phi(\mathbf{1})) = 1$. This extends, by the cartesian decomposition, to a positive linear map $\mathcal{B}(\mathbf{H}) \to \mathcal{B}(\mathbf{K})$, where $\mathcal{B}(\mathbf{H})$ is the space of all bounded linear operators on $\mathbf{H}$. Conversely, any positive linear map $\phi : \mathcal{B}(\mathbf{H}) \to \mathcal{B}(\mathbf{H})$ determines a state ω on $\mathfrak{F}(\mathbf{H}) \times \mathfrak{F}(\mathbf{K})$ via

$$\omega(xy) := \mathrm{Tr}\,(\phi(P_x)P_y) = \langle \phi(P_x)y, y \rangle.$$

where P_x is the orthogonal projection operator determined by $x \in \mathbf{H}$.

Thus, the set of influence-free states on $\mathfrak{F}(\mathbf{H}) \times \mathfrak{F}(\mathbf{K})$ is affinely isomorphic to the space of positive linear maps on $\mathcal{L}(\mathbf{K})$, normalized as above. Suppose now that $\mathbf{H}$ and $\mathbf{K}$ are *complex* Hilbert spaces with $\dim(\mathbf{H}) = \dim(\mathbf{K}) < \infty$. For simplicity, we may assume $\mathbf{H} = \mathbf{K}$. In this setting, one can represent influence-free states on $\mathfrak{F}(\mathbf{H}) \times \mathfrak{F}(\mathbf{K})$ by operators on $\mathbf{H} \otimes \mathbf{H}$, thanks to the following useful observation:

LEMMA 180. *Let $\mathbf{H}$ be a finite-dimensional complex Hilbert space. For any linear map $\phi : \mathcal{L}(\mathbf{H}) \to \mathcal{L}(\mathbf{H})$, there exists a unique operator $W = W_\phi$ on $\mathbf{H} \otimes \mathbf{H}$ such that, for all $x, y, u, v \in \mathbf{H}$, $\langle \phi(P_x)y, y \rangle = \langle Wx \otimes y, x \otimes y \rangle$. Conversely, every* *operator* W *on* $\mathbf{H} \otimes \mathbf{H}$ *arises in this way from a unique linear map* $\phi : \mathcal{B}(\mathbf{H}) \to \mathcal{B}(\mathbf{H})$.

Proof. For any linear operator on $\mathcal{L}(\mathbf{H})$, the quantity $\langle \phi(P_x)y, y\rangle$ is bi-quadratic in x and y. Polarizing twice, we see that ϕ is uniquely determined by the form $(x, u, y, v) \mapsto \langle \phi(x \odot u)y, v\rangle$, where $x \odot u$ is the skew-projection $y \mapsto \langle y, u\rangle x$. Note that this is linear in x and y, conjugate-linear in u and v. Accordingly, there is a unique *sesquilinear* form Φ on $\mathbf{H} \otimes \mathbf{H}$ satisfying $\Phi(x \otimes y, u \otimes v) := \langle \phi(x \odot u)y, v\rangle$. By the Riesz representation theorem, there is a unique operator W_ϕ on $\mathbf{H} \otimes \mathbf{H}$ such that $\Phi(\tau_1, \tau_2) = \langle W_\phi \tau_1, \tau_2\rangle$ for all tensors $\tau_1, \tau_2 \in \mathbf{H} \otimes \mathbf{H}$. Setting $\tau_1 = x \otimes y$ and $\tau_2 = u \otimes v$ gives the result. ■

In combination with Example 179, this immediately yields the following "unentangled Gleason theorem" [Kläy *et al.*, 1987] (see also [Fuchs, 2002; Wallach, 2000]):

COROLLARY 181 [Kläy, Randall and Foulis, 1987]. *Let* $\mathbf{H}$ *be a finite-dimensional complex Hilbert space. For every influence-free state* ω *on* $\mathfrak{F}(\mathbf{H}) \times \mathfrak{F}(\mathbf{H})$, *there exists an operator* $W = W^*$ *on* $\mathbf{H}$ *with* $\omega(xy) = \langle Wx \otimes y, x \otimes y\rangle$ *for all unit vectors* $x, y \in \mathbf{H}$.

Evidently, the operator W must be *positive on pure tensors* (POPT), in that $\langle Wx \otimes y, x \otimes y\rangle \geq 0$ for all $x, y \in \mathbf{H}$. However, W need not be positive:

EXAMPLE 182. Let S be the unitary operator on $\mathbf{H} \otimes \mathbf{H}$ (uniquely) defined by $S(x \otimes y) = y \otimes x$ for all unit vectors $x, y \in \mathbf{H}$. Then S is POPT, since $\langle Sx \otimes y, x \otimes y\rangle = \langle y \otimes x, x \otimes y\rangle = \langle y, x\rangle\langle x, y\rangle = |\langle x, y\rangle|^2$. But S is certainly not positive. Indeed, if $\tau = x \otimes y - y \otimes x$, then $S\tau = -\tau$, whence $\langle S\tau, \tau\rangle = -\|\tau\|^2$.

The question now arises: when is the POPT operator W_ϕ arising from a positive linear map $\phi : \mathcal{L}(\mathbf{H}) \to \mathcal{L}(\mathbf{H})$ in fact *positive* on $\mathbf{H} \otimes \mathbf{H}$? Recall that a linear map $\phi : \mathcal{L}(\mathbf{H}) \to \mathcal{L}(\mathbf{H})$ is *completely positive* (CP) iff the map $\phi \otimes \mathrm{Id} : \mathcal{L}(\mathbf{H} \otimes \mathbf{K}) \to \mathcal{L}(\mathbf{H} \otimes \mathbf{K})$ remains positive for all Hilbert spaces $\mathbf{K}$. The following well-known result is due independently to Choi [Choi, 1975] and Hellwig and Kraus [Hellwig and Kraus, 1969]:

THEOREM 183. *Let* $W = W_\phi$ *be the operator associated with the linear map* $\phi : \mathcal{L}(\mathbf{H}) \to \mathcal{L}(\mathbf{H})$ *as in Proposition 4.1. Then* W *is positive iff* ϕ *is completely positive.*

Remark: Since non-completely positive "states" will not comport happily with the usual quantum-mechanical coupling via the tensor product, we have here, perhaps, a reason to reject the non-quantum mechanical states on a tensor product of frame manuals as "unphysical". This line of thought is further explored in the paper [Barnum *et al.*, 2005].

6 SYMMETRIC TEST SPACES

Quantum test spaces are marked by a particularly high degree of symmetry. If $\mathfrak{F}(\mathbf{H})$ is the frame manual of a Hilbert space $\mathbf{H}$, then any two tests (that is, frames) E and F have the same cardinality, and any bijection $E \to F$ extends uniquely to a unitary operator on $\mathbf{H}$, which in turn defines a symmetry of $\mathfrak{F}$.

In this section, we study abstract test spaces that are in various degrees homogeneous with respect to a given group action. Most of what follows derives from the papers [Foulis, 2000; Foulis and Wilce, 2000; Wilce, 1997a; Wilce, 2005c; Quan and Wilce, 2008].

6.1 G-Test Spaces

If G is a group, a *G-set* is a set X equipped with an *action* of G on X, that is, a homomorphism $\phi : G \to S(X)$, where $S(X)$ is the group of all bijections on X. We generally suppress reference to ϕ in the notation, writing $\phi(\alpha)(x)$ more briefly as αx, for $\alpha \in G$ and $x \in X$. If X and Y are two G-sets, we say that a mapping $f : X \to Y$ is *G-equivariant* iff $g(\alpha x) = \alpha f(x)$ for every $x \in X$ and every $\alpha \in G$. The *orbit* of an element x of a G-set X under G is the set

$$Gx := \{ \alpha x \mid \alpha \in G \}.$$

The set of distinct orbits partition X. More generally, the orbit of a set $A \subseteq X$ is the collection $\{\alpha A | \alpha \in G\}$. We say that a G-set X is *transitive* iff it has only one orbit, i.e., $\forall x, y \in X$, there exists some $\alpha \in X$ with $\alpha x = y$. The *stabilizer* of $x \in X$ is the subgroup $G_x \leq G$ consisting of all $\alpha \in G$ with $\alpha x = x$. Notice that if $y = \alpha x$, then $G_y = \alpha G_x \alpha^{-1}$.

DEFINITION 184. A *symmetry* of a test space $(X, \mathfrak{A})$ is a bijection $\alpha : X \to X$ such that, for all $E \in \mathfrak{A}$, $\alpha E \in \mathfrak{A}$ and $\alpha^{-1} E \in \mathfrak{A}$. The group of all symmetries of $(X, \mathfrak{A})$ will be denoted by $S(X, \mathfrak{A})$. By an action of a group G on $\mathfrak{A}$, I mean an action of G on X by symmetries of $\mathfrak{A}$, i.e., a homomorphism $\phi : G \to S(X, \mathfrak{A})$.

Clearly, any symmetry of $\mathfrak{A}$ takes events to events, and respects both orthogonality and perspectivity. Hence, if G acts on $\mathfrak{A}$, it also acts (by appropriate automorphisms) on the lattice $\mathcal{S}$ of all $\mathfrak{A}$-supports, and on the logic $\Pi(\mathfrak{A})$. Note, too, that, for any $x, y \in X$ and any symmetry α of $\mathfrak{A}$, $x \perp \alpha y$ iff $\alpha^{-1} x \perp y$. Hence, $\alpha(x^{\perp}) = (\alpha x)^{\perp}$; more generally, $\alpha(A^{\perp}) = (\alpha A)^{\perp}$ for all events (indeed, for all *subsets* of X). Thus, α takes $\perp$-closed subsets of X to $\perp$-closed subsets, and thus defines an automorphism of the ortholattice $\mathcal{C}(X, \perp)$.

DEFINITION 185. Let G be a group. A *G-test space* a test space $\mathfrak{A}$ equipped with a fixed G-action. $\mathfrak{A}$ is *symmetric* iff

(i) G acts transitively on $\mathfrak{A}$, and

(ii) the stabilizer, G_E, of any test $E \in \mathfrak{A}$ acts transitively on E.

We shall say that $\mathfrak{A}$ is *fully symmetric* iff all tests have the same size, and any bijection between two tests is effected by some element of G. If this element is always unique, then we shall say that $\mathfrak{A}$ is *strongly* symmetric.

EXAMPLES 186. (i) As noted above, the test space of frames of a Hilbert space $\mathbf{H}$ is strongly symmetric with respect to $\mathbf{H}$'s unitary group $U(\mathbf{H})$. On the other hand, the projective test space of $\mathbf{H}$ is fully, but not strongly symmetric with respect to $U(\mathbf{H})$, since a bijection between two maximal orthogonal sets of one-dimensional projections determines a unitary operator only up to a choice of phase factors.

(ii) The Fano plane test space described earlier (52, Remark) is strongly symmetric with respect to its full symmetry group, i.e., the collineation group of the Fano plane. More generally, any finite projective plane furnishes an example of a strongly symmetric test space. 'For another source of finite examples, let X be the set of edges of a platonic solid; for each vertex v, let E_v be the set of edges meeting at that vertex, and let $\mathfrak{A}$ be the collection of all sets of the form E_v as v ranges over the vertices of the solid. The test space $(X, \mathfrak{A})$ is strongly symmetric with respect to the group of symmetries of the solid.

Constructing Symmetric Test Spaces

All symmetric test spaces can be recovered from group-theoretic data, as instances of the following construction.

Let E_o be a set, regarded as the outcome-set for some "standard" experiment on a system of interest. Let H be a group acting transitively on E_o, reflecting some physical symmetries of this system under which E_o is invariant. Suppose we believe that the system is invariant under a larger symmetry group G, of which H is a subgroup. We should like to enlarge our repertoire of tests by considering, roughly speaking, the orbit of E_o under this enlarged group.

To accomplish this, fix an outcome $x_o \in E_o$, and let H_{x_o} denote the stabilizer of x_o in H. Let K be any subgroup of G such that $K \cap H = H_{x_o}$. (Think of this as consisting of symmetries in G under which we expect the outcome x_o to be invariant.) Let $X = G/K$, the space of left K-cosets in G. There is a natural H-equivariant injection $i : E \to X$ given by $\sigma x \mapsto \sigma K$, where $\sigma \in H$. Thus, identifying E_o with its image under this injection, we may regard E_o as a subset of X, invariant under H. Now let $\mathfrak{A}(E_o)$ denote the orbit of E_o under the action of G — that is, let

$$\mathfrak{A} = \{\alpha E_o | \alpha \in G\}.$$

Clearly, G acts transitively on $\mathfrak{A}$, and the stabilizer G_{E_o} of E_o, as it contains H, acts transitively on E_o. It follows that the stabilizer of any test $E \in \mathfrak{A}$ acts transitively on E. Thus, $\mathfrak{A}$ is a G-symmetric test space. If we take H to act as the full symmetric group S_E of all bijections on E, the resulting symmetric test space $\mathfrak{A}$ will be fully symmetric. It will be *strongly* symmetric iff, in addition, the only element of G fixing every outcome in E is the identity element.

Conversely, given a G-symmetric test space $\mathfrak{A}$, choose any test $E_o \in \mathfrak{A}$ and any outcome $x_o \in E_o$; setting $H = G_{E_o}$ and $K = G_{x_o}$ (the stabilizers, respectively, of E_o and x_o in G), the preceding construction reproduces $\mathfrak{A}$.

DEFINITION 187. The G-test space $\mathfrak{A}$ constructed above from the classical H-set E_o and the subgroup K, will be called the *G-expansion* of E_o based on K, and will be denoted by $\mathfrak{A}_{G,K}(E_o)$.

Remarks:

(a) Notice that, in the construction $\mathfrak{A}_{G,K}(E_o)$, we can begin with purely group theoretic data. Indeed, if G is a group and H, K are subgroups of G, set $X = G/K$ and let $E = \{\eta K | \eta \in K\} \subseteq X$. Let $\mathfrak{A} = \{\alpha E | \alpha \in G\}$. Then $\mathfrak{A}$ is a G-symmetric test space. Every G-symmetric test space has this form.

(b) The possible G-symmetric test spaces extending the classical H-symmetric test space E_o in 187 are parametrized by the subgroups $K \leq G$ with $K \cap H = H_{x_o}$. If K_1 and K_2 are two such subgroups with $K_1 \leq K_2$, the natural surjection $\phi : G/K_1 \to G/K_2$ induces a surjective outcome-preserving interpretation $\mathfrak{A}_{G,K_1}(E_o) \to \mathfrak{A}_{G,K_2}(E_o)$. The smallest possible choice for K is H_{x_o} itself. In this case, the orbit of E_o in $X = G/K$ *partitions* X. In other words, the test space $\mathfrak{A}_{G,H_{x_o}}(E_o)$ is semi-classical — a horizontal sum of copies of E_o. Choosing a larger subgroup K will in general produce a non-semiclassical test space.

We now characterize the orthogonality relation on a G-symmetric test space. If $\mathfrak{A}$ is G-symmetric and $x_o \in E \in \mathfrak{A}$ are given, let $x_\alpha = \alpha x_o$ and $E_\alpha = \alpha E_o$ for all $\alpha \in G$.

LEMMA 188. *Let $\mathfrak{A}$ be G-symmetric, let $x_o \in E \in \mathfrak{A}$ be given, and let $K = G_{x_o}$ and $H = G_E$, as above. Then, for all $\alpha, \beta \in G$, $x_\alpha \perp x_\beta$ iff $\beta^{-1}\alpha \in K(H \setminus K)K$.*

Proof. Since $x_\alpha \perp x_\beta$ iff $x_{\beta^{-1}\alpha} \perp x_o$, we need only show that $x_\alpha \in x_o^\perp$ iff $\alpha \in K(H \setminus K)K$. Suppose first that $\alpha = \beta\sigma\gamma$ where $\beta, \gamma \in K$ and $\sigma \in H \setminus K$. Then $x_o \perp \sigma x_o$, so $x_o = \beta x_o \perp \beta\sigma x_o = \beta\sigma\gamma x_o = \alpha x_o$.

Conversely, suppose $x_\alpha \perp x_o$. Then $x_\alpha \neq x_o$, and there exists some $E = E_\beta \in \mathfrak{A}$ with $x_o, x_\alpha \in E_\beta$. It follows that there exist $\sigma, \sigma' \in H$ with (i) $x_\alpha = \beta\sigma x_o$ and (ii) $x_o = \beta\sigma' x_o$. From (ii), we have $\beta\sigma' \in K$, whence, $\beta \in K\sigma'^{-1}$. Now (i) requires that $x_\alpha = \beta\sigma x_o \neq x_o$, so $\sigma'^{-1}\sigma \in H \setminus K$. We also have from (i) that $(\beta\sigma)^{-1}\alpha \in K$, whence, $\alpha \in \beta\sigma K \subseteq K\sigma'^{-1}\sigma K \subseteq K(H \setminus K)K$. ■

6.2 *Fully Symmetric Test Spaces*

If $\mathfrak{A}$ is *fully* G-symmetric, then G acts transitively on each of the sets $\mathcal{E}_k$ of k-element events. To see this, suppose $A, B \in \mathcal{E}_k$: choose tests $E \supseteq A$ and $F \supseteq B$ and a bijection $f : A \to B$. Since $|E| = |F|$, we can extend f to a bijection $\overline{f} : E \to F$; by assumption, this is induced by a group element $\alpha \in G$. But then $\alpha A = B$.

If X is a G-set and $A \subseteq X$, we denote by F_A the *fixing subroup* of A, that is, the subgroup consisting of all elements α of G such that $\alpha x = x$ for every $x \in A$. (If $A = \{x\}$, this is just the stabilizer of x.) The following observation is very simple, but also very useful:

LEMMA 189 Pivoting Lemma. *Let $\mathfrak{A}$ be fully G-symmetric. If $A \subseteq E \cap F$, where $E, F \in \mathfrak{A}$, then there exists some $\gamma \in F_A$ with $F = \gamma E$.*

Proof. Since $\mathfrak{A}$ is G-symmetric, $|E| = |F|$. Choose a bijection $f : E \to F$ that fixes each $x \in A$, and extend this to an element of G. ■

.

The following theorem gives a sharp characterization of algebraicity for fully symmetric test spaces, in terms of the fixing subgroups of complementary events.

THEOREM 190. *Let $\mathfrak{A}$ be a fully-symmetric G-test space. Choose and fix $E \in \mathfrak{A}$. If $A \subseteq E$, write A' for $E \setminus A$, and let F_A be the subgroup of G fixing each $x \in A$. Then $\mathfrak{A}$ is algebraic iff, for every $A \subseteq E$, $F_A F_{A'} = F_{A'} F_A$.*

Proof. ($\Rightarrow$) Suppose $\mathfrak{A}$ is algebraic. Let $A \subseteq E$ and $A' = E \setminus A$. It is sufficient to show that $F_A F_{A'} \subseteq F_{A'} F_A$. If $\alpha \in F_A$ and $\alpha' \in F_{A'}$, we obtain a "hook" of events $\alpha A' \mathsf{co} A \mathsf{co} A' \mathsf{co} \alpha' A$:

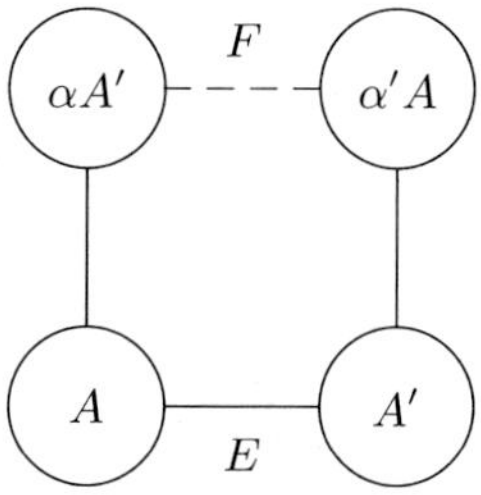

Since $\mathfrak{A}$ is algebraic, $\alpha A' \mathsf{co} \alpha' A$. Let $\alpha A' \cup \alpha' A =: F \in \mathfrak{A}$. Since $|\alpha A'| = |A'|$ and $|\alpha' A| = |A|$, and since every bijection $E \to F$ extends to an element of G, we can find $\beta \in G$ with $\beta x = \alpha x$ for every $x \in A'$ and $\beta x = \alpha' x$ for every $x \in A$. Then $\alpha^{-1}\beta \in F_{A'}$ and $\beta^{-1}\alpha' \in F_A$, whence, $\alpha^{-1}\alpha' = (\alpha^{-1}\beta)(\beta^{-1}\alpha') \in F_{A'} F_A$. Since $\alpha^{-1}\alpha'$ is an arbitrary element of $F_{A'} F_A$, we have $F_A F_{A'} \subseteq F_{A'} F_A$.

($\Leftarrow$) Now suppose that $F_A F_{A'} = F_{A'} F_A$ for every $A \subseteq E$. To show $\mathfrak{A}$ is algebraic, it is sufficient to consider configurations of the form $A \mathsf{co} A' \mathsf{co} B \mathsf{co} C$, with $A \subseteq E$

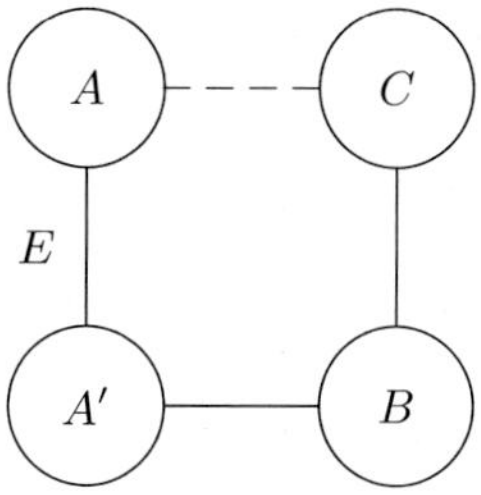

(any other hook in $\mathcal{E}$ being a translate of one of these). We wish to show that $A \mathsf{co} C$. Now, by Lemma 189, $B = \alpha' A$ for some $\alpha' \in F_{A'}$, and $C = \beta A'$ for some $\beta \in F_B$. But $F_B = F_{\alpha' A} = \alpha' F_A \alpha'^{-1} \subseteq F_{A'} F_A F_{A'}$. Since $F_{A'} F_A = F_A F_{A'}$, we have $F_{A'} F_A F_{A'} \subseteq F_A F_{A'}$. Thus, $\beta \in F_B \Rightarrow \beta = \alpha\alpha''$ where $\alpha \in F_A$ and $\alpha'' \in F_{A'}$. But then $C = \beta A' = \alpha A'$ — whence, indeed, $A \mathsf{co} C$. ■

EXAMPLE 191. As an illustration of the preceding result, let $G = U(\mathbf{H})$, the unitary group of a Hilbert space $\mathbf{H}$, and let E be an orthonormal basis for $\mathbf{H}$. If $A \subseteq E$, let $\langle A \rangle$ be the subspace spanned by A. Then F_A is the group of unitaries of the form $W = \mathbf{1}_{\langle A \rangle} \oplus U$, where $\mathbf{1}_{\langle A \rangle}$ is the identity operator on $\langle A \rangle$ and U is any unitary operator on $\langle A \rangle^{\perp}$. Likewise, $F_{A'}$ consists of unitaries of the form $W' = V \oplus \mathbf{1}_{\langle A \rangle^{\perp}}$, V a unitary on $\langle A \rangle$. Since $WW' = W'W$ for any two such W and W', we have $F_A F_{A'} = F_{A'} F_A$.

Recall that a test space $\mathfrak{A}$ is *chain connected* iff every pair of tests is linked by a finite sequence of consecutively overlapping tests. By Zorn's lemma, every test in $\mathfrak{A}$ is contained in a maximal chain-connected subset $\mathfrak{C}$ of $\mathfrak{A}$, called the *chain component* of $\mathfrak{A}$. Note that if $\mathfrak{C}_1$ and $\mathfrak{C}_2$ are distinct chain components, then $X_1 = \bigcup \mathfrak{C}_1$ and $X_2 = \bigcup \mathfrak{C}_2$ are disjoint. Thus, every test space is the horizontal sum of its chain-connected components.

PROPOSITION 192. *Let $\mathfrak{A}$ be a fully G-symmetric test space. Let $x_o \in E_o \in \mathfrak{A}$, let $K = G_{x_o}$, $H = G_{E_o}$ be the stabilizers of x_o and E_o, respectively, and let $\langle H, K \rangle$ be the subgroup of G generated by H and K. Then $\mathfrak{A}$ is chain-connected iff it is fully $\langle H, K \rangle$- symmetric.*

Proof. Let $\mathfrak{A}_o$ be the chain component of E_o. Then obviously $\mathfrak{A}_o$ is both H-invariant and K-invariant, whence, $\langle H, K \rangle$-invariant. Hence, if $\mathfrak{A}$ is $\langle H, K \rangle$ transitive (much less fully symmetric), $\mathfrak{A}_o = \mathfrak{A}$, whence, $\mathfrak{A}$ is connected. For the converse, let $\mathfrak{B}$ denote the orbit of E_o under $\langle H, K \rangle$. Suppose that $E = \alpha E_o \in \mathfrak{B}$, where $\alpha \in \langle H, K \rangle$, and let $F \in \mathfrak{A}$ with $E \cap F \neq \emptyset$. Let $x \in E \cap F$. Then we have $x = \alpha \sigma x_o$ for some $\sigma \in H$ (since, by assumption, any permutation of E_o can be effected by an element of H). Let $\beta := \alpha \sigma \in \langle H, K \rangle$. Then, by the Pivoting Lemma, $F = \gamma E$ for some $\gamma \in G_x = \beta K \beta^{-1}$. But $\beta K \beta^{-1} \subseteq \langle H, K \rangle$. Thus, any test intersecting a test in $\mathfrak{B}$, is again in $\mathfrak{B}$. In particular, $\mathfrak{B}$ is chain-connected. If $\mathfrak{A}$ is chain-connected, therefore, $\mathfrak{B} = \mathfrak{A}$. This shows that $\mathfrak{A}$ is $\langle H, K \rangle$-transitive. That it is *fully* $\langle H, K \rangle$-transitive follows from the fact that H acts on E_o as the latter's full permutation group (since $\mathfrak{A}$ is, by assumption, fully G-transitive). ■

Ortho-symmetric test spaces

We now isolate a particularly strong symmetry condition, enjoyed by the frame manual of a Hilbert space, that, among other things, forces a test space to be both algebraic and regular.

DEFINITION 193. A G test space $\mathfrak{A}$ is *ortho-symmetric* iff for all events A and B of $\mathfrak{A}$

(i) $A \sim B \Rightarrow |A| = |B|$, and

(ii) for every bijection $f : A \to B$, there exists some $\alpha \in G$ with $\alpha x = f(x)$ for all $x \in A$ and $\alpha x = x$ for all $x \in A^{\perp}$.

If $\mathbf{H}$ is a Hilbert space, then two perspective events A and B are simply orthonormal bases for some closed subspace $\mathbf{M}$ of $\mathbf{H}$. Any bijection $f : A \to B$

extends uniquely to a unitary operator $U : \mathbf{M} \to \mathbf{M}$; this in turn extends uniquely to a unitary operator on $\mathbf{H}$ that fixes $\mathbf{M}^{\perp}$ pointwise, namely, $U \otimes \mathbf{1}_{\mathbf{M}^{\perp}}$. Thus, $\mathfrak{F}(\mathbf{H})$ is ortho-symmetric with respect to the unitary group on $\mathbf{H}$.

Orthosymmetric test spaces satisfy many of the regularity requirements considered thus far:

LEMMA 194. *Let $\mathfrak{A}$ be an ortho-symmetric G-test space. Then*

(a) $\mathfrak{A}$ is symmetric.

(b) $\mathfrak{A}$ is algebraic.

(c) $\mathfrak{A}$ is regular.

(d) $\forall p \in \Pi(\mathfrak{A})$, $\mathfrak{A}_p$ is ortho-symmetric under its stabilizer $G_p \leq G$.

Proof. That $\mathfrak{A}$ is symmetric is clear. To see that $\mathfrak{A}$ is algebraic, let $A \sim B$ and $B\mathsf{co}C$. Then there exists some $\alpha \in G$ with $\alpha B = A$ and $\alpha x = x$ for all $x \in A^{\perp}$ — including all $x \in C$. Hence, $C = \alpha C \mathsf{co} \alpha B = A$. To see that $\mathfrak{A}$ is regular, suppose that $A \sim B$; again select α such that $B = \alpha A$ and $gx = x$ for all $x \in A^{\perp}$. Then $\alpha(A^{\perp}) = (\alpha A)^{\perp} = B^{\perp}$. To prove (d), suppose A and B are perspective events of $\mathfrak{A}_p$. By Lemma 142, it follows that A and B are also perspective as events of $\mathfrak{A}$; hence, there is a unique $\alpha \in G$ with $B = \alpha A$ and $\alpha C = C$ for any $C \perp A$. It follows that $\alpha \in G_{p'}$, whence, $\alpha \in G_p$. ■

The foregoing result notwithstanding, orthosymmetric test spaces can be fairly pathological. For instance, the test space

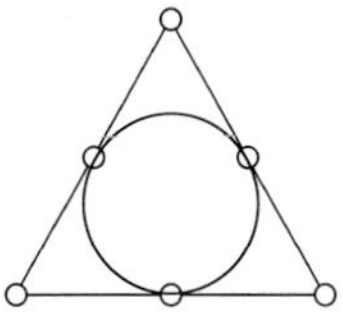

is ortho-symmetric with respect to its automorphism group, but is not ortho-coherent.

6.3 Systems of Imprimitivity

In this section, we discuss how a test space can be enlarged so as to accommodate a larger symmetry group, using the classical theory of induced group actions. This is related to, but not identical with, the construction of induced unitary representations in quantum mechanics. The definitions in this section are in a sense dual to those discussed by Randall and Foulis in [Randall and Foulis, 1978]; for a comparison of the two, see [Wilce, 1997a]. A result more general than Theorem

198 below can be found in [Foulis, 2000]. (See also [Gudder, 1971] for a discussion of induced representations in the context of orthomodular lattices.)

LEMMA 195. *Let $\mathfrak{A}$ be a transitive G-test space.*

(a) If $\mathfrak{A}$ is algebraic and $p \in \Pi$ is G-invariant, then $X_p \in \mathcal{Z}(\mathfrak{A})$.

(b) If $\mathfrak{A}$ is locally finite, then every G-invariant subset of X is a support. Consequently, every G-invariant support is complemented.

Proof. (a) If $p = p(A)$ with $A \subseteq E \in \mathfrak{A}$ then for any $F = \alpha E \in \mathfrak{A}$,

$$F \cap X_p = \alpha E \cap X_p = \alpha(E \cap X_p) = \alpha(A) \in \mathfrak{A}_p.$$

Hence, $\mathfrak{A}_{X_p} \subseteq \mathfrak{A}_p$. Hence, X_p is a support (since $\mathfrak{A}_p$ is irredundant) and central.

(b) Notice first that a non-empty G-invariant set S must intersect every operation of $\mathfrak{A}$. Now, suppose S is a G-invariant subset of X. Given tests $E, F \in \mathfrak{A}$ with $S \cap E \subseteq S \cap F$, let $\alpha \in G$ be such that $\alpha F = E$. Then

$$\alpha(S \cap F) \subseteq \alpha S \cap E \subseteq S \cap E.$$

Thus, α induces an injection $S \cap F \to S \cap E$. Thus, the two sets have the same cardinality. Being finite, and one containing the other, they must then be equal. The second statement is immediate, since the complement of an invariant set is also invariant. ■

The following is straightforward:

THEOREM 196. *Let $\mathfrak{A}$ be transitive and algebraic, and let $\mathbf{1} = p_1 \oplus \cdots \oplus p_n$ be a partition of $\mathbf{1}$ in $\Pi(\mathfrak{A})$ by invariant propositions. Then $\mathfrak{A} = \mathfrak{A}_{p_1} \oplus \cdots \oplus \mathfrak{A}_{p_n}$, and G acts transitively on each summand $\mathfrak{A}_{p_i}$.*

Let G and H be groups, with $H \leq G$, and let X be an H-set. A *G- extension* of X is an H-equivariant injection $\eta : X \to Y$, Y a G-space. A universal, or *free* G-extension can be constructed as follows [Foulis and Wilce, 2000]. Define an equivalence relation on $G \times_H X$ by setting $(\alpha, x) \equiv (\beta, y)$ iff $\beta^{-1}\alpha = \eta \in H$ and $\eta x = y$. The quotient set $G \times_H X := (G \times H)/\equiv$ is then a G-set under the action $\alpha[\beta, x] = [\alpha\beta, x]$. Denoting the equivalence class of (α, x) by $[\alpha, x]$, we have $[\alpha\eta, x] = [\alpha, \eta x]$ $\forall \eta \in H$. In particular, the map $x \mapsto [e, x]$ is H-equivariant. Identifying x with $[e, x]$, we may regard X as a subset of $G \times_H X$ invariant under the action of H. It is easy to show that if $p : X \to Z$ is any H-equivariant map into a G-space Z, then p can be extended uniquely to a G-equivariant map $f : G \times_H X \to Z$ via $f([\alpha, x]) = \alpha p(x)$. (This is an abstract version of the so-called *Frobenius reciprocity*; cf [Foulis and Wilce, 2000]).

The orbit of X in $G \times_H X$ in fact partitions the latter. Conversely, given a G-space Y and an invariant partition E of Y on which G acts transitively, there is a unique H-equivariant bijection $Y \simeq G \times_H X$ where $X \in E$ and $H = G_X$, the stabilizer of X in G (namely, $[\alpha, x] \mapsto \alpha x$). Such an invariant, transitive partition of Y is called a *system of imprimitivity* for Y. Any subset of Y, the orbit of which is a partition of Y (and hence, a system of imprimitivity), is called a *set of imprimitivity* for Y.

DEFINITION 197. Let $\mathfrak{A}$ be an H-test space. If $H \leq G$, the *free G-extension* of $\mathfrak{A}$ is

$$G \times_H \mathfrak{A} := \left\{ \bigcup_{\alpha \in G} [\alpha, E_\alpha] \mid E_\alpha \in \mathfrak{A} \;\&\; \forall \eta \in H, E_{\alpha\eta} = \eta^{-1} E_\alpha]. \right\}$$

Any mapping $\alpha \mapsto E_\alpha$ with $E_{\alpha\eta} = \eta^{-1} E_\alpha$ yields an element of $G \times_H \mathfrak{A}$. In particular, we may select E_e arbitrarily and define $E_\alpha = \tau(\alpha) E_e$ where τ is a *twist* [Foulis and Wilce, 2000], i.e., a mapping $\tau : G \to H$ with $\tau(1) = 1$ and $\tau(\alpha\eta) = \eta^{-1}\tau(\alpha)$ for all g. Twists are abundant: any normalized cross-section $\sigma : G/H \to G$ of the canonical surjection $G \to G/H$ defines a twist via the formula $\tau(\alpha) = \alpha^{-1}\sigma(\alpha H)$. Thus, $G \times_H \mathfrak{A}$ contains sufficiently many sets to cover $G \times_H X$. Moreover, X (which we identify with the set $[e, X] = \{[e, x] | x \in X\}$) is a support — indeed, a central support — of $G \times_H \mathfrak{A}$ and $(G \times_H \mathfrak{A})|_X$ is isomorphic to $\mathfrak{A}$.

Let π be the H-equivariant interpretation $G \times_H \mathfrak{A}$ to $\mathfrak{A}$ sending $[\alpha, x]$ to $\{\alpha x\}$ if $\alpha \in H$, and to the empty set otherwise.

THEOREM 198. *Let G and H be as above. If H acts on $\mathfrak{A}$, G acts on $\mathcal{B}$, and $\phi : \mathcal{B} \to \mathfrak{A}$ is an H-equivariant interpretation, then there exists a unique G-equivariant interpretation $\widehat{\phi} : \mathcal{B} \to G \times_H \mathfrak{A}$ with $\pi \circ \widehat{\phi} = \phi$*

Proof. Let $Z = \cup\mathcal{B}$. The only possible G- equivariant map $\widehat{\phi} : Z \to 2^{G \times_H X}$ such that $\phi(z) \cap X = \phi(z)$ for all $z \in Z$ is given by

$$\widehat{\phi}(z) := \bigcup_{\alpha \in G} [\alpha, \phi(\alpha^{-1} z)].$$

For, if $[\alpha, x] \in \widehat{\phi}(z)$, then $x \in \alpha^{-1}\widehat{\phi}(z) = \widehat{\phi}(\alpha^{-1}z)$, whence $x \in \widehat{\phi}(\alpha^{-1}z) \cap X = \phi(\alpha^{-1}z)$, whence $\alpha x \in \alpha\phi(\alpha^{-1}z)$. Conversely,

$$[\alpha, \phi(\alpha^{-1}z)] = \alpha(\widehat{\phi}(\alpha^{-1}z) \cap X) = \widehat{\phi}(z) \cap \alpha X \subseteq \widehat{\phi}(z).$$

We now show that the map $\widehat{\phi}$ defined above is indeed an interpretation. It is enough to verify, first, that $\widehat{\phi}(E)$ belongs to $G \times_H \mathfrak{A}$ for every operation E of $\mathfrak{A}$, and second, that $z \perp z'$ entails $\widehat{\phi}(z) \cap \widehat{\phi}(z') = \emptyset$ for all z, z' in Z. To this end note that

$$\widehat{\phi}(E) = \bigcup_{z \in E} \bigcup_{\alpha \in G} [\alpha, \phi(\alpha^{-1} z)] = \bigcup_{\alpha \in G} [\alpha, \phi(\alpha^{-1} E)]$$

Let $E_\alpha = \phi(\alpha^{-1}E)$: Then $E_{\alpha\eta} = \phi(\eta^{-1}g^{-1}E) = \eta^{-1}E_\alpha$. Thus, $\widehat{\phi}(E) = \bigcup_{\alpha \in G} \alpha E_\alpha$ indeed belongs to $G \times_H \mathfrak{A}$. Now, notice that since ϕ is a morphism, $z \perp z'$ implies that $\phi(\alpha^{-1}z) \perp \phi(\alpha^{-1}z')$, whence, $\widehat{\phi}(z) \cap \widehat{\phi}(z') \cap X = \emptyset$. Since $\{ [\alpha, X] \mid \alpha \in G \}$ partitions $G \times_H X$ and since $\widehat{\phi}$ is G- equivariant, $\widehat{\phi}(z) \cap \widehat{\phi}(z') = \emptyset$. ■

DEFINITION 199. A *set of imprimitivity* for a G-test space $\mathfrak{A}$ is a central support $S \in \mathcal{Z}(\mathfrak{A})$ such that the orbit of S under G partitions $X = \bigcup \mathfrak{A}$.

Notice that such a set S is also a set of imprimitivity for the G-set X. Hence, $X = G \times_H S$ where H is the stabilizer of S in $\mathfrak{A}$. The following may be regarded as an imprimitivity theorem for finite G-test spaces.

COROLLARY 200. *Let S be a set of imprimitivity for a G-test space $\mathfrak{A}$, where G is a finite group. Then $\mathfrak{A} \simeq G \times_H \mathfrak{A}|_S$.*

Proof. Let $\phi : S \to \mathcal{E}(\mathfrak{B})$ be any H-equivariant interpretation from $\mathfrak{A}|_S$ to a G-test space $\mathfrak{B}$. Since $X = G \times_H S$ and $\mathcal{E}(\mathfrak{B})$ is a G- set, there is a unique extension of ϕ to a G-equivariant mapping $\overline{\phi} : X \to \mathcal{E}(\mathfrak{B})$, namely, for $\alpha x \in \alpha S$, $\overline{\phi}(\alpha x) = \alpha\phi(x)$. (Since ϕ is H-equivariant, this is well-defined: If $\alpha x = \alpha' y$ for $x, y \in S$, then $\alpha^{-1}\alpha' x = y$, so $\alpha^{-1}\alpha' S \cap S \neq \emptyset$, whence $\alpha^{-1}\alpha' S = S$, i.e., $\alpha^{-1}\alpha' \in H$. Thus, $\alpha^{-1}\alpha\phi(x) = \phi(\alpha^{-1}\alpha x) = \phi(y)$, since ϕ is H-equivariant.) Now, since S is central, $\mathfrak{A} \simeq \bigoplus_{\alpha \in H} \mathfrak{A}|_{\alpha S}$, and $\overline{\phi} = \bigoplus_{\alpha \in G} \alpha \circ \phi \circ \alpha^{-1}$, which is clearly an interpretation.Thus, $\mathfrak{A}$ enjoys the same universal property attributed by Theorem 198 to $G \times_H \mathfrak{A}|_S$; hence, the two are isomorphic via a unique G-equivariant interpretation. ■

Remarks: (a) There is a construction, due to Foulis and Randall [Randall and Foulis, 1978], that is in a sense dual to that of $G \times_H \mathfrak{A}$. Let $G \times_H X$ be the free G-expansion of the set $X = \bigcup \mathfrak{A}$, as described above. Identifying $x \in X$ with $[e, x] \in G \times_H X$, we view X as an H-invariant subset of $G \times_H X$. In particular, every test $E \in \mathfrak{A}$ may be regarded as a subset of $G \times_H X$. Define

$$G \times^H \mathfrak{A} \;=\; \{\, gE \subseteq G \times_H X \mid E \in \mathfrak{A} \,\}.$$

It is not hard to see that, as a collection of sets, $G \times^H \mathfrak{A}$ is a horizontal sum of copies of $\mathfrak{A}$, parametrized by G/H. It is interesting to note that, in the special case in which $\mathfrak{A}$ is a classical test space, say $\mathfrak{A} = \{E_o\}$, the test space $G \times^H \mathfrak{A}$ is just the (semi-classical) test space $\mathfrak{A}_{G,K}(E_o)$ of Definition 187, where $K = H_{x_o}$.

(b) It should be understood that, when applied to the the frame manual of a Hilbert space carrying a unitary representation of a group H, the foregoing free extension construction does not yield the frame manual of the Hilbert space of the induced unitary representation. (Indeed, our "imprimitivity theorem" yields something more akin to a covariant superselection rule than to the covariant observable associated with a system of imprimitivity in the sense of Mackey [Mackey, 1963].)

7 TOPOLOGICAL CONSIDERATIONS

In this final section, we discuss how the theory of test spaces can be topologized. There are at least two compelling reasons to undertake this exercise. First, the basic quantum test space, the frame manual $\mathfrak{F}(\mathbf{H})$ of a Hilbert space $\mathbf{H}$, *has* a significant topological structure, as does its logic, the projection lattice $L(\mathbf{H})$. Secondly, continuity assumptions are both very natural and very powerful in the context of axomatic physics — and, indeed, have played a role in both earlier [Zierler, 1961] and more recent [Holland, 1995; Hardy, 2001] efforts to axiomatize quantum theory. This suggests that it may be fruitful to study test spaces endowed with *a priori* topological structure. Here, I shall summarize what is presently known about topological test spaces, topological orthoalgebras, and the relationship between the two. The material for this section derives from [Wilce, 2005a; Wilce, 2005b; Wilce, 2005c], and represents quite recent work — with many loose ends.

7.1 Topological Test Spaces

We begin by asking, what ought we to mean by a topological test space? Among several possible definitions that suggest themselves, the following one, while not the most general that might be considered, seems in practice not too constraining — and is, as we'll presently see, fruitful.

DEFINITION 201. A *topological test space* is a test space $(X, \mathfrak{A})$, where X is a Hausdorff space and the relation $\perp$ is closed in the product topology on $X \times X$.

EXAMPLES 202. (i) Let $\mathbf{H}$ be a Hilbert space. Let S be the unit sphere of $\mathbf{H}$, in any topology making the inner product continuous. Then the test space $(S, \mathfrak{F})$ defined above is a topological test space, since the orthogonality relation is closed in S^2.

(ii) More generally, suppose that X is Hausdorff and that $(X, \mathfrak{A})$ is locally finite and supports a set Γ of *continuous* probability weights that are $\perp$-separating in the sense that $p \not\perp q$ iff $\exists \omega \in \Gamma$ with $\omega(p) + \omega(q) > 1$. Then $\perp$ is closed in X^2, so again $(X, \mathfrak{A})$ is a topological test space.

(iii) A *topological OML*, or TOML, in the sense of [Cho and Greechie, 1993], is an orthomodular lattice equipped with a Hausdorff topology making the lattice operations and the orthocomplementation continuous. If L is any TOML, the mapping $\phi : L^2 \to L^2$ given by $\phi(p, q) = (p, p \wedge q')$ is continuous, and $\perp = \phi^{-1}(\Delta)$ where Δ is the diagonal of L^2. Since L is Hausdorff, Δ is closed, whence, so is $\perp$. Hence, the test space $(L \setminus \{0\}, \mathfrak{A}_L)$ of orthopartitions of unity in L is topological.

The following two lemmas collect some elementary, but important, observations about topological test spaces.

DEFINITION 203. Let $(X, \mathfrak{A})$ be a topological test space. A set $A \subseteq X$ is *totally non-orthogonal* iff it contains no two orthogonal elements.

LEMMA 204. *Let $\mathfrak{A}$ be a topological test space with outcome set X. Then*

(a) Each outcome $x \in X$ has a totally non-orthogonal open neighborhood.

(b) If X is compact, then all pairwise orthogonal sets are finite, and of uniformly bounded size. In particular, $\mathfrak{A}$ is of finite rank.

Proof. (a) Since $(x,x) \notin \perp$, and $\perp$ is closed in X^2, there exist open sets V and W about x with $V \times W \cap \perp = \emptyset$. Taking $U = V \cap W$, we have that $U \times U \cap \perp = \emptyset$, which is to say, U is totally non-orthogonal.

(b) By part (a), every point $x \in X$ is contained in some totally non- orthogonal open set. Since X is compact, a finite number of these, say $U_1, ..., U_n$, cover X. A pairwise orthogonal set $D \subseteq X$ can meet each U_i at most once; hence, $|D| \leq n$. ■

LEMMA 205. *Let $(X, \mathfrak{A})$ be a topological test space. Then*

(a) For every set $A \subseteq X$, $A^\perp$ is closed.

(b) Each pairwise orthogonal subset of X is discrete

(c) Each pairwise orthogonal subset of X is closed.

Proof. (a) Let $y \in X \setminus x^\perp$. Then $(x,y) \notin \perp$. Since the latter is closed, there exist open sets $U, V \subseteq X$ with $(x,y) \in U \times V$ and $(U \times V) \cap \perp = \emptyset$. Thus, no element of V is orthogonal to any element of U; in particular, we have $y \in V \subseteq X \setminus x^\perp$. Thus, $X \setminus x^\perp$ is open, i.e., $x^\perp$ is closed. It now follows that for any set $A \subseteq X$, the set $A^\perp = \bigcap_{x \in A} x^\perp$ is closed.

(b) Let D be pairwise orthogonal. Let $x \in D$: by part (a), $X \setminus x^\perp$ is open, whence, $\{x\} = D \cap (X \setminus x^\perp)$ is relatively open in D. Thus, D is discrete.

(c) Now suppose D is pairwise orthogonal, and let $z \in \overline{D}$: if $z \notin D$, then for every open neighborhood U of z, $U \cap D$ is infinite; hence, we can find distinct elements $x, y \in D \cap U$. Since D is pairwise orthogonal, this tells us that $(U \times U) \cap \perp \neq \emptyset$. But then (x,x) is a limit point of $\perp$. Since $\perp$ is closed, $(x,x) \in \perp$, which is a contradiction. Thus, $z \in D$, i.e., D is closed. ■

A Topology for Events

The collection of all closed subsets of a topological space X carries a natural topology, called the *Vietoris* topology. This is the weakest topology making the set $[U] = \{F \in 2^X | F \cap U \neq \emptyset\}$ open whenver $U \subseteq X$ is open, and closed whenever U is closed. Equivalently, the Vietoris topology is generated by open sets of the form $[U]$ and $(U) = [U^c]^c = \{F \in 2^X | F \subseteq U\}$, with U ranging over open subsets of X. It is not hard to see that a basis for the resulting topology consists of sets of the form

$$\langle U_1, ..., U_n; V \rangle := [U_1] \cap \cdots \cap [U_n] \cap (V).$$

Evidently, $\langle U_1, ..., U_n; V \rangle$ is the collection of all closed sets F meeting each of the open sets U_1, and contained in the open set V. Equipped with the Vietoris topology, the collection of closed subsets of X is commonly denoted by 2^X, and called the *hyperspace* of X. [30] For later reference, we collect here some basic facts about hyperspaces. Proofs can be found in, e.g., [Illones and Nadler, 1999].

[30] This usage courts trouble, of course, since 2^X is also used for the power set of X. I'll rely on context to distinguish between the two.

PROPOSITION 206. *Let X be any topological space, and let 2^X be the collection of closed subsets of X, with the Vietoris topology. Then*

(a) The union of a Vietoris-compact set of compact sets, is compact;

(b) If X is compact, so is 2^X, and vice versa;

(c) The operation $\bigcup : 2^X \times 2^X \to 2^X$ of taking unions of closed sets, is continuous

(d) If X and Y are two spaces, then the mapping $\pi : 2^X \times 2^Y \to 2^{X \times Y}$ given by $\pi(A, B) = A \times B$, is continuous.

If $\mathfrak{A}$ be a topological test space with outcome space X. It follows from parts (b) and (c) of Lemma 205 that every event of $\mathfrak{A}$ is a closed, discrete subset of X. Thus, we can construe the set $\mathcal{E}$ of events as a subspace of 2^X of all closed subsets of X.

Remark: The empty set is isolated in 2^X — and hence, in $\mathcal{E}$ — since $\emptyset = [\emptyset]$. Many authors omit $\emptyset$ from 2^X, but for our purposes, it is more convenient to include it.

207 Standard Neighborhoods. For locally finite topological test spaces, the Vietoris topology on the space of events has a particularly nice description. Suppose A is a finite event: By Part (a) of Lemma 204, for each point $x \in A$ we can find a totally non-orthogonal open neighborhood U_x. Since X is Hausdorff and A is finite, we can choose these to be disjoint from one another. The set

$$\mathcal{V} = \langle U_x, x \in A \rangle \cap \mathcal{E}$$

is a Vietoris open neighborhood of A in $\mathcal{E}$. An event B belonging to $\mathcal{V}$ is contained in $\bigcup_{x \in A} U_x$ and meets each U_x in at least one point; however, being pairwise orthogonal, B can meet each U_x *at most* once. Thus, B selects *exactly one* point from each of the open sets U_x — in particular, $|B| = |A|$.

Since the totally non-orthogonal sets form a basis for the topology on X, open sets of the form just described form a basis for the Vietoris topology on $\mathcal{E}$. We shall refer to these sets as *standard* neighborhoods.

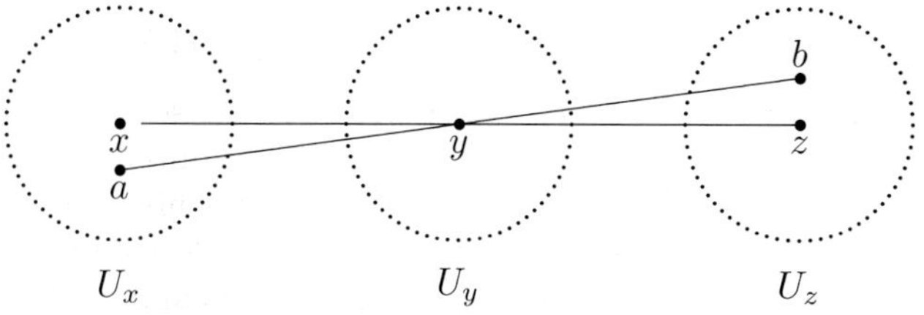

A standard open neighborhood for a finite event $A = \{x, y, z\}$, and a second event $\{a, y, b\}$ in this neighborhood.

An immediate consequence of these remarks is the following:

PROPOSITION 208. *Let $(X, \mathfrak{A})$ be locally finite. Then the set $\mathcal{E}_n$ of all events of a given cardinality n is clopen in $\mathcal{E}(X, \mathfrak{A})$.*

Remark: If $\mathcal{V}$ is a standard open neighborhood of a test $E \in \mathfrak{A}$, then we an regard $\mathcal{V}$ as a test space in its own right. The considerations above show that this is UDF. In this sense, then, any locally finite topological test space is *locally UDF.*

THE CLOSURE OF A TOPOLOGICAL TEST SPACE

If $(X, \mathfrak{A})$ is a topological test space, let $\overline{\mathfrak{A}}$ denote the closure of $\mathfrak{A}$ in 2^X. Note that elements of $\overline{\mathfrak{A}}$ are (closed) subsets of X, so we can interpret $(X, \overline{\mathfrak{A}})$ as a test space.

LEMMA 209. *Let $(X, \mathfrak{A})$ be any topological test space, and let $E \in \overline{\mathfrak{A}}$. Then E is pairwise orthogonal (with respect to the orthogonality induced by $\mathfrak{A}$).*

Proof. Let x and y be two distinct points of E. Let U and V be disjoint neighborhoods of x and y respectively, and let $(E_\lambda)_{\lambda \in \Lambda}$ be a net of closed sets in $\mathfrak{A}$ converging to E in the Vietoris topology. Since $E \in [U] \cap [V]$, we can find $\lambda_{U,V} \in \Lambda$ such that $E_\lambda \in [U] \cap [V]$ for all $\lambda \geq \lambda_{U,V}$. In particular, we can find $x_{\lambda_{U,V}} \in E_{\lambda_{U,V}} \cap U$ and $y_{\lambda_{U,V}} \in E_{\lambda_{U,V}} \cap V$. Since U and V are disjoint, $x_{\lambda_{U,V}}$ and $y_{\lambda_{U,V}}$ are distinct, and hence, — since they belong to a common test E_λ — orthogonal. This gives us a net $(x_{\lambda_{U,V}}, y_{\lambda_{U,V}})$ in $X \times X$ converging to (x, y) and with $(x_{\lambda_{U,V}}, y_{\lambda_{U,V}}) \in \perp$. Since $\perp$ is closed, $(x, y) \in \perp$, i.e., $x \perp y$. ■

It follows that the orthogonality relation on X induced by $\overline{\mathfrak{A}}$ is the same as that induced by $\mathfrak{A}$, and, in particular, is closed. Thus, $(X, \overline{\mathfrak{A}})$ is a topological test space. We shall now show that, if $(X, \mathfrak{A})$ is locally finite, then $(X, \overline{\mathfrak{A}})$ has the same states as $(X, \mathfrak{A})$.

We need a preliminary lemma. Let $\mathcal{F}_n$ denote the collection of all non-empty finite subsets of X having n or fewer elements, understood as a subspace of 2^X. Let $q : X^n \to \mathcal{F}_n$ be the natural surjection $q(x_1, ..., x_n) = \{x_1, ..., x_n\}$.

LEMMA 210. *Let X be Hausdorff. Then for every n,*

(a) q is an open continuous mapping.

(b) $\mathcal{F}_n$ is closed in 2^X.

(c) If $f : X \to \mathbb{R}$ is continuous, then so is the mapping $\widehat{f} : \mathfrak{F}_n \to \mathbb{R}$ given by

$$\widehat{f}(A) := \sum_{x \in A} f(x).$$

Proof. (a) Let $U_1, ..., U_k$ be open subsets of X. Then

$$q(U_1 \times \cdots \times U_k) = \langle U_1, ..., U_n \rangle \cap F_k(X),$$

so q is an open mapping. Also, if $\{U_1, ..., U_k\}$ is pairwise disjoint, so that $\langle U_1, ..., U_k\rangle \cap F_k(X)$ is a basic open set in $F_k(X)$, then

$$q^{-1}(\langle U_1, ..., U_n\rangle \cap F_k(X)) = \bigcup_{\sigma}(U_{\sigma(1)} \times \cdots \times U_{\sigma(n)})$$

where σ runs over all permutations of $\{1, 2, ..., k\}$; thus, q is continuous.

(b) Let F be a closed set of cardinality greater than n. Let $x_1, ..., x_{n+1}$ be distinct elements of F, and let $U_1, ..., U_n$ be pairwise disjoint open sets with $x_i \in U_i$ for each $i = 1, ..., n$. Then no closed set in $\mathcal{U} := [U_1] \cap \cdots \cap [U_n]$ has fewer than $n+1$ points — i.e, $\mathcal{U}$ is an open neighborhood of F disjoint from $\mathcal{F}_n$. This shows that $2^X \setminus \mathcal{F}_n$ is open, i.e., $\mathcal{F}_n$ is closed.

(c) By part (a), $\mathfrak{F}_n$ is the quotient space of X^n induced by the surjection $q : (x_1, ..., x_n) \mapsto \{x_1, ..., x_n\}$. The mapping $\overline{f} : X^n \to \mathbb{R}$ given by $(x_1, ..., x_n) \mapsto \sum_{i=1}^n f(x_i)$ is plainly continuous; hence, so is $\hat{f}$. ■

PROPOSITION 211. *Let $(X, \mathfrak{A})$ be a rank-n (respectively, n-uniform) test space. Then $\overline{\mathfrak{A}}$ is also a rank-n (respectively, n-uniform) test space having the same continuous states as $\mathfrak{A}$.*

Proof. If $\mathfrak{A}$ is rank-n, then $\mathfrak{A} \subseteq \mathfrak{F}_n$. Since the latter is closed, $\overline{\mathfrak{A}} \subseteq \mathfrak{F}_n$ also. If $\mathfrak{A}$ is n-uniform and $E \in \overline{\mathfrak{A}}$, then any net $E_\lambda \to E$ is eventually in bijective correspondence with E, by Proposition 208; hence, $\overline{\mathfrak{A}}$ is also n-uniform. Finally, every continuous state on $(X, \mathfrak{A})$ lifts to a continuous state on $\overline{\mathfrak{A}}$ by Lemma 210 (c). ■

COROLLARY 212. *Let $\mathfrak{A}$ be a topological test space of finite rank, supporting a semi-unital set of* continuous *states. Then $\mathfrak{A}$ can be extended to an algebraic topological test space $\mathfrak{B}$ having the same outcomes and continuous states as $\mathfrak{A}$, that is closed in 2^X.*

Proof. Recall that any test space having a semi-unital set of states is pre-algebraic. Let $\mathfrak{A}_o = \mathfrak{A}$, and define a sequence of test spaces $\mathfrak{A}_\alpha$ by transfinite recursion, setting $\mathfrak{A}_{\alpha+1} = \overline{\langle \mathfrak{A}_\alpha \rangle}$, where the closure is taken in 2^X. Lemma 107 and Corollary 110, together with Proposition 211, imply that $\mathfrak{A}_{\alpha+1}$ has the same continuous states as $\mathfrak{A}_i$, and hence, is again pre-algebraic. If α is a limit ordinal, let $\mathfrak{A}_\alpha = \bigcup_{\beta<\alpha} \mathfrak{A}_\beta$. This again supports the same continuous states as $\mathfrak{A}$. Since each $\mathfrak{A}_\alpha \subseteq 2^X$, the sequence $(\mathfrak{A}_\alpha)$ stabilizes, with limit $\mathfrak{B}$ having the desired properties. ■

Dense semi-classical subspaces

We now show that every member of a very large and natural class of test spaces, which includes the frame manuals of infinite-dimensional Hilbert spaces, has a dense, semi-classical subspace. Since every state on a (locally countable) semi-classical test space is an average, or mixture, of dispersion- free states, this can be interpreted as saying the given test space has an *approximate* "non-contextual hidden-variables" interpretation. This generalizes the results of Meyer [Meyer,

1999] and Clifton and Kent [Clifton-Kent, 2000], who establish substantially the same thing in the special context of quantum states.[31]

LEMMA 213. *Let X be any Hausdorff (indeed, T_1) space, and let $U \subseteq X$ be a dense open set. Then $(U) = \{F \in 2^X | F \subseteq U\}$ is a dense open set in 2^X.*

Proof. Since sets of the form $\langle U_1, ..., U_n \rangle$, $U_1, ..., U_n$ open in X, form a basis for the Vietoris topology on 2^X, it will suffice to show that $(U) \cap \langle U_1, ..., U_n \rangle \neq 0$ for all choices of non-empty opens $U_1, ..., U_n$. Since U is dense, we can select for each $i = 1, ..., n$ a point $x_i \in U \cap U_i$. The finite set $F := \{x_1, ..., x_n\}$ is closed (since X is T_1), and by construction lies in $(U) \cap \langle U_1, ..., U_n \rangle$. ■

COROLLARY 214. *Let $(X, \mathfrak{A})$ be any topological test space with X having no isolated points, and let E be any test in $\mathfrak{A}$. Then open set $(E^c) = [E]^c$ of tests disjoint from E is dense in $\mathfrak{A}$.*

Proof. Since E is a closed set, its complement E^c is an open set; since E is discrete and includes no isolated point, E^c is dense. The result follows from the preceding Lemma. ■

THEOREM 215. *Let $(X, \mathfrak{A})$ be a topological test space with X (and hence, $\mathfrak{A}$) second countable, and without isolated points. Then there exists a countable, pairwise-disjoint sequence $E_n \in \mathfrak{A}$ such that (i) $\{E_n\}$ is dense in $\mathfrak{A}$, and (ii) $\bigcup_n E_n$ is dense in X.*

Proof. Since it is second countable, $\mathfrak{A}$ has a countable basis of open sets $\mathcal{W}_k$, $k \in \mathbb{N}$. Selecting an element $F_k \in \mathcal{W}_k$ for each $k \in \mathbb{N}$, we obtain a countable dense subset of $\mathfrak{A}$. We shall construct a countable dense pairwise-disjoint subsequence $\{E_j\}$ of $\{F_k\}$. Let $E_1 = F_1$. By Corollary 214, $[E_1]^c$ is a dense open set; hence, it has a non-empty intersection with $\mathcal{W}_2$. As $\{F_k\}$ is dense, there exists an index $k(2)$ with $E_2 := F_{k(2)} \in \mathcal{W}_2 \cap [E_1]^c$. We now have $E_1 \in W_1$, $E_2 \in \mathcal{W}_2$, and $E_1 \cap E_2 = \emptyset$. Now proceed recursively: Since $[E_1]^c \cap [E_2]^c \cap \cdots \cap [E_j]^c$ is a dense open and $\mathcal{W}_{j+1}$ is a non-empty open, they have a non-empty intersection; hence, we can select $E_{j+1} = F_{k(j+1)}$ belonging to this intersection. This will give us a test belonging to $\mathcal{W}_{j+1}$ but disjoint from each of the pairwise disjoint sets $E_1, ..., E_j$. Thus, we obtain a sequence $E_j := F_{k(j)}$ of pairwise disjoint tests, one of which lies in each non-empty basic open set $\mathcal{W}_j$ — and which are, therefore, dense.

For the second assertion, notice that for each open set $U \subseteq X$, $[U]$ is a non-empty open in $\mathfrak{A}$, and hence contains some E_j. But then $E_j \cap U \neq \emptyset$, whence, $\bigcup_j E_j$ is dense in X. ■

[31]Indeed, the problem of extending these results to arbitrary test spaces played a role in motivating the theory of topological test spaces.

7.2 Topological Orthoalgebras

If $\mathfrak{A}$ is a topological test space, and the space $\mathcal{E}$ of $\mathfrak{A}$-events has its relative Vietoris topology, as discussed above, then we can endow the logic $\Pi = \Pi(X, \mathfrak{A})$ of $\mathfrak{A}$ with the quotient topology induced by the canonical surjection $p : \mathcal{E} \to \Pi$. In this section, we establish a set of conditions on $(X, \mathfrak{A})$ sufficient to make Π, in this quotient topology, a topological orthoalgebra in the sense of the following

DEFINITION 216. A *topological orthoalgebra* (hereafter: TOA) is an orthoalgebra $(L, \oplus)$ equipped with a topology making the relation $\perp \subseteq L \times L$ closed, and the mappings $\oplus : \perp \to L$ and $' : L \to L$, continuous.

LEMMA 217. *Let $(L, \oplus)$ be a topological orthoalgebra. Then*

(a) The order relation $\leq$ is closed in $L \times L$

(b) L is a Hausdorff space.

(c) The mapping $\ominus : \leq \to L$ is continuous.

Proof. For (a), notice that $a \leq b$ iff $a \perp b'$. Thus, $\leq = f^{-1}(\perp)$ where $f : L \times L \to L \times L$ is the continuous mapping $f(a, b) = (a, b')$. Since $\perp$ is closed, so is $\leq$. That L is Hausdorff now follows easily.[32] Finally, since $b \ominus a = (b' \oplus a)'$, and $\oplus$ and $'$ are both continuous, $\ominus$ is also continuous. ■

EXAMPLES 218. (i) Any product of discrete orthoalgebras, with the product topology, is a TOA.

(ii) Another source of examples are *topological orthomodular lattices* (TOMLs) [Cho and Greechie, 1993]. A TOML is an orthomodular lattice equiped with a Hausdorff topology making both the lattice operations and the orthocomplementation continuous. If L is a TOML and $a, b \in L$, then $a \perp b$ iff $a \leq b'$ iff $a = a \wedge b'$. This is obviously a closed relation, since L is Hausdorff and both $\wedge$ and $'$ are continuous. Thus, every TOML may be regarded as a TOA. However, there are simple and important examples of lattice-ordered TOAs that are not TOMLs — in particular, we have the following:

(iii) Let $\mathbf{H}$ be a Hilbert space, and let $L = L(\mathbf{H})$ be the space of projection operators on $\mathbf{H}$, with its operator-norm topology. As multiplication is jointly continuous, the relation $P \perp Q$ iff $PQ = QP = 0$ is closed. Since addition and subtraction are continuous, the partial operation $P, Q \mapsto P \oplus Q := P + Q$ is continuous on $\perp$, as is the operation $P \mapsto P' := \mathbf{1} - P$. Thus, $L(\mathbf{H})$ is a lattice-ordered topological orthoalgebra. It is not, however, a topological lattice. Indeed, if Q is a non-trivial projection, choose unit vectors x_n not lying in $\text{ran}\,(Q)$, but converging to a unit vector in $x \in \text{ran}\,(Q)$. If P_n is the projection generated by x_n and P, that generated by x, then $P_n \to P$. But $P_n \wedge Q = 0$, while $P \wedge Q = P$.

[32] Indeed, suppose $a \neq b$. Then either $a \not\leq b$ or $b \not\leq a$. Without loss of generality, suppose the former. Then $(a, b) \notin \leq$. Since $\leq$ is closed, there exist open sets U and V in L with $(a, b) \in U \times V$ and $U \times V \cap \leq = \emptyset$. But then $a \in U$, $b \in V$, and $U \times V = \emptyset$.

We can also endow $L(\mathbf{H})$ with the relative strong operator topology.[33] Multiplication is jointly SOT-continuous for operators of norm ≤ 1, the same arguments as given above show that this topology also makes $L(\mathbf{H})$ a lattice-ordered TOA, but not a topological lattice.

Remark: Topologically, projection lattices and TOMLs are strikingly different. As shown by Tae-Hae Cho and R. Greechie, any compact TOML is totally disconnected ([Cho and Greechie, 1993], Lemma 3). In strong contrast to this, if $\mathbf{H}$ is finite dimensional, then $L(\mathbf{H})$ is compact, but the set of projections of a given dimension in $L(\mathbf{H})$ is a manifold.

COMPACT ORTHOALGEBRAS

The structure of *compact* TOAs is particularly tractable. It is a standard fact ([Johnstone, 1983], Corollary VII.1.3) that any ordered topological space with a closed order is isomorphic to a closed subspace of a cartesian power of the real unit interval $[0, 1]$ in its product order and topology. It follows that such a space L is *topologically order-complete*, meaning that any upwardly-directed net in L has a supremum, to which it converges. Applied to a compact TOA, this yields the following completeness result:

LEMMA 219. *Any compact TOA L is orthocomplete. Moreover, if $A \subseteq L$ is jointly orthogonal, the net of finite partial sums of A converges topologically to $\bigoplus A$.*

We are going to show (Theorem 223 and Proposition 225 below) that any compact TOA that any compact *regular* TOA, and likewise, any compact TOA with an isolated zero, is atomistic. If L is any orthoalgebra, let

$$\mathbf{M}(L) := \{(a, c, b) \in L \times L \times L | c \leq a, \ c \leq b, \text{ and } a \perp (b \ominus c)\}.$$

In other words, $(a, c, b) \in \mathbf{M}(L)$ iff $(a \ominus c, c, b \ominus c)$ is a Mackey decomposition for a and b.

LEMMA 220. *For any TOA L, the relation $\mathbf{M}(L)$ is closed in $L \times L \times L$.*

Proof. Just note that $\mathbf{M}(L) = (\geq \times L) \cap (L \times \leq) \cap (\mathrm{Id} \times \ominus)^{-1}(\perp)$. Since the relations $\leq$ and $\perp$ are closed and $\ominus : \leq \to L$ is continuous, this also is closed. ■

Since lattice-ordered TOAs need not be topological lattices, the following is noteworthy:

PROPOSITION 221. *A compact Boolean topological orthoalgebra is a topological lattice, and hence, a compact topological Boolean algebra.*

Proof. If L is Boolean, then $\mathbf{M}(L)$ is, up to a permutation, the graph of the mapping $a, b \mapsto a \wedge b$. Thus, by Lemma 220, $\wedge$ has a closed graph. Since L is

[33] Or, equivalently, the relative weak operator topology; these coincide for projections: [Conway, 1990], Section 2.8, Exercise 4.

compact, this suffices to show that $\wedge$ is continuous.[34] It now follows from the continuity of $'$ that $\vee$ is also continuous. ■

Remarks:

(a) The compactness assumption in Proposition 221 is quite necessary. John Harding (personal communication) has constructed a non- compact Boolean TOA that is not a topological Boolean algebra.

(b) Every compact topological Boolean algebra has the form 2^E, where E is a set and 2^E has the product topology [Johnstone, 1983]. In particular, every compact Boolean algebra is atomistic. This will be useful below.

For any orthoalgebra L, let Comp(L) be the set of all compatible pairs in L, and for any fixed $a \in L$, let Comp(a) be the set of elements compatible with a.

PROPOSITION 222. *Let L be a compact TOA. Then*

(a) Comp(L) is closed in $L \times L$;

(b) For every $a \in L$, Comp(a) is closed in L;

(c) The closure of a pairwise compatible set in L is pairwise compatible;

(d) A maximal pairwise compatible set in L is closed.

Proof. (a) Comp$(L) = (\pi_1 \times \pi_3)(\mathbf{M}(L))$. Since $\mathbf{M}(L)$ is closed, and hence compact, and $\pi_1 \times \pi_3$ is continuous, Comp(L) is also compact, hence closed. For (b), note that Comp$(a) = \pi_1(\text{Comp}(L) \cap (L \times \{a\}))$. As Comp$(L)$ is closed, so is Comp$(L) \cap (L \times \{a\})$; hence, its image under π_1 is also closed (remembering here that L is compact). For (c), suppose $M \subseteq L$ is pairwise compatible. Then $M \times M \subset \text{Comp}(L)$. By part (a), Comp$(L)$ is closed, so we have

$$\overline{M} \times \overline{M} \subseteq \overline{M \times M} \subseteq \text{Comp}(L),$$

whence, $\overline{M}$ is again pairwise compatible. For (d), if M is maximally pairwise compatible, then since $M \subseteq \overline{M}$ and $\overline{M}$ is also pairwise compatible, $M = \overline{M}$. ■

There exist (non-orthocoherent) orthoalgebras in which Comp$(L) = L \times L$ (for instance, the Fano test space of 52). However, in an OML, Comp$(L) = \mathbf{C}(L)$, the center of L. Thus, from part (a) of Proposition 222 we have the result (not hard to prove directly [Cho and Greechie, 1993]) that the center of a compact TOML is a compact Boolean algebra. In fact, we get more than this. Recall that an orthoalgebra regular iff every pairwise compatible subset is contained in a Boolean sub-orthoalgebra. Many orthoalgebras that arise in practice, including all lattice-ordered orthoalgebras, are regular. Recall that a *block* in an orthoalgebra is

[34] Recall here that if X and Y are compact spaces, and the graph G_f of $f : X \to Y$ is closed, then f is continuous. Indeed, let $F \subseteq Y$ be closed. Then $f^{-1}(F) = \pi_1((X \times F) \cap G_f)$, where π_1 is projection on the first factor. Since X and Y are compact, π_1 sends closed sets to closed sets.

a maximal Boolean sub-orthoalgebra. In a regular orthoalgebra, this is the same thing as a maximal pairwise compatible set.

THEOREM 223. *Let L be a compact, regular TOA. Then*

(a) *Every block of L is a compact Boolean algebra, as is the center of L;*

(b) *L is atomistic .*

Proof. (a) If L is regular, then a block of L is the same thing as a maximal pairwise compatible set. It follows from part (d) of Proposition 222 that every block is closed in L, and hence compact. It is not hard to show that in a regular TOA the center is the intersection of the blocks. Thus we also have that $\mathbf{C}(L)$ is also closed, hence compact. Proposition 221 now supplies the result.

To prove (b), suppose $a \in L$. By Zorn's Lemma, there is some block $B \subseteq L$ with $a \in B$. Since B is a compact Boolean algebra, it is complete and atomistic ; hence, a can be written as the join, $\bigvee_B A$, of a set A of atoms in B. Equivalently, $a = \bigvee_B \{\bigoplus F | F \subseteq A, F \text{ finite}\}$. By Lemma 219, L is orthocomplete, hence, $\bigoplus A = \bigvee_L \{\bigoplus F | F \subseteq A,\ F \text{ finite}\}$ also exists, and is the limit of the partial sums $\bigoplus F$, $F \subseteq A$ finite. Since each partial sum lies in B, and B is closed, $\bigoplus A \in B$. It follows that $\bigoplus A = a$. It remains to show that every atom of B is an atom of L. Suppose that b is an atom of B and that $x \in L$ with $0 < x \leq b$. Since B is Boolean, every $y \in B$ satisfies either $b \leq y$ or $y \leq b'$; thus, either $x \leq y$ or $y \leq x$. In particular, x is compatible with every element of B. Since a block in a regular orthoalgebra is a maximal pairwise compatible set, $x \in B$, whence, $x = b$. ■

TOAs WITH ISOLATED ZERO

In [Cho and Greechie, 1993], it is established that any TOML with an isolated point is discrete. In particular, a *compact* TOML with an isolated point is finite. This does not hold for lattice-ordered TOAs generally. Indeed, if $\mathbf{H}$ is a finite-dimensional Hilbert space, then $L(\mathbf{H})$ is a compact lattice-ordered TOA in which 0 is isolated. On the other hand, as I'll now show, a compact TOA with isolated zero does have quite special properties.

To begin with, call an open set in a TOA L *totally non-orthogonal* if it contains no two orthogonal elements. Since $\perp$ is a relatively closed orthogonality relation on $L \setminus \{0\}$, and since the latter set is open (L being Hausdorff), we have the following

LEMMA 224. *Every non-zero element of a TOA has a totally non-orthogonal open neighborhood.*

PROPOSITION 225. *Let L be a compact TOA with 0 isolated. Then*

(a) *L is atomistic and of finite height;*

(b) *The set of atoms of L is open.*

Proof. (a) We first show that there is a finite upper bound on the size of a pairwise orthogonal set. Since 0 is isolated in L, $L \setminus \{0\}$ is compact. By Lemma 224, we can cover $L \setminus \{0\}$ by finitely many totally non-orthogonal open sets $U_1, ..., U_n$. A pairwise- orthogonal subset of $L \setminus \{0\}$ can meet each U_i at most once, and so, can have at most n elements. Now given a finite chain $x_1 < x_2 < ... < x_m$ in L, we can construct a pairwise orthogonal set $y_1, ..., y_{m-1}$ defined by $y_1 = x_1$ and $y_k = x_k \ominus y_{k-1}$ for $k = 2, ..., m-1$. Hence, $m - 1 \leq n$, so $m \leq n+1$. This shows that L has finite height, from which it follows that L is atomistic .

(b) Note that if A and B are any closed subsets of L, then $(A \times B) \cap \perp$ is a closed, hence compact, subset of $\perp$. Since $\oplus$ is continuous on $\perp$, the set

$$A \oplus B := \{a \oplus b | a \in A, b \in B \text{ and } a \perp b\} = \oplus((A \times B) \cap \perp)$$

is compact, hence closed. The set of non-atoms is precisely $(L \setminus \{0\}) \oplus (L \setminus \{0\})$. Since 0 is isolated, $(L \setminus \{0\})$ is closed. Thus, the set of non-atoms is closed. ■

Remark: Both the statements and proofs of Lemma 224 and Part (a) of Proposition 230 apply verbatim to any topological orthoposet, i.e., any ordered space having a closed order and equipped with a continuous orthocomplementation.

If a belongs to the center of a TOA L, then $[0, a] \times [0, a'] \subseteq \perp$. Hence, the natural isomorphism $\phi : [0, a] \times [0, a'] \to L$ given by $(x, y) \mapsto x \oplus y$ is continuous. If L is compact, then so are $[0, a]$ and $[0, a']$; hence, ϕ is also an homeomorphism. Since the center of an orthoalgebra is a Boolean sub- orthoalgebra of L, and since a Boolean algebra of finite height is finite, Proposition 225 has the following

COROLLARY 226. *Let L be a compact TOA with 0 isolated. Then the center of L is finite. In particular, L decomposes, both algebraically and topologically, as the product of finitely many compact irreducible TOAs.*

7.3 Logics of Topological Test Spaces

We now return to the question: when is the logic of a topological test space, in the quotient topology, a topological orthoalgebra? For the balance of this section, $(X, \mathfrak{A})$ is a topological *algebraic* test space, and $\mathcal{E} = \mathcal{E}(\mathfrak{A})$ is understood as having its relative Vietoris topology as a subspace of 2^X.

LEMMA 227. *Suppose $\mathcal{E}$ is closed in 2^X. Then*

(a) The orthogonality relation $\perp_{\mathcal{E}}$ on $\mathcal{E}$ is closed in $\mathcal{E}^2$.

(b) The mapping $\cup : \perp_{\mathcal{E}} \to \mathcal{E}$ is continuous

Proof. The mapping $\mathcal{E}^2 \to 2^{[X]}$ given by $(A, B) \mapsto A \cup B$ is continuous; hence, if $\mathcal{E}$ is closed in $2^{[X]}$, then so is the set $\mathbf{C} := \{(A, B) \in \mathcal{E}^2 | A \cup B \in \mathcal{E}\}$ of *compatible* pairs of events. It will suffice to show that the set $\mathbf{O} := \{(A, B) \in \mathcal{E} | A \subseteq B^{\perp}\}$ is also closed, since $\perp = \mathbf{C} \cap \mathbf{O}$. But $(A, B) \in \mathbf{O}$ iff $A \times B \subseteq \perp$, i.e., $\mathbf{O} = \pi^{-1}((\perp)) \cap \mathcal{E}$

where $\pi : 2^X \times 2^X \to 2^{X \times X}$ is the product mapping $(A, B) \mapsto A \times B$. As observed in section 1 (Proposition 206 (d)), this mapping is continuous, and since $\perp$ is closed in $2^{X \times X}$, so is $(\perp)$ in $2^{X \times X}$. Statement (b) now follows from the Vietoris continuity of $\cup$ (Proposition 206 (c)). ■

Remarks: We did not use the hypothesis that $\mathcal{E}$ be closed in 2^X in showing that the relation $\mathbf{O}$ is closed. If $(X, \mathfrak{A})$ is *coherent*, then $\mathbf{O} = \perp$, so in this case, the hypothesis can be avoided altogether. On the other hand, if X is compact and $\mathfrak{A}$ is closed, then $\mathcal{E}$ will also be compact and hence, closed: by 206 (b), To 2^X is compact; hence, so is the closed set $(E) = \{A \in 2^X | A \subseteq E\}$ for each $E \in \mathfrak{A}$. The mapping $2^X \to 2^{2^X}$ given by $E \mapsto (E)$ is easily seen to be continuous. Since $\mathfrak{A}$ is closed, hence compact, in 2^X, it follows that $\{(E) | E \in \mathfrak{A}\}$ is a compact subset of 2^{2^X}. By 206 (a), then, $\mathcal{E} = \bigcup_{E \in \mathfrak{A}} (E)$ is compact, hence closed, in 2^X.)

In order to apply Lemma 227 to show that $\perp \subseteq \Pi^2$ is closed and $\oplus : \perp \to \Pi$ is continuous, we'd like to the canonical surjection $p : \mathcal{E} \to \Pi$ be be an open mapping. The following condition is sufficient to secure this, along with the continuity of the orthocomplementation $' : \Pi \to \Pi$.

DEFINITION 228. Call a topological test space $(X, \mathfrak{A})$ is *stably complemented* iff for any open set $\mathcal{U}$ in $\mathcal{E}$, the set $\mathcal{U}^{\mathsf{co}}$ of events complementary to events in $\mathcal{U}$ is again open.

As we'll show in Section 7.4, the frame manual of a finite-dimensional Hilbert space is stably complemented.

LEMMA 229. *Let $(X, \mathfrak{A})$ be a topological test space, and let $p : \mathcal{E} \to \Pi$ be the canonical quotient mapping (with Π having the quotient topology). Then the following are equivalent:*

(a) $(X, \mathfrak{A})$ is stably complemented

(b) The mapping $p : \mathcal{E} \to \Pi$ is open and the mapping $' : \Pi \to \Pi$ is continuous.

Proof. Suppose first that $(X, \mathfrak{A})$ is stably complemented, and let $\mathcal{U}$ be an open set in $\mathcal{E}$. Then

$$\begin{aligned} p^{-1}(p(\mathcal{U})) &= \{A \in \mathcal{E} | \exists B \in \mathcal{U} A \sim B\} \\ &= \{A \in \mathcal{E} | \exists C \in \mathcal{U}^{\mathsf{co}} A \mathsf{co} C\} \\ &= \left(\mathcal{U}^{\mathsf{co}}\right)^{\mathsf{co}} \end{aligned}$$

which is open. Thus, $p(\mathcal{U})$ is open. Now note that $' : \Pi \to \Pi$ is continuous iff, for every open set $V \subseteq \Pi$, the set $V' = \{p' | p \in V\}$ is also open. But $p^{-1}(V') = (p^{-1}(V))^{\mathsf{co}}$: since p is continuous and $(X, \mathfrak{A})$ is stably complemented, this last is open. Hence, V' is open. For the converse, note first that if $'$ is continuous, it is also open (since $a'' = a$ for all $a \in \Pi$). Now for any open set $\mathcal{U} \subseteq \mathcal{E}$, $\mathcal{U}^{\mathsf{co}} = p^{-1}(p(\mathcal{U})')$, which is open, since p and $'$ are continuous open mappings. ■

PROPOSITION 230. *Let $(X, \mathfrak{A})$ be a stably complemented algebraic test space with $\mathcal{E}$ closed in 2^X. Then Π is a topological orthoalgebra.*

Proof. Continuity of $'$ has already been established. We show first that $\perp \subseteq \Pi^2$ is closed. If $(a, b) \notin \perp$, then for all $A \in p^{-1}(a)$ and $B \in p^{-1}(b)$, $(A, B) \notin \perp_{\mathcal{E}}$. The latter is closed, by Part (a) of Lemma 227. Hence, we can find Vietoris-open neighborhoods $\mathcal{U}$ and $\mathcal{V}$ of A and B, respectively, with $(\mathcal{U} \times \mathcal{V}) \cap \perp_{\mathcal{E}} = \emptyset$. Since p is open, $U := p(\mathcal{U})$ and $V := p(\mathcal{V})$ are open neighborhoods of a and b with $(U \times V) \cap \perp = \emptyset$. To establish the continuity of $\oplus : \perp \to \Pi$, let $a \oplus b = c$ and let $A \in p^{-1}(a), B \in p^{-1}(B)$ and $C \in p^{-1}(c)$ be representative events. Note that $A \perp B$ and $A \cup B = C$. Let W be an open set containing c: then $\mathcal{W} := p^{-1}(W)$ is an open set containing C. By Part (b) of Lemma 227, the mapping $\cup : \perp_{\mathcal{E}} \to \mathcal{E}$ is continuous; hence, we can find open sets $\mathcal{U}$ about A and $\mathcal{V}$ about B with $A_1 \cup B_1 \in \mathcal{W}$ for every $(A_1, B_1) \in (\mathcal{U} \times \mathcal{V}) \cap \perp_{\mathcal{E}}$. Now let $U = p(\mathcal{U})$ and $V = p(\mathcal{V})$: these are open neighborhoods of a and b, and for every $a_1 \in U$ and $b_1 \in V$ with $a_1 \perp b_1$, $a_1 \oplus b_1 \in p(p^{-1}(W)) = W$ (recalling here that p is surjective). Thus, $(U \times V) \cap \perp \subseteq \oplus^{-1}(W)$, so $\oplus$ is continuous. ■

7.4 Topological G-Test Spaces

We conclude this section by considering topological G-test spaces associated with a compact group G. We shall see that, in the presence of sufficient symmetry, the topological hypotheses of Proposition 230 are automatically satisfied.

PROPOSITION 231. *Let $(X, \mathfrak{A})$ be a G-test space, where G is a compact group acting continuously on X. Then the natural action of G on $\mathcal{E}(X, \mathfrak{A})$ is likewise continuous.*

Proof. Let $\mathcal{E}_k$ denote the space of k-element events of $(X, \mathfrak{A})$: this is G-invariant and, as remarked above, clopen in $\mathcal{E}$. Thus, it is sufficient to show that G's action on $\mathcal{E}_k$ is continuous. By *Lemma 210 (a)*, the canonical surjection $q : X^k \to F_k$, where F_k is the set of all finite subsets of X having k or fewer elements, is continuous and open. Giving X^k the natural diagonal G-action, q is equivariant. It follows easily that the action of G on $F_k(X)$ — and hence, on any invariant subset of $F_k(X)$, e.g., $\mathcal{E}_k$ — is also continuous. ■

Let G be a group and let H and K be closed subgroups of G. As noted above (Remark (a) following Definition 189), we can construct from this data a G-symmetric test space $(X, \mathfrak{A})$, where $X = G/K$; letting $x_o = K \in X$, and writing x_α for $\alpha x_o = \alpha K$ for $\alpha \in G$, we set $E_o = H x_o \subseteq X$, and take $\mathfrak{A}$ be the G-orbit of the set E_o, i.e., $\mathfrak{A} = \{\alpha E_o | \alpha \in G\}$.

THEOREM 232. *Let G be a compact topological group, and let H and K be closed subgroups. Let $(X, \mathfrak{A})$ be the G-symmetric test space constructed from this data as discussed above. Let $X = G/K$ have the quotient topology. Then $(X, \mathfrak{A})$ is a topological test space iff $H \setminus K$ is closed in G.*

Proof. Note first that since G is compact and Hausdorff, $X = G/K$ is likewise Hausdorff. It remains to show that the orthogonality relation on X is closed in $X \times X$ iff $H \setminus K$ is closed in G. Since G is compact, the closed subgroups K and H are compact. If $H \setminus K$ is closed, then certainly so is $K(H \setminus K)K$ (as this is the image of the compact set $K \times (H \setminus K) \times K$ under the continuous mapping $(\alpha, \beta, \gamma) \mapsto \alpha\beta\gamma$). Thus, so is the set $\{(\alpha, \beta) | \beta^{-1}\alpha \in K(H \setminus K)K\}$. Finally, since G is compact, the image of this set under the quotient mapping $(\alpha, \beta) \mapsto (x_\alpha, y_\beta)$ is closed. But, by Lemma 188, this image is just the orthogonality relation on X. For the converse, suppose $\perp$ is closed. Then so is $K(H \setminus K)K$, again by *Lemma 188*. It follows that $(H \setminus K)$ is likewise closed. For suppose $\eta_i \to \eta$ in H, with $\eta_i \notin K$. If $\eta \in H \cap K$, then we have $\eta^{-1}\eta_i\eta \to \eta$ and $\eta^{-1}\eta_i\eta \in K(H \setminus K)K$, whence, $\eta \in K(H \setminus K)K$. Thus, we can find $\phi, \psi \in K$ and $\eta' \in H \setminus K$ with $\eta = \phi\eta'\psi$. Then $\eta' = \phi^{-1}\eta\psi^{-1} \in K$, a contradiction. ■

Notice that the condition that $H \setminus K$ be closed will certainly hold if H is discrete. This is the case, for instance, for the frame test space of a Hilbert space $\mathbf{H}$ with respect to $U(n)$, since here the stabilizer of an orthonormal basis E is isomorphic to the group of permutations of E.

THEOREM 233. *Let $(X, \mathfrak{A})$ be a fully G-symmetric topological test space, with G compact. Then $(X, \mathfrak{A})$ is stably complemented, and $\mathcal{E}$ is closed in 2^X.*

Proof. As $(X, \mathfrak{A})$ has finite rank n, and as each set $\mathcal{E}_k$ of k-element events ($k = 0, ..., n$) is clopen in $\mathcal{E}$, it suffices to show that, for every $k = 0, ..., n$, if $\mathcal{U}$ is open in $\mathcal{E}_k$, then $\mathcal{U}^{\mathsf{co}}$ is open in $\mathcal{E}_{n-k}$. As observed above, the mapping $G \to \mathcal{E}_k$ given by $\alpha \mapsto \alpha A$ is continuous and open for each $A \in \mathcal{E}_k$. Thus, if $\mathcal{U}$ is an open neighborhood of an event $A \in \mathcal{E}_k$, then the set $U = \{\alpha \in G | \alpha A \in \mathcal{U}\}$ is open in G. Let $B\mathsf{co}A$. Then for every $\alpha \in U$, $\alpha B \mathsf{co} \alpha A \in \mathcal{U}$, i.e., $\alpha B \in \mathcal{U}^{\mathsf{co}}$. In other words, the open set $U \cdot B = \{\alpha B | \alpha \in U\}$ about B is contained in $\mathcal{U}^{\mathsf{co}}$. Thus, $\mathcal{U}^c$ is open in $\mathcal{E}_{n-k}$. It remains to show that $\mathcal{E}$ is closed in 2^X. It will suffice to show that each clopen set $\mathcal{E}_k$ is closed in $\mathcal{F}_k(X)$ (since, by *Lemma 210 (b)*, the latter is closed in 2^X). Suppose, then, that A_i is a net in $\mathcal{E}_k$ converging in $\mathcal{F}_k(X)$ to a set A. Since G acts transitively on $\mathcal{E}_k$, we can find a net α_i in G with $A_i = \alpha_i A_o$, where A_o is some arbitrary "base" event in $\mathcal{E}_k$. Since G is compact, we can choose a convergent sub-net $\alpha_{i'} \to \alpha \in G$. By the continuity of the map $G \to \mathcal{F}_k(X)$ given by $\alpha \mapsto \alpha A_o$, we have $A_{i'} = \alpha_{i'} A_o \to \alpha A_o \in \mathcal{E}$, in the latter's Vietoris topology. Since 2^X is Hausdorff, it follows that $A = \alpha A_o \in \mathcal{E}$. ■

Thus, the topological assumptions of Proposition 230 are automatically satisfied for any fully symmetric test space of a compact topological group. If $(X, \mathfrak{A})$ is algebraic, its follows that its logic $L = \Pi(X, \mathfrak{A})$ is a compact TOA with isolated zero — hence, in particular, that L is atomistic. Indeed, the atoms of L are precisely the points of the form $p(\{x\})$, where $x \in X$. It is easy to see that G continues to act on L by continuous automorphisms, and that the atoms of L form a transitive G-space.

ACKNOWLEDGEMENTS

It is a great pleasure to thank D. J. Foulis, from whom I learned the better part of the theory outlined in this Chapter. I have also benefited from the opportunity to present pieces of this material at various conferences and workshops. I would like particularly to thank Howard Barnum, Bob Coecke, Chris Fuchs, Matt Lieffer, and Sonja Smets for organizing several of these. Thanks are due also to Anatolij Dvurecenskij and Frank Valkenborgh for comments on an earlier manuscript containing parts of Sections 1–3. I have been far from scrupulous about citations; but the reader will not go far wrong in assuming that every result not otherwise attributed is due to Foulis and Randall. Or, at any rate, every correct result: for the errors, I must claim sole credit.

BIBLIOGRAPHY

[Aerts, 1982] Aerts, D., Description of many separated physical entities without the paradoxes encountered in quantum mechanics, Foundations of Physics **12** (1982), 1131-1170.

[Amemiya and Araki, 1967] Amemiya, I., and Araki, H., *A Remark on Piron's Paper*, Publications of the Research Institute of Mathematical Sciences, Kyoto University, A **2** (1967), 423-429.

[Barnum *et al.*, 2005] Barnum, H., Fuchs, C., Renes, J., and Wilce, A., Influence-free states on compound quantum systems, preprint (arXiv:quant-ph/0507108).

[Beltrametti and Cassinelli, 1981] Beltrametti, E., and Cassinelli, G., *The Logic of Quantum Mechanics*, Addison-Wesley, 1981.

[Bennett and Foulis, 1990] Bennett, M. K., and Foulis, D. J., Superposition in quantum and classical mechanics, *Foundations of Physics* **20** (1990) 733-744.

[Bennett and Foulis, 1998] Bennett, M. K., and Foulis, D. J., A generalized Sasaki projection for effect algebras, *Tatra Mountains Mathematical Publications* **15** (1998), 55-66.

[Bennett and Foulis, 1994] Foulis, D. J., and Bennett, M. K., Effect algebras and unsharp quantum logis, Foundations of Physics **24** (1994) 1325-1346.

[Birkhoff and von Neumann, 1936] Birkhoff, G., and von Neumann, J., The logic of quantum mechanics, *Annals of Mathematics* **37** (1936), 823-843

[Bruns and Harding, 2000] Bruns, G., and Harding, J., The algebraic theory of orthomodular lattices, in Coecke et al (eds.), **Current Research in Operational Quantum Logic**, Kluwer, 2000.

[Berge, 1989] Berge, C., **Hypergraphs**, North-Holland, 1989.

[Bunce and Maitland-Wright, 1992] Bunce, L. J., and Maitland-Wright, J. D., The Mackey-Gleason problem, *Bulletin of the American Mathematical Society* **26** (1992) 288-293.

[Cho and Greechie, 1993] Cho, T. H., and Greechie, R. J., Profinite orthomodular lattices, *Proceedings of the American Mathematical Society* **118** (1993), 1053-1060.

[Choi, 1975] Choi, M-D., Completely positive maps on complex matrices, *Linear Algebra and its Applications* **10** (1975) 285-290.

[Christensen, 1982] Christensen, E., Measures on projections and physical states, *Communications in Mathematical Physics* **86**, 529-538, 1982.

[Clifton-Kent, 2000] Simulating quantum mechanics with non-contextual hidden variables, *The Proceedings of the Royal Society of London A* **456** (2000), 2101-2114.

[Coecke *et al.*, 2000] Coecke, B., Moore, D., and Wilce, A., *Operational Quantum Logic: an overview*, in Coecke, B., et al., eds., **Current Research in Operational Quantum Logic**, Kluwer, 2000.

[Cohen and Svetlichny, 1987] Cohen, D., and Svetlichny, G., Minimal supports in Quantum Logics, *International Journal of Theoretical Phsics* **27** (1987) 435-450.

[Conway, 1990] Conway, J. H., *A Course in Functional Analysis*, Springer Verlag, 1990.

[Cook, 1985] Cook, T. Banach spaces of weights on test spaces, *Int. J. Theor. Phys.* **24** (1985) 1113-1131.

[Cooke and Hilgeviirde, 1981] Cooke, R., and Hilgevoorde, J., A new approach to equivalence in quantum logic, in E. Beltrametti and B. van Fraassen (eds.), *Current Issues in Quantum Logic*, Plenum, 1981.

[Dacey, 1968] Dacey, J. R., **Orthomodular spaces**, Ph. D. Dissertation, University of Massachusetts, Amherst, 1968

[D'Andrea and De Lucia, 1991] D'Andrea, A., and De Lucia, P., The Brooks-Jewett theorem on an orthomodular lattice, *J. Math. Anal. Appl.* **154** (1991), 507–522.

[De Simone and Navara, 2001] De Simone, A.; Navara, M., Yosida-Hewitt and Lebesgue decompositions of states on orthomodular posets, *Journal of Mathematical Analysis and Applications* **255** (2001) 74-104.

[De Simone and Navara, 2002] De Simone, A., and Navara, M., On the Yosida-Hewitt decomposition and Rttimann decomposition of states, *Sci. Math. Japon.* **56** (2002), 49-62.

[Eilers and Horst, 1975] Eilers, M., and Horst, E., The theorem of Gleason for nonseparable Hilbert spaces, *International Journal of Theoretical Phsics* **13** (1975), 419-424.

[Einstein *et al.*, 1935] Einstein, A., Podolsky, B., and Rosen, N., Can quantum mechanical description of reality be considered complete? *Physcial Review* **47** (1935) 777-780.

[Feldman and Wilce, 1998] Feldman, D. V., and Wilce, A., Abelian extensions of quantum logics, *International Journal of Theoretical Physics* **37** (1998), 39-43.

[Foulis, 2000] Foulis, D. J., Representations of groups on unigroups, in B. Coecke et al. (eds.), **Current Research in Operational Quantum Logic**, Kluwer, 2000.

[Foulis *et al.*, 1996] Foulis, D. J., Greechie, R. J., and Bennett, M. K.,, Test groups and effect algebras, International Journal of Theoretical Phsics **35** (1996) 1117-1140.

[Foulis *et al.*, 1992] Foulis, D. J., Greechie, R. J., and Rüttimann, G. T., Filters and supports in orthoalgebras, *Int. J. Theor. Phys* **31** (1992), 789-807.

[Foulis *et al.*, 1993] Foulis, D. J., Greechie, R. J., and Rüttimann, G. T., *Logico-Algebraic Structures II: Supports on Test Spaces*, International Journal of Theoretical Phsics **32** (1993) 1675-1690.

[Foulis *et al.*, 1985] Foulis, D. J., Piron, C., and Randall, C. H., Realism, operationalism and quantum mechanics, *Foundations of Physics* **13** (1985) 813-842.

[Foulis and Randall, 1971] Foulis, D. J., and Randall, C. H., Lexicographic orthogonality, *Journal of Combinatorial Theory* **11** (1971), 157-162.

[Foulis and Randall, 1981a] Foulis, D.J., and Randall, C. H., What are quantum logics and what ought they to be? In E. Beltrametti and B. van Fraassen (eds.), **Current Issues in Quantum Logic**, Plenum, 1981.

[Foulis and Randall, 1981b] Foulis, D., J., and Randall, C. H., Empirical logic and tensor products, in H. Neumann (ed.), **Interpretations and Foundations of Quantum Theory**, B. I., Wissenschaft, 1981.

[Foulis and Wilce, 2000] Foulis, D. J., and Wilce, A., Free extensions of group actions, induced representations, and the foundations of physics, in Coecke et al, eds., **Current Research in Operational Quantum Logic**, Kluwer, 2000.

[Fuchs, 2002] Fuchs, C. A., *Quantum mechanics as quantum information (and only a little more)*, quant-ph/0205039.

[Gaudard, 1977] Gaudard, M., *The Prior in Bayesian Inference: a non-classical approach*, Ph. D. Dissertation, University of Massachusetts/Amherst, 1977.

[Gleason, 1957] Gleason, A., Measures on the closed subspaces of a Hilbert space, *Journal of Mathematics and Mechanics* **6** (1957) 885-893.

[Goldblatt, 1984] Goldblatt, R., Orthomodularity is not elementary, *Journal of Symbolic Logic* **49** (1984), 401-404.

[Golfin, 1987] Golfin, A., **Representations and Products of Lattices**, Ph. D. Dissertation, University of Massachusetts, Amherst, 1987.

[Greechie, 1966] Greechie, R. J., *Orthomodular lattices*, Ph. D. Dissertation, University of Florida, Gainsville, 1966.

[Greechie, 1968] Greechie, R. J., On the structure of orthomodular lattices satisfying the chain condition, *Journal of Combinatorial Theory* **4** (1968), 210-218.

[Greechie, 1971] Greechie, R. J., Orthomodular lattices admitting no states, J. Comb. Theory **10** (1971), 119-132.

[Greechie, 1974] Greechie, R. J., Some results from the combinatorial approach to quantum logic, *Synthese* **29** (1974), 113-127.
[Greechie *et al.*, 1995] Greechie, R. J., Foulis, D. J., and Pulmannová, S., The center of an effect algebra, *Order* **12** (1995), 91-106.
[Gudder, 1971] Gudder, S., Representations of groups by automorphisms of orthomodular lattices and posets, *Canadian Journal of Mathematics* **23** (1971), 659-673.
[Gudder, 1988] Gudder, S., **Quantum Probability**, Academic Press, San Diego 1988.
[Habil, 1995] Habil, E., Brooks-Jewett and Nikodým convergence theorems for orthoalgebras that have the weak subsequential interpolation property, *Internat. J. Theoret. Phys.* **34** (1995) 465-491
[Halmos, 1957] Halmos, P., *Introduction to Hilbert Space and the Theory of Spectral Multiplicity*, Chelsea, 1957.
[Harding and Navara, 2000] Harding, J., and Navara, M., Embeddings into orthomodular lattices with given centers, state spaces and automorphism groups, *Order* **17** (2000), 239-254.
[Hardy, 2001] Hardy, L., Quantum theory from five reasonable axioms, preprint (quant-ph/0101012) 2001
[Hellwig and Kraus, 1969] K. E. Hellwig and K. Kraus, Pure operations and measurements, *Communications in Mathematical Physics* **11** (1969), 214-220 .
[Holland, 1995] Holland, S., Orthomodularity in infinite dimensions: a theorem of M. P. Solèr, *Bulletin of the American Mathematical Society* **32** (1995), 205-328.
[Illones and Nadler, 1999] Illones and Nadler, **Hyperspaces**, S. Dekker, 1999.
[Ischi, 2002] Ischi, B., The property lattice for independent quantum systems, *Reports on Mathematical Physics* **50** (2002), 155-165.
[Ischi, 2005] Ischi, B., Orthocomplementation and compound systems, preprint (quant-ph/0410085).
[Johnstone, 1983] Johnstone, P. T., **Stone Spaces**, Cambridge, 1983
[Kalmbach, 1983] Kalmbach, G., **Orthomodular lattices**, Academic Press, 1983.
[Keller, 1980] Keller, H. A., Ein Nichtklassischer Hilbertscher Raum, *Mathematische Zeitschrift* **172** (1980) 41-49.
[Kochen and Specker, 1967] Kochen, S., and Specker, E., The problem of hidden variables in quantum mechanics, *Journal of Mathematics and Mechanics* **17**, 59-87, 1967.
[Kläy *et al.*, 1987] Kläy, M., Randall, C. H., and Foulis, D. J., Tensor products and probability weights, International Journal of Theoretical Phsics **26** (1987), 199-219.
[Lock, P., 1981] Lock, P., Categories of Manuals, Ph.D. University of Massachusetts, Amherst, 1981.
[Lock, R., 1981] Lock, R., Constructing the Tensor Product of Generalized Sample Spaces, Ph. D. Dissertation, University of Massachusetts, Amherst, 1981.
[Mackey, 1957] Mackey, G., Quantum mechanics in Hilbert space, *American Mathematical Monthly* **64:2** (1957) 45-57.
[Mackey, 1963] Mackey, G., **Mathematical Foundations of Quantum Mechanics**, W. A. Benjamin, 1963.
[Mayet, 1998] Mayet, R., Some characterizations of the underlying division ring of a Hilbert lattice by automorphisms, International Journal of Theoretical Phsics **37** (1998), 109-114.
[Meyer, 1999] Meyer, D., Finite precision measurement nullifies the Kochen-Specker theorem. *Physical Review Letters* **83** (1999), 3751-3754.
[Nachbin, 1965] Nachbin, L., **Topology and Order**, van Nostrand, 1965.
[Obeid, 1990] Obeid, Mustafa, *Pastings and Centeria of Orthoalgebras*, Ph. D. Dissertation, Kansas State University, 1990.
[Navara, 1992] Navara, M., Independence of automorphism group, center and state space of quantum logics, *Int. J. Theor. Phys.* 31 (1992) 925-935.
[Nishimura, 1995] Nishimura, H., Manuals in orthogonal categories, International Journal of Theoretical Phsics **34** (1995), 211-228, and Empirical sets, Ibid., 229-252.
[Pitowsky, 2005] Pitowsky, I., Quantum mechanics as a probability theory, preprint (quant-ph 0510095), 2005.
[Piron, 1964] Piron, C., Axiomatique quantique, *Helvetica Physica Acta* **37** 439-468, 1964.
[Piron, 1976] Piron, C., *Foundations of Quantum Mechanics*, W. A. Benjamin, 1976
[Prestel, 1995] Prestel, A., On Solèr's characterization of Hilbert spaces, *Manuscripta Mathematica* **86**, 225-238, 1995.

[Pták, 1983] Pták, P., Logics with given centers and state spaces, *Proc. Amer. Math. Soc.* **88** (1983), 106-109.

[Pulmannová and Wilce, 1995] Pulmannová, S., and Wilce, A., Representations of D-posets, International Journal of Theoretical Phsics **34** (1995), 1689-1696

[Quan and Wilce, 2008] Quan Tran and Wilce, A., "Covariance an quantum logic", International Journal of Theoretical Physics **47** (2008), 15-25.

[Randall, 1966] Randall, C. H., *A Mathematical Foundation for Empirical Science*, Ph. D. dissertation, Rensselaer Polytchnic Institute, 1966.

[Randall and Foulis, 1970] Randall, C. H., and Foulis, D. J., An approach to empirical logic, *American Mathematical Monthly* **71** (1970) 363-374.

[Randall and Foulis, 1976] Randall, C. H., and Foulis, D. J., A mathematical settting for inductive reasoning, in W. L. Harper and C. A. Hooker (eds.), *Foundations of Probability Theory and Statistical Theories of Science*, D. Reidel, Dordrecht, 1976

[Randall and Foulis, 1978] Randall, C. H., and Foulis, D. J., The operational approach to quantum mechanics, in Hooker ed., *The Logico-Algebraic Approach to Quantum Mechanics*, D. Riedel, 1978.

[Randall and Foulis, 1983a] Randall, C. H., and Foulis, D. J., Properties and operational propositions in quantum mechanics, *Foundations of Physics***13** (1983) 843-857.

[Randall and Foulis, 1983b] Randall, C. H., and Foulis, D.J., A mathematical language for quantum physics, in C. Gruber, C. Piron, T. M. Tâm and R. Weill (eds.), *Les fondements de la mécanique quantique* AVCP, Lausanne, 1983.

[Randall *et al.*, 1973] Randall, C. H., Foulis, D. J., and Janowitz, M. F., Orthomodular generalizations of homogeneous Boolean algebras, *Journal of the Australian Mathematical Society* **15** (1973), 94-104.

[Shmuely, 1974] Shmuely, Z., The structure of Galois connections, *Pacific Journal of Mathematics* **54** (1974), 209-225.

[Schultz, 1974] Shultz, F. W., A characterization of state spaces of orthomodular lattices, *J. Combin. Theory Ser. A* **17** (1974), 317–328.

[Solèr, 1995] Solèr, M. P., Characterization of Hilbert space by orthomodular spaces, *Communications in Algebra* **23**, 219-243, 1995.

[Varadarajan, 1985] Varadarajan, V. S., *The Geometry of Quantum Theory*, 2nd ed., Springer Verlag, 1985

[von Neumann, 1932] Von Neumann, J., *Mathematische Grundlagen der Quantenmechanik*, Springer, Berlin, 1932 (Reissued in English translation as *M*athematical Foundations of Quantum Mechanics by Princeton University Press, 1957).

[Wallach, 2000] Wallach, N. R., *An Unentangled Gleason's Theorem*, preprint (quant-ph/0002058)

[Wilce, 1990] Wilce, A., Tensor products of frame manuals, International Journal of Theoretical Phsics **29** (1990), 805-814.

[Wilce, 1992] Wilce, A., Tensor products in generalized measure theory, International Journal of Theoretical Phsics **31** (1992) 1915-1928

[Wilce, 1995] Wilce, A., Spaces of vector-valued weights on test spaces, preprint

[Wilce, 1997a] Wilce, A., Symmetric test spaces, in S. P. Hotaling and A. R. Pirich (Eds.), *Proceedings of SPIE* **3076** (1997), 111-130.

[Wilce, 1997b] Wilce, A., Pull-backs and product tests, *Helvetica Physica Acta* **70** (1997), 803-812.

[Wilce, 1998] Wilce, A., Perspectivity and congruence in partial abelian Semigroups, *Mathematica Slovaca* **47**, 1998, 117–135.

[Wilce, 2000] Wilce, A., On generalized Sasaki projections, *International Journal of Theoretical Physics* **39** (2000), 969-974.

[Wilce, 2002] Wilce, A., Quantum logic and quantum probability, *Stanford Encyclopedia of Philosophy*, Feb. 2002

[Wilce, 2005a] Wilce, A., Topological test spaces, *International Journal of Theoretical Physics* **44** (2005), 1227-1238.

[Wilce, 2005b] Wilce, A., Compact orthoalgebras, *Proceedings of the American Mathematical Society* **133** (2005), 2911-2920.

[Wilce, 2005c] Wilce, A., Symmetry and Topology in Quantum Logic, *International Journal of Theoretical Physics* **44** (2005), 2265-2278.

[Wright, 1977] Wright, R., The structure of projection-valued states, *International Journal of Theoretical Physics* **16** (1977) 567-573.

[Wright, 1978a] Wright, R., Spin manuals, in R. Marlow (ed.), *Mathematical Foundations of Quantum Mechanics*, Academic Press, 1978

[Wright, 1978b] Wright, R., The state of the pentagon, in R. Marlow (ed.), *Mathematical Foundations of Quantum Mechanics*, Academic Press, 1978

[Yeadon, 1983] Yeadon, F. J., Measures on projections in W^*-algebras of type II_1, *Bulletin of the London Mathematical Society* **15**, 139-145, 1983.

[Zierler, 1961] Zierler, N., Axioms for non-relativistic quantum mechanics, *Pacific Journal of Mathematics* **11** (1961), 1151-1169.

Edited by K. Engesser, D. M. Gabbay and D. Lehmann

CONTEXTS IN QUANTUM, CLASSICAL AND PARTITION LOGIC

Karl Svozil

It is not enough to have no concept,
one must also be incapable of expressing it.
Karl Kraus

But no sooner do we depart from sense and instinct to follow the light of a superior principle, to reason, meditate, and reflect on the nature of things, but a thousand scruples spring up in our minds concerning those things which before we seemed fully to comprehend. Prejudices and errors of sense do from all parts discover themselves to our view; and, endeavouring to correct these by reason, we are insensibly drawn into uncouth paradoxes, difficulties, and inconsistencies, which multiply and grow upon us as we advance in speculation, till at length, having wandered through many intricate mazes, we find ourselves just where we were, or, which is worse, sit down in a forlorn Scepticism
George Berkeley

1 MOTIVATION

In what follows, the term *context* refers to a maximal collection of co-measurable observables "bundled together" to form a "quasi-classical mini-universe" within some "larger" nonclassical structure. Similarly, the contexts of an observable are often defined as maximal collections of mutually co-measurable (compatible) observables which are measured or at least could in principle be measured alongside of this observable [Bohr, 1949; Bell, 1966; Heywood and Redhead, 1983; Redhead, 1990]. Quantum mechanically, this amounts to a formalization of contexts by Boolean subalgebras of Hilbert lattices [Svozil, 2005c; Svozil, 2005d], or equivalently, to maximal operators (e.g., Ref. [von Neumann, 1932, Sec. II.10, p. 90], English translation in Ref. [von Neumann, 1955, p. 173], Ref. [Kochen and Specker, 1967, § 2], Ref. [Neumark, 1954, pp. 227,228], and Ref. [Halmos, 1974, § 84]).

In classical physics, contexts are rather unrevealing, as all classical observables are in principle co-measurable, and there is only a single context which comprises the entirety of observables. Indeed, that two or more observables may not be co-measurable; i.e., operationally obtainable simultaneously, and thus may belong to different, distinct contexts, did not bother the classical mind until around 1920. This situation has changed dramatically with the emergence of quantum mechanics, and in particular with the discovery of complementarity and value indefiniteness. Contexts are the building blocks of quantum logics; i.e., the pastings of a continuity of contexts form the Hilbert lattices.

We shall make use of algebraic formalizations, in particular logic. Quantum logic is about the relations and operations among statements referring to the quantum world. As quantum physics is an extension of classical physics, so is quantum logic an extension of classical logic. Classical physics can be extended in many mindboggling, weird ways. The question as to why Nature "prefers" the quantum mindboggling way over others appears most fascinating to the open mind. Before understanding some of the issues, one has to review classical as well as quantum logic and some of its doubles.

Logic will be expressed as algebra. That is an approach which can be formalized. Other approaches, such as the widely held opportunistic belief that something is true because it is useful might also be applicable (for instance in acrimonious divorces), though less formalized. Some of the material presented here has already been published elsewhere [Svozil, 1998], in particular the partition logic part [Svozil, 2005b], or the section on quantum probabilities [Svozil and Tkadlec, 1996]. Here we emphasize the importance of the notion of *context*, which may serve as a unifying principle for all of the logics discussed.

2 CLASSICAL CONTEXTS

Logic is an ancient philosophical discipline. Its algebraization started in the mid-nineteenth century with Boole's *Laws of Thought* [Boole, 1958]. In what follows, Boole's approach, in particular to probability theory, is reviewed.

2.1 Boolean algebra

A Boolean algebra $\mathfrak{B}$ is a set endowed with two binary operations $\wedge$ (called "and") and $\vee$ (called "or"), as well as a unary operation " $'$ " (called "complement" or "negation"). It also contains two elements 1 (called "true") and 0 (called "false"). These entities satisfy associativity, commutativity, the absorption law and distributivity. Every element has a unique complement.

A typical example of a Boolean algebra is set theory. The operations are identified with the set theoretic intersection, union, and complement, respectively. The implication relation is identified with the subset relation.

2.2 Classical contexts as classical logics

A classical Boolean algebra is the representation of all possible "propositions" or "knowables." Every knowable can be combined with every other one by the standard logical operations "and" and "or." Operationally, all knowables are in principle knowable simultaneously. Stated differently: within the Boolean "universe," the knowables are all consistently co-knowable. In this sense, classical contexts

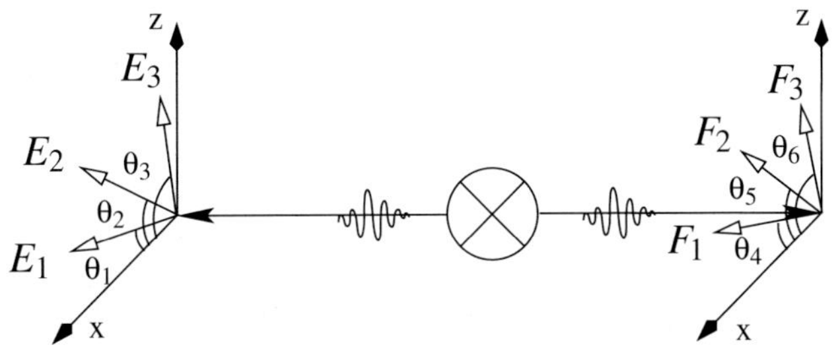

Figure 1. Measurements of E_1, E_2, E_3 on the "left," and F_1, F_2, F_3 on the "right" hand side, along directions θ_i.

coincide with the collection of all possible observables, which are expressed by Boolean algebras. Thus, classical contexts can be identified with the respective classical logics.

2.3 Classical probabilities

Classical probabilities and joint probabilities can be represented as points of a *convex polytope* spanned by all possible "extreme cases" of the classical Boolean algebra; more formally: by all two-valued measures on the Boolean algebra. Two-valued measures, also called dispersionless measures or valuations, acquire only the values "0" and "1," interpretable as falsity and truth, respectively. If some events are independent, then their joint probability $pq\cdots$ can be expressed as the product of their individual probabilities p, q, ...

The associated *correlation polytope* [Pitowsky, 1989b; Pitowsky, 1989a; Pitowsky, 1991; Pitowsky, 1994; Pitowsky and Svozil, 2001] (see also Refs. [Froissart, 1981; Cirel'son (=Tsirel'son), 1980; Cirel'son (=Tsirel'son), 1993]) is spanned by a convex combination of vertices, which are vectors of the form $(p, q, \ldots, pq, \ldots)$, where the components are the individual probabilities of independent events which take on the values 0 and 1, together with their joint probabilities, which are the products of the individual probabilities. The polytope faces impose "inside–outside" distinctions. The associated inequalities must be obeyed by all classical probability distributions; they are bounds on classical (joint) probabilities termed *"conditions of possible experience"* by Boole [Boole, 1958; Boole, 1862].

Two-event "1–1" case

Let us demonstrate the bounds on classical probabilities by the simplest nontrivial example of two propositions; e.g.,

$E \equiv$ *"a particle detector aligned along direction* **a** *clicks,"* and
$F \equiv$ *"a particle detector aligned along direction* **b** *clicks."*

Consider also the joint proposition

	E	F	$E \wedge F \equiv E \cdot F$
1	0	0	0
2	0	1	0
3	1	0	0
4	1	1	1

(a)

	full facet inequality
1	$pq \geq 0$
2	$p \geq pq$
3	$p \geq pq$
4	$pq \geq p + q - 1$

(b)

Table 1. Construction of the correlation polytope for two events: (a) the four possible cases are represented by the truth table, whose rows can be interpreted as three-dimensional vectors forming the vertices of the correlation polytope; (b) the resulting four faces of the polytope are characterized by half-spaces which are obtained by solving the hull problem. .

> $E \wedge F \equiv$ *"the two particle detectors aligned along directions* **a** *and* **b** *click."*

The notation "1–1" alludes to the experimental setup, in which the two events are registered by detectors located at two "adjacent sites." For multiple direction measurements, see Fig. 1.

There exist four possible cases, enumerated in Table 1(a). The correlation polytope in this case is formed by interpreting the rows as vectors in three-dimensional vector space. Four cases, interpretable as truth assignments or two-valued measures, correspond to the four vectors $(0, 0, 0)$, $(0, 1, 0)$, $(1, 0, 0)$, and $(1, 1, 1)$. The correlation polytope for the probabilities p, q and the joint probabilities pq of an occurrence of E, F, and both $E\&F$

$$(p, q, pq) = \kappa_1(0, 0, 0) + \kappa_2(0, 1, 0) + \kappa_3(1, 0, 0) + \kappa_4(1, 1, 1) = (\kappa_3 + \kappa_4, \kappa_2 + \kappa_4, \kappa_4)$$

is spanned by the convex sum $\kappa_1 + \kappa_2 + \kappa_3 + \kappa_4 = 1$ of these four vectors, which thus are vertices of the polytope. κ_i can be interpreted as the normalized weight for event i to occur. The configuration is drawn in Figure 2.

By the Minkoswki-Weyl representation theorem (e.g, Ref. [Ziegler, 1994, p.29]), every convex polytope has a dual (equivalent) description: either as the convex hull of its extreme points (vertices); or as the intersection of a finite number of half-spaces. Such facets are given by linear inequalities, which are obtained from the set of vertices by solving the so called *hull problem*. The inequalities coincide with Boole's "conditions of possible experience." The hull problem is algorithmically solvable but computationally hard [Pitowsky, 1990].

In the above example, the "conditions of possible experience" are given by the inequalities enumerated in Table 1b). One of their consequences are bounds on joint occurrences of events. Suppose, for example, that the probability of a click in detector aligned along direction **a** is 0.9, and the probability of a click in the second detector aligned along direction **b** is 0.7. Then inequality 4 forces us to accept that the probability that both detector register clicks cannot be smaller than $0.9 + 0.7 - 1 = 0.6$. If, for instance, somebody comes up with a

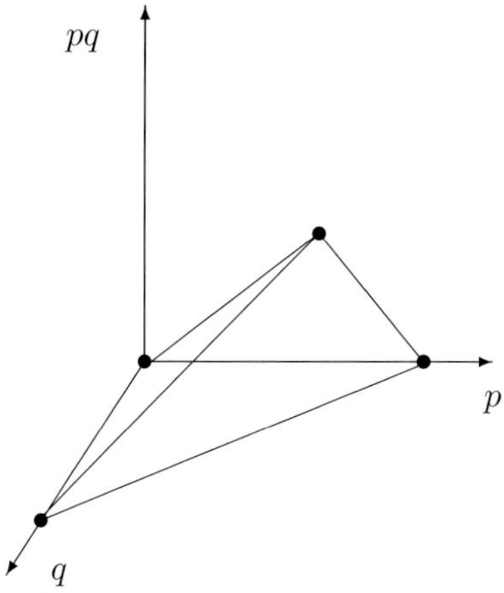

Figure 2. The correlation polytope for two events. The vertices are $(0,0,0)$, $(0,1,0)$, $(1,0,0)$, and $(1,1,1)$. The four faces of the polytope are characterized by the inequalities in Table 1(b).

joint probability of 0.4, we would know that this result is flawed, possibly by fundamental measurement errors, or by cheating.

Four-event "2–2" case

A configuration discussed in quantum mechanics is one with four events grouped into two equal parts E_1, E_2 and F_1, F_2. There are 2^4 different cases of occurrence or nonoccurrence of these four events enumerated in Table 2.

By solving the hull problem, one obtains a set of conditions of possible experience which represent the bounds on classical probabilities enumerated in Table 3. For historical reasons, the bounds 17-18, 19-20, 21-22, and 23-24 are called the Clauser-Horne inequalities [Clauser and Horne, 1974; Clauser and Shimony, 1978]. They are equivalent (up to permutations of p_i, q_i), and are the only additional inequalities structurally different from the two-event "1–1" case.

Six event "3–3" case

A similar calculation [Pitowsky and Svozil, 2001] for six events $E_1, E_2, E_3, F_1, F_2, F_3$ depicted in Fig. 1 yields an additional independent [Colins and Gisin, 2004; Sliwa, 2003] inequality for their probabilities $p_1, p_2, p_3, q_1, q_2, q_3$ and their joint probabilities of the type

$$p_1q_1 + p_2q_2 + p_1q_3 + p_2q_1 + p_2q_2 - p_2q_3 + p_3q_1 - p_3q_2 \leq p_1 + 2q_1 + q_2.$$

	E_1	E_2	F_1	F_2	E_1F_1	E_1F_2	E_2F_1	E_2F_2
1	0	0	0	0	0	0	0	0
2	0	0	0	1	0	0	0	0
3	0	0	1	0	0	0	0	0
4	0	0	1	1	0	0	0	0
5	0	1	0	0	0	0	0	0
6	0	1	0	1	0	0	0	1
7	0	1	1	0	0	0	1	0
8	0	1	1	1	0	0	1	1
9	1	0	0	0	0	0	0	0
10	1	0	0	1	0	1	0	0
11	1	0	1	0	1	0	0	0
12	1	0	1	1	1	1	0	0
13	1	1	0	0	0	0	0	0
14	1	1	0	1	0	1	0	1
15	1	1	1	0	1	0	1	0
16	1	1	1	1	1	1	1	1

Table 2. Construction of the correlation polytope for four events. The 16 possible cases are represented by the truth table, whose rows can be interpreted as three-dimensional vectors forming the vertices of the correlation polytope; (b) the resulting four faces of the polytope are characterized by half-spaces which are obtained by solving the hull problem.

3 QUANTUM CONTEXTS

Omniscience in a classical sense is no longer possible for quantum systems. Some of the reasons are: (i) quantum complementarity and, algebraically associated with it, the breakdown of distributivity; (ii) the impossibility to consistently assign truth and falsity for all observables simultaneously and, associated with it, the nonexistence of two-valued measures on even finite subsets of Hilbert logics; and (iii) the alleged randomness of certain single outcomes.

3.1 Hilbert lattices as quantum logics

Quantum logic has been introduced by Garrett Birkhoff and John von Neumann [von Neumann, 1932; Birkhoff and von Neumann, 1936; Mackey, 1957; Jauch, 1968; Pták and Pulmannová, 1991] in the thirties. They organized it *top-down*, starting from the Hilbert space formalism of quantum mechanics. Certain entities of Hilbert spaces are identified with propositions, partial order relations and lattice operations. These relations and operations are identified with the logical implication relation and operations such as "and," "or," and the negation. Thereby, as we shall see, the resulting logical structures are "nonclassical," in particular "nonboolean."

Kochen and Specker [Kochen and Specker, 1965b; Kochen and Specker, 1965a] suggested to consider only relations and operations among compatible, co-measurable observables; i.e., within Boolean subalgebras, which will be identified with blocks

	full facet inequality	inequality for $p_1 = p_2 = q_1 = q_2 = \frac{1}{2}$
1	$p_1q_1 \geq 0$	$p_1q_1 \geq 0$
2	$p_1q_2 \geq 0$	$p_1q_2 \geq 0$
3	$p_2q_1 \geq 0$	$p_2q_1 \geq 0$
4	$p_2q_2 \geq 0$	$p_2q_2 \geq 0$
5	$p_1 \geq p_1q_1$	$\frac{1}{2} \geq p_1q_1$
6	$p_1 \geq p_1q_2$	$\frac{1}{2} \geq p_1q_2$
7	$q_1 \geq p_1q_1$	$\frac{1}{2} \geq p_1q_1$
8	$q_1 \geq p_1q_2$	$\frac{1}{2} \geq p_1q_2$
9	$p_2 \geq p_2q_1$	$\frac{1}{2} \geq p_2q_1$
10	$p_2 \geq p_2q_2$	$\frac{1}{2} \geq p_2q_2$
11	$q_2 \geq p_2q_1$	$\frac{1}{2} \geq p_2q_1$
12	$q_2 \geq p_1q_2$	$\frac{1}{2} \geq p_2q_2$
13	$p_1q_1 \geq p_1 + q_1 - 1$	$p_1q_1 \geq 0$
14	$p_1q_2 \geq p_1 + q_2 - 1$	$p_1q_2 \geq 0$
15	$p_2q_1 \geq p_2 + q_1 - 1$	$p_2q_1 \geq 0$
16	$p_2q_2 \geq p_2 + q_2 - 1$	$p_2q_2 \geq 0$
17	$0 \geq p_1q_1 + p_1q_2 + p_2q_1 - p_2q_2 - p_1 - q_1$	$1 \geq +p_1q_1 + p_1q_2 + p_2q_1 - p_2q_2$
18	$p_1q_1 + p_1q_2 + p_2q_1 - p_2q_2 - p_1 - q_1 \geq -1$	$p_1q_1 + p_1q_2 + p_2q_1 - p_2q_2 \geq 0$
19	$0 \geq p_1q_1 + p_1q_2 - p_2q_1 + p_2q_2 - p_1 - q_2$	$1 \geq +p_1q_1 + p_1q_2 - p_2q_1 + p_2q_2$
20	$p_1q_1 + p_1q_2 - p_2q_1 + p_2q_2 - p_1 - q_2 \geq -1$	$p_1q_1 + p_1q_2 - p_2q_1 + p_2q_2 \geq 0$
21	$0 \geq p_1q_1 - p_1q_2 + p_2q_1 + p_2q_2 - p_2 - q_1$	$1 \geq p_1q_1 - p_1q_2 + p_2q_1 + p_2q_2$
22	$p_1q_1 - p_1q_2 + p_2q_1 + p_2q_2 - p_2 - q_1 \geq -1$	$p_1q_1 - p_1q_2 + p_2q_1 + p_2q_2 \geq 0$
23	$0 \geq -p_1q_1 + p_1q_2 + p_2q_1 + p_2q_2 - p_2 - q_2$	$1 \geq -p_1q_1 + p_1q_2 + p_2q_1 + p_2q_2$
24	$-p_1q_1 + p_1q_2 + p_2q_1 + p_2q_2 - p_2 - q_2 \geq -1$	$-p_1q_1 + p_1q_2 + p_2q_1 - p_2q_2 \geq 0$

Table 3. Construction of the correlation polytope for four events. The 24 faces of the polytope spanned by the vertices corresponding to the rows enumerated in Table 2. The bounds 17-18, 19-20, 21-22, and 23-24 are the Clauser-Horne inequalities.

and contexts of Hilbert lattices. Nevertheless, some of their theorems formally take into account ensembles of contexts [Kochen and Specker, 1967] for which a multitude of incompatible observables contribute.

If theoretical physics is assumed to be a faithful representation of our experience, such an "empirical," "operational" [Bridgman, 1927; Bridgman, 1934; Bridgman, 1952] logic derives its justification by the phenomena themselves. In this sense, one of the main justifications for quantum logic is the construction of the logical and algebraic order of events based on empirical findings.

Definition

The dimensionality of the Hilbert space for a given quantum system depends on the number of possible mutually exclusive outcomes. In the spin–$\frac{1}{2}$ case, for example, there are two outcomes "up" and "down," associated with spin state measurements along arbitrary directions. Thus, the dimensionality of Hilbert space needs to be two.

generic lattice	order relation	"meet"	"join"	"complement"
propositional calculus	implication $\rightarrow$	disjunction "and" $\wedge$	conjunction "or" $\vee$	negation "not" $\neg$
"classical" lattice of subsets of a set	subset $\subset$	intersection $\cap$	union $\cup$	complement
Hilbert lattice	subspace relation $\subset$	intersection of subspaces $\cap$	closure of linear span $\oplus$	orthogonal subspace $\perp$
lattice of commuting {noncommuting} projection operators	$E_1E_2 = E_1$	E_1E_2 $\{\lim_{n\to\infty}(E_1E_2)^n\}$	$E_1 + E_2 - E_1E_2$	orthogonal projection

Table 4. Comparison of the identifications of lattice relations and operations for the lattices of subsets of a set, for experimental propositional calculi, for Hilbert lattices, and for lattices of commuting projection operators.

Then the following identifications can be made. Table 4 lists the identifications of relations of operations of classical Boolean set-theoretic and quantum Hillbert lattice types.

- Any closed linear subspace of — or, equivalently, any projection operator on — a Hilbert space corresponds to an elementary proposition. The elementary "true"–"false" proposition can in English be spelled out explicitly as

  > "The physical system has a property corresponding to the associated closed linear subspace."

- The logical "and" operation is identified with the set theoretical intersection of two propositions "$\cap$"; i.e., with the intersection of two subspaces. It is denoted by the symbol "$\wedge$". So, for two propositions p and q and their associated closed linear subspaces $\mathfrak{M}_p$ and $\mathfrak{M}_q$,

$$\mathfrak{M}_{p\wedge q} = \{x \mid x \in \mathfrak{M}_p,\ x \in \mathfrak{M}_q\}.$$

- The logical "or" operation is identified with the closure of the linear span "$\oplus$" of the subspaces corresponding to the two propositions. It is denoted by the symbol "$\vee$". So, for two propositions p and q and their associated closed linear subspaces $\mathfrak{M}_p$ and $\mathfrak{M}_q$,

$$\mathfrak{M}_{p\vee q} = \mathfrak{M}_p \oplus \mathfrak{M}_q = \{x \mid x = \alpha y + \beta z,\ \alpha, \beta \in \mathbb{C},\ y \in \mathfrak{M}_p,\ z \in \mathfrak{M}_q\}.$$

  The symbol $\oplus$ will used to indicate the closed linear subspace spanned by two vectors. That is,

$$u \oplus v = \{w \mid w = \alpha u + \beta v,\ \alpha, \beta \in \mathbb{C},\ u, v \in \mathfrak{H}\}.$$

Notice that a vector of Hilbert space may be an element of $\mathfrak{M}_p \oplus \mathfrak{M}_q$ without being an element of either $\mathfrak{M}_p$ or $\mathfrak{M}_q$, since $\mathfrak{M}_p \oplus \mathfrak{M}_q$ includes all the vectors in $\mathfrak{M}_p \cup \mathfrak{M}_q$, as well as all of their linear combinations (superpositions) and their limit vectors.

- The logical "not"-operation, or "negation" or "complement," is identified with operation of taking the orthogonal subspace "$\perp$". It is denoted by the symbol "$\,'\,$". In particular, for a proposition p and its associated closed linear subspace $\mathfrak{M}_p$, the negation p' is associated with

$$\mathfrak{M}_{p'} = \{x \mid (x, y) = 0,\ y \in \mathfrak{M}_p\},$$

where (x, y) denotes the scalar product of x and y.

- The logical "implication" relation is identified with the set theoretical subset relation "$\subset$". It is denoted by the symbol "$\rightarrow$". So, for two propositions p and q and their associated closed linear subspaces $\mathfrak{M}_p$ and $\mathfrak{M}_q$,

$$p \rightarrow q \Longleftrightarrow \mathfrak{M}_p \subset \mathfrak{M}_q.$$

- A trivial statement which is always "true" is denoted by 1. It is represented by the entire Hilbert space $\mathfrak{H}$. So,

$$\mathfrak{M}_1 = \mathfrak{H}.$$

- An absurd statement which is always "false" is denoted by 0. It is represented by the zero vector 0. So,

$$\mathfrak{M}_0 = 0.$$

Diagrammatical representation, blocks, complementarity

Propositional structures are often represented by Hasse and Greechie diagrams. A *Hasse diagram* is a convenient representation of the logical implication, as well as of the "and" and "or" operations among propositions. Points " $\bullet$ " represent propositions. Propositions which are implied by other ones are drawn higher than the other ones. Two propositions are connected by a line if one implies the other. Atoms are propositions which "cover" the least element 0; i.e., they lie "just above" 0 in a Hasse diagram of the partial order.

A much more compact representation of the propositional calculus can be given in terms of its *Greechie diagram* [Greechie, 1971]. In this representation, the emphasis is on Boolean subalgebras. Points " $\circ$ " represent the atoms. If they belong to the same Boolean subalgebra, they are connected by edges or smooth curves. The collection of all atoms and elements belonging to the same Boolean subalgebra is called *block*; i.e., every block represents a Boolean subalgebra within a nonboolean structure. The blocks can be joined or pasted together as follows.

- The tautologies of all blocks are identified.

- The absurdities of all blocks are identified.
- Identical elements in different blocks are identified.
- The logical and algebraic structures of all blocks remain intact.

This construction is often referred to as *pasting* construction. If the blocks are only pasted together at the tautology and the absurdity, one calls the resulting logic a *horizontal sum*.

Every single block represents some "maximal collection of co-measurable observables" which will be identified with some quantum *context*. Hilbert lattices can be thought of as the pasting of a continuity of such blocks or contexts.

Note that whereas all propositions within a given block or context are co-measurable; propositions belonging to different blocks are not. This latter feature is an expression of complementarity. Thus from a strictly operational point of view, it makes no sense to speak of the "real physical existence" of different contexts, as knowledge of a single context makes impossible the measurement of all the other ones.

Einstein-Podolski-Rosen (EPR) type arguments [Einstein *et al.*, 1935] utilizing a configuration sketched in Fig. 1 claim to be able to infer two different contexts counterfactually. One context is measured on one side of the setup, the other context on the other side of it. By the uniqueness property [Svozil, 2005d; Svozil, 2005a] of certain two-particle states, knowledge of a property of one particle entails the certainty that, if this property were measured on the other particle as well, the outcome of the measurement would be a unique function of the outcome of the measurement performed. This makes possible the measurement of one context, as well as the simultaneous counterfactual inference of another, mutual exclusive, context. Because, one could argue, although one has actually measured on one side a different, incompatible context compared to the context measured on the other side, if on both sides the same context *would be measured*, the outcomes on both sides *would be uniquely correlated*. Hence measurement of one context per side is sufficient, for the outcome could be counterfactually inferred on the other side.

As problematic as counterfactual physical reasoning may appear from an operational point of view even for a two particle state, the simultaneous "counterfactual inference" of three or more blocks or contexts fails because of the missing uniqueness property [Svozil, 2005a] of quantum states.

As a first example, we shall paste together observables of the spin one-half systems. We have associated a propositional system

$$L(\mathbf{a}) = \{0, E, E', 1\},$$

corresponding to the outcomes of a measurement of the spin states along some arbitrary direction $\mathbf{a}$. If the spin states would be measured along a different spatial direction, say $\mathbf{b} \neq \pm\mathbf{a}$, an identical propositional system

$$L(\mathbf{b}) = \{0, F, F', 1\}$$

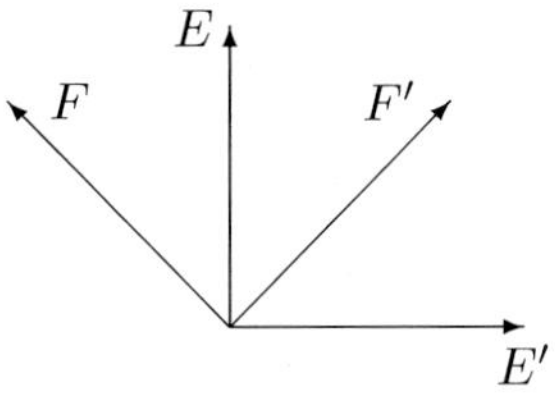

Figure 3. Two-dimensional configuration of spin 1/2 state measurements along two directions **a** and **b**.

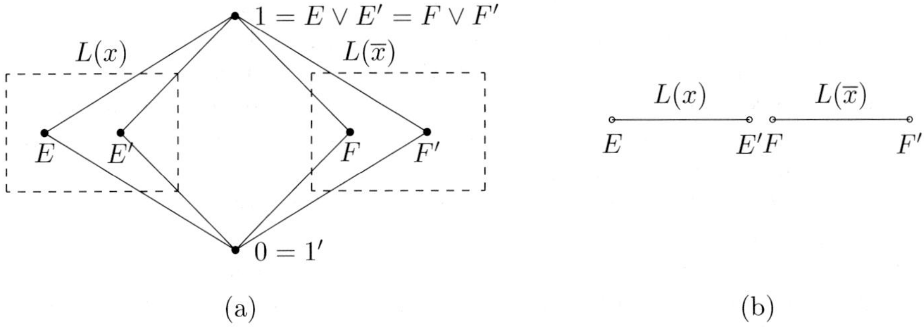

Figure 4. (a) Hasse diagram of the "Chinese lantern" form obtained by the pasting of two spin one-half propositional systems $L(x)$ and $L(\overline{x})$ which are noncomeasurable. The resulting logical structure is a modular orthocomplemented lattice $L(x) \oplus L(\overline{x}) = MO_2$. The blocks (without $0, 1$) are indicated by dashed boxes. (b) Greechie diagram of the configuration depicted in (a).

would have resulted, with the propositions E and F explicitly expressed before. The two-dimensional Hilbert space representation of this configuration is depicted in Figure 3.

$L(\mathbf{a})$ and $L(\mathbf{b})$ can be joined by pasting them together. In particular, we identify their tautologies and absurdities; i.e., 0 and 1. All the other propositions remain distinct. We then obtain a propositional structure

$$L(\mathbf{a}) \oplus L(\mathbf{b}) = MO_2$$

whose Hasse diagram is of the "Chinese lantern" form and is drawn in Figure 4(a). The corresponding Greechie Diagram is drawn in Figure 4(b). Here, the "O" stands for *orthocomplementation,* expressing the fact that for every element there exists an orthogonal complement. The term "M" stands for *modularity*; i.e., for all $x \rightarrow b$, $x \vee (a \wedge b) = (x \vee a) \wedge b$. The subscript "2" stands for the pasting of two Boolean subalgebras 2^2. Since all possible directions $\mathbf{a} \in \mathbb{R}^3$ form a continuum, the Hilbert lattice is a continuum of pastings of subalgebras of the form $L(\mathbf{a})$.

The propositional system obtained is not a classical Boolean algebra, since the distributive laws are not satisfied; i.e.,

$$\begin{array}{ccc} F \vee (E \wedge E') & \stackrel{?}{=} & (F \vee E) \wedge (F \vee E') \\ F \vee 0 & \stackrel{?}{=} & 1 \wedge 1 \\ F & \neq & 1, \\ F \wedge (E \vee E') & \stackrel{?}{=} & (F \wedge E) \vee (F \wedge E') \\ F \wedge 1 & \stackrel{?}{=} & 0 \vee 0 \\ F & \neq & 0. \end{array}$$

Notice that the expressions can be easily evaluated by using the Hasse diagram 4(a): For any a, b, $a \vee b$ is just the least element which is connected by a and b; $a \wedge b$ is just the highest element connected to a and b. Intermediates which are not connected to both a and b do not count. That is,

$a \vee b$ is called a least upper bound of a and b. $a \wedge b$ is called a greatest lower bound of a and b.

MO_2 is a specific example of an algebraic structure which is called a *lattice*. Any two elements of a lattice have a least upper and a greatest lower bound satisfying the commutative, associative and absorption laws.

Nondistributivity is the algebraic expression of nonclassicality, but what is the algebraic reason for nondistributivity? It is, heuristically speaking, scarcity, the lack of necessary algebraic elements to "fill up" all propositions necessary to obtain one and the same result in both ways as expressed by the distributive law.

3.2 Quantum contexts as blocks

All that is operationally knowable for a given quantized system is a *single block* representing co-measurable observables. Thus, single blocks or, in another terminology, maximal Boolean subalgebras of Hilbert lattices, will be identified with quantum contexts. As Hilbert lattices are pastings of a continuity of blocks or contexts, contexts are the building blocks of quantum logics.

A quantum context can equivalently be formalized by a single (nondegenerate) "maximal" self-adjoint operator C, such that all commuting, compatible co-measurable observables are functions thereof. (e.g., Ref. [von Neumann, 1932], Sec. II.10, p. 90, English translation p. 173; Ref. [Kochen and Specker, 1967], § 2; Ref. [Neumark, 1954], pp. 227,228; Ref. [Halmos, 1974], § 84). Note that mutually commuting opators have identical pairwise orthogonal sets of eigenvectors (forming an orthonormal basis) which correspond to pairwise orthogonal projectors adding up to unity. The spectral decompositions of the mutually commuting opators thus contain sums of identical pairwise orthogonal projectors.

Thus the "maximal" self-adjoint operator C has a spectral decomposition into some complete set of orthogonal projectors E_i which correspond to elementary "yes"-"no" propositions in the Von Neumann-Birkhoff type sense [von Neumann, 1932; Birkhoff and von Neumann, 1936]. That is, $C = \sum_{i=1}^{n} c_i E_i$ with mutually different c_i and $\sum_{i=1}^{n} E_i = \mathbb{I}$. In n dimensions, contexts can be viewed as n-pods spanned by the n orthogonal vectors corresponding to the projectors $E_1, E_2, \cdots, E_n$. As there exist many such representations with many different sets of coefficients c_i, "maximal" operator are not unique.

An observable belonging to two or more contexts is called *link observable*. Contexts can thus be depicted by Greechie diagrams [Greechie, 1971], consisting of *points* which symbolize observables (representable by the spans of vectors in n-dimensional Hilbert space). Any n points belonging to a context; i.e., to a maximal set of co-measurable observables (representable as some orthonormal basis of n-dimensional Hilbert space), are connected by *smooth curves*. Two smooth curves may be crossing in common *link observables*. In three dimensions, smooth curves and the associated points stand for tripods. Still another compact representation is in terms of Tkadlec diagrams [Tkadlec, 2000], in which points represent complete tripods and smooth curves represent single legs interconnecting them.

In two dimensional Hilbert space, interlinked contexts do not exist, since every context is fixed by the assumption of one property. The entire context is just this property, together with its negation, which corresponds to the orthogonal ray (which spans a one dimensional subspace) or projection associated with the ray corresponding to the property.

The simplest nontrivial configuration of interlinked contexts exists in three-dimensional Hilbert space. Consider an arrangement of five observables A, B, C, D, K with two systems of operators $\{A, B, C\}$ and $\{D, K, A\}$, the contexts, which are interconnected by A. Within a context, the operators commute and the associated observables are co-measurable. For two different contexts, operators outside the link operators do not commute. A is a link observable. This propositional structure (also known as L_{12}) can be represented in three-dimensional Hilbert space by two tripods with a single common leg. Fig. 5 depicts this configuration in three-dimensional real vector space, as well as in the associated Greechie and Tkadlec diagrams. The operators B, C, A and D, K, A can be identified with the projectors corresponding to the two bases

$$\begin{aligned} B_{B-C-A} &= \{(1,0,0)^T, (0,1,0)^T, (0,0,1)^T\}, \\ B_{D-K-A} &= \{(\cos\varphi, \sin\varphi, 0)^T, (-\sin\varphi, \cos\varphi, 0)^T, (0,0,1)^T\}, \end{aligned}$$

(the superscript "T" indicates transposition). Their matrix representation is the dyadic product of every vector with itself.

Physically, the union of contexts $\{B, C, A\}$ and $\{D, K, A\}$ interlinked along A does not have any direct operational meaning; only a single context can be measured along a single quantum at a time; the other being irretrievably lost if no reconstruction of the original state is possible. Thus, in a direct way, testing the value of observable A against different contexts $\{B, C, A\}$ and $\{D, K, A\}$ is

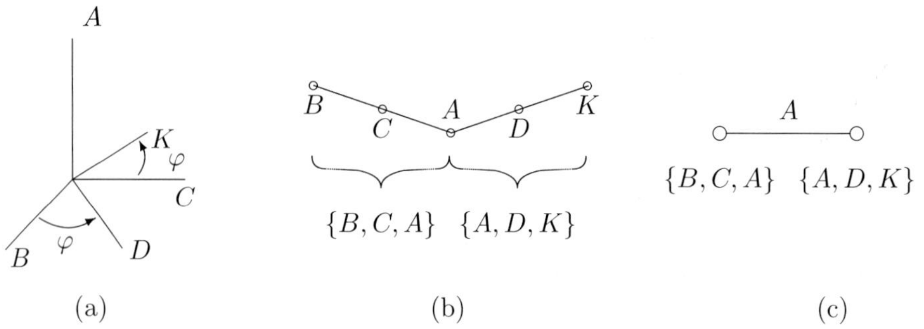

Figure 5. Three equivalent representations of the same geometric configuration: (a) Two tripods with a common leg; (b) Greechie (orthogonality) diagram: points stand for individual basis vectors, and orthogonal tripods are drawn as smooth curves; (c) Tkadlec diagram: points represent complete tripods and smooth curves represent single legs interconnecting them.

metaphysical.

It is, however, possible to counterfactually retrieve information about the two different contexts of a single quantum indirectly by considering a singlet state $|\Psi_2\rangle = (1/\sqrt{3})(|+-\rangle + |-+\rangle - |00\rangle)$ via the "explosion view" Einstein-Podolsky-Rosen type of argument depicted in Fig. 1. Since the state is form invariant with respect to variations of the measurement angle and at the same time satisfies the uniqueness property [Svozil, 2005a], one may retrieve the first context $\{B, C, A\}$ from the first quantum and the second context $\{D, K, A\}$ from the second quantum. (This is a standard procedure in Bell type arguments with two spin one-half quanta.)

More tightly interlinked contexts such as $\{A, B, C\} - \{C, D, E\} - \{E, F, A\}$, whose Greechie diagram is a triangle with the edges A, C and E, or $\{A, B, C\} - \{C, D, E\} - \{E, F, G\} - \{G, H, A\}$, whose Greechie diagram is a quadrangle with the edges A, C, E and G, cannot be represented in Hilbert space and thus have no realization in quantum logics. The five contexts $\{A, B, C\} - \{C, D, E\} - \{E, F, G\} - \{G, H, I\} - \{I, J, A\}$ whose Greechie diagrams is a pentagon with the edges A, C, E, G and I have realizations in $\mathbb{R}^3$ [Svozil and Tkadlec, 1996].

3.3 Probability theory

Kochen-Specker theorem

Quantum logics of Hilbert space dimension greater than two have not a single two-valued state interpretable as consistent, overall truth assignment [Specker, 1960]. This is the gist of the beautiful construction of Kochen and Specker [Kochen and Specker, 1967]. For similar theorems, see Refs. [Zierler and Schlessinger, 1965; Alda, 1980; Alda, 1981; Kamber, 1964; Kamber, 1965]. As a result of the nonexistence of two-valued states, the classical strategy to construct probabilities by a convex combination of all two-valued states fails entirely.

One of the most compact and comprehensive versions of the Kochen-Specker proof by contradiction in three-dimensional Hilbert space $\mathbb{R}^3$ has been given by Peres [Peres, 1991]. (For other discussions, see Refs. [Stairs, 1983; Redhead, 1990; Jammer, 1992; Brown, 1992; Peres, 1991; Peres, 1993; Zimba and Penrose, 1993; Clifton, 1993; Mermin, 1993; Svozil and Tkadlec, 1996].) Peres' version uses a 33-element set of lines without a two-valued state. The direction vectors of these lines arise by all permutations of coordinates from

$$(0,0,1),\ (0,\pm 1,1),\ (0,\pm 1,\sqrt{2}),\quad \text{and}\quad (\pm 1,\pm 1,\sqrt{2}).$$

These lines can be generated (by the "nor"-operation between nonorthogonal propositions) by the three lines [Svozil and Tkadlec, 1996]

$$(1,0,0),\ (1,1,0),\ (\sqrt{2},1,1).$$

Note that as three arbitrary but mutually nonorthogonal lines generate a dense set of lines [Havlicek and Svozil, 1996], it can be expected that any such triple of lines (not just the one explicitly mentioned) generates a finite set of lines which does not allow a two-valued probability measure.

The way it is defined, this set of lines is invariant under interchanges (permutations) of the x_1, x_2 and x_3 axes, and under a reversal of the direction of each of these axes. This symmetry property allows us to assign the probability measure 1 to some of the rays without loss of generality. Assignment of probability measure 0 to these rays would be equivalent to renaming the axes, or reversing one of the axes.

The Greechie diagram of the Peres configuration is given in Figure 6 [Svozil and Tkadlec, 1996]. For simplicity, 24 points which belong to exactly one edge are omitted. The coordinates should be read as follows: $\bar{1} \to -1$ and $2 \to \sqrt{2}$; e.g., $1\bar{1}2$ denotes $\mathrm{Sp}(1,-1,\sqrt{2})$. Concentric circles indicate the (non orthogonal) generators mentioned above.

Let us prove that there is no two-valued probability measure [Svozil and Tkadlec, 1996; Tkadlec, 1998]. Due to the symmetry of the problem, we can choose a particular coordinate axis such that, without loss of generality, $P(100) = 1$. Furthermore, we may assume (case 1) that $P(21\bar{1}) = 1$. It immediately follows that $P(001) = P(010) = P(102) = P(\bar{1}20) = 0$. A second glance shows that $P(20\bar{1}) = 1$, $P(1\bar{1}2) = P(112) = 0$.

Let us now suppose (case 1a) that $P(201) = 1$. Then we obtain $P(\bar{1}12) = P(\bar{1}\bar{1}2) = 0$. We are forced to accept $P(110) = P(1\bar{1}0) = 1$ — a contradiction, since (110) and $(1\bar{1}0)$ are orthogonal to each other and lie on one edge.

Hence we have to assume (case 1b) that $P(201) = 0$. This gives immediately $P(\bar{1}02) = 1$ and $P(211) = 0$. Since $P(01\bar{1}) = 0$, we obtain $P(2\bar{1}\bar{1}) = 1$ and thus $P(120) = 0$. This requires $P(2\bar{1}0) = 1$ and therefore $P(12\bar{1}) = P(121) = 0$. Observe that $P(210) = 1$, and thus $P(\bar{1}2\bar{1}) = P(\bar{1}21) = 0$. In the following step, we notice that $P(10\bar{1}) = P(101) = 1$ — a contradiction, since (101) and $(10\bar{1})$ are orthogonal to each other and lie on one edge.

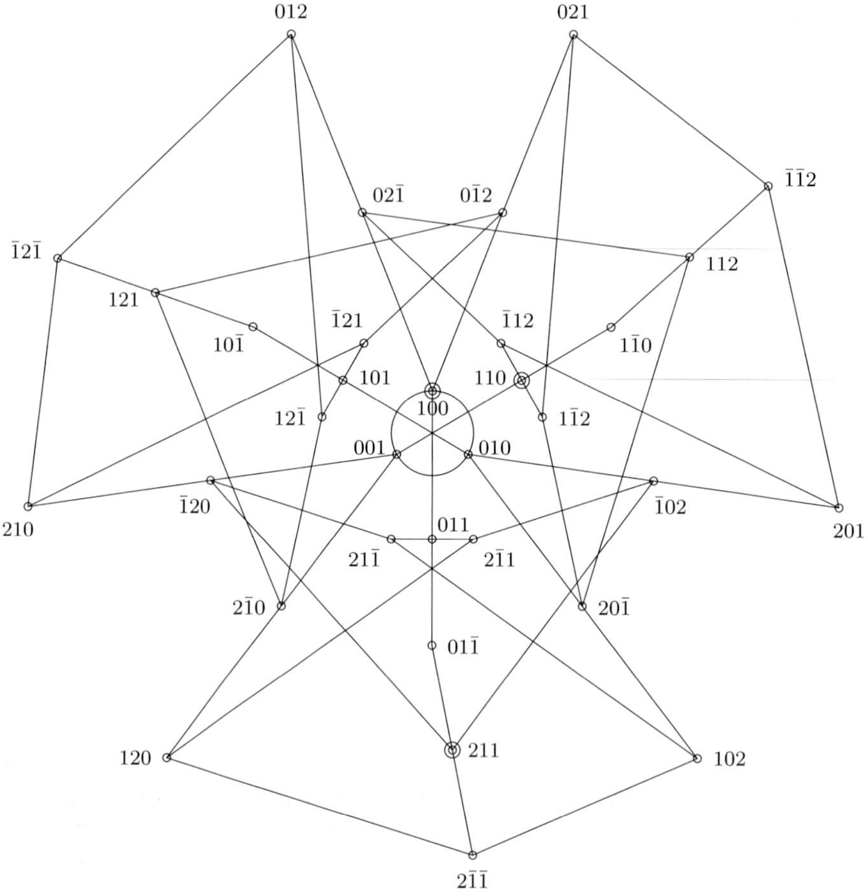

Figure 6. Greechie diagram of a finite subset of the continuum of blocks or contexts embeddable in three-dimensional real Hilbert space without a two-valued probability measure [Svozil and Tkadlec, 1996, Figure 9].

Thus we are forced to assume (case 2) that $P(2\bar{1}1) = 1$. There is no third alternative, since $P(011) = 0$ due to the orthogonality with (100). Now we can repeat the argument for case 1 in its mirrored form.

The most compact way of deriving the Kochen-Specker theorem in four dimensions has been given by Cabello [Cabello *et al.*, 1996; Cabello, 2000]. It is depicted in Fig. 7.

Gleason's derivation of the Born rule

In view of the nonexistence of classical two-valued states on even finite superstructures of blocks or contexts associated with quantized systems, one could still resort to classicality *within* blocks or contexts. According to Gleason's theorem, this is exactly the route, the *"via regia,"* to the quantum probabilities, in particular to the Born rule.

According to the *Born rule*, the expectation value $\langle A \rangle$ of an observable A is the trace of ρA; i.e., $\langle A \rangle = \text{tr}(\rho A)$. In particular, if A is a projector E corresponding to an elementary yes-no proposition *"the system has property Q,"* then $\langle E \rangle = \text{tr}(\rho E)$ corresponds to the probability of that property Q if the system is in state ρ. The equations $\rho^2 = \rho$ and $\text{tr}(\rho^2) = 1$ are only valid for pure states, because ρ is not an projector and thus idempotent for mixed states.

It is still possible to ascribe a certain degree of classical probabilistic behaviour to a quantum logic by considering its block superstructure. Due to their Boolean algebra, blocks are "classical mini-universes." It is one of the mindboggling features of quantum logic that it can be decomposed into a pasting of blocks. Conversely, by a proper arrangement of "classical mini-universes," quantum Hilbert logics can be obtained. This theme is used in quantum probability theory, in particular by the Gleason and the Kochen-Specker theorems. In this sense, Gleason's theorem can be understood as the functional analytic generalization of the generation of all classical probability distributions by a convex sum of the extreme cases.

Gleason's theorem [Gleason, 1957; Dvurečenskij, 1993; Cooke *et al.*, 1985; Peres, 1993; Hrushovski and Pitowsky, 2004; Richman and Bridges, 1999] is a derivation of the Born rule from fundamental assumptions about quantum probabilities, guided by the quasi–classical; i.e., Boolean, sub-parts of quantum theory. Essentially, the main assumption required for Gleason's theorem is that *within* blocks or contexts, the quantum probabilities behave as classical probabilities; in particular the sum of probabilities over a complete set of mutually exclusive events add up to unity. With these quasi–classical provisos, Gleason proved that there is no alternative to the Born rule for Hilbert spaces of dimension greater than two.

3.4 *Quantum violations of classical probability bounds*

Due to the different form of quantum correlations, which formally is a consequence of the different way of defining quantum probabilities, the constraints on classical probabilities are violated by quantum probabilities. Quantitatively, this can be

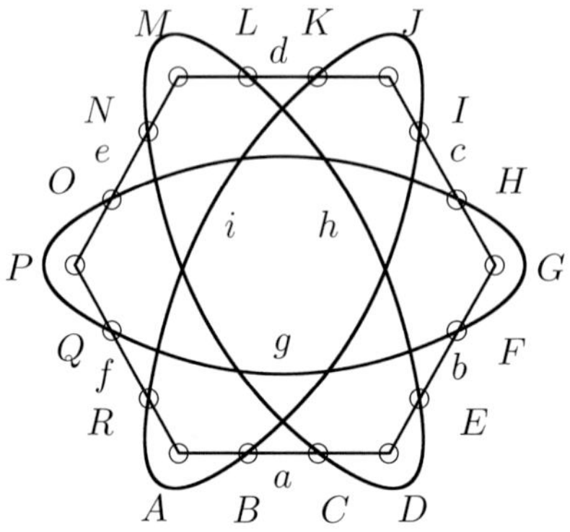

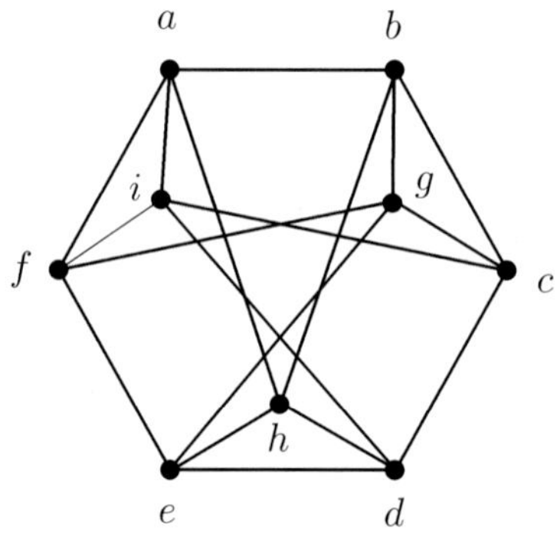

Figure 7. Greechie diagram of a finite subset of the continuum of blocks or contexts embeddable in four-dimensional real Hilbert space without a two-valued probability measure [Cabello *et al.*, 1996; Cabello, 2000]. The proof of the Kochen-Specker theorem uses nine tightly interconnected contexts $a = \{A, B, C, D\}$, $b = \{D, E, F, G\}$, $c = \{G, H, I, J\}$, $d = \{J, K, L, M\}$, $e = \{M, N, O, P\}$, $f = \{P, Q, R, A\}$, $g = \{B, I, K, R\}$, $h = \{C, E, L, N\}$, $i = \{F, H, O, Q\}$ consisting of the 18 projectors associated with the one dimensional subspaces spanned by $A = (0, 0, 1, -1)$, $B = (1, -1, 0, 0)$, $C = (1, 1, -1, -1)$, $D = (1, 1, 1, 1)$, $E = (1, -1, 1, -1)$, $F = (1, 0, -1, 0)$, $G = (0, 1, 0, -1)$, $H = (1, 0, 1, 0)$, $I = (1, 1, -1, 1)$, $J = (-1, 1, 1, 1)$, $K = (1, 1, 1, -1)$, $L = (1, 0, 0, 1)$, $M = (0, 1, -1, 0)$, $N = (0, 1, 1, 0)$, $O = (0, 0, 0, 1)$, $P = (1, 0, 0, 0)$, $Q = (0, 1, 0, 0)$, $R = (0, 0, 1, 1)$. (a) Greechie diagram representing atoms by points, and contexts by maximal smooth, unbroken curves. (b) Dual Tkadlec diagram representing contexts by filled points, and interconnected contexts are connected by lines. (Duality means that points represent blocks and maximal smooth curves represent atoms.) Every observable proposition occurs in exactly two contexts. Thus, in an enumeration of the four observable propositions of each of the nine contexts, there appears to be an *even* number of true propositions. Yet, as there is an odd number of contexts, there should be an *odd* number (actually nine) of true propositions.

investigated [Filipp and Svozil, 2004a] by substituting the classical probabilities by the quantum ones; i.e.,

$$\begin{array}{lcl} p_1 & \rightarrow & q_1(\theta) = \frac{1}{2}\left[\mathbb{I}_2 + \sigma(\theta)\right] \otimes \mathbb{I}_2, \\ p_3 & \rightarrow & q_3(\theta) = \mathbb{I}_2 \otimes \frac{1}{2}\left[\mathbb{I}_2 + \sigma(\theta)\right], \\ p_{ij} & \rightarrow & q_{ij}(\theta,\theta') = \frac{1}{2}\left[\mathbb{I}_2 + \sigma(\theta)\right] \otimes \frac{1}{2}\left[\mathbb{I}_2 + \sigma(\theta')\right], \end{array}$$

with $\sigma(\theta) = \begin{pmatrix} \cos\theta & \sin\theta \\ \sin\theta & -\cos\theta \end{pmatrix}$, where θ is the relative measurement angle in the x–z-plane, and the two particles propagate along the y-axis, as depicted in Fig. 1.

The quantum transformation associated with the Clauser-Horne inequality for the 2–2 case is given by

$$\begin{array}{rcl} O_{22}(\alpha,\beta,\gamma,\delta) & = & q_{13}(\alpha,\gamma) + q_{14}(\alpha,\delta) + q_{23}(\beta,\gamma) - q_{24}(\beta,\delta) - q_1(\alpha) - q_3(\gamma) \\ & = & \frac{1}{2}\left[\mathbb{I}_2 + \sigma(\alpha)\right] \otimes \frac{1}{2}\left[\mathbb{I}_2 + \sigma(\gamma)\right] + \frac{1}{2}\left[\mathbb{I}_2 + \sigma(\alpha)\right] \otimes \frac{1}{2}\left[\mathbb{I}_2 + \sigma(\delta)\right] \\ & & +\frac{1}{2}\left[\mathbb{I}_2 + \sigma(\beta)\right] \otimes \frac{1}{2}\left[\mathbb{I}_2 + \sigma(\gamma)\right] - \frac{1}{2}\left[\mathbb{I}_2 + \sigma(\beta)\right] \otimes \frac{1}{2}\left[\mathbb{I}_2 + \sigma(\delta)\right] \\ & & -\frac{1}{2}\left[\mathbb{I}_2 + \sigma(\alpha)\right] \otimes \mathbb{I}_2 - \mathbb{I}_2 \otimes \frac{1}{2}\left[\mathbb{I}_2 + \sigma(\gamma)\right], \end{array}$$

where α, β, γ, δ denote the measurement angles lying in the x–z-plane: α and β for one particle, γ and δ for the other one. The eigenvalues are

$$\lambda_{1,2,3,4}(\alpha,\beta,\gamma,\delta) = \frac{1}{2}\left(\pm\sqrt{1 \pm \sin(\alpha-\beta)\sin(\gamma-\delta)} - 1\right)$$

yielding the maximum bound $\|O_{22}\| = \max_{i=1,2,3,4}\lambda_i$. Note that for the particular choice of parameters $\alpha = 0, \beta = 2\theta, \gamma = \theta, \delta = 3\theta$ adopted in [Cabello, 2004; Filipp and Svozil, 2004b], one obtains $|O_{22}| = \frac{1}{2}\left\{\left[(3-\cos 4\theta)/2\right]^{1/2} - 1\right\} \leq \frac{1}{2}\left(\sqrt{2}-1\right)$, as compared to the classically allowed bound from above 0.

3.5 Interpretations

The nonexistence of two-valued states on the set of quantum propositions (of greater than two-dimensional Hilbert spaces) interpretable as truth assignments poses a great challenge for the interpretation of quantum logical propositions, relations and operations, as well as for quantum mechanics in general. At stake is the meaning and physical co-existence of observables which are not co-measurable. Several interpretations have been proposed, among them contextuality, as well as the abandonment of classical omniscience and realism discussed below.

Contextuality

Contextuality abandons the context independence of measurement outcomes [Bell, 1966; Heywood and Redhead, 1983; Redhead, 1990] by supposing that it is wrong to assume (cf. Ref. [Bell, 1966], Sec. 5) that the result of an observation is independent of what observables are measured alongside of it. Bell [Bell, 1966, Sec. 5] states that the "*... result of an observation may reasonably depend not* *only on the state of the system ... but also on the complete disposition of the apparatus."* Note also Bohr's remarks [Bohr, 1949] about *"the impossibility of any sharp separation between the behavior of atomic objects and the interaction with the measuring instruments which serve to define the conditions under which the phenomena appear."*

Contextuality might be criticized as an attempt to maintain omniscience and omni-realism even in view of a lack of consistently assignable truth values on quantum propositions. Omniscience or omni-realism is the belief that *"all observables exist even without being experienced by any finite mind."* Contextuality supposes that an *"observable exists without being experienced by any finite mind, but it may have different values, depending on its context."*

So far, despite some claims to have measured contextuality, there is no direct experimental evidence. Some experimental findings inspired by Bell-type inequalities [Aspect *et al.*, 1981; Aspect *et al.*, 1982; Weihs *et al.*, 1998], the Kochen-Specker theorem [Simon *et al.*, 2000; Hasegawa *et al.*, 2006] as well as the Greenberger-Horne-Zeilinger theorem [Pan *et al.*, 2000] measure incompatible contexts one after another; i.e., temporally sequentially, and not simultaneously. Hence, different contexts can only be measured on different particles. A more direct test of contextuality might be an EPR configuration of two quanta in three-dimensional Hilbert space interlinked in a single observable, as discussed above.

Abandonment of classical omniscience

As has been pointed out already, contextuality might be criticized for its presumption of quantum omniscience; in particular the supposition that a physical system, at least in principle, is capable of "carrying" all answers to any classically retrievable question. This is true classically, since the classical context is the entirety of observables. But it need not be true for other types of (finite) systems or agents. Take for example, a refrigerator. If it is automated in a way to tell you whether or not there is enough milk in it, it will be at a complete loss at answering a totally different question, such as if there is enough oil in the engine of your car. It is a matter of everday experience that not all agents are prepared to give answers to all perceivable questions.

Nevertheless, if one forces an agent to answer a question it is incapable to answer, the agent might throw some sort of "fair coin" — if it is capable of doing so — and present random answers. This scenario of a context mismatch between preparation and measurement is the basis of quantum random number generators [id Quantique, 2004] which serve as a kind of "quantum random oracle" [Calude, 2004; Calude and Dinneen, 2005]. It should be kept in mind that randomness, at least algorithmically [Chaitin, 1990; Chaitin, 1987; Calude, 2002], does not come "for free," thus exhibiting an amazing capacity of single quanta to support random outcomes. Alternatively, the unpredictable, erratic outcomes might, in the context translation [Svozil, 2004] scenario, be due to some stochasticity originating from the interaction with a "macroscopic" measurement apparatus, and the undefined.

One interpretation of the impossibility to operationalize more than a single context is the abandonment of classical omniscience: in this view, whereas it might be meaningful theoretically and formally to study the entirety of the context superstructure, only a single context operationally exists. Note that, in a similar way as *retrieving* information from a quantized system, the only information *codable* into a quantized system is given by a single block or context. If the block contains n atoms corresponding to n possible measurement outcomes, then the information content is a $n$it [Zeilinger, 1999; Donath and Svozil, 2002; Svozil, 2002]. The information needs not be "located" at a particular particle, as it can be "distributed" over a multi–partite state. In this sense, the quantum system could be viewed as a kind of (possibly nonlocal) *programmable integrated circuit,* such as a *field programmable qate array* or an *application specific integrated circuit.*

Quantum observables make only sense when interpreted as a function of some context, formalized by either some Boolean subalgebra or by the maximal operator. It is useless in this framework to believe in the existence of a single isolated observable devoid of the context from which it is derived. In this holistic approach, isolated observables separated from its missing contexts do not exist.

Likwise, it is wrong to assume that all observables which could in principle ("potentially") have been measured, also co-exist, irrespective of whether or not they have or could have been actually measured. Realism in the sense of *"co-measurable entities sometimes exist without being experienced by any finite mind"* might still be assumed for a *single* context, in particular the one in which the system was prepared.

Subjective idealism

Still another option is subjective idealism, denying the "existence" of observables which could in principle ("potentially") have been measured, but actually have not been measured: in this view, it is wrong to assume that [Stace, 1949]

> *"entities sometimes exist without being experienced by any finite mind."*

Indeed, Bekeley states [Berkeley, 1710],

> *"For as to what is said of the absolute existence of unthinking things without any relation to their being perceived, that seems perfectly unintelligible. Their esse [[to be]] is percepi [[to be perceived]], nor is it possible they should have any existence out of the minds or thinking things which perceive them."*

With this assumption, the Bell, Kochen-Specker and Greenberger-Horne-Zeilinger theorems and similar have merely theoretical, formal relevance for physics, because they operate with unobservable physical "observables" and entities or with counterfactuals which are inferred rather than measured.

4 AUTOMATA AND GENERALIZED URN LOGIC

The following quasi–classical logics take up the notion of contexts as blocks representing Boolean subalgebras and the pastings among them. They are quasi–classical, because unlike quantum logics they possess sufficiently many two-valued states to allow embeddings into Boolean algebras.

4.1 Partition logic

The empirical logics (i.e., the propositional calculi) associated with the generalized urn models suggested by Ron Wright [Wright, 1978; Wright, 1990], and automaton logics (APL) [Svozil, 1993; Schaller and Svozil, 1996; Dvurečenskij *et al.*, 1995; Calude *et al.*, 1997; Svozil, 1998] are equivalent (cf. Refs. [Svozil, 1998, p.145] and [Svozil, 2005b]) and can be subsumed by partition logics. The logical equivalence of automaton models with generalized urn models suggests that these logics are more general and "robust" with respect to changes of the particular model than could have been expected from the particular instances of their first appearance.

Again the concept of context or block is very important here. Partition logics are formed by pasting together contexts or blocks based on the *partitions of a set of states.* The contexts themselves are derived from the input/output analysis of experiments.

4.2 Generalized urn models

A generalized urn model $\mathcal{U} = \langle U, C, L, \Lambda \rangle$ is characterized as follows. Consider an ensemble of balls with black background color. Printed on these balls are some color symbols from a symbolic alphabet L. The colors are elements of a set of colors C. A particular ball type is associated with a unique combination of mono-spectrally (no mixture of wavelength) colored symbols printed on the black ball background. Let U be the set of ball types. We shall assume that every ball contains just one single symbol per color. (Not all types of balls; i.e., not all color/symbol combinations, may be present in the ensemble, though.)

Let $|U|$ be the number of different types of balls, $|C|$ be the number of different mono-spectral colors, $|L|$ be the number of different output symbols.

Consider the deterministic "output" or "lookup" function $\Lambda(u, c) = v$, $u \in U$, $c \in C$, $v \in L$, which returns one symbol per ball type and color. One interpretation of this lookup function Λ is as follows. Consider a set of $|C|$ eyeglasses build from filters for the $|C|$ different colors. Let us assume that these mono-spectral filters are "perfect" in that they totally absorb light of all other colors but a particular single one. In that way, every color can be associated with a particular eyeglass and vice versa.

When a spectator looks at a particular ball through such an eyeglass, the only operationally recognizable symbol will be the one in the particular color which is transmitted through the eyeglass. All other colors are absorbed, and the symbols printed in them will appear black and therefore cannot be differentiated from the black background. Hence the ball appears to carry a different "message" or symbol, depending on the color at which it is viewed. This kind of "complementarity" has been used for a demonstration of quantum cryptography [Svozil, 2006].

An empirical logic can be constructed as follows. Consider the set of all ball types. With respect to a particular colored eyeglass, this set disjointly "decays" or gets partitioned into those ball types which can be separated by the particular color of the eyeglass. Every such partition of ball types can then be identified with a Boolean algebra whose atoms are the elements of the partition. A pasting of all of these Boolean algebras yields the empirical logic associated with the particular urn model.

Consider, for the sake of demonstration, a single color and its associated partition of the set of ball types (ball types within a given element of the partition cannot be differetiated by that color). In the generalized urn model, an element a of this partition is a set of ball types which corresponds to an elementary proposition

"the ball drawn from the urn is of the type contained in a."

4.3 Automaton models

A (Mealy type) automaton $\mathcal{A} = \langle S, I, O, \delta, \lambda \rangle$ is characterized by the set of states S, by the set of input symbols I, and by the set of output symbols O. $\delta(s, i) = s'$ and $\lambda(s, i) = o$, $s, s' \in S$, $i \in I$ and $o \in O$ represent the transition and the output functions, respectively. The restriction to Mealy automata is for convenience only.

In the analysis of a *state identification problem*, a typical automaton experiment aims at an operational determination of an *unknown initial state* by the input of some symbolic sequence and the observation of the resulting output symbols. Every such input/output experiment results in a state partition in the following way. Consider a particular automaton. Every experiment on such an automaton which tries to solve the initial state problem is characterized by a set of input/output symbols as a result of the possible input/output sequences for this experiment. Every such distinct set of input/output symbols is associated with a set of initial automaton states which would reproduce that sequence. This state set may contain one or more states, depending on the ability of the experiment to separate different initial automaton states. A partitioning of the automaton states is obtained if one considers a single input sequence and the variety of all possible output sequences (given a particular automaton). Stated differently: given a set of inputs, the set of initial automaton states "break down" into disjoint subsets associated with the possible output sequences. (All elements of a subset yield the same output on the same input.)

This partition can then be identified with a Boolean algebra, with the elements of the partition interpreted as atoms. By pasting the Boolean algebras of the "finest" partitions together one obtains an empirical partition logic associated with the particular automaton. (The converse construction is also possible, but not unique; see below.)

For the sake of simplicity, we shall assume that every experiment just deals with a single input/output combination. That is, the finest partitions are reached already after the first symbol. This does not impose any restriction on the partition logic, since given any particular automaton, it is always possible to construct another automaton with exactly the same partition logic as the first one with the above property.

More explicitly, given any partition logic, it is always possible to construct a corresponding automaton with the following specification: associate with every element of the set of partitions a single input symbol. Then take the partition with the highest number of elements and associate a single output symbol with any element of this partition. (There are then sufficient output symbols available for the other partitions as well.) Different partitions require different input symbols; one input symbol per partition. The output function can then be defined by associating a single output symbol per element of the partition (associated with a particular input symbol). Finally, choose a transition function which completely looses the state information after only one transition; i.e., a transition function which maps all automaton state into a single one.

A typical proposition in the automaton model refers to a partition element a containing automaton states which cannot be distinguished by the analysis of the strings of input and output symbols; i.e., it can be expressed by

"the automaton is initially in a state which is contained in a."

4.4 Contexts

In the generalized urn model represent everything that is knowable by looking in only a single color. For automata, this is equivalent to considering only a single string of input symbols. Formally, this amounts to the identification of blocks with contexts, as in the quantum case.

4.5 Proof of logical equivalence of automata and generalized urn models

From the definitions and constructions mentioned in the previous sections it is intuitively clear that, with respect to the empirical logics, generalized urn models and finite automata models are equivalent. Every logic associated with a generalized urn model can be interpreted as an automaton partition logic associated with some (Mealy) automaton (actually an infinity thereof). Conversely, any logic associated with some (Mealy) automaton can be interpreted as a logic associated with some generalized urn model (an infinity thereof). We shall proof these claims by explicit construction. Essentially, the lookup function Λ and the output function λ will be identified. Again, the restriction to Mealy automata is for convenience only. The considerations are robust with respect to variations of finite input/output automata.

Direct construction of automaton models from generalized urn models

In order to define an APL associated with a Mealy automaton $\mathcal{A} = \langle S, I, O, \delta, \lambda\rangle$ from a generalized urn model $\mathcal{U} = \langle U, C, L, \Lambda\rangle$, let $u \in U$, $c \in C$, $v \in L$, and $s, s' \in S$, $i \in I$, $o \in O$, and assume $|U| = |S|$, $|C| = |I|$, $|L| = |O|$. The following identifications can be made with the help of the bijections t_S, t_I and t_O:

$$\begin{aligned}
&t_S(u) = s,\; t_I(c) = i,\; t_O(v) = o,\\
&\delta(s,i) = s_i \quad \text{for fixed } s_i \in S \text{ and arbitrary } s \in S,\; i \in I,\\
&\lambda(s,i) = t_O\left(\Lambda(t_S^{-1}(s), t_I^{-1}(i))\right).
\end{aligned}$$

More generally, one could use equivalence classes instead of a bijection. Since the input-output behavior is equivalent and the automaton transition function is trivially $|L|$-to-one, both entities yield the same propositional calculus.

Direct construction of generalized urn models from automaton models

Conversely, consider an arbitrary Mealy automaton $\mathcal{A} = \langle S, I, O, \delta, \lambda\rangle$ and its associated propositional calculus APL.

Just as before, associate with every single automaton state $s \in S$ a ball type u, associate with every input symbol $i \in I$ a unique color c, and associate with every output symbol $o \in O$ a unique symbol v; i.e., again $|U| = |S|$, $|C| = |I|$, $|L| = |O|$. The following identifications can be made with the help of the bijections τ_U, τ_C and τ_L:

$$\tau_U(s) = u,\; \tau_C(i) = c,\; \tau_L(o) = v,\; \Lambda(u,c) = \tau_L(\lambda(\tau_U^{-1}(u), \tau_C^{-1}(c))).$$

A comparison yields

$$\tau_U^{-1} = t_S,\; \tau_C^{-1} = t_I,\; \tau_L^{-1} = t_O.$$

Schemes using dispersion-free states

Another equivalence scheme uses the fact that both automaton partition logics and the logic of generalized urn models have a separating (indeed, full) set of dispersion-free states. Stated differently, given a finite atomic logic with a separating set of states, then the enumeration of the complete set of dispersion-free states enables the explicit construction of generalized urn models and automaton logics whose logic corresponds to the original one.

This can be achieved by "inverting" the set of two-valued states as follows. (The method is probably best understood by considering the examples below.) Let us start with an atomic logic with a separating set of states.

(i) In the first step, every atom of this lattice is labeled by some natural number, starting from "1" to "n", where n stands for the number of lattice atoms. The set of atoms is denoted by $A = \{1, 2, \ldots, n\}$.

(ii) Then, all two-valued states of this lattice are labeled consecutively by natural numbers, starting from "m_1" to "m_r", where r stands for the number of two-valued states. The set of states is denoted by $M = \{m_1, m_2, \ldots, m_r\}$.

(iii) Now partitions are defined as follows. For every atom, a set is created whose members are the numbers or "labels" of the two-valued states which are "true" or take on the value "1" on this atom. More precisely, the elements $p_i(a)$ of the partition $\mathcal{P}_j$ corresponding to some atom $a \in A$ are defined by

$$p_i(a) = \{k \mid m_k(a) = 1, \; k \in M\}\,.$$

The partitions are obtained by taking the unions of all p_i which belong to the same subalgebra $\mathcal{P}_j$. That the corresponding sets are indeed partitions follows from the properties of two-valued states: two-valued states (are "true" or) take on the value "1" on just one atom per subalgebra and ("false" or) take on the value "0" on all other atoms of this subalgebra.

(iv) Let there be t partitions labeled by "1" through "t". The partition logic is obtained by a pasting of all partitions $\mathcal{P}_j$, $1 \le j \le t$.

(v) In the following step, a corresponding generalized urn model or automaton model is obtained from the partition logic just constructed.

(a) A generalized urn model is obtained by the following identifications (see also [Wright, 1978, p. 271]).

- Take as many ball types as there are two-valued states; i.e., r types of balls.
- Take as many colors as there are subalgebras or partitions; i.e., t colors.
- Take as many symbols as there are elements in the partition(s) with the maximal number of elements; i.e., $\max_{1\le j\le t} |\mathcal{P}_j| \le n$. To make the construction easier, we may just take as many symbols as there are atoms; i.e., n symbols. (In some cases, much less symbols will suffice). Label the symbols by v_l. Finally, take r "generic" balls with black background. Now associate with every measure a different ball type. (There are r two-valued states, so there will be r ball types.)
- The ith ball type is painted by colored symbols as follows: Find the atoms for which the ith two-valued state m_i is 1. Then paint the symbol corresponding to every such lattice atom on the ball, thereby choosing the color associated with the subalgebra or partition the atom belongs to. If the atom belongs to more than one subalgebra, then paint the same symbol in as many colors as there are partitions or subalgebras the atom belongs to (one symbol per subalgebra).

This completes the construction.

(b) A Mealy automaton is obtained by the following identifications (see also [Svozil, 1993, pp. 154–155]).

- Take as many automaton states as there are two-valued states; i.e., r automaton states.
- Take as many input symbols as there are subalgebras or partitions; i.e., t symbols.
- Take as many output symbols as there are elements in the partition(s) with the maximal number of elements (plus one additional auxiliary output symbol "$*$", see below); i.e., $\max_{1\leq j\leq t}|\mathcal{P}_j| \leq n+1$.
- The output function is chosen to match the elements of the state partition corresponding to some input symbol. Alternatively, let the lattice atom $a_q \in A$ must be an atom of the subalgebra corresponding to the input i_l. Then one may choose an output function such as

$$\lambda(m_k, i_l) = \begin{cases} a_q & \text{if } m_k(a_q) = 1 \\ * & \text{if } m_k(a_q) = 0 \end{cases}$$

  with $1 \leq k \leq r$ and $1 \leq l \leq t$. Here, the additional output symbol "$*$" is needed.
- The transition function is r–to–1 (e.g., by $\delta(s,i) = s_1$, $s, s_1 \in S$, $i \in I$), i.e., after one input the information about the initial state is completely lost.

This completes the construction.

Example 1: The generalized urn logic L_{12}

In what follows we shall illustrate the above constructions with a couple of examples. First, consider the generalized urn model

$$\langle \{u_1, \ldots, u_5\}, \{\text{red}, \text{green}\}, \{1, \ldots, 5\}, \Lambda\rangle$$

with Λ listed in Table 5(a).

The associated Mealy automaton can be directly constructed as follows. Take $t_S = t_O = \text{id}$, where id represents the identity function, and take $t_I(\text{red}) = 0$ and $t_I(\text{green}) = 1$, respectively. Furthermore, fix a (five$\times$two)-to-one transition function by $\delta(.,.) = 1$. The transition and output tables are listed in Table 5(b). Both empirical structures yield the same propositional logic L_{12} which is depicted in Fig. 5(b).

Example 2: The automaton partition logic L_{12}

Let us start with an automaton whose transition and output tables are listed in Table 5(b) and indirectly construct a logically equivalent generalized urn model by using dispersion-free states. The first thing to do is to figure out all dispersion-free

ball type	red	green
1	1	3
2	1	4
3	2	3
4	2	4
5	5	5

(a)

state	δ					λ				
	1	2	3	4	5	1	2	3	4	5
0	1	1	1	1	1	1	1	2	2	5
1	1	1	1	1	1	3	4	3	4	5

(b)

Table 5. (a) Ball types in Wright's generalized urn model [Wright, 1990] (cf. also [Svozil, 1998, p.143ff]). (b) Transition and output table of an associated automaton model.

	colors									
ball type	c_1 "red"					c_2 "green"				
1	*	*	*	*	5	*	*	*	*	5
2	*	2	*	*	*	*	*	*	4	*
3	*	2	*	*	*	*	*	3	*	*
4	1	*	*	*	*	*	*	*	4	*
5	1	*	*	*	*	*	*	3	*	*

Table 6. Representation of the sign coloring scheme Λ. "*" means no sign at all (black) for the corresponding atom.

states of L_{12} depicted in Fig. 5(b). There are five of them, which we might write in vector form; i.e., in lexicographic order:

$$m_1 = (0,0,0,0,1),\ m_2 = (0,1,0,1,0),\ m_3 = (0,1,1,0,0),$$
$$m_4 = (1,0,0,1,0),\ m_5 = (1,0,1,0,0).$$

Now define the following generalized urn model as follows. There are two subalgebras with the atoms $1,2,5$ and $3,4,5$, respectively. Since there are five two-valued measures corresponding to five ball types. They are colored according to the coloring rules defined above. and Λ as listed in Table 6.

Example 3: generalized urn model of the Kochen-Specker "bug" logic

Another, less simple example, is a logic which is already mentioned by Kochen and Specker [Kochen and Specker, 1967] (this is a subgraph of their Γ_1) whose automaton partition logic is depicted in Fig. 8. (It is called "bug" by Professor Specker [Specker, 1999] because of the similar shape with a bug.) There are 14 dispersion-free states which are listed in Table 7(a). The associated generalized urn model is listed in Table 7(b).

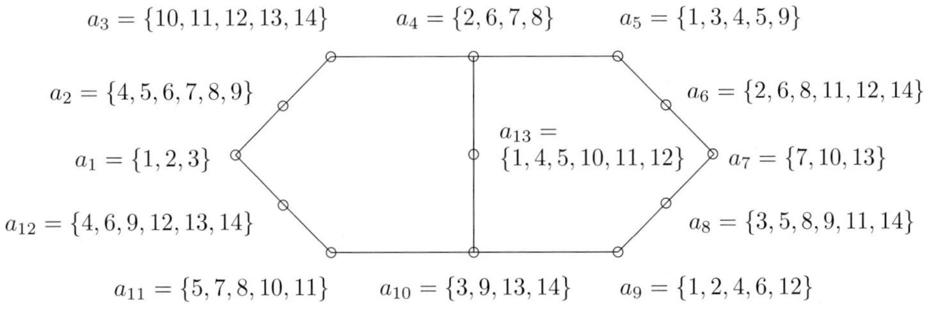

Figure 8. Greechie diagram of automaton partition logic with a nonfull set of dispersion-free measures.

m_r and ball type	(a) lattice atoms													(b) colors						
	a_1	a_2	a_3	a_4	a_5	a_6	a_7	a_8	a_9	a_{10}	a_{11}	a_{12}	a_{13}	c_1	c_2	c_3	c_4	c_5	c_6	c_7
1	1	0	0	0	1	0	0	0	1	0	0	0	1	1	5	5	9	9	1	13
2	1	0	0	1	0	1	0	0	1	0	0	0	0	1	4	6	9	9	1	4
3	1	0	0	0	1	0	0	1	0	1	0	0	0	1	5	5	8	10	3	10
4	0	1	0	0	1	0	0	0	1	0	0	1	1	2	5	5	9	9	12	13
5	0	1	0	0	1	0	0	1	0	0	1	0	1	2	5	5	8	11	11	13
6	0	1	0	1	0	1	0	0	1	0	0	1	0	2	4	6	9	9	12	4
7	0	1	0	1	0	0	1	0	0	0	1	0	0	2	4	7	7	11	11	4
8	0	1	0	1	0	1	0	1	0	0	1	0	0	2	4	6	8	11	11	4
9	0	1	0	0	1	0	0	1	0	1	0	1	0	2	5	5	8	10	12	10
10	0	0	1	0	0	0	1	0	0	0	1	0	1	3	3	7	7	11	11	13
11	0	0	1	0	0	1	0	1	0	0	1	0	1	3	3	6	8	11	11	13
12	0	0	1	0	0	1	0	0	1	0	0	1	1	3	3	6	9	9	12	13
13	0	0	1	0	0	0	1	0	0	1	0	1	0	3	3	7	7	10	13	10
14	0	0	1	0	0	1	0	1	0	1	0	1	0	3	3	6	8	10	12	10

Table 7. (a) Dispersion-free states of the Kochen-Specker "bug" logic with 14 dispersion-free states and (b) the associated generalized urn model (all blank entries "$*$"have been omitted).

4.6 Probability theory

The probability theory of partition logics is based on a full set of state, allowing to define probabilities via the convex sum of those states. This is essentially the same procedure as for classical probabilities. In the same way, bounds on probabilities can be found through the computation of the faces of correlation polytopes.

Consider, as an example, a logic already discussed. Its automaton partition logic is depicted in Fig. 8. The correlation polytope of this lattice consists of 14 vertices listed in Table 7, where the 14 rows indicate the vertices corresponding to the 14 dispersion-free states. The columns represent the partitioning of the automaton states. The solution of the hull problem yields the equalities

$$
\begin{aligned}
1 &= P_1 + P_2 + P_3 = P_4 + P_{10} + P_{13},\\
1 &= P_1 + P_2 - P_4 + P_6 + P_7 = -P_2 + P_4 - P_6 + P_8 - P_{10} + P_{12},\\
1 &= P_1 + P_2 - P_4 + P_6 - P_8 + P_{10} + P_{11},\\
0 &= P_1 + P_2 - P_4 - P_5 = -P_1 - P_2 + P_4 - P_6 + P_8 + P_9.
\end{aligned}
$$

The operational meaning of $P_i = P_{a_i}$ is "the probability to find the automaton in state a_i." The above equations are equivalent to all probabilistic conditions on the contexts (subalgebras) $1 = P_1 + P_2 + P_3 = P_3 + P_4 + P_5 = P_5 + P_6 + P_7 = P_7 + P_8 + P_9 = P_9 + P_{10} + P_{11} = P_4 + P_{10} + P_{13}$.

Let us now turn to the joint probability case. Notice that formally it is possible to form a statement such as $a_1 \wedge a_{13}$ (which would be true for measure number 1 and false otherwise), but this is not operational on a single automaton, since no experiment can decide such a proposition on a single automaton. Nevertheless, if one considers a "singlet state" of two automata which are in an unknown yet identical initial state, then an expression such as $a_1 \wedge a_{13}$ makes operational sense if property a_1 is measured on the first automaton and property a_{13} on the second automaton. Indeed, all joint probabilities $a_i \wedge a_j \wedge \ldots a_n$ make sense for n-automaton singlets.

5 SUMMARY

Regarding contexts; i.e., the maximum collection of co-measurable observables, three different cases have been discussed. The first, classical case, is characterized by omniscience. Within the classical framework, all observables form a *single* context, and everything that is in principle knowable is also knowable simultaneously. Classical probability can be based upon the convex combinations of all two-valued states. Fig. 9 depicts a "mind map" representing the use of contexts to build up logics and construct probabilities.

In the generalized urn or automaton cases, if one sticks to the rules — that is, if one does not view the object unfiltered or "screw the automaton box open" — omniscience is impossible and a quasi–classical sort of complementarity emerges: depending on the color (or input string) chosen, one obtains knowledge of a particular observable or context. All other contexts are hidden to the experimenter

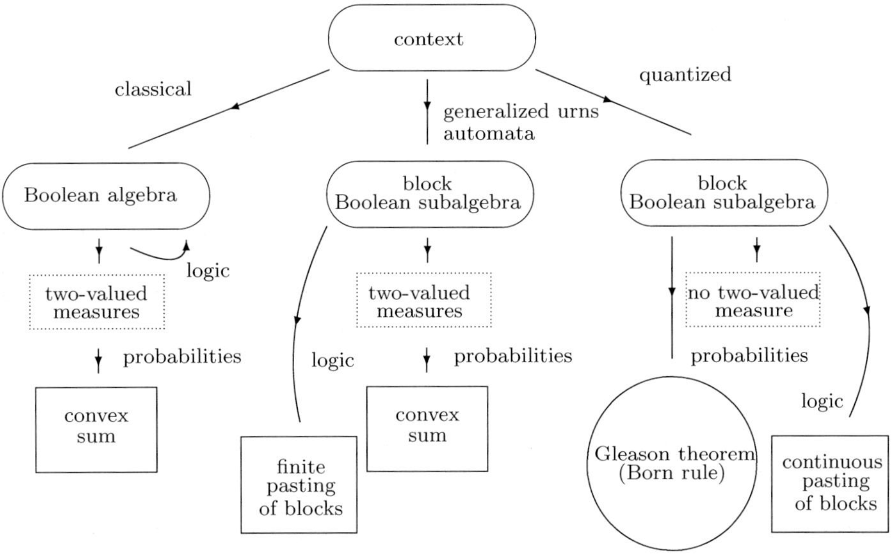

Figure 9. "Mind map" representing the use of contexts to build up logics and construct probabilities.

unable to lift the bounds of one color filter or one input sequence. A system science issue is emerging here; namely the question of how intrinsic observers perform inside of a given system [Svozil, 1993; Svozil, 1994]. The situation resembles quantum mechanics even more if reversible systems are considered; where an experiment can be "undone" only by investing all the information gained from previous experiments (without being able to copy these)[Greenberger and YaSin, 1989; Herzog *et al.*, 1995]. All incompatible blocks or contexts are pasted together to form the partition logic. These pasting still allow a sufficient number of two-valued states for the construction of probabilities based upon the convex combinations thereof.

In the quantum case, the Hilbert lattices can formally be thought of as pastings of a continuum of blocks or contexts, but the mere assumption of the physical existence — albeit inaccessible to an intrisic observer — of even a finite number of contexts yields a complete contradiction. In view of this, one can adopt at least two interpretations: that an observable depends on its context; or that more than one context for quantum systems has no operational meaning. The former view has been mentioned by Bell (and also by Bohr to some degree), and can be subsumed by the term "contextuality." To the author, contextuality is the last resort of a realism which is inclined to maintain "a sort of" classical omniscience, even in view of the Kochen-Specker and Bell-type theorems.

The latter viewpoint — that quantum systems do not encode more than a single context — abandons omniscience, but needs to cope with the fact that it *is* indeed possible to measure different contexts; even if there is a mismatch between the preparation and the measurement context. It has been proposed that in these cases the measurement apparatus "translates" one context into the other at the prize of randomizing the measurement result [Svozil, 2004]. This *context translation principle* could be tested by changing the measurement apparatus' ability of translation.

All in all, contexts seem to be an exciting subject. The notion may become more useful and relevant, as progress is made towards a better comprehension of the quantum world and its differences with respect to other classical and quasi–classical systems.

BIBLIOGRAPHY

[Alda, 1980] V. Alda. On 0-1 measures for projectors i. *Aplik. mate.*, **25**, 373–374 (1980).

[Alda, 1981] V. Alda. On 0-1 measures for projectors ii. *Aplik. mate.*, **26**, 57–58 (1981).

[Aspect *et al.*, 1981] Alain Aspect, Philippe Grangier, and Gérard Roger. Experimental tests of realistic local theories via Bell's theorem. *Physical Review Letters*, **47**, 460–463 (1981). `http://dx.doi.org/10.1103/PhysRevLett.47.460`

[Aspect *et al.*, 1982] Alain Aspect, Philippe Grangier, and Gérard Roger. Experimental realization of Einstein-Podolsky-Rosen-Bohm *gedankenexperiment*: A new violation of Bell's inequalities. *Physical Review Letters*, **49**, 1804–1807 (1982). `http://dx.doi.org/10.1103/PhysRevLett.49.91`

[Bell, 1966] John S. Bell. On the problem of hidden variables in quantum mechanics. *Reviews of Modern Physics*, **38**, 447–452 (1966). Reprinted in [Bell, 1987, pp. 1-13]. `http://dx.doi.org/10.1103/RevModPhys.38.447`

[Bell, 1987] John S. Bell. *Speakable and Unspeakable in Quantum Mechanics.* Cambridge University Press, Cambridge (1987).

[Berkeley, 1710] George Berkeley. *A Treatise Concerning the Principles of Human Knowledge* (1710). `http://www.gutenberg.org/etext/4723`

[Birkhoff and von Neumann, 1936] Garrett Birkhoff and John von Neumann. The logic of quantum mechanics. *Annals of Mathematics*, **37**(4), 823–843 (1936).

[Bohr, 1949] Niels Bohr. Discussion with Einstein on epistemological problems in atomic physics. In P. A. Schilpp, editor, *Albert Einstein: Philosopher-Scientist*, pages 200–241. The Library of Living Philosophers, Evanston, Ill. (1949). `http://www.emr.hibu.no/lars/eng/schilpp/Default.html`

[Boole, 1862] George Boole. On the theory of probabilities. *Philosophical Transactions of the Royal Society of London*, **152**, 225–252 (1862).

[Boole, 1958] George Boole. *An investigation of the laws of thought.* Dover edition, New York (1958).

[Bridgman, 1927] Percy W. Bridgman. *The Logic of Modern Physics.* New York (1927).

[Bridgman, 1934] Percy W. Bridgman. A physicist's second reaction to Mengenlehre. *Scripta Mathematica*, **2**, 101–117, 224–234 (1934). Cf. R. Landauer [Landauer, 1994].

[Bridgman, 1952] Percy W. Bridgman. *The Nature of Some of Our Physical Concepts.* Philosophical Library, New York (1952).

[Brown, 1992] Harvey R. Brown. Bell's other theorem and its connection with nonlocality, part 1. In A. van der Merwe, F. Selleri, and G. Tarozzi, editors, *Bell's Theorem and the Foundations of Modern Physics*, pages 104–116, Singapore (1992). World Scientific.

[Cabello *et al.*, 1996] Adán Cabello, José M. Estebaranz, and G. García-Alcaine. Bell-Kochen-Specker theorem: A proof with 18 vectors. *Physics Letters A*, **212**(4), 183–187 (1996). `http://dx.doi.org/10.1016/0375-9601(96)00134-X`

[Cabello, 2000] Adán Cabello. Kochen-Specker theorem and experimental test on hidden variables. *International Journal of Modern Physics*, **A 15**(18), 2813–2820 (2000). `http://dx.doi.org/10.1142/S0217751X00002020`

[Cabello, 2004] Adán Cabello. Proposed experiment to test the bounds of quantum correlations. *Physical Review Letters*, **92**, 060403 (2004). `http://dx.doi.org/10.1103/PhysRevLett.92.060403`

[Calude and Dinneen, 2005] Cristian S. Calude and Michael J. Dinneen. Is quantum randomness algorithmic random? a preliminary attack. In S. Bozapalidis, A. Kalampakas, and G. Rahonis, editors, *Proceedings 1st International Conference on Algebraic Informatics*, pages 195–196. Aristotle University of Thessaloniki (2005).

[Calude *et al.*, 1997] Cristian Calude, Elena Calude, Karl Svozil, and Sheng Yu. Physical versus computational complementarity I. *International Journal of Theoretical Physics*, **36**(7), 1495–1523 (1997).

[Calude, 2002] Cristian Calude. *Information and Randomness—An Algorithmic Perspective.* Springer, Berlin, second edition (2002).

[Calude, 2004] Cristian S. Calude. Algorithmic randomness, quantum physics, and incompleteness. In M. Margenstern, editor, *Proceedings of the Conference "Machines, Computations and Universality" (MCU'2004)*, pages 1–17. Lectures Notes in Comput. Sci. 3354, Springer, Berlin (2004).

[Chaitin, 1987] Gregory J. Chaitin. *Algorithmic Information Theory.* Cambridge University Press, Cambridge (1987).

[Chaitin, 1990] Gregory J. Chaitin. *Information, Randomness and Incompleteness.* World Scientific, Singapore, second edition (1990). This is a collection of G. Chaitin's early publications.

[Cirel'son (=Tsirel'son), 1980] Boris S. Cirel'son (=Tsirel'son). Quantum generalizations of Bell's inequality. *Letters in Mathematical Physics*, **4**, 93–100 (1980).

[Cirel'son (=Tsirel'son), 1993] Boris S. Cirel'son (=Tsirel'son). Some results and problems on quantum Bell-type inequalities. *Hadronic Journal Supplement*, **8**, 329–345 (1993).

[Clauser and Horne, 1974] J. F. Clauser and M. A. Horne. Experimental consequences of objective local theories. *Physical Review D*, **10**, 526–535 (1974). `http://dx.doi.org/10.1103/PhysRevD.10.526`

[Clauser and Shimony, 1978] J. F. Clauser and A. Shimony. Bell's theorem: experimental tests and implications. *Rep. Prog. Phys.*, **41**, 1881–1926 (1978). `http://dx.doi.org/10.1088/0034-4885/41/12/002`

[Clifton, 1993] Rob Clifton. Getting contextual and nonlocal elements–of–reality the easy way. *American Journal of Physics*, **61**, 443–447 (1993).

[Colins and Gisin, 2004] Daniel Colins and Nicolas Gisin. A relevant two qbit Bell inequality inequivalent to the CHSH inequality. *J. Phys. A: Math. Gen.*, **37**, 1775–1787 (2004). `http://dx.doi.org/10.1088/0305-4470/37/5/021`

[Cooke *et al.*, 1985] R. Cooke, M. Keane, and W. Moran. An elementary proof of Gleason's theorem. *Math. Proc. Camb. Soc.*, **98**, 117–128 (1985).

[Donath and Svozil, 2002] Niko Donath and Karl Svozil. Finding a state among a complete set of orthogonal ones. *Physical Review A (Atomic, Molecular, and Optical Physics)*, **65**, 044302 (2002). `http://dx.doi.org/10.1103/PhysRevA.65.044302`

[Dvurečenskij *et al.*, 1995] Anatolij Dvurečenskij, Sylvia Pulmannová, and Karl Svozil. Partition logics, orthoalgebras and automata. *Helvetica Physica Acta*, **68**, 407–428 (1995).

[Dvurečenskij, 1993] Anatolij Dvurečenskij. *Gleason's Theorem and Its Applications.* Kluwer Academic Publishers, Dordrecht (1993).

[Einstein *et al.*, 1935] Albert Einstein, Boris Podolsky, and Nathan Rosen. Can quantum-mechanical description of physical reality be considered complete? *Physical Review*, **47**, 777–780 (1935). `http://dx.doi.org/10.1103/PhysRev.47.777`

[Filipp and Svozil, 2004a] Stefan Filipp and Karl Svozil. Generalizing Tsirelson's bound on bell inequalities using a min-max principle. *Physical Review Letters*, **93**, 130407 (2004). `http://dx.doi.org/10.1103/PhysRevLett.93.130407`

[Filipp and Svozil, 2004b] Stefan Filipp and Karl Svozil. Testing the bounds on quantum probabilities. *Physical Review A (Atomic, Molecular, and Optical Physics)*, **69**, 032101 (2004). `http://dx.doi.org/10.1103/PhysRevA.69.032101`

[Froissart, 1981] M. Froissart. Constructive generalization of Bell's inequalities. *Nuovo Cimento B*, **64**(2), 241–251 (1981).

[Gleason, 1957] Andrew M. Gleason. Measures on the closed subspaces of a Hilbert space. *Journal of Mathematics and Mechanics*, **6**, 885–893 (1957).

[Greechie, 1971] J. R. Greechie. Orthomodular lattices admitting no states. *Journal of Combinatorial Theory*, **10**, 119–132 (1971).

[Greenberger and YaSin, 1989] Daniel B. Greenberger and A. YaSin. "Haunted" measurements in quantum theory. *Foundation of Physics*, **19**(6), 679–704 (1989).

[Halmos, 1974] Paul R.. Halmos. *Finite-dimensional vector spaces.* Springer, New York, Heidelberg, Berlin (1974).

[Hasegawa *et al.*, 2006] Yuji Hasegawa, Rudolf Loidl, Gerald Badurek, Matthias Baron, and Helmut Rauch. Quantum contextuality in a single-neutron optical experiment. *Physical Review Letters*, **97**(23), 230401 (2006). `http://dx.doi.org/10.1103/PhysRevLett.97.230401`

[Havlicek and Svozil, 1996] Hans Havlicek and Karl Svozil. Density conditions for quantum propositions. *Journal of Mathematical Physics*, **37**(11), 5337–5341 (November 1996).

[Herzog *et al.*, 1995] Thomas J. Herzog, Paul G. Kwiat, Harald Weinfurter, and Anton Zeilinger. Complementarity and the quantum eraser. *Physical Review Letters*, **75**(17), 3034–3037 (1995). `http://dx.doi.org/10.1103/PhysRevLett.75.3034`

[Heywood and Redhead, 1983] Peter Heywood and Michael L. G. Redhead. Nonlocality and the Kochen-Specker paradox. *Foundations of Physics*, **13**(5), 481–499 (1983).

[Hooker, 1975] Clifford Alan Hooker. *The Logico-Algebraic Approach to Quantum Mechanics. Volume I: Historical Evolution.* Reidel, Dordrecht (1975).

[Hrushovski and Pitowsky, 2004] Ehud Hrushovski and Itamar Pitowsky. Generalizations of Kochen and Specker's theorem and the effectiveness of Gleason's theorem. *Studies in History and Philosophy of Science Part B: Studies in History and Philosophy of Modern Physics*, **35**(2), 177194 (2004). `http://dx.doi.org/10.1016/j.shpsb.2003.10.002`

[id Quantique, 2004] id Quantique. Quantis - quantum random number generators (2004). `http://www.idquantique.com`

[Jammer, 1992] Max Jammer. John Steward Bell and the debate on the significance of his contributions to the foundations of quantum mechanics. In A. van der Merwe, F. Selleri, and G. Tarozzi, editors, *Bell's Theorem and the Foundations of Modern Physics*, pages 1–23. World Scientific, Singapore (1992).

[Jauch, 1968] J. M. Jauch. *Foundations of Quantum Mechanics.* Addison-Wesley, Reading, MA. (1968).

[Kamber, 1964] Franz Kamber. Die Struktur des Aussagenkalküls in einer physikalischen Theorie. *Nachr. Akad. Wiss. Göttingen*, **10**, 103–124 (1964).

[Kamber, 1965] Franz Kamber. Zweiwertige Wahrscheinlichkeitsfunktionen auf orthokomplementären Verbänden. *Mathematische Annalen*, **158**, 158–196 (1965).

[Kochen and Specker, 1965a] Simon Kochen and Ernst P. Specker. The calculus of partial propositional functions. In *Proceedings of the 1964 International Congress for Logic, Methodology and Philosophy of Science, Jerusalem*, pages 45–57, Amsterdam (1965). North Holland. Reprinted in [Specker, 1990, pp. 222–234].

[Kochen and Specker, 1965b] Simon Kochen and Ernst P. Specker. Logical structures arising in quantum theory. In *Symposium on the Theory of Models, Proceedings of the 1963 International Symposium at Berkeley*, pages 177–189, Amsterdam (1965). North Holland. Reprinted in [Specker, 1990, pp. 209–221].

[Kochen and Specker, 1967] Simon Kochen and Ernst P. Specker. The problem of hidden variables in quantum mechanics. *Journal of Mathematics and Mechanics*, **17**(1), 59–87 (1967). Reprinted in [Specker, 1990, pp. 235–263].

[Landauer, 1994] Rolf Landauer. Advertisement for a paper I like. In John L. Casti and J. F. Traub, editors, *On Limits*, page 39. Santa Fe Institute Report 94-10-056, Santa Fe, NM (1994). `http://www.santafe.edu/research/publications/workingpapers/94-10-056.pdf`

[Mackey, 1957] George W. Mackey. Quantum mechanics and Hilbert space. *Amer. Math. Monthly, Supplement*, **64**(8), 45–57 (1957).

[Mermin, 1993] N. D. Mermin. Hidden variables and the two theorems of John Bell. *Reviews of Modern Physics*, **65**, 803–815 (1993). `http://dx.doi.org/10.1103/RevModPhys.65.803`

[Neumark, 1954] M. A. Neumark. Principles of quantum theory. In Klaus Matthes, editor, *Sowjetische Arbeiten zur Funktionalanalysis. Beiheft zur Sowjetwissenschaft*, volume 44, pages 195–273. Gesellschaft für Deutsch-Sowjetische Freundschaft, Berlin (1954).

[Pan *et al.*, 2000] J.-W. Pan, D. Bouwmeester, M. Daniell, H. Weinfurter, and A. Zeilinger. Experimental test of quantum nonlocality in three-photon Greenberger-Horne-Zeilinger entanglement. *Nature*, **403**, 515–519 (2000).

[Peres, 1991] Asher Peres. Two simple proofs of the Kochen–Specker theorem. *Journal of Physics*, **A24**, L175–L178 (1991). Cf. [Peres, 1993, pp. 186-200].

[Peres, 1993] Asher Peres. *Quantum Theory: Concepts and Methods.* Kluwer Academic Publishers, Dordrecht (1993).

[Pitowsky and Svozil, 2001] Itamar Pitowsky and Karl Svozil. New optimal tests of quantum nonlocality. *Physical Review A (Atomic, Molecular, and Optical Physics)*, **64**, 014102 (2001). http://dx.doi.org/10.1103/PhysRevA.64.014102

[Pitowsky, 1989a] Itamar Pitowsky. From George Boole to John Bell: The origin of Bell's inequality. In M. Kafatos, editor, *Bell's Theorem, Quantum Theory and the Conceptions of the Universe*, pages 37–49. Kluwer, Dordrecht (1989).

[Pitowsky, 1989b] Itamar Pitowsky. *Quantum Probability—Quantum Logic.* Springer, Berlin (1989).

[Pitowsky, 1990] Itamar Pitowsky. The physical Church-Turing thesis and physical computational complexity. *Iyyun*, **39**, 81–99 (1990).

[Pitowsky, 1991] Itamar Pitowsky. Correlation polytopes their geometry and complexity. *Mathematical Programming*, **50**, 395–414 (1991).

[Pitowsky, 1994] Itamar Pitowsky. George Boole's 'conditions od possible experience' and the quantum puzzle. *The British Journal for the Philosophy of Science*, **45**, 95–125 (1994). http://dx.doi.org/10.1093/bjps/45.1.95

[Pták and Pulmannová, 1991] Pavel Pták and Sylvia Pulmannová. *Orthomodular Structures as Quantum Logics.* Kluwer Academic Publishers, Dordrecht (1991).

[Redhead, 1990] Michael Redhead. *Incompleteness, Nonlocality, and Realism: A Prolegomenon to the Philosophy of Quantum Mechanics.* Clarendon Press, Oxford (1990).

[Richman and Bridges, 1999] Fred Richman and Douglas Bridges. A constructive proof of Gleason's theorem. *Journal of Functional Analysis*, **162**, 287–312 (1999). http://dx.doi.org/10.1006/jfan.1998.3372

[Schaller and Svozil, 1996] Martin Schaller and Karl Svozil. Automaton logic. *International Journal of Theoretical Physics*, **35**(5), 911–940 (May 1996).

[Simon *et al.*, 2000] C. Simon, M. Zukowski, H. Weinfurter, and A. Zeilinger. Feasible "Kochen-Specker experiment with single particles. *Physical Review Letters*, **85**, 1783–1786 (2000). http://dx.doi.org/10.1103/PhysRevLett.85.1783

[Sliwa, 2003] Cezary Sliwa. Symmetries of the Bell correlation inequalities. *Physics Letters A*, **317**, 165–168 (2003). http://dx.doi.org/10.1016/S0375-9601(03)01115-0

[Specker, 1960] Ernst Specker. Die Logik nicht gleichzeitig entscheidbarer Aussagen. *Dialectica*, **14**, 175–182 (1960). Reprinted in [Specker, 1990, pp. 175–182]; English translation: *The logic of propositions which are not simultaneously decidable*, reprinted in [Hooker, 1975, pp. 135-140].

[Specker, 1990] Ernst Specker. *Selecta.* Birkhäuser Verlag, Basel (1990).

[Specker, 1999] Ernst Specker. private communication (1999).

[Stace, 1949] Walter Terence Stace. The refutation of realism. In Herbert Feigl and Wilfrid Sellars, editors, *Readings in philosophical analysis.* Appleton–Century–Crofts, New York (1949). previously published in *Mind* **53**, 1934.

[Stairs, 1983] Allen Stairs. Quantum logic, realism, and value definiteness. *Philosophy of Science*, **50**, 578–602 (1983).

[Svozil and Tkadlec, 1996] Karl Svozil and Josef Tkadlec. Greechie diagrams, nonexistence of measures in quantum logics and Kochen–Specker type constructions. *Journal of Mathematical Physics*, **37**(11), 5380–5401 (November 1996). http://dx.doi.org/10.1063/1.531710

[Svozil, 1993] Karl Svozil. *Randomness & Undecidability in Physics.* World Scientific, Singapore (1993).

[Svozil, 1994] Karl Svozil. Extrinsic-intrinsec concept and complementarity. In H. Atmanspacker and G. J. Dalenoort, editors, *Inside versus Outside*, pages 273–288. Springer-Verlag, Heidelberg (1994).

[Svozil, 1998] Karl Svozil. *Quantum Logic.* Springer, Singapore (1998).

[Svozil, 2002] Karl Svozil. Quantum information in base n defined by state partitions. *Physical Review A (Atomic, Molecular, and Optical Physics)*, **66**, 044306 (2002). http://dx.doi.org/10.1103/PhysRevA.66.044306

[Svozil, 2004] Karl Svozil. Quantum information via state partitions and the context translation principle. *Journal of Modern Optics*, **51**, 811–819 (2004). http://dx.doi.org/10.1080/09500340410001664179

[Svozil, 2005a] Karl Svozil. Are simultaneous bell measurements possible? *New J. Phys.*, **8**, 39 (2005). http://dx.doi.org/10.1088/1367-2630/8/3/039

[Svozil, 2005b] Karl Svozil. Logical equivalence between generalized urn models and finite automata. *International Journal of Theoretical Physics*, **44**, 745–754 (2005). http://dx.doi.org/10.1007/s10773-005-7052-0

[Svozil, 2005c] Karl Svozil. Noncontextuality in multipartite entanglement. *J. Phys. A: Math. Gen.*, **38**, 5781–5798 (2005). http://dx.doi.org/10.1088/0305-4470/38/25/013

[Svozil, 2005d] Karl Svozil. On counterfactuals and contextuality. In Andrei Khrennikov, editor, *AIP Conference Proceedings 750. Foundations of Probability and Physics-3*, pages 351–360, Melville, NY (2005). American Institute of Physics. http://dx.doi.org/10.1063/1.1874586

[Svozil, 2006] Karl Svozil. Staging quantum cryptography with chocolate balls. *American Journal of Physics*, **74**(9), 800–803 (2006). http://dx.doi.org/10.1119/1.2205879

[Tkadlec, 1998] Josef Tkadlec. Greechie diagrams of small quantum logics with small state spaces. *International Journal of Theoretical Physics*, **37**(1), 203–209 (Jan 1998). http://dx.doi.org/10.1023/A:1026646229896

[Tkadlec, 2000] Josef Tkadlec. Diagrams of Kochen-Specker type constructions. *International Journal of Theoretical Physics*, **39**(3), 921–926 (March 2000). http://dx.doi.org/10.1023/A:1003695317353

[von Neumann, 1932] John von Neumann. *Mathematische Grundlagen der Quantenmechanik.* Springer, Berlin (1932). English translation in [von Neumann, 1955].

[von Neumann, 1955] John von Neumann. *Mathematical Foundations of Quantum Mechanics.* Princeton University Press, Princeton (1955).

[Weihs *et al.*, 1998] Gregor Weihs, Thomas Jennewein, Christoph Simon, Harald Weinfurter, and Anton Zeilinger. Violation of Bell's inequality under strict einstein locality conditions. *Phys. Rev. Lett.*, **81**, 5039–5043 (1998). http://dx.doi.org/10.1103/PhysRevLett.81.5039

[Wright, 1978] Ron Wright. The state of the pentagon. A nonclassical example. In A. R. Marlow, editor, *Mathematical Foundations of Quantum Theory*, pages 255–274. Academic Press, New York (1978).

[Wright, 1990] Ron Wright. Generalized urn models. *Foundations of Physics*, **20**, 881–903 (1990).

[Zeilinger, 1999] Anton Zeilinger. A foundational principle for quantum mechanics. *Foundations of Physics*, **29**(4), 631–643 (1999). http://dx.doi.org/10.1023/A:1018820410908

[Ziegler, 1994] Günter M. Ziegler. *Lectures on Polytopes.* Springer, New York (1994).

[Zierler and Schlessinger, 1965] Neal Zierler and Michael Schlessinger. Boolean embeddings of orthomodular sets and quantum logic. *Duke Mathematical Journal*, **32**, 251–262 (1965).

[Zimba and Penrose, 1993] Jason Zimba and Roger Penrose. On Bell non-locality without probabilities: more curious geometry. *Studies in History and Philosophy of Modern Physics*, **24**(5), 697–720 (1993).

Edited by K. Engesser, D. M. Gabbay and D. Lehmann

NONMONOTONICITY AND HOLICITY IN QUANTUM LOGIC

Kurt Engesser, Dov Gabbay and Daniel Lehmann

INTRODUCTION

The question whether we need a 'new logic' in order to reason properly in quantum theory is asked frequently. Do we have to depart from classical logic in building 'quantum logic' and, if so, how? The answer that most physicists give to this question is that we do not. In fact, physicists put quantum mechanics to good use in an unprecedentedly successful way, and in this do they not use classical logic? In [Popper, 1968] Popper denies any need to depart from classical logic in order to reason properly in quantum mechanics.

Why did the question arise at all? The question of 'the logic of quantum mechanics' was in the scientific literature first raised by Birkhoff and von Neumann in their seminal 1936 paper [Birkhoff and von Neumann, 1936]. Their chief motivation for trying to discover the 'logic of quantum mechanics' was the fact that they considered the novel features of quantum mechanics such as the uncertainty relations to be logical in nature. Since these features are not reflected in classical logic, there is, according to Birkhoff-von Neumann, a need to construct a (logical) 'calculus' in which they are actually represented.

Later on it was Putnam, Finkelstein and others who put forward a view of quantum logic which for some time attracted considerable attention. Central to this paradigm is the idea that logic may be empirical. Putnam and his followers argued that the role of logic in quantum mechanics was similar to that of geometry in the theory of relativity. In the theory of relativity, Euclidean geometry, which in Newtonian physics was still considered a priori, had to be revised on empirical grounds. In quantum mechanics, Putnam argued, it is (classical) logic that needs revision on empirical grounds. Similar to the way the theory of relativity teaches us the 'real' geometry quantum mechanics would teach us the 'real' logic. This is undoubtedly an attractive idea which, however, we will not pursue here. We refer the interested reader to Bacciagaluppi's chapter "Is Logic Empirical?" in this Handbook.

Rather we adopt a different attitude, which is already implicit in the Birkhoff-von Neumann paper. In the Introduction they write: "The object of the present paper is to discover what logical structure one may hope to find in physical theories which, like quantum mechanics, do not conform to classical logic".

In fact, this is the task we pose ourselves: searching for logical structures in quantum mechanics. The procedure is this. We take a close look at Hilbert space and, as a result, identify and study certain logical structures implicit in Hilbert space. We then pursue the question whether these logical structures represent essential features of the formalism of quantum mechanics. Can the structures found shed light on the formalism of quantum mechanics? We think that they in fact can. We think that the science of logic can detect and study logical structures in the formalism of quantum mechanics which are relevant to the understanding of the formalism itself. Are there any guidelines that may help in searching for these structures? Are there any traits of quantum mechanics itself that could suggest certain directions of investigation? Let us speculate about this.

As a good starting point we may look at the relationship between classical and quantum mechanics. We may start by analysing the way how quantum mechanics departs from classical mechanics. Given that quantum mechanics, as is often claimed more or less vaguely, does not conform to classical logic, then it is reasonable to ask how the transition from classical to quantum mechanics is reflected in the logical structures we are looking for.

There are various ways of viewing the relationship between classical and quantum mechanics. Since in classical mechanics we have no uncertainty relations, it is the uncertainty relations that are often regarded as constituting the essential difference. Another crucial difference concerns the role of measurement. In classical mechanics a measurement does not involve a change of the state of the system measured. The fact that in quantum mechanics measurement does, in general, involve such a change of state is undoubtedly an essential difference between classical and quantum mechanics. Classical mechanics is often considered to be a limiting case of quantum mechanics as Newtonian mechanics is a limiting case of the theory of relativity. We may ask the question how these observations are reflected in the logical structures we may find. What are uncertainty relations from the logical point of view? We will see that, logically, the presence of uncertainty relations is reflected as nonmonotonicity of the logical structures implicit in the formalism of quantum mechanics. There is, however, a general feeling expressed in a vast body of literature, popular scientific and seriously scientific or philosophical alike, that the fairly obvious differences between classical and quantum mechanics mentioned above are not the whole story. Rather the general impression seems to be that the way how quantum mechanics departs from classical mechanics touches on deeper ground. Take the famous Einstein-Podolsky-Rosen (EPR) argument put forward in their famous paper entitled "Can the quantum-mechanical description of reality be considered complete?" [Einstein *et al.*, 1935]. In the EPR argument the term 'element of reality' plays a crucial role. EPR take it for granted that (physical) reality is to be viewed as consisting of separate 'elements of reality'. And, in fact, once this fragmenting view of reality is accepted, it is hard to avoid the EPR conclusion that quantum mechanics does not provide a complete description of physical reality. Therefore, if quantum mechanics is in fact a complete description of physical reality as seems to be generally assumed nowadays, then something must be wrong with this view of reality. It seems that the way quantum mechanics departs from classical mechanics is of an even more profound nature than the way the theory of relativity departs from classical mechanics. In the latter case we 'just' have to abandon our view of space and time. In the case of quantum mechanics it seems that we have to abandon our view of the very nature of reality. This is all pervading the literature on the foundations of quantum mechanics be it popular scientific or seriously philosophical. It is the intuition of oneness, interconnectedness and wholeness, which is prevalent in Eastern thought for instance, that finds strong support in quantum mechanics.

How can we, at the level of logic, reflect the shift in our perception of reality which is forced upon us in the transition from classical mechanics to quantum mechanics? A possible answer is this. Classical mechanics and classical logic conform to each other and the view of reality that underlies classical mechanics also underlies classical logic. If, as seems to be the case, our 'classical' view of reality is to be revised in quantum mechanics, we must ask the question whether logic, i.e. the quantum logic to be constructed, can account for quantum mechanics if it does not reflect this shift. We will in Section 4 describe a way of departing from classical logic for the sake of quantum logic which may be regarded as reflecting this intuition.

Most, though not all of the material presented in this chapter was first published in a less condensed form in our monograph [Engesser and Gabbay, 2002].

1 STRUCTURE OF THIS CHAPTER

The core logical structure of this chapter is what we call a *holistic logic.* This concept reflects two intuitions that inevitably arise in connection with quantum mechanics, namely the intuition of nonmonotonicity as expressed by the uncertainty relations and the intuiton of holicity , which pervades the literature on the foundations of quantum mechanics. This concept is introduced and studied extensively in Section 4, the core section of this chapter. Section 2 entitled "Basics of Nonmonotonic Logic" and Section 3 entitled "Consequence Revision Systems" provide the background for Section 4. In Section 6 entitled "Towards Hilbert Space" we study the connection between the concept of a holistic logic and that of a Hilbert space. In particular, we prove a representation theorem for holistic logics in terms of Hilbert spaces. Section 5 entitled "Some Hilbert Space Theory" provides the mathematical machinery needed for the investigations in Section 6.

2 BASICS OF NONMONOTONIC LOGIC

2.1 What is nonmonotonic logic?

Classical logic is monotonic. Given a set Σ of assumptions and a formula α such that $\Sigma \vdash \alpha$. If we add more assumptions to Σ so as to get Σ^* we will still have

$\Sigma^* \vdash \alpha$. 'More information' cannot invalidate inferences drawn on the basis of 'less information'. This is what monotonicity means. In the last decades, logicians studied modes of reasoning that do not have this property. In these so-called *non-monotonic logics* 'old inferences' may be invalidated by 'new information'. What reason can there be for this phenomenon? One reason is incomplete information. This is for instance the case in common sense reasoning. If we view a common sense reasoner's activity as 'jumping to conclusions' on the basis of certain 'pieces of information', it seems quite natural that certain of his conclusions cannot be maintained in the light of additional information.

Another source of nonmonotonicity is perfect introspection of the (reasoning) agent. Imagine a reasoner, i.e. an agent who can infer propositions from sets of assumptions. Suppose, moreover, this reasoner has an additional ability. Namely assume that whenever he can, in his system of reasoning, infer a certain proposition α from a certain set Σ of assumptions, he can, in the same system, infer the proposition saying "I can infer α" denoted by $I\alpha$ and whenever he cannot infer α from Σ he can infer the proposition "I cannot infer α", i.e. $\neg I\alpha$. The former capability is called positive introspection, the latter is called negative introspection. Assume a consistent agent having both capabilities. We give an informal argument to the effect that such a reasoner cannot be monotonic. So assume he is monotonic. Given a set Σ of assumptions and let α be a proposition the reasoner cannot infer from Σ. By negative introspection he can then infer $\neg I\alpha$. Assume that α is consistent with Σ and can be consistently added to Σ. Then the agent *can* infer α from $\Sigma \cup \{\alpha\}$ and thus by (positive) introspection he can infer $I\alpha$ from the enlarged set of assumptions. Since he is assumed to be monotonic, he can still infer $\neg I\alpha$. But this would mean that he is inconsistent. It follows that he cannot be monotonic.

2.2 *Consequence relations*

A consequence relation is a binary relation between formulas satisfying certain intuitive conditions we expect logical consequence to satisfy. We assume the (full) language $\mathcal{L}$ of propositional logic. We state some minimal conditions a consequence relation is supposed to satisfy. The following are the minimal conditions as suggested by Gabbay in [Gabbay, 1985]. The reader may verify that the classical consequence relation $\vdash$ in fact satisfies these conditions.

Reflexivity

$$\alpha \mid\!\sim \alpha$$

Cut

$$\frac{\alpha \wedge \beta \mid\!\sim \gamma,\ \alpha \mid\!\sim \beta}{\alpha \mid\!\sim \gamma}$$

Restricted Monotonicity

$$\frac{\alpha \mid\!\sim \beta,\ \alpha \mid\!\sim \gamma}{\alpha \wedge \beta \mid\!\sim \gamma}$$

As observed in [Kraus *et al.*, 1990], any consequence relation satisfying the above conditions has the following property *AND*:

$$\frac{\alpha \mid\!\sim \beta, \alpha \mid\!\sim \gamma}{\alpha \mid\!\sim \beta \wedge \gamma}$$

In nonmonotonic logic we of course do not insist on the requirement of monotonicity.

2.3 Semantics of nonmonotonic logic

How can nonmonotonic consequence relations be presented? As to this problem a breakthrough was achieved in the seminal paper by Kraus-Lehmann-Magidor (KLM) [Kraus *et al.*, 1990]. Namely, it was shown by KLM that certain semantic structures which have become known as KLM models are suitable for this purpose. We present here a slight modification of the original KLM structures introduced in [Gabbay, 1996].

DEFINITION 1.

- A Scott model for Fml is any function $s : Fml \to \{0, 1\}$.
- A $GKLM$ (Generalised Kraus–Lehmann–Magidor) model is a structure of the form $\langle S, <, l \rangle$, where S is a non-empty set, $<$ is a binary relation on S and l is a function associating with each $t \in S$ a set of Scott models $l(t)$. The model is required to satisfy the smoothness condition stated in the next definition.

DEFINITION 2. Let $\mathcal{M} = \langle S, <, l \rangle$ be a structure as described in the last definition. Let $t \in S$ and α a formula. Then define the satisfaction relation $t \models \alpha$ as follows:

- $t \models \alpha$ iff for all $s \in l(t)$ we have $s(\alpha) = 1$
- Let $A \subset S$. We say that t is $<$-minimal in A iff for all $t' \in A$ such that $t' < t$ we have $t' = t$. We say that A is smooth iff for every $t \in A$, either t is minimal in A or for some $s \in A$, $s < t$ and s is minimal in A.
- Let $[\alpha] = \{t \in S \mid t \models \alpha\}$. We say that $\mathcal{M}$ is smooth iff for all α, $[\alpha]$ is smooth.
- For a smooth model $\mathcal{M}$ we define the consequence relation $\mid\!\sim_{\mathcal{M}}$ as follows: $\alpha \mid\!\sim_{\mathcal{M}} \beta$ iff for all t minimal in $[\alpha]$, we have $t \models \beta$.
- Given a consequence relation $\mid\!\sim$ and a smooth model $\mathcal{M}$. We say $\mathcal{M}$ is a model for $\mid\!\sim$ iff $\mid\!\sim = \mid\!\sim_{\mathcal{M}}$.

We cannot motivate the above definitions here. We leave it by reporting that $GKLM$ models have turned out to be extremely suited for presenting nonmonotonic consequence relations semantically. The concept has a long history taking its origin in investigations on the semantics of conditionals.

2.4 Uncertainty relations and nonmonotonicity

The reader may ask the question why we want to consider nonmonotonic logics in our study of quantum logic. The answer is that nonmonotonicity is an essential feature of quantum mechanics. We encounter nonmonotonic (logical) systems in nature so to speak. Imagine a (quantum) physical system $\mathcal{S}$ and let A and B be two observables pertaining to $\mathcal{S}$. Suppose we measure A. Through measurement we get a certain value, say μ. Viewing a measurement as a sort of proof we have then 'proved' the proposition $A = \mu$. This is then a fact about the system $\mathcal{S}$. Now assume we measure B. Again, we get a value, say λ, and we have proved the proposition $B = \lambda$, another fact about the system $\mathcal{S}$. We are now, used to classical physics and classical logic as we are, inclined to say that we now know the facts $A = \mu$ and $B = \lambda$ and that subsequent measurements could only confirm these facts. This, however, is according to quantum mechanics not neccessarily the case. Namely, if there exists an uncertainty relation between A and B, a subsequent measurement of A may yield a value different from μ. The measurement of B can invalidate the result of the measurement of A and vice versa. This is by Heisenberg's uncertainty principle for instance the case if observable A is momentum and observable B is position. From the point of view of logic this is nonmonotonicity. It is via the uncertainty relations that nonmonotonicity enters quantum logic.

3 CONSEQUENCE REVISION SYSTEMS

3.1 Formal motivation: the Lindenbaum algebra viewed as an operator algebra

In order to motivate the concepts we are going to introduce we start with an observation from classical logic. We denote the language by Fml and the classical consequence relation by $\vdash$. Given a formula α, we may form a new consequence relation $\vdash_\alpha$ as follows: $\beta \vdash_\alpha \gamma$ iff $\alpha \wedge \beta \vdash \gamma$. We get a class of consequence relations $\mathcal{C} = \{\vdash_\alpha \mid \alpha \in Fml\}$. By the deduction theorem of classical logic we have $\beta \vdash_\alpha \gamma$ iff $\vdash_\alpha \beta \to \gamma$ for all $\vdash_\alpha \in \mathcal{C}$. We say that $\to$, i.e. material implication, is an internalising connective for $\mathcal{C}$. Again, given $\alpha \in Fml$ and $\vdash_\beta \in \mathcal{C}$, we may form the consequence relation $\vdash_{\alpha\wedge\beta}$. Thus every $\alpha \in Fml$ induces an operator $\overline{\alpha} : \mathcal{C} \to \mathcal{C}$. We have $\overline{\alpha} = \overline{\beta}$ iff α and β are classically equivalent. It is readily verified that the class of operators is partially ordered by: $\overline{\alpha} \leq \overline{\beta}$ iff $\alpha \vdash \beta$. Moreover, it is routine to verify that this structure forms a Boolean algebra isomorphic to the Lindenbaum algebra of classical logic. This is our motivating example of what we will call a *consequence revision system*. Its main ingredients are a *class of consequence relations* $\mathcal{C}$, a function $F : Fml \times \mathcal{C} \to \mathcal{C}$ and a connective which is an internalising connective for all consequence relations of $\mathcal{C}$. In this case this is material implication. We have $\vdash_\alpha \beta \to \gamma$ iff $\beta \vdash_\alpha \gamma$ for any α. The structure of interest is the triple $\mathcal{L} = \langle \mathcal{C}, \mathcal{F}, \to \rangle$.

There is a straightforward generalisation of the above consideration. We could have started with any consistent set of formulas Σ and the consequence relation $\vdash_\Sigma$ defined by: $\alpha \vdash_\Sigma \beta$ iff $\Sigma \cup \{\alpha\} \vdash \beta$ and would by the same procedure as above have arrived at the structure $\mathcal{L}_\Sigma = \langle \mathcal{C}_\Sigma, \mathcal{F}_\Sigma, \rightarrow \rangle$. Note that, by the deduction theorem of classical logic, material implication is still the internalising connective in this more general case.

3.2 The concept of a consequence revision system

We will, in this chapter, be concerned with classes of consequence relations and must therefore consider conditions these consequence relations are supposed to satisfy. These conditions go beyond those stated so far. We will see in Section 6 that all these conditions are satisfied in Hilbert space.

We denote the universal (inconsistent) universal consequence relation by 0. We assume that for the consequence relations we consider this is equivalent to the existence of a formula α such that $\mid\sim \alpha$ and $\mid\sim \neg\alpha$. Any class of consequence relations considered is assumed to contain 0. That means we assume for any $\mid\sim \neq 0$ that for no $\alpha \in Fml$ we have $\mid\sim \alpha$ and $\mid\sim \neg\alpha$. If $\alpha \mid\sim \beta$ and $\beta \mid\sim \alpha$ we write $\alpha \equiv \beta$. Given a class $\mathcal{C}$ of consequence relations. Then we write $\alpha \mid\sim_\mathcal{C} \beta$ iff $\alpha \mid\sim \beta$ for every $\mid\sim \in \mathcal{C}$. We say $\alpha \equiv_\mathcal{C} \beta$ if $\alpha \mid\sim_\mathcal{C} \beta$ and $\beta \mid\sim_\mathcal{C} \alpha$. For the concept of a consequence revision system we impose in addition to the conditions of the last section the following conditions on the consequence relations.

$$\alpha \equiv \neg\neg\alpha$$

$$\top \equiv \alpha \vee \neg\alpha$$

$$\bot \equiv \alpha \wedge \neg\alpha$$

$$\alpha \wedge \beta \mid\sim \alpha$$

$$\mid\sim \alpha \text{ and } \mid\sim \beta \text{ implies } \mid\sim \alpha \wedge \beta$$

$$\alpha \wedge \beta \mid\sim \beta$$

$$\alpha \mid\sim \alpha \vee \beta$$

$$\beta \mid\sim \alpha \vee \beta$$

$$\mid\sim \alpha \vee \neg\alpha$$

$$\alpha \mid\sim \top$$

$$\bot \mid\sim \alpha$$

$$\neg(\alpha \wedge \beta) \equiv \neg\alpha \vee \neg\beta$$

$$\neg(\alpha \vee \beta) \equiv \neg\alpha \wedge \neg\beta$$

The conditions we imposed so far are 'local' in the sense that they are imposed separately on every single consequence relation belonging to the class considered. We, moreover, impose the following conditions which have a global character in the sense that they are related to the class $\mathcal{C}$ as a whole.

$$\frac{\alpha \mid\sim_{\mathcal{C}} \gamma, \beta \mid\sim_{\mathcal{C}} \gamma}{\alpha \vee \beta \mid\sim_{\mathcal{C}} \gamma}$$

$$\frac{\alpha \mid\sim_{\mathcal{C}} \beta}{\neg\beta \mid\sim_{\mathcal{C}} \neg\alpha}$$

Let us now define the key concept of a consequence revision system.

DEFINITION 3. Let $\mathcal{C}$ be a class of consequence relations over Fml satisfying the conditions described. Let F be a function

$$F : Fml \times \mathcal{C} \to \mathcal{C}.$$

We say that F is an action on $\mathcal{C}$ iff for every $\mid\sim \in \mathcal{C}$ and $\alpha, \beta \in Fml$ the following conditions are satisfied.

$$\text{(i) } F(\top, \mid\sim) = \mid\sim$$

$$\text{(ii) } F(\alpha, \mid\sim) = 0 \text{ iff } \mid\sim \neg\alpha$$

$$\text{(iii) } F(\beta, F(\alpha, \mid\sim)) = F(\alpha, \mid\sim) \text{ iff } \alpha \mid\sim \beta$$

If F is an action on $\mathcal{C}$, we call the pair $\langle \mathcal{C}, F\rangle$ a *consequence revision system* (CRS).

Note that by $\mid\sim \alpha$ we mean $\top \mid\sim \alpha$. For a given class $\mathcal{C}$ of consequence relations call the formulas α and β $\mathcal{C}$-equivalent, in symbols $\alpha \equiv_{\mathcal{C}} \beta$, if for every $\mid\sim \in \mathcal{C}$ we have $\alpha \mid\sim \beta$ and $\beta \mid\sim \alpha$.

Remark: We are aware of the fact that the way we use the term revision in the above definition does not fully capture the way it is used in traditional revision theory (see for instance [Alchourrón *et al.*, 1985]). If at all, the action of formulas on consequence relations as defined above represents a simple type of revision. Condition (*ii*) above says that given a consequence relation $\mid\sim$ and a formula α which is inconsistent with $\mid\sim$ then the result of 'revising' $\mid\sim$ by α is the inconsistent consequence relation. The corresponding case in traditional revision theory is that of a theory T and a formula α inconsistent with T. The result of revising T by α usually denoted by $T * \alpha$ is then, according to traditional revision theory, not necessarily the inconsistent theory. Since, however, in our most important examples, namely those arising from Hilbert spaces, we are concerned with a process which, in the intuitive sense, deserves to be called revision, we freely use the term revision. Every $\alpha \in Fml$ induces a (revision) operator on $\mathcal{C}$

$$\overline{\alpha} : \mathcal{C} \to \mathcal{C}$$

via

$$\overline{\alpha} \mid\!\sim =: F(\alpha, \mid\!\sim)$$

For $\overline{\alpha} \mid\!\sim$ we will also write $\mid\!\sim_\alpha$.

Denote the class of these operators by $\overline{Fml}$. We have $\overline{\alpha} = \overline{\beta}$ iff $\alpha \equiv_{\mathcal{C}} \beta$.

LEMMA 4. *For any $\alpha \in Fml$ we have $\overline{\alpha} \circ \overline{\alpha} = \overline{\alpha}$.*

Proof. By *Reflexivity* we have $\alpha \mid\!\sim \alpha$ for every $\mid\!\sim \in \mathcal{C}$. Then the claim follows by condition (iii) of the definition of an action. ■

LEMMA 5. *Let $\langle \mathcal{C}, F\rangle$ be a CRS. Then for any $\mid\!\sim \in \mathcal{C}$ the following conditions are equivalent*

- *(i)* $\mid\!\sim \alpha$
- *(ii)* $\mid\!\sim_\alpha = \mid\!\sim$
- *(iii) There exists a* $\mid\!\sim_1 \in \mathcal{C}$ *such that* $\mid\!\sim_{1,\alpha} = \mid\!\sim$

Proof. For the equivalence of (i) and (ii) observe first that $\mid\!\sim_{\top,\alpha} = \mid\!\sim_\alpha$. By condition (iii) of the definition of an action we have that $\top \mid\!\sim \alpha$ iff $\mid\!\sim_\alpha = \mid\!\sim_{\top,\alpha} = \mid\!\sim_\top = \mid\!\sim$. Clearly, (ii) implies (iii). In order to show that (iii) implies (ii) suppose $\mid\!\sim_{1,\alpha} = \mid\!\sim$. Note that by *Reflexivity* we have $\alpha \mid\!\sim_1 \alpha$. Then it follows by condition (iii) of the definition of an action that $\mid\!\sim_\alpha = \mid\!\sim$. ■

LEMMA 6. $\alpha \mid\!\sim \beta$ *iff* $\mid\!\sim_\alpha \beta$,

Proof. Suppose $\alpha \mid\!\sim \beta$. By condition (iii) of the definition of an action this is equivalent to $\mid\!\sim_{\alpha,\beta} = \mid\!\sim_\alpha$. By (i) of the above lemma this means that $\mid\!\sim_\alpha \beta$. ■

It follows by the above two lemmas that $\mid\!\sim_\alpha = \mid\!\sim_\beta$ implies $\alpha \equiv \beta$, i.e. $\alpha \mid\!\sim \beta$ and $\beta \mid\!\sim \alpha$. We see that $\overline{\alpha} = \overline{\beta}$ iff $\alpha \equiv_{\mathcal{C}} \beta$.

DEFINITION 7. Let $\langle \mathcal{C}, F\rangle$ be a CRS. Then define the proposition $[\alpha]$ by

$$[\alpha] =: \{\mid\!\sim \mid\ \mid\!\sim \alpha\}$$

We denote the class of propositions of $\langle \mathcal{C}, F\rangle$ by Prop.

It is routine to verify the statements made in the following lemma.

LEMMA 8. *Let* $\langle \mathcal{C}, F \rangle$ *be a* CRS. *Then*

$$\overline{\alpha} \leq \overline{\beta} \text{ iff } [\alpha] \subset [\beta]$$

$$\overline{\alpha} = \overline{\beta} \text{ iff } [\alpha] = [\beta]$$

$$\alpha \mathrel{|\!\sim_{\mathcal{C}}} \beta \text{ iff } [\alpha] \subset [\beta]$$

$$\alpha \equiv_{\mathcal{C}} \beta \text{ iff } [\alpha] = [\beta]$$

$$\overline{\alpha} \leq \overline{\beta} \text{ iff } \overline{\neg\beta} \leq \overline{\neg\alpha}$$

$$[\alpha] \subset [\beta] \text{ iff } [\neg\beta] \subset [\neg\alpha]$$

The conditions we imposed on the consequence relations guarantee that the following holds.

PROPOSITION 9. *For any* CRS *both* $\langle \overline{Fml}, \leq \rangle$ *and* $\langle Prop, \subset \rangle$ *are lattices. For* $\overline{\alpha}, \overline{\beta} \in \overline{Fml}$ *and* $[\alpha], [\beta] \in Prop$ *the greatest lower bounds are* $\overline{\alpha \wedge \beta}$ *and* $[\alpha \wedge \beta]$ *respectively. The lowest upper bounds are given by* $\overline{\alpha \vee \beta}$ *and* $[\alpha \vee \beta]$ *respectively. The unit and the zero element are given by* $[\top]$ *and* $[\bot]$ *respectively.*

Given a CRS $\langle \mathcal{C}, F \rangle$. Then define unary operations
$^* : \overline{Fml} \to \overline{Fml}$ and $^* : Prop \to Prop$ as follows.

$$\overline{\alpha}^* =: \overline{\neg\alpha}$$

and

$$[\alpha]^* =: [\neg\alpha]$$

Note that in view of Lemma 8 these operations are well defined. Moreover, we define a mapping $\psi : \overline{Fml} \to Prop$ by

$$\psi(\overline{\alpha}) = [\alpha]$$

again, by Lemma 8 this mapping is well defined. It is routine to verify the following proposition which bears an analogy to the well known fact that in Hilbert space the lattice of projections and the lattices of closed subspaces are isomorphic (orthomodular) lattices.

PROPOSITION 10. *Let* $\langle \mathcal{C}, F \rangle$ *be a* CRS. *Then*

- $\langle \overline{Fml}, \leq, ^* \rangle$ *and* $\langle Prop, \subset, ^* \rangle$ *are ortholattices.*
- ψ *is an isomorphism between ortholattices.*

We now define the concept of an *internalising connective* , which provides the link between the object level and the meta level. Whenever we use the term connective we mean a connective definable by the usual propositional connectives in the following sense. We say that $\alpha \rightsquigarrow \beta$ is definable if there exists a formula of propositional logic $\varphi(p,q)$ with exactly two propositional variables such that the formula $\alpha \rightsquigarrow \beta$ is the result of uniformly substituting α in φ for p and β for q. We say that φ defines $\rightsquigarrow$. Given two connectives $\rightsquigarrow_1$ and $\rightsquigarrow_2$ defined by φ_1 and φ_2 respectively. Then we say that $\rightsquigarrow_1$ and $\rightsquigarrow_2$ are classically equivalent if φ_1 and φ_2 are classically equivalent. Consider for instance the Sasaki hook $\rightsquigarrow_s$ which is defined by $\varphi(p,q) =: \neg p \vee (p \wedge q)$. This says that $\alpha \rightsquigarrow_s \beta$ is just short for $\neg\alpha \vee (\alpha \wedge \beta)$. The Sasaki hook is classically equivalent to material implication.

DEFINITION 11. Let $\mid\!\sim$ be a consequence relation and $\rightsquigarrow$ a connective such that $\alpha \mid\!\sim \beta$ iff $\mid\!\sim \alpha \rightsquigarrow \beta$. Then we say that $\rightsquigarrow$ is an internalising connective for $\mid\!\sim$. Given a *CRS* $\langle \mathcal{C}, F\rangle$. Then we say that $\rightsquigarrow$ is an internalising connective for $\langle \mathcal{C}, F\rangle$ iff $\rightsquigarrow$ is an internalising connective for all $\mid\!\sim \in \mathcal{C}$.

PROPOSITION 12. *Let $\langle \mathcal{C}, F\rangle$ be a CRS and let $\rightsquigarrow$ be an internalising connective for $\langle \mathcal{C}, F\rangle$. Then the following holds.*

- *(i) $\alpha \mid\!\sim (\beta \rightsquigarrow \gamma)$ iff $\beta \mid\!\sim_\alpha \gamma$*
- *(ii)$\{\mid\!\sim \mid \alpha \mid\!\sim \beta\}$ is a proposition, namely $[\alpha \rightsquigarrow \beta]$*

Proof. By Lemma 6 we have $\alpha \mid\!\sim (\beta \rightsquigarrow \gamma)$ iff $\mid\!\sim_\alpha (\beta \rightsquigarrow \gamma)$. Since $\rightsquigarrow$ is internalising, this is equivalent to $\beta \mid\!\sim_\alpha \gamma$. This proves (i).
(ii) follows from the fact that $\rightsquigarrow$ is internalising. ∎

Note that in view of the above we can in case we have an internalising connective $\rightsquigarrow$ describe the process of revision simply as follows. Revise the consequence relation $\mid\!\sim$ by α so as to get $\mid\!\sim_\alpha$. Then γ can be proved from β in $\mid\!\sim_\alpha$ iff $\beta \rightsquigarrow \gamma$ can be proved from α in $\mid\!\sim$.

Given a class of consequence relations $\mathcal{C}$ and two connectives $\rightsquigarrow_1$ and $\rightsquigarrow_2$. We then say that $\rightsquigarrow_1$ and $\rightsquigarrow_2$ are $\mathcal{C}$-equivalent iff for all formulas $\alpha, \beta \in Fml$ we have $\alpha \rightsquigarrow_1 \beta \equiv_{\mathcal{C}} \alpha \rightsquigarrow_2 \beta$.

LEMMA 13. *Let $\langle \mathcal{C}, F\rangle$ be a CRS. Then any two internalising connectives for $\langle \mathcal{C}, F\rangle$ are $\mathcal{C}$-equivalent.*

Proof. Let $\rightsquigarrow_1$ and $\rightsquigarrow_2$ be two internalising connectives for $\langle \mathcal{C}, F\rangle$. By symmetry it suffices to prove that $\alpha \rightsquigarrow_1 \beta \mid\!\sim_{\mathcal{C}} \alpha \rightsquigarrow_2 \beta$. So let $\mid\!\sim$ be any element of $\mathcal{C}$ such that $\mid\!\sim \alpha \rightsquigarrow_1 \beta$. Since $\rightsquigarrow_1$ is internalising, we have $\alpha \mid\!\sim \beta$ and, since $\rightsquigarrow_2$ is internalising, $\mid\!\sim \alpha \rightsquigarrow_2 \beta$. ∎

The above lemma says that the action 'determines' the internalising connective modulo $\mathcal{C}$-equivalence. The next lemma states a sort of converse for this, namely that the internalising connective 'determines' the action.

LEMMA 14. *Let $\langle \mathcal{C}, F_1 \rangle$ and $\langle \mathcal{C}, F_2 \rangle$ be CRS and let $\rightsquigarrow$ be a connective which is internalising for both. Then we have $F_1 = F_2$.*

The concept of an orthomodular lattice is a dominant concept in virtually all approaches to quantum logic. It is the quantum logical counterpart of the concept of a Boolean algebra in classical logic. In our approach, this fact is highlighted by Theorem 16. For its proof we need the following theorem of Mittelstaedt, the proof of which can for instance be found in [Engesser *et al.*, 2008].

THEOREM 15 Mittelstaedt. *Let L be an orthocomplemented lattice with orthocomplementation $^{\perp}$. Then L is orthomodular if there exists a conditional $S(A, B)$ such that the following conditions are satisfied.*

$$(i)\ A \wedge S(A,B) \leq B$$

$$(ii)\ A \wedge C \leq B \text{ implies } A^{\perp} \vee (A \wedge C) \leq S(A,B)$$

A conditional satisfying the above conditions is unique, namely

$$S(A,B) = A^{\perp} \vee (A \wedge B).$$

L is a Boolean algebra if the above conditions are satisfied by 'material implication', i.e. $S(A,B) = A^{\perp} \vee B$.

We denote $A^{\perp} \vee (A \wedge B)$ by $A \rightsquigarrow_s B$.

THEOREM 16. *Let $\langle \mathcal{C}, F \rangle$ be a CRS such that for any $\mid\!\sim\, \in \mathcal{C}$, $\mid\!\sim \alpha \rightsquigarrow_s \beta$ implies $\alpha \mid\!\sim \beta$ and let $\rightsquigarrow$ be an internalising connective for $\langle \mathcal{C}, F \rangle$. Then $\langle \overline{Fml}, \leq, ^* \rangle$ and thus $\langle Prop, \subset, ^* \rangle$ are orthomodular lattices and $\rightsquigarrow$ is $\mathcal{C}$-equivalent to $\rightsquigarrow_s$.*
If $\rightsquigarrow_s$ is $\mathcal{C}$-equivalent to $\rightarrow$, i.e. material implication, then the above lattices are Boolean algebras.

Proof. In view of Proposition 10 it suffices to prove orthomodularity. We first show that for any $\mid\!\sim\, \in \mathcal{C}$

$$(1)\ \alpha \wedge (\alpha \rightsquigarrow \beta) \mid\!\sim \beta$$

By Lemma 6 it suffices to show that $\mid\!\sim_{\alpha\wedge(\alpha\rightsquigarrow\beta)} \beta$. By the same lemma we get $\mid\!\sim_{\alpha\wedge(\alpha\rightsquigarrow\beta)} \alpha \wedge (\alpha \rightsquigarrow \beta)$ and thus $\mid\!\sim_{\alpha\wedge(\alpha\rightsquigarrow\beta)} \alpha$ and $\mid\!\sim_{\alpha\wedge(\alpha\rightsquigarrow\beta)} \alpha \rightsquigarrow \beta$. Moreover, since $\rightsquigarrow$ is internalising, we have $\mid\!\sim_{\alpha\wedge(\alpha\rightsquigarrow\beta),\alpha} \beta$. But $\mid\!\sim_{\alpha\wedge(\alpha\rightsquigarrow\beta),\alpha} = \mid\!\sim_{\alpha\wedge(\alpha\rightsquigarrow\beta)}$, since $\mid\!\sim_{\alpha\wedge(\alpha\rightsquigarrow\beta)} \alpha$. Now (1) is proved.
It follows that

$$(2)\ \overline{\alpha} \wedge \overline{\alpha \rightsquigarrow \beta} \leq \overline{\beta}$$

We now prove that the operator $\overline{\alpha \rightsquigarrow \beta}$ has the following property.

$$(3)\ \overline{\alpha} \wedge \overline{\beta} \leq \overline{\gamma} \text{ implies } \overline{\alpha \rightsquigarrow_s \beta} \leq \overline{\alpha \rightsquigarrow \gamma}.$$

For this we have to use that every $\mid\!\sim\in\mathcal{C}$ satisfies *Cut*. Assume $\overline{\alpha}\wedge\overline{\beta}\leq\overline{\gamma}$ and let $\mid\!\sim\in\mathcal{C}$ be such that $\mid\!\sim\alpha\rightsquigarrow\beta$. We then have $\alpha\wedge\beta\mid\!\sim\gamma$ and, since $\rightsquigarrow$ is internalising, $\alpha\mid\!\sim\beta$. Then we get, using *Cut*, that $\alpha\mid\!\sim\gamma$ and again, since $\rightsquigarrow$ is internalising, $\mid\!\sim(\alpha\rightsquigarrow\gamma)$. Thus $\overline{\alpha\rightsquigarrow\beta}\leq\overline{\alpha\rightsquigarrow\gamma}$.
Now, by the hypothesis, $\mid\!\sim(\alpha\rightsquigarrow_s\beta$ implies $\alpha\mid\!\sim\beta$ and thus, since $\rightsquigarrow$ is internalising, $\mid\!\sim(\alpha\rightsquigarrow\beta)$. This means $\overline{\alpha\rightsquigarrow_s\beta}\leq\overline{\alpha\rightsquigarrow\beta}$. By transitivity we have $\overline{\alpha\rightsquigarrow_s\beta}\leq\overline{\alpha\rightsquigarrow\gamma}$.
We have proved that, if $\overline{\alpha}\wedge\overline{\beta}\leq\overline{\gamma}$, then $\mid\!\sim\alpha\rightsquigarrow_s\beta$ implies $\mid\!\sim\alpha\rightsquigarrow\gamma$ for any $\mid\!\sim\in\mathcal{C}$, which means $\overline{\alpha\rightsquigarrow_s\beta}\leq\overline{\alpha\rightsquigarrow\gamma}$. We now get by (2), (3) and Mittelstaedt's Theorem 15 that $\langle\overline{Fml},\leq,^*\rangle$ and thus $\langle Prop,\subset,^*\rangle$ are orthomodular and, moreover, $\overline{\alpha\rightsquigarrow\beta}=\overline{\alpha}^*\vee(\overline{\alpha}\wedge\overline{\beta})$. From this it follows that $\rightsquigarrow$ and $\rightsquigarrow_s$ are $\mathcal{C}$-equivalent.
That the lattices under consideration are Boolean if $\rightsquigarrow_s$ is $\mathcal{C}$-equivalent to material implication, again, follows by Mittelstaedt's theorem. ■

Remark: Note that in the above proof two 'logical' properties of the class $\mathcal{C}$ play a crucial role in establishing the fact that the lattices $\langle\overline{Fml},\leq,^*\rangle$ and $\langle Prop,\subset\ ^*\rangle$ have the algebraic property of being orthomodular. The first 'logical' property is that an internalising connective having a certain property exists for $\langle\mathcal{C},F\rangle$. This property of an action can, as we will see, be viewed as a generalisation of the property that the deduction theorem holds. The second crucial property is that all consequence relations of $\mathcal{C}$ satisfy *Cut*.
For the purposes of this chapter we introduce the following notion of a *logic*.

DEFINITION 17. Let $\langle\mathcal{C},F\rangle$ be a *CRS* and $\rightsquigarrow$ an internalising connective for $\langle\mathcal{C},F\rangle$. Then call the triple $\mathcal{L}=\langle\mathcal{C},F,\rightsquigarrow\rangle$ a logic.

We may thus interpret the above theorem as essentially saying that for a *CRS* to become a logic (with $\rightsquigarrow_s$ as its internalising connective), it is necessary that the lattice of operators $\langle\overline{Fml},\leq,^*\rangle$ and thus the lattice of propositions $\langle Prop,\subset,^*\rangle$ have the algebraic property of being orthomodular.
Given a consequence relation $\mid\!\sim$, then define $C(\mid\!\sim)=:\{\alpha\mid\;\mid\!\sim\alpha\}$. We have the

PROPOSITION 18. *Let* $\mathcal{L}=\langle\mathcal{C},F,\rightsquigarrow\rangle$ *be a logic. Given* $\mid\!\sim_1,\mid\!\sim_2\in\mathcal{C}$. *Then* $C(\mid\!\sim_1)=C(\mid\!\sim_2)$ *iff* $\mid\!\sim_1=\mid\!\sim_2$.

Proof. Suppose $\mathcal{C}(\mid\!\sim_1)=\mathcal{C}(\mid\!\sim_2)$ and let $\alpha\mid\!\sim_1\beta$. It follows, since $\rightsquigarrow$ is internalising that $\mid\!\sim_1(\alpha\rightsquigarrow\beta)$ and thus by the hypothesis $\mid\!\sim_2(\alpha\rightsquigarrow\beta)$. Again, since $\rightsquigarrow$ is internalising, we get $\alpha\mid\!\sim_2\beta$, thus $\mid\!\sim_1\subset\mid\!\sim_2$. By symmetry we also get the other inclusion. ■

3.3 Classical logic revisited

Let us now return to our motivating example from classical logic and look at it from the point of view of the framework developed in the last subsection. Let $\Sigma\subset Fml$ be any consistent set of formulas. Define the class $\mathcal{C}_{\Sigma,\alpha}$ of consequence relations as follows. For a given formula α, define $\vdash_{\Sigma,\alpha}$ by:

$$\beta\vdash_{\Sigma,\alpha}\gamma\text{ iff }\Sigma\cup\{\alpha\wedge\beta\}\vdash\gamma$$

Moreover, define $\mathcal{C}_\Sigma = \{\vdash_{\Sigma,\alpha} \mid \alpha \in Fml\}$ and the function $\mathcal{F}_\Sigma : Fml \times \mathcal{C}_\Sigma \to \mathcal{C}_\Sigma$ by $\mathcal{F}_\Sigma(\alpha, \vdash_{\Sigma,\alpha}) \vdash_{\Sigma,\alpha\wedge\beta}$. It is immediately verified, using familiar facts of classical logic such as the deduction theorem, that consequence relations as defined above satisfy all the conditions we imposed and that $\langle \mathcal{C}_\Sigma, \mathcal{F}_\Sigma \rangle$ is a CRS. We have

$$\vdash_{\Sigma,\alpha} = \vdash_{\Sigma,\beta} \text{ iff } \Sigma \vdash \alpha \leftrightarrow \beta$$

THEOREM 19. *$\mathcal{L}_\Sigma = \langle \mathcal{C}_\Sigma, \mathcal{F}_\Sigma, \to \rangle$ is a logic. The lattice of operators $\mathcal{O}_{\mathcal{L}_\Sigma}$ and thus the lattice of propositions $\mathcal{P}_{\mathcal{L}_\Sigma}$ are Boolean algebras isomorphic to the Lindenbaum algebra $\mathcal{B}(\Sigma)$.*

Proof. For the first part of our claim we need to prove that $\to$ is an internalising connective for $\langle \mathcal{C}_{L,\Sigma}, \mathcal{F}_{L,\Sigma} \rangle$. But this is exactly what the deduction theorem says:

$$\Sigma \cup \{\alpha\} \vdash (\beta \to \gamma) \text{ iff } \Sigma \cup \{\alpha \wedge \beta\} \vdash \gamma$$

It follows from the fact that $\to$ is internalising and Theorem 16 that the lattices under consideration are Boolean algebras. Moreover, it is straightforward to prove that the following function $\varphi : \mathcal{O}_{\mathcal{L}_\Sigma} \to \mathcal{B}(\Sigma)$ is well defined and is an isomorphism

$$\varphi(\overline{\alpha}) = [\alpha]_\Sigma,$$

where $[\alpha]_\Sigma$ denotes the (unique) element of the Lindenbaum algebra $\mathcal{B}(\Sigma)$ to which α belongs. ■

Note that this way of establishing the well known fact that the Lindenbaum algebra is a Boolean algebra is radically different from the usual method.

4 HOLISTIC LOGICS: EVERYTHING IS ENCODED IN EVERYTHING

4.1 What is a physical state from the logical point of view?

In classical physics, the concept of a state is, from the logical point of view, unproblematic. Logically, in classical physics a state is a complete classical theory. It can be identified with the set of all physical statements *true* about the system. In this sense the state of the system at a certain point in time fully contains all the information about the system. What in classical mechanics is particularly convenient is the fact that once we know the momenta and the positions of the particles constituting the system, we know all relevant physical properties. Therefore, from the logical point of view, a state can be described by a single proposition, namely by the proposition specifying all values of the momenta and positions at a given time. From this we can then compute (deduce) the values of all relevant physical quantities. This is what in classical mechanics is known as *phase space*. So, the logical analogue of the concept of a state in classical mechanics is that of a *complete classical theory*.

Why can't we represent the state of a quantum system analogously, namely by the set of those propositions that are true in this state? In quantum mechanics things aren't that simple. Given a state x in quantum mechanics and a proposition of the form $A = \mu$, where A is an observable and μ a real number. Suppose we perform a measurement of the observable A in x. Then the following three cases may occur. First, the probability to measure μ is 1 in which case we may reasonably say that $A = \mu$ is true (in state x). Second, the probability to get μ as a result of measurement may be 0. In this case we may reasonably say that $A = \mu$ is false or, equivalently, $\neg(A = \mu)$ is true. In quantum mechanics there is, however, a third case which marks the difference with classical mechanics. Namely, the probability to get μ may be greater than zero and smaller than one. Let us for the moment call these propositions contingent with respect to x. It is then obviously insufficient to represent the state x by the set of those propositions that are true in x because this does not give us any information about the contingent propositions and their probabilities. It seems that a proper representation of a quantum state must specify probabilities. In a purely logical treatment of the concept of a physical state we should, however, try to avoid specifying probabilities.

We may think of the contingent propositions as *coming true* rather than *being true* in the following sense. Let us think of a proposition α of the form $A = \mu$ as a projection in a Hilbert space H , namely as the projection corresponding to the eigenspace for eigenvalue μ of observable A. This eigenspace is the set $FP(\alpha)$ of fixed points of the projection α. We may say then that α *is* true in x if $x \in FP(\alpha)$ and x is false in x if $x \in FP(\neg\alpha)$. Otherwise, i.e. in case that $\alpha(x) = y \neq x$ and $\alpha(x) \neq 0$ we may say that α *comes* true in x. Thus α *comes* true in x if it *is* true in $\alpha(x)$. Hence the representation of the state x must give us information not just on what is true in x but about what comes true in x. Thus in the quantum case it is the coming true of a proposition that replaces or generalises the being true of a proposition in classical physics and classical logic. This is, in the quantum case, the dynamic analogue of the static concept of being true in classical physics and classical logic. However, coming true in x involves a different state which in turn must be specified. Thus, intuitively, we must require the logical entity representing a state x as also specifying other states, namely all those states in which a proposition *is* true when it *comes* true in state x.

Technically speaking, it is as follows. The logical entity representing a (physical) state x in classical logic, namely a complete theory, contains all propositions true in state x. When, however, we are concerned with propositions that act on the state x or, as we said, have the property of coming true rather than being true in state x, then the logical representation of x must encode all propositions that come true at x. In other words, the logical representation of a state x must encode the action of the propositions on x. The action of a proposition on x, however, yields a new state y, and therefore the state x must encode other states. So we inevitably hit here on the phenomenon of *encodedness of states in other states* which will play a dominant role in our study of *holistic logics* introduced in this chapter. We will see that, there, a state is a logical entity that encodes *other states* and also itself.

4.2 The concept of a holistic logic

Let us start from our motivating Example 3.1. In that example the consequence relations cannot be 'characterised' by a single formula, i.e. given any $\vdash_\alpha$, then there exists no formula β such that β is provable in $\vdash_\alpha$ and only in $\vdash_\alpha$. We take this observation as a motivation for studying logics in which every consequence relation has a 'characterising' formula.

DEFINITION 20. Given a logic $\mathcal{L} = \langle \mathcal{C}, F, \rightsquigarrow \rangle$ in the sense of Definition 17. Consider the following conditions.

- For any non-zero $\mid\!\sim_0 \in \mathcal{C}$ there exists a formula $\sigma_{\mid\sim_0}$ such that $\mid\!\sim \sigma_{\mid\sim_0}$ iff $\mid\!\sim = \mid\!\sim_0$. We call σ a pointer to $\mid\!\sim_0$.
- For every $\mid\!\sim \in \mathcal{C}$ there exist a formula α such that neither $\mid\!\sim \alpha$ nor $\mid\!\sim \neg\alpha$.

We call $\mathcal{L}$ a (non-degenerate) holistic logic if both of the above conditions are satisfied. We call $\mathcal{L}$ degenerate holistic if the first condition is satisfied but not the second. We call $\mathcal{L}$ totally degenerate holistic if the first condition is satisfied but for no consequence relation $\mid\!\sim$ does there exist an α such that neither $\mid\!\sim \alpha$ nor $\mid\!\sim \neg\alpha$.

Remarks: Any two pointers σ_1 and σ_2 to the same consequence relation are equivalent, i.e. $[\sigma_1] = [\sigma_2]$ We assume the consequence relation referred to later to be non-zero, i.e. consistent without explicit mentioning.

Intuitively, the second condition says that every consequence relation $\mid\!\sim$ must be genuinely revisable, i.e. we assume that there exists a formula α such that $\mid\!\sim_\alpha$ is consistent and distinct from $\mid\!\sim$. It follows that a (non-degenerate) holistic logic has at least two consequence relations. We will always use the term 'holistic' in the sense of 'non-degenerate holistic' except in the theorem which we call the limiting case theorem. In the case of a totally degenerate holistic logic there is no genuine revision at all.

In the next subsections we state some salient properties of holistic logics.

4.3 Everything is Encoded in Everything

The findings of this subsection have given rise to the term 'holistic'. We will see that, roughly, in a holistic logic every consequence relation is 'encoded' in every other consequence relation and that it also 'encodes' itself.

DEFINITION 21. Let $\mathcal{L}$ be a holistic logic and let $\mid\!\sim_1$ and $\mid\!\sim_2$ be two consequence relations of $\mathcal{L}$ with pointers σ_1 and σ_2 respectively. Then we say that $\mid\!\sim_1$ and $\mid\!\sim_2$ are orthogonal if $\mid\!\sim_1 \neg\sigma_2$ and $\mid\!\sim_2 \neg\sigma_1$.

Actually it suffices to require one of the two conditions. It can then be proved using the second of the global conditions we impose on the consequence relations that the relation of orthogonality is symmetric.

The following lemma follows from the definition of a pointer and that of a consequence revision systems.

LEMMA 22. *Let $\mathcal{L} = \langle \mathcal{C}, F, \rightsquigarrow \rangle$ be a holistic logic, $\mid\!\sim_1, \mid\!\sim_2 \in \mathcal{C}$. If $\mid\!\sim_1$ and $\mid\!\sim_2$ are not orthogonal, then $\mid\!\sim_{1,\sigma_2} = \mid\!\sim_2$ and vice versa. If they are orthogonal we have $\mid\!\sim_{1,\sigma_2} = 0$ and vice versa.*

DEFINITION 23. Let $\mathcal{L}$ be a holistic logic. We call a family $(\mid\!\sim_i)_{i \in I}$ of pairwise orthogonal consequence relations a basis of $\mathcal{L}$ if for any $\mid\!\sim$ of $\mathcal{L}$ there exists a basis consequence relation $\mid\!\sim_j$ not orthogonal to $\mid\!\sim$. We call $\mathcal{L}$ finite-dimensional iff it admits a finite basis. If $\mathcal{L}$ is finite-dimensional we say it has dimension n if it admits a basis of n elements and no basis of fewer elements.

LEMMA 24. *Let $\mathcal{L} = \langle \mathcal{C}, F, \rightsquigarrow \rangle$ be a holistic logic and $\mid\!\sim \in \mathcal{C}$ with pointer σ. Then we have*

- *(i) $\mid\!\sim \alpha$ iff $[\sigma \rightsquigarrow \alpha] = [\top]$ and thus $[\neg(\sigma \rightsquigarrow \alpha)] = [\bot]$*
- *(ii) $\not\mid\!\sim \alpha$ iff $[\sigma \rightsquigarrow \alpha] = [\neg\sigma]$ and thus $[\neg(\sigma \rightsquigarrow \alpha)] = [\sigma]$*

Remark: Note that by the above lemma we have that $\mid\!\sim \alpha$ iff $\mid\!\sim \sigma \rightsquigarrow \alpha$ and $\not\mid\!\sim \alpha$ iff $\mid\!\sim \neg(\sigma \rightsquigarrow \alpha)$. We may therefore view the formula $\sigma \rightsquigarrow \alpha$ as expressing provability of α at the object level and the formula $\neg(\sigma \rightsquigarrow \alpha)$ as expressing the unprovability of α at the object level. In particular we have 'provability of unprovability' in the sense that if $\mid\!\sim$ cannot prove α, then it can prove that it cannot prove α.

Proof. (i) For the direction from left to right suppose $\mid\!\sim \alpha$ and note that for any $\mid\!\sim_1$ orthogonal to $\mid\!\sim_1$ we have $\mid\!\sim_1 \sigma \rightsquigarrow \alpha$, see Lemma 22. If $\not\mid\!\sim_1$ is non-orthogonal to $\mid\!\sim$, we have by Lemma 22 that $\mid\!\sim \sigma \rightsquigarrow \alpha$. Thus $[\sigma \rightsquigarrow \alpha] = \mathcal{C} = [\top]$. The other direction is obvious.
(ii) Suppose that $\not\mid\!\sim \alpha$. Then, again, we have for every $\mid\!\sim_1$ orthogonal to $\mid\!\sim$ that $\mid\!\sim_1 \sigma \rightsquigarrow \alpha$. But if $\mid\!\sim_1$ is not orthogonal to $\mid\!\sim$, $\mid\!\sim_1 \sigma \rightsquigarrow \alpha$ cannot hold, since this would imply $\mid\!\sim \alpha$ contrary to the hypothesis. Thus $[\sigma \rightsquigarrow \alpha] = [\neg\sigma]$. The other direction is obvious. ■

Remark: Note that the propositions $[\sigma], [\neg\sigma], [\top], [\bot]$ do not depend on the pointer σ, since any two pointers are equivalent. They form a Boolean algebra in a natural way.

PROPOSITION 25. *Assume the hypotheses of the last lemma and let φ and ψ have the form $\varphi = \sigma \rightsquigarrow \alpha$ or $\varphi = \neg(\sigma \rightsquigarrow \alpha)$ and $\psi = \sigma \rightsquigarrow \beta$ or $\psi = \neg(\sigma \rightsquigarrow \beta)$. Then we have*

- *(i) $\mid\!\sim \neg\varphi$ iff $\not\mid\!\sim \varphi$*
- *(ii) $\mid\!\sim \varphi \wedge \psi$ iff $\mid\!\sim \varphi$ and $\mid\!\sim \psi$*
- *(iii) $\mid\!\sim \varphi \vee \psi$ iff $\mid\!\sim \varphi$ or $\mid\!\sim \psi$*
- *(iv) $\mid\!\sim \varphi \rightarrow \psi$ iff $\not\mid\!\sim \varphi$ or $\mid\!\sim \psi$*
- *(v) $\varphi \mid\!\sim \psi$ iff $\mid\!\sim \varphi \rightarrow \psi$*

- (vi) $\mid\!\sim \varphi \rightarrow \psi$ *iff* $\mid\!\sim \varphi \rightsquigarrow \psi$

Proof. For (i) we reason as follows. Let φ have the form $\varphi = \sigma \rightsquigarrow \alpha$ and suppose $\mid\!\sim \neg\varphi$. Then, clearly, $\not\mid\!\sim \varphi$. Now suppose $\not\mid\!\sim \varphi$. By Lemma 24 we then have $[\neg\varphi] = [\sigma]$. It follows that $\mid\!\sim \neg\varphi$.

(ii) is a general property of the consequence relations considered. (iii) and (iv) follow from (i) and the definition of the connectives $\vee$ and $\rightarrow$. In order to see that (v) holds recall that $\varphi \mid\!\sim \psi$ means that $\mid\!\sim_\varphi \psi$ and observe that by Lemma 24 $\mid\!\sim_\varphi = \mid\!\sim$, namely if $\mid\!\sim \varphi$, or $\mid\!\sim_\varphi = 0$, namely if $\not\mid\!\sim \varphi$. (vi) follows from (v) and the fact that $\rightsquigarrow$ is an internalising connective. ■

THEOREM 26. *Let $\mathcal{L}$ be a holistic logic and $\mid\!\sim_1$ and $\mid\!\sim_2$ two non-orthogonal consequence relations with pointers σ_1 and σ_2 respectively. Then we have*

- (i) $\alpha \mid\!\sim_1 \beta$ *iff* $\mid\!\sim_2 \sigma_1 \rightsquigarrow (\alpha \rightsquigarrow \beta)$
- $(ii$ $\alpha \not\mid\!\sim_1 \beta$ *iff* $\mid\!\sim_2 \sigma_1 \rightsquigarrow \neg(\sigma_1 \rightsquigarrow (\alpha \rightsquigarrow \beta))$

By symmetry the claim also holds if we interchange the indices 1 *and* 2.

Proof. (i) Recall that $\mid\!\sim_1$ and $\mid\!\sim_2$ are non-orthogonal iff $\mid\!\sim_{2_{\sigma_1}} = \mid\!\sim_1$ (and vice versa). We have $\alpha \mid\!\sim_2 \beta$ iff $\mid\!\sim_2 \alpha \rightsquigarrow \beta$, since $\rightsquigarrow$ is internalising. $\alpha \mid\!\sim_2 \beta$ is thus equivalent to $\mid\!\sim_{2_{\sigma_1}} \alpha \rightsquigarrow \beta$. This is the case iff $\sigma_1 \mid\!\sim_2 \alpha \rightsquigarrow \beta$, which is equivalent to $\mid\!\sim_2 \sigma_1 \rightsquigarrow (\alpha \rightsquigarrow \beta)$.

(ii) Note that $\alpha \not\mid\!\sim_1 \beta$ is by 'provability of unprovability' equivalent to $\mid\!\sim_1 \neg(\sigma_1 \rightsquigarrow (\alpha \rightsquigarrow \beta)$ and apply (i). ■

The above theorem says that non-orthogonal consequence relations of a holistic logic are 'encoded' in each other. This fact is the motivation for calling these structures holistic.

THEOREM 27. *Let $\mathcal{L}$ be a holistic logic and $\mid\!\sim_1$ and $\mid\!\sim_2$ be any consequence relations. Denote for convenience of notation the respective pointers by σ_x and σ_y. Suppose $\mid\!\sim_3$ is non-orthogonal to both $\mid\!\sim_1$ and $\mid\!\sim_2$ and denote its pointer by σ_z. Then we have*

- $\alpha \mid\!\sim_2 \beta$ *iff* $\mid\!\sim_1 \sigma_z \rightsquigarrow (\sigma_y \rightsquigarrow (\alpha \rightsquigarrow \beta))$
- $\alpha \not\mid\!\sim_2 \beta$ *iff* $\mid\!\sim_1 \sigma_z \rightsquigarrow (\sigma_y \rightsquigarrow \neg(\sigma_y \rightsquigarrow (\alpha \rightsquigarrow \beta)))$

Proof. Repeated application of Theorem 26. ■

The significance of Theorem 27 is this. For those holistic logics that are presented by a Hilbert space (see Section 6) it is true that for any two consequence relations there exists a third consequence relation (superposition) which is non-orthogonal to both. This means that in such systems *any* two consequence relations encode each other. Loosely speaking, every single consequence relation 'mirrors' the whole system. For those readers familiar with Leibniz's *Monadology* this may come as a striking resemblance. In fact, there are several such parallels between our theory of holistic logics and Leibniz's famous metaphysical treatise. We elaborated on this in the form of a Platonic dialogue in our monograph [Engesser *et al.*, 2008].

4.4 Self-referential soundness and completeness

In this section we study self-referential soundness and completeness in holistic logics. This notion was, essentially, first introduced by Smullyan in [Smullyan, 1987] and [Smullyan, 1992] for modal systems. We will prove that the consequence relations of a holistic logic are self-referentially sound and complete and –apart from the classical limiting case– nonmonotonic.

We now define a meta language in which we can talk about provability. Intuitively, $DER(\alpha, \beta)$ means "β is derivable from α in $\mid\!\sim$".

DEFINITION 28.

- (i) If α, β are formulas of the object language, then $DER(\alpha, \beta) \in ML$.
- If α is a formula of the object language and $\varphi \in ML$, then $DER(\alpha, \varphi) \in ML$ and $DER(\varphi, \alpha) \in ML$.
- If $\varphi, \psi \in ML$, then $DER(\varphi, \psi) \in ML$.
- If $\varphi, \psi \in ML$, so are $\neg\varphi$ and $\varphi \wedge \psi$, $\varphi \vee \psi$, $\varphi \rightarrow \psi$, where $\vee$ and $\rightarrow$ are defined as usual in terms of $\neg$ and $\wedge$.

We will use the following abbreviations:

$$PROV\alpha =: DER(\top, \alpha)$$

$$CON\alpha =: \neg PROV \neg\alpha$$

$$EQUIV(\alpha, \beta) =: DER(\alpha, \beta) \wedge DER(\beta, \alpha)$$

We now define a natural translation of the meta language ML into the object language. We assume that we have a logic $\mathcal{L} = \langle \mathcal{C}, F, \rightsquigarrow \rangle$. The following definitions are relative to a fixed $\mid\!\sim \in \mathcal{C}$ having a pointer σ to itself. Since any two pointers are equivalent, they do not 'depend' on the pointer chosen.

DEFINITION 29. Let σ be a pointer to $\mid\!\sim$. Define the translation $'$ as follows.

- (i) If $\varphi = DER(\alpha, \beta)$ where α and β are formulas of the object language, $\varphi' =: \sigma \rightsquigarrow (\alpha \rightsquigarrow \beta)$.
- (ii) If $\varphi = DER(\alpha, \psi)$, where α is a formula of the object language and $\psi \in ML$, then $\varphi' =: \sigma \rightsquigarrow (\alpha \rightsquigarrow \psi')$; analogously for the case $DER(\psi, \alpha)$.
- (iii) If $\varphi = DER(\psi, \rho)$ with $\psi, \rho \in ML$, $\varphi' =: \sigma \rightsquigarrow (\psi' \rightsquigarrow \rho')$.

- (iv) If $\varphi = \neg\psi$, $\varphi' =: \neg(\sigma \rightsquigarrow \psi')$.
- (v) If $\varphi = \psi \wedge \rho$, $\varphi' =: \psi' \wedge \rho'$.

We now define the notion of truth for ML in a natural way. This definition of truth is in the spirit of what Smullyan calls a self-referential interpretation in the above mentioned books. The essential feature of Smullyan's notion of self-referential truth is this. Given a modal system M with the modal operator $\Box$. Then we say that a formula of the form $\Box A$ is (self-referentially) true with respect to M iff A is provable in M.

DEFINITION 30.

- (i) If $\varphi = DER(\alpha, \beta)$, where α, β are formulas of the object language, then TRUE φ iff $\alpha \mathrel{|\!\sim} \beta$.
- (ii) If $\varphi = DER(\alpha, \psi)$, where α is a wff of the object language, then TRUE φ iff $\alpha \mathrel{|\!\sim} \psi'$; analogously for the case $DER(\psi, \alpha)$.
- (iii) If $\varphi = DER(\psi, \rho)$ for $\psi, \rho \in ML$, then TRUE φ iff $\psi' \mathrel{|\!\sim} \rho'$.
- (iv) If $\varphi = \neg\psi$, then TRUE φ iff not TRUE ψ.
- (v) If $\varphi = \psi \wedge \rho$, then TRUE φ iff TRUE ψ and TRUE ρ.

THEOREM 31. *Let* $\mathcal{L} = \langle \mathcal{C}, F, \rightsquigarrow \rangle$ *be a holistic logic,* $\mathrel{|\!\sim} \in \mathcal{C}$. *Then we have for any* $\mathrel{|\!\sim} \in \mathcal{C}$ *and any* $\varphi \in ML$

$$TRUE\ \varphi \text{ iff } \mathrel{|\!\sim} \varphi'$$

The above theorem expresses *self-referential soundness and completeness* of the consequence relations of a holistic logic. The fact that $\mathrel{|\!\sim} \varphi'$ implies TRUE φ expresses self-referential soundness and the fact that TRUE φ implies $\mathrel{|\!\sim} \varphi'$ expresses self-referential completeness.

Proof. By induction on the construction of the formulas of ML.
(i) Case $\varphi = DER(\alpha, \beta)$. By definition $TRUE$ φ means $\alpha \mathrel{|\!\sim} \beta$. This means $\mathrel{|\!\sim} \alpha \rightsquigarrow \beta$, which is equivalent to $\mathrel{|\!\sim} \sigma \rightsquigarrow (\alpha \rightsquigarrow \beta)$. But this says that $\mathrel{|\!\sim} \varphi'$.
(ii) Case $\varphi = DER(\alpha, \psi)$. Suppose $TRUE$ φ. By definition this says $\alpha \mathrel{|\!\sim} \psi'$ or equivalently $\mathrel{|\!\sim} \sigma \rightsquigarrow (\alpha \rightsquigarrow \psi')$. But this is exactly what $\mathrel{|\!\sim} \varphi'$ means.
(iii) The case $\varphi = DER(\psi, \rho)$ is proved analogously to that of (ii).
(iv) Case $\varphi = \neg\psi$. $TRUE$ φ means that not TRUE ψ. By the induction hypothesis this is equivalent to $\not\mathrel{|\!\sim} \psi'$. This is by Lemma 24 the case iff $\mathrel{|\!\sim} \neg(\sigma \rightsquigarrow \psi')$. But this says $\mathrel{|\!\sim} \varphi'$.
(v) Case $\varphi\psi \wedge \rho$. We have by definition TRUE φ iff TRUE ψ and TRUE ρ. The latter is by the induction hypothesis equivalent to "$\mathrel{|\!\sim} \varphi'$ and $\mathrel{|\!\sim} \rho'$" which in turn is equivalent to $\mathrel{|\!\sim} \varphi'$. ∎

Remark: Inspecting the translation of the metalanguage into the object language, we may view the metalanguage as a 'sublanguage' of the object language. The peculiar feature of this 'sublanguage' is that it contains a 'proof operator' $\Box$, namely $\Box\alpha =: \sigma \rightsquigarrow \alpha$, as opposed to 'proof predicates' which we have in other languages.

Instead of explicitly defining a metalanguage we could have proceeded as follows. We could from the outset have confined ourselves to defining a sublanguage MSL of the object language doing the job of the metalanguage. In this case the expression $DER(\alpha, \beta)$ would be an *abbreviation* for $\sigma \rightsquigarrow (\alpha \rightsquigarrow \beta)$. We could then have defined TRUE φ directly for the formulas of MSL, i.e. for object formulas. Theorem 31 would then read: TRUE φ iff $\mid\!\sim \varphi$.

The notion of self-referentiality thus becomes fully analogous to that introduced by Smullyan in connection with self-application of modal systems, where the modal operator $\Box$ plays the role of a proof operator.

Example: Let us consider an example and let us for the sake of illustration verify the truth of the claim made in the above theorem directly. Let α be an object formula and consider the following meta-statement

$$\varphi = PROV\alpha \rightarrow CON\alpha$$

Its translation is

$$\varphi' = (\sigma \rightsquigarrow (\top \rightsquigarrow \alpha)) \rightarrow \neg(\sigma \rightsquigarrow (\sigma \rightsquigarrow (\top \rightsquigarrow \neg\alpha)))$$

Let us first verify that $TRUE\ \varphi$ implies $\mid\!\sim \varphi'$. Assume that not $TRUE\ PROV\alpha$. This means that $TRUE\neg PROV\alpha$, which says that $\not\mid\!\sim \alpha$. By Lemma 24 we have $[\neg(\sigma \rightsquigarrow (\top \rightsquigarrow \alpha))] = [\sigma]$. Thus $[\varphi'] = [\sigma \vee ...]$ and we have $\mid\!\sim \varphi'$.
Now assume $TRUE\ CON\alpha$, i.e. $\not\mid\!\sim \neg\alpha$ and thus $\not\mid\!\sim \top \rightsquigarrow \neg\alpha$, hence $\not\mid\!\sim \sigma \rightsquigarrow (\top \rightsquigarrow \neg\alpha)$. In this case we have by Lemma 24 $[\neg(\sigma \rightsquigarrow (\sigma \rightsquigarrow (\top \rightsquigarrow \neg\alpha)))] = [\sigma]$. Thus $[\varphi'] = [... \vee \sigma]$ and we have $\mid\!\sim \varphi'$.
Let us now verify that $\mid\!\sim \varphi'$ implies $TRUE\ \varphi$. So assume $\mid\!\sim \varphi'$. $[\neg(\sigma \rightsquigarrow (\sigma \rightsquigarrow (\top \rightsquigarrow \alpha)))]$ equals either $[\bot]$ or $[\sigma]$. In the first case we have $\mid\!\sim \alpha$. Since $\mid\!\sim$ is assumed to be consistent, we have $\not\mid\!\sim \neg\alpha$, which means $TRUE\ CON\alpha$. But this says that $TRUE\ \varphi$.
In the second case we have $\not\mid\!\sim \alpha$ and thus not $TRUE\ PROV\alpha$ in which case again $TRUE\ \varphi$.

Some examples of true meta-statements

PROPOSITION 32. *The following meta-statements are true and thus their translations are provable.*

- $\varphi_1 = PROV\varphi \leftrightarrow PROVPROV\varphi$
- $\varphi_2 = \neg PROV\varphi \leftrightarrow PROV\neg PROV\varphi$
- $\varphi_3 = CON\varphi \rightarrow \neg EQUIV(\varphi, \neg PROV\varphi)$

- $\varphi_4 = (CON\varphi \wedge \neg PROV\varphi) \rightarrow (PROV\neg PROV\varphi \wedge \neg DER(\varphi, \neg PROV\varphi)$
- $\varphi_5 = PROV\varphi \leftrightarrow EQUIV(\varphi, \neg PROV\bot)$
- $\varphi_6 = PROV\neg PROV\bot$
- $\varphi_6 = (PROV\varphi \wedge DER(\varphi, \psi)) \rightarrow PROV\psi$

Comment: The above claims are immediate consequences of Lemma 24. The reader should note that by self-referential completeness the consequence relation 'knows' the facts expressed by the above meta-statements.

Intuitively, φ_1 expresses 'provability of provability': φ is provable iff it is provable that φ is provable.

φ_2 expresses 'provability of unprovability': φ is not provable iff its unprovability can be proved.

φ_3 says that if φ is consistent, then it is not equivalent to its 'own unprovability'. This says that the consequence relations of a holistic logic do not admit Gödel fixed points.

φ_4 says the following. Suppose φ is consistent and not provable. Then we know that its unprovability can be proved. What φ_4 says is that, however, its unprovability cannot be proved 'from φ'. So, φ_4 can be rephrased as follows. If φ is consistent and not provable, then its unprovability can be proved but not from φ. In the non-degenerate case, φ_4 says in particular that the consequence relation is nonmonotonic, since in the non-degenerate case we assume it to have an object formula which is consistent and not provable.

φ_5 says that φ is provable iff it is equivalent to the consistency of the consequence relation.

φ_6 says that the consequence relation can prove its consistency.

φ_7 expresses modus ponens for meta-statements. In a sense, the consequence relation can 'justify' the logical rule of modus ponens. Normally, logical rules such as modus ponens are justified at the meta level as preserving truth. The intuitive meaning of φ_7 is that holistic logics can *prove* its own rules (at the object level).

The case of a complete classical theory

Recall the definition of $\mathcal{L}_\Sigma = \langle \mathcal{C}_\Sigma, \mathcal{F}_\Sigma, \rightarrow\rangle$ from the motivating Example 3.1. We have the

PROPOSITION 33. *Let Σ be a consistent set of formulas. Then $\mathcal{L}_\Sigma = \langle \mathcal{C}_\Sigma, \mathcal{F}_\Sigma, \rightarrow\rangle$ is holistic iff Σ is a complete classical theory. In this case $\mathcal{L}_\Sigma$ is (totally) degenerate. It has dimension* 1 *and we have $\mathcal{C} = \{\vdash_\Sigma, 0\}$ and $\mathcal{F}_\Sigma(\alpha, \vdash_\Sigma) = \vdash_\Sigma$ if $\alpha \in \Sigma$, else* 0.

Proof. Observe that for any α such that neither $\Sigma \vdash \neg\alpha$ nor $\Sigma \vdash \alpha$, $\vdash_\Sigma$ is a proper subset of $\vdash_\Sigma, \alpha$. So in this case $\vdash_\Sigma$ cannot have a pointer. It follows that $\vdash_\Sigma$ can have a pointer only if for every α either $\Sigma \vdash \alpha$ or $\Sigma \vdash \neg\alpha$, i.e. Σ is a complete theory. In fact, in this case any formula α such that $\Sigma \vdash \alpha$ is a pointer to $\vdash_\Sigma$ ■

4.5 No windows theorems

The local no windows theorem

Given a holistic logic $\mathcal{L} = \langle \mathcal{C}, F, \rightsquigarrow \rangle$ and $\mid\!\sim \in \mathcal{C}$ with pointer σ. Then we define $\Sigma_{\mid\sim} =: \{\alpha \mid \mid\!\sim \alpha\}$, to be the *local theory* of $\mid\!\sim$. We denote by Σ_g its *global theory*, i.e. $\Sigma_g =: \{\alpha \mid \mid\!\sim \alpha \text{ for all } \mid\!\sim \in \mathcal{C}\}$.

LEMMA 34. *Let $\mathcal{L} = \langle \mathcal{C}, F, \rightsquigarrow \rangle$ be a holistic logic and $\mid\!\sim \in \mathcal{C}$ with pointer σ. Suppose $\mid\!\sim \alpha$. Then $\sigma \rightsquigarrow \alpha \in \Sigma_g$.*

Proof. Let $\mid\!\sim_1 \in \mathcal{C}$. Suppose $\mid\!\sim_1$ is orthogonal to $\mid\!\sim$. Then $\mid\!\sim_{1\sigma} = 0$ and, clearly, $\sigma \mid\!\sim_1 \alpha$. Hence $\mid\!\sim_1 \sigma \rightsquigarrow \alpha$. Suppose $\mid\!\sim_1$ is not orthogonal to $\mid\!\sim$. In this case we have $\mid\!\sim_{1_\sigma} = \mid\!\sim$. Thus $\mid\!\sim_{1_\sigma} \alpha$. It follows that $\sigma \mid\!\sim_1 \alpha$ which means $\mid\!\sim_1 \sigma \rightsquigarrow \alpha$. We have proved that $\sigma \rightsquigarrow \alpha \in \Sigma_g$. ∎

In the sequel we will use the terminology 'the connective $\rightsquigarrow$ is classically equivalent to material implication'. By this we simply mean that $\alpha \rightsquigarrow \beta$ is an abbreviation for a formula which is classically equivalent to $\alpha \rightarrow \beta$. For instance the Sasaki hook $\rightsquigarrow_s$ has this property because $\alpha \rightsquigarrow_s \beta =: \neg\alpha \vee (\alpha \wedge \beta)$ is classically equivalent to $\alpha \rightarrow \beta$.

LEMMA 35. *Let $\mathcal{L} = \langle \mathcal{C}, F, \rightsquigarrow \rangle$ be a holistic logic and $\mid\!\sim \in \mathcal{C}$ with pointer σ. Suppose $\rightsquigarrow$ is classically equivalent to $\rightarrow$, i.e. material implication. Assume that $\Sigma_g \cup \{\sigma\}$ is classically consistent. Then we have $\mid\!\sim \alpha$ iff $\Sigma_g \cup \{\sigma\} \vdash \alpha$.*

Proof. For the direction from left to right note that $\mid\!\sim \alpha$ implies by Lemma 34 that $\sigma \rightsquigarrow \alpha \in \Sigma_g$ and, since $\rightsquigarrow$ is assumed to be classically equivalent to $\rightarrow$, we have $\Sigma_g \cup \{\sigma\} \vdash \alpha$.
For the other direction suppose $\Sigma_g \cup \{\sigma\} \vdash \alpha$ and assume $\not\mid\!\sim \alpha$. We have $\Sigma_g \cup \{\sigma\} \vdash \sigma \rightarrow \alpha$. On the other hand we have by 'provability of unprovability' $\mid\!\sim \neg(\sigma \rightsquigarrow \alpha)$ and thus, by the direction already proved, $\Sigma_g \cup \{\sigma\} \vdash \neg(\sigma \rightsquigarrow \alpha)$ and thus, since $\rightsquigarrow$ is classically equivalent to $\rightarrow$ $\Sigma_g \cup \{\sigma\} \vdash \neg(\sigma \rightarrow \alpha)$

$\Sigma_g \cup \{\sigma\}$ would thus be classically inconsistent contrary to the hypothesis. It follows that $\mid\!\sim \alpha$. ∎

We call the following theorem the (local) no windows theorem because it is reminiscent of what Leibniz in his *Monadology* says about the monads: "The monads have no windows". Again, the interested reader is referred to our monograph [Engesser *et al.*, 2008].

THEOREM 36. *Let $\mathcal{L} = \langle \mathcal{C}, F, \rightsquigarrow \rangle$ be a non-degenerate holistic logic. Suppose $\rightsquigarrow$ is classically equivalent to material implication. Let $\mid\!\sim \in \mathcal{C}$ with pointer σ. Then $\Sigma_g \cup \{\sigma\}$ is classically inconsistent. Thus, $\Sigma_{\mid\sim}$ is classically inconsistent.*

Proof. Let σ be any pointer with corresponding $\mid\!\sim \in \mathcal{C}$. Assume that $\Sigma_g \cup \{\sigma\}$ is classically consistent. Let α be such that $\not\mid\!\sim \neg\alpha$ and $\not\mid\!\sim \alpha$. By the hypothesis of non-degeneracy such a formula exists. Then we have by 'provability of unprovability' and nonmonotonicity

$$(1) \mid\!\sim \neg(\sigma \rightsquigarrow \alpha)$$

$$(2)\ \alpha \not\mid\!\sim \neg(\sigma \rightsquigarrow \alpha)$$

We have by Lemma 35

$$(3)\ \Sigma_g \cup \{\sigma\} \vdash \neg(\sigma \rightsquigarrow \alpha)$$

and thus by classical logic

$$(4)\ \Sigma_g \cup \{\sigma\} \vdash \alpha \rightarrow \neg(\sigma \rightsquigarrow \alpha)$$

Since $\rightsquigarrow$ is classically equivalent to $\rightarrow$, it follows that

$$(5)\ \Sigma_g \cup \{\sigma\} \vdash \alpha \rightsquigarrow \neg(\sigma \rightsquigarrow \alpha)$$

Again, by Lemma 35 we get

$$(6) \mid\!\sim \alpha \rightsquigarrow \neg(\sigma \rightsquigarrow \alpha)$$

and thus

$$(7)\ \alpha \mid\!\sim \neg(\sigma \rightsquigarrow \alpha)$$

But (7) contradicts (2). It follows that $\Sigma_g \cup \{\sigma\}$ is classically inconsistent. ■

The global no windows theorem

We now restrict ourselves to the case of a finite-dimensional holistic logic. In this case we can sharpen the no windows theorem so as to get a Kochen-Specker type result as a special case.

LEMMA 37. *Let $\mathcal{L}$ be any logic and α such that $\not\mid\!\sim \alpha$ for every $\mid\!\sim \neq 0$. Then $\alpha \rightsquigarrow \bot \in \Sigma_g$.*

Proof. Given any $\mid\!\sim \in \mathcal{C}$ and α as in the hypothesis. We claim that $\mid\!\sim_\alpha = 0$. For otherwise we would have $\mid\!\sim_\alpha \alpha$ with $\mid\!\sim_\alpha \neq 0$ contrary to the hypothesis. Thus $\mid\!\sim_\alpha \bot$, which means $\alpha \mid\!\sim \bot$ and thus $\mid\!\sim \alpha \rightsquigarrow \bot$. We have proved that $\alpha \rightsquigarrow \bot \in \Sigma_g$. ■

The following theorem is a summary of previous results and, moreover, contains the strengthened version of the no windows theorem.

THEOREM 38. *Let $\mathcal{L} = \langle \mathcal{C}, \mathcal{F}, \rightsquigarrow \rangle$ be a non-degenerate holistic logic. Suppose that $\rightsquigarrow$ is classically equivalent to $\rightarrow$, i.e. material implication. Then we have the following*

(i) Every consistent $\mid\!\sim \in \mathcal{C}$ is nonmonotonic.

(ii) For any $\mid\!\sim \in \mathcal{C}$, $\Sigma_{\mid\sim}$ is classically inconsistent.

(iii) If $\mathcal{L}$ is finite dimensional, then Σ_g is classically inconsistent. In fact, it contains a classical contradiction.

Proof. (i) and (ii) summarise results proved earlier.

As to (iii) let $(\mid\!\sim_i), i = 1, .., n$ a basis with pointers σ_i. The local no windows Theorem 36 tells that $\Sigma \cup \{\sigma_i\}$ is classically inconsistent, $i = 1, ..., n$. This means that

$$\Sigma_g \vdash \neg\sigma_i,\ i = 1, ..., n.$$

Therefore

$$\Sigma_g \vdash \bigwedge_i \neg\sigma_i$$

For any $\mid\!\sim \neq 0$ we have by the definition of a basis

$$\not\mid\!\sim \bigwedge_i \neg\sigma_i$$

For otherwise $\mid\!\sim$ would be orthogonal to all elements of the basis contrary to the definition of a basis. It follows by Lemma 37 that

$$\bigwedge_i \neg\sigma_i \rightsquigarrow \bot \in \Sigma_g$$

Thus

$$\Sigma_g \vdash \bigwedge_i \neg\sigma_i \rightsquigarrow \bot$$

and, since $\rightsquigarrow$ is classically equivalent to $\rightarrow$,

$$\Sigma_g \vdash \bigwedge_i \neg\sigma_i \rightarrow \bot$$

It follows that

$$\Sigma_g \vdash \bot$$

Thus Σ_g is classically inconsistent. Then there exists a finite set $\{\alpha_1, ..., \alpha_n\} \subset \Sigma_g$ which is classically inconsistent. Since Σ_g is closed under conjunctions we have

$$\bigwedge_i \alpha_i \in \Sigma_g$$

But this conjunction is a classical contradiction. ■

4.6 Limiting case theorem

In this section we prove a limiting case theorem for holistic logics. Let us start with the following observation.

PROPOSITION 39. *Given a consequence revision system* $\langle \mathcal{C}, F \rangle$. *Suppose that all revision operators commute. Then every* $\mid\!\sim \in \mathcal{C}$ *is monotonic.*

Proof. Assume that all operators commute and let $\mid\!\sim \in \mathcal{C}$. Assume $\mid\!\sim \beta$. This means $\mid\!\sim_\beta = \mid\!\sim$. Now let α be any formula. Note that $\mid\!\sim_{\alpha,\beta} \beta$. The above notation means that $\mid\!\sim$ is first revised by α and then by β. Since the revision operators corresponding to α and β commute, we have $\mid\!\sim_{\beta,\alpha} = \mid\!\sim_{\alpha,\beta}$. But $\mid\!\sim_{\beta,\alpha} = \mid\!\sim_\alpha$. It follows that $\mid\!\sim_\alpha \beta$. This says that $\alpha \mid\!\sim \beta$. We have proved that $\mid\!\sim$ is monotonic. ■

It follows from the above lemma that in a consequence revision system containing nonmonotonic consequence relations we have non-commuting revision operators, i.e. 'uncertainty relations'.

LEMMA 40. *Let* $\mathcal{L} = \langle \mathcal{C}, F, \rightsquigarrow \rangle$ *be a holistic logic such that every* $\mid\!\sim \in \mathcal{C}$ *is monotonic. Then for every* $\mid\!\sim \in \mathcal{C}$ *we have* $\mid\!\sim = \vdash_{\Sigma_{\mid\sim}}$, *i.e.* $\mathcal{L}$ *is totally degenerate.*

Proof. Suppose that $\mid\!\sim \in \mathcal{C}$ is monotonic. Let α be such that $\not\mid\!\sim \alpha$. Hence $\mid\!\sim_\alpha \neq \mid\!\sim$. By 'provability of unprovability' we have that that $\mid\!\sim \neg(\sigma \rightsquigarrow \alpha)$ and by monotonicity $\alpha \mid\!\sim \neg(\sigma \rightsquigarrow \alpha)$. Since $[\neg(\sigma \rightsquigarrow \alpha)] = \{\mid\!\sim, 0\}$, we have $\mid\!\sim_\alpha = 0$. But this says that $\mid\!\sim \neg\alpha$. It follows that for any α we have either $\mid\!\sim \alpha$ or $\mid\!\sim \neg\alpha$. We say that $\mid\!\sim$ is complete as a consequence relation.

Recall that by $\Sigma_{\mid\sim}$ we denote the set $\{\alpha \mid \mid\!\sim \alpha\}$. We have proved that for any α we have $\alpha \in \Sigma_{\mid\sim}$ or $\neg\alpha \in \Sigma_{\mid\sim}$. We prove, moreover, that $\Sigma_{\mid\sim}$ has the property that $(\alpha \to \beta) \in \Sigma_{\mid\sim}$ iff not $\alpha \in \Sigma_{\mid\sim}$ or $\beta \in \Sigma_{\mid\sim}$. It follows that $\Sigma_{\mid\sim}$ is a complete theory. Suppose that $(\alpha \to \beta) \in \Sigma_{\mid\sim}$. This says that $\neg(\alpha \wedge \neg\beta) \in \Sigma_{\mid\sim}$. By the property already proved this is equivalent to $\not\mid\!\sim \alpha \wedge \neg\beta$. This is by a general condition imposed on the consequence relation considered the case iff $\not\mid\!\sim \alpha$ or $\not\mid\!\sim \neg\beta$ which in turn is equivalent to not $\alpha \in \Sigma_{\mid\sim}$ or $\beta \in \Sigma_{\mid\sim}$.

We now need to prove that $\mid\!\sim = \vdash_{\Sigma_{\mid\sim}}$. For this we need to see that $\alpha \mid\!\sim \beta$ iff $\mid\!\sim \neg\alpha$ or $\mid\!\sim \beta$. We have $\alpha \mid\!\sim \beta$ iff $\mid\!\sim_\alpha \beta$. Note that $\mid\!\sim_\alpha = 0$, which means $\mid\!\sim \neg\alpha$ or $\mid\!\sim_\alpha = \mid\!\sim$. $\mid\!\sim_\alpha \beta$ therefore holds iff $\mid\!\sim \neg\alpha$, which says that not $\alpha \in \Sigma_{\mid\sim}$ or $\mid\!\sim \beta$, which means that $\beta \in \Sigma_{\mid\sim}$ ■

The following theorem is the limiting case theorem for holistic logics.

THEOREM 41. *Let* $\mathcal{L} = \langle \mathcal{C}, F, \rightsquigarrow \rangle$ *be a holistic logic. Then the following conditions are equivalent.*

(i) Every $\mid\!\sim \in \mathcal{C}$ *is monotonic.*

(ii) For every $\mid\!\sim \in \mathcal{C}$ *we have* $\mid\!\sim = \vdash_{\Sigma_{\mid\sim}}$.

(iii) All operators commute.

Proof. (i) implies $(ii$ by Lemma 40. That (ii) implies (iii) from the fact that if (ii) holds, then the operation is trivial in the sense that $F(\alpha, \mid\!\sim) = \mid\!\sim$ or $F(\alpha, \mid\!\sim) = 0$. From Proposition 39 we see that (iii) implies (i). ■

In the limit we have in particular monotonicity , classical consistency and thus *models*, a 'reality outside the logic'.

5 SOME HILBERT SPACE THEORY

For this section we refer the reader also to the chapter "Solèr's Theorem" by A. Prestel of this Handbook, in particular for the definition of an orthomodular space.

In quantum mechanics we are (primarily) concerned with infinite-dimensional Hilbert spaces. We define, for historical reasons, a *classical Hilbert lattice* to be a lattice isomorphic to the lattice of closed subspaces of an infinite-dimensional Hilbert space over the real numbers, the complex numbers or the quaternions.

In this section we give a characterisation of classical Hilbert lattices among ortholattices which was first presented in [Engesser, 2000]. For this purpose we need, apart from Piron's theorem, two deep theorems of modern Hilbert space theory, namely the theorems of Solèr and Mayet stated below. Mayet's theorem heavily relies on a theorem of Wigner, which we state too. In his pioneering paper [Keller, 1980], Keller settled a long standing question, namely the question whether every infinite-dimensional orthomodular space is already a Hilbert space. Keller's construction of a counter example settled the question in the negative. This, however, posed another problem, namely the problem of characterising those orthomodular spaces that are in fact Hilbert spaces. This problem was solved by Maria Pia Solèr in [Solèr, 1995].

THEOREM 42 Solèr. *Let $\langle H, \langle , \rangle\rangle$ be an orthomodular space over K and let $c \in K$. Suppose there exists an infinite family $(x_i)_{i \in I}$ of pairwise orthogonal elements of H such that for all $i \in I$, $\langle x_i, x_i\rangle = c$. Then K must be the (skew-) field of the real numbers, the complex numbers or the quaternions and H is an infinite-dimensional Hilbert space.*

This way of stating Solèr's theorem is due to Holland, see [Holland, 1995].

DEFINITION 43. Let H_1 and H_2 be two orthomodular spaces and $\sigma : H_1 \to H_2$ be a bijective map. We say that σ is a semi-unitary map iff the following conditions are satisfied.

- For any $x, y \in H_1$, $\sigma(x + y) = \sigma(x) + \sigma(y)$.
- There exists an automorphism ρ of K such that, for any $\lambda \in K$ and any $x \in H_1$, we have $\sigma(\lambda x) = \rho(\lambda)(\sigma x)$.
- There exists $\lambda_\sigma \in K$ such that, for any $x, y \in H_1$, we have $\langle\sigma(x), \sigma(y)\rangle = \rho(\langle x, y\rangle)\lambda_\sigma$.

If, moreover, we have $\rho = id_K$ and $\lambda_\sigma = 1$, we say that σ is unitary.

THEOREM 44 Wigner. *Let H_1 and H_2 be orthomodular spaces of dimension at least 3. Then every ortholattice isomorphism $f : Sub(H1) \to Sub(H_2)$ is induced by some semi-unitary map.*

We need the following result by Mayet which, essentially, is a consequence of Wigner's theorem.

THEOREM 45 Mayet. *Let H be an orthomodular space of dimension at least 3 and let $X \in Sub(H)$ of dimension at least 2. Let f be an automorphism of $Sub(H)$ whose restriction to $[0, X]$ is the identical map. Then there exists a unique unitary operator σ on H inducing f such that the restriction of σ to X is the identical map.*

Solèr's theorem characterises Hilbert spaces among orthomodular spaces. We are interested in a characterisation of classical Hilbert lattices among ortholattices. For this we need Piron's Representation Theorem.

THEOREM 46 Piron. *An ortholattice $\mathcal{L}$ of height ≥ 4 is a Hilbert lattice iff it is atomistic, complete, irreducible, orthomodular and satisfies the covering property.*

For more more on this see [Holland, 1995]. The characterisation we give is in terms of a symmetry property, see [Engesser, 2000].

For a given ortholattice L we call two atoms σ_1 and σ_2 orthogonal if $\sigma_1 \leq \sigma_2^\perp$. This relation is readily seen to be symmetric.

DEFINITION 47. Let L be a complete ortholattice and let $\Delta = (\sigma_i)_{i \in I}$ be an infinite pairwise orthogonal family of atoms of L. We say that L satisfies the symmetry property (synonymously: is symmetric) with respect to Δ iff the following holds. For any permutation $f : I \to I$ there exists an ortholattice automorphism ρ_f of L with the following properties.

- ρ_f extends f, i.e. $\rho_f(\sigma_i) = \sigma_{f(i)})$ for any $i \in I$.
- If the set J of those elements of I which are left fixed by f is non-empty, ρ_f induces the identical map on $[0, A]$, where A denotes the least upper bound of the family $(\sigma_j)_{j \in J}$.

We say that L is symmetric iff there exists an infinite pairwise orthogonal family Δ of atoms of L such that L is symmetric with respect to Δ.

We have the following characterisation theorem.

THEOREM 48 Engesser. *A Hilbert lattice is a classical Hilbert lattice iff it is symmetric.*

Proof. Let us first verify that for a given infinite-dimensional classical Hilbert space H, $Sub(H)$ is symmetric. To see this consider an orthonormal basis $(x_i)_{i \in I}$ of H. Then the family of one-dimensional subspaces $(\langle x_i \rangle)_{i \in I}$ is an infinite orthogonal system of atoms of $Sub(H)$. Let $f : I \to I$ be any permutation of I. Recall that $x = \sum_{i \in I} \langle x, x_i \rangle x_i$. Define the map φ_f as follows. For $x = \sum_{i \in I} \langle x, x_i \rangle x_i$ put

$\varphi_f(x) =: \sum_{i \in I} \langle x, x_{f^{-1}(i)} \rangle x_i$. φ is well defined. For any $i \in I$ we have $\varphi_f(x_i) = x_{f(i)}$. Moreover, φ_f is unitary, since for any $x, y \in H$ we have by Parseval's identity $\langle \varphi_f(x), \varphi_f(y) \rangle = \sum_{i \in I} \langle x, x_{f^{-1}(i)} \rangle \overline{\langle y, x_{f^{-1}(i)} \rangle} = \sum_{i \in I} \langle x, x_i \rangle \overline{\langle y, x_i \rangle} = \langle x, y \rangle$. Suppose $\{i \mid f(i) = i\}$ is non-empty and denote by X the smallest closed subspace containing $\{x_i \mid f(i) = i\}$. X is the smallest closed subspace containing $\{\langle x_i \rangle \mid f(i) = i\}$ and φ_f induces the identity on X. For the latter claim observe that φ_f induces the identity on the subspace spanned by $\{x_i \mid f(i) = i\}$ and and X is the closure of that subspace. Since φ_f is continuous, it induces the identity on X too. φ_f thus induces an ortholattice automorphism ρ_f on $Sub(H)$ such that for any $i \in I$, $\rho_f(\langle x_i \rangle) = \langle x_{f(i)} \rangle$. Clearly, ρ_f induces the identical map on $[0, X]$. Thus symmetry of $Sub(H)$ is proved.

For the other direction note that the symmetry property implies infinite height. By Piron's Representation Theorem it therefore suffices to show that any orthomodular space H such that $Sub(H)$ has the symmetry property is an infinite-dimensional classical Hilbert space. So let $(\langle x_i \rangle)_{i \in I}$ be an infinite orthogonal family with respect to which $Sub(H)$ is symmetric. Let $i_0 \in I$. For any $j \in I, i_0 \neq j$ consider the permutation f_j of I defined as follows.

$$f_j(i_0) = j,\ f_j(j) = i_0, f_j(i) = i \text{ else.}$$

Denote by X the smallest closed closed subspace of X containing $\langle x_i \rangle$ for all $i \in I$. X is infinite - dimensional. By symmetry there exists an automorphism ρ_j of $Sub(H)$ inducing the identity on $[0, X]$ such that for all $i \in I$, $\rho_j(\langle x_i \rangle) = \langle x_{f_j(i)} \rangle$. So, by Mayet's theorem, ρ_j is induced by some unitary map φ_j. Put $y_j =: \varphi_j(x_{i_0})$ for $j \neq i_0$ and $y_{i_0} = x_{i_0}$. Then, since φ_j is unitary, the family $(y_j)_{j \in I}$ is a family as required in Solèr's theorem. It follows by Solér's theorem that H must be an infinite-dimensional classical Hilbert space. ■

As a corollary we get the following theorem, which gives another characterisation of Hilbert spaces among orthomodular spaces.

THEOREM 49. *Let $\langle H, \langle . \rangle \rangle$ be an orthomodular space over K. Then the following conditions are equivalent.*

- *There exists an infinite family $(x_i)_{i \in I}$ of pairwise orthogonal elements of H and a non-zero $c \in H$ such that for all $i \in I$ we have $\langle x_i, x_i \rangle = c$.*
- *$Sub(H)$ is symmetric.*
- *H is an infinite–dimensional classical Hilbert space.*

6 TOWARDS HILBERT SPACE

Let us in this section come to an idea which is central to the enterprise of this chapter. In the Introduction we speculated about what logic can do about quantum mechanics. We said that it would not be our aim to find a new deductive system especially suited for reasoning in quantum mechanics, which does not mean that we would not consider this a reasonable enterprise. Rather we said that we should look for logical structures implicit in the formalism of quantum mechanics which could prove useful in the task of trying to understand this very formalism. In this section we ask the question "Can we cast light on the core concept of the formalism, namely that of a Hilbert space, using these structures?" This is part of what in the literature on the foundations of quantum mechanics is sometimes called the representation enterprise.

6.1 Presenting holistic logics

Let H be an orthomodular space. $Sub(H)$ denotes its the set of closed subspaces of H. We know that $\langle Sub(H), \subset, ^{\perp} \rangle$ is an orthomodular lattice. Recall that $^{\perp}$ means orthogonal complement formation. We will, as we did earlier, use capital letters $A, B, ...$ for closed subspaces and, if there is no danger of confusion, for the corresponding projections. Moreover, we use the symbols for Boolean connectives in connection with closed subspaces, i.e we write $A \wedge B$ for $A \cap B$ and we denote the smallest closed subspace containing the closed subspaces A and B by $A \vee B$.

Let Fml be the a language propositional logic. Let $\Psi : Fml \to Sub(H)$ be a surjective function such that $\Psi(\neg\alpha) = \Psi(\alpha)^{\perp}$ and $\Psi(\alpha \wedge \beta)) = \Psi(\alpha) \wedge \Psi(\beta)$. Denote the projection corresponding to $\Psi(\alpha)$ by A. Let $x \in H$. Then we define the consequence relation $\vdash_x$ by

$$\alpha \vdash_{x,\Psi} \beta \text{ iff } Ax \in \Psi(\beta).$$

We will simply write $\vdash_x$ if Ψ is clear from the context. Note that $\vdash_x$ depends only on the ray of x, i.e. $\vdash_{x_1} = \vdash_{x_2}$ iff the one dimensional subspace $\langle x_1 \rangle$ generated by x_1 is equal to the one dimensional subspace $\langle x_2 \rangle$ generated by x_2.
Given an orthomodular space H and a function Ψ as described above, we define

$$\mathcal{C}_{H,\Psi} =: \{\vdash_x | \, x \in H\}.$$

Let us now define a function that will turn out to be an action on $\mathcal{C}_{H,\Psi}$. Define $\mathcal{F}_{H,\Psi} : Fml \times \mathcal{C}_{H,\Psi} \to \mathcal{C}_{H,\Psi}$ by

$$\mathcal{F}_{H,\Psi}(\alpha, \vdash_x) =: \vdash_{Ax}.$$

Note that $\mathcal{F}_{H,\Psi}$ is well defined, since $\langle x_1 \rangle = \langle x_2 \rangle$ implies $\langle Ax_1 \rangle = \langle Ax_2 \rangle$.
Recall that the Sasaki hook $\rightsquigarrow_s$ is the connective defined as follows: $\alpha \rightsquigarrow \beta =: \neg\alpha \vee (\alpha \wedge \beta$. The following lemma generalises an observation made by Hardegree in [Hardegree, 1974] in connection with Hilbert spaces.

LEMMA 50. *Let H be an orthomodular space, $x \in H$, $A, B \in Sub(H)$. Then $Ax \in B$ iff $x \in A^{\perp} \vee (A \wedge B)$*

Proof. First note that the closed subspaces $A^{\perp}$ and $A \wedge B$ are orthogonal. Then we have $A^{\perp} \vee (A \wedge B) = A^{\perp} \oplus (A \wedge B)$.

For the direction from left to right let $x = y + z$ be the unique decomposition of x with respect to A and $A^\perp$, i.e. $y \in A$ and $z \in A^\perp$. We have $Ax = y$. The hypothesis says that $y \in B$. Thus $y \in A \wedge B$. It follows that $x \in A^\perp \oplus (A \wedge B)$.

For the direction from right to left observe $A^\perp \oplus (A \wedge B)$ is again an orthomodular space with the Hermitian form properly restricted. We have thus, in addition to the above decomposition, a decomposition $x = y_1 + z_1$ with $y_1 \in A$ and $z_1 \in A \wedge B$. Since the decomposition is unique we have $y = y_1$ and $z = z_1$. It follows that $Ax = y = y_1 \in B$. ■

THEOREM 51. *Let H be an orthomodular space and Ψ a function as described above. Then $\mathcal{L}_{H,\Psi} =: \langle \mathcal{C}_{H,\Psi}, \mathcal{F}_{H,\Psi}, \rightsquigarrow_s \rangle$ is a holistic logic. All consequence relations satisfy the conditions we impose on consequence relations with the possible exception of Cut and Cautious Monotonicity. In case H is a Hilbert space all conditions are satisfied.*

The following proof is in case that H is a Hilbert space. Cut and Cautious Monotonicity work in the Hilbert space case only.

Proof. We first need to verify the conditions imposed on the elements of $\mathcal{C}$. This is routine for the most part.
Reflexivity is a consequence of the fact that for $x \in \Psi(\alpha)$ we have $Ax = x$.
Let us first verify *Cut*. So let $x \in H$ and assume $\alpha \wedge \beta \vdash_x \gamma$ and $\alpha \vdash_x \beta$. $\alpha \vdash_x \beta$ says that $\Psi(\alpha)_x \in \Psi(\beta)$. Moreover, from the above assumptions it follows that $\Psi(\alpha \wedge \beta)_x = \Psi(\alpha)_x$. By the hypothesis we have $\Psi(\alpha \wedge \beta)_x \in \Psi(\gamma)$ and thus $\Psi(\alpha)_x \in \Psi(\gamma)$. But this means that $\alpha \vdash_x \gamma$. Thus, *Cut* is verified.
We now verify *Restricted Monotonicity*. Assume $\alpha \vdash_x \beta$ and $\alpha \vdash_x \gamma$. It follows that $\Psi(\alpha)_x = \Psi(\alpha \wedge \beta)_x$ and, since by the hypothesis we have $\Psi(\alpha)_x \in \Psi(\gamma)$, we see that $\Psi(\alpha \wedge \beta)_x \in \Psi(\gamma)$, which says that $\alpha \vdash_x \gamma$. Thus *Restricted Monotonicity* is verified.
In order to verify the other conditions use that by definition we have $\Psi(\alpha \wedge \beta) = \Psi(\alpha) \wedge \Psi(\beta)$ and $\Psi(\neg\alpha) = \Psi(\alpha)^\perp$ and elementary Hilbert space theory.

For the first global condition for instance suppose $\alpha \mid\!\sim_{\mathcal{C}_{H,\Psi}} \gamma$ and $\beta \mid\!\sim_{\mathcal{C}_{H,\Psi}} \gamma$. This means $\Psi(\alpha) \subset \Psi(\gamma)$ and $\Psi(\beta) \subset \Psi(\gamma)$. It is then elementary Hilbert space theory that $\Psi(\alpha \vee \beta) \subset \Psi(\gamma)$. But this says that $\alpha \vee \beta \mid\!\sim_{\mathcal{C}_{H,\Psi}} \gamma$.

We now prove that $\mathcal{F}_{H,\Psi}$ is an action on $\mathcal{C}$. Condition *(i)* in the definition of an action is obvious, see Definition 3. Consider condition *(ii)* in the definition of an action. Suppose $\vdash_x \neg\alpha$. This is equivalent to $x \in \Psi(\alpha)^\perp$, which is the case iff $Ax = 0$. But this means $\vdash_{Ax} = \mathcal{F}_{H,\Psi}(\alpha, \vdash_x) = 0$.
As to condition *(iii)* in the definition of an action let $\mathcal{F}_{H,\Psi}(\beta, (\mathcal{F}_{H,\Psi}(\alpha, \vdash_x)) = \mathcal{F}_{H,\Psi}(\alpha), \vdash_x)$. This is the case iff $BAx = Ax$, which is equivalent to $Ax \in \Psi(\beta)$. But this says that $\alpha \vdash_x \beta$.

We still need to prove that $\rightsquigarrow_s$ is internalising for $\mathcal{C}$. Suppose $\alpha \vdash_x \beta$. By definition this means that $Ax \in \Psi(\beta)$. By Lemma 50 this is the case iff $x \in \neg A \vee (A \wedge B)$. But this says $\vdash_x \alpha \rightsquigarrow_s \beta$.

We still need to see that $\mathcal{L}_{H,\Psi}$ is holistic. We need to show that any $\vdash_x$ has a pointer. Since Ψ is assumed to be surjective, it is easily seen that such a pointer is given by any formula σ such that $\Psi(\sigma) = \langle x \rangle$. ■

We call a logic of the above form an *orthomodular space logic.* In case case H is a classical Hilbert space we call $\mathcal{L}_{H,\Psi}$ a *Hilbert space logic.*

6.2 Classical inconsistency in Hilbert space logics

Let us recall here a phenomenon first observed by Kochen and Specker, namely that Birkhoff-von Neumann quantum logic is in a sense 'classically inconsistent'. We will see that this phenomenon is not accidental. In fact, it is a consequence of the no windows theorem for holistic logics.

We start with the following simple observation. We denote by H_n an n-dimensional Hilbert space. Let x_1, x_2 be non-orthogonal and non-colinear vectors of H_2. Let Fml be the language of propositional logic and consider a Hilbert space logic $\mathcal{L}_{H_2,\Psi_0}$ such that for the propositional variables p_1, p_2 we have $\Psi_0(p_i) = \langle x_i \rangle$, $i = 1, 2$. Consider the formula $\phi = \phi_1 \wedge \phi_2 \wedge \phi_3 \wedge \phi_4$ such that

$$\phi_1 = p_1 \vee p_2$$

$$\phi_2 = \neg p_1 \vee p_2$$

$$\phi_3 = p_1 \vee \neg p_2$$

$$\phi_4 = \neg p_1 \vee \neg p_2$$

It is easily seen that ϕ is a classical contradiction which is provable in all consequence relations of $\mathcal{L}_H, \Psi_0$.

PROPOSITION 52. *ϕ is a classical contradiction and for all consequence relations $\mid\!\sim$ of $\mathcal{L}_{H_2}, \Psi_0$ we have $\mid\!\sim \phi$*

In the case of three dimensional Hilbert space H the above result is more difficult to establish. In [Kochen and Specker, 1967] Kochen and Specker presented a classical tautology the negation of which is provable in all consequence relations of a Hilbert space logic presented by a three dimensional Hilbert space.

PROPOSITION 53. *There exists a classical contradiction α and a Hilbert space logic $\mathcal{L}_{H_3,\Psi}$ such that $\mid\!\sim \alpha$ for all $\mid\!\sim$ of $\mathcal{L}$.*

Remark: The formula presented by Kochen and Specker contains 117 variables. It represents the full space under a certain 'valuation' of these variables. It is important to note that this valuation is such that only compatible elements of $Sub(H_3)$ are combined by the connectives. The following theorem predicts the existence of a classical contradiction which is a 'quantum tautology'. It does, however, not capture the additional property of the Kochen-Specker formula just mentioned.

THEOREM 54. *Let H be a finite dimensional orthomodular space and $dim H \geq 2$. Let $\mathcal{L}_{H,\psi}$ be a logic presented by H. Then Σ_g is classically inconsistent.*

COROLLARY 55. *Under the above hypotheses there exists a classical tautology* ϕ *such that for all* $x \in H$ *we have* $\vdash_x \neg\phi$.

Proof. The claim is an immediate consequence of Theorem 51 and the global no windows theorem 38. Recall that the Sasaki hook is classically equivalent to material implication and note that holistic logics presented by finite-dimensional orthomodular spaces are finite dimensional as holistic logics. ■

6.3 Symmetry and Hilbert space presentability: A representation theorem

In this subsection we are looking for properties characterising Hilbert space logics. To pose the problem more precisely, let us introduce the following terminology. Given a logic $\mathcal{L} = \langle \mathcal{C}, F, \rightsquigarrow \rangle$, a Hilbert space H and a function $\Psi \to Sub(H)$ such that $\mathcal{L} = \mathcal{L}_{H,\Psi}$. Then we say that $\mathcal{L}$ is *presented* by H via Ψ. We say that $\mathcal{L}$ is presentable by H if there exists a function Ψ such that $\mathcal{L}$ is presented by H via Ψ. It is our aim to characterise the logics presentable by some Hilbert space H. In other words, we are looking for necessary and sufficient conditions for a logic $\mathcal{L}$ to be presentable by some Hilbert space H. We will see that, besides some natural logical conditions, there are two properties essential for the characterisation we have in mind. The first property is *holicity.* The second essential property is a symmetry property. We will call it the *symmetry property.* These two properties play a vital role in our characterisation of Hilbert space logics.

LEMMA 56. *Let* $\mathcal{L} = \langle \mathcal{C}, F, \rightsquigarrow_s \rangle$ *be a holistic logic. Then* $\langle Fml, \leq, ^* \rangle$ *and thus* $\langle Prop, \subset, ^* \rangle$ *are orthomodular, atomistic and irreducible lattices.*

Proof. We have orthomodularity by the fact that $\mathcal{L}$ is a logic with the Sasaki hook as its internalising connective and Theorem 16. As to atomicity observe that the atoms of $\langle Prop, \subset, ^* \rangle$ are of the form $[\sigma_{\mid\sim}]$.

For irreducibility we need to prove that the centre of that lattice consists of truth and falsity only. For this it suffices to prove that for every proposition $[\alpha]$ not representing truth or falsity there exists an atom $[\sigma_{\mid\sim}]$ such that $[\alpha]$ and $[\sigma_{\mid\sim}]$ are not compatible. In the special case of a pointer $\sigma_{\mid\sim}$ and a formula α compatibility says that $[\sigma_{\mid\sim}] \subset [\alpha]$ or $[\sigma_{\mid\sim}] \subset [\neg\alpha]$. Since $\mathcal{L}$ is non-trivial, for a given formula α there exists a $\mid\sim_o \in \mathcal{C}$ such that neither $[\sigma_{\mid\sim_0}] \subset [\alpha]$ nor $[\sigma_{\mid\sim_0}] \subset [\neg\alpha]$ and thus $[\alpha]$ and $[\sigma_{\mid\sim_0}]$ are not compatible. ■

DEFINITION 57. Let $\mathcal{L} = \langle \mathcal{C}, F, \rightsquigarrow \rangle$ be a logic.

- We say that $\mathcal{L}$ has the upward finiteness property, in brief the uf-property, iff the following holds: Given a set Σ of formulas. Then there exists a formula ψ such that $\sigma \mid\sim_{\mathcal{C}} \psi$ for every $\sigma \in \Sigma$ and the following condition is satisfied. For any formula ρ such that $\sigma \mid\sim_{\mathcal{C}} \rho$ for every $\sigma \in \Sigma$, we have $\psi \mid\sim_{\mathcal{C}} \rho$.
- We say that $\mathcal{L}$ has the downward finiteness property, in brief the df-property iff the following holds:

Given a set Σ of formulas. Then there exists a formula χ such that $\chi \mid\!\sim_C \sigma$ for every $\sigma \in \Sigma$ and the following condition is satisfied. For any formula ρ such that $\rho \mid\!\sim_C \sigma$ for every $\sigma \in \Sigma$, we have $\rho \mid\!\sim_C \chi$.

- In case that $\mathcal{L}$ is holistic we say that $\mathcal{L}$ has the covering property iff the following condition is satisfied. Given a formula α and $\mid\!\sim \in \mathcal{C}$ such that $\not\mid\!\sim \alpha$. Then for any formula ρ such that $\alpha \mid\!\sim_C \rho$ and $\rho \mid\!\sim_C \alpha \vee \sigma_{\mid\sim}$ we have $\rho \equiv_C \alpha \vee \sigma_{\mid\sim}$ or $\rho \equiv_C \alpha$

Intuitively we may think of the formulas ψ and χ in the above definition of playing the role of 'infinite disjunction' and 'infinite conjunction' of the formulas of Σ. The properties defined above are such that the following lemma holds.

LEMMA 58. *Let $\mathcal{L} = \langle \mathcal{C}, F, \rightsquigarrow_s \rangle$ be a holistic logic having the df, uf and the covering properties. Then the lattices $\langle \overline{F}ml, \leq, ^* \rangle$ and thus $\langle Prop, \subset, ^* \rangle$ are orthomodular, atomic, irreducible, complete lattices having the covering property.*

Recall the following observation already made in the proof of Theorem 48.

Let H be a Hilbert space and let $(x_i)_{i \in I}$ be a complete orthonormal system of H. Then any permutation of the system $(x_i)_{i \in I}$, more precisely any permutation of the index set I, induces a unique unitary transformation on H and thus an automorphism of the lattice $Sub(H)$. This fact reflects a symmetry property of Hilbert spaces and in view of Solér's theorem seems to be at the heart of the concept of a Hilbert space. It is the above fact that serves us as a motivation for the concept of a symmetric logic which we will study in the sequel.

DEFINITION 59. Let $\mathcal{L}$ be a holistic logic having the properties in the last lemma. Let $\Delta = (\mid\!\sim_i)_{i \in I}$ be an infinite family of consequence relations of $\mathcal{L}$ with the following properties

(i) For $i \neq j$, $\mid\!\sim_i$ and $\mid\!\sim_j$ are orthogonal.

(ii) For any consequence relation $\mid\!\sim$ of $\mathcal{L}$ there exists an $i_0 \in I$ such that $\mid\!\sim$ and $\mid\!\sim_{i_0}$ are not orthogonal.

Then we call Δ a basis for $\mathcal{L}$.

Remark: Intuitively, we may think of a basis Δ of a holistic logic $\mathcal{L}$ as follows. Given any consequence relation $\mid\!\sim$ of $\mathcal{L}$. Then there exists a member of Δ in which $\mid\!\sim$ is encoded. The system Δ may thus be viewed as containing the 'whole information' of $\mathcal{L}$.

DEFINITION 60. Let $\mathcal{L}$ be a logic as in the last definition and let $\Delta = (\mid\!\sim_i)_{i \in I}$ be a basis for $\mathcal{L}$. We say that $\mathcal{L}$ satisfies the symmetry condition with respect to Δ iff the following holds. Let $f : I \rightarrow I$ be any permutation of the index set I. Then there exists an automorphism φ_f of the algebra of propositions of $\mathcal{L}$ (and thus of th algebra of operators) such that

- $\varphi_f([\sigma_i]) = [\sigma_{f(i)}]$, where $(\sigma_i)_{i \in I}$ is any family such that σ_i is a pointer to $\mid\!\sim_i$.

- If the subset $J \subset I$ of those elements of I that are left fixed by f is non-empty, then φ_f induces the identity on $[0, A]$, where A is the smallest proposition containing $[\sigma_j]$ for all $j \in J$.

We say that $\mathcal{L}$ satisfies the (synonymously: is symmetric) iff there exists a basis Δ for $\mathcal{L}$ such that $\mathcal{L}$ is symmetric with respect to Δ.

Recall the notation $[0, A]$. It is the set of all propositions smaller than or equal to A equipped with a lattice structure in a natural way. The next theorem is our *Representation Theorem.* The proof we give can be simplified by making use of the Theorem 48 characterising classical Hilbert lattices. Essentially, we repeat the argument in the proof of 48 so as to make this section as self-contained as possible.

THEOREM 61. *Let* $\mathcal{L} = \langle \mathcal{C}, F, \leadsto_s \rangle$ *be a logic. Then the following conditions are equivalent.*

(i) $\mathcal{L}$ *is symmetric.*

(ii) There exists an infinite-dimensional classical Hilbert Space H *presenting* $\mathcal{L}$.

Proof. Most of the work has been done in the proof of Theorem 48. We therefore give just a sketch of the proof.

For the direction from (ii) to (i) assume that there exists an infinite-dimensional classical Hilbert space H and a (surjective) function Ψ such that $\mathcal{L} = \mathcal{L}_{H,\Psi}$. We need to verify that $\mathcal{L}$ is symmetric. Let $(x_i)_{i \in I}$ be a complete orthonormal system of H. Then $\Delta = (\vdash_{x_i})_{i \in I}$ is a basis for $\mathcal{L}$. Observe that the lattice of propositions of $\mathcal{L}$ and $Sub(H)$ are isomorphic in a canonical way, namely via $[\alpha] \mapsto \Psi(\alpha)$. Then, for the proof of the symmetry of $\mathcal{L}$, essentially, use the argument establishing the symmetry of $Sub(H)$ in the proof of Theorem 48.

For the other direction it is routinely verified that the conditions imposed on a symmetric logic guarantee that its lattice of propositions $Prop_{\mathcal{L}}$ has infinite height and also satisfies the other hypotheses of Piron's representation Theorem 46. Therefore there exists an orthomodular space H and an isomorphism $\Phi : Prop_{\mathcal{L}} \rightarrow Sub(H)$. Moreover, this lattice is symmetric in the sense of Definition 47. It then follows by Theorem 48 that H is an infinite-dimensional classical Hilbert space. We then need to see that H presents $\mathcal{L}$. For this define the function Ψ. Define $\Psi : Fml \rightarrow Sub(H)$ by $\Psi(\alpha) = \Phi([\alpha])$ and routinely verify the following:

1. $\mathcal{C} = \mathcal{C}_{H,\Psi}$

2. If $\mid\!\sim = \vdash_x$, then for any α, $\mid\!\sim_\alpha = \mathcal{F}_{H,\Psi}(\alpha, \vdash_x)$

∎

BIBLIOGRAPHY

[Alchourrón *et al.*, 1985] C. Alchourrón, P. Gärdenfors, D. Makinson. On the logic of theory change: Partial meet contractions and their associated revision functions. *Journal of Symbolic Logic* **50**, pp. 185-205, 1985

[Antoniou, 1997] G. Antoniou. *Nonmonotonic Reasoning*, MIT Press, 1997

[Birkhoff and von Neumann, 1936] G. Birkhoff, J. von Neumann. The logic of quantum mechanics. *Annals of Mathematics* **37**, pp. 823-843, 1936

[Boolos, 1993] G. Boolos. *The Logic of Provability*, Cambridge University Press, 1993.

[Bohm, 1980] D. Bohm. *Wholeness and the Implicate Order*, Routledge, 1980

[Bub, 1999] J. Bub. *Interpreting the quantum world*, Cambridge University Press, 1999

[Dalla Chiara, 1986] M.L. Dalla Chiara. Quantum Logic. In Gabbay and Guenthner (eds.) *Handbook of Philosophical Logic* Vol. *III*, pp. 427-469, 1986. Revised version in *Handbook of Philosophical Logic*, Second edition, Volume **6**, pp. 129-228, Kluwer, 2001.

[Dalla Chiara, 1977] M.L. Dalla Chiara. Quantum logic and physical modalities. *J. Philosophical Logic*, **6**, 391-404, 1977

[Dalla Chiara, 2004] M.L. Dalla Chiara, R. Giuntini, R. Greechie. *Reasoning in Quantum Theory*, Kluwer, 2004

[Einstein *et al.*, 1935] A. Einstein, B. Podolsky, N. Rosen. Can the quantum mechanical description of physical reality be considered complete? *Physical Review* **47**, pp. 777-780, 1935

[Engesser, 2000] K. Engesser. Characterisation of classical Hilbert lattices. In P. Hitzler and G. Kalmbach (eds.) *Begabtenförderung im MINT- Bereich*, Band **5**, pp. 1-8, Aegis-Verlag, 2000

[Engesser *et al.*, 2008] K. Engesser, D.M. Gabbay, D. Lehmann. *A New Approach to Quantum Logic*, College Publications, 2008

[Engesser and Gabbay, 2002] K. Engesser, D.M. Gabbay, Quantum logic, Hilbert space, revision theory. *Artificial Intelligence* **136**, pp. 61-100, 2002

[Gabbay, 1996] D.M. Gabbay. *Labelled Deductive Systems.* Clarendon Press, Oxford, 1996

[Gabbay, 1999] D. M. Gabbay. *Fibring Logics.* Oxford University Press, 1999

[Gabbay, 1985] D.M. Gabbay. Theoretical Foundations for nonmonotonic reasoning in expert systems. In K.R. Apt (ed.) *Proceedings NATO Advanced Study Institute on Logics and Models of Concurrent Systems*, pp. 439-457, Springer-Verlag, Berlin, 1985.

[Goldblatt, 1974] R.H. Goldblatt. Semantic analysis of orthologic. *J. Philosophical Logic*, **3**, pp. 19-35, 1974

[Halmos, 1957] P. Halmos. *Introduction to Hilbert Space and the Theory of Spectral Multiplicity*, 2nd edition, Chelsea, New York, 1957

[Hardegree, 1974] G.M. Hardegree. The Conditional in Quantum Logic. *Synthese*, **29**, pp. 63-80, 1974

[Holland, 1995] S.S. Holland. Orthomodularity in Infinite Dimensions, A Theorem of M. Solèr. *Bulletin of the American Mathematical Society*, **32**, pp. 205-234, 1995

[Kalmbach, 1983] G. Kalmbach. *Orthomodular Lattices. London Math. Soc. Monographs*, Vol. **18** Academic Press, London and New York, 1983

[Katsuno and Sato, 1995] H. Katsuno and K. Sato. A unified view of consequence relation, belief revision and conditional logic. In G. Crocco, L. Farinas del Cerro, and A. Herzig, editors, *Conditionals : From Philosophy to Computer Science*, pp. 33-66, Oxford University Press, 1995.

[Kochen and Specker, 1967] S. Kochen, E. Specker. The problem of hidden variables in quantum mechanics. *Journal of Mathematics and mechanics* **17**, pp. 59-87, 1967

[Keller, 1980] H. A. Keller. Ein nichtklassischer Hilbertscher Raum. *Mathematische Zeitschrift*, **172**, 41-49, 1980.

[Kraus *et al.*, 1990] S. Kraus, D. Lehmann, and M. Magidor. nonmonotonic reasoning, preferential models and cumulative logics. *Artificial Intelligence*, **44**, pp. 167-207, 1990

[Lehmann, 2001] D. Lehmann. Nonmonotonic Logic and Semantics, *Journal of Logic and Computation*, **11**, pp. 229-256, 2001

[Lehmann *et al.*, 2006] D. Lehmann, K. Engesser, D.M. Gabbay. Algebras of Measurements: the Logical Structure of Quantum Mechanics. *International Journal of Theoretical Physics* **45**, 698-723, 2006

[Lehmann and Magidor, 1992] D. Lehmann, M. Magidor. What does a conditional knowledge base entail? *Artificial Intelligence*, **55**, pp. 1-60, 1992

[Makinson and Gärdenfors, 1991] D. Makinson and P. Gärdenfors. Relation between the logic of theory change and nonmonotonic logic. In *The Logic Theory of Change*, A. Fuhrmann and H. Morreau, eds. pp. 185–205. Lecture Notes in AI, **465**, Springer Verlag, 1991.

[Mayet, 1998] R. Mayet. Some Characterizations of the Underlying Division Ring of a Hilbert Lattice by Automorphisms. *International Journal of Theoretical Physics*, **37**, pp. 109-114, 1998

[Mittelstaedt, 1978] P. Mittelstaedt. *Quantum Logic.* D. Reidel Publishing Company, 1978

[Moore, 1985] R.C. Moore. Semantical Considerations on Nonmonotonic Logic. *Artificial Intelligence* **25**, pp. 75-94, 1985

[Piron, 1976] C. Piron. *Foundations of quantum mechanics.* W.A. Benjamin, 1976

[Popper, 1968] K. Popper. Birkhoff and von Neumann's Interpretation of Quantum Mechanics. *Nature*, **219**, pp. 682-705, 1968

[Prestel, 2008] A. Prestel. On Solèr's Theorem. This volume.

[Putnam, 1969] H. Putnam. Is logic empirical? in R.S. Cohen and M.W. Wartofsky (eds), *Boston Studies in the Philosophy of Science*, Vol **5**, Reidel-Dordrecht, pp. 216-241, 1969,

[Savile, 2000] A. Savile. *Leibniz and the Monadology.* Routledge, 2000

[Smullyan, 1987] R.M. Smullyan. *Forever Undecided.* Oxford University Press, 1987

[Smullyan, 1992] R.M. Smullyan. *Gödel's Incompleteness Theorems.* Oxford University Press, 1992

[Solèr, 1995] M.P. Solèr. Characterization of Hilbert spaces with orthomodular spaces. *Communications in Algebra* **23**, pp. 219-234, 1995

[von Neumann, 1932] J. von Neumann. *Mathematische Grundlagen der Quantenmechanik*, Berlin, Springer, 1932

Edited by K. Engesser, D. M. Gabbay and D. Lehmann

A QUANTUM LOGIC OF DOWN BELOW

P. D. Bruza, D. Widdows and J. H. Woods

1 INTRODUCTION

The logic that was purpose-built to accommodate the hoped-for reduction of arithmetic gave to language a dominant and pivotal place. Flowing from the founding efforts of Frege, Peirce, and Whitehead and Russell, this was a logic that incorporated proof theory into syntax, and in so doing made of grammar a senior partner in the logicistic enterprise. The seniority was reinforced by soundness and completeness metatheorems, and, in time, Quine would quip that the "grammar [of logic] is linguistics on purpose" [Quine, 1970, p. 15] and that "logic chases truth up the tree of grammar" [Quine, 1970, p. 35]. Nor was the centrality of syntax lost with the Gödel incompleteness results, which, except for the arithmeticization of syntax, would have been impossible to achieve.

Logic's preoccupation with language is no recent thing. In Aristotle's logic of the syllogism, the target properties of necessitation and syllogistic entailment are properties of sentences or sets of sentences of Greek. Only with the likes of Peirce and Frege is the rejection of natural language explicit, each calling for a logic whose properties would attach to elements of artificial languages, and — after Tarski — to such elements in semantic relation to non-linguistic set theoretic structures.

It is hardly surprising that mathematical logic should have given such emphasis to language, given that the motivating project of logic was to facilitate the reduction of arithmetic to an obviously analytic discipline. Still, it is also worthy of note that the historic role of logic was to lay bare the logical structure of human reasoning. Aristotle is clear on this point. The logic of syllogisms would serve as the theoretical core of a wholly general theory of real-life, two-party argumentation. Even here, the centrality of language could not be ignored. For one thing, it was obvious that real-life argumentation is transacted in speech. For a second, it was widely held (and still is) that reasoning is just soliloquial argumentation (just as argumentation is held to be reasoning in public — out loud, so to speak). Given these purported equivalences, reasoning too was thought of as linguistic.

It is convenient to date the birth of modern mathematical logic from the appearance in 1879 of Frege's great book on the language of logic, *Begriffsschrift*. It is easy to think of logic as having a relatively unfettered and richly progressive course ever since, one in which even brutal setbacks could be celebrated as triumphs of metalogic. There is, however, much of intervening importance from 1904 onwards, what with developments in intuitionist, modal, many-valued and relevant logics, which in retrospect may seem to presage crucial developments in the second half of that century. Suffice it here to mention Hintikka's seminal work on epistemic logic [Hintikka, 1962], which is notable in two important respects. One is the introduction of agents as load-bearing objects of the logic. The other is the influence that agents are allowed to have on what the theory is prepared to count as its logical truths. The logical truths of this system include its indefensible sentences, where these in turn include sentences which it would be self-defeating for an agent to utter (e.g., "I can't speak a single word of English"). It is easy to see that Hintikka here allows for a sentence to be a truth of logic if its negation is pragmatically inconsistent. To this extent, the presence of agents in his logic occasions the pragmaticization of its semantics.

Agents now enter logic with a certain brisk frequency. They are either expressly there or are looming forces in theories of belief dynamics and situation semantics, in theories of default and non-monotonic reasoning, and in the incipient stirrings of logics of practical reasoning. Notable as these developments are, they all lie comfortably within the embrace of the linguistic presumption. Agents may come or go in logic, but whether here or there, they are, in all that makes them of interest to logicians, manipulators of language. What is more, notwithstanding the presence of agents, these were logics that took an interest in human reasoning rather than human *reasoners*. This made for a fateful asymmetry in which what human reasoners are like (or should be) is read off from what human reasoning is like (or should be).

It may be said, of course, that this is exactly the wrong way around, that what reasoning is (or should be) can only be read off from what reasoners are (and can be). Such a view one finds, for example in [Gabbay and Woods, 2001] and [Gabbay and Woods, 2003b], among logicians, and, also in the social scientific literature [Simon, 1957; Stanovich, 1999; Gigerenzer and Selten, 2001b]. Here the leading idea of the "new logic" is twofold. First, that logic's original mission as a theory of human reasoning should be re-affirmed. Second, that a theory of human reasoning must take empirical account of what human reasoners are like – what they are interested in and what they are capable of.

It is easy to see that the human agent is a cognitive being, that human beings have a drive to know. They desire to know what to believe and what to do. And since, whatever else it is, reasoning is an aid to cognition, a theory of human reasoning must take into account how beings like us operate as cognitive systems. Here, too, the empirical record is indispensable. It is the first point of contact between logic and cognition. In this way symbolic inference becomes "married" to computations through state (dimensional) spaces motivated from cognition which may open the door the large-scale operational symbolic inference systems. The logicians Barwise and Seligman have advocated such a marriage between logic and cognition [Barwise and Seligman, 1997, p.234]. This bears in an important way on what we have been calling the linguistic presumption. For if the empirical record is anything to go on, much of the human cognitive project is sublinguistic, and inaccessible to introspection. This, the cognition of "down below", carries consequences for the new logic. If logic is to attend to the cognizing agent, it must take the cognizer as he comes, warts and all. Accordingly, a theory of human reasoning must subsume a logic of down below.

These days, the logic of down below appears to have a certain memetic status. It is an idea whose time has come. In addition to the work of Gabbay and Woods, the idea is independently in play in a number of recent writings. In [Churchland, 1989; Churchland, 1995] we find a connectionist approach to subconscious abductive processes (cf. [Burton, 1999]). In a series of papers, Horgan and Tiensen develop a rules without representation (RWR) framework for cognitive modeling [Horgan and Tienson, 1988; Horgan and Tienson, 1989; Horgan and Tienson, 1990; Horgan and Tienson, 1992; Horgan and Tienson, 1996; Horgan and Tienson, 1999b; Horgan and Tienson, 1999a] (Cf. [Guarini, 2001]). Other non-representational orientations include [Wheeler, 2001; Sterelny, 1990; Brooks, 1991; Globus, 1992; Shannon, 1993; Thelen and Smith, 1993; Wheeler, 1994; Webb, 1994; Beer, 1995] (Cf. [Wimsatt, 1986] and [Clark, 1997]). A neural symbolic learning systemic framework is developed in [d'Avila Garcez *et al.*, 2002; d'Avila Garcez and Lamb, 2004] and extended to abductive environments in [Gabbay and Woods, 2005, ch. 6]. Bruza and his colleagues advance a semantic space framework [Bruza *et al.*, 2004; Bruza and Cole, 2005b; Bruza *et al.*, 2006].

The present chapter is offered as a contribution to the logic of down below. In the section to follow, we attempt to demonstrate that the nature of human agency necessitates that there actually be such a logic. The ensuing sections develop the suggestion that cognition down below has a structure strikingly similar to the physical structure of quantum states. In its general form, this is not an idea that originates with the present authors. It is known that there exist mathematical models from the cognitive science of cognition down below that have certain formal similarities to quantum mechanics. We want to take this idea seriously. We will propose that the subspaces of von Neumann-Birkhoff lattices are too crisp for modelling requisite cognitive aspects in relation to subsymbolic reasoning. Instead, we adopt an approach which relies on projections into nonorthogonal density states. The projection operator is motivated from cues which probe human memory.

2 AGENCY

In this section our task is to orient the logic of down below by giving an overview of salient features of individual cognitive agency. Investigations of non-monotonic reasoning (NMR) have successfully provided an impressive symbolic account of human practical reasoning over the last two and half decades. The symbolic characterization of practical reasoning, however, is only part of the picture. Gärdenfors [Gärdenfors, 2000, p. 127] argues that one must go under the symbolic level of cognition. In this vein, he states, "...information about an object may be of two kinds: *propositional* and *conceptual*. When the new information is propositional, one learns new *facts* about the object, for example, that x is a penguin. When the new information is conceptual, one *categorizes* the object in a new way, for example, x is *seen as* a penguin instead of as just a bird". Gärdenfors' mention of "conceptual" refers to the conceptual level of a three level model of cognition [Gärdenfors, 2000]. How information is represented varies greatly across the different levels. The sub-conceptual level is the lowest level within which information is carried by a connectionist representation. Within the uppermost level information is represented symbolically. It is the intermediate, *conceptual level*, or *conceptual space*, which is of particular relevance to this account. Here properties and concepts have a geometric representation in a dimensional space. For example, the property of "redness" is represented as a convex region in a tri-dimensional space determined by the dimensions hue, chromaticity and brightness. The point left dangling for the moment is that representation at the conceptual level is rich in associations, both explicit and implicit. We speculate that the dynamics of associations are primordial stimuli for practical inferences drawn at the symbolic level of cognition. For example, it seems that associations and analogies generated within conceptual space play an important role in hypothesis generation. Gärdenfors ([Gärdenfors, 2000], p48) alludes to this point when he states, "most of scientific theorizing takes place within the conceptual level."

Gärdenfors' conjecture receives strong endorsement from an account of practical reasoning developed in [Gabbay and Woods, 2003a; Gabbay and Woods, 2005], in which reasoning on the ground is understood to function under economic constraints. In this essay, our own point of departure is that subsymbolic reasoning is valuable to human agents precisely for the economies it achieves. It will help to place this assumption in its proper context by giving a brief overview of our approach to cognitive agency.

A Hierarchy of Agency Types

It is useful to repeat the point that since reasoning is an aid to cognition, a logic, when conceived of as a theory of reasoning, must take this cognitive orientation deeply into account. Accordingly, we will say that a *cognitive system* is a triple of a cognitive agent, cognitive resources, and cognitive target performed in real time. (See here [Norman, 1993; Hutchins, 1995].) Correspondingly, a logic of a cognitive system is a principled description of conditions under which agents deploy resources in order to perform cognitive tasks. Such is a practical logic when the agent it describes is a *practical agent.*

A practical logic is but an instance of a more general conception of logic. The more general notion is reasoning that is target-motivated and resource-dependent. Correspondingly, a logic that deals with such reasoning is a Resource-Target Logic (*RT*-logic). In our use of the term, a practical logic is a *RT*-logic relativized to practical agents.

How agents perform is constrained in three crucial ways: in what they are disposed towards doing or have it in mind to do (i.e., their *agendas*); in what they are capable of doing (i.e., their *competence*); and in the means they have for converting competence into performance (i.e., their *resources*). Loosely speaking, agendas here are programmes of action, exemplified by belief-revision and belief-update, decision-making and various kinds of case-making and criticism transacted by argument. [1]

Agency-type is set by two complementary factors. One is the degree of command of resources an agent needs to advance or close his (or its) agendas. For cognitive agendas, three types of resources are especially important. They are (1) *information*, (2) *time*, and (3) *computational capacity*. The other factor is the height of the cognitive bar that the agent has set for himself. Seen this way, agency-types form a hierarchy H partially ordered by the relation C of commanding-greater-resources-in-support-of-higher-goals-than. H is a poset (a partially ordered set) fixed by the ordered pair $\langle C, X \rangle$ of the relation C on the unordered set of agents X.

Human agency divides roughly into the individual and the institutional. By comparison, *individual* agency ranks low in H. For large classes of cases, individuals perform their cognitive tasks on the basis of less information and less time than they might otherwise like to have, and under limitations on the processing and manipulating of complexity. Even so, paucity must not be confused with scarcity. There are lots of cases in which an individual's resources are adequate for the attainment of the attendant goal. In a rough and ready way, we can say that the comparative modesty of an agent's cognitive goals inoculates him against cognitive-resource scarcity. But there are exceptions, of course.

Institutional entities contrast with human agents in all these respects. A research group usually has more information to work with than any individual, and more time at its disposal; and if the team has access to the appropriate computer networks, more fire-power than most individuals even with good PCs. The same is true, only more so, for agents placed higher in the hierarchy — for corporate actors such as NASA, and collective endeavours such as particle physics since 1970. Similarly, the cognitive agendas that are typical of institutional agents are by and large stricter than the run-of-the-mill goals that motivate individual agents. In most things, NASA aims at stable levels of scientific confirmation, but, for individuals the defeasibly plausible often suffices for local circumstances.

These are vital differences. Agencies of higher rank can afford to give maximization more of a shot. They can wait long enough to make a try for total information, and they can run the calculations that close their agendas both powerfully and precisely. Individual agents stand conspicuously apart. He must do his business with the information at hand, and, much of the time, sooner rather than later. Making do in a timely way with what he knows now is not just the only chance of achieving whatever degree of cognitive success is open to him as regards the agenda at hand; it may also be what is needed in order to avert unwelcome disutilities, or even death. Given the comparative humbleness of his place in H, the human individual is frequently faced with the need to practise cognitive economies. This is certainly so when either the loftiness of his goal or the supply of drawable resources create a cognitive strain. In such cases, he must turn *scantness*

[1] Agendas are discussed at greater length in [Gabbay and Woods, 2002].

to *advantage*. That is, he must (1) deal with his resource-limits and in so doing (2) must do his best not to kill himself. There is a tension in this dyad. The paucities with which the individual is chronically faced are often the natural enemy of getting things right, of producing accurate and justified answers to the questions posed by his agenda. And yet, not only do human beings contrive to get most of what they do right enough not to be killed by it, they also in varying degrees prosper and flourish. This being so, we postulate for the individual agent *slight-resource adjustment strategies (SRAS),* which he uses to advantage in dealing with the cognitive limitations that inhere in the paucities presently in view. We make this assumption in the spirit of Simon [1957] and an ensuing literature in psychology and economics. At the heart of this approach is the well-evidenced fact that, for ranges of cases, "fast and frugal" is almost as good as full optimization, and at much lower cost [Gigerenzer and Selten, 2001a]. We shall not take time here to detail the various conditions under which individuals extract outcome economies from resource limitations and target modesty, but the examples to follow will give some idea of how these strategies work.

Although resource-paucity should not be equated with resource-scarcity, it remains the case that in some sense practical agents operate at a cognitive disadvantage. It is advisable not to make too much of this. What should be emphasized is that in relation to the cognitive standards that an institutional agent might be expected to meet, the resources available to a practical agent will typically not enable him (or it) to achieve that standard. Whether this constitutes an unqualified disadvantage depends on the nature of the task the individual has set for himself and the cognitive resources available to him. For a practical agent to suffer an unqualified disadvantage, two factors must intersect in the appropriate way: his resources must be inadequate for the standard he should hit, in relation to a goal that has reasonably been set for him. So, the measure of an agent's cognitive achievement is a function of three factors: his cognitive *goal*; the *standard* required (or sufficient) for achieving that goal; and the cognitive *wherewithal* on which he can draw to meet that standard.

In discharging his cognitive agendas, the practical agent tends to set goals that he can attain and to be stocked with the wherewithal that makes attainment possible (and frequent). In the matter of both goals set and the execution of standards for meeting them, the individual is a satisficer rather than an optimizer. There are exceptions, of course; a working mathematician won't have a solution of Fermat's Last Theorem unless he has a full-coverage proof that is sound (and, as it happens, extremely long).

The tendency to satisfice rather than maximize (or optimize) is not what is *distinctive* of practical agency. This is a point to emphasize. In most of what they set out to do and end up achieving, institutional agents exhibit this same favoritism. What matters — and sets them apart from the likes of us — is not that they routinely optimize but that they satisfice against loftier goals and tougher standards.

Slight-resource Adjustment Strategies

Slight-resource adjustment strategies lie at the crux of the economy of effort, as Rescher calls it [Rescher, 1996, p.10]. They instantiate a principle of least effort, and they bear on our tendency to minimize the expenditure of cognitive assets.[2]We note here some examples.

Statistical studies such as opinion polls always give results to within a given level of confidence (e.g., "these predictions are valid to within $\pm 3\%$ with 95% confidence"), and part of the science of statistics lies in making reliable statements of this nature given the size of sample taken. Medical tests are often only correct to a known precision, and given the fequency of false-positives, the result of a positive test-result is often a further round of more reliable but more invasive tests.

It may be tempting to presume that such knowledge-constrained strategies are mainly confined to empirical or practical sciences, but this is far from the case. For example, mathematics is full of rules-of-thumb and famous theorems that reduce difficult problems to easy ones. These begin for many early students with the familiar division rules, such as "if a number ends in a 2 or a 5, it is divisible by 2 or 5", or the more complex "if the alternating sum of the digits of a number is divisible by 11, the number itself is divisible by 11". Such results do not produce the quotient of the division, but they may tell the student whether such a computation is worth the trouble if the goal is to end up with a whole number. More advanced division properties are embodied in results such as *Fermat's Little Theorem*, which states that if p is prime and $1 \leq a \leq p$, then $a^{p-1} \cong 1 \pmod{p}$. Like many important theorems, this only gives necessary but not sufficient conditions for a statement (in this case, the statement "p is prime") to be true. However, if this necessary condition holds for enough values of a, we may conclude that p is *probably* prime, which is in fact a strong enough guarantee for some efficient encryption algorithms. Even in mathematics, often regarded as the most exact and uncompromising discipline, short-cuts that are close enough are not only important, they are actively sought after.

2.1 Hasty Generalization

Individual cognitive agents are hasty generalizers, otherwise known as *thin-slicers*. Hasty generalization is a *SRAS*. In standard approaches to fallacy theory and theories of statistical inference, hasty generalization is a blooper; it is a serious sampling error. This is the correct assessment if the agent's objective is to find a sample that is guaranteed to raise the conditional probability of the generalization, and to do so in ways that comport with the theorems of the applied mathematics of chance. Such is an admirable goal for agents who have the time and know-how to construct, or find samples that underwrite such guarantees. But as J.S. Mill shrewdly observed, human individuals often lack the wherewithal for constructing these inferences. The business of sample-to-generalization induction often exceeds

[2]See here the classic work of George Zipf. [Zipf, 1949].

the resources of individuals and is better left to institutions. (See [Woods, 2004].) A related issue, even supposing that the requisitely high inductive standards are meetable in a given situation in which a practical agent finds himself, is whether it is necessary or desirable for him (or it) to meet that standard. Again, it depends on what the associated cognitive goal is. If, for example, an individual's goal is to have a reasonable belief about the leggedness of ocelots is, rather than to achieve the highest available degree of scientific certainty about it, it would suffice for him to visit the ocelot at the local zoo, and generalize hastily "Well, I see that ocelots are four-legged".

2.2 Generic Inference

Often part of what is involved in a human reasoner's facility with the one-off generalization is his tendency to eschew generalizations in the form of universally quantified conditional propositions. When he generalizes hastily the individual agent is often making a *generic* inference. In contrast to universally quantified conditional propositions, a generic claim is a claim about what is characteristically the case. "For all x, if x is a ocelot, then x is four-legged" is one thing; "Ocelots are four-legged" is quite another thing [Krifka *et al.*, 1995]. The first is felled by any true negative instance, and thus is *brittle.* The second can withstand multiples of true negative instances, and thus is *elastic.* There are significant economies in this. A true generic claim can have up to lots of true negative instances. So it is true that ocelots are four-legged, even though there are up to lots of ocelots that aren't four-legged. The economy of the set-up is evident: With generic claims, it is unnecessary to pay for every exception. One can be wrong in particular without being wrong in general.

Generic claims are a more affordable form of generalization than the universally quantified conditional. This is part of what explains their dominance in the generalizations that individual agents tend actually to make (and to get right, or some near thing). It must not be thought, however, that what constitutes the rightness (or some near thing) of an individual's hasty generalizations is that when he generalizes thus he generalizes to a generic claim. Although part of the story, the greater part of the rightness of those hasty generalizations arises from the fact that, in making them, an individual typically has neither set himself, nor met, the standard of inductive strength. This, together with our earlier remarks about validity, is telling. Given the cognitive goals typically set by practical agents, validity and inductive strength are typically not appropriate (or possible) standards for their attainment. This, rather than computational costs, is the deep reason that practical agents do not in the main execute systems of deductive or inductive logic as classically conceived.

2.3 *Natural Kinds*

Our adeptness with generic inference and hasty generalization is connected to our ability to recognize *natural kinds* [Krifka *et al.*, 1995, pp.63–95]. Natural kinds have been the object of much metaphysical skepticism of late [Quine, 1969], but it is a distinction that appeals to various empirical theorists. The basic idea is evident in concepts such as *frame* [Minsky, 1975], *prototype* [Rosch, 1978], *script* [Schank and Abelson, 1977] and *exemplar* [Smith and Medin, 1981]. It is possible, of course, that such are not a matter of metaphysical unity but rather of perceptual and conceptual organization.

It goes without saying that even when the goal is comparatively modest — say, what might plausibly be believed about something at hand — not every hasty generalization that could be made comes anywhere close to hitting even that target. The (defeasible) rule of thumb is this: The hasty generalizations that succeed with these more modest goals are by and large those we actually draw in actual cognitive practice. We conjecture that the comparative success of such generalizations is that they generalize to generic propositions, in which the process is facilitated by the agent's adeptness in recognizing natural kinds. In section 5, we discuss the extent to which a quantum logical framework provides a more useful model for adapting to natural kinds than either Boolean set theory or taxonomy.

2.4 *Consciousness*

A further important respect in which individual agency stands apart from institutional agency is that human agents are conscious. (The consciousness of institutions, such as it may be figuratively speaking, supervenes on the consciousness of the individual agents who constitute them.) Consciousness is both a resource and a limitation. Consciousness has a narrow bandwidth. This makes most of the information that is active in a human system at a time consciously unprocessible at that time. In what the mediaevals called the *sensorium* (the collective of the five senses operating concurrently), there exist something in excess of 10 million bits of information per second; but fewer than 40 bits filter into consciousness at those times. Linguistic agency involves even greater informational entropy. Conversation has a bandwidth of about 16 bits per second.[3]

The narrow bandwidth of consciousness bears on the need for cognitive economy. It helps elucidate what the scarcity of information consists in. We see it explained that at any given time the human agent has only slight information by the fact that if it is consciously held information there is a bandwidth constraint which

[3][Zimmermann, 1989]. Here is John Gray on the same point: "If we do not act in the way we think we do, the reason is partly to do with the bandwidth of consciousness — its ability to transmit information measured in terms of bits per second. This is much too narrow to be able to register the information we routinely receive and act on. As organisms active in the world, we process perhaps 14 million bits of information per second. The bandwidth of consciousness is around eighteen bits. This means that we have conscious access to about a millionth of the information we daily use to survive" [Gray, 2002, p. 66].

regulates its quantity. There are also devices that regulate consciously processible information as to *type*. A case in point is informational relevance. When H.P. Grice issued the injunction, "Be relevant", he left it undiscussed whether such an imperative could in fact be honoured or ignored by a conscious act of will. There is evidence that the answer to this question is "No"; that, in lot's of cases, the mechanisms that steer us relevantly in the transaction of our cognitive tasks, especially those that enable us to discount or evade irrelevance, are automatic and pre-linguistic [Gabbay and Woods, 2003a]. If there is marginal capacity in us to heed Grice's maxim by consciously sorting out relevant from irrelevant information, it is likely that these informational relevancies are less conducive to the closing of cognitive agendas than the relevancies that operate "down below". Thus vitally relevant information often can't be processed consciously, and much of what can is not especially vital.[4]

Consciousness can claim the distinction of being one of the toughest problems, and correspondingly, one of the most contentious issues in the cognitive sciences. Since the agency-approach to logic subsumes psychological factors, it is an issue to which the present authors fall heir, like it or not. Many researchers accept the idea that information carries negative entropy, that it tends to impose order on chaos.[5] If true, this makes consciousness a thermodynamically expensive state to be in, since consciousness is a radical suppressor of information. Against this are critics who abjure so latitudinarian a conception of information [Hamlyn, 1990] and who remind us that talk about entropy is most assured scientifically for closed systems (and that ordinary individual agents are hardly *that*).

The grudge against promiscuous "informationalism", in which even physics goes digital [Wolfram, 1984], is that it fails to explain the distinction between energy-to-energy transductions and energy-to-information transformations [Tallis, 1999, p. 94]. Also targeted for criticism is the view that consciousness arises from or inheres in neural processes. If so, "[h]ow does the energy impinging on the nervous system become transformed into consciousness?" [Tallis, 1999, p. 94].

In the interests of economy, we decline to join the metaphysical fray over consciousness. The remarks we have made about consciousness are intended not as advancing the metaphysical project but rather as helping characterize the economic limitations under which individual cognitive agents are required to perform.

Consciousness is tied to a family of cognitively significant issues. This is reflected in the less than perfect concurrence among the following pairs of contrasts:

1. conscious v unconscious processing

[4]Consider here taxonomies of vision in which implicit perception has a well-established place [Rensink, 2000].

[5]Thus Colin Cherry: "In a descriptive sense, entropy is often referred to as a 'measure of disorder' and the Second Law of Thermodynamics as stating that 'systems can only proceed to a state of increased disorder; as time passes, entropy can never decrease.' The properties of a gas can change only in such a way that our knowledge of the positions and energies of the particles lessens; randomness always increases. In a similar descriptive way, information is contrasted, as bringing order out of chaos. Information, then is said to be 'like' negative energy" [Cherry, 1966, p. 215].

2. controlled v automatic processing
3. attentive v inattentive processing
4. voluntary v involuntary processing
5. linguistic v nonlinguistic processing
6. semantic v nonsemantic processing
7. surface v depth processing

What is striking about this septet of contrasts is not that they admit of large intersections on each side, but rather that their concurrence is approximate at best. For one thing, "tasks are never wholly automatic or attentive, and are always accomplished by mixtures of automatic and attentive processes" [Shiffrin, 1997, p. 50]. For another, "depth of processing does not provide a promising vehicle for distinguishing consciousness from unconsciousness (just as depth of processing should not be used as a criterial attribute for distinguishing automatic processes ..." [Shiffrin, 1997, p. 58]). Indeed "[s]ometimes parallel processing produces an advantage for automatic processing, but not always .... Thoughts high in consciousness often seem serial, probably because they are associated with language, but at other times consciousness seems parallel ..." [Shiffrin, 1997, p. 62].

It is characteristic of agents of all types to adjust their cognitive targets upwards as the cognitive resources for attaining them are acquired. A practical agent may take on commitments previously reserved for agents of higher rank if, for example, he is given the time afforded by a tenured position in a university, the information stored in the university's library and in his own PC, and the fire-power of his university's mainframe. In like fashion, institutional agents constantly seek to expand their cognitive resources (while driving down the costs of their acquisition, storage and deployment), so that even more demanding targets might realistically be set. Accordingly, agents tend toward the enhancement of cognitive assets when this makes possible the realization of cognitive goals previously unattainable (or unaffordable). Asset enhancement is always tied to rising levels of cognitive ambition. In relation to cognitive tasks adequately performed with present resources, an interest in asset enhancement is obsessive beyond the range of what would count as natural and proportionate improvements upon what is already adequately dealt with.

2.5 Subsymbolic reasoning

Practical reasoning is reasoning performed by practical agents, and is therefore subject to economic constraints. In this connection, we advance the following conjecture: It may well be that because such associations are formed below the symbolic level of cognition, significant cognitive economy results. This is not only interesting from a cognitive point of view, but also opens the door to providing a computationally tractable practical reasoning systems, for example, operational abduction to drive scientific discovery in biomedical literature [Bruza *et al.*, 2004; Bruza *et al.*, 2006]

The appeal of Gärdenfors' cognitive model is that it allows inference to be considered not only at the symbolic level, but also at the conceptual (geometric) level. Inference at the symbolic level is typically a linear, deductive process. Within a conceptual space, inference takes on a decidedly associational character because associations are often based on similarity (e.g., semantic or analogical similarity), and notions of similarity are naturally expressed within a dimensional space. For example, Gärdenfors states that a more natural interpretation of "defaults" is to view them as "relations between concepts".[6] This is a view which flows into the account which follows: the strength of associations between concepts change dynamically under the influence of context. This, in turn, influences the defaults haboured within the symbolic level of cognition.

It is important to note the paucity of representation at the symbolic level and reflect how symbolic reasoning systems are hamstrung as a result. In this connection, Gärdenfors ([Gärdenfors, 2000, p. 127]) states, " ..information about categorization can be quite naturally transfered to propositional information: categorizing x as an emu, for example, can be expressed by the proposition "x is an emu". This transformation into the propositional form, however, tends to suppress the internal *structure* of concepts. Once one formalizes categorizations of objects by *predicates* in a first-order language, there is a strong tendency to view the predicates as primitive, atomic notions and to forget that there are rich relations among concepts that disappear when put into standard logical formalism."

The above contrast between the conceptual and symbolic levels raises the question as to what are the implications for providing an account of practical reasoning. Gärdenfors states that concepts generate "expectations that result in different forms of *non-monotonic reasoning*", which are summarized as follows:

Change from a general category to a subordinate

When shifting from a basic category, e.g., "bird" to a subordinate category, e.g., "penguin", certain default associations are given up (e.g., "Tweety flies"), and new default properties may arise (e.g., "Tweety lives in Antarctica").

Context effects

The context of a concept triggers different associations that "lead to non-monotonic inferences". For example, *Reagan* has default associations "Reagan is a president", "Reagan is a republican" etc., but *Reagan* seen in the context of *Iran* triggers associations of "Reagan" with "arms scandal", etc.

[6] In the theory of Gabbay and Woods, default reasoning is a core slight-resource compensation strategy.

The effect of contrast classes

Properties can be relative, for example, "a tall Chihuahua is not a tall dog" ([Gärdenfors, 2000, p. 119]),. In the first contrast class "tall" is applied to Chihuahuas and the second instance it is applied to dogs in general. Contrast classes generate conceptual subspaces, for example, skin colours form a subspace of the space generated by colours in general. Embedding into a subspace produces non-monotonic effects. For example, from the fact that x is a white wine and also an object, one cannot conclude that x is a white object (as it is yellow).

Concept combination

Combining concepts results in non-monotonic effects. For example, *metaphors* ([Gärdenfors, 2000, p. 130]), Knowing that something is a lion usually leads to inferences of the form that it is alive, that it has fur, and so forth. In the combination, *stone lion*, however, the only aspect of the object that is lion-like is its shape. One cannot conclude that a stone lion has the other usual properties of a lion, and thus we see the non-monotonicity of the combined concept.

An example of the non-monotonic effects of concept combination not involving metaphor is the following: *A guppy is not a typical pet, nor is guppy is a typical fish, but a guppy is a typical pet fish.*

In short, concept combination leads to conceptual change. These correspond to revisions of the concept and parallel belief revisions modelled at the symbolic level, the latter having received thorough examination in the artificial intelligence literature.

The preceding characterization of the dynamics of concepts and associated non-monotonic effects is intended to leave the impression that a lot of what happens in connection with practical reasoning takes place within a conceptual (geometric) space, or a space of down-below. What is more, this impression may provide a foothold towards realizing genuine operational systems. This would require that at least three issues be addressed. The first is that a computational variant of the conceptual level of cognition is necessary. Secondly, the non-monotonic effects surrounding concepts would need to be formalized and implemented. Thirdly, the connection between these effects and NMR at the symbolic level needs to be specified. This account will cover aspects related to the first two of these questions. Computational approximations of conceptual space will be furnished by semantic space models which are emerging from the fields of cognition and computational linguistics. Semantic space models not only provide a cognitively motivated basis to underpin human practical reasoning, but from a mathematical perspective, they show a marked similarity with quantum mechanics (QM) [Aerts and Czachor, 2004]. This introduces the tantalizing and unavoidably speculative prospect of formalizing aspects of human practical reasoning via QM.

3 SEMANTIC SPACE: COMPUTATIONAL APPROXIMATIONS OF CONCEPTUAL SPACE

To illustrate how the gap between cognitive knowledge representation and actual computational representations may be bridged, the Hyperspace Analogue to Language (HAL) semantic space model is employed [Lund and Burgess, 1996; Burgess *et al.*, 1998]. HAL produces representations of words in a high dimensional space that seem to correlate with the equivalent human representations. For example, "...simulations using HAL accounted for a variety of semantic and associative word priming effects that can be found in the literature...and shed light on the nature of the word relations found in human word-association norm data" [Lund and Burgess, 1996]. Given an n-word vocabulary, HAL computes an $n \times n$ matrix constructed by moving a window of length l over the corpus by one word increment ignoring punctuation, sentence and paragraph boundaries. All words within the window are considered as co-occurring with the last word in the window with a strength inversely proportional to the distance between the words. Each row i in the matrix represents accumulated weighted associations of word i with respect to other words which preceded i in a context window. Conversely, column i represents accumulated weighted associations with words that appeared after i in a window. For example, consider the text "President Reagan ignorant of the arms scandal", with $l = 5$, the resulting HAL matrix H would be:

	arms	ig	of	pres	reag	scand	the
arms	0	3	4	1	2	0	5
ig	0	0	0	4	5	0	0
of	0	5	0	3	4	0	0
pres	0	0	0	0	0	0	0
reag	0	0	0	5	0	0	0
scand	5	2	3	0	1	0	4
the	0	4	5	2	3	0	0

Table 1. A simple semantic space computed by HAL

If word precedence information is considered unimportant the matrix $S = H + H^T$ denotes a symmetric matrix in which $S[i, j]$ reflects the strength of association of word i seen in the context of word j, irrespective of whether word i appeared before or after word j in the context window. The column vector S_j represents the strengths of association between j and other words seen in the context of the sliding window: the higher the weight of a word, the more it has lexically co-occurred with j in the same context(s). For example, table 2 illustrates the vector representation for "Reagan" taken from a matrix S computed from a corpus of 21578 Reuters[7]

[7]The Reuters-21578 collection is standard test collection used for research into automatic text classification.

```
def calculate_hal(documents, n)
  HAL = 2DArray.new()
  for d in documents {
    for i in 1 ..  d.len {
      for j in max(1,i-n) ..  i-1 {
        HAL[d.word(i),d.word(j)] += n+1-(i-j)
  }}}
  return HAL
end
```

Figure 1. Algorithm to compute the HAL matrix for a collection of documents. It is assumed that the documents have been pruned of stop words and punctuation.

president (5259), administration (2859), trade (1451), house (1426), budget (1023), congress (991), bill (889), tax (795), veto (786), white (779), japan (767), senate (726), iran (687), billion (666), dlrs (615), japanese (597), officials (554), arms (547), tariffs (536) ...

Table 2. Example representation of the word "Reagan"

news feeds taken from the year 1988. (The weights in the table are not normalized). Highly weighted associations reflect Reagan in his presidential role dealing with congress, tax, vetoes etc. In addition, the more highly weighted association reflect a default-like character, e.g., "president" and "administration". Associations with lower weights seem to reflect the trade war with Japan ("japan", "tariffs") and the Iran-contra scandal ("Iran", "arms"). In other words, the representation of Reagan represents a mixture of different "senses" of Reagan. This facet is intuitively similar to the QM phenomenon of a particle being in a state of superposition.

HAL is an exemplar of a growing ensemble of computational models emerging from cognitive science, which are generally referred to as *semantic spaces* [Lund and Burgess, 1996; Burgess *et al.*, 1998; Lowe, 2000; Lowe, 2001; Landauer and Dumais, 1997; Landauer *et al.*, 1998; Patel *et al.*, 1997; Schütze, 1998; Levy and Bullinaria, 1999; Sahlgren, 2002]. Even though there is ongoing debate about specific details of the respective models, they all feature a remarkable level of compatibility with a variety of human information processing tasks such as word association. Semantic spaces provide a geometric, rather than propositional, representation of knowledge. They can be considered to be approximations of conceptual space proposed by Gärdenfors [Gärdenfors, 2000], and of reasoning down below as proposed by [Gabbay and Woods, 2003a; Gabbay and Woods, 2005].

Within a conceptual space, knowledge has a dimensional structure. For example, the property colour can be represented in terms of three dimensions: hue, chromaticity, and brightness. Gärdenfors argues that a property is represented as a convex region in a geometric space. In terms of the example, the property "red" is a convex region within the tri-dimensional space made up of hue, chromaticity and brightness. The property "blue" would occupy a different region of this space. A domain is a set of integral dimensions in the sense that a value in one dimension(s) determines or affects the value in another dimension(s). For example, the three dimensions defining the colour space are integral since the brightness of a colour will affect both its saturation (chromaticity) and hue. Gärdenfors extends the notion of properties into concepts, which are based on domains. The concept "apple" may have domains taste, shape, colour, etc. Context is modelled as a weighting function on the domains, for example, when eating an apple, the taste domain will be prominent, but when playing with it, the shape domain will be heavily weighted (i.e., it's roundness). One of the goals of this article is to provide both a formal and operational account of this weighting function.

Observe the distinction between representations at the symbolic and conceptual levels. At the symbolic level "apple" can be represented as the atomic proposition $apple(x)$. However, within a conceptual space (conceptual level), it has a representation involving multiple inter-related dimensions and domains. Colloquially speaking, the token "apple" (symbolic level) is the tip of an iceberg with a rich underlying representation at the conceptual level. Gärdenfors points out that the symbolic and conceptual representations of information are not in conflict with each other, but are to be seen as "different perspectives on how information is described".

Barwise and Seligman [Barwise and Seligman, 1997] also propose a geometric foundation to their account of inferential information content via the use of real-valued state spaces. In a state space, the colour "red" would be represented as a point in a tri-dimensional real-valued space. For example, brightness can be modelled as a real-value between white (0) and black (1). Integral dimensions are modelled by so called observation functions defining how the value(s) in dimension(s) determine the value in another dimension. Observe that this is a similar proposal, albeit more primitive, to that of Gärdenfors as the representations correspond to points rather than regions in the space.

Semantic space models are also an approximation of Barwise and Seligman state spaces whereby the dimensions of the space correspond to words. A word j is a point in the space. This point represents the "state" in the context of the associated text collection from which the semantic space was computed. If the collection changes, the state of the word may also change. Semantic space models, however, do not make provision for integral dimensions. An important intuition for the following is the state of a word in semantic space is tied very much with its "meaning", and this meaning is context-sensitive. Further, context-sensitivity will be realized by state changes of a word.

In short, HAL, and more generally semantic spaces, are a promising, pragmatic means for knowledge representation based on text. They are computational approximations, albeit rather primitively, of Gärdenfors' conceptual space. Moreover, due to their cognitive track record, semantic spaces would seem to be a fitting foundation for considering realizing computational variants of human reasoning.

Finally, it has been shown that a semantic space is formally a density matrix, a notion from QM [Aerts and Czachor, 2004; Bruza and Cole, 2005a]. This opens the door to exploring further connections with QM.

4 BRIDGING SEMANTIC SPACE AND QUANTUM MECHANICS

HAL exemplifies how a semantic space model assigns each word in a given vocabulary a point in a finite dimensional vector space. Lowe [Lowe, 2001] formalizes semantic space models as a quadruple $\langle A, B, F, M\rangle$ where

- B is a set of m basis elements
- A is a function which maps the co-occurrence frequencies between words in a vocabulary V and the basis elements so each $w \in V$ is represented by a vector $(A(b_1, w), \ldots, A(b_m, w))$
- F is a function which maps pairs of vectors onto continuous valued quantity. The interpretation of F is often "semantic similarity" between the two vectors in question.
- M is a transformation which takes one semantic space and maps it into another, for example via dimensional reduction

A semantic space[8] S is an instance of the range of the function A. That is, S is a $m \times n$ matrix where the columns $\{1, \ldots, n\}$ correspond to vector representations of words. A typical method for deriving the vocabulary V is to tokenize the corpus from which the semantic space is computed and remove non information bearing words such as "the", "a", etc. The letters u, v, w will be used to identify individual words.

The interpretation of the basis elements corresponding to the rows $\{1 \ldots m\}$ depends of the type of semantic space in question. For example, table 3 illustrates that HAL produces a square matrix in which the rows are also interpreted as representations of terms from the vocabulary V. In contrast, a row in the semantic space models produced by Latent Semantic Analysis [Landauer *et al.*, 1998] corresponds to a text item, for example, a whole document, a paragraph, or even a fixed window of text, as above. The value $S[t, w] = x$ denotes the salience x of word w in text t. Information-theoretic approaches are sometimes use to compute salience. Alternatively, the (normalized) frequency of word w in context t can be used.

For reasons of a more straightforward embedding of semantic space into QM, we will focus on square, symmetric semantic spaces ($m = n$). The following draws from [van Rijsbergen, 2004].

A word w is represented as a column vector in S:

---

[8]Bear in mind that the term "space" should not be interpreted as a "vector space". This unfortunate blurring between "matrix" and "space" in the technical sense occurs because "semantic space" is a term from the cognitive science literature.

(1) $$|w\rangle = \begin{pmatrix} w_1 \\ \vdots \\ w_n \end{pmatrix}$$

The notation on the LHS is called a *ket*, and originates from quantum physicist Paul Dirac. Conversely, a row vector $v = (v_1, \ldots, v_n)$ is denoted by the *bra* $\langle v|$.

Multiplying a ket by a scalar α is as would be expected:

(2) $$\alpha|w\rangle = \begin{pmatrix} \alpha w_1 \\ \vdots \\ \alpha w_n \end{pmatrix}$$

Addition of vectors $|u\rangle + |v\rangle$ is also as one would expect. In Dirac notation, the scalar product of two n-dimensional real[9] valued vectors u and v produces a real number:

(3) $$\langle u|v\rangle = \sum_{i=1}^{n} u_i v_i$$

The outer product $|u\rangle\langle u|$ produces a $n \times n$ symmetric matrix. Vectors u and v are *orthogonal* iff $\langle u|v\rangle = 0$. Scalar product allows the length of a vector to be defined: $\|u\| = \sqrt{\langle u|u\rangle}$. A vector $|u\rangle$ can be normalized to unit length ($\|u\| = 1$) by dividing each of its components by the vector's length: $\frac{1}{\|u\|}|u\rangle$.

A Hilbert space is a complete[10] inner product space. In the formalization to be presented in ensuing sections, a semantic space S is an n-dimensional real-valued Hilbert space using Euclidean scalar product as the inner product.

A Hilbert space allows the state of a quantum system to be represented. It is important to note that a Hilbert space is an *abstract* state space meaning QM does not prescribe the state space of specific systems such as electrons. This is the responsibility of a physical theory such as quantum electrodynamics. Accordingly, it is the responsibility of semantic space theory to offer the specifics: In a nutshell, a ket $|w\rangle$ describes the state of "meaning" of a word w. It is akin to a particle in QM. The state of a word changes due to context effects in a process somewhat akin to quantum collapse. This in turn bears on practical inferences drawn due to context effects of word seen together with other words as described above.

In QM, the state can represent a superposition of potentialities. By way of illustration consider the state σ of a quantum bit, or *qubit* as:

(4) $$|\sigma\rangle = \alpha|0\rangle + \beta|1\rangle$$

[9]QM is founded on complex vector spaces. We restrict our attention to finite vector spaces of real numbers.

[10]The notion of a "complete" vector space should not be confused with "completeness" in logic. The definition of a completeness in a vector space is rather technical, the details of which are not relevant to this account.

where $\alpha^2 + \beta^2 = 1$. The vectors $|0\rangle$ and $|1\rangle$ represent the characteristic states, or *eigenstates*, of "off" and "on". Eigenstates are sometimes referred to as *pure*, or *basis* states. They can be pictured as defining orthogonal axes in a 2- D plane:

(5) $$\alpha|0\rangle = \begin{pmatrix} 0 \\ 1 \end{pmatrix}$$

and

(6) $$\alpha|1\rangle = \begin{pmatrix} 1 \\ 0 \end{pmatrix}$$

The state σ is a linear combination of eigenstates. Hard though it is to conceptualize, the linear combination allows the state of the qubit to be a mixture of the eigenstates of being "off" and "on" at the same time.

In summary, a quantum state encodes the probabilities of its measurable properties, or eigenstates. The probability of observing the qubit being off (i.e., $|0\rangle$ is α^2). Similarly, β^2 is the probability of observing it being "on".

The above detour into QM raises questions in relation to semantic space. What does it mean that a word is a superposition? What are the eigenstates of a word?

4.1 Superposed and eigenstates of a word meaning

Consider the following traces of text from the Reuters-21578 collection:

- *President Reagan was ignorant about much of the Iran arms scandal*
- *Reagan says U.S to offer missile treaty*
- *Reagan seeks more aid for Central America*
- *Kemp urges Reagan to oppose stock tax.*

Each of these is a window which HAL will process accumulating weighted word associations in relation to the word "Reagan". In other words, included in the HAL vector for "Reagan" are associations dealing with the Iran-contra scandal, missile treaty negotiations with the Soviets, stock tax etc. The point is when HAL runs over the full collection, the vector representation for "Reagan"is a mixture of eigenstates, whereby an eigenstate corresponds to a particular "sense", or "characteristic meaning" of the concept "Reagan". For example, Reagan, in the political sense, in the sense dealing with the Iran-Contra scandal, etc. The senses of a concept are equivalent of the eigenstates of a particle in QM [Bruza and Cole, 2005a; Aerts *et al.*, 2005; Widdows and Peters, 2003].

Consider once again the HAL matrix H computed from the text "President Reagan ignorant of the arms scandal". As mentioned before, $S = H + H^T$ is a real symmetric matrix. Consider a set of y text windows of length l which are centred around a word w. Associated with each such text window $j, 1 \leq j \leq y$, is a semantic space S_j. It is assumed that the semantic space is n-dimensional, whereby the n dimensions correspond to a fixed vocabulary V as above. The semantic space around word w, denoted by S_w, can be calculated by the sum:

$$S_w = \sum_{j=1}^{y} S_j \tag{7}$$

The above formula provides a toehold for computing a semantic space in terms of a sum of semantic spaces; each constituent semantic space corresponding to a specific sense of the concept w. By way of illustration, Let the concept w be "Reagan" and assume there are a total of y traces centred on the word "Reagan", x of which deal with the Iran-contra issue. These x traces can be used to construct a semantic space using equation 7. Let S_i denote this semantic space. Its associated probability $p_i = \frac{x}{y}$. Assume the concept w has m senses. As each sense i represents a particular state of w, each can be represented as a semantic space S_i with an associated probability.

$$S_w = p_1 S_1 + \ldots + p_m S_m \tag{8}$$

where $p_1 + \ldots + p_m = 1$.

This formula expresses that the semantic space around a concept w can be conceived of as a linear combination of semantic spaces around senses of w. The formula is intuitively close to an analogous formula from QM whereby a density matrix can be expressed as a probability mixture of density matrices [Barndorff-Nielsen *et al.*, 2003, p. 778]. A density matrix corresponding to a superposed state can be expressed as a weighted combination of density matrices corresponding to basis states. There is no requirement that the state vectors of the pure states are orthogonal to one another. This is a very important point. Intuitively, it is unrealistic to require the senses of a concept to be orthogonal. For this reason, the term "sense" will be used to denote the basis state of a word meaning, rather than "eigenstate", because, in QM, eigenstates are assumed to be mutually orthogonal.

The connection between the notions of semantic space and density matrix have been detailed elsewhere [Aerts and Czachor, 2004; Bruza and Cole, 2005a]. As mentioned in the introduction, there are various semantic space models presented in the literature. Each will involve a different rendering as density matrix. The method adopted in this account rests on the intuition the ket $|e_i\rangle$ in each semantic space S_i of equation 8 corresponds to a state vector representing a sense of concept w. A density matrix ρ_i can be formed by the product $|e_i\rangle\langle e_i|$. Building on this, a density matrix ρ_w corresponding to the semantic space S_w can be constructed as follows.

$$\rho_w = p_1\rho_1 + \ldots + p_m\rho_m \tag{9}$$

Importantly, no assumption of orthogonality has been made.

This approach to representing a semantic space in a state contrasts approaches using the spectral decomposition of the semantic space [Aerts and Czachor, 2004;

Aerts and Gabora, 2005]. As the semantic space S_w is a symmetric matrix, the spectral decomposition of singular value decomposition (SVD) allows S_w to be reconstructed, where $k \leq n$:

$$\begin{aligned} S_w &= \sum_{i=1}^{k} |e_i\rangle d_i \langle e_i| \\ &= \sum_{i=1}^{k} d_i |e_i\rangle\langle e_i| \\ &= d_1|e_1\rangle\langle e_1| + \ldots + d_k|e_k\rangle\langle e_k| \end{aligned}$$

This equation parallels the one given in equation 9. The singular values d_i relate to the probabilities of the associated eigenvectors (eigenstates in QM terminology). Each eigenstate $|e_i\rangle$ contributes to the linear combination via the density matrix $|e_i\rangle\langle e_i|$. The eigenstates $|e_i\rangle$ of S_w should ideally correspond to the senses of word w. Unfortunately, this does not bear out in practice. A fundamental problem is that the eigenstates $|e_i\rangle$ computed by SVD are orthogonal, and in reality the senses of a word w need not be. (See [Bruza and Cole, 2005a] for more details).

4.2 The collapse of meaning in the light of context

We continue by connecting the above development of quantum mechanics in semantic space to Gärdenfors' views on the interaction of context and the meaning of concepts. He states, "The starting point is that, for some concepts, the meaning of the concept is determined by the *context* in which it occurs" [Gärdenfors, 2000, p.119]. Context effects manifest in relation to contrast classes. In the introduction, the Chihuahua showed how property tall is relative, "a tall Chihuahua is not a tall dog". He also illustrates how contrast classes manifest in word combinations. Consider, "red" in the following combinations, "red book", "red wine", "red hair", "red skin", "red soil". Gärdenfors argues contrast classes generate conceptual subspaces, for example, skin colours form a subspace of the space generated by colours in general. In other words, each of the combinations involving "red" results in a separate subspace representing the particular quality of "red", for example, the quality of "red" would actually be "purple" when "red" is seen in the context of "wine".

The collapse of word meaning can be thought of in terms of the quantum collapse of the particle but with an important difference: The collapse due to context may not always result in a basis state because the context may not be sufficient to fully resolve the sense in question. By way of illustration, consider "Reagan" in the context of "Iran". For the purposes of discussion, assume there are two possible senses. The first deals with the Iran-contra scandal, and the other deals with hostage crisis at the American embassy in Teheran. The distinction between a measurement due to context and a physical measurement possibly has its roots in human memory. Matrix models of human memory also contain the notion of superimposed memory states, and it has been argued, "The superposition of memory traces in a vector bundle resulting from a memory retrieval has often been considered to be a noisy signal that needs to be 'cleaned up' [i.e., full collapse onto a basis state as in QM]. The point we make here is that this is *not necessarily so* and that the superposition of vectors [after retrieval] is a powerful process that adds to the flexibility of memory processes." (Emphasis ours) [Wiles *et al.*, 1994].

This distinction requires a less stringent notion of collapse as maintained within QM. Consider a concept w considered in the light of some context, for example, other words. The context is denoted generically by X. The effect of context X is brought about by a projection operator P_x. Assuming the density matrix ρ_w corresponding to a concept w, the collapse of meaning in the light of context X is characterized by the following equation:

$$(10) \qquad P_x \rho_w = p \rho_w^x$$

where p denotes the probability of collapse and ρ_w^x is the state of w after the "collapse" of its meaning.

In terms of QM, ρ_w is an "observable" meaning an observable physical quantity. An observable is represented by a self-adjoint operator. As ρ_w is a real symmetric matrix, it is therefore also a self-adjoint operator. This is consistent with the second axiom of QM [Byron and Fuller, 1992]. Even though this equation has the form of an eigenvalue problem, the value p is not an eigenvalue. It is a theorem that the eigenstates of a self adjoint operator belonging to different eigenvalues must be orthogonal, a requirement which is too strong for word meanings as was motivated earlier. Nevertheless, it will be be shown later that p derives from the geometry of the space as do eigenvalues.

The previous equation is also consistent with the third axiom of QM as the result of "measurements of the observable [ρ_w]" is an element of "the spectrum of the operator". In our case, the spectrum is specified by the probability mixture given in equation 9, but more of the flexibility of this equation is exploited than is the case in QM. The key to this flexibility revolves around the fact that the sum of density matrices is a density matrix. By way of illustration, equation 9 can be equivalently written as the probability mixture:

$$\rho_w = p_1 \rho_1 + p_2 \rho_2$$

where $p_1 + p_2 = 1$. Let ρ_1 correspond to the state of "Reagan" in the context of "Iran" and ρ_2 the state of "Reagan" in all other contexts. Assume, that "Reagan" is seen in the context of "Iran". The projection operator P_x collapses ρ_w onto ρ_1 with probability p_1. Unlike, QM, the state ρ_1 is not a basis state but corresponds to a partially resolved sense. Let the Iran-contra sense be denoted $|c\rangle$ and the Iranian embassy hostage crisis be denoted $|h\rangle$. In the light of this example, the density matrix corresponding to the state after collapse due to "Iran" would be of the form $\rho_1 = p_c|c\rangle\langle c| + p_h|h\rangle\langle h|$, where $p_c + p_h = 1$.

It has been argued in [Bruza and Cole, 2005a] that in terms of this running example many would assume the "Iran-contra" sense of "Reagan" when "Reagan" is seen in the context of "Iran". This phenomenon may have its roots in cognitive economy. Full resolution requires processing, and to avoid this processing, humans "guess" the more likely sense (In the example, p_c happens to be substantially greater than p_h). In other words, we cautiously put forward the conjecture that collapse of meaning and abductive processes go hand in hand to fully resolve the sense, i.e., collapse onto a basis state. Even though "full" collapse eventually results, the process is not direct as is the the case of the collapse in QM.[11].

The running example reveals something of the nature of the projection operator P_x. If P_x is orthogonal to a sense $|e_i\rangle$ represented by the density matrix $\rho_i = |e_i\rangle\langle e_i|$, then P_x projects this sense onto the zero vector $|0\rangle$. (Note the corresponding density matrix is $|0\rangle\langle 0|$). If the projection P_x is not orthogonal to a sense $|e_i\rangle$, then it has the effect of retrieving those senses out of the combination expressed in equation 9. This is not unlike the notion of a cue which probes human memory. Cues can be used to access memory in two ways; via *matching* or *retrieval* processes. Matching entails the "comparison of the test cue(s) with the information stored in memory" [Humphreys *et al.*, 1989, p 41.]. This process measures the similarity of the cue(s) and the memory representation. The output of this process is a scalar quantity (i.e., a single numeric value representing the degree or strength of the match). Memory tasks which utilise this access procedure include recognition and familiarity tasks. Retrieval involves the "recovery of qualitative information associated with a cue" [Humphreys *et al.*, 1989, p 141.]. This information is modelled as a vector of feature weights. Retrieval tasks include free recall, cued-recall, and indirect production tasks.

The intuition we will attempt to develop is that collapse of word meaning due to context is akin to a cued-recall retrieval operation driven by the projector P_x on a given density matrix corresponding to the state of a word meaning. The probability of collapse p is a function of the scalar quantity resulting from matching.

In the matrix model of memory [Humphreys *et al.*, 1989] , memory representations can include items, contexts or, combinations of items and contexts (associations). Items can comprise stimuli, words, or concepts. Each item is modelled as a vector of feature weights. Feature weights are used to specify the degree to which certain features form part of an item. There are two possible levels of vector representation for items. These include:

- modality specific peripheral representations (e.g., graphemic or phonemic representations of words)
- modality independent central representations (e.g., semantic represenatations of words)

In our case, our discussion will naturally focus on the latter due to assumption that semantic spaces deliver semantic representations of words. For example, the "Rea-

[11]For a more detailed discussion of how the logic of abduction engages with the cognitive economy of practical agency, see [Gabbay and Woods, 2005]. For the link between abduction and semantic space, see [Bruza *et al.*, 2006]

gan" vector $|r\rangle$ from the semantic space S_r illustrates a "modality independent central representation".

Context can be conceptualised as a mental representation (overall holistic picture) of the context in which items, or events have occurred. (e.g., "Reagan" in the context of "Iran"). Context is also modelled as a vector of feature weights. Following from this, context is X is assumed to be represented by a ket $|x\rangle$. In the case of the running example, the "Iran" vector $|i\rangle$ drawn from the semantic space S_i could be employed as a context vector.

Memories are associative by nature and unique representations are created by combining features of items and contexts. Several different types of associations are possible [Humphreys *et al.*, 1989]. The association of interest here is a two way association between a word $|w\rangle$ and a context $|x\rangle$. In the matrix model of memory, an association between context and a word is represented by an outer product; $|w\rangle\langle x|$. Seeing a given word (a target) in the context of other words (cue) forms an association which probes memory. Observe with respect to the running example how the probe $|r\rangle\langle i|$ embodies both the cue of the probe "Iran" and the target "Reagan".

In the light of the above brief digression into a matrix model of human memory, one possibility is to formalize the projector P_x as the probe $|w\rangle\langle x|$. The object being probed is a density matrix which is not a superposition of memory traces but of semantic spaces hinged around a particular word or concept. Equation 8 and its density matrix equivalent (equation 9) reflect this superposition, however in this case the traces, in their raw form, are windows of text.

In short, viewing the collapse of meaning in terms of retrieval and matching processes in memory refines the collapse equation 10 as follows. Let $|w\rangle$ be a target concept and $|x\rangle$ be the context. Firstly, collapse of meaning is characterized by projecting the probe into the memory corresponding to the state of the target word w. The collapse equates with retrieving a new state of meaning reflecting the change of meaning of w in light of the context.

(11) $$P_x \rho_w = |w\rangle\langle x|\rho_w = e_w^x$$

The probability p of collapse is assumed to be a function[12] of the match between the probe and the memory:

(12) $$p = f(\langle x|\rho_w|w\rangle)$$

Motivating the collapse of meaning by means of the matrix model of memory introduces a deviation from orthodox QM. After application of the probe $|w\rangle\langle x|$, the the state after the collapse, denoted ρ_w^x is not guaranteed to be density matrix. This deviation from orthodox QM is not solely a technical issue. It may well be that there are different qualities of probe. For example, "Reagan" in the context of "Iran" would intuitively involve a projection of the global "Reagan"

[12]Further research is needed to provide the specifics of this function which will take into account issues such as decay processes in memory.

semantic space onto a subspace dealing with "Iran". On the other hand, consider "lion" in the context of "stone". In this case, the result after the application of the context would seem to be considerably outside the "lion" space as a "stone lion" does not share many of the attributes of a living one. It would seem, then, a projection operator is not the appropriate mechanism, but rather a more general linear operator which can project "outside" the space. In the latter case, equating P_x with the probe $|w\rangle\langle x|$ is arguably justified as such probes in the matrix model of memory briefly described earlier are transformations of the space, rather than projections into it. An alternative view is that "stone lion" is a result of concept combination and mechanisms other than projection operators are required to suitably formalize it. For example, Aerts and Gabora [Aerts and Gabora, 2005] resort to tensor products for concept combination. These are slippery issues requiring a clean distinction between context effects and concept combination. More research is needed to clarify these issues in relation to a logic of down below.

It remains to provide a characterization of P_x as an orthodox projector as typified by the "Reagan in the context of "Iran" example. In order to do this, the senses $B = \{|e_1\rangle, \ldots, |e_m\rangle\}$ are assumed to form a basis. (The assumption here is linear independence, which is a weaker assumption than assuming the $|e_i\rangle$'s are mutually orthogonal, i.e., an orthonomal basis as is commonly seen in orthodox QM). The set B represents the basis of the space S_w in relation to ρ_w. Let $B_x = \{|x_1\rangle, \ldots, |x_r\rangle\}$ and $B_y = \{|y_1\rangle, \ldots, |y_{m-r}\rangle\}$ such that $B_x \cup B_y = B$. The set B_x is the basis of the subspace S_x due to context X. The complementary space is denoted S_y. By way of illustration in terms of the running example, $B_x = \{|x_1\rangle, |x_2\rangle\}$ would corresponds to the two senses of "Reagan" in the context of "Iran" previously introduced as $|e_c\rangle$ and $|e_h\rangle$. Though complementary spaces, S_x and S_y are *not* assumed to be orthogonal. Consequently, the projection operator P_x is "oblique" rather than orthogonal. Once again, this is a deviation from orthodox QM, but nevertheless faithful to the underlying intuition behind projection operators. As stated earlier, the projection operator P_x "retrieves" the relevant senses out of the probability mixture (equation 9), that is $P_x|x_i\rangle = |x_i\rangle$, for $x_i \in \{|x_1\rangle, \ldots, |x_r\rangle\}$. These are the so called *fixed points* of the projector P_x. As a consequence, the density matrix form of the fixed points also holds as $P_x(|x_i\rangle\langle x_i|) = (P_x|x_i\rangle)\langle x_i| = |x_i\rangle\langle x_i|$. This establishes that P_x will retrieve the density matrix form of the relevant senses expressed in equation 9.

$B_{n\times m}$ is an $n \times m$ matrix with columns

$$[|x_1\rangle|x_2\rangle\cdots|x_r\rangle||y_1\rangle|y_2\rangle\cdots|y_{m-r}\rangle] \quad = \quad [X_{n\times r}|Y_{n\times(m-r)}]$$

The projection operator P_x retrieves those fixed points relevant to the context. All other senses are projected onto the zero vector $|0\rangle$:

$$\begin{aligned} P_xB &= P_x[X|Y] \\ &= [P_xX|P_xY] \\ &= [X|0] \end{aligned}$$

For the case $m = n$, the matrix B has an inverse B^{-1} so the makeup of the required projection operator is given by:

$$P_x = [X|0]B^{-1} = B \begin{pmatrix} I_r & 0 \\ 0 & 0 \end{pmatrix} B^{-1} \qquad (13)$$

With an eye on operational deployment on a large scale, a simple algorithmic construction of P_x is based on the intuition that those senses which are not orthogonal to the cue should be retrieved from the linear combination of m senses (equation 9):

$$B_x = \{|e_i\rangle | \langle x|\rho_i|x\rangle > 0, 1 \leq i \leq m\}$$

(Recall that $\rho_i = |e_i\rangle\langle e_i|$). In terms of the running example, $B_x = \{|e_c\rangle, |e_h\rangle\}$, the two senses relevant to "Reagan" seen in the context of "Iran".

The scalar $\langle x|\rho_i|x\rangle$ decomposes as follows:

$$\begin{aligned} \langle x|\rho_i|x\rangle &= \langle x|(|e_i\rangle\langle e_i|)|x\rangle \\ &= (\langle x|e_i\rangle)^2 \\ &= \cos^2\theta_i \\ &= a_i \end{aligned}$$

where $\cos\theta_i$ is the angle between $|x\rangle$ and $|e_i\rangle$. In the second last line the equivalence between Euclidean scalar product and cosine was employed due to the vectors being normalized to unit length. This value reflects how much the given sense is being activated to the level a_i by the cue $|x\rangle$. Stated otherwise, a_i reflects the strength with which the sense ρ_i is aligned with the cue $|x\rangle$. All senses $|e_i\rangle$ in the basis B_x will have a positive activation value a_i. By appropriately scaling the values a_i, the effect of projector P_x can now be expressed as a probability mixture:

$$P_x\rho_w = p_1\rho_1 + \ldots + p_r\rho_r \qquad (14)$$

where $\rho_i = |e_i\rangle\langle e_i|$, for all $|e_i\rangle \in B_x$ and $p_1 + \ldots + p_r = 1$. The import of the last equation is that the effect of the projector P_x results in a density matrix.

4.3 *The probability of collapse*

It is illustrative to examine how in the light of the running example the scalar value resulting from the matching process determines the probability of collapse (equation 12). First, the effect of the cue "Iran" via the context vector $|i\rangle$ is shown. The "memory" to be probed derives from the target "Reagan" and is denoted by the density matrix ρ_r.

$$\begin{aligned} \langle i|\rho_r &= \langle i|(p_1\rho_1 + \ldots + p_m\rho_m) \\ &= p_1\langle i|\rho_1 + \ldots + p_m\langle i|\rho_m \end{aligned}$$

Recall that each of the m constituent density matrices ρ_i derives from a particular sense of "Reagan" denoted e_i. Therefore the previous equation can be written as,

$$\begin{aligned}\langle i|\rho_r &= p_1\langle i|(|e_1\rangle\langle e_1|) + \ldots + p_m\langle i|(|e_m\rangle\langle e_m|)\\ &= p_1(\langle i|e_1\rangle)\langle e_1| + \ldots + p_m(\langle i|e_m\rangle)\langle e_m|\\ &= p_1\cos\theta_1\langle e_1| + \ldots + p_m\cos\theta_m\langle e_m|\end{aligned}$$

The salient facet of the last line is those senses that are not orthogonal to the context vectors will be retrieved ($\cos\theta_i > 0$)and will contribute to the probability of collapse. This accords with the intuitions expressed in the previous section. In the running example, these senses were denoted $|e_c\rangle$ and $|e_h\rangle$. So,

$$\langle i|\rho_r = p_c\cos\theta_c\langle e_c| + p_h\cos\theta_h\langle e_h|$$

A second aspect of the matching is post multiplying with the target vector "Reagan", denoted $|r\rangle$:

$$\begin{aligned}(p_c\cos\theta_c\langle e_c| + p_h\cos\theta_h\langle e_h|)|r\rangle &= p_c\cos\theta_c(\langle e_c|r\rangle) + p_h\cos\theta_h(\langle e_h|r\rangle)\\ &= p_c\cos\theta_c\cos\psi_c + p_h\cos\theta_h\cos\psi_h\\ &= p_c m_c + p_h m_h\end{aligned}$$

The angles $\cos\psi$ reflects how strongly the the sense correlates with the given target. It can be envisaged as a measure of significance of the given sense with the target $|r\rangle$. The scores due to matching of the probe with memory are reflected by the scalars m_c and m_h. These are modified by associated probabilities of the respective senses. Finally, the two terms are added to return a single scalar. The probability of collapse is assumed to be a function of this value.

4.4 Summary

The preceding development has centred around providing an account of the collapse of meaning in the light of context. It is important that the formalization rests on non-orthogonal density matrices, which is in contrast to the orthogonal approach used in the SCOP model [Aerts and Gabora, 2005]. The approach presented here draws inspiration from a cue which probe human memory and describes collapse of meaning in terms of memory cues. The notion of a "probe" is not foreign to QM. The most useful probes of the various wave functions of atoms and molecules are the various forms of spectroscopy. In spectroscopy , an atom or molecule starting with some wave function (represented by a density matrix) is probed with light, or some other particle. The light interacts with the molecule and leaves it in another state. This process is analogous to the probing of memory just described. Chemical physics also shares another similarity with our account in the sense that the underlying density matrices cannot be assumed to be orthogonal. Nonorthogonal density matrix perturbation theory has arisen to deal with nonorthogonal density matrices and may turn out to be a relevant area for formalizing additional aspects of a logic of "down below". The analogy should be mindfully employed, however. Human memory is a vast topic abundant with texture and nuance, not to mention strident debate. However we feel investigations into the memory literature can bear further fruit in relation to a QM inspired account of a logic of "down below". The matrix model of memory described above has been extended to provide an account of analogical mapping [Wiles *et al.*, 1994]. In our opinion, it is reasonable to assume that analogical reasoning has roots in subsymbolic logic. Dunbar [Dunbar, 1999] concludes from cognitive studies that scientists frequently resort to analogies when there is not a straightforward answer to their current problem. Therefore, analogical reasoning sometimes plays a crucial role in hypothesis formation which is fundamental to abduction[Gabbay and Woods, 2005, Chapter 7]. Reasoning, then, becomes highly confounded with memory processes. Consider the "Tweety" example described earlier. When one learns that "Tweety is a penguin", it is debatable whether any reasoning takes place at all. We would argue that the example can be explained in terms of probes to memory and the associated dynamics of defaults emerge out of context effects. We have argued such probes bear a striking similarity to quantum collapse.

5 QUANTUM LOGIC AND CONCEPTUAL GENERALIZATION

A proposal for reasoning at the subsymbolic level must give an account for how conceptual structures may arise from perceptual observations. For example, in *Word and Object*, Quine [Quine, 1960, p. 25] famously challenged philosophers to give an account for how a hearer might reliably deduce that a speaker who utters the word "gavagai" upon seeing a rabbit actually means "rabbit", instead of "part of a rabbit", or a member of some other class such as "rabbit or guppy", or even "rabbit or Reagan". In other words, how might a conceptual logic give rise to a recognition and representation of natural kinds, in such a way that this logic is cognitively beneficial?

It is known that some logics are more amenable to inductive learning than others, and that direct adherence to the Boolean distributive law effectively prevents the sort of smoothing or closure operations that may lead to the formation of natural kinds (see [Widdows and Higgins, 2004]). For example, since Boolean logic is modelled on set theory and the union of the set of rabbits and the set of frogs is a perfectly well-formed set, the concept "rabbit or guppy" is as natural as the concept of "rabbit" in Boolean logic. At the other extreme, a single-inheritance taxonomic logic (based, for example, on phylogenetic inheritance) may overgeneralize by assuming that the disjunction of "rabbit" and "guppy" must be the lowest common phylogenetic ancestor "vertebrate". This would lead also to unfortunate consequences, such as the presumption that, since a rabbit makes a good pet for a child and a guppy makes a good pet for a child, any vertebrate makes a good pet for a child.

Compared with the discrete extreme of Boolean classification, and the opposite extreme of a single-inheritance taxonomy, the vector lattice of quantum logic presents and attractive middle ground. There are distinctly well-formed concepts represented by lines and planes, there is a natural closure or smoothing operator defined by the linear span of a set, and there is a scope for multiple inheritance (since a line is contained is many different planes and an m-dimensional subspace is contained in many $m+n$-dimensional subspaces). Some practical evidence for the usefulness of the linear span as a disjunction of two concept vectors was provided by the experiments in [Widdows, 2003], in which the removal of a pair of concepts using negated quantum disjunction proved greatly more effective than Boolean negation at the task of removing unwanted keywords and their synonyms. The argument that projection onto subspaces of a vector space can be used as a solution to the age-old problem of learning from incomplete experience has been made one of the mainstays of Latent Semantic Analysis, by [Landauer and Dumais, 1997] and others.

It should also be noted that the use of a pure quantum logic for concept generalization in semantic space leads to problems of its own, as one would expect with any attempt to apply such a simple mathematical model to a wholesale description of language. In particular, quantum disjunctions may often overgenerate, because of the nature of the linear independence and the operation of taking the linear span. In practice, vectors that are very close to one another in semantic space may still be linearly independent, and will thus generate a large subspace that does not reflect that fact that the vectors were in fact drawn from a small region of this subspace. This danger is illustrated in Figure 2, which depicts two groups of three vectors in a 3-dimensional vector space. In the left hand picture, the vectors A, B and C are orthogonal and can be used to generate the whole of the space. The vectors D, E and F, far from being orthogonal, have high mutual similarity. However, since these vectors are still linearly independent, they can still be used to generate the whole of the space. In other words, the quantum disjunctions $A \vee B \vee C$ and $D \vee E \vee F$ are identical. This seems quite contrary to intuition, which would suggest that the concept $D \vee E \vee F$ should be much more specific than the concept $A \vee B \vee C$. A practical drawback of this overgeneration is that a search engine that used quantum disjunction too liberally would be likely to generate results that would only be judged relevant by users willing for their queries to be extrapolated considerably.

There is a natural way to fix this problem in the formalism, and it bears an interesting relation to the observation that non-orthogonal vectors and subspaces give rise to subtly related non-commuting density matrices. In the diagram, the vector E lies nearly but not quite upon the line from D to F. To simplify the description of the local situation, a reasonable approximation would be to represent E by its projection onto the subspace $D \vee F$. This would amount to making the assertion "E is between D and F", which might not be *exact*, but from a human standpoint is certainly *true*. To generalize from this example, it would be reasonable to say that a vector B can be *approximately derived from* a set $A_1, \ldots, A_j$ if distance between B and the projection of B onto $A_1 \vee \ldots \vee A_j$ is small. Defining 'small' in practice is a subtle challenge, and to some extent is

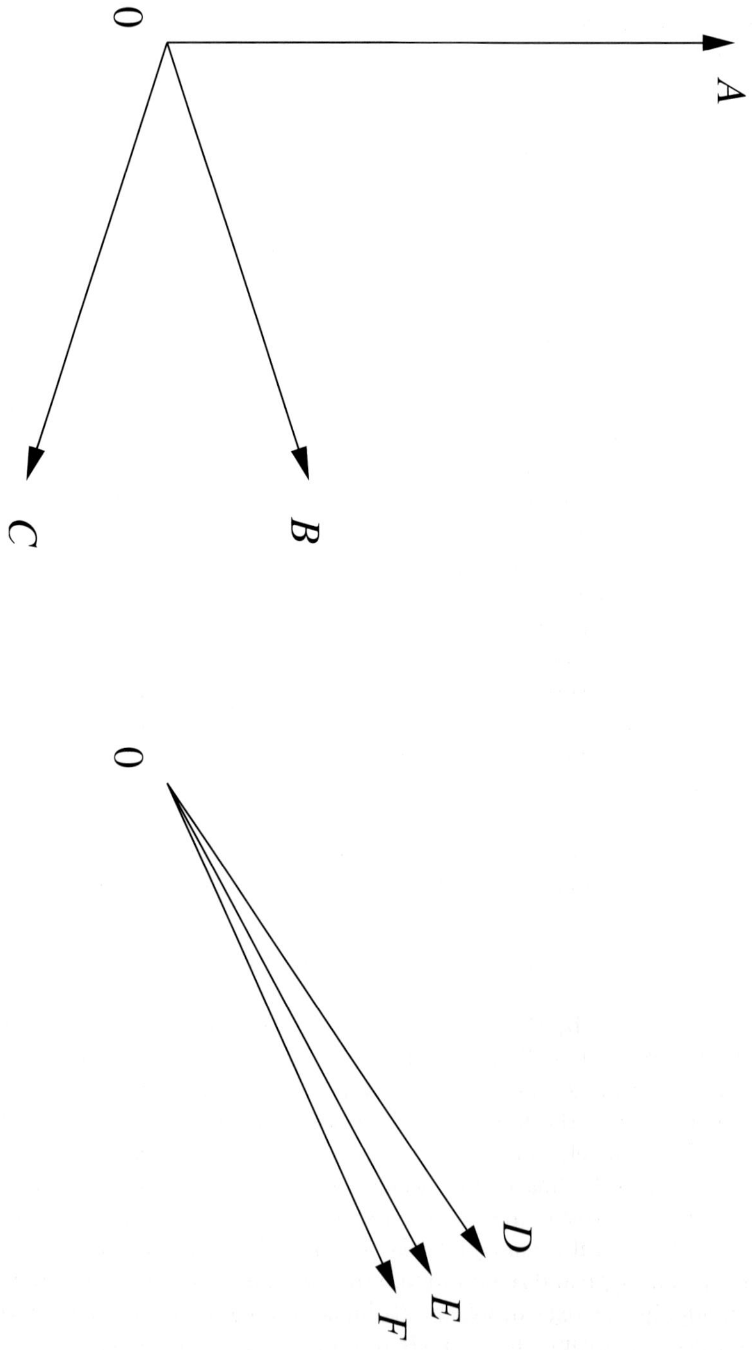

Figure 2. Orthogonal vectors in 3-space compared with 3 similar vectors.

in the eye of the beholder — to some, *Liszt* is adequately represented as being somewhere between *Beethoven* and *Brahms*, but to some, his music has special qualities independently of both.

Such a discussion suggest that one of the requirements for a working quantum logic of semantic space is the ability to model automatically the "natural dimension" of a sample of points. This problem will have many variants, and different solutions will be appropriate for different users. However, there are some general techniques such as the Isomap algorithm [Tenenbaum *et al.*, 2000] that provide dimensional decompositions of this sort, even for samples of points taken from nonlinear submanifolds of vector spaces. From a cognitive point of view, such a dimensional simplification is to be expected and indeed preferred. From microscopic observation and subsequent progress in chemistry and physics, we know that the surface of a wooden tabletop is a complex 3-dimensional structure, which may have a detailed fractal surface and according to some physical theories may consist of particles that need several more dimensions to be represented correctly. However, even to those humans who are well versed in such scientific precision, the tabletop is for all practical purposes a 2-dimensional structure, and you can certainly describe the whereabouts of any perceptual object on the table at a relevant scale of reference by giving two coordianates.

The challenges for adapting the vector space model to describing semantics and perception do not end here, of course. Many of the vector space axioms (such as the underlying assumption that vectors form a commutative group under addition) are seriously off the mark when viewed from a cognitive perspective. The purpose of this discussion is not to convince the reader that these problems have been completely solved, but that the immediate drawbacks of a naive implementation of quantum logic and concept formation in semantic space can be anticipated by a more careful consideration of the cognitive and logical goals of the system, whereupon plausible solutions can be found using existing mathematics.

6 SUMMARY AND CONCLUSIONS

A logic that is shaped by the empirical make-up of reasoning agents is subject to the same experimental challenges and limitations that affect the investigation of human subjects quite generally. The interior of the atom is, in well-known ways, difficult to access, but the interior of the reasoning agent throws up accessibility difficulties of an entirely different order. Experimental psychology, to take the most obvious example, has had to learn how to flourish despite the collapse of behaviourism and introspectivism. A great part of its success, such as it is, is owed to the skill with which it organizes its theoretical outputs around strongly plausible conjectures. In a rough and ready way, conjecturing is what one does in the absence of observation. In this regard, we are reminded of the grand conjecture with which Planck launched quantum theory itself, an idea whose immediate import in 1900 was the unification of the laws of black body radiation. In taking a quantum approach to the logic of reasoning down below, two sources of conjecture merge.

In probing the down below, we conjecture in ways that incorporate the conjectures of quantum mechanics. Of course, by now there is ample empirical confirmation of quantum theory, as well as encouraging empirical support for cognitive psychology in some of its manifestations, but neither of those desirable outcomes would have been possible without the founding conjectures around which the original theories organized themselves. It is the same way with the logic of down below. A reliable empirical understanding of it has no chance of occurring spontaneously. It must be preceded by theoretical speculation. Eddington once quipped that theories are "put-up jobs", anticipating Quine's crack that theories are "free for the thinking up". These, of course, are jokes. The fact is that the practice of scientific conjecture is respectable to the degree that it conforms to the canons of abductive logic. One of the marks of an abductively successful conjecture is its *narrative coherence* with what is known observationally about the subject in question. [Thagard, 1989],[Gabbay and Woods, 2005]. Smooth narratives identify possible scenarios. This is how we find ourselves positioned here. We have sketched what we take to be a coherent narrative of the quantum character of reasoning down below. To the extent that we have succeeded in this, we have outlined a possible theory for such reasoning. What remains now is to sort out ways in which the theory might be made responsive to observational test. Initial steps in this direction have been made in the realm of text-mining and search, a field which benefits from the comparative ease of empirical measurement and hypothesis testing. Whether the theory provides a truly useful model of cognitive processes will require different observational methods.

ACKNOWLEDGMENTS

The first author was supported in part by the Australian Research Council Discovery Project DP0773341. The third author thanks Dov Gabbay and Kent Peacock for helpful advice, and, for its financial support, the Engineering and Physical Sciences Research Council of the United Kingdom. Thanks to Ian Turner for helpful comments of a technical nature.

BIBLIOGRAPHY

[Aerts and Czachor, 2004] D. Aerts and M. Czachor. Quantum Aspects of Semantic Analysis and Symbolic Artificial Intelligence. *Journal of Physics A-Mathematical and General*, 37:L123–L132, 2004. http://uk.arxiv.org/abs/quant-ph/0309022.

[Aerts and Gabora, 2005] D. Aerts and L. Gabora. A Theory of Concepts and Their Combinations II: A Hilbert Space Representation. *Kybernetes*, 34:176–205, 2005. http://uk.arxiv.org/abs/quant-ph/0402205.

[Aerts *et al.*, 2005] D. Aerts, J. Broekaert, and L. Gabora. A Case for Applying an Abstracted Quantum Formalism to Cognition. In M.H. Bickhard and R. Campbell, editors, *Mind in Interaction*. John Benjamins: Amsterdam, 2005.

[Barndorff-Nielsen *et al.*, 2003] O.E. Barndorff-Nielsen, R.D. Gill, and P.E Jupp. On quantum statistical inference. *Journal of the Royal Statistical Society*, 65(4):775–816, 2003.

[Barwise and Seligman, 1997] J. Barwise and J. Seligman. *Information flow: The logic of distributed systems.* Cambridge University Press, 1997.

[Beer, 1995] R.D. Beer. Comptuational and dynamical languages for autonomous agents. In R. Port and T. van Gelder, editors, *Mind as Motion: Explorations in the Dynamics of Cognition*, pages 121–147. Cambridge, MA: MIT Press/Bradford Books, 1995.

[Brooks, 1991] R.A. Brooks. Intelligence without representation. *Artificial Intelligence*, 47:139–159, 1991.

[Bruza and Cole, 2005a] P. Bruza and R. Cole. Quantum Logic of Semantic Space: An Exploratory Investigation of Context Effects in Practical Reasoning . In S. Artemov, H. Barringer, A. S. d'Avila Garcez, L. C. Lamb, and J. Woods, editors, *We Will Show Them: Essays in Honour of Dov Gabbay*, volume 1, pages 339–361. London: College Publications, 2005.

[Bruza and Cole, 2005b] P.D. Bruza and R.J. Cole. A Bare Bones Approach to Literature-Based Discovery: An analysis of the Raynaud's/fish-oil and migraine-magnesium discoveries in semantic space. In A. Hoffman, H. Motoda, and T. Scheffer, editors, *Discovery Science, 8th International Conference, DS 2005, Singapore, October 8-11, 2005, Proceedings*, volume 3735 of *Lecture Notes in Artificial Intelligence*, pages 84–98. Springer, 2005.

[Bruza *et al.*, 2004] P.D. Bruza, D.W. Song, and R.M. McArthur. Abduction in semantic space: Towards a logic of discovery. *Logic Journal of the IGPL*, 12(2):97–110, 2004.

[Bruza *et al.*, 2006] P.D. Bruza, R.J. Cole, Z. Abdul Bari, and D. Song. Towards operational abduction from a cognitive perspective. In L. Magnani, editor, *Abduction and Creative Inferences in Science*, Logic Journal of the IGPL. Oxford University Press, 2006.

[Burgess *et al.*, 1998] C. Burgess, K. Livesay, and K. Lund. Explorations in context space: words, sentences, discourse. *Discourse Processes*, 25(2&3):211–257, 1998.

[Burton, 1999] Robert G. Burton. A neurocomputational approach to abduction. *Mind*, 9:257–265, 1999.

[Byron and Fuller, 1992] F.W Byron and R.W. Fuller. *The Mathematics of Classical and Quantum Physics.* Dover Publications, 1992.

[Cherry, 1966] C. Cherry. *On Human Communication.* Cambridge, MA: MIT Press, 1966.

[Churchland, 1989] Paul Churchland. *A Neurocomputational Perspective: The nature of mind and the structure of science.* Cambridge, MA: MIT Press, 1989.

[Churchland, 1995] Paul M. Churchland. *The Engine of Reason, The Seat of the Soul.* Cambridge, MA: The MIT Press, 1995.

[Clark, 1997] A. Clark. *Being There: Putting Brain, Body and World Together Again.* Cambridge, MA: MIT Press/Bradford Books, 1997.

[d'Avila Garcez and Lamb, 2004] A.S. d'Avila Garcez and L.C. Lamb. Reasoning about time and knowledge in neural-symbolic learning systems. In S. Thrum and B. Schoelkopk, editors, *Advances in Neural Information Processing Systems 16: Proceedings of the NIPS 2003 Conference*, Vancouver, BC, 2004. Cambridge, MA: MIT Press.

[d'Avila Garcez *et al.*, 2002] A.S. d'Avila Garcez, K. Broda, and Dov M. Gabbay. *Neural-Symbolic Learning Systems: Foundations and Applications.* Berlin: Springer-Verlag, 2002.

[Dunbar, 1999] K. Dunbar. How scientists build models invivo science as a window on the scientific mind. In L. Magnani, editor, *Model-Based Reasoning in Scientific Discovery*, pages 85–99. Kluwer Academic/Plenum Publishers, 1999.

[Gabbay and Woods, 2001] Dov M. Gabbay and John Woods. The new logic. *Logic Journal of the IGPL*, 9:157–190, 2001.

[Gabbay and Woods, 2002] Dov M. Gabbay and John Woods. Formal approaches to practical reasoning: A survey. In Dov M. Gabbay, Ralph H. Johnson, Hans Jürgen Ohlbach, and John Woods, editors, *Handbook of the Logic of Argument and Inference: The Turn Towards the Practical*, volume 1 of *Studies in Logic and Practical Reasoning*, pages 445–478. Amsterdam: North-Holland, 2002.

[Gabbay and Woods, 2003a] Dov M. Gabbay and John Woods. *Agenda Relevance: A Study in Formal Pragmatics*, volume 1 of *A Practical Logic of Cognitive Systems.* Amsterdam: North-Holland, 2003.

[Gabbay and Woods, 2003b] Dov M. Gabbay and John Woods. Normative models of rationality: The disutility of some approaches. *Logic Journal of IGPL*, 11:597–613, 2003.

[Gabbay and Woods, 2005] Dov M. Gabbay and John Woods. *The Reach of Abduction: Insight and Trial*, volume 2 of *A Practical Logic of Cognitive Systems.* Amsterdam: North-Holland, 2005.

[Gärdenfors, 2000] P. Gärdenfors. *Conceptual Spaces: The Geometry of Thought.* MIT Press, 2000.

[Gigerenzer and Selten, 2001a] G. Gigerenzer and R. Selten, editors. *Bounded Rationality: The Adaptive Toolbox.* Cambridge, MA: MIT Press, 2001.

[Gigerenzer and Selten, 2001b] G. Gigerenzer and R. Selten. Rethinking rationality. In *Bounded Rationality: The Adaptive Toolbox*, pages 1–12. Cambridge, MA: MIT Press, 2001.

[Globus, 1992] G. Globus. Towards a non-computational cognitive neuroscience. *Journal of Cognitive Neuroscience*, 4:299–310, 1992.

[Gray, 2002] John Gray. *Straw Dogs: Thoughts on Humans and Other Animals.* London: Granta, 2002.

[Guarini, 2001] Marcello Guarini. A defence of connectionism against the "SYNTACTIC" argument. *Synthese*, 128:287–317, 2001.

[Hamlyn, 1990] D.W. Hamlyn. *In and Out of the Black Box.* Oxford: Basil Blackwell, 1990.

[Hintikka, 1962] Jaakko Hintikka. *Knowledge and Belief.* Ithaca, NY: Cornell University Press, 1962.

[Horgan and Tienson, 1988] T. Horgan and J. Tienson. Settling into a new paradigm. *Southern Journal of Philosophy*, 26:97–113, 1988. *Connectionism and the Philosophy of Mind: Proceedings of the 1987 Spindel Conference*, special supplement.

[Horgan and Tienson, 1989] T. Horgan and J. Tienson. Representations without rules. *Philosophical Topics*, 17:147–174, 1989.

[Horgan and Tienson, 1990] T. Horgan and J. Tienson. Soft laws. In Peter A. French, Theodore E. Uehling, Jr., and Howard K. Wettstein, editors, *The Philosophy of the Human Sciences*, volume 15 of *Midwest Studies in Philosophy*, pages 256–279. Notre Dame, IN: University of Notre Dame Press, 1990.

[Horgan and Tienson, 1992] T. Horgan and J. Tienson. Cognitive systems as dynamical systems. *Topoi*, 11:27–43, 1992.

[Horgan and Tienson, 1996] Terence Horgan and John Tienson. *Connectionism and the Philosophy of Psychology.* Cambridge, MA: MIT Press, 1996.

[Horgan and Tienson, 1999a] T. Horgan and J. Tienson. Authors' replies. *Acta Analytica*, 22:275–287, 1999.

[Horgan and Tienson, 1999b] T. Horgan and J. Tienson. Short précis of *Connectionism and the Philosophy of Psychology. Acta Analytica*, 22:9–21, 1999.

[Humphreys *et al.*, 1989] M.S. Humphreys, J.D. Bain, and R. Pike. Different ways to cue a coherent memory system: A theory for episodic, semantic and procedural tasks. *Psychological Review*, 96:208–233, 1989.

[Hutchins, 1995] E. Hutchins. *Cognition in the Wild.* Cambridge, MA: MIT Press, 1995.

[Krifka *et al.*, 1995] Manfred Krifka, Francis Jeffry Pelletier, Gregory N. Carlson, Alice ter Meulen, Godehard Link, and Germano Chierchia. Genericity: An introduction. In Gregory N. Carlson and Francis Jeffry Pelletier, editors, *The Generic Book*, pages 1–124. Chicago, IL: The Univeristy of Chicago Press, 1995.

[Landauer and Dumais, 1997] T.K. Landauer and S.T. Dumais. A solution to Plato's problem: The latent semantic analysis theory of acquisition, induction and representation of knowledge. *Psychological Review*, 104:211–240, 1997.

[Landauer *et al.*, 1998] T.K. Landauer, P.W. Foltz, and D. Laham. An introduction to latent semantic analysis. *Discourse Processes*, 25(2&3):259–284, 1998.

[Levy and Bullinaria, 1999] J.P. Levy and J.A. Bullinaria. Learning lexical properties from word usage patterns: Which context words should be used? In R.F. French and J.P. Sounge, editors, *Connectionist Models of Learning, development and Evolution: Proceedings of the Sixth Neural Computation and psychology Workshop*, pages 273–282. Springer, 1999.

[Lowe, 2000] W. Lowe. What is the dimensionality of human semantic space? In *Proceedings of the 6th Neural Computation and Psychology workshop*, pages 303–311. Springer Verlag, 2000.

[Lowe, 2001] W. Lowe. Towards a theory of semantic space. In J. D. Moore and K. Stenning, editors, *Proceedings of the Twenty-Third Annual Conference of the Cognitive Science Society*, pages 576–581. Lawrence Erlbaum Associates, 2001.

[Lund and Burgess, 1996] K. Lund and C. Burgess. Producing high-dimensional semantic spaces from lexical co-occurrence. *Behavior Research Methods, Instruments & Computers*, 28(2):203–208, 1996.

[Minsky, 1975] Marvin Minsky. Frame-system theory. In R.C. Schank and B.L. Nash-Webber, editors, *Interdisciplinary Workshop on Theoretical Issues in Natural Language Processing.* Cambridge, MA: M.I.T Press, 1975. Preprints of a conference at MIT, June 1975. Reprinted in P.N. Johnson-Laird and P.C. Wason, editors. *Thinking: Readings in Cognitive Science*, Cambridge: Cambridge University Press 1977; pp. 355–376.

[Norman, 1993] D.A. Norman. *Things That Make Us Smart: Defending Human Attributes in the Age of the Machine.* Reading, MA: Addison-Wesley, 1993.

[Patel *et al.*, 1997] M. Patel, J.A. Bullinaria, and J.P. Levy. Extracting semantic representations from large text corpora. In R.F. French and J.P. Sounge, editors, *Connectionist Models of Learning, Development and Evolution: Proceedings of the Fourth Neural Computation and Psychology Workshop*, pages 199–212. Springer, 1997.

[Quine, 1960] W.V. Quine. *Word and Object.* Cambridge, MA and New York: MIT Press and John Wiley, 1960.

[Quine, 1969] W.V. Quine. Natural kinds. In Nicholas Rescher, editor, *Essays in Honor of Carl G. Hempel*, pages 5–23. Dordrecht: Reidel, 1969.

[Quine, 1970] W.V. Quine. *Philosophy of Logic.* Englewood Cliffs, NJ: Prentice-Hall, 1970.

[Rensink, 2000] R. Rensink. Visual sensing without seeing. *Psychological Science*, 15:27–32, 2000.

[Rescher, 1996] Nicholas Rescher. *Priceless Knowledge? Natural Science in Economic Perspective.* Lanham, MD: Rowman and Littlefield, 1996.

[Rosch, 1978] Eleanor Rosch. Principles of categorization. In Eleanor Rosch and B.B. Lloyd, editors, *Cognition and Categorization*, pages 27–48. Hillsdale, NJ: Erlbaum, 1978.

[Sahlgren, 2002] M. Sahlgren. Towards a Flexible Model of Word Meaning. Paper presented at the AAAI Spring Symposium 2002, March 25-27, Stanford University, Palo Alto, California, USA, 2002.

[Schank and Abelson, 1977] Roger Schank and Robert Abelson. *Scripts, Plans, Goals and Understanding: An Inquiry into Human Knowledge Structures.* Hillsdale, NJ: Lawrence Erlbaum Associates, 1977.

[Schütze, 1998] H. Schütze. Automatic word sense discrimination. *Computational Linguistics*, 24(1):97–124, 1998.

[Shannon, 1993] B. Shannon. *The Representation and the Presentational: An Essay on Cognition and the Study of Mind.* New York and London: Harvester Wheatsheaf, 1993.

[Shiffrin, 1997] Richard M. Shiffrin. Attention, automatism and consciousness. In Jonathan D. Cohen and Jonathan W. Schooler, editors, *Scientific Approaches to Consciousness*, pages 49–64. Mahwah, NJ: Erlbaum, 1997.

[Simon, 1957] H.A. Simon. *Models of Man.* New York: John Wiley, 1957.

[Smith and Medin, 1981] Edward E. Smith and Douglas L. Medin. *Categories and Concepts.* Cambridge, MA: Harvard University Press, 1981.

[Stanovich, 1999] Keith A. Stanovich. *Who is Rational? Studies of Individual Differences in Reasoning.* Mahwah, NJ: Erlbaum, 1999.

[Sterelny, 1990] K. Sterelny. *The Representation Theory of Mind.* Oxford: Blackwell, 1990.

[Tallis, 1999] Raymond Tallis. *The Explicit Animal: A Defence of Human Consciousness.* London: Macmillan and New York: Martin's Press, 2nd edition, 1999.

[Tenenbaum *et al.*, 2000] Joshua B. Tenenbaum, Vin de Silva, and John C. Langford. A global geometric framework for nonlinear dimensionality reduction. *Science*, 290(5500):2319–2323, December 2000.

[Thagard, 1989] Paul Thagard. Explanatory coherence. *Behavioral and Brain Sciences*, 12:381–433, 1989.

[Thelen and Smith, 1993] E. Thelen and I.B. Smith. *A Dynamic Systems Approach to the Development of Cognition and Action.* Cambridge, MA: MIT Press, 1993.

[van Rijsbergen, 2004] C.J. van Rijsbergen. *The Geometry of Information Retrieval.* Cambridge University Press, 2004.

[Webb, 1994] B. Webb. Robotic experiments in cricket phonotaxis. In D. Cliff, P. Husbands, J.A. Meyer, and S. Wilson, editors, *From Animals to Animats 3: Proceedings of the Third International Conference on Simulation of Adaptive Behavior*, pages 45–54. Cambridge, MA: MIT Press/Bradford Books, 1994.

[Wheeler, 1994] M. Wheeler. From activation to activity: Representation, computation and the dynamics of neural network control systems. *Artificial Intelligence and Simulation of Behaviour Quarterly*, 87:36–42, 1994.

[Wheeler, 2001] M. Wheeler. Two threats to representation. *Synthese*, 129:211–231, 2001.

[Widdows and Higgins, 2004] D. Widdows and M. Higgins. Geometric ordering of concepts, logical disjuntion, and learning by induction. In *Compositional Connectionism in Cognitive Science*, 2004. AAAI Fall Symposium Series.

[Widdows and Peters, 2003] D. Widdows and S. Peters. Word vectors and quantum logic: experiments with negation and disjunction. In *Proceedings of Mathematics of Language 8*, pages 141–154, 2003.

[Widdows, 2003] Dominic Widdows. Orthogonal negation in vector spaces for modelling word-meanings and document retrieval. In *Proceedings of the 41st Annual Meeting of the Association for Computational Linguistics (ACL)*, Sapporo, Japan, 2003.

[Wiles *et al.*, 1994] J. Wiles, G.S. Halford, J.E.M. Stewart, M.S. Humphreys, J.D. Bain, and W.H. Wilson. Tensor Models: A creative basis for memory and analogical mapping. In T. Dartnall, editor, *Artificial Intelligence and Creativity*, pages 145–159. Kluwer Academic Publishers, 1994.

[Wimsatt, 1986] W. Wimsatt. Forms of aggregativity. In A. Donagan, N. Perovich, and M. Wedin, editors, *Human Nature and Natural Knowledge*, pages 259–293. Dordrecht: Reidel, 1986.

[Wolfram, 1984] Stephen Wolfram. Computer softwear in science and mathematics. *Scientific American*, 251:188, September 1984.

[Woods, 2004] John Woods. *The Death of Argument: Fallacies in Agent-Based Reasoning.* Dordrecht and Boston: Kluwer, 2004.

[Zimmermann, 1989] Manfred Zimmermann. The nervous system and the context of information theory. In R.F. Schmidt and G. Thews, editors, *Human Physiology*, pages 166–175. Berlin: Springer-Verlag, 2nd edition, 1989. Marguerite A. Biederman-Thorson, translator.

[Zipf, 1949] George K. Zipf. *Human Behavior and the Principle of Least Effort.* Cambridge, MA: Addison Wesley, 1949.

Edited by K. Engesser, D. M. Gabbay and D. Lehmann

A COMPLETENESS THEOREM OF QUANTUM SET THEORY

Satoko Titani

1 INTRODUCTION

In [1936], Birkhoff and von Neumann introduced a new mathematical formulation of quantum physics in the language of lattice, where the structure of propositions was represented as the algebraic structure of lattices and considered as logic.

By the formulation, quantum propositions are represented as projections on a Hilbert space H, and quantum physical system is represented as complete orthomodular lattice $P(H)$ consisting of all projections on a Hilbert space H, or equivalently, consisting of all closed subspaces of H. We call the logic which corresponds to complete orthomodular lattices a *quantum logic*.

In Titani [1999], we formulated a lattice valued logic corresponding to general complete lattices. A complete lattice is an object of mathematics which is developed in the classical set theory based on the classical logic. We fix a universe V of the classical set theory, and construct a lattice valued universe V^L in V. The lattice valued logic is a logic in the universe V^L. Let $(L, \leqslant, \wedge, \vee)$ be a complete lattice in the universe V of classical set theory. The relation $\leqslant$ is represented as an operator $\rightarrow$ on L :

$$(a \rightarrow b) = \begin{cases} 1 & \text{if } a \leqslant b \\ 0 & \text{otherwise,} \end{cases} \tag{1}$$

where $1, 0$ are the top and bottom of the lattice. We introduced the logical operator $\rightarrow$ corresponding to the algebraic operator $\rightarrow$. Either algebraically or logically, the operator $\rightarrow$ is called a *basic implication.* Thus, primitive logical symbols of the lattice valued logic are $\rightarrow$, $\wedge$, $\vee$, $\neg$ $\forall x$, and $\exists x$. The completeness of the lattice valued logic was proved by Takano [Takano, 2002].

In the present paper, first we formulate a quantum logic QL as the lattice valued logic with additional logical operator $^\perp$ and logical axioms.

Takeuti proposed a quantum set theory developed in $P(H)$-valued universe $V^{P(H)}$ in [Takeuti, 1978] and [Takeuti, 1981], where he defined an implication in terms of operators $\wedge$, $\vee$, $^\perp$. We denote his implication by $\rightarrow_T$ to distinguish from the basic implication, and call it *local implication* :

$$\varphi \rightarrow_T \psi \overset{\text{def}}{\Longleftrightarrow} \varphi^\perp \vee (\psi \wedge \varphi).$$

The corresponding operator $\rightarrow_T$ on an orthomodular lattice, defined by $(a \rightarrow_T b) \overset{\text{def}}{=} a^{\perp} \vee (a \wedge b)$ is an implication in the sense that :

$$a \wedge (a \rightarrow_T b) \leqslant b. \tag{2}$$

However,

$$a \wedge b \leqslant c \Longrightarrow a \leqslant (b \rightarrow_T c) \tag{3}$$

does not hold generally. (3) holds if a, b are compatible, i.e. $a \underset{\circ}{\downarrow} b$ (cf. Definition 4). On the other hand,

$$a \wedge b \leqslant c \Longrightarrow a \leqslant (b \rightarrow c) \tag{4}$$

holds if a is global (i.e. 1 or 0). We use both of $\rightarrow$ and $\rightarrow_T$ to develop a quantum set theory. $\rightarrow$ is global and $\rightarrow_T$ is local, in the sense that:

$$(a \rightarrow b) = \bigvee \{c \in \{1, 0\} \mid a \wedge c \leqslant b\}, \quad (a \rightarrow_q b) = \bigvee \{c \mid a \underset{\circ}{\downarrow} c,\ a \wedge c \leqslant b\}.$$

REMARK 1. On a complete orthomodular lattice Q, we could interprete the basic implication of QL as $\rightarrow_z$ defined by

$$(a \rightarrow_z b) = \bigvee \{c \in Z \mid a \wedge c \leqslant b\},$$

where Z is a complete Boolean sub-algebra of the center of Q, consisting of all elements which are distributive over arbitrary joins in Q. The center of $P(H)$ is $\{0, 1\}$, and $\rightarrow_{\{0,1\}}$ coincides with our basic implication $\rightarrow$.

REMARK 2. We could formulate the quantum logic which is equivalent to QL, by introducing a modal operator $\Box$ interpreted as :

$$\Box a = \begin{cases} 1 & \text{if } a = 1 \\ 0 & \text{otherwise,} \end{cases} \tag{5}$$

as a primitive operator, instead of the basic implication $\rightarrow$. For, the basic implication $\rightarrow$ on a complete orthomodular lattice can be defined in terms of $\Box$, $\wedge$, $\vee$ and $^{\perp}$:

$$(a \rightarrow b) = \Box(a \rightarrow_T b) \ ;$$

and the modal operator $\Box$ is defined by the basic implication $\rightarrow$:

$$\Box a = (a \rightarrow a) \rightarrow a.$$

We formulated a lattice valued set theory LZFZ in [Titani, 1999] as the lattice valued logic with ZF-type non-logical axioms A1–A11 (in Theorem 24), and proved a completeness theorem of LZFZ, where " a formula φ of LZFZ is valid" means that "" $[\![\varphi]\!] = 1$ on V^L for all complete lattice L" is provable in ZFC". Thus, the completeness theorem of LZFZ asserts that:

$$\begin{aligned}&\text{ZFC} \vdash \text{“} [\![\varphi]\!] = 1 \text{ in } V^L \text{ for all complete lattice } L\text{”}\\ &\Longrightarrow \text{LZFZ} \vdash \varphi.\end{aligned}$$

In the present paper, we formulate a quantum set theory QZFZ as quantum logic QL with the nonlogical axioms A1– A11 of LZFZ, and prove the "completeness theorem" of quantum set theory QZFZ (Theorem 28) :

$$\begin{aligned}&\text{ZFC} \vdash \text{“} [\![\varphi]\!] = 1 \text{ in } V^Q \text{ for all complete orthomodular lattice } Q\text{”}\\ &\Longrightarrow \text{QZFZ} \vdash \varphi.\end{aligned}$$

We use several notations of implication. $\rightarrow$ and $\rightarrow_T$ are used as the logical and algebraic operators, $\Rightarrow$ is a formal symbol which constructs sequents. Furthermore, we use long arrow $\Longrightarrow$ as the implication in meta-language.

2 COMPLETE ORTHOMODULAR LATTICES

A logical system of quantum physics is represented as a complete lattice $P(H)$ consisting of all closed subspaces of a Hilbert space H. $P(H)$ satisfies the following conditions (C), (P) and (A).

(C) There exists a unary operation $c \mapsto c^\perp$ such that

- (C_1) $c^{\perp\perp} = c$
- (C_2) $c \vee c^\perp = 1,$ and $c \wedge c^\perp = 0$
- (C_3) $b \leqslant c \Longrightarrow c^\perp \leqslant b^\perp$ for every element b, c.

(P) If b, c are elements such that $b \leqslant c$, then the sub-lattice generated by $\{b, b^\perp, c, c^\perp\}$ is distributive.

(A) If $b \neq c$ and $b \leqslant c$, one say that c *covers* b when

$$b \leqslant x \leqslant c \Longrightarrow x = b \ \text{ or } \ x = c.$$

An element which covers 0 is called an *atom.*

- (A_1) If b is an element different from 0, there exists an atom p such that $p \leqslant b$.
- (A_2) If p is an atom and if $p \wedge b = 0$, then $p \vee b$ covers b.

DEFINITION 3. A lattice satisfying (C) is called an *ortho-lattice*. A lattice satisfying (C) and (P) is called an *orthomodular lattice*. A complete lattice satisfying (C), (P), (A) is called a *proposition system*. Top and bottom elements of a lattice will be denoted by 1 and 0, respectively.

DEFINITION 4. Elements b, c of a complete orthomodular lattice Q are said to be *compatible*, in symbols $b \underset{\circ}{\rfloor} c$, if the sublattice generated by $\{b, b^{\perp}, c, c^{\perp}\}$ is distributive. For a subset C of Q and $b \in \mathrm{Q}$,

$$b \underset{\circ}{\rfloor} C \overset{\text{def}}{\iff} \forall c \in C\, (b \underset{\circ}{\rfloor} c).$$

THEOREM 5. [Piron, 1976] *For elements b, c of a complete orthomodular lattice, the following conditions are equivalent.*

(1) *b, c are compatible*

(2) $(b \wedge c) \vee (b^{\perp} \wedge c) \vee (b \wedge c^{\perp}) \vee (b^{\perp} \wedge c^{\perp}) = 1$

(3) $(b \wedge c) \vee (b^{\perp} \wedge c) = c$

(4) $(b \vee c^{\perp}) \wedge c = b \wedge c$

THEOREM 6. [Piron, 1976] *In a complete orthomodular lattice* Q, *if $b \in \mathrm{Q}$, $C \subset \mathrm{Q}$ and $b \underset{\circ}{\rfloor} C$, then*

$$\bigvee_{c \in C} (b \wedge c) = b \wedge (\bigvee C), \qquad \bigwedge_{c \in C} (b \vee c) = b \vee (\bigwedge C).$$

THEOREM 7. [Piron, 1976] *In an orthomodular lattice* Q, *if $b \in \mathrm{Q}$, $C \subset \mathrm{Q}$ and $b \underset{\circ}{\rfloor} C$, then*

$$b \underset{\circ}{\rfloor} \bigvee C \quad \textit{and} \quad b \underset{\circ}{\rfloor} \bigwedge C.$$

DEFINITION 8. [Takeuti, 1981] In an orthomodular lattice,

$$(a \to_T b) \overset{\text{def}}{=} a^{\perp} \vee (a \wedge b).$$

THEOREM 9. *In an orthomodular lattice, if $a \underset{\circ}{\rfloor} c$, then*

$$c \leqslant (a \to_T b) \iff a \wedge c \leqslant b.$$

Proof. Assume $a \mathrel{\underset{\circ}{\rfloor}} c$. Then $c = (a^{\perp} \wedge c) \vee (a \wedge c)$. Since $a \mathrel{\underset{\circ}{\rfloor}} a^{\perp}$ and $a \mathrel{\underset{\circ}{\rfloor}} (a \wedge b)$,

$$\begin{aligned} c \leqslant (a \rightarrow_T b) &\implies a \wedge c \leqslant a \wedge (a^{\perp} \vee (a \wedge b)) \\ &\implies a \wedge c \leqslant (a \wedge a^{\perp}) \vee (a \wedge b)) \leqslant b \\ a \wedge c \leqslant b &\implies a \wedge c \leqslant a \wedge b \\ &\implies c = (a^{\perp} \wedge c) \vee (a \wedge c) \leqslant a^{\perp} \vee (a \wedge b) \end{aligned}$$

■

On a complete orthomodular lattice, operations $\rightarrow$, $\neg$, $\Box$ and $\Diamond$ are defined by

$$(a \rightarrow b) \stackrel{\text{def}}{=} \bigvee \{ x \in \mathbf{2} \mid a \wedge x \leqslant b \} = \begin{cases} 1, & \text{if } a \leqslant b, \\ 0, & \text{otherwise,} \end{cases}$$

$$\neg a \stackrel{\text{def}}{=} (a \rightarrow 0) = \bigvee \{ x \in \mathbf{2} \mid a \wedge x \leqslant 0 \} = \begin{cases} 1, & \text{if } a = 0, \\ 0, & \text{otherwise.} \end{cases}$$

$$\Box a \stackrel{\text{def}}{=} ((a \rightarrow a) \rightarrow a) = (1 \rightarrow a) = \begin{cases} 1 & \text{if } a = 1 \\ 0 & \text{otherwise.} \end{cases}$$

$$\Diamond a \stackrel{\text{def}}{=} \neg \Box \neg a = \begin{cases} 1 & \text{if } a \neq 0 \\ 0 & \text{otherwise.} \end{cases}$$

As immediate consequents of the definitions, we have:

THEOREM 10. *In a complete orthomodular lattice,*

(1) $\neg 0 = 1$; $\neg 1 = 0$

(2) $a \wedge \neg a = 0$; $a \leqslant \neg\neg a$

(3) $(a \rightarrow b) \leqslant (\neg b \rightarrow \neg a)$

(4) $\neg(a \vee b) = \neg a \wedge \neg b$; $\neg a \vee \neg b \leqslant \neg(a \wedge b)$

(5) $(a \vee b)^{\perp} = a^{\perp} \wedge b^{\perp}$; $(a \wedge b)^{\perp} = a^{\perp} \vee b^{\perp}$

(6) $(\Box a)^{\perp} = \neg(\Box a)$

(7) $\Box a \leqslant a$; $\Box\Box a = \Box a$

(8) $\neg a = \Box \neg a$

(9) *If* $\Box a \leqslant b$, *then* $\Box a \leqslant \Box b$

(10) $a \leqslant \Diamond a$

(11) *If* $a \leqslant \Box b$ *then* $\Diamond a \leqslant \Box b$

(12) $\Box a \wedge \bigvee_k b_k = \bigvee_k (\Box a \wedge b_k)$; $a \wedge \bigvee_k \Box b_k = \bigvee_k (a \wedge \Box b_k)$

(13) $\Box a \vee \bigwedge_k b_k = \bigwedge_k (\Box a \vee b_k)$; $a \vee \bigwedge_k \Box b_k = \bigwedge_k (a \vee \Box b_k)$

(14) $\Box a \vee \neg \Box a = 1$

(15) *If* $a \wedge \Box c \leqslant b$, *then* $\neg b \wedge \Box c \leqslant \neg a$.

(16) $(a \to b) = \bigvee \{c \in \mathrm{Q} \mid c = \Box c,\ a \wedge c \leqslant b\}$

(17) $\Box(\bigwedge_k a_k) = \bigwedge_k \Box a_k$

(18) $\bigwedge_k \Box a_k = \Box \bigwedge_k \Box a_k$

(19) $\bigvee_k \Box a_k = \Box \bigvee_k \Box a_k$

(20) $\Box(a \to b) = (a \to b)$.

(21) *If* $\Box a \wedge b \leqslant c$ *then* $\Box a \leqslant (b \to c)$

(22) $((\Box a \wedge b) \to c) = (\Box a \to (b \to c))$

(23) $\Diamond \bigvee_k a_k = \bigvee_k \Diamond a_k$

(24) $\Diamond(\Box a \wedge b) = \Box a \wedge \Diamond b$

3 QUANTUM LOGIC QL

Quantum logic QL is a logic representing the structure of complete orthomodular lattice. QL is formulated as a lattice valued logic with an additional logical operator $^\perp$ and additional logical axioms. Here, we consider only $=$ and $\in$ as predicate symbols.

3.1 Language

Primitive symbols are :

(1) free variables : $a_1, a_2, \cdots$

(2) bound variables : $x_1, x_2, \cdots$

(3) constants : $c_1, c_2, \cdots$

(4) predicate symbols : $=$, $\in$,

(5) logical symbols : $\wedge$, $\vee$, $\to$, $\neg$, $^\perp$, $\forall$, $\exists$.

(6) parentheses : (,), [,].

Free variables and constants are called *terms*, and denoted by $t_1, t_2, \cdots$. *Atomic formula* of QL are expressions of the form $t_1 = t_2$ or $t_1 \in t_2$ with terms t_1, t_2. *Formulas* of QL are constructed from atomic formulas, by using the logical symbols. To denote formulas, we use

$$\varphi, \psi, \xi, \cdots, \ \varphi(a), \cdots .$$

DEFINITION 11. $\Box$-*closed formulas* are defined inductively :

(1) A formula of the form $\varphi \to \psi$ or $\neg\varphi$ is $\Box$-closed.

(2) If formulas φ and ψ are $\Box$-closed, then $\varphi \wedge \psi$, $\varphi \vee \psi$ and $\varphi^{\perp}$ are $\Box$-closed.

(3) If a formula $\varphi(a)$ is a $\Box$-closed formula with free variable a, then $\forall x\varphi(x)$ and $\exists x\varphi(x)$ are $\Box$-closed.

(4) $\Box$-closed formulas are only those obtained by (1)–(3).

$\Gamma, \Delta, \Pi, \Lambda, \cdots$ will be used to denote finite sequences of formulas ; $\overline{\varphi}, \overline{\psi}, \cdots$ to denote $\Box$-closed formulas ; and $\overline{\Gamma}, \overline{\Delta}, \overline{\Pi}, \overline{\Lambda}, \cdots$ to denote finite sequences of $\Box$-closed formulas.

A formal expression of the form $\Gamma \Rightarrow \Delta$ is called a *sequent*.

3.2 Inference rules

Beginning sequents of QL

Every proof starts with sequents of the following form (E), (C) or (P), which are called *logical axioms*.

E : $\varphi \Rightarrow \varphi$

C (Orthogonal complements):

C1 $\varphi \Rightarrow \varphi^{\perp\perp}, \quad \varphi^{\perp\perp} \Rightarrow \varphi$

C2 $\Rightarrow \varphi \vee \varphi^{\perp}, \quad \varphi \wedge \varphi^{\perp} \Rightarrow$

C3 $(\varphi \to \psi) \Rightarrow (\psi^{\perp} \to \varphi^{\perp})$

P (Orthomodularity): $(\varphi \to \psi), \psi \Rightarrow \psi \wedge \varphi, \psi \wedge \varphi^{\perp}$.

Structural rules:

Thinning : $\dfrac{\Gamma \Rightarrow \Delta}{\varphi, \Gamma \Rightarrow \Delta}$ $\dfrac{\Gamma \Rightarrow \Delta}{\Gamma \Rightarrow \Delta, \varphi}$

Contraction : $\dfrac{\varphi, \varphi, \Gamma \Rightarrow \Delta}{\varphi, \Gamma \Rightarrow \Delta}$ $\dfrac{\Gamma \Rightarrow \Delta, \varphi, \varphi}{\Gamma \Rightarrow \Delta, \varphi}$

Interchange : $\dfrac{\Gamma, \varphi, \psi, \Pi \Rightarrow \Delta}{\Gamma, \psi, \varphi, \Pi \Rightarrow \Delta}$ $\dfrac{\Gamma \Rightarrow \Delta, \varphi, \psi, \Lambda}{\Gamma \Rightarrow \Delta, \psi, \varphi, \Lambda}$

Cut : $\dfrac{\Gamma \Rightarrow \overline{\Delta}, \varphi \quad \varphi, \Pi \Rightarrow \Lambda}{\Gamma, \Pi \Rightarrow \overline{\Delta}, \Lambda}$ $\dfrac{\Gamma \Rightarrow \Delta, \varphi \quad \varphi, \overline{\Pi} \Rightarrow \Lambda}{\Gamma, \overline{\Pi} \Rightarrow \Delta, \Lambda}$

$$\dfrac{\Gamma \Rightarrow \Delta, \overline{\varphi} \quad \overline{\varphi}, \Pi \Rightarrow \Lambda}{\Gamma, \Pi \Rightarrow \Delta, \Lambda}$$

Logical rules:

$\neg$: $\dfrac{\Gamma \Rightarrow \overline{\Delta}, \varphi}{\neg\varphi, \Gamma \Rightarrow \overline{\Delta}}$ $\dfrac{\Gamma \Rightarrow \Delta, \overline{\varphi}}{\neg\overline{\varphi}, \Gamma \Rightarrow \Delta}$ $\dfrac{\varphi, \overline{\Gamma} \Rightarrow \overline{\Delta}}{\overline{\Gamma} \Rightarrow \overline{\Delta}, \neg\varphi}$ $\dfrac{\overline{\varphi}, \Gamma \Rightarrow \Delta}{\Gamma \Rightarrow \Delta, \neg\overline{\varphi}}$

$\wedge$: $\dfrac{\varphi, \Gamma \Rightarrow \Delta}{\varphi \wedge \psi, \Gamma \Rightarrow \Delta}$ $\dfrac{\Gamma \Rightarrow \overline{\Delta}, \varphi \quad \Gamma \Rightarrow \overline{\Delta}, \psi}{\Gamma \Rightarrow \overline{\Delta}, \varphi \wedge \psi}$

$\dfrac{\psi, \Gamma \Rightarrow \Delta}{\varphi \wedge \psi, \Gamma \Rightarrow \Delta}$ $\dfrac{\Gamma \Rightarrow \Delta, \overline{\varphi} \quad \Gamma \Rightarrow \Delta, \overline{\psi}}{\Gamma \Rightarrow \Delta, \overline{\varphi} \wedge \overline{\psi}}$

$\vee$: $\dfrac{\varphi, \overline{\Gamma} \Rightarrow \Delta \quad \psi, \overline{\Gamma} \Rightarrow \Delta}{\varphi \vee \psi, \overline{\Gamma} \Rightarrow \Delta}$ $\dfrac{\Gamma \Rightarrow \Delta, \varphi}{\Gamma \Rightarrow \Delta, \varphi \vee \psi}$

$\dfrac{\overline{\varphi}, \Gamma \Rightarrow \Delta \quad \overline{\psi}, \Gamma \Rightarrow \Delta}{\overline{\varphi} \vee \overline{\psi}, \Gamma \Rightarrow \Delta}$ $\dfrac{\Gamma \Rightarrow \Delta, \psi}{\Gamma \Rightarrow \Delta, \varphi \vee \psi}$

$\rightarrow$: $\dfrac{\Gamma \Rightarrow \overline{\Delta}, \varphi \quad \psi, \overline{\Pi} \Rightarrow \Lambda}{(\varphi \rightarrow \psi), \Gamma, \overline{\Pi} \Rightarrow \overline{\Delta}, \Lambda}$ $\dfrac{\varphi, \overline{\Gamma} \Rightarrow \overline{\Delta}, \psi}{\overline{\Gamma} \Rightarrow \overline{\Delta}, (\varphi \rightarrow \psi)}$ $\dfrac{\overline{\varphi}, \Gamma \Rightarrow \Delta, \overline{\psi}}{\Gamma \Rightarrow \Delta, (\overline{\varphi} \rightarrow \overline{\psi})}$

$\forall$:
$$\frac{\varphi(t), \Gamma \Rightarrow \Delta}{\forall x \varphi(x), \Gamma \Rightarrow \Delta}$$
where t is any term

$$\frac{\Gamma \Rightarrow \overline{\Delta}, \varphi(a)}{\Gamma \Rightarrow \overline{\Delta}, \forall x \varphi(x)} \qquad \frac{\Gamma \Rightarrow \Delta, \overline{\varphi}(a)}{\Gamma \Rightarrow \Delta, \forall x \overline{\varphi}(x)}$$
where a is a free variable which does not occur in the lower sequent.

$\exists$:
$$\frac{\varphi(a), \overline{\Gamma} \Rightarrow \Delta}{\exists x \varphi(x), \overline{\Gamma} \Rightarrow \Delta} \qquad \frac{\overline{\varphi}(a), \Gamma \Rightarrow \Delta}{\exists x \overline{\varphi}(x), \Gamma \Rightarrow \Delta}$$
where a is a free variable which does not occur in the lower sequent.

$$\frac{\Gamma \Rightarrow \Delta, \varphi(t)}{\Gamma \Rightarrow \Delta, \exists x \varphi(x)}$$
where t is any term

We use the following abbreviations :

$$\Box\varphi \overset{\text{def}}{\Longleftrightarrow} (\varphi \to \varphi) \to \varphi, \qquad \Diamond\varphi \overset{\text{def}}{\Longleftrightarrow} \neg\Box\neg\varphi,$$

$$\varphi \leftrightarrow \psi \overset{\text{def}}{\Longleftrightarrow} (\varphi \to \psi) \wedge (\psi \to \varphi).$$

If $\Gamma \Rightarrow \Delta$ is provable in QL, then we write

$$\text{QL} \vdash \Gamma \Rightarrow \Delta,$$

where QL may be omitted if it is obvious. $\vdash \varphi \Leftrightarrow \psi$ means that "$\vdash \varphi \Rightarrow \psi$ and $\vdash \psi \Rightarrow \varphi$".

REMARK 12. The quantum logic is also formulated as a sequential system, using the compatibility instead of $\Box$-closedness, by Kodera [Kodera and Titani, submitted].

THEOREM 13. *For formulas of* φ, ψ, ξ *of* QL,

(1) $\vdash \varphi \wedge \psi, \Gamma \Rightarrow \Delta$ *if and only if* $\vdash \varphi, \psi, \Gamma \Rightarrow \Delta$

(2) $\vdash \Gamma \Rightarrow \Delta, \varphi \vee \psi$ *if and only if* $\vdash \Gamma \Rightarrow \Delta, \varphi, \psi$

(3) $\vdash \Box\varphi \Rightarrow \varphi$

(4) $\vdash \varphi \Rightarrow \Diamond\varphi$

(5) $\vdash \overline{\Gamma} \Rightarrow \overline{\Delta}, \varphi$ *if and only if* $\vdash \overline{\Gamma} \Rightarrow \overline{\Delta}, \Box\varphi$

(6) *If* φ *is* $\Box$*-closed, then* $\vdash \varphi \Leftrightarrow \Box\varphi$

(7) $\vdash \varphi, \overline{\Gamma} \Rightarrow \overline{\Delta}$ *if and only if* $\vdash \Diamond\varphi, \overline{\Gamma} \Rightarrow \overline{\Delta}$

(8) $\vdash \varphi \wedge (\varphi \to \psi) \Rightarrow \psi$

(9) $\vdash \varphi \wedge \neg\varphi \Rightarrow$

(10) $\vdash \varphi, \overline{\Gamma} \Rightarrow \overline{\Delta}, \psi$ *implies* $\vdash \neg\psi, \overline{\Gamma} \Rightarrow \overline{\Delta}, \neg\varphi$

(11) $\vdash \varphi \Rightarrow \neg\neg\varphi$; $\vdash \Box\varphi \Leftrightarrow \neg\neg\Box\varphi$;

(12) $\vdash \neg(\varphi \vee \psi) \Leftrightarrow (\neg\varphi \wedge \neg\psi)$

(13) $\vdash (\neg\varphi \vee \neg\psi) \Rightarrow \neg(\varphi \wedge \psi)$

(14) $\vdash (\Box\varphi \rightarrow \psi) \Leftrightarrow (\Box\varphi \rightarrow \Box\psi) \Leftrightarrow (\neg\Box\varphi \vee \Box\psi)$

(15) $\vdash \Box\varphi \wedge \exists x\psi(x) \Leftrightarrow \exists x(\Box\varphi \wedge \psi(x))$;
$\vdash \varphi \wedge \exists x\Box\psi(x) \Leftrightarrow \exists x(\varphi \wedge \Box\psi(x))$

(16) $\vdash \quad \Rightarrow \Box\varphi \vee \neg\Box\varphi$

(17) $\vdash [(\varphi \wedge \Box\xi) \rightarrow \psi] \Rightarrow [(\neg\psi \wedge \Box\xi) \rightarrow \neg\varphi]$

(18) $\vdash (\Box\varphi)^{\perp} \Leftrightarrow \neg\Box\varphi$

(19) $\vdash (\varphi \rightarrow \Box\psi) \Rightarrow (\Diamond\varphi \rightarrow \Box\psi)$

(20) $\vdash \Diamond(\Box\varphi \wedge \psi) \Rightarrow \Box\varphi \wedge \Diamond\psi$

(21) $\vdash \forall x\Box\varphi(x) \Leftrightarrow \Box\forall x\varphi(x)$

(22) $\vdash \exists x\Diamond\varphi(x) \Leftrightarrow \Diamond\exists x\varphi(x)$

(23) *If* $A(\varphi)$ *is a formula with subformula* φ *and* $\vdash \varphi \Leftrightarrow \psi$, *then*

$$\vdash A(\varphi) \Leftrightarrow A(\psi).$$

Proofs are in Appendix A.

4 QUANTUM SET THEORY QZFZ

4.1 Q-valued universe V^{Q}

In what follows, Q denotes a complete orthomodular lattice unless otherwise mentioned. Q-valued universe V^{Q} is constructed by transfinite induction :

$$\begin{aligned} V_\alpha^{\mathrm{Q}} &= \{u \mid \exists\beta<\alpha\, \exists\mathcal{D}u \subset V_\beta^{\mathrm{Q}}(u : \mathcal{D}u \rightarrow \mathrm{Q})\}, \\ V^{\mathrm{Q}} &= \bigcup_{\alpha\in On} V_\alpha^{\mathrm{Q}}. \end{aligned}$$

The least α such that $u \in V_\alpha^{\mathrm{Q}}$ is called the *rank* of u.

Formulas of quantum set theory are constructed from atomic formulas of the form $t_1 = t_2$ or $t_1 \in t_2$ by logical operators $\wedge$, $\vee$, $\neg$, $^\perp$, $\rightarrow$, $\forall$, $\exists$. Atomic formulas are interpreted in V^{Q} as

$$\begin{aligned} [\![u = v]\!] &= \bigwedge_{x \in \mathcal{D}u} (u(x) \rightarrow [\![x \in v]\!]) \wedge \bigwedge_{x \in \mathcal{D}v} (v(x) \rightarrow [\![x \in u]\!]), \\ [\![u \in v]\!] &= \bigvee_{x \in \mathcal{D}v} (v(x) \wedge [\![u = x]\!]). \end{aligned}$$

Logical operators $\wedge$, $\vee$, $\neg$, $\rightarrow$, $\forall x$, $\exists x$ and $^\perp$ are interpreted as the corresponding operators on Q, and

$$[\![\Box\varphi]\!] = [\![(\varphi \rightarrow \varphi) \rightarrow \varphi]\!] = \Box[\![\varphi]\!] = \begin{cases} 1 & \text{if } [\![\varphi]\!] = 1 \\ 0 & \text{otherwise.} \end{cases}$$

We say an element p of Q is $\Box$-*closed* if $p = \Box p$. Obviously we have:

LEMMA 14. [Titani, 1999] *For a formula φ of* QZFZ *and $u, v \in V^{\mathrm{Q}}$,*

(1) *If φ is $\Box$-closed then $[\![\varphi]\!]$ is $\Box$-closed in* Q.

(2) $[\![u = v]\!]$ *is $\Box$-closed.*

(3) $[\![u = v]\!]$ *is distributive over arbitrary join in* Q :

$$(\bigvee_k b_k) \wedge [\![u = v]\!] = \bigvee_k (b_k \wedge [\![u = v]\!]).$$

(4) $[\![u = v]\!] = [\![v = u]\!]$

(5) *If $x \in \mathcal{D}u$ then $u(x) \leqslant [\![x \in u]\!]$.*

(6) $[\![u = u]\!] = 1$

LEMMA 15. *For $u, v, w \in V^{\mathrm{Q}}$,*

(1) $[\![u = v \wedge v = w]\!] \leqslant [\![u = w]\!]$

(2) $[\![u = v \wedge v \in w]\!] \leqslant [\![u \in w]\!]$

(3) $[\![u = v \wedge w \in v]\!] \leqslant [\![w \in u]\!]$

Proof. (1) We proceed by induction. Assume that $u, v, w \in V^{\mathrm{Q}}_\alpha$. If $x \in \mathcal{D}u$ and $y \in \mathcal{D}v$, then $x, y \in V^{Q}_{<\alpha}$ and

$$[\![u = v]\!] \wedge u(x) \leqslant (u(x) \rightarrow [\![x \in v]\!]) \wedge u(x) \leqslant [\![x \in v]\!].$$

Hence, by using Lemma 14,

$$\begin{aligned}
[\![u=v \wedge v=w]\!] \wedge u(x) &\leqslant [\![v=w]\!] \wedge \bigvee_{y\in\mathcal{D}v} [\![x=y]\!] \wedge v(y) \\
&\leqslant \bigvee_{y\in\mathcal{D}v} [\![x=y]\!] \wedge [\![v=w]\!] \wedge v(y) \\
&\leqslant \bigvee_{y\in\mathcal{D}v} [\![x=y]\!] \wedge \bigvee_{z\in\mathcal{D}w} ([\![y=z]\!] \wedge w(z)) \\
&\leqslant \bigvee_{y\in\mathcal{D}v} \bigvee_{z\in\mathcal{D}w} [\![x=y \wedge y=z]\!] \wedge w(z).
\end{aligned}$$

By induction hypothesis,

$$\begin{aligned}
&\leqslant \bigvee_{z\in\mathcal{D}w} [\![x=z]\!] \wedge w(z) \\
&\leqslant [\![x\in w]\!].
\end{aligned}$$

Since $[\![u = v \wedge v = w]\!]$ is $\Box$-closed, $[\![u = v \wedge v = w]\!] \leqslant \bigwedge_{x\in\mathcal{D}u}(u(x) \to [\![x \in w]\!])$. Similarly, we have $[\![u=v \wedge v=w]\!] \leqslant \bigwedge_{z\in\mathcal{D}w}(w(z) \to [\![z\in u]\!])$. Hence,

$$[\![u=v \wedge v=w]\!] \leqslant [\![u=w]\!].$$

(2) and (3) follows from Lemma 14 and (1). ■

As imediate consequents of Lemma 15, we have:

THEOREM 16. *For* $u, v, w \in V^{\mathrm{Q}}$,

(1) $[\![u=v]\!] \leqslant [\![v=w \leftrightarrow u=w]\!]$

(2) $[\![u=v]\!] \leqslant [\![v\in w \leftrightarrow u\in w]\!]$

(3) $[\![u=v]\!] \leqslant [\![w\in v \leftrightarrow w\in u]\!]$

(4) $[\![u=v]\!] \leqslant [\![\varphi(u) \leftrightarrow \varphi(v)]\!]$ *for a formula* $\varphi(a)$.

DEFINITION 17. $\forall x\in u\varphi(x)$ and $\exists x\in u\varphi(x)$ are usual abbreviations, i.e.

$$\forall x\in u\,\varphi(x) \overset{\text{def}}{\iff} \forall x(x\in u \to \varphi(x)), \quad \exists x\in u\,\varphi(x) \overset{\text{def}}{\iff} \exists x(x\in u \wedge \varphi(x)).$$

THEOREM 18. *For a formula* $\varphi(a)$ *and* $u\in V^{\mathrm{Q}}$,

(1) $[\![\forall x\in u\varphi(x)]\!] = \bigwedge_{x\in\mathcal{D}u}[\![x\in u \to \varphi(x)]\!]$

(2) $[\![\exists x\in u\varphi(x)]\!] = \bigvee_{x\in\mathcal{D}u}[\![x \in u \wedge \varphi(x)]\!]$

Proof.

(1) $[\![\forall x(x \in u \to \varphi(x))]\!] \leqslant \bigwedge_{x \in \mathcal{D}u} [\![x \in u \to \varphi(x)]\!]$ is obvious.

Proof of ($\geqslant$) : Let $y \in V^{\mathrm{Q}}$. $\bigwedge_{x \in \mathcal{D}u} [\![x \in u \to \varphi(x)]\!]$ is $\square$-closed, and

$$
\begin{aligned}
(\bigwedge_{x \in \mathcal{D}u} [\![x \in u \to \varphi(x)]\!]) \wedge [\![y \in u]\!] \\
&= (\bigwedge_{x \in \mathcal{D}u} [\![x \in u \to \varphi(x)]\!]) \wedge (\bigvee_{x' \in \mathcal{D}u} [\![y = x']\!] \wedge u(x')) \\
&\leqslant \bigvee_{x' \in \mathcal{D}u} ([\![x' \in u \to \varphi(x')]\!] \wedge [\![y = x']\!] \wedge [\![x' \in u]\!]) \\
&\leqslant [\![\varphi(y)]\!]
\end{aligned}
$$

Hence,

$$\bigwedge_{x \in \mathcal{D}u} [\![x \in u \to \varphi(x)]\!]) \leqslant [\![\forall x(x \in u \to \varphi(x))]\!].$$

(2) Since $[\![x \in u]\!] \leqslant \bigvee_{x' \in \mathcal{D}u} [\![x = x']\!]$ and $[\![x = x']\!]$ is $\square$-closed for $x' \in \mathcal{D}u$,

$$
\begin{aligned}
[\![\exists x(x \in u \wedge \varphi(x))]\!] &\leqslant \bigvee_{x \in V^{\mathrm{Q}}} \bigvee_{x' \in \mathcal{D}u} ([\![x = x']\!] \wedge [\![x \in u \wedge \varphi(x)]\!]) \\
&\leqslant \bigvee_{x' \in \mathcal{D}u} [\![x' \in u \wedge \varphi(x')]\!].
\end{aligned}
$$

■

DEFINITION 19. $u \in V^{\mathrm{Q}}$ is said to be *definite* if $u(x) = [\![x \in u]\!]$ for all $x \in \mathcal{D}u$.

THEOREM 20. *For any* $u \in V^{\mathrm{Q}}$ *there exists a definte* $v \in V^{\mathrm{Q}}$ *such that*

$$[\![u = v]\!] = 1.$$

Proof. For $u \in V^{\mathrm{Q}}$, let

$$\mathcal{D}v = \mathcal{D}u, \quad v(x) = [\![x \in u]\!].$$

Then v is definte and $[\![u = v]\!] = 1$. ■

In what follows we may assume that each $u \in V^{\mathrm{Q}}$ is definite. For a formula $\varphi(a)$ and definite $u \in V^{\mathrm{Q}}$,

(1) $[\![\forall x \in u \varphi(x)]\!] = \bigwedge_{x \in \mathcal{D}u} (u(x) \to [\![\varphi(x)]\!])$

(2) $[\![\exists x \in u\varphi(x)]\!] = \bigvee_{x \in \mathcal{D}u} u(x) \wedge [\![\varphi(x)]\!]$

For convenience, we write $x \overset{\Box}{\in} y$ instead for $\Box(x \in y)$:

$$x \overset{\Box}{\in} y \overset{\text{def}}{\Longleftrightarrow} \Box(x \in y)$$

DEFINITION 21. x is said to be *global* if $\forall t \in x(t \overset{\Box}{\in} x)$, in symbols $\mathrm{Gl}(x)$:

$$\mathrm{Gl}(x) \overset{\text{def}}{\Longleftrightarrow} \forall t \in x(t \overset{\Box}{\in} x)$$

Let V be a universe of ZFC in which Q-valued universe V^{Q} is constructed. Sublattice $\mathbf{2}$ ($= \{1, 0\}$) of Q is a Boolean algebra, and the universe V is isomorphic to $V^{\mathbf{2}}$.

DEFINITION 22. Elements of the sub-universe $V^{\mathbf{2}} \subset V^{\mathrm{Q}}$ are called *check sets* in V^{Q}.

If u is a set in V, $\check{u} \in V^{\mathrm{Q}}$ defined by

$$\begin{cases} \mathcal{D}\check{u} = \{\check{x} \mid x \in u\} \\ \check{u}(\check{x}) = 1 \end{cases}$$

is a check set. "x is a check set ", in symbols $\mathrm{ck}(x)$, can be expressed in the language of quantum set theory by $\overset{\Box}{\in}$-recursion (Theorem 37), as

$$\mathrm{ck}(x) \Longleftrightarrow \forall t[t \in x \leftrightarrow (t \overset{\Box}{\in} x) \wedge \mathrm{ck}(t)].$$

Embedding $\mathrm{I} : \mathbf{2} \to \mathrm{Q}$ induces an embedding $V \to V^{\mathrm{Q}}$ by

$$u \mapsto \check{u} \in V^{\mathbf{2}} \subset V^{\mathrm{Q}} \quad (u \in V).$$

Then, obvioully

$$u = v \Longleftrightarrow [\![\check{u} = \check{v}]\!] = 1, \quad u \in v \Longleftrightarrow [\![\check{u} \in \check{v}]\!] = 1,$$

and generally, we have the following theorem.

THEOREM 23. *If $\varphi(u_1, \cdots, u_n)$ is a bounded sentence of* ZFC *with constants $u_1, \cdots, u_n$ in V, then*

$$\varphi(u_1, \cdots, u_n) \textit{ holds in } V \Longleftrightarrow [\![\varphi(\check{u}_1, \cdots, \check{u}_n)]\!] = 1 \textit{ in } V^{\mathrm{Q}}.$$

4.2 Non-logical axioms of QZFZ

For any complete orthomodular lattice Q, $(V^{\mathrm{Q}}, [\![\ \ ]\!])$ is a model of lattice valued set theory in [Titani, 1999], and also a model of quantum logic QL, since the logical axioms (C) and (P) of QL represent the axioms (C) and (P) of orthomodular lattice. Thus, we adopt the quantum logic QL and the non-logical axioms A1–A11 of lattice valued set theory as those of quantum set theory.

THEOREM 24. [Titani, 1999] *The following A1–A11 are valid, that is,* $[\![\mathrm{A1}]\!] = \cdots = [\![\mathrm{A11}]\!] = 1$ *have truth value 1 in* $(V^{\mathrm{Q}}, [\![\ \ ]\!])$ *for any complete orthomodular lattice* Q.

A1 (Equality) $\forall u \forall v\, [u = v \wedge \varphi(u) \rightarrow \varphi(v)]$.

A2 (Extensionality) $\forall u, v\, [\forall x (x \in u \leftrightarrow x \in v) \rightarrow u = v]$.

A3 (Pairing) $\forall u, v \exists z \forall x (x \in z \leftrightarrow (x = u \vee x = v))$.

The set z *satisfying* $\forall x[x \in z \leftrightarrow (x = u \vee x = v)]$ *is denoted by* $\{u, v\}$.

A4 (Union) $\forall u \exists z \forall x [x \in z \leftrightarrow \exists y \in u (x \in y)]$.

The set z *satisfying* $\forall x[x \in z \leftrightarrow \exists y \in u (x \in y)]$ *is denoted by* $\bigcup u$.

A5 (Power set) $\forall u \exists z \forall x (x \in z \leftrightarrow x \subset u)$, *where* $x \subset u \overset{def}{\Longleftrightarrow} \forall y (y \in x \rightarrow y \in u)$.

The set z *satisfying* $\forall x (x \in z \leftrightarrow x \subset u)$ *is denoted by* $\mathcal{P}(u)$.

A6 (Infinity) $\exists u \left[\exists x (x \overset{\square}{\in} u) \wedge \forall x \overset{\square}{\in} u \exists y \overset{\square}{\in} u (x \overset{\square}{\in} y) \right]$.

A7 (Separation) $\forall u \exists v \forall x [x \in v \leftrightarrow x \in u \wedge \varphi(x)]$.

The set v *satisfying* $\forall x[x \in v \leftrightarrow x \in u \wedge \varphi(x)]$ *is denoted by* $\{x \in u \mid \varphi(x)\}$.

A8 (Collection) $\forall u \exists v \left[\forall x \in u \exists y \varphi(x, y) \rightarrow \forall x \in u \exists y \overset{\square}{\in} v \varphi(x, y) \right]$.

A9 ($\in$-induction) $\forall x\, [\forall y \in x \varphi(y)) \rightarrow \varphi(x)] \rightarrow \forall x \varphi(x)$.

A10 (Zorn) $\exists x (x \in u) \wedge \mathrm{Gl}(u) \wedge \forall v\, [Chain(v, u) \rightarrow \bigcup v \in u] \rightarrow \exists z Max(z, u)$,

$$\text{where } Chain(v, u) \overset{def}{\Longleftrightarrow} v \subset u \wedge \forall x, y \in v (x \subset y \vee y \subset x),$$

$$Max(z, u) \overset{def}{\Longleftrightarrow} z \in u \wedge \forall x \in u (z \subset x \rightarrow z = x).$$

A11 ($\Diamond$) $\forall u \exists z \forall t\,[t \in z \leftrightarrow\ \Diamond(t \in u)\,]$, *where* $\Diamond\varphi \overset{def}{\Longleftrightarrow} \neg\Box\neg\varphi$.

The set z *satisfying* $\forall t[t \in z \leftrightarrow \Diamond(t \in u)\,]$ *is denoted by* $\Diamond u$.

Proof. See Appendix B. ■

DEFINITION 25. *Quantum set theory* QZFZ is the quantum logic QL with non-logical axioms A1–A11 in Theorem 24. That is, quantum set theory QZFZ is obtained from lattice valued set theory LZFZ by adding logical operator $^{\perp}$ and logical axioms (C1), (C2), (C3), (P).

" Sequent $\Gamma \Rightarrow \Delta$ is provable in QZFZ " is written as

$$\mathrm{QZFZ} \vdash \Gamma \Rightarrow \Delta,$$

where QZFZ will be omitted if it is obvious, and QZFZ $\vdash\ \Rightarrow \varphi$ is shortened as $\vdash \varphi$. Even " $\vdash$ " will be omitted if it is obvious.

DEFINITION 26. Formulas φ and ψ are said to be *compatible* , in symbols $\varphi \downarrow_{\circ} \psi$, if $\varphi \to (\varphi \wedge \psi) \vee (\varphi \wedge \psi^{\perp})$.

$$\varphi \downarrow_{\circ} \psi \overset{\text{def}}{\Longleftrightarrow} \varphi \to (\varphi \wedge \psi) \vee (\varphi \wedge \psi^{\perp})$$

By Theorem 5, $[\![\varphi \downarrow_{\circ} \psi]\!] = 1$ in V^{Q} if and only if $[\![\varphi]\!] \downarrow_{\circ} [\![\psi]\!]$ in Q.

G. Takeuti developed a quantum set theory in $V^{P(H)}$, where he used an implication defined in terms of $\vee$, $\wedge$, and $^{\perp}$. We denote his implication by $\to_{\mathrm{T}}$ with subscript $\perp$ to distinguish from the basic implication $\to$. Implication $\to_T$ is defined in QZFZ by

$$\varphi \to_T \psi \overset{\text{def}}{\Longleftrightarrow} \varphi^{\perp} \vee (\varphi \wedge \psi),$$

and called a *local implication.* Equality $\underset{T}{=}$ and membership relation $\underset{T}{\in}$ corresponding to the quantum implication are defined in QZFZ by

$$u \underset{T}{=} v \overset{\text{def}}{\Longleftrightarrow} \forall x(x \in u \to_T x \underset{T}{\in} v) \wedge \forall x(x \in v \to_T x \underset{T}{\in} u),$$

$$u \underset{T}{\in} v \overset{\text{def}}{\Longleftrightarrow} \exists x(x \in v \wedge u \underset{T}{=} x).$$

By equality axiom we have the following lemma.

LEMMA 27. $\vdash u = v \Rightarrow u \underset{T}{=} v, \quad \vdash u \in v \Rightarrow u \underset{T}{\in} v.$

5 A COMPLETENESS THEOREM OF QZFZ

By "a sentence φ of QZFZ is valid" we mean that ""$[\![\varphi]\!] = 1$ on V^{Q} for all complete orthomodular lattice Q" is provable in ZFC", i.e.

$$\mathrm{ZFC} \vdash \text{“ } [\![\varphi]\!] = 1 \text{ in } V^{\mathrm{Q}} \text{ for all complete orthomodular lattice Q ”}$$

In this section, we prove a completeness theorem of QZFZ in that sense of validity:

THEOREM 28. *For a sentence φ of* QZFZ,

$$\begin{aligned} \mathrm{ZFC} \quad &\vdash \quad \text{“ } [\![\varphi]\!] = 1 \textit{ in } V^{\mathrm{Q}} \textit{ for all complete orthomodular lattice } \mathrm{Q} \text{ ”} \\ &\Longrightarrow \quad \mathrm{QZFZ} \vdash \varphi, \end{aligned}$$

where V^{Q} is the Q*-valued universe constructed in* ZFC.

5.1 *Well-Founded Relations in* QZFZ

Any formula with two free variables determines a binary relation. For a binary relation $A(x, y)$, we use the following abbreviations:

$$x \in \mathrm{Dom}\, A \overset{\mathrm{def}}{\Longleftrightarrow} \exists y A(x, y), \quad x \in \mathrm{Rge}\, A \overset{\mathrm{def}}{\Longleftrightarrow} \exists y A(y, x),$$

$$x \in \mathrm{Fld}\, A \overset{\mathrm{def}}{\Longleftrightarrow} \exists y \left[A(x, y) \vee A(y, x)\right].$$

DEFINITION 29. A binary relation $\prec$ is said to be *well-founded* if the following conditions are satisfied.

WF1 $\forall x, y \neg(x \prec y \wedge y \prec x)$,

WF2 $\forall x \in \mathrm{Fld}(\prec)[\forall y \prec x \varphi(y) \to \varphi(x)] \to \forall x \in \mathrm{Fld}(\prec) \varphi(x)$,

WF3 $\forall x \exists y \forall z \prec x\, (z \in y)$.

In view of the axiom A9 ($\in$-induction), the membership relation $\in$ is a well-founded relation, and so is $\overset{\square}{\in}$.

Singlton $\{x\}$ and ordered pair $\langle x, y\rangle$ are defined as usual:

$$\{x\} \overset{\mathrm{def}}{=} \{x, x\}, \quad \langle x, y\rangle \overset{\mathrm{def}}{=} \{\{x\}, \{x, y\}\}$$

Then

$$\vdash x \in \{y\} \Leftrightarrow x = y, \quad \vdash \langle x, y\rangle = \langle x', y'\rangle \Leftrightarrow x = x' \wedge y = y'.$$

DEFINITION 30. A binary relation $F(x, y)$ is said to be *global*, if

$$\forall x, y\, [F(x, y) \to \Box F(x, y)]\,;$$

and a global relation $F(x, y)$ is *functional*, if

$$\forall x, y, y'\, [F(x, y) \wedge F(x, y') \to y = y']\,.$$

For a global functional relation F, we write $F(x) = y$ instead for $F(x, y)$. If F is a global functional relation and $\prec$ is a well-founded relation, then $F_{\prec u}$ is defined for each set $u \in \mathrm{Fld}(\prec)$ by

$$F_{\prec u} \overset{\text{def}}{=} \{\langle x, y\rangle \mid F(x, y) \wedge \Diamond(x \prec u)\,\}.$$

$F_{\prec u}$ is a global set of QZFZ, i.e. $\vdash \exists x(F_{\prec u} \in x) \wedge \mathrm{Gl}(F_{\prec u})$, by WF3, A11($\Diamond$) and A8(Collection).

The following recursion principle is justified in QZFZ, as usual.

THEOREM 31 Recursion Principle in QZFZ. *Let $\prec$ be a well founded relation and $H(a, b)$ be a global functional relation such that $\forall x \exists y H(x, y)$. Then there exists a unique global functional relation F such that*

$$\mathrm{Dom}\, F = \mathrm{Fld}(\prec) \;\wedge\; \forall x \in \mathrm{Fld}(\prec)\, (F(x) = H(F_{\prec x}))\,.$$

Proof. Outline of the proof is in Appendix C. ■

DEFINITION 32. We define the formula $\mathrm{Ord}(\alpha)$ ("α is an ordinal") in QZFZ by $\in$-recursion :

$$\begin{aligned}
\mathrm{Ord}(\alpha) &\overset{\text{def}}{\Longleftrightarrow} \mathrm{Gl}(\alpha) \wedge \mathrm{Tr}(\alpha) \wedge \forall \beta \in \alpha\, [\mathrm{Gl}(\beta) \wedge \mathrm{Tr}(\beta)\,], \quad \text{where} \\
&\qquad \mathrm{Gl}(\alpha) \overset{\text{def}}{\Longleftrightarrow} \forall \beta(\beta \in \alpha \to \beta \overset{\Box}{\in} \alpha) \\
&\qquad \mathrm{Tr}(\alpha) \overset{\text{def}}{\Longleftrightarrow} \forall \beta, \gamma(\beta \in \alpha \wedge \gamma \in \beta \to \gamma \in \alpha) \\
\mathrm{On} &\overset{\text{def}}{=} \{\alpha \mid \mathrm{Ord}(\alpha)\}
\end{aligned}$$

The following lemma is an immediate consequence of the definition of Ord.

LEMMA 33.

(1) $\vdash \mathrm{Ord}(\alpha) \wedge \beta \in \alpha \Rightarrow \mathrm{Ord}(\beta)$

(2) $\vdash \mathrm{Gl}(X) \wedge \forall x \in X\, \mathrm{Ord}(x) \Rightarrow \mathrm{Ord}(\bigcup X)$

DEFINITION 34. A global well founded relation $\prec$ is called a *well order* on u if

$$(\mathrm{Fld}(\prec) = u) \wedge (\prec \text{ is transitive}) \wedge (\prec \text{ is extensional}), \quad \text{where}$$

$$\prec \text{ is transitive} \overset{\text{def}}{\Longleftrightarrow} \forall x, y, z[(x \prec y) \wedge (y \prec z) \rightarrow (x \prec z)]$$

$$\prec \text{ is extensional} \overset{\text{def}}{\Longleftrightarrow} \forall x, y[x, y \in u \wedge \forall z(z \prec x \leftrightarrow z \prec y) \rightarrow x = y].$$

THEOREM 35 in QZFZ. *Every global set can be well-ordered, i.e. for every global set u, there exists a well-order $\prec$ on u.*

Proof. Suppose $\mathrm{Gl}(u)$, and let

$$P \overset{\text{def}}{=} \{\langle v, w\rangle \mid \mathrm{Gl}(v) \wedge \mathrm{Gl}(w) \wedge v \subset u \wedge (w \text{ is a well-order on } v)\},$$

and let $\langle v, w\rangle \prec \langle v', w'\rangle$ mean that $w = w' \lceil v$ and v is an initial w'-section of v', i.e.

$$\langle v, w\rangle \prec \langle v', w'\rangle \overset{\text{def}}{\Longleftrightarrow} (v \subset v') \wedge (w = w' \cap (v \times v)) \wedge (v \times (v' - v) \subset w').$$

Furthermore, let

$$\mathcal{I} \overset{\text{def}}{=} \{I \subset P \mid \forall p, q \in I[\,(p \prec q) \vee (p = q) \vee (q \prec p)\,] \wedge$$
$$\forall p, q \in P[(p \in I) \wedge (q \prec p) \rightarrow (q \in I)]\}.$$

Then

$$(\mathcal{I}' \subset \mathcal{I}) \wedge \forall I, I' \in \mathcal{I}'\,[(I \subset I') \vee (I' \subset I)] \Rightarrow \bigcup \mathcal{I}' \in \mathcal{I}.$$

By using GA10(Zorn), there exists a maximal $I_0 \in \mathcal{I}$. Set

$$v_0 = \bigcup\{v \mid \langle v, w\rangle \overset{\Box}{\in} I_0\}, \quad w_0 = \bigcup\{w \mid \langle v, w\rangle \overset{\Box}{\in} I_0\}$$

It suffices to show that $\langle v_0, w_0\rangle \in P$ and $v_0 = u$.

Lemma 35-1 If $\langle v, w\rangle, \langle v', w'\rangle \in P$, then

$$\vdash \langle v, w\rangle \prec \langle v', w'\rangle \wedge x \in v \wedge \langle y, x\rangle \in w' \Rightarrow y \in v.$$

Proof. Let Ψ be the formula $\langle v, w\rangle \prec \langle v', w'\rangle \wedge x \in v \wedge \langle y, x\rangle \in w'$. Ψ is $\Box$-closed, and by WF1,

$$\begin{aligned} \vdash \Psi \wedge \neg(y \in v) \quad &\Rightarrow \quad \langle x, y\rangle \in w' \\ &\Rightarrow \quad \bot \end{aligned}$$

Hence, $\vdash \Psi \Rightarrow \neg\neg(y \in v)$. Since $\vdash \mathrm{Gl}(v) \wedge \neg\neg(y \in v) \Rightarrow y \in v$, $\vdash \Psi \Rightarrow y \in v$. ■

Lemma 35-2 $\vdash \langle v, w\rangle \overset{\Box}{\in} I_0 \wedge x \in v \wedge \langle y, x\rangle \in w_0 \Rightarrow \langle y, x\rangle \in w.$

Proof. Let Ψ be the formula $\langle v, w\rangle \overset{\Box}{\in} I_0 \wedge x \in v \wedge \langle y, x\rangle \in w_0$.

$$\begin{aligned} \vdash \Psi \quad &\Rightarrow \quad \exists \langle v', w'\rangle \overset{\Box}{\in} I_0[\langle y, x\rangle \in w' \wedge x \in v \wedge \langle v, w\rangle \overset{\Box}{\in} I_0] \\ &\Rightarrow \quad y \in v \quad \text{by using Lemma 35-1} \\ &\Rightarrow \quad \langle y, x\rangle \in w. \end{aligned}$$

■

$\mathrm{Gl}(v_0)$, $\mathrm{Gl}(w_0)$ and $v_0 \subset u$ are obvious. In order to prove $\langle v_0, w_0\rangle \in P$, it is to prove that "$w_0$ is a well order on v_0".

WF1 "$\vdash \langle x, y\rangle \in w_0 \wedge \langle y, x\rangle \in w_0 \Rightarrow \bot$" is obvious.

WF2 $\forall x \in v_0[\forall y(\langle y, x\rangle \in w_0 \to \varphi(y)) \to \varphi(x)] \to \forall x \in v_0 \varphi(x)$

Proof. Let $\Psi(x)$ be formula $\forall y(\langle y, x\rangle \in w_0 \to \varphi(y)) \to \varphi(x)$. By Lemma 35-2, $x \in v \wedge \langle v, w\rangle \overset{\Box}{\in} I_0 \Rightarrow [\langle y, x\rangle \in w_0 \to \langle y, x\rangle \in w]$. Hence,

$$\begin{aligned} \vdash x \in v_0 \wedge \Psi(x) \wedge \langle v, w\rangle \overset{\Box}{\in} I_0 \quad &\Rightarrow \quad \Psi(x) \wedge [\langle y, x\rangle \in w_0 \to \langle y, x\rangle \in w] \\ &\Rightarrow \quad \forall y(\langle y, x\rangle \in w \to \varphi(y)) \to \varphi(x) \\ &\Rightarrow \quad \varphi(x) \end{aligned}$$

■

WF3 $\forall x \exists y \forall z(\langle z, x\rangle \in w_0 \to z \overset{\Box}{\in} y)$ is obvious.

Transitivity of w_0 is obvious, and extensionality of w_0 on v_0 follows from Lemma 35-2. Therefore, w_0 is a well order, and $\langle v_0, w_0\rangle \in P$.

Now we prove that $u = v_0$. Assume that $x \in u - v_0$ and let

$$v_1 = v_0 \cup \{x\}, \quad w_1 = w_0 \cup (v_0 \times \{x\}).$$

Then it is straightforward to prove that

$$\langle v_0, w_0\rangle \prec \langle v_1, w_1\rangle \in P,$$

and this contradicts the maximality of I_0. Therefore, $u = v_0$. ■

THEOREM 36 in QZFZ. *If u is a global set and $\prec$ is a well-order on u, then $\langle u, \prec\rangle$ is isomorphic to an ordinal $\langle \alpha, \in\rangle$, i.e. there exists ρ such that*

$$\exists \rho\big[(\rho : u \to \alpha) \wedge \rho(u) = \alpha \wedge \forall x, y \in u(x \prec y \leftrightarrow \rho(x) \in \rho(y)) \wedge$$
$$\forall x, y \in u(x = y \leftrightarrow \rho(x) = \rho(y))\big].$$

Proof. We define ρ by $\prec$-recursion :

$$\rho(x) = \bigcup\{\rho(y)+1 \mid y \prec x\}, \quad \text{where } \rho(y)+1 = \rho(y) \cup \{\rho(y)\}.$$

It is easy to see by WF2 ($\prec$-induction) that $\forall x \in u\, \mathrm{Ord}(\rho(x))$, and

$$\forall x \in u \forall t \in \rho(x) \exists y \prec x (t = \rho(y)).$$

Set $\alpha = \{\rho(x) \mid x \in u\}$. Then $\mathrm{Ord}(\alpha)$, and $\langle u, \prec\rangle$ is isomorphic to $(\alpha, \in)$. ■

5.2 *Check sets*

The notion of check set is defined in QZFZ, by $\overset{\Box}{\in}$-recursion:

$$\mathrm{ck}(x) \overset{\text{def}}{\Longleftrightarrow} \forall t \left(t \in x \leftrightarrow t \overset{\Box}{\in} x \wedge \mathrm{ck}(t) \right).$$

That is, set

$$H(u,v) \overset{\text{def}}{\Longleftrightarrow} v = \{t \mid \langle t,t\rangle \in u\}.$$

H is a global functional relation such that $\forall u \exists v H(u,v)$. Let $\prec$ be $\overset{\Box}{\in}$. $\prec$ is a well founded relation. Since $\forall x\, [x \in \mathrm{Fld}(\prec)]$, there exists a unique global functional relation $C(x,y)$ such that

$$\forall x\, [x \in \mathrm{Dom}(C) \wedge C(x) = H(C_{\prec x})],$$

by Recursion Principle (Theorem 31).

THEOREM 37. *The followings are provable in* QZFZ.

(1) $y \in C(x) \leftrightarrow (y \overset{\Box}{\in} x) \wedge C(y,y)$

(2) $C(x,x) \leftrightarrow \forall t [t \in x \leftrightarrow (t \overset{\Box}{\in} x \wedge C(t,t))]$

(3) $C(x) = CC(x)$

Proof. (1) and (2) are immediate results of the definition of C. (3) is proved by :

$$y \in CC(x) \Leftrightarrow [y \overset{\Box}{\in} C(x) \wedge C(y,y)] \Leftrightarrow [y \overset{\Box}{\in} x \wedge C(y,y)] \Leftrightarrow y \in C(x)$$

■

$\mathrm{ck}(x)$ ("x is a *check set*") is defined as $C(x,x)$, i.e.

$$\mathrm{ck}(x) \overset{\text{def}}{\Longleftrightarrow} x = C(x) \Longleftrightarrow \forall t \left(t \in x \leftrightarrow t \overset{\Box}{\in} x \wedge \mathrm{ck}(t) \right).$$

The class of check sets will be denoted by W:

$$x \in W \overset{\text{def}}{\Longleftrightarrow} \mathrm{ck}(x).$$

5.3 *The model* W *of* ZFC *in* QZFZ

An interpretation of ZFC in QZFZ is obtained by relativizing the range of quantifiers to check sets. In this section we will show that

$$\text{QZFZ} \vdash \text{“the class } W \text{ of check sets is a model of ZFC”}, \quad \text{where}$$

$$W = \{x \mid \text{ck}(x)\}, \quad \text{ck}(x) \Leftrightarrow \forall t[t \in x \leftrightarrow (t \overset{\Box}{\in} x \wedge \text{ck}(t))].$$

We denote quantifiers relativized on check sets by $\forall^W, \exists^W$, i.e.

$$\forall^W x\varphi(x) \overset{\text{def}}{\Longleftrightarrow} \forall x(\text{ck}(x) \to \varphi(x))$$

$$\exists^W x\varphi(x) \overset{\text{def}}{\Longleftrightarrow} \exists x(\text{ck}(x) \wedge \varphi(x)).$$

For a formula φ of ZFC, φ^W is the formula obtained from φ by replacing all quantifiers $\forall x$, $\exists x$, by $\forall^W x$, $\exists^W x$, respectively.

LEMMA 38. *The following (1)–(9) are provable in* QZFZ, *for any formula* φ *of* ZFC.

(1) $\forall^W x, y(x \in y \to x \overset{\Box}{\in} y)$

(2) $\forall^W x_1 \cdots x_n[\varphi^W(x_1, \cdots, x_n) \to \Box\varphi^W(x_1, \cdots, x_n)]$

(3) $\forall^W x(\forall^W y(y \in x \to \varphi^W(y)) \to \varphi^W(x)) \to \forall^W x\varphi^W(x)$

(4) $\forall\alpha[\text{Ord}(\alpha) \leftrightarrow \text{ck}(\alpha) \wedge \text{Ord}^W(\alpha)]$

(5) $\text{ck}(\emptyset)$, *where* $\emptyset$ *is the empty set.*

(6) $\forall^W x, y[\text{ck}(\{x, y\}) \wedge \text{ck}(\bigcup x) \wedge \text{ck}(\{z \in x \mid \Box\varphi(z)\})]$

(7) *The set of natural numbers* ω *is defined by recursion as follows:*

$$\begin{aligned} \text{Suc}(x) &\overset{def}{\Longleftrightarrow} [x = \emptyset \vee \exists z(x = z + 1)], \ \textit{where} \ z + 1 = z \cup \{z\}, \\ \text{HSuc}(x) &\overset{def}{\Longleftrightarrow} [\text{Suc}(x) \wedge \forall y \in x\, \text{HSuc}(y)], \quad \textit{and} \\ \omega &\overset{def}{=} \{x : \text{HSuc}(x)\}. \end{aligned}$$

Then $\exists x(\omega \in x) \wedge \text{Ord}(\omega) \wedge \forall^W n \in \omega[n = \emptyset \vee \exists^W m \in n(n = m + 1)]$.

Proof.

(1) By the definition of $\mathrm{ck}(y)$, $\vdash \mathrm{ck}(x) \wedge \mathrm{ck}(y) \wedge x \in y \;\rightarrow\; x \overset{\Box}{\in} y$.

(2) By induction on complexity of φ. If φ has no logical symbol, then φ is of the form $x = y$ or $x \in y$. $\forall^W x, y[x = y \rightarrow \Box(x = y)]$ since $x = y$ is $\Box$-closed, and $\forall^W x, y[x \in y \rightarrow \Box(x \in y)]$ by (1). Here we prove only the case that φ is of the form $\exists x \psi(x, x_1, \cdots, x_n)$, since the other cases are similar. Let $\vdash \mathrm{ck}(x_1) \wedge \cdots \wedge \mathrm{ck}(x_n)$. By using induction hypothesis,

$$\vdash \psi^W(x, x_1, \cdots, x_n) \wedge \mathrm{ck}(x) \Rightarrow \Box\left(\mathrm{ck}(x) \wedge \psi^W(x, x_1, \cdots, x_n)\right).$$

Hence, by Theorem 13(21),

$$\vdash \exists^W x \psi^W(x, x_1, \cdots, x_n) \;\Rightarrow\; \Box \exists^W x \psi^W(x, x_1, \cdots, x_n).$$

(3) Let $\Psi(x)$ be formula $\mathrm{ck}(x) \rightarrow \varphi^W(x)$. Then, using $\in$-induction, we have

$$\begin{aligned} \vdash \forall^W x[\forall^W y(y \in x \rightarrow \varphi^W(y)) \;&\rightarrow\; \varphi^W(x)] \\ &\Rightarrow\; \forall x[\forall y(y \in x \rightarrow \Psi(y)) \rightarrow \Psi(x)] \\ &\Rightarrow\; \forall x \Psi(x) \\ &\Rightarrow\; \forall^W x \varphi^W(x). \end{aligned}$$

(4) $\vdash \mathrm{Ord}\,\alpha \wedge \beta \in \alpha \Rightarrow \mathrm{Ord}\,\beta$ by the definition of Ord. Set

$$\Psi(\alpha) \overset{\mathrm{def}}{\Longleftrightarrow} (\mathrm{Ord}(\alpha) \rightarrow \mathrm{ck}(\alpha)).$$

$$\begin{aligned} \vdash \mathrm{Ord}\,\alpha \wedge \forall \beta \in \alpha \Psi(\beta) \;&\Rightarrow\; \mathrm{Gl}(\alpha) \wedge \forall \beta \in \alpha[\mathrm{Ord}(\beta) \wedge \Psi(\beta)] \\ &\Rightarrow\; \mathrm{Gl}(\alpha) \wedge \forall \beta \in \alpha\, \mathrm{ck}(\beta) \\ &\Rightarrow\; \mathrm{ck}(\alpha) \end{aligned}$$

Since $\forall \beta \in \alpha \Psi(\beta)$ is $\Box$-closed, $\vdash \forall \beta \in \alpha \Psi(\beta) \Rightarrow \Psi(\alpha)$. Therfore, by $\in$-induction, $\forall \alpha \Psi(\alpha)$. That is,
$\vdash \forall \alpha[\mathrm{Ord}(\alpha) \rightarrow \mathrm{ck}(\alpha)]$.

$\vdash \forall \alpha[\mathrm{Ord}(\alpha) \leftrightarrow \mathrm{ck}(\alpha) \wedge \mathrm{Ord}^W(\alpha)]$ is straightforward.

(5) $\vdash \mathrm{ck}(\emptyset)$ follows from:

$$\begin{aligned} \vdash x \in \emptyset \;&\Rightarrow\; \neg(x = x) \\ &\Rightarrow\; x \overset{\Box}{\in} \emptyset \wedge \mathrm{ck}(x). \end{aligned}$$

(6)

$$\vdash \mathrm{ck}(x) \wedge \mathrm{ck}(y) \wedge (z \in \{x, y\}) \quad \Leftrightarrow \quad \mathrm{ck}(x) \wedge \mathrm{ck}(y) \wedge (z = x \vee z = y)$$
$$\Rightarrow \quad \mathrm{ck}(z) \wedge z \overset{\Box}{\in} \{x, y\}.$$

$$\vdash \mathrm{ck}(x) \wedge z \in \bigcup x \quad \Rightarrow \quad \mathrm{ck}(x) \wedge \exists t \in x (z \in t)$$
$$\Rightarrow \quad \exists t[\mathrm{ck}(t) \wedge t \overset{\Box}{\in} x \wedge z \in t]$$
$$\Rightarrow \quad \mathrm{ck}(z) \wedge z \overset{\Box}{\in} \bigcup x.$$

$$\vdash \mathrm{ck}(x) \wedge t \in \{z \in x \mid \Box\varphi(z)\} \quad \Rightarrow \quad \mathrm{ck}(t) \wedge t \overset{\Box}{\in} x \wedge \Box\varphi(t)$$
$$\Rightarrow \quad \mathrm{ck}(t) \wedge t \overset{\Box}{\in} \{z \in x \mid \Box\varphi(t)\}.$$

(7) (a) $\vdash \forall x(\mathrm{HSuc}(x) \to \mathrm{ck}(x))$.

Proof. Let $\Psi(x) \overset{\text{def}}{\Longleftrightarrow} [\,\mathrm{HSuc}(x) \to \mathrm{ck}(x)\,]$. Then

$$\vdash \forall y \in x \Psi(y) \to \Psi(x).$$

Therefore, $\vdash \forall x \Psi(x)$ by $\in$-induction. ■

(b) $\vdash$"ω is a set", i.e. $\vdash \exists x(\omega \in x)$.

Proof. Let

$$U_x \overset{\text{def}}{=} \bigcup \{C(\mathcal{P}(U_y)) \mid y \overset{\Box}{\in} x\},$$

where $C(\mathcal{P}(X)) = \{Y \subset X \mid \mathrm{ck}(Y)\}$ (cf. Theorem 37). By Axiom A6(Infinity), $\vdash \exists u[\,\exists x(x \overset{\Box}{\in} u) \wedge \forall x \overset{\Box}{\in} u \exists y \overset{\Box}{\in} u(x \overset{\Box}{\in} y)\,])$. Assume

$$(a \overset{\Box}{\in} u) \wedge \forall x \overset{\Box}{\in} u \exists y \overset{\Box}{\in} u(x \overset{\Box}{\in} y),$$

and we show $\vdash \omega \in \mathcal{P}(U_u)$.

(i) $\vdash \mathrm{ck}(U_x)$ and

(ii) $\vdash \mathrm{Tr}(U_x)$, i.e. $\vdash \forall y \in U_x (y \subset U_x)$. Because:

$$\vdash \forall z \in x\, \mathrm{Tr}(U_z) \wedge y \in U_x \wedge t \in y$$
$$\Rightarrow \forall z \in x\, \mathrm{Tr}(U_z) \wedge \exists z \in x[\,y \in C(\mathcal{P}(U_z))\,] \wedge\ t \in y\ \wedge \mathrm{ck}(y)$$
$$\Rightarrow \exists z[\,z \in x \wedge \mathrm{Tr}(U_z) \wedge \mathrm{ck}(t) \wedge t \in U_z]$$
$$\Rightarrow \exists z[z \in x \wedge \mathrm{ck}(t) \wedge t \subset U_z]$$
$$\Rightarrow t \in U_x$$

$$\therefore \quad \vdash \forall z \in x\, \mathrm{Tr}(U_z) \Rightarrow \mathrm{Tr}(U_x).$$

Therefore, $\vdash \forall x\, \mathrm{Tr}(U_x)$ by $\in$-induction.

(iii) $\emptyset \in U_u$, since $\vdash \mathrm{ck}(\emptyset) \wedge a \overset{\Box}{\in} u \wedge \emptyset \subset U_a$.

(iv) $\beta \in U_u \Rightarrow \beta + 1 \in U_u$.

$$\begin{aligned}
\because)\quad \beta \in U_u \quad &\Rightarrow \quad \mathrm{ck}(\beta) \wedge \exists x \overset{\Box}{\in} u(\beta \subset U_x) \wedge \forall x \overset{\Box}{\in} u \exists y \overset{\Box}{\in} u(x \overset{\Box}{\in} y) \\
&\Rightarrow \quad \mathrm{ck}(\beta) \wedge \exists x, y[x \overset{\Box}{\in} u \wedge y \overset{\Box}{\in} u \wedge x \overset{\Box}{\in} y \wedge \beta \subset U_x] \\
&\Rightarrow \quad \exists x, y[\mathrm{ck}(\beta) \wedge y \overset{\Box}{\in} u \wedge x \overset{\Box}{\in} y \wedge \beta \subset U_x] \\
&\Rightarrow \quad y \overset{\Box}{\in} u \wedge \beta \in U_y \\
&\Rightarrow \quad y \overset{\Box}{\in} u \wedge \beta \cup \{\beta\} \subset U_y, \quad \text{sinse } \mathrm{Tr}(U_y) \\
&\Rightarrow \quad \beta \cup \{\beta\} = \beta + 1 \in U_u
\end{aligned}$$

It follows that $\vdash \omega \subset U_u$, that is, $\omega = \{x \in U_u \mid \mathrm{HSuc}(x)\} \in \mathcal{P}(U_u)$. ■

It is easy to see that

$$\vdash \mathrm{Gl}(\omega) \wedge \mathrm{Tr}(\omega) \wedge \forall \alpha \in \omega[\mathrm{Gl}(\alpha) \wedge \mathrm{Tr}(\alpha)].$$

That is, $\vdash \mathrm{Ord}(\omega)$.

$\vdash \forall^W n \in \omega(n = \emptyset \vee \exists^W m \in n(n = m + 1))$ is obvious.

■

THEOREM 39 Interpretation of ZFC in QZFZ. *If φ is a theorem of* ZFC, *then φ^W is provable in* QZFZ.

Proof. For a formula $\varphi(a_1, \cdots, a_n)$ of ZFC, if $u_1, \cdots, u_n \in W$, then $\varphi^W(u_1, \cdots, u_n)$ is $\Box$-closed by Lemma 38(2), i.e.

$$\mathrm{QZFZ} \vdash \forall^W x_1, \cdots, x_n(\varphi^W \to \Box\varphi^W).$$

QZFZ $\vdash \Box\varphi \vee \neg\Box\varphi$, and $\Box\varphi$ is distributive over all $\vee$ and $\exists$. Hence, $\Box$-closed formulas form a system of classical logic, i.e. W is a model of classical logic. Now it suffices to show that QZFZ $\vdash A^W$ for each nonlogical axiom A of ZFC. In what follows in this proof, $\vdash \Psi$ means QZFZ $\vdash \Psi$ as above.

$\vdash$ (Equality axiom)^W and $\vdash$ (Extensionality)^W are obvious.

$\vdash$ (Pairing)^W: By Lemma 38(6),

$$\vdash \mathrm{ck}(u) \wedge \mathrm{ck}(v) \Rightarrow \mathrm{ck}(\{u, v\}) \wedge \forall^W x(x \in \{u, v\} \leftrightarrow x = u \vee x = v).$$

$\vdash$ (Union)^W: Similarly.

$\vdash$ (Power set)^W: $\vdash \mathrm{ck}(C(\mathcal{P}(u)))$ and

$$\vdash \forall^W u, x[x \in C(\mathcal{P}(u)) \leftrightarrow \forall^W t(t \in x \to t \in u)],$$

where $C(\mathcal{P}(u)) = \{x \in \mathcal{P}(u) \mid \mathrm{ck}(x)\}$.

$\vdash$ ($\in$ -induction)^W: By Lemma 38(3).

$\vdash$ (Separation)^W: If $\vdash \mathrm{ck}(u)$, then $\vdash \mathrm{ck}(\{x \in u \mid \varphi^W(x)\})$ by Lemma 38(6), and

$$\vdash \forall^W u, x \left[x \in \{z \in u \mid \varphi^W(z)\} \leftrightarrow x \in u \wedge \varphi^W(x)\right].$$

$\vdash$ (Collection)^W: Suppose $\vdash \mathrm{ck}(u) \wedge \forall^W x \in u \exists^W y \varphi^W(x, y)$. By GA8,

$$\vdash \exists v \forall x \in u \exists y \overset{\Box}{\in} v(\mathrm{ck}(y) \wedge \varphi^W(z, y)).$$

Since $\vdash y \overset{\Box}{\in} v \wedge \mathrm{ck}(y) \to y \in C(v) \wedge \mathrm{ck}(C(v))$, where $y \in C(v) = y \overset{\Box}{\in} v \wedge \mathrm{ck}(y)$ (cf. Theorem 37), we have

$$\vdash \exists^W v \forall^W x \in u \exists^W y \in v \varphi^W(z, y).$$

$\vdash$ (Infinity)^W: By Lemma 38(7).

$\vdash$ (Choice)^W, i.e. $\vdash \forall^W u \exists^W f \forall^W x \in u[x \neq \emptyset \to \exists!^W y \in x(\langle x, y\rangle \in f)]$, where $x \neq \emptyset$ stands for $\exists^W y(y \in x)$. Assume $\mathrm{ck}(u)$ and apply A10(Zorn) to the set of partial choice functions on u :

$$\mathcal{I} = \{f \mid \exists I \subset u\,[\,\mathrm{ck}(I) \wedge f : I \to \bigcup I \wedge \forall x \in I(f(x) \in x)\,]\,\}.$$

Let f_0 be a maximal element of $\mathcal{I}$. Then $f_0 : u \to \bigcup u$ such that $\forall x \in u(f_0(x) \in x)$. ■

5.4 *Power set* $\mathcal{P}(1)$ *of 1 in* QZFZ

The power set $\mathcal{P}(1)$ of 1 ($= \{\emptyset\}$) is a global set and a complete lattice with respect to inclusion $\subset$. Namely, if $\{x_\alpha\}_{\alpha \in K} \subset \mathcal{P}(1)$, where index set K is a check set, then the supremum $\bigcup_{\alpha \in K} x_\alpha$ and the infimum $\bigcap_{\alpha \in K} x_\alpha$ of $\{x_\alpha\}_{\alpha \in K}$ are elements of $\mathcal{P}(1)$. Let

$$a^\perp = \{x \in 1 \mid (0 \in a)^\perp\} \quad \text{for} \quad a \in \mathcal{P}(1).$$

Then $^\perp$ is an operator on $\mathcal{P}(1)$ satisfying conditions (C) of ortholattice, and $(\mathcal{P}(1), \subset, \bigcup, \bigcap, {}^\perp)$ is a complete orthomodular lattice with top $1 = \bigcup \mathcal{P}(1)$ and bottom $0 = \bigcap \mathcal{P}(1)$.

Since $\mathcal{P}(1)$ is a global set, there exists a check set $\mathcal{Q}$ and a bijection ρ such that $\rho : \mathcal{P}(1) \to \mathcal{Q}$ by Theorem 36.

DEFINITION 40. We fix a check set $\mathcal{Q}$ and a bijection ρ such that

$$\rho : \mathcal{P}(1) \to \mathcal{Q},$$

and define $\leqslant$, $\bigwedge$, $\bigvee$, $^\perp$, $\to$ on $\mathcal{Q}$ so that $\mathcal{Q}$ is a complete orthomodular lattice isomorphic to $\mathcal{P}(1)$, i.e.

$$\begin{aligned}
p \leqslant q &\overset{\text{def}}{\iff} \rho^{-1}(p) \subset \rho^{-1}(q) \\
\bigwedge_\alpha p_\alpha &\overset{\text{def}}{=} \rho(\bigcap_\alpha \rho^{-1}(p_\alpha)) \\
\bigvee_\alpha p_\alpha &\overset{\text{def}}{=} \rho(\bigcup_\alpha \rho^{-1}(p_\alpha)) \\
p^\perp &\overset{\text{def}}{=} \rho(\{x \in 1 \mid (0 \in \rho^{-1}(p))^\perp\}) \\
(p \to q) &\overset{\text{def}}{=} \rho(\{x \in 1 \mid \rho^{-1}(p) \subset \rho^{-1}(q)\})
\end{aligned}$$

Then $(\mathcal{Q}, \leqslant, \bigwedge, \bigvee, ^\perp)$ is a complete orthomodular lattice in W.

DEFINITION 41. For a sentence φ, let

$$|\varphi| \overset{\text{def}}{=} \rho(\{x \in 1 \mid \varphi\}).$$

LEMMA 42. *If φ is a sentence of* QZFZ, *then*

$$|\varphi| \in \mathcal{Q} \quad \textit{and} \quad \text{QZFZ} \vdash \varphi \Leftrightarrow 0 \in \rho^{-1}(|\varphi|).$$

Proof. QZFZ $\vdash \varphi \Leftrightarrow 0 \in \{x \in 1 \mid \varphi\} \Leftrightarrow 0 \in \rho^{-1}|\varphi|$. ■

LEMMA 43. *For $p, q \in \mathcal{Q}$,*

$$\begin{aligned}
\text{QZFZ} &\vdash 0 \in \rho^{-1}(p^\perp) \Leftrightarrow (0 \in \rho^{-1}(p))^\perp, \\
\text{QZFZ} &\vdash 0 \in \rho^{-1}(p \to q) \Leftrightarrow [\rho^{-1}(p) \subset \rho^{-1}(q)]
\end{aligned}$$

Proof. By the definitions of $^\perp$ and $\to$. ■

5.5 *$\mathcal{Q}$-valued universe $W^{\mathcal{Q}}$ in* QZFZ

Complete orthomodular lattice $\mathcal{Q}$ in W represents the truth value set of QZFZ, and the relation $\prec$ defined by $\alpha \prec \beta \overset{\text{def}}{\iff} (\alpha, \beta \in \text{On} \wedge \alpha \in \beta)$ is a well founded relation with $\mathrm{Fld}(\prec) = \mathrm{On}$. Thus, recursive definition on On is justified in QZFZ, by Recursion Principle. $\mathcal{Q}$-valued universe $W^{\mathcal{Q}}$ is constructed in W by $\in$-recursion.

$$\begin{aligned} W^{\mathcal{Q}}_{\alpha} &= \{u \in W \mid \left(\exists\beta\in\alpha\exists\mathcal{D}u\subset W^{\mathcal{Q}}_{\beta}(\mathrm{Gl}(\mathcal{D}u)\wedge u:\mathcal{D}u\to\mathcal{Q})\right)^{W}\} \\ W^{\mathcal{Q}} &= \bigcup_{\alpha\in\mathrm{On}} W^{\mathcal{Q}}_{\alpha} \end{aligned}$$

Truth value of φ in $W^{\mathcal{Q}}$ will be denoted by $[\![\varphi]\!]^{W}$. Atomic relations $=$ and $\in$ are interpreted in $W^{\mathcal{Q}}$, as

$$\begin{aligned} [\![x=y]\!]^{W} &= \bigwedge_{t\in\mathcal{D}x}(x(t)\to[\![t\in y]\!]^{W})\wedge\bigwedge_{t\in\mathcal{D}y}(y(t)\to[\![t\in x]\!]^{W}) \\ [\![x\in y]\!]^{W} &= \bigvee_{t\in\mathcal{D}y}[\![x=t]\!]^{W}\wedge y(t). \end{aligned}$$

Logical operations $\wedge$, $\vee$, ${}^{\perp}$, $\to$, $\neg, \forall$, $\exists$ are interpreted as the corresponding operations on $\mathcal{Q}$. Then every sentence in $W^{\mathcal{Q}}$ has its truth value in $\mathcal{Q}$. We will show that $\mathrm{QZFZ}\vdash[\![\varphi]\!]^{W}=|\varphi|$.

DEFINITION 44. For $x\in W^{\mathcal{Q}}$, define $F(x)$ by

$$F(x)=\{F(t)\mid t\in\mathcal{D}x\ \wedge\ 0\in\rho^{-1}(x(t))\}.$$

LEMMA 45. $\mathrm{QZFZ}\vdash\ \forall u\exists x\in W^{\mathcal{Q}}(F(x)=u)$.

Proof. By $\in$-induction. Let $\Psi(u,\alpha)\overset{\text{def}}{\Longleftrightarrow}\exists x\in W^{\mathcal{Q}}_{\alpha}(u=F(x))$. Then by using A8 (Collection) we have

$$\exists\alpha\in\mathrm{On}[\forall v\in u\exists\beta\Psi(v,\beta))\ \to\ \forall v\in u\exists\beta\overset{\square}{\in}\alpha\exists y\in W^{\mathcal{Q}}_{\beta}(v=F(y))].$$

Let

$$\begin{cases} \mathcal{D}x = W^{\mathcal{Q}}_{\alpha} \\ x(y)=\rho(\{t\in 1\mid F(y)\in u\}) \end{cases}$$

Then $x\in W^{\mathcal{Q}}$ and $F(x)=u$, i.e.

$$\forall v\in u\exists\beta\Psi(v,\beta))\ \to\ \exists\alpha\Psi(u,\alpha)$$

Hence, $\forall u\exists x(F(x)=u)$. ■

THEOREM 46. *For every sentence* φ *of* QZFZ,

$$\mathrm{QZFZ}\vdash[\![\varphi]\!]^{W}=|\varphi|.$$

Proof. It suffices to prove that for every formula $\varphi(a_1, \cdots, a_n)$ of QZFZ, and for $x_1, \cdots x_n \in W^{\mathcal{Q}}$,

$$\mathrm{QZFZ} \vdash [\![\varphi(x_1, \cdots, x_n)]\!]^W = |\varphi(F(x_1), \cdots, F(x_n))|,$$

by induction on the complexity of φ.

(1) $\mathrm{QZFZ} \vdash |F(x) = F(y)| = [\![x = y]\!]^W, \qquad \mathrm{QZFZ} \vdash |F(x) \in F(y)| = [\![x \in y]\!]^W.$

$\because$) By induction.

$$\begin{aligned} \mathrm{QZFZ} \vdash 0 \in \rho^{-1}([\![x = y]\!]^W), F(t) \in F(x) &\Rightarrow t \in \mathcal{D}x \wedge 0 \in \rho^{-1}(x(t)) \\ &\Rightarrow 0 \in \rho^{-1}[\![t \in y]\!]^W \\ &\Rightarrow F(t) \in F(y) \end{aligned}$$

Since $0 \in \rho^{-1}([\![x = y]\!]^W)$ is $\Box$-closed, we have

$$\begin{aligned} \mathrm{QZFZ} \vdash 0 \in \rho^{-1}[\![x = y]\!]^W &\Rightarrow F(x) = F(y) \\ &\Rightarrow 0 \in \rho^{-1}|F(x) = F(y)| \end{aligned}$$

$$\begin{aligned} \mathrm{QZFZ} \vdash F(x) = F(y) \wedge t \in \mathcal{D}x \wedge 0 \in \rho^{-1}(x(t)) &\Rightarrow F(t) \in F(y) \\ &\Rightarrow \exists s \in \mathcal{D}y[F(t) = F(s) \wedge 0 \in \rho^{-1}y(s))]. \end{aligned}$$

$$\therefore \ \mathrm{QZFZ} \vdash F(x) = F(y) \Rightarrow 0 \in \rho^{-1}[\![x = y]\!]^W.$$

$$\begin{aligned} \mathrm{QZFZ} \vdash 0 \in \rho^{-1}[\![x \in y]\!]^W &\Leftrightarrow \exists t \in \mathcal{D}y[0 \in \rho^{-1}([\![x = t]\!]^W \wedge y(t))] \\ &\Leftrightarrow F(x) = F(t) \in F(y) \end{aligned}$$

(2) If φ is of the form $\varphi_1 \wedge \varphi_2$, then

$$\begin{aligned} \mathrm{QZFZ} \vdash 0 \in \rho^{-1}|\varphi \wedge \psi| &\Leftrightarrow \varphi \wedge \psi \\ &\Leftrightarrow 0 \in \rho^{-1}|\varphi| \wedge 0 \in \rho^{-1}\psi \\ &\Leftrightarrow 0 \in \rho^{-1}|\varphi| \cap \rho^{-1}\psi \\ &\Leftrightarrow 0 \in \rho^{-1}|\varphi \wedge \psi| \end{aligned}$$

(3) If φ is of the form $\varphi_1 \vee \varphi_2$, then similarly to (2).

(4) If φ is of the form $\varphi_1^\perp$, then by Lemma 43,

$$\begin{aligned} \mathrm{QZFZ} \vdash 0 \in \rho^{-1}|\varphi_1^\perp| &\Leftrightarrow \varphi_1^\perp \\ &\Leftrightarrow (0 \in \rho^{-1}|\varphi_1|)^\perp \\ &\Leftrightarrow 0 \in \rho^{-1}(|\varphi_1|^\perp) \end{aligned}$$

(5) If φ is of the form $\varphi_1 \to \varphi_2$, then similarly

$$\begin{aligned} \text{QZFZ} \vdash 0 \in \rho^{-1}|\varphi_1 \to \varphi_2| &\Leftrightarrow \varphi_1 \to \varphi_2 \\ &\Leftrightarrow 0 \in \rho^{-1}|\varphi_1| \to 0 \in \rho^{-1}|\varphi_2| \\ &\Leftrightarrow \rho^{-1}|\varphi_1| \subset \rho^{-1}|\varphi_2| \\ &\Leftrightarrow 0 \in \rho^{-1}(|\varphi_1| \to |\varphi_2|) \end{aligned}$$

(6) If $\varphi(x_1, \cdots, x_n)$ is of the form $\forall x \psi(x, x_1, \cdots, x_n)$, then using Lemma 45

$$\begin{aligned} \text{QZFZ} \vdash 0 \in \rho^{-1}[\![\varphi]\!]^W &\Leftrightarrow 0 \in \rho^{-1}(\bigwedge_x [\![\psi(x, x_1, \cdots, x_n)]\!]^W) \\ &\Leftrightarrow \forall x(\psi(F(x), F(x_1), \cdots, F(x_n)) \\ &\Leftrightarrow \forall z \psi(z, F(x_1), \cdots, F(x_n)). \end{aligned}$$

where Lemma 45 was used.

(7) If φ is of the form $\exists x \psi(x, x_1, \cdots, x_n)$, then similar to (6).

■

COROLLARY 47. QZFZ $\vdash \Box\varphi \Leftrightarrow$ "$[\![\varphi]\!]^W = 1$ *in* $W^{\mathcal{Q}}$ ".

5.6 *Proof of a completeness theorem of* QZFZ

Suppose that φ is a sentence of QZFZ and

$$\text{ZFC} \vdash \text{“ } [\![\varphi]\!] = 1 \text{ in } V^{\mathrm{Q}} \text{ for all complete orthomodular lattice Q ”}.$$

Then, since W is a model of ZFC and $\mathcal{Q}$ is a complete orthomodular lattice in W,

$$\text{QZFZ} \vdash [\![\varphi]\!]^W = 1 \text{ in } W^{\mathcal{Q}} .$$

Hence, by Corollary 47, QZFZ $\vdash \varphi$. Therefore, our completeness theorem of QZFZ (Theorem 28) has proved. That is, for a sentence φ of QZFZ,

$$\begin{aligned} \text{ZFC} \quad &\vdash \quad \text{“ } [\![\varphi]\!] = 1 \text{ in } V^{\mathrm{Q}} \text{ for all complete orthomodular lattice Q ”} \\ &\Longrightarrow \quad \text{QZFZ} \vdash \varphi. \end{aligned}$$

A PROOF OF THEOREM 13

Some obvious parts in the following proof figures will be skipped.

(1) $\vdash \varphi \wedge \psi, \Gamma \Rightarrow \Delta$ if and only if $\vdash \varphi, \psi, \Gamma \Rightarrow \Delta$

Proof. $\vdash \varphi, \psi \Rightarrow \varphi \wedge \psi$ by $\wedge$-right. Hence, by using Cut,

$$\vdash \varphi \wedge \psi, \Gamma \Rightarrow \Delta \quad \text{implies} \quad \vdash \varphi, \psi, \Gamma \Rightarrow \Delta.$$

The converse follows from $\wedge$-left as usual. ∎

(2) $\vdash \Gamma \Rightarrow \Delta, \varphi \vee \psi$ if and only if $\vdash \Gamma \Rightarrow \Delta, \varphi, \psi$

Proof. Similar to (1). ∎

(3) $\vdash \Box\varphi \Rightarrow \varphi$

Proof.

$$\dfrac{\dfrac{\varphi \Rightarrow \varphi}{\Rightarrow (\varphi \to \varphi)} \qquad \varphi \Rightarrow \varphi}{(\varphi \to \varphi) \to \varphi \Rightarrow \varphi}$$

∎

(4) $\vdash \varphi \Rightarrow \Diamond\varphi$

Proof. $\neg\varphi$ and $\Box\neg\varphi$ are $\Box$-closed. Hence,

$$\dfrac{\dfrac{\dfrac{\vdots}{\Box\neg\varphi \Rightarrow \neg\varphi} \qquad \dfrac{\vdots}{\neg\varphi, \varphi \Rightarrow}}{\Box\neg\varphi, \varphi \Rightarrow}}{\varphi \Rightarrow \neg\Box\neg\varphi}$$

∎

(5) $\vdash \overline{\Gamma} \Rightarrow \overline{\Delta}, \varphi$ if and only if $\vdash \overline{\Gamma} \Rightarrow \overline{\Delta}, \Box\varphi$

Proof.

If-part is obvious by (3). The converse follows from the fact:

$$\dfrac{\dfrac{\overline{\Gamma} \Rightarrow \overline{\Delta}, \varphi}{(\varphi \to \varphi), \overline{\Gamma} \Rightarrow \overline{\Delta}, \varphi}}{\overline{\Gamma} \Rightarrow \overline{\Delta}, (\varphi \to \varphi) \to \varphi}$$

∎

(6) If φ is $\Box$-closed, then $\vdash \Box\varphi \Leftrightarrow \varphi$.

Proof. By (5). ∎

(7) $\vdash \varphi, \overline{\Gamma} \Rightarrow \overline{\Delta}$ if and only if $\vdash \Diamond\varphi, \overline{\Gamma} \Rightarrow \overline{\Delta}$

Proof. If-part is obvious by (4). The converse follows from:

$$\cfrac{\cfrac{\cfrac{\varphi, \overline{\Gamma} \Rightarrow \overline{\Delta}}{\overline{\Gamma} \Rightarrow \overline{\Delta}, \neg\varphi}}{\overline{\Gamma} \Rightarrow \overline{\Delta}, \Box\neg\varphi}}{\neg\Box\neg\varphi, \overline{\Gamma} \Rightarrow \overline{\Delta}}$$

∎

(8) $\vdash \varphi \wedge (\varphi \to \psi) \Rightarrow \psi$

Proof.

$$\cfrac{\cfrac{\varphi \Rightarrow \varphi \quad \psi \Rightarrow \psi}{(\varphi \to \psi), \varphi \Rightarrow \psi}}{\varphi \wedge (\varphi \to \psi) \Rightarrow \psi}$$

∎

(9) $\vdash \varphi \wedge \neg\varphi \Rightarrow$

Proof. By $\neg$-left and (1). ∎

(10) $\vdash \varphi, \overline{\Gamma} \Rightarrow \psi, \overline{\Delta}$ implies $\vdash \neg\psi, \overline{\Gamma} \Rightarrow \neg\varphi, \overline{\Delta}$

Proof. By $\neg$-left and then $\neg$-right, where we use the $\Box$-closedness of $\neg\varphi$ and $\overline{\Gamma}$. ∎

(11) $\vdash \varphi \Rightarrow \neg\neg\varphi$; $\vdash \Box\varphi \Leftrightarrow \neg\neg\Box\varphi$;

Proof. Since $\neg\varphi$ is $\Box$-closed, by $\neg$-right,

$$\cfrac{\cfrac{\varphi \Rightarrow \varphi}{\neg\varphi, \varphi \Rightarrow}}{\varphi \Rightarrow \neg\neg\varphi}$$

By the $\Box$-closedness of $\Box\varphi$, $\vdash \neg\neg\Box\varphi \Rightarrow \Box\varphi$. ∎

(12) $\vdash \neg(\varphi \vee \psi) \Leftrightarrow (\neg\varphi \wedge \neg\psi)$

Proof. Since $\neg(\varphi \vee \psi)$ is $\Box$-closed, $\vdash \neg(\varphi \vee \psi) \Rightarrow \neg\varphi$.

$$\because)\quad \dfrac{\dfrac{\dfrac{\varphi \Rightarrow \varphi}{\varphi \Rightarrow \varphi \vee \psi}}{\neg(\varphi \vee \psi), \varphi \Rightarrow}}{\neg(\varphi \vee \psi) \Rightarrow \neg\varphi}$$

Similarly, $\vdash \neg(\varphi \vee \psi) \Rightarrow \neg\psi$. Hence, $\vdash \neg(\varphi \vee \psi) \Rightarrow (\neg\varphi \wedge \neg\psi)$. The converse is proved, by using the fact that $\neg\varphi \wedge \neg\psi$ is $\Box$-closed, i.e.

$$\dfrac{\dfrac{\dfrac{\vdots}{\neg\varphi \wedge \neg\psi, \varphi \Rightarrow} \qquad \dfrac{\vdots}{\neg\varphi \wedge \neg\psi, \psi \Rightarrow}}{\neg\varphi \wedge \neg\psi, \varphi \vee \psi \Rightarrow}}{\neg\varphi \wedge \neg\psi \Rightarrow \neg(\varphi \vee \psi)}$$

■

(13) $\vdash \neg\varphi \vee \neg\psi \Rightarrow \neg(\varphi \wedge \psi)$.

Proof. Similar to the first part of (12). ■

(14) $\vdash (\Box\varphi \to \psi) \Leftrightarrow (\Box\varphi \to \Box\psi) \Leftrightarrow (\neg\Box\varphi \vee \Box\psi)$

Proof. Since the first $(\Leftrightarrow)$ is obvious from (5), we prove only the second $(\Leftrightarrow)$.

$$\dfrac{\dfrac{\dfrac{\vdots}{\Box\varphi \to \Box\psi, \Box\varphi \Rightarrow \Box\psi}}{\Box\varphi \to \Box\psi \Rightarrow \Box\psi, \neg\Box\varphi}}{\Box\varphi \to \Box\psi \Rightarrow \neg\Box\varphi \vee \Box\psi}$$

$$\dfrac{\dfrac{\dfrac{\Box\psi \Rightarrow \Box\psi}{\Box\varphi, \Box\psi \Rightarrow \Box\psi} \qquad \dfrac{\vdots}{\neg\Box\varphi, \Box\varphi \Rightarrow \Box\psi}}{\neg\Box\varphi \vee \Box\psi, \Box\varphi \Rightarrow \Box\psi}}{\neg\Box\varphi \vee \Box\psi \Rightarrow (\Box\varphi \to \Box\psi)}$$

■

(15) $\vdash \Box\varphi \wedge \exists x\psi(x) \Leftrightarrow \exists x(\Box\varphi \wedge \psi(x))$; $\vdash \varphi \wedge \exists x\Box\psi(x) \Leftrightarrow \exists x(\varphi \wedge \Box\psi(x))$

Proof.

$$\dfrac{\dfrac{\dfrac{\dfrac{\vdots}{\Box\varphi, \psi(a) \Rightarrow \Box\varphi \wedge \psi(a)}}{\Box\varphi, \psi(a) \Rightarrow \exists x(\Box\varphi \wedge \psi(x))}}{\Box\varphi, \exists x\psi(x) \Rightarrow \exists x(\Box\varphi \wedge \psi(x))}}{\Box\varphi \wedge \exists x\psi(x) \Rightarrow \exists x(\Box\varphi \wedge \psi(x))}$$

$$\dfrac{\dfrac{\dfrac{\Box\varphi \Rightarrow \Box\varphi}{\Box\varphi \wedge \psi(a) \Rightarrow \Box\varphi} \quad \dfrac{\vdots}{\Box\varphi \wedge \psi(a) \Rightarrow \exists x\psi(x)}}{\Box\varphi \wedge \psi(a) \Rightarrow \Box\varphi \wedge \exists x\psi(x)}}{\exists x(\Box\varphi \wedge \psi(x)) \Rightarrow \Box\varphi \wedge \exists x\psi(x)}$$

Similarly, $\vdash (\varphi \wedge \exists x\Box\psi(x)) \Leftrightarrow \exists x(\varphi \wedge \Box\psi(x))$. ■

(16) $\vdash \Box\varphi \vee \neg\Box\varphi$

Proof. Obvious. ■

(17) $\vdash [(\varphi \wedge \Box\xi) \to \psi] \Rightarrow [(\neg\psi \wedge \Box\xi) \to \neg\varphi]$

Proof.

$$\dfrac{\dfrac{\dfrac{\dfrac{\dfrac{\vdots}{\varphi, \Box\xi \Rightarrow \varphi \wedge \Box\xi} \quad \dfrac{\vdots}{\psi, \neg\psi \Rightarrow}}{(\varphi \wedge \Box\xi \to \psi), \neg\psi, \varphi, \Box\xi \Rightarrow}}{\vdots}}{(\varphi \wedge \Box\xi \to \psi), \neg\psi \wedge \Box\xi \Rightarrow \neg\varphi}}{(\varphi \wedge \Box\xi \to \psi) \Rightarrow (\neg\psi \wedge \Box\xi) \to \neg\varphi}$$

■

(18) $\vdash (\Box\varphi)^{\perp} \Leftrightarrow \neg\Box\varphi$

Proof. $\Rightarrow \Box\varphi, (\Box\varphi)^{\perp}$ and $\Box\varphi, (\Box\varphi)^{\perp} \Rightarrow$ are axioms (C2), and $(\Box\varphi)^{\perp}$ is $\Box$-closed. Hence,

$$\vdash \neg\Box\varphi \Rightarrow (\Box\varphi)^{\perp} \quad \text{and} \quad \vdash (\Box\varphi)^{\perp} \Rightarrow \neg\Box\varphi.$$

■

(19) $\vdash (\varphi \to \Box\psi) \Rightarrow (\Diamond\varphi \to \Box\psi)$

Proof. By (7). ■

(20) $\vdash \Diamond(\Box\varphi \wedge \psi) \Rightarrow \Box\varphi \wedge \Diamond\psi$

Proof. By (4) and (7). ■

(21) $\vdash \forall x\Box\varphi(x) \Leftrightarrow \Box\forall x\varphi(x)$

Proof. ($\Rightarrow$) is obvious. Proof of ($\Leftarrow$) is:

$$\dfrac{\dfrac{\vdots}{\Box\forall x\varphi(x) \Rightarrow \Box\varphi(a)}}{\Box\forall x\varphi(x) \Rightarrow \forall x\Box\varphi(x)}$$

■

(22) $\vdash \exists x\Diamond\varphi(x) \Leftrightarrow \Diamond\exists x\varphi(x)$

Proof. Similar to (21). ■

(23) If $A(\varphi)$ is a formula with subformula φ and $\vdash \varphi \Leftrightarrow \psi$, then

$$\vdash A(\varphi) \Leftrightarrow A(\psi).$$

Proof. By induction on the complexity of $A(\varphi)$. ■

B PROOF OF THEOREM 24

In this section it will be proved that each of nonlogical axioms A1–A11 of lattice valued set theory has truth value 1 in Q-valued universe V^{Q}.

A1 (Equality) For any formula $\varphi(a)$ of QZFZ and $u, v \in V^{\mathrm{Q}}$,

$$[\![u = v \wedge \varphi(u)]\!] \leqslant [\![\varphi(v)]\!].$$

Proof. By Theorem 16(4). ■

A2 (Extensionality) $\forall x(x \in u \leftrightarrow x \in v) \rightarrow u = v$.

Proof. $[\![\forall x(x \in u \leftrightarrow x \in v)]\!] = [\![u = v]\!]$ by Theorem 18. ■

A3 (Pairing) $\forall u, v \exists z \forall x[\, x \in z \leftrightarrow x = u \vee x = v\,]$.

Proof. For $u, v \in V^{\mathrm{Q}}$ define z by

$$\begin{cases} \mathcal{D}z = \{u, v\} \\ z(t) = 1 \quad \text{for } \ t \in \mathcal{D}z \end{cases}$$

Then $[\![x \in z]\!] = \bigvee_{t \in \mathcal{D}z} [\![x = t]\!] \wedge z(t) = [\![x = u]\!] \vee [\![x = v]\!]$.
Therefore, $[\![\forall x(x \in z \leftrightarrow x = u \vee x = v)]\!] = 1$. ■

A4 (Union) $\forall u \exists v \forall x[x \in v \leftrightarrow \exists y(y \in u \wedge x \in y)]$.

Proof. For $u \in V^{\mathrm{Q}}$ defined v by

$$\begin{cases} \mathcal{D}v = \bigcup_{y \in \mathcal{D}u} \mathcal{D}y \\ v(x) = [\![\exists y(y \in u \wedge x \in y)]\!]. \end{cases}$$

Then, by Theorem 18,

$$\begin{aligned} [\![\exists y(y \in u \wedge x \in y)]\!] &= \bigvee_{y \in \mathcal{D}u} [\![y \in u]\!] \wedge [\![x \in y]\!] \\ &= \bigvee_{y \in \mathcal{D}u} [\![y \in u]\!] \wedge [\![x \in y]\!] \wedge \bigvee_{x' \in \mathcal{D}y} [\![x = x']\!] \\ &= \bigvee_{y \in \mathcal{D}u,\ x' \in \mathcal{D}y} [\![x = x']\!] \wedge [\![x' \in y \wedge y \in u]\!] \\ &= [\![x \in v]\!] \end{aligned}$$

■

A5 (Power set) $\forall u \exists v \forall x[x \in v \leftrightarrow x \subset u]$, where $x \subset u \overset{\text{def}}{\Longleftrightarrow} \forall t(t \in x \to t \in u)$.

Proof. Let $u \in V^{\mathrm{Q}}_{\alpha}$. For every $x \in V^{\mathrm{Q}}$, define x^* by

$$\begin{cases} \mathcal{D}x^* = \mathcal{D}u \\ x^*(t) = [\![x \subset u \wedge t \in x]\!]. \end{cases}$$

Since

$$[\![x \subset u \wedge t \in x]\!] \leqslant [\![t \in u]\!] \leqslant \bigvee_{t' \in \mathcal{D}u} [\![t = t']\!],$$

we have

$$[\![x \subset u \wedge t \in x]\!] \leqslant \bigvee_{t' \in \mathcal{D}u} [\![t = t' \wedge x \subset u \wedge t' \in x]\!] \leqslant [\![t \in x^*]\!].$$

It follows that for every $x \in V^{\mathrm{Q}}$ there exists $x^* \in V^{\mathrm{Q}}_{\alpha}$ such that $[\![x \subset u]\!] \leqslant [\![x = x^*]\!]$. Now we define v by

$$\begin{cases} \mathcal{D}v = \{x \in V^{\mathrm{Q}}_{\alpha} \mid \mathcal{D}x = \mathcal{D}u\} \\ v(x) = [\![x \subset u]\!]. \end{cases}$$

Then

$$[\![\forall x(x \in v \leftrightarrow x \subset u)]\!] = 1.$$

■

Definition 4.1 For each set x we define $\check{x} \in V^{\mathrm{Q}}$ by

$$\begin{cases} \mathcal{D}\check{x} = \{\check{t} \mid t \in x\} \\ \check{x}(\check{t}) = 1. \end{cases}$$

$\check{x}$ is called the *check set associated with* x. For check sets $\check{x}, \check{y}$, we have

$$[\![\check{x} = \check{y}]\!] = \begin{cases} 1 & \text{if} \quad x = y \\ 0 & \text{if} \quad x \neq y \end{cases} \quad ; \quad [\![\check{x} \in \check{y}]\!] = \begin{cases} 1 & \text{if} \quad x \in y \\ 0 & \text{if} \quad x \notin y. \end{cases}$$

A6 (Infinity) $\exists u \left[\exists x(x \overset{\Box}{\in} u) \wedge \forall x \overset{\Box}{\in} u \exists y \in u (x \overset{\Box}{\in} y) \right].$

Proof. $\check{\omega}$ associated with the set ω of all natural numbers satisfies

$$[\![\exists x(x \overset{\Box}{\in} \check{\omega}) \wedge \forall x \overset{\Box}{\in} \check{\omega} \exists y \overset{\Box}{\in} \check{\omega} (x \overset{\Box}{\in} y))]\!] = 1.$$

■

A7 (Separation) $\forall u \exists v \forall x [\, x \in v \leftrightarrow x \in u \wedge \varphi(x) \,].$

Proof. For a given $u \in V^{\mathrm{Q}}$ define v by

$$\begin{cases} \mathcal{D}v = \mathcal{D}u \\ v(x) = [\![x \in u \wedge \varphi(x)]\!] \end{cases}$$

Then

$$[\![\forall x (x \in v \leftrightarrow x \in u \wedge \varphi(x))]\!] = 1.$$

■

A8 (Collection) $\forall u \exists v \left[\forall x \in u \exists y \varphi(x, y) \rightarrow \forall x \in u \exists y \overset{\Box}{\in} v \varphi(x, y) \right].$

Proof. Let

$$p \;=\; [\![\forall x \in u \exists y \varphi(x, y))]\!] = \bigwedge_{x \in \mathcal{D}u} ([\![x \in u]\!] \rightarrow \bigvee_{y} [\![\varphi(x, y)]\!]).$$

Since Q is a set, there exists an ordinal $\alpha(x)$ such that

$$p \wedge [\![x \in u]\!] \leqslant \bigvee_{y \in V^{\mathrm{Q}}_{\alpha(x)}} [\![\varphi(x, y)]\!]$$

for each $x \in \mathcal{D}u$. By using the axiom of collection externally, there exists an ordinal α such that

$$p \wedge [\![x \in u]\!] \leqslant \bigvee_{y \in V_\alpha^{\mathrm{Q}}} [\![\varphi(x,y)]\!] \quad \text{for all } x \in \mathcal{D}u.$$

Now we defined v by

$$\begin{cases} \mathcal{D}v = V_\alpha^{\mathrm{Q}} \\ v(y) = 1 \end{cases}$$

Then

$$p \wedge [\![x \in u]\!] \leqslant \bigvee_{y \in \mathcal{D}v} [\![y \overset{\square}{\in} v \wedge \varphi(x,y)]\!] = [\![\exists y \overset{\square}{\in} v \varphi(x,y)]\!] \quad \text{for all } x \in \mathcal{D}u.$$

Since $p = \square p$, we have

$$p \leqslant [\![\forall x (x \in u \rightarrow \exists y \overset{\square}{\in} v \varphi(x,y)]\!].$$

■

A9 (∈-induction) $\forall x \, [\forall y (y \in x \rightarrow \varphi(y)) \rightarrow \varphi(x)] \rightarrow \forall x \varphi(x)$.

Proof. Let $p = [\![\forall x \, (\forall y (y \in x \rightarrow \varphi(y)) \rightarrow \varphi(x)]\!]$. We prove

$$p \leqslant [\![\forall x \varphi(x)]\!] = \bigwedge_{x \in V^{\mathrm{Q}}} [\![\varphi(x)]\!]$$

by induction on the rank of x. Let $x \in V_\alpha^{\mathrm{Q}}$. Since $p \leqslant [\![\varphi(y)]\!]$ for all $y \in \mathcal{D}x \subset V_{<\alpha}^{\mathrm{Q}}$ by induction hypothesis,

$$p \wedge [\![y \in x]\!] \leqslant [\![\varphi(y)]\!] \text{ for all } y \in \mathcal{D}x.$$

Hence, by using $p = \square p$, we have

$$p \leqslant [\![\forall y (y \in x \rightarrow \varphi(y))]\!].$$

It follows that $p \leqslant [\![\forall x \varphi(x)]\!]$. ■

DEFINITION 48. *Restriction* $u \restriction p$ *of* $u \in V^{\mathrm{Q}}$ *by* $p \in \mathrm{Q}$ *is defined inductively by*

$$\begin{cases} \mathcal{D}(u \restriction p) = \{x \restriction p \mid x \in \mathcal{D}u\} \\ (u \restriction p)(x \restriction p) = \bigvee \{u(x') \wedge p \mid x' \in \mathcal{D}u, \; x \restriction p = x' \restriction p\} \text{ for } x \in \mathcal{D}u. \end{cases}$$

If $u \in V_\alpha^{\mathrm{Q}}$ then $u \restriction p \in V_\alpha^{\mathrm{Q}}$, and we have

LEMMA 49. *Let* $u, x \in V^{\mathrm{Q}}$, $p, q \in \mathrm{Q}$, *and* $p = \square p$. *Then*

(1) $p \leqslant [\![u = u \restriction p]\!]$
(2) $[\![x \in u \restriction p]\!] = [\![x \in u]\!] \wedge p$
(3) $(u \restriction q) \restriction p = u \restriction (p \wedge q)$.

Proof. We proceed by induction on the rank of u,

(1) : For $x \in \mathcal{D}u$,

$$\begin{aligned} p \wedge u(x) &\leqslant (u \restriction p)(x \restriction p) \wedge [\![x = x \restriction p]\!] \leqslant [\![x \in u \restriction p]\!] \\ (u \restriction p)(x \restriction p) &= \bigvee_{x' \in \mathcal{D}u,\, x \restriction p = x' \restriction p} u(x') \wedge p \wedge [\![x = x' = x \restriction p]\!] \\ &\leqslant [\![x \restriction p \in u]\!]. \end{aligned}$$

Therefore, $p \leqslant [\![u = u \restriction p]\!]$.

(2) : By (1) and Theorem 16,

$$[\![x \in u]\!] \wedge p \quad \leqslant \quad [\![x \in u \restriction p]\!].$$

($\leqslant$) follows from the fact that $x'' \restriction p = x' \restriction p$ implies $p \leqslant [\![x'' = x']\!]$:

$$\begin{aligned} [\![x \in u \restriction p]\!] &= \bigvee_{x' \in \mathcal{D}u} [\![x = x' \restriction p]\!] \wedge \bigvee_{x'' \in \mathcal{D}u,\, x'' \restriction p = x' \restriction p} u(x'') \wedge p \\ &\leqslant [\![x \in u]\!] \wedge p \end{aligned}$$

(3) : $\mathcal{D}((u \restriction q) \restriction p) = \mathcal{D}(u \restriction (q \wedge p))$, by the induction hypothesis. Since p is $\Box$-closed, $(\bigvee_{x'} (u(x') \wedge q)) \wedge p = \bigvee_{x'} (u(x') \wedge q \wedge p)$. Therefore,

$$((u \restriction q) \restriction p)((x \restriction q) \restriction p) = (u \restriction (q \wedge p))(x \restriction (q \wedge p))$$

■

A10 (Zorn) $\exists x (x \in u) \wedge \mathrm{Gl}(u) \wedge \forall v [\mathrm{Chain}(v, u) \rightarrow \bigcup v \in u] \rightarrow \exists z \mathrm{Max}(z, u)$, where

$$\begin{aligned} \mathrm{Gl}(u) &\overset{\text{def}}{\Longleftrightarrow} \forall x (x \in u \rightarrow x \overset{\Box}{\in} u), \\ \mathrm{Chain}(v, u) &\overset{\text{def}}{\Longleftrightarrow} v \subset u \wedge \forall x, y (x, y \in v \rightarrow x \subset y \vee y \subset x), \\ \mathrm{Max}(z, u) &\overset{\text{def}}{\Longleftrightarrow} z \in u \wedge \forall x (x \in u \wedge z \subset x \rightarrow z = x). \end{aligned}$$

Proof. For $u \in V_\alpha^{\mathrm{Q}}$, let

$$p = [\![\mathrm{Gl}(u) \wedge \forall v (\mathrm{Chain}(v, u) \rightarrow \bigcup v \in u)]\!].$$

By using Zorn's lemma externally, let U be a maximal subset of V_α^{Q} such that

$$\forall x, y \in U\big(\llbracket x \in u \wedge \exists t(t \in x) \wedge y \in u \wedge \exists t(t \in y)\rrbracket \wedge p \leqslant \llbracket x \subset y \vee y \subset x\rrbracket\big)$$

Define v by

$$\begin{cases} \mathcal{D}v = U \\ v(x) = p \wedge \llbracket x \in u \wedge \exists t(t \in x)\rrbracket. \end{cases}$$

Then it suffices to show that $p \leqslant \llbracket \mathrm{Max}(\bigcup v, u)\rrbracket$. Since $p = \Box p$ and $p \wedge v(x) \leqslant \llbracket x \in u\rrbracket$ for all $x \in \mathcal{D}v$, we have $p \leqslant \llbracket v \subset u\rrbracket$. By the definition of v, $p \leqslant \llbracket \mathrm{Chain}(v, u)\rrbracket$. Therefore, $p \leqslant \llbracket \bigcup v \in u\rrbracket$. Now we prove the maximality of $\bigcup v$ in u, i.e.

$$p \wedge \llbracket x \in u \wedge \bigcup v \subset x\rrbracket \leqslant \llbracket x \subset \bigcup v\rrbracket \quad \text{for} \quad x \in \mathcal{D}u.$$

Let $x \in \mathcal{D}u$ and $r = p \wedge \llbracket x \in u \wedge \bigcup v \subset x\rrbracket$. Then r is $\Box$-closed and $r \leqslant \llbracket x = x \restriction r\rrbracket$ by Lemma 49. Besides, for each $y \in U$,

$$\begin{aligned} \llbracket y \in u \wedge \exists t(t \in y) \ &\wedge \ (x \restriction r) \in u \wedge \exists t(t \in x \restriction r)\rrbracket \wedge p \\ &\leqslant \ \llbracket y \in v\rrbracket \wedge r \\ &\leqslant \ \llbracket y \subset \bigcup v \subset x\rrbracket \wedge \llbracket x = x \restriction r\rrbracket \\ &\leqslant \ \llbracket y \subset x \restriction r\rrbracket \\ &\leqslant \ \llbracket y \subset x \restriction r \ \vee \ x \restriction r \subset y\rrbracket. \end{aligned}$$

Therefore, $x \restriction r \in U$. It follows that :

$$\begin{aligned} r \wedge x(t) \ &\leqslant \ \llbracket x = x \restriction r \wedge x \in u \wedge t \in x\rrbracket \wedge p \\ &\leqslant \ \llbracket x = x \restriction r \wedge x \restriction r \in u \wedge \exists t(t \in x \restriction r)\rrbracket \wedge p \\ &\leqslant \ \llbracket x = x \restriction r\rrbracket \wedge v(x \restriction r) \\ &\leqslant \ \llbracket x \in v\rrbracket \leqslant \llbracket x \subset \bigcup v\rrbracket. \end{aligned}$$

Consequently, we have $r \leqslant \llbracket x \subset \bigcup v\rrbracket$. That is, $\llbracket \mathrm{Zorn}\rrbracket = 1$. ■

A11 ($\Diamond$) $\forall u \exists v \forall x(x \in v \leftrightarrow \Diamond(x \in u))$, where $\Diamond\varphi \overset{\mathrm{def}}{\iff} \neg\Box\neg\varphi$.

Proof. For a given $u \in V^{\mathrm{Q}}$, defined v by

$$\begin{cases} \mathcal{D}v = \mathcal{D}u \\ v(x) = \llbracket \Diamond(x \in u)\rrbracket. \end{cases}$$

By using Theorem 10,

$$\begin{aligned}[\![\Diamond(x \in u)]\!] &= \Diamond \bigvee_{x' \in \mathcal{D}u} [\![x = x']\!] \wedge u(x') \\ &\leqslant \bigvee_{x' \in \mathcal{D}u} [\![x = x']\!] \wedge [\![\Diamond(x' \in u)]\!] = [\![x \in v]\!].\end{aligned}$$

Hence $[\![\forall x(x \in v \leftrightarrow \Diamond(x \in u))]\!] = 1$. ■

C OUTLINE OF A PROOF OF THEOREM 31(RECURSION PRINCIPLE)

Theorem 31 (Recursion Principle)
Let $\prec$ be a well founded relation and H be a global functional relation such that $\forall x \exists y H(x, y)$. Then there exists a unique global functional relation F such that

$$\mathrm{Dom}\, F = \mathrm{Fld}(\prec) \ \wedge \ \forall x\, [x \in \mathrm{Fld}(\prec) \rightarrow (F(x) = H(F_{\prec x}))],$$

where $F_{\prec x} = \{\langle t, s\rangle \in F \mid \Diamond(t \prec x)\}$.

Proof. Set

$$\begin{aligned} S_x &\stackrel{\text{def}}{=} \{y \mid \Diamond(y \prec x)\} \\ F_n(f) &\stackrel{\text{def}}{\Longleftrightarrow} f \text{ is a global functional relation, i.e.} \\ & \qquad \forall x, y\, [f(x,y) \rightarrow \Box f(x,y)] \wedge \forall x, y, y' [f(x,y) \wedge f(x,y') \rightarrow y = y'] \\ R(f) &\stackrel{\text{def}}{\Longleftrightarrow} F_n(f) \wedge [\mathrm{Dom}\, f \subset \mathrm{Fld}(\prec)] \wedge \\ & \qquad \forall x [x \in \mathrm{Dom}\, f \rightarrow (S_x \subset \mathrm{Dom}\, f) \wedge f(x) = H(f_{\prec x})] \\ T(f,x) &\stackrel{\text{def}}{\Longleftrightarrow} R(f) \wedge S_x \subset \mathrm{Dom}\, f \ \wedge \forall f' [R(f') \wedge S_x \subset \mathrm{Dom}(f') \rightarrow f \subset f'] \end{aligned}$$

If there exists a unique f such that $T(f, x)$, i.e. $\forall x \in \mathrm{Fld}(\prec) \exists! f T(f, x)$, then the desired functional relation F is obtained by

$$F(x,y) \stackrel{\text{def}}{\Longleftrightarrow} \exists f [T(f,x) \ \wedge \ y = H(f_{\prec x})].$$

S_x is a global set by WF3 in Definition 29. If there exists f such that $T(f, x)$, then uniqueness of f follows from

$$\forall f, f' \forall x \big[R(f) \wedge R(f') \wedge (x \in \mathrm{Dom}\, f \cap \mathrm{Dom}\, f') \rightarrow f(x) = f'(x)\big],$$

which is proved by $\prec$-induction.

Existence of f is proved again by $\prec$-induction : Assuming

$$x \in \mathrm{Fld}(\prec) \wedge \forall y (y \prec x \rightarrow \exists! f T(f, y)),$$

put

$$f_0 \stackrel{\text{def}}{=} \bigcup\{f \mid \exists y(\Diamond(y \prec x) \wedge T(f,y))\} \cup \{\langle y, H(f \restriction S_y)\rangle \mid \Diamond(y \prec x) \wedge T(f,y)\}$$

Then $T(f_0, x)$ ■

BIBLIOGRAPHY

[Birkhoff and von Neumann, 1936] Garrett Birkhoff and John von Neumann, The logic of Quantum Mechanics, *Ann. Math.*, 37, 823, 1936.

[Grayson, 1975] Robin J. Grayson, *A sheaf approach to models of set theory.* M.Sc. thesis. Oxford 1975.

[Halmos, 1951] Paul R. Halmos, *Introduction to Hilbert Space*, Chelsea Publishing Company, 1951.

[Kodera and Titani, submitted] Heiji Kodera and Satoko Titani, The equivalence of two sequential calculi of quantum logic. Submitted.

[von Neumann, 1955] John von Neumann, *Mathematical Foundation of Quantum Mechanics*, Princeton University Press, 1955.

[Piron, 1976] Constantin P. Piron, *Foundations of Quantum Physics*, W.A. Benjamin, Inc. 1976.

[Takano, 2002] Michio Takano, Strong Completeness of Lattice Valued Logic, *Archive for Mathematical Logic*, 41 (2002) 497-505.

[Takeuti, 1978] Gaisi Takeuti, *Two Applications of Logic to Mathematics*, Iwanami and Princeton University Press, Tokyo and Princeton (1978).

[Takeuti, 1981] Gaisi Takeuti, Quantum Set Theory, *Current Issues in Quantum Logic*, eds. E.Beltrametti and B.C.van Frassen, Plenum,New York (1981) pp.303-322

[Titani, 1999] Satoko Titani, Lattice Valued Set Theory, *Archive for Mathematical Logic* 38-6(1999) pp.395-421.

INDEX